TEACHER WRAPAROUND EDITION

Glencoe

PHYSICAL SCIENCE

GLENCOE

McGraw-Hill

New York, New York   Columbus, Ohio   Mission Hills, California   Peoria, Illinois

GLENCOE Physical Science

Student Edition

Teacher Wraparound Edition

Laboratory Manual: SE

Laboratory Manual: TE

Study Guide: SE

Study Guide: TE

Teaching Transparency Package

Section Focus Transparency Package

Science Integration Transparency Package

Assessment

Performance Assessment

Computer Test Bank

MindJogger Videoquiz

English/Spanish Audiocassettes

Interactive Videodisc Program

Glencoe/McGraw-Hill

A Division of The **McGraw·Hill** *Companies*

Send all inquiries to:

Glencoe/McGraw-Hill
8787 Orion Place
Columbus, OH 43240
ISBN 0-02-827566-7

Printed in the United States of America.
4 5 6 7 8 9 10   027/046   04 03 02 01

Teacher Guide

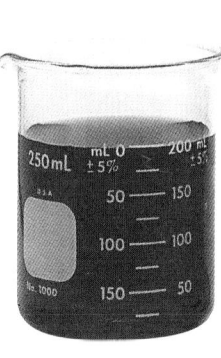

About Our Authors

CHARLES W. McLAUGHLIN has more than 25 years of experience as a high school chemistry teacher and is currently the coordinator of science education for the St. Joseph, Missouri, school district. He holds a Ph.D. in chemistry from the University of Nebraska. The National Science Teachers Association has presented Dr. McLaughlin with three national awards for innovative science education. He also has received the Presidential Award for Science Teaching.

MARILYN THOMPSON is currently teaching and pursuing graduate work in science education at the University of Kansas. Her career has included teaching physics, chemistry, and physical science while serving as science department chair at Center High School in Kansas City, Missouri. Ms. Thompson received her B.A. degree in chemistry from Carleton College and her M.A. degree in science curriculum and instruction from the University of Kansas. She has conducted summer workshops and courses in the physical sciences for K–12 teachers. Ms. Thompson is a member of NSTA and AAPT.

Table of Contents

Student Edition

Table of Contents

Table of Contents

Glencoe Physical Science
The National Science Education Standards

The *National Science Education Standards*, published by the National Research Council and representing the contribution of thousands of educators and scientists, offer a comprehensive vision of a scientifically literate society. The standards not only describe what students should know but also offer guidelines for science teaching and assessment. Bill Aldridge, a Glencoe author, helped originate this standards initiative and served on its Chair's Advisory Committee. If you are using, or plan to use, the standards to guide changes in your science curriculum, you can be assured that *Glencoe Physical Science* aligns with the *National Science Education Standards*.

Glencoe Physical Science is an example of how Glencoe's commitment to effective science education is changing the materials used in science classrooms today. More than just a collection of facts in a textbook, *Glencoe Physical Science* is a program that provides numerous opportunities for students, teachers, and school districts to meet the *National Science Education Standards*.

Content Standards
The accompanying table shows the close alignment between *Glencoe Physical Science* and the content standards. The integration included in *Glencoe Physical Science* allows students to discover concepts within each of the content standards, giving them opportunities to make connections between the science disciplines. Our hands-on activities and inquiry-based lessons reinforce the science processes emphasized in the standards.

Teaching Standards
Alignment with the *National Science Education Standards* requires much more than alignment with the outcomes in the content standards. The way in which concepts are presented is critical to effective learning. The teaching

standards within the *National Science Education Standards* recommend an inquiry-based program facilitated and guided by teachers. *Glencoe Physical Science* provides such opportunities through activities and discussions that allow students to discover by inquiry critical concepts and apply the knowledge they've constructed to their own lives.

Throughout the program, students are building critical skills that will be available to them for lifelong learning. *The Teacher Wraparound Edition* helps you make the most of every instructional moment. It offers an abundance of effective strategies and suggestions for guiding students as they explore science.

Assessment Standards
The assessment standards are supported by many of the components that make up the *Glencoe Physical Science* program. *The Teacher Wraparound Edition* and *Teacher Classroom Resources* provide multiple chances to assess students' understanding of important concepts as well as their ability to perform a wide range of skills. Ideas for portfolios, performance activities, written reports, and other assessment activities accompany every lesson. Rubrics and performance task assessment lists can be found in Glencoe's Professional Series booklet *Performance Assessment in the Science Classroom*.

Program Coordination
The scope of the content standards requires students to meet the outcomes over the course of their education. The correlation on the following pages demonstrates the close alignment of this course of *Glencoe Physical Science* with the content standards.

Correlation of *Glencoe Physical Science* to the National Science Standards, Content Standards, Grades 5–12

Content Standard	Page Numbers
(UCP) UNIFYING CONCEPTS AND PROCESSESS	
1. Systems, order, and organization	6-17, 20-23, 33-53, 68-69, 110-116, 128, 141-144, 162-174, 180-203, 214-221, 224-227, 246-263, 270-291, 298-308, 310-311, 314-320, 330-353, 360-369, 373-381, 388-391, 396-403, 416-418, 424-434, 442-445, 448-459, 466, 470, 472-477, 498-513, 516-521, 528-529, 536-543, 546-551, 558-581, 592-617, 624-639, 648-665, 674-677, 704-705, 712-719
2. Evidence, models, and explanation	12, 16-19, 68-69, 73, 76-87, 94-117, 124-144, 152-158, 162-174, 180-203, 214-239, 246-263, 270-291, 298-320, 330-353, 360-381, 396-405, 416-434, 442-459, 466-487, 498-521, 528-551, 558-581, 592-617, 624-641, 648-665, 674-693, 700-719
3. Change, constancy, and measurement	20-23, 26-27, 34-47, 64-87, 94-117, 124-145, 152-158, 162-174, 181-183, 186-203, 224-239, 254-263, 279-281, 298-308, 332-336, 338-353, 363-364, 380-381, 389, 398, 404-405, 417-421, 424-429, 432-434, 442-459, 466, 470, 472-485, 498-521, 528-543, 546-551, 558-571, 574-581, 592-617, 624-639, 648-663, 674-677, 679-693, 700-702, 704-708, 714-717
4. Evolution and equilibrium	68, 78, 99, 426, 593, 640
5. Form and function	34-53, 70-71, 76-77, 80, 82, 97-105, 128, 139, 143-144, 154-158, 162-174, 180-203, 217, 230-231, 234-239, 254-263, 273, 284-291, 298-302, 304-308, 312-313, 319-320, 330-353, 360-381, 388-405, 419-423, 432-434, 444, 446-447, 451-458, 466-485, 504-521, 529-552, 558-581, 592-617, 624-641, 648-665, 683, 686-693, 700-719
(A) SCIENCE AS INQUIRY	
1. Abilities necessary to do scientific inquiry	26-27, 46-47, 58-59, 70-71, 106-107, 132-133, 160-161, 184-185, 232-233, 262-263, 282-283, 310-311, 352-353, 380-381, 402-403, 430-431, 454-455, 486-487, 502-503, 542-543, 580-581, 612-613, 638-639, 664-665, 684-685, 718-719
2. Understandings about scientific inquiry	7, 22, 25, 35-37, 40-45, 48-53, 58-59, 65-67, 69, 74-75, 80, 87, 95-96, 99, 103, 117, 127, 130-131, 141-145, 158, 169, 182-183, 186-188, 192-193, 196-197, 200, 202-203, 215, 220, 226, 230, 239, 248, 254, 257, 261, 271, 274, 290-291, 300-301, 307-308, 316-318, 320, 335, 338-339, 348, 361, 363, 365, 377, 389-391, 394, 396-401, 404-405, 418, 426-428, 445, 457, 469, 501, 504, 509, 513-515, 541, 549-550, 578, 604, 606-607, 615-617, 629, 636, 640-641, 657, 662, 675-676, 680, 682, 702, 706, 717
(B) PHYSICAL SCIENCE (Grades 5–8)	
1. Properties and changes of properties in matter	221, 224-231, 254-263, 284-285, 288-289, 291, 329-353, 392-394, 466, 470, 680-682
2. Motions and forces	63-117, 127-128, 181, 591-597, 600-613, 628-637
3. Transfer of energy	8, 124-125, 128-131, 152-158, 172-174, 370-371, 456-459, 498, 529-535, 544-545, 614-617, 700-705, 707-708, 716
(B) PHYSICAL SCIENCE (Grades 9–12)	
1. Structure of atoms	269-291
2. Structure and properties of matter	221, 224-231, 254-263, 284-285, 288-289, 291, 298-308, 329-353, 361-369, 373-381, 392-394, 466, 470, 674-677, 689-690
3. Chemical reactions	441-459, 478-487
4. Motions and forces	63-117, 127-128, 181, 591-597, 600-613, 628-637
5. Conservation of energy and increase in disorder	128-129, 701-703
6. Interactions of energy and matter	498-521, 528-552, 558-582, 591-607
(C) LIFE SCIENCE (Grades 5–8)	
1. Structure and function in living systems	6-7, 155, 169, 182, 189, 305-306, 368, 373-381, 400, 434, 457, 473, 476, 481, 507-508, 566-567, 692-693
2. Reproduction and heredity	375, 377, 692-693
3. Regulation and behavior	169, 305-306, 378-379, 457, 476, 492-493
4. Populations and ecosystems	146, 378-379
5. Diversity and adaptations of organisms	85, 104-105, 146, 155, 163, 169-170, 182, 196-197, 378-379, 434, 457
(C) LIFE SCIENCE (Grades 9–12)	
1. The cell	305-306, 373, 375, 692-693
2. Molecular basis of heredity	375, 377, 692-693
3. Biological evolution	19, 104-105, 692-693

4. Interdependence of organisms	378-379
5. Matter, energy, and organization in living systems	130, 155, 169-170, 189, 281, 305-306, 364-365, 368-371, 373-381, 457, 476, 481, 507-508, 531, 540-541, 566-567, 676
6. Behavior of organisms	146, 170, 307, 378-379

(D) EARTH AND SPACE SCIENCE (Grades 5–8)

1. Structure of the Earth system	69, 229, 336-337, 446-447, 533, 625, 708
2. Earth's history	68-69, 281, 625, 683
3. Earth in the solar system	34, 108-109, 154, 572-573, 690-691, 714-715

(D) EARTH AND SPACE SCIENCE (Grades 9–12)

1. Energy in the Earth system	125, 142, 154, 160-161, 172-174, 446-447, 499, 533, 714-717
2. Geochemical cycles	258, 412-413
3. Origin and evolution of the Earth system	68-69, 572-573, 690-691
4. Origin and evolution of the universe	17, 572-573, 690-691

(E) SCIENCE AND TECHNOLOGY

1. Abilities of technological design	26-27, 46-47, 70-71, 106-107, 132-133, 160-161, 184-185, 232-233, 262-263, 282-283, 310-311, 352-353, 380-381, 402-403, 430-431, 454-455, 486-487, 502-503, 542-543, 580-581, 612-613, 638-639, 664-665, 684-685, 718-719
2. Understandings about science and technology	7, 18-19, 41, 80, 103, 108-109, 142, 169, 172-174, 196-197, 202, 230, 252-253, 257, 276-277, 290, 307, 335, 338-339, 377, 399, 428, 457, 469, 509, 514-515, 544-545, 549, 572-573, 578, 604, 629, 640-641, 646-667, 676, 692-693, 702

(F) SCIENCE IN PERSONAL AND SOCIAL PERSPECTIVES (Grades 5–8)

1. Personal health	76-77, 196-197, 202, 222-223, 252-253, 307, 312-313, 335, 370-371, 422-423, 428, 446-447, 457, 478-479, 514-515, 578, 598-599, 629, 676, 692-693, 710-711
2. Populations, resources, and environments	138-140, 142,172-174, 202, 222-223, 252-253, 307, 312-313, 370-371, 422-423, 446-447, 457, 478-479, 544-545, 598-599, 710-711
3. Natural hazards	76-77, 138-140, 172-174, 202, 222-223, 252-253, 307, 312-313, 370-371, 422-423, 446-447, 478-479, 598-599, 710-711
4. Risks and benefits	76-77, 80, 138-140, 172-174, 222-223, 312-313, 370-371, 404-405, 422-423, 446-447, 478-479, 544-545, 598-599, 702, 710-711
5. Science and technology in society	7, 18-19, 41, 52-53, 76-77, 80, 108-109, 138-140, 142, 172-174, 196-197, 202, 222-223, 230, 252-253, 257, 276-277, 307, 312-313, 335, 338-339, 370-371, 377, 399, 404-405, 422-423, 428, 446-447, 457, 469, 478-479, 509, 514-515, 544-545, 549, 572-573, 578, 598-599, 604, 629, 640-641, 662, 666-667, 676, 692-693, 702, 710-711

(F) SCIENCE IN PERSONAL AND SOCIAL PERSPECTIVES (Grades 9–12)

1. Personal and community health	76-77, 80, 138-140, 196-197, 202, 222-223, 252-253, 307, 312-313, 335, 370-371, 422-423, 428, 446-447, 457, 478-489, 514-515, 578, 598-599, 629, 666-667, 676, 692-693, 710-711
2. Population growth	138-140, 172-174, 222-223, 307, 370-371, 446-447
3. Natural resources	138-140, 142, 172-174, 202, 222-223, 252-253, 307, 312-313, 370-371, 446-447, 478-479, 544-545, 598-599, 710-711
4. Environmental quality	138-140, 172-174, 202, 222-223, 252-253, 307, 312-313, 370-371, 422-423, 446-447, 457, 478-479, 544-545, 598-599, 710-711
5. Natural and human-induced hazards	76-77, 80, 138-140, 172-174, 202, 222-223, 252-253, 307, 312-313, 370-371, 422-423, 446-447, 457, 478-479, 598-599, 676, 710-711
6. Science and technology in local, national, and global challenges	7, 18-19, 41, 52-53, 76-77, 80, 108-109, 138-140, 142, 172-174, 196-197, 202, 222-223, 230, 252-253, 257, 276-277, 307, 312-313, 335, 338-339, 370-371, 377, 399, 404-405, 422-423, 428, 446-447, 457, 469, 478-479, 509, 514-515, 544-545, 549, 572-573, 578, 598-599, 604, 629, 640-641, 662, 666-667, 676, 692-693, 702, 710-711

(G) HISTORY AND NATURE OF SCIENCE

1. Science as a human endeavor	28, 81, 88, 94-95, 103-105, 108-112, 118, 127, 159, 172-174, 196-197, 199, 202, 204, 229-230, 236-237, 240, 272, 284-285, 292, 309, 348, 382, 399, 406, 422-423, 442-443, 488, 529, 548-550, 572-573, 582, 605-606, 618, 628, 633, 640, 646-668, 675, 689, 710-711, 720
2. Nature of scientific knowledge	6-17, 20-25, 270-273, 284-285
3. Historical perspectives	81, 94-95, 108-112, 127, 146, 196-197, 199, 202, 204, 229-230, 236-237, 271-273, 284-285, 348, 399, 442-443, 529, 548-549, 605-606, 628, 633, 640, 659, 675, 689

Responding to Changes in
Science Education

The need for new directions in science education

By today's projections, seven out of every ten American jobs will be related to science, mathematics, or electronics by the year 2000. And according to the experts, if students haven't grasped the fundamentals—they probably won't go further in science and may not have a future in a global job market. Studies also reveal that students are avoiding taking "advanced" science classes.

The time for action is now!

In the past decade, educators, public policy makers, corporate America, and parents have recognized the need for reform in science education. These groups have united in a call to action to solve this national problem. As a result, three important projects have published reports to point the way for America:

- **Project on National Science Standards ...** by the National Research Council.
- **Benchmarks for Science Literacy ...** by the American Association for the Advancement of Science.
- **Scope, Sequence, and Coordination of Secondary School Science ...** by the National Science Teachers Association (NSTA).

Together, these reports spell out unified guiding principles for new directions in U.S. science education. *Glencoe Physical Science* is based on these guiding principles:

- developing scientific, technological, and mathematical literacy in all students.
- educating students to use scientific principles and processes appropriately in making personal decisions.
- helping students to experience the richness and excitement of knowing about and understanding the natural world.
- teaching students how to engage intelligently in public discourse and debate about matters of scientific and technological concern.

Glencoe Physical Science answers the challenge!

At Glencoe Publishing, we believe that *Glencoe Physical Science* will help you bring science reform to the front lines—the classrooms of America. But more importantly, we believe it will help students succeed in science so that they will want to continue learning about science through high school and into adulthood.

In response to the goals of science curriculum reform, *Glencoe Physical Science* promotes:

- solid, accurate content,
- a thematic orientation,
- hands-on activities that promote a constructivist approach through more student-planned activities,
- frequent practice of science process skills,
- integration of science concepts across the curriculum,
- numerous opportunities for performance assessment, and
- decision making and problem solving.

GLENCOE PHYSICAL SCIENCE is loaded with activities.

You'll choose from dozens of activities in the *Student Edition.* These activities are easy to set up and manage and will allow you to teach using a hands-on, inquiry-based approach to learning.

MiniLABS and **Activities** involve students in learning and applying scientific methods to practice thinking skills and construct science concepts. They provide an engaging, diverse, active program. MiniLABS require a minimum of equipment, and students may take responsibility for organization and execution.

Activities develop and reinforce or restructure concepts as well as develop the ability to use process skills. Activity formats are structured to guide students to make their own discoveries. Students collect real evidence and are encouraged through open-ended questions to reflect and reformulate their ideas based on this evidence.

In each chapter, there is one open-ended Activity called **Design Your Own Experiment** that gives students a broad topic to explore and guides them with leading questions to the point where they can determine the direction of the investigation themselves. Students are asked to brainstorm hypotheses, make a decision to investigate one that can be tested, plan procedures, and in the end, think about why their hypotheses were supported or not.

GLENCOE PHYSICAL SCIENCE integrates science and math.

Mathematics is a tool that all students, regardless of their career goals, will use throughout their lives. *Glencoe Physical Science* provides opportunities to hone mathematics skills while learning about the natural world. **Using Math, Practice Problems, MiniLABS,** and **Activities** offer numerous options for practicing math, including making and using tables and graphs, measuring in SI, and calculating. **Across the Curriculum** strategies in the *Teacher Wraparound Edition* provide additional connections between science and mathematics.

GLENCOE PHYSICAL SCIENCE integrates the sciences for understanding.

No subject exists in isolation. At appropriate points in *Glencoe Physical Science*, attention is called to other sciences in the *Student Edition,* in the *Teacher Wraparound Edition,* and in the *Teacher Classroom Resources.*

Connect to... margin features relate basic questions in physics, chemistry, Earth science, and life science to one another. Suggested answers to these questions are contained in the *Teacher Wraparound Edition.* Opportunities for **Integration** of science topics are highlighted throughout the *Student Edition.* Further explanation of how to weave the **Integration** topic into your lesson is provided in the *Teacher Wraparound Edition.* Look for this logo  for ideas on how to make the sciences connect for your students. *Science Integration Activities* are full-page activities in the *Teacher Classroom Resources* that relate physics, chemistry, Earth science, and life science to each other. *Science Integration Transparencies* also relate the sciences to one another using full-color illustrations that can be used with an overhead projector. These opportunities for integrating the sciences will help your students discover the science behind things they see every day.

It's time for new directions in science education.

The need for new directions in science education has been established by the experts. America's students must prepare themselves for the high-tech jobs of the future. We at Glencoe believe that *Glencoe Physical Science* answers this challenge. We believe *Glencoe Physical Science* will assist you better in preparing your students for a lifetime of science learning.

Themes & Scope & Sequence

Our society is becoming more aware of the interrelationship of the disciplines of science. It is also necessary to recognize the precarious nature of the stability of some systems and the ease with which this stability can be disturbed so that the system changes. For most people, then, the ideas that unify the sciences and make connections between them are the most important.

Themes provide a construct for unifying the sciences. Woven throughout *Glencoe Physical Science,* themes integrate facts and concepts. Like story lines, themes provide "big ideas" that link the science disciplines. Themes are an important part of any teaching strategy because they help students see the importance of truly understanding concepts rather than simply memorizing isolated facts. While there are many possible themes around which to unify science, we have chosen four: Energy, Systems and Interactions, Scale and Structure, and Stability and Change.

Energy

Energy is a central concept of the physical sciences that pervades the biological and Earth sciences. In physical terms, energy is the ability of an object to change itself or its surroundings—the capacity to do work. In chemical terms, it forms the basis of reactions between compounds. In biological terms, it gives living systems the ability to maintain themselves, to grow, and to reproduce. Energy sources are crucial in the interactions among science, technology, and society.

Systems and Interactions

A system can be incredibly tiny, such as an atom's nucleus and electrons; extremely complex, such as an ecosystem; or unbelievably large, as the stars in a galaxy. By defining the boundaries of the system, one can study the interactions among its parts. The interactions may be a force of attraction between the positively charged nucleus and negatively charged electron. In an ecosystem, on the other hand, the interactions may be between the predator and its prey, or among the plants and animals.

Scale and Structure

Used as a theme, "structure" emphasizes the relationship among different structures. "Scale" defines the focus of the relationship. As the focus is shifted from a system to its components, the properties of the structure may remain constant. In other systems, an ecosystem for example, which includes a change in scale from interactions between prey and predator to the interactions among systems inside an animal, the structure changes drastically. In *Glencoe Physical Science,* the authors have tried to stress how we know what we know and why we believe it to be so. Thus, explanations remain on the macroscopic level until students have the background needed to understand how the microscopic structure operates.

Stability and Change

A system that is stable is constant. Often the stability is the result of a system being in equilibrium. If a system is not stable, it undergoes change. Changes in an unstable system may be characterized as trends (position of falling objects), cycles (the motion of planets around the sun), or irregular changes (radioactive decay).

Theme Development

These four major themes are developed within the student material and discussed throughout the *Teacher Wraparound Edition.* Each chapter of *Glencoe Physical Science* incorporates at least one of these themes. These themes are interwoven throughout each level and are developed as appropriate to the topic presented.

The *Teacher Wraparound Edition* includes a **Theme Connection** section for each unit opener. This section provides a framework for integrating concepts discussed in the unit opener. Each chapter opener includes a **Theme Connection** section to explain the chapter's themes and to point out the major chapter concepts supporting those themes. Throughout the chapters, **Theme Connections** show specifically how a topic in the *Student Edition* relates to the themes.

Theme connections stressed in each chapter of *Glencoe Physical Science* are indicated in the chart on page 13T.

GLENCOE PHYSICAL SCIENCE Chapters	Themes			
	Scale and Structure	Energy	Stability and Change	Systems and Interactions
1 The Nature of Science				✔
2 Physical Science Methods	✔			
3 Exploring Motion and Forces			✔	
4 Acceleration and Momentum			✔	
5 Energy		✔		
6 Using Thermal Energy		✔		
7 Machines—Making Work Easier				✔
8 Solids, Liquids, and Gases		✔		
9 Classification of Matter			✔	
10 Atomic Structure and the Periodic Table	✔			
11 Chemical Bonds			✔	
12 Elements and Their Properties	✔			
13 Organic and Biological Compounds	✔			
14 Useful Materials	✔			
15 Solutions				✔
16 Chemical Reactions			✔	
17 Acids, Bases, and Salts				✔
18 Waves and Sound		✔		
19 Light		✔		
20 Mirrors and Lenses				✔
21 Electricity		✔		
22 Magnetism and Its Uses				✔
23 Electronics and Computers				✔
24 Radioactivity and Nuclear Reactions			✔	
25 Energy Sources		✔		

Resources for Different Needs

In addition to the wide array of instructional options provided in the student and teacher editions, *Glencoe Physical Science* also offers an extensive list of support materials and program resources. Some of these materials offer alternative ways of enriching or extending your science program, others provide tools for reinforcing and evaluating student learning, while still others will help you directly in delivering instruction. You won't have time to use them all, but the ones you use will help you save the time you have.

Hands-On Activities

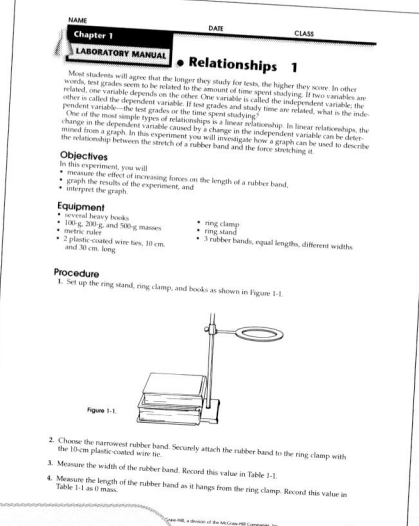

If you want more hands-on options, the *Lab Manual* offers you one or more additional labs per chapter. Each lab is complete with setup diagrams, data tables, and space for student responses.

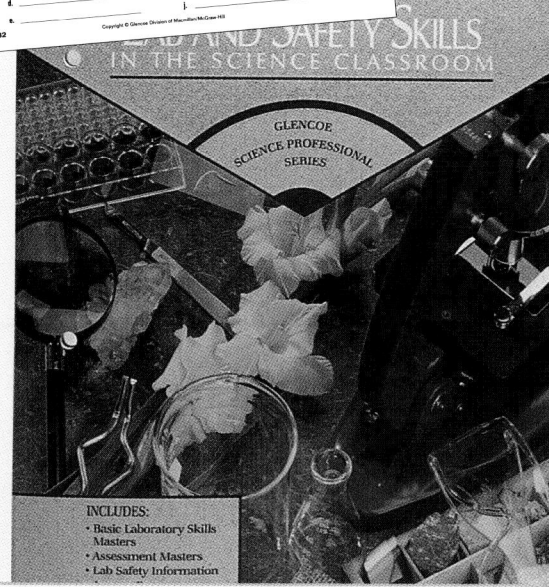

Reinforce basic lab and safety skills, including graphing and measurement, with activities from the *Lab and Safety Skills* book.

LABORATORY MANUAL
TEACHER EDITION

Glencoe

PHYSICAL SCIENCE

Includes:
- 50 additional Physical Science lab activities
- Detailed safety guidelines
- Answer pages

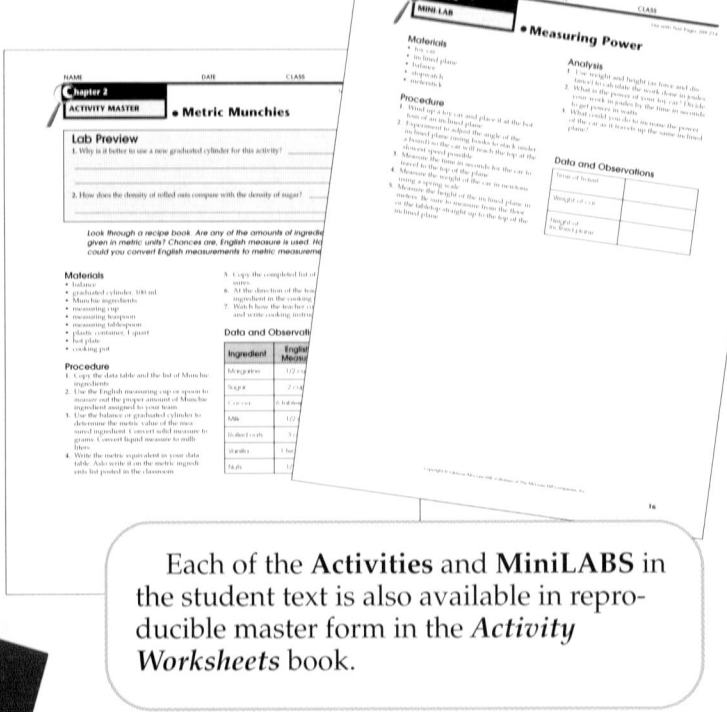

Each of the **Activities** and **MiniLABS** in the student text is also available in reproducible master form in the *Activity Worksheets* book.

ACTIVITY WORKSHEETS

Glencoe

PHYSICAL SCIENCE

SCIENCE INTEGRATION ACTIVITIES

Glencoe

PHYSICAL SCIENCE

Includes:

• Twenty-five activities that integrate the physical sciences with other sciences
• Answers to questions

Science Integration Activities are laboratory activities that relate Earth and life science to specific physical science chapters.

Reinforcement and Enrichment

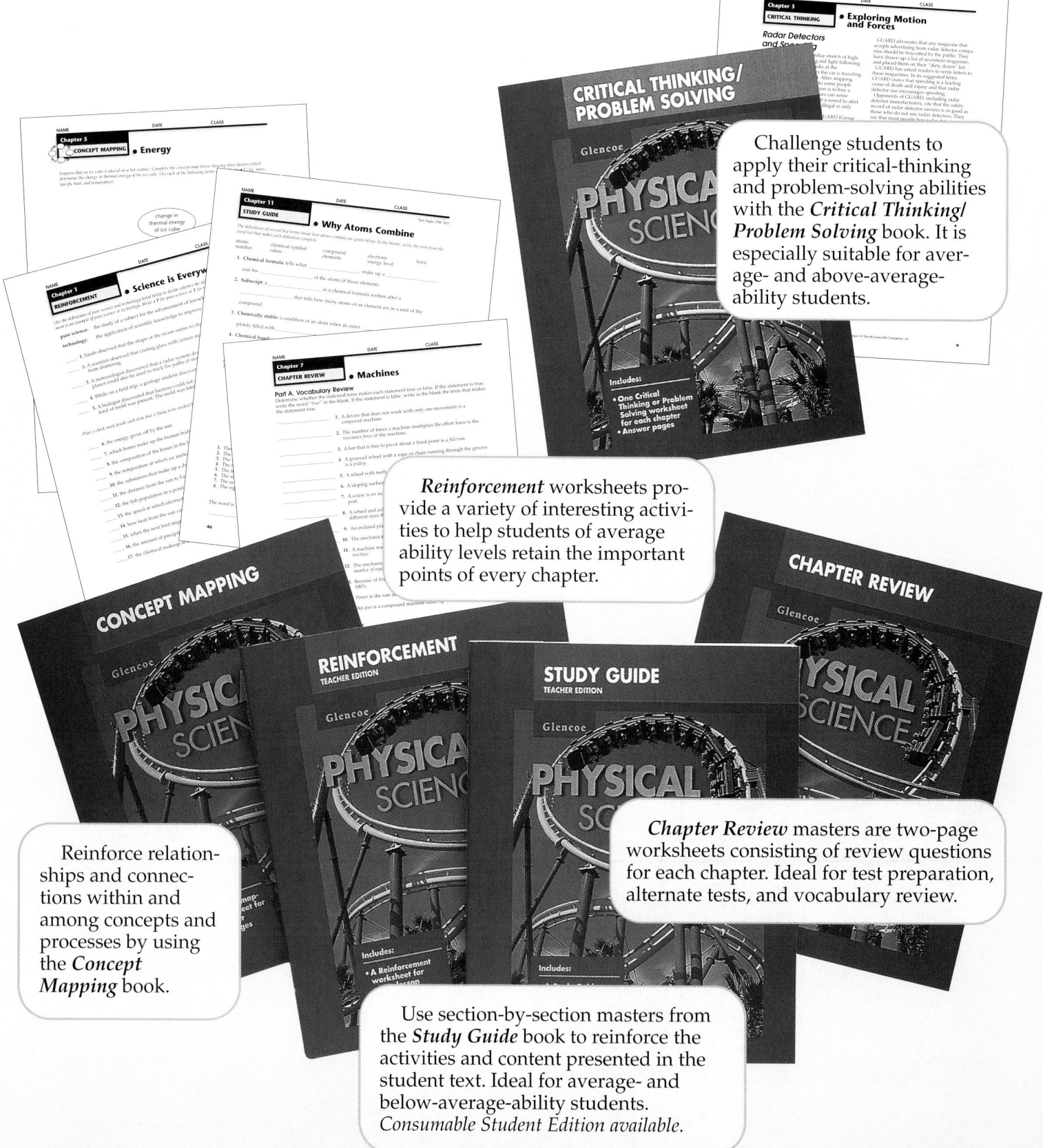

Challenge students to apply their critical-thinking and problem-solving abilities with the *Critical Thinking/ Problem Solving* book. It is especially suitable for average- and above-average-ability students.

Reinforcement worksheets provide a variety of interesting activities to help students of average ability levels retain the important points of every chapter.

Chapter Review masters are two-page worksheets consisting of review questions for each chapter. Ideal for test preparation, alternate tests, and vocabulary review.

Reinforce relationships and connections within and among concepts and processes by using the *Concept Mapping* book.

Use section-by-section masters from the *Study Guide* book to reinforce the activities and content presented in the student text. Ideal for average- and below-average-ability students. *Consumable Student Edition available.*

Enrichment worksheets challenge your students of above-average ability to design, interpret, and research scientific topics based on the text in each lesson.

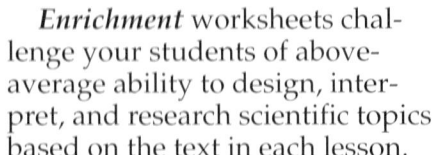

Science and Society Integration worksheets encourage further involvement with science as it applies to society.

The *Cross-curricular Integration* book includes interdisciplinary worksheets that emphasize learning by doing and provide valuable insight into the connection physical science has with other disciplines.

Technology Integration masters explore recent developments in science and technology or explain how familiar machines, tools, or systems work.

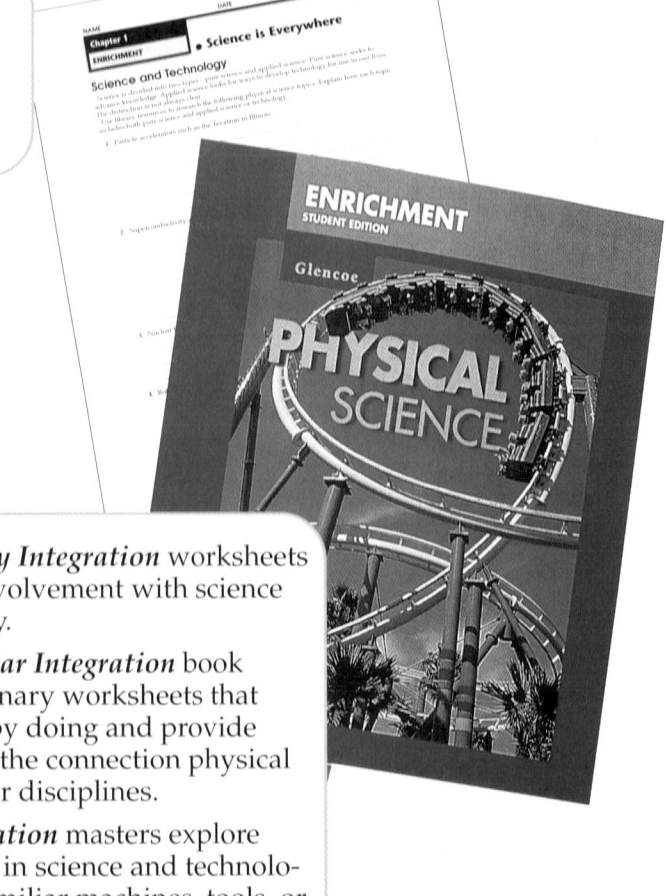

Explore past and present contributions to science of individuals and societies of various cultural backgrounds through the readings and activities in the *Multicultural Connections* book.

Assessment

Assess student learning and performance through a variety of questioning strategies and formats in the *Assessment* book. Test items cover activity procedures and analysis as well as science concepts within the four pages of reproducible masters available for each chapter.

ASSESSMENT
CHAPTER AND UNIT TESTS

Glencoe

PHYSICAL SCIENCE

PERFORMANCE ASSESSMENT
IN THE SCIENCE CLASSROOM

GLENCOE
SCIENCE PROFESSIONAL
SERIES

INCLUDES:
• Rationale for Assessing Performance
• Strategies for Implementation

PERFORMANCE TASK ASSESSMENT LIST

Using Math in Science

Element	Points Possible	Earned Assessment
		Self / Teacher's

Understanding the Problem

1. The problem is clearly defined by being re-stated.
2. Given information is identified.
3. Information that must be assumed is listed.
4. Information that must be obtained is listed.
5. A clear diagram is drawn that shows the important elements of the problem.

Solving the Problem

6. The algebraic formula(s) for this problem is listed.
7. The formula(s) is rearranged correctly to solve for the unknown quantity.
8. Appropriately labeled values are put in the final formula.
9. Appropriate arithmetic operations are used accurately.
10. All values are labeled.
11. Reasoning can be easily followed by the sequence of arithmetic operations.
12. The appropriate number of significant figures is used.
13. Scientific notation is correctly used for very large or very small numbers.
14. The answer is correct and labeled correctly.
15. The answer is appropriate according to the assumptions and reasoning used.

Communicating the Result

16. A clear, concise statement of the problem, the strategy for the solution, and the answer are made. Math vocabulary is used correctly.
17. A labeled diagram is used to support the written statement.

Total

Performance Assessment in the Science Classroom provides classroom assessment lists and scoring rubrics for a variety of assessment strategies, including group assessment, oral presentations, modeling, and writing in science.

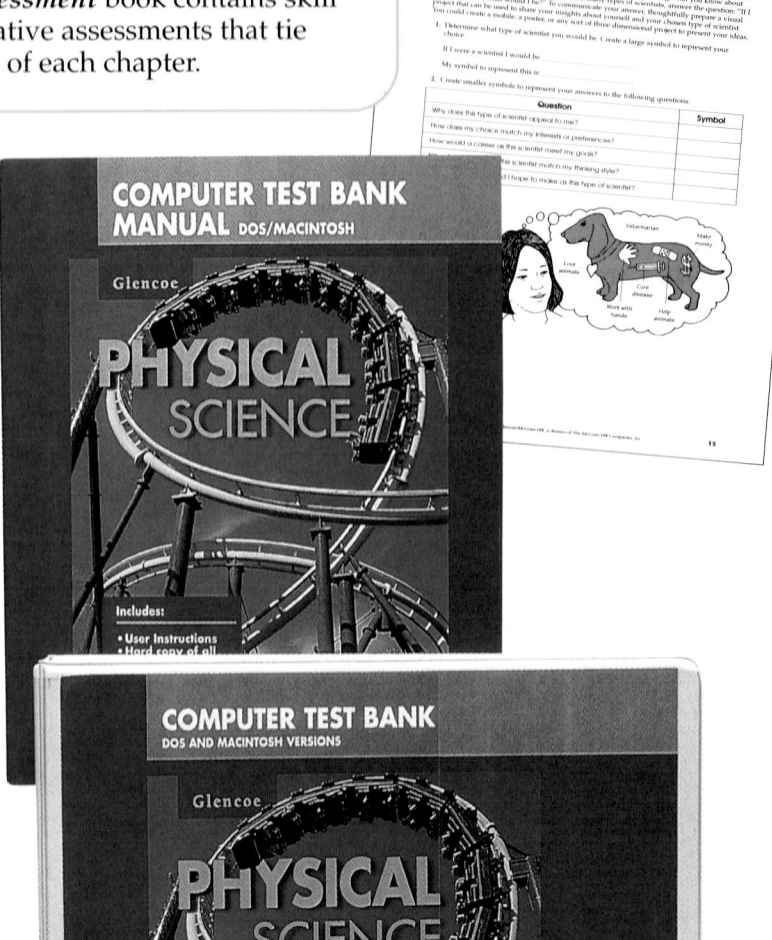

The *Performance Assessment* book contains skill assessments and summative assessments that tie together major concepts of each chapter.

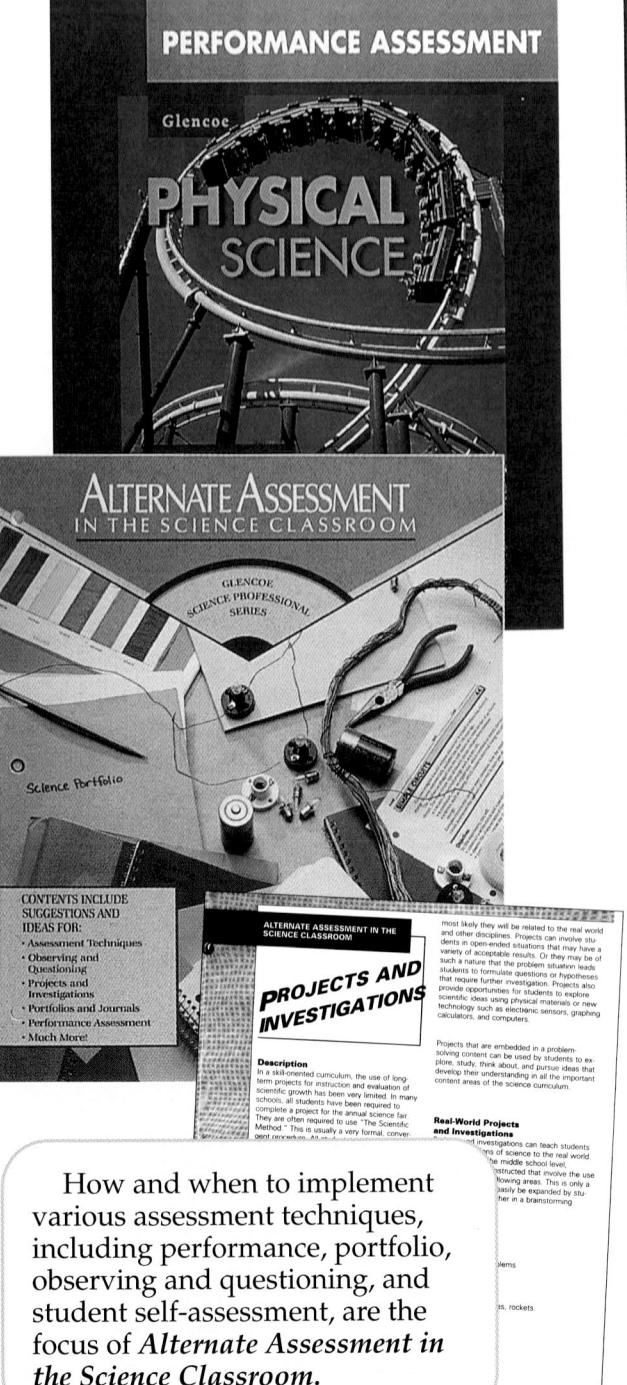

How and when to implement various assessment techniques, including performance, portfolio, observing and questioning, and student self-assessment, are the focus of *Alternate Assessment in the Science Classroom.*

Computer Test Banks, available in IBM and Macintosh versions, provide the ultimate flexibility in designing and creating your own test instruments. Select test items from two different levels of difficulty, or write and edit your own.

Teaching Resources

Chapter-by-chapter *Lesson Plans* will help you organize your lessons more efficiently.

LESSON PLANS
INCLUDING BLOCK SCHEDULING

Glencoe

PHYSICAL SCIENCE

Includes:
- Complete Lesson Plans for every lesson in the student text

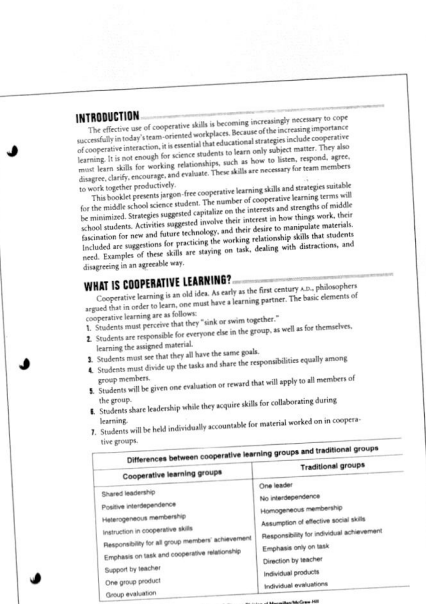

INTRODUCTION

The effective use of cooperative skills is becoming increasingly necessary to cope successfully in today's team-oriented workplaces. Because of the increasing importance of cooperative interaction, it is essential that educational strategies include cooperative learning. It is not enough for science students to learn only subject matter. They also must learn skills for working relationships, such as how to listen, respond, agree, disagree, clarify, encourage, and evaluate. These skills are necessary for team members to work together productively.

This booklet presents jargon-free cooperative learning skills and strategies suitable for the middle school science student. The number of cooperative learning terms will be minimized. Strategies suggested capitalize on the interests and strengths of middle school students. Activities suggested involve their interest in how things work, their fascination for new and future technology, and their desire to manipulate materials. Included are suggestions for practicing the working relationship skills that students need. Examples of these skills are staying on task, dealing with distractions, and disagreeing in an agreeable way.

WHAT IS COOPERATIVE LEARNING?

Cooperative learning is an old idea. As early as the first century A.D., philosophers argued that in order to learn, one must have a learning partner. The basic elements of cooperative learning are as follows:

1. Students must perceive that they "sink or swim together."
2. Students are responsible for everyone else in the group, as well as for themselves, learning the assigned material.
3. Students must see that they all have the same goals.
4. Students must divide up the tasks and share the responsibilities equally among group members.
5. Students will be given one evaluation or reward that will apply to all members of the group.
6. Students share leadership while they acquire skills for collaborating during learning.
7. Students will be held individually accountable for material worked on in cooperative groups.

Differences between cooperative learning groups and traditional groups	
Cooperative learning groups	**Traditional groups**
Shared leadership	One leader
Positive interdependence	No interdependence
Heterogeneous membership	Homogeneous membership
Instruction in cooperative skills	Assumption of effective social skills
Responsibility for all group members' achievement	Responsibility for individual achievement
Emphasis on task and cooperative relationship	Emphasis only on task
Support by teacher	Direction by teacher
One group product	Individual products
Group evaluation	Individual evaluations

Copyright © Glencoe Division of Macmillan/McGraw-Hill

COOPERATIVE LEARNING
IN THE SCIENCE CLASSROOM

GLENCOE
SCIENCE PROFESSIONAL
SERIES

INCLUDES:
- Strategies for Implementing and Evaluating Cooperative Learning Activities in Science
- Blackline Masters and Facsimiles

Cooperative Learning in the Science Classroom contains background information, strategies, and practical tips for using cooperative learning techniques whenever you do activities from *Glencoe Physical Science.*

Enhance your presentation of science concepts with the color transparency packages. The *Teaching Transparency Package* includes two full-color transparencies per chapter and a **Study Guide** booklet with reproducible student worksheets and blackline reproductions for each of the program's full-color transparencies. The *Section Focus Transparency Package* includes one full-color transparency for each section of the student text and a **Study Guide** booklet with reproducible student worksheets. The *Science Integration Transparency Package* includes one full-color transparency per chapter and a **Study Guide** booklet with student worksheets.

Help your Spanish-speaking students get more out of your science lessons by reproducing pages from the *Spanish Resources* book. In addition to a complete English-Spanish glossary, the book contains translations of all objectives, key terms, activities, and main ideas for each chapter of the student text.

the Technology Connection

Glencoe Physical Science offers a wide selection of video and audio products to stimulate and challenge your students.

Interactive videos that provide a fun way for your students to review chapter concepts make up the *MindJogger Videoquiz* series.

The *Glencoe Interactive Videodisc Program* includes interactive lessons that illustrate and reinforce major physical science concepts.

The *English/Spanish Audiocassettes* summarize the content of the Student Edition in both English and Spanish.

The *Secret of Life* videotapes produced by WGBH in Boston can demonstrate the myriad ways in which our expanding knowledge of DNA and heredity have affected all areas of biology and created many ethical dilemmas.

Lab Partner Software provides spreadsheet and graphing tools to record and display data from lab activities.

The *Infinite Voyage* (videotapes and videodiscs) is 20 programs from the award-winning PBS series, covering physical sciences, biology, ecology and environmental education, archaeology, and ancient history. An *Instructor's Guide* accompanies each videotape.

Program Resources

Glencoe is proud to offer quality educational technology from the *National Geographic Society.* These interactive learning tools remove the walls from your classroom and transport your students to the far corners of the world. Award-winning titles from National Geographic give students exciting, easy-to-use resources. With the click of a mouse, students can explore the world—and all that's in it—through video, photographs, sound, and text. Also available from Glencoe is the *Science and Technology Videodisc Series (STVS),* based on the Mr. Wizard television program. Both National Geographic Society and STVS products have been correlated to each chapter of *Glencoe Physical Science.*

NATIONAL GEOGRAPHIC SOCIETY

- STV (Single-Topic Videodiscs)
    - Atmosphere
    - Water
    - Human Body Volumes 1, 2, & 3
    - Plants
    - The Cell
    - Rain Forest
    - Solar System
    - Restless Earth
    - Biodiversity

National Geographic Society
- GTV (Group-Topic Videodiscs)
    - Planetary Manager

24T

The *Science and Technology Videodisc Series* is a seven-disc series that contains more than 280 full-motion video reports. Reports cover a broad spectrum of topics relating to current research in various science fields, innovations in technology, and science and society issues. In addition to reinforcing science concepts, the videoreports are ideal for illustrating science methods, laboratory techniques, and careers in science.

National Geographic Society
Newton's Apple (Videodisc)
  Physical Sciences
  Life Sciences

National Geographic Society
CD-ROM
  NGS Picture Show

Constructivism

Strategies suggested in *Glencoe Physical Science* support a constructivist approach to science education. The role of the teacher is to provide an atmosphere in which students design and direct activities. To develop the idea that science investigation is not made up of closed-end questions, the teacher should ask guiding questions and be prepared to help his or her students draw meaningful conclusions when their results do not match predictions. Through the numerous activities, cooperative learning opportunities, and a variety of critical thinking exercises in *Glencoe Physical Science,* you can feel comfortable taking a constructivist approach to science in your classroom.

Activities

A constructivist approach to science is rooted in an activities-based plan. Students must be provided with sensorimotor experiences as a base for developing abstract ideas. *Glencoe Physical Science* utilizes a variety of learning-by-doing opportunities. **MiniLABS** allow students to consider questions about the concepts to come, make observations, and share prior knowledge. **MiniLABS** require a minimum of equipment, and students may take responsibility for organization and execution.

Design Your Own Experiments develop and reinforce or restructure concepts as well as develop the ability to use process skills. **Design Your Own Experiment** formats guide students to make their own discoveries. Students collect real evidence and are encouraged through open-ended questions to reflect and reformulate their ideas based on this evidence.

Cooperative Learning

Cooperative learning adds the element of social interaction to science learning. Group learning allows students to verbalize ideas, and encourages the reflection that leads to active construction of concepts. It allows students to recognize the inconsistencies in their own perspectives and the strengths of others'. By presenting the idea that there is no one, "ready-made" answer, all students may gain the courage to try to find a viable solution. **Cooperative Learning** strategies appear in the *Teacher Wraparound Edition* margins whenever appropriate.

And More...

Flex Your Brain—a self-directed, critical-thinking matrix is introduced in Chapter 1, "The Nature of Science." This activity, referenced wherever appropriate in the **Teacher Wraparound Edition** margins, assists students in identifying what they already know about a subject, then in developing independent strategies to investigate further.

Students are encouraged to discover the pleasure of solving a problem through a variety of features. The **Science and Society** features in each chapter invite students to confront real-life problems. **Explore the Issue** or **Explore the Technology** questions encourage students to reflect on issues related to technology and society. The **Skill Handbook** gives specific examples to guide students through the steps of developing thinking and process skills. **Developing Skills** and **Thinking Critically** sections of the **Chapter Review** allow the teacher to assess and reward successful thinking skills.

Developing and Applying
Thinking Skills

Science is not just a collection of facts for students to memorize. Rather, it is a process of applying those observations and intuitions to situations and problems, formulating hypotheses, and drawing conclusions. This interaction of the thinking process with the content of science is the core of science and should be the focus of science study. Students, much like scientists, will plan and conduct research or experiments based on observations, evaluate their findings, and draw conclusions to explain their results. This process then begins again, using the new information for further investigation.

Basic Process Skills

Observing

The most basic process is observing. Through observation—seeing, hearing, touching, smelling, tasting—the student begins to acquire information about an object or event. Observation allows a student to gather information regarding size, shape, texture, or quantity of an object or event.

Classifying

One of the simplest ways to organize information gathered through observation is by classifying. Classifying involves the sorting of objects according to similarities or differences.

Using Numbers

Numbers are used to quantify information, including variables and measurements. Quantified information is useful for making comparisons and classifying data or objects.

Communicating

Communicating information is an important part of science. Once all the information is gathered, it is necessary to organize the observations so that the findings can be considered and shared by others. Information can be presented in tables, charts, a variety of graphs, or models.

Developing Thinking Skills

Measuring

Measuring is a way of quantifying information in order to classify or order it by magnitude, such as area, length, volume, or mass. Measuring can involve the use of instruments and the necessary skills to manipulate them.

Inferring

Inferences are logical conclusions based on observations and are made after careful evaluation of all the available facts or data. Inferences are a means of explaining or interpreting observations and are based on making judgments. Therefore, inferences can be invalid.

Predicting

Predicting involves suggesting future events based on observations and inferences about current events. Reliable predictions are based on making accurate observations and measurements and interpreting them accurately.

Using Space/Time Relationships

This process skill involves describing the spatial relationships of objects and how those relationships change with time. For example, a student may be required to describe the motion, direction, symmetry, or shape of an object or objects.

Sequencing

Sequencing involves arranging objects or events in a particular order and may imply a hierarchy or a chronology. Developing a sequence involves the skill of identifying relationships among objects or events. A sequence may be in the form of a numbered list or a series of objects or ideas with directional arrows.

Comparing and Contrasting

Comparing is a way of identifying similarities among objects or events, while contrasting identifies their differences. Comparing and contrasting provides a way of describing and evaluating what is known about something.

Recognizing Cause and Effect

Recognizing cause and effect involves observing actions or events and making logical inferences about why they occur. Recognizing cause and effect can lead to further investigation to isolate a specific cause of a particular event.

Interpreting Scientific Illustrations

Illustrations provide examples that clarify difficult concepts or give additional information about the topic being studied. Interpreting illustrations involves processing visual information into a conceptual construct.

Integrated Process Skills

Interpreting Data

Interpreting data involves synthesizing information (collected data) to make generalizations about the problem under study and apply those generalizations to new problems. Interpreting data may therefore involve many of the other process skills, such as predicting, inferring, classifying, and using numbers.

Defining Operationally

Forming operational definitions involves stating the meaning of something in terms of its function. Operational definitions are formed through observation.

Experimenting

Experimenting involves gathering data to test a hypothesis. For data to be reliable, the experiment must be conducted under controlled conditions with a limited number of variables.

Controlling Variables

Controlling variables requires understanding and regulating all the possible variables that might affect the outcome of an experiment. Failure to control variables can lead to unreliable results.

Formulating Hypotheses

Formulating a hypothesis involves making observations followed by some kind of inference as to what those observations mean. The hypothesis is then tested via experimentation to determine its validity.

Formulating Models

A model is a way to concretely represent abstract ideas or relationships. Models can also be used to predict the outcome of future events or relationships. Models may be expressed physically, verbally, or mentally.

Concept Mapping

Concept maps are visual representations or graphic organizers of relationships among concepts. Concept mapping may involve sequencing temporal events in which a final outcome is observed (events chain), or in which the last event relates back to the beginning of

Developing Thinking Skills

the sequence (cycle map). Hierarchies can be represented by network tree concept maps. Central and supporting ideas that are nonhierarchical can be represented by spider concept maps.

Making and Using Tables

Making tables involves understanding cause and effect and relationships among ideas and representing those relationships in a tabular format.

Making and Using Graphs

Making graphs involves interpreting data and representing those data in a visual format.

Interaction of content and process

Glencoe Physical Science encourages the interaction between science content and thinking processes. We've known for a long time that hands-on activities are a way of providing a bridge between science content and student comprehension. *Glencoe Physical Science* encourages the interaction between content and thinking processes by offering literally hundreds of hands-on activities that are easy to set up and do. In the student text, the **MiniLABS** require students to make observations and collect and record a variety of data. Full-page **Activities** connect hands-on experiences to the content information.

At the end of each chapter, students use thinking processes as they complete **Developing Skills, Thinking Critically,** and **Performance Assessment** questions.

Skill Handbook

The **Skill Handbook** provides the student with another opportunity to practice the thinking processes relevant to the material they are studying. The **Skill Handbook** also provides examples of the processes that students may refer to as they do the **Skillbuilders, Activities,** and **MiniLABS.**

Thinking Skills in the *Teacher Wraparound Edition*

These processes are also featured throughout the *Teacher Wraparound Edition.* In the margins are suggestions for students to write in a journal. Keeping a journal encourages students to communicate their ideas, a key process in science.

Flex Your Brain is a self-directed activity designed to assist students in developing thinking processes as they investigate content areas. Suggestions for using this decision-making matrix are in the margins of the *Teacher Wraparound Edition* whenever appropriate for the topic. Further discussion of **Flex Your Brain** can be found in the *Activity Worksheets* booklet of the *Teacher Classroom Resources.*

Thinking Skills Map

On page 30T, you will find a Thinking Skills Map that indicates how frequently thinking skills are encouraged and developed in *Glencoe Physical Science.*

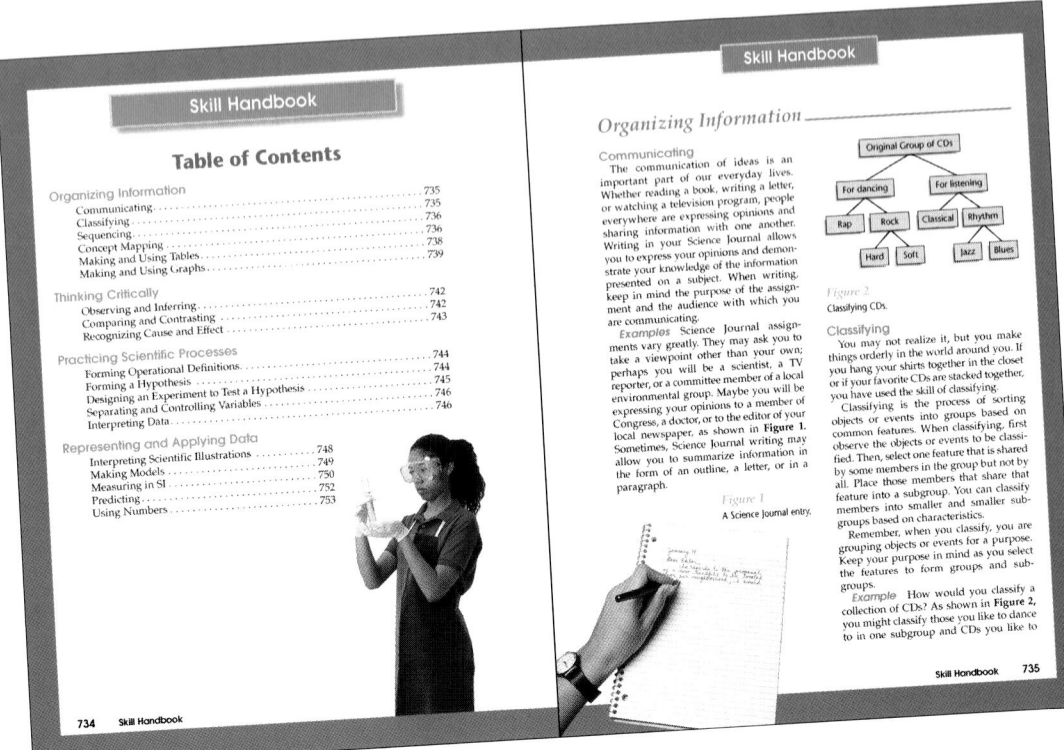

Developing Thinking Skills

Thinking Skills

Chapters

Thinking Skills	1	2	3	4	5	6	7	8	9	10	11	12	13	14	15	16	17	18	19	20	21	22	23	24	25
ORGANIZING INFORMATION																									
Communicating	✓	✓	✓	✓	✓	✓	✓	✓		✓	✓	✓	✓	✓	✓	✓	✓	✓	✓	✓	✓	✓	✓		✓
Classifying		✓	✓			✓	✓	✓	✓	✓	✓	✓	✓	✓			✓	✓			✓			✓	✓
Sequencing						✓	✓						✓				✓	✓			✓	✓		✓	
Concept Mapping	✓	✓	✓	✓	✓	✓	✓	✓	✓	✓	✓	✓	✓	✓	✓	✓	✓	✓	✓	✓	✓	✓	✓	✓	✓
Making and Using Tables		✓	✓	✓	✓	✓			✓	✓	✓	✓		✓	✓	✓		✓	✓	✓	✓	✓	✓		
Making and Using Graphs		✓	✓		✓		✓	✓		✓			✓	✓		✓					✓			✓	
THINKING CRITICALLY																									
Observing and Inferring	✓	✓	✓	✓	✓	✓	✓	✓	✓	✓	✓	✓	✓	✓	✓	✓	✓	✓	✓	✓	✓	✓	✓	✓	✓
Comparing and Contrasting		✓	✓	✓	✓	✓	✓		✓	✓	✓	✓			✓	✓	✓	✓	✓	✓	✓	✓			✓
Recognizing Cause and Effect		✓	✓	✓	✓	✓	✓	✓				✓	✓			✓	✓	✓	✓	✓	✓	✓	✓	✓	✓
Forming Operational Definitions		✓		✓	✓	✓	✓	✓		✓	✓	✓		✓		✓	✓	✓	✓	✓	✓			✓	✓
PRACTICING SCIENTIFIC PROCESSES																									
Forming a Hypothesis	✓		✓	✓	✓	✓	✓	✓		✓	✓	✓		✓		✓	✓	✓	✓	✓	✓	✓			
Designing an Experiment to Test a Hypothesis	✓		✓	✓	✓	✓				✓	✓	✓		✓		✓	✓		✓						
Separating and Controlling Variables	✓	✓	✓	✓	✓	✓		✓					✓			✓			✓						
Interpreting Data	✓	✓	✓	✓		✓	✓	✓	✓		✓	✓	✓	✓	✓	✓	✓	✓	✓	✓	✓	✓	✓	✓	✓
REPRESENTING AND APPLYING DATA																									
Interpreting Scientific Illustrations						✓	✓				✓	✓	✓			✓								✓	
Making Models		✓	✓	✓		✓	✓	✓		✓	✓	✓	✓	✓			✓	✓			✓	✓	✓	✓	✓
Measuring in SI		✓	✓	✓	✓	✓	✓	✓			✓			✓	✓	✓	✓	✓		✓	✓				
Predicting	✓					✓			✓			✓		✓	✓	✓			✓	✓	✓			✓	✓
Using Numbers		✓	✓	✓	✓		✓							✓			✓	✓	✓	✓	✓		✓		

Flex Your Brain

A key element in the coverage of problem-solving and critical-thinking skills in *Glencoe Physical Science* is a critical-thinking matrix called **Flex Your Brain.**

Flex Your Brain provides students with an opportunity to explore a topic in an organized, self-checking way, and then identify how they arrived at their responses during each step of their investigation. The activity incorporates many of the skills of critical thinking. It helps students to consider their own thinking and learn about thinking from their peers.

Where is Flex Your Brain found?

Chapter 1, "The Nature of Science," is an introduction to the topics of critical thinking and problem solving. **Flex Your Brain** accompanies the text section in the introductory chapter. A worksheet for **Flex Your Brain** appears on page 5 of the *Activity Worksheets* book of the *Teacher Resources.* This version provides spaces for students to write in their responses.

In the *Teacher Wraparound Edition,* suggested topics are given in each chapter for the use of *Flex Your Brain.* You can photocopy the worksheet master from the *Teacher Resources.*

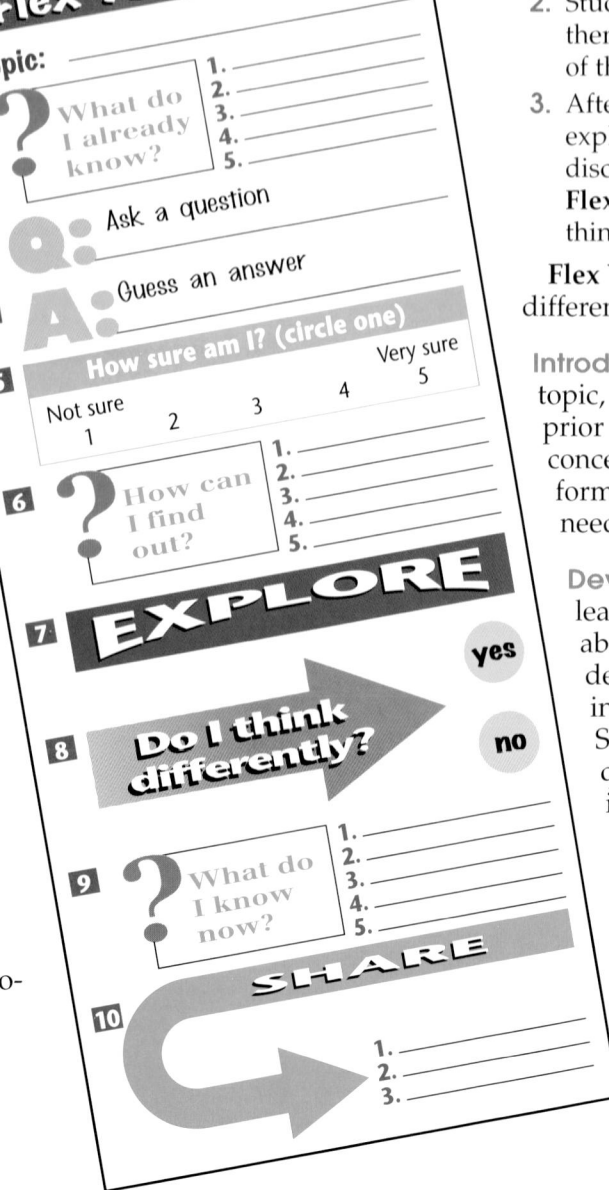

Using Flex Your Brain

Flex Your Brain can be used as a whole-class activity or in cooperative groups, but is primarily designed to be used by individual students within the class. There are three basic steps.

1. Teachers assign a class topic to be investigated using **Flex Your Brain.**
2. Students use **Flex Your Brain** to guide them in their individual explorations of the topic.
3. After students have completed their explorations, teachers guide them in a discussion of their experiences with **Flex Your Brain,** bridging content and thinking processes.

Flex Your Brain can be used at many different points in the lesson plan.

Introduction: Ideal for introducing a topic, **Flex Your Brain** elicits students' prior knowledge and identifies misconceptions, enabling the teacher to formulate plans specific to student needs.

Development: Flex Your Brain leads students to find out more about a topic on their own, and develops their research skills while increasing their knowledge. Students actually pose their own questions to explore, making their investigations relevant to their personal interests and concerns.

Review and Extension: Flex Your Brain allows teachers to check student understanding while allowing students to explore aspects of the topic that go beyond the material presented in class.

Concept Maps

In science, concept maps make abstract information concrete and useful, improve retention of information, and show students that thought has shape.

Concept maps are visual representations or graphic organizers of relationships among particular concepts. Concept maps can be generated by individual students, small groups, or an entire class. *Glencoe Physical Science* develops and reinforces four types of concept maps—the **network tree, events chain, cycle concept map,** and **spider concept map**—that are most applicable to studying science. Examples of the four types and their applications are shown on this page and page 33T. Students can learn how to construct each of these types of concept maps by referring to the **Skill Handbook.**

Building Concept Mapping Skills

The **Developing Skills** section of the **Chapter Review** provides opportunities for practicing concept mapping. A variety of concept mapping approaches is used in almost every chapter. Students may be directed to make a specific type of concept map and be provided the terms to use. At other times, students may be given only general guidelines. For example, concept terms to be used may be provided and students required to select the appropriate model to apply, or vice versa. Finally, students may be asked to provide both the terms and type of concept map to explain relationships among concepts. When students are given this flexibility, it is important for you to recognize that, while sample answers are provided, student responses may vary. Look for the conceptual strength of student responses, not absolute accuracy. You'll notice that most network tree maps provide connecting words that explain the relationships between concepts. We recommend that you not require all students to supply these words, but many students may be challenged by this aspect.

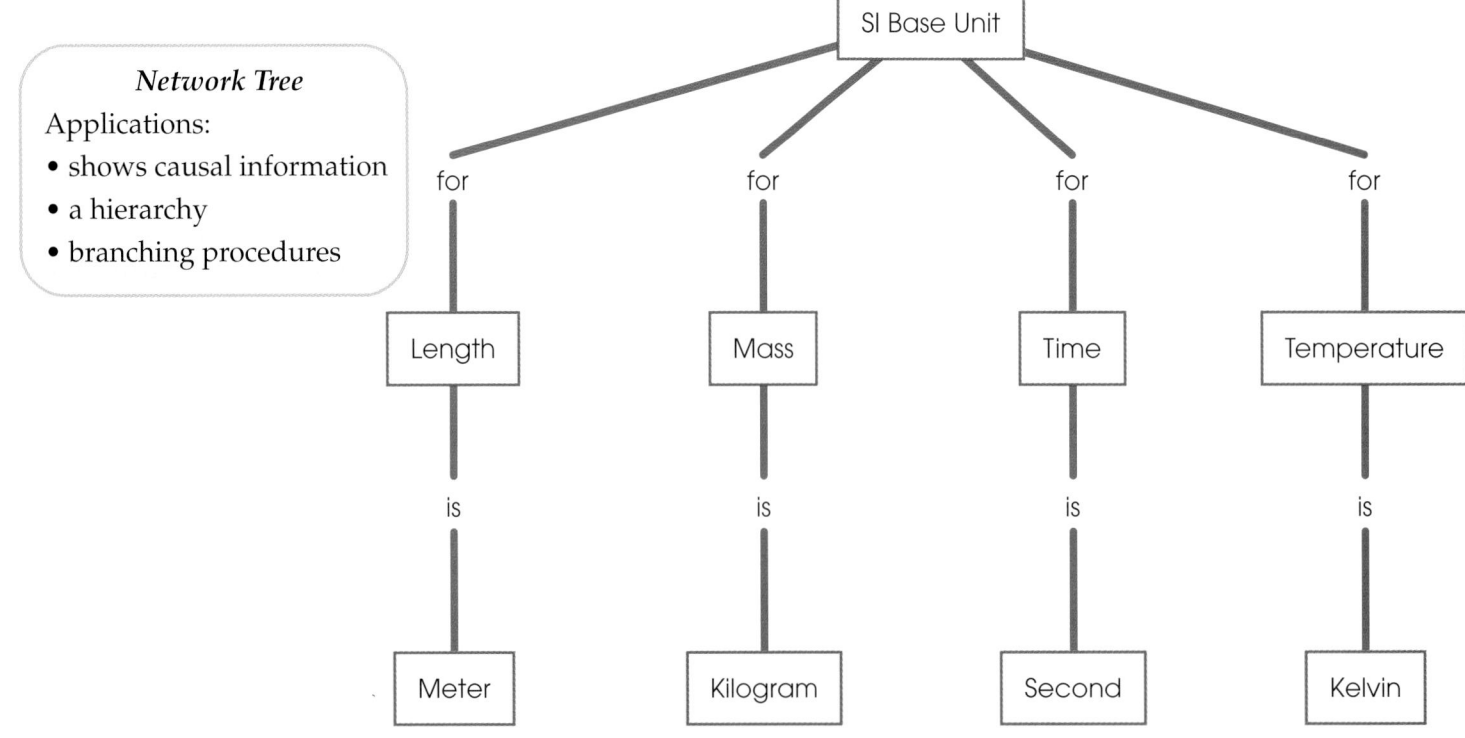

Network Tree
Applications:
• shows causal information
• a hierarchy
• branching procedures

Concept Mapping Booklet

The *Concept Mapping* book of the *Teacher Classroom Resources*, too, provides a developmental approach for students to practice concept mapping.

As a teaching strategy, generating concept maps can be used to preview a chapter's content by visually relating the concepts to be learned and allowing the students to read with purpose. Using concept maps for previewing is especially useful when there are many new key science terms for students to learn. As a review strategy, constructing concept maps reinforces main ideas and clarifies their relationships. Construction of concept maps using cooperative learning strategies as described in this Teacher Guide will allow students to practice both interpersonal and process skills.

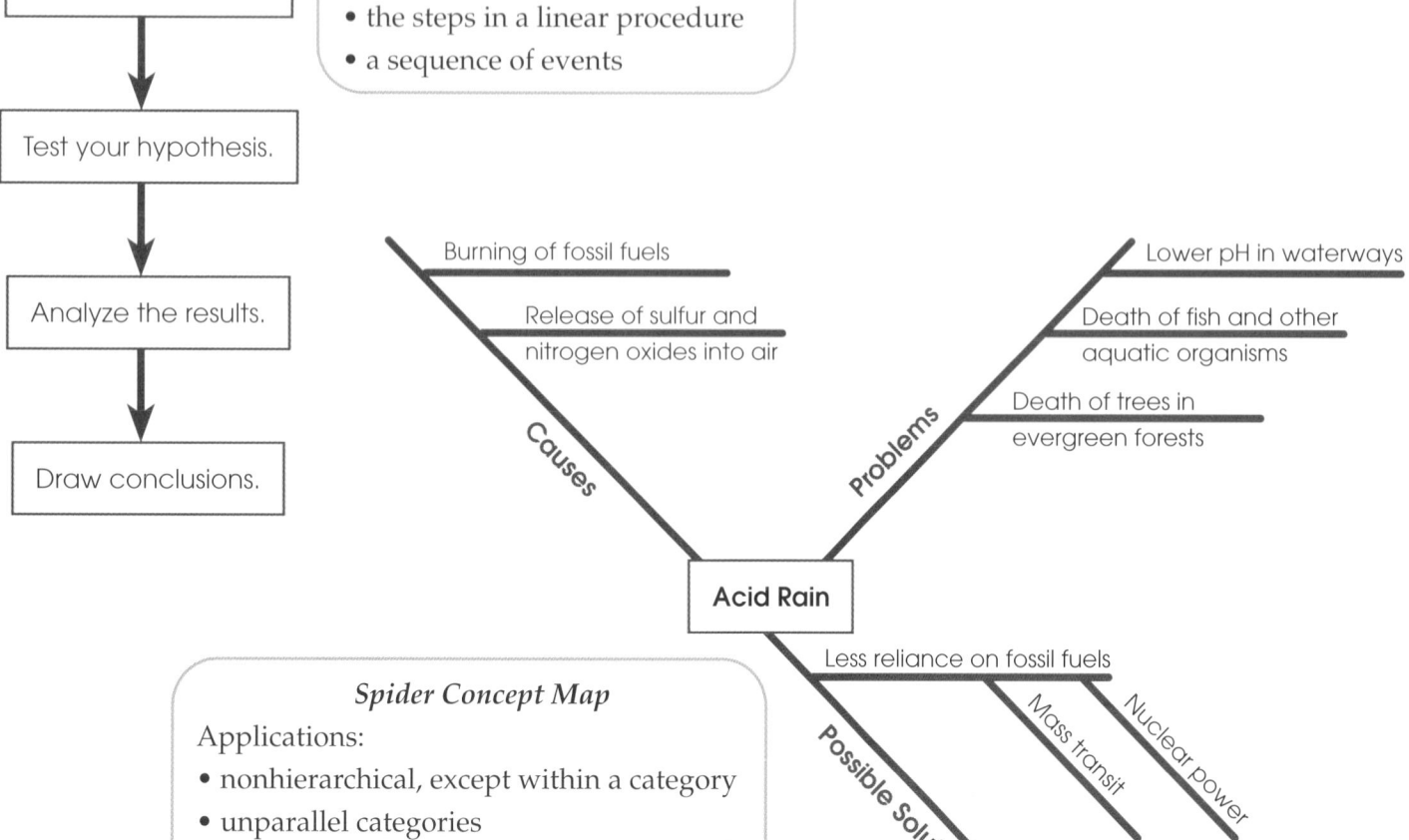

CO_2 + H_2O — Sunlight/Chlorophyll

Respiration — Energy released — Energy stored — Photosysthesis

O_2 + $C_6H_{12}O_6$

Cycle Concept Map

Application:
- shows how a series of events interacts to produce a set of results again and again

Initiating Event

Determine the problem.

Make a hypothesis.

Test your hypothesis.

Analyze the results.

Draw conclusions.

Events Chain

Applications:
- describes the stages of a process
- the steps in a linear procedure
- a sequence of events

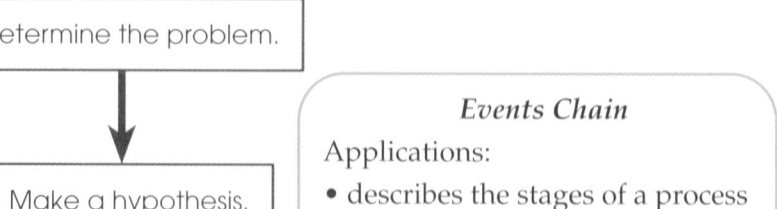

Burning of fossil fuels

Release of sulfur and nitrogen oxides into air

Lower pH in waterways

Death of fish and other aquatic organisms

Death of trees in evergreen forests

Causes

Problems

Acid Rain

Less reliance on fossil fuels

Mass transit

Nuclear power

Possible Solutions

Coal scrubbers

Spider Concept Map

Applications:
- nonhierarchical, except within a category
- unparallel categories
- brainstorming

Planning Your Course

Glencoe Physical Science provides flexibility in the selection of topics and content that allows teachers to adapt the text to the needs of individual students and classes. In this regard, the teacher is in the best position to decide what topics to present, the pace at which to cover the content, and what material to give the most emphasis. To assist the teacher in planning the course, a planning guide has been provided.

Glencoe Physical Science may be used in a full-year, two-semester course that is comprised of 180 periods of approximately 45 minutes each. This type of schedule is represented in the table under the heading of Single-Class Scheduling. As an alternative, the table also outlines a plan for block scheduling, which is described in the next section.

Block Scheduling

To build flexibility into the curriculum, many schools are introducing a block scheduling approach. Block scheduling often involves covering the same information in fewer days with longer class periods. Thus, a course that lasts an entire year may take only one semester under a block scheduling approach. This type of approach allows curriculum supervisors and teachers to tailor the curriculum to meet students' needs while achieving local and/or state curriculum goals. Long, concentrated periods of study can facilitate the learning of complex material. Furthermore, students may be able to take a wider variety of coursework under a block scheduling plan than under a traditional full-year plan, thus enriching their high-school experience and giving them a broader foundation for college-level work.

If you follow a block schedule, you may want to consider either combining lessons or eliminating certain topics and spending more time on the topics you do cover. *Glencoe Physical Science* provides the flexibility that allows you to tailor the program to your needs. *Glencoe Physical Science* also provides a wide variety of support materials that will help you and your students whether you follow a block schedule or full-year schedule. A block schedule may enable you to complete some of the longer **Activities** in one class period. Have students fill out the accompanying *Activity Worksheets* from the *Teacher Classroom Resources* as they complete the activities. The *Teacher Classroom Resources* provide a wealth of materials that provide classroom management support.

In the table shown here, it is assumed that for block scheduling the course will be taught for one semester and include 90 periods of approximately 90 minutes each.

Please remember that the planning guide is provided as an aid in planning the best course for your students. You should use the planning guide in relation to your curriculum and the ability levels of the classes you teach, the materials available for activities, and the time allotted for teaching.

Planning Guide for GLENCOE PHYSICAL SCIENCE — Scheduling

Unit	Chapter/Section	Single-Class (180 days)	Block (90 days)
UNIT 1	**Physical Science Basics**	14	7
1	THE NATURE OF SCIENCE	5	2
1-1	Science Is Everywhere	1	$\frac{1}{2}$
1-2	Finding Out	1	$\frac{1}{2}$
1-3	Science and Society: Getting Real with Special Effects	1	$\frac{1}{2}$
1-4	Exploring Science	1	
	Chapter Review	1	$\frac{1}{2}$
2	PHYSICAL SCIENCE METHODS	9	5
2-1	Standards of Measurement	2	1
2-2	Using SI Units	2	1
2-3	Graphing	3	2
2-4	Science and Society: SI for All?	1	$\frac{1}{2}$
	Chapter Review	1	$\frac{1}{2}$
UNIT 2	**Energy and Motion**	35	18
3	EXPLORING MOTION AND FORCES	7	4
3-1	Motion and Speed	1	$\frac{1}{2}$
3-2	Velocity and Acceleration	1	$\frac{1}{2}$
3-3	Science and Society: A Crash Course in Safety	$\frac{1}{2}$	$\frac{1}{2}$
3-4	Connecting Motion with Forces	2	1
3-5	Gravity—A Familiar Force	$1\frac{1}{2}$	1
	Chapter Review	1	$\frac{1}{2}$
4	ACCELERATION AND MOMENTUM	7	4
4-1	Accelerated Motion	1	1
4-2	Projectile and Circular Motion	2	1
4-3	Science and Society: Sending up Satellites	1	$\frac{1}{2}$
4-4	Action and Reaction	2	1
	Chapter Review	1	$\frac{1}{2}$

Planning Guide for GLENCOE PHYSICAL SCIENCE — Scheduling

Unit	Chapter/Section	Single-Class (180 days)	Block (90 days)
5	ENERGY	7	4
5-1	Energy and Work	2	1
5-2	Temperature and Heat	1	1
5-3	Science and Society: Thermal Pollution: Waste You Can't See	1	$1\frac{1}{2}$
5-4	Measuring Thermal Energy	2	1
	Chapter Review	1	$1\frac{1}{2}$
6	USING THERMAL ENERGY	7	3
6-1	Thermal Energy on the Move	2	1
6-2	Using Heat to Stay Warm	$1\frac{1}{2}$	$\frac{1}{2}$
6-3	Using Heat to Do Work	$1\frac{1}{2}$	$\frac{1}{2}$
6-4	Science and Society: Energy from the Oceans	1	$\frac{1}{2}$
	Chapter Review	1	$\frac{1}{2}$
7	MACHINES—MAKING WORK EASIER	7	3
7-1	Why We Use Machines	$\frac{1}{2}$	
7-2	The Simple Machines	$2\frac{1}{2}$	$1\frac{1}{2}$
7-3	Science and Society: Mending with Machines	1	
7-4	Using Machines	2	1
	Chapter Review	1	$\frac{1}{2}$
UNIT 3	**The Nature of Matter**	35	18
8	SOLIDS, LIQUIDS, AND GASES	10	4
8-1	Matter and Temperature	1	$\frac{1}{2}$
8-2	Science and Society: Fresh Water—Will There Be Enough?	1	$\frac{1}{2}$
8-3	Changes in State	2	1
8-4	Behavior of Gases	3	$1\frac{1}{2}$
8-5	Uses of Fluids	2	
	Chapter Review	1	$\frac{1}{2}$

Planning Guide

Planning Guide for GLENCOE PHYSICAL SCIENCE — Scheduling

Unit	Chapter/Section	Single-Class (180 days)	Block (90 days)
	17-4 Acids, Bases, and Salts	3	$1\frac{1}{2}$
	Chapter Review	1	$\frac{1}{2}$
UNIT 6	**Waves, Light, and Sound**	**18**	**11**
18	**WAVES AND SOUND**	6	4
	18-1 Characteristics of Waves	2	1
	18-2 The Nature of Sound	1	1
	18-3 Science and Society: Using Sound Advice in Medicine	1	$\frac{1}{2}$
	18-4 Music to Your Ears	1	1
	Chapter Review	1	$\frac{1}{2}$
19	**LIGHT**	6	4
	19-1 Electromagnetic Radiation	1	1
	19-2 Light and Color	1	$\frac{1}{2}$
	19-3 Science and Society: Battle of the Bulbs	1	$\frac{1}{2}$
	19-4 Wave Properties of Light	2	$1\frac{1}{2}$
	Chapter Review	1	$\frac{1}{2}$
20	**MIRRORS AND LENSES**	6	3
	20-1 The Optics of Mirrors	1	$\frac{1}{2}$
	20-2 The Optics of Lenses	2	1
	20-3 Optical Instruments	$\frac{1}{2}$	$\frac{1}{2}$
	20-4 Science and Society: The Hubble Space Telescope	$\frac{1}{2}$	
	20-5 Applications of Light	1	$\frac{1}{2}$
	Chapter Review	1	$\frac{1}{2}$
UNIT 7	**Electricity and Energy Resources**	**35**	**14**
21	**ELECTRICITY**	8	4
	21-1 Electric Charge	1	$\frac{1}{2}$
	21-2 Science and Society: To Burn or Not	1	
	21-3 Electric Current	$1\frac{1}{2}$	1

Planning Guide for GLENCOE PHYSICAL SCIENCE — Scheduling

Unit	Chapter/Section	Single-Class (180 days)	Block (90 days)
	21-4 Electrical Circuits	$2\frac{1}{2}$	$1\frac{1}{2}$
	21-5 Electrical Power and Energy	1	$\frac{1}{2}$
	Chapter Review	1	$\frac{1}{2}$
22	**MAGNETISM AND ITS USES**	9	4
	22-1 Characteristics of Magnets	2	1
	22-2 Uses of Magnetic Fields	2	1
	22-3 Producing Electric Current	3	1
	22-4 Science and Society: Superconductivity	1	$\frac{1}{2}$
	Chapter Review	1	$\frac{1}{2}$
23	**ELECTRONICS AND COMPUTERS**	5	3
	23-1 Semiconductor Devices	1	1
	23-2 Radio and Television	1	$\frac{1}{2}$
	23-3 Microcomputers	1	$\frac{1}{2}$
	23-4 Science and Society: Computer Crimes	1	$\frac{1}{2}$
	Chapter Review	1	$\frac{1}{2}$
24	**RADIOACTIVITY AND NUCLEAR REACTIONS**	7	3
	24-1 Radioactivity	1	$\frac{1}{2}$
	24-2 Nuclear Decay	1	$\frac{1}{2}$
	24-3 Detecting Radioactivity	1	
	24-4 Nuclear Reactions	2	1
	24-5 Science and Society: Using Nuclear Reactions in Medicine	1	$\frac{1}{2}$
	Chapter Review	1	$\frac{1}{2}$
25	**ENERGY SOURCES**	6	
	25-1 Fossil Fuels	1	
	25-2 Nuclear Energy	1	
	25-3 Science and Society: Nuclear Waste and NIMBY	1	
	25-4 Alternative Energy Sources	2	
	Chapter Review	1	

Using Technology in the Classroom

Technology helps you adapt your teaching methods to the needs of your students. Glencoe classroom technology products provide many pathways to help you match students' different learning styles. To make your lesson planning easier, all of the technology products listed below are correlated to the student text.

Videodiscs

Glencoe's *Integrated Science Videodiscs* are designed to be used interactively in the classroom. Barcodes in this book allow you to step through the programs and pause to discuss and answer on-screen questions. Barcodes for the following videodiscs also appear throughout this teacher edition: *Mr. Wizard's Science and Technology Videodisc, Infinite Voyage, Newton's Apple,* and *National Geographic Society.*

MindJogger Videoquiz

A videoquiz for each chapter can be used to assess prior knowledge or review content before an exam. Student teams work cooperatively to answer three rounds of questions posed by the video host. Each round requires higher-level thinking skills than the previous round.

Glencoe Interactive CD-ROMs

The *Glencoe Physical Science* CD-ROM is correlated to the Student Edition.

You can:
- use a CD-ROM interaction as a whole class presentation;
- allow 2-3 small groups to use the CD-ROMs at once;
- rotate student groups through a single computer station;
- place the materials in a computer lab; or
- use CD-ROMs as a library resource.

Glencoe also offers three National Geographic Society CD-ROMs. These image-rich, reference CD-ROMs are faithful to the Society's long history of excellence in science teaching and journalism.

Computer Test Bank

Glencoe's Test Generator for Macintosh and for DOS makes creating, editing, and printing tests quick and easy. You also can edit questions or add your own favorite questions and graphics.

English/Spanish Audiocassettes

Audio chapter summaries in English and in Spanish are a way for auditory learners, lower-level readers, and LEP students to review key chapter concepts. Students can listen individually during class or check out tapes and use them at home. You may find them useful for reviewing the chapter as you plan lessons.

Glencoe Software Support Hotline **1-800-437-3715**

Should you encounter any difficulty when setting up or running Glencoe software, contact the Software Support Center at Glencoe Publishing between 8:30 a.m. and 4:30 p.m. Eastern Time.

Using the Internet

If you're already familiar with the Internet, skip to the sites listed at the bottom of this page. If you need some tips on how to get started, keep reading.

The Internet is an enormous reference library and a communication tool. You can use it to retrieve information quickly from computers around the globe. Like any good reference, it has an index so you can locate the right piece of information. An Internet index entry is called a *Universal Resource Locator*, or URL. Here's an example:

http://www.glencoe.com/intro/index.html

The first part of the URL tells the computer how to display the information. The second part, after the double slash, names the organization and the computer where the information is stored. The part after the first single slash tells the computer which directory to search and which file to retrieve. File locations change frequently. If you can't find what you're looking for, use the first part of the address only, for ex-

ample, http://www.glencoe.com/ from the URL shown, and follow links to what you need.

The World Wide Web

The World Wide Web (WWW), a subset of the Internet, began in 1992. Unlike regular text files, web files can have links to other text files, images, and sound files. By clicking on a link, you can see or hear the linked information.

How do I get access?

To use the Internet, you need a computer, a modem, a telephone line, and a connection to the Internet. If your school doesn't have a connection, contact your local public library or a university; they often give free access to students and educators.

CAUTION: Contents may shift!

The sites referenced in Glencoe's Internet Connections are not under Glencoe's control. Therefore,

Glencoe can make no representation concerning the content of these sites. Extreme care has been taken to list only reputable links by using educational and government sites whenever possible. Internet searches have been used that return only sites that contain no content apparently intended for mature audiences.

Where to Start

The brief list of science Internet sites below may prove useful. You can also find Internet addresses throughout this book correlated to selected features.

Useful tools for searching the Internet include:
http://www.yahoo.com/search.html
http://www.altavista.digital.com and
http://www.msn.com/access/
    allinone/hv1

Science Internet Site	Description
http://ericir.syr.edu/	**Ask ERIC**, an ask-the-expert service for K-12 teachers
http://www.enc.org/	**The Eisenhower National Clearinghouse for Math and Science**, instructional materials
http://medinfo.wustl.edu/~ysp/MSN/	**The Mad Scientist Information Network**, an ask-the-expert service for science students

For more information about Glencoe Technology, call Customer Service at **1-800-334-7344.**

Cultural Diversity

American classrooms reflect the rich and diverse cultural heritage of the American people. Students come from different ethnic backgrounds and different cultural experiences into a common classroom that must assist all of them in learning. The diversity itself is an important focus of the learning experience.

Diversity can be repressed, creating a hostile environment; ignored, creating an indifferent environment; or appreciated, creating a receptive and productive environment. Responding to diversity and approaching it as a part of every curriculum is challenging to a teacher, experienced or not. The goal of science is understanding. The goal of multicultural education is to promote the understanding of how people from different cultures approach and solve the basic problems all humans have in living and learning.

Cultural Diversity in Science

Every culture has utilized science to explore fundamental questions, meet challenges, and address human needs. The growth of knowledge in science has advanced along culturally diverse roots. No single culture has a monopoly on the development of scientific knowledge. The history of science is rich with the contributions and accomplishments of men and women from diverse cultural backgrounds. A brief look at some of the milestones in science reveals this diversity. Cultural groups on all continents have devised methods for measuring time and space. For example, the Mayan civilization, which flourished in what is now southern Mexico, Guatemala, and Belize, developed sophisticated and accurate mathematical systems and calendars. Lewis Latimer, an African-American draftsman and inventor, contributed to the development of the electric lightbulb and drew the plans that resulted in the 1876 patent of the telephone. Today, individuals from diverse cultural and ethnic backgrounds continue to make significant contributions to science. The **People and Science** feature in the *Student Edition* of *Glencoe Physical Science* highlights some of these individuals and their contributions.

The scientific enterprise is a human enterprise that utilizes science processes (observing and inferring, classifying, and so on) to discover, explore, and explain the environment. Consequently, students must experience and understand that all people, regardless of the cultural orientation, have made major contributions to science. *Glencoe Physical Science* addresses this issue. In addition to the individuals highlighted in the *Student Edition*, the **Cultural Diversity** sections of the *Teacher Wraparound Edition* provide information about people and groups who have traditionally been misrepresented or omitted. The intent is to build awareness and appreciation for the global community in which we all live.

Goals of Multicultural Education

The *Glencoe Physical Science Teacher Classroom Resources* include a *Multicultural Connections* booklet that offers additional opportunities to integrate multicultural materials into the curriculum. By providing these opportunities, *Glencoe Physical Science* is helping to meet four major goals of multicultural education:

1. promoting the strength and value of cultural diversity
2. promoting human rights and respect for those who are different from oneself
3. promoting social justice and equal opportunity for all people
4. promoting equity in the distribution of power among groups

These goals are accomplished when students see others like themselves in positive settings in their textbooks. They also are accomplished when students realize that major contributions in science are the result of the work of hundreds of individuals from diverse backgrounds. *Glencoe Physical Science* has been designed to enable you, the science teacher, to achieve these goals.

Books that provide additional information on multicultural education are

Atwater, Mary, et al. *Multicultural Education: Inclusion of All.* Athens, Georgia: University of Georgia Press, 1994.

Banks, James A. (with Cherry A. McGee Banks). *Multicultural Education: Issues and Perspectives.* Boston: Allyn and Bacon, 1989.

Selin, Helaine. *Science Across Cultures: An Annotated Bibliography of Books on Non-Western Science, Technology, and Medicine.* New York: Garland, 1992.

Meeting Individual Needs

Each student brings his or her own unique set of abilities, perceptions, and needs into the classroom. It is important that the teacher try to make the classroom environment as receptive to these differences as possible and to ensure a good learning environment for all students.

It is important to recognize that individual learning styles are different and that learning style does not reflect a student's ability level. While some students learn primarily through visual or auditory senses, others are kinesthetic learners and do best if they have hands-on exploratory interaction with materials. Some students work best alone and others learn best in a group environment. While some students seek to understand the "big picture" in order to deal with specifics, others need to understand the details first in order to put the whole concept together.

In an effort to provide all students with a positive science experience, this text offers a variety of ways for students to interact with materials so that they can utilize their preferred method of learning the concepts. The variety of approaches allows students to become familiar with other learning approaches, as well.

Ability Levels

The activities are broken down into three levels to accommodate all student ability levels. *Glencoe Physical Science Teacher Wraparound Edition* designates the activities as follows:

L1 activities are basic activities designed to be within the ability range of all students. These activities reinforce the concepts presented.

L2 activities are application activities designed for students who have mastered the concepts presented. These activities give students an opportunity for practical application of the concepts presented.

L3 activities are challenging activities designed for the students who are able to go beyond the basic concepts presented. These activities allow students to expand their perspectives on the basic concepts presented.

The chart on pages 42T-43T gives tips you may find useful in structuring the learning environment in your classroom to meet students' special needs. In addition, **Inclusion Strategies** that address special needs are provided in the bottom margins of the *Teacher Wraparound Edition.*

Limited English Proficiency

In providing for the student with limited English proficiency, the focus needs to be on overcoming a language barrier. Once again, it is important not to confuse ability in speaking/reading English with academic ability or "intelligence." In general, the best method for dealing with LEP, variations in learning styles, and ability levels is to provide all students with a variety of ways to learn, apply, and be assessed on the concepts. Look for this symbol **LEP** in the teacher margin for specific strategies for students with limited English proficiency.

Learning Styles

We at Glencoe believe it is our responsibility to provide you with a program that allows you to apply diverse instructional strategies to a population of students with diverse learning styles. Several learning styles are emphasized in the *Teacher Wraparound Edition:* Kinesthetic, Visual-Spatial, Logical-Mathematical, Interpersonal, Intrapersonal, Linguistic, and Auditory-Musical. A student with a kinesthetic style learns from touch, movement, and manipulating objects. Visual-spatial learners respond to images and illustrations. Using numbers and reasoning are characteristics of the logical-mathematical learner. A student with an interpersonal style has confidence in social settings, while a student with an intrapersonal style may prefer to learn on his or her own. Linguistic learning involves the use and understanding of words. Finally, auditory-musical learning involves listening to the spoken word and to tones and rhythms.

Any student may display any or all of these styles depending on the learning situation. The *Student Edition* and *Teacher Wraparound Edition* provide a number of strategies for encouraging students with diverse learning styles. These include **Using Math, Activities, MiniLABS,** and **Science Journal** features. The *Teacher Wraparound Edition* contains **Visual Learning** and **Science Journal,** as well as a number of other strategies. Look for this logo **LS** that identifies strategies for different learning styles.

Meeting Individual Needs

	Description	Sources of Help/Information
Learning Disabled	All learning-disabled students have an academic problem in one or more areas, such as academic learning, language, perception, social-emotional adjustment, memory, or attention.	*Journal of Learning Disabilities* *Learning Disability Quarterly*
Behaviorally Disordered	Children with behavior disorders deviate from standards or expectations of behavior and impair the functioning of others and themselves. These children may also be gifted or learning-disabled.	*Exceptional Children* *Journal of Special Education*
Physically Challenged	Children who are physically disabled fall into two categories—those with orthopedic impairments and those with other health impairments. Orthopedically impaired children have the use of one or more limbs severely restricted, so the use of wheelchairs, crutches, or braces may be necessary. Children with other health impairments may require the use of respirators or other medical equipment.	Batshaw, M.L. and M.Y. Perset. *Children with Handicaps: A Medical Primer.* Baltimore: Paul H. Brooks, 1981. Hale, G. (Ed.). *The Source Book for the Disabled.* New York: Holt, Rinehart & Winston, 1982. *Teaching Exceptional Children*
Visually Impaired	Children who are visually disabled have partial or total loss of sight. Individuals with visual impairments are not significantly different from their sighted peers in ability range or personality. However, blindness may affect cognitive, motor, and social development, especially if early intervention is lacking.	*Journal of Visual Impairment and Blindness* *Education of Visually Handicapped* American Foundation for the Blind
Hearing Impaired	Children who are hearing impaired have partial or total loss of hearing. Individuals with hearing impairments are not significantly different from their hearing peers in ability range or personality. However, the chronic condition of deafness may affect cognitive, motor, and social development if early intervention is lacking. Speech development also is often affected.	*American Annals of the Deaf* *Journal of Speech and Hearing Research* *Sign Language Studies*
Limited English Proficiency	Multicultural and/or bilingual children often speak English as a second language or not at all. The customs and behavior of people in the majority culture may be confusing for some of these students. Cultural values may inhibit some of these students from full participation.	*Teaching English as a Second Language Reporter* R.L. Jones (Ed.). *Mainstreaming and the Minority Child.* Reston, VA: Council for Exceptional Children, 1976.
Gifted	Although no formal definition exists, these students can be described as having above-average ability, task commitment, and creativity. Gifted students rank in the top five percent of their class. They usually finish work more quickly than other students, and are capable of divergent thinking.	*Journal for the Education of the Gifted* *Gifted Child Quarterly* *Gifted Creative/Talented*

Strategies

Tips for Instruction
With careful planning, the needs of all students can be met in the science classroom.

1. Provide support and structure; clearly specify rules, assignments, and duties.
2. Practice skills frequently. Use games and drills to help maintain student interest.
3. Allow students to record answers on tape and allow extra time to complete tests and assignments.
4. Provide outlines or tape lecture material.
5. Pair students with peer helpers, and provide class time for pair interaction.

1. Provide a clearly structured environment with regard to scheduling, rules, room arrangement, and safety.
2. Clearly outline objectives and how you will help students obtain objectives.
3. Reinforce appropriate behavior and model it for students.
4. Do not expect immediate success. Instead, work for long-term improvement.
5. Balance individual needs with group requirements.

1. Openly discuss with the student any uncertainties you have about when to offer aid.
2. Ask parents or therapists and students what special devices or procedures are needed, and if any special safety precautions need to be taken.
3. Allow physically disabled students to do everything their peers do, including participating in field trips, special events, and projects.
4. Help nondisabled students and adults understand physically disabled students.

1. Help the student become independent. Modify assignments as needed.
2. Teach classmates how to serve as guides.
3. Limit unnecessary noise in the classroom.
4. Provide tactile models whenever possible.
5. Describe people and events as they occur in the classroom.
6. Provide taped lectures and reading assignments.
7. Team the student with a sighted peer for laboratory work.

The *Student Edition* and *Teacher Wraparound Edition* contain a variety of strategies for addressing the individual learning styles of students:

- Both structured and open-ended **Activities**
- Hands-on **MiniLABS**
- **Enrichment** activities
- **Visual Learning** strategies
- **Across the Curriculum** strategies

The following Glencoe products also can help you tailor your instruction to meet the individual needs of your students:

- **Study Guide**
- **Teaching Transparencies**
- **Section Focus Transparencies**
- **Concept Mapping**
- **MindJogger Videoquiz**
- **English/Spanish Audiocassettes**
- **Critical Thinking/Problem Solving**
- **Cooperative Learning Resource Guide**
- Various multimedia products, including **Interactive Videodisc Program, The Secret of Life Series, The Infinite Voyage Series, Science and Technology Videodisc Series, National Geographic Society Series**

1. Seat students where they can see your lip movements easily, and avoid visual distractions.
2. Avoid standing with your back to the window or light source.
3. Using an overhead projector allows you to maintain eye contact while writing.
4. Seat students where they can see speakers.
5. Write all assignments on the board, or hand out written instructions.
6. If the student has a manual interpreter, allow both student and interpreter to select the most favorable seating arrangements.

1. Remember, students' ability to speak English does not reflect their academic ability.
2. Try to incorporate the student's cultural experience into your instruction. The help of a bilingual aide may be effective.
3. Include information about different cultures in your curriculum to help build students' self-image. Avoid cultural stereotypes.
4. Encourage students to share their cultures in the classroom.

1. Make arrangements for students to take selected subjects early and to work on independent projects.
2. Let students express themselves in art forms such as drawing, creative writing, or acting.
3. Make public services available through a catalog of resources, such as agencies providing free and inexpensive materials, community services and programs, and people in the community with specific expertise.
4. Ask "what if" questions to develop high-level thinking skills. Establish an environment safe for risk taking.
5. Emphasize concepts, theories, ideas, relationships, and generalizations.

Assessment

What criteria do you use to assess your students as they progress through a course? Do you rely on formal tests and quizzes? To assess students' achievement in science, you need to measure not only their knowledge of the subject matter, but also their ability to handle apparatus; to organize, predict, record, and interpret data; to design experiments; and to communicate orally and in writing. *Glencoe Physical Science* has been designed to provide you with a variety of assessment tools, both formal and informal, to help you develop a clearer picture of your students' progress.

Performance Assessment

Performance assessments are becoming more common in today's schools. Science curricula are being revised to prepare students to cope with change and with futures that will depend on their abilities to think, learn, and solve problems. Although learning fundamental concepts will always be important in the science curriculum, the concepts alone are no longer sufficient in a student's scientific education.

Defining Performance Assessment

Performance assessment is based on judging the quality of a student's response to a performance task. A performance task is constructed to require the use of important concepts with supporting information, work habits important to science, and one or more of the elements of scientific literacy. The performance task attempts to put the student in a "real-world" context so that the class learning can be put to authentic uses.

Performance Assessment Is Also a Learning Activity

Performance assessment is designed to improve the student. While a traditional test is designed to take a snapshot of what the student knows, the performance task involves the student in work that actually makes the learning more meaningful and builds on the student's knowledge and skills. As a student is engaged in a performance task with performance assessment lists and examples of excellent work, both learning and assessment are occurring.

Performance Task Assessment Lists

Performance task assessment lists break the assessment criteria into several well-defined categories. Possible points for each category are assigned by the teacher. Both the teacher and the student assess the work and assign the number of points earned. The teacher is scoring not only the quality of the product, but also the quality of the student's self-assessment.

Besides using the performance task assessment list to help guide his or her work and then to self-assess it, the student needs to study examples of excellent work. These examples could come from published sources, or they could be students' work. These examples could include the same type of product for different topics, but they would all be rated excellent using the assessment list for that product type.

Performance Assessment in *GLENCOE PHYSICAL SCIENCE*

Many performance task assessment lists can be found in Glencoe's *Performance Assessment in the Science Classroom.* These lists were developed for the summative and skill performance tasks in the *Performance Assessment* book that accompanies the *Glencoe Physical Science* program. The *Performance Assessment* book contains a mix of skill assessments and summative assessments that tie together major concepts of each chapter. The lists can be used to support **MiniLABS, Activities,** and the **Performance Assessment** problems in the **Chapter Review.**

Glencoe's *Alternate Assessment in the Science Classroom* provides additional background and examples of performance assessment. Activity sheets in the *Activity Worksheets* book provide yet another vehicle for formal assessment of student products. The **MindJogger Videoquiz** series offers interactive videos that provide a fun way for your students to review chapter concepts. You can extend the use of the videoquizzes by implementing them in a testing situation. Questions are at three difficulty levels: basic, intermediate, and advanced.

Assessment and Grading

Assessment is giving the learner feedback on the individual elements of his or her performance or product. Therefore, assessment provides specific information on strengths and weaknesses and allows the student to set targets for improvement. Grading is an act to evaluate the overall quality of the performance or product according to some norms of quality. From the performance task assessment list, the student can see the quality of the pieces. The points on an assessment list can be summed and an overall grade awarded. The grade alone is not very helpful to the student, but it does describe overall quality of the performance or product. The information from both the assessment list and the grade should be reported to the student, parents, and other audiences.

Assessing Student Work with Rubrics

A rubric is a set of descriptions of the quality of a process and/or a product. The set of descriptions includes a continuum of quality from excellent to poor. Rubrics for various types of assessment products are in the Glencoe Professional Development Series booklet *Performance Assessment in the Science Classroom.*

When to Use the Performance Task Assessment List and When to Use the Rubric

Performance task assessment lists are used for all or most performance tasks. If a grade is also necessary, it can be derived from the total points earned on the assessment list. Rubrics are used much less often. Their best use is to help students periodically assess the overall quality of their work. After making a series of products, the student is asked how he or she is doing overall on one of these types of products. With reference to the standards of quality set for that product at that grade level, the student can decide where on the continuum of quality his or her work fits. The student is asked to assign a rubric score and explain why the score was chosen. Because the student has used a performance task assessment list to examine the elements of each of the products, he or she can justify the rubric score. Experience with performance task assessment lists comes before use of a rubric.

Group Performance Assessment

Recent research has shown that a cooperative learning environment improves student learning outcomes for students of all ability levels. *Glencoe Physical Science* provides many opportunities for cooperative learning and, as a result, many opportunities to observe group work processes and products. *Glencoe Physical Science: Cooperative Learning Resource Guide* provides strategies and resources for implementing and evaluating group activities. In cooperative group assessment, all members of the group contribute to the work process and the products it produces. For example, if a mixed-ability, four-member laboratory work group conducts an activity, you can use a rating scale or checklist to assess the quality of both group interaction and work skills. An example, along with information about evaluating cooperative work, is provided in the booklet *Alternate Assessment in the Science Classroom.*

The Science Journal

A science journal is intended to help the student organize his or her thinking. It is not a lecture or laboratory notebook. It is a place for students to make their thinking explicit in drawings and writing. It is the place to explore what makes science fun and what makes it hard.

Portfolios: Putting It All Together

The portfolio should help the student see the "big picture" of how he or she is performing in gaining knowledge and skills and how effective his or her work habits are. The portfolio is a way for students to see how the individual performance tasks done during the year fit into a pattern that reveals the overall quality of their learning. The process of assembling the portfolio should be both integrative (of process and content) and reflective. The performance portfolio is not a complete collection of all worksheets and other assignments for a grading period. At its best, the portfolio should include integrated performance products that show growth in concept attainment and skill development.

The Portfolio: Criteria for Success

To be successful, a science portfolio should

1. improve the student's performance in science.
2. promote the student's skills of self-assessment and goal setting.
3. promote a sense of ownership and pride of accomplishment in the student.
4. be a reasonable amount of work for the teacher.
5. be highly valued by the teacher receiving the portfolio the next year.
6. be useful in parent conferences.

The Portfolio: Its Contents

Evidence of the student's growth in the following five categories should be included.

1. Range of thinking and creativity
2. Use of scientific method
3. Inventions and models
4. Connections between science and other subjects
5. Readings in science

For each of the five categories, students should make indexes that describe what they have selected for their portfolios and why. From their portfolio selections, have students pick items that show the quality of their work habits. Finally, have each student write a letter to next year's science teacher introducing him- or herself to the teacher and explaining how the work in this section shows what the student can do.

Opportunities for Using Science Journals and Portfolios

Glencoe Physical Science presents a wealth of opportunities for performance portfolio development. Each chapter in the student text contains enrichment activities and connections with life, society, and literature. Each of the student activities results in a product. A mixture of these products can be used to document student growth during the grading period. Descriptions, examples, and assessment criteria for portfolios are discussed in *Alternate Assessment in the Science Classroom.* Glencoe's *Performance Assessment in the Science Classroom* contains even more information on using science journals and making portfolios. Performance task assessment lists and rubrics for both journals and portfolios are found there.

Content Assessment

While new and exciting performance skill assessments are emerging, paper-and-pencil tests are still a mainstay of student evaluation. Students must learn to conceptualize, process, and prepare for traditional content assessments. Presently and in the foreseeable future, students will be required to pass pencil-and-paper tests to exit high school and to enter college, trade schools, and other training programs.

Glencoe Physical Science contains numerous strategies and formative checkpoints for evaluating student progress toward mastery of science concepts. Throughout the chapters in the student text, **Section Wrap-up** questions and application tasks are presented. This spaced review process helps build learning bridges that allow all students to confidently progress from one lesson to the next.

For formal review that precedes the written content assessment, *Glencoe Physical Science* presents a three-page **Chapter Review** at the end of each chapter. By evaluating the student responses to this extensive review, you can determine whether any substantial reteaching is needed.

For the formal content assessment, a two-page review and a four-page chapter test are provided for each chapter in the *Review and Assessment* book. Using the review in a whole-class session, you can correct any misperceptions and provide closure for the text. If your individual assessment plan requires a test that differs from the chapter test in the resource package, customized tests can be produced easily using the *Computer Test Bank.*

Tech-Prep Education

What Is Tech Prep?

Tech prep is a rigorous and focused program of study that aims to create a workforce in the United States that is technically literate. It is designed to prepare students enrolled in a general curriculum for the demands of further education or for employment by providing them with essential academic and technical foundations, along with problem-solving, group-process, and lifelong-learning skills. These goals can be achieved by integrating vocational study with higher-level academic study.

What Are the Characteristics of the Tech Prep Curriculum?

In 1990, Congress passed the Carl D. Perkins Vocational and Applied Technology Act to set aside funds for the development and administration of Tech Prep programs. The criteria outlined in Title III of the Perkins Act specify that tech prep programs take place during the last two years of high school, followed by two years of post-secondary occupational education, and that this education culminate in a certificate or associate degree. The Secretary's Commission on Achieving Necessary Skills (SCANS), an arm of the United States Department of Labor, published a report in June 1991 that outlined several competencies that characterize successful workers. The Tech Prep curriculum seeks to address these competencies, which include the

- ability to use resources productively.
- ability to use interpersonal skills effectively, including fostering teamwork, teaching others, serving customers, leading, negotiating, and working well with individuals from culturally diverse backgrounds.
- ability to acquire, evaluate, interpret, and communicate data and information.
- ability to understand social, organizational, and technological systems.
- ability to apply technology to specific tasks.

How Does Tech Prep Relate to the Middle School Curriculum?

Tech prep provides an alternative to the college-prep course of study that leads to a career goal.

Because of this, tech prep is available to all students whether they are taking a college prep, vocational, or general course of study. The middle school years provide an opportunity to identify those students who might benefit from a tech-prep curriculum once they reach high school. They also provide an opportunity to introduce unfamiliar students to technological applications leading to career opportunities, or to provide practice for students who already have some knowledge of how technology is used. Young children are fascinated by the technology that surrounds them, and they are especially quick to learn how to use and apply it. Take advantage of their enthusiasm by preparing them for a possible tech-prep education when they are in middle school.

How Does *GLENCOE PHYSICAL SCIENCE* Address Tech Prep Issues?

Glencoe Physical Science helps you develop scientific and technological literacy in your students through a variety of performance-based activities that emphasize problem solving, critical thinking skills, and teamwork. In the *Student Edition*, **Problem Solving, Using Technology,** and **Design Your Own Experiment** provide opportunities for practical applications of concepts. **Using Computers** and **Using Math** features give additional practice in computer and math applications. **Unit Projects** provide ideas for activities that can be used as extended projects or for science fairs.

Collecting weather data using computers

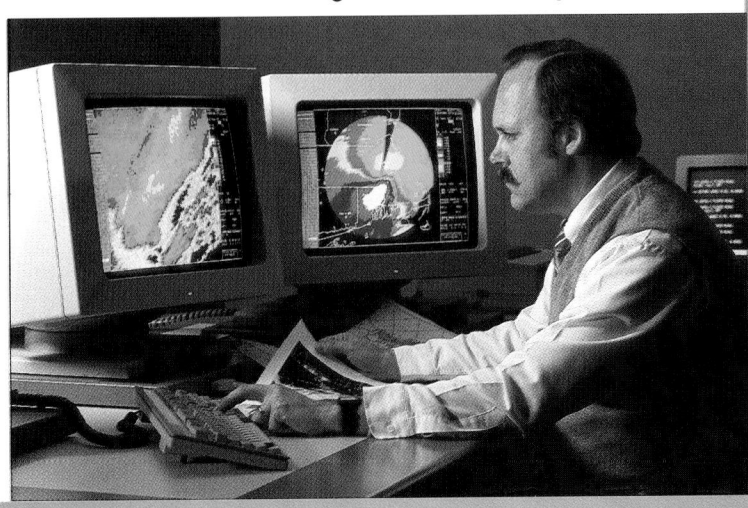

Cooperative Learning

What Is Cooperative Learning?

In cooperative learning, students work together in small groups to learn academic material and interpersonal skills. Group members each learn that they are responsible for accomplishing an assigned group task as well as for learning the material. Cooperative learning fosters academic, personal, and social success for all students.

Recent research shows that cooperative learning results in

- development of positive attitudes toward science and toward school.
- lower drop-out rates for at-risk students.
- building respect for others regardless of race, ethnic origin, or sex.
- increased sensitivity to and tolerance of diverse perspectives.

Establishing a Cooperative Classroom

Cooperative groups in the high school usually contain from two to five students. Heterogeneous groups that represent a mixture of abilities, genders, and ethnicity expose students to ideas different from their own and help them learn to work with different people.

Initially, cooperative learning groups should work together for only a day or two. After the students are more experienced, they can work with a group for longer periods of time. It is important to keep groups together long enough for each group to experience success and to change groups often enough that students have the opportunity to work with a variety of students.

Students must understand that they are responsible for group members learning the material. Before beginning, discuss the basic rules for effective cooperative learning: (1) listen while others are speaking, (2) respect other people and their ideas, (3) stay on tasks, and (4) be responsible for your own actions.

The *Teacher Wraparound Edition* uses the code **COOP LEARN** at the end of activities and teaching ideas where cooperative learning strategies are useful. For additional help, refer again to these pages of background information on cooperative learning.

Using Cooperative Learning Strategies

The *Cooperative Learning Resource Guide* of the *Teacher Classroom Resources* provides help for selecting cooperative learning strategies, as well as methods for troubleshooting and evaluation.

During cooperative learning activities, monitor the functioning of groups. Praise group cooperation and good use of interpersonal skills. When students are having trouble with the task, clarify the assignment, reteach, or provide background as needed. Answer questions only when no students in the group can.

Evaluating Cooperative Learning

At the close of the lesson, have groups share their products or summarize the assignment. You can evaluate group performance during a lesson by frequently asking questions to group members picked at random or having each group take a quiz together. You might have all students write papers and then choose one at random to grade. Assess individual learning by your traditional methods.

Managing Activities

In the Science Classroom

Glencoe Physical Science engages students in a variety of experiences to provide all students with an opportunity to learn by doing. The many hands-on activities throughout *Glencoe Physical Science* require simple, common materials, making them easy to set up and manage in the classroom.

MiniLAB

MiniLABS are intended to be short and occur many times throughout the text. The integration of these activities with the core material provides for thorough development and reinforcement of concepts.

Design Your Own Experiment

What makes science exciting to students is working in the lab, observing natural phenomena, and tackling concrete problems that challenge them to find their own answers. *Glencoe Physical Science* provides more than the same "cookbook" activities you've seen hundreds of times before. Students work cooperatively to develop their own experimental designs in **Design Your Own Experiments.** They discover firsthand that developing procedures for studying a problem is not as hard as they thought it might be. Watch your students grow in confidence and ability as they progress through the self-directed **Design Your Own Experiments.**

Preparing Students for Open-Ended Lab Experiences

To prepare students for the **Design Your Own Experiments,** you should follow the guidelines in the *Teacher Wraparound Edition,* especially in the sections titled Possible Procedures and Teaching Strategies. In these sections, you will be given information about what demonstrations to do and what questions to ask students so that they will be able to design their experiment. Your introduction to a **Design Your Own Experiment** will be very different from traditional activity introductions in that it will be designed to focus students on the problem and stimulate their thinking without giving them directions for how to set up their experiment. Different groups of students will develop alternative hypotheses and alternative procedures. Check their procedures before they begin. In contrast to some "cookbook" activities, there may not be just one right answer. Finally, students should be encouraged to use questions that come up during the activity in designing a new experiment.

If you feel that your students are not prepared to begin designing experiments, you may want to practice some paper-and-pencil lab designs using the following format: Select a simple scenario and ask students to design an experiment that will test a hypothesis they make about the problem presented in the reading. Give them a set of questions that will lead them to an appropriate experimental design.

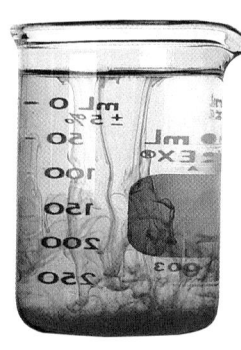

Laboratory Safety

Safety is of prime importance in every classroom. However, the need for safety is even greater when science is taught. The activities in *Glencoe Physical Science* are designed to minimize dangers in the laboratory. Even so, there are no guarantees against accidents. Careful planning and preparation as well as being aware of hazards can keep accidents to a minimum. Numerous books and pamphlets are available on laboratory safety with detailed instructions on preventing accidents. In addition, the *Glencoe Physical Science* program provides safety guidelines in several forms. The *Lab and Safety Skills* booklet contains detailed guidelines, in addition to masters you can use to test students' lab and safety skills. The *Student Edition* and *Teacher Wraparound Edition* provide safety precautions and symbols designed to alert students to possible dangers. Know the rules of safety and what common violations occur. Know the **Safety Symbols** used in this book. A safety symbol chart is on page 731. Know where emergency equipment is stored and how to use it. Practice good laboratory housekeeping and management to ensure the safety of your students.

It is most important to use safe laboratory techniques when handling all chemicals. Many substances may appear harmless but are, in fact, toxic, corrosive, or very reactive. Always check with the manufacturer or with Flinn Scientific, Inc., (312) 879-6900. Chemicals should never be ingested. Be sure to use proper techniques to smell solutions or other agents. Always wear safety goggles and an apron. The following general cautions should be used.

1. Poisonous/corrosive liquid and/or vapor. Use in the fume hood. Examples: *acetic acid, hydrochloric acid, ammonia hydroxide, nitric acid.*

2. Poisonous and corrosive to eyes, lungs, and skin. Examples: *acids, limewater, iron(III) chloride, bases, silver nitrate, iodine, potassium permanganate.*

3. Poisonous if swallowed, inhaled, or absorbed through the skin. Examples: *glacial acetic acid, copper compounds, barium chloride, lead compounds, chromium compounds, lithium compounds, cobalt(II) chloride, silver compounds.*

4. Always add acids to water, never the reverse.

5. When sulfuric acid or sodium hydroxide is added to water, a large amount of thermal energy is released. Sodium metal reacts violently with water. Use extra care when handling any of these substances.

Unless otherwise specified, solutions are prepared by adding the solid to a small amount of distilled water and then diluting with water to the volume listed. For example, to make a $0.1M$ solution of aluminum sulfate, dissolve 34.2 g of $Al_2(SO_4)_3$ in a small amount of distilled water and dilute to a liter with water. If you use a hydrate that is different from the one specified in a particular preparation, you will need to adjust the amount of the hydrate to obtain the required concentration.

Chemical
Storage & Disposal

General Guidelines

Be sure to store all chemicals properly. The following are guidelines commonly used. Your school, city, county, or state may have additional requirements for handling chemicals. It is the responsibility of each teacher to become informed as to what rules or guidelines are in effect in his or her area.

1. Separate chemicals by reaction type. Strong acids should be stored together. Likewise, strong bases should be stored together and should be separated from acids. Oxidants should be stored away from easily oxidized materials, and so on.

2. Be sure all chemicals are stored in labeled containers indicating contents, concentration, source, date purchased (or prepared), any precautions for handling and storage, and expiration date.

3. Dispose of any outdated or waste chemicals properly, according to accepted disposal procedures.

4. Do not store chemicals above eye level.

5. Wood shelving is preferable to metal. All shelving should be firmly attached to the wall and should have antiroll edges.

6. Store only those chemicals that you plan to use.

7. Hazardous chemicals require special storage containers and conditions. Be sure to know what those chemicals are and the accepted practices for your area. Some substances must even be stored outside the building.

8. When working with chemicals or preparing solutions, observe the same general safety precautions that you would expect from students. These include wearing an apron and goggles. Wear gloves and use the fume hood when necessary. Students will want to do as you do whether they admit it or not.

9. If you are a new teacher in a particular laboratory, it is your responsibility to survey the chemicals stored there and to be sure they are stored properly or disposed of. Consult the rules and laws in your area concerning what chemicals can be kept in your classroom. For disposal, consult up-to-date disposal information from the state and federal governments.

Disposal of Chemicals

Local, state, and federal laws regulate the proper disposal of chemicals. These laws should be consulted before chemical disposal is attempted. Although most substances encountered in middle school science can be flushed down the drain with plenty of water, it is not safe to assume that this is always true. It is recommended that teachers who use chemicals consult the following books from the National Research Council:

Prudent Practices for Handling Hazardous Chemicals in Laboratories. Washington, DC: National Academy Press, 1981.

Prudent Practices for Disposal of Chemicals from Laboratories. Washington, DC: National Academy Press, 1983.

These books are useful and still in print, although they are several years old. Current laws in your area would, of course, supersede the information in these books.

Disclaimer

Glencoe Publishing Company makes no claims to the completeness of this discussion of laboratory safety and chemical storage. The material presented is not all-inclusive, nor does it address all of the hazards associated with handling, storage, and disposal of chemicals, or with laboratory management.

Using a Graphing Calculator

What is it?

What does it do?

How is it going to help me learn math?

These are just a few of the questions many students ask themselves when they first see a graphing calculator. Some students may think, "Oh, no! Do we *have* to use one?" while others may think, "All right! We get to use these neat calculators!" There are as many thoughts and feelings about graphing calculators as there are students, but one thing is for sure: a graphing calculator can help you learn mathematics and science.

So what is a graphing calculator? Very simply, it is a calculator that draws graphs. This means that it will do all of the things that a "regular" scientific calculator will do, *plus* it will draw graphs of equations. This capability is helpful as you analyze graphs.

But a graphing calculator can do more than just calculate and draw graphs. For example, you can program it or work with data to make statistical graphs and perform computations. If you need to generate random numbers, you can do that on the graphing calculator. You can even draw and manipulate geometric figures on some graphing calculators. It's really a very powerful tool—so powerful that it is often called a pocket computer.

As you may have noticed, graphing calculators have some keys that other calculators do not. These instructions are for the Texas Instruments TI-82 and TI-83 calculators. Most of the keystrokes for the TI-82 and TI-83 are the same. The keys on the bottom half of the TI-82 and TI-83 are those found on scientific calculators. The keys located just below the screen are the graphing keys. You will also notice the up, down, left, and right arrow keys. These allow you to move the cursor around on the screen, to "trace" graphs that have been plotted, and to choose items from the menus. The other keys located on the top half of the calculator access the special features such as statistical computations and programming features.

A few of the keystrokes that can save you time when using the graphing calculator are listed below.

- The commands above the calculator keys are accessed with the `2nd` or `ALPHA` key. On a TI-82, the `2nd` key and the commands accessed by it are blue and the `ALPHA` key and its commands are gray. On a TI-83, the `2nd` key and its commands are yellow and the `ALPHA` and its commands are green.

- `2nd` [ENTRY] copies the previous calculation so you can edit and use it again.

- Pressing `ON` while the calculator is graphing stops the calculator from completing the graph.

- `2nd` [QUIT] will return you to the home (or text) screen.

- `2nd` [A-LOCK] locks the `ALPHA` key, which is like pressing "shift lock" or "caps locks" on a typewriter or computer. The result is that all letters will be typed and you do not have to repeatedly press the `ALPHA` key. (This is handy for programming.) Stop typing letters by pressing `ALPHA` again.

- `2nd` [OFF] turns the calculator off.

Some commonly used mathematical functions are shown in the table below. As with any scientific calculator, the graphing calculator observes the order of operations.

Mathematical Operation	Examples	Keys	Display
evaluate expressions	Find 2 + 5.	2 [+] 5 [ENTER]	2+5 7
exponents	Find 3^5.	3 [^] 5 [ENTER]	3^5 243
multiplication	Evaluate 3(9.1 + 0.8).	3 [×] [(] 9.1 [+] .8 [)] [ENTER]	3(9.1+.8) 29.7
roots	Find $\sqrt{14}$.	[2nd] [√] 14 [ENTER]	√14 3.741657387
opposites	Enter −3.	[(−)] 3	−3

Graphing on the TI-82 or TI-83

Before graphing, we must instruct the calculator how to set up the axes in the coordinate plane. To do this, we define a **viewing window.** The viewing window for a graph is the portion of the coordinate grid that is displayed on the graphics screen of the calculator. The viewing window is written as [left, right] by [bottom, top] or [Xmin, Xmax] by [Ymin, Ymax]. A viewing window of [−10, 10] by [−10, 10] is called the **standard viewing window** and is a good viewing window to start with to graph an equation. The standard viewing window can be easily obtained by pressing [ZOOM] 6. Try this. Move the arrow keys around and observe what happens. You are seeing a portion of the coordinate plane that includes the region from −10 to 10 on the x-axis and from −10 to 10 on the y-axis. Move the cursor, and you can see the coordinates of the point for the position of the cursor.

Any viewing window can be set manually by pressing the [WINDOW] key. The window screen will appear and display the current settings for your viewing window. Using the arrow and [ENTER] keys, move the cursor to edit the window settings. Xscl and Yscl refer to the x-scale and y-scale. This is the number of units between tick marks on the x- and y-axes. Xscl=1 means that there will be a tick mark for every unit of one along the x-axis.

Programming on the TI-82 or TI-83

The TI-82 and TI-83 have programming features that allow us to write and execute a series of commands to perform tasks that may be too complex or cumbersome to perform otherwise. Each program is given a name. Commands begin with a colon (:), which the calculator enters automatically, followed by an expression or an instruction. Most of the features of the calculator are accessible from program mode.

When you press [PRGM], you see three menus: EXEC, EDIT, and NEW. EXEC allows you to execute a stored program, EDIT allows you to edit or change a program, and NEW allows you to create a program. The following tips will help you as you enter and run programs on the TI-82.

- To begin entering a new program, press [PRGM] [▶] [▶] [ENTER].

- After a program is entered, press [2nd] [QUIT] to exit the program mode and return to the home screen.

- To execute a program, press [PRGM]. Then use the down arrow key to locate the program name and press [ENTER], or press the number or letter next to the program name.

- If you wish to edit a program, press [PRGM] [▶] and choose the program from the menu.

- To immediately re-execute a program after it is run, press [ENTER] when "Done" appears on the screen.

- To stop a program during execution, press [ON].

While a calculator cannot do everything, it can make some things easier. To prepare for whatever lies ahead, you should try to learn as much as you can. The future will definitely involve technology, and using a graphing calculator is a good start toward becoming familiar with technology. Who knows? Maybe one day you will be designing the next satellite, building the next skyscraper, or helping students learn mathematics and science with the aid of a graphing calculator!

Activity Materials

Get the materials you need quickly and easily!

Glencoe and Science Kit, Inc. have teamed up to make materials selection for *Glencoe Physical Science* easier with an activity materials folder. This folder contains two convenient ways to order materials and equipment for the program: the **Activity Plan Checklist** and the **Activity Materials List** master.

Call Science Kit at 1-800-828-7777 to get your folder.

Materials Support Provided By:
Science Kit® & Boreal®Laboratories
Your Classroom Resource
777 East Park Drive
Tonawanda, NY 14150-6784
800-828-7777
Fax 716-874-9572

Non-Consumables

Item	Activity	MiniLAB	Explore
Apron	8-1, 9-2, 11-2, 12-1, 13-1, 14-1, 15-1, 15-2, 16-1, 16-2, 17-1, 17-2	331, 342, 470	329
Balance	2-1, 3-2, 5-2, 7-2	44, 443, 701	465
Bar magnet	22-1	627	623
Batteries, (2) C or D cells		603	
Batteries, dry-cell, 6-V	21-2, 22-2, 23-1	630	
Beaker(s), 250-mL	3-2, 6-1, 8-2, 14-2, 15-2, 16-2, 21-1	260, 331, 367, 470	297, 415, 465, 673
Beaker(s), 100-mL	25-1	419, 701	329
Board, wooden, 100 cm long		201	
Book, old		97	
Books		201, 566	123, 151, 179
Bottle, dropper		443	
Building bricks (2)	4-2		
Burner, laboratory	11-2	153, 331	329, 359
Calculators		661	
Clay, modeling	19-2, 24-2, 25-1	248, 715	269, 673
Clips, alligator		616	
Cloth or yarn, small piece	6-2		
Coins (quarter, nickel, dime)	7-2		
Compact disc (CD)			527
Compass, magnetic		630	
Conductivity tester	12-1	468	
Container, large, plastic	19-2	715	527
Cooking pot	2-1		
Cork	4-1		
Cup(s), foam, large		532	415
Cup(s), foam, small	5-2		93
Cup(s), foam with lid		48, 135	
Cup(s), paper	8-1		
Cup(s), plastic	3-2		63, 245, 415
Cup(s), short, opaque		548	
Diffraction grating	19-1		
Diode	23-1		
Dominoes		690	
Dropper(s)	8-1, 13-1, 17-1, 17-2	225, 235, 342, 367, 441, 484, 566	
Dynamics carts (2)	4-2		
Egg carton, modified	11-1		
Electric toy with battery		616	
Evaporating dish, small	9-2, 14-1		
Filters, plastic, red and green		539	
Filters, 2 polarizing	20-2		
Flashlight		559	527

Equipment List

Non-Consumables

Item	Activity	MiniLAB	Explore
Flask, 250-mL	6-2		
Forceps	25-1		
Funnel, plastic	21-1		465
Galvanometer or milliammeter	22-1	616	
Glass prism			527
Goggles, safety	1-1, 3-1, 8-1, 8-2, 9-2, 11-2, 12-1, 12-2, 13-1, 14-1, 14-2, 15-1, 15-2, 16-1, 16-2, 17-1, 17-2, 18-1, 20-2, 25-1	331, 342, 367, 470	245, 329, 359, 441
Graduated cylinder, 10-mL	9-2, 13-1, 15-1, 16-1	367	
Graduated cylinder, 100-mL	2-1, 5-2, 8-1, 16-2, 17-1, 17-2	44, 223, 701	465
Granite, coarse-grained	9-1		
Hair curlers, set, electric	5-2		
Hammer	12-1, 14-2		
Hand lens, powerful	9-2		647
Hardware (small nails, screws, paper clips, ball bearings)			269
Hot plate	2-1, 8-2, 9-2, 14-1, 15-1, 15-2, 16-2	367	
Jar, glass, 1-gallon capacity		248	
Lamp with bare bulb	23-1		
LEDs, 2 different colors			151
Lightbulb(s)		603	
Lightbulb(s), clear, tubular, and socket	19-1		
Lightbulb(s), incandescent and fluorescent, same wattage		532	
Lightbulb(s), low voltage	22-2		
Lightbulb(s), small with sockets	21-2	630	
Light source	19-2, 20-2		
Magnetic board, about 20 cm × 27 cm	10-1		
Magnetic tape, rubber	10-1		
Magnets, 16 small, disk-type	23-2		
Marbles	11-1		673
Markers, colored	2-2, 8-2, 10-1		
Marking pen	7-1, 13-2, 25-2	452	
Mass, 100-g		81, 627	
Measuring cup, English units	2-1, 14-2		
Measuring spoons, teaspoon and tablespoon	2-1		
Metal object such as a hanger or spoon		506	
Meterstick	3-1, 4-2, 5-1, 7-1, 8-2, 18-1, 21-1	74, 201	179
Microwave oven		393	
Microwave-safe bowl, about 400-mL capacity		393	
Mirrors, concave, large		559	
Mirrors, flat	20-1		557
Model car, small		74	

Item	Activity	MiniLAB	Explore
Model car, wind-up type		201	
Mortar and pestle		419	
Nail, large	22-2		
Pan, shallow		277	
Paper clips, metal	3-1, 13-2, 14-2, 20-1, 21-2, 22-2	627	329, 623
Pencil(s)	7-1, 19-2	44, 74, 288, 500, 569, 715	415
Pencil(s), colored	19-1	539	
Pennies, copper	14-1, 24-1	548, 566	63, 441
Pens		288	
Pinch clamps (2)	6-2		
Pipet		484	
Plastic bottle, 2-L, clean, clear, empty, with cap	4-1	235	
Plastic container, 1-quart	2-1		
Plastic freezer bag containing tagged items	9-1		
Plastic pipes, various lengths	18-2		
Polystyrene sheets, thin	12-2		
Power supply, AC, low-voltage (6-V)	22-2, 23-1		
Power supply, with dimmer switch	19-1		
Protractor	20-1		
PVC pipe or broom handles, (2) 1-m lengths		190	
Radio(s)		520, 655	
Resistor, 1000 ohms	23-1		
Ring stand(s)	1-1, 15-2		
Ring stand(s), with clamp		627	
Ring stand(s), with ring	3-2, 5-1, 21-1	701	
Rope, 1-m		190	
Rubber bands, long	4-2, 18-2	616	
Rubber bands, small	25-1	14	
Rubber tubing	6-2, 21-1		
Ruler, metric	1-1, 6-1, 7-2, 18-2, 23-2	101, 164, 201, 500, 559, 569	151, 497, 699
Sandpaper, coarse-grained, 1 sheet		81	
Saucer			441, 591
Scissors	1-1, 2-2, 3-2, 12-2, 17-1, 21-2, 22-1, 22-2	97	699
Spatula	12-1	153	
Spiral spring, coiled	18-1		
Spoon, large		393, 559	
Spoon, small, plastic	16-2		
Spring scale		201	
Stapler	3-1		
Stick, wooden	8-1	715	
Stirring rod, copper wire	15-1		
Stirring rod(s), glass	6-1	153, 260, 367, 419, 484	
Stopper, 2-hole, rubber, medium	5-1		
Stopper, 2-hole, rubber with glass tubing	6-2		
Stopwatch or timer with second hand	1-1, 3-1, 11-2, 17-1, 18-1, 21-1	48, 74, 153, 164, 201, 225, 367, 419, 532	673

Equipment List

Non-Consumables

Item	Activity	MiniLAB	Explore
Straight pins			699
String	1-1, 2-2, 3-1, 3-2, 5-1, 8-2	112, 164, 506	
Support rod, 30-cm	5-1		
Support rod clamp, right angle	5-1		
Syringe, large, plastic	6-2		
"T," glass or plastic	6-2		
Television or monitor			5
Test tube, 1 medium		367	
Test tube(s), several large, identical	11-2, 15-1, 16-1, 16-2, 17-1, 17-2	342	297, 359
Test tube(s), with stopper	13-1	569	
Test-tube holder, wire	11-2, 15-1	367	359
Test-tube rack	15-1, 16-1, 16-2		
Thermal mitt	2-1, 8-2, 9-2, 11-2, 15-1, 19-1, 19-2, 20-2	153, 367	
Thermometer(s), Celsius	5-2, 6-1, 8-2, 15-1, 15-2	48, 135, 367, 532, 701	151
Thread	4-1	627	
Thumbtacks		153	
Tongs	6-1, 8-2, 14-1, 16-1	331	329
Twist ties	25-1		
Video camera			5
Videotapes			5
Washers, 2 identical, metal		101	
Water faucet		306	
Weights, 3 small, identical	1-1		
Window fan			699
Wire, 4-cm-long pieces		331	
Wire gauze		701	
Wire, insulated, 32-gauge	22-2		
Wire, insulated	22-1, 23-1	603, 616, 630	
Wool			591
Wool or fur, small piece		306	

Consumables

Item	Activity	MiniLAB	Explore
Aluminum foil, heavy duty	9-1, 14-2, 21-2, 25-1	164, 701	591
Balloon(s), long, slender		112	93
Balloon(s), small, inflated		306	591
Balloon(s), (2) round, medium	8-2		

Consumables

Item	Activity	MiniLAB	Explore
Binder-hole reinforcing stickers	23-2		
Bread, small piece			359
Candle(s)		701	213
Candy, chocolate-center, green and red	10-2		
Candy, peanut-center, green and red	10-2		
Cardboard tube	22-1		
Chalk, small piece, finely crushed	9-1		
Charcoal, powdered			415
Chocolate, small piece		164	
Cloth (heavy) or foam pads		520	
Coat hangers, wire		164	
Cocoa	2-1		
Copper foil	9-1		
Cornstarch		484	
Crayons	14-2	539	
Cream of tartar, 5 mL		393	
Dishwashing liquid soap			527
Drinking straw(s), plastic	14-2, 22-2	112	
Filter paper		223	465
Flour		277, 393	
Food coloring	8-1		
Fruit juices, different kinds		484	
Glue		164	
Gravel		248	415
Gumdrops, small, soft, unsugared	12-2	363	
Index cards, small		452, 559, 569	63
Index cards, 20 small, unlined		14	
Light cooking oil, 5 mL		393	
Litmus paper		223	
Marble chips, 5.0 g			465
Margarine	2-1		
Matches	6-2		213
Metals, small pieces	16-1		
Milk	2-1		
Miscellaneous construction objects such as empty paper towel or toilet paper rolls, glue, cellophane tape, aluminum foil, empty cardboard milk cartons	25-2		
Musical greeting card			647
Newspaper, sheets	3-2		
Note cards, 2 different colors	13-2		
Nuts	2-1		
Orange juice		468	
Paper, cardboard sheet (22 cm × 28 cm), thin, marked with rectangles	17-1		

Consumables

Item	Activity	MiniLAB	Explore
Paper, card stock	12-2, 23-2		
Paper, graph	5-1, 5-2, 24-1	48	
Paper, notebook, small piece		627	63, 359
Paper, poster-sized	7-1		
Paper, red 1-cm circles	10-1		
Paper, sheet, 20 cm × 28 cm (8 1/2" × 11")	7-2, 19-2		
Paper, sheet(s), plain white	10-1, 17-1	81, 97, 500	151, 699
Paper, sheets of various types	3-1		
Paper, strips of different types	3-2		
Paper, tissue, small piece(s)		101	591
Paper towels	14-2		441
Pebbles, small		248	245
Pepper			591
pH paper	17-1		
Pipe cleaners	14-2		
Plaster of paris	14-2		
Plastic food wrap	25-1	532, 566	441
Poster board	25-2	164	93
Raisins		363	
Rolled oats	2-1		
Salt, table	6-1, 9-1, 11-2, 15-2	393	441, 591
Sand	3-2, 16-2	135, 248, 277	63, 415
Sand, fine-grained (20 g)			245
Soap, small plug		443	
Soda water, 50 mL			465
Soft drinks, clear, 2 different kinds	17-2		
Soil		248	
Solder, small piece	9-1		
Spaghetti, thin	12-2		
Steel wool (without soap)			297
Sugar, cubes		419	
Sugar, granulated	2-1, 9-1, 11-2, 15-1		
Tape, masking	2-2, 3-2, 4-2, 20-1, 20-2	112	
Tape, transparent	3-1, 20-1, 21-2, 23-2	593, 627	93
Toothpicks	12-2	363	269
Vanilla flavoring	2-1		
Vinegar		223	441
Water, distilled	15-1, 15-2	342	329
Wax		153	
Wooden splint	16-1, 16-2		

Chemical Supplies

Item	Activity	MiniLAB	Explore
Acetic acid, dilute	17-1		
Aluminum	12-1		
Ammonia-based household cleaner, solution		468	
Baking soda (NaHCO$_3$)	9-1, 9-2		
Borax solution, 4%	8-1		
Calcium carbonate, 10 g crushed			245
Carbon	12-1		
Copper(II) sulfate			329
Ethanol	13-1		
Hydrochloric acid, dilute	9-2, 16-1, 17-1		
Hydrogen peroxide, 3%	16-2		
Iodine solution		484	
Iron(III) chloride solution		443	
Magnesium	12-1		
Manganese dioxide	16-2		
Methyl alcohol		367	
Nitric acid, dilute	14-1		
Phenolphthalein indicator solution, 1%	17-2, 25-1		
Polyvinyl alcohol (PVA) solution, 4%	8-1		
Potassium permanganate	13-1	260	
Rubbing alcohol		225	
Salicylic acid		367	
Silver nitrate solution		342	
Sodium chloride		342	329
Sodium hydrogen sulfite		260	
Sodium hydroxide	13-1, 14-1, 17-2, 25-1	443	
Strontium chloride			329
Sulfur	12-1		
Sulfuric acid, concentrated		367	
Tin	12-1		
Zinc, 30-mesh	14-1		

References

Supplier Addresses

Software Distributors

(AIT) Agency for Instructional Technology
Box A
Bloomington, IN 47402-0120

Aims Media
9710 Desoto Avenue
Chatsworth, CA 91311-4409

Aquarium Instructional
P.O. Box 128
Indian Rocks Beach, FL 34635

Cambridge Development Lab (CDL)
1696 Massachusetts Avenue
Cambridge, MA 02138

(Classroom Consortia Media Inc.)
Gemstar
P.O. Box 050228
Staten Island, NY 10305

COMPress
P.O. Box 102
Wentworth, NH 03282

Computer Software

Earthware Computer Services
P.O. Box 30039
Eugene, OR 97403

Educational Activities, Inc.
1937 Grand Avenue
Baldwin, NY 11510

Educational Materials and Equipment
Company (EME)
P.O. Box 2805
Danbury, CT 06813-2805

Focus Media, Inc.
839 Stewart Avenue
P.O. Box 865
Garden City, NY 11530

Human Relations Media (HRM)
175 Tompkins Avenue
Pleasantville, NY 10570

IBM Educational Systems
Department PC
4111 Northside Parkway
Atlanta, GA 30327

J & S Software
135 Haven Avenue
Port Washington, NY 11050

McGraw-Hill Webster Division
1221 Avenue of the Americas
New York, NY 10020

Microphys
1737 W. Second Street
Brooklyn, NY 11223

MicroPower and Light Co.
Suite 120
12820 Hillcrest Road 219
Dallas, TX 75230

Minnesota Educational Computing
Corporation (MECC)
3490 Lexington Avenue N.
Saint Paul, MN 55126

Queue, Inc.
562 Boston Avenue
Bridgeport, CT 06610

Texas Instruments, Data Systems Group
P.O. Box 1444
Houston, TX 77251

Ventura Educational System
3440 Brokenhill Street
Newbury Park, CA 91320

Audiovisual Distributors

Aims Media
9710 Desoto Avenue
Chatsworth, CA 91311-4409

BFA Educational Media
468 Park Avenue S.
New York, NY 10016

Churchill Films
662 N. Robertson Blvd.
Los Angeles, CA 90069

Coronet/MTI Film and Video
Distributors of LCA
108 Wilmot Road
Deerfield, IL 60015

CRM Films
2233 Faraday Avenue
Suite F
Carlsbad, CA 92008

Diversified Education Enterprise
725 Main Street
Lafayette, IN 47901

Encyclopedia Britannica Educational
Corp. (EBEC)
310 S. Michigan Avenue
Chicago, IL 60604

Focus Media, Inc.
839 Stewart Avenue
P.O. Box 865
Garden City, NY 11530

Hawkill Associates, Inc.
125 E. Gilman Street
Madison, WI 53703

(HRM) Human Relations Media
175 Tompkins Avenue
Pleasantville, NY 10570

Indiana University
Audiovisual Center
Bloomington, IN 47405-5901

Journal Films, Inc.
930 Pitner Avenue
Evanston, IL 60202

Lumivision
1490 Lafayette
Suite 305
Denver, CO 80218

National Earth Science Teachers
c/o Art Weinle
733 Loraine
Grosse Point, MI 48230

National Geographic Society Educational
Services
17th and "M" Streets, NW
Washington, DC 20036

Science Software Systems
11890 W. Pico Blvd.
Los Angeles, CA 90064

Time-Life Videos
Time and Life Building
1271 Avenue of the Americas
New York, NY 10020

Universal Education & Visual Arts
(UEVA)
100 Universal City Plaza
Universal City, CA 91608

U.S. Department of Energy
DOE 535
Washington, DC

Video Discovery
1515 Dexter Avenue N.
Suite 400
Seattle, WA 98109

Glencoe

PHYSICAL SCIENCE

A GLENCOE PROGRAM

Glencoe Physical Science

Student Edition
Teacher Wraparound Edition
Study Guide SE and TE
Reinforcement SE and TE
Enrichment SE and TE
Concept Mapping
Critical Thinking/Problem Solving
Activity Worksheets
Chapter Review
Chapter Review Software
Lab Manual SE and TE
Science Integration Activities
Cross-Curricular Integration
Science and Society Integration

Transparency Packages:
    Teaching Transparencies
    Section Focus Transparencies
    Science Integration Transparencies

Technology Integration
Multicultural Connection
Performance Assessment
Review and Assessment
Lesson Plans
Spanish Resources
MindJogger Videoquizzes and
    Teacher Guide
English/Spanish Audiocassettes
CD-ROM Multimedia System
Interactive Videodisc Program
Computer Testbank—
    DOS and Macintosh Versions

The Glencoe Science Professional Development Series
    Performance Assessment in the Science Classroom
    Lab and Safety Skills in the Science Classroom
    Cooperative Learning in the Science Classroom
    Alternate Assessment in the Science Classroom
    Exploring Environmental Issues

Cover Photograph: The roller coaster on the cover was photographed in Phoenix, Arizona by Ken Ross/FPG International.

Glencoe/McGraw-Hill

A Division of The McGraw·Hill Companies

Send all inquiries to:
Glencoe/McGraw-Hill
8787 Orion Place
Columbus, OH 43240

ISBN 0-02-827567-5
Printed in the United States of America.
8 9 10 11 12 071/046 05 04 03 02 01 00

Authors

Charles W. McLaughlin has more than 25 years of experience as a high school chemistry teacher and is currently the coordinator of science education for the St. Joseph, Missouri school district. He holds a Ph.D. in chemistry from the University of Nebraska. The National Science Teachers Association has presented Dr. McLaughlin with three national awards for innovative science education, and he received a Presidential Award for Science Teaching.

Marilyn Thompson is currently teaching and pursuing graduate work in science education at the University of Kansas. Her career has included teaching physics, chemistry and physical science while serving as science department chair at Center High School in Kansas City, Missouri. Ms. Thompson received her B.A. degree in chemistry from Carleton College and her M.A. degree in science curriculum and instruction from the University of Kansas. She has conducted several summer workshops and courses in the physical sciences for K-12 teachers. Ms. Thompson is a member of NSTA and AAPT.

Contributing Writers

Linda Barr
Freelance Writer
Westerville, Ohio

Dan Blaustein
Science Teacher and Author
Evanston, Illinois

Pam Bliss
Freelance Science Writer
Grayslake, Illinois

Mary Dylewski
Freelance Science Writer
Houston, Texas

Nancy Ross-Flanigan
Syndicated Science Reporter
Detroit, Michigan

Helen Frensch
Freelance Writer
Santa Barbara, California

Steve Glazer
Freelance Writer
Champaign, Illinois

Rebecca Johnson
Freelance Science Writer
and Author
Sioux Falls, South Dakota

Devi Mathieu
Freelance Science Writer
Sebastopol, California

Patricia West
Freelance Writer
Oakland, California

Consultants

Physics

Donald Bord, Ph.D.
Professor of Physics
University of Michigan-Dearborn
Dearborn, Michigan

Patrick Hamill, Ph.D.
Professor of Physics
San Jose State University
San Jose, California

Chemistry

Teresa Anne McCowen
Senior Lecturer
Butler University
Indianapolis, Indiana

Robert C. Smoot, M.S.
Rollins Teacher Fellow in Science
McDonough School
Owings Mills, Maryland

Multicultural

Karen Muir, Ph.D.
Lead Instructor
Department of Social and Behavioral Sciences
Columbus State Community College
Columbus, Ohio

Safety

Peggy W. Holliday
Science Education Specialist
NIOSH Certified Lab Trainer
Raleigh, North Carolina

Reviewers

Al Alvarez
Curriculum Consultant
San Antonio, TX

James E. Arnett III
South Doyle Middle School
Knoxville, Tennessee

Robert W. Avakian
Alamo Junior High School
Midland, Texas

D. Vance Baugher
Hannibal High School
Hannibal, Missouri

Anthony J. DiSipio, Jr.
Octorara Intermediate School
Atglen, Pennsylvania

Lorena R. Farrar
Westwood Junior High
Dallas, Texas

Elizabeth Hezel
St. Mary's School
Canandaigua, New York

Bettye Lou Jerrel
Evansville-Vanerburgh
  School Corporation
Evansville, Indiana

Ron Kinoshita
Gilroy Unified School District
Gilroy, California

Denise E. McCarthy
Ben Franklin Junior High
Fargo, North Dakota

Donna McLeish
Vigo County School District
Terre Haute, Indiana

Jim Nelson
Orange County Public Schools
Orlando, Florida

Jo Ann Pelkki
Saginaw City Schools
Saginaw, Michigan

Curtis A. Spenrath
Goose Creek Consolidated
  Independent School District
Baytown, Texas

Lisa Jean Wakeland
Chesterfield County
  Chester Middle School
Chester, Virginia

Raquél S. Williams
West End High School
Birmingham, Alabama

Contents

UNIT 1

UNIT 2

Contents

Contents

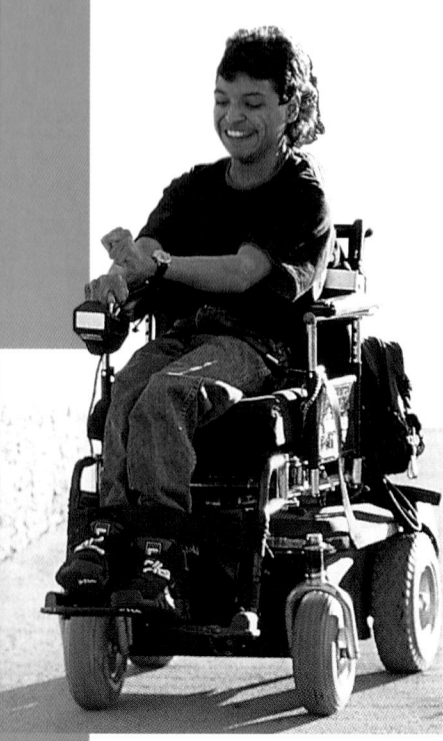

UNIT 3

The Nature of Matter   210

Contents

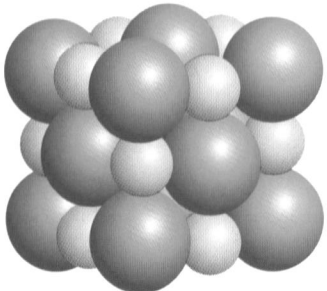

Kinds of Substances   326

U N I T 4

Contents

Contents

Interactions of Matter 412

Contents

Contents

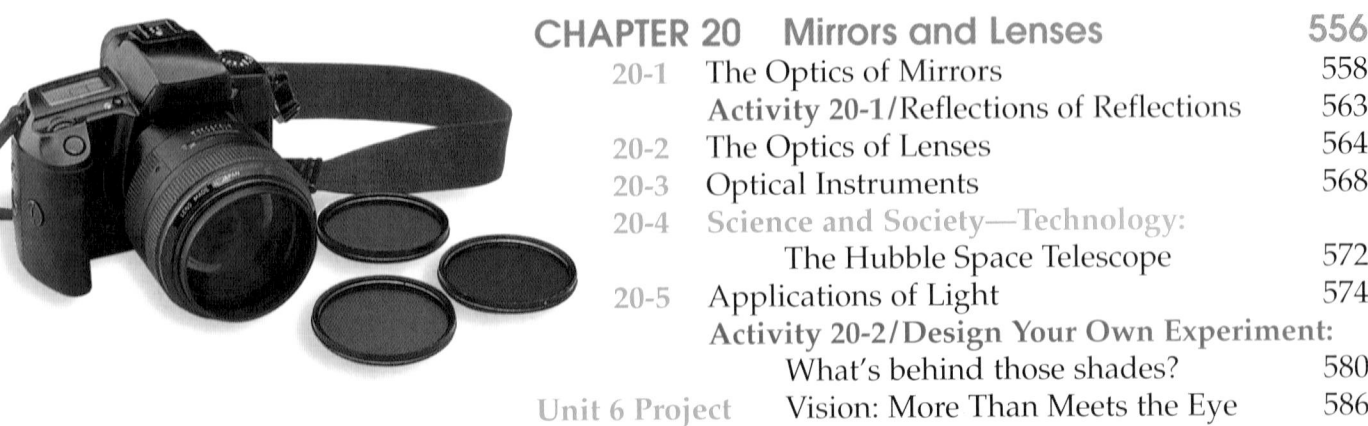

Electricity and Energy Resources   588

UNIT 7

Contents

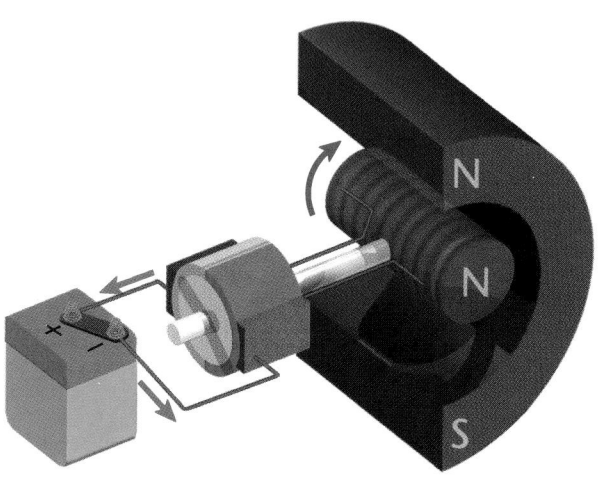
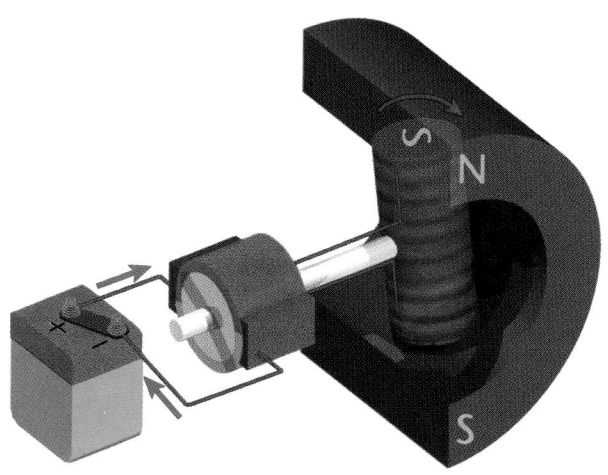

Contents

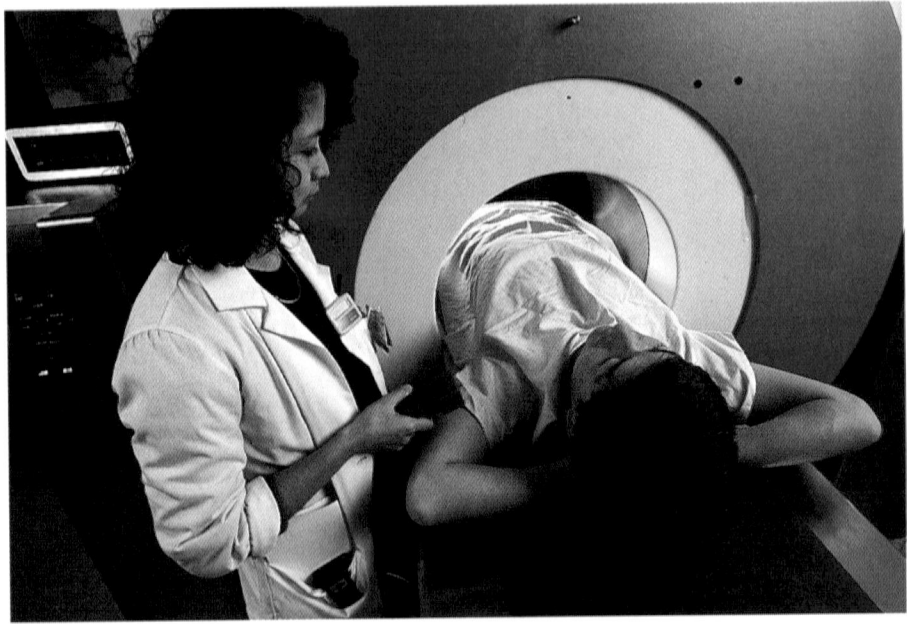

Contents

Appendices

Skill Handbook

Activities

Activities

Explore Activities

MiniLABs

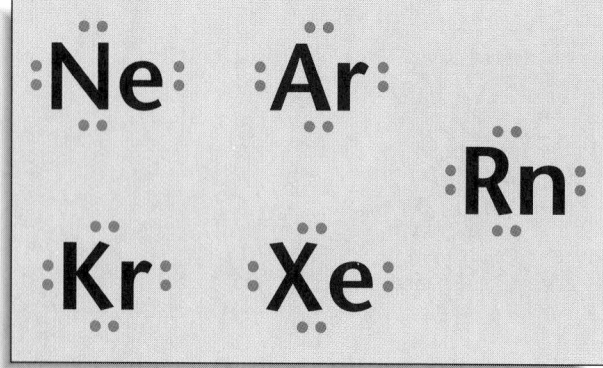

MiniLABs

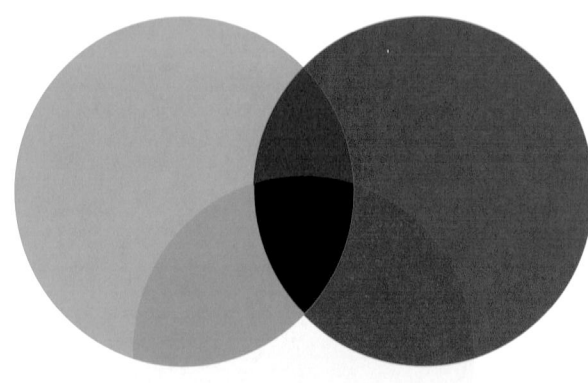

Problem Solving

Using Technology

Skill Builders

ORGANIZING INFORMATION

THINKING CRITICALLY

DESIGNING AN EXPERIMENT

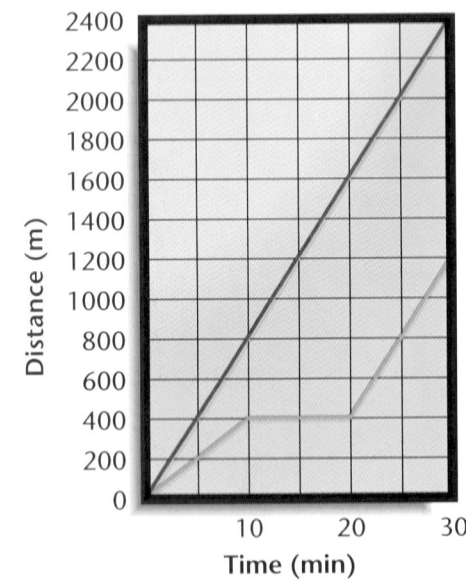

GRAPHICS

People and Science

Science Connections

Science and Art

Science and History

Science and Literature

UNIT 1

Physical Science Basics

In Unit 1, students briefly explore the content of physical science. Students are also introduced to the important role of problem solving as an activity of science. The identification of problems and the choosing of strategies to solve them are related to scientific method and experimental design. The unit closes with a discussion on the importance of measurements in science, SI measurements, and graphing as a useful means of displaying and interpreting data.

CONTENTS

UNIT 1

Unit Contents

Science at Home

Home Measurement Have students keep a log of the types of measurements, measuring instruments, and units of measurement that they use at home or that they observe being used in various occupations. After one week, have students present their findings to the class. [L1]

Physical Science Basics

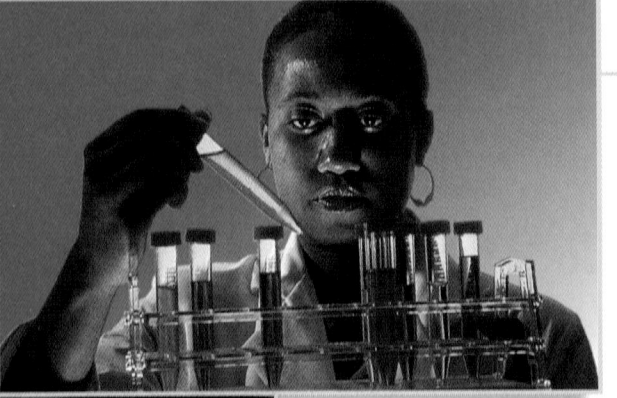

What's Happening Here?

Anxiously waiting to see photographic images appear on developing film may seem to be far removed from learning science. Yet the chemical reactions, solutions, and techniques involved all depend on important scientific principles. Within this unit, you will observe that studying science allows you to find out about new ideas and to understand and measure what has been around you all your life. Begin this study, and see what develops.

Science Journal

The light in the photo developing lab looks mysterious, but it is used for an important reason. Find out the difference in the developing process for black-and-white and color photographs. Compare the two processes in your Science Journal.

3

Theme Connection

Systems and Interactions/Scale and Structure Unit 1 provides an overview of how physical systems are studied scientifically using observations, inferences, and data collection by means of measuring and interpreting data.

Cultural Diversity

Navigating the Globe Many early cultures used observations and measurements of the sun, planets, and the night sky to navigate over sea and land. Have interested students list specific techniques and inventions used by various cultures. L1

3

Chapter Organizer

Section	Objectives/Standards	Activities/Features
Chapter Opener		**Explore Activity:** Learn how developing skills and knowing how to apply them is necessary in both daily life and the study of science. p. 5
1-1 Science is Everywhere (1 session, ½ block)*	1. **Compare** and **contrast** pure science and technology. 2. **Define** *physical science.* 3. **Discuss** some of the topics covered in physical science. National Content Standards: (5-8) UCP1, A2, B3, C1, E2, F5, G2; (9-12) UCP1, A2, E2, F6, G2	**Using Technology:** A Sticky Story, p. 7 **Science Journal,** p. 9 **Skill Builder:** Observing and Inferring, p. 9
1-2 Finding Out (1 session, ½ block)*	1. **Distinguish** between problems and exercises. 2. **Evaluate** approaches to solving problems. 3. **Compare** and **contrast** hypothesis, theory, and scientific law. National Content Standards: (5-8) UCP1, UCP2, G2; (9-12) UCP1, UCP2, D4, G2	**MiniLAB:** How can still pictures appear to be moving? p. 14 **Connect to Earth Science,** p. 12 **Problem Solving:** Can you put that in writing? p. 13 **Science Journal,** p. 17 **Skill Builder:** Outlining, p. 17
1-3 Science and Society (1 session, ½ block)*	1. **Discuss** the ways technology has made movie special effects more realistic. 2. **List** the areas of science that are drawn upon for movie special effects. National Content Standards: (5-8) UCP2, E2, F5; (9-12) UCP2, C3, E2, F6	**Connect to Life Science,** p. 19 **Explore the Technology,** p. 19
1-4 Exploring Science (1 session)*	1. **Understand** the importance of following guidelines in doing experiments. 2. **Define** *constant, independent variable,* and *dependent variable.* 3. **Understand** laboratory safety rules. National Content Standards: (5-8) UCP1, UCP3, A1, A2, E1, G1, G2; (9-12) UCP1, UCP3, A1, A2, E1, G1, G2	**Using Math,** pp. 22, 25 **Skill Builder:** Hypothesizing, p. 25 **Activity 1-1:** Identifying and Controlling Variables, pp. 26-27 **People and Science:** On the Job, p. 28

* A complete Planning Guide that includes block scheduling is provided on pages 32T-35T.

Activity Materials

Explore	Activities	MiniLAB
page 5 video camera	pages 26-27 string, identical weights, metric ruler, ring stands, scissors, stopwatch or timer, safety goggles	page 14 12 small, unlined index cards; rubber band

Need Materials? Call Science Kit (1-800-828-7777).

Teacher Classroom Resources

Reproducible Masters	Transparencies	Teaching Resources
Study Guide, p. 5 Reinforcement, p. 5 Enrichment, p. 5	Section Focus Transparency 1, What Do Scientists Do?	Physical Science CD-ROM Spanish Resources English/Spanish Audiocassettes Cooperative Learning Resource Guide Lab Partner Lab and Safety Skills Lesson Plans
Study Guide, p. 6 Reinforcement, p. 6 Enrichment, p. 6 Activity Worksheets, p. 7 Concept Mapping, pp. 7-8 Cross-Curricular Integration, p. 5 Critical Thinking/Problem Solving, p. 7	Section Focus Transparency 2, Science Is Problem Solving Science Integration Transparency 1, Using the Best Graph Teaching Transparency 1, Flex Your Brain	**Assessment Resources** Chapter Review, pp. 5-6 Assessment, pp. 5-8 Performance Assessment in the Science Classroom (PASC) MindJogger Videoquiz Alternate Assessment in the Science Classroom Performance Assessment Chapter Review Software Computer Test Bank
Study Guide, p. 7 Reinforcement, p. 7 Enrichment, p. 7	Section Focus Transparency 3, Special Effects	
Study Guide, p. 8 Reinforcement, p. 8 Enrichment, p. 8 Activity Worksheets, pp. 5-6 Lab Manual 1, Relationships Multicultural Connections, p. 5 Science Integration Activity 1, p. 1, Cabbage Patch Chemist Science and Society Integration, p. 5	Section Focus Transparency 4, Designing Experiments Teaching Transparency 2, Safety Symbols	

GLENCOE TECHNOLOGY

The following multimedia resources are available from Glencoe.

Science and Technology Videodisc Series (STVS)
Physics
   Wind Engineering
   Computer Graphics
   Computerized Star Imaging
Chemistry
   Glass Making for Science
   Fire-Resistant Clothing
Earth & Space
   Modeling Black Holes
   Computerized Weather Forecasting

The Infinite Voyage Series
Unseen Worlds
The Search for Ancient Americans
The Great Dinosaur Hunt
Miracles by Design

Physical Science CD-ROM

Key to Teaching Strategies

The following designations will help you decide which activities are appropriate for your students.

L1 Level 1 activities should be within the ability range of all students, including those with learning difficulties.

L2 Level 2 activities should be within the ability range of the average to above-average student.

L3 Level 3 activities are designed for the ability range of above-average students.

LEP LEP activities should be within the ability range of Limited English Proficiency students.

LS These activities are designed to address different learning styles.

COOP LEARN Cooperative Learning activities are designed for small group work.

P These strategies represent student products that can be placed into a best-work portfolio.

Teacher Classroom Resources

This is a representation of key blackline masters available in the Teacher Classroom Resources.

Teaching Aids

Section Focus Transparencies

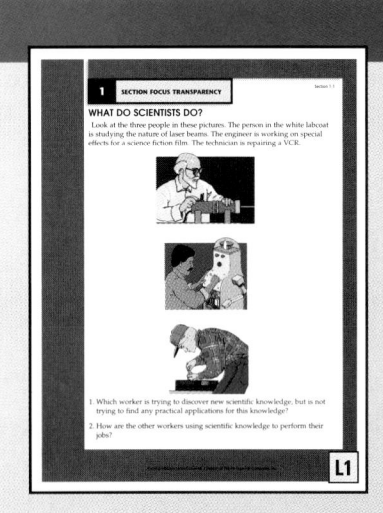

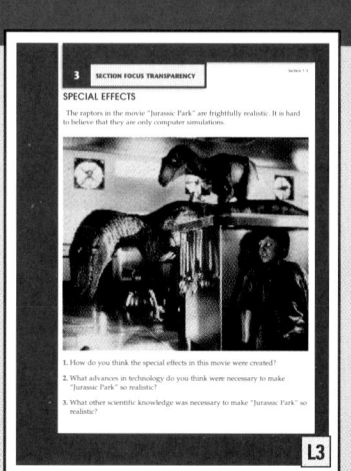

Science Integration Transparencies

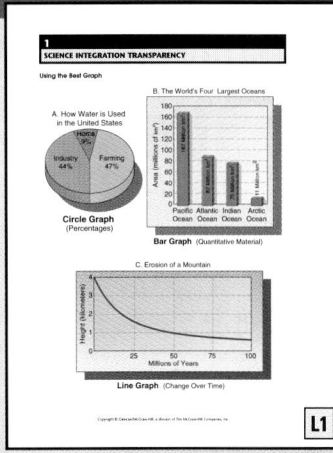

Teaching Transparencies

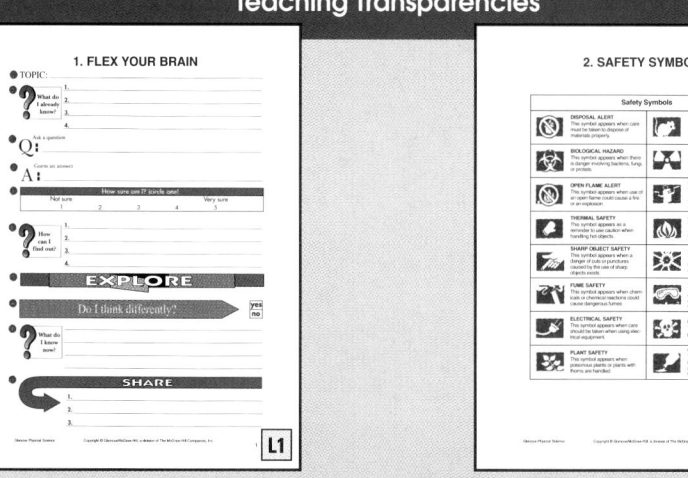

Meeting Different Ability Levels

Study Guide

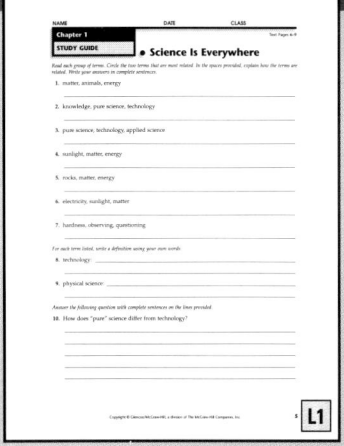

Reinforcement

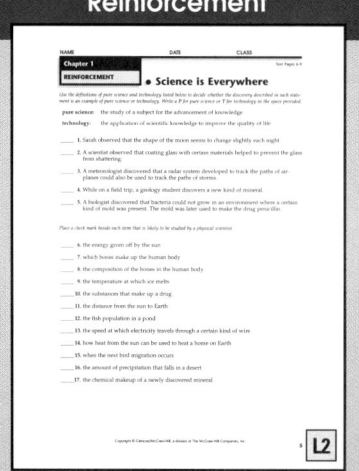

Enrichment Worksheets

Hands-On Activities

Science Integration Activity

Cabbage Patch Chemists

L1

Lab Manual

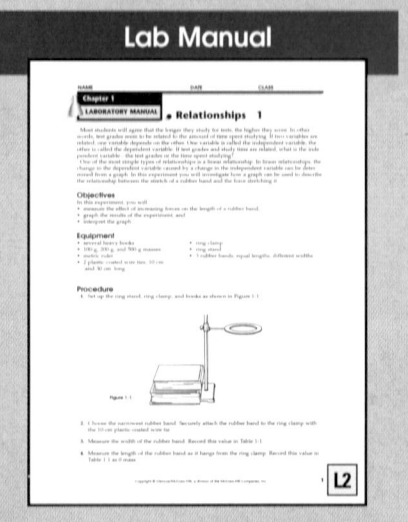

Relationships   1

L2

Assessment

Performance Assessment

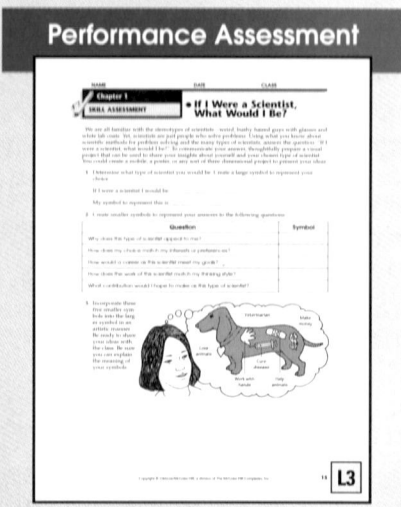

If I Were a Scientist, What Would I Be?

L3

Enrichment and Application

Critical Thinking/ Problem Solving

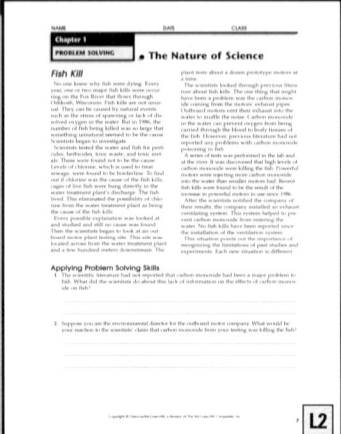

The Nature of Science

Fish Kill

L2

Cross-Curricular Integration

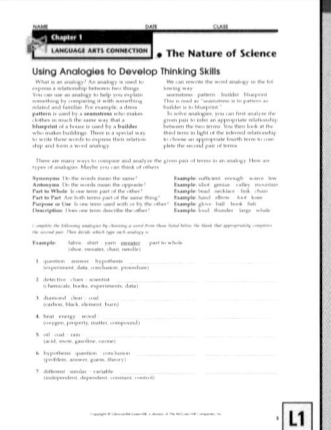

Using Analogies to Develop Thinking Skills

L1

Science and Society Integration

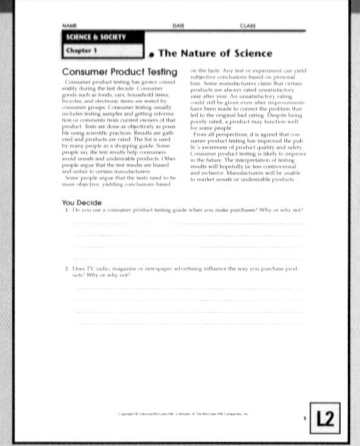

Consumer Product Testing

L2

Multicultural Connections

Walter E. Massey— Scientist and Role Model

L2

Concept Mapping

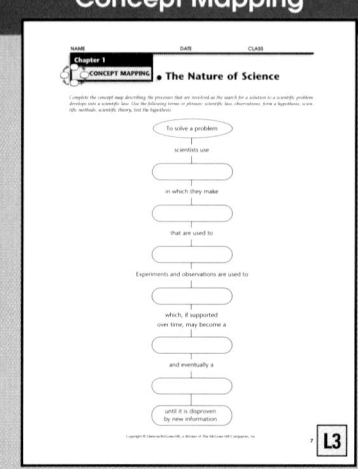

The Nature of Science

L3

Chapter 1

The Nature of Science

CHAPTER OVERVIEW

Section 1-1 In this section the relationship between science and technology is discussed. The scope of topics studied in the physical sciences is presented.

Section 1-2 This section distinguishes problems from exercises. Hypotheses, theories, and scientific laws are compared.

Section 1-3 Science and Society Specific examples of new technology, materials, and computer graphics used for movie making are introduced.

Section 1-4 The design of experiments is explained with emphasis on the control of variables. Lab safety is also emphasized.

Chapter Vocabulary

technology	experiment
physical science	control
model	constant
observation	independent variable
hypothesis	dependent variable
theory	
scientific law	

Theme Connection

Systems and Interactions Discoveries in pure science (research) can interact to solve specific problems or to create new materials or products (technology).

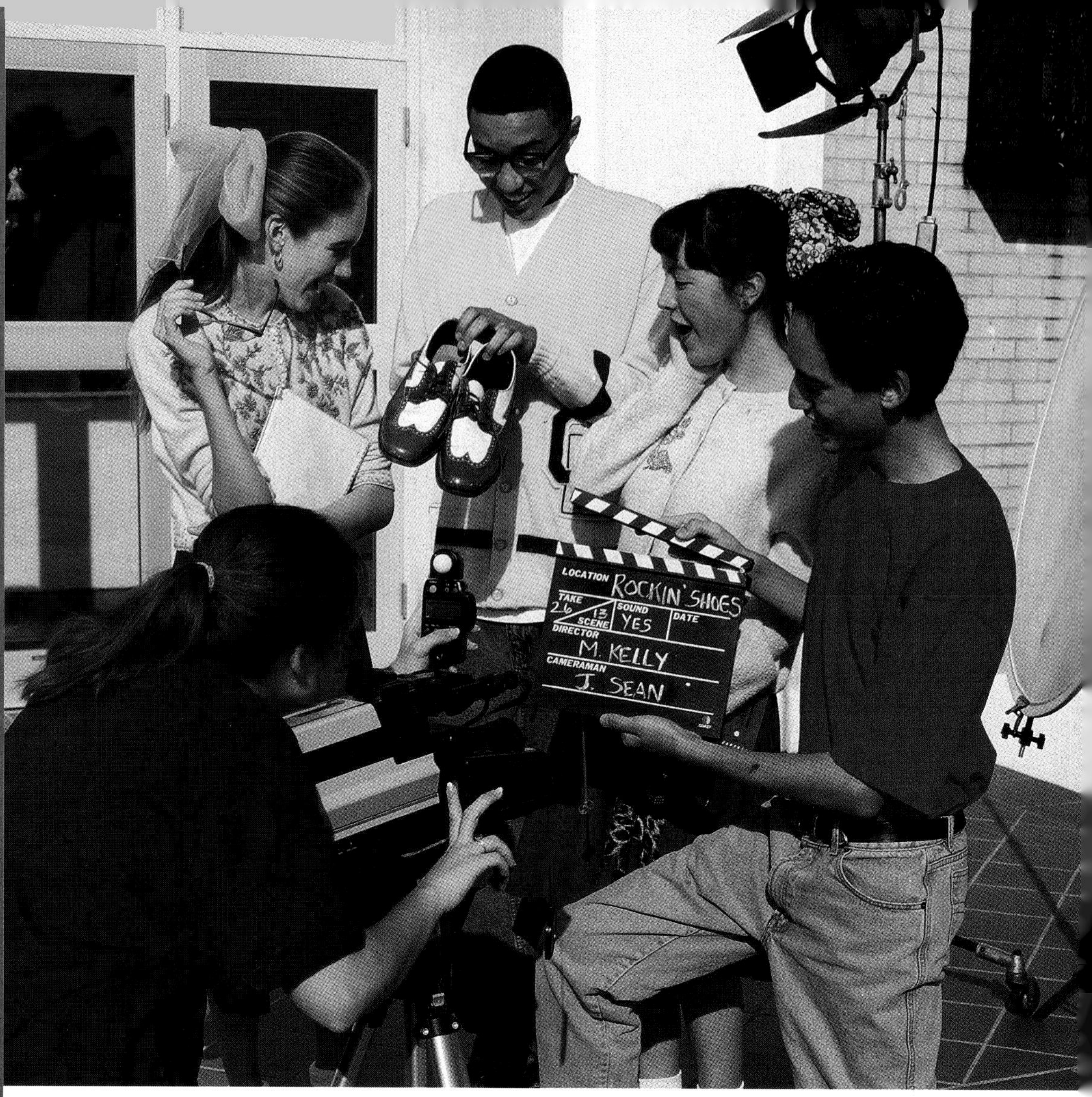

Previewing the Chapter

Section 1-1 Science Is Everywhere
- ▶ Behind the Scenes
- ▶ Science—Pure and Applied
- ▶ A Guidance System

Section 1-2 Finding Out
- ▶ Detecting Clues . . . Solving Problems
- ▶ Thinking Critically
- ▶ Scientific Methods

Section 1-3 Science and Society Technology: Getting Real with Special Effects

Section 1-4 Exploring Science
- ▶ Scientifically Tested
- ▶ Play It Safe
- ▶ Exploring Your World

Learning Styles

Look for the following logo for strategies that emphasize different learning modalities. LS

Kinesthetic	Explore, p. 5; Reteach, p. 8; MiniLAB, p. 14
Visual-Spatial	Visual Learning, pp. 7, 11, 12, 14, 21; Across the Curriculum, p. 23
Interpersonal	Concept Development, p. 11; Activity, p. 22
Logical-Mathematical	Across the Curriculum, p. 13; Reteach, p. 16; Activity 1-1, pp. 26-27
Linguistic	Science Journal, p. 9; Across the Curriculum, p. 22

LS

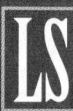

The Nature of Science

Is science really everywhere you look? Something as common as videotaping a school event would not be possible without the application of many scientific principles and discoveries. From receiving, recording, and transmitting light and sound to the nerve impulses that tell your finger to press the *Record* button, you can notice many ways that science can be put to work for you.

EXPLORE ACTIVITY

Learn how developing skills and knowing how to apply them is necessary in both daily life and the study of science.

1. Obtain a video camera from your home or school. Be sure you have permission and know how to use the camera correctly.
2. Record several short school or class activities.
3. Watch your video with your classmates, and observe any problems with the taping. Were some events better taped than others?

Observe: In your Science Journal, record any problem present with the taping. Explain how developing certain skills could make taping more effective.

Previewing Science Skills

▶ In the Skill Builders, you will **observe and infer, outline**, and **map concepts**.

▶ In the Activity, you will **observe, predict**, and **control variables**.

▶ In the MiniLABs, you will **measure in SI, control variables**, and **observe**.

5

Assessment Planner

Portfolio
Refer to page 29 for suggested items that students might select for their portfolios.

Performance Assessment
See page 29 for additional Performance Assessment options.
Skill Builders, pp. 9, 17, 25
MiniLAB, p. 14
Activity 1-1, pp. 26-27

Content Assessment
Section Wrap-ups, pp. 9, 17, 19, 25
Chapter Review, pp. 29-31
Mini Quizzes, pp. 16, 24

Group Assessment
Opportunities for group assessment occur with Cooperative Learning Strategies and Flex Your Brain Activities.

EXPLORE ACTIVITY

Purpose
IS **Kinesthetic** Use this activity to help students begin to think about the connections between science and technology. **L1** **LEP** **COOP LEARN**

Preparation
Obtain several video cameras. Be sure students understand the importance of proper and responsible operation of the cameras.

Materials
video cameras, videotapes, television or monitor

Teaching Strategies
• Depending on the camera features, a VCR may be necessary for viewing the tapes.
• In reviewing the tapes, be sure discussion is focused on the tapes themselves and not on the students involved.

Observe
Students will probably notice several problems, such as focus problems when the camera is moved quickly, or excessive camera movement. Students should predict that the problems can be solved or improved by examining the problem and deciding on an approach to solving it.

✓ Assessment

Performance After students decide on problem-solving techniques, have them again tape an event, using the techniques to minimize problems. Use the Performance Task Assessment List for Video in **PASC,** p. 81. **P**

Prepare

Section Background

Discoveries in scientific research often lead to advances in technology.

1 Motivate

Bellringer

 Before presenting the lesson, display **Section Focus Transparency 1** on the overhead projector. Assign the accompanying **Focus Activity** worksheet. L1 LEP

Demonstration

Show sections of several videotapes to the classes. Ask them to write down their favorite special effects and infer how science is involved in them.

2 Teach

GLENCOE TECHNOLOGY

 Videodisc

The Infinite Voyage Series: Unseen Worlds

Chapter 1
Technology Reconstructs Egyptian Mummies

Chapter 2
Supercomputer Models: Photosynthesis, Space Exploration, Weather

---

Science Words

technology
physical science

Objectives

* Compare and contrast pure science and technology.
* Define *physical science.*
* Discuss some of the topics covered in physical science.

Behind the Scenes

What was the best movie you saw last year? Was it an action/adventure film, a comedy, a thriller, or a romance? As you watched, did you ever wish you could sneak behind the scenes to see how the movie was made? You might be surprised to see that it isn't just the actors, producer, director, and crew that contribute to what you see on the screen. Science, too, plays a starring role, as shown in **Figure 1-1.**

Lights, Camera, Action!

Before a movie makes it to your local theater, every moment of every scene is captured on film through the lens of a motion picture camera. How do moviemakers use lenses to bring about the feeling of being close up or to produce panoramic shots of breathtaking scenery? The science of optics holds the answers.

An on-screen explosion, a lifelike robotic dinosaur, and heart-stopping sound effects all rely on scientific principles. None of these effects is possible without an understanding of chemical reactions, mechanics, hydraulics, electronics, and acoustics.

Science—Pure and Applied

It's not hard to see how a knowledge of chemical reactions would help someone figure out how to make a mock explosion in a movie, but not all connections between scientific knowledge and its practical uses are so obvious. For example, can you see any connection between moviemaking and the details of how human vision works? Read on!

A Great Vision

The human eye receives images in the form of light waves, which are translated into nerve impulses. The impulses are sent to the brain, which converts the information back into images and then interprets those images. It takes the eye

Figure 1-1
See science in action behind the scenes on a movie set. *How do you think science is used on this set?*

 INTEGRATION
Life Science

Program Resources

 Reproducible Masters
Study Guide, p. 5 L1
Reinforcement, p. 5 L1
Enrichment, p. 5 L3

Transparencies
Section Focus Transparency 1 L1

one-quarter of a second to receive and transmit the information for a single image. Now, here's where moviemaking comes in. Movie cameras take a series of still images at a rate of 24 per second. Projectors show the film at the same speed. But, because the human eye can't process more than four separate images per second, the images on the screen seem to blend together, creating the optical illusion of continuous motion.

Do you think the scientists who first studied human vision had motion pictures in mind? Certainly not. They just wanted to understand how the eye works. They were engaged in what is called *pure science.* But their pure science discoveries eventually made it possible for others to invent motion pictures, eyeglasses, contact lenses, and other visually related products we enjoy and depend on every day. These are examples of *applied science.*

USING TECHNOLOGY

A Sticky Story

Visit any office where people pass paperwork back and forth, and you'll probably see notes written on colored slips of paper that can be stuck on and easily peeled off. The story of how those peel-and-stick notepads came to be invented could itself be the subject of a movie. The sound track would be supplied by the choir at the church where chemist Art Fry sang on Sundays. Fry always marked his place in the hymnbook with little scraps of paper. Sometimes the scraps fell out, and he couldn't find the right hymn when it came time to sing.

He remembered an adhesive developed a few years earlier by another scientist, Dr. Spencer Silver. Silver had thought the adhesive was useless because it wasn't strong enough to hold things together permanently. But sitting there in church, Fry suddenly saw thousands of uses for notepads with "temporarily permanent adhesive," as he called it. It took months of experimenting to get the formula just right, but when the sticky notepads finally were offered to the public, they were an instant hit.

Think Critically:
Use the above example to explain how pure science and applied science can be used together to solve problems.

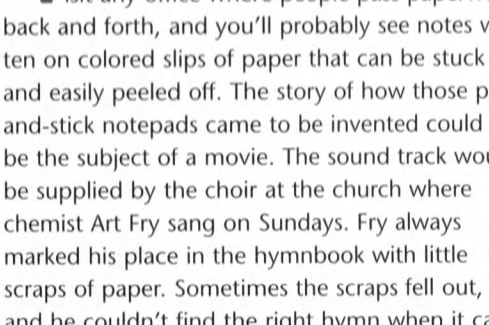
Peel-and-stick notes

INTEGRATION
Life Science

As a movie can be out of focus, so can vision. Unfocused vision most commonly results from irregularities in the shape of the cornea. These vision problems are usually taken care of by corrective lenses, although new surgical methods are being used more frequently.

Visual Learning

Figure 1-1 How do you think science is used on this set? *Student answers may include use of light, sound, and electricity.* **LEP** **LS**

GLENCOE TECHNOLOGY

 **Videodisc**
The Infinite Voyage Series: Miracles by Design
Chapter 1
Auto Racing: Advancement in the Field of Metallurgy

Chapter 9
"Smart Materials" and Their Construction

USING TECHNOLOGY

Think Critically
The pure science involved in developing the adhesive itself is applied (technology) to solve the problem of non-stick paper.

3 Assess

Check for Understanding

Ask questions 1-2 and the Think Critically question in the Section Wrap-up.

Reteach

IKS Kinesthetic Have students bring in examples of packaging that show technological improvements such as twist-off bottle caps, tab-open cans, twist-tie and self-sealing food storage bags, and plastic foam containers. Have them compare the benefits and possible drawbacks of such new products.

Extension

For students who have mastered this section, use the **Reinforcement** and **Enrichment** masters.

8

Something New

Pure science is the gathering of new information or the discovery of a new relationship or fact. Pure science adds to the body of scientific knowledge but does not have to have practical uses. Not every scientific discovery will have common, everyday applications, but even those that do not may open up whole new fields of investigation.

Technology is the practical use of scientific information. Where does pure science leave off and technology begin? It's not always easy to tell. Scientific discoveries may lead to technical innovations which, in turn, may lead to further discoveries. Both pure science, for the advancement of knowledge, and applied science, or technology, are considered important parts of science as a whole.

A Guidance System

When you're looking for a specific movie at the video store, there is a logical system to guide you. New releases are in one section; action/adventure films, comedies, dramas, and musicals all have their own places. When you go to the grocery store with a shopping list, you know that certain foods are grouped together and that the aisles are arranged in a particular order. But suppose your mission was not to find a movie or a can of peaches, but to find out all you could about how cameras work. What sort of system would help you in your search through the vast world of science?

Major Categories

Your first step would be to narrow your search to one of the main categories of science: life science, which deals with living organisms; Earth science, which investigates Earth and space; or physical science, which deals with matter and energy. Physical science topics can be grouped into chemistry and physics. Each of these subdivisions also can be divided into more specific branches of science. How would you decide which branches of science would best help you understand how a camera works?

Physical Science

This textbook will be your guide through the major topics of physical science. **Physical science** is the study of matter and energy. Every measurable thing in the universe is either matter or energy. Plants and animals, rocks and

Figure 1-2
Lightning, thunder, and motion are examples of energy. Trees, rain, and clouds are all examples of matter.

Integrating the Sciences

Earth Science Ask students to make a list of all of the technologies that are used to make weather forecasts. The list should include measuring systems such as weather balloons, wind gauges, Doppler radar, etc. Also included would be the various communications systems that are used, such as satellite imaging microwave transmissions, newspapers, radio, and TV. **L1**

Use **Study Guide,** p. 5, as you teach this lesson.

clouds, eggs and elephants are all examples of matter. Lightning, thunder, and motion are all examples of energy, as shown in **Figure 1-2.**

In physical science, you'll study what makes up matter—its composition. You'll also learn about what matter is like and how it behaves. Hardness and taste are two examples of properties of matter that should already be familiar to you.

As you study physical science, you'll learn how matter and energy are related. You'll discover how energy is transferred through matter, as when the sound from the CD player in **Figure 1-3** reaches your ears. You'll find out how refrigerators stay cold and why an insulated jacket keeps you warmer than a thin shirt.

As you learn about sound and light, you'll discover the scientific side of music, photography, and other everyday activities. The study of electricity will help you understand what keeps your television and compact disc player running. In the study of force and motion, you'll learn why acorns fall to Earth and why satellites remain in orbit.

As you learn about physical science, notice how things and ideas are organized into systems and how different parts of a system interact. Your observations will lead you to ask questions. The rest of this chapter will help you learn how to go about answering those questions.

Figure 1-3
Everyday activities have scientific sides.

Section Wrap-up

Review

1. What is the difference between pure science and technology? Give an example of each.

2. Define *physical science*.

3. **Think Critically:** Pose three questions that you think could be answered by using physical science.

Skill Builder
Observing and Inferring
Bring in a newspaper or magazine article related to physical science. Summarize the article and explain whether it deals with a scientific discovery or a technological development. If you need help, refer to Observing and Inferring in the **Skill Handbook.**

Science Journal

Imagine that you are writing a screenplay about the invention of peel-and-stick notepads. Describe three scenes that would show audiences how pure science discoveries led to the invention.

Skill Builder
Articles collected by students will vary. Make sure that students correctly identify the content as a scientific discovery or as a technological development.

**Assessment**

Performance Have students use their articles to create a bulletin board that relates pure science to technology. Use the Performance Task Assessment List for Bulletin Board in **PASC,** p. 59.

4 Close

Divide the class into two groups. Ask students to consider the photo of the CD player. One group should list types of energy changes. The other group should list the different technologies in the CD player. **L1** **LEP** **COOP LEARN**

Section Wrap-up

Review

1. Science is a study solely for the advancement of knowledge. Technology is the application of scientific knowledge to solve everyday problems. The study of the electrical nature of matter is science, whereas the development of an electric generator is technology.

2. the study of matter and energy

3. **Think Critically** Student responses will vary. Examples: (1) What causes tides? (2) How fast does sound travel in air? (3) Why might fuses be used in an electrical circuit?

Science Journal

LS **Linguistic** Answers will vary, but will likely include supposed failures of the glue because it doesn't hold heavy objects.

GLENCOE TECHNOLOGY

CD-ROM
Physical Science CD-ROM
Have students perform the interactive exploration for Chapter 1 to reinforce important chapter concepts and thinking processes.

Prepare

Section Background

Problem solving is an important skill. The techniques used are the same as those used to develop and test hypotheses by scientific experimentation.

Preplanning

Gather unlined 3" × 5" note cards for the MiniLAB.

1 Motivate

Bellringer

🕹 Before presenting the lesson, display **Section Focus Transparency 2** on the overhead projector. Assign the accompanying **Focus Activity** worksheet. L1 LEP

Tying to Previous Knowledge

Have students recall how they solved the problem of finding different classrooms when they first started school in this building. Point out that they will learn other methods of problem solving in this section.

1•2 Finding Out

Science Words

model
observation
hypothesis
theory
scientific law

Objectives

- Distinguish between problems and exercises.
- Evaluate approaches to solving problems.
- Compare and contrast hypothesis, theory, and scientific law.

Detecting Clues . . . Solving Problems

Science involves a lot more than collecting facts and sorting things into categories. The real business of science is solving problems—some that have everyday applications and others that are more abstract. In this section, you'll explore ways of solving problems.

At your after-school job at the local movie theater, the boss calls you aside and asks for some help. The projector shown in **Figure 1-4** isn't working right, and he can't figure out why. Could something be wrong with the lens that focuses the images on the screen? Is the film running too slowly or too quickly? What about the light source that shines through the film to project it onto the screen? Has the light burned out? Is there something wrong with the photoelectric cell that "reads" the sound track from the film and creates an electrical signal that is amplified and sent to speakers in the theater?

What is a problem?

Your boss has a *problem*. You've probably used that word many times without stopping to think exactly what it means. A problem is a situation in which something appears to be missing. A problem is not the same thing as an exercise. In an exercise, the steps required to find the solution are usually obvious. For example, finding the average weight of five classmates is an exercise if you know how to calculate averages. If you don't know how to find averages, then you do have a problem.

Figure 1-4

Most problems, such as what is wrong with this broken projector, can be solved by following a series of logical steps.

10   Chapter 1 The Nature of Science

Program Resources

📁 **Reproducible Masters**
Study Guide, p. 6 L1
Reinforcement, p. 6 L1
Enrichment, p. 6 L3
Cross-Curricular Integration, p. 5
Activity Worksheets, pp. 5, 7 L1
Critical Thinking/Problem Solving, p. 7
Concept Mapping, p. 7

 Transparencies
Section Focus Transparency 2 L1
Teaching Transparency 1 L1
Science Integration Transparency 1

The broken projector is a problem, and because your boss asked you for help, it's now *your* problem. Simply observing that the projector isn't working won't help you find a solution. You must look for more clues before you can figure out what to do. Solving a problem involves finding missing information, but it's not always clear what kind of information is needed. Where would you start?

Patterns

As **Figure 1-5** shows, finding patterns can help solve problems. Does the projector work sometimes but not always? Does it stop working after it has been running for a while? To answer questions such as these, you need to observe the projector over time. You may need to take measurements. For example, you could use your watch to measure how long it takes from the time you turn on the projector until something goes wrong. Would one measurement be enough to show a pattern? No. You would need to make many identical measurements before you could detect a pattern.

Solving Problems

The problem of the broken projector has many possible solutions. An obvious one is to check that the projector is plugged in. Then, you might check to make sure the projector is turned on. After ruling out those easy solutions, how would you proceed?

Before you can solve a problem, you need to understand exactly what the problem is. That may sound obvious, but people often have trouble solving problems because they don't know where to start. Sometimes, that's because they haven't defined the problem clearly enough.

As you attack the movie-projector problem, you begin to look it over and ask the manager a lot of questions. Does it make any strange noises? Is a burning smell obvious when it runs? Your questions and observations will help you pinpoint the specific problem. You'll progress from knowing only that the projector isn't working to knowing, for example, that the mechanism that moves the film is broken. The more precisely you can define the problem, the less time you'll spend looking for solutions.

Figure 1-5

Searching for patterns is a good problem-solving strategy. *What patterns could help you solve this problem?*

2 Teach

Inquiry Question

Early cartoons required 16 drawings for each second of action. **How many sketches would be required to make a 5-minute cartoon?** *16 sketches/s × 60 s/min × 5 min = 4800 sketches*

Concept Development

LS Interpersonal Have teams of students select a sport, board game, or video game and explain how it poses a problem to be solved. Have teams discuss how the various game moves can be used as strategies to solve the problem posed by the game.

 Use these program resources as you teach this lesson.

Cross-Curricular Integration, p. 5

Critical Thinking/Problem Solving, p. 7

Concept Mapping, p. 7

GLENCOE TECHNOLOGY

Videodisc

The Infinite Voyage Series: The Search for Ancient Americans

Chapter 7

The Anasazi: Unraveling the Puzzle of Chaco Canyon

Visual Learning

Figure 1-5 What patterns could help you solve this problem? *Answers may include identifying edge pieces and pieces of a certain color.* **LEP** **LS**

Visual Learning

Figure 1-6 The person at the computer is using a computer simulation. Today, most airline companies train pilots on complicated, realistic simulators. **What are some of the advantages of this training?** *The pilots can practice takeoffs and landings and also practice emergency procedures without risking having an accident.* **LEP** **LS**

CONNECT TO
EARTH SCIENCE

Answer

1. Use past knowledge to make a guess: other formations seemed to coincide with season changes and the sun's changes and sun's position.
2. Look for patterns: Stonehenge slabs, when viewed at certain angles, lined up with the sun's position on winter solstice and spring equinox.
3. Develop a model: some ancient societies in England, called Druids, followed rituals based on Earth and sun positions.
4. Break the problem into smaller problems: the scientists examined various portions of the Stonehenge formations and used aerial views to see the overall formations.

Use **Science Integration Transparency 1** as you teach this lesson.

Figure 1-6

Computer models are programs that predict how different designs behave under various conditions.

CONNECT TO
EARTH SCIENCE

The famous stone slabs in England called Stonehenge may represent an ancient calendar. *Research* how scientists used the four problem-solving strategies to decide possible meanings of Stonehenge.

Plan a Strategy

Once you've defined the problem, look for answers in a systematic way. Proceed in logical steps, starting with what you know. Keep track of your steps so that if an idea doesn't work, you can move on to something else instead of trying the same thing again. Here are some ways that will help you solve problems.

• *Use what you know about the problem to predict a solution, and try it.* If your effort fails, think about why it didn't work. Then make another prediction and try that. Keep eliminating possibilities until you find the one that works. You've probably used this method to search for something you've lost—such as last night's homework. Think about when you last saw the homework, and review where you've been since then. This process helps you to direct your search. Maybe you remember that you went to the kitchen for a snack after you finished your homework, so you look there. If you don't find it, you look in the next logical place—the chair where you sat to have your snack—and so on.

• *Look for patterns that will help you make predictions about the problem.* Suppose you occasionally break out in a rash. If you pay attention to what you eat, touch, and wear every day, you may find a pattern that helps you discover what's causing the rash. Putting information into a table or graph or making a drawing sometimes can help you find patterns.

• *Develop a model.* When your problem deals with something complicated or difficult to see, it may help to develop a model. A **model** is an idea, system, or structure that represents whatever you're trying to explain. The model is never exactly like the thing being explained, but it is similar enough to allow comparisons.

Before starting to shoot a movie, filmmakers sometimes build models to get an idea of how particular scenes will look. Models are used in scientific investigations, too. To find out

12 Chapter 1 The Nature of Science

Community Connection

Auto Mechanic One of the greatest problem solvers in your area is the local auto mechanic. Visit the mechanic and ask him or her what steps that he or she would take to fix a car that won't start.

how the shape of an airplane affects performance, scientists and engineers make model airplanes of different shapes. These models are based on theories that describe how moving air behaves. Engineers then test the different models under various conditions to find out which shape works best. They may also create different computer models, such as those shown in **Figure 1-6.**

• *Break the problem down into smaller, simpler problems.* Sometimes, it's hard to see what needs to be done when the problem is complicated. Look for ways to solve it step-by-step.

As you search for solutions, keep thinking about what you first knew about the problem and what you've learned from each problem-solving attempt. When you find a solution, think again about the problem and ask yourself whether your solution makes sense.

Thinking Critically

Imagine that you're in charge of sound effects for a school play. The script calls for a barking dog in one scene; a squawking bird in another; an alarm clock, a siren, a scream, and a car crash in others. You carefully tape-record each sound and splice them together on one tape, in the order the script calls for. But at dress rehearsal, everything goes wrong. When it's time for the alarm clock, you press the *Play* button on your tape recorder and get the sound of a squawking bird. Instead of the barking dog,

Problem Solving

Can you put that in writing?

Your report is due tomorrow morning and it has to be neat and readable. You enter the report into your computer and then give the command to print. Nothing happens. No, this is not a nightmare. It is really happening, and you need help. Can you apply problem-solving steps to analyze the problem and solve it?

Solve the Problem:

1. As you gather information, look for any patterns. Has anyone else had problems with the printer? Have you had the same problem before?
2. Plan a strategy. What does a printer need to operate? How are those needs supplied to the printer? Can you check out each of those needs one at a time?
3. Once you think that you have enough information to solve the problem, make a prediction that you can test. The prediction can be simple. Is the printer plugged in? Is it turned on? If these predictions when tested don't solve your problem, you can still use them to guide you to another prediction.

Think Critically:

If the printer is receiving power, what else could be causing the problem? Think about how the printer receives instructions to print. What might you predict about the computer and printer connection?

Problem Solving

Solve the Problem

1. If no one else has had this happen with the printer, then you have a problem.
2. The printer needs to be plugged in, turned on, and connected to the computer. Additionally, the printer should have paper and should not be jammed. Each of these items can be checked out one at a time.
3. Hypotheses should reflect a possible solution and be testable.

Think Critically

Student hypotheses may include that the printer cable is bad or that a connection on the printer or computer is loose.

GLENCOE TECHNOLOGY

◉ Videodisc
STVS: Earth & Space
Disc 3, Side 1
Modeling Black Holes (Ch. 12)

Disc 3, Side 2
Computerized Weather Forecasting (Ch. 3)

Across the Curriculum

Mathematics Point out that patterns can be used to solve sequence problems. Have students discuss patterns in the following sequences and predict the missing integer.

0, 1, 2, 3, 4, ... *next integer, 5*
0, 1, 4, 9, 16, ... *square of an integer, 25*
0, 1, 1, 2, 3, 5, 8, ... *two previous terms added,*
13 L1 LS

Integrating the Sciences

Earth Science Have students identify and discuss patterns that occur in the Earth sciences. These patterns may include crystal structures or cycles such as days, lunar months, years, tides, and phases of the moon. Ask students how these patterns can be used to identify materials or predict events. L1

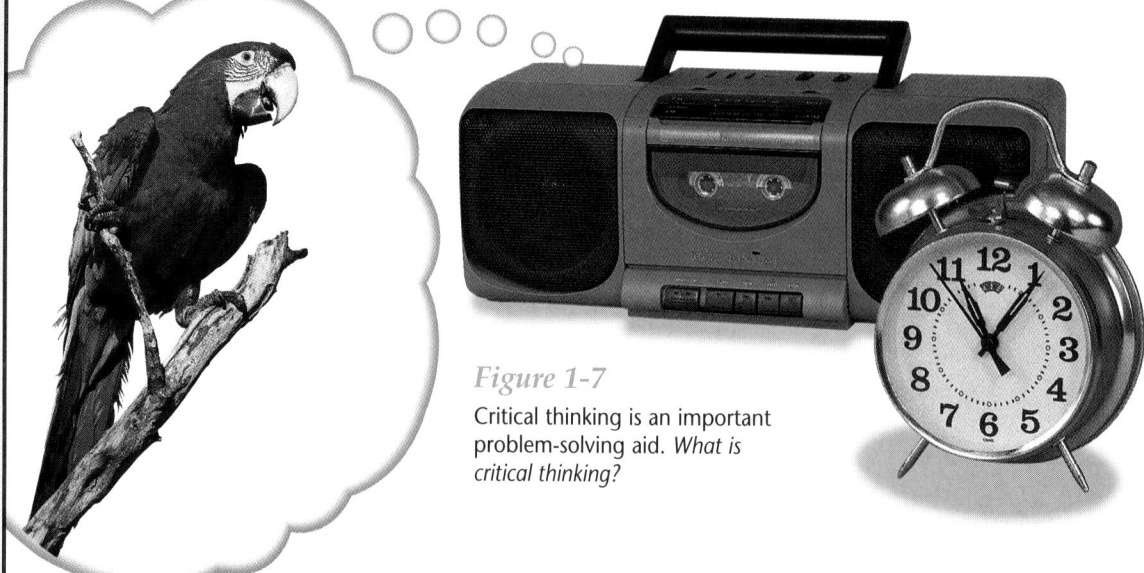

Figure 1-7

Critical thinking is an important problem-solving aid. *What is critical thinking?*

MiniLAB

How can still pictures appear to be moving?

Procedure

1. Obtain about a dozen small, unlined index cards. Place a rubber band tightly around one end of the stack of cards.
2. Using simple figures, draw a starting scene on the first card. On the next card, draw the same scene with a slight change in position to indicate motion. Repeat this step until all the cards are used.

Analysis

1. Grasp the end of the stack that has the rubber band on it, and flip the cards quickly. Record your observations in your Science Journal.
2. If you flip the cards more slowly, how does the smoothness of the apparent motion change?

you get a car crash. What's going on? Did you mix up the order of the sounds when you made the tape? Was there a script change that nobody told you about? Did someone play a prank on you and switch your tape for another one?

After thinking about it, you decide that you may have mixed up the order of sounds on your tape. How did you arrive at this conclusion? Without being aware of it, you probably used some aspect of critical thinking. Read on to find out more.

What is critical thinking?

Critical thinking is a process that uses certain skills to solve a problem. Let's see how you may have used critical thinking to solve the case of the scrambled sounds. First, you identified the problem—the wrong order of sounds—by comparing the sounds coming from your tape recorder to the sounds called for in the script. Next, you may have separated important information from unimportant information by deciding that it wasn't likely that the script had been changed at the last minute without your knowledge. Finally, you examined your assumption that you had spliced the sounds together in the correct order. You looked back at the script, thought about how rushed you were when you made the tape, and concluded that you made a mistake. You probably went one step further and analyzed your conclusion. You asked yourself if you had ever made similar mistakes when you were in a hurry. If the answer was yes, then you may have solved the problem.

Flex Your Brain, the series of steps outlined on the next page, is an activity that will help you to think about and examine how you think. Flex Your Brain is a way to keep your thinking on track when you are investigating a topic.

Cultural Diversity

Shamanic Medicine Principles of science are used to solve problems throughout the world. In the rain forests of Latin America, much of the healing is done by shamans, who are local people who use native plants for medicines. Today, researchers are working with shamans and investigating the medicinal value of the plants. Perhaps the cure to some diseases can be found in these plants.

You start with what you know about a topic and move on to new conclusions and new awareness. You end by reviewing and discussing the steps you took.

Flex Your Brain

1 Topic: _____

2 ? **What do I already know?**
1. _____
2. _____
3. _____
4. _____
5. _____

3 **Q:** Ask a question
_____

4 **A:** Guess an answer
_____

5 **How sure am I? (circle one)**

Not sure				Very sure
1	2	3	4	5

6 ? **How can I find out?**
1. _____
2. _____
3. _____
4. _____
5. _____

7 **EXPLORE**

8 **Do I think differently?** → yes no

9 ? **What do I know now?**
1. _____
2. _____
3. _____
4. _____
5. _____

10 **SHARE**
1. _____
2. _____
3. _____

1 Fill in the topic.

2 Jot down what you already know about the topic.

3 Using what you already know (step 2), form a question about the topic. Are you unsure about one of the items you listed? Do you want to know more? Do you want to know what, how, or why? Write down your question.

4 Guess an answer to your question. In the next few steps, you will be exploring the reasonableness of your answer. Write down your guesses.

5 Circle the number in the box that matches how sure you are of your answer in step 4. This is your chance to rate your confidence in what you've done so far and, later, to see how your level of sureness affects your thinking.

6 How can you find out more about your topic? You might want to read a book, ask an expert, or do an experiment. Write down ways you can use to find out more.

7 Make a plan to explore your answer. Use the resources you listed in step 6. Then, carry out your plan.

8 Now that you've explored, go back to your answer in step 4. Would you answer differently?

9 Considering what you learned in your exploration, answer your question again, adding new things you've learned. You may completely change your answer.

10 It's important to be able to talk about thinking. Choose three people to tell about how you arrived at your response in every step. For example, don't just read what you wrote down in step 2. Try to share how you thought of those things.

Activity

To acquaint students with this strategy, divide the class into small groups and have them explore a topic using the Flex Your Brain worksheet. Students should have some familiarity with the topic chosen. Possible topics might include cacti, volcanoes, robots, or topics from a list the class brainstorms. See steps on the student page for a more detailed description of the Flex Your Brain process. A reproducible version of the Flex Your Brain activity with space for students to write in their responses can be found on page 5 in the **Activity Worksheets** booklet in the **Teacher Classroom Resources.**

Use **Teaching Transparency 1** as you teach this lesson.

Use the **Study Guide,** p. 6, as you teach this lesson.

3 Assess

Check for Understanding

❓ FLEX Your Brain

Use the Flex Your Brain activity to have students explore SCIENTIFIC METHODS.

Activity Worksheets, p. 5

Reteach

LS **Logical-Mathematical** Have each student prepare an outline of a scientific method. Have them compare these outlines with those of other students and discuss any differences.

Extension

For students who have mastered this section, use the **Reinforcement** and **Enrichment** masters.

4 Close

•MINI•QUIZ•

Use the Mini Quiz to check for understanding.

1. A(n) _____ is a testable prediction about the answer to a problem. *hypothesis*

2. A(n) _____ is an explanation based on many observations supported by experimental results. *theory*

3. A(n) _____ is a "rule of nature." *scientific law*

Figure 1-8

You can learn about gravity by observing how different objects fall.

16

Scientific Methods

As you puzzled over the broken movie projector or the scrambled sounds on the tape, you planned a strategy and went through a series of steps to solve the problem. That's how scientists have approached scientific problems since the time of the founders of modern science. The series of steps scientists use are called scientific methods. You'll be using these methods in investigations throughout this textbook.

The first step in a scientific method is **observation,** which is using your senses to gather information. Observations can be used to study what is happening in **Figure 1-8.** In science, tools such as microscopes, rulers, and clocks can help make your observations more precise.

Good observations lead to testable predictions about how to solve a problem or explain how something works. In science, a testable prediction is called a **hypothesis.** When studying falling bodies, you can observe repeatedly that objects fall with increasing speed. You could come up with a hypothesis that the farther something falls, the greater the speed it attains.

A hypothesis can be tested by conducting experiments and making further observations. If your first hypothesis is not supported, does that mean you are a failure as a scientist? No. Even today, in modern laboratories where top scientists use state-of-the-art equipment, science advances by trial and error.

Test the Hypothesis

If one hypothesis turns out to be unsupported, another one can be proposed and tested. It's never possible to prove that a hypothesis is absolutely right. However, you can keep ruling out possibilities until you settle on one you think is most supported. It may take many experiments and many different kinds of data to thoroughly test a hypothesis. The more results you obtain that support the hypothesis, the more confident you can be that it's supported.

Even when you're confident that your hypothesis is confirmed, other scientists may disagree with you. In science, differences of opinions and even conflicting results from similar experiments are normal. Scientists are trained to question everything and always to look for other possible explanations.

Theories and Laws

Scientists use the information they gather during experiments to form a theory. A **theory** is an explanation based on many observations supported by experimental results. A theory is the most logical explanation of why things work the way they do. As new information is collected, a theory may need to be revised or discarded and replaced with another theory. An example of theory revision is shown in **Figure 1-9.**

Scientists also state scientific laws. A **scientific law** is a rule of nature that sums up related observations and experimental results to describe a pattern in nature. For example, scientists have observed for years that matter is never created or destroyed in chemical changes. This fact is stated as the law of conservation of mass. Generally, laws can be used to predict what will happen in a given situation, but they don't explain why. Theories can serve as explanations of laws. Like theories, laws can be changed or discarded if new observations show them to be incorrect.

Whether you're puzzling over a physical science mystery in the classroom or just trying to figure out where you left your homework, you can use what you've learned about problem solving to lead you to a solution.

Figure 1-9

Theories of the origin of the universe have changed over the years and will continue to change as new discoveries are made.

Section Wrap-up

Review

1. Explain the difference between an exercise and a problem. Give an example of each.

2. Compare and contrast these terms: *hypothesis, theory, scientific law.*

3. **Think Critically:** Develop a model to describe how you decide what to wear to school every day. Could someone who doesn't know you use the model to predict what you'll wear tomorrow?

Skill Builder
Outlining
Think of a problem you recently solved or tried to solve. Outline the steps you followed in searching for the solution. If you need help, refer to Outlining in the **Skill Handbook.**

Science Journal

In your Science Journal, describe exactly how you could solve the problem of making sure you are awake at a certain time each day. Then, describe how this problem can become an exercise.

Section Wrap-up

Review

1. With an exercise, the method for reaching the solution is known; solving a crossword puzzle. With a problem, the method of solution must be developed; fixing a flat tire.

2. A hypothesis is a testable prediction, based on the best available knowledge. A theory is an explanation based on observations supported by test results. A scientific law is a rule of nature that sums up related observations and experimental results.

3. **Think Critically** Student responses will vary, but could include such considerations as weather conditions, school dress codes, and current fashion trends. A stranger should be able to use a well-conceived model to make fairly accurate predictions.

Science Journal

You could buy an alarm clock, set the time, and set the alarm. The first time you did this, you would be solving a problem. To reset it every night is just an exercise.

Skill Builder
Student outlines should reflect steps and strategies in solving a particular problem.

Assessment

Performance To determine if students can effectively use problem-solving strategies, present a new problem, such as the classroom door would not unlock this morning, for them to solve. Use the Performance Task Assessment List for Making Observations and Inferences in **PASC,** p. 17.

Prepare

Section Background

Technology, such as special materials, extensive modeling, and advanced computer graphics, has greatly improved movie special effects.

1 Motivate

Bellringer

Before presenting the lesson, display **Section Focus Transparency 3** on the overhead projector. Assign the accompanying **Focus Activity** worksheet. L1 LEP

Tying to Previous Knowledge

Ask students to list special materials they have seen in the movies, such as a window made with sugar-glass.

2 Teach

Discussion

Ask students whether they think that technology in one area, such as the movies, has applications in other areas, such as transportation.

3 Assess

Check for Understanding

? FLEX Your Brain

Use the Flex Your Brain activity to have students explore CLAYMATION.

 **Activity Worksheets,** p. 5

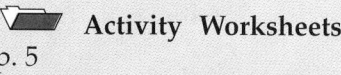

TECHNOLOGY:
1•3 Getting Real with Special Effects

Objectives

* Discuss the ways technology has made movie special effects more realistic.
* List the areas of science that are drawn upon for movie special effects.

Figure 1-10

Technology has helped make movie special effects more realistic.

Technology Makes It Real

Have you ever watched an old horror movie and laughed at how unrealistic the monsters seemed? Maybe there were also scenes in which the cars, planes, ships, and even whole cities were obviously toys or poorly made models. In your mind, compare those scenes to a recent science fiction or horror film you've seen. Even though you knew that much of what you were seeing on the screen was achieved through special effects, didn't it all seem realistic?

Thanks to technological advances in everything from materials to computers, special effects in today's movies are more convincing than ever, as shown in **Figure 1-10**. Even the simplest props, such as fake rocks and windows that an actor can safely jump through, are improvements over the old versions. In old movies, rocks were made from papier mâché. Today, polyurethane foam can be molded, cut, or carved into realistic-looking boulders. In a method similar to that used to build surfboards and some automobile bodies, fiberglass is used to create lightweight imitation rocks. At one time, fake glass was made from hardened sugar water. This candy glass, as it was called, would melt under bright lights or when water was splashed on it. Now, most fake glass in the movies is made of plastic that snaps into pieces on impact.

Program Resources

📁 **Reproducible Masters**
Study Guide, p. 7 L1
Reinforcement, p. 7 L1
Enrichment, p. 7 L3

🔦 **Transparencies**
Section Focus Transparency 3

Figure 1-11
Computers have improved the quality of movement in modern animation films.

Help by Computer

As you can see in **Figure 1-11,** computers have revolutionized special effects in all sorts of films—from cartoons to action adventures. You've probably seen animated films that use a process called claymation, in which figures are photographed one frame at a time and moved slightly between frames. Old attempts at claymation looked jerky and unnatural, but a technique called go-motion, in which computers move cameras and creatures at the same time, gives a much smoother effect.

Section Wrap-up

Review

1. Describe three ways that technological advances have made movie special effects more realistic.
2. Other than making movies more realistic, what are some advantages of using improved materials to make movie props?

inter*NET* CONNECTION Animation companies are using a technique called *motion capture* to produce more realistic animation. Visit the Chapter 1 Internet Connection at Glencoe Online Science, **www.glencoe.com/sec/science/physical,** for a link to more information about animation using motion capture.

1-3 Getting Real with Special Effects   19

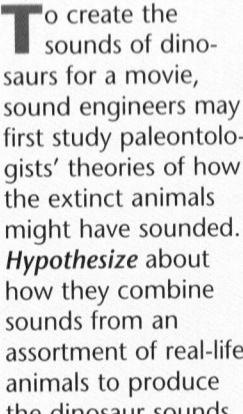

CONNECT TO

LIFE SCIENCE

To create the sounds of dinosaurs for a movie, sound engineers may first study paleontologists' theories of how the extinct animals might have sounded. *Hypothesize* about how they combine sounds from an assortment of real-life animals to produce the dinosaur sounds.

SCIENCE & SOCIETY

Use the **Study Guide,** p. 7, as you teach this lesson.

GLENCOE TECHNOLOGY

 Videodisc

The Infinite Voyage Series: The Great Dinosaur Hunt

Chapter 5

Dinomation International: Building Dinosaurs

Reteach

Ask a computer programmer to speak to your class about how computers are used to model items, colorize black-and-white films, and edit what appears on film.

Extension

For students who have mastered this section, use the **Reinforcement** and **Enrichment** masters.

CONNECT TO

LIFE SCIENCE

Answer Students will probably notice that animals similar in size make sounds similar in pitch. Animal sounds similar in type and pitch can be blended to simulate other animal sounds.

4 Close

Discussion

Ask students to give examples of how technology has made movies safer for the actors.

Section Wrap-up

Review

1. Better, more accurate models; new materials for props; and computer graphics are used.
2. New techniques and safer materials have reduced serious injuries.

Prepare

Section Background

Galileo is credited with establishing a scientific method in which explanations of phenomena are verified by controlled observations.

Preplanning

Assemble possible materials for Activity 1-1.

1 Motivate

Bellringer

 Before presenting the lesson, display **Section Focus Transparency 4** on the overhead projector. Assign the accompanying **Focus Activity** worksheet. L1 LEP

Tying to Previous Knowledge

Have students recall the meaning of the term *hypothesis*. Point out that in this section they will learn a standard method that is used to test a hypothesis.

2 Teach

Concept Development

Point out that experiments are conducted to confirm a hypothesis or to confirm a null hypothesis, that is, the hypothesis that the independent variable in the experiment does not affect the dependent variable.

1•4 Exploring Science

Science Words

experiment
control
constant
independent variable
dependent variable

Objectives

• Understand the importance of following guidelines in doing experiments.
• Define *constant, independent variable,* and *dependent variable.*
• Understand laboratory safety rules.

Scientifically Tested

Anyone who performs in a play or a film can be called an actor, whether or not that person makes a living by acting. A person who plays the piano doesn't have to sell out concert halls to be called a pianist. If you solve problems or find new facts by following the steps described in Section 1-2, are you a scientist?

What do scientists do? Scientists, such as the one shown in **Figure 1-12,** gather information from observations, look for patterns, form hypotheses, test their hypotheses with experiments, analyze the results, draw conclusions, and communicate results. The main work of scientists revolves around experiments—designing them, performing them, studying their outcomes, and deciding whether more experiments need to be done.

An **experiment** is an organized procedure for testing a hypothesis. When scientists conduct experiments, they usually are seeking new information. Classroom experiments often demonstrate and verify information that already is known but may be new to you. When doing an experiment, it is important to follow certain guidelines to reduce the chance of reaching wrong conclusions.

Popcorn on Ice

Look at **Figure 1-13.** Another problem has come up at the theater where you work. The popcorn just isn't popping as it should. Every batch has lots of unpopped kernels. You did so well solving the projector problem that the boss wants you to tackle this one. This time, you're going to approach the problem as a scientist would. You're going to experiment.

Figure 1-12

Scientists test their own hypotheses and check the work of other scientists.

20

Program Resources

Reproducible Masters
Study Guide, p. 8 L1
Reinforcement, p. 8 L1
Enrichment, p. 8 L3
Activity Worksheets, pp. 5-7 L1
Science Integration Activities, pp. 1-2
Multicultural Connections, pp. 5-6
Science and Society Integration, p. 5
Lab Manual 1

Transparencies
Section Focus Transparency 4 L1
Teaching Transparency 2 L1

Be in Control

You remember hearing somewhere that storing popcorn in a freezer makes it pop better. So, you try storing a package of popcorn in the theater's freezer for a day or two. Then, you pop the corn and count the unpopped kernels. Can you draw a meaningful conclusion from this experiment? No, because you don't have anything to compare your results with. Maybe you'd get the same number of unpopped kernels with corn stored at room temperature. You don't know because you didn't test any unfrozen popcorn.

To draw a conclusion, you often need a **control**—a standard for comparison. In an experiment, a control shows that your result is related to the condition you're testing and not to some other condition.

Making Changes

To find out whether storing popcorn in the freezer makes a difference, you need to test popcorn stored at room temperature as a control. All other conditions should be the same for both batches of popcorn. The popcorn should be the same brand, equally fresh, stored for the same period of time, and popped in the same corn-popping machine. Only one factor, the place of storage, should differ between batches. By changing only one factor at a time, while keeping everything else the same, you can find out the effect of that factor. If you change more than one factor at a time, you will not be able to sort out the effects of each change. All of the other factors—brand of popcorn, freshness, storage period, and popping conditions—are constants. A **constant** is a factor that doesn't vary in an experiment.

Can you draw conclusions after popping just one bag of popcorn stored each way? What if one of the bags happened to be defective? To be sure of your conclusions, you should repeat the experiment several times with different bags of popcorn, storing half the bags in the freezer and half at room temperature.

Figure 1-13

If there is a problem with popping this popcorn, an experiment can be conducted to help find out what's wrong. *What are some constants in such an experiment?*

1-4 Exploring Science   21

What factors would be constants in this experiment? *The constants may include brand, freshness, amount of corn to be popped, and the time and method of cooking, with the addition of similar storage conditions.*

Activity

LS **Interpersonal** Give student teams a large index card on which to write a problem for an experiment. Have teams exchange their cards. Each team generates a hypothesis for the question they now hold and writes it on the card. Again, they exchange the cards. Ask each team to devise an experiment based on the question and hypothesis on the card that they now have. **L1** **COOP LEARN**

USING MATH

Answer In both algebra and science, a variable represents something that can change.

USING MATH

In algebra, a variable is defined as a symbol that is used to represent an unspecified number. How is this definition similar to the definitions of independent and dependent variables?

Figure 1-14

Keeping accurate records is an important part of scientific experimentation.

Going Further

If your experiment shows that storing popcorn in the freezer makes it pop better, you may want to do another experiment to find out whether the length of time the popcorn is stored makes a difference. In the second experiment, all the popcorn would be stored in the freezer but for different lengths of time. Length of time would be the **independent variable,** the factor adjusted by the experimenter. The measure of popping success—the number of unpopped kernels—would be the dependent variable. A **dependent variable** is a factor whose value depends upon the value of the independent variable. What factors would be constants in this experiment?

Watch Your Steps

As shown in **Figure 1-14,** keeping accurate records for an experiment will help you to draw conclusions and decide what to do next in your investigation. It will also allow others to duplicate your investigation. Although scientists agree that there is no one way to solve a problem, their scientific methods often include the following steps.

- *Determine the problem.* What do you want to find out?

- *Make a hypothesis.* What prediction do you want to test?

- *Test your hypothesis.* What steps can you take to reach a conclusion about your hypothesis? What measurements should you record?

- *Analyze the results.* What happens during your experiment?

- *Draw conclusions.* Do your observations and data suggest that your hypothesis is supported? If not, do you think your hypothesis should be changed, or does your experimental procedure need adjustment?

Across the Curriculum

Language Arts Have students look up the word *experiment* in a dictionary. Students should find that it is related to the Latin root *experiri* (to try). Have students explain how the meaning of the word reflects its root. **LS**

As you've seen, scientists use logical methods that can lead to new discoveries and solutions to many problems—just as you did with the popcorn problem. As you did, they record their results so that they can properly analyze them. And, they probably like good popcorn, too. So, are you a scientist? In the sense that you use the same methods that scientists use—yes, you are.

Play It Safe

In movies, the chase scenes, car crashes, and fiery explosions add to the excitement. Filming these scenes can be dangerous, but stunt people take many precautions to minimize their chances of getting hurt. They wear fireproof clothing and protective pads and use props that can be smashed over someone's head without hurting the person. The cars used in crash scenes are reinforced with steel bars to protect the drivers. Special seat belts, air bags, and padding also are used.

While doing science experiments, you'll be asked to follow safety rules, too. There are risks involved in handling glass, hot objects, sharp objects, and certain chemicals. Show proper respect for these new situations, and follow all safety rules, shown in **Figure 1-15,** below and on the next page.

Figure 1-15

The safety rules depicted here and on the next page are essential parts of working in a laboratory.

A First, know what you're supposed to do before you begin. Read and follow all directions carefully, and pay close attention to any caution statements.

B Dress appropriately to reduce the chance of accidents. Avoid clothing with large, floppy sleeves. Remove dangling jewelry, and tie back or cover long hair.

1-4 Exploring Science   23

Teacher F.Y.I.

Alfred Nobel, the man who left his fortune to fund the Nobel prizes, was the inventor of dynamite.

Use **Teaching Transparency 2** as you teach this lesson.

Use these program resources as you teach this lesson.
Study Guide, p. 8
Science Integration Activities, pp. 1-2
Multicultural Connections, pp. 5-6
Science and Society Integration, p. 5
Lab Manual 1

Across the Curriculum

Art Safety symbols are icons—images that convey information. Have students design an icon to alert students to the laboratory safety hazards of long hair, clothing with floppy sleeves, or jewelry. L1
LEP LS

Inclusion Strategies

Learning Disabled Have students make a chart showing the major safety rules that should be followed while conducting a lab. Their charts should illustrate the rules and be displayed all year in the classroom.

? FLEX Your Brain

Use the Flex Your Brain activity to have students explore SAFETY ICONS.

📁 **Activity Worksheets,** p. 5

Reteach

Discussion Ask students, "How many of you have heard the statement that hot water freezes faster than cold water?" Ask students to devise an experiment to find out. Make sure that students use the same volume of water, the same size and kind of container, and the same freezer. Both samples should be placed in the freezer at the same time.

Extension

📁 For students who have mastered this section, use the **Reinforcement** and **Enrichment** masters.

4 Close

•MINI•QUIZ•

Use the Mini Quiz to check students' recall of chapter content.

1. **What is an experiment?** *an organized procedure for testing a hypothesis*

2. **In an experiment a(n) _____ is used as a standard for comparison.** *control*

3. **The _____ is the factor that is adjusted by the experimenter.** *independent variable*

C Think about what you're doing. Handle glass carefully, and don't touch hot objects with your bare hands. To be on the safe side, treat all equipment as if it were hot. Wear safety glasses and a protective apron.

D Be sure you understand the safety symbols explained in Appendix B on page 731. They will be used throughout the textbook to alert you to possible laboratory dangers and precautions to take.

E Your workspace should contain only materials needed. Keep other objects out of the way. Arrange your equipment so that you won't have to reach over burners or across equipment that could be knocked over.

F Watch closely when your teacher demonstrates how to handle equipment, how to dispose of materials, how to clean up spills, and any other safety practices. Make sure you know the location and proper operation of safety equipment, such as fire extinguishers.

Exploring Your World

You've seen how scientists make observations, look for patterns, experiment, and draw conclusions. Every day, you can use the same methods to investigate what's going on around you and to safely solve problems for yourself, your friends, family, and community, as shown in **Figure 1-16**.

Figure 1-16

Safety precautions reduce the risk of injury whether on a movie set or in the lab.

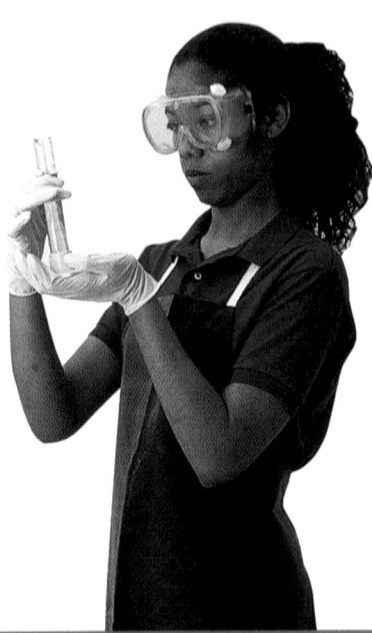

Section Wrap-up

Review

1. What is the function of a control in an experiment?

2. You are doing an experiment to find out how the temperature of water affects the rate at which sugar dissolves. What factors should remain constant? What factors will vary?

3. **Think Critically:** Design an experiment to determine the fastest route from your home to school.

Skill Builder

Hypothesizing

A chemist made some observations about a new type of plastic that may biodegrade in acidic soils. State a possible hypothesis about these findings. How could your hypothesis be tested? If you need help, refer to Hypothesizing in the **Skill Handbook.**

USING MATH

Select one or more classmates to try the popcorn experiment described in this chapter at home. Use microwave popcorn, with at least five bags stored in the freezer and an equal number stored at room temperature. For popcorn stored each way, calculate the average number of kernels left after popping.

Section Wrap-up

Review

1. A control is used as a basis for comparison with other data.
2. Constants: quantity of water, quantity of sugar; variable: temperature of water
3. **Think Critically** Student responses will vary, but should include at least one constant—mode of transportation—and one variable—routes taken.

USING MATH

This activity will have students calculating simple averages, where the average is the sum divided by the number of trials (5). Be sure brands, popping time, and other factors remain constant.

Skill Builder

Possible Hypothesis: The plastic material will degrade faster as the soil it is in becomes more acidic. *Testing:* Bury equal-sized pieces of plastic the same depth in the same soil under measured variable acid conditions. After a constant time, examine the plastic for signs of decay.

✓Assessment

Performance Have students list the constants, dependent variable, and independent variables for the Skill Builder. Use the Performance Task Assessment List for Making Observations and Inferences in **PASC,** p. 17.
P

PREPARATION

Purpose
LS **Logical-Mathematical** Students will test a hypothesis by controlling the variables in an experiment. **L1**

COOP LEARN

Process Skills
observing, predicting, communicating, forming a hypothesis, designing an experiment, interpreting data, controlling the variables in an experiment

Time
30 minutes

Materials
Students may choose other methods of suspending the weights.

Alternate Materials Metal washers or hex nuts are easier to tie to the strings and are less likely to fly off than bolts.

Safety Precautions
Warn students not to swing the weights so hard that they fly off or tip the ring stands.

Possible Hypotheses
Some groups will suggest that the most important variable is the weight. Others will vary the string length. A few groups may decide to change the distance between the strings.

📁 Activity Worksheets, pp. 5-7

Activity 1-1

Design Your Own Experiment
Identifying and Controlling Variables

When you have a problem to solve, a logical plan often works best. Suppose that at the theater concession stand, all the instructions for popping popcorn are followed, but the popcorn still burns. What might be the cause? Maybe not enough oil was used, the popcorn itself did not contain the right amount of moisture, the temperature was too high, or the popping time was too long. You have to be sure that you change only one variable at a time. Here's another problem to solve by controlling variables.

PREPARATION

Problem
In what way does one swinging weight influence another swinging weight?

Form a Hypothesis
Choose a variable and make a hypothesis about the relationship between the variable and how one swinging weight might affect another swinging weight.

Possible variables include string length, distance between weights, amount of weight, and distance of the swing.

Objectives
- Observe the movements of one swinging weight, and notice how these movements affect the movement of another swinging weight.
- Accurately measure changes of the variable.

Possible Materials
- string
- identical weights
- metric ruler
- ring stands
- scissors
- stopwatch or timer

Safety Precautions 🥽 ✋
Refer to Appendix B on page 731 for an explanation of safety symbols.

PLAN THE EXPERIMENT

1. As a group, agree upon and write out your hypothesis statement, clearly stating which variable you will be testing.
2. As a group, decide on the specific steps that you will use to test your hypothesis. Describe each step carefully.
3. Use the suggested materials to build a suspension setup similar to that shown. Your setup will differ according to the variable you have chosen to test. Remember to keep all other possible variables constant.
4. In your Science Journal, design a data table that can be used to record data and observations.

Check the Plan
1. Read over your steps to make sure that they are in a logical order, and test your hypothesis.
2. Will you want to test the variable more than once?
3. *Make sure your teacher approves your plan and that you have included any changes suggested in the plan.*

DO THE EXPERIMENT

1. With the suspended weights starting in a stationary position, pull one back a measured distance and release it. As it swings, observe the effect on the other weight that is attached to the string.
2. Be sure to make all time and distance measurements accurately and record them in your data table.

Analyze and Apply
1. **Compare** the findings about your variable with those of other groups that checked the same variable.
2. Was your hypothesis supported? Explain.
3. **Compare** your results with those of groups that tested different variables. Which variable seemed to have the most effect on the other weight? Explain.

PLAN THE EXPERIMENT

Possible Procedures
After setting up the materials as determined by the hypothesis, students should swing one weight and observe and take careful measurements of what happens to the other weight.

Teaching Strategies
Troubleshooting Make sure that students change only one variable at a time. Also suggest that students measure the time for ten swings rather than the time for one swing.

DO THE EXPERIMENT

Expected Outcome
The swinging weight will transfer some energy to the other weight. As the weights swing, the distance from vertical will decrease due to friction. If the length of the string varies, longer strings take more time. If weights vary, there is more effect on the lighter weight. If distance between weights varies, the closer they are, the more effect there is.

Analyze and Apply
1. Students will find that other lab groups will have similar results.
2. Answers will vary. Students should justify their answers.
3. The string length had the largest effect on the results. The horizontal distance between the strings had the least effect.

Go Further

With a third weight added, a swinging weight will transfer some of its energy into both of the other weights.

✓ Assessment

Oral Ask students to predict if the experiment would work for 1-kg masses. Have two students build the large-scale system and then put it in motion. Use the Performance Task Assessment List for Carrying Out a Strategy and Collecting Data in **PASC**, p. 25. **P**

Background

Science teacher Priscilla King believes in involving her students in hands-on projects. She says, "Kids seem to be able to learn more easily and remember what they learn when they do something like building a model of cells with yarn, paper, and other art materials." Several science museums in the United States share her philosophy. Their exhibits encourage visitors to push, pull, and prod to discover the how and why of science. The Exploratorium in San Francisco, California, is one of the leading hands-on museums. Its hundreds of interactive displays let visitors touch a tornado, look inside an eye, blow giant soap bubbles, and build a bridge they can walk across.

Teaching Strategies

Challenge your students to think of a tradition in their own culture or a culture they know about and to tell what principles of science are involved in that tradition. Traditions might include a cut Christmas tree illustrating capillary action or a spinning Hanukkah dreidel demonstrating centrifugal force.

Career Connection

Career Path High school science teachers usually earn a bachelor's or master's degree in their chosen area of science while completing an approved teacher training program.

People and Science

PRISCILLA KING, *Science Teacher*

On the Job

Q How do the science classes you teach on the Navajo reservation differ from those in an urban school?

A I try to build bridges between science and Navajo traditions. For example, when we're studying stars, I invite an elder from the tribe to present Navajo myths about the creation of the constellations. In a botany class, I help the students collect plants to dye wool for a Navajo rug. Instead of dissecting a frog, my class may butcher a sheep to study the muscles, organs, and bones. Then, in the Navajo way, we make a stew to eat in class and dry some meat to preserve it. Another project might be hunting a deer, studying its body, then tanning the hide. There are scientific principles to learn in every step of that process.

Q Is there a reason, besides the long Navajo tradition of sheep herding, for choosing a sheep over a frog for dissection?

A Yes. The Navajo believe that sheep are a part of our food source but that killing a frog, which is not a usual food source, requires a ceremony to reestablish harmony among living things.

Personal Insights

Q Does scientific problem solving also take a local twist in your classes?

A Yes. Because some of the students don't have running water in their homes, I show them how to test their water source and figure out how to improve its safety. Also, we do an experiment with yucca, a common plant in our area. Students make a shampoo from the yucca root and compare its cleansing properties with those of commercial shampoo.

Q How do you maintain your Navajo heritage?

A I continue to speak the language and keep my Navajo name, Nangisbah, which was given to me by my medicine-man grandfather when I was six years old. That name means "warrior." I guess it fits me because I stand up for what I think is right, even in small ways, like making sure the girls' volleyball team I coach has the same privileges as the football team.

Career Connection

Talk to several teachers in your school. Find out how they relate their subject to students' lives. Then make a poster to show what you learn.

- **Safety Consultant**
- **Curriculum Specialist**

For More Information

Students could write to these organizations:

American Federation of Teachers
555 New Jersey Avenue, NW
Washington, DC 20001

National Education Association
1201 16th Street, NW
Washington, DC 20036

Review

Summary

1-1: Science Is Everywhere

1. Pure science involves gathering information or the discovery of a pattern. When that information is applied to some use, it becomes technology.
2. Physical science is the study of matter and energy.

1-2: Finding Out

1. A problem is a situation that seems to be missing some information. An exercise is a situation where known steps need to be taken to find a solution.
2. Solving problems involves looking for patterns, making predictions, and testing the predictions. Often, a model can be developed that helps to solve a problem.
3. A hypothesis is a testable prediction. A theory, based on many observations, is the most logical explanation of why things work. A scientific law is a summary of many experimental results that describes a pattern in nature.

1-3: Science and Society: Getting Real with Special Effects

1. New technology, new materials, and computer assistance have helped to make some events and objects in movies seem much more realistic than they have seemed in the past.

1-4: Exploring Science

1. An experiment is an organized procedure for testing a hypothesis.
2. Typically, an experiment will have a control, which is a standard for comparison, and two types of variables. Independent variables are adjusted to different values by the experimenter. Dependent variables have values that change with a change in an independent variable.
3. Tested experiments can be done safely when everyone is aware of and follows standard safety warnings.

Key Science Words

a. constant
b. control
c. dependent variable
d. experiment
e. hypothesis
f. independent variable
g. model
h. observation
i. physical science
j. scientific law
k. technology
l. theory

Reviewing Vocabulary

Match each phrase with the correct term from the list of Key Science Words.

1. the study of matter and energy
2. applying science to solve problems
3. an idea, system, or structure that can be used to solve a problem
4. a testable prediction
5. the use of human senses to gather information
6. an organized method of testing a hypothesis
7. describes but doesn't explain a pattern in nature
8. a factor that doesn't change in an experiment
9. a standard for comparison
10. factor in an experiment that is adjusted by the experimenter

Summary

Have students read the summary statements to review the major concepts of the chapter.

Reviewing Vocabulary

1. i	**6.** d
2. k	**7.** j
3. g	**8.** a
4. e	**9.** b
5. h	**10.** f

✔ Assessment

Portfolio Encourage students to place in their portfolios one or two items of what they consider to be their best work. Examples include:
- Performance Assessment, p. 5
- MiniLAB Assessment, p. 14
- Skill Builder Assessment, p. 25 **P**

Performance Additional performance assessments may be found in **Performance Assessment** and **Science Integration Activities**. Performance Task Assessment Lists and rubrics for evaluating these activities can be found in Glencoe's **Performance Assessment in the Science Classroom.**

GLENCOE TECHNOLOGY

▭ MindJogger Videoquiz

Chapter 1 Have students work in groups as they play the Videoquiz game to review key chapter concepts.

Chapter 1 Review

Checking Concepts

1. b	**6.** c
2. c	**7.** d
3. a	**8.** b
4. b	**9.** b
5. b	**10.** a

Understanding Concepts

11. With an exercise, you know exactly what you are trying to solve and the method to use to reach the solution. Solving a crossword puzzle is an example of an exercise. A problem involves a great deal of uncertainty. A crossword puzzle would be a problem if the clues were not numbered.

12. A hypothesis, a theory, and a scientific law all deal with using information to answer a question or solve a problem. A hypothesis is a testable prediction based on the best information available. A theory is an explanation that has been tested and supported by results. A scientific law is a statement that describes a pattern in nature.

13. A written record of observations and data allows another person to duplicate the experiment exactly and to compare the results with those of the original experiment.

14. Student responses may vary, but should include four of the safety precautions included in the chapter.

15. Student responses will vary, but most students will realize that the use of models and computer graphics can be important.

Thinking Critically

16. A pure scientist might ask about the sun's size, composition, age, temperature,

---

Checking Concepts

Choose the word or phrase that completes the sentence or answers the question.

1. The study of science for the sole purpose of advancing our knowledge is called _____ science.
   a. experimental   c. hypothetical
   b. pure   d. technological
2. Physical science would involve the study of all of the following except _____ .
   a. the melting point of wax
   b. the composition of wax
   c. the behavior of bees producing wax
   d. energy released by burning wax
3. A(n) _____ is used as a standard for comparison in an experiment.
   a. control   c. independent variable
   b. theory   d. constant
4. A _____ is an explanation supported by experimental results.
   a. conclusion   c. scientific law
   b. theory   d. model
5. Advances in technology have helped to make modern movies _____.
   a. unsafe to make   c. look jerky
   b. more realistic   d. appear unnatural
6. Which of these is not a safety rule to be followed in every laboratory?
   a. Follow all directions carefully.
   b. Clean up all spills immediately.
   c. Wear rubber-soled shoes at all times.
   d. Know the location of the fire extinguisher.
7. An experiment is used for testing a _____.
   a. model   c. law
   b. theory   d. hypothesis
8. The _____ describe(s) the steps followed in conducting an experiment.
   a. problem   c. conclusion
   b. procedure   d. data
9. A scientific _____ is sometimes called a rule of nature.
   a. theory   c. model
   b. law   d. hypothesis

---

10. Which is the final step in an experiment?
    a. Reach a conclusion.
    b. State the problem.
    c. Set up a procedure.
    d. Record data.

Understanding Concepts

Answer the following questions in your Science Journal using complete sentences.

11. Why is it usually easier to find the solution to an exercise than it is to find the solution to a problem? Include examples in your answer.

12. Describe how a hypothesis, a theory, and a scientific law are related.

13. Discuss the importance of recording all observations and data when conducting a science experiment.

14. List four safety precautions to be followed in the science laboratory.

15. Describe a special effect that you remember from a movie, and explain how science may have been used to bring about the effect.

Thinking Critically

16. In a study of the sun, what questions might an applied scientist ask? A pure scientist? How might the two sets of questions be related?

17. What aspects of the following items would a physical scientist be interested in?
    a. an electric guitar
    b. a piece of coal

---

movements, and methods of energy production. An applied scientist might ask how the sun's energy can be harnessed for use on Earth. The applied scientist must use the pure scientist's findings.

17. Examples could include (a) how the length and thickness of the strings or the shape, composition, and structure of the guitar body affect the sound; (b) the chemical composition of the coal, its hardness and color, how hot it must be before it burns.

18. a. Break it down into simpler problems.
    b. Look for patterns.
    c. Create and study models.

19. *Constants:* volume of water, mass of sugar; *Control:* water at same temperature; *Independent variable:* water temperature; *Dependent variable:* time required for sugar to dissolve

20. A scientist is able to eliminate an incorrect hypothesis as a possible solution to a problem.

18. Which problem-solving approach would you use to solve the following problems? Explain your choice in each case.
    a. solving a complex word problem in math
    b. figuring out a secret code
    c. predicting how a skyscraper would be affected by an earthquake
19. Design an experiment to determine how the temperature of water affects the time it takes sugar to dissolve in it. Identify the variables, constants, and control, if any.
20. Explain this statement: Scientists often learn as much from an incorrect hypothesis as they do from one that is correct.

Developing Skills

If you need help, refer to the **Skill Handbook**.

21. **Comparing and Contrasting:** Compare and contrast the methods used to complete an exercise and to solve a problem.
22. **Hypothesizing:** Propose a hypothesis to explain why a balloon filled with air weighs more than a deflated balloon.
23. **Interpreting Scientific Illustrations:** The following are safety symbols used in two different experiments. Using Table B-2 on page 731 in Appendix B, write what safety precaution is indicated by each symbol.
    a.
    b.
24. **Using Variables, Constants, and Controls:** Do some objects fall faster than others? Design an experiment to find out. State your hypothesis and describe your procedure for testing it. Identify your variables, constants, and controls.

25. **Concept Mapping:** Fill in the following events chain concept map to show the steps in organizing a scientific experiment. Use the following terms: *draw conclusions, make a hypothesis, determine the problem, analyze the results,* and *test your hypothesis.*

Initiating event

```
┌─────────────────────────────┐
│   Determine the problem.     │
└─────────────────────────────┘
              │
              ▼
┌─────────────────────────────┐
│     Make a hypothesis.       │
└─────────────────────────────┘
              │
              ▼
┌─────────────────────────────┐
│     Test your hypothesis.    │
└─────────────────────────────┘
              │
              ▼
┌─────────────────────────────┐
│     Analyze the results.     │
└─────────────────────────────┘
              │
              ▼
┌─────────────────────────────┐
│      Draw conclusions.       │
└─────────────────────────────┘
```

Performance Assessment

1. **Making and Using a Classification System:** One way to solve a problem is to look for patterns within a large system. When you enter a store that sells recorded music, what patterns can you see that help the overall organization of the music in the store? Explain.
2. **Designing an Experiment:** In Activity 1-1, you carefully controlled and identified variables to solve a problem. Assume you are testing two different ink pens to see which one writes for the longer time. What are some important constants to consider?
3. **Poster:** Make a poster that illustrates laboratory safety techniques.

Developing Skills

21. **Comparing and Contrasting** The methods used to solve a problem are developed as the solution is sought. The methods used to complete an exercise have already been tested.
22. **Hypothesizing** The inflated balloon contains something that the deflated balloon does not—air—and it has weight.
23. **Interpreting Scientific Illustrations** (a) an open flame; safety goggles should be worn; chemicals used are poisonous; the chemicals are caustic to the skin. (b) a possible electric hazard; the equipment being used is hot; safety goggles should be worn; the chemicals used are potentially explosive.
24. **Using Variables, Constants, and Controls** Students should describe an experiment in which they measure the time it takes for objects having different physical characteristics (mass, shape, size) to fall the same distance. The distance will be constant. Variables will consist of the differences in the objects tested.
25. **Concept Mapping** See student page.

Performance Assessment

1. The store separates music by type (rock, R&B, classical, etc.). Then the items are arranged by artist, then by media (tapes are separated from CDs). This system makes it quite simple to find specific items. Use the Performance Task Assessment List for Making and Using a Classification System in **PASC,** p. 49. [P]

2. Things to consider include the length of line and the thickness of lines. Use the Performance Task Assessment List for Designing an Experiment in **PASC,** p. 23. [P]
3. Posters should reflect safety rules from Section 1-4. Use the Performance Task Assessment List for Poster in **PASC,** p. 73. [P]

Chapter Organizer

Section	Objectives/Standards	Activities/Features
Chapter Opener		Explore Activity: Design your own measuring device. p. 33
2-1 Standards of Measurement (2 sessions, 1 block)*	1. **Define** standard of measurement. 2. **Identify** the need for standards of measurement. 3. **Name** the prefixes used in SI and indicate what multiple of ten each represents. National Content Standards: (5-8) UCP1, UCP3, UCP5, A2, D3; (9-12) UCP1, UCP3, UCP5, A2	Connect to Earth Science, p. 36 Science Journal, p. 36 Skill Builder: Sequencing, p. 36 Activity 2-1: Metric Munchies, p. 37
2-2 Using SI Units (2 sessions, 1 block)*	1. **Identify** SI units and symbols for length, volume, mass, density, time, and temperature. 2. **Define** derived unit. 3. **Convert** a measurement among related SI units. National Content Standards: (5-8) UCP1, UCP3, UCP5, A1, A2, E1, E2, F5; (9-12) UCP1, UCP3, UCP5, A1, A2, E1, E2, F6	Using Math: Converting Meters to Centimeters, p. 40 Using Technology: Space Spheres, p. 41 Using Math: Converting Centimeters to Liters, p. 42 Problem Solving: Verifying Volumes: Finding the Best Value, p. 43 MiniLAB: What is the density of a pencil? p. 44 Using Math, p. 45 Skill Builder: Concept Mapping, p. 45 Activity 2-2: Setting High Standards...For Measurement, pp. 46-47
2-3 Graphing (3 sessions, 2 blocks)*	1. **Identify** three types of graphs and explain the correct use of each type. 2. **Distinguish** between dependent and independent variables. 3. **Interpret** graphs. National Content Standards: (5-8) UPC1, UCP5, A2; (9-12) UCP1, UCP5, A2	MiniLAB: How can a graph help you observe change? p. 48 Using Computers, p. 51 Skill Builder: Making and Using Graphs, p. 51
2-4 Science and Society (1 session, ½ block)*	1. **Analyze** the advantages and disadvantages of universal use of SI measurements. 2. **Give examples** of SI units already commonly used in the United States. National Content Standards: (5-8) UCP1, UCP5, A1, A2, F5; (9-12) UCP1, UCP5, A1, A2, F6	Science Journal, p. 53 Explore the Issue, p. 53 Science and Art: Cubism, p. 54

Activity Materials

Explore	Activities	MiniLABs
page 33 device for measuring length	page 37 balance, 100-mL graduated cylinder, 1/2 cup margarine, 2 cups sugar, 6 T cocoa, 1/2 cup milk, 3 cups rolled oats, 1 t vanilla, 1/2 cup nuts, measuring cup, measuring teaspoon, measuring tablespoon, 1-quart plastic container, hot plate, cooking pot, oven mitt pages 46-47 string, scissors, marking pen, masking tape, miscellaneous objects for standards	page 44 balance with masses, pencil, water, 100-mL graduated cylinder page 48 thermometer, plastic foam cup with lid, water, hot plate, clock or watch with second hand

Need Materials? Call Science Kit (1-800-828-7777).   * A complete Planning Guide that includes block scheduling is provided on pages 32T-35T.

Teacher Classroom Resources

Reproducible Masters	Transparencies	Teaching Resources
Study Guide, p. 9 Reinforcement, p. 9 Enrichment, p. 10 Activity Worksheets, pp. 8-9 Multicultural Connections, p. 7 Concept Mapping, pp. 9-10 Critical Thinking/Problem Solving, p. 8	Section Focus Transparency 5, Measurements in Football Teaching Transparency 3, SI Units and Prefixes	Physical Science CD-ROM Spanish Resources English/Spanish Audiocassettes Cooperative Learning Resource Guide Lab Partner Lab and Safety Skills Lesson Plans
Study Guide, p. 10 Reinforcement, p. 10 Enrichment, p. 10 Activity Worksheets, pp. 10-11, 12 Lab Manual 2, No Need to Count Your Pennies Science Integration Activity 2, Measuring Up to Standards Technology Integration, pp. 7-8	Section Focus Transparency 6, Speedometers	**Assessment Resources** Chapter Review, pp. 7-8 Assessment, pp. 9-12 Performance Assessment in the Science Classroom (PASC) MindJogger Videoquiz Alternate Assessment in the Science Classroom
Study Guide, p. 11 Reinforcement, p. 11 Enrichment, p. 11 Activity Worksheets, p. 13 Lab Manual 3, Viscosity	Section Focus Transparency 7, Average Daily Temperatures Science Integration Transparency 2, Graphing a Greenhouse Gas Teaching Transparency 4, Types of Graphs	Performance Assessment Chapter Review Software Computer Test Bank
Study Guide, p. 12 Reinforcement, p. 12 Enrichment, p. 12 Science and Society Integration, p. 6 Cross-Curricular Integration, p. 6	Section Focus Transparency 8, Olympic Performers	**Key to Teaching Strategies**

Key to Teaching Strategies

The following designations will help you decide which activities are appropriate for your students.

L1 Level 1 activities should be within the ability range of all students, including those with learning difficulties.

L2 Level 2 activities should be within the ability range of the average to above-average student.

L3 Level 3 activities are designed for the ability range of above-average students.

LEP LEP activities should be within the ability range of Limited English Proficiency students.

LS These activities are designed to address different learning styles.

COOP LEARN Cooperative Learning activities are designed for small group work.

P These strategies represent student products that can be placed into a best-work portfolio.

GLENCOE TECHNOLOGY

The following multimedia resources are available from Glencoe.

Science and Technology Videodisc Series (STVS)
Chemistry
  Losing Weight by Design
Earth & Space
  Mapping with a Rifle
  Map Science
  Charting Air Space

The Infinite Voyage Series
Sail On, Voyager

National Geographic Society Series
STV: Solar System

Physical Science CD-ROM

Teacher Classroom Resources

This is a representation of key blackline masters available in the Teacher Classroom Resources.

Teaching Aids

Section Focus Transparencies

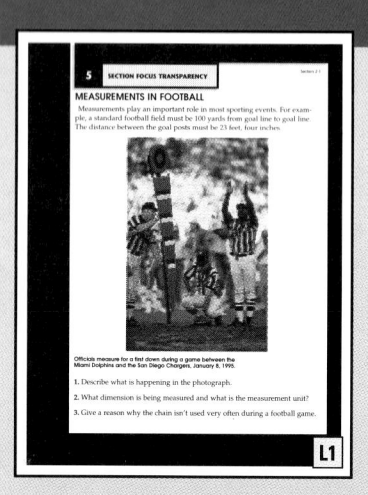

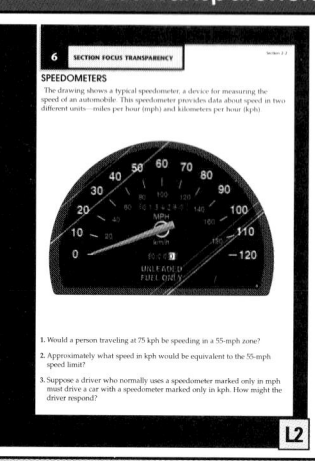

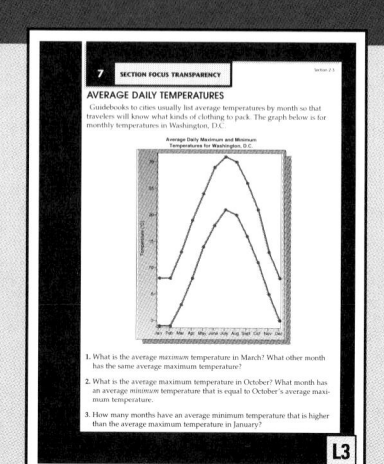

Science Integration Transparencies

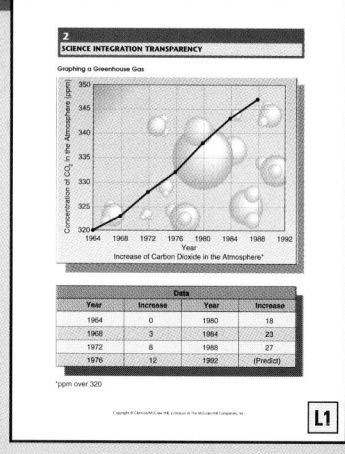

Teaching Transparencies

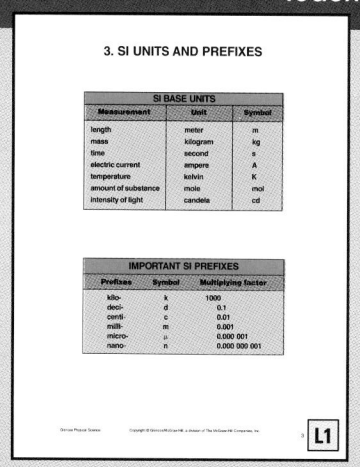

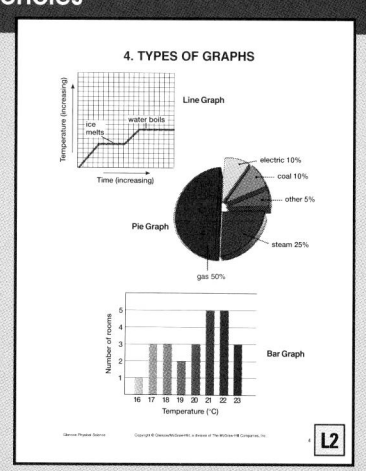

Meeting Different Ability Levels

Study Guide

Reinforcement

Enrichment Worksheets

Chapter 2 Physical Science Methods

Hands-On Activities

Science Integration Activity

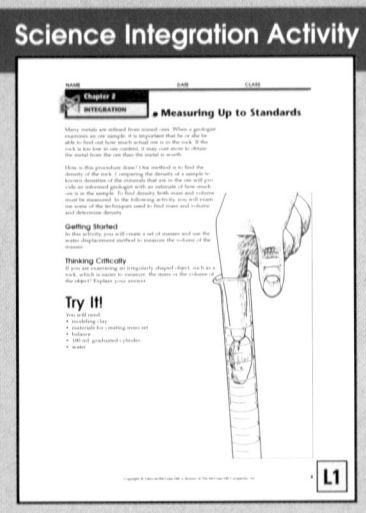

Measuring Up to Standards

L1

Lab Manual

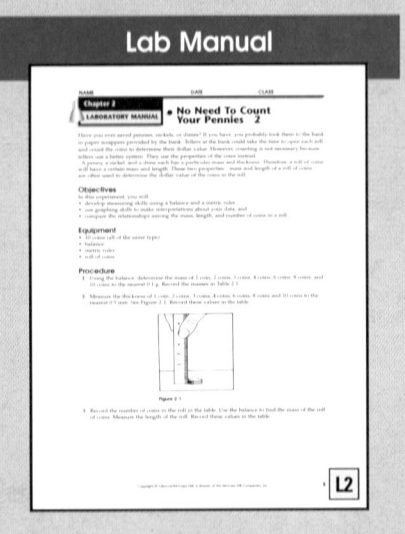

No Need To Count Your Pennies 2

L2

Assessment

Performance Assessment

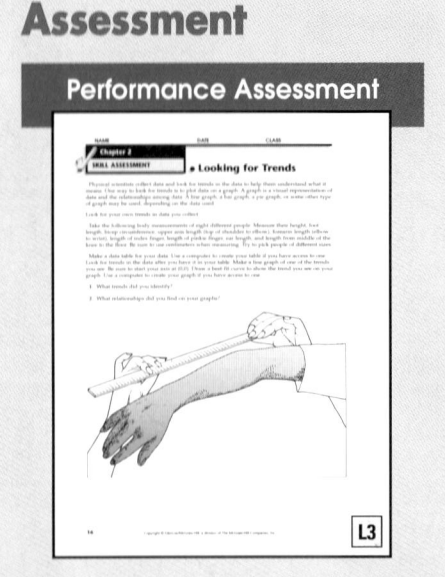

Looking for Trends

L3

Enrichment and Application

Critical Thinking/ Problem Solving

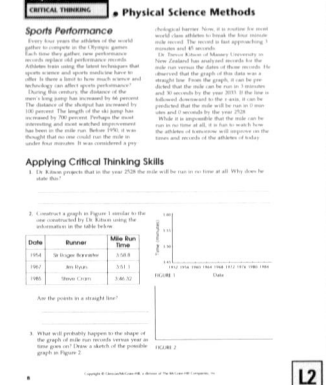

Physical Science Methods

L2

Cross-Curricular Integration

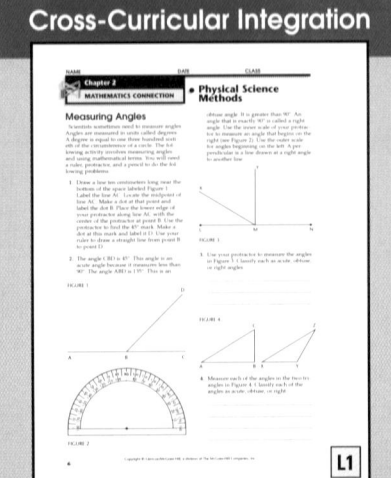

Physical Science Methods

L1

Science and Society Integration

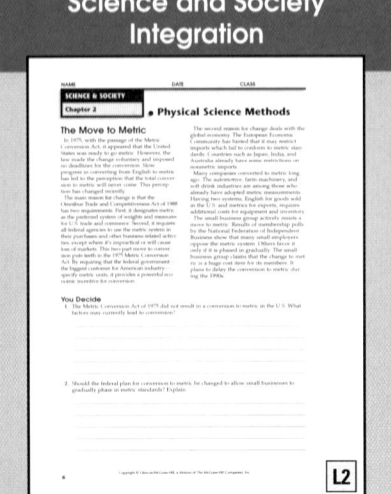

Physical Science Methods

L2

Technology Integration

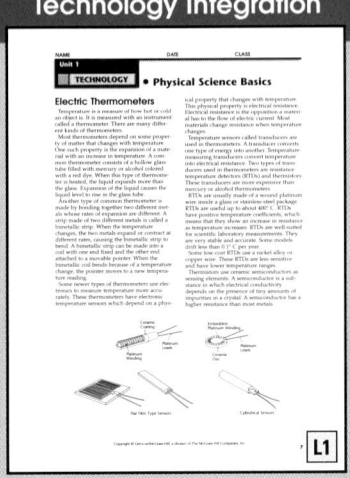

Physical Science Basics

L1

Multicultural Connections

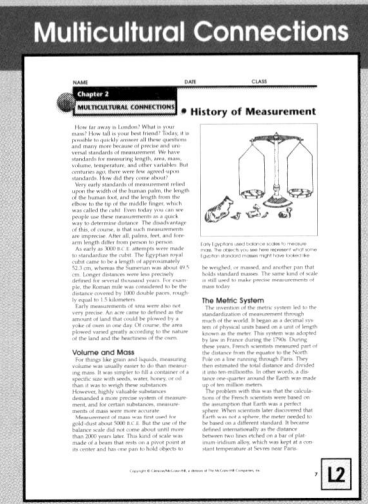

History of Measurement

L2

Concept Mapping

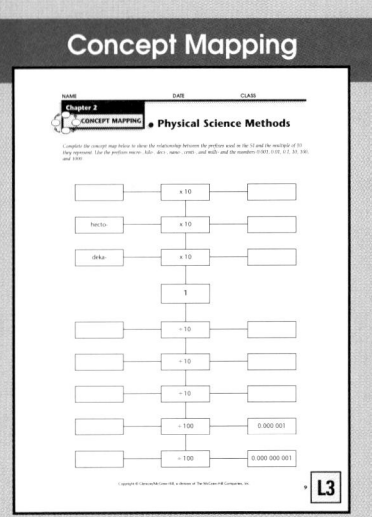

Physical Science Methods

L3

32D

Physical Science Methods

CHAPTER OVERVIEW

Section 2-1 This section introduces the SI system as a standardized decimal system of measurement that is being used by the scientific community and by most countries.

Section 2-2 The SI prefixes and base units for length, mass, time, and temperature are introduced in this section. The relationships among length, volume, mass, and density are discussed.

Section 2-3 Three types of graphs and their specific uses are presented as ways of displaying information.

Section 2-4 Science and Society Students are asked to evaluate the impact of converting the United States to SI.

Chapter Vocabulary

standard	kilogram
SI	density
meter	time
volume	second
derived unit	kelvin
liter	graph
mass	

Theme Connection

Scale and Structure This chapter introduces the SI system of measurement and graphing as two methods of gaining and interpreting information. In developing the SI system, stress how base units are related to each other and how the same set of prefixes is used for all units.

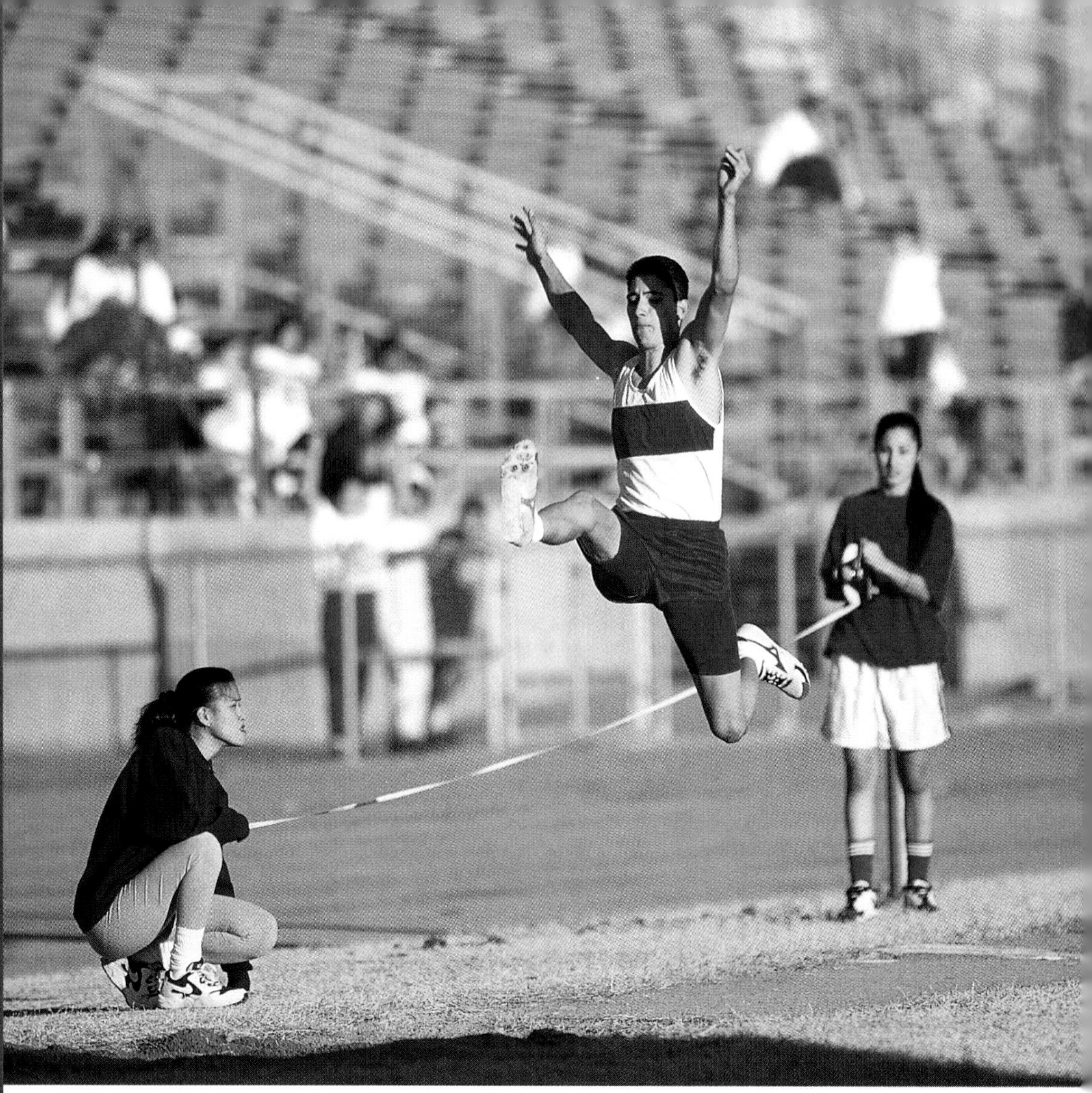

Previewing the Chapter

Section 2-1 Standards of Measurement
  ▶ Units and Standards
  ▶ Measurement Systems
  ▶ Measuring to the Moon

Section 2-2 Using SI Units
  ▶ SI Units and Symbols
  ▶ Length
  ▶ Volume
  ▶ Mass

  ▶ Density
  ▶ Time and Temperature

Section 2-3 Graphing
  ▶ Using Graphs
  ▶ Line Graphs
  ▶ Bar Graphs
  ▶ Circle Graphs

Section 2-4 Science and Society
Issue: SI for All?

32

Physical Science Methods

How are the winners in each event determined in the track meet pictured to the left? By carefully measuring and comparing quantities such as length and time, winners of sports events are decided. To be of value, the measurements must relate to a standard with which everyone agrees. Standard measurements are especially important in scientific experiments. In the chapter that follows, you will see how some of the measurement and graphing methods used in physical science are the same as those used to keep sports results consistent and fair.

EXPLORE ACTIVITY

Design your own measuring device.

1. Pick something in your classroom to use as a tool for measuring length. It might be a notebook, a pencil, a hand, or any other convenient device.
2. Working with a partner, measure a distance in the room with your measuring device. Record your measurement and make up a name for your measuring unit.
3. Now, have your partner measure the same distance, first using his or her own measuring device, then using yours. Record measures with proper unit names.

Observe: In your Science Journal, explain why you think it might be important to have standard, well-defined units to make measurements.

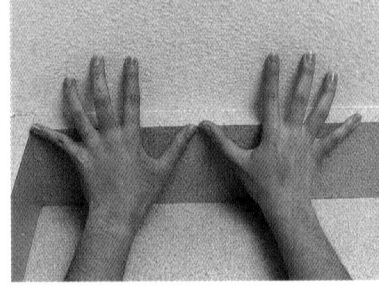

Previewing Science Skills

▶ In the Skill Builders, you will **sequence, map concepts,** and **make and use graphs.**

▶ In the Activities, you will **observe, collect and organize data,** and **measure in SI.**

▶ In the MiniLABs, you will **measure in SI, make inferences,** and **make and use graphs.**

33

Purpose

IS **Interpersonal** Students will model and experiment with a classroom object as a measuring device. This activity serves as an introduction to measurement, an important method in physical science. **L1** **LEP** **COOP LEARN**

Preparation

Gather a collection of objects students could use for their measuring devices.

Materials

Do not make standard devices such as metersticks available.

Teaching Strategies

• Have all students measure at least two common objects, such as the length of the chalk tray and the width of a student desk, for comparisons.

Observe

Only measurements made with the same device can be easily compared. Standard units of measurement make it possible to compare measurements by different people in different locations.

▼ **Assessment**

Oral Have students explain the limitations of the measurements they made. For example, they might have trouble measuring between whole units accurately or comparing consistently between groups using a different device. Use the Performance Task Assessment List for Making Observations and Inferences in **PASC**, p. 17. **P**

Assessment Planner

Portfolio
Refer to page 55 for suggested items that students might select for their portfolios.

Performance Assessment
See page 55 for additional Performance Assessment options.
Skill Builders, pp. 36, 45, 51
MiniLABs, pp. 44, 49
Activities 2-1, p. 37; 2-2, pp. 46-47

Content Assessment
Section Wrap-ups, pp. 36, 45, 51, 53
Chapter Review, pp. 55-57
Mini Quizzes, pp. 36, 45, 51

Group Assessment
Opportunities for group assessment occur with Cooperative Learning Strategies and Flex Your Brain Activities.

Prepare

Section Background

One major reason that science has progressed is because of its reliance on quantitative observations. The SI system is used by the scientific community to collect and communicate these observations.

Preplanning

Collect ingredients and clean equipment for Activity 2-1.

1 Motivate

Bellringer

 Before presenting the lesson, display **Section Focus Transparency 5** on the overhead projector. Assign the accompanying **Focus Activity** worksheet. L1 LEP

GLENCOE TECHNOLOGY

 **Videodisc**

The Infinite Voyage: Sail On, Voyager
Chapter 1
Preparation for the Grand Tour

Chapter 11
Voyager 2, Neptune

 **CD-ROM**

Physical Science CD-ROM
Have students perform the interactive exploration for Chapter 2 to reinforce important chapter concepts and thinking processes.

2•1 Standards of Measurement

Science Words

standard
SI

Objectives

• Define standard of measurement.
• Identify the need for standards of measurement.
• Name the prefixes used in SI and indicate what multiple of ten each represents.

Units and Standards

Measuring is an important skill. In order for a measurement to be useful, a measurement standard must be used. A **standard** is an exact quantity that people agree to use for comparison. When all measurements are made using the same standard, the measurements can be compared to each other.

Look at **Figure 2-1.** Suppose you and a friend want to make some measurements to find out whether a desk will fit through a doorway. You have no ruler, so you decide to use your hands as measuring tools. Using the width of his or her hands, your friend measures the doorway and says it is 8 hands wide. Using the width of your hands, you measure the desk and find it is 7¾ hands wide. Will the desk fit through the doorway? You can't be sure. What if your hands are wider than your friend's hands? Then, the distance equal to 7¾ of your hands might be greater than the distance equal to 8 of your friend's hands.

What did you forget to do? Even though you both used hands to measure, you didn't check to see whether your hands are the same width as your friend's hands. In other words, you didn't use a measurement standard, so you can't compare the measurements.

Figure 2-1

Hands are a convenient measuring tool, but using them can lead to misunderstanding.

Program Resources

 Reproducible Masters
Study Guide, p. 9 L1
Reinforcement, p. 9 L1
Enrichment, p. 9 L3
Activity Worksheets, pp. 8-9 L1
Critical Thinking/Problem Solving, p. 8
Multicultural Connection, pp. 7-8
Concept Mapping, p. 9

Transparencies
Section Focus Transparency 5 L1
Teaching Transparency 3

Measurement Systems

Suppose the label on a ball of string indicates that the length of the string in the ball is 150. Can you tell how much string is in the ball? No. It could be 150 feet, 150 meters, or 150 of some unit you've never heard of. In order for a measurement to make sense, it must include a number and a unit.

Your family may buy lumber by the foot, milk by the gallon, and potatoes by the pound. These measurement units are part of the English system of measurement, which is most commonly used in the United States. Most other nations use a system of measurement based on multiples of ten. The first such system of measurement, called the metric system, was devised by a group of scientists in the late 1700s.

Table 2-1

Common SI Prefixes		
Prefix	Symbol	Multiplying Factor
kilo-	k	1000
deci-	d	0.1
centi-	c	0.01
milli-	m	0.001
micro-	μ	0.000 001
nano-	n	0.000 000 001

Figure 2-2

The standard kilogram mass, composed of a platinum-iridium alloy, is kept at the International Bureau of Weights and Measures in Sèvres, France. *What is the purpose of a standard?*

International System of Units

In 1960, an improved version of the metric system was devised. Known as the International System of Units, this system is often abbreviated *SI*, from the French *Le Système Internationale d'Unités*. SI is the standard system of measurement used worldwide. All **SI** standards are universally accepted and understood by scientists. The standard kilogram is shown in **Figure 2-2**.

In SI, each type of measurement has a base unit, such as the meter, which is the base unit of length. In the next section, you will learn the base units in SI used to measure length, mass, time, temperature, volume, and density.

The SI system is easy to use because it is based on powers of ten. Prefixes are used with the names of the base units to indicate what power of ten should be used with the base unit. For example, the prefix *kilo-* means 1000, so a *kilometer* is 1000 meters. The most frequently used prefixes are shown in **Table 2-1.**

Inclusion Strategies

Gifted Have gifted students research the standardization of currency. When did it occur? Why? They can compare the need for standard units of measurement discussed in the chapter to the need for standard currency. [L3]

Gifted Students can use centimeter-graph paper to draw a scaled floor plan of their classroom, home, or school. [L3]

CONNECT TO
EARTH SCIENCE

See p. 36.
Answer

$$1.5 \times 10^{11}\,\text{m} \times \frac{1\,\text{km}}{1000\,\text{m}} =$$

$$1.5 \times 10^8\,\text{km}$$

Visual Learning

Figure 2-2 What is the purpose of a standard? *to have an exact quantity that people agree to use for comparison*
LEP [LS]

3 Assess

Check for Understanding

? Flex Your Brain

Use the Flex Your Brain activity to have students explore MEASUREMENT SYSTEMS.

📁 **Activity Worksheets,** p. 5

Reteach

[LS] **Linguistic** Write the terms *decade, century,* and *millennium* on the chalkboard and have students identify them as 10-, 100-, and 1000-year periods. Ask the students to use the metric prefixes to define a year in terms of a decade (decidecade), century (centicentury), and millennium (millimillennium). Ask students to interpret the meaning of the *i* in the three metric prefixes.

Extension

📁 For students who have mastered this section, use the **Reinforcement** and **Enrichment** masters.

INTEGRATION
Earth Science

The distance from the moon to Earth is not constant because the moon's orbit is elliptical. At its closest point to Earth (perigee), it is about 363 300 km away, and at its farthest point from Earth (apogee), it is about 405 500 km away. The circumference of the moon is about 11 000 km, making it about 27% the size of Earth.

4 Close

•MINI•QUIZ•

Use the Mini Quiz to check students' recall of chapter content.

1. **A standard is an exact quantity that people agree to use for _____ .** *comparison*

2. **_____ is the standard system of measurement used worldwide.** *SI*

3. **SI is easy to use because it is based on powers of _____ .** *ten*

Section Wrap-up

Review

1. 0.1 m = 1 dm, 1000 m = 1 km

2. They would not have to convert measurements and prices in international trade. Tools and equipment sizes would also be standard.

3. **Think Critically** 16 km

Science Journal

Accept all reasonable examples, such as 12 fluid ounces and 355 mL for a soft drink.

INTEGRATION
Earth Science

CONNECT TO
EARTH SCIENCE

The standard measurement for the distance from Earth to the sun is called the *astronomical unit,* or AU. This distance is about 150 000 000 000 (1.5×10^{11}) m. In your Science Journal, *calculate* what one AU would equal in km.

Measuring to the Moon

Even with dependable SI units for measurement standards, it is difficult to imagine measuring large distances such as that from Earth to the moon. Obviously, it would not be practical to use a meterstick or gigantic tape measure to find this distance. However, this distance has been measured. How was this done? Lasers produce extremely focused beams of light that are often used to measure both large and small distances. Scientists have determined that light travels a distance of 299 792 500 meters in every second through space. If you know how fast something travels and how long it travels, then the distance it travels can be calculated.

Mirrors on the Moon

U.S. astronauts set up light reflectors on the moon during an exploration trip in the 1970s. Pulses of laser light originating on Earth were aimed at these reflectors. The time for the light to travel from Earth to the moon and back was measured. Knowing this time and the speed of light, the distance from Earth to the moon was calculated to be 378 000 000 meters.

Astronomers have developed large units of measurement based on the speed of light for use in measuring vast distances between objects in space. A light-year is the distance light travels in one year. One light-year equals about 9.5 trillion kilometers. The nearest star is 4.2 light-years away.

Section Wrap-up

Review

1. In SI, the base unit of length is the meter. What would you call 0.1 meter? 1000 meters?

2. Why might it be desirable for scientists and manufacturers in the United States and other countries to use the same measurement standard?

3. **Think Critically:** Deimos is a small, natural satellite that orbits Mars. At one point, its diameter is 16 000 meters. How many kilometers is this?

Skill Builder
Sequencing

Using the information in **Table 2-1,** sequence the following units in order from largest to smallest: centigram, gram, milligram, kilogram, decigram. If you need help, refer to Sequencing in the **Skill Handbook.**

Science Journal

Examine the labels of at least five different food or drink products. In your Science Journal, report the metric and English equivalents of the weights or volumes of these products. Write a statement about how you feel about the use of SI units.

Skill Builder

kilogram, gram, decigram, centigram, milligram

Assessment

Process Ask students which has more mass: a 1-g vitamin pill or a 1-mg pill. Use the Performance Task Assessment List for Making Observations and Inferences in **PASC,** p. 17. **P**

Activity 2-1

Metric Munchies

Look through a recipe book. Are any of the amounts of ingredients stated in metric units? Chances are, English measure is used. How could you convert English measurements to metric measurements?

Problem
How do kitchen measurements compare with metric measurements?

Materials
- balance
- graduated cylinder, 100-mL
- munchie ingredients
- measuring cup
- measuring teaspoon
- measuring tablespoon
- plastic container, 1-quart
- hot plate
- cooking pot
- oven mitt

Procedure
1. Copy the data table into your Science Journal.
2. Use the English-measure cup or spoon to measure out the amount of munchie ingredient assigned to your team.
3. Use the balance or graduated cylinder to determine the metric equivalent of the measured ingredient. Convert solid measurements to grams. Convert liquid measurements to milliliters.
4. Write the metric equivalent in your data table. Also write it on the metric ingredients list posted in the classroom.
5. At the direction of the teacher, place your ingredient in the cooking pot.
6. Watch as the teacher cooks the munchies, then write cooking instructions.

Analyze
1. The volume ratio of sugar to oatmeal is 2 to 3. What is the mass ratio?
2. Which recipe, English or metric, requires the use of the greatest number of measuring devices?
3. Which kind of measure tends to be more accurate, volume or mass?

Conclude and Apply
4. English-measured recipes tend to use whole numbers and simple fractions. **Explain** how you could simplify the metric recipe.
5. **Predict** how kitchen equipment would change if all recipes were metric.
6. **Explain** benefits and problems in changing all recipes to metric.

Data and Observations
Sample Data

Ingredient	English Measure	Metric
Margarine	1/2 cup	114 g
Sugar	2 cups	432 g
Cocoa	6 tablespoons	39 g
Milk	1/2 cup	118 mL
Rolled oats	3 cups	258 g
Vanilla	1 teaspoon	5 mL
Nuts	1/2 cup	52 g

2-1 Standards of Measurement **37**

---

5. A gram scale and volume measure in milliliters would be added.
6. Measuring quantities by mass will simplify many measurements. However, kitchens will need to be equipped with balances capable of measuring grams. Accept other answers dealing with such concerns as publishing new cookbooks and assessing nutritional values in metric serving sizes.

✔ Assessment
Performance To further assess students' understanding of the metric system, see Performance Assessment question 1 on page 57. Use the Performance Task Assessment List for Making Observations and Inferences in **PASC**, p. 17. **P**

Activity 2-1

Purpose
IS **Interpersonal** Students will compare advantages and disadvantages of converting from English to metric measure. **COOP LEARN** **L1**

Process Skills
measuring in SI, interpreting data, comparing and contrasting, observing, using numbers, forming operational definitions

Time
40 minutes

Safety Precautions
Use clean kitchen pots and utensils—not laboratory glassware—because the product is edible.

📁 **Activity Worksheets**, pp. 5, 8-9

Teaching Strategies
Troubleshooting To cook the munchies, you will need a large saucepan, a large spoon for mixing, a hot plate or burner, a small spoon, and a roll of waxed paper. Allow ample room for students to view the preparation.
- Provide each team with an assigned ingredient and the appropriate measuring equipment.
- Have students put the first five ingredients into the pot. Heat slowly to a boil, then boil for 4 minutes while mixing constantly. Remove from heat and add the rest of the ingredients. Immediately portion onto waxed paper using the small spoon. You should have enough for the class.

Answers to Questions
1. roughly 2 to 1
2. The English recipe requires at least three. Metric could be done with two.
3. mass
4. Round numbers off to the nearest 5 or 10.

Prepare

Section Background

The SI system defines seven base units. In this section, four base units—those that measure length, mass, time, and temperature—are introduced. These units are necessary to describe motion and energy, the subjects of the next six chapters.

Preplanning

• To prepare for the MiniLAB, obtain several 100-mL graduated cylinders.

• Collect string, scissors, pens, tape, and standard objects for Activity 2-2.

1 Motivate

Bellringer

Before presenting the lesson, display **Section Focus Transparency 6** on the overhead projector. Assign the accompanying **Focus Activity** worksheet. L1 LEP

Tying to Previous Knowledge

Logical-Mathematical Have students use Tables A-1 through A-5 on pages 727-729 to associate familiar metric units with products or uses such as milligrams (vitamin pills), watts (lightbulbs), liters (soda bottles), and volts (batteries). Point out that these units are associated with the metric system of measurement that students will be learning more about in this section. L1

Science Words

   meter
   volume
   derived unit
   liter
   mass
   kilogram
   density
   time
   second
   kelvin

Objectives

• Identify SI units and symbols for length, volume, mass, density, time, and temperature.
• Define *derived unit*.
• Convert a measurement among related SI units.

2•2 Using SI Units

SI Units and Symbols

Every type of quantity measured in SI has a base unit and a symbol for that unit. These names and symbols are shown in **Table 2-2**. All other SI units can be derived from these seven base units.

Table 2-2

SI Base Units		
Quantity Measured	**Unit**	**Symbol**
Length	Meter	m
Mass	Kilogram	kg
Time	Second	s
Electric current	Ampere	A
Temperature	Kelvin	K
Amount of substance	Mole	mol
Intensity of light	Candela	cd

Length

The word *length* is used in many different ways. For example, the length of a novel is the number of pages or words it contains. In scientific measurement, length is the distance between two points. That distance may be the diameter of a period on this page or the distance from Earth to the moon. The SI base unit of length is the **meter** (m). A baseball bat is about 1 meter long. Metric rulers and metersticks are used to measure length.

Recall that measurement in SI is based on powers of ten. The prefix *deci-* means 1/10, so a decimeter is one-tenth of a meter. Similarly, a centimeter (cm) is one-hundredth of a meter. **Figure 2-3** shows how a meter and a yard compare. The diameter of a shirt button is about 1 cm.

$$100 \text{ cm} = 10 \text{ dm} = 1 \text{ m}$$

Centimeters can be divided into smaller units called millimeters (mm). A millimeter is 1/1000 of a meter. One tooth along the edge of a postage stamp is about 1 mm long.

Program Resources

Reproducible Masters
Study Guide, p. 10 L1
Reinforcement, p. 10 L1
Enrichment, p. 10 L3
Science Integration Activities, pp. 3-4
Activity Worksheets, pp. 10-11, 12
Lab Manual 2
Technology Integration, pp. 7-8

 **Transparencies**
Section Focus Transparency 6 L1

Figure 2-3

A meter is slightly (about 1/10) longer than a yard; 100 meters is slightly longer than a football field. *Would you expect your best time for a 100-m dash to be slightly more or less than your time for a 100-yard dash?*

Meterstick
(1 meter)

Yardstick
(1 yard)

Choosing a Unit

As in **Figure 2-4**, the size of the unit you use to measure with will depend on the size of the item being measured. For example, you would use the centimeter to measure the length of your pencil and the meter to measure the length of your classroom. What unit would you use to measure the distance from your home to school? You would probably want to use a unit larger than a meter. The kilometer (km), which is 1000 meters, is used to measure long distances. One kilometer is about ten football fields long.

Suppose you know the length of something in meters and want to change, or *convert*, the measurement to centimeters. Because 1 m = 100 cm, you can convert from meters to centimeters simply by multiplying by 100. Because SI is based on powers of ten, any unit can be converted to a related

1 cm

Figure 2-4

The size of the object being measured determines which unit you should use. *What unit would you use to measure your height?*

1 mm

2-2 Using SI Units **39**

2 Teach

Activity

LS **Visual-Spatial** Distribute metric rulers and have students identify the millimeter, centimeter, and decimeter markings on the ruler. Have them measure common objects. As they are measuring, have each of them choose an object that best represents a centimeter. Allow them to make estimations and use their rulers to check their estimations. **L1**

Revealing Preconceptions

Students may believe that the SI system is more precise than the English system because it is used by scientists. Point out that both systems can yield equally precise measurements. For instance, a micrometer used by machinists can measure the diameters of bolts to the nearest 0.0001 inch. Scientists use SI because it is easier to use than the English system.

Visual Learning

Figure 2-3 Would your best time for a 100-m dash be slightly less or slightly more than your time for a 100-yard dash? *Your time for 100 m would be slightly more.*

Figure 2-4 What unit would you use to measure your height? *You could use either meters or centimeters.* **LEP** **LS**

GLENCOE TECHNOLOGY

 Videodisc
STVS: Earth and Space
Disc 3, Side 2
Charting Air Space (Ch. 22)

Inquiry Questions

- **How many millimeters are there in 2.5 meters?** *2500 millimeters*

- **How many meters are there in 650 millimeters?** *0.65 meter*

- **How many centimeters are there in 0.464 meter?** *46.4 centimeters*

- **How many centimeters are there in a decimeter?** *10 cm* **How many cubic centimeters are there in a cubic decimeter?** *1000 cm³* **A cubic decimeter is the same as a liter. How many cubic centimeters are there in a liter?** *1000 cm³*

Visual Learning

Figure 2-5 How many millimeters are in 10 cm? *There are 100 mm in 10 cm.*
LEP LS

Practice Problem Answers

1. 253.8 km = 25 380 000 cm

2. 75 cm = 7.5 dm
   75 cm = 0.75 m

Teacher F.Y.I.

- There are prefixes that represent 10 and 100, namely *deka-* and *hecto-*. However, they are rarely used.

- In SI measurements that have values greater than 9999 or less than 0.999, groups of three integers to the left and to the right of the decimal point are separated by spaces, not commas, because a comma is used to represent the decimal point in some European countries. Thus, 12345.6789 cm is expressed 12 345.678 9 cm. Numbers of only four digits have neither a comma nor a space.

Figure 2-5

One centimeter contains ten millimeters. *How many millimeters are in ten centimeters?*

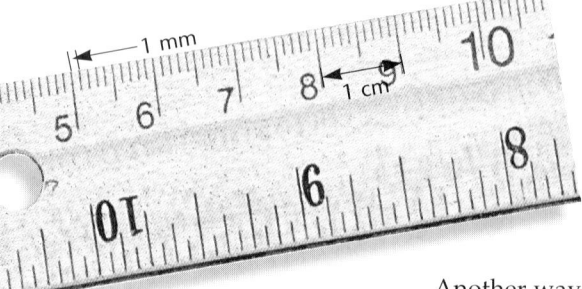

unit by multiplying or dividing it by the appropriate power of ten. **Figure 2-5** shows units of length on a meterstick. If you follow these rules, conversion will be easy.

- A measurement consists of two parts: a number and a unit label.
- To convert from larger to smaller units, multiply.
- To convert from smaller to larger units, divide.

For example, to convert 532 cm to meters, you divide because you're converting from smaller (cm) to larger (m) units. Because 100 cm = 1 m, divide 532 by 100. So, 532 cm = 5.32 m.

Another way to convert units is to multiply by a ratio that equals one. For example, 1 meter = 100 centimeters. Therefore,

$$\frac{1 \text{ meter}}{100 \text{ centimeters}} = \frac{100 \text{ centimeters}}{100 \text{ centimeters}} = 1$$

To convert 532 cm to meters, write

$$532 \text{ cm} \times \frac{1 \text{ m}}{100 \text{ cm}} = \frac{532 \text{ m}}{100} = 5.32 \text{ m}$$

The *cm* label is canceled because it appears in both the numerator and the denominator. Try the practice problems below.

USING MATH

Converting Meters to Centimeters

Example Problem:
How many centimeters are in 1.98 meters?

Problem-Solving Steps:
1. What is known?
   1 m = 100 cm
   You are converting from larger to smaller units.
2. When converting from larger to smaller units, multiply.
3. **Solution:** Multiply 1.98 by 100.   1.98 m = 198 cm

Practice Problems
1. How many centimeters are in 253.8 km?
Strategy Hint: First, change kilometers to meters; then, change meters to centimeters.
2. A bookshelf is 75 cm wide. How many decimeters is this? How many meters would this be?
Strategy Hint: Remember the meanings of the prefixes.

40   Chapter 2 Physical Science Methods

Across the Curriculum

Math The SI prefix *nano-* is used for tiny measurements. The wavelength of red light in a compact disc player may be around 710 nanometers. Have students find out what *nano-* means and find the wavelength of red light in meters. *The prefix nano- means one-billionth. The red light would be 710 billionths of a meter.* L2
LS

As you move ahead in this section, you will learn about several different types of measurements. Keep in mind that the conversion rules you have used here can be used with any base unit in SI because the system is based on powers of ten. The prefixes remain the same no matter what base unit you may be using.

Volume

The amount of space occupied by an object is called its **volume.** If you want to know the volume of a solid object, such as a building brick, you measure its length, width, and height, and multiply the three numbers together. For the brick, your measurements would be in centimeters, and the volume would be expressed in cubic centimeters (cm^3). For a larger object, such as a truck, your measurements might be in meters and the volume in cubic meters (m^3).

Width of sphere = 0.01 mm

USING TECHNOLOGY

Space Spheres

The space shuttle *Challenger* served as the manufacturing site for one of the latest standard or reference materials produced by the National Bureau of Standards. The reference material is a polystyrene sphere that measures 10 micrometers across. The head of a pin could hold 18 000 of these spheres. The spheres will be packaged in a 5-mL vial containing about 30 million spheres in water. These spheres can be used as a reference for manufacturers and researchers who need to calibrate instruments to check particle size in products such as cosmetics, paint pigments, flour, toner used in photocopiers, and so on. The spheres can also be used to improve microscopic measurements made in areas such as medicine and electronics.

Weightless Manufacturing

The *Challenger* was chosen as the manufacturing site because of its weightless environment. Spheres produced by conventional processes on Earth float or sink during their formation. This results in spheres with a variation in diameter that is too great for use as a standard. Weightlessness allows the production of spheres that are uniform in size and shape.

Think Critically:

In your Science Journal, explain why particle size is important in the products listed.

USING TECHNOLOGY

Have students use a magnifying glass or microscope to observe the tiny spheres in cosmetic powder or flour.

Think Critically
All products mentioned need a uniform, small-particle size for consistent performance or application.

Demonstration

LS Logical-Mathematical Have students use metersticks to measure the circumference of a round object. Have them repeat the measurements with a metric tape measure. Ask them to explain why the measurements made with the tape measure are more accurate. **L1** **LEP**

GLENCOE TECHNOLOGY

 **Videodisc**
STVS: Earth and Space
Disc 3, Side 2
Mapping with a Rifle (Ch. 20)

Map Science (Ch. 21)

NATIONAL GEOGRAPHIC SOCIETY

 Videodisc
STV: Solar System
The Big Picture
Unit 1, Side 1
The Big Picture

00691-11815

Cultural Diversity

Social Time The passage of time may be subject to accurate measurement by a variety of instruments, but social time varies a great deal from region to region and culture to culture. People's ideas of punctuality vary from culture to culture. Americans are concerned even with anticipated lateness, calling ahead to apologize because they think they are going to be late. In other cultures, such as many Latin American and Arab cultures, one may be an hour or more late without an apology being made or expected.

The pace of life varies, too. A study that compared accuracy of bank clocks and the average walking speed in six societies found that Japan came in first and the United States came in second in time preoccupation in both measures of social time.

 Use **Science Integration Activities,** pp. 3-4, as you teach this lesson.

Practice Problem Answer

1. 1.45 L × 1000 cm³/1 L = 1450 cm³ of milk

Problem Solving

Solve the Problem

1. 2-L bottle = 2000 mL
6 cans × 355 mL/can = 2130 mL = 2.130 L

2. \$1.30/2 L = \$0.65/L for 2-L bottle
\$2.50/2.13 L = \$1.17/L for cans
The 2-L bottle is the best buy.

 The trace function can be used to determine that 30.8 is the *X* value that corresponds to a *Y* value of 40. Thus, 30 bottles can be purchased.

GLENCOE TECHNOLOGY

Videodisc
STVS: Chemistry
Disc 2, Side 2
Losing Weight by Design (Ch. 19)

Student Text Question
See p. 43

How many grams are there in 1 kilogram? *There are 1000 g in 1 kg.*

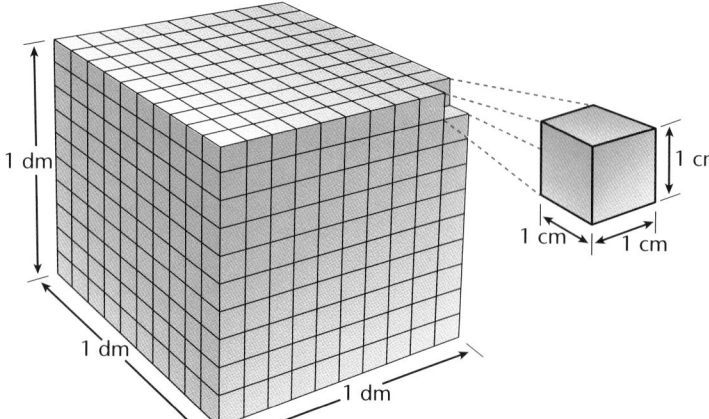

Figure 2-6
The large cube has a volume of 1 cubic decimeter (dm³), which is equivalent to a liter. *How many cubic centimeters (cm³) are in the large cube?*

Measuring Volume

No simple tool can be used to measure the volume of an object directly. As you saw in the example with the brick on the previous page, volume units are obtained by combining other SI units of length. Units obtained by combining SI units are called **derived units.**

How do you measure the volume of a liquid? A liquid has no sides to measure. In measuring a liquid's volume, you indicate the capacity of a container that holds that amount of liquid. Liquid volumes are sometimes expressed in cubic centimeters, as in doses of medicine. The most common units for expressing liquid volumes are liters and milliliters. A **liter** occupies the same volume as a cubic decimeter (dm³). That is, a liter is the same volume as a cube that is 1 dm (10 cm) on each side, as in **Figure 2-6.** A liter is slightly larger than a quart. The

USING MATH

Converting Centimeters to Liters

Example Problem:
How many liters of gasoline are in 538 cm³?

Problem-Solving Steps:

1. What is known?
    1 cm³ = 1 mL and 1 L = 1000 mL;
    therefore, 1 L = 1000 cm³
    You are converting from smaller units to larger units.

2. When converting from smaller to larger units, divide.

3. **Solution:** Divide 538 by 1000.   538 cm³ = 0.538 L

Practice Problem

1. How many cubic centimeters are in 1.45 L of milk?

Across the Curriculum

Math In many countries, gasoline is sold in units of liters. If one liter is approximately the same volume as one quart, find the approximate cost of one gallon of gasoline if it costs 60 cents per liter. L1 IS

\$0.60/1 L × 4 L/1 gal = \$2.40/gal

liter is not an SI unit, but it is used frequently with that system. Study the problems on page 42.

One liter (L) is equal to 1000 milliliters (mL). A cubic decimeter (dm^3) is equal to 1000 cubic centimeters (cm^3). So, because $1 L = 1 dm^3$, $1 mL = 1 cm^3$.

Suppose you wanted to convert a measurement in liters to cubic centimeters. The same rules you used for converting length can be used for any SI unit.

Mass

A table-tennis ball and a golf ball have about the same volume. But if you pick them up, you notice a difference. The golf ball has more mass. **Mass** is a measurement of the matter in an object. The SI unit of mass is the **kilogram** (kg). For measuring objects of small mass, grams (g) or milligrams are used. How many grams are in 1 kilogram?

In the laboratory, mass is measured with a balance. There are several different types of balances, but they all operate on the same principle. You use something of known mass to balance something else of unknown mass.

Density

If you were to take a cube of polished aluminum and a cube of silver the same size, they would look similar, and they would have the same volume. But the cube of silver would have more mass. The mass and volume of an object can be used to find the density of the material it is made of. **Density** is the mass per unit volume of a material.

Problem Solving

Finding the Best Value

Suppose you are in charge of supplying soft drink for a picnic. Two-L bottles cost $1.30 each, and six-packs of 355-mL cans cost $2.50 per pack. Assuming cups are provided, how will you decide which package is a more economical way to purchase the soft drink?

Solve the Problem:
1. **Determine how many liters of soft drink are in a 2-L bottle and how many are in a six-pack of soda cans.**
2. **Calculate the price per liter of soda for each type of packaging. Which is the best buy?**

GRAPHING CALCULATOR

You can find the total cost of any number of 2-L bottles of soda by the equation $Y = \$1.30 \times X$, where Y is the total cost, $1.30 is the cost of one bottle, and X is the number of bottles purchased. Graph this equation using your graphing calculator, and use your graph to determine how many 2-L bottles can be purchased if you have $40.00 to spend.

3 Assess

Flex Your Brain

Use the Flex Your Brain activity to have students explore DENSITY.

Activity Worksheets, p. 5

Reteach

Visual-Spatial Use the following diagram to illustrate the mechanics of converting decimal units.

k---[h–]-[dk–]-(base)--d----c----m—
| | | | | | |
k---[h–]-[dk–]-(base)--d----c----m—

Locate the original unit on the top line, and draw an arrow from it to the desired unit on the bottom line. For example, to convert 3.46 m to centimeters, locate m on the top line and draw an arrow to cm on the bottom line.

km - [hm] - [dkm] - - m - - -dm - - -cm - - -mm
| | | | | | |
km - [hm] - [dkm] - - m - - -dm - - -cm - - -mm

To convert, move the decimal point *two* places to the *right*. Thus, 3.46 m = 346 cm. To express 350 g as kilograms, locate g on the top line and connect it by an arrow to kg on the line below.

kg - [hg] - [dkg] - - g - - -dg - - -cg - - -mg
| | | | | | |
kg - [hg] - [dkg] - - g - - -dg - - -cg - - -mg

The diagram indicates that the decimal point must be moved *three* places to the *left*. Thus, 350 g = 0.350 kg.

Extension

For students who have mastered this section, use the **Reinforcement** and **Enrichment** masters.

Across the Curriculum

Math A vitamin capsule contains 200 milligrams of vitamin C. **How many capsules could be made from 1.0 kilogram of vitamin C?** Hint: Convert 1.0 kilogram to milligrams.

1.0 kilogram = 1000 grams = 1 000 000 milligrams.

Therefore, 1 000 000 milligrams × 1 capsule/200 milligrams = 5000 capsules.

What is the area of a rectangular tabletop that measures 150 cm by 200 cm? *30 000 cm²* **What is the area of the table in square meters?** *30 000 cm² = 30 000 cm² × 1 m²/10 000 cm² = 3 m²* [L1] [LS]

43

MiniLAB

What is the density of a pencil?

Procedure

1. Use a balance to measure the mass of a pencil in grams.
2. Put 90 mL of water into a 100-mL graduated cylinder. Lower the pencil, eraser end down, into the cylinder. Continue to push the pencil point down until it is completely underwater, but be sure your finger is not also submerged. Read the new volume to the nearest tenth of a milliliter.

Analysis

1. Calculate the pencil's density by dividing its mass by the change in volume of the water level when the pencil was completely underwater.
2. Is the density of the pencil greater or less than the density of water? How do you know?

Table 2-3

Material	Density (g/cm³)	Material	Density (g/cm³)
Hydrogen	0.000 09	Quartz	2.6
Oxygen	0.0013	Aluminum	2.7
Cork	0.24	Iron	7.9
Water	1.0	Copper	8.9
Glue	1.27	Lead	11.3
Sugar	1.6	Mercury	13.6
Table salt	2.2	Gold	19.3

Densities of Some Materials at 20°C

Like volume, density is a derived unit. You can find the density of an object by dividing its mass by its volume. For example, the density of an object having a mass of 10 g and a volume of 2 cm³ is 5 g/cm³. This value is expressed as 5 grams per cubic centimeter. Notice that both the mass and volume units are used to express density. **Table 2-3** lists the densities of some familiar materials.

Time and Temperature

It is often necessary to keep track of how long it takes for something to happen, or whether something heats up or cools down. These measurements involve time and temperature. **Figure 2-7** shows some different kinds of clocks.

Time is the interval between two events. The SI unit for time is the **second.** In the laboratory, you will use a stopwatch or a clock with a second hand to measure time.

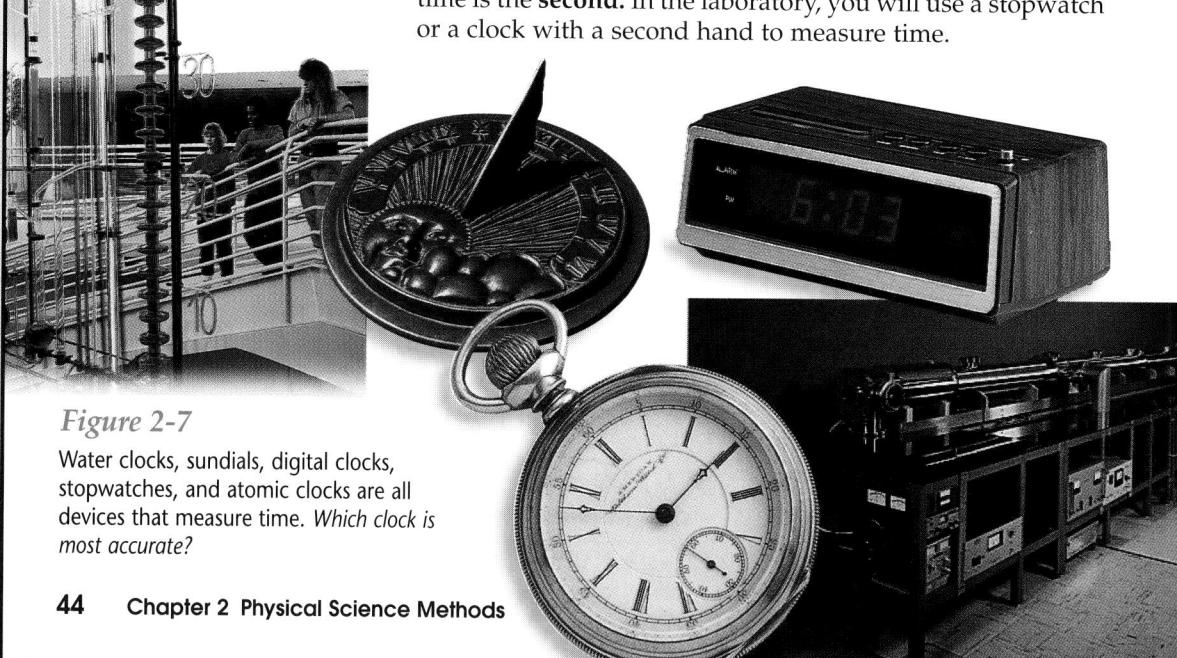

Figure 2-7

Water clocks, sundials, digital clocks, stopwatches, and atomic clocks are all devices that measure time. *Which clock is most accurate?*

44 Chapter 2 Physical Science Methods

What's Hot and What's Not

You will learn the scientific meaning of temperature in a later chapter. For now, you can think of temperature as a measure of how hot or how cold something is. The temperature of a material is measured with a thermometer.

Look at **Figure 2-8**. For most scientific work, temperature is measured on the Celsius (C) scale. On this scale, the freezing point of water is zero degrees (0°C), and the boiling point of water is one hundred degrees (100°C). Between these points, the scale is divided into 100 equal divisions. Each one represents 1 Celsius degree. On the Celsius scale, average human body temperature is 37°C, and a typical room temperature may be between 20°C and 25°C.

The SI unit of temperature is the **kelvin** (K). Zero on the Kelvin scale (0 K) is the coldest possible temperature, also known as *absolute zero*. That is −273°C, which is 273 degrees below the freezing point of water.

Most laboratory thermometers are marked only with the Celsius scale. Because the divisions on the two scales are the same size, the Kelvin temperature can be found by adding 273 to the Celsius reading. So, on the Kelvin scale, water freezes at 273 K and boils at 373 K. Notice that degree symbols are not used with the Kelvin scale.

Figure 2-8

These three thermometers illustrate the range of temperatures between the freezing point and boiling point of water. Although the numbers on the scales are different, they measure the same temperature range. *How do a Celsius and a Fahrenheit degree compare?*

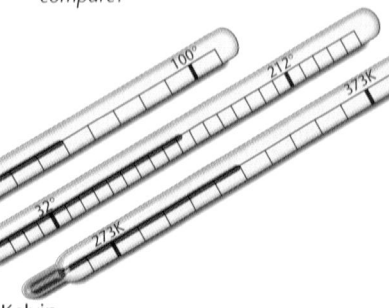

Celsius Fahrenheit Kelvin

Section Wrap-up

Review

1. Make the following conversions.
   a. 100 cm to meters
   b. 2.3 dm^3 to liters
   c. 27°C to K

2. Explain why density is a derived unit.

3. **Think Critically:** Chemists sometimes use density to identify a sample. What is the density of an unknown metal that has a mass of 178.0 g and a volume of 20.0 mL? Use **Table 2-3** to help you predict the possible identity of this metal.

Skill Builder
Concept Mapping

Make a network tree concept map to show the SI base units used to measure length, mass, time, and temperature. If you need help, refer to Concept Mapping in the **Skill Handbook.**

USING MATH

You probably know how much you weigh in pounds. Calculate your mass in kilograms. HINT: At Earth's surface, an object weighing 1 pound has a mass of 0.45 kg.

---

4 Close

•MINI•QUIZ•

Use the Mini Quiz to check students' recall of chapter content.

1. In SI, the meter is the base unit of _____ . *length*

2. Units of m^3 are used in what type of measurement? *volume*

3. _____ is the amount of matter in an object. *Mass*

Section Wrap-up

Review

1. a. 1 cm = 0.01 m; 100 cm = 1 m
   b. 1 dm^3 = 1 L; 2.3 dm^3 = 2.3 L
   c. 27°C + 273 = 300 K

2. Density units are obtained by dividing the SI base unit for mass by the SI derived unit for volume.

3. **Think Critically** 178.0 g/20.0 mL = 8.9 g/mL. Copper

USING MATH

LS **Logical-Mathematical** To find your mass, multiply your weight expressed in pounds by the factor 0.454 kg/1 lb. For example,

70 lb × 0.45 kg\1 lb = (70 × 0.45) kg = 32 kg

Skill Builder

```
            SI Base Unit
         /    |     |     \
       for   for   for    for
        |     |     |      |
     Length  Mass  Time  Temperature
        |     |     |      |
        is    is    is     is
        |     |     |      |
      Meter Kilogram Second Kelvin
```

✓ Assessment

Performance Use the network tree to assess the student's ability to organize information in the format of a concept map. Use the Performance Task Assessment List for Concept Map in **PASC**, p. 89. P

45

Activity
2-2

PREPARATION

Purpose

LS **Interpersonal** Students will design and carry out an experiment to show the necessary components of an acceptable measurement system. **COOP LEARN** **L1**

Process Skills

measuring in SI, collecting and organizing data, making and using tables, separating and controlling variables, communicating, forming operational definitions, making models, using numbers, classifying, observing and inferring

Time

One class period to brainstorm, and one-half to one class period to complete the activity and summarize results.

Materials

Have various colors of string available, if possible.

Possible Hypotheses

Students may hypothesize that using a defined measurement standard will make it possible for other students to measure objects consistently.

📁 **Activity Worksheets,** pp. 10-11

PLAN THE EXPERIMENT

Possible Procedures

Students may choose any device as their standard, such as a piece of chalk, a paper clip, a book, etc. They may mark the units on their string with a marker or tape. They should plan to try several different-sized scale divisions of the base unit to measure halves, quarters, and tenths of units.

Design Your Own Experiment

Setting High Standards. . . for Measurement

To develop the International System, people had to agree on set standards and basic definitions of scale. If you had to develop a new measurement system, people would have to agree with your new standards and definitions. In this activity, your team will use string to devise and test its own SI (String International) system for measuring length.

PREPARATION

Problem

What are the requirements for designing a new measurement system using string?

Form a Hypothesis

Based on your knowledge of measurement standards and systems, state a hypothesis about the relationship between measurement standards and consistency of measurements.

Objectives

- Design an experiment that involves devising and testing your own measurement system for length.
- Measure various objects with the string measurement system.

Possible Materials

- string
- scissors
- marking pen
- masking tape
- miscellaneous objects for standards

Teaching Strategies

Use heavy string to make handling easy. Be sure metersticks and other standard measuring devices are NOT available to students. Encourage students to use common classroom objects as standards.

PLAN THE EXPERIMENT

1. As a group, agree upon and write out the hypothesis statement.
2. As a group, list the steps that you need to take to test your hypothesis. Be specific, describing exactly what you will do at each step.
3. Make a list of the materials that you will need to complete your experiment.
4. Design a data table in your Science Journal so it is ready to use as your group collects data.

Check the Plan

1. As you read over your experimental plan, be sure you have chosen an object in your classroom to serve as a standard. Keep in mind, it should be in the same size range as what you will measure.

2. Consider how you will mark scale divisions on your string. Plan to use different pieces of string to try different-sized scale divisions of the base unit.
3. What will your new unit of measurement be called? Come up with an abbreviation for your unit. Will you name the smaller scale divisions, too?
4. What objects will you measure with your new measuring unit? Be sure to include objects longer and shorter than your string. Will you measure each object more than once to test consistency?
5. *Make sure your teacher approves your plan and that you have included any changes suggested in the plan.*

DO THE EXPERIMENT

1. Carry out the experiment as planned.
2. Record observations that you make, and complete the data table in your Science Journal. Be sure to include a unit with your measurements.

Analyze and Apply

1. Which of your string scale systems will provide the most accurate measurement of small objects? **Explain.**

2. How did you record measurements that were between two whole numbers of your units? Are any of your scale divisions easy to report as decimal numbers? Why or why not?
3. When sharing your results with other groups, why is it important for them to know what you used as a standard?
4. **Infer** how it is possible for different numbers to represent the same length of an object.

Go Further

If you were to design a new measurement system for mass, what parts of your experimental plan would be the same as they were for this length-measuring system? What parts would be different?

Expected Outcome

Students will devise a tested measuring system that other groups can use to make consistent measurements. They may encounter errors due to stretching of the string and estimating between units.

Analyze and Apply

1. The system with the smallest divisions will measure small objects most accurately.
2. As fractions or decimals. If a group divided their string into ten parts, it would be easiest to express measurements as decimal numbers.
3. The measurements must have a label, or unit, reflecting what the standard is. A lone number is meaningless.
4. When the size of the unit of measurement varies, the numerical value attached to the measurement must also be different.

Go Further

Answers may include that this system must also be based on a standard, but the means of measurement would be different.

Assessment

Portfolio Tape a sample of your measuring string to an explanation of your standard and measuring system. Discuss any problems you solved in the process of developing this system. Include this paper in your portfolio. Use the Performance Task Assessment List for Writing in Science in **PASC,** p. 87. **P**

Prepare

Section Background

Line graphs are the most important type of graph in science because some can be analyzed to provide equations that relate the information being displayed.

Preplanning

Gather thermometers and foam cups for the MiniLAB.

1 Motivate

Bellringer

Before presenting the lesson, display **Section Focus Transparency 7** on the overhead projector. Assign the accompanying **Focus Activity** worksheet. L1 LEP

Tying to Previous Knowledge

LS Interpersonal Assign students to cut graphs from newspapers and magazines and bring them to class. Using the Numbered Heads Together strategy, have each group devise a classification system for the graphs. L1

2 Teach

Visual Learning

Figure 2-9 Where is the girl and dog walking fastest? *They are walking fastest in last segment of the graph.* **Figure 2-10** Which graph shows a bigger temperature change over the same 25-minute time period? *Both graphs show the same temperature change.* LEP LS

Science Words

graph

Objectives

- Identify three types of graphs and explain the correct use of each type.
- Distinguish between dependent and independent variables.
- Interpret graphs.

MiniLAB

How can a graph help you observe change?

Procedure
1. Place a thermometer in a plastic foam cup of hot, but not boiling, water.
2. Measure and record the temperature every 30 seconds for 5 minutes.
3. Repeat the experiment with freshly heated water. This time, cover the cup with a plastic lid.

Analysis
1. Make a line graph of the changing temperature from step 2, showing time on the x-axis and temperature on the y-axis. Then plot the changing temperature from step 3 on the same graph.
2. Use the graph to describe what happens during cooling in each cup.

2•3 Graphing

Using Graphs

Often, it is helpful to be able to illustrate what happens during the course of an experiment. This can be done with a graph. A **graph** is a visual display of information or data. **Figure 2-9** is a graph that shows a girl walking her dog.

Graphs are useful for displaying information in business, sports, and many everyday situations. Different kinds of graphs—line, bar, and circle—are appropriate for displaying different types of information. It is important to use the correct kind of graph for the data you are presenting.

Figure 2-9

This graph tells the story of the motion that takes place when a girl takes her dog for an 8-minute walk. Study each segment of the graph. Use the drawings above each one to help you understand what the shape of that segment tells you about the motion of the girl and dog during that time interval. *Where are the girl and dog moving fastest?*

Distance from home (m)

Time (minutes)

Program Resources

 Reproducible Masters
Study Guide, p. 11 L1
Reinforcement, p. 11 L1
Enrichment, p. 11 L3
Activity Worksheets, p. 13 L1
Lab Manual 3

Transparencies
Section Focus Transparency 7 L1
Science Integration Transparency 2
Teaching Transparency 4

Line Graphs

Line graphs are used to show trends or how the data change over time. Suppose you want to show how the temperature of a room changes after you switch on the heat one chilly morning. Taking temperature readings in the room every 5 minutes, you might collect information that looks like that shown in **Table 2-4.** If you look closely at the data, you can see how temperature changed over time. But the relationship is easier to see in the graphs shown in **Figure 2-10.** Both graphs show that temperature increased for the first 15 minutes, then stayed constant. Do the graphs tell you anything about what made the temperature change?

Table 2-4

Room Temperature	
Time*	Temperature (°C)
0	16
5	17
10	19
15	20
20	20
25	20

*minutes after turning on heat

Figure 2-10

Both of the graphs below show the same information from **Table 2-4.** Notice, however, that the temperature scales along the vertical axes are different. *Which graph shows a bigger temperature change over the same 25-minute time period?*

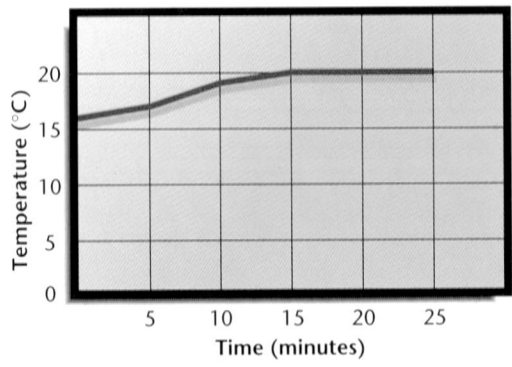

Ⓐ In this graph, each temperature interval represents 5°C. Because the temperatures measured were between 16 and 20°C, the portion of the graph below 16°C is not even used.

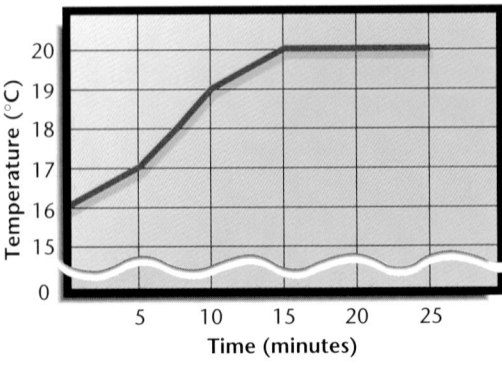

Ⓑ In this graph, the break in the vertical axis between the 0 and the 15 means that numbers in this range have been left out to save space. Each temperature mark represents 1°C. The temperature range graphed covers about 5 degrees instead of the 20-degree range covered in **Figure 2-10A.** Because there is more space between the numbers on this vertical axis, the same change in temperature looks larger.

2-3 Graphing   **49**

Use **Teaching Transparency 4** as you teach this lesson.

3 Assess

Check for Understanding

❓ Flex Your Brain

Use the Flex Your Brain activity to have students explore GRAPHING.

📁 **Activity Worksheets,** p. 5

Reteach

🔲 **Visual-Spatial** Stick ten birthday cake candles in holders in a long piece of plastic foam. Light the second candle and let it burn for only 5 seconds. Light the remaining candles in turn, letting the third candle burn for 10 seconds, the fourth for 15 seconds, etc. Remove the candles from the holders, clip their wicks, and place them side by side, bases aligned, on an overhead projector. Ask students to discuss what the silhouette displays.

Extension

📁 For students who have mastered this section, use the **Reinforcement** and **Enrichment** masters.

Graphs Show Change

In the example in **Figure 2-10,** two things are changing, or varying—time and temperature. Time is the independent variable. Its value does not depend on changes in the value of the other variable, temperature. Temperature is the dependent variable. Its value depends on changes in the time. In a line graph, the dependent variable always is plotted on the vertical y-axis, and the independent variable is plotted on the horizontal x-axis.

Bar Graphs

A bar graph is useful for comparing information collected by counting. For example, suppose you measured the temperature in every classroom in your school and organized your data in a table like **Table 2-5.** You could show these data in a bar graph like the one shown in **Figure 2-11.** The height of each bar corresponds to the number of rooms at a particular temperature.

In a line graph, adjacent points are related with a straight or curving line. In a bar graph, the bars are not connected.

Figure 2-11

The height of each bar corresponds to the number of rooms at a particular temperature.

Table 2-5

Temperature of Classrooms	
Temp. (°C)	Number of Classrooms
16	1
17	3
18	3
19	2
20	3
21	5
22	5
23	3

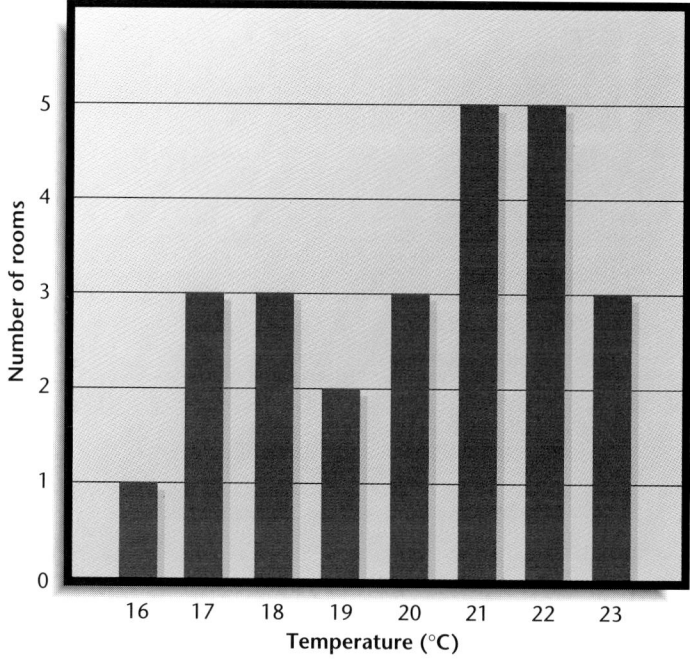

Science Journal

🔲 **Logical-Mathematical** Take a class survey of the months in which each student's birthday occurs and tally this on the board. Have students record this information in their Science Journal and have them design both a bar graph and a circle graph that shows this information. L1

Circle Graphs

A circle graph, or pie graph, is used to show how some fixed quantity is broken down into parts. The circular pie represents the total. The slices represent the parts and usually are represented as percentages of the total.

Figure 2-12 illustrates how a circle graph could be used to show the percent of buildings in a neighborhood using each of a variety of heating fuels. You can easily see that more buildings use gas heat than any other kind of system. What else does the graph tell you?

When you use graphs, think carefully about the conclusions you can draw from them. Can you infer cause and effect from looking at a graph? How might the scale of the graph affect your conclusions? Is any information missing or improperly connected?

Figure 2-12

A circle graph shows the different parts of a whole quantity.

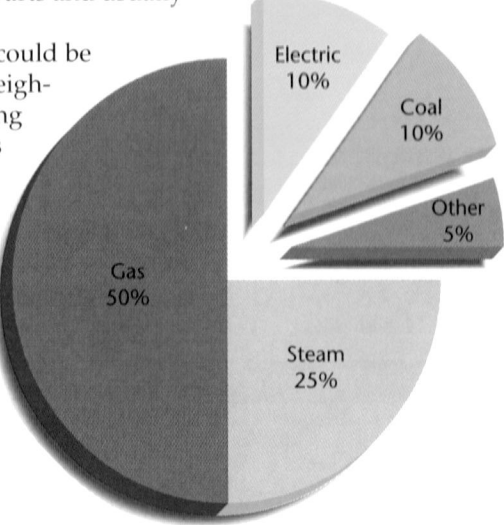

Electric 10%

Coal 10%

Other 5%

Steam 25%

Gas 50%

Section Wrap-up

Review

1. Your class does an experiment to show how the volume of a gas changes with changes in temperature. You are asked to make a line graph of the results. What are the dependent and independent variables?

2. Explain why points are connected in a line graph, but not in a bar graph.

3. **Think Critically:** A survey shows that, in your neighborhood, 75 people ride the bus; 45 drive their own cars; 15 carpool; 9 walk, ride bikes, or ride motorcycles; and the rest use different methods on different days to get to work. What kind of graph would be best for displaying these results at a neighborhood meeting? Draw it.

Skill Builder
Making and Using Graphs
Find a graph in a newspaper or magazine. Identify the kind of graph you found and write an explanation of what the graph shows. If you need help, refer to Making and Using Graphs in the **Skill Handbook.**

> **Using Computers**
>
> **Spreadsheet** Some computer programs make creating data tables and making graphs an easier task. Use a spreadsheet and a graphing program to make a data table and a line graph of the data you collected in the MiniLAB on page 48.

✓ **Assessment**

Process If students found line graphs, have them identify the dependent and independent variables. For pie graphs, have them check to see that the percentages total 100 percent. For bar graphs, have them explain why the bars are not connected. Use the Performance Task Assessment List for Written Summary of a Graph in **PASC,** p. 41. **P**

4 Close

•MINI•QUIZ•

Use the Mini Quiz to check students' recall of chapter content.

1. In a line graph, the _____ variable is plotted on the vertical, *y*-axis. *dependent*

2. _____ graphs are useful for showing information collected by counting. *Bar*

3. The slices of a pie graph usually are represented as _____ of the total. *percentages*

Section Wrap-up

Review

1. Independent variable is temperature; dependent variable is volume.

2. Line graphs show trends, and the spaces between measured points have meaning. Bar graphs show information collected by counting, and the spaces between bars do not represent any information.

3. **Think Critically** This information would best be represented by either a bar graph or a circle graph.

Using Computers

The computer graph will have the same basic shape as the student-drawn graph. However, students may get a better idea of the best-fit curve from the computer-drawn graph.

Prepare

Section Background

Other SI units are defined by adding prefixes to these base units or are derived from base units.

1 Motivate

Bellringer

 Before presenting the lesson, display **Section Focus Transparency 8** on the overhead projector. Assign the accompanying **Focus Activity** worksheet. L1 LEP

Tying to Previous Knowledge

Have students recall SI measurements that they are familiar with. Point out that these measurements are also familiar to citizens of most countries of the world.

2 Teach

Content Background

Inform students that the Metric Conversion Act of 1975 outlined a method by which the voluntary metrification of the United States would take place.

3 Assess

Check for Understanding

? Flex Your Brain

Use the Flex Your Brain activity to have students explore SI USE IN THE UNITED STATES.

52

ISSUE:

2•4 SI for All?

Objectives

- Analyze the advantages and disadvantages of universal use of SI measurements.
- Give examples of SI units already commonly used in the United States.

SI Use in the United States

Can you think of any situations, other than in science classes, where you see SI units in use? Athletes compete on courses that are measured in meters, medicine is sold in milligrams and milliliters, and some soft drinks are sold in 2-L bottles. Imported equipment is manufactured to metric specifications, as shown in **Figure 2-13**. However, in the United States, other units of measure are still common. Carpenters still buy lumber measured in feet and inches, farmers measure their land in acres and their crops in bushels, fabric is sold by the yard, and highway signs give distances in miles and speed limits in miles per hour. Sometimes both systems are used, as shown in **Figure 2-14**.

A History of Conflict

For nearly 100 years, advocates of the metric system and the newer SI units have argued for widespread adoption of the system in the United States. Only a few countries in the world have not adopted SI as the official system of measurement. But opponents to changing the measurement system have argued just as vigorously against it.

Figure 2-13

The metric system is used universally, so when working on imported equipment, it is necessary to use metric tools.

52 Chapter 2 Physical Science Methods

Program Resources

 Reproducible Masters
Study Guide, p. 12 L1
Reinforcement, p. 12 L1
Enrichment, p. 12 L3
Science and Society Integration, p. 6
Cross-Curricular Integration, p. 6

Transparencies
Section Focus Transparency 8 L1

2 Points of View

▶ Drawbacks of Adopting SI

Many citizens resist the switch to SI units because they have grown up using such units as feet, pounds, and gallons, and they feel more comfortable using them. Some representatives of industry point out that such a changeover would require them to replace or convert their machinery and tools to the appropriate SI dimensions—a costly process.

▶ Benefits of Adopting SI

Citizens in favor of adopting SI argue that because SI is based on powers of ten, conversions and calculations between different units are easier. For example, do you know how many cubic inches are in a fluid ounce? It is much easier to recall that one cubic centimeter is equivalent to one milliliter of fluid. Some business people also point out that one-time costs of making industrial machinery compatible with SI will easily be offset by the time and money saved by having products automatically ready for trade with other countries, nearly all of which use SI.

Figure 2-14

In order to acquaint people in the United States with metric measurements, many lengths and package net contents are given in both English and SI units.

Section Wrap-up

Review

1. Write a list of different units used to measure length in the United States. Then show the calculations necessary to convert from one unit to another. Do the same thing using SI units for length. Which is easier?

2. Write a one-paragraph essay summarizing your position on the adoption of SI in the United States.

Explore the Issue

Suppose that on the first day of the new year, the whole United States is to begin using only SI units. In your Science Journal, explain what changes would need to be made by that time. For whom would the change be most difficult or costly? Who would benefit most from this change? Explain your answers clearly.

Science Journal

In your Science Journal, explain your opinions about adopting SI in the United States.

SCIENCE & SOCIETY

2-4 SI for All? **53**

53

Source

de la Croix, Horst, and Richard G. Tansey. *Gardener's Art Through the Ages.* New York: Harcourt Brace College Pubs., 1990.

Biography

Pablo Picasso was born in 1881 in Spain but spent much of his adult life in France. He "evolved" through various artistic periods during his long life. He painted somber Realist works, dabbled a bit in an Impressionistic phase, went through his famous Blue Period, and experimented with the spatial relationships of Cubism. In the late 1920s, his style of painting might have been classified as contemporary Expressionism.

Background

- The Post-Impressionist Paul Cézanne is thought to have had a strong influence on the development of the Cubist school.
- Many Cubist works were still lifes of tabletops, instruments, glassware, and bottles. Often, words, or fragments of words and numbers, were included in the art.
- Collages were an important Cubist medium. Bits of paper, oilcloth, actual objects, newspaper clippings, and so on were included in Cubist collages.

Teaching Strategies

LS Visual-Spatial Students' analyses of Picasso's musicians might include that the *Accordionist* is very abstract, fairly monotone, and involves many overlapping planes, whereas the *Three Musicians*, while still a Cubist piece, is quite colorful, not as abstract, appears much flatter than the other work, and makes use of predominantly small, rectangular shapes.

54

inter NET
CONNECTION

Visit the Chapter 2 Internet Connection at Glencoe Online Science, **www.glencoe.com/sec/ science/physical,** for a link to more information about Cubism.

Cubism

In this chapter, you learned about standard methods of measurement. What do you think might happen if some people decided to use different tools or units of measurement? Would their methods be understood and accepted by all? Probably not. In a similar vein, various artists in the early 1900s decided to challenge the Impressionist school of art that emphasized light, atmosphere, and perspective, and formed a new school of art called Cubism.

Cubists concerned themselves with the representation of form in their paintings and sculptures. They strove to use cylinders, spheres, and cones to create a three-dimensional perspective on a flat surface. This use of illusion gave the works an abstract look.

Some art critics link the Cubist movement to the physical world proposed by physicist Albert Einstein at the time. These critics interpret time as an actual dimension in the artwork. Objects in Cubist works, they argue, are not seen at one specific moment. Rather, the different views shown would have been seen by the artist at different times. The result is a piece of art as seen in both time and space.

Analytical Cubism

Analytical Cubism is the early phase of Cubism. It lasted only two years, from 1910 to 1912. Mentally, Analytical Cubists divided their subjects into numerous planes. When arranged on the canvas, these planes interlocked and overlapped to produce an abstract view of reality. Strong contrasts were made using only blacks, whites, grays, and sometimes browns.

Synthetic Cubism

The second phase of Cubism—Synthetic Cubism—gained popularity about 1912. The nearly one-color, abstract Analytical Cubist works of art gave way to more colorful works with a limited number of views of the subject. These Synthetic Cubist works of art also used a broader variety of textures and shapes.

54 Chapter 2 Physical Science Methods

Classics

Other classic artists who belonged to the Cubist movement include Georges Braque and Fernand Léger of France and Juan Gris of Spain.

Other Works

Other classic works by Pablo Picasso include *Still Life with Chair Caning* and *Guernica*. The former work, a collage, marked the end of the analytical phase of Cubism. *Guernica* was a black, white, and gray protest of the bombing in the Spanish Civil War.

Summary

2-1: Standards of Measurement

1. A standard of measurement is an exact quantity that people agree to use as a basis of comparison.
2. When a standard of measurement is established, all measurements are compared to the same exact quantity—the standard. Therefore, all measurements can be compared with one another.
3. In SI, prefixes are used to make the base units larger or smaller by powers of ten. The most common prefixes and their values are: *kilo-*, 1000; *deci-*, 0.1; *centi-*, 0.01; *milli-*, 0.001; *micro-*, 0.000 001; and *nano-*, 0.000 000 001.

2-2: Using SI Units

1. The most commonly used units in SI and their symbols include: length—meter, m; volume—cubic decimeter, dm^3; mass—kilogram, kg; time—second, s.
2. Any SI unit can be converted to any other related SI unit by multiplying or dividing by the appropriate multiple of ten.

2-3: Graphing

1. Line graphs show continuous changes between related variables. Bar graphs are used to show data collected by counting. Circle or pie graphs show how a fixed quantity can be broken into parts.
2. In a line graph, the independent variable is always plotted on the horizontal x-axis; the dependent variable is always plotted on the vertical y-axis.

2-4: Science and Society: SI for All?

1. There are many benefits and drawbacks to the adoption of SI.
2. SI units are already in wide use on consumer goods in the United States.

Key Science Words

a. density
b. derived unit
c. graph
d. kelvin
e. kilogram
f. liter
g. mass
h. meter
i. second
j. SI
k. standard
l. time
m. volume

Reviewing Vocabulary

Match each phrase with the correct term from the list of Key Science Words.

1. the modern version of the metric system
2. the amount of space occupied by an object
3. an agreed-upon quantity to be used for comparison
4. the amount of matter in an object
5. obtained by combining SI units
6. a visual display of data
7. the SI unit of mass
8. the SI unit of length
9. a metric unit of volume
10. mass per unit volume

Chapter 2

Review

Summary

Have students read the summary statements to review the major concepts of the chapter.

Reviewing Vocabulary

1. j	**6.** c
2. m	**7.** e
3. k	**8.** h
4. g	**9.** f
5. b	**10.** a

Assessment

Portfolio Encourage students to place in their portfolios one or two items of what they consider to be their best work. Examples include:

- Flex Your Brain, p. 43
- MiniLAB graph and conclusions, p. 49
- Activity 2-2 Measuring system and answers, pp. 46-47 **P**

Performance Additional performance assessments may be found in **Performance Assessment** and **Science Integration Activities.** Performance Task Assessment Lists and rubrics for evaluating these activities can be found in Glencoe's **Performance Assessment in the Science Classroom.**

GLENCOE TECHNOLOGY

MindJogger Videoquiz

Chapter 2 Have students work in groups as they play the Videoquiz game to review key chapter concepts.

Chapter 2 Review

Checking Concepts

1. c   5. d   8. b
2. b   6. a   9. b
3. a   7. c   10. d
4. d

Understanding Concepts

11. SI measurements are used in science classes because they are used throughout the world and allow scientists to compare measurements.

12. Yes. Volume and mass in SI are both based on units of ten. Conversion between two different units of length can be made by multiplying or dividing by the proper power of ten.

13. Line graphs consist of data plotted on a grid within a set of axes. Line graphs are best used to show a continuous change in data that involves two variables.

    Bar graphs consist of a pair of axes and a series of bars. Bar graphs are best used to show data collected by counting.

    Circle graphs consist of a circle divided into segments. Circle graphs are best used to show the different parts that make up a whole quantity.

14. A line graph; time would be the independent variable and temperature would be the dependent variable.

15. Advantages: SI measurements are based on powers of ten, and the same prefixes are used for all types of measurement, thus making conversion easy; people in most nations measure in SI units. Disadvantages: Conversion would involve considerable expense; people are reluctant to change from a system they are familiar with.

Checking Concepts

Choose the word or phrase that completes the sentence or answers the question.

1. A meaningful measurement must have _____.
   a. a number only
   b. a unit only
   c. a number and a unit
   d. a prefix and a unit

2. The _____ is an example of an SI unit.
   a. foot   c. pound
   b. second   d. gallon

3. The system of measurement used by scientists around the world is the _____.
   a. SI   c. English system
   b. Standard system   d. Kelvin system

4. SI is based on _____.
   a. inches   c. English units
   b. powers of five d. power of ten

5. The SI prefix that means 1/1000 is _____.
   a. kilo-   c. centi-
   b. nano-   d. milli-

6. The symbol for deciliter is _____.
   a. dL   c. dkL
   b. dcL   d. Ld

7. The symbol for _____ is µg.
   a. nanogram   c. microgram
   b. kilogram   d. milligram

8. _____ is the distance between two points.
   a. Volume   c. Mass
   b. Length   d. Density

9. Which of the following is not a derived unit?
   a. cubic decimeter   c. cubic centimeter
   b. meter   d. grams per milliliter

10. 1000 mL is equal to all of the following except _____.
    a. 1 L   c. 1 dm^3
    b. 100 cL   d. 1 cm^3

Understanding Concepts

Answer the following questions in your Science Journal using complete sentences.

11. Why are SI measurements used in science classes?

12. Explain how to convert from one length measurement in SI to another. Can volume or mass measurements be converted in the same way? Why or why not?

13. Describe three different types of graphs and the types of data best displayed by each.

14. Suppose you set a glass of water in direct sunlight for two hours and measure its temperature every 10 minutes. What type of graph would you use to display your data? What would the dependent variable be? The independent variable?

15. What are some advantages and disadvantages of adopting SI for use in the United States?

Thinking Critically

16. Make the following conversions.
    a. 1500 mL to liters
    b. 2 km to cm
    c. 5.8 dg to mg
    d. 22°C to kelvin

17. Standards of measurement used during the Middle Ages were often based on such things as the length of the king's arm. What arguments would you use to convince people of the need for a different system of standard measurements?

18. List the SI units of length you would use to express the following. Refer to **Table 2-1** on page 35.
    a. the diameter of an atom
    b. the width of your classroom
    c. the width of a pencil lead
    d. the length of a sheet of paper
    e. the distance to the moon

Thinking Critically

16. **a.** 1.5 L
    b. 200 000 cm
    c. 580 mg
    d. 295 K

17. Student responses should include the importance of being able to compare measurements.

18. **a.** nanometer, nm
    b. meter, m
    c. millimeter, mm
    d. centimeter, cm
    e. kilometer, km

19. **a.** 7.5 g/cm^3
    b. 2.5 g/mL
    c. 9 kg/L

20. A line graph; time in seconds would be the independent variable and distance would be the dependent variable.

56

19. Determine the density of each of the following objects.
    a. mass = 15 g, volume = 2 cm³
    b. mass = 200 g, volume = 80 mL
    c. mass = 1.8 kg, volume = 0.2 L

20. Suppose you measure the distance a go-cart travels every 5 seconds for a 60-second period. What kind of graph would you use to display your data? What would the dependent and independent variables be?

Developing Skills

If you need help, refer to the **Skill Handbook**.

21. **Comparing and Contrasting:** Consider the base units and conversion procedures in SI and the English system. Compare and contrast the ease with which conversions can be made among units within each system.

22. **Hypothesizing:** A metal sphere is found to have a density of 5.2 g/cm³ at 25°C and a density of 5.1 g/cm³ at 50°C. Propose a hypothesis to explain this observation. How could you easily test your hypothesis?

23. **Concept Mapping:** Complete the Concept Map for metric measurements using the following terms: *trends or change over time, bar, circle, information collected by counting, line, proportional parts of a whole.*

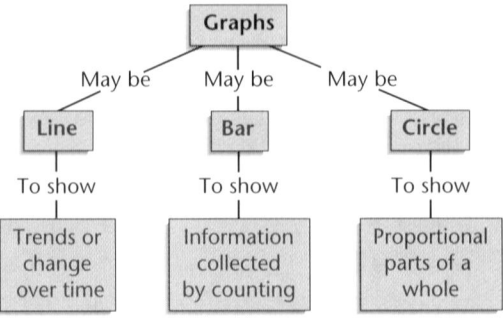

24. **Making and Using Graphs:** Using the data in **Table 2-3** on page 44, graph the densities of the following materials: water, sugar, salt, iron, copper, lead, and gold. Use the proper type of graph and let each unit on your graph represent 1.0 g/cm³. Then, answer the following questions:
    a. Why did you choose this type of graph?
    b. What would be the mass of 5 cm³ of sugar?
    c. If you had a 1-g sample of each material shown in the graph, which sample would have the greatest volume?
    d. How could your graph be made more accurate?

25. **Measuring in SI:** Determine the mass, volume, and density of your textbook, a container of milk, and an air-filled balloon. Make your measurements in SI units using the appropriate measuring tool and the quickest, most accurate method possible.

Performance Assessment

1. **Making Observations and Inferences:** In Activity 2-1, suppose you wanted to make five times that much food. Compare the difficulty of multiplying the English measurements five times with multiplying the metric measurements five times.

2. **Designing an Experiment:** Explain how you could improve the accuracy of the measuring system you used in Activity 2-2 for marking the scale divisions.

3. **Using Math in Science:** Find the SI equivalents of familiar measurements such as
    a. your weight in newtons (1 lb = 4.5 N)
    b. your mass in kilograms (1 kg = 2.2 lb)
    c. volume of 1.5 gallons of milk in liters (1 gal = 3780 mL)

Developing Skills

21. **Comparing and Contrasting** Students should name and compare the base units of measurement of length, mass, volume, temperature, and time for the two systems. The fact that SI measurements are based on powers of ten should be contrasted with the lack of any consistent or logical base in our system. The ease of converting in SI should be contrasted with the difficulty in our system.

22. **Hypothesizing** The metal expands when heated. The hypothesis can be tested by measuring the mass and volume of the ball at the two temperatures.

23. **Concept Mapping** See student page.

24. **Making and Using Graphs**

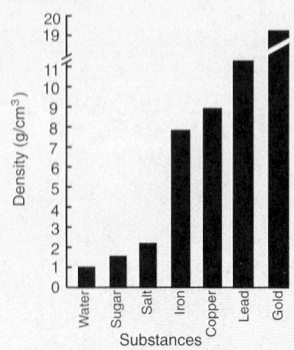

    a. A bar graph was chosen because the data consist of discrete bits of information.
    b. mass = density × volume
    mass of sugar = 1.6 g/cm³ × 5 cm³
    = 8 g
    c. water
    d. The graph could be made more accurate by using units smaller than 1 g/cm³.

25. **Measuring in SI** Student answers will vary.

Performance Assessment

1. The English system will sometimes produce fractional amounts that are more difficult to measure, particularly if 1/3 is involved. Use the Performance Task Assessment list for Using Math in Science in **PASC**, p. 29. **P**

2. Student answers may include that when you tape the scale divisions onto the string, decide on a consistent edge or area of the tape to make your exact division mark. Have the entire class agree to use the same marking system. Use the Performance Task Assessment list for Using Math in Science in **PASC**, p. 29. **P**

3. **a.** Answers will vary (wt in lbs) (4.5 N/1 lb); **b.** Answers will vary (wt in lbs)/2.2 lb/kg; **c.** (1.5 gal) (3780 $\frac{mL}{gal}$) /1000 $\frac{mL}{L}$ = 5.670 L

Use the Performance Task Assessment list for Using Math in Science in **PASC**, p. 29. **P**

Objectives

IS **Interpersonal** Students will investigate the Celsius and Fahrenheit temperature scales and measure and graph their relationship. In so doing, students will develop a means of converting from one scale to the other. **L1** **COOP LEARN**

Summary

- This project is designed to support the introductory concepts in Unit 1 that deal with problem solving, measuring, and graphing.

- Students are already familiar with the Fahrenheit scale for temperature measurement. In this project, they will simultaneously measure the Celsius and Fahrenheit temperatures of heated water as it cools. Students will then graph these data and use the graph to compare the two temperature scales. By analyzing the data, students will determine the relationship of the two temperature scales and develop a means of converting from one scale to the other.

- In this introductory project, students will also learn the importance of clearly communicating their findings.

Time Required

This project can be done after problem solving, graphing, and measuring have been discussed in Chapters 1 and 2. One class period will be required to gather the temperature data. Another period may be required to explain and demonstrate graphing techniques. Actual graphing, making posters, and collecting photos can be done outside of class. Another class period may be required for sharing of the displays.

UNIT PROJECT 1

Think Like A Scientist

Your gym locker won't open. You can't decide what to eat for lunch. You need to explain to a friend how to get to your house. Every day you have problems that you need to solve and ideas that you need to communicate. Problem solving and communicating are not always easy to do. However, your ability to describe events or materials that you come in contact with can be improved by using measurements. In this project, begin to understand the importance of scientific processes by combining all three of these skills: problem solving, communication, and measurement.

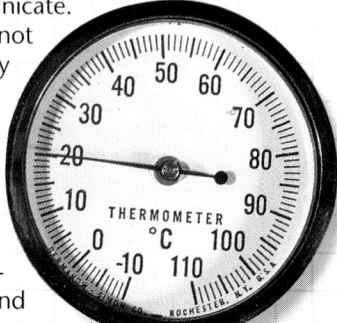

Two Languages for Temperature

At the present time, the world uses primarily two systems for measuring temperature. The Fahrenheit and Celsius systems were both named after the scientists who developed these now-familiar temperature scales. Most countries around the world use the Celsius scale. Scientists also use the Celsius scale. In the United States, however, temperature is reported in degrees Fahrenheit. Because two temperature scales are commonly used, it is important to be able to convert, or translate, from one to the other.

Purpose

In this experiment, work with a partner to show the conversion of temperature measurements between the two systems.

Materials

- a Fahrenheit thermometer
- a Celsius thermometer
- a beaker of water
- a hot plate
- clamps for the thermometers

Procedure

1. Use the hot plate to heat a quantity of water in a beaker.

2. When the water begins to boil, turn off the power to the hot plate.

3. Carefully suspend both of your thermometers near each other in the hot water.

4. As the temperature begins to drop, record the temperature readings, taken at the same time, from each of the thermometers. Be sure to label each of your paired readings as either °C or °F.

5. Collect a total of ten separate paired temperature readings. You may place the beaker and thermometers in a refrigerator to make the water cool faster.

Using Your Research

To help you compare the two scales, use a technique scientists frequently use—prepare a graph. With your partner, set up a graph with °C on the vertical axis and °F on the horizontal axis.

Even though you may not have measured it, label your vertical axis so that it includes 0°C. Make a line graph that plots each of your pairs of temperature readings. What value on the Fahrenheit axis matches the 0°C reading? This value is the amount that each Celsius degree is offset from each Fahrenheit measurement.

Next, select any ten-degree change on the Celsius axis. Use the line that you made to determine how many degrees the Fahrenheit scale would change in that interval. Using these values, determine what ratio compares one Celsius degree to one Fahrenheit degree.

Use this information to show how to change a Fahrenheit measurement to a Celsius value. With your partner, make a large poster of your graph and use your poster to explain the differences in the two systems and how to convert from one system to the other.

Go Further

Find photographs of different scenes that show people or events that are in situations of obvious temperature ranges. Examples might include a sunny beach and a snow scene. Attach these to your graph with questions relating the temperature in each scene to the graph.

Another temperature scale used commonly by scientists is Kelvin. Find out how and why the Kelvin scale was developed and how it compares to the Celsius scale. Design a method to convert Celsius to Kelvin.

59

Preparation

Obtain an electric hot plate, a beaker, and a Celsius and a Fahrenheit thermometer for each group. For graphs and posters, have rulers, graph paper, and poster board or butcher paper available.

Teaching Strategies

- As an introduction to the project, use a term or phrase from a foreign language, and ask students to translate the term into English. Explain that, for efficient communication, some study and experience are necessary. Relate this example to the study and experience needed to make temperature conversions.

- Caution students to use care not to touch any hot materials unless a thermal mitt is used.

- Because this project takes place early in the year, use this opportunity to set standards for neat, organized reporting of data.

- Emphasize the need to label graph axes with units that are of proper and equal size.

- Students' graphs should indicate that a 0°C reading corresponds to 32°F and that one Celsius degree is 1.8 (9/5) times larger than a Fahrenheit degree.

Go Further

- For student posters, have several old travel magazines available for students to use as sources for photos that represent extremes in temperature.

- The Kelvin scale was developed by Lord Kelvin, an English physicist, in the 1800s. This scale uses degrees that are the same size as Celsius degrees. The zero point is the temperature where, theoretically, no molecular motion exists, and there is thus no temperature.

- 0 K equals –273°C, which is called absolute zero. Students may want to investigate the concept of absolute zero and the effects of extreme cold on matter.

- You may want to have students use their graphs to convert the temperature in the daily weather forecast, in a recipe, or in some other source from Fahrenheit to Celsius.

References

Asimov, Isaac. *The Measure of the Universe.* New York: Harper and Row Publishers, 1983.

Kurtz, V. Ray. *Metrics for Elementary and Middle Schools.* Washington, DC: National Education Association, 1978.

Youden, W. J. *Experimentation and Measurement.* Washington, DC: NSTA, 1989.

UNIT 2

Energy and Motion

In Unit 2, students are introduced to kinematics, the study of motion, and dynamics, the study of the causes of motion. Newton's three laws of motion are introduced to encapsulate the relationship between motion and force. From forces, students then move on to the concept of energy, the various types of energy, and the law of conservation of energy.

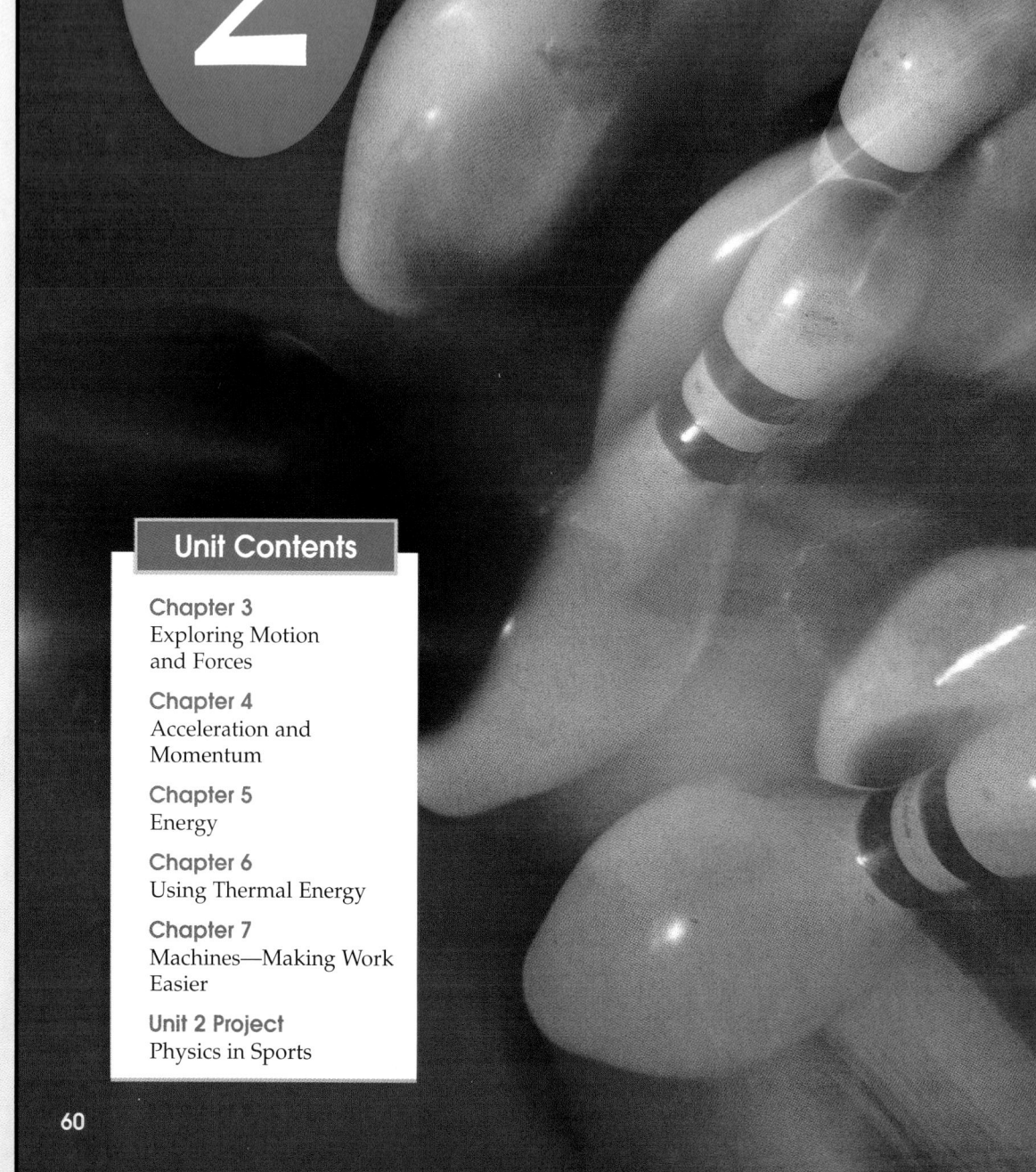

CONTENTS

Unit Contents

60

Science at Home

Sports Events Have students monitor TV and newspaper articles on sports events and collect verbs and adjectives used to describe the action. **L1**

Measuring Motion Have students monitor TV and newspapers to find measurements of motion. Have them categorize the types of motion and the units used to measure motion. **L1**

Energy and Motion

What's Happening Here?

A bowling ball strikes the pins and scatters them in many directions. Some pins fall down, some fly through the air, and some may remain standing. What determines a pin's motion? You will discover that all motions are the result of forces. The forces acting on each bowling pin determine whether it will move and the direction of the movement. Forces may also be at work when an object is standing still, like the rock balanced in the small photo above. In this unit you will study the connections among force, motion, and energy.

Science Journal

The rock in the photograph above looks as if it will topple over any minute, but it has been there for thousands of years. In your Science Journal, describe how forces are involved in producing the formation. What force keeps it standing? What force will eventually make it fall?

61

Chapter Organizer

Section	Objectives/Standards	Activities/Features
Chapter Opener		**Explore Activity:** Find out about physics for one cent. p. 63
3-1 Motion and Speed (1 session, ½ block)*	1. **Define** *speed* as a rate. 2. **Perform** calculations involving speed, time, and distance. 3. **Interpret** distance-time graphs. **National Content Standards:** (5-8) UCP1-UCP5, A1, A2, B2, D1, D2, E1; (9-12) UCP1-UCP5, A1, A2, B4, D3, E1	**Using Math,** p. 65 **Using Math:** Calculating Speed, p. 66 **Using Math:** Calculating Time from Speed, p. 67 **Using Computers,** p. 69 **Skill Builder:** Comparing and Contrasting, p. 69 **Activity 3-1:** Design a Slow Flyer, pp. 70-71
3-2 Velocity and Acceleration (1 session, ½ block)*	1. **Compare** and **contrast** speed, velocity, and acceleration. 2. **Calculate** acceleration. **National Content Standards:** (5-8) UCP2, UCP3, A2, B2; (9-12) UCP2, UCP3, A2, B4	**MiniLAB:** How can you describe the motion of a car? p. 74 **Using Math:** Calculating Acceleration, p. 75 **Science Journal,** p. 75 **Skill Builder:** Making and Using Graphs, p. 75
3-3 Science and Society (½ session, ½ block)*	1. **Analyze** the motion that takes place in a car crash. 2. **Evaluate** the effects of wearing seat belts during a car crash. 3. **State** an informed opinion about whether laws should require people to wear seat belts. **National Content Standards:** (5-8) UCP2, UCP3, UCP5, B2, F1, F3-F5; (9-12) UCP2, UCP3, UCP5, B4, F1, F5, F6	**Explore the Issue,** p. 77
3-4 Connecting Motion with Forces (2 sessions, 1 block)*	1. **Recognize** different examples of forces. 2. **Identify** cause-and-effect relationships between force and changes in velocity. 3. **Give** examples of the effects of inertia. 4. **State** Newton's first law of motion. **National Content Standards:** (5-8) UCP2-UCP5, A2, B2, E2, F4, F5, G1, G3; (9-12) UCP2-UCP5, A2, B4, E2, F1, F5, F6, G1, G3	**Using Technology:** Air Bags, p. 80 **MiniLAB:** How does friction act as a force? p. 81 **Science Journal,** p. 82 **Skill Builder:** Recognizing Cause and Effect, p. 82
3-5 Gravity—A Familiar Force (1½ sessions, 1 block)*	1. **Give** examples of the effects of gravity. 2. **Relate** gravitational force to mass and distance. 3. **Distinguish** between mass and weight. **National Content Standards:** (5-8) UCP2, UCP3, A2, B2, C5, G1; (9-12) UCP2, UCP3, A2, B4, G1	**Problem Solving:** An Experiment for the Shuttle, p. 84 **Connect to Life Science,** p. 85 **Science Journal,** p. 86 **Skill Builder:** Observing and Inferring, p. 86 **Activity 3-2:** Balancing Forces Against Gravity, p. 87 **People and Science:** Derique, Juggler, p. 88

Activity Materials

Explore	Activities	MiniLABs
page 63 plastic drinking cup, water, sand, index card, penny, small piece of paper	pages 70-71 stopwatch or timer with second hand, meterstick or metric tape measure, goggles, string, paper sheets of various types, transparent tape, paper clips, stapler page 87 selection of different types of paper strips, masking tape, string, beaker, ring stand and ring, sheets of newspaper, scissors, plastic cup, 250 mL of sand, balance	page 74 small model car, stopwatch or timer with second hand, pencils, metric tape measure page 81 sheet of plain white paper, 100-g mass, sheet of coarse-grained sandpaper, meterstick

Need Materials? Call Science Kit (1-800-828-7777).   * A complete Planning Guide that includes block scheduling is provided on pages 32T-35T.

Teacher Classroom Resources

Reproducible Masters	Transparencies	Teaching Resources
Study Guide, p. 13 **Reinforcement**, p. 13 **Enrichment**, p. 13 **Activity Worksheets**, pp. 14-15 **Science and Society Integration**, p. 7	**Section Focus Transparency 9**, The Tortoise and the Hare **Science Integration Transparency 3**, Earth's Moving and Shaping Forces **Teaching Transparency 5**, Distance-Time Graph	**Physical Science CD-ROM** **Glencoe Physical Science Interactive Videodisc** **Spanish Resources** **English/Spanish Audiocassettes** **Cooperative Learning Resource Guide** **Lab Partner** **Lab and Safety Skills** **Lesson Plans**
Study Guide, p. 14 **Reinforcement**, p. 14 **Enrichment**, p. 14 **Activity Worksheets**, p. 18 **Lab Manual 4**, Speed and Acceleration **Cross-Curricular Integration**, p. 7	**Section Focus Transparency 10**, A Roller Coaster Ride	
Study Guide, p. 15 **Reinforcement**, p. 15 **Enrichment**, p. 15 **Critical Thinking/Problem Solving**, p. 9	**Section Focus Transparency 11**, Why Wear a Seatbelt?	**Assessment Resources** **Chapter Review**, pp. 9-10 **Assessment**, pp. 20-23 **Performance Assessment in the Science Classroom (PASC)** **MindJogger Videoquiz** **Alternate Assessment in the Science Classroom** **Performance Assessment** **Chapter Review Software** **Computer Test Bank**
Study Guide, p. 16 **Reinforcement**, p. 16 **Enrichment**, p. 16 **Activity Worksheets**, p. 19	**Section Focus Transparency 12**, Motion in Outer Space **Teaching Transparency 6**, Newton's First Law	
Study Guide, p. 17 **Reinforcement**, p. 17 **Enrichment**, p. 17 **Activity Worksheets**, pp. 16-17 **Multicultural Connections**, p. 9 **Science Integration Activity 3**, Down to Earth(worms) **Concept Mapping**, pp. 11-12	**Section Focus Transparency 13**, Astronauts on the Moon	

Key to Teaching Strategies

The following designations will help you decide which activities are appropriate for your students.

L1 Level 1 activities should be within the ability range of all students, including those with learning difficulties.

L2 Level 2 activities should be within the ability range of the average to above-average student.

L3 Level 3 activities are designed for the ability range of above-average students.

LEP LEP activities should be within the ability range of Limited English Proficiency students.

LS These activities are designed to address different learning styles.

COOP LEARN Cooperative Learning activities are designed for small group work.

P These strategies represent student products that can be placed into a best-work portfolio.

GLENCOE TECHNOLOGY

The following multimedia resources are available from Glencoe.

Science and Technology Videodisc Series (STVS)
Physics
   Laminar Flow over Airplane Wings
Earth & Space
   Moving Continents

National Geographic Society Series
STV: Restless Earth
Newton's Apple: Physical Sciences

Glencoe Physical Science Interactive Videodisc
Motion

Physical Science CD-ROM

Teacher Classroom Resources

This is a representation of key blackline masters available in the Teacher Classroom Resources.

Teaching Aids

Section Focus Transparencies

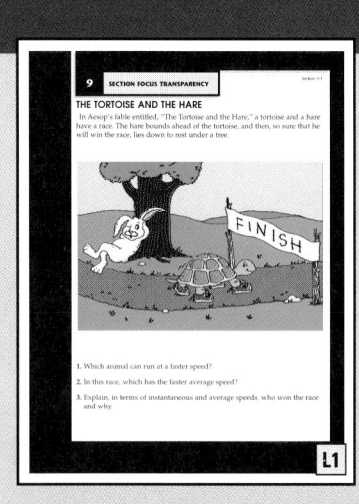

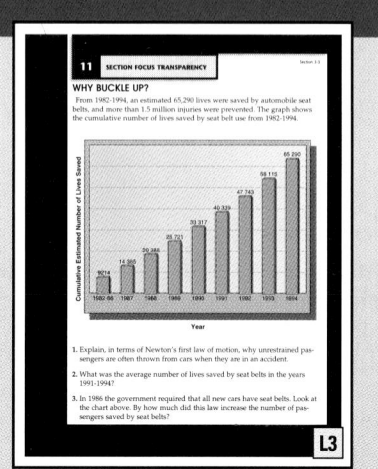

Science Integration Transparencies

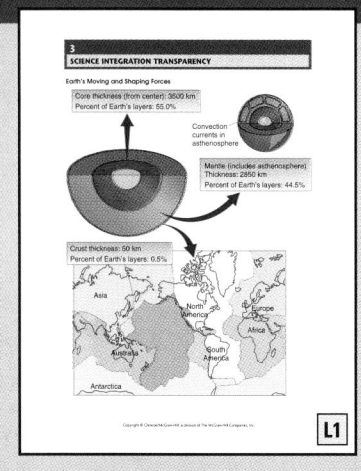

Teaching Transparencies

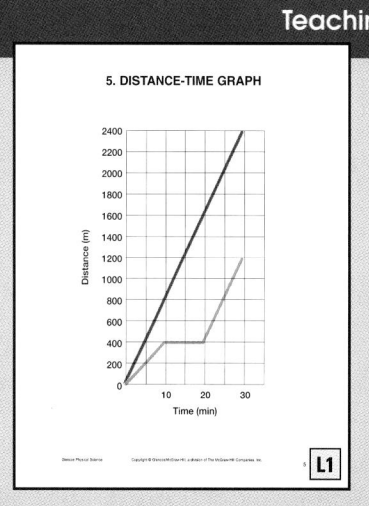

Meeting Different Ability Levels

Study Guide

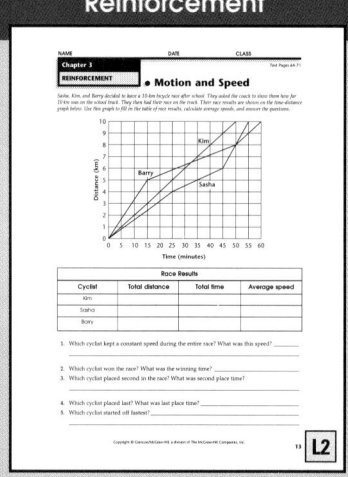

Reinforcement

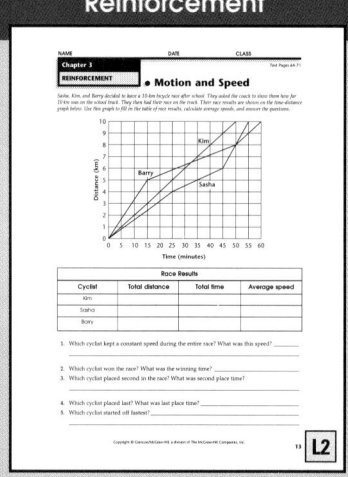

Enrichment Worksheets

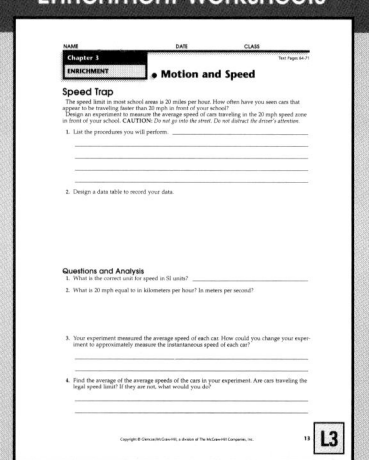

Hands-On Activities

Science Integration Activity

L1

Lab Manual

L2

Assessment

Performance Assessment

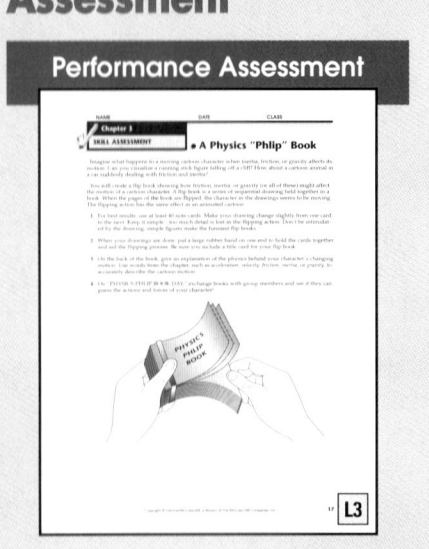

L3

Enrichment and Application

Critical Thinking/ Problem Solving

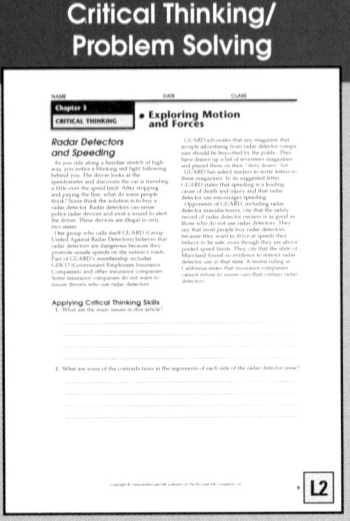

L2

Cross-Curricular Integration

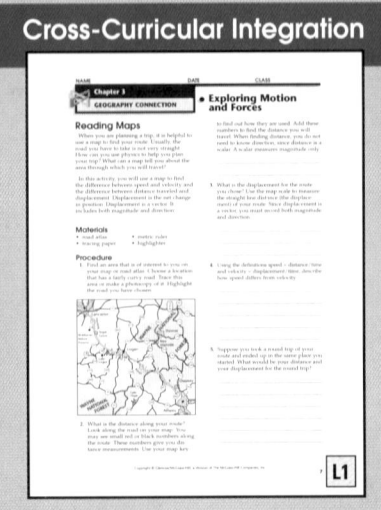

L1

Science and Society Integration

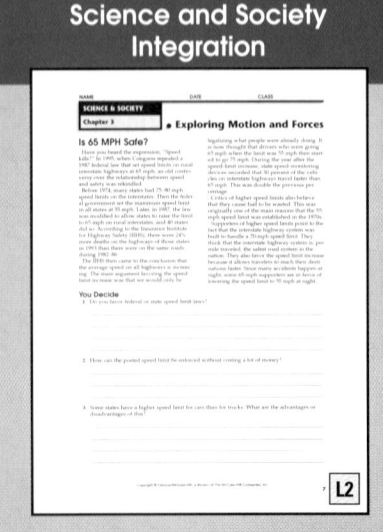

L2

Multicultural Connections

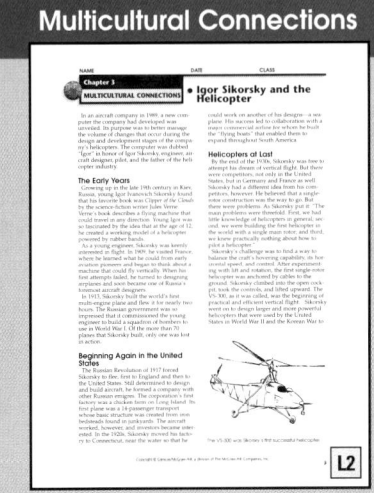

L2

Concept Mapping

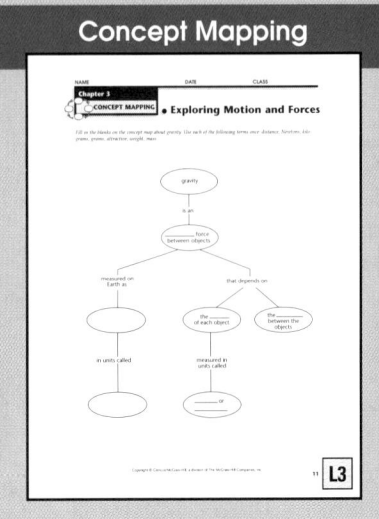

L3

Exploring Motion and Forces

CHAPTER OVERVIEW

Section 3-1 This section introduces the concept of an object's speed as the rate of change in its position. Instantaneous speed, constant speed, average speed, and distance-time graphs are discussed.

Section 3-2 Velocity and speed are compared. The concept of acceleration is introduced as a rate of change in velocity.

Section 3-3 Science and Society This section describes how seat belts reduce injuries from car crashes and asks students to form an opinion about the mandatory use of seat belts.

Section 3-4 The concept of force and the property of inertia are described, followed by a discussion of friction and Newton's first law of motion.

Section 3-5 The concept of an object's weight is developed, and the operation of scales is explained.

Chapter Vocabulary

speed	balanced
instantaneous	forces
speed	net force
constant speed	inertia
average speed	friction
velocity	gravity
acceleration	weight
force	

Theme Connection

Stability and Change A theme that emerges in this chapter is that objects change their positions when unbalanced forces act upon them. A net force acting on an object can be inferred from changes in its motion.

Previewing the Chapter

62

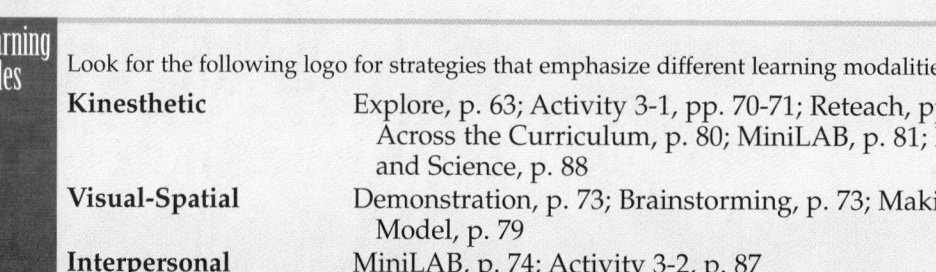

Learning Styles

Look for the following logo for strategies that emphasize different learning modalities. **LS**

Kinesthetic Explore, p. 63; Activity 3-1, pp. 70-71; Reteach, pp. 75, 77; Across the Curriculum, p. 80; MiniLAB, p. 81; People and Science, p. 88

Visual-Spatial Demonstration, p. 73; Brainstorming, p. 73; Making a Model, p. 79

Interpersonal MiniLAB, p. 74; Activity 3-2, p. 87

Intrapersonal Activity, p. 77

Logical-Mathematical Activity, p. 65; Reteach, pp. 68, 77, 85

Linguistic Using Science Terms, p. 66; Revealing Preconceptions, p. 84; Science Journal, p. 84

Chapter 3

Exploring Motion and Forces

Why is a water slide so exciting for many people? You may have enjoyed sliding down the slippery slope and splashing into the water at the bottom. What makes amusement park rides so much fun is their creative use of some basic laws of physics. In this chapter, you'll begin to explore the important connection between motion and forces that explains many familiar observations. The activity below illustrates this connection.

EXPLORE ACTIVITY

Find out about physics for one cent.

1. Use a plastic drink cup that is about two-thirds full of water or sand to anchor it. Lay a flat index card over the top of the cup.
2. Place a penny on the card, centered over the cup.
3. With a quick flick of your finger, give the card a horizontal push. What happens?
4. Crumple up a small piece of paper to about the size of a marble. Place it on the card and flick the card away. Record your observations.

Observe: In your Science Journal, explain why you think the penny and the crumpled paper behaved as they did.

Previewing Science Skills

▶ In the Skill Builders, you will **compare and contrast, make and use graphs,** and **observe and infer.**

▶ In the Activities, you will **measure, control variables,** and **experiment.**

▶ In the MiniLABs, you will **calculate, observe,** and **interpret.**

63

Prepare

Section Background

- Because an object's motion is a change in its position, the term *position* must be understood. An object's position at any moment is relative to, or measured from, some arbitrary reference point that is stationary.

- The *slope* at any point on a distance-time graph is equivalent to the instantaneous speed of the object at that time and location. If a segment of a graph is a straight line, the speed is constant and the slope is the average velocity during the interval.

Preplanning

To prepare for Activity 3-1, obtain the materials listed on page 70.

1 Motivate

Bellringer

Before presenting the lesson, display **Section Focus Transparency 9** on the overhead projector. Assign the accompanying **Focus Activity** worksheet. L1 LEP

Tying to Previous Knowledge

Ask students to recall that the units marked on a speedometer are miles per hour and kilometers per hour—quantities introduced in Chapter 2.

Science Words

- speed
- instantaneous speed
- constant speed
- average speed

Objectives

- Define *speed* as a rate.
- Perform calculations involving speed, time, and distance.
- Interpret distance-time graphs.

Understanding Speed

When something moves, it changes position. It travels from one place to another, even if only for a brief time. Think about an ice skater. If asked to describe the motion of speed skater Bonnie Blair in **Figure 3-1,** you could say something like "leaning forward with repetitive gliding strokes." How would you describe the motion of a person on a swing?

Motion and Position

You don't always have to see something move to know that motion has taken place. For example, suppose you look out a window and see a mail truck parked next to a mailbox. One minute later, you look out again and see the same truck parked down the street from the mailbox. Although you didn't observe the motion, you know the truck moved because its position relative to the mailbox has changed.

Thus, motion can be described as a change in position. To know whether the position of something has changed, you need a reference point. In the case of the mail truck, the mailbox was a reference point. You can also use a reference point to get a rough idea of how far the truck moved. However, you don't know how fast the truck moved to reach its new position.

Motion and Time

Descriptions of motion often include speed—how fast something moves. If you think of motion as a change in position, then speed is an expression of how much time it takes for that change in position to occur. Any change over time is called a rate. **Speed,** then, is the rate of change in position. Speed can also be described as simply a "rate of motion."

There are several useful measures of speed. The speedometer in a car shows instantaneous speed. **Instantaneous speed** is the rate of

Figure 3-1

Some speed skaters reach a speed of 56 km (35 mi)/h. Their uniforms help to maximize their speed.

64

Program Resources

Reproducible Masters

Study Guide, p. 13 L1
Reinforcement, p. 13 L1
Enrichment, p. 13 L3
Activity Worksheets, pp. 5, 14, 15 L1
Science and Society Integration, p. 7 L1

Transparencies

Section Focus Transparency 9 L1
Teaching Transparency 5 L1
Science Integration Transparency 3 L2

motion at any given instant. At the moment the picture in **Figure 3-2** was taken, the car was traveling at a speed of 80 km/h. On a highway, a car may travel at the same speed for a fairly long period of time. A speed that does not vary is called a **constant speed.**

Figure 3-2

The speedometer of a car shows how fast the car is moving at any given instant. *Would the car's instantaneous speed remain the same or change if you measure it frequently during a rush-hour drive?*

Changing Speed

Much of the time, the speeds you deal with are not constant. Think about riding your bicycle for a distance of 5 km. As you start out, your speed increases from 0 km/h to, say, 20 km/h. You slow down to 12 km/h as you pedal up a steep hill and speed up to 35 km/h going down the other side of the hill. You stop for a red light, speed up again, and move at a constant speed for a while. As you near the end of the trip, you slow down and then stop. Checking your watch, you find that the trip took 15 minutes, or one-quarter of an hour. How would you express your speed on such a trip? Would you use your fastest speed, your slowest speed, or some speed in between the two?

In cases where rate of motion varies a great deal, the best way to describe speed is to use average speed. **Average speed** is the total distance traveled divided by total time of travel. On the trip just described, your average speed was 5 km divided by ¼ hour, or 20 km/h.

Figure 3-3

A runner's speed can change from moment to moment during a race.

USING MATH

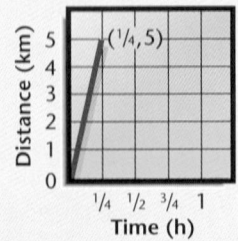

A line graph provides a "picture" of the bicyclist traveling 5 km in ¼ hour.

The *slope,* or steepness, of the line represents the average speed. Suppose another bicyclist traveled 5 km in ½ hour. Compared to the first graph, is this more steep or less steep? Is the average speed greater or less than the first speed?

2 Teach

Activity

LS **Logical-Mathematical** Review the concept of rate by having students measure their breathing rates. Discuss other time rates, such as heart rate (pulse), growth rates, and interest rates. Discuss the units that describe these rates and point out similarities in the units.

Use **Study Guide,** p. 13 as you teach this lesson.

USING MATH

The second graph is less steep than the first one, and the average speed is less than the first speed.

Student Text Question

How would you express your speed on such a trip? *as average speed*

Would you use your fastest speed, your slowest speed, or some speed in between the two? *some speed between the two*

GLENCOE TECHNOLOGY

 Videodisc

Glencoe Physical Science Interactive Videodisc

Side 1, Lesson 1

Distance and Displacement

14-1470

1473-2365

 CD-ROM

Physical Science CD-ROM

Have students perform the interactive exploration for Chapter 3 to reinforce important chapter concepts and thinking processes.

Content Background

Show students that time units really do emerge from the equation $t = d/v$. That is,

$$m \div m/s = m \times s/m = s$$

Emphasize that the vertical axis in **Figure 3-4** represents the distance each swimmer has covered from the beginning of the workout, and the horizontal axis represents the time that has elapsed since the start of the workout.

Revealing Preconceptions

Students may interpret the height of a graph as its slope. The steepness of the straight line of a distance-time graph is the average speed; the height of the line is the distance the object has traveled.

Using Science Terms

LS **Linguistic** Students may have difficulty with the term *relative* used in describing an object's position. Explain that the term means "dependent upon" or "with reference to."

66

Calculating Speed

How could you find out who the fastest runner in your school is? One way would be to get all the students together to run in a giant race. However, this isn't very practical. A better way would be to have each student run a certain distance and to time each runner, as shown in **Figure 3-3** on page 65. The runner with the shortest time is the fastest student.

The relationship between distance, speed, and time is shown in the equation $d = v \times t$, where d = distance, v = speed, and t = time. If you know the distance and time, you can rewrite the equation as

$$v = \frac{d}{t}$$

to calculate the average speed.

The problems below will give you practice in calculating the average speed.

Graphing Speed

A distance-time graph makes it possible to "see" the motion of an object over a period of time. For example, the graph in **Figure 3-4** shows how two swimmers performed during a 30-minute workout. The smooth red line represents the

USING MATH

Calculating Speed

Example Problem:

Your neighbor says she can skate at a speed of 4 m/s. To see if you can skate faster, you have her time you as you skate as fast as you can for 100 m. Your time is 20 s. Who skates faster?

Problem-Solving Steps:

1. What is known?
   distance, d = 100 m; time, t = 20 s
2. What is unknown? average speed, v
3. Use the equation $v = \dfrac{d}{t}$
4. **Solution:** $v = \dfrac{d}{t}$

$$v = \frac{100 \text{ m}}{20 \text{ s}} = 5 \text{ m/s}$$

Your neighbor's speed is 4 m/s; you skate faster.

Practice Problem

Florence Griffith Joyner set a world record by running 200 m in 21.34 s. What was her average speed?

Strategy Hint: In what units will your answer be given?

Inclusion Strategies

Gifted Have students investigate the speeds at which various organisms such as ants, wasps, turtles, horses, leopards, ostriches, and falcons move. Ask them to present the information as a bar graph on a poster. **L3**

motion of a swimmer who swam 800 m during each 10-minute period. Her speed was constant at 80 m/min.

The blue line represents the motion of a second swimmer, who did not swim at a constant speed. She covered 400 m during the first 10 minutes of her workout. Then she rested for the next 10 minutes. During this time, her speed was 0 m/min. The slope of the graph over the next 10 minutes shows that she swam faster than before and covered 800 m. What total distance did she cover? What was her average speed for the 30-minute period? The problems below will give you practice in calculating the time during which a motion occurs.

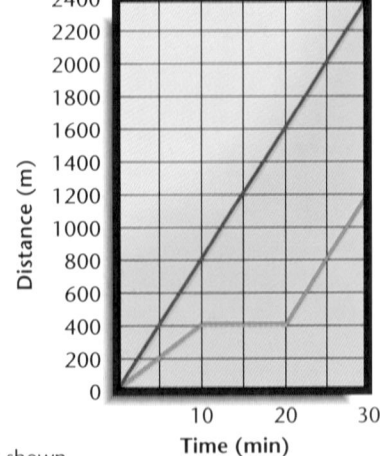

Figure 3-4

The distance-time graph shown here makes it possible to visualize the motion of the swimmers. *Which person swam the farthest during the 30-minute workout?*

USING MATH

Calculating Time from Speed

Example Problem:

Sound travels at a speed of 330 m/s. If a lightning bolt strikes the ground 1 km away from you, how long will it take for the sound to reach you?

1. What is known?   distance, $d = 1$ km; speed, $v = 330$ m/s

2. Use the equation $d = v \times t$. Since you know d and v, rewrite the equation as $t = \dfrac{d}{v}$ to find the time t.

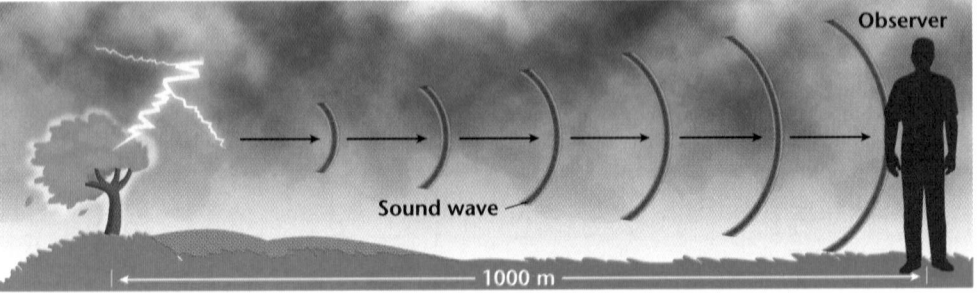

Observer

Sound wave

1000 m

3. Solution: $t = \dfrac{d}{v}$ $t = \dfrac{1 \text{ km}}{330 \text{ m/s}} = \dfrac{1000 \text{ m}}{330 \text{ m/s}} = 3.03$ s

Practice Problem

The world's fastest passenger elevator operates at an average speed of about 10 m/s. If the 60th floor is 219 m above the first floor, how long does it take the elevator to go from the first floor to the 60th floor?

Strategy Hint: Rearrange the equation.

3-1 Motion and Speed   **67**

Visual Learning

Figure 3-5 Are modern-day Europe, Asia, and North America found in the northern or southern portion of Pangaea? *in the northern portion* B. In which direction is India moving in this diagram? *northward* C. Will the present arrangement of the continents change in the future? Explain. *Yes, because the continents are in constant motion; the motion is so slow that we're not aware of it.* **LEP** **LS**

3 Assess

Check for Understanding

? FLEX Your Brain

Use the Flex Your Brain activity to have students explore SPEED. **L1**

📁 **Activity Worksheets,** p. 5

Reteach

LS Logical-Mathematical Have students calculate the average speed of a wind-up or battery-operated toy car using metersticks and a wall clock.

Extension

📁 For students who have mastered this section, use the **Reinforcement** and **Enrichment** masters.

INTEGRATION
Earth Science

Earth's Crust—Moving Right Along

Can you think of something that is moving so slowly you cannot watch its motion directly, yet you can see evidence of its motion over long periods of time? As you look around the surface of Earth from year to year, the basic structure of our planet seems the same. Yet, if you examined geological evidence of what Earth's surface looked like over hundreds of millions of years, you would see an ever-changing place, as shown in **Figure 3-5.**

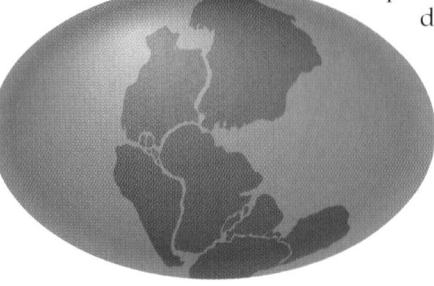

250 million years ago

Figure 3-5

About 250 million years ago, the arrangement of the continents formed the supercontinent called Pangaea. *Are modern-day Europe, Asia and North America found in the northern or southern portion of Pangaea?*

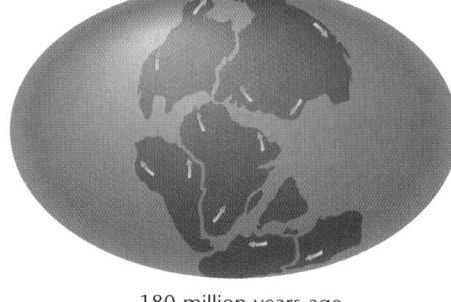

180 million years ago

 About 180 million years ago, Pangaea began to separate into smaller pieces.

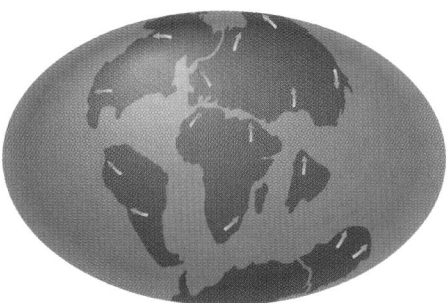

66 million years ago

B Continental movement continued. *In which direction is India moving in this diagram?*

Present

C *Will the present arrangement of the continents change in the future? Explain.*

How can continents move?

You may have studied about the theory of plate tectonics, which suggests that Earth's crust and solid upper mantle, shown in **Figure 3-6** as the lithosphere, form huge sections called plates. If you compare Earth to an egg, these plates are about as thick as the eggshell. The plates slide around slowly

68 Chapter 3 Exploring Motion and Forces

Integrating the Sciences

Earth Science In the early 1900s, Alfred Wegener, a German meteorologist, proposed that all Earth's continents were once part of a single landmass called *Pangaea*, meaning "all Earth." Evidence supporting this theory includes similarities in fossils, glacial sediments, rock layers, and the ocean floor between continents, thought once to have been directly connected in a landmass.

on a putty-like layer of partly molten mantle called the asthenosphere. A variety of geological changes, such as the formation of mountain ranges, earthquakes, and volcanic eruptions, can occur when these moving plates interact in different ways.

Measuring Speeds of Tectonic Movements

In most of the examples in this chapter, we were measuring speed based on traveling an easily measurable distance in a few seconds or minutes. The speeds of tectonic movements are best described in distance per year. In California, the movement along the San Andreas fault has an average drift speed of about 2 cm per year. The Australian plate's movement is one of the fastest, pushing Australia north at an average speed of about 17 cm per year. **Figure 3-5** illustrates the slow, long-term movement of the continents.

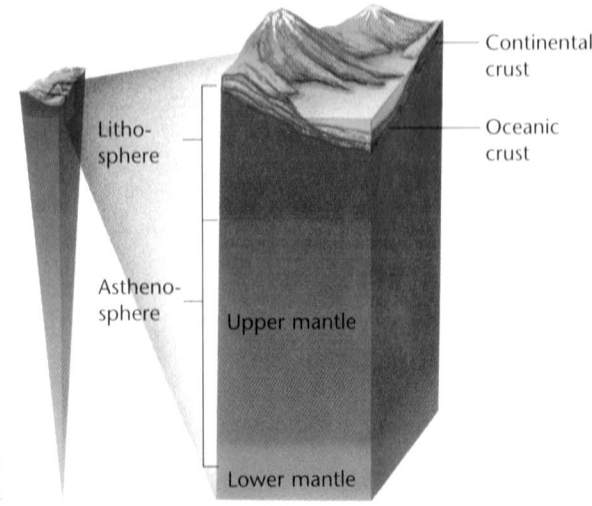

Figure 3-6

The diagram shows the different layers of Earth's crust and mantle.

Section Wrap-up

Review

1. What units would you use to describe the speed of a car? Would you use different units for the speeds of runners in a neighborhood race? Explain.

2. In a skateboarding marathon, the winner covered 435 km in 36.75 h. What was the winner's average speed?

3. **Think Critically:** Make a distance-time graph for a 2-hour car trip. The car covered 50 km in the first 30 minutes, stopped for 30 minutes, and covered 60 km in the final 60 minutes. Note the three graph segments. Which graph segment slopes the most? Which one does not slope? What was the car's average speed?

Skill Builder
Comparing and Contrasting
Compare and contrast the motions of a person on a swing and a person riding a merry-go-round. If you need help, refer to Comparing and Contrasting in the **Skill Handbook.**

> **Using Computers**
>
> **Spreadsheet** Given the information in the graph in **Figure 3-4**, use a computer to construct a data table listing the distance and time measurements from which the graph was made. Use a spreadsheet program to make your data table and use it, if possible, to re-create the graphs.

Skill Builder
Both riders experience repetitive motions along the path of an arc. The arc of the person on a merry-go-round horse is up and down; the arc of a person on a swing is wider. In addition to the up-and-down motion, the merry-go-round rider experiences continuous motion around a wide circle.

 **Assessment**

Content Have students relate the formula $v = d/t$ to a portion of the concepts in a concept map. Use the Performance Task Assessment List for Concept Map in **PASC,** p. 89.

P

4 Close

•MINI•QUIZ•

Use the Mini Quiz to check students' recall of chapter content.

1. **What is motion?** *a change in position*

2. **What two quantities do you need to know to calculate average speed?** *distance and time*

3. **What does a flat line on a segment of a distance-time graph tell you about the average speed for that time period?** *No change in distance occurs; therefore, average speed is zero.*

Section Wrap-up

Review

1. km/h; yes, m/s; the neighborhood race would involve short distances and short periods of time

2. 11.8 km/h

3. **Think Critically** The graph segment of the first 30 minutes slopes the most; the segment for the next 30 minutes doesn't slope; the average speed is 55 km/h.

Using Computers		
Time (min)	Distance (red)	Distance (blue)
0	0	0
5	400	200
10	800	400
15	1200	400
20	1600	400
25	2000	800
30	2400	1200

Activity 3-1

PREPARATION

Purpose

 Kinesthetic Students will examine a variety of variables included in the measurement of speed in a glider designed to fly as slowly as possible.

L1 LEP

Process Skills

observing, measuring, separating and controlling variables, communicating, forming a hypothesis, classifying, comparing and contrasting, recognizing cause and effect, designing an experiment, interpreting data, making models

Time

40 minutes

Materials

Conduct the activity in a large, open space.

Safety Precautions

Be sure the flight test area is cleared of people and breakable objects. Have all students wear safety goggles.

Possible Hypotheses

Most students will hypothesize that having a wide surface area and light weight would probably be beneficial for a long flight time.

 **Activity Worksheets,** pp. 5, 14, 15

Activity 3-1

Design Your Own Experiment
Design a Slow Flyer

How do heavy jet planes fly high above Earth's surface? Their shape, wing design, and high thrust all contribute to their flying capabilities. Engineers use many physics principles to design flying machines to move fast, glide, or maneuver quickly.

PREPARATION

Problem

Can you design planes with specific flying behaviors from materials such as paper and tape? These behaviors can include speed and ability to glide for long distances.

Form a Hypothesis

Write a hypothesis about what types of design features will help a paper glider fly as slowly as possible. One important design feature will be the size and shape of the wings.

Objectives

• Design a paper glider to fly as slowly as possible.
• Measure distance and time and calculate your glider's speed.

Materials

• stopwatch or timer with second hand
• meterstick or metric tape measure
• goggles
• string
• paper sheets of various types
• transparent tape
• paper clips and stapler

Safety Precautions 👓

Make sure you have a clear area to launch your glider; do not throw it toward another person.

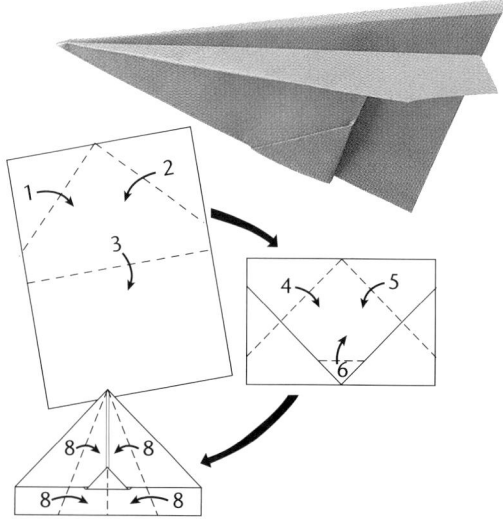

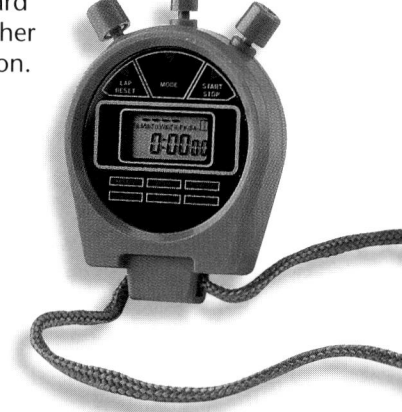

Across the Curriculum

Math Students can compare their average flying speed with the plane's weight and the surface area of its wings. Have them estimate the surface area by tracing the plane's wings onto graph paper (or making the plane out of graph paper) and counting the squares covered by the wing. Have students determine the area of the wings in square centimeters. **L1 LS**

PLAN THE EXPERIMENT

1. When you plan your experiment, decide how you can measure the speed of a paper glider.
2. In constructing your glider, be sure to consider its balance and stability.
3. Think about how the way you launch your glider will affect its speed.
4. If you need a data table, design one in your Science Journal so it is ready to use as your group collects data.

Check the Plan

1. Be sure you have considered all the factors that can affect glider speed.
2. Will you have more than one trial run with each design?
3. What measurements will you need to make to determine the speed of your glider? Who will make the measurements?

4. Will the data be summarized in a graph?
5. Be sure to make adjustments to your glider after your first trials to try to improve its performance. It will be best to adjust one factor at a time.
6. *Make sure your teacher approves your plan and that you have included any changes suggested in the plan.*

DO THE EXPERIMENT

1. Carry out the experiment as planned.
2. Be sure to enter all useful data in your Science Journal so you can make effective adjustments to your glider.

Analyze and Apply

1. **Compare** your results with those of other groups.
2. **Calculate** your glider's maximum, minimum, and average speeds.

3. If your glider travels a curved path, which distance measurement will give the slowest speed calculation—along the curved path or along the straight line between the starting and landing points?
4. What factors affected your glider's flight? **Predict** how you could change some of these factors.

Go Further

What types of contests exist for full-sized gliders piloted by people? Investigate and write a brief report that summarizes your findings.

✓ Assessment

Oral Ask students to explain why average speed is easier to determine and why the speed varies during flight. Use the Performance Task Assessment List for Making Observations and Inferences in **PASC**, p. 17.

 P

Go Further

The most common contests are races against time. Contestants follow a course and may fly as far as 1000 km.

PLAN THE EXPERIMENT

Possible Procedures

Students should realize that by measuring the linear flight distance and total time, they can calculate average speed.

For each design, students should average speeds measured in two or more trials.

Teaching Strategies

Each team will design and test a glider to compete with other teams. All teams should contribute to devising a set of rules to make this a fair contest—flight location and distance, launching methods, assignment of judges, etc.

DO THE EXPERIMENT

Expected Outcome

Most groups will find that lightweight gliders with wings having a wide surface area will fly at the slowest average speed.

Analyze and Apply

1. Students can write their results on the board to compare them.
2. Actual speed is likely to change during travel. Have students calculate the average speed by dividing distance traveled by the time of flight.
3. The curved-path distance is greater than a corresponding straight path. The straight-path distance measurement will result in a slower average speed.
4. Predictions will vary, but weight, balance, shape, and type of paper used can be important factors to experiment with.

Prepare

Section Background

Velocity and acceleration are both rates of motion. The slope of a velocity-time graph represents acceleration.

Preplanning

- Obtain carts and stopwatches for the MiniLAB. Ask students who have digital wristwatches with stopwatches to wear them the day of the activity.
- Obtain several wind-up or battery-operated toy cars.

1 Motivate

Bellringer

Before presenting the lesson, display **Section Focus Transparency 10** on the overhead projector. Assign the accompanying **Focus Activity** worksheet. L1 LEP

Tying to Previous Knowledge

Have a student explain the phrase "stepping on the gas." Relate the function of the accelerator and the motion of a car.

Science Words

velocity
acceleration

Objectives

- Compare and contrast speed, velocity, and acceleration.
- Calculate acceleration.

Figure 3-7
Knowing the speed and direction of a tornado can be important for safety.

3•2 Velocity and Acceleration

Velocity and Speed

You turn on the radio and hear the tail end of a news story about a tornado sighting in a storm, such as the one in **Figure 3-7**. The storm, moving at a speed of 60 km/h, has just left a town 10 km north of your location. Should you be worried? Unfortunately, you don't have enough information. Knowing only the speed of the storm isn't much help. Speed only describes how fast something is moving. You also need to know the direction the storm is moving. In other words, you need to know the velocity of the storm. **Velocity** describes both *speed* and *direction* of an object.

Picture a motorcycle racing down the highway at 100 km/h. It meets another motorcycle going 100 km/h in the opposite direction. The speeds of the motorcycles are the same, but their velocities are different because the motorcycles are not moving in the same direction.

You learned earlier that speed isn't always constant. Like speed, velocity may also change. Unlike speed, the velocity of an object can change even if the speed of the object remains constant! For example, if a car goes around a curve in the road, its direction changes. Even if the speed remains constant, the velocity changes because the direction changes.

Acceleration

At the starting line of a drag strip, the driver idles the dragster's engine. When the starting signal flashes, the driver presses the gas pedal.

The car leaps forward, as shown in **Figure 3-8A**, moving faster and faster until it crosses the finish line. Then, in **Figure 3-8B**, the driver releases a drag chute and the car rapidly slows down and comes to a stop. While the car gains speed, it is accelerating. Strange as it may seem, the car is also accelerating as it slows down. How is this possible?

72 Chapter 3 Exploring Motion and Forces

Program Resources

 **Reproducible Masters**
Study Guide, p. 14 L1
Reinforcement, p. 14 L1
Enrichment, p. 14 L3
Activity Worksheets, pp. 5, 18 L1
Lab Manual 4 L2
Cross-Curricular Integration, p. 7 L1

Transparencies
Section Focus Transparency 10 L1

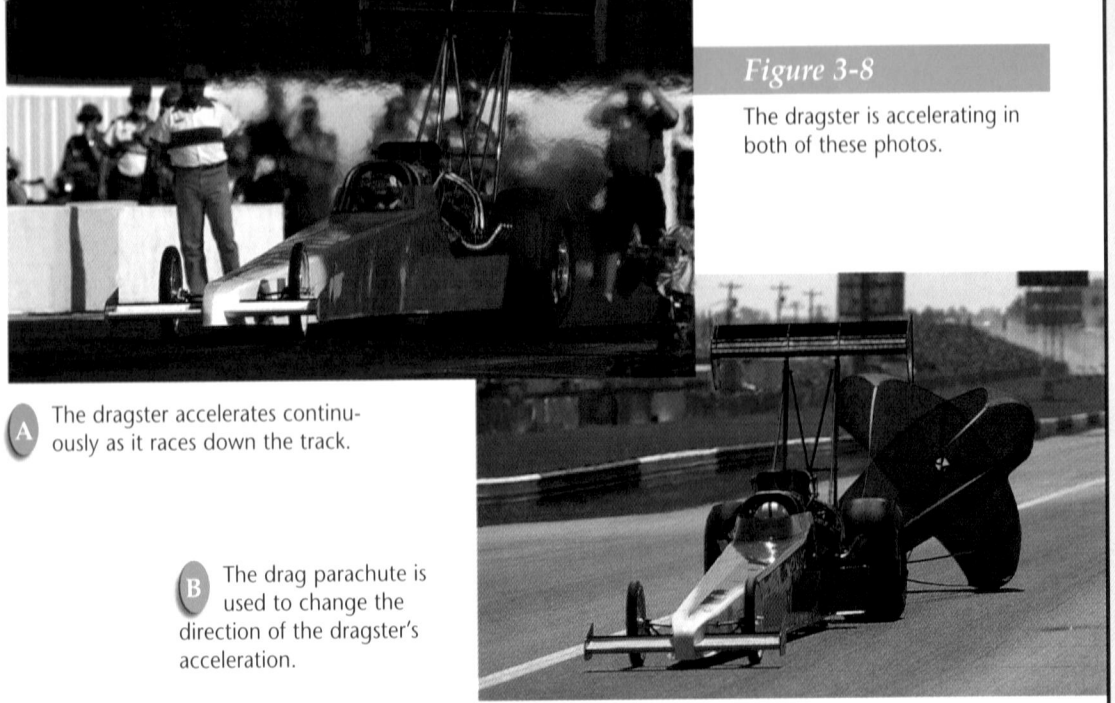

Figure 3-8

The dragster is accelerating in both of these photos.

A The dragster accelerates continuously as it races down the track.

B The drag parachute is used to change the direction of the dragster's acceleration.

What is acceleration?

Acceleration is the rate of change of velocity. Because velocity includes both speed and direction, if either one changes, velocity will change. In other words, acceleration can occur through a change in speed or a change in direction.

If an object travels in a straight line and the directions of acceleration and velocity occur along the same line, as with the dragster, then acceleration is just the rate of change of speed. If the acceleration is in the same direction as the velocity, then the dragster *speeds up.* Its acceleration is positive. If they are in opposite directions, then the dragster *slows down.* Its acceleration is negative. These principles are illustrated in **Figure 3-9.**

The amount of acceleration depends on both the change in velocity and the time interval. The *time interval* is the amount of time that passed while the change in velocity was taking place. The acceleration will be large if the change in velocity is large; it will also be large if the change in velocity occurs in a small time interval.

Figure 3-9

The diagram indicates how positive and negative acceleration are defined. *Explain these two definitions.*

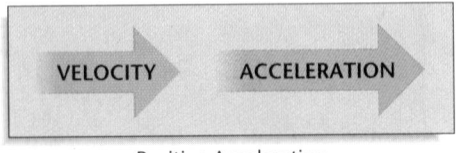

Positive Acceleration

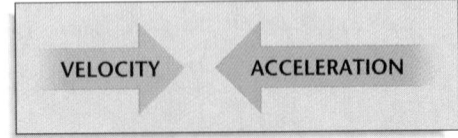

Negative Acceleration

---

2 Teach

Brainstorming

LS **Visual-Spatial** Ask students to give ways of describing direction, such as uphill, downhill, right, left, or compass points.

Demonstration

LS **Visual-Spatial** Place a wind-up or battery-operated toy car on the floor and release it first in one direction and then in the opposite. Emphasize that the average speed in both directions was the same. Ask students whether speed indicates how fast and/or where something is going. Have them explain how they would describe direction.
LEP

📂 Use **Study Guide,** p. 14 as you teach this lesson.

📂 Use Lab 4 in the **Lab Manual** as you teach this lesson.

GLENCOE TECHNOLOGY

💿 **Videodisc**
Glencoe Physical Science Interactive Videodisc
Side 1, Lesson 1
Velocity

3739-3875
Acceleration as a Change in Velocity

4870-7230

Visual Learning

Figure 3-9 Explain these two definitions. *Positive acceleration occurs when acceleration and velocity are in the same direction. When acceleration and velocity are in opposite directions, negative acceleration occurs.* **LEP** **LS**

Theme Connection

Stability and Change In the discussion of acceleration, the stability and change theme should be emphasized. Students likely have a preconception that acceleration applies only to increasing speed, so it should be emphasized that decreasing speed and changing direction are also considered acceleration.

MiniLAB

Purpose

LS **Interpersonal** Students will calculate the average speed and velocity of a cart's motion. **L1** **COOP LEARN**

Materials

cart or toy car, stopwatch, meterstick

Teaching Strategies

Troubleshooting Try to use carts or cars that will travel approximately in a straight line when pushed gently.

Analysis

1. Average speed = distance/time. Velocity should be speed in a given direction (north, southeast, etc.).

2. The speed would be the same, but the direction would be different.

3. The car accelerated when it began to move and decelerated as it slowed and came to rest. It may also have changed direction slightly. Push it harder to cause a greater acceleration.

✔ Assessment

Process Have students use their observations to hypothesize why the cart eventually comes to a rest. Use the Performance Task Assessment List for Formulating a Hypothesis in PASC, p. 21. **P**

📁 **Activity Worksheets, p. 18**

Practice Problem Answer

See p. 75.
7.3 m/s^2

MiniLAB

How can you describe the motion of a car?

Procedure

1. Select a line or place a pencil on the floor to mark your starting point.
2. Beginning at the starting line, give your car a gentle push forward. At the same time, start your stopwatch.
3. Stop timing exactly when the car comes to a complete stop. Mark the spot at the front of the car with another pencil.
4. Record the time for the entire trip. Measure the distance to the nearest tenth of a centimeter and convert it to meters.

Analysis

1. Calculate the average speed during your car's trip. What was the car's average velocity?
2. How would the velocity reading differ if you repeated your experiment in exactly the same way in the opposite direction?
3. Describe any acceleration you observed the car undergo. How would you cause a greater acceleration?

Figure 3-10

Horses often tire near the end of a race. *What shape would the graph have if Secretariat had slowed down near the finish line?*

Visual Learning

Figure 3-10 What shape would the graph have if Secretariat had slowed down near the finish line? *The right-hand portion of the graph would begin to flatten out to the right.*

LEP **LS**

Calculating Acceleration

To calculate average acceleration, divide the change in velocity by the time interval. To find the *change in velocity*, subtract the initial velocity (starting velocity), v_i, from the final velocity, v_f.

$$a = \frac{v_f - v_i}{t} = \frac{\Delta v}{t}$$

The symbol Δ is the Greek letter *delta* and stands for "change in."

When calculating acceleration, be sure to include all proper units and algebraic signs. The unit for velocity is meters/second (m/s) and the unit for time is seconds (s). Thus, the unit for acceleration is meters/second/second. This unit is usually written as m/s^2 and is read as "meters per second squared" or "meters per second per second."

You can see from the equation that acceleration will be positive if an object is speeding up and negative if an object is slowing down. In a horse race, for example, the outcome is often determined by whether a horse is speeding up or slowing down during the stretch run near the end of the race. When Secretariat set the speed record for the Kentucky Derby, **Figure 3-10,** he ran each quarter mile faster than the previous one. The graph in **Figure 3-10** illustrates Secretariat's motion.

When you work on acceleration problems, be careful to correctly identify the initial velocity and the final velocity. If

you confuse them, the acceleration will have the wrong sign (positive or negative). The problems below provide practice in calculating acceleration.

USING MATH

Calculating Acceleration

Example Problem:

A car's velocity changes from 0 m/s to 30 m/s 10 seconds later. Calculate the car's average acceleration.

1. What is known?   initial velocity, v_i = 0 m/s; final velocity, v_f = 30 m/s; time, t = 10 s.

2. Use the equation $a = \dfrac{v_f - v_i}{t}$

3. **Solution:** $a = \dfrac{v_f - v_i}{t}$ $a = \dfrac{30 \text{ m/s} - 0 \text{ m/s}}{10 \text{ s}} = \dfrac{30 \text{ m/s}}{10 \text{ s}} = 3 \text{ m/s}^2$

The car's average acceleration is 3 m/s^2.

Practice Problem

As a roller coaster starts down a hill, its speed is 10 m/s. Three seconds later, its speed is 32 m/s at the bottom of the hill. What is the roller coaster's acceleration?
Strategy Hint: Find the change in velocity by subtracting initial velocity from final velocity.

Section Wrap-up

Review

1. One jet plane is flying east at 880 km/h, and another plane is traveling north at 880 km/h. Do they have the same velocities? The same speeds? Explain.

2. A swimmer speeds up from 1.1 m/s to 1.3 m/s during the last 20 s of the workout. What is the acceleration during this interval?

3. **Think Critically:** Describe three different ways to change your velocity when you're riding a bicycle.

Science Journal

In your Science Journal, explain why we have speed limits on streets and highways rather than velocity limits.

Skill Builder

Making and Using Graphs
Decide on a style of graph to show how velocity changes over time. Explain your choice. Show the graph. If you need help, refer to Making and Using Graphs in the **Skill Handbook**.

Science Journal

Speed is only the rate at which we change our position. Velocity includes direction, too. So, going around a curve might violate a velocity limit.

Skill Builder

Students should select a line graph.

Assessment

Performance   Ask students to describe and then sketch a velocity-time graph for a bicycle ride through a hilly park. Use the Performance Task Assessment List for Graph from Data in **PASC**, p. 39. **P**

3 Assess

Check for Understanding

? FLEX Your Brain

Use the Flex Your Brain activity to have students explore ACCELERATION. **L1**

Activity Worksheets, p. 5

Reteach

LS Kinesthetic Have students demonstrate three ways to accelerate while walking (speeding up, slowing down, or changing direction).

Extension

For students who have mastered this section, use the **Reinforcement** and **Enrichment** masters.

4 Close

·MINI·QUIZ·

Use the Mini Quiz to check students' recall of chapter content.

1. **What is velocity?** *both speed and direction*

2. **What is acceleration?** *rate of change of velocity*

3. **What is the unit of acceleration?** *m/s^2*

Section Wrap-up

Review

1. No, they are traveling in different directions. The speeds are the same; they will travel the same distance in the same time.

2. 0.01 m/s^2

3. **Think Critically** Answers may include: apply brakes, pedal faster, change directions, etc.

Prepare

Section Background

Inertia is the tendency of an object to resist any change in its motion. It accounts for unbelted car passengers crashing into windshields and backs of seats. This concept will be developed in Section 3-4.

1 Motivate

Tying to Previous Knowledge

Have students discuss commercials that advocate seat-belt use.

Bellringer

Before presenting the lesson, display **Section Focus Transparency 11** on the overhead projector. Assign the accompanying **Focus Activity** worksheet. L1 LEP

2 Teach

Content Background

Emphasize that wearing seat belts increases the stopping time and spreads the force over a larger area.

3 Assess

Check for Understanding

? FLEX Your Brain

Use the Flex Your Brain activity to explore REQUIRING SEAT BELTS IN SCHOOL BUSES. L1 P

📁 **Activity Worksheets,** p. 5

SCIENCE & SOCIETY

ISSUE:
3•3 A Crash Course in Safety

Objectives

- Analyze the motion that takes place in a car crash.
- Evaluate the effects of wearing seat belts during a car crash.
- State an informed opinion about whether laws should require people to wear seat belts.

Do you usually wear your seat belt when you travel in a car? Through experiments called crash tests, shown in **Figure 3-11,** and by studying actual car collisions, scientists have learned what happens to people in accidents. They have determined that many serious injuries and deaths can be prevented by the use of relatively inexpensive seat belts that cross a passenger's shoulder, chest, and lap. Many states have responded by requiring that drivers and front-seat passengers wear these restraining belts. Although the belts save lives, some citizens argue that seat-belt use should be a personal decision and not required by law.

What happens in a crash?

When a car traveling about 50 km/h collides head-on with something solid, the car crumples, slows down, and stops within approximately 0.1 second. Any passenger not wearing a seat belt continues to move forward at the same speed the car was traveling. Within about 0.02 second (one-fiftieth of a second) after the car stops, unbelted persons slam into the dashboard, steering wheel, windshield, or the backs of the front seats, as in **Figure 3-11.** They are traveling at the car's original speed of 50 km/h—about the same speed they would reach falling from a three-story building.

The person putting on a seat belt in **Figure 3-12** will be attached to the car and will slow down as the car slows down. The force needed to slow a person from 50 km/h to zero in 0.1 second is equal to 14 times that person's weight.

Figure 3-11

In crash tests, researchers put life-like dummies in cars and crash the cars into each other or into crash barriers.

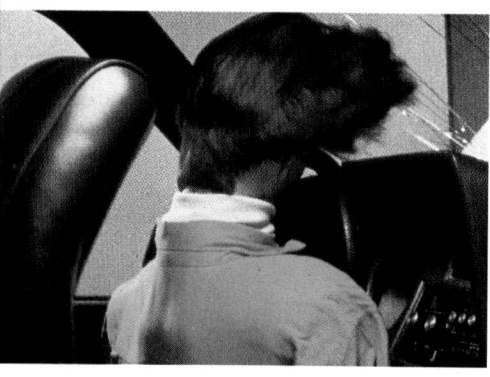 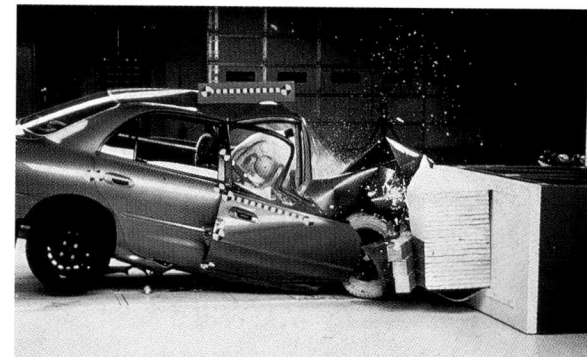

Program Resources

📁 **Reproducible Masters**
Study Guide, p. 15 L1
Reinforcement, p. 15 L1
Enrichment, p. 15 L3
Critical Thinking/Problem Solving, p. 9 L1

📦 **Transparencies**
Section Focus Transparency 11 L1

The belt "gives" a little as it restrains the person, increasing the time it takes to slow him or her down and spreading out the force so it's not focused on only one part of the body.

2 Points of View

Making Laws to Save Lives and Money

Car safety experts say that about half the people who die in car crashes would survive if they wore seat belts. Thousands of others would suffer fewer serious injuries. Those in favor of seat-belt laws maintain that it is a matter of public health and safety, similar to other laws aimed at preventing accidents and injuries. These preventable injuries and deaths are sad events for families involved and costly for individuals, insurance companies, and governmental agencies that must pay treatment bills.

Making Our Own Choices

Some people resent laws that restrict individual choices. They think everyone should be free to decide whether or not to use seat belts. A person who violates other traffic laws, such as speed limits, endangers other innocent people. However, a person who chooses not to wear a seat belt endangers only himself or herself. Therefore, the individual should be free to make the choice without being penalized.

Figure 3-12

Drivers who are concerned about their personal safety wear seat belts.

Section Wrap-up

Review

1. Describe what happens to a person who's not wearing a seat belt during a car crash.

2. Describe how seat belts protect passengers and drivers during accidents.

Explore the Issue

Should states create laws that make seat-belt use mandatory? If you were a member of your state legislature, how would you vote? Support your position.

SCIENCE & SOCIETY

Community Connection

Seat-Belt Survey Have students survey seat-belt use by interviewing at least five people in each of three age divisions using the question, "Do you Always, Usually, Seldom, or Never wear seat belts?" Be sure to tell them that they may not ask people who have already participated in the survey with another student. Pool the class results, and find the percent in each category. L2

Science Journal

Make a List Have students make a list of safety features they can identify in a typical modern car. Be sure to have students look at both the inside and outside of the car. L1 P

Reteach

Demonstration Attach a hard-boiled egg to the middle of a skateboard with a small bit of clay. Starting about 2 m from a wall, thrust the skateboard toward the wall. Attach another egg to the skateboard with several rubber bands and repeat. Have students compare the effects of the crashes on both eggs. LS

Extension

For students who have mastered this section, use the **Reinforcement** and **Enrichment** masters.

4 Close

Activity

LS **Intrapersonal** Divide students into groups of four. Have them pretend that they are a state legislative committee discussing a seat-belt law. Ask them to try to reach a consensus. A spokesperson from each group could report the group's decision to the class. L1

Section Wrap-up

Review

1. A person not wearing a seat belt in a head-on crash continues to move forward at the same speed the car was moving.

2. The belt holds people in place (they become "part" of the car), it gives a little, increasing the time it takes for them to come to rest, and it spreads out the force of the impact over more of the body area so it's not concentrated at one point.

Explore the Issue

Accept all reasonable, supported answers.

Prepare

Section Background

Quantities that require a direction to fully describe them, such as velocity and acceleration, are called *vectors*. Vectors are represented by small arrows.

Preplanning

Obtain plain white paper, sheets of coarse sandpaper, and 100-g masses or their equivalent for the MiniLAB on p. 81.

1 Motivate

Bellringer

Before presenting the lesson, display **Section Focus Transparency 12** on the overhead projector. Assign the accompanying **Focus Activity** worksheet. **L1** **LEP**

? Flex Your Brain

Use the Flex Your Brain activity to have students explore FORCE.

Activity Worksheets, p. 5

Visual Learning

Figure 3-14 What happens if the forces on her become unbalanced? *She will move in the same direction as the net force.* **LEP** **LS**

3•4 Connecting Motion with Forces

Science Words
force
balanced forces
net force
inertia
friction

Objectives
* Recognize different examples of forces.
* Identify cause-and-effect relationships between force and changes in velocity.
* Give examples of the effects of inertia.
* State Newton's first law of motion.

Figure 3-13
The force exerted by the bat sends the ball flying.

What is a force?

Push a door open. Stretch a rubber band. Squeeze a piece of clay. Slide a book across a table. In each case, you are applying a force. A **force** is a push or a pull one body exerts on another. Sometimes, the effects of a force are obvious, as when a moving car crashes into a stationary object, such as a tree. Other forces aren't as noticeable. Can you feel the force the floor exerts on your feet?

List all the forces you might exert or encounter in a typical day. Think about actions such as pushing, pulling, stretching, squeezing, bending, and falling.

Effects of Forces on Objects

In your list, what happens to the objects that have forces exerted on them? If an object is moving, does the force change the object's velocity? Think of a swinging bat hitting a softball, as in **Figure 3-13**. The ball's velocity certainly changes upon impact.

Balanced Forces

Force does not always change velocity. **Figure 3-14** illustrates a game of tug-of-war with a dog. You plant yourself firmly and lean back to push against the ground, causing the ground to push back on you. Your dog does the same. If you don't move forward or backward, the force of the dog pulling you forward must be balancing the force of the ground pushing you back. Forces on an object that are equal in size and opposite in direction are called **balanced forces.**

Unbalanced Forces

Now what happens if your feet hit a slippery spot on the ground? Your feet slip, and the ground can't exert as much force back on you. The forces of the dog pulling you forward and the ground pushing you back become unbalanced, and there is a net force on you. A **net force** on an object *always* changes the velocity of the object. When the dog pulls you forward with more force than the ground pushes back on you, you accelerate in the direction of the greater force.

78   Chapter 3 Exploring Motion and Forces

Program Resources

 **Reproducible Masters**
Study Guide, p. 16 **L1**
Reinforcement, p. 16 **L1**
Enrichment, p. 16 **L3**
Activity Worksheets, pp. 5, 19 **L1**

 Transparencies
Teaching Transparency 6 **L1**
Section Focus Transparency 12 **L1**

Remember that velocity involves both speed and direction. A net force acting on an object will change its speed, direction, or both. In the tug-of-war, the net force on you causes both your speed and direction to change.

Figure 3-14

When the forces on the girl are balanced, she does not move. *What happens if the forces on her become unbalanced?*

Inertia and Mass

Picture a hockey puck sliding across the ice as in **Figure 3-15.** Its velocity hardly changes until it hits something, such as the wall, the net, or a player's stick. The velocity of the puck is constant, and its acceleration is zero until it hits something that alters its speed or direction.

The sliding puck demonstrates the property of inertia. **Inertia** is the tendency of an object to resist any change in its motion. If an object is moving, it will keep moving at the same speed and in the same direction unless an unbalanced force acts on it. In other words, the velocity of the object remains constant unless a force changes it. If an object is at rest, it tends to remain at rest. Its velocity is zero unless a force makes it move.

Would you expect that a bowling ball would have the same inertia as a table-tennis ball? Why would there be a difference? The more mass an object has, the greater its inertia is. Recall that mass is the

Figure 3-15

The velocity and acceleration of a hockey puck are constantly changing during a game.

2 Teach

Making a Model

[LS] **Visual-Spatial**   Introduce forces as arrows. Sketch two arrows representing the force needed to lift a book and a TV. Explain that the length of the arrow indicates the size of the force. The direction of the arrow indicates the direction of the force. Encourage volunteers to try to sketch arrows to represent the forces in the lists they were asked to make on page 78. These are commonly called vectors. [LEP]

Use **Study Guide,** p. 16 as you teach this lesson.

 Videodisc

Newton's Apple: Physical Sciences
Frisbee Physics
Chapter 2, Side A
Introduction

13812-15304
Frisbees and the laws of physics

16504-16974

Student Text Question

Would you expect that a bowling ball would have the same inertia as a table-tennis ball? *no*

Why would there be a difference? *because the bowling ball has much more mass and its motion is harder to change*

USING TECHNOLOGY

Air Bags

Seat belts are designed to reduce the effect of inertia in a crash by holding the passengers in place. However, in a high-speed crash, belts provide limited protection to the head and upper body. These parts of the body can be protected by air bags, which provide an instantaneous cushion for the head and upper body at the time of impact.

Air bags are designed to be used in addition to seat belts. An air-bag system consists of one or more crash sensors, an ignitor and gas generator, and an inflatable nylon bag. If a car hits something with speed in excess of 15 to 20 km/h, impact sensors trigger the flow of electric current to an ignitor. The ignitor causes an explosive chemical reaction to occur, producing harmless nitrogen gas. The nitrogen gas propels the air bag from its storage compartment just in time to absorb the inertia of the occupants. The bag then immediately deflates so it will not interfere with the driver.

Some people should not sit in a seat protected by an air bag. Children and small adults have been injured or killed by the force of the expanding bag.

interNET
CONNECTION

Visit the Chapter 3 Internet Connection at Glencoe Online Science, **www.glencoe.com/sec/ science/physical**, for a link to more information about air bags.

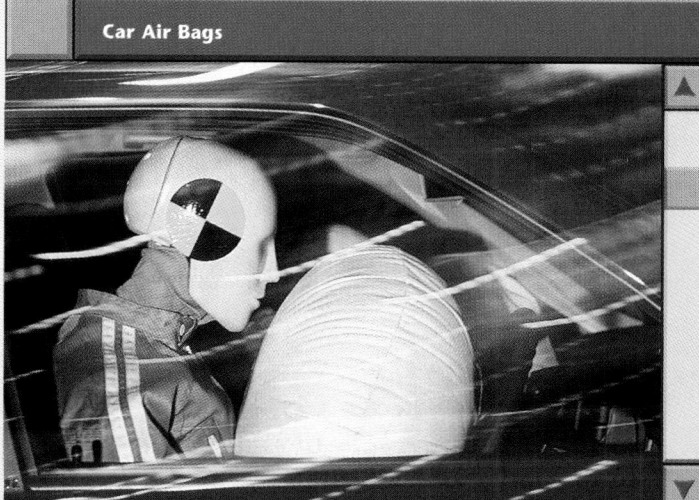

Car Air Bags

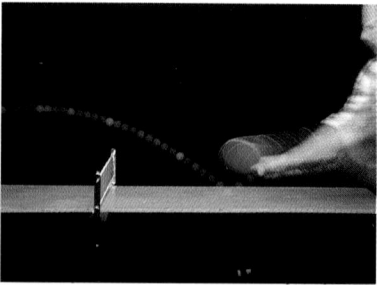

Figure 3-16

Besides the paddle, what else changes the ball's velocity?

amount of matter in an object, and a bowling ball certainly contains more matter than a table-tennis ball. So the bowling ball would have greater inertia than the table-tennis ball. You wouldn't change the velocity of a bowling ball very much by swatting it with a paddle, but you could easily change the velocity of the table-tennis ball in **Figure 3-16.** Because the bowling ball has greater inertia, a much greater force would be needed to change its velocity.

Across the Curriculum

Physical Education Before a runner is given credit for a record in outdoor track, officials carefully analyze the conditions under which the event was held. Ask students to speculate what some of the factors they consider might be. How does this relate to forces? **LEP** **LS**

Newton's First Law

Forces change the motion of an object in very specific ways—so specific that Sir Isaac Newton (1642-1727) was able to state laws that describe the effects of forces.

Newton's first law of motion is that an object moving at a constant velocity keeps moving at that velocity unless a net force acts on it. If an object is at rest, it stays at rest unless a net force acts on it. Does this sound familiar? It is the same as the earlier discussion of inertia. Thus, you'll understand why this law is sometimes called the *law of inertia*. You've probably seen—and felt—this law at work without even knowing it. In **Figure 3-17**, you can see this law in action.

Figure 3-17

You can observe many common examples of Newton's first law in action.

A The young man is pushing the hand truck at a constant velocity. The boxes are also moving at the same constant velocity. There is no net force acting on either the hand truck or the boxes.

B The hand truck bumps into the curb and stops. However, the boxes keep moving forward even though no one has pushed them. Their inertia keeps them moving forward. The boxes also begin to fall because the force of gravity pulls them downward.

C The boxes finally come to rest on the ground. *Are the forces acting on them balanced or unbalanced?*

3-4 Connecting Motion with Forces   81

MiniLAB

How does friction act as a force?

Procedure

1. Place a sheet of plain white paper near the edge of a flat surface, then set a 100-g mass on the paper about 7 cm from the far end of the paper.
2. Grip the end of the paper near the table's edge and give it a quick, smooth, downward yank. What happens?
3. Replace the paper with a sheet of coarse sandpaper, rough side up, and repeat the procedure. Observe what happens to the mass.

Analysis

1. How do you interpret the different results?
2. How can Newton's first law help explain your observations?

MiniLAB

Purpose

Kinesthetic Students will observe and infer how friction affects a model of the familiar tablecloth trick. **L1**

LEP **COOP LEARN**

Materials

a sheet of smooth paper or slick piece of cloth, a 100-g mass or another unbreakable object with a smooth bottom surface, piece of coarse sandpaper, table

Teaching Strategies

Troubleshooting It is important to give the paper a quick, downward yank.

Analysis

1. The net force due to friction is greater with the sandpaper than with the smooth paper. The friction between the mass and the sandpaper provided a strong enough force to cause the mass to move with the paper.
2. The mass tended to remain at rest due to its inertia until the force of friction was great enough to cause it to move.

Assessment

Oral Ask students to predict the results of using other materials in this experiment, such as plastic, construction paper, or a fuzzy towel. Discuss qualities of materials that seem to cause friction. Use the Performance Task Assessment List for Formulating a Hypothesis in **PASC**, p. 21. **P**

Activity Worksheets, p. 19

Visual Learning

Figure 3-16 Besides the paddle, what else changes the ball's velocity? *the impact of the ball on the table and the table on the ball, and the force of gravity on the ball*

Figure 3-17 C. Are the forces acting on them balanced or unbalanced? *balanced*

Figure 3-18 See p. 82. **How is friction helping this soccer player?** *Friction keeps her left foot from sliding or slipping as she kicks the ball.*

Science Journal

Use with Science Journal entry on page 82. Answers will vary, but students should describe specific relationships with examples relating to their sport.

82

3 Assess

Check for Understanding

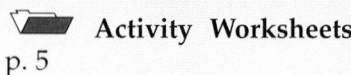

Use the Flex Your Brain activity to explore INERTIA. L2

📁 **Activity Worksheets,** p. 5

Reteach

Tie a string around a book and suspend it. Cut the string. Before the string was cut, the book was at rest. After the string was cut, it accelerated—a net force was acting on it. **LEP**

Extension

📁 For students who have mastered this section, use the **Reinforcement** and **Enrichment** masters.

4 Close

•MINI•QUIZ•

Use the Mini Quiz to check students' recall of chapter content.

1. **What are balanced forces?** *forces equal in size and opposite in direction*
2. **What is inertia?** *the tendency of an object to resist any change in motion*
3. **What is friction?** *the force that opposes motion between two touching surfaces*

Section Wrap-up

Review

1. Rosin reduces slipping. They are taking advantage of friction.
2. The jet; inertia is related to mass, not speed.
3. **Think Critically** catching a football, sliding into base, or diving

Friction

You've just learned that inertia causes an object that is moving at constant velocity to keep moving at that velocity unless a net force acts on it. But you know that if you slide a book across a long table, it eventually slows down and stops. Why does it stop?

An unseen force is acting between the book and the table. The force is friction. **Friction** is the force that opposes motion between two surfaces that are touching each other. Would you expect more friction between an oily floor and a slick, leather shoe sole or between a rough sidewalk and the bottom of a tennis shoe? The amount of friction depends on two factors—the kinds of surfaces and the force pressing the surfaces together.

Life Without Friction

If there were no friction, your life would be much different. You wouldn't be able to walk or hold things between your fingers. Your shoes would fall off. Friction between the soles of your shoes and the floor makes it possible for you to walk. You can hold something with your fingers because of friction. Shoelaces remain tied because of friction. **Figure 3-18** shows the effect of friction on a soccer player.

As you complete this section, you should be more aware that force and motion are part of everything you do and everything that happens around you.

Figure 3-18

Soccer requires players to run, jump, pivot, and kick. *How is friction helping this soccer player?*

Section Wrap-up

Review

1. Before dancing on a smooth wooden floor, ballet dancers sometimes put a sticky powder called rosin on their shoe soles. Why? What force are they taking advantage of?
2. Explain which has greater inertia, a speeding car or a jet airplane sitting on a runway.
3. **Think Critically:** Think of and describe three examples from sports in which a force changes the velocity of an object or a person.

Skill Builder

Recognizing Cause and Effect
Explain what happens to your body in terms of inertia, friction, and forces when you slip and fall on an icy sidewalk. If you need help, refer to Recognizing Cause and Effect in the **Skill Handbook.**

Science Journal

Inertia plays an important role in most sports. In your Science Journal, write a paragraph describing the role of inertia in your favorite sport.

Skill Builder

With little friction between your feet and the ice, your feet keep moving and slip out from under you when you exert a force by stepping forward. Inertia keeps your feet and legs sliding forward and out from under you. Gravity causes you to fall to the sidewalk.

✓ Assessment

Performance Have students prepare a diagram showing the forces acting on the person who slips and falls. The diagram should show both horizontal and vertical forces. Use the Performance Task Assessment List for Idea Organizer in **PASC,** p. 93.

Gravity— A Familiar Force

Gravitational Force

A diver, poised high above the water in **Figure 3-19,** jumps up in the air and then hurtles toward the water. Why does the diver plunge downward instead of flying off into space? Seeing objects fall is a common experience. You may not have thought about it, but when you see something fall, you're watching the effects of the force of gravity.

Every object in the universe exerts a force on every other object. That force is **gravity.** Often, the force is too slight to notice. For example, there's a gravitational force of attraction between your hand and your pencil, but when you let go of your pencil, it doesn't stay in your hand. The gravitational force of Earth is much stronger than the gravitational force of your hand, so the pencil falls toward Earth.

What determines the force of gravity?

The amount of gravitational force between objects depends on two things— their masses and the distance between them. The masses of your hand and your pencil are small, so the force of attraction between them is weak. The mass of Earth is large, so the gravitational force between Earth and your pencil is strong.

The amount of Earth's gravitational force acting on an orbiting satellite depends on how close the satellite is to Earth. The closer it is to Earth, the stronger the pull of gravity. The farther away it is, the less effect gravity has on it. Thus, gravity will exert a stronger pull on a satellite in a lower orbit and a weaker pull on a satellite in a higher orbit.

Science Words

gravity
weight

Objectives

- Give examples of the effects of gravity.
- Relate gravitational force to mass and distance.
- Distinguish between mass and weight.

Figure 3-19

As the diver falls toward the water, she is under the influence of gravity and accelerates downward.

3-4 Connecting Motion with Forces **83**

Prepare

Section Background

Newton was the first to describe the gravitational force between two objects quantitatively. The force is directly proportional to the product of the two masses and inversely proportional to the square of the distance between their centers. If the distance between two objects is doubled, the force between them will fall to one-quarter of the value it was before the objects were moved.

Preplanning

Gather the materials for Activity 3-2. Be sure to select several different types of paper. The plastic cup should be large enough to hold the amount of sand it takes to tear the paper. Provide an effective method for cleanup.

1 Motivate

Bellringer

 Before presenting the lesson, display **Section Focus Transparency 13** on the overhead projector. Assign the accompanying **Focus Activity** worksheet. L1 LEP

? Flex Your Brain

Use the Flex Your Brain activity to have students explore GRAVITY.

📁Activity Worksheets, p. 5

Student Text Question

Why does the diver plunge downward instead of flying off into space? *The force acting on her is gravity, which pulls her downward.*

Program Resources

📁 **Reproducible Masters**
Study Guide, p. 17 L1
Reinforcement, p. 17 L1
Enrichment, p. 17 L3
Activity Worksheets, pp. 5, 16, 17 L1
Concept Mapping, pp. 11, 12 L1
Science and Integration Activities,
   pp. 5, 6 L2
Multicultural Connections, pp. 9, 10 L1

📠 **Transparencies**
Section Focus Transparency 13 L1

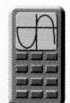

Problem Solving

An Experiment for the Shuttle

NASA was having a competition to select several student experiments for scientists to conduct while the space shuttle orbited Earth. Ellen decided to enter an experiment about how the time spent on a shuttle flight would affect the masses of the astronauts. The mass of each astronaut would be determined before and after the mission. During the flight, the astronauts would have to list the foods and liquids they consumed and eliminated, and the masses of each. Ellen's experiment required that the masses of the food and liquid packets be measured and labeled to the nearest 0.01 gram. It also required the astronauts to report on their daily activities.

Solve the Problem:

1. Design a data table that can be used to record the required information for one astronaut for a period of five days.
2. How would the mass of any material on Earth compare with its mass during a shuttle mission?

GRAPHING CALCULATOR

Gravity acting on mass produces weight. You can find your weight on Earth in newtons by the equation $Y = 0.23 \times X$, where *Y* is your weight in newtons, and *X* is your weight in pounds. Graph this equation using your graphing calculator, and use your graph to determine your weight in newtons.

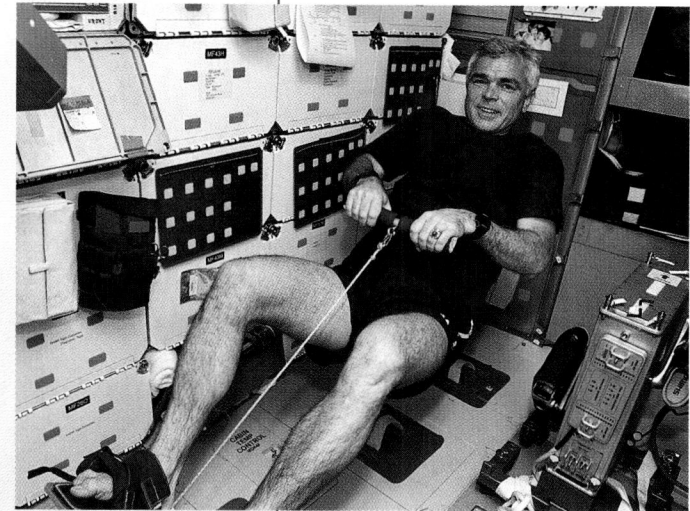

Weight

The measure of the force of gravity on an object is the object's **weight.** The term *weight* is most often used in reference to the gravitational force between Earth and a body at Earth's surface. Weight isn't the same as mass, but the two are related. Recall from Chapter 2 that mass is the amount of matter in an object. The greater an object's mass, the stronger the gravitational force on it. In other words, the more mass an object has, the more it weighs.

Mass is measured in grams (g) and kilograms (kg). Weight, which is a force, is measured in units called newtons (N). A kilogram of mass on Earth's surface weighs 9.8 N. On Earth, a cassette tape weighs about 0.5 N, a backpack full of books weighs about 40 N, and a wide-bodied jumbo jet weighs about 3.4 million N.

Table 3-1 shows how an object's weight depends upon the object's location.

Table 3-1

Weight on Earth (lb)	Weight on Other Bodies in the Solar System (lb)				
	Moon	Venus	Mars	Jupiter	Saturn
75	12.5	67.5	28.5	190.5	87.0
100	16.7	90.0	38.0	254.0	116.0
125	20.8	112.5	47.5	317.5	145.0
150	25.0	135.0	57.0	381.0	174.0
2000	333.3	1800.0	760.0	5080.0	2320.0

Weight—It Depends on Where You Are

The gravitational force an object exerts on other objects is related to its mass. Earth exerts a stronger gravitational force on an object at its surface than the moon does on an object at its surface because Earth has more mass. Because of the moon's weaker gravitational force, a person weighing about 480 N on Earth would weigh only about 80 N on the moon. Does this mean that the person would have less mass on the moon than on Earth? No. Unlike weight, mass doesn't change with changes in gravity. Because mass remains constant, the astronaut in **Figure 3-20** would also have the same inertia on the moon as on Earth.

Figure 3-20

Astronauts jump higher on the moon than they can on Earth because of differences in gravity.

Measuring Forces

Scales use the principle of balanced forces to measure how much something weighs. To see the principle work, hang a rubber band from a hook or a nail. Then, attach something heavy to the lower end of the rubber band and see how much the band stretches. If you attach something even heavier, will the band stretch more? Yes, the band stretches until it is exerting an upward force equal to the weight of the object hung. Thus, the length of the rubber band is a measure of the force on it.

A scale works something like the rubber band. When you step on a bathroom scale, the force of your body stretches a spring inside the scale. The stretched spring pulls on levers,

CONNECT TO

LIFE SCIENCE

Research health problems encountered by astronauts after extended periods of reduced gravity.

3-5 Gravity—A Familiar Force   85

Integrating the Sciences

Earth Science On ocean beaches, the water level rises and falls regularly each day—a phenomenon commonly called tides. Have students find out what causes tides. If coastal newspaper or electronic resources are available, you could also have students find the high-tide and low-tide times for a certain location for that day. *As Earth rotates, different areas of water are closest to the moon and are pulled more strongly by the moon's gravity. This causes a net force on the ocean's water, and the water flows in that direction.* L2

Use the Mini Quiz to check students' recall of chapter content.

1. **What is the force of gravity?** *the force every object exerts on every other object*

2. **What is weight?** *the measure of the force of gravity on an object*

3. **Weight is measured in what units?** *newtons*

4. **What instrument is used to measure weight?** *a scale*

Section Wrap-up

Review

1. The mass is less.

2. The masses are small and the gravitational forces weak. You feel the gravitational attraction between the pencil and Earth and the friction between your fingers and the pencil.

3. **Think Critically** The satellite's velocity and inertia make it move in a straight line. However, gravity pulls the satellite toward Earth and makes it move in a curved path. The satellite stays in orbit because of the balance created by its own velocity and the force of gravity.

Science Journal

Encourage creativity, but also encourage scientific accuracy. **P**

Figure 3-21

A scale measures the force of gravity on the object or person being weighed.

A When you stand on a bathroom scale, the downward force of gravity—your weight—is balanced by the upward force exerted by the spring inside the scale.

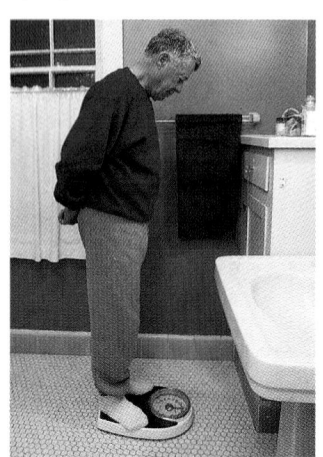

B A system of levers and a spring are used to indicate the person's weight.

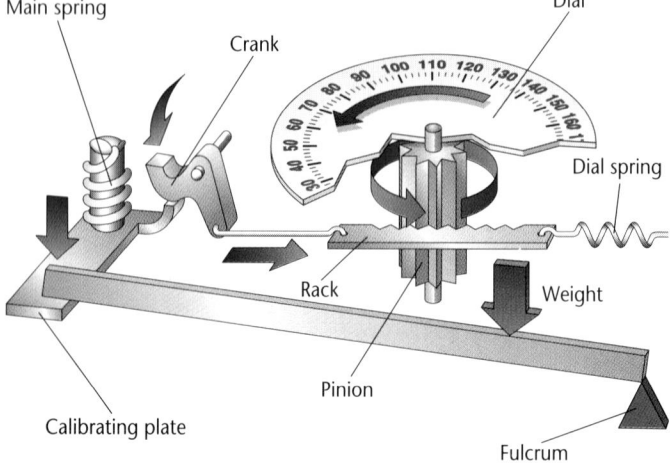

Main spring · Crank · Dial · Dial spring · Rack · Weight · Pinion · Calibrating plate · Fulcrum

as shown in **Figure 3-21,** until the upward force of the scale equals the downward force of your weight. The dial on the scale, which is marked off in units of weight, moves until the spring stops stretching. The number showing on the dial indicates your weight.

Section Wrap-up

Review

1. Arriving on a newly discovered Earth-sized planet, you find that you weigh one-third as much as on Earth. Is the planet's mass greater or less than Earth's?

2. Why don't you feel the gravitational force between your hand and a pencil? What do you feel?

3. **Think Critically:** Use Newton's first law and the concept of gravity to explain how a satellite orbits Earth.

Skill Builder
Observing and Inferring
Select three objects in the room and, without touching them, try to guess which weighs the most. What clues help you to make your guesses? How can you tell if you guessed right? If you need help, refer to Observing and Inferring in the **Skill Handbook.**

Science Journal

In your Science Journal, write a short story describing what a certain activity would be like if the gravitational force were half as strong. For example, you could write about playing sports, constructing a building, or riding a bicycle.

Skill Builder
Students might infer that larger objects are heavier. They might also consider the density of the object. They can determine whether they are correct by weighing each object.

✓ Assessment

Oral Check the understanding of mass and weight by asking students how the mass and weight of the three objects would change on the moon. *Mass would not. The weight would be 1/6 of the value on Earth.* Use the Performance Task Assessment List for Making Observations and Inferences in **PASC,** p. 17. **P**

Activity 3-2

Balancing Forces Against Gravity

Imagine yourself hanging by your arms from a tree limb. Why do you eventually get tired and let go? When your arms get tired, they can't exert an upward force on your body to balance the force of gravity pulling your body down (your weight).

Problem
Determine the strengths of different types of paper.

Materials
- selection of different types of paper strips
- masking tape
- string
- beaker
- ring stand and ring
- sheets of newspaper
- scissors
- plastic cup
- sand, 250 mL
- balance

Procedure
1. Prepare a table with the headings *sample, mass of sand,* and *weight of sand.*
2. Place the ring stand as shown. Select a paper strip and record its identity in the data table. Cut the strip as shown. Tape one side of the strip to the ring and tape the other side to the cup.
3. Slowly pour sand into the cup until the paper strip begins to tear.
4. Use the balance to measure the combined mass of the cup and sand. Record this measurement and return the sand to its container.

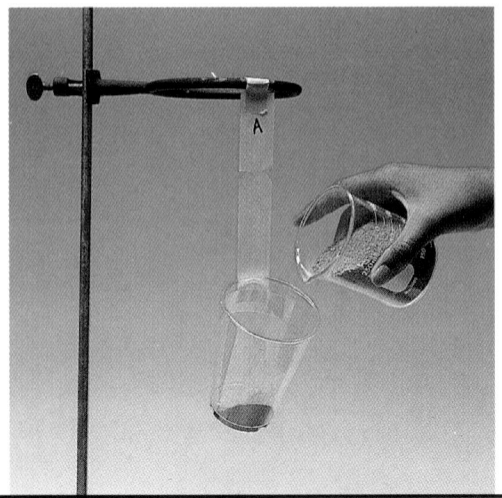

5. Repeat steps 2, 3, and 4 using different types of paper strips.
6. Using the formula below, calculate and record the weights of the cup and sand for each paper strip tested.

Weight = mass (in kg) × 9.80 N/kg

Data and Observations   Sample Data

Sample	Mass of Sand (g)	Weight of Sand (N)
a	85	0.83
b	44	0.43
c	82	0.80

Analyze
1. The ability of a material to resist tearing is called shear strength. What did this investigation show about the shear strengths of different types of paper?
2. Why would it be incorrect to refer to shear strength in grams or kilograms?

Conclude and Apply
3. This investigation tests the shear strength in one direction only. How could you find out whether the shear strength depends on the direction in which the paper is torn?
4. **Predict** whether this investigation would work as well if water were used in place of sand.
5. Why is gravity a good choice of force to use for measuring the strength of different materials?

3-5 Gravity—A Familiar Force   **87**

✓ Assessment

Oral Ask students to explain why it is harder to hang from a tree by one arm as opposed to two. Use the Performance Task Assessment List for Making Observations and Inferences in **PASC**, p. 17. **P**

Background

• Here's how juggler Derique McGee suggests that you get started in learning to juggle: "Hold your arms straight out in front of you and bent comfortably. Toss a juggling ball from one hand to the other at about eye level 50 times or so, just to get the feel of it. Then take a ball in each hand. Toss the first one and watch for the moment when it starts to drop. Then toss the second one. It takes a while to get used to the motion because there's nothing in real life that requires that kind of hand-to-hand movement."

Teaching Strategies

Kinesthetic Ask students to rank the following sets of objects from easiest to hardest to juggle. Have them give reasons for the order they decide upon.

1. three torches
2. three bouncy balls
3. a Ping-Pong ball, a basketball ball, and an ax
4. three bowling balls
5. three beanbags

Answer: 5, 2, 1, 4, 3. Reasons: Beanbags won't roll away when dropped; objects of different sizes and weights are hardest to juggle.

Career Connection

Career Path Possible career directions vary; people with good vision, eye-hand coordination, and spatial perception include astronauts, airplane pilots, and physical therapists.

People and Science

DERIQUE, *Juggler*

On the Job

Q Can the principles of juggling be explained in words?

A You can think of a box with an X drawn inside it. A juggling pattern forms the X, with a ball going from a lower corner diagonally to an upper corner. Then the ball drops. The left hand throws from the left bottom corner of the box to the right side. The balls just miss one another.

Q What do you do when gravity, the juggler's "downfall," strikes?

A I accept that important law of physics and juggling: What goes up must come down. To overcome this, I trap the escaping juggling ball between my feet, jump up, release it and get it in the air again.

Q You sometimes juggle atop a unicycle. Does staying on a unicycle involve any other scientifically based skills?

A Yes, it's called counterbalancing. I have to keep shifting my weight from side to side and front to back.

Personal Insights

Q What made you want to become a juggler?

A In high school, I wasn't good enough for regular organized sports, but I wanted to be part of a group, so I took clown classes. When my group did shows for kids in hospitals, I discovered how great it feels to make people forget their troubles for a while. Overcoming gravity comes from the heart.

Career Connection

Performing arts and other occupations often require skills involving eye-hand coordination, spatial relationships, and overcoming forces through manipulation of objects. Choose an occupation that involves one of these skills and write a want ad for a newspaper for that job. Consider the following careers:

• **Astronaut**   • **Physical Therapist**

For More Information

Students can investigate through library research the basic training and background needed to become a physical therapist. They can also contact a local physical therapy clinic and interview a licensed therapist. During the interview, the student might ask what role attitude and humor play in the progress that some patients make.

Students can also contact the local chapter of the American Physical Therapists Association.

Chapter 3 Review

Summary

3-1: Motion and Speed

1. Motion is a change in position of a body. Speed is the rate at which a body changes position.
2. Average speed is the ratio of distance traveled to time and describes motion, even if speed varies.

3-2: Velocity and Acceleration

1. Velocity describes the speed and direction of a moving body.
2. Acceleration is the rate of change in velocity.

3-3: Science and Society: A Crash Course in Safety

1. Seat belts reduce injuries by limiting impacts and spreading out the force over more of the body.
2. Some people believe that seat belts should be required by law; others don't.

3-4: Connecting Motion with Forces

1. A force is a push or a pull one body exerts on another. Balanced forces acting on a body do not change the motion of the body. Unbalanced forces result in a net force, which always changes the motion of a body.
2. Inertia explains why a massive, fast-moving bowling ball is more difficult to stop than a table-tennis ball at the same speed.
3. Newton's first law says an object's motion will not change unless a net force acts on it.

3-5: Gravity—A Familiar Force

1. Gravity causes planets to orbit the sun and people to remain on Earth's surface.

2. The gravitational force between two objects depends on their masses and the distance between them.
3. Mass is the amount of matter in an object. Weight is the force of gravity on that mass.

Key Science Words

a. acceleration
b. average speed
c. balanced forces
d. constant speed
e. force
f. friction
g. gravity
h. inertia
i. instantaneous speed
j. net force
k. speed
l. velocity
m. weight

Reviewing Vocabulary

Match each phrase with the correct term from the list of Key Science Words.

1. rate of change in position
2. speed that does not change
3. rate of change in velocity
4. a push or pull exerted on an object
5. type of force that changes the motion of an object
6. tendency of an object to resist change in motion
7. a force that opposes motion between surfaces
8. force exerted by every object in the universe on every other object
9. measure of the force of gravity on an object
10. rate of motion at a given point in time

Chapter 3

Review

Summary

Have students read the summary statements to review the major concepts of the chapter.

Reviewing Vocabulary

1. k	6. h
2. d	7. f
3. a	8. g
4. e	9. m
5. j	10. i

✓ Assessment

Portfolio Encourage students to place in their portfolios one or two items of what they consider to be their best work. Examples include:

- Developing Skills concept map, p. 91
- Flex Your Brain, p. 76
- Science Journal, pp. 77 and 86 **P**

Performance Additional performance assessments may be found in **Performance Assessment** and **Science Integration Activities.** Performance Task Assessment Lists and rubrics for evaluating these activities can be found in Glencoe's **Performance Assessment in the Science Classroom.**

GLENCOE TECHNOLOGY

 Videodisc

Glencoe Physical Science
Interactive Videodisc

Use the videodisc *Motion* to review motion, speed, velocity, and acceleration and some of the forces that affect motion.

MindJogger Videoquiz

Chapter 3 Have students work in groups as they play the Videoquiz game to review key chapter concepts.

Chapter 3 Review

Checking Concepts

1. a	**6.** a
2. b	**7.** b
3. b	**8.** c
4. b	**9.** b
5. d	**10.** d

Understanding Concepts

11. An object has moved if its position relative to a reference point has changed. To calculate its speed, you would need to know the distance it moved and the amount of time it took to move that distance. To determine its velocity, you need to know the direction in which it traveled.

12. Any change in the speed and/or direction of a moving object is accelerated motion. Therefore, if the car changes direction by turning a corner, the car is accelerating, even if its speed does not change.

13. Friction acts only when two surfaces are in contact. Gravity acts over distances. The texture of surfaces affects friction, but not gravity.

14. Weight is a measure of the force of gravity on a body. The more mass a body has, the more it will weigh. Mass and weight differ in that the mass of a body never changes due to its position. The weight of a body is not constant; it varies with location. The weight of a body depends upon the distance of the body from Earth.

15. Inertia resists any change in motion. Because of its inertia, a body at rest tends to remain at rest, and a body in motion tends to keep moving in a straight line at a constant velocity.

Checking Concepts

Choose the word or phrase that completes the sentence or answers the question.

1. The best way to describe the rate of motion of an object that changes speed several times is to calculate the object's _____.
a. average speed   c. instantaneous speed
b. constant speed   d. variable speed

2. Which of the following is a force?
a. inertia   c. acceleration
b. friction   d. velocity

3. The unit for _____ is m/s^2.
a. weight   c. inertia
b. acceleration   d. velocity

4. Which of the following is not used in calculating acceleration?
a. initial velocity   c. time interval
b. average speed   d. final velocity

5. A body accelerates if it _____.
a. speeds up   c. changes direction
b. slows down   d. all of these

6. The gravitational force between two objects depends on their _____.
a. masses   c. shapes
b. velocities   d. volumes

7. _____ acts only between surfaces that are in contact.
a. Inertia   c. Gravity
b. Friction   d. A net force

8. An object's weight is directly related to its _____.
a. volume   c. mass
b. velocity   d. shape

9. An object of large mass has _____ than an object of small mass.
a. less inertia   c. less weight
b. more inertia   d. greater acceleration

10. A constant velocity means acceleration is _____.
a. positive   c. increasing
b. negative   d. zero

Understanding Concepts

Answer the following questions in your Science Journal using complete sentences.

11. Can you tell an object has moved if you do not see it move? What information would you need to calculate the object's speed? Its velocity?

12. Explain how it is possible for an automobile traveling at constant speed to be accelerating.

13. Friction and gravity are both forces. Describe at least two differences between them.

14. Compare and contrast mass and weight.

15. Describe some common effects of inertia.

Thinking Critically

16. Explain why a fast-moving freight train cannot be stopped quickly.

17. A cyclist must travel 800 km. How many days will the trip take if the cyclist travels 8 h per day at an average speed of 16 km/h?

18. A satellite's original velocity is 10 000 m/s. After one minute, it is 5000 m/s. What is the satellite's acceleration?

19. A cyclist leaves home and rides due east for a distance of 45 km. She returns home on the same bike path. If the entire trip takes 4 h, what is her average speed?

20. The return trip of the cyclist in question 19 took 30 min longer than her trip east, although her total time was still 4 h. What was her velocity in each direction?

Thinking Critically

16. Because of its large mass, the train has a huge amount of inertia. The braking force of the train is limited, so it requires time to overcome a large inertia.

17. Known information: distance = 800 km; speed = 16 km/h; daily travel time = 8 h
Unknown information: Time in days
Formula to use: $t = d/v$

Solution: $t = 800$ km/16 km per h
$= 50$ h
Days $= 50$ h/8 hours per day
$= 6\ 1/4$ days

18. Known information:
$v_i = 10\ 000$ m/s; $v_f = 5000$ m/s
time interval = 1 minute or 60 s

Unknown information: acceleration
Formula to use: $a = (v_f - v_i)/t$
Solution:
$a = (5000$ m/s $- 10\ 000$ m/s$)/60$s
$= -83.3$ m/s^2

Developing Skills

If you need help, refer to the **Skill Handbook.**

21. **Measuring in SI:** Which of the following represents the greatest speed: 20 m/s, 200 cm/s, or 0.2 km/s? HINT: Express all three in meters/second and then compare them.

22. **Observing and Inferring:** A car sits motionless on a hill. What forces are acting on the car? Are the forces balanced or unbalanced? Explain how you inferred your answers.

23. **Making and Using Tables:** The four cars shown in the table were traveling at the same speed, and the brakes were applied in all four cars at the same instant.

Car	Mass	Stopping Distance
A	1000 kg	80 m
B	1250 kg	100 m
C	1500 kg	120 m
D	2000 kg	160 m

What is the relationship between the mass of a car and its stopping distance? How do you account for this relationship?

24. **Making and Using Graphs:** The following data were obtained for two runners.

	Sally	Alonzo
Time	**Distance**	**Distance**
1 s	2 m	1 m
2 s	4 m	2 m
3 s	6 m	2 m
4 s	8 m	4 m

Make a distance-time graph that shows the motion of both runners. What is the average speed of each runner? What is the instantaneous speed of each runner 1 s after he or she starts? Which runner stops briefly? During what time interval do Sally and Alonzo run at the same speed?

25. **Concept Mapping:** Make a network tree concept map that defines the three kinds of speed described in this chapter.

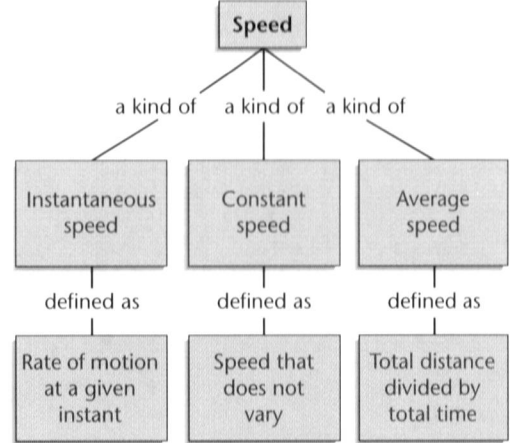

Performance Assessment

1. **Making Observations and Inferences:** In Activity 3-1 on page 70, you designed an experiment to find out how slowly you could make a glider fly. After observing the glider flights, you can make inferences about how glider design relates to glider speed. Redesign your glider to see how fast you can make it fly. What changes would you make? Try it and calculate your glider's speed.

2. **Data Table:** Use the procedure in Activity 3-2 on page 87 to test the shear strength of other materials such as aluminum foil, waxed paper, or human hair. Report your findings in a table.

3. **Formulating a Hypothesis:** Oils and waxes are lubricants, materials that are used to reduce the friction between surfaces. Form a hypothesis about which lubricant would work best on (a) a squeaky hinge and (b) a sticking dresser drawer.

2 seconds for a 1-s period. Between 3 and 4 seconds, the two runners have the same speed, 2 m/s. This can be seen by noting the identical slopes of the graphs during this interval, or it can be calculated from the data table.

25. **Concept Mapping** See annotations on reduced student page. P

Performance Assessment

1. The glider would probably have a long, thin design. Use the Performance Task Assessment List for Making Observations and Inferences in **PASC**, p. 17. P

2. Answers will reflect materials used. Use the Performance Task Assessment List for Data Table in **PASC**, p. 37. P

3. Oil will work better on a squeaky metal hinge and wax will work better on the wooden dresser drawer. Use the Performance Task Assessment List for Formulating a Hypothesis in **PASC**, p. 21. P

19. Known information:
    distance = 90 km;
    time = 4 h
    Unknown information: average speed
    Formula to use: $v = d/t$
    Solution: $v = 90$ km/4 h = 22.5 km/h

20. trip west—2.25 h; trip east—1.75 h
    $v = 45$ km/1.75 h = 25.7 km/h east
    $v = 45$ km/2.25 h = 20 km/h west

Developing Skills

21. **Measuring in SI** In meters/second, the three speeds are: 20 m/s, 2 m/s, and 200 m/s. The third speed—0.2 km/s— is the greatest.

22. **Observing and Inferring** Gravity and friction act on the car. Gravity tends to pull the car down the hill, while friction of the hand brake prevents the car from rolling down the hill. Because the car is at rest, no net force is acting on it. Thus, the logical inference is that the forces are balanced.

23. **Making and Using Tables** The greater the mass of the car, the greater its stopping distance. The greater the mass of the car, the greater its inertia, and therefore, the harder it is to change the motion of the car and bring it to rest. Assuming equal braking forces, the most massive car, D, will take the greatest amount of time to come to rest, and will therefore travel the greatest distance.

24. **Making and Using Graphs** Average speed:
    Sally, 8 m/4 s = 2 m/s
    Alonzo, 4 m/4 s = 1 m/s
    Instantaneous speed:
    Sally, 2 m/1 s = 2 m/s
    Alonzo, 1 m/1 s = 1 m/s
    Alonzo stops briefly after

Chapter Organizer

Section	Objectives/Standards	Activities/Features
Chapter Opener		**Explore Activity:** What propels a rocket upward? p. 93
4-1 **Accelerated Motion** (1 session, 1 block)*	1. **Explain** how force, mass, and acceleration are related. 2. **Compare** the rates at which different objects fall. 3. **Observe** the effects of air resistance. **National Content Standards:** (5-8) UCP2-UCP5, A2, B2, G1, G3; (9-12) UCP2-UCP5, A2, B4, G1, G3	**Connect to Life Science,** p. 95 **Using Math:** Calculating Force, p. 95 **MiniLAB:** How can air resistance change the acceleration of a falling object? p. 97 **Using Math,** p. 99 **Skill Builder:** Comparing and Contrasting, p. 99
4-2 **Projectile and Circular Motion** (2 sessions, 1 block)*	1. **Explain** why things that are thrown or shot follow a curved path. 2. **Compare** motion in a straight line with circular motion. 3. **Define** *weightlessness*. **National Content Standards:** (5-8) UCP2, UCP3, UCP5, A1, A2, B2, C5, E1, E2, G1; (9-12) UCP2, UCP3, UCP5, A1, A2, B4, C3, E1, E2, G1	**MiniLAB:** Do projectiles fall more slowly than dropped objects? p. 101 **Using Technology:** Roller Coaster Physics, p. 103 **Science Journal,** p. 105 **Skill Builder:** Making and Using Tables, p. 105 **Activity 4-1:** Ready...Set...Accelerate! pp. 106-107
4-3 **Science and Society** (1 session, ½ block)*	1. **Explain** how satellites are placed into orbit around Earth. 2. **Give examples** of how satellites can be used. **National Content Standards:** (5-8) UCP2, UCP3, B2, D3, E2, F5, G1, G3; (9-12) UCP2, UCP3, B4, E2, F6, G1, G3	**Explore the Technology,** p. 109
4-4 **Action and Reaction** (2 sessions, 1 block)*	1. **Analyze** action and reaction forces. 2. **Calculate** momentum. 3. **Explain** conservation of momentum. **National Content Standards:** (5-8) UCP1-UCP3, A2, B2, G1, G3; (9-12) UCP1-UCP3, A2, B4, G1, G3	**MiniLAB:** How can you deliver a payload with a balloon rocket? p. 112 **Problem Solving:** An Icy Challenge, p. 115 **Science Journal,** p. 116 **Skill Builder:** Hypothesizing, p. 116 **Activity 4-2:** A Massive Problem, p. 117 **People and Science:** Sammy Chan, Pool Player, p. 118

* A complete Planning Guide that includes block scheduling is provided on pages 32T-35T.

Activity Materials

Explore	Activities	MiniLABs
page 93 small sheets of poster board; tape; small polystyrene cup; long, slender balloon	pages 106-107 clean, empty 2-L plastic bottle with cap; cork or other small object that floats in water; thread; water page 117 2 dynamics carts, long rubber band, 2 building bricks, meterstick, masking tape	page 97 discarded book, piece of paper, scissors page 101 flat ruler, 2 identical metal washers, small piece of tissue paper page 112 long, slender balloon; drinking straw; 8-m piece of string; tape; small piece of candy

Need Materials? Call Science Kit (1-800-828-7777).

Teacher Classroom Resources

Reproducible Masters	Transparencies	Teaching Resources
Study Guide, p. 18 Reinforcement, p. 18 Enrichment, p. 18 Activity Worksheets, p. 24 Cross-Curricular Integration, p. 8	Section Focus Transparency 14, Downhill Skiing	Glencoe Physical Science Interactive Videodisc Physical Science CD-ROM Spanish Resources English/Spanish Audiocassettes Cooperative Learning Resource Guide Lab Partner Lab and Safety Skills Lesson Plans
Study Guide, p. 19 Reinforcement, p. 19 Enrichment, p. 19 Activity Worksheets, pp. 20-21, 25 Lab Manual 5, Projectile Motion Science Integration Activity 4, Going in Circles Concept Mapping, pp. 13-14	Section Focus Transparency 15, Playing Volleyball Science Integration Transparency 4, Weightlessness in Orbit Teaching Transparency 7, Motion of a Projectile	**Assessment Resources** Chapter Review, pp. 11-12 Assessment, pp. 24-27 Performance Assessment in the Science Classroom (PASC) MindJogger Videoquiz Alternate Assessment in the Science Classroom Performance Assessment Chapter Review Software Computer Test Bank
Study Guide, p. 20 Reinforcement, p. 20 Enrichment, p. 20	Section Focus Transparency 16, Satellite Missions	
Study Guide, p. 21 Reinforcement, p. 21 Enrichment, p. 21 Activity Worksheets, pp. 22-23, 26 Lab Manual 6, Conservation of Motion Lab Manual 7, Velocity and Momentum Multicultural Connections, p. 11 Science and Society Integration, p. 8 Critical Thinking/Problem Solving, p. 10	Section Focus Transparency 17, Preventing Automobile Accidents Teaching Transparency 8, Newton's Third Law	

GLENCOE TECHNOLOGY

The following multimedia resources are available from Glencoe.

Science and Technology Videodisc Series (STVS)
Physics
  Science of Bowling
  Safer Roads

National Geographic Society Series
Newton's Apple: Physical Sciences

Glencoe Physical Science Interactive Videodisc
Motion
Machines and Forces

Physical Science CD-ROM

Key to Teaching Strategies

The following designations will help you decide which activities are appropriate for your students.

L1 Level 1 activities should be within the ability range of all students, including those with learning difficulties.

L2 Level 2 activities should be within the ability range of the average to above-average student.

L3 Level 3 activities are designed for the ability range of above-average students.

LEP LEP activities should be within the ability range of Limited English Proficiency students.

LS These activities are designed to address different learning styles.

COOP LEARN Cooperative Learning activities are designed for small group work.

P These strategies represent student products that can be placed into a best-work portfolio.

Teacher Classroom Resources

This is a representation of key blackline masters available in the Teacher Classroom Resources.

Teaching Aids

Section Focus Transparencies

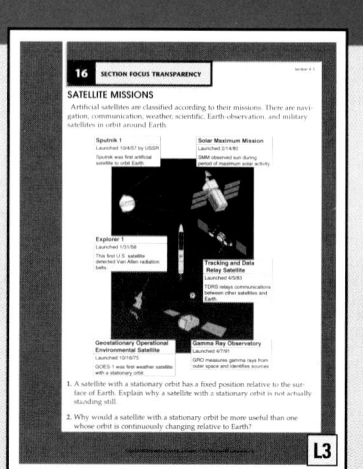

Science Integration Transparencies

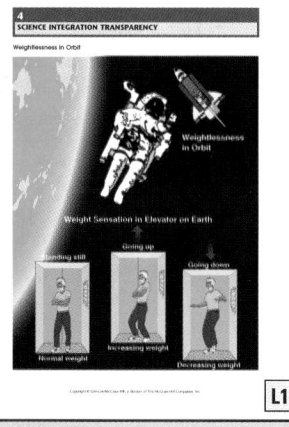

Teaching Transparencies

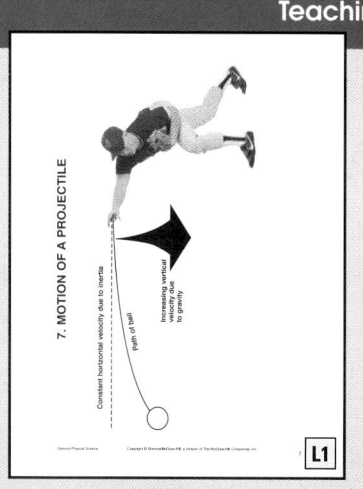

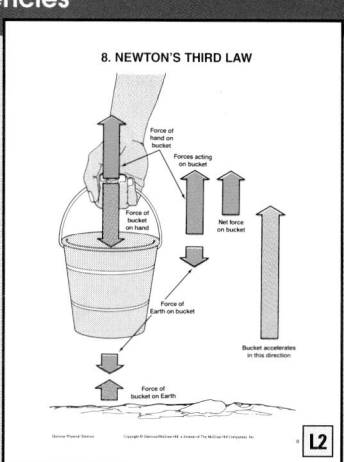

Meeting Different Ability Levels

Study Guide

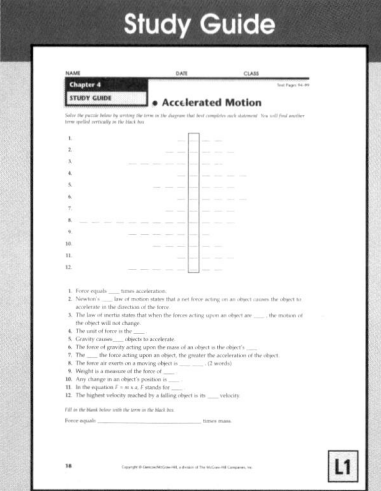

Reinforcement

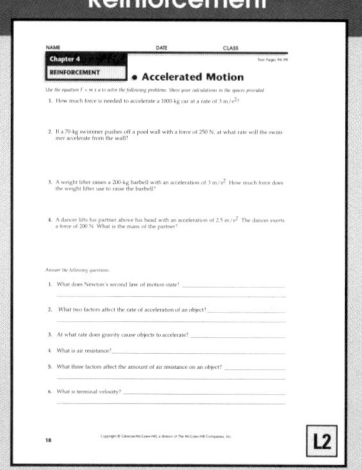

Enrichment Worksheets

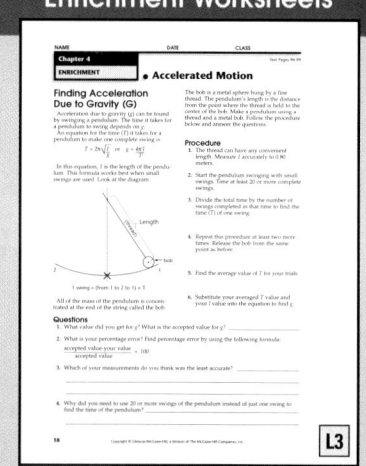

Chapter 4 Acceleration and Momentum

Hands-On Activities

Science Integration Activity

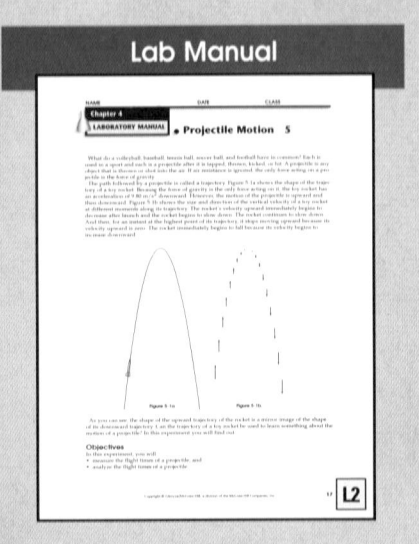

L1

Lab Manual

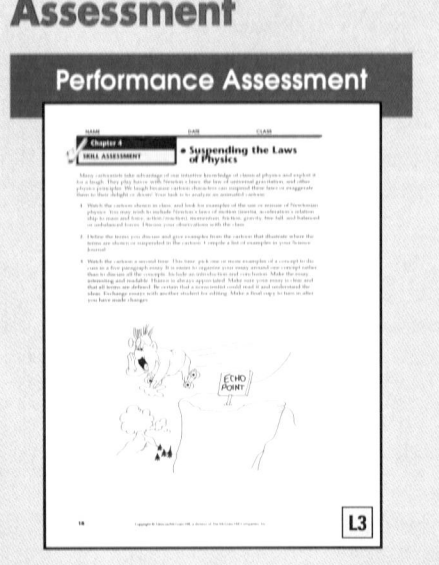

L2

Assessment

Performance Assessment

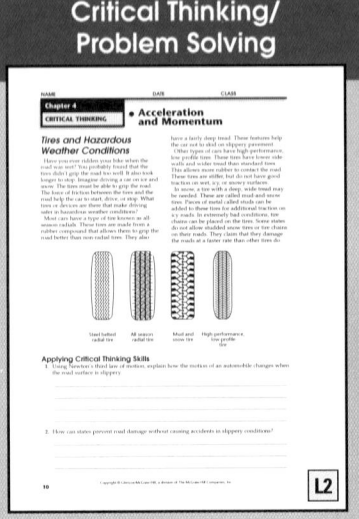

L3

Enrichment and Application

Critical Thinking/Problem Solving

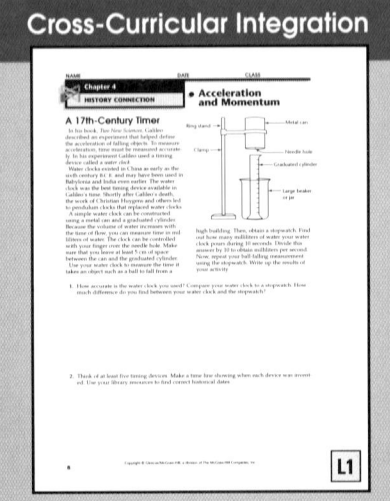

L2

Cross-Curricular Integration

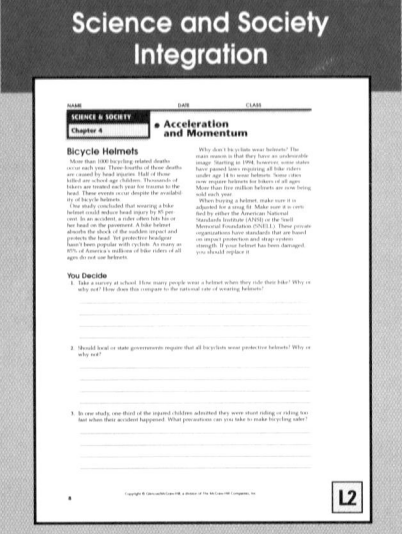

L1

Science and Society Integration

L2

Multicultural Connections

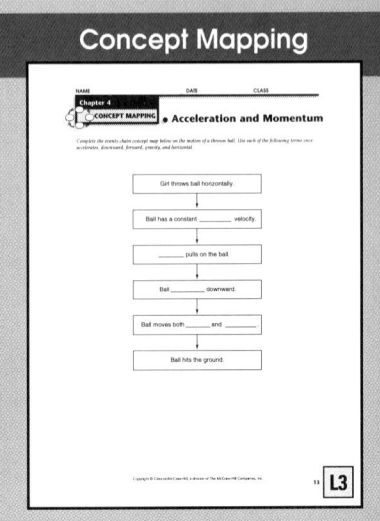

L2

Concept Mapping

L3

Acceleration and Momentum

CHAPTER OVERVIEW

Section 4-1 This section develops Newton's second law of motion. Falling objects and the effect of air resistance on their motion are discussed.

Section 4-2 This section develops the concepts of projectile motion and circular motion. Centripetal acceleration and centripetal force are introduced, and weightlessness is discussed.

Section 4-3 Science and Society The science and technology involved in launching artificial satellites are presented. A variety of uses of artificial satellites is also discussed.

Section 4-4 Newton's third law of motion is presented, with a discussion of momentum and the law of conservation of momentum.

Chapter Vocabulary

Newton's second law
  of motion

air resistance

terminal velocity

projectile

centripetal acceleration

centripetal force

artificial satellite

Newton's third law
  of motion

momentum

law of conservation
  of momentum

Theme Connection

Stability and Change Newton established his three laws of motion by considering common patterns of change in all motion. These laws can be used to analyze and predict changes in the motion of objects. Stress the importance of having only three laws to explain the motion of almost all objects.

Previewing the Chapter

92

Learning Styles

Look for the following logo for strategies that emphasize different learning modalities.

Kinesthetic	Explore, p. 93; MiniLAB, pp. 97, 112; People and Science, p. 118
Visual-Spatial	Demonstration, pp. 95, 103, 111; Activity 4-1, p. 106; Close, p. 109; Reteach, p. 115; Visual Learning, pp. 95, 96, 98, 101, 102, 104, 108, 111, 113, 114, 116
Interpersonal	MiniLAB, p. 101; Activity 4-2, p. 117
Logical-Mathematical	Across the Curriculum, p. 95; Reteach, pp. 98, 104; Making a Model, p. 113; People and Science, p. 118
Linguistic	Science Journal, p. 112

Chapter 4

Acceleration and Momentum

When you watch a televised image of a rocket or space shuttle being launched, do you wonder how such a massive object is accelerated nearly straight up away from Earth? The exhaust gases escaping from the bottom of the rocket play a major role in this launch. Although advanced technology is used to control the motion of the rocket, a few basic physics principles can help you understand more about such events.

EXPLORE ACTIVITY

What propels a rocket upward?
1. Make a model rocket launcher by rolling and taping a piece of poster board into a cylinder with a diameter a few centimeters wider than a polystyrene cup. Cut a door large enough to insert an inflated balloon and your hand (as shown).
2. Inflate a long, slender balloon. Pinch the end closed and insert the top of the balloon into the launcher door. Place a polystyrene cup, open end down, over the top of the balloon as your rocket.
3. To launch your rocket, let the balloon go. Repeat steps 1 and 2. Record your observations and a sketch in your Science Journal.

Observe: What provides the force that pushes the rocket up? How is this like a real rocket?

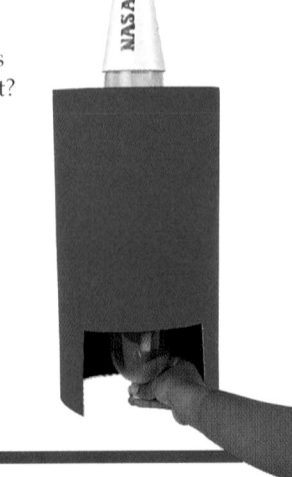

Previewing Science Skills

► In the Skill Builders, you will **compare and contrast, make and use tables,** and **hypothesize.**

► In the Activities, you will **hypothesize, observe, collect and organize data,** and **analyze.**

► In the MiniLABs, you will **experiment, observe and infer,** and **hypothesize.**

93

Prepare

Section Background

Because of gravitational attraction, an object near Earth's surface falls with an acceleration of about 9.8 m/s² in a vacuum. However, the observed acceleration of an object is usually less because of the effects of air resistance.

Preplanning

Have students bring in old, hardcover books for the Mini-LAB on page 97.

1 Motivate

Bellringer

✎ Before presenting the lesson, display **Section Focus Transparency 14** on the overhead projector. Assign the accompanying **Focus Activity** worksheet. L1 LEP

Tying to Previous Knowledge

Have students recall from Section 3-5 the weight of a kilogram on Earth's surface. Tell them that they will learn why this is so in this section.

Student Text Question

What effect would using a slightly heavier tennis ball have on the maximum acceleration? *The maximum acceleration would be less due to the greater mass of the tennis ball.*

4•1 Accelerated Motion

Science Words

Newton's second law of motion
air resistance
terminal velocity

Objectives

• Explain how force, mass, and acceleration are related.
• Compare the rates at which different objects fall.
• Observe the effects of air resistance.

Newton's Second Law

With Newton's first law, you learned how to describe the motion of a speeding sports car or a stationary hockey puck. You also learned that the motion of an object only changes if a net force acts on it, such as the brakes of a car or a fast-moving hockey stick. In the activity on page 93, you observed a net force acting on your balloon "rocket." As you study this chapter, you will find out some reasons why things move in the ways they do.

Study the photo of the tennis players in **Figure 4-1.** Each player can use a racket to produce an acceleration of the tennis ball. What effect would using a slightly heavier tennis ball have on the maximum accelerations? You may not realize it, but in arriving at your answer, you have taken into account Newton's second law of motion. From experience, you know that mass, force, and acceleration are somehow related. **Newton's second law of motion** says that a net force acting on an object causes the object to accelerate in the direction of the force. The acceleration is determined by the size of the force and the mass of the object. A larger force acting on an object causes a greater acceleration. A larger mass requires a greater force than a smaller mass would require to achieve the same acceleration.

Figure 4-1

It is a definite advantage to be able to cause large accelerations of a tennis ball without losing placement control. *Which player is likely to apply the greater force on the ball? Why?*

Program Resources

📁 **Reproducible Masters**
Study Guide, p. 18 L1
Reinforcement, p. 18 L1
Enrichment, p. 18 L3
Cross-Curricular Integration, p. 8
Activity Worksheets, p. 24 L1

📦 **Transparencies**
Section Focus Transparency 14 L1

The Force Equation

Newton's second law can be expressed in equation form as:

$$force = mass \times acceleration$$
$$F = ma$$

Mass is expressed in kilograms and acceleration is expressed in m/s^2. Thus, force is expressed in units of $kg \cdot m/s^2$.

You learned in Chapter 3 that one newton is the standard unit for measuring force. A newton is the amount of force needed to accelerate an object with a mass of 1 kg at an acceleration of 1 m/s^2. In other words, $1 N = 1 kg \cdot m/s^2$.

Study the following Example Problem, then do the Practice Problems that follow.

USING MATH

Calculating Force

Example Problem:

How much force is needed to accelerate a 70-kg rider and her 200-kg motorcycle at 4 m/s^2?

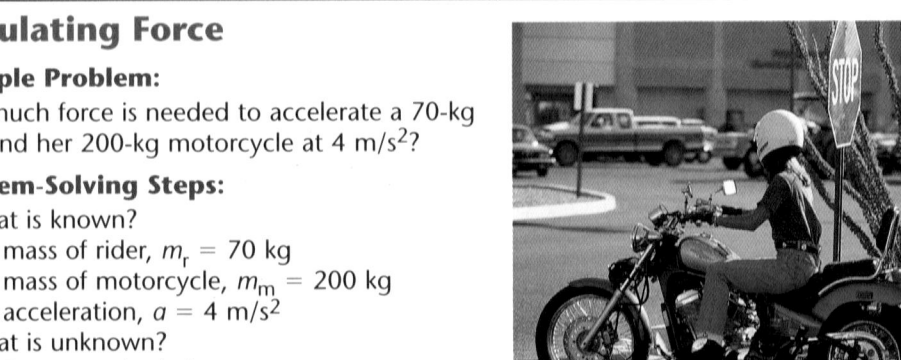

Problem-Solving Steps:

1. What is known?
   mass of rider, m_r = 70 kg
   mass of motorcycle, m_m = 200 kg
   acceleration, a = 4 m/s^2
2. What is unknown?
   Force required, F
3. Choose the equation, $F = ma$.

4. **Solution:**
   The total mass, m, is 70 kg + 200 kg = 270 kg
   F = 270 kg $\times$ 4 m/s^2
   = 1080 kg $\cdot$ m/s^2 = 1080 N
   A force of 1080 N is needed.

Practice Problems

1. It takes a force of 3000 N to accelerate an empty 1000-kg car at 3 m/s^2. If a 160-kg wrestler is inside the car, how much force will be needed to produce the same acceleration?
Strategy Hint: What units combine to make a newton?
2. A 63-kg skater pushes off from a wall with a force of 300 N. What is the skater's acceleration?
Strategy Hint: Rearrange the equation.

4-1 Accelerated Motion   95

Falling Objects

It is hard to believe, but, if you dropped a bowling ball and a marble from a bridge at the same time, they'd both splash into the water at almost the same instant. (As you read further, you'll find out why they don't hit the water at exactly the same instant.) This means their accelerations would be just about the same. Would you have expected the bowling ball to hit the water sooner because it has more mass? It's true that the force of gravity would be greater on the bowling ball because of its larger mass. But the larger mass also gives the bowling ball more inertia, so more force is needed to change its velocity. The marble has a much smaller mass than the bowling ball, but its inertia also is much less. **Figure 4-2** shows the falling motion of two balls revealed by high-speed photography. The blue ball is more massive than the green one, but you can see that they fall at the same rate.

Acceleration Caused by Gravity

Near Earth's surface, gravity causes all falling objects to accelerate at 9.8 m/s^2. Does the number 9.8 seem familiar? When you studied the relationship between mass and weight, you learned that any object with a mass of 1 kg weighs 9.8 N on Earth. Now, you'll find out why.

Any force can be calculated using the equation:

$$F = m \times a.$$

The weight of an object, W, is the force of gravity acting on its mass. So, we can substitute and write:

$$W = m \times a.$$

Acceleration due to gravity is 9.8 m/s^2, therefore:

$$W = m \times 9.8 \text{ m/s}^2.$$

This means that a mass of 1 kg weighs 9.8 kg · m/s^2, or 9.8 N. You could calculate your weight in newtons if you knew your mass. For example, a person with a mass of 50 kg would have a weight of 490 N.

Remember that this discussion is concerned only with *falling* objects. This refers to an object that is dropped from some height and allowed to fall freely. As the object is released, the only downward force acting on it is gravity. The situation changes for an object that is *thrown* downward. In this case, the object is affected by gravity *and* the downward force of the throwing hand. Therefore, the object's downward acceleration would be greater than 9.8 m/s^2.

Figure 4-2

As the photograph shows, the rate of acceleration of a falling body is not affected by the mass of the body. *What effect does inertia have on the falling bodies?*

Use these program resources as you teach this lesson.
Study Guide, p. 18
Cross-Curricular Integration, p. 8

Air Resistance

Acceleration due to gravity (g) is the same for all objects, regardless of mass. This means that if no force other than gravity is present, all objects accelerate at 9.8 m/s^2. Think about that for a minute. Does a leaf accelerate as fast as an acorn? Does a feather fall as fast as a penny?

What would happen if you took two identical sheets of paper, crumpled one into a ball, and dropped both sheets at the same time, as shown in **Figure 4-3?** If your answer is that the crumpled ball would fall faster than the flat sheet, you are correct. But this behavior does not agree with what you've just learned. How can this disagreement be explained?

The only explanation for the behavior of the two sheets of paper is that some force is at work in addition to gravity. Anything that moves in Earth's atmosphere is affected by air resistance. **Air resistance** is the force air exerts on a moving object. This force acts in the opposite direction to that of the object's motion. In the case of a falling object, air resistance pushes up as gravity pulls down.

Figure 4-3

Gravity and air resistance, the two forces acting on the paper, are invisible. *How do we know these forces exist?*

MiniLAB

How can air resistance change the acceleration of a falling object?

Procedure
1. Obtain a discarded book. Cut a piece of paper the same size as the cover of the book.
2. Predict whether the book or the paper will hit the floor first if you drop them the same way and at the same time.
3. Try dropping them side by side with the greatest surface area facing down. Which object has the greater acceleration?
4. Next, devise several methods to make the book and paper fall at nearly the same acceleration.

Analysis
1. If the acceleration of gravity is the same for all objects, how can you explain your results?
2. Describe the methods you discovered for making the book and paper fall at nearly the same acceleration. Analyze how they work.

MiniLAB

Purpose
LS **Kinesthetic** Students will observe how air resistance can affect the acceleration of a falling object. **L1** **LEP**
COOP LEARN

Materials
old, hardcover book; scissors; piece of paper

Teaching Strategies
- Suggest that materials be dropped from waist height.
- Have extra paper available. Students may wish to crumple it in their experiment.

Analysis
1. Because the book is more massive than the paper, it has greater inertia and resists the upward push of air resistance more effectively.
2. Students may place the paper on top of the book or in the book before they are dropped. They may crumple the paper into a dense wad. These methods reduce the surface area acted upon by air resistance.

✔Assessment

Oral Have students explain and discuss how this experiment would differ if performed on the moon, where there is no air resistance and about one-sixth of the gravity as on Earth's surface. *The paper and book would fall with equal acceleration, but the acceleration rate would be less than on Earth.* Use the Performance Task Assessment List for Making Observations and Inferences in **PASC**, p. 17.
P

Activity Worksheets, pp. 5, 24

Air resistance

Figure 4-4

The Frisbee is designed to use air resistance to help it soar and maneuver as it moves through the air. *How does its shape produce different effects of horizontal and vertical air resistance?*

Figure 4-5

The snowflakes (A) and the hailstones (B) are both solid forms of water.

A

B

The amount of air resistance on an object depends on the speed, size, shape, and density of the object. It is air resistance that helps the Frisbee in **Figure 4-4** stay aloft for several seconds. The larger the surface area of the object, the greater the amount of air resistance on it. This is why feathers, leaves, and sheets of paper fall more slowly than pennies, acorns, and crumpled balls of paper. The snowflakes and hailstones shown in **Figure 4-5** fall at different speeds. A snowflake may follow a long and winding path to the ground. Once a hailstone is large enough, it falls straight down. Why do the shapes and masses of the snowflakes and hailstones cause them to have different accelerations?

98   Chapter 4 Acceleration and Momentum

Integrating the Sciences

Life Science **Some small insects can fall hundreds of meters and walk away unharmed after hitting the ground. Can you explain this phenomenon in terms of air resistance and terminal velocity?** *If the insect has a low weight and a relatively large surface area,* *then air resistance can overcome the force of gravity before the insect reaches a high velocity. This is its terminal velocity—the highest velocity the falling insect will reach. The insect is falling slowly enough when it lands that it won't be harmed.*

Terminal Velocity

As an object falls through air, air resistance gradually increases until it balances the pull of gravity. According to the law of inertia, when the forces acting on an object are balanced, the motion of the object will not change. When this happens, the falling object will stop accelerating. It will continue to fall, but at a constant, final velocity like the parachutist in **Figure 4-6.** This **terminal velocity** is the highest velocity that will be reached by a falling object.

Think about the things you know about moving objects. You know that a leaf falls more slowly than an acorn. A rock that is dropped from a height of 10 m is traveling faster when it hits the ground than a rock dropped from 1 m. You will learn more about moving things as you continue reading this chapter.

Air resistance

Gravity

Figure 4-6

Air resistance acts on the parachute, allowing the parachutist to fall at a terminal velocity that is slow enough to allow a safe landing.

Section Wrap-up

Review

1. A weightlifter raises a 440-kg barbell with an acceleration of 2 m/s². How much force does the weightlifter exert on the barbell?

2. A softball is larger and more massive than a baseball. Use Newton's second law to explain why players can't throw or hit a softball as far as a baseball.

3. **Think Critically:** Use what you have learned about falling objects, air resistance, and terminal velocity to explain why a person can parachute safely to Earth from a high-flying airplane.

USING MATH

You apply a force of 50 N to lift a package with a mass of 4 kg. Calculate the resulting rate of acceleration of the package.

Skill Builder

Comparing and Contrasting
Use a balance to find the masses of an uninflated balloon and a balloon filled with air. Drop the two balloons from the same height at the same time, and compare and contrast the rates at which they fall. If you need help, refer to Comparing and Contrasting in the **Skill Handbook.**

4-1 Accelerated Motion   99

Skill Builder

Students should set up an experiment where they can measure the speed at which the balloons fall. Distance should be measured in meters and time in seconds.

 **Assessment**

Oral Ask students to explain why the rates differ. They should note that, although the masses are nearly the same, the difference in surface area causes the force of air resistance to differ. Use the Performance Task Assessment List for Analyzing the Data in **PASC**, p. 27. **P**

Prepare

Section Background

The motion of a projectile can be explained by considering its vertical and horizontal motions independently. The constant horizontal speed and constantly increasing speed downward due to its weight cause the projectile to follow a curved path called a parabola.

Preplanning

- For the MiniLAB, gather flat rulers and metal washers.
- For Activity 4-1, obtain empty 2-L plastic bottles, corks, and thread.

1 Motivate

Bellringer

Before presenting the lesson, display **Section Focus Transparency 15** on the overhead projector. Assign the accompanying **Focus Activity** worksheet. [L1] [LEP]

Tying to Previous Knowledge

Have students recall the motion of a hockey puck using Newton's first law from Section 3-4 and the motion of a falling ball described in the previous section of this chapter. Tell them that this lesson will help them explain the motion of a pop fly to center field.

4•2 Projectile and Circular Motion

Science Words
- projectile
- centripetal acceleration
- centripetal force

Objectives
- Explain why things that are thrown or shot follow a curved path.
- Compare motion in a straight line with circular motion.
- Define *weightlessness*.

Projectiles

Have you noticed that nearly all the moving objects described in this unit so far have been moving in straight lines? But that's not the only kind of motion you know about. Skateboarders wheel around in circles, cars go around hairpin curves, and rockets shoot upward and curve back to Earth. How do the laws of motion account for these kinds of motion?

If you've ever played darts, thrown a ball, or shot an arrow from a bow, you have probably noticed that these objects didn't travel in straight lines. They start off straight, but they curve downward. That's why dart players and archers learn to aim above their targets.

Anything that's thrown or shot through the air is called a **projectile.** Because of Earth's gravitational pull and their own inertia, projectiles follow a curved path. They have both horizontal and vertical velocities. The basketball in **Figure 4-7** is a projectile.

Horizontal Motion

When you throw a ball, like the pitcher in **Figure 4-8,** the force from your hand makes the ball move forward. It gives the ball *horizontal motion*, that is, motion parallel to Earth's surface. Once you let go of the ball, no other force accelerates it forward, so its horizontal velocity is constant if you ignore air resistance.

Figure 4-7

A basketball becomes a projectile when it is shot toward the basket. *Describe the path the ball takes after it leaves the shooter's hands.*

Program Resources

 Reproducible Masters
Study Guide, p. 19 [L1]
Reinforcement, p. 19 [L1]
Enrichment, p. 19 [L3]
Activity Worksheets, pp. 20-21, 25 [L1]
Science Integration Activity, pp. 7-8
Concept Mapping, pp. 13-14
Lab Manual 5

Transparencies
Section Focus Transparency 15 [L1]
Teaching Transparency 7 [L1]
Science Integration Transparency 4

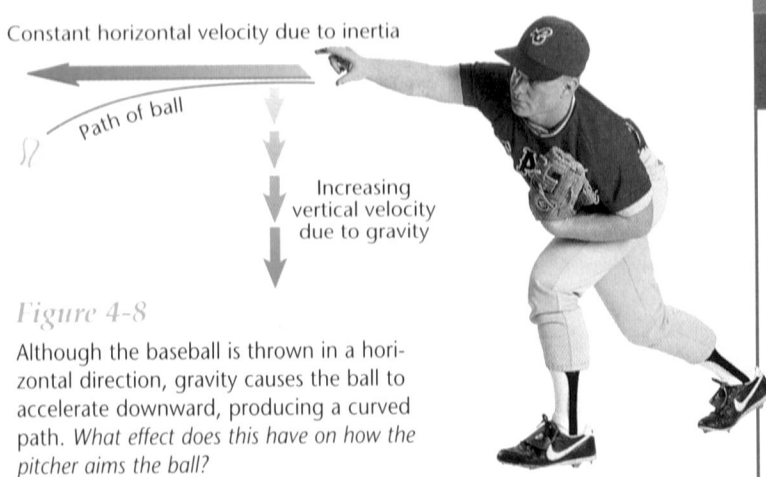

Constant horizontal velocity due to inertia

Path of ball

Increasing vertical velocity due to gravity

Figure 4-8

Although the baseball is thrown in a horizontal direction, gravity causes the ball to accelerate downward, producing a curved path. *What effect does this have on how the pitcher aims the ball?*

Vertical Motion

When you let go of the ball, something else happens, as well. Gravity starts pulling it downward, giving it *vertical motion*, or motion perpendicular to Earth's surface. Now the ball has constant horizontal velocity but increasing vertical velocity. Gravity exerts an unbalanced force on the ball, changing the direction of its path from forward only to forward and downward. The ball's horizontal and vertical motions are completely independent of each other.

If you throw a ball horizontally from shoulder height, will it take longer to hit the ground than a ball you simply drop from the same height? Surprisingly, both will hit the ground at the same time. If you have a hard time believing this, **Figure 4-9** may help. The two balls have the same acceleration due to gravity, 9.8 m/s^2 downward.

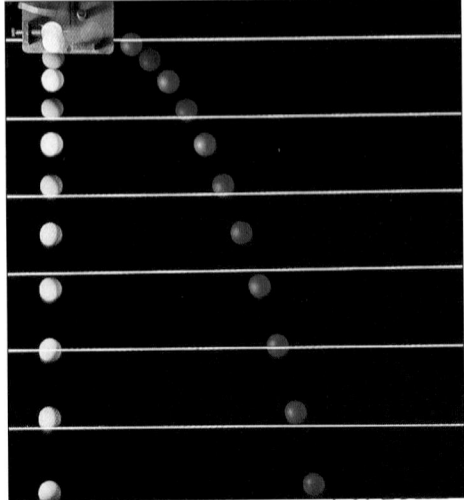

Figure 4-9

The two balls in the photograph were released at the same time. Although the red ball has horizontal velocity, gravity causes both balls to accelerate downward at the same rate.

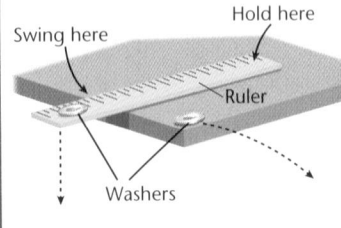
4-2 Projectile and Circular Motion   101

101

Motion Along Curves

Recall that acceleration is a rate of change in velocity caused by a change in speed, in direction, or both. Now, picture a bicycle moving at a constant speed along the westbound straightaway of an oval track. Because its speed is constant in a straight line, the bicycle is not accelerating. However, when the bicycle enters a curve, even if its speed does not change, it is accelerating because its direction is changing. The change in the direction of the velocity is toward the center of the curve. Acceleration toward the center of a curved or circular path is called **centripetal acceleration.** The bicycles and their riders in **Figure 4-10** experience centripetal acceleration and lean toward the inside of the curve. The word *centripetal* means "toward the center."

Figure 4-10

These bicycle riders are experiencing centripetal acceleration.

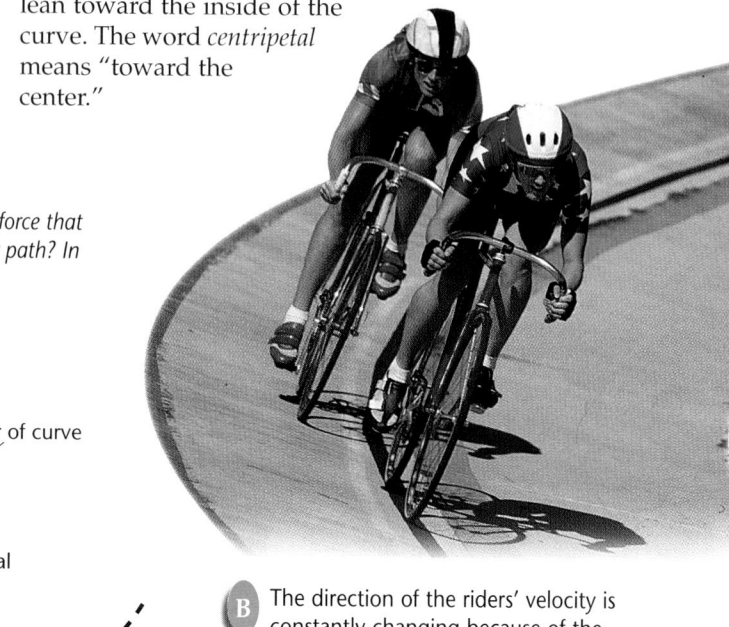

A *What is providing the centripetal force that helps these riders follow a circular path? In what direction are they accelerating?*

Center of curve

Centripetal force

Bicycle direction

B The direction of the riders' velocity is constantly changing because of the influence of centripetal force.

Centripetal Force

In order for the bicycle to be accelerating, some unbalanced force must be acting on it in a direction toward the center of the curve. That force is a centripetal force. **Centripetal force** is a force acting toward the center of a curved or circular path.

When a car rounds a sharp curve on a highway, the centripetal force is the friction between the tires and the road surface. But if the road is icy or wet and the tires lose their grip, the centripetal force may not be enough to overcome the car's inertia. The car would then keep moving in a straight line in the direction it was traveling at the spot where it lost traction.

Inclusion Strategies

Learning Disabled Make sure students realize that as an object moves in a circle, the *change* in its direction of motion is always toward the center of the circle, while the direction of its motion is always perpendicular to the radius of the circle. To help students understand the difference, have them visualize moving a heavy brick along a big circle. Each time they move the brick, they must slide it forward just a little and then twist it a bit so it fits along the circle. The twist, which represents the change in direction of the brick, is always toward the center of the circle. The brick itself is always aligned along the circle itself, which means it is always perpendicular to the radius of the circle.

Weightlessness in Orbit

Maybe you've seen pictures of astronauts floating inside the space shuttle with various pieces of equipment suspended in midair beside them. Any item that is not fastened down in the shuttle will float around and pose a possible hazard for the astronauts and their equipment. The astronauts and their belongings are experiencing the sensation of weightlessness.

But to be truly weightless, the astronauts would have to be free from the effects of gravity. Orbiting 400 km above Earth, the shuttle and everything inside it still respond to those effects. In fact, gravity keeps the shuttle in orbit.

Free-Falling

So what does it really mean to say that something is weightless in orbit? Think about how the weight of an object is measured. When you place an object on a scale, gravity causes the object to

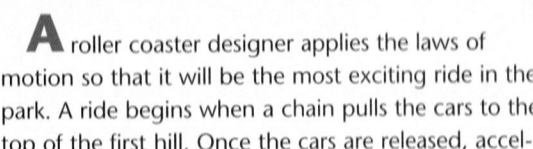

USING TECHNOLOGY

Roller Coaster Physics

A roller coaster designer applies the laws of motion so that it will be the most exciting ride in the park. A ride begins when a chain pulls the cars to the top of the first hill. Once the cars are released, acceleration increases until all of the cars are headed downward.

The debate on which seat is the scariest continues to rage, and the answer is, "that depends." As the cars descend, their speed increases. The rear car starts down the slope at a much greater speed than the front car, thus giving the passengers the sense of being hurled over the edge. At the bottom of the hill, it is a different story. When the change in direction from down to up occurs, the front car will be going fastest and its passengers will experience the greater forces. As the cars pop over the top of the hill, the passengers in the rear car may experience a considerable force, resulting in the sensation of being thrown free.

As ride technology improves, roller coasters get larger and faster. The Magnum XL 200 at Cedar Point in Sandusky, Ohio, has a first hill 61 m (201 feet) high, reaches a speed of 112 km/h, and covers 1.6 km (5106 feet) of track in 2½ minutes.

Riders experience rapid acceleration

Think Critically:

Describe the roller-coaster design that would result in the greatest sensations for the passengers.

USING TECHNOLOGY

Have the students share their scariest roller-coaster ride. Have them explain which seat they chose and why.

Think Critically
The first hill should be high with a steep down-slope. Subsequent hills, though not as high, should be steep, and there should be tight radius curves.

Demonstration

 Visual-Spatial Place a small hole in the bottom of a foam plastic cup. While holding the hole shut, fill the cup with water. Hold the cup as high as you can and drop it into a bucket placed directly beneath it on the floor. Have students observe that no water ran from the cup as it fell because both the cup and water were in free-fall. **LEP**

Revealing Preconceptions

Point out that weightlessness is the sensation felt in free-fall when there is no support force under your body. The word *weightless* really refers to the state of an object so far from any other object that the gravitational force would be negligible.

GLENCOE TECHNOLOGY

Videodisc

Glencoe Physical Science Interactive Videodisc
Side 1, Lesson 1
Free-fall

8141-10601

Use these program resources as you teach this lesson.
Teaching Transparency 7
Science Integration Transparency 4

Visual Learning

Figure 4-11 Are objects in Earth orbit really weightless? Explain. *No, they only appear to be weightless. They are in free-fall.* **LEP** **LS**

INTEGRATION
Life Science

One way to overcome weightlessness in space is to build a large, doughnut-shaped ring called a torus. If the torus is set spinning in outer space, the centripetal force on someone standing near the outer wall of the torus would cause a feeling similar to standing on the floor on Earth.

3 Assess

Check for Understanding

FLEX Your Brain

Use the Flex Your Brain activity to have students explore PROJECTILE MOTION.

📁 **Activity Worksheets,** p. 5

Reteach

LS **Logical-Mathematical** Have students place tracing paper over Figure 4-2 and mark the vertical positions of one of the balls. Then have them imagine that the ball was also moving sideways with a constant speed and change their drawing to account for this.

Extension

📁 For students who have mastered this section, use the **Reinforcement** and **Enrichment** masters.

Figure 4-11

The space shuttle and everything in it are in free-fall around Earth, thus producing apparent weightlessness. *Are objects in orbit around Earth really weightless? Explain.*

push down on the scale. In turn, the scale pushes back on the object with the same force. The dial on the scale measures the amount of upward force needed to offset gravity.

Now suppose that the scale is falling, being pulled downward by gravity at the same rate as the object being weighed. The scale couldn't push back on the object, so its dial would read zero. The object would seem to be weightless. This is what is happening in a space shuttle orbiting Earth. The shuttle and everything in it, including the astronaut working in **Figure 4-11,** are all "falling" toward Earth at exactly the same rate of acceleration. When an object is influenced only by gravity, it is said to be in free-fall. An orbiting space shuttle and all its contents are in free-fall around Earth.

INTEGRATION
Life Science

Effects of Weightlessness

As you move about in your daily routines, the resistance provided by gravity helps you exercise your body. What happens to the physical condition of astronauts who experience the sensation of weightlessness for extended periods of time? When Russian cosmonauts experienced weightlessness for more than 200 consecutive days, they developed health problems. Health tests performed on American astronauts have also shown that some bone and muscle deterioration occurs during long periods of weightlessness.

104 Chapter 4 Acceleration and Momentum

📁 Use these program resources as you teach this lesson.
Study Guide, p. 19
Science Integration Activity, pp. 7-8
Concept Mapping, pp. 13-14
Lab Manual 5

Exercising in Space

Flight doctors have developed special exercise programs for astronauts to reduce the health problems related to weightlessness. Several activities, such as the one in **Figure 4-12,** are helpful. They include isometric exercises, in which muscles push against other muscles in the body. For example, you can feel some resistance if you place your left palm against your right palm and push your palms together for 10 seconds.

Aboard the space shuttle, astronauts also walk on a unique treadmill for 15 to 30 minutes each day. Special attachments to their arms and legs make their muscles work even harder. An air duct circulates air to help dry off their sweat. They can even listen to music or gaze out the window into space while they exercise.

Figure 4-12

Astronauts don't have much room to exercise inside the shuttle. Here astronaut Ellen Baker exercises aboard Columbia while in orbit.

Section Wrap-up

Review

1. Use a diagram similar to **Figure 4-8** to show why a dart player has to aim above the target to hit the bull's-eye. In your diagram, show the forces acting on the dart, the dart's path, and the two kinds of motion involved.

2. A child is swinging a yo-yo in a circle. What provides the force to keep the yo-yo going in a circle? What is the force called? What happens if the string breaks?

3. **Think Critically:** Does the mass of an astronaut change when he or she is in orbit around Earth? Does an astronaut's weight change?

Skill Builder
Making and Using Tables

Make a table showing important characteristics of projectile motion, circular motion, and free-fall. Table headings should include: *Kind of Motion, Shape of Path,* and *Laws or Forces Involved.* You may add other headings. If you need help, refer to Making and Using Tables in the **Skill Handbook.**

Science Journal

Write a paragraph describing a situation in which you experienced something close to free-fall or a feeling of weightlessness. Think about amusement park rides, elevators, athletic events, or even movie scenes.

Skill Builder

Tables should be set up as shown here. Accept any correct data students include to complete the table.

Important Characteristics of Three Kinds of Motion		
Kind of Motion	Shape of Path	Laws or Forces
Projectile		
Circular		
Free fall		

 **Assessment**

Performance Have students add a column entitled "Practical Examples." Use the Performance Task Assessment List for Data Table in **PASC,** p. 37. [P]

4 Close

Section Wrap-up

Review

1. Refer to Figure 4-8 in the student text. Students' diagrams should show the dart being launched on a slightly upward path, so that the downward curve of the dart caused by gravity brings the dart to the target bull's-eye.

2. The string provides the centripetal force that keeps the yo-yo moving in a circular path. If the string breaks, the yo-yo will fly off in a straight line in the direction it was moving when the string broke.

3. **Think Critically** No; mass is independent of gravity. Yes; weight changes because the force of gravity is reduced.

Science Journal

Accept all reasonable answers. Roller coasters, free-fall rides, jumping out of a tree, and pole vaulting are all examples of activities involving near free-fall situations. [P]

PREPARATION

Purpose
LS **Visual-Spatial** Students will make a model of a simple accelerometer and use it for determining the direction of acceleration in various movements. **L1** **LEP** **COOP LEARN**

Process skills
observing and inferring, recognizing cause and effect, designing an experiment, communicating, making and using tables, comparing and contrasting, forming operational definitions, forming a hypothesis, separating and controlling variables, making models

Time
30 minutes

Materials
For safety considerations, you may wish to punch holes in the corks before class to facilitate tying thread to them. After the thread has been tied to the corks, students can adjust the thread to make the cork float near the top of the upside-down container by leaving some of the thread outside when the lid is tightened.

Possible Hypotheses
When motion starts in a forward, backward, left, or right direction, the acceleration is in the direction of the motion. Acceleration in circular motion is toward the center of the circle.

📁 **Activity Worksheets,** pp. 5, 20-21

PLAN THE EXPERIMENT

Possible Procedures
• Students should construct an accelerometer similar to the one shown in the

---

Activity 4-1

Design Your Own Experiment
Ready...Set... Accelerate!

Imagine the changes in motion you would experience while riding a bicycle straight ahead and then going around a corner. As you started out, would you be accelerating forward? When you whip around a corner at constant speed, you are accelerating because your direction is changing. But in which direction is the acceleration?

PREPARATION

Problem
How can you use an accelerometer to determine the direction of acceleration in various types of motion?

Form a Hypothesis
Develop a hypothesis about the direction of acceleration in simple movements such as moving forward, backward, left, right, and in a circle.

Objectives
• Construct a simple accelerometer to use in determining the direction of acceleration in various movements.
• Observe the accelerometer as you accelerate in different directions.

Possible Materials
• clean, empty, 2-L plastic bottle with cap
• cork or other small object that floats in water
• thread
• water

Safety Precautions 🧤 🥽
Protect clothing and eyes. Provide enough floor space for students to safely spin around.

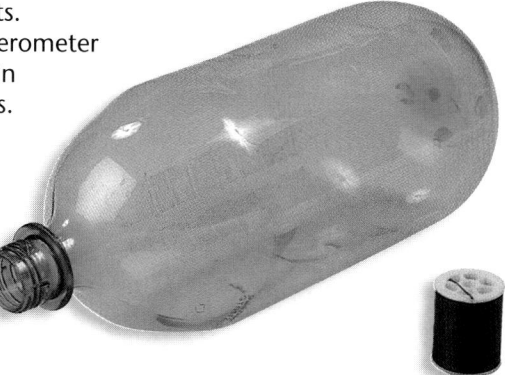

---

photo. They may try to analyze a variety of motions, but be sure they include circular motion.
• Pour water into the bottle until it is 3/4 full, tie the thread to the cork, and leave the thread long enough so it hangs outside the bottle. This allows the length to be adjusted so the cork floats at the surface of the water.

Teaching Strategies
Troubleshooting It is important that students observe the motion of the cork just as *acceleration is beginning*. This allows them to observe the motion due to the initial applied force, before other forces come into play. The cork indicates the direction of acceleration because the water, which is more dense and has greater inertia than the cork, undergoes what we commonly

PLAN THE EXPERIMENT

1. Be sure your group has agreed upon a hypothesis statement.
2. Sketch the design for your accelerometer. See the photo and the list of suggested materials for some ideas.
3. As a group, make a detailed list of the steps you will take to test your hypothesis.
4. Gather the materials you will need to carry out your experiment.

Check the Plan

1. What will you use as your floating device in the accelerometer?
2. How long will your thread that anchors the floating device be?

3. What types of motion will you try to use your accelerometer to analyze? Be sure to include circular motion.
4. Be sure you observe the motion of your floating device when the accelerated motion begins.
5. *Make sure your teacher approves your plan and that you have included any changes suggested in the plan.*

DO THE EXPERIMENT

1. Carry out the experiment as planned.
2. Be sure to write down any observations that you make while doing your experiment. You might also include suggestions for improving your accelerometer.

Analyze and Apply

1. How did the floater act when you moved your accelerometer to the left, to the right, forward, and backward? What other motions did you try?

2. Did you try to **observe** any vertical motions with your accelerometer? If not, try it now. Was it effective? Why or why not?
3. What happened in the accelerometer when you held it while rotating in a circle? In what direction is this centripetal acceleration directed? What is providing the centripetal force?

4-2 Projectile and Circular Motion   **107**

Go Further

Think of a familiar carnival or amusement park ride. If you were carrying your accelerometer on this ride, what changes would you observe? Write a descriptive paragraph about your predictions. What industrial uses do you think more accurate accelerometers might have?

---

call *whiplash* and pushes the cork away. For example, when the bottle is accelerated to the right, the force of the water on the left side increases and forces the cork to the right in the direction of the acceleration.

DO THE EXPERIMENT

Expected Outcome

Students will observe that the cork moves in the direction of acceleration.

Analyze and Apply

1. The floater moved in the same direction as acceleration.
2. Generally, it would not be effective. Buoyancy of the cork would tend to affect any downward acceleration force.
3. The cork moved toward the center of the circular motion—toward the person holding the bottle. The centripetal acceleration is directed toward the person applying the spinning motion. The centripetal force is provided by the water as it undergoes rapid circular movement.

Go Further

On a roller coaster, for example, an accelerometer would show rapid changes in acceleration during the entire ride, going down and up hills and around curves. In industry, accelerometers could be used in developing guided missiles, space vehicles, transportation equipment, and industrial machinery. They are useful in measuring the effects of vibration, shock, acceleration, and braking.

 **Assessment**

Content Have students sketch the trip they must take in a car to travel to school. Ask them to predict what their accelerometer would indicate as they make various accelerations on this trip (speeding up, slowing down, changing directions). Use the Performance Task Assessment List for Making Observations and Inferences in **PASC**, p. 17.
P

Prepare

Section Background

The speed required for a satellite to move in an elliptical orbit is more than 30 000 km/h, which is greater than for a circular orbit.

1 Motivate

Bellringer

Before presenting the lesson, display **Section Focus Transparency 16** on the overhead projector. Assign the accompanying **Focus Activity** worksheet. [L1] [LEP]

Tying to Previous Knowledge

Before beginning this lesson, ask students where they have heard the phrase "via satellite."

2 Teach

Visual Learning

Figure 4-13 How is his idea similar to the way satellites are put into orbit today? *Newton's idea involved boosting the satellite into the air, but cannonballs couldn't be shot far enough or fast enough to reach orbital velocity. Today, rockets are used to propel satellites after they have been lifted to the proper orbital altitude.*

Figure 4-15 Why are communication satellites placed in geostationary orbits? *This keeps them at a fixed position above Earth for better transmission and reception of signals.* [LEP] [LS]

TECHNOLOGY:
4•3 Sending up Satellites

Science Words

artificial satellite

Objectives

- Explain how satellites are placed into orbit around Earth.
- Give examples of how satellites can be used.

Figure 4-13

Newton's idea of using cannons to launch satellites was based on the technology of the time. He proposed that if a cannonball were launched with large enough speed, the curve of its trajectory would match Earth's curve and the ball would orbit Earth at a constant height. *How is his idea similar to the way satellites are put into orbit today?*

What is an Earth satellite?

When watching a sports program, newscast, or weather report, have you heard the announcer state that a report was received "via satellite"? You may know that Earth's moon is a natural satellite, but the moon was not the source of these reports. What other satellites are orbiting Earth? **Artificial satellites** are human-made devices that orbit Earth for a specific purpose. Once in orbit, their motion and behavior are like those of natural satellites such as Earth's moon.

Launching Artificial Satellites

Newton's original idea of launching a satellite by blasting it horizontally from a mountaintop, illustrated in **Figure 4-13,** did not turn out to be practical, but in the 1950s, the developing field of rocketry did make launching satellites possible. A multistage rocket system typically boosts the satellite to the desired height of the orbit and then fires again to accelerate the satellite to the speed required to stay in orbit. This speed for circular orbits is around 8 km/s, or about 29 000 km/h.

In 1957, the former Soviet Union launched the first artificial Earth satellite. The satellite, called Sputnik, was a metallic instrument module having a mass of only 84 kg and a diameter of 60 cm. It orbited Earth about once every 90 minutes. Literally thousands of satellites have now been placed into orbit, including some that have orbited Mars, Venus, and Earth's moon.

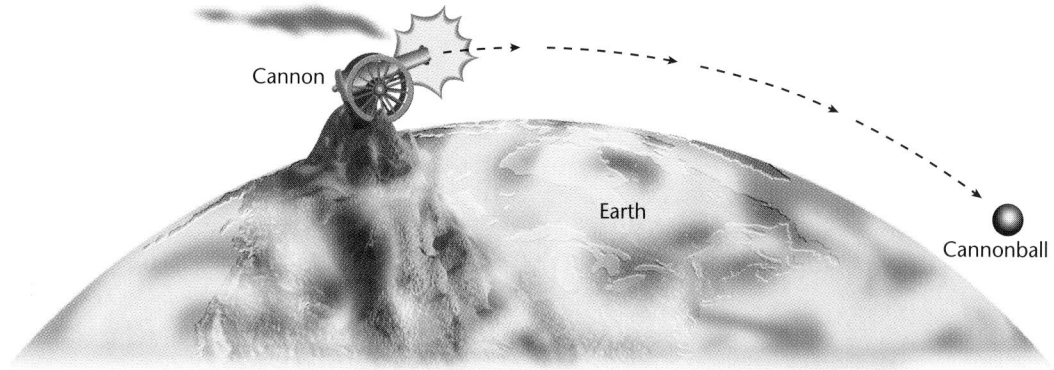

Cannon

Earth

Cannonball

108 Chapter 4 Acceleration and Momentum

Program Resources

Reproducible Masters
Study Guide, p. 20 [L1]
Reinforcement, p. 20 [L1]
Enrichment, p. 20 [L3]

Transparencies
Section Focus Transparency 16 [L1]

How We Use Satellites

Artificial satellites, such as the one in **Figure 4-14,** serve a variety of different purposes. Communications and weather satellites are most familiar. These are usually geostationary satellites, which are put into orbit with a speed that matches the movement of Earth as it spins on its axis. Thus, as **Figure 4-15** shows, they appear to be stationary high above a given location on Earth. Communication satellites serve as receivers and transmitters to relay TV and radio signals around the world. Weather satellites use different photographic techniques to monitor weather patterns and ground temperatures.

Many satellites are used by the military to monitor actions in other countries—tiny details can be photographed from hundreds of kilometers above Earth.

Falling Satellites

Artificial satellites cannot orbit forever. Even in high orbits, small amounts of air resistance gradually cause a satellite to lose energy, allowing Earth's gravity to pull it lower. As a satellite is pulled downward into the denser part of Earth's atmosphere, it usually burns up in the extreme heat generated by atmospheric friction.

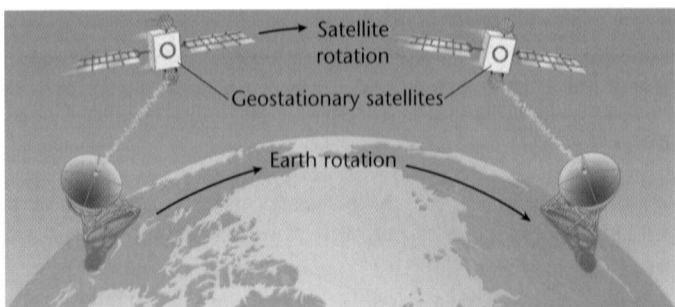

Section Wrap-up

Review

1. For what purposes are artificial satellites used?
2. Why are geostationary satellites often useful for communications and monitoring weather?

 Visit the Chapter 4 Internet Connection at Glencoe Online Science, **www.glencoe.com/sec/science/physical,** for a link to more information about satellites.

SCIENCE & SOCIETY

Figure 4-14

Satellites can remain in orbit performing their tasks for many years.

Figure 4-15

If you look up at a geostationary satellite, it appears to be hovering motionless overhead. *Why are communication satellites placed in geostationary orbits?*

3 Assess

Check for Understanding

Ask students what would happen if a satellite gradually slowed below the velocity required for orbit. *Without a high enough velocity, it will return to Earth.*

Reteach

Recall **Figure 4-13.** Tell students that a satellite must rise more than 200 km above Earth's surface before it enters orbit. **LEP**

Extension

For students who have mastered this section, use the **Reinforcement** and **Enrichment** masters.

4 Close

LS **Visual-Spatial** Use a rotating globe and a small ball to model the behavior of satellites and geostationary satellites.

Section Wrap-up

Review

1. Artificial satellites may be used for communications, weather monitoring, and military investigations.
2. Geostationary satellites would allow measurements or communications to be made constantly over a desired region.

Prepare

Section Background

- The momentum of an object is equal to the product of its mass and its instantaneous velocity. The momentum of a system is the sum of all the momenta of the parts of the system.
- The law of conservation of momentum states that if there is no external, net force acting on an object or a system, the momentum of the object or system remains the same.

Preplanning

- Obtain balloons, fishing line, plastic drinking straws, and tape for the MiniLAB.
- For Activity 4-2, gather 2 dynamics carts, a long rubber band, 2 bricks, a meterstick, and masking tape for each group.

1 Motivate

Bellringer

Before presenting the lesson, display **Section Focus Transparency 17** on the overhead projector. Assign the accompanying **Focus Activity** worksheet. L1 LEP

Tying to Previous Knowledge

Have students recall the motion of an untied, inflated balloon. Point out that the cause of the motion is much like that which causes a rocket to move. Tell students they will learn more about rocket propulsion in this section.

Science Words

Newton's third law of motion
momentum
law of conservation of momentum

Objectives

- Analyze action and reaction forces.
- Calculate momentum.
- Explain conservation of momentum.

Figure 4-16

The boy's leap will fall short of the dock. *What didn't he think about before he jumped?*

110

4•4 Action and Reaction

Newton's Third Law

A girl blows up a balloon and releases it, watching it dart away on a zigzag course. A young person enjoys bouncing on a trampoline. A boy leaping from a boat toward land falls in the water when the boat scoots away from the dock. As different as these examples of motion may seem, they all illustrate one point: forces always act in pairs, called action-reaction pairs.

Newton's third law of motion describes action-reaction pairs this way: When one object exerts a force on a second object, the second one exerts a force on the first that is equal in size and opposite in direction. A less formal way of saying the same thing is "to every action force there is an equal and opposite reaction force."

Action-Reaction Pairs

Let's look at two examples of the action-reaction pairs described earlier. When a person jumps on a trampoline, he or she exerts a downward force on the trampoline. The trampoline then exerts an equal force upward, sending the person high into the air.

In **Figure 4-16,** when the boy leaps from the boat, the boat exerts a force on his feet, moving him forward. His feet exert an equal and opposite force on the boat, sending it backward.

Program Resources

📁 **Reproducible Masters**
Study Guide, p. 21 L1
Reinforcement, p. 21 L1
Enrichment, p. 21 L3
Science and Society Integration, p. 8
Activity Worksheets, pp. 22-23, 26 L1
Lab Manual 6, 7
Critical Thinking/Problem Solving, p. 10
Multicultural Connections, pp.11-12

📽 **Transparencies**
Section Focus Transparency 17 L1
Teaching Transparency 8 L1

Figure 4-17

As the swimmer's hands and feet push against the water, the water pushes back, moving the swimmer forward.

Forces Acting on Different Objects

Newton's third law can be used to explain how the swimmer in **Figure 4-17** moves through the water. With each stroke, the swimmer's arm exerts a force on the water. The water pushes back on the swimmer with an equal force in the opposite direction. But if the forces are equal, how can the swimmer move forward? It's possible because the forces are acting on different things. The "action" force acts on the water; the "reaction" force acts on the swimmer. The swimmer, having less mass than the pool full of water, accelerates more than the water.

An important point to keep in mind when dealing with Newton's third law is that action-reaction forces always act on *different* objects. Thus, even though the forces may be equal, they are not balanced. In the case of the swimmer, water pushes her forward, overcoming the friction, or drag, she encounters. Thus, a net force, or unbalanced force, acts on the swimmer, and a change in motion can take place. The force pairs involved in passing a basketball will cause the wheelchair in **Figure 4-18** to roll backward.

Figure 4-18

The action of passing the basketball forward will propel the player slightly backward. *Would this be an advantage or disadvantage for the player?*

Demonstration

Visual-Spatial Have a student measure the lengths of three rubber bands. Suspend a weight from one rubber band and have a student measure its new length. Remove and link an identical rubber band to the first with a wire tie from a plastic food-storage bag. Ask students to predict by how much each rubber band will elongate if you suspend a weight from the second rubber band. Demonstrate the elongation of each rubber band, and have a student measure the length of each rubber band. Have students compare the elongations with the elongation of the single rubber band. Link a third rubber band to the chain in a similar fashion and repeat the questioning and the demonstration. Ask students to summarize their observations.

Visual Learning

Figure 4-16 What didn't he think about before he jumped? *His feet exert an equal and opposite force on the boat, sending it backward away from the dock.*

Figure 4-18 Would this be an advantage or a disadvantage for the player? *A disadvantage because the player and his wheelchair will be propelled backward slightly every time the player passes the ball. However, all players will face this same problem.* LEP

Community Connection

Safe Structures Civil engineers, especially those responsible for designing safe roads, bridges, and building codes, must be aware of the importance of the effects of balanced and unbalanced forces. Invite a civil engineer from your community to discuss how these principles are used in designing safe, efficient communities.

Rocket Propulsion

Rockets work on a principle similar to that which causes a balloon to move when air escapes. In the rocket, burning fuel produces hot gases that push against the inside of the rocket and escape out the bottom. The upward push of gases propels the rocket.

Figure 4-19

The "action force" of the expanding gases pushes the rocket upward. The "reaction force" of the rocket pushes the exhaust gases out.

MiniLAB

How can you deliver a payload with a balloon rocket?

Procedure

1. Make a guiding line for your balloon rocket by threading an 8-m string through a drinking straw. Tie each end to a chair and stretch it taut between the chairs.
2. Blow up a balloon and pinch the end. Tape a very small object to the balloon in a location that will interfere as little as possible with the rocket's flight.
3. With the straw at one end of the string, use tape to attach the inflated balloon to the straw.
4. Release the balloon. Measure to the nearest centimeter how far your rocket travels.

Analysis

1. What propels the balloon forward? How does this demonstrate Newton's third law?
2. Do you think your rocket would travel farther if it were not carrying the object? What changes could you make to allow your balloon to travel farther? Try it.

Action Force
pushes rocket up

Reaction Force
pushes gases out

Momentum—Mass on the Move

If a toy truck rolls toward you, you can easily stop it with your hand. However, such a tactic would not work with a full-sized truck, even if it were moving at the same speed as the toy truck. It takes more force to stop the full-sized truck because it has more momentum than the toy truck has. **Momentum** is a property a moving object has because of its mass and velocity. Momentum can be calculated with this equation, in which p represents momentum.

$$momentum = mass \times velocity$$

$$p = mv$$

The unit for momentum is kg · m/s. Notice that momentum has a direction because velocity has a direction. Notice also that if an object is not moving, it has zero momentum.

Science Journal

LS **Linguistic** Have students use Newton's third law to describe the similarities between a human walking on a sidewalk and a fish swimming in the ocean. *A human exerts a force on the sidewalk, and the reaction force of the sidewalk on the person moves the person forward. A fish pushes on the water, and the reaction force of the water propels the fish forward.* **L1**

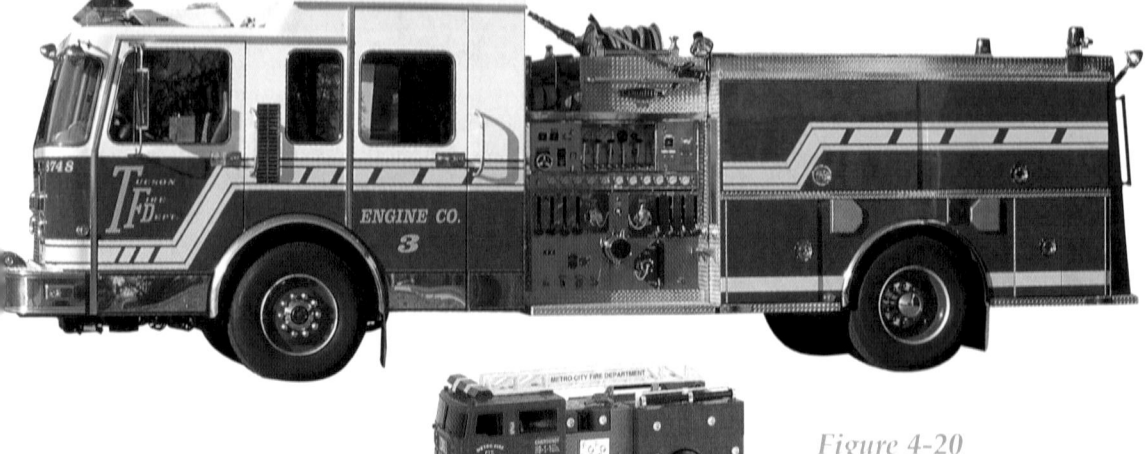

Figure 4-20

Because of its much greater mass, the real truck has more momentum than the toy truck, even when the two trucks are moving at the same speed. *What would happen in a collision between a large truck and a small truck?*

The two trucks pictured in **Figure 4-20** may have the same velocity, but the full-sized truck has more momentum because of its greater mass. A speeding bullet has a large momentum because of its high velocity, even though its mass is small. A lumbering elephant may have a low velocity, but because of its large mass, it has a large momentum. The bumper cars shown in **Figure 4-21** are an example of how rapidly changing momentums can be fun.

Figure 4-21

Bumper cars are a favorite amusement-park ride for many people. *What factors could cause the momentum of one bumper car to be different from that of another bumper car?*

4-4 Action and Reaction   113

 Use these program resources as you teach this lesson.
Multicultural Connections, pp. 11-12
Lab Manual 7

Use **Teaching Transparency 8** as you teach this lesson.

The pool player is ready to start a game. *What is the total momentum of all the balls on the table?*

Conserving Total Momentum

The momentum of an object doesn't change unless its mass, or velocity, or both, change. But momentum can be transferred from one object to another. Consider the game of pool shown in **Figure 4-22.** Before the game starts, all the balls are motionless. Therefore, the total momentum of the balls is zero. There can be no momentum because none of the balls has a velocity.

What happens when a cue ball rolling across a pool table hits the group of balls that is standing still? At first, the rolling ball has momentum and the motionless balls do not. When the cue ball collides with the balls that were at rest, all the balls start moving. They gain momentum. The cue ball slows down and loses momentum. If you were to measure the total momentum of all the balls before and after the collision, it would be the same. The momentum the group of balls gains is equal to the momentum that the cue ball loses. Total momentum is conserved—it doesn't change.

A The player strikes the cue ball and sends it hurtling toward the group of balls at the other end of the table. The cue ball is the only ball moving; thus, it is the only ball with momentum. It carries the total momentum of all the balls.

B After being struck by the cue ball, the other balls scatter around the table; some hit other balls, some hit the edge of the table, and others may hit nothing before stopping. At any instant, the total momentum of all the balls is equal to the momentum the cue ball had before it struck the rest of the balls.

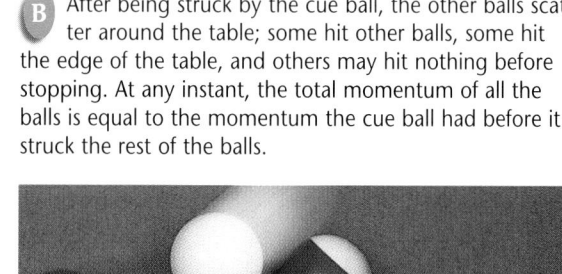

The **law of conservation of momentum** states that the total amount of momentum of a group of objects does not change unless outside forces act on the objects. After the collision, the balls on the pool table eventually slow down and stop rolling. What outside force makes that happen?

With Newton's third law and conservation of momentum, you can explain many types of motion that may seem complicated at first. Bouncing on a trampoline, knocking down bowling pins with a ball, and tackling a football player are a few examples. Think about how you would explain these and other examples of motion.

C All the balls have come to a stop and the player is ready to take the next shot. The total momentum of all the balls is zero again.

 **Problem Solving**

An Icy Challenge

Suppose you are playing hockey on a frozen pond with some friends when one of them presents a challenge. From the middle of the frozen pond, you are challenged to find a way to get to the ground at the edge of the pond without pushing or walking with your legs or arms. Assume the pond is nearly frictionless. What do you have with you that will help you get off of the pond?

Solve the Problem:

1. How does the statement that the pond is nearly frictionless affect your approach to solving this problem?
2. What is your momentum as you are standing in the middle of the frozen pond?

Think Critically:

1. Describe your plan for getting to the edge of the pond. Be sure to describe any forces involved in your plan.
2. What motions would take place and why would they occur?
3. How would you change your plan if you almost made it to the edge?

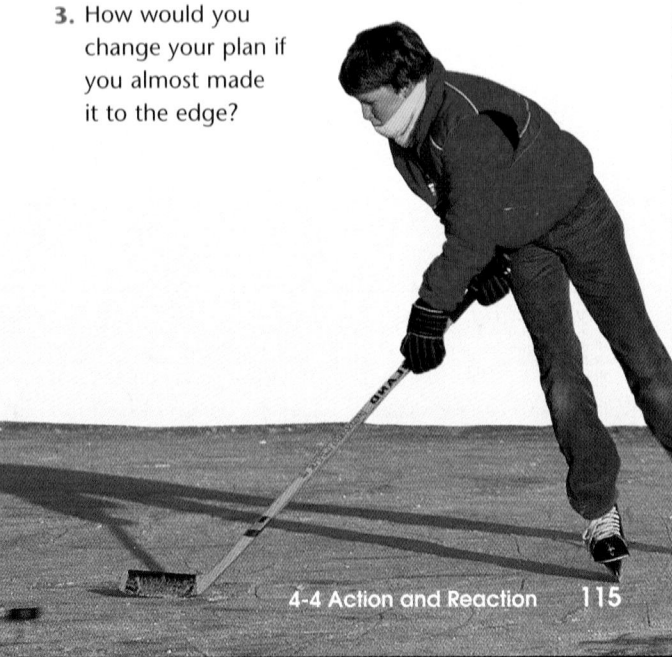

4-4 Action and Reaction   115

Use these program resources as you teach this lesson.
Critical Thinking/Problem Solving, p. 10
Lab Manual 6
Study Guide, p. 21
Science and Society Integration, p. 8

Figure 4-23

A tackle being made in football is a good example of momentum being transferred and conserved. *If a football player wanted to increase his momentum, how could he do it?*

4 Close

Most sports involve many rapid changes in momentum. The momentums of the ball carrier and the tacklers in **Figure 4-23** change when the tackle is made. Players who are moving have momentum, and the equipment they use—bats, balls, rackets, and pucks—also have their own momentum when they are put into play. In every case, the law of conservation of momentum is in effect.

Section Wrap-up

Review

1. A boater tries to jump a few feet from a boat to land. Instead, he lands in the water. Explain why.

2. Compare the momentums of a 50-kg dolphin swimming 16.4 m/s and a 6300-kg elephant walking 0.11 m/s.

3. **Think Critically:** Some ballet directors assign larger dancers to perform slow, graceful steps and smaller dancers to perform quick movements. Does this plan make sense? Why?

Skill Builder
Hypothesizing
You are a crane operator using a wrecking ball to demolish an old building. You can choose to use a 100-kg ball or a 150-kg ball. Which ball would knock the walls down faster? Which ball would be easier for you to control? Explain. If you need help, refer to Hypothesizing in the **Skill Handbook.**

Science Journal

In your Science Journal, use the law of conservation of momentum to explain the results of a particular collision you have witnessed. For example, think of games, sports, or amusement park rides or contests.

Activity 4-2

A Massive Problem

Some or all of the momentum of a billiard ball transfers to a second ball when they collide. What would happen if the second ball were a bowling ball?

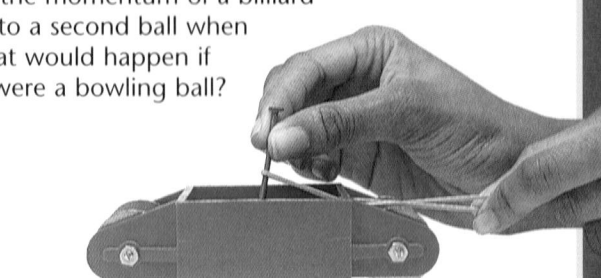

Problem
How does mass affect momentum when two objects collide?

Materials
- dynamics carts (2)
- long rubber band
- building bricks (2)
- meterstick
- masking tape

Safety Precautions 🥽
Be sure that there is a clear path between the carts when they are released.

Procedure
1. Prepare a data table like the one shown.
2. Attach a long rubber band to the two dynamics carts. Move the carts apart until the rubber band is taut, but not stretched.
3. Mark the halfway point between the two carts with a piece of tape. Lay the meterstick on the table with the 50-cm mark beside the tape.
4. Pull the carts apart until the ends of the rubber band line up with the ends of the meterstick.
5. Release both carts at the same time so that they gain equal and opposite momentums. Observe and record the point along the meterstick where the carts collide and where the carts stop.
6. Place a brick on cart A and repeat steps 4 and 5.
7. Add a second brick to cart A and repeat steps 4 and 5.

Analyze
1. From the data, **conclude** which cart was moving faster.
2. How can both carts have equal and opposite momentums if they are traveling at different speeds?
3. How does the addition of mass to one cart affect the location of the collision point? How does it affect what happens to the carts after the collision?

Conclude and Apply
4. If one cart were released slightly before the other, how would the momentums of the two carts be affected?
5. Suppose momentum is said to be negative (−) in one direction and positive (+) in the opposite direction. If two carts with equal momentums are traveling in opposite directions, **predict** the total momentum of the two-cart system.

Data and Observations
Sample Data

	Collision Point	Distance Traveled		Location after Collision	
		Cart A	Cart B	Cart A	Cart B
Trial 1	50 cm	50 cm	50 cm	20 cm	80 cm
Trial 2	40 cm	40 cm	60 cm	15 cm	77 cm
Trial 3	20 cm	20 cm	80 cm	12 cm	68 cm

4-4 Action and Reaction   **117**

Purpose
IS **Interpersonal** Students will compare the roles of mass and velocity in the momentum of an object. **L1** **LEP** **COOP LEARN**

Process Skills
observing, separating and controlling variables, forming operational definitions, recognizing cause and effect, making and using tables, comparing and contrasting, interpreting data, measuring in SI, using numbers

Time
30 minutes

📁 **Activity Worksheets,** pp. 5, 22-23

Teaching Strategies
Troubleshooting The lab setup requires about 2 m of counter or table space.

- Keep extra rubber bands handy. If the extra-long type is not available, link together several shorter ones.
- Lighter carts than those pictured can be used but may not survive the collisions without damage.
- If appropriate, the lab can lead to a discussion of momentum within a stationary or a moving frame of reference. (What would be the momentum of the carts if this lab were conducted inside a moving railroad car?)

Answers to Questions
1. Because both carts traveled for the same period of time, the cart that traveled farther had the higher average velocity.
2. The faster cart has a lower mass.
3. The collision point will be closer to the origin of cart A. Cart A will move less distance from the collision than will cart B.
4. The cart released first will gain the greater momentum.
5. The total momentum is zero.

✔ Assessment
Oral Ask students to discuss the question about the bowling ball in the introductory paragraph. Use the Performance Task Assessment List for Making Observations and Inferences in **PASC**, p. 17. **P**

Background

- The popularity of pool used to be bad news for elephants. In 1920, 10 000 elephants were killed so that their ivory tusks could be made into pool balls. Each tusk yielded five balls. Today, most pool balls are made of cast phenolic-resin. However, manufacturers still favor ivory as a component of pool cues because ivory expands and contracts at the same rate as the wood in the cues. The use of ivory may also be phased out in cues, possibly by substituting a paper-based epoxy resin.

- Where did the game of pool originate? France, England, China, Italy, and Spain have all been named as possible sites, but that question remains unanswered. The earliest known written reference to the game in Europe is from the 15th century.

Teaching Strategies

IS Logical-Mathematical Pool is a game that can be enjoyed by many people who have physical limitations. For example, the Silver Spokes players, a group in Phoenix, Arizona, play pool from wheelchairs. Challenge students to think of ways to adapt the game to accommodate players with various physical problems.

IS Kinesthetic A good pool player puts chalk on the tip of his or her cue. Bring in a cue and some chalk and have students practice chalking up. Challenge them to figure out the reason for doing this. *The friction provided by the rough chalk keeps the cue tip from slipping off the cue ball.* L1

SAMMY CHAN, *Pool Player*

On the Job

Q Mr. Chan, what scientific principles do you take into consideration as you plan shots?

A I often have to figure out angles and hit the ball a little off center. A difference of just millimeters in where I hit the ball will widen or narrow the angle and make all the difference in a shot. Often, I rehearse a shot in my mind, visualizing a clock face on the cue ball. This gives me a specific spot to aim for.

Q What does planning well result in?

A Sometimes, I set up three balls at the end of the table, one by each pocket and one in the middle. If I hit the cue ball just right, I can sink the left-hand ball, curve around the middle ball, and sink the right-hand ball.

Q Why is chalk put on the tip of a pool cue?

A Chalk roughens the tip of the cue and increases friction when the tip of the pool cue hits the shiny, smooth surface of the cue ball. Even weather affects a game. The felt on the table swells in hot, humid weather and becomes flatter and tighter in cold weather.

Personal Insights

Q Outside of the fun of the game of pool, why are you attracted to it?

A Pool is a game of discipline and planning. Actually, it's a lot like chess because you have to plan ahead. I never make a shot without planning my next several shots. I want to know exactly where the cue ball will end up after a shot.

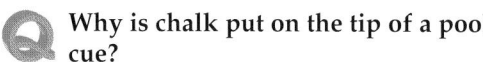

Career Connection

Think about science-related careers that require good hand-and-eye coordination and planning. Investigate your choices in the *United States Government Handbook of Careers,* available in any public library.

Choose someone in your community to interview who is involved in one of these careers and present a summary of the interview in class. Some of these careers might be:

- **Hygienist**
- **Mechanical Engineer**
- **Civil Engineer**

118 Chapter 4 Acceleration and Momentum

Career Connection

Career Path Professional pool players play competitively, and many own or manage pool halls. Some teach other aspiring players.

- High School: Geometry

For More Information

Students can contact the United States Pool Players Association for more information about the game.

Chapter 4 Review

Summary

4-1: Accelerated Motion

1. According to Newton's second law, a net force acting on an object causes the object to accelerate in the direction of the force. The size of the acceleration depends on the strength of the force and the mass of the object.
2. Near Earth's surface, gravity causes falling objects to accelerate at a rate of 9.8 m/s². Ignoring air resistance, all objects accelerate at this rate, regardless of mass.
3. Air resistance acts in the direction opposite to that in which the object is moving.

4-2: Projectile and Circular Motion

1. Objects thrown or shot through the air are called projectiles. All projectiles have both horizontal and vertical velocities. If air resistance is ignored, the horizontal velocity is constant; the vertical velocity, which is affected by gravity, increases.
2. When an object moves along a circular path, it is accelerated toward the center of the circle.
3. When an object is influenced only by gravity, it is said to be in free-fall. Objects in free-fall can be considered weightless.

4-3: Science and Society: Sending up Satellites

1. Artificial satellites are placed in Earth's orbit for communication, weather-monitoring, scientific, and military purposes.
2. Rockets are used to carry satellites up to the desired orbit height and to then give them the proper orbital velocity.

4-4: Action and Reaction

1. Forces always act in pairs. The forces in an action-reaction pair are always equal in size and opposite in direction.
2. All moving objects have momentum. The momentum of an object is the product of its mass and velocity.
3. The total momentum of a set of objects is conserved unless a net force acts on the set.

Key Science Words

a. air resistance
b. artificial satellite
c. centripetal acceleration
d. centripetal force
e. law of conservation of momentum
f. momentum
g. Newton's second law of motion
h. Newton's third law of motion
i. projectile
j. terminal velocity

Reviewing Vocabulary

Match each phrase with the correct term from the list of Key Science Words.

1. deals with action-reaction forces
2. force that opposes the motion of a falling object near Earth's surface
3. an object that is thrown through the air
4. acceleration toward the center of a circle
5. product of an object's mass and velocity
6. describes the effect of a net force on an object
7. causes circular motion
8. achieved when acceleration due to gravity is balanced by air resistance
9. describes the unchanging nature of the total momentum of a set of objects
10. human-made device that orbits Earth

Review

Summary

Have students read the summary statements to review the major concepts of the chapter.

Reviewing Vocabulary

1. h	**6.** g
2. a	**7.** d
3. i	**8.** j
4. c	**9.** e
5. f	**10.** b

✓ Assessment

Portfolio Encourage students to place in their portfolios one or two items of what they consider to be their best work. Examples include:
- Connect to Life Science, p. 95
- Science Journal, p. 105
- Problem Solving, p. 115 **P**

Performance Additional performance assessments may be found in **Performance Assessment** and **Science Integration Activities**. Performance Task Assessment Lists and rubrics for evaluating these activities can be found in **Glencoe's Performance Assessment in the Science Classroom**.

GLENCOE TECHNOLOGY

⊙ Videodisc

Glencoe Physical Science
Interactive Videodisc
Use videodisc lesson 2 *Machines and Forces* to review the principles of mass and acceleration.

MindJogger Videoquiz

Chapter 4 Have students work in groups as they play the Videoquiz game to review key chapter concepts.

Chapter 4 Review

Checking Concepts

1. b   6. c
2. d   7. a
3. b   8. d
4. c   9. b
5. d   10. a

Understanding Concepts

11. The greater the mass of an object, the greater the force of gravity that acts on it and the greater its weight.

12. The rate at which gravity causes a falling object to accelerate is independent of the mass of the object. Although the force of gravity acting on a large mass at the same distance from Earth is greater than for a small mass, the greater inertia of the larger mass requires that more force be exerted on it to change its motion.

13. The forces in an action-reaction pair act on different objects; therefore, they cannot be balanced by each other.

14. Although the spaceship is kept from flying off into space by Earth's gravity, the spaceship and everything in it are in free-fall around Earth. This condition produces the impression of weightlessness because there is no upward force to act against gravity.

15. A marble rolling across a table has no net force acting on it to cause it to change its motion. Once the marble rolls off the edge of the table, gravity is the only force acting on it, and the marble becomes a projectile. Because the marble has a constant horizontal velocity and a constant vertical acceleration downward, the path of the marble is curved.

Checking Concepts

Choose the word or phrase that completes the sentence or answers the question.

1. A net force acting on a moving object causes the object to _____.
   a. fall   c. stop
   b. accelerate   d. curve

2. _____ is the force of gravity on an object.
   a. Mass   c. Centripetal force
   b. Momentum   d. Weight

3. Which of these opposes acceleration due to gravity?
   a. momentum   c. reaction force
   b. air resistance   d. terminal velocity

4. According to Newton's second law, _____ equals mass times acceleration.
   a. gravity   c. force
   b. momentum   d. weight

5. What force causes a leaf to fall more slowly than a penny?
   a. gravity   c. inertia
   b. momentum   d. air resistance

6. Which best illustrates Newton's third law?
   a. projectile motion   c. rocket propulsion
   b. circular motion   d. centripetal force

7. The _____ velocity of a projectile is considered to be constant.
   a. horizontal   c. accelerated
   b. circular   d. vertical

8. An object in free-fall can be considered _____.
   a. moving horizontally   c. motionless
   b. heavy   d. weightless

9. _____ is reached when air resistance and force due to gravity are equal.
   a. Negative acceleration
   b. Terminal velocity
   c. Centripetal acceleration
   d. Weightlessness

10. Which of the following does not affect the amount of air resistance that acts on an object?
   a. mass   c. shape
   b. size   d. speed

Understanding Concepts

Answer the following questions in your Science Journal using complete sentences.

11. On Earth, why does an object of large mass weigh more than an object of smaller mass?

12. Explain why gravity does not cause a falling object of large mass to accelerate at a faster rate than a falling object of smaller mass.

13. If the forces in an action-reaction pair are equal in size and opposite in direction, why aren't they balanced forces?

14. A spaceship orbiting Earth is held in its orbit by Earth's gravity. Yet, astronauts in the spaceship are said to be weightless. Explain.

15. Explain why a marble moves in a straight line as it rolls across the table but follows a curved path once it rolls off the table.

Thinking Critically

16. What force is exerted on a 1000-kg car accelerating at a rate of 15 m/s²?

17. The motion of a 12-kg object is opposed by a 30-N force of friction. At what rate does friction slow the object down?

18. You are asked to design a winding mountain road. What force must you try to increase in designing this road? How might you do this?

19. A 4-kg bowling ball rolling at 6 m/s collides head-on with an identical, motionless bowling ball. If the first ball is moving forward at 2 m/s right after the collision, what is the speed and direction of the second ball?

Thinking Critically

16. $F = m \times a$
$$= 1000 \text{ kg} \times 15 \text{ m/s}^2$$
$$= 15\ 000 \text{ N}$$

17. $F = m \times a$
$$a = \frac{F}{m}$$
$$= \frac{30 \text{ N}}{12 \text{ kg}} = \frac{30 \text{ kg} \bullet \text{m/s}^2}{12 \text{ kg}}$$
$$= 2.5 \text{ m/s}^2$$

18. Centripetal force must be made as large as possible. This force is affected by the force of friction between the road surface and the tires and by the shape of the curves. Therefore, you could use a slightly rougher material for the road surface to increase friction. The road surface could also be angled upward in the outside parts of the curves.

20. The moon has no atmosphere and its gravity is about one-sixth as strong as that of Earth. Considering these factors, how would the motions of objects near the moon be different from the same motions near Earth?

Developing Skills

If you need help, refer to the **Skill Handbook**.

21. **Interpreting Data:** The table contains data about four objects dropped to the ground from the same height at the same time.
    a. Which object falls fastest? Slowest?
    b. On which object is the force of gravity greatest?
    c. Is air resistance stronger on A or B?
    d. Which object is probably largest in size? Explain your reasoning.

Time of Fall for Dropped Objects

Object	Mass	Time of Fall
A	5.0 g	2.0 s
B	5.0 g	1.0 s
C	30.0 g	0.5 s
D	35.0 g	1.5 s

22. **Observing and Inferring:** A girl weighing 360 N steps into a motionless elevator at the ground floor. She steps onto a scale and remains on the scale while the elevator accelerates, then moves at a constant speed, and finally slows to a stop at the 120th floor. During that time, the readings on the scale range from a high of 365 N to a low of 355 N. Infer what the reading was on the ground floor, on the way up, stopping, and stopped on the 120th floor. Why do these readings change?

23. **Recognizing Cause and Effect:** When using a high-pressure hose, why is it necessary for firefighters to grip the hose strongly and plant their feet firmly?

24. **Concept Mapping:** Make two events chains showing what happens when a rolling ball (Ball One) hits a resting ball (Ball Two). Use the phrases: *gains momentum, rests, rolls more slowly or stops, hits Ball Two, rolls, loses momentum, is hit by Ball One, starts rolling.*

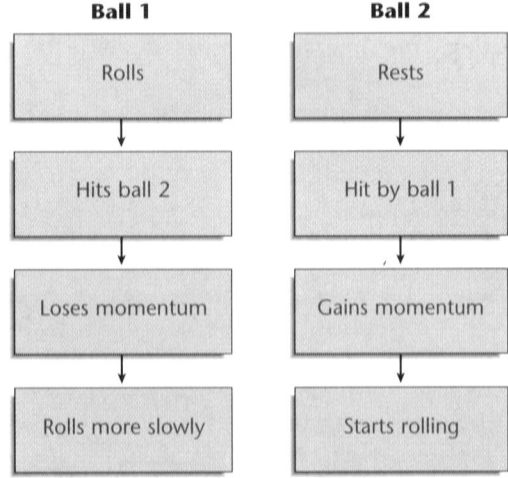

Ball 1

- Rolls
- Hits ball 2
- Loses momentum
- Rolls more slowly

Ball 2

- Rests
- Hit by ball 1
- Gains momentum
- Starts rolling

Performance Assessment

1. **Designing an Experiment:** Review the MiniLAB on page 112. Devise a way to make the rocket travel farther by using two balloons. Use physics terms to explain why this could be effective.

2. **Analyzing the Data:** Look back at your results from Activity 4-2 on page 117. Suppose you have cart A carrying a load that makes it twice as massive as cart B. Under what conditions could both carts have the same momentum?

3. **Poster:** Make a poster showing how a rocket engine works. Use Newton's third law to explain why a rocket lifts off.

Chapter 4 Review

19. initial momentum = final momentum
    $p_i = p_f$
    before collision, $p_i =$
    $m_1 v_1 + m_2 v_2$
    $= (4 \text{ kg} \cdot 6 \text{ m/s}) +$
    $(4 \text{ kg} \cdot 0 \text{ m/s})$
    $= 24 \text{ kg} \cdot \text{m/s}$
    after collision, $p_f =$
    $m_1 v_1 + m_2 v_2 =$
    $24 \text{ kg} \cdot \text{m/s}$
    $v_2 = (p_f - m_1 v_1)/m_2$
    $= [24 - (4 \text{ kg} \cdot 2 \text{ m/s})]/$
    4 kg
    $= 4 \text{ m/s}$
    Direction is forward, in the same direction as the first ball.

20. Because of the moon's smaller gravity, acceleration of a falling object would be smaller on the moon than on Earth. However, without air resistance a falling object would accelerate at the same rate until it struck the moon's surface.

Developing Skills

21. **Interpreting Data**
    a. Object C falls fastest, object A slowest
    b. Object D c. Object A
    d. Object D because it has more mass but falls slowly

22. **Observing and Inferring** As the elevator begins to go up, the scale must exert an upward force greater than the girl's weight to produce a net upward force. As the elevator slows at the top, the scale must exert a force less than the girl's weight to produce a net downward force. Any time the elevator is not accelerating, the scale must produce a force equal to the girl's weight.

23. **Recognizing Cause and Effect** The force of the water may knock them down (Newton's third law).

24. See student page.

Performance Assessment

1. The rocket is propelled forward by the force of the air molecules on the balloons. Using two balloons doubles the force. Also, because momentum is conserved, the momentum of the escaping air equals the momentum of the rocket. Use the Performance Task Assessment List for Designing an Experiment in **PASC,** p. 23. **P**

2. Cart B must have twice the velocity of cart A. Use the Performance Task Assessment List for Analyzing the Data in **PASC,** p. 27. **P**

3. Burning fuel exerts forces on the interior of the rocket. The interior of the rocket exerts equal and opposite forces on the burning fuel. The upward force of expanding gases lifts the rocket. Use the Performance Task Assessment List for Poster in **PASC,** p. 73. **P**

Chapter Organizer

Section	Objectives/Standards	Activities/Features
Chapter Opener		**Explore Activity:** Think about the forces and energy involved as you try this physical workout. p. 123
5-1 **Energy and Work** (2 sessions, 1 block)*	1. **Distinguish** between kinetic and potential energy. 2. **Recognize** that energy is conserved when changing from one form to another. 3. **Compare** the scientific meaning of *work* with its everyday meaning. **National Content Standards:** (5-8) UCP1-UCP3, UCP5, A1, A2, B2, B3, E1, G1, G3; (9-12) UCP1-UCP3, UCP5, A1, A2, B4, B5, D1, E1, G1, G3	**Connect to Earth Science,** p. 125 **Using Math:** Calculating Work, p. 127 **Connect to Life Science,** p. 129 **Using Math,** p. 130 **Science Journal,** p. 131 **Skill Builder:** Comparing and Contrasting, p. 131 **Activity 5-1:** Swinging Energy, pp. 132-133
5-2 **Temperature and Heat** (1 session, 1 block)*	1. **Recognize** the difference between the motion of an object and the motion of the particles that make up the object. 2. **Contrast** heat and temperature. 3. **Explain** what determines the thermal energy of a sample of matter. **National Content Standards:** (5-8) UCP2, UCP3; (9-12) UCP2, UCP3	**MiniLAB:** Can shaking increase the temperature of sand? p. 135 **Problem Solving:** A Cold Cup of Hot Chocolate ...and a Warm Soft Drink, p. 136 **Science Journal,** p. 137 **Skill Builder:** Concept Mapping, p. 137
5-3 **Science and Society** (1 session, ½ block)*	1. **Identify** some causes and effects of thermal pollution. 2. **Discuss** possible solutions for thermal pollution problems. **National Content Standards:** (5-8) UCP2, UCP3, UCP5, F2-F5; (9-12) UCP2, UCP3, UCP5, F1-F6	**Explore the Issue,** p. 140 **Using Computers,** p. 140
5-4 **Measuring Thermal Energy** (2 sessions, 1 block)*	1. **Define** *specific heat.* 2. **Calculate** changes in thermal energy. **National Content Standards:** (5-8) UCP1-UCP3, UCP5, A2, C4, C5, F2, F5, G3; (9-12) UCP1-UCP3, UCP5, A2, C5, D1, F3, F6, G3	**Using Technology:** Infrared Weather Eyes, p. 142 **Using Math:** Calculating Changes in Thermal Energy, p. 143 **Connect to Chemistry,** p. 144 **Using Math,** p. 144 **Skill Builder:** Interpreting Data, p. 144 **Activity 5-2:** A Hot Topic, p. 145 **Science & History:** Inuit Clothing: Life on Top of the World, p. 146 **Science Journal,** p. 146

* A complete Planning Guide that includes block scheduling is provided on pages 32T-35T.

Activity Materials

Explore	Activities	MiniLAB
page 123 4 books	pages 132-133 ring stand and ring, right-angle support rod clamp, 30-cm support rod, 2-hole medium rubber stopper, 1m of string, metersticks, graph paper page 145 set of electric hair curlers, 2 Celsius thermometers, plastic foam cup, graph paper, balance, graduated cylinder	page 135 sand, plastic cup with tight-fitting lid, thermometer

Need Materials? Call Science Kit (1-800-828-7777).

Teacher Classroom Resources

Reproducible Masters	Transparencies	Teaching Resources

Reproducible Masters

Study Gui~~de~~
Reinforce
Enrichme
Activity W
Lab Manu~~a~~
  Pendulu~~m~~
Lab Manua
Science Inte
  on the Riv
Cross-Curric
Science and

Study Guide,
Reinforcement
Enrichment, p.
Activity Worksh
Multicultural C

Study Guide, p. 2
Reinforcement, p.
Enrichment, p. 24
Critical Thinking/P

Study Guide, p. 25
Reinforcement, p. 25
Enrichment, p. 25
Activity Worksheets,
Lab Manual 10, Heat
Concept Mapping, pp

Transparencies

Section Focus Transparency 18, Work It Out
Science Integration Transparency 5, Energy Changes in Earth's Yearly Voyage
Teaching Transparency 9, Kinetic and Potential Energy

~~S~~ction Focus Transparency 19, How Hot Is It?

~~Secti~~on Focus Transparency 20, Keep ~~You~~r Cool

~~Section~~ Focus Transparency 21, Why ~~Does~~ This Heat Faster?
~~Teaching~~ Transparency 10, Calorimeter ~~Applic~~ations

Teaching Resources

Physical Science CD-ROM
Spanish Resources
English/Spanish Audiocassettes
Cooperative Learning Resource Guide
Lab Partner
Lab and Safety Skills
Lesson Plans

Assessment Resources

Chapter Review, pp. 13-14
Assessment, pp. 28-31
Performance Assessment in the Science Classroom (PASC)
MindJogger Videoquiz
Alternate Assessment in the Science Classroom
Performance Assessment
Chapter Review Software
Computer Test Bank

Overlaid note card — MINI·QUIZ:

MINI·QUIZ

Use the Mini Quiz to check students' recall of chapter content.

1. **What is energy?** *the ability to cause change*
2. **What is the unit of energy?** *joule*
3. **What is potential energy?** *stored energy*
4. **What is kinetic energy?** *energy in the form of motion*
5. **What is work?** *the transfer of energy through motion*
6. **What unit measure work?** *joule*

Section Wrap-up

Review
1. When you drop the ba~~ll~~ its potential energy ~~is~~ changing to kinetic e~~n~~ergy. When the ball hi~~ts~~ the floor and bounces u~~p~~ward, kinetic energy ~~is~~ changing back to pote~~n~~tial energy. The potent~~ial~~ energy becomes less w~~ith~~ each bounce. When t~~he~~ ball stops bouncing, ~~the~~ energy has been tra~~ns~~formed into heat, sou~~nd~~ etc.

2. $W = F \times d$
   $$F = \frac{W}{d} = \frac{1470 \text{ J}}{20 \text{ m}}$$
   $$= 73.5 \text{ N}$$

3. **Think Critically** It is ~~mo~~re available from Glencoe.
   energy,
   burned

Key to Teaching Strategies

The following designations will help you decide which activities are appropriate for your students.

L1 Level 1 activities should be within the ability range of all students, including those with learning difficulties.

L2 Level 2 activities should be within the ability range of the average to above-average student.

L3 Level 3 activities are designed for the ability range of above-average students.

LEP LEP activities should be within the ability range of Limited English Proficiency students.

LS These activities are designed to address different learning styles.

COOP LEARN Cooperative Learning activities are designed for small group work.

P These strategies represent student products that can be placed into a best-work portfolio.

The following m~~aterials~~ ~~a~~re available from Glencoe.

Science and Technology Series (STVS)
Physics
  Seismic Simulator
  Images of Heat
Chemistry
  Solar Food Dryer
Ecology
  Greenhouse Effect

The Infinite Voyage Series
A Taste of Health
Living with Disaster

National Geographic Society Series
GTV: Planetary Manager
STV: Atmosphere

Physical Science CD-ROM

Teacher Classroom Resources

This is a representation of key blackline masters available in the Teacher Classroom Resources.

Teaching Aids

Section Focus Transparencies

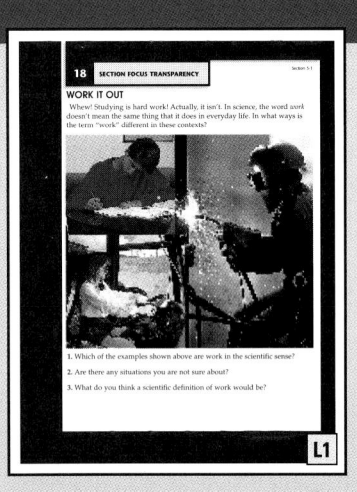

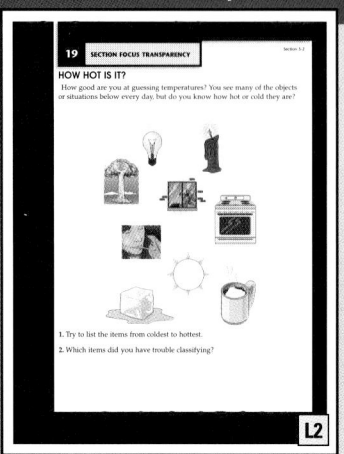

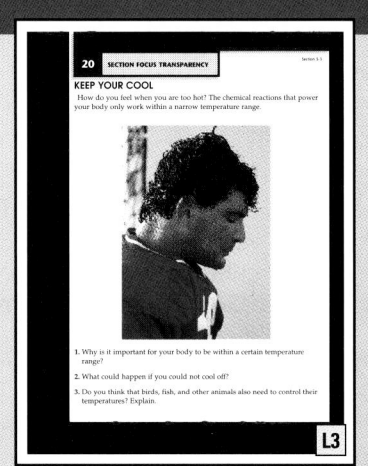

Science Integration Transparencies

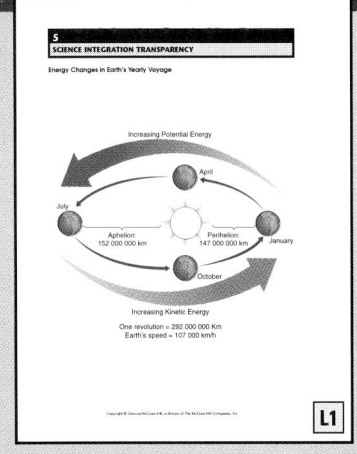

Teaching Transparencies

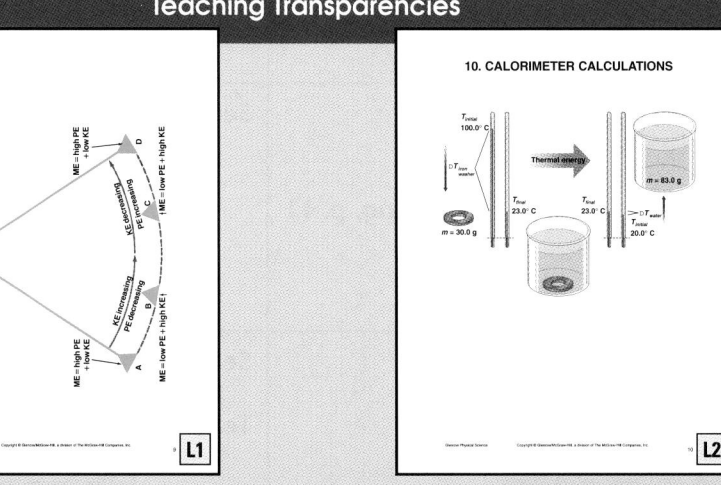

Meeting Different Ability Levels

Study Guide

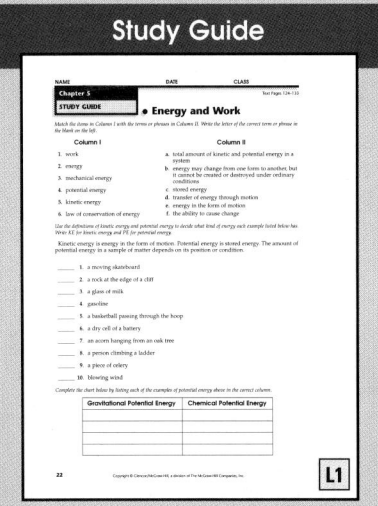

Reinforcement

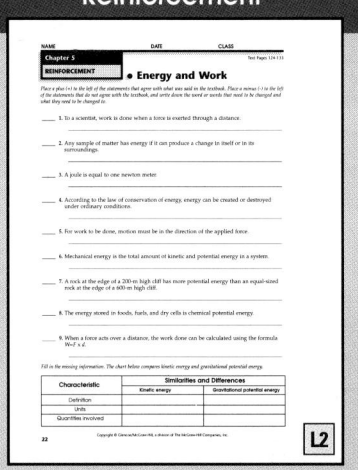

Enrichment Worksheets

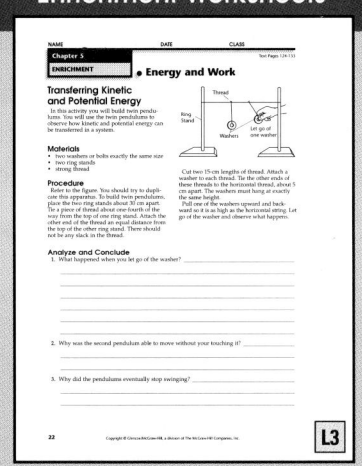

Hands-On Activities

Science Integration Activity

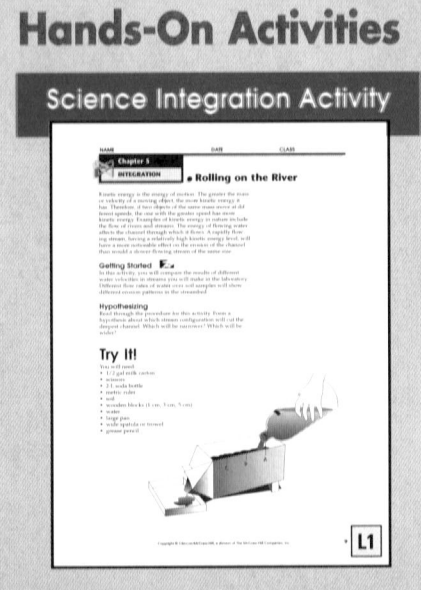

L1

Lab Manual

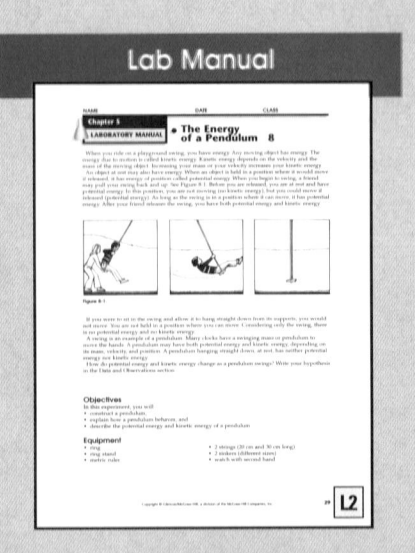

L2

Assessment

Performance Assessment

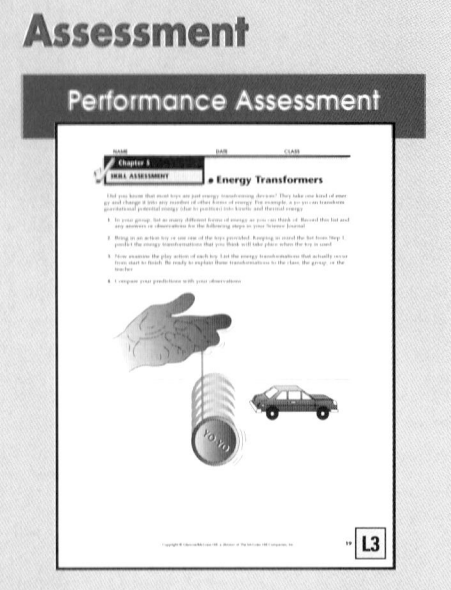

L3

Enrichment and Application

Critical Thinking/Problem Solving

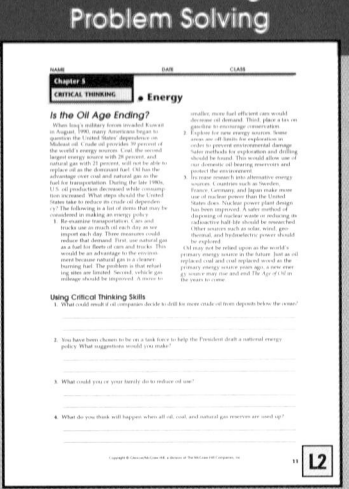

L2

Cross-Curricular Integration

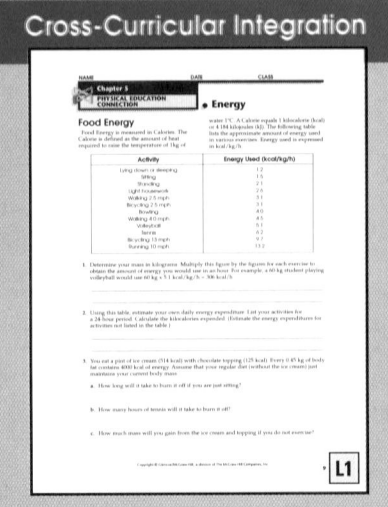

L1

Science and Society Integration

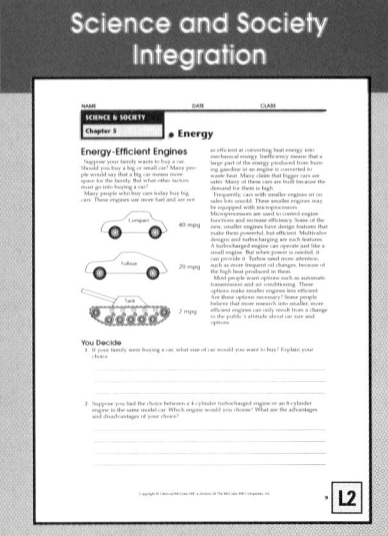

L2

Multicultural Connections

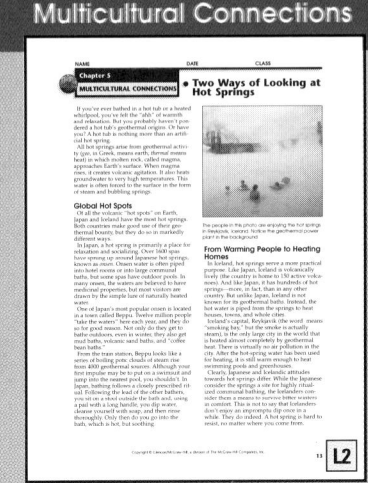

L2

Concept Mapping

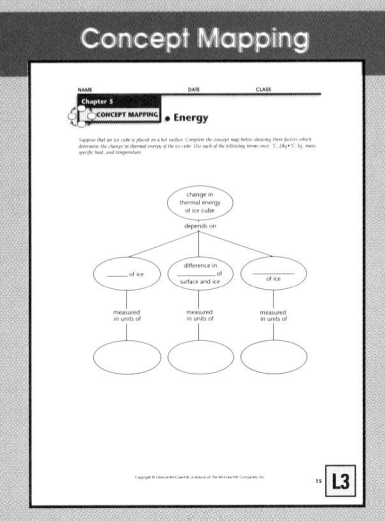

L3

Energy

CHAPTER OVERVIEW

Section 5-1 Kinetic and potential energies are discussed qualitatively. Work is introduced as the means of transferring energy through motion. Mechanical energy in a swing is used to introduce the law of conservation of energy.

Section 5-2 This section presents the relationship between the temperature of a material and the average kinetic energy of its particles. The difference between thermal energy and heat is discussed.

Section 5-3 Science and Society The concept of thermal pollution is introduced. Two points of view are presented regarding the desirability of regulating industry to reduce the pollution.

Section 5-4 Specific heat is introduced as a means of measuring changes in thermal energy.

Chapter Vocabulary

energy	temperature
kinetic energy	thermal
potential	energy
energy	heat
work	thermal
mechanical	pollution
energy	specific heat
law of conser-	
vation of energy	

Theme Connection

Energy The theme of this chapter is energy and its use in describing how matter behaves. Energy as the ability to cause a change is introduced and should be emphasized in each section. Mechanical energy is discussed and is used to introduce the concept of thermal energy at the microscopic level.

Previewing the Chapter

Section 5-1   Energy and Work
▶ What is energy?
▶ What is work?
▶ Conservation of Energy
▶ The Human Body—Balancing the Energy Equation

Section 5-2   Temperature and Heat
▶ Temperature

▶ Thermal Energy
▶ Heat

Section 5-3   Science and Society
Issue: Thermal Pollution: Waste You Can't See

Section 5-4   Measuring Thermal Energy
▶ Specific Heat
▶ Calculating Thermal Changes

122

Learning Styles

Look for the following logo for strategies that emphasize different learning modalities. **LS**

Kinesthetic	Explore, p. 123; MiniLAB, p. 135
Visual-Spatial	Demonstration, pp. 125, 127, 128; Reteach, p. 130; Visual Learning, pp. 125, 127, 129, 130, 134, 136, 138, 144
Interpersonal	Activity 5-1, pp. 132-133
Intrapersonal	Science Journal, p. 127; Activity, p. 139
Logical-Mathematical	Inclusion Strategies, p. 128; Activity 5-2, p. 145
Linguistic	Across the Curriculum, pp. 125, 126, 143; Close Activity, p. 140

Chapter

5

Energy

How many ways do you use your own energy every day? Every move you make requires the release of energy stored in your body. Some tasks, such as exercising on a stair climber, require more energy than others. In this chapter, find out about several different forms of energy and how they can be measured. In the activity below, observe how dependent you are on a ready source of energy.

EXPLORE ACTIVITY

Think about the energy involved as you try this workout.

1. Stand and hold your arms with your hands together, palms up, extended at waist level. Have a classmate stack books on your hands.
2. Raise the books to shoulder level, then lower them. Now try lifting them over your head.
3. Have your classmate add two more books to your pile. Repeat the lifting exercise.
4. Hold the books at shoulder level until your arms get tired.

Observe: Which activities require the most energy? Describe them in your Science Journal.

Previewing Science Skills

► In the Skill Builders, you will **compare and contrast, map concepts,** and **interpret data.**

► In the Activities, you will **measure in SI, observe and infer, collect and interpret data, make and use graphs,** and **hypothesize.**

► In the MiniLAB, you will **measure in SI, observe and infer,** and **hypothesize.**

123

EXPLORE ACTIVITY

Purpose
IS Kinesthetic Use the activity to introduce students to the concepts of work and energy. Tell them that they will learn more about these concepts and how they are related in this chapter. **L1** **LEP**

Preparation
Use sets of similar books.

Materials
four books per pair of students

Teaching Strategies
• **CAUTION:** *Students with health problems should limit participation in this activity.*
• Have students raise or lower the books each time at a slow, constant speed.
• During this activity, ask students to state when work is done and when work is increased.

Observe
Lifting the books over the head requires the most energy.

✔ Assessment

Process Assign Problem Solving Teams to conduct this activity quantitatively. Have students measure the weight of the books or calculate the weight from mass measurements of the books. Use the Performance Task Assessment List for Making Observations and Inferences in **PASC,** p. 17. **P**

Assessment Planner

Portfolio
Refer to page 147 for suggested items that students might select for their portfolios.

Performance Assessment
See page 147 for additional Performance Assessment options.
Skill Builders, pp. 131, 137, 144
MiniLAB, p. 135
Activities 5-1, pp. 132, 133; 5-2, p. 145

Content Assessment
Section Wrap-ups, pp. 131, 137, 140, 144
Chapter Review, pp. 147-149
Mini Quizzes, pp. 131, 137

Group Assessment
Opportunities for group assessment occur with Cooperative Learning Strategies and Flex Your Brain Activities.

Section 5•1

Prepare

Section Background

- Energy is defined as the ability of an object or phenomenon to cause a change. A moving baseball has energy because you can see and feel the change in motion of your hand as you catch it. Light is a form of energy because it can change pigments in photographic film and in the retina of the eye and can cause sunburn.

- An object can be described as having kinetic energy, a measure of its motion, and potential energy, a measure of its position or condition. Mechanical energy for an object is the sum of its kinetic and potential energies.

Preplanning

To prepare for Activity 5-1, obtain ring stands, rings, support rods and clamps, rubber stoppers, string, metersticks, and graph paper.

1 Motivate

Bellringer

 Before presenting the lesson, display **Section Focus Transparency 18** on the overhead projector. Assign the accompanying **Focus Activity** worksheet. L1 LEP

Tying to Previous Knowledge

Have students recall the conservation of momentum law from Section 4-4 and give examples. Allow them to explain the concept of conservation. Tell them that in this section they will learn about another conservation law, that of energy.

Science Words

- energy
- kinetic energy
- potential energy
- work
- mechanical energy
- law of conservation of energy

Objectives

- Distinguish between kinetic and potential energy.
- Recognize that energy is conserved when changing from one form to another.
- Compare the scientific meaning of *work* with its everyday meaning.

What is energy?

You have seen several examples of energy being used today. Just about everything you see or do involves energy. Energy is a bit mysterious. You can't smell it. In most cases, you can't even see it. Light is one form of energy you can see and without it, you wouldn't be able to see anything. You can't see electricity, but you can see its effects in a glowing lightbulb, and you can feel its effects in the heat produced by the coils of a toaster. You can't see the energy in the food in **Figure 5-1,** but you can see and feel its effects when your muscles use that energy to move.

Energy—An Agent of Change

If a baseball flies through the air and shatters a window, it certainly changes the window! When an object is able to change its environment, we say the object has energy. The baseball had energy and did work on the window by causing the window to move. In summary, the use of energy involves change.

Scientists have some difficulty defining energy because it exists in so many different forms. Some of these forms include radiant, electrical, chemical, thermal, and nuclear energy. Traditionally, energy has been defined as the ability to do work—to cause something to move. But when work is performed, there is always a change. This connection offers a useful general definition. **Energy** is the ability to cause change.

Figure 5-1

A nutritious, well-balanced lunch provides the body with "fuel." Energy stored in food is converted to a form that the body can use.

124 Chapter 5 Energy

Program Resources

📁 **Reproducible Masters**
Study Guide, p. 22 L1
Reinforcement, p. 22 L1
Enrichment, p. 22 L3
Cross-Curricular Integration, p. 9
Science and Society Integration, p. 9
Activity Worksheets, pp. 27-28 L1
Science Integration Activities, pp. 9-10
Lab Manual 8, 9

🔊 **Transparencies**
Section Focus Transparency 18 L1
Science Integration Transparency 5
Teaching Transparency 9

Kinetic Energy: On the Move

Usually, when you think of energy, you think of action—of some motion taking place. **Kinetic energy** is energy in the form of motion. A spinning bicycle wheel, the sprinting cross-country runners in **Figure 5-2,** and a flying plastic disk all have kinetic energy. How much? That depends on the mass and velocity of the moving object.

The greater the mass a moving object has, the more kinetic energy it has. Similarly, the greater an object's velocity, the more kinetic energy it has. The truck traveling at 100 km/h in **Figure 5-3** has more kinetic energy than the motorcycle traveling at the same speed. However, the motorcycle has more kinetic energy than an identical motorcycle moving at 80 km/h.

Potential Energy: Ready and Waiting

Energy doesn't have to involve motion. Even motionless, any sample of matter may have stored energy that gives it the potential to cause change if certain conditions are met. **Potential energy** is stored energy. The amount of potential energy a sample of matter has depends on its position or its condition.

A flowerpot sitting on a second-floor windowsill has potential energy due to its position. If something knocks it off the windowsill, gravity will cause it to fall toward the ground. As it falls, its potential energy changes to kinetic energy.

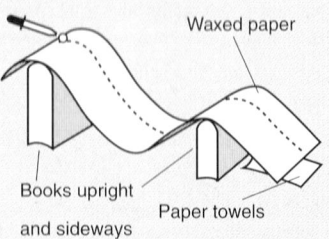

Figure 5-2

The runners are using energy at a high rate. *What types of activities require only a little energy?*

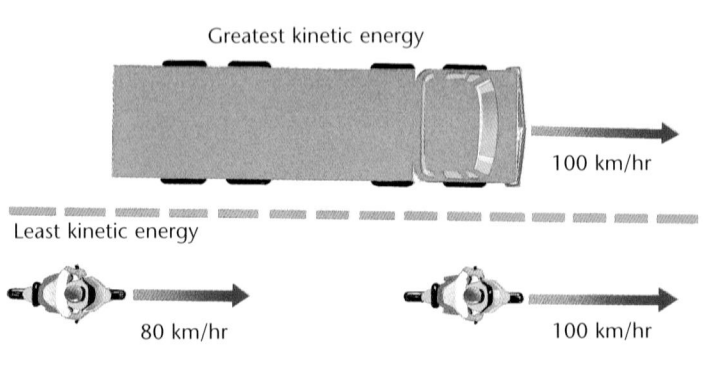

Greatest kinetic energy

100 km/hr

Least kinetic energy

80 km/hr   100 km/hr

Figure 5-3

The kinetic energy of each vehicle is different because kinetic energy depends on mass and velocity.

CONNECT TO
EARTH SCIENCE

Fast-flowing rivers and slow-moving glaciers both contain kinetic energy. A rock balanced high on a hill contains potential energy. *Hypothesize* other examples of kinetic and potential energy in nature.

5-1 Energy and Work   125

Figure 5-4

Objects that can fall have one form of potential energy.

The potential energy of the flowerpot in **Figure 5-4** is related to its distance above the ground. The greater its height, the greater its potential energy. A flowerpot sitting on a fifth-floor ledge has more potential energy than one sitting on a lower ledge.

If a flowerpot falls, the force of gravity accelerates it downward. The higher the flowerpot, the greater its final velocity will be. Thus, a flowerpot falling from a higher floor will have a higher velocity and more kinetic energy when it hits the ground than a similar pot falling from a lower floor.

You have learned about kinetic energy and some forms of potential energy. Other forms of potential energy will be discussed in later chapters of this book. In the following paragraphs, find out how the concepts of energy and work are related.

What is work?

To most people, the word *work* means something they do to earn money. In that sense, work can be anything from filling fast-food orders or loading trucks to teaching or doing office work at a desk. The scientific meaning of work is more specific. **Work** is the transfer of energy through motion. In order for work to take place, a force must be exerted through a distance.

Figure 5-5

Work is done on the books when they are being lifted from the floor. The young man is not performing work on his books.

Calculating Work

In which case would you do more work—lifting a pack of gum from the floor to waist level or lifting a pile of books through the same distance? Would more work be done if the young woman in **Figure 5-5** lifted the books from the floor to her waist or if she raised them all the way over her head? The amount of work done depends on two things: the amount of force exerted and the distance over which the force is applied.

When a force acts over a distance *in the direction of an object's motion*, the work done can be calculated as

$$Work = force \times distance$$

$$W = F \times d$$

Work, like energy, is measured in joules. The joule is named for the British scientist, James Prescott Joule. One joule is equal to a newton-meter (N · m), which is the amount of work done when a force of one newton acts through a distance of one meter. The stockperson in **Figure 5-6** is doing work. In the problems below, calculate the work done.

Figure 5-6
Stocking shelves is an important job in grocery stores. *How would you determine the amount of work done in stocking the shelves?*

USING MATH

Calculating Work

Example Problem:
A student's full backpack weighs 30 N. She lifts it from the floor to a shelf 1.5 m high. How much work is done on the pack full of books?

Problem-Solving Steps:
1. What is known?   Backpack weight, $F = 30$ N; distance, $d = 1.5$ m
2. What is unknown?   Work, W
3. Use the equation $W = F \times d$.
4. **Solution:** $W = 30$ N $\times 1.5$ m
   $= 45$ N · m
   $= 45$ J

Practice Problems
1. A carpenter lifts a 45-kg beam 1.2 m. How much work is done on the beam?
Strategy Hint: The joule is a derived unit. What other units make it up?
2. A dancer lifts a 400-N ballerina overhead a distance of 1.4 m and holds her there for several seconds. How much work is done during the lift? During the time the ballerina is held overhead?
Strategy Hint: Work requires a force to be exerted over a distance.

There are two factors to keep in mind when deciding whether work is being done: something has to move, and the motion must be *in the same direction* as the applied force. If you pick up a pile of books from the floor, work is done on the books. They move upward, in the direction of the applied force. If you hold the books in your arms, no work is done on the books. Some upward force is still being applied (to keep the books from falling), but no movement is taking place. Even if you carry the books across the floor at a constant speed, no work is done on the books. The force being applied to the books is still upward, or vertical, but your motion across the floor is sideward, or horizontal. In the following paragraphs, you will examine the energy changes that result from the motion of a swing.

Figure 5-7

The rider experiences energy changes as the swing moves.

KE increasing
PE decreasing

KE decreasing
PE increasing

A   B   C   D

A At the rider's highest point, the mechanical energy is entirely potential.

B As she falls toward the bottom of the path, the rider accelerates and gains kinetic energy. Because the rider is not as high, her potential energy decreases by the same amount.

C As it rises toward the opposite side, the swing begins to slow down and lose kinetic energy. As it gains height, it also increases in potential energy.

D As the rider reaches the peak and prepares to fall again in the opposite direction, the mechanical energy of the swing is entirely potential.

128   Chapter 5 Energy

Conservation of Energy

Perhaps you have ridden on a playground swing like the person in **Figure 5-7.** Try to remember what it was like swinging back and forth, high and low. Now think about the energy changes involved in such a ride.

The ride starts with a push to get you moving—to give you some kinetic energy. As the swing rises, kinetic energy changes to potential energy of position. At the top of each swing, potential energy is greatest. Then, as the swing moves downward, potential energy changes to kinetic energy. At the bottom of each swing, kinetic energy is greatest and potential energy is at its minimum.

As the swing continues to move back and forth, energy is converted from kinetic to potential to kinetic, over and over again. Taken together, the potential and kinetic energy of the swing make up its mechanical energy. **Mechanical energy** is the total amount of kinetic and potential energy in a system.

Conserving Energy—A Natural Law

Scientists have learned that in any given situation, energy may change from one form to another, but the total amount of energy remains constant. In other words, energy is conserved. This fact is recognized as a law of nature. According to the **law of conservation of energy,** energy may change form but it cannot be created or destroyed under ordinary conditions. This law applies to closed systems, in which energy cannot enter or leave the system.

Suppose the law of conservation of energy is applied to the swing. Would you expect the swing to continue moving back and forth forever? You know this doesn't happen. The swing slows down and comes to a stop. Where does the energy go?

If you think about it, friction and air resistance are constantly acting on the swing and rider just as they are also acting on the skateboarder in **Figure 5-8.** These unbalanced forces cause some of the mechanical energy of the swing to change to thermal energy—heat. With every pass of the swing, the temperatures of the swing, the rider, and the air around them go up a little bit. So the mechanical energy of the swing isn't lost, it is converted to thermal energy. Thermal energy is discussed in the next section.

CONNECT TO

LIFE SCIENCE

A roller coaster lifts you high above the ground and also moves you at high speeds. Your body experiences many changes in kinetic and potential energy. *Hypothesize* the point at which your body has the most kinetic energy.

Figure 5-8

Gravity provides the force to move the skateboarder down the ramp. *What energy changes are taking place?*

Inquiry Questions

- **Why is the first hill of a roller coaster ride the highest?** *The total mechanical energy of the ride is mostly due to the potential energy of the roller coaster at the top of this hill. The contribution to this total due to kinetic energy is very small at this point. If a later hill were higher, the mechanical energy needed would be larger than the original mechanical energy, and the coaster could not reach the top.*

CONNECT TO

LIFE SCIENCE

Answer at the lowest point on the ride

Visual Learning

Figure 5-8 **What energy changes are taking place?** *From potential energy at the top of the ramp to kinetic energy at the bottom.* LEP LS

NATIONAL GEOGRAPHIC SOCIETY

Videodisc
GTV: Planetary Manager
Energy

47474

47902

47903

47904

INTEGRATION
Life Science

The Human Body—Balancing the Energy Equation

What forms of energy discussed in this chapter can you find in the human body? With your right hand, reach up and feel your left shoulder. Even with that simple action, you participate in the conversion of stored potential energy within your body to the kinetic energy of your moving arm as you do work on it. Did your shoulder feel warm to your hand? Some of the potential energy stored in your body is used to maintain a nearly constant internal temperature. Some of this energy is also converted to the excess heat that your body gives off to its surroundings. Even the person resting in **Figure 5-9** requires energy conversions.

Figure 5-9
Even a person at rest is using energy. *Describe several ways that energy is being used in her body.*

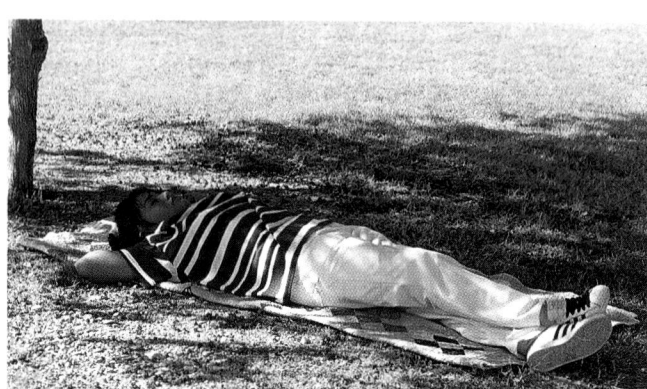

The complex chemical and physical processes of your body also obey the laws of physics, including the law of conservation of energy. In your body, stored energy (such as that found in fat) is lost when work is done or heat is lost by your body to your surroundings. To maintain a healthy weight, you must have a proper balance between energy taken in and energy lost from your body as work or heat.

Food—Our Chemical Potential Energy

What did you eat for breakfast this morning? Your body has been busy chemically changing your food into molecules that can combine with oxygen and be used as fuel. Even if you did not eat breakfast this morning, your body converts energy stored in fat for its immediate needs until you eat again. You are probably familiar with the food Calorie, a unit used by nutritionists to measure how much energy we get from specific foods. One Calorie is actually equivalent to a kilocalorie, or about 4180 joules. Every gram of fat a person consumes can supply nine Calories of energy. Carbohydrates and proteins each supply about four Calories of energy per gram.

Why does your body need such an energy supply? Food supplies the fuel used to maintain a nearly constant body temperature, to help your organs function, and to do work as you move your body. **Table 5-1** shows the amount of energy used in doing various activities. The energy needs of an individual are based upon factors such as body size, age,

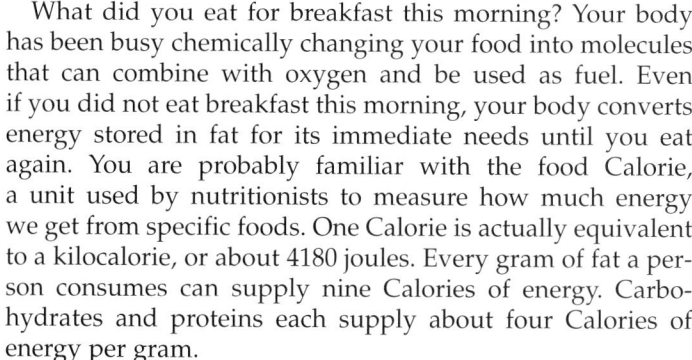

USING MATH

A small bag of potato chips has about 270 Calories. Using **Table 5-1**, calculate how many minutes you would have to walk to "burn off" these Calories.

Table 5-1

Calories Used in 1 Hour			
Type of Activity	Body Frames		
	Small	Medium	Large
Sleeping	48	56	64
Sitting	72	84	96
Eating	84	98	112
Standing	96	112	123
Walking	180	210	240
Playing tennis	380	420	460
Bicycling fast	500	600	700
Running	700	850	1000

gender, heredity, and level of daily activity. In this section, you have learned about the sources and uses of energy and the different forms it takes.

Section Wrap-up

Review

1. Imagine that you're standing on a stepladder and you drop a basketball. The first bounce will be highest. Each bounce after that will be lower until the ball stops bouncing. Describe the energy changes that take place, starting with dropping the ball.

2. A game-show contestant won a prize by pushing a bowling ball 20 m using his nose. The amount of work done was 1470 J. How much force did the person exert on the ball?

3. **Think Critically:** Much discussion has focused on the need to drive more efficient cars and use less electricity. If the law of conservation of energy is true, why are people concerned about energy usage?

Skill Builder
Comparing and Contrasting
Compare and contrast the everyday meaning of work with the scientific definition of that term. Give examples of work in the everyday sense that would not be considered work in the scientific sense. If you need help, refer to Comparing and Contrasting in the **Skill Handbook.**

Science Journal

Your body used energy to move you into the room today. Where did this energy come from? In your Science Journal, write a paragraph describing where you acquired this energy. Trace it back through as many transformations as you can.

Skill Builder
The everyday meaning of work would include any time effort was given. The scientific meaning of work would require a force being exerted through a distance. Examples of work in the everyday sense that would not meet scientific meaning are: Jim worked hard reading his English assignment, or Jennifer worked on holding the garage door open.

 Assessment

Oral Perform several tasks, such as pulling on a locked door, lifting a chair, or dropping a book. Ask the students if work is being done and to explain why or why not. Use the Performance Task Assessment List for Making Observations and Inferences in **PASC,** p. 17. **P**

4 Close

•MINI•QUIZ•

Use the Mini Quiz to check students' recall of chapter content.
1. **What is energy?** *the ability to cause change*
2. **What is the unit of energy?** *joule*
3. **What is potential energy?** *stored energy*
4. **What is kinetic energy?** *energy in the form of motion*
5. **What is work?** *the transfer of energy through motion*
6. **What unit measures work?** *joule*

Section Wrap-up

Review

1. When you drop the ball, its potential energy is changing to kinetic energy. When the ball hits the floor and bounces upward, kinetic energy is changing back to potential energy. The potential energy becomes less with each bounce. When the ball stops bouncing, its energy has been transformed into heat, sound, etc.

2. $W = F \times d$
$F = \dfrac{W}{d} = \dfrac{1470 \text{ J}}{20 \text{ m}}$
$= 73.5 \text{ N}$

3. **Think Critically** It is not energy, but the fuels burned to supply our energy that are in short supply.

Science Journal
Your energy came largely from the food you ate. The animals you have eaten consumed plants or other animals. The plants used energy from the sun to grow.

PREPARATION

Purpose
LS **Interpersonal** Construct a pendulum to compare the exchange of potential and kinetic energy. **L1** **COOP LEARN**

Process Skills
measuring, collecting and organizing data, observing and inferring, communicating, classifying, making and using tables, comparing and contrasting, recognizing cause and effect, forming operational definitions, forming a hypothesis, designing an experiment, using numbers, and separating and controlling variables

Time
one class period for setting up lab and devising measurement methods, and one-half class period for post-lab discussion

Materials
Have additional materials available (string, extra ring stands, banner paper, and masking tape).

Possible Hypotheses
Students may suggest that the crossarm's interference halfway up the length of the string will cause the maximum height to decrease by one half. Alternatively, they may suggest a doubling of the maximum height. The correct hypothesis is that the shape of the pendulum's path will change, but the maximum height on the opposite end should still be close to the original height because energy will be conserved.

📂 **Activity Worksheets,** pp. 5, 27-28

Activity 5-1

Design Your Own Experiment
Swinging Energy

Imagine yourself swinging in a swing. What would happen if a friend tried to stop you by grabbing the swing's chains halfway up as you passed the lowest point? Would you come to a complete stop or continue rising to your previous maximum height?

PREPARATION

Problem
How could you use a model to answer the questions in the swing situation described above?

Form a Hypothesis
Examine the pictured apparatus. How is it similar to the situation in the introductory paragraph? Hypothesize what will happen to the pendulum's motion and final height if its swing is interrupted by the crossarm.

Objectives
* Construct a pendulum to compare the exchange of potential and kinetic energy when a swing is interrupted.
* Measure starting and ending heights of the pendulum.

Possible Materials
* ring stand and ring
* support rod clamp, right angle
* support rod, 30 cm
* rubber stopper, 2-hole, medium
* string (1 m)
* metersticks
* graph paper

Safety Precautions
Be sure the base is heavy enough or well anchored so the apparatus will not tip over.

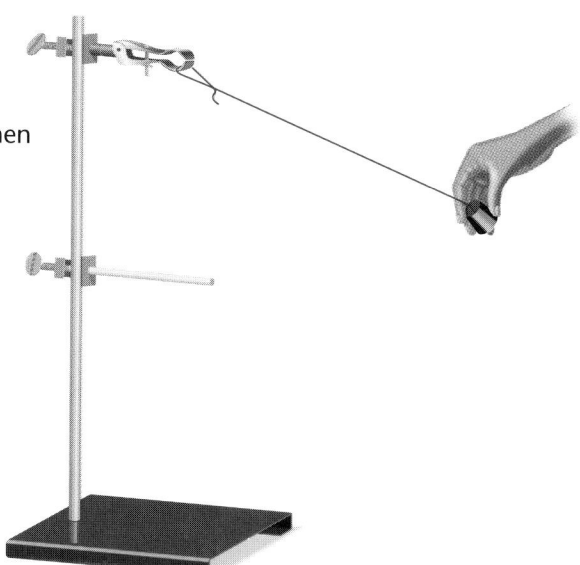

PLAN THE EXPERIMENT

Possible Procedures
One possible set of procedures might be to start the stopper from progressively lower heights and record the maximum height reached on the opposite side of the crossarm. Alternatively, the same height could be used with the crossarm moved to progressively lower or higher levels.

Teaching Strategies
* **Troubleshooting** Students may need some help developing measurement systems. A piece of banner paper could be taped to a wall, with the pendulum placed directly in front of it. One person could view the pendulum at eye level and mark the highest points on the paper. Or tie a piece of string at the height of the crossarm between two ring stands to make the starting and ending heights easier to spot.

PLAN THE EXPERIMENT

1. As a group, write your hypothesis and list the steps that you will need to take to test your hypothesis. Be specific. Also, list the materials you will need.
2. If you need a data table, design one in your Science Journal.

Check the Plan

1. Set up the apparatus as shown in the diagrams. Be sure the crossarm intersects the pendulum string.
2. Devise a way to measure the starting and ending heights of the pendulum. Record your starting and ending heights in a data table.
3. Decide how you will release the stopper from the same height each time.
4. Be sure you try starting your swing both above and below the height of the crossarm.

How many times should you repeat each type of swing?

5. *Make sure your teacher approves your plan and that you have included any changes suggested in the plan.*

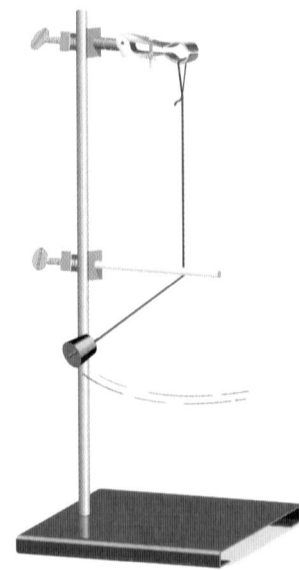

DO THE EXPERIMENT

1. Carry out the approved experiment as planned.
2. While the experiment is going on, write down any observations that you make and complete the data table in your Science Journal.

Analyze and Apply

1. When the stopper is released from the same height as the crossarm, is the ending height of the stopper exactly the same as its starting height? **Communicate** by explaining why or why not.
2. **Predict** what will happen when the starting height is higher than the crossarm.
3. **Analyze** the energy transfers. At what point along a single swing does the stopper have the greatest kinetic energy? The greatest potential energy?

5-1 Energy and Work   **133**

Go Further

What happens if the mass of the stopper is increased? Try it. What effect would increased rider mass have on the person grabbing the swing?

- Remind students that changing the release height affects the pendulum's maximum P.E., changing the maximum K.E., too.

DO THE EXPERIMENT

Expected Outcome

Students will observe that even with the crossarm interrupting the swing, the approximate original maximum height will be reached due to the conservation of energy. They will infer that conservation of energy has occurred as potential energy is transferred to kinetic energy and back again. They will also observe how air friction affects the motion and duration of the pendulum.

Analyze and Apply

1. Air resistance and friction slow the stopper and take some energy out of the system, so it cannot reach the starting height.
2. If released higher than the crossarm, the stopper wraps the string around the crossarm until it stops and falls.
3. K.E. is greatest at the bottom and lowest at the top. P.E. is greatest at the top and lowest at the bottom.

Go Further

Increasing the mass of the stopper increases the maximum K.E. and P.E. The person grabbing the swing would have to pull harder to try to stop the rider.

Assessment

Process When dropped from the same height, a pendulum without interference from a crossarm will remain in motion longer than a pendulum with an arm. Ask students to discuss and explain this. Some energy transfers into the crossarm. Use the Performance Task Assessment List for Making Observations and Inferences in **PASC**, p. 17. **P**

Prepare

Section Background

- The average K.E. of the particles in a material is directly related to the absolute temperature of the material. Comparing the temperatures of two samples of a material compares the average K.E. of their particles.
- The thermal energy of a material is the total kinetic and potential energies of all of its particles and depends on the amount of material. Heat is the thermal energy transferred between objects at different temperatures.

Preplanning

To prepare for the MiniLAB, obtain dry sand and a plastic cup with lid for each student.

1 Motivate

Bellringer

Before presenting the lesson, display **Section Focus Transparency 19** on the overhead projector. Assign the accompanying **Focus Activity** worksheet. [L1] [LEP]

Tying to Previous Knowledge

Have students recall the definition of energy from Section 5-1 and list and discuss situations that indicate heat is a kind of energy. Tell them they will be learning more about heat in this section.

Visual Learning

Figure 5-10 **How does the particle motion in A differ from that in B?** *Particle motion in A is slower.* [LEP] [IS]

Science Words

temperature
thermal energy
heat

Objectives

- Recognize the difference between the motion of an object and the motion of the particles that make up the object.
- Contrast heat and temperature.
- Explain what determines the thermal energy of a sample of matter.

Figure 5-10

As the crucible is heated, the kinetic energy of the particles that make up the crucible increases. *How does the particle motion in (A) differ from that in (B)?*

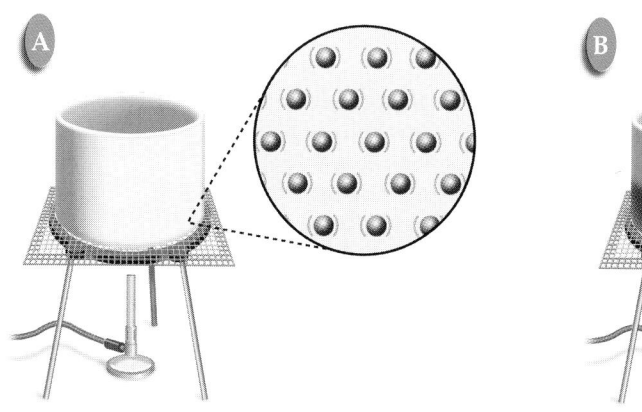
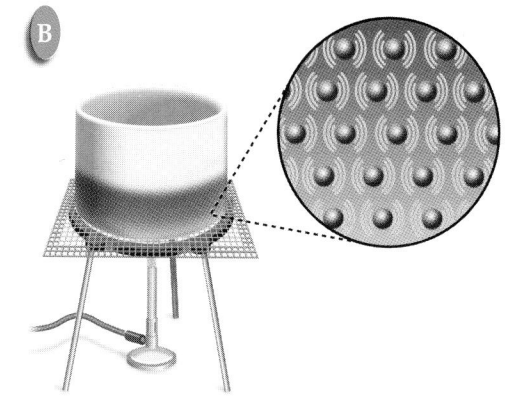

5•2 Temperature and Heat

Temperature

What do you know about temperature? When the air temperature outside is high, you probably describe the weather as hot. Ice cream, which has a lower temperature, feels cold. The words *hot* and *cold* are commonly used to describe the temperature of a material. Although not very scientific, these terms can be useful. Just about everyone understands that *hot* indicates high temperature and *cold* indicates low temperature.

When most people think of temperature, they automatically think of heat, also. This association makes sense, because heat and temperature are related. However, they are not the same. To understand the relationship between temperature and heat, you need to know about matter.

Matter in Motion

All matter is made up of particles so small that you can't see them, even with an ordinary microscope. The particles that make up any object are constantly moving, even if the object itself appears perfectly still. Everything you can think of—the book on your desk, the shoe on your foot, even the foot inside your shoe—is made up of moving particles.

You know that moving things have kinetic energy. Because the particles that make up matter are in constant motion, you can conclude that they have kinetic energy. The faster the particles move, the more kinetic energy they have. **Figure 5-10** illustrates particle movement in heated and unheated objects. **Temperature** is a measure of the average kinetic energy of the

134 Chapter 5 Energy

Program Resources

📁 **Reproducible Masters**
Study Guide, p. 23 [L1]
Reinforcement, p. 23 [L1]
Enrichment, p. 23 [L3]
Activity Worksheets, p. 31 [L1]
Multicultural Connections, pp. 13-14

🔖 **Transparencies**
Section Focus Transparency 19 [L1]

particles in a sample of matter. As the particles in an object move faster and their average kinetic energy becomes greater, the temperature of the object rises. Similarly, as the particles in an object move more slowly, their average kinetic energy decreases. Which particles are moving faster, those in a cup of hot tea or those in the same amount of iced tea?

Thermal Energy

If you place an ice-cold teaspoon on top of the scoop of ice cream in **Figure 5-11,** nothing will happen. But suppose you place a hot teaspoon on the ice cream. Now the ice cream under the spoon starts to melt. The hot spoon causes the ice cream to change because it transfers energy to the ice cream. Where does this energy come from? The spoon isn't moving, and its position is the same as that which the cold spoon had. So the energy is not due to the motion or the position of the spoon. Instead, it is related to the temperature of the spoon.

The change in the ice cream is caused by the flow of thermal energy from the spoon. Like temperature, thermal energy is related to the energy of the particles that make up matter. **Thermal energy** is the total energy of the particles in a material. This total includes both kinetic and potential energy. The kinetic energy is due to vibrations and movement within and between particles. The potential energy is determined by forces that act within or between the particles.

Suppose you stack two hot teaspoons at the same temperature on a scoop of ice cream. What will happen? The two spoons will melt more ice cream than the single spoon did in the same amount of time. The stacked spoons have twice as much mass and, therefore, twice as many moving particles. The more mass a material has at the same temperature, the more thermal energy it has.

Figure 5-11

The rate of melting depends on the amount of thermal energy present.

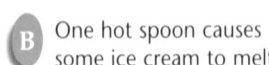

A An ice-cold spoon has no effect on the ice cream.

B One hot spoon causes some ice cream to melt.

C Two hot spoons cause more ice cream to melt.

MiniLAB

Can shaking increase the temperature of sand?

Procedure

1. Pour sand into a plastic cup (with a well-fitting lid) until the sand is deep enough to just cover the bulb of a thermometer.
2. Measure and record the temperature of the sand to the nearest tenth of a degree. Remove the thermometer.
3. Place the lid snugly on the cup. While holding the lid firmly in place, shake the cup vigorously for several minutes.
4. Stop shaking, remove the lid, and immediately use the thermometer to measure the temperature of the sand.

Analysis

1. What effect did shaking have on the temperature of the sand?
2. Was energy created? Explain why or why not.

5-2 Temperature and Heat   **135**

Student Text Question

Which particles are moving faster, those in a cup of hot tea or those in the same amount of iced tea? *the particles in the hot tea*

 # Problem Solving

A Cold Cup of Hot Chocolate . . . and a Warm Soft Drink

How can two liquids originally at very different temperatures become the same temperature? While working at a restaurant, Jonas was clearing the table where two customers had been sitting. One had been served a cup of hot chocolate and the other had been served a cold soft drink. When he picked up the remaining portions of their drinks, Jonas noticed that the hot chocolate and the cold soft drink were now both at the same temperature.

Solve the Problem:

1. Assuming both drinks had the same mass, how did the thermal energy of the two drinks compare when the customers were first served the drinks?

2. Describe the heat transfer that took place to bring both drinks to room temperature.

Think Critically:

Heat will flow between two objects only if they are not at the same temperature.

1. Where does the heat lost by the hot chocolate go?

2. If heat flows in only one direction, how can both hot and cold liquids reach room temperature as they sit on a table?

Different kinds of matter have different thermal energies, even when mass and temperature are the same. For example, a 5-g sample of sand has a different thermal energy than a 5-g sample of pudding at the same temperature. This difference is due mainly to the ways in which the particles of the materials are bound together.

It is important to remember that the thermal energy of a material depends on the *total energy of its particles*. The kinetic energy of the object itself has no effect on its thermal energy. For example, at 20°C, a golf ball has the same thermal energy whether it's sitting on the ground or speeding through the air.

Heat

What would happen if you pressed your left hand against a cool tile wall? Your hand would feel cooler. Its temperature would have decreased. If you then touched the same spot on the wall with your right hand, the spot wouldn't feel as cool as it did when you originally touched it. The temperature of the spot would have increased when you touched it with your left hand. Energy flowed from your warm left hand to the cool tile. **Heat** is the thermal (or internal) energy that flows from something with a higher temperature to something with a lower temperature. Remember that, in most cases, heat flows from warmer to cooler materials—

---

not from cooler to warmer. In **Figure 5-12**, the heat flows are making one hand warmer and one hand colder.

The next time you listen to a weather report, think about the difference between temperature and heat. And when you drop an ice cube into your drink, think about how the cooling occurs. Does the melting ice cause the liquid to cool? Or does heat flowing from the liquid to the ice cool the drink and cause the ice to melt?

Like work, heat is measured in joules and involves transfer of energy. Heat is energy transferred between objects at different temperatures. Work is energy transferred when a force acts over a distance.

Figure 5-12

Heat flows from hot chocolate to your hand, making your hand feel hot. Heat flows from your hand to a glass of iced tea, making your hand feel cold. *Why do people wear gloves in cold weather?*

Section Wrap-up

Review

1. In terms of thermal energy, why does a bottle of soda left in the sun have a higher temperature than one left in an ice chest?

2. Which has more thermal energy, a 5-kg bowling ball that has been resting on a hot driveway for 4 hours on a 35°C day, or the same bowling ball rolling down a lane in an air-conditioned bowling alley?

3. **Think Critically:** Using your knowledge of heat, explain what happens when you heat a pan of soup on the stove, then put some leftover warm soup in the refrigerator.

Science Journal

Use physics terms and concepts to explain why you take your temperature by placing a thermometer under your tongue and waiting several minutes. Write your explanation in your Science Journal.

Skill Builder
Concept Mapping

Create a network tree that shows how the following words and phrases are related: *energy, potential energy of particles, energy transfer (higher to lower temperature), heat, kinetic energy of particles,* and *thermal energy.* If you need help, refer to Concept Mapping in the **Skill Handbook.**

4 Close

•MINI•QUIZ•

Use the Mini Quiz to check students' recall of chapter content.

1. **What is thermal energy?** *the total energy of the particles that make up a material*

2. **What is heat?** *energy that flows from something at a higher temperature to something at a lower one*

Section Wrap-up

Review

1. The soda left in the sun absorbs more energy, and the average K.E. of its particles goes up.

2. The ball that rested in the sun for several hours has more thermal energy.

3. **Think Critically** Energy from the stove burner transfers to the soup. Heat will flow from the warm soup to the air in the refrigerator.

Science Journal

You begin with the thermometer at a temperature below 98.6°F (average human temperature). If your body is warmer than this, the heat from your body will transfer to the thermometer. By placing the thermometer under your tongue, you seal it off from outside air. You wait several minutes for the energy conversion to be complete and for the system to equilibrate.

Skill Builder

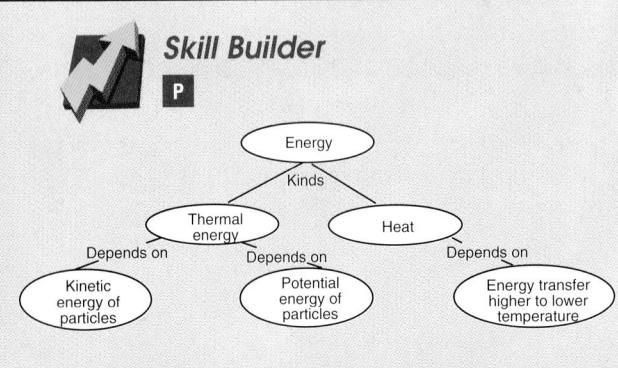

Assessment

Performance Have students shake hands with students near them. Ask them to use specific terms and phrases from this Skill Builder to explain why some hands feel warm and some feel cold. **CAUTION:** *Have students wash their hands after the activity.* Use the Performance Task Assessment List for Making Observations and Inferences in **PASC**, p. 17.

SCIENCE & SOCIETY
Section 5•3

Prepare

Section Background

Point out that pollution does not just manifest itself as solid forms, i.e., solid waste and particulates in the air. Excess heat from machines, buildings, and other sources contribute to the problem of thermal pollution.

1 Motivate

Bellringer

Before presenting the lesson, display **Section Focus Transparency 20** on the overhead projector. Assign the accompanying **Focus Activity** worksheet. L1 LEP

2 Teach

Visual Learning

Figure 5-13 How might this affect the ecology of the river? *Increased water temperature and lowered oxygen level. Some water-dwelling organisms can't live in warmer water or in water with lower oxygen levels. The food chain will be disrupted when some organisms die. The warm water may eventually kill plant and animal life.*

Figure 5-14 Why are they designed to be wide and open at the top? *To facilitate evaporation and cooling of the water.* LEP IS

SCIENCE & SOCIETY

5•3

ISSUE:

Thermal Pollution: Waste You Can't See

Science Words

thermal pollution

Objectives

- Identify some causes and effects of thermal pollution.
- Discuss possible solutions for thermal pollution problems.

Where does it come from?

If you have been in a city on a warm day, surrounded by cars, large buildings, and concrete, you may have noticed that it seemed to be hotter there than in the suburbs, where there's less traffic and less activity. Much of the energy used in everyday life—electrical, chemical, mechanical, radiant, or nuclear energy—ends up as waste thermal energy that is given off to the surroundings. The heat removed from air-conditioned buildings and vehicles is released to the outside air, just as heat is released into a kitchen by a refrigerator.

The level of waste heat can reach unhealthy levels. **Thermal pollution** occurs when waste heat significantly changes the temperature of the environment. It is a particular problem in areas where power plants and factories use water to cool their buildings and equipment, warming the water in the process. If the warmed water is dumped into a river, lake, or ocean, as in **Figure 5-13,** the added heat lowers the dissolved oxygen content and may cause serious problems for the plants and animals living there.

The added heat accelerates chemical-biological processes and can alter the reproductive processes of the water plants and animals. In extreme cases, major fish kills can result.

Figure 5-13

Hot wastewater is discharged into a river. *How might this affect the ecology of the river?*

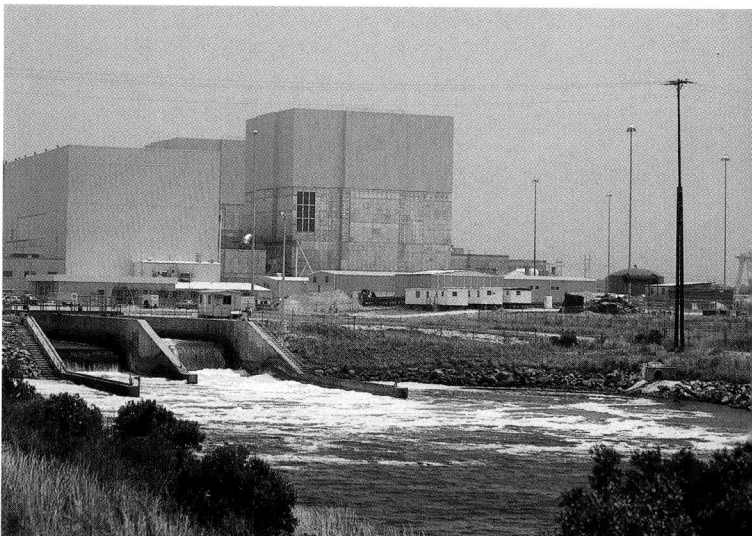

Program Resources

Reproducible Masters

Study Guide, p. 24 L1
Reinforcement, p. 24 L1
Enrichment, p. 24 L3
Critical Thinking/Problem Solving, p. 11

Transparencies

Section Focus Transparency 20 L1

2 Points of View

▶ Too many regulations for industry?

Thermal pollution can be reduced by releasing small amounts of warm water mixed with plenty of cooler water. But in order to cool the water, factories and power plants must build devices such as cooling towers. In these large structures, illustrated in **Figure 5-14,** water is cooled by fans or evaporation. Although the cooling towers are effective, they are also expensive. Businesses may be able to recover these costs by charging more for their products and services. However, if competition in the industry is so strong that a company can't raise prices, the company runs the risk of losing money and perhaps going out of business.

Some people argue that the government passes unnecessary laws and sets tight regulations that require businesses to build costly cooling mechanisms. They think that the effects of thermal pollution on the environment are not severe enough to justify raising the costs of operation, which then increases the prices paid by consumers. In their view, the government is creating regulations that are wasteful and unnecessary.

▶ Cooling Costs: It's Worth the Effort

Typically, the temperature of water discharged from an electric power plant ranges from 5 to 11 Celsius degrees above the temperature of the waterway that is receiving the discharge.

Because it is possible to cool water before releasing it back into a water system, some people argue that it is our responsibility to prevent damage to the environment from thermal pollution. They emphasize that the physical, chemical, and biological effects of a temperature rise can upset delicate aquatic ecosystems. They are concerned that if thermal pollution is not controlled, it will kill water plants and animals, destroy water-based ecosystems, and ruin recreational waterways for people who enjoy fishing, swimming, and boating.

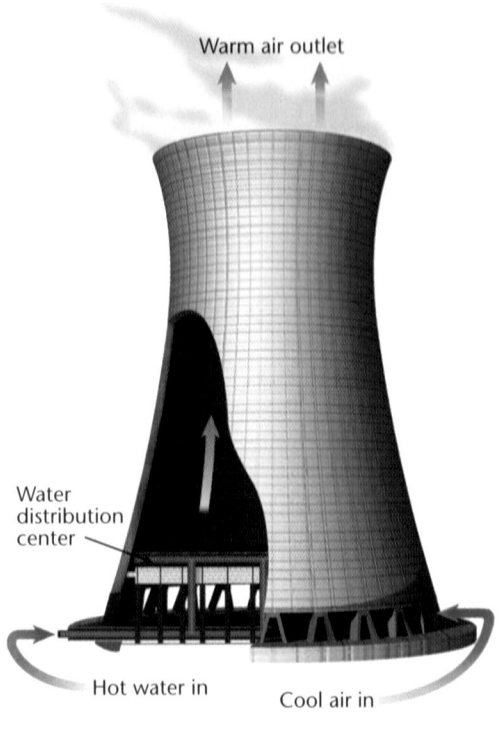

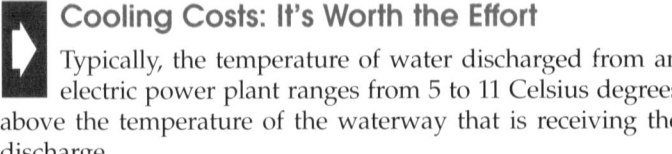

Warm air outlet

Water distribution center

Hot water in

Cool air in

Figure 5-14

Cooling towers can cool large volumes of hot wastewater. *Why are they designed to be wide and open at the top?*

Activity

LS **Intrapersonal** Have students observe the effects of thermal pollution by adding cool, warm, and hot water to three small plants over a period of several days. Ask them to describe what condition would provide these water temperatures. Warm and hot water should cause the plants to wilt and/or die. Rain and streams provide cool water; industry may provide the warm and hot water. **L1**

GLENCOE TECHNOLOGY

🔘 **Videodisc**
STVS: Ecology
Disc 6, Side 2
Greenhouse Effect (Ch. 14)

‖‖‖‖‖‖‖‖‖‖‖

3 Assess

Check for Understanding

? FLEX Your Brain

Use the Flex Your Brain activity to have students explore THERMAL POLLUTION.

🗂 **Activity Worksheets,** p. 5

Reteach

For students who don't understand that thermal pollution is not just a water problem, have them observe several house plants that have been placed next to heat generating objects (overhead projectors, lab incubators or ovens, etc.). Ask them to expand this list to sources outside of the classroom and building. **LEP**

4 Close

Figure 5-15
For healthy fish to exist in a waterway, the water temperature must not exceed harmful levels.

Many people argue that we can develop creative uses for the warm water. This would reduce or eliminate the need to dump the warm water into streams, rivers, or other waterways. Instead, the extra thermal energy in the water can be used to heat greenhouses or other buildings. These people argue that the costs associated with minimizing thermal pollution are a necessary part of an industrial society. In their view, we need to be willing to pay the costs so that people, such as the person fishing in **Figure 5-15,** can continue to enjoy the benefits.

Using Computers

Word Processing
Your representative in Congress will soon vote on a bill to require companies to reduce thermal pollution below specified levels. Write a two-paragraph letter to her explaining why she should or should not vote for the bill.

SCIENCE & SOCIETY

Section Wrap-up

Review

1. Thermal pollution may encourage the growth of certain water plants and slow or destroy the growth of others. Why might this be a problem?

2. A company plans to build several greenhouses and to heat them using waste heat from cooling water. What factors should the company consider when making its plan?

Explore the Issue

A company wants to build a factory by a river that runs through your city, creating new jobs but also generating waste thermal energy. What steps would you take to find out whether the environment will be damaged? Should the company be allowed to build its factory?

Measuring Thermal Energy

Specific Heat

If you have jumped into a swimming pool or lake on a hot summer day, did you find the water surprisingly cold? Even though lots of radiant energy has been transferred to the water from the sun, the temperature of the water is still cooler than that of the surroundings.

Different materials need different amounts of heat to produce similar changes in their temperatures. The materials have different specific heats. The **specific heat** (C) of a material is the amount of energy it takes to raise the temperature of 1 kg of the material 1 kelvin. Specific heat is measured in joules per kilogram per kelvin [J/(kg · K)]. **Table 5-2** shows the specific heats of some familiar materials. How does the specific heat of water compare with the specific heats of the other materials?

Heat Absorption

When you see how high the specific heat of water is compared to other materials, do you understand why one of the persons in **Figure 5-16** is testing the water before going in? The materials around the water heat up much faster than the water itself, so the water is cold compared with its surroundings.

Science Words
specific heat

Objectives
- Define *specific heat*.
- Calculate changes in thermal energy.

Table 5-2

Specific Heat of Some Common Materials (J/(kg · K))	
Water	4184
Alcohol	2450
Aluminum	920
Carbon (graphite)	710
Sand	664
Iron	450
Copper	380
Silver	235

Figure 5-16

Because of its high specific heat, water warms up more slowly than its surroundings.

Prepare

Section Background

The specific heat of a material can be used to quantitatively describe a change in its thermal energy. Specific heat is defined as the amount of energy needed to change the temperature of one kilogram of a material by one kelvin. Its unit is J/(kg • K).

Preplanning

Gather thermometers, foam cups, graduated cylinders, and balances for Activity 5-2. Have students bring in electric hair curlers.

1 Motivate

Bellringer

Before presenting the lesson, display **Section Focus Transparency 21** on the overhead projector. Assign the accompanying **Focus Activity** worksheet. [L1] [LEP]

Tying to Previous Knowledge

Have students recall the metric units and, where appropriate, the instruments used to measure energy, mass, and temperature. Tell students that they will use this knowledge to learn more about changes in thermal energy as something is heated or cooled.

Student Text Question

How does the specific heat of water compare with the specific heats of other materials? *It is greater.*

Program Resources

Reproducible Masters
Study Guide, p. 25 [L1]
Reinforcement, p. 25 [L1]
Enrichment, p. 25 [L3]
Activity Worksheets, pp. 29-30 [L1]
Concept Mapping, p. 15
Lab Manual 10

Transparencies
Section Focus Transparency 21 [L1]
Teaching Transparency 10

2 Teach

Content Background

Although the absolute value of the temperature of an object is different on the Celsius and Kelvin scales, the change in thermal energy is the same for 1°C or 1 K. So, specific heat can have °C or K in the denominator interchangeably.

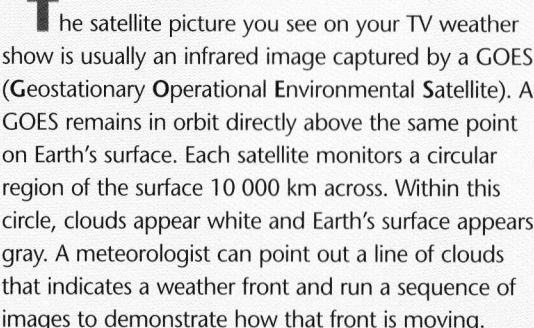

USING TECHNOLOGY

Infrared Weather Eyes

The satellite picture you see on your TV weather show is usually an infrared image captured by a GOES (Geostationary Operational Environmental Satellite). A GOES remains in orbit directly above the same point on Earth's surface. Each satellite monitors a circular region of the surface 10 000 km across. Within this circle, clouds appear white and Earth's surface appears gray. A meteorologist can point out a line of clouds that indicates a weather front and run a sequence of images to demonstrate how that front is moving.

Warm objects give off more infrared radiation than cool objects. The satellite's sensors measure many bands of infrared radiation. One band indicates moisture in the air. Dry air appears black and moist air appears in varying shades of gray. Because bad weather often breaks out when dry air intrudes into moist air, these water-vapor images can be used to predict storms. Other bands of radiation reveal information about wind speed and direction, rainfall, fog, snow, and ice. Newer satellites may one day provide data for making accurate global weather forecasts 30 to 90 days in advance.

interNET CONNECTION

Visit the Chapter 5 Internet Connection at Glencoe Online Science, **www.glencoe.com/sec/science/physical**, for a link to more information about weather satellites.

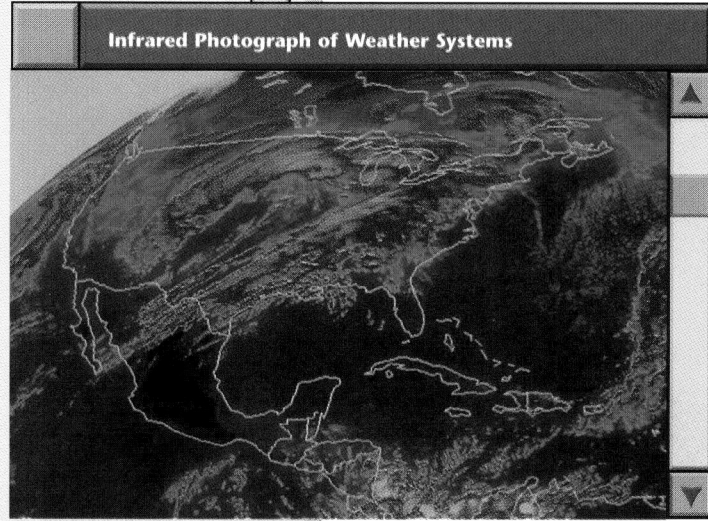

Infrared Photograph of Weather Systems

Water, alcohol, and other materials with a high specific heat can absorb a lot of energy with little change in temperature. The specific heats of different substances depend on the chemical makeup of the substances.

Calculating Thermal Changes

Changes in thermal energy cannot be measured directly, but they can be calculated. Specific heat can be used to determine changes in thermal energy. For example, you can place a heated mass or object in a calorimeter such as the one shown in **Figure 5-17.** By measuring the temperature increase in the water in the calorimeter, you can determine the change in thermal energy. Now, suppose you take a 32-g silver spoon from a pot of water at a temperature of 60°C and allow it to cool to room temperature, which is 20°C. You have enough information to find out the

Use these program resources as you teach this lesson.
Study Guide, p. 25
Concept Mapping, p. 15

change in the thermal energy of the spoon using the equation below.

$$\text{Change in thermal energy} = \text{mass} \times \text{Change in temperature} \times \text{specific heat}$$

$$Q = m \times \Delta T \times C$$

The symbol Δ (delta) means "change," so ΔT is the change in temperature. "Change" is included in Q, which is the variable for energy change.

$$\Delta T = T_{final} - T_{initial}$$

When ΔT is positive, Q is also positive; the object has increased in temperature and gained thermal energy. When ΔT is negative, Q is also negative; the object has lost thermal energy and decreased in temperature. The problems below will give you practice in calculating changes in thermal energy.

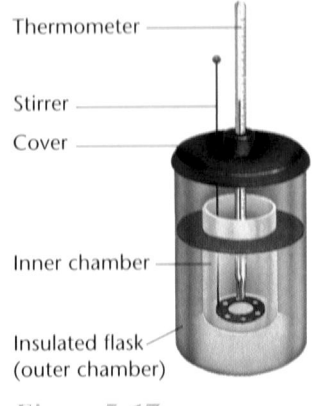
Thermometer
Stirrer
Cover
Inner chamber
Insulated flask (outer chamber)

Figure 5-17
Devices like this simple calorimeter are used to measure thermal energy transfer.

USING MATH

Calculating Changes in Thermal Energy

Example Problem:
A 32-g silver spoon cools from 60°C to 20°C. What is the change in its thermal energy?

Problem-Solving Steps:
1. What is known?
   Mass of spoon, m = 32.0 g = 0.0320 kg
   The spoon is made of silver.
   The specific heat, C, of silver is 235 J/(kg · K).
   Initial temperature, $T_{initial}$ = 60.0°C + 273 = 333K
   Final temperature, T_{final} = 20.0°C + 273 = 293K
2. What is unknown?
   Change in thermal energy, Q
3. Use the equation $Q = m \times \Delta T \times C$.
4. **Solution:** $Q = m \times (T_{final} - T_{initial}) \times C$ $\quad \Delta T = T_{final} - T_{initial}$
   = 0.0320 kg × (293K − 333K) × 235 J/(kg · K)
   = −301 J
   The spoon loses 301 J of thermal energy as it cools.

Practice Problems
1. Calculate the change in thermal energy when 230 g of water warms from 12°C to 90°C.
   Strategy Hint: Will your answer be a positive or negative number?
2. A 45-kg brass sculpture gains 180 480 J of thermal energy as its temperature increases from 28°C to 40°C. What is the approximate specific heat of brass?
   Strategy Hint: What is the unknown information?

5-4 Measuring Thermal Energy   **143**

Across the Curriculum

History Have students research the experiments conducted by James Prescott Joule that quantitatively relate thermal energy and work. L2 LS

Practice Problem Answers
1. $T_{initial}$ = 12°C + 273
      = 285 K
   T_{final} = 90°C + 273 = 363 K
   $Q = \Delta T \times m \times C$
      = (363 K − 285 K)
        × 0.23 kg
        × 4190 J/(kg • K)
      = 75 000 J
2. $T_{initial}$ = 28°C + 273
      = 301 K
   T_{final} = 40°C + 273 = 313 K
   $Q = \Delta T \times m \times C$
   $C = \dfrac{Q}{\Delta T \times m}$
   $= \dfrac{180\ 480\ \text{J}}{(313\ \text{K} - 301\ \text{K}) \times 45\ \text{kg}}$
   $= \dfrac{180\ 480\ \text{J}}{540\ \text{kg} \bullet \text{K}}$
      = 334 J/(kg • K) P

3 Assess

Check for Understanding

? FLEX Your Brain

Use the Flex Your Brain activity to have students explore SPECIFIC HEAT.

Activity Worksheets, p. 5

Reteach
Ask students to look at **Table 5-2.** The values of the specific heats are ranked from high to low. Have the students think about cooking food on top of a stove. Explain that their objective is to heat the food as quickly as possible. Ask them whether they would use an aluminum pan or a copper pan to cook the food quickly.

Extension
For students who have mastered this section, use the **Reinforcement** and **Enrichment** masters.

143

Visual Learning

Figure 5-18 Why are elephants' ears so large? *So they can provide adequate cooling. The large surface area carries heat away from the large body.* LEP IS

CONNECT TO

CHEMISTRY

Answer If her earlobe remained cooler than the earring, her earlobe must have a higher specific heat. P

4 Close

Discussion

Ask students if they have any evidence that thermal energy is also conserved.

Section Wrap-up

Review

1. The sand heats more quickly than the water because it has a lower specific heat than water.
2. $Q = \Delta T \times m \times C$
   $Q = -65°C \times 0.055\ kg \times 450\ J/(kg \bullet K)$
   $= -1609\ J$
3. **Think Critically** Coolants should have a high specific heat so they can absorb a lot of energy without drastic temperature changes.

USING MATH

$Q = mC\Delta T$, solving for ΔT,
$\Delta T = Q/(m \bullet C)$
$\Delta T = 9 \times 10^6\ J/(50\ kg) \bullet 4190\ J/kg \bullet K$
$\Delta T = 9 \times 10^6/209\ 500$
$\Delta T = 42.95°C$

CONNECT TO

CHEMISTRY

While Karen was blow-drying her hair, she noticed that her silver earring had become uncomfortably hot. If her earring was about the same mass as her earlobe, ***infer*** which of the two had the higher specific heat.

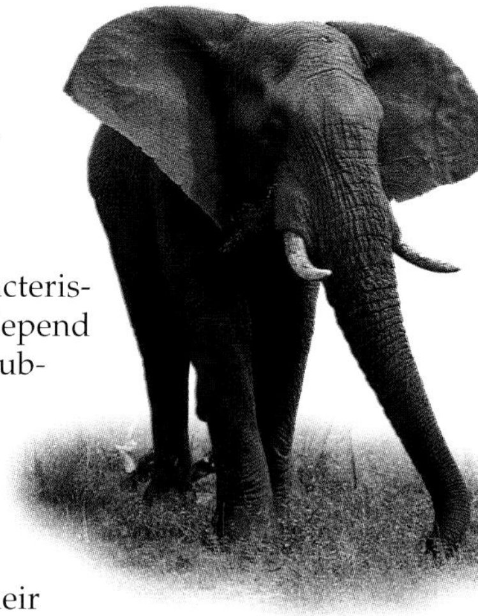

Figure 5-18
Different animals use different methods to cool themselves. *Why are elephants' ears so large?*

The thermal energy characteristics of complex systems depend on the specific heats of the substances involved and the masses and shapes of the systems. For example, as shown in **Figure 5-18,** elephants fan their ears to dissipate heat and lower their body temperature. Without this ability, the large mass, shape, and high specific heat of the materials in an elephant's body would keep its temperature too high.

With your new understanding of specific heat, think about temperature changes you've observed in objects around you. Now you have an idea about why a kettle of water takes so long to boil and why a sandy beach heats up quickly on a sunny day.

Section Wrap-up

Review

1. A bucket of sand and a bucket of water are side by side in direct sunlight. Which warms faster? Why?
2. Use **Table 5-2** on page 141 to calculate the change in thermal energy when a 55-g iron nail cools from 90°C to 25°C.
3. **Think Critically:** Water and a liquid called ethylene glycol are used in automobile radiators to keep the engine from overheating. Explain whether you would rather have a coolant with a high or a low specific heat.

Skill Builder

Interpreting Data

Equal amounts of iron, water, and sand, all at the same initial temperature, were placed in an oven and heated briefly. Use the data in **Table 5-2** on page 141 to match each final temperature with the appropriate material: 31°C, 5°C, and 46°C. If you need help, refer to Interpreting Data in the **Skill Handbook.**

USING MATH

Suppose you are on a diet that recommends intake of 9 000 000 J (more than 2000 Calories) each day. If you add this amount of heat to 50 kg of water, how much would the water's temperature rise?

Skill Builder

5°C = water, 31°C = sand, 46°C = iron

Assessment

Process Two students heated two equal masses of water and alcohol on a hot plate for 10 minutes. Assuming neither one boils, which would have the higher temperature? Why? The alcohol; because it has a lower specific heat. Use the Performance Task Assessment List for Making Observations and Inferences in **PASC,** p. 17. P

A Hot Topic

Heat can often be used to change the shape of something. If you have ironed clothes or curled your hair, you have used heat to help you accomplish the task. How much heat is required to do these things? In this activity, you will use your understanding of specific heat to find out how much heat is released from an electric hair curler.

Problem
How much thermal energy is transferred by an electric hair curler?

Materials

- set of electric hair curlers
- thermometers, Celsius (2)
- plastic foam cup
- graph paper
- balance
- graduated cylinder

Procedure

1. Prepare a data table like the one shown.
2. Measure the mass of the plastic cup. Fill the cup about halfway with cool water and measure its mass again. Find the mass of the water in kilograms.
3. Carefully remove a heated curler from its heating unit. Stand the curler, open end up, on the table and pour in 2 mL of cool water. **CAUTION:** *Don't touch any hot metal parts when handling heated curlers—wear gloves if they are available.*
4. Using two thermometers, measure and record the temperature of the water in the cup and the temperature of the water in the curler.
5. Carefully lower the curler into the cup of water. Measure and record the water temperature every minute for 5 minutes. **CAUTION:** *Turn the heating unit off immediately after completing the activity.*
6. Prepare a time-temperature graph, with time along the *x*-axis and temperature along the *y*-axis. Plot both the heating and cooling curves.

Data and Observations   Sample Data

Time in Minutes	Water Temperature (°C) in curler	in cup
0	78	21
1	73	24
2	67	26
3	64	28
4	61	30
5	60	31

Analyze

1. What happened to the water in the cup? Explain in terms of heat.
2. What was the change in temperature of the water in the cup over the 5-minute interval?
3. Using the formula below, find the change in thermal energy (*Q*), in joules, of the water in the cup.
   $$Q = m \times \Delta T \times C$$

Conclude and Apply

4. Does the curler still have thermal energy available to continue heating the water after 5 minutes? How do you know?
5. **Compare** your results with those of your classmates. Do some curlers provide more heat than others? Which curlers provide the most heat?
6. How do the starting temperatures of all the curlers **compare?**
7. Which statement is more accurate: "Some curlers are hotter than others," or "Some curlers contain more thermal energy than others"? **Communicate** your answer in a brief oral explanation.

5-4 Measuring Thermal Energy   **145**

The following is a graph using sample data for step 6 in the Procedure. Students should obtain similar graphs.

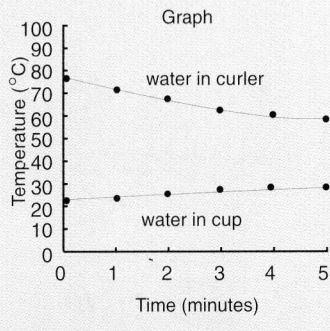

Activity 5-2

Purpose

IS **Logical-Mathematical** Measure the heat delivered by a hair curler over a 5-minute period.
L1 **LEP** **COOP LEARN**

Process Skills
measuring, using numbers, and interpreting

Time
40 minutes

📁 **Activity Worksheets,** pp. 5, 29-30

Teaching Strategies
Troubleshooting Do not allow the water in the cup to overflow into the curler.

- One set of hair curlers will serve the entire class. The type with a hard plastic outer covering is recommended.
- Heat the set of curlers to maximum temperature prior to class. Warn students not to touch metal interiors of the cylinders.
- Assign different-sized curlers to different teams.

Answers to Questions

1. The water became hotter; its temperature rose. Heat produced by electricity was transferred to and stored in the curler. The stored heat was transferred to the water in the cup.
2. probably about 10°C
3. Most results will be in the 2000 J to 7000 J range.
4. Yes. The temperature inside the curler is still higher than that of the surrounding water.
5. Yes. The larger curlers provide the most heat.
6. The starting temperatures are all nearly the same.
7. "Hotter" refers to temperature, or average kinetic energy of the particles. "Contain more thermal energy" refers to total energy, both kinetic and potential, and is more accurate.

✔ Assessment

Content Ask students to explain why hot water bottles are used to warm people in bed. Use the Performance Task Assessment List for Making Observations and Inferences in **PASC**, p. 17. **P**

Sources

- Cheney, Theodore A. Rees. *Living in Polar Regions.* Franklin Watts: New York, 1987.
- Hahn, Elizabeth. *Native American People: The Inuit.* Rourke Publications: Vero Beach, FL, 1990.
- Newman, Sherlee. *The Inuits.* Franklin Watts: New York, 1993.

Background

- Wearing clothing and using shelter are the two principal ways humans adapt to the environment. New types of clothing fibers and internal heating systems in houses and buildings have reduced, to a certain degree, the importance of warm clothing in areas that experience cold weather. However, from experience, most people recognize that wearing layers of clothes is the best way to conserve heat and stay warm in winter. The air trapped between the layers is an insulator that reduces the flow of heat away from the body.

- The importance of layering clothes and conserving energy is not lost on the Inuit. The Inuit live in extreme cold for most of the year, an environment that many people might describe as unpleasant. Also called the "Land of the Midnight Sun," the Arctic is an area of extremes. The sun shines nearly all day and night for several months of summer. The reverse is true in the winter, when the sun shines very little for two to three months at a time.

Teaching Strategies

- Initiate a discussion with your class about clothing and cold weather. Discuss some of the standard clothing items people wear to keep warm in winter, such

Inuit Clothing: Life on Top of the World

Anyone who has ever experienced the stinging cold of sub-zero weather knows this fact well: The trick to winter survival is to prevent your body's thermal energy from being lost to the environment. For the Inuit people of Greenland and the northernmost parts of North America, surviving in bitterly cold weather for a large part of the year is a way of life.

Insulating with Clothing

With its frigid, sub-zero temperatures and its icy, snow-covered landscape, the Arctic is one of the harshest places on Earth. Perhaps more than anything else, Inuit clothing and the way it is worn allows its wearers to survive comfortably in the far north.

Although technology and other cultural influences have changed the basic way of life of the Inuit, many of their historical practices still exist. This is particularly true in their clothing. Traditionally, skins and furs from seal, caribou, fox, and polar bear were used to fashion all articles of clothing, including coats, hats, mittens, shirts, boots, pants, and socks.

In cold winter weather, the Inuit wear two suits of clothing. The inner layer of clothing is worn with the fur against the skin, and the outer layer is worn with the fur facing out. The use of fur and two layers of clothing illustrates a knowledge of how to conserve thermal energy and stay warm in cold weather. Animal fur is excellent for preventing thermal energy loss. It acts as an insulator, which is a material that prevents heat transfer. The long, thick hairs trap pockets of air that are warmed when placed close to the body. Also, the hairs themselves are hollow, so the air inside them is another barrier to the flow of heat away from the body. An additional layer of insulation is also formed between the two layers of clothing.

Science Journal

Write a short story or a non-fiction piece about either the coldest or warmest weather you've experienced in your life. Describe the temperature and other environmental conditions you observed. How were you able to adapt to the weather conditions? Explain how your clothing choices allowed you to remain comfortable.

as sweaters, hats, gloves, and heavy coats. Point out to students how each of the clothing items relates to the concept of conserving thermal energy.

- After students have read the feature, discuss some of the subsistence and cultural practices of the Inuit. In particular, discuss the ways people in the Arctic adapt to cold weather. Besides clothing, include in your discussion some information about Inuit shelter, food choices, recreational activities, and hunting practices. Point out how

all these cultural practices relate to the harsh environmental conditions that the Inuit endure.

Other Works

Additional books about the Inuit:

- Meyer, Carolyn. *Eskimos: Growing Up in a Changing Culture.* Atheneum: New York, 1977.
- Olinski, Alice. *The Eskimo: The Inuit and Yupik People.* Childrens Press: Chicago, 1985.

Summary

5-1: Energy and Work

1. Energy is the ability to cause change. Energy may be in the form of motion (kinetic energy) or it may be stored (potential energy).
2. Energy exists in many different forms, and it can change from one form to other forms with no loss of total energy.
3. Work is the transfer of energy through motion. Work is done only when force produces motion in the direction of the force.

5-2: Temperature and Heat

1. All matter is made up of tiny particles that are in constant motion. This motion is not related to or dependent on the motion of the object.
2. Heat and temperature are related, but they are not the same. The temperature of a material is a measure of the average kinetic energy of the particles that make up the material. Heat is the energy that flows from a warmer to a cooler material.
3. The thermal energy of a material is the total energy—both kinetic and potential—of the particles that make up the material.

5-3: Science and Society: Thermal Pollution: Waste You Can't See

1. When waste heat finds its way into rivers, lakes, and oceans, it can damage or destroy the plants and animals living there.
2. Government and industry continue to debate whether the benefits of regulating thermal pollution are worth the costs to society.

5-4: Measuring Thermal Energy

1. Different materials have different heat capacities, or specific heats.
2. The specific heat of a material can be used to calculate changes in the thermal energy of the material.

Key Science Words

a. energy
b. heat
c. kinetic energy
d. law of conservation of energy
e. mechanical energy
f. potential energy
g. specific heat
h. temperature
i. thermal energy
j. thermal pollution
k. work

Reviewing Vocabulary

Match each phrase with the correct term from the list of Key Science Words.

1. the ability to cause change
2. energy of motion
3. Under normal conditions, energy cannot be created or destroyed.
4. energy required to raise the temperature of 1 kg of a material 1 K
5. total energy of the particles in a material
6. the transfer of energy through motion
7. transfer of energy from warmer to cooler materials
8. stored energy
9. measure of the average kinetic energy of the particles in a material
10. the total amount of kinetic and potential energy of an object

Chapter 5

Review

Summary

Have students read the summary statements to review the major concepts of the chapter.

Reviewing Vocabulary

1. a	6. k
2. c	7. b
3. d	8. f
4. g	9. h
5. i	10. e

✔ Assessment

Portfolio Encourage students to place in their portfolios one or two items of what they consider to be their best work. Examples include:

- Skill Builder concept map, p. 137
- Practice Problems calculations, p. 143
- Connect to Chemistry answer, p. 144 **P**

Performance Additional performance assessments may be found in **Performance Assessment** and **Science Integration Activities.** Performance Task Assessment Lists and rubrics for evaluating these activities can be found in Glencoe's **Performance Assessment in the Science Classroom.**

GLENCOE TECHNOLOGY

▭ MindJogger Videoquiz

Chapter 5 Have students work in groups as they play the Videoquiz game to review key chapter concepts.

Checking Concepts

1. b	**6.** b
2. a	**7.** a
3. a	**8.** c
4. c	**9.** c
5. c	**10.** d

Understanding Concepts

11. Gravitational potential energy is the energy an object has due to its position above Earth's surface. Another type of potential energy is the chemical energy of food stored in the body. Kinetic energy is energy of motion.

12. At the top of a swing, potential energy is maximum; as the pendulum falls from this position, potential energy changes to kinetic energy. At the bottom of a swing, kinetic energy is maximum. As the pendulum moves up the other side of its swing, kinetic energy changes to potential energy. The total mechanical energy of the pendulum decreases with each swing as it is converted to thermal energy.

13. Thermal energy is the total kinetic and potential energy of all the particles that make up a material. Temperature is a measure of the average kinetic energy of the particles of a material. Heat is the transfer of energy due to a difference in temperature.

14. Copper has a higher specific heat and thus required more heat to raise its temperature.

15. Benefits—reduce the amount of warm water released into the environment, thus reducing the threat to the ecosystem. Drawbacks—expensive to build, requires additional energy to power the fans if used in cooling process.

148

Checking Concepts

Choose the word or phrase that completes the sentence or answers the question.

1. The basic SI unit of energy is the _____.
   a. kilogram   c. newton
   b. joule   d. kelvin

2. If the velocity of an object increases, the _____ of the object will also increase.
   a. kinetic energy   c. specific heat
   b. mass   d. potential energy

3. Which of these is not used to calculate change in thermal energy?
   a. volume   c. specific heat
   b. temperature change   d. mass

4. The _____ of a material is a measure of the average kinetic energy of its particles.
   a. potential energy   c. temperature
   b. thermal energy   d. specific heat

5. Which of these does not represent work done on a rock in the scientific sense?
   a. lifting a rock   c. holding a rock
   b. throwing a rock   d. dropping a rock

6. The total amount of kinetic and potential energy in a closed system is called _____.
   a. specific heat   c. stored energy
   b. mechanical energy   d. temperature

7. As the temperature of a material increases, the average _____ of its particles increases.
   a. kinetic energy   c. specific heat
   b. potential energy   d. mass

8. Kinetic energy is directly related to _____.
   a. volume   c. mass
   b. force   d. position

9. The _____ of an object depends upon its position.
   a. kinetic energy   c. potential energy
   b. thermal energy   d. temperature

10. Materials with a high _____ absorb a lot of energy with little change in temperature.
   a. potential energy
   b. kinetic energy
   c. velocity
   d. specific heat

Understanding Concepts

Answer the following questions in your Science Journal using complete sentences.

11. Describe two types of potential energy. How does this energy differ from kinetic energy?

12. Describe the energy changes in a swinging pendulum. Explain how energy is conserved, even as a pendulum slows down.

13. Compare and contrast heat, temperature, and thermal energy.

14. A copper bowl and a silver bowl of equal mass were heated from 27°C to 100°C. Which required more heat? Explain.

15. Describe some possible benefits and drawbacks of building large cooling towers to reduce thermal pollution from a power plant.

Thinking Critically

16. While performing a chin-up, Carlos raises himself 0.8 m. How much work does Carlos, who weighs 600 N, accomplish doing a chin-up?

17. A football player picks up a football, runs with it, and then throws it to a teammate. Describe the work done on the ball.

18. Why does air feel cold on a windy winter day and hot on a calm summer day?

19. How much thermal energy does 420 g of liquid water gain when it is heated from its freezing point to its boiling point?

20. If 50 g of water and 50 g of sand each absorb 200 J of solar energy, what will be the temperature change of each material?

Thinking Critically

16. $W = F \times d$
   $= 600 \text{ N} \times 0.8 \text{ m}$
   $= 480 \text{ J}$

17. Work is done on the ball when the player picks it up and when the ball is thrown to a teammate. No work is done on the ball when the player carries it.

18. Cold air feels cold because the particles of air have lower average kinetic energy than the skin, regardless of whether the

Developing Skills

If you need help, refer to the **Skill Handbook.**

21. **Comparing and Contrasting:** Compare and contrast heat and work.

22. **Hypothesizing:** Suppose you are sitting beside an outdoor swimming pool when someone dives into the pool and splashes water on you. Your body jumps or tenses in reaction to the cold water. Propose a hypothesis to explain why, after a few seconds, the water on your skin no longer feels cold.

23. **Concept Mapping:** Below is a concept map of the energy changes of a gymnast bouncing on a trampoline. Complete the map by indicating the type of energy—kinetic, potential, or both—the gymnast has at each of the following stages in his or her path:
    a. halfway up;
    b. the highest point;
    c. halfway down;
    d. the lowest point, just before hitting the trampoline.

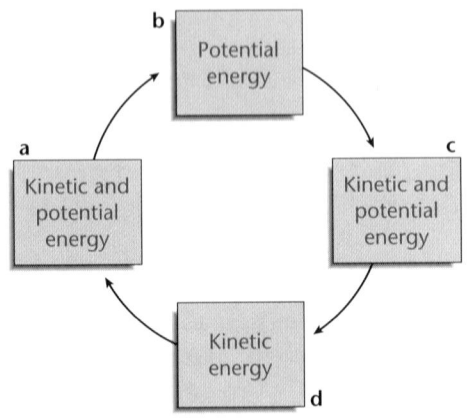

24. **Interpreting Data:** A young man with a medium-size body frame wants to burn at least 400 Calories by exercising for an hour. Using **Table 5-1** on page 131, decide which activities will enable him to reach his goal.

25. **Using Numbers:** A non-SI unit often used to measure thermal energy is the Calorie, which is equal to 4.18 J. The numbers of Calories needed to raise the temperatures of 1-kg samples of three different materials 1°C are given below. Convert each value to joules.

Specific Heats (Non-SI Units)

Material	Calories
wood	550
glass	120
mercury	33

Performance Assessment

1. **Making a Graph:** Choose a playground activity or amusement park ride that you enjoy. Placing time on the horizontal axis, sketch a graph that shows how your kinetic and potential energies change during the activity or ride.

2. **Designing an Experiment:** In Activity 5-2 on page 145, you determined the amount of thermal energy stored in a heated hair curler. Design an experiment to reheat the rollers to their original temperature without using the base of the hair curler apparatus.

3. **Poster:** Make a poster showing sources of energy other than burning fossil fuels (gas and oil). List the advantages and disadvantages of each. Indicate the resource that you think is the best and explain why.

---

air is moving or not. Warm air feels warm because the particles of air have higher average kinetic energy.

19. $Q = \Delta T \times m \times C$
    $= 100°C \times 0.420 \text{ kg} \times 4190 \text{ J}/(\text{kg} \cdot \text{K})$
    $= 175\ 980 \text{ J}$

20. $Q = \Delta T \times m \times C$; solving for ΔT,
    $\Delta T = Q/(m \cdot C)$
    $$\Delta T_{water} = \frac{(200 \text{ J})}{(0.05\text{kg}) \cdot [4190 \text{ J}/(\text{kg} \cdot \text{K})]}$$
    $\Delta T_{water} = 0.95C°$
    $$\Delta T_{sand} = \frac{(200 \text{ J})}{(0.05\text{kg}) \cdot [664 \text{ J}/(\text{kg} \cdot \text{K})]}$$
    $\Delta T_{sand} = 6.0C°$

Developing Skills

21. **Comparing and Contrasting** Both heat and work involve the transfer of energy, and both are measured in joules. Energy is transferred as heat when two materials are at different temperatures. Energy is transferred as work when a force acts through a distance.

22. **Hypothesizing** At first, the water feels cold because its temperature is lower than that of your body. Heat flows from your body to the water, raising its temperature so it doesn't feel cold.

23. **Concept Mapping** See student page.

24. **Interpreting Data** Playing tennis, bicycling fast, or running will let him reach his goal.

25. **Using Numbers**
    Wood: 420 cal × 4.18 J/cal = 1756 J
    Glass: 120 cal × 4.18 J/cal = 502 J
    Mercury: 33 cal × 4.18 J/cal = 138 J

Performance Assessment

1. Graphs will vary depending upon the activity or ride chosen by each student. They should have time on the horizontal axis and energy on the vertical axis. Use the Performance Task Assessment List for Graph from Data in **PASC,** p. 39. **P**

2. Students will need to put the curlers in an environment with enough thermal energy to transfer to the curlers. For example, they might boil or bake them.

Use the Performance Task Assessment List for Designing an Experiment in **PASC,** p. 23. **P**

3. Posters could show resources such as the sun (solar energy), the atom (nuclear energy), water (hydroelectric and tidal power), the wind (driving windmills), and steam or heat from below Earth's surface (geothermal energy). Use the Performance Task Assessment List for Poster in **PASC,** p. 73. **P**

Chapter Organizer

Section	Objectives/Standards	Activities/Features
Chapter Opener		**Explore Activity:** Measure the effects of thermal energy on the move. p. 151
6-1 **Thermal Energy on the Move** (2 sessions, 1 block)*	1. **Compare** and **contrast** the transfer of thermal energy by conduction, convection, and radiation. 2. **Differentiate** between conductors and insulators. 3. **Explain** how insulation affects the transfer of energy. **National Content Standards: (5-8) UCP2, UCP3, UCP5, A1, A2, B3, C1, C5, D3, E1, G1; (9-12) UCP2, UCP3, UCP5, A1, A2, D1, E1, G1**	**MiniLAB:** How is heat transferred by conduction? p. 15 **Problem Solving:** The Warm House, p. 156 **Using Math,** p. 158 **Skill Builder:** Using Variables, Constants, and Controls, p. 158 **People and Science:** Magali Regis, Architect and Solar Cooking Expert, p. 159 **Activity 6-1:** Creating Convection Currents, pp. 160-161
6-2 **Using Heat to Stay Warm** (1½ sessions, ½ block)*	1. **Describe** three types of conventional heating systems. 2. **Explain** how solar energy can be used to heat buildings. 3. **Explain** the differences between passive and active solar heating systems. **National Content Standards: (5-8) UCP1-UCP3, UCP5, C5; (9-12) UCP1-UCP3, UCP5**	**Connect to Life Science,** p. 163 **MiniLAB:** How can the sun be used to warm food? p. 164 **Science Journal,** p. 165 **Skill Builder:** Making and Using Tables, p. 165
6-3 **Using Heat to Do Work** (1½ sessions, ½ block)*	1. **Describe** how internal combustion engines and external combustion engines work. 2. **Explain** how a heat mover can transfer thermal energy in a direction opposite to that of its natural movement. **National Content Standards: (5-8) UCP1-UCP3, UCP5, A2, C1, C3, C5, E2; (9-12) UCP1-UCP3, UCP5, A2, C5, E2**	**Using Technology:** Cooling Crystals, p. 169 **Science Journal,** p. 170 **Connect to Earth Science,** p. 170 **Skill Builder:** Concept Mapping, p. 170 **Activity 6-2:** The Four-Stroke Engine, p. 171
6-4 **Science and Society** (1 session, ½ block)*	1. **Explain** how differences in ocean temperatures can be used to operate a heat engine that changes thermal energy to mechanical energy. 2. **Discuss** the advantages and disadvantages of ocean thermal energy conversion. **National Content Standards: (5-8) UCP1-UCP3, UCP5, B3, E2, F2-F5, G1; (9-12) UCP1-UCP3, UCP5, D1, E2, F2-F6, G1**	**Explore the Technology,** p. 174

Activity Materials

Explore	Activities	MiniLABs
page 151 lamp, thermometer, metric ruler, book, sheet of paper	pages 160-161 colored ice cubes, warm fresh water, salt, tongs, metric ruler, thermometer, 250-mL beaker, stirring rod page 171 matches, 250-mL flask, glass or plastic "T", 2 pinch clamps, large plastic syringe, 3 pieces of rubber tubing, small piece of cloth or yarn, 2-holed rubber stopper with glass tubing	page 153 5 small thumbtacks, wax, long metal spatula or fork used for outdoor grilling, oven mitt, lab burner, heat-resistant glass stirring rod, clock or watch with second hand, burner, container for melting wax page 164 aluminum foil, large piece of poster board, glue, metric ruler, string, wire coat hangers, small piece of chocolate, clock or watch

Need Materials? Call Science Kit (1-800-828-7777).   * A complete Planning Guide that includes block scheduling is provided on pages 32T-35T.

Teacher Classroom Resources

Reproducible Masters	Transparencies	Teaching Resources
Study Guide, p. 26 Reinforcement, p. 26 Enrichment, p. 26 Activity Worksheets, pp. 32-33, 36 Lab Manual 11, Conduction of Heat Cross-Curricular Integration, p. 10	Section Focus Transparency 22, Solar Radiation and Earth	Glencoe Physical Science Interactive Videodisc Physical Science CD-ROM Spanish Resources English/Spanish Audiocassettes Cooperative Learning Resource Guide Lab Partner Lab and Safety Skills Lesson Plans
Study Guide, p. 27 Reinforcement, p. 27 Enrichment, p. 27 Activity Worksheets, p. 37 Lab Manual 12, Specific Heats of Metals Science Integration Activity 6, Canned Sunshine Concept Mapping, pp. 17-18 Science and Society Integration, p. 10	Section Focus Transparency 23, Energy Consumption in the United States Teaching Transparency 11, Heating Systems	**Assessment Resources** Chapter Review, pp. 15-16 Assessment, pp. 32-35 Performance Assessment in the Science Classroom (PASC) MindJogger Videoquiz Alternate Assessment in the Science Classroom
Study Guide, p. 28 Reinforcement, p. 28 Enrichment, p. 28 Activity Worksheets, pp. 34-35 Lab Manual 13, Thermal Energy from Foods Multicultural Connections, p. 15 Critical Thinking/Problem Solving, p. 12 Technology Integration, pp. 9-10	Section Focus Transparency 24, Steam Engines Science Integration Transparency 6, Beating the Heat with Evaporation Teaching Transparency 12, Four-Stroke Engine	Performance Assessment Chapter Review Software Computer Test Bank
Study Guide, p. 29 Reinforcement, p. 29 Enrichment, p. 29	Section Focus Transparency 25, Temperatures in the Ocean	

Key to Teaching Strategies

The following designations will help you decide which activities are appropriate for your students.

L1 Level 1 activities should be within the ability range of all students, including those with learning difficulties.

L2 Level 2 activities should be within the ability range of the average to above-average student.

L3 Level 3 activities are designed for the ability range of above-average students.

LEP LEP activities should be within the ability range of Limited English Proficiency students.

LS These activities are designed to address different learning styles.

COOP LEARN Cooperative Learning activities are designed for small group work.

P These strategies represent student products that can be placed into a best-work portfolio.

GLENCOE TECHNOLOGY

The following multimedia resources are available from Glencoe.

Science and Technology Videodisc Series (STVS)
Physics
   Images of Heat
   Solar Food Dryer
Chemistry
   Fire Safety Tests
   Fire-Resistant Clothing
Animals
   Sea Urchins and Power Plants

Glencoe Physical Science Interactive Videodisc
Behavior of Gases

Physical Science CD-ROM

National Geographic Society Series
STV: Atmosphere

Teacher Classroom Resources

This is a representation of key blackline masters available in the Teacher Classroom Resources.

Teaching Aids

Section Focus Transparencies

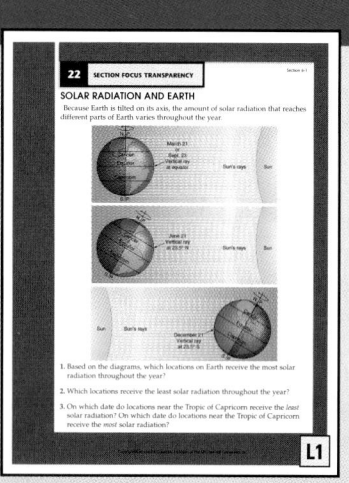

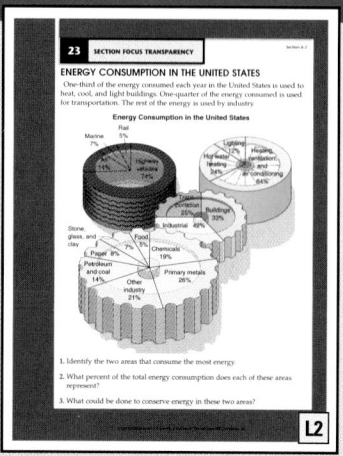

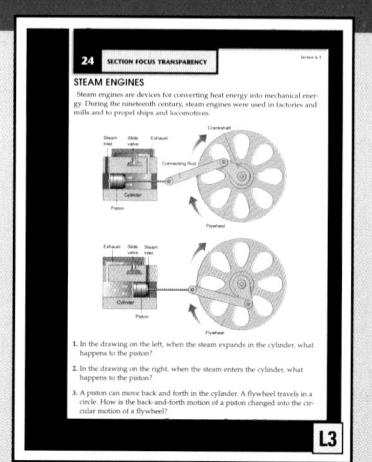

Science Integration Transparencies

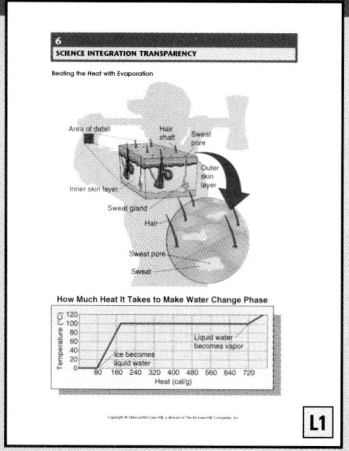

Teaching Transparencies

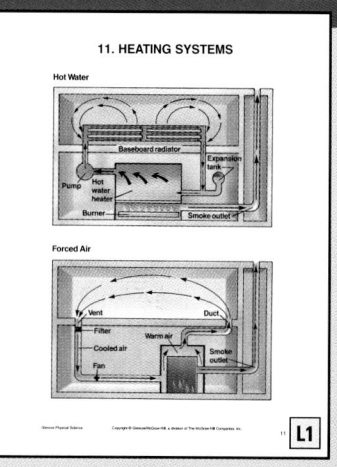

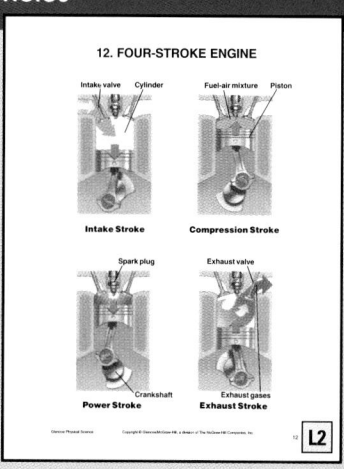

Meeting Different Ability Levels

Study Guide

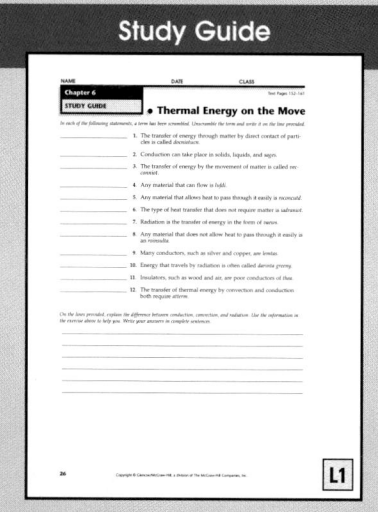

Reinforcement

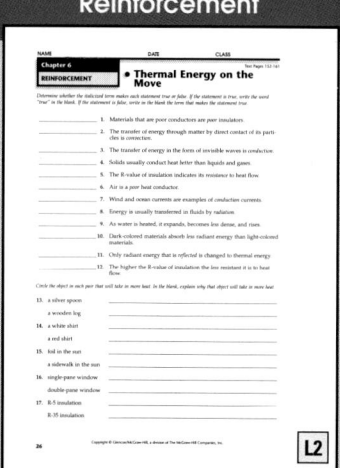

Enrichment Worksheets

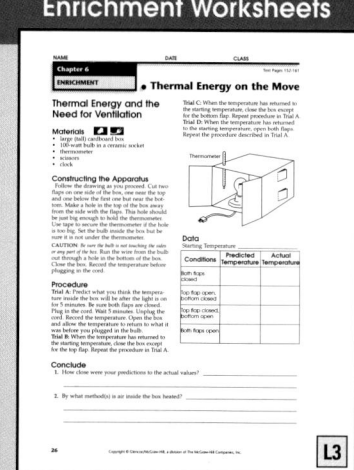

Hands-On Activities

Science Integration Activity

L1

Lab Manual

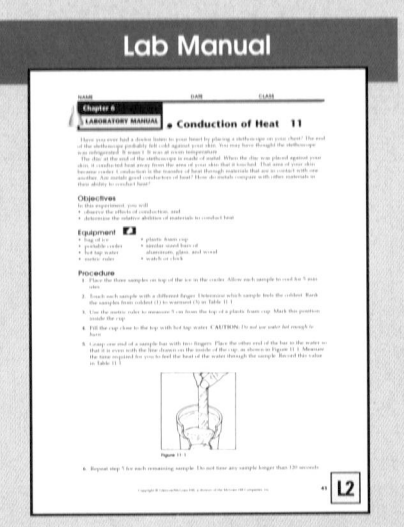

L2

Assessment

Performance Assessment

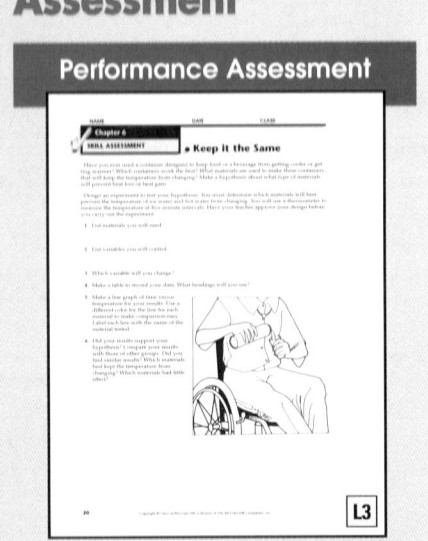

L3

Enrichment and Application

Critical Thinking/ Problem Solving

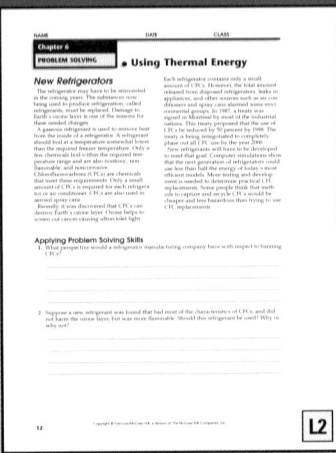

L2

Cross-Curricular Integration

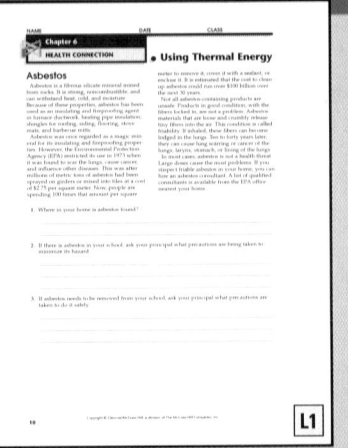

L1

Science and Society Integration

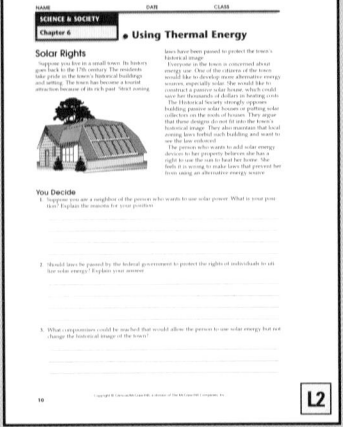

L2

Technology Integration

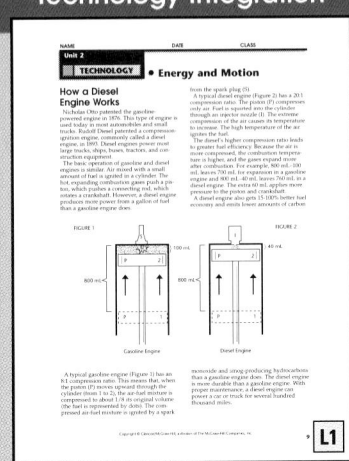

L1

Multicultural Connections

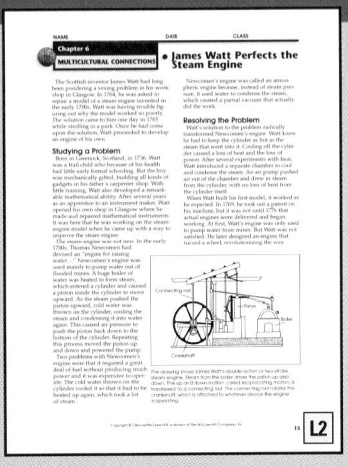

L2

Concept Mapping

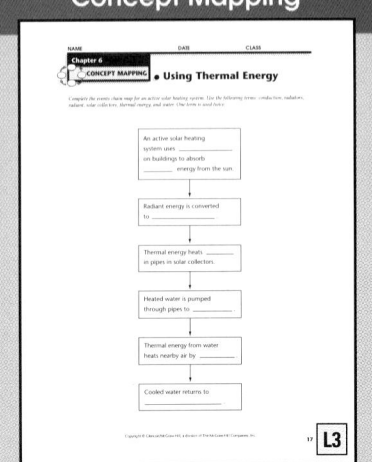

L3

Using Thermal Energy

CHAPTER OVERVIEW

Section 6-1 The three types of thermal energy transfer—conduction, convection, and radiation—are discussed in this section. Thermal conductors and insulators are compared and contrasted.

Section 6-2 This section discusses the three primary types of home-heating systems in terms of energy conversions and thermal energy transfers. Passive and active solar heating methods are introduced.

Section 6-3 Heat engines are introduced as devices that change thermal energy into mechanical energy. Internal and external combustion engines, refrigerators, and heat pumps are discussed.

Section 6-4 Science and Society The vast energy resources available in the oceans are presented. The use of ocean thermal energy conversion to drive turbines is discussed.

Chapter Vocabulary

conduction	heat engine
fluid	combustion
convection	internal
radiation	combustion
insulator	engine
radiator	external
solar energy	combustion
solar	engine
collector	heat mover
ocean thermal	
energy conversion	

Theme Connection

Energy The chapter discusses the transfer, production, and uses of thermal energy. Energy conversions and conservation should be stressed in each section.

Previewing the Chapter

Section 6-1 Thermal Energy on the Move
- ► Conduction
- ► Convection
- ► Radiation
- ► Reducing Thermal-Energy Flow

Section 6-2 Using Heat to Stay Warm
- ► Conventional Heating Systems
- ► Solar Heating

Section 6-3 Using Heat to Do Work
- ► Heat Engines
- ► Heat Movers
- ► Sweat—The Human Coolant

**Section 6-4 Science and Society
Technology: Energy from the Oceans**

Look for the following logo for strategies that emphasize different learning modalities. **LS**

Learning Styles	
Kinesthetic	Explore, p. 151; Inclusion Strategy, p. 155; Activity 6-1, pp. 160-161; Activity 6-2, p. 171
Visual-Spatial	Visual Learning, pp. 152, 153, 155, 157, 162, 166, 168, 169, 172; MiniLAB, p. 153; Reteach, p. 173
Interpersonal	Reteach, p. 169; Close, p. 174
Logical-Mathematical	People and Science, p. 159; MiniLAB, p. 164
Linguistic	Across the Curriculum, p. 154; Reteach, p. 163; Science Journal, p. 165

Using Thermal Energy

Is the process of keeping something warm similar to the process of keeping something cool? The hot-dog vendor and the ice-cream vendor shown in this photograph are both making use of methods that block the transfer of thermal energy. In the activity below, measure the effects of thermal-energy flows.

EXPLORE ACTIVITY

Measure the effects of thermal energy on the move.

1. Turn on a lamp with a bare lightbulb. Do not touch the bulb with your hand or any other object.
2. For one minute, hold a thermometer steadily 5 cm from the lightbulb, at the level of the filament. Record the temperature.
3. Repeat step 2 at distances of 10 cm, 15 cm, and 20 cm. At what distance is there no measurable temperature change after one minute?
4. With the thermometer at the 10-cm position, place a book between the thermometer and the lamp and repeat the measurement. Also, try this with a piece of paper instead of a book.

Observe: How does distance affect the temperature change caused by thermal energy on the move? How effective were the barriers? Explain your answers in your Science Journal.

Previewing Science Skills

▶ In the Skill Builders, you will **use variables, constants, and controls; make and use tables;** and **map concepts.**

▶ In the Activities, you will **observe, collect** and **organize data, model, analyze,** and **infer.**

▶ In the MiniLABs, you will **analyze** and **hypothesize.**

151

EXPLORE ACTIVITY

Purpose
LS Kinesthetic Use the Explore activity to introduce students to the concept of heat transfer by convection and radiation. **L1** **LEP**

Preparation
You might want to have volunteers bring in small desk lamps that use incandescent lightbulbs. Obtain several electrical extension cords (if needed) and wide masking tape.

Materials
lamp and bare lightbulb, metric ruler, thermometer, book, paper

Teaching Strategies
Safety Precautions Hot lightbulbs should not be touched by people, thermometers, or paper. Secure extension cords with strong tape to avoid tripping.

Observe
The measured temperature decreases as the distance from the bulb increases. The book should block transfer of energy better than the paper.

✓ Assessment

Performance Have students design an experiment to test the effectiveness of a variety of barriers to heat transfer. Their designs should control variables such as the distance to the thermometer, the placement of the barrier, and the time of exposure. Cardboard frames used for overhead transparencies can be used to mount different blocking materials, such as paper, clear plastic wrap, and foil. Use the Performance Task Assessment List for Designing an Experiment in **PASC**, p. 23. **P**

Section Background

Materials such as metals, with many free electrons that easily transport kinetic energy, are good thermal conductors. Materials that lack this property are good insulators.

Preplanning

- To prepare for Activity 6-1, make several trays of colored ice cubes.
- For the MiniLAB, find a long metal utensil used for grilling, wax, and thumbtacks.

1 Motivate

Bellringer

 Before presenting the lesson, display **Section Focus Transparency 22** on the overhead projector. Assign the accompanying **Focus Activity** worksheet. L1 LEP

2 Teach

Visual Learning

Figure 6-1 What would happen if you used a wooden spoon? *Thermal energy would not be conducted through the spoon, and the end not in contact with the heat source would remain cool.* **What if the spoon were bent?** *It still would conduct heat.* LEP IS

Student Text Question

Why do you think cooking pots are made of metal? *Metals are good conductors of thermal energy.* **What are the handles usually made of?** *plastic or wood (poor thermal energy conductors)*

Science Words

conduction
fluid
convection
radiation
insulator

Objectives

- Compare and contrast the transfer of thermal energy by conduction, convection, and radiation.
- Differentiate between conductors and insulators.
- Explain how insulation affects the transfer of energy.

6•1 Thermal Energy on the Move

Conduction

Thermal energy travels as heat from a material at higher temperature to a material at lower temperature. If you pick up an ice cube, heat from your hand transfers to the ice, causing the ice to melt. If you pick up a hot spoon, the heat from the spoon moves to your hand, perhaps causing you to drop the spoon. How does thermal energy move from place to place?

One way is by conduction. **Conduction** is the transfer of energy through matter by direct contact of particles. Recall that all matter is made up of tiny particles that are in constant motion. The temperature of a material is a measure of the average kinetic energy of its particles.

Transfer by Collisions

Energy is transferred when particles moving at different speeds bump into each other. When faster-moving particles collide with slower-moving particles, some of the momentum of the faster-moving particles passes along to the slower-moving particles. The faster particles slow down and the slower particles speed up.

Heat may be transferred by conduction through a given material or from one material to another. Think about what happens when one end of a metal spoon is placed in boiling water. Heat from the water is transferred to the spoon. The end of the spoon in the water becomes hotter than the other end of the spoon. But, eventually, the entire spoon becomes hot, as in **Figure 6-1.**

Figure 6-1

Part of the spoon is heated by contact with the hot water (A). Heat is transferred through the metal spoon, particle by particle, until the entire spoon is hot (B). *What would happen if you used a wooden spoon? What if the spoon were bent?*

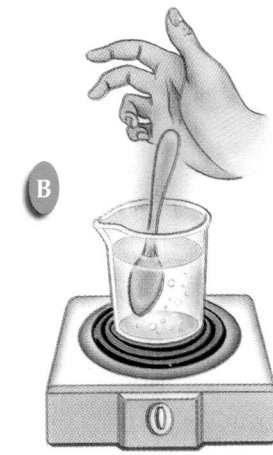

Program Resources

 Reproducible Masters
Study Guide, p. 26 L1
Reinforcement, p. 26 L1
Enrichment, p. 26 L3
Cross-Curricular Integration, p. 10
Activity Worksheets, pp. 5, 32–33, 36 L1
Lab Manual 11

Transparencies
Section Focus Transparency 22 L1

Figure 6-2

Use of these kitchen items is related to their heat-conducting properties. *Which items conduct heat well? Which ones block it?*

Conduction takes place in solids, liquids, and gases. Because their particles are packed closer together, solids usually conduct heat better than liquids or gases. However, some solids, such as the pots in **Figure 6-2,** conduct heat better than others. Many metals have loosely held electrons that move around easily and transfer kinetic energy to nearby particles more efficiently. Silver, copper, and aluminum are good heat conductors, while wood, plastic, glass, and fiberglass are poor conductors of heat. Why do you think cooking pots are made of metal? What are the handles usually made of?

Convection

Liquids and gases differ from solids because they flow. Any material that flows is a **fluid.** The most important way thermal energy is transferred in fluids is by convection. **Convection** is the transfer of energy by the bulk movement of matter. How does this differ from conduction? In conduction, energy moves from particle to particle, but the particles themselves remain approximately in place. In convection, fluid particles move from one location to another, carrying energy with them.

When heat is added to a fluid, the particles begin to move faster, just as the particles of a solid do. However, the particles of a fluid have more freedom to move. Therefore, they move farther apart, or expand. The hot-air balloon in **Figure 6-3** shows the expansion of a heated fluid.

Figure 6-3

Heat from the flame causes the air inside the balloon to expand.

153

Figure 6-4
Only the radiant energy that is absorbed is converted to thermal energy.

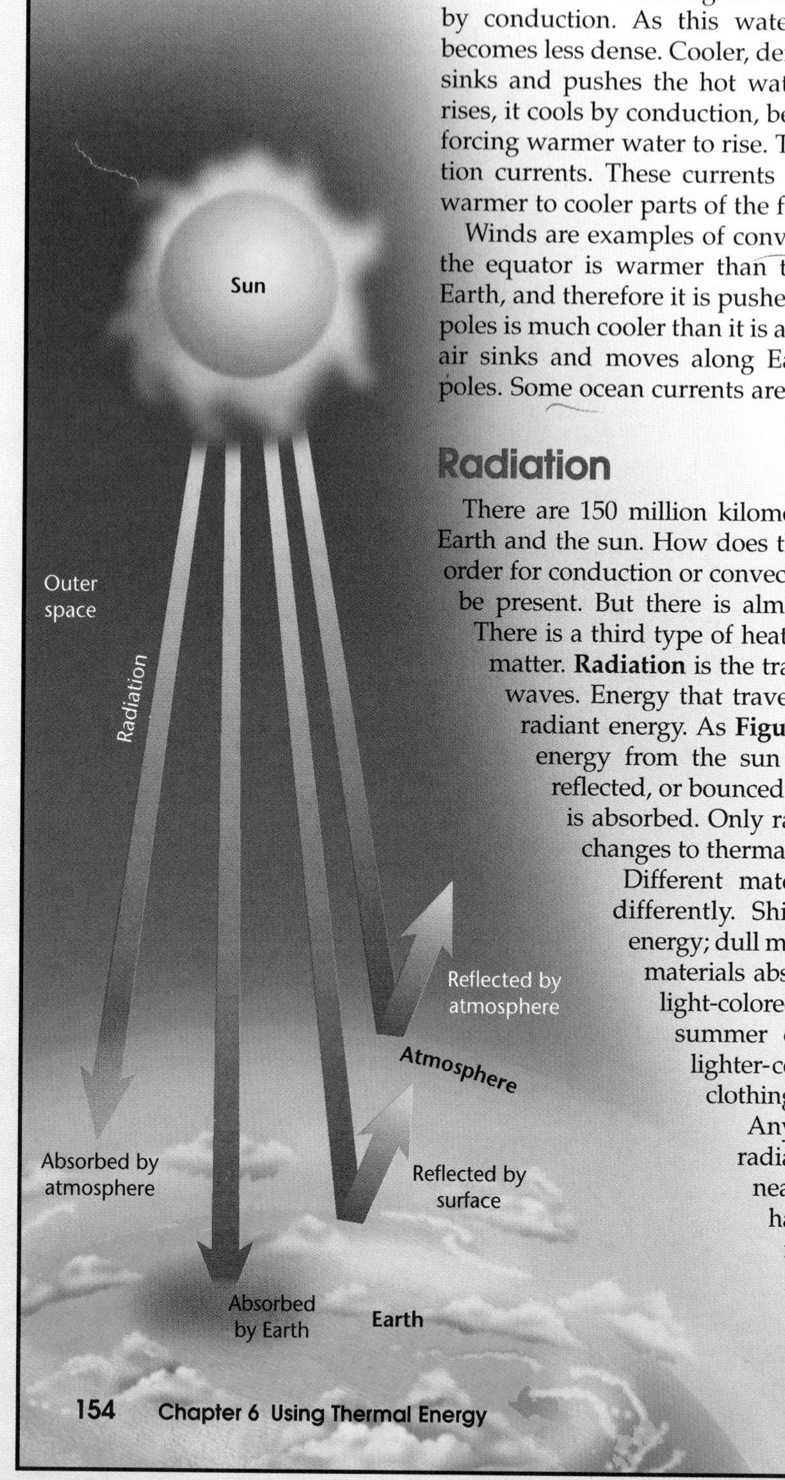

Transfer by Currents

Now think about what happens when a pot of water is heated. The stove burner heats the bottom of the pot by conduction. Water touching the bottom of the pot is also heated by conduction. As this water is heated, it expands and becomes less dense. Cooler, denser water at the top of the pot sinks and pushes the hot water upward. As the hot water rises, it cools by conduction, becomes more dense, and sinks, forcing warmer water to rise. This movement creates convection currents. These currents transfer thermal energy from warmer to cooler parts of the fluid.

Winds are examples of convection currents. The air above the equator is warmer than the air at any other place on Earth, and therefore it is pushed upward. However, air at the poles is much cooler than it is anywhere else. This cold, dense air sinks and moves along Earth's surface away from the poles. Some ocean currents are also convection currents.

Radiation

There are 150 million kilometers of empty space between Earth and the sun. How does thermal energy reach Earth? In order for conduction or convection to take place, matter must be present. But there is almost no matter in outer space. There is a third type of heat transfer that does not require matter. **Radiation** is the transfer of energy in the form of waves. Energy that travels by radiation is often called radiant energy. As **Figure 6-4** illustrates, once radiant energy from the sun reaches Earth, some of it is reflected, or bounced back, toward space, and some is absorbed. Only radiant energy that is absorbed changes to thermal energy.

Different materials absorb radiant energy differently. Shiny materials reflect radiant energy; dull materials absorb it. Dark-colored materials absorb more radiant energy than light-colored materials. This explains why summer clothing is usually made of lighter-colored materials, and winter clothing is darker colored.

Any object warmer than 0 K emits radiation. If you hold your hand near a lighted electric bulb, your hand feels the heat. The bulb's radiation is converted to thermal energy as it is absorbed by your hand.

Sun

Outer space

Radiation

Reflected by atmosphere

Atmosphere

Reflected by surface

Absorbed by atmosphere

Absorbed by Earth   Earth

Across the Curriculum

Language Arts Have students look up the words *conduction, convection,* and *radiation* in a dictionary. Students should find that the words come from the Latin roots *conducere* (to carry), *convehere* (to bring together), and *radius* (ray), respectively. Ask volunteers to explain how the meaning of each word reflects its root. [L1] [IS]

Reducing Thermal-Energy Flow

When the weather is cold, doesn't it feel good to wrap yourself up in a thick sweater or curl up under a thick quilt? And in hot weather, wouldn't you rather have juice from a bottle that has been inside a picnic cooler than one that has been sitting out on a table?

What do these situations have in common? In each case, some method is used to reduce the flow of heat by conduction, convection, or radiation. In the first case, the material of the sweater or quilt traps your body heat and keeps it from escaping to the open air. In the second case, the material of the picnic cooler keeps heat from flowing to the juice. Similarly, the feathers of the birds in **Figure 6-5A** help to restrict heat flow away from the bird's body.

Insulators

Good **insulators** do not allow heat to move easily through them. Some insulators, such as wood, plastic, glass, and fiberglass, have already been identified as being poor conductors.

Gases, such as air, are excellent insulators. Several types of insulating material contain many tiny pockets of trapped air. These pockets restrict the formation of convection currents, and the trapped air is a good insulator. Plastic foam is a type of insulation commonly used in beverage cups and picnic coolers. This foam is mostly tiny pockets of trapped air. Down jackets, such as the one shown in **Figure 6-5B**, and quilts are stuffed with tiny feathers or fibers that trap air.

Buildings are insulated to keep warm air inside in cold weather and outside in hot weather. In the United States, about ten percent of all energy produced is used to heat buildings, so you can see why it is important to prevent as much heat loss as possible. Later in this chapter, you will learn how air conditioners are used to remove heat from building interiors. Insulation reduces the amount of heat that flows into a cooled building from the outside, keeping the inside of the building cooler.

Building insulation is usually made of some fluffy material, such as fiberglass, cellulose, or treated paper. The insulation is packed into outer walls and attics, where it reduces heat flow.

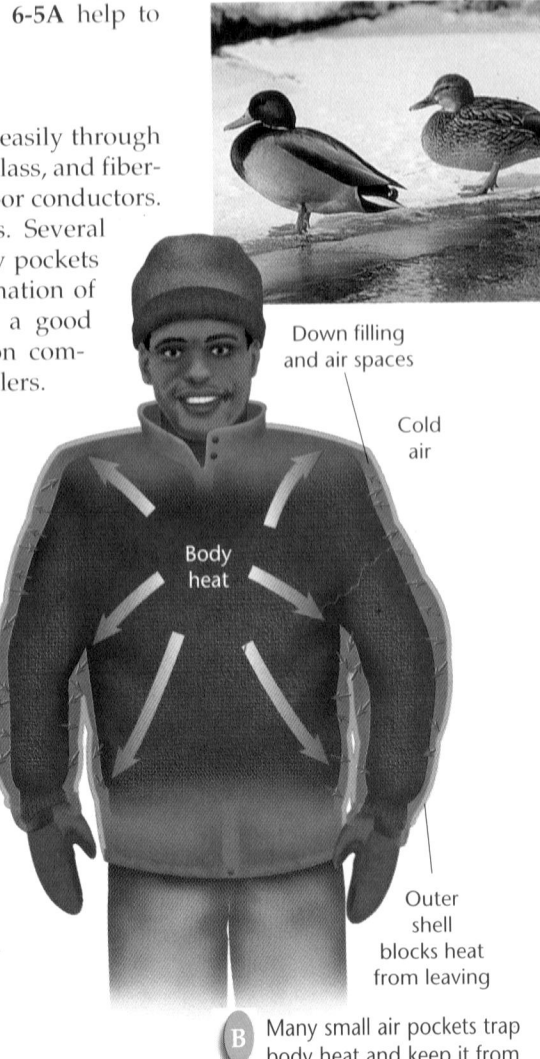

Figure 6-5

Although people use coats and birds have feathers, the effect is the same—to reduce thermal transfer.

A Feathers protect ducks by insulating them against loss of body heat.

Down filling and air spaces

Cold air

Body heat

Outer shell blocks heat from leaving

B Many small air pockets trap body heat and keep it from escaping.

6-1 Thermal Energy on the Move   155

Revealing Preconceptions

Some students may think that materials are intrinsically cold or hot. For example, ask them why bathroom tiles feel cold to bare feet but a shaggy rug on top of the tiles feels warm. The discussion may reveal that students believe tiles are always cold and rugs are always warm. The feelings that the students experience are correct; however, both materials are at the same temperature. Make sure students are aware that the tile feels cool because it absorbs heat from a bare foot and cools the foot. The trapped air among the fibers of the rug provides insulation that dramatically reduces heat conduction and, therefore, reduces heat loss.

Inquiry Question

Why won't you get burned if you grab the hot handle of a pan with a dry dishcloth, but you might if you grab the handle with a wet dishcloth? *The tiny spaces among the fibers of a dry dishcloth are filled with air, which is a poor conductor of heat. However, the spaces among the fibers of a wet dishcloth are filled with water, which conducts heat more readily.*

Use these program resources as you teach this lesson.
Study Guide, p. 26
Cross-Curricular Integration, p. 10
Lab Manual 11

Inclusion Strategies

Learning Disabled Students can test various brands of polystyrene foam cups from fast-food restaurants for their effectiveness. They should place the same amount of the same temperature hot or cold water in the various cups. A thermometer should be placed in the water and the top should be covered with plastic wrap. Students should record the temperature after 5 and then 10 minutes. **LS**

Problem Solving

Solve the Problem

1. More heat is lost through the attic because of thermal updrafts.
2. The insulation and thermal windows all contain pockets of trapped air. These pockets reduce the transfer of thermal energy because air is a poor conductor of heat.

Think Critically

The insulation may save money by preventing energy loss and therefore saving on fuel consumption.

Content Background

Students should also be made aware that the *R*-value of a wall or ceiling made of several layered materials is the sum of the individual *R*-values for each material. For example, a wall made of 1-cm layers of plasterboard, brick, and wood siding would have an *R*-value of 1.03 (0.35 + 0.08 + 0.60).

Problem Solving

The Warm House

Mary and a group of her friends volunteered to help build a house with the Habitat for Humanity organization.

One day, the group went to the new house to install insulation. They wanted to make sure that the house was properly constructed to withstand the cold winters in the area.

The group worked on all the exterior walls to see that they were properly insulated. They used 1"-thick foam insulation backed with aluminum foil. During the next week, they went up to the attic to install two layers of 12"-thick fiberglass insulation between the rafters. Afterward, they checked to see that all the windows had stickers indicating that they were double-paned thermal windows.

Solve the Problem:

1. Why did the Habitat group install more insulation in the attic than in the walls?
2. What do the foam insulation, the fiberglass insulation, and the thermal windows have in common that helps them to prevent the loss of heat?

Think Critically:

The insulation in the home that the group worked on was expensive. Why might they think that it was a good investment or not?

156

Insulation Ratings

To help consumers understand the energy-transfer qualities of the many materials used in building construction, a rating system has been established in which each material is given an *R*-value. The *R* indicates resistance to heat flow. **Table 6-1** shows the *R*-values for some common building materials. The *R*-value of a material is a measure of the resistance that a 1 m x 1 m square slab of the material has to heat flow per centimeter of the material's thickness.

To have a well-insulated house, materials with a total *R*-value of at least 19 should be used in the outer walls and those with total *R*-values of 30-44 should be used in the roof or ceilings. Higher *R*-values are needed for roofs and ceilings because the warmest air inside the house is carried upward by convection currents. The *R*-values needed for comfort also depend on the local climate.

Across the Curriculum

Life Science Ask the following questions: **Why do birds fluff their feathers and mammals fluff their fur to keep warm?** *The fluffing increases the amount of air trapped in the feathers or fur, which increases their insulating properties.* **Does blubber have a high or low *R*-value? Explain.** *Blubber has a high R-value; it acts as an insulator in Arctic mammals.*

Even though glass is a good insulator, windows do transfer thermal energy. A single pane of glass has an *R*-value of only 1, but using double-pane windows, such as the one in **Figure 6-6**, reduces heat loss considerably because a thin sandwich of air is trapped between two panes of glass. The air is an excellent insulator, and the narrow space restricts the formation of convection currents.

Some double-pane windows have a higher *R*-value because air is replaced with a harmless, colorless gas that is a better insulator. Argon is the gas most frequently used in windows because it is nonreactive and inexpensive. These windows are especially useful in cold climates or where temperature extremes can be uncomfortable or even dangerous.

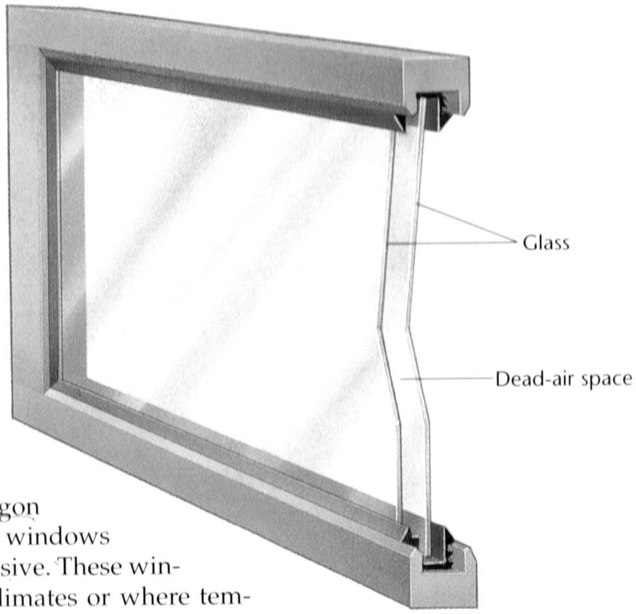

Adding insulation and specially equipped windows are not the only ways to reduce heat loss from buildings. Many buildings, especially older ones, have cracks and gaps around windows and doors through which heat escapes. These heat leaks can be filled with caulking, putty, or weather stripping to reduce energy loss and to lower fuel bills.

Glass

Dead-air space

Figure 6-6

The air trapped in this double-pane window serves as a layer of insulation.

Table 6-1

R-Values of Various Materials*

Material	R-Value
Brick	0.08/cm
Plasterboard (drywall)	0.35/cm
Stucco	0.08/cm
Wood siding	0.60/cm
Air space	1.82-3.56/cm
Fiberglass	1.22/cm
Loose cellulose	1.46/cm
Aluminum siding	0.01/cm
Loose foam	1.89/cm
Loose vermiculite	1.09/cm

*The total *R*-value is the product of the *R*-value and the thickness of the layer in centimeters.

6-1 Thermal Energy on the Move   157

- **What is the *R*-value of a wall made of bricks that are 9 cm thick?** R-*value* = 0.72; 9 cm × 0.08/cm = 0.72
- **How thick would a solid brick wall have to be to have an *R*-value of 19?** 19 ÷ 0.08/cm = 240 cm

Visual Learning

See p. 158.

Figure 6-7 How does each of these features block the transfer of heat? *Plastic and the vacuum space are poor conductors of thermal energy. The reflecting surface reflects thermal energy back toward the contents of the container.* LEP

LS

Student Text Question

See p. 158.

What other ways can you think of? *Large trees block thermal radiation from the sun.*

3 Assess

Check for Understanding

? FLEX Your Brain

Use the Flex Your Brain activity to have students explore INSULATORS.

📁 **Activity Worksheets,** p. 5

Reteach

Explain how a vacuum bottle insulates any material it contains.

Extension

📁 For students who have mastered this section, use the **Reinforcement** and **Enrichment** masters.

4 Close

Use the Mini Quiz to check students' recall of chapter content.

1. _____ is the transfer of energy through matter by direct contact of particles. *Conduction*

2. _____ is the transfer of energy in the form of waves. *Radiation*

Section Wrap-up

Review

1. Conduction transfers thermal energy by direct contact; convection by matter moving from one location to another; radiation in the form of waves.

2. Heat is not transferred easily through a poor conductor, which is characteristic of a good insulator.

3. Wood and plastic are poor conductors.

4. **Think Critically** The plastic foam contains hundreds of tiny air bubbles that act as insulation to keep the food warm.

USING MATH

30/1.22/cm = 24.6 cm

Skill Builder

Example Experiment

Hypothesize Different materials do not insulate equally.

Objective Compare the abilities of plastic foam pellets, shredded newspaper, and crumpled plastic bags to keep an ice cube from melting.

Procedure Place three similarly sized ice cubes, one each, into three plastic sandwich bags. Center each of these bags into identical empty soup cans

and pack each with a different insulating material. Place all three cans in direct sunlight. After 30 minutes remove the bags and measure the amount of melted ice water in each bag.

Controls size of ice cubes, size of cans, amount of insulating material, time exposed to the sun

Figure 6-7

The plastic container, vacuum layer, and reflecting surfaces all add to the insulating properties of vacuum bottles. *How does each of these features block the transfer of heat?*

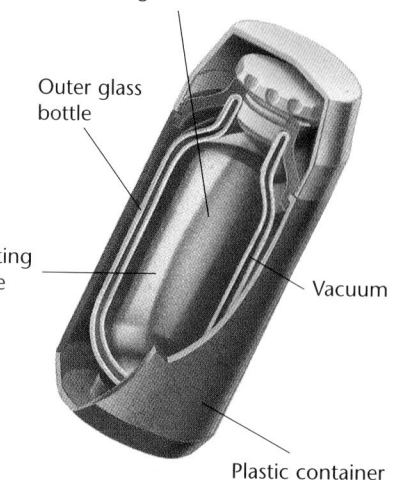

Inner glass bottle

Outer glass bottle

Reflecting surface

Vacuum

Plastic container

You may have used a vacuum bottle, like the one in **Figure 6-7,** to carry cold or hot liquids for lunch. This bottle is similar to a double-pane window. It has a double glass wall with a vacuum between the layers to prevent heat transfer by convection. One side of each layer is coated with aluminum to reduce heat transfer by radiation.

Now that you know about how thermal energy is transferred, think about things you can do to stay cooler or warmer. You can sit under a shady umbrella or put on a sweater. You can wear light-colored clothing or close a window. You can weatherproof your home by adding insulation and eliminating heat leaks around windows and doors. What other ways can you think of?

Section Wrap-up

Review

1. Name and describe three methods of thermal energy transfer.

2. Explain why poor conductors of heat are good insulators of heat.

3. Why do many cooking pots and pans have wooden or plastic handles?

4. **Think Critically:** For many years, many fast-food restaurants served their hot sandwiches in plastic foam containers instead of wrapping them in less expensive paper. Why was plastic foam used?

Skill Builder

Using Variables, Constants, and Controls
Design an experiment to find out which material makes the best insulation: plastic foam pellets, shredded newspaper, or crumpled plastic bags. Remember to state your hypothesis and indicate what factors must be held constant. If you need help, refer to Using Variables, Constants, and Controls in the **Skill Handbook.**

USING MATH

Suppose you were planning to use fiberglass insulation to achieve a total *R*-value of 30. Use the information in **Table 6-1** to calculate how thick (cm) the fiberglass layer must be.

Assessment

Performance After students complete the experiment, ask them to describe how they would improve the experimental design. Use the Performance Task Assessment List for Assessing a Whole Experiment and Planning the Next Experiment in **PASC,** p. 33.

People and Science

MAGALI REGIS, *Architect and Solar Cooking Expert*

On the Job

Q Ms. Regis, you teach people in Haiti how to build and use solar cookers. Why is this important?

A One of the reasons we're trying to introduce solar cooking is because of the terrible deforestation problems Haiti faces. The Haitian people cook with charcoal, which is made from wood. And people have to pay for charcoal. It's not cheap. Sunlight is free. So solar cookers save them money. That means they can eat better food, or maybe send an extra child to school.

Q How does a solar box cooker work?

A You have a box with a glass cover on top that lets in sunlight. The box is well sealed, so heat cannot escape. You put your food in a thin-walled black pot in the center of the box. The inside of the box is lined with a reflecting material like aluminum foil. The box becomes an oven that can be used to bake bread, or cook cornmeal, rice, and vegetables.

Personal Insights

Q Are there other solar cookers?

A Another type is the parabolic cooker, which is more like the burner on a stove. It has the shape of a satellite dish and is lined with reflective material. It focuses sunlight on a point just above the center of the dish, where you place your pot. It produces temperatures high enough for boiling and frying.

Q How did you become involved in teaching solar cooking?

A To me, solar technology is the way to go to help reduce the need to burn fuels. I first saw a solar cooker during a visit to an organic farm in Oregon—we cooked on it almost every day during the summer.

Career Connection

Architects design buildings and supervise their construction. Use the yellow pages of the phone book to find an architect in your area. Interview him or her to find out how architects are trained and what their work is like. Ask how the need for heating and air conditioning influences building design.

- **Architect**
- **Building Contractor**

- Solar technology experts include nonscientists who operate all kinds of solar devices. Solar experts also include highly educated engineers and physicists who develop new solar technologies.

- International aid workers include teachers, engineers, farmers, and doctors who help people whose lives are affected by poverty, war, drought, or famine. The International Red Cross and the Peace Corps are agencies that train and send out aid workers.

For More Information

- **Aprovecho Institute**
  80574 Hazelton
  Cottage Grove, OR 97424

Develops solar cookers and high-efficiency stoves and teaches people how to use them.

- **Solar Cookers International**
  1724 11th Street
  Sacramento, CA 95814

Distributes simple plans for making solar cookers with inexpensive materials.

People and Science

Background

- More than 15% of Haiti's people have left the countryside and moved to cities, partly because deforestation has led to large-scale erosion. Much of the land can no longer be cultivated. Haitian families spend as much as one-third of their income for cooking fuel.

- Fuelwood scarcity is a problem in Africa, Asia, and the Middle East, as well as Haiti. Charcoal is used in urban areas in developing countries because it is lighter than wood and easier to transport. But converting wood to charcoal uses up more than half the energy contained in the wood, speeding the loss of trees. Solutions include planting fast-growing trees in areas where fuel is needed; developing and using fuel-efficient stoves that can be made easily with cheap, local materials; and using alternative fuels, such as solar energy.

Teaching Strategies

- Review with students the basic principles involved in solar-cooking design: absorption, reflection, and insulation.

LS **Logical-Mathematical** Challenge students to develop short activities that demonstrate the principles involved in designing and using solar cookers. Activities could answer questions, such as: What materials best absorb heat? What materials trap heat? What materials reflect heat?

Career Connection

- An architect must have a high school diploma and complete a course of study at an architectural school. The profession involves technical skills, including drafting, as well as creative ability.

159

Activity 6-1

PREPARATION

Purpose

LS **Kinesthetic** Recognize the cause and effect in relationships between convection currents and fluid density. **L1**

LEP **COOP LEARN**

Process Skills

recognizing cause and effect, observing and inferring, classifying, making and using tables, comparing and contrasting, forming operational definitions, forming a hypothesis, designing an experiment, separating and controlling variables, interpreting data, making models, measuring in SI

Time

two 45-minute class periods

Materials

Make plenty of ice cubes colored with food coloring (four per group). Allow about 70 g of salt to each team. Most of the salt will dissolve with considerable stirring.

Safety Precautions

Caution students to place thermometers far from the edge of their tables. Remind them to stir with stirring rods, not thermometers.

Possible Hypotheses

Student hypotheses may reflect that the cold water from the melting ice cube will be more dense than the warm fresh water and will therefore sink. The cold water may be less dense than the warm salt water, so the meltwater will stay on top.

📁 **Activity Worksheets,** pp. 5, 32–33

Activity
6-1

Design Your Own Experiment
Creating Convection Currents

What happens when you use ice cubes to cool a drink? After a while, the drink feels cooler than before you added the ice. What actually happens to the particles that make up the ice cube when it melts?

PREPARATION

Problem
How does convection aid in the cooling of a drink when ice is added?

Form a Hypothesis
Write a hypothesis describing what you think will happen when a colored ice cube melts in a sample of warm fresh water and in a sample of warm salt water.

Objectives
- Design an experiment to explore the relationships between convection currents and fluid density.
- Observe what happens when a colored ice cube melts in warm fresh water and salt water.

Possible Materials
- colored ice cubes
- warm, fresh water
- salt
- tongs
- a ruler
- thermometer
- beaker, 250-mL
- stirring rod

Safety Precautions 🧪👓
Be sure to keep thermometers away from the table's edge. Stir with stirring rods only, not thermometers.

160   Chapter 6 Using Thermal Energy

Inclusion Strategies

Visually Impaired Place students who have difficulty reading thermometers and recording results in charge of formulating and evaluating the hypothesis.

PLAN THE EXPERIMENT

1. Remember that the temperature will vary with the depth of the water.
2. Avoid creating convection currents by physically disturbing the water.
3. If you need a data table, design one in your Science Journal.

Check the Plan

1. Be sure your data table has space to record the water temperature in salt water and in fresh water at the surface, the bottom, and at one (or more) measured depth(s) in between. Will you measure these temperatures before adding ice, after the ice has been melting for a few minutes, or both?
2. The photo at the right shows a way to place the colored ice cube into the water samples while keeping the water as still as possible.

3. How will you make a saturated solution of warm salt water? Will you do the experiment the same way with the fresh water and salt water?
4. Do you have to run any tests more than one time?
5. In addition to your temperature measurements, how will you record your observations of what is happening?
6. *Make sure your teacher approves your plan and that you have included any changes suggested in the plan.*

DO THE EXPERIMENT

1. Carry out the approved experiment as planned.
2. During the experiment, write down any observations that you make and complete the data table in your Science Journal.

Analyze and Apply

1. As the ice melts, **compare** what happens to the colored

meltwater in the samples of fresh water and salt water.
2. Do you observe convection currents in this activity? Explain.
3. Of the three liquids involved in this activity (warm fresh water; warm salt water; and cold colored meltwater), **predict** which is the most dense.

Go Further

Predict what would happen to the coloring and temperature of the liquid if you left the beaker containing fresh water and a colored ice cube untouched for 30 minutes. Test your prediction.

Go Further

After a while, the temperature throughout the liquid will become nearly the same and the food coloring will be mixed evenly in the solution.

 Assessment

Process Ask students if this activity could be used as a model of convection currents in a large lake. Have them explain why the water is colder at the bottom of a lake in the summertime. Use the Performance Task Assessment List for Making Observations and Inferences in **PASC**, p. 17. **P**

Possible Procedures

Students should measure the temperatures at the three depths both before adding the ice and after the ice has been melting for a few minutes. They should use tongs to gently lower the ice cube into the water. They should prepare a data table for recording their temperature measurements and their observations.

Teaching Strategies

- Remind students that less dense materials will float on more dense materials.
- Students may need help setting up their data table and deciding at what depths to make their measurements.

DO THE EXPERIMENT

Expected Outcome

The colored meltwater will be more dense than warm fresh water, and the colored water will sink. The colored meltwater will be less dense than the warm salt water, so it will float.

Analyze and Apply

1. In fresh water, the cold meltwater streams to the bottom and the warm water moves to the top, resulting in a convection current. In salt water, the colored meltwater stays on top.
2. Yes, the movement of water between areas of different density causes a convection current.
3. Salt water is the most dense because cold meltwater sinks in fresh water but floats in the salt water.

Prepare

Section Background

- The most common types of heating systems are hot water, steam, and forced hot air.
- In homes containing passive or active solar heating, it is used primarily in conjunction with conventional heating systems or to supplement heating water for washing and bathing.

Preplanning

Gather some glue, aluminum foil, poster board, string, coat hangers, and chocolate for the MiniLAB.

1 Motivate

Bellringer

Before presenting the lesson, display **Section Focus Transparency 23** on the overhead projector. Assign the accompanying **Focus Activity** worksheet. L1 LEP

Tying to Previous Knowledge

Have students identify the energy sources and the type of energy used in their home-heating systems. From a class poll, rank the energy sources from most used to least used.

Visual Learning

Figure 6-8 How does the warm air near the radiator get to other parts of the room? *by convection* LEP IS

6•2 Using Heat to Stay Warm

Science Words
radiator
solar energy
solar collector

Objectives
- Describe three types of conventional heating systems.
- Explain how solar energy can be used to heat buildings.
- Explain the differences between passive and active solar heating systems.

Figure 6-8

The design of this radiator permits large quantities of air to be heated by contact with its surface. *How does the warm air near the radiator get to other parts of the room?*

Figure 6-9

In this heating system, water is heated by a furnace. The hot water is pumped to the radiators to heat the rooms of the house.

Conventional Heating Systems

What is the climate like where you live? Is it cold part of the year or is it warm all year-round? Sometimes, even in warm climates, the weather can be cold enough that buildings need to be heated. For this reason, most buildings have some kind of heating system.

All heating systems must have a source of energy, such as fuel or electricity. The simplest type of heating system is one in which fuel, such as wood, coal, or gas, is burned right in the area to be heated. The energy released by the burning fuel is transferred to the surrounding air by conduction, convection, and radiation.

Radiator Systems

Many heating systems use radiators to transfer energy. The **radiator** shown in **Figure 6-8** is a device with a large surface area designed to heat the air near it by conduction. Convection currents then circulate the heat to all parts of the room.

In some heating systems, radiators are heated by electricity. However, most systems are designed so that fuel is burned in a furnace, and the heat is transported to radiators throughout the building. In the system shown in

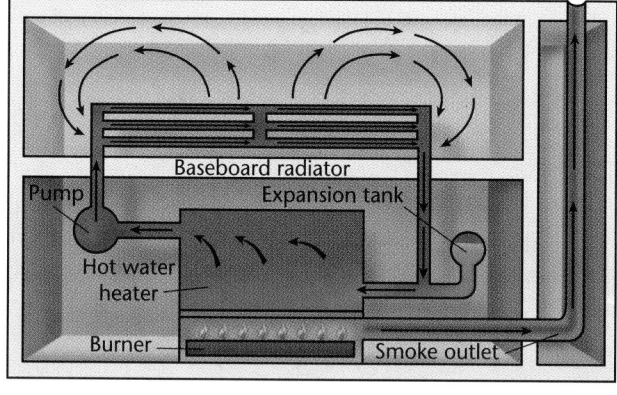

Baseboard radiator
Pump
Expansion tank
Hot water heater
Burner
Smoke outlet

Program Resources

📁 **Reproducible Masters**
Study Guide, p. 27 L1
Reinforcement, p. 27 L1
Enrichment, p. 27 L3
Science and Society Integration, p. 10
Activity Worksheets, pp. 5, 37 L1
Science Integration Activities, pp. 11–12
Lab Manual 12
Concept Mapping, p. 17

🔦 **Transparencies**
Section Focus Transparency 23 L1
Teaching Transparency 11 L1

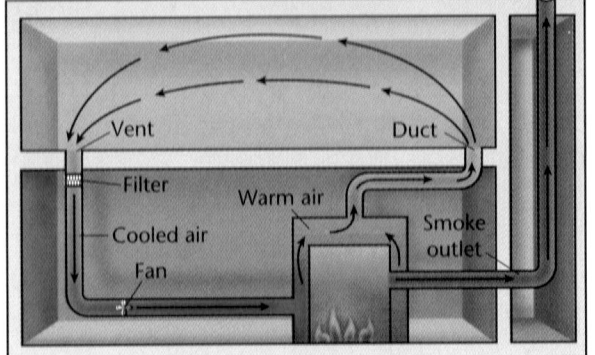

Figure 6-10

In a forced-air heating system, air heated by the furnace is used to heat the rooms of the house.

Figure 6-9, the furnace uses the energy to heat water, which is then pumped through pipes to the radiators. The cooler water is returned to the furnace to be heated again.

In a similar type of heating system, the furnace heats water to its boiling point, producing steam. The steam travels through insulated pipes to the radiators. As it cools, the steam condenses to water, which returns to the furnace. Steam-heating systems use about one-fiftieth as much water as hot-water systems, but the pipes and furnace need special insulation to keep the steam from condensing before it reaches the radiators.

Forced-Air Systems

In the forced-air system shown in **Figure 6-10,** energy released in the furnace heats air. A blower forces the heated air through a system of large pipes, called ducts, to openings, called vents, in each room. In the rooms, the warm air circulates by convection. Cooler air returns through other vents and ducts to the furnace to be reheated.

Electrical Heating Systems

Some buildings are heated entirely by electricity. Heating coils are enclosed within floors or ceilings and are heated by electrical energy. Nearby air is heated by conduction, and people and materials in the room are also warmed by radiation. Such systems, sometimes called radiant electric heating systems, provide even heating but may be too expensive to operate because of the cost of electricity in some areas.

Solar Heating

If you have ever gotten into a car that has been sitting in direct sunlight for any length of time, you know that energy from the sun can be changed to thermal energy. Energy from the sun is known as **solar energy.** Because solar energy is free, the idea of using it to heat buildings is especially appealing.

6-2 Using Heat to Stay Warm   **163**

Cultural Diversity

Central Heating The Chinese were the first to use central heating for their homes. Instead of building fires inside the house, they would build one outside and heat air to create a convection current that would move into the house. The Inuit reverse the concept in their igloos. By digging the entranceway lower than the inside of the igloo, they create a natural pressure barrier to keep cold air out. Since cold air sinks to the lowest point, the entranceway fills with cold air from the outside and then the flow stops. The warmer air inside the igloo stays in because it would have to go down through colder air to escape.

Purpose

LS **Logical-Mathematical** Students will design and build a solar cooker to melt a piece of chocolate. **L2**

COOP LEARN

Materials

glue, aluminum foil, strips of poster board about 30 cm × 1 m (stiff enough to hold a smooth curve), string, wire coat hangers, chocolate, watch, ruler

Teaching Strategies

- The arc will be closer to a parabola than a circle.
- The performance of the solar cooker will be affected by many factors: curvature and placement of the chocolate, time of day, weather, etc.

Analysis

1. Radiant energy from the sun is heating the chocolate most effectively.
2. adjust the placement of the chocolate, change the curvature, etc.

✓ Assessment

Performance Ask students to design a way to use soft wax to find the hottest areas in the solar cooker. The melting order of equal masses of wax placed in different locations would show the location of the hottest areas. Use the Performance Task Assessment List for Designing an Experiment in **PASC**, p. 23. **P**

📁 **Activity Worksheets,** pp. 5, 37

Student Text Question

Why do you think it's painted black? *Black absorbs more radiant energy than lighter colors absorb.*

Mini**LAB**

How can the sun be used to warm food?

Procedure

1. Make a solar cooker by gluing aluminum foil to a large piece of poster board.
2. Bend the board into a U-shape so that the ends of the board are nearly parallel, but about 30 cm apart. Use string stretched between its corners to hold the board in this shape.
3. Use wire coat hangers to make a mount that will hold the board with the center of the U pointing directly at the sun.
4. Predict where the hottest region of your solar cooker is. From another string bridge connecting the edges of the board, dangle a piece of chocolate into the hottest area.
5. Check the chocolate every few minutes to see if it begins to melt. You might wish to try other spots, as well.

Analysis

1. Which energy-transfer process is heating your chocolate the most? Explain.
2. How might you make the cooker more effective?

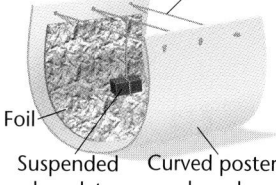

String holding poster board in curved shape

Foil

Suspended chocolate

Curved poster board

Figure 6-11

This passive solar heating system changes the sun's energy to thermal energy.

Passive Solar Heating

Two basic types of systems are used to capture solar energy and convert it to thermal energy to heat buildings. The first type is passive solar heating. Passive systems use no fans or mechanical devices to transfer heat from one area to another. Some materials in the system absorb radiant energy during the day, convert it to thermal energy, and radiate the thermal energy after dark.

As shown in **Figure 6-11,** a home with a passive solar heating system usually has a wall of large windows on the south side to receive maximum sunlight. The other exterior walls are heavily insulated and have few windows. During the day, sunlight passes through the windows and is absorbed by materials such as water or concrete. The radiant energy is converted to thermal energy, which is stored in the absorbing material. Later, when the house begins to cool, the stored energy warms the rooms.

Active Solar Heating

As you may have guessed, the second type is called an active solar heating system. Most active solar heating systems include **solar collectors,** devices that absorb radiant energy from the sun. The collectors are usually installed on the roof or south side of a building. The radiant energy is used to heat water or air that is then pumped throughout the house.

Figure 6-12 shows a home with one type of active solar collector. The metal plate absorbs radiant energy from the sun. Why do you think it's painted black? The glass or plastic sheets reduce energy loss due to convection. Water-filled pipes are located just beneath the metal plate. Radiant energy is absorbed and converted to thermal energy to heat the water in the pipes. A pump circulates the heated water to radiators in the system. Cooled water is returned to the collector to be reheated. Some systems have large, insulated tanks for storing heated water to be used as needed.

Community Connection

Energy Efficiency Invite a representative from a utility company or a commercial company to discuss the advantages and disadvantages of various heating and cooling system alternatives. Have them address initial cost, energy efficiency, and potential long-term maintenance requirements. They might also discuss the importance of insulation.

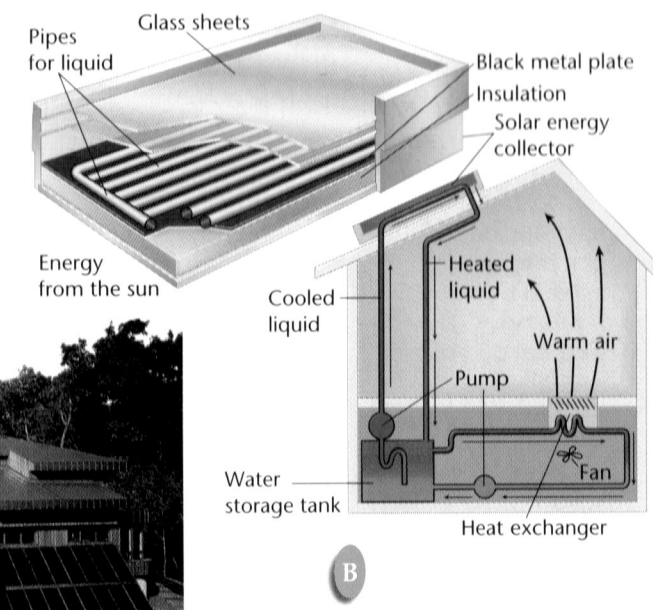

Figure 6-12

The house below is heated by an active solar heating system. Metal plates on the roof absorb heat from the sun. The operation of the system is illustrated in B.

A

Glass sheets

Pipes for liquid

Black metal plate
Insulation
Solar energy collector

Energy from the sun

Cooled liquid

Heated liquid

Warm air

Pump

Water storage tank

Fan

Heat exchanger

B

Section Wrap-up

Review

1. What are the main differences among electrical, radiator, and forced-air heating systems?

2. Compare and contrast active and passive solar heating systems.

3. **Think Critically:** Suppose you are an architect from a cold climate who has been asked to design a building for Tucson, Arizona, where the temperature averages 22°C and can reach 50°C in summer. What do you know about keeping buildings warm that will help a building stay cool in hot weather?

Science Journal

In your Science Journal, describe and diagram the heating system in your home, apartment, or school. Explain how conduction, convection, and radiation are involved.

Skill Builder
Making and Using Tables

In a table, organize information about the kinds of heating systems discussed in this section. Include any type of information you think is important. If you need help, refer to Making and Using Tables in the **Skill Handbook.**

Skill Builder

Type of Heating System	Specific Name	Heating Method
Conventional	Radiator	Conduction/convection
	Forced-air	Convection
	Radiant electric	Conduction/radiation/convection
Solar	Passive	Radiation
	Active	Radiation/conduction

✓ Assessment

Performance Have students use their table to analyze the type of heating system used in at least two buildings other than their home or school. Use the Performance Task Assessment List for Making Observations and Inferences in **PASC,** p. 17. **P**

---

4 CLOSE

Prepare

Section Background

By reversing the processes of a heat engine, heat can be transferred from a material at a lower temperature to a material at a higher temperature. Work must be done on a system to reverse the natural flow of heat.

Preplanning

For Activity 6-2, make sure the materials listed are available in your supply room.

1 Motivate

Bellringer

 Before presenting the lesson, display **Section Focus Transparency 24** on the overhead projector. Assign the accompanying **Focus Activity** worksheet. L1 LEP

Tying to Previous Knowledge

Point out to students that just as they have the ability to lift and move objects, heat also can do work.

Visual Learning

Figure 6-13 List at least one advantage and one disadvantage of one of these engine types. *four cylinder: advantage—consumes less energy, disadvantage—produces less power; six cylinder: advantage—produces more power, disadvantage—consumes more energy* LEP [S]

Science Words

heat engine
combustion
internal combustion engine
external combustion engine
heat mover

Objectives

- Describe how internal combustion engines and external combustion engines work.
- Explain how a heat mover can transfer thermal energy in a direction opposite to that of its natural movement.

Figure 6-13

Most car engines have either four or six cylinders. *List at least one advantage and one disadvantage of one of these engine types.*

Heat Engines

Heat engines are devices that convert thermal energy into mechanical energy. Heat engines burn fuel in a process called **combustion,** which means rapid burning. The two main classes of heat engines are based on where combustion happens—inside the engine or outside the engine.

Internal Combustion Engines

In an **internal combustion engine,** fuel burns inside the engine in chambers called cylinders. Gasoline and diesel engines such as those used in cars and trucks are examples of internal combustion engines.

Figure 6-13A shows what the cylinders look like inside the gasoline engine of a car. Each cylinder has at least two openings that open or close with valves. A piston inside each cylinder moves up and down, turning a crankshaft. The motion of the crankshaft is transferred to the wheels of the car through a series of moving parts. The wheels exert a force on the road through the tires. The equal and opposite force of the road on the tires pushes the car forward.

Each up or down movement of the piston is called a stroke. An automobile engine is called a four-stroke engine because the piston makes four strokes in each cycle. Follow the steps of the four-stroke cycle in **Figure 6-13B–E.**

Some engines have fuel injectors instead of carburetors. In fuel-injected engines, only air enters the cylinder during the intake stroke. During the compression stroke, fine droplets of fuel are injected directly into the compressed air in the cylinder. The other steps are the same as in an engine with a carburetor.

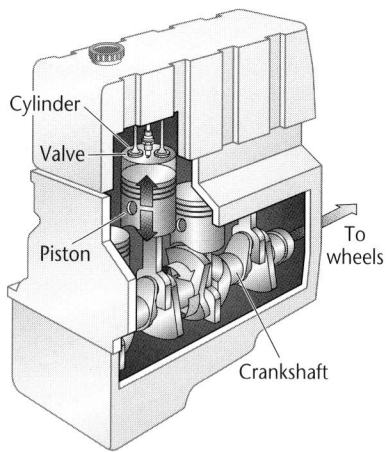

Cylinder
Valve
Piston
To wheels
Crankshaft

 A typical arrangement of cylinders and pistons in a gasoline engine is shown.

166 Chapter 6 Using Thermal Energy

Program Resources

📁 **Reproducible Masters**
Study Guide, p. 28 L1
Reinforcement, p. 28 L1
Enrichment, p. 28 L3
Activity Worksheets, pp. 5, 34–35 L1
Multicultural Connections, pp. 15–16
Lab Manual 13
Technology Integration, pp. 9–10
Critical Thinking/Problem Solving, p. 12

📽 **Transparencies**
Section Focus Transparency 24 L1
Teaching Transparency 12 L1
Science Integration Transparency 6

In a diesel engine, fuel also is injected into compressed air in the cylinder. But the engine has no spark plugs. The fuel-air mixture is compressed so much that it becomes hot enough to ignite without a spark.

In an internal combustion engine, only part of the thermal energy produced by burning fuel is converted to mechanical energy. The rest is left over as waste thermal energy due to friction and the heating of engine parts. Gasoline engines convert only about 12 percent of the chemical potential energy in the fuel to mechanical energy. Diesel engines convert up to 25 percent of the fuel's potential energy to mechanical energy.

External Combustion Engines

In an **external combustion engine,** fuel is burned outside the engine. In old-fashioned steam engines such as those used to power early locomotives, fuel was burned to boil water in a chamber outside the engine. The resulting steam passed through a valve in the engine, where it pushed a piston. The motion of the piston was transferred to the wheels of the locomotive.

Intake valve   Cylinder   Fuel-air mixture   Piston   Spark plug   Exhaust valve

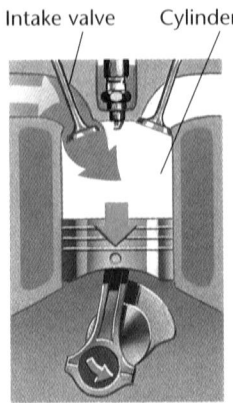 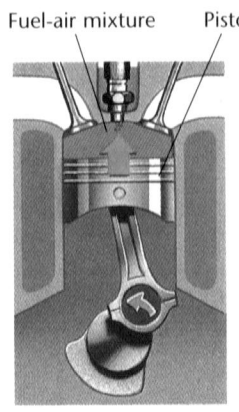 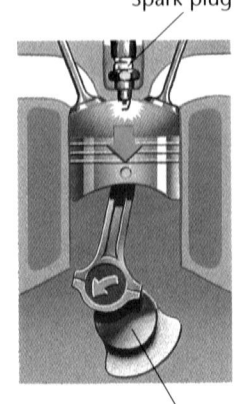 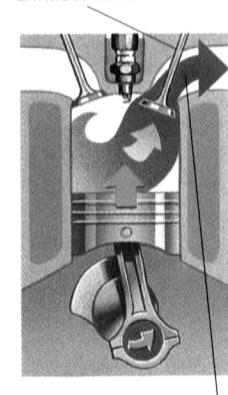

Crankshaft   Exhaust gases

B *Intake stroke.* In another part of the engine called the carburetor, gasoline is broken up into fine droplets and mixed with air. In the cylinder, the intake valve opens and the piston moves downward, drawing the fuel-air mixture into the cylinder.

C *Compression stroke.* The intake valve closes, and the piston moves up. The fuel-air mixture is squeezed, or compressed, into a smaller space.

D *Power stroke.* When the piston is almost at the top of the cylinder, a spark plug produces a hot spark that ignites the fuel-air mixture. As the mixture burns, hot gases expand, forcing the piston down. Energy is transferred from the piston to the wheels of the car through the crankshaft and other moving parts.

E *Exhaust stroke.* The piston moves up again, forcing the waste products from burning the fuel-air mixture out the exhaust valve.

Content Background

- In engines with several cylinders, the piston in each cylinder goes through a four-stroke cycle. However, at any given time, the pistons in the various cylinders will be in different positions.

- As more air is pumped into a bicycle tire, the tire becomes warm. The air temperature rises as work is done to compress air inside the tire. As a gas expands, it cools. A pressurized can of paint cools as the paint is released.

GLENCOE TECHNOLOGY

 Videodisc

STVS: Chemistry
Disc 2, Side 1
Fire Safety Tests (Ch. 6)

Fire-Resistant Clothing (Ch. 7)

Use these program resources as you teach this lesson.
Technology Integration, pp. 9-10
Study Guide, p. 28
Multicultural Connections, pp. 15-16

Use **Teaching Transparency 12** as you teach this lesson.

Making a Model

Picture a heat engine as a system that takes in a quantity of heat at a higher temperature and produces work and a smaller quantity of heat at a lower temperature as exhaust.

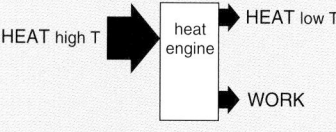

Running the system in reverse and doing work on the heat engine reverses the flow of heat, as in a refrigerator or heat pump.

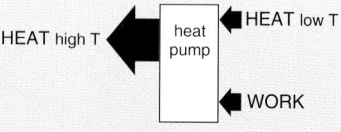

INTEGRATION
Life Science

Sweat glands exist in the dermis, the inner layer of skin lying below the epidermis. About 2 m² of skin, which has a mass of about 2.7 kg (6 pounds), covers an average mature human body.

Modern steam engines, such as those used in electrical power stations, don't have pistons. Instead, steam is directed onto huge, fanlike devices called turbines. The steam pushes against the turbine blades, which then rotate rapidly to turn a cylindrical shaft, thus producing mechanical energy.

Heat Movers

When you put warm food into a refrigerator, the food gets cooler. Where does the thermal energy in the food go? Feel the back of the refrigerator, and you'll notice that it's warm. But you know that heat always flows from warmer to cooler areas. So how can heat flow from the cool refrigerator to the warm room? It can't unless work is done. The energy to do the work comes from the electricity that powers the refrigerator.

A refrigerator is an example of a heat mover. A **heat mover** is a device that removes thermal energy from one location and transfers it to another location at a different temperature. As shown in **Figure 6-14**, refrigerators use the process of evaporation to remove heat from the food inside. A liquid is pumped through coils inside the refrigerator. In most cooling systems, liquid Freon is used because it evaporates at low temperatures. As the liquid evaporates, it absorbs heat, cooling the inside of the refrigerator. The Freon gas is then pumped to a compressor on the outside of the refrigerator. Compressing the Freon causes its temperature to rise above room temperature. So, the Freon loses heat to the air around it and condenses again. The excess heat is transferred into the room, sometimes with the help of fans.

An air conditioner is another kind of heat mover. It removes thermal energy from a warm house and transfers it to the even warmer outdoor surroundings. Refrigerators and air conditioners are heat engines working in reverse —they use mechanical energy to move thermal energy from cooler to warmer areas.

A heat pump is a special two-way heat mover. In warm weather, it operates like an air conditioner. In cold weather, it removes thermal energy from cool outside air and transfers it to the inside of the house.

Figure 6-14

In a heat mover such as a refrigerator, thermal energy is removed from one location and is released in another location. *Where does the heat that is removed from food in a refrigerator go?*

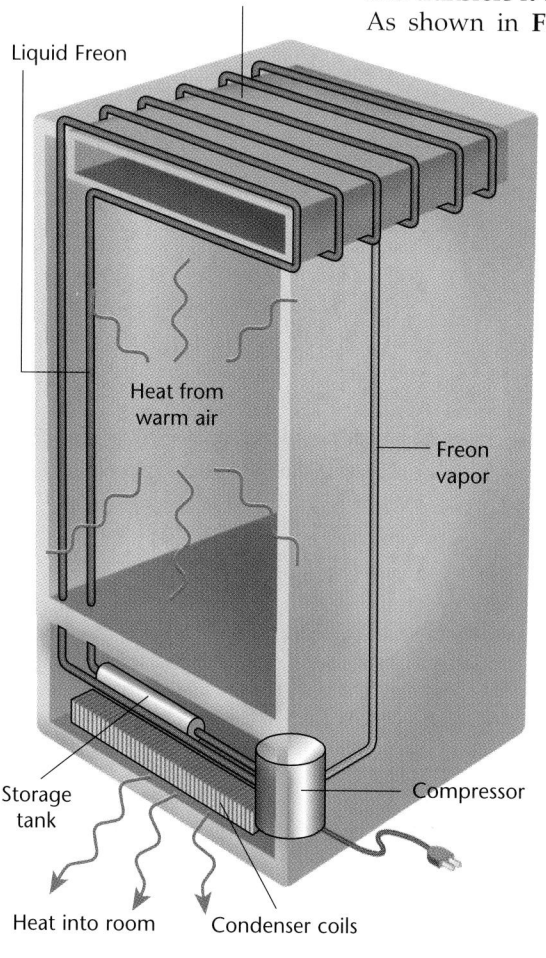

Freezer unit
Liquid Freon
Heat from warm air
Freon vapor
Storage tank
Compressor
Heat into room
Condenser coils

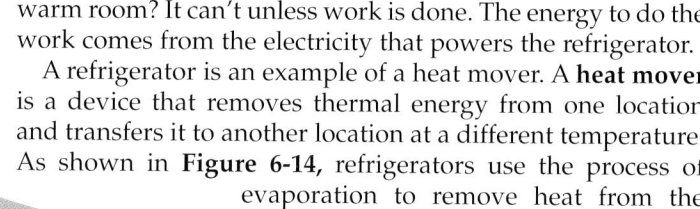
Use these program resources as you teach this lesson.
Critical Thinking/Problem Solving, p. 12
Lab Manual 13

Use **Science Integration Transparency 6** as you teach this lesson.

INTEGRATION
Life Science

Sweat—The Human Coolant

After exercising on a warm day, you may feel hot and be drenched with sweat. But if you were to take your temperature, you would probably find it close to your normal body temperature of 37°C. How does your body stay cool in hot weather?

In reading about refrigerators, you learned that a refrigerant such as Freon removes heat from food as it changes from a liquid to a gas. This process, called evaporation, is also important in maintaining a healthy internal body temperature. When a liquid changes to a gas, energy is absorbed from the liquid's surroundings. Heat is removed from your body as your perspiration evaporates, making you cooler.

USING TECHNOLOGY

Cooling Crystals

Do you know the coldest temperature something could actually become? Recall that zero on the Kelvin scale is called absolute zero, the point at which all particle motion stops. Physicists are trying to bring small samples of matter to the coldest temperature possible. Some samples have been chilled to 180 billionths of a degree above absolute zero (−273.15°C).

Physicists think about extremely low temperatures in terms of slowness of particle vibrations rather than as chilliness. They can bring matter to such low temperatures by using magnetic fields or special lasers to decrease the particle vibrations.

Property Changes

Many properties of matter change at temperatures near absolute zero. Studying materials at extremely cold temperatures may better explain magnetism and the behavior of the particles in atoms.

Researchers at the National Institute of Standards and Technology hope that reaching such temperatures will improve the accuracy of atomic clocks, which depend upon natural vibrations of atoms to keep time.

Think Critically:

Do you think super-cold freezers could be used to bring matter to a temperature of absolute zero? Why or why not?

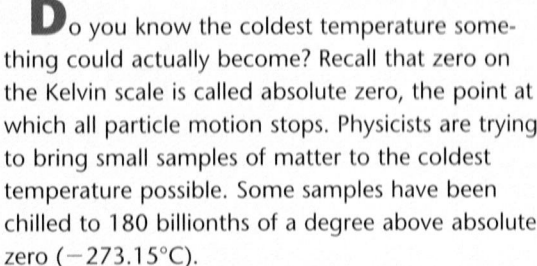

Sodium Atoms Trapped and Cooled by Laser Light

USING TECHNOLOGY

Although absolute zero has never been observed, it is the temperature at which a substance would have zero thermal energy. This temperature is equivalent to 0 K, −273.15°C, and −459.67°F

Think Critically

No, super-cold freezers would have to be at a temperature below absolute zero to remove thermal energy from even a slightly warmer substance. Sub-absolute zero temperatures are not possible.

3 Assess

Check for Understanding

?FLEX Your Brain

Use the Flex Your Brain activity to have students explore INTERNAL COMBUSTION ENGINES.

📁 **Activity Worksheets,** p. 5

Reteach

IS Interpersonal Have students make flash cards with the names of the strokes of the four-stroke cycle. On the reverse side of each card, describe the stroke. Partners can then test each other by shuffling the cards and sequencing them both by name and description. Students can alter the cards to represent the four-stroke cycle of a diesel engine. **L1** **LEP** **COOP LEARN**

Extension

📁 For students who have mastered this section, use the **Reinforcement** and **Enrichment** masters.

Visual Learning

See p. 170.
Figure 6-15 Would there be more evaporation on a hot, dry day or a hot, humid day? *hot, dry day* **LEP**

4 Close

170

The energy lost by your body becomes part of the thermal energy of your evaporated sweat.

Why do humid days feel hotter than dry days?

Have you ever wondered why humid days feel warmer than dry days at the same temperature? On humid days, more water vapor is in the air around you. Your sweat doesn't evaporate as quickly. In fact, some of this water vapor actually condenses and releases energy to your skin. Condensation, the change from a gas to a liquid, is the opposite of evaporation. Evaporation is a cooling process, and condensation is a warming process.

The dog in **Figure 6-15** is cooling herself by panting. Because dogs only sweat between their toes, they pant to cause evaporation from their respiratory systems to stay cool.

Figure 6-15

Evaporation is helping to keep this dog cool. *Would there be more evaporation on a hot, dry day or a hot, humid day?*

Section Wrap-up

Review

1. How do diesel and gasoline engines differ?

2. In a heat engine, what happens to the thermal energy that is converted from chemical energy?

3. **Think Critically:** Explain whether it is a good idea to cool your house by leaving the refrigerator door open.

Skill Builder
Concept Mapping

Make a cycle concept map to show the steps in one cycle of a four-stroke internal combustion engine. If you need help, refer to Concept Mapping in the **Skill Handbook.**

Science Journal

Only a small percentage of the chemical potential energy of gasoline is actually used to move your car. In your Science Journal, describe the evidence that much of the energy is wasted.

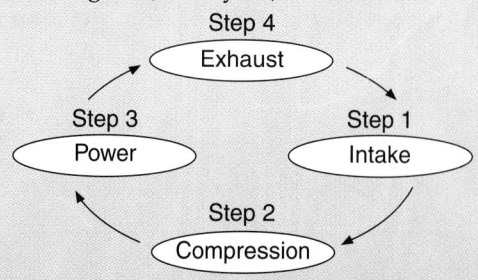

Activity 6-2

The Four-Stroke Engine

As you examine the piston movements in a four-stroke engine, can you identify and detect changes in kinetic and potential energy?

Problem
How do the kinetic and potential energies change during a four-stroke cycle?

Materials
- matches
- flask, 250-mL
- large syringe, plastic
- rubber tubing (3 pieces)
- small piece of cloth or yarn
- rubber stopper, 2-hole, with glass tubing
- glass or plastic "T"
- pinch clamps (2)

Procedure
1. Set up a model four-stroke engine as shown.
2. Prepare a data table like the one below.
3. With all clamps closed, ignite a small piece of cloth or yarn and drop it into the flask. **CAUTION:** *Exercise caution when using open flames.* Allow the flask, representing the carburetor, to fill with smoke.
4. With the piston inside the syringe, open the intake clamp and pull the piston down.
5. Close the intake clamp and push the piston up into the syringe.
6. Release the piston and observe what happens.
7. Open the exhaust clamp and push the piston up into the syringe.
8. Close the exhaust clamp and repeat steps 4-7.

Analyze
1. What happens to the smoke particles when the intake clamp is closed and the piston moves into the syringe?
2. What happens when the piston is released?
3. What happens to the smoke particles when the exhaust clamp is opened and the piston moves into the syringe?

Conclude and Apply
4. In this model, **explain** what the smoke represents in steps 4, 5, and 7.
5. **Infer** the best instant to explode a fuel-air mixture.
6. **Identify** in which stroke kinetic energy is changed to potential energy.

Data and Observations

Step	Stroke	Observations	Sample Data
4	intake	intake clamp open; piston moves out of syringe; smoke from carburetor fills syringe	
5	compression	both clamps closed; piston moves into syringe; particles of smoke squeezed together	
6	power	both clamps closed; pressure of gases squeezed at top of syringe forces piston to move out	
7	exhaust	exhaust clamp open; piston moves into syringe; smoke is pushed out through exhaust clamp	

5. when the piston is at the top of its stroke and the fuel-air mixture is most compressed
6. Kinetic energy is changed to potential energy during the compression stroke. Potential energy is changed to kinetic energy during the power stroke.

Assessment

Process Have students compare and contrast their model engines to the internal combustion engine diagrammed in **Figure 6-13.** Use the Performance Task Assessment List for Science Journal in **PASC,** p. 103. **P**

Activity 6-2

Purpose
LS Kinesthetic Construct and analyze a model of a four-stroke engine. **L2 COOP LEARN**

Process Skills
modeling, communicating, sequencing, comparing and contrasting, interpreting scientific illustrations, recognizing cause and effect

Time
30 minutes

Safety Precautions
- Test the small pieces of cloth or yarn in advance to be sure they are a small enough size to burn under control.
- Insert glass tubing prior to the activity. This will save time and increase safety.

Activity Worksheets, pp. 5, 34–35

Teaching Strategies
Alternate Materials Torch paper can be used to generate smoke. Otherwise, you should use paper towels. Soft rubber tubing and strong pinch clamps are needed to avoid gas leaks.
- Students should prepare by reading pages 166-167 about internal combustion engines.
- Use the activity to discuss the use of models to study specific parts of a complex problem.

Answers to Questions
1. The smoke particles are compressed—squeezed together.
2. It moves downward in the syringe.
3. Smoke particles are pushed out through the clamp by the piston.
4. In steps 4 and 5, the smoke represents the fuel-air mixture; in step 7, it represents exhaust gases.

171

Prepare

Section Background

Although not practical for inland areas, OTEC has potential for many urban coastal areas. Around the United States, OTEC could prove beneficial for Hawaii, Florida, and Puerto Rico, as they are the regions that have a significant thermal gradient between tropical surface waters and deeper waters. It has been estimated that using OTEC in Puerto Rico as a supplement to burning oil could save more than 60 000 barrels of oil each day.

1 Motivate

Bellringer

Before presenting the lesson, display **Section Focus Transparency 25** on the overhead projector. Assign the accompanying **Focus Activity** worksheet. L1 LEP

2 Teach

Visual Learning

Figure 6-16 Explain the effects they have on the rocks and sand along the shoreline. *Rocks are eroded and sand is moved great distances.* LEP LS

Content Background

The OTEC idea was first proposed by d'Arsonval in 1880. In 1930, the first working plant was developed at Matanzas Bay, Cuba.

SCIENCE & SOCIETY

6•4 TECHNOLOGY: Energy from the Oceans

Science Words
ocean thermal energy conversion

Objectives
- Explain how differences in ocean temperatures can be used to operate a heat engine that changes thermal energy to mechanical energy.
- Discuss the advantages and disadvantages of ocean thermal energy conversion.

Figure 6-16

Ocean waves and tides carry large amounts of energy. *Explain the effects they have on the rocks and sand along the shoreline.*

Oceans—Natural Reservoirs of Energy

Earth's dwindling supply of fossil fuels is making it necessary for researchers to develop ways to make use of natural, renewable energy sources. One of the largest energy sources covers nearly three-fourths of Earth's surface—the oceans. With such a large surface area and depths up to 11 000 m, the oceans have the ability to absorb radiant energy from the sun and store it as thermal energy. In addition to thermal energy, the oceans also contain a vast amount of mechanical energy. You can see the effects of this energy in **Figure 6-16.**

Ocean waters are in constant motion. The mechanical energy of moving water can be used to rotate turbines and generate electricity, as in hydroelectric plants in rivers. Several plants already make use of moving water resulting from tides, and scientists are working on ways to use the energy in ocean currents and waves.

172

Program Resources

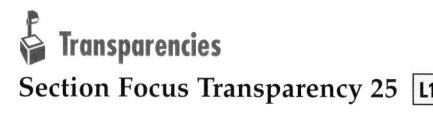

📁 **Reproducible Masters**
Study Guide, p. 29 L1
Reinforcement, p. 29 L1
Enrichment, p. 29 L3

🔦 **Transparencies**
Section Focus Transparency 25 L1

Ocean Thermal Energy Conversion

Where do you think ocean temperatures are the highest and the lowest? In tropical or subtropical regions, there can be a more than 20°C difference between warm surface waters and cold bottom waters. Energy researchers are trying to improve methods of extracting and using energy stored in warm surface waters.

Ocean thermal energy conversion (OTEC) is a process that uses heat engines to convert differences in ocean water temperature into mechanical energy to drive turbines. An OTEC heat engine uses heat from warm surface water to vaporize a working fluid with a low boiling point, such as ammonia. **Figure 6-17** illustrates how an OTEC heat engine works. How is this similar to the refrigeration process described in the last section?

Figure 6-17

This OTEC heat engine uses ammonia as a working fluid. The turbine is driven by the ammonia vapor and is connected to a generator to produce electricity. The warm water is drawn from the ocean surface; the cold water from a depth of 1000 m.

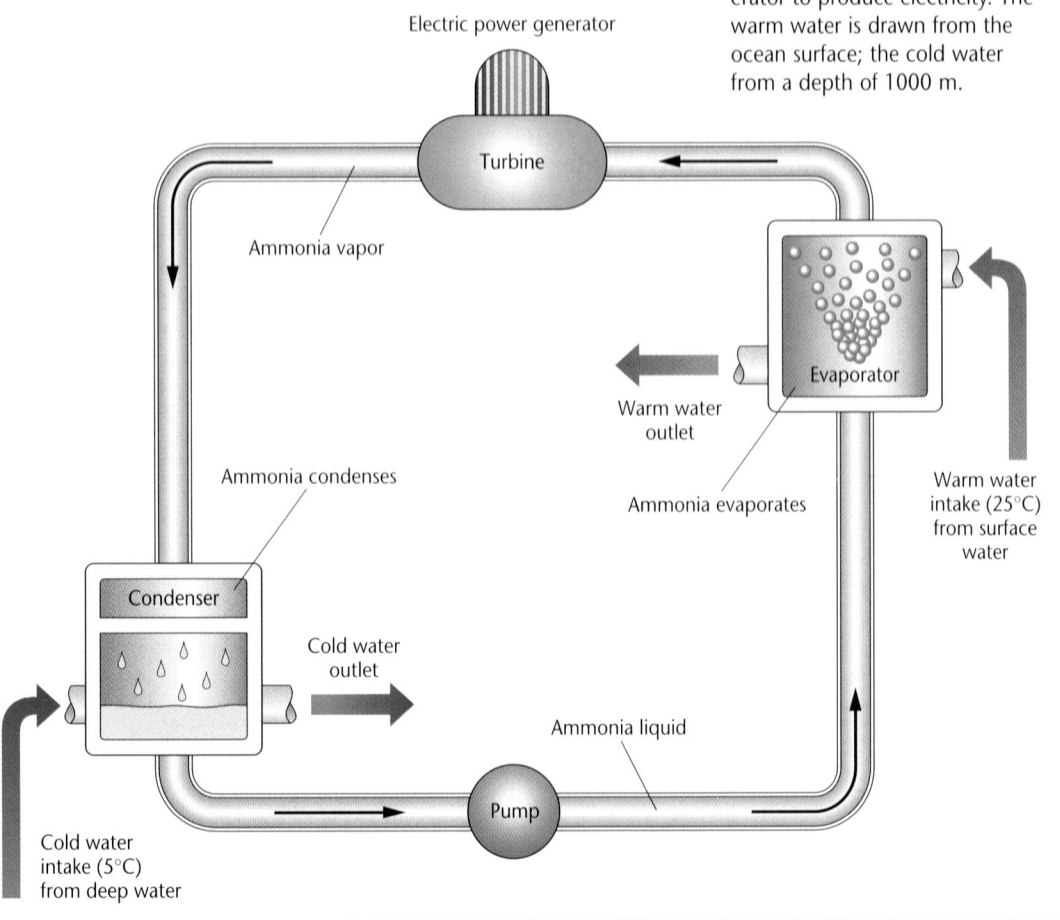

Electric power generator

Turbine

Ammonia vapor

Evaporator

Warm water outlet

Ammonia evaporates

Warm water intake (25°C) from surface water

Ammonia condenses

Condenser

Cold water outlet

Ammonia liquid

Cold water intake (5°C) from deep water

Pump

Teacher F.Y.I.

The highest ocean tides occur at the Bay of Fundy in New Brunswick, Canada, averaging more than 14 m high.

Student Text Question

How is this similar to the refrigeration process described in the last section? *The same principles of evaporation and condensation are used to move thermal energy from one location to another.*

3 Assess

Check for Understanding

?FLEX Your Brain

Use the Flex Your Brain activity to have students explore OTEC.

▱ **Activity Worksheets,** p. 5

Reteach

⬚ **Visual-Spatial** Ask students to look back at **Figure 6-14,** the diagram of a refrigerator. Have them compare and contrast this process to the OTEC process shown in **Figure 6-17.** Note that the same principles of evaporation and condensation are used to move thermal energy from one location to another, but the OTEC process harnesses this thermal energy to drive a turbine and produce electricity. **LEP**

Extension

▱ For students who have mastered this section, use the **Reinforcement** and **Enrichment** masters.

4 Close

LS Interpersonal Use a globe to have students describe the types of locations that would likely benefit most from the development of OTEC plants. Note that islands and coastal regions in tropical areas would likely have exposure to ocean water with a large temperature gradient.

Section Wrap-up

Review

1. Heat from warm surface water is used to vaporize a fluid with a low boiling point, such as ammonia. As this vapor expands, it is used to rotate a turbine. The vapor is then recondensed by cold bottom water to complete the cycle.

2. Advantages may include having a renewable energy source for generating electricity without polluting the atmosphere. Disadvantages may include disruptions due to temperature changes in the marine environment and inefficiency due to pumping large amounts of water.

Is OTEC practical?

The construction and operation of several small plant designs have shown that OTEC is possible. Improved designs will likely increase the efficiency of such plants. They require a lot of energy to operate because large amounts of water must be pumped from ocean depths.

Additional study must be done to determine possible environmental effects of pumping so much cold water to the surface in warm-water regions. The environment marine organisms are accustomed to, especially with respect to dissolved oxygen and nutrient levels, could be altered by temperature changes. The main attraction of OTEC lies in the fact that radiant energy supplied by the sun to heat the surface water is a dependable and free source of energy. OTEC plants may be land-based or built on floating ocean platforms.

Figure 6-18
This land-based OTEC plant uses a system of pipes and pumps to transport water from the nearby ocean.

Section Wrap-up

Review

1. Explain how temperature differences at different depths of the ocean provide a possible energy source.

2. What do you think are the possible advantages and disadvantages of developing OTEC as an energy source?

 Visit the Chapter 6 Internet Connection at Glencoe Online Science, **www.glencoe.com/sec/science/physical,** for a link to more information about OTEC.

SCIENCE & SOCIETY

174   Chapter 6 Using Thermal Energy

Review

Summary

6-1: Thermal Energy on the Move
1. Thermal energy can be transferred by conduction, convection, and radiation. Unlike radiation, conduction and convection can occur only when matter is present.
2. Conductors allow heat to flow easily. Insulators resist heat flows.

6-2: Using Heat to Stay Warm
1. Heating systems are generally identified by the medium that transfers the thermal energy. The three most common media are hot water, steam, and air.
2. A solar heating system converts radiant energy from the sun to thermal energy.
3. Passive solar systems do not have devices to transfer heat from one part of the system to another. Active solar systems use fans or pumps to serve this purpose.

6-3: Using Heat to Do Work
1. Heat engines are devices that convert thermal energy produced by burning fuel into mechanical energy. In an internal combustion engine, fuel is burned inside the engine. In an external combustion engine, fuel is burned outside the engine.
2. Heat movers move thermal energy from one place and release it in another place.
3. Sweating helps humans cool their bodies through evaporation.

6-4: Science and Society: Energy from the Oceans
1. Differences in ocean temperatures can be used to operate a heat engine.
2. Additional research needs to be done to improve the efficiency and evaluate the environmental effects of OTEC.

Key Science Words

a. combustion
b. conduction
c. convection
d. external combustion engine
e. fluid
f. heat engine
g. heat mover
h. insulator
i. internal combustion engine
j. ocean thermal energy conversion
k. radiation
l. radiator
m. solar collector
n. solar energy

Reviewing Vocabulary

Match each phrase with the correct term from the list of Key Science Words.

1. energy transfer by direct contact of particles
2. energy transfer by movement of particles from place to place
3. energy transfer by electromagnetic waves
4. material that resists the flow of heat
5. device that absorbs the sun's radiant energy
6. device that converts thermal energy into mechanical energy
7. rapid burning
8. a device that transfers thermal energy from one place to another
9. type of engine in which fuel is burned inside the engine
10. a process that uses a difference in water temperatures to operate a heat engine

Summary

Have students read the summary statements to review the major concepts of the chapter.

Reviewing Vocabulary

1. b	6. f
2. c	7. a
3. k	8. g
4. h	9. i
5. m	10. j

✔ Assessment

Portfolio Encourage students to place in their portfolios one or two items of what they consider to be their best work. Examples include:
- Connect to Life Science, p. 163
- Science Journal, p. 165
- Science Journal, p. 170 **P**

Performance Additional performance assessments may be found in **Performance Assessment** and **Science Integration Activities**. Performance Task Assessment Lists and rubrics for evaluating these activities can be found in Glencoe's **Performance Assessment in the Science Classroom**.

GLENCOE TECHNOLOGY

 Videodisc

Glencoe Physical Science
Interactive Videodisc
Use the videodisc lesson *Behavior of Gases* to review the principles of heat transfer.

MindJogger Videoquiz

Chapter 6 Have students work in groups as they play the Videoquiz game to review key chapter concepts.

Checking Concepts

1. b	**6.** b
2. a	**7.** c
3. d	**8.** a
4. a	**9.** d
5. a	**10.** a

Understanding Concepts

11. Some of the energy of the burner top is transferred by conduction to the particles of the pot bottom, which transfers energy to the soup. As the bottom portion of the soup warms, the cooler, denser liquid sinks, forcing the warmer liquid to rise. This movement transfers heat throughout the liquid by convection.

12. Convection currents are created when one part of a fluid is heated. The cooler, denser fluid moves down and forces the warmer, less dense fluid to rise. Winds and ocean currents are convection currents.

13. Darker materials absorb radiation more effectively. When layers of clothing are worn, a thin layer of insulating air will be present between each layer of clothing.

14. In passive solar heating, radiant energy is directed onto a material that will absorb the energy, convert it to thermal energy, and transfer it to the surroundings as heat. In active solar heating, the thermal energy is transferred by the use of mechanical devices. Both systems are based on the idea that radiant energy is changed to thermal energy when it is absorbed.

15. The mechanical energy of tides can be used to generate electricity. The temperature difference between warm surface water and cold bottom water can be used to drive a heat engine.

Chapter 6 Review

Checking Concepts

Choose the word or phrase that completes the sentence or answers the question.

1. Which is not a method of heat transfer?
a. conduction   c. radiation
b. insulation   d. convection

2. In _____, fuel is burned inside chambers called cylinders.
a. internal combustion engines
b. external combustion engines
c. heat pumps
d. steam engines

3. Waste gases are removed during the _____ of a four-stroke engine.
a. power stroke   c. compression stroke
b. intake stroke   d. exhaust stroke

4. Which material is a poor insulator of heat?
a. aluminum   c. air
b. feathers   d. plastic

5. A _____ is an example of a heat mover.
a. refrigerator   c. combustion engine
b. steam engine   d. four-stroke engine

6. In which of these forms is water not a fluid?
a. liquid water   c. water vapor
b. ice   d. steam

7. Heat can move easily through a good _____.
a. insulator   c. conductor
b. carburetor   d. collector

8. Which of these does not require the presence of particles of matter?
a. radiation   c. convection
b. conduction   d. combustion

9. In order for radiant energy to change to thermal energy, it must be _____.
a. reflected   c. convected
b. conducted   d. absorbed

10. A _____ changes thermal energy into mechanical energy.
a. heat engine   c. solar collector
b. refrigerator   d. conductor

Understanding Concepts

Answer the following questions in your Science Journal using complete sentences.

11. Describe all of the ways in which energy is transferred while a pot of soup is heated on an electric stove. Indicate how each type of energy transfer takes place.

12. Explain how convection currents are created. Describe two examples of convection currents that occur in nature.

13. Why is winter clothing generally darker in color than summer clothing? Explain why wearing two or three layers of clothing helps to keep you warmer in cold weather than does one thick layer.

14. Compare and contrast passive and active solar heating. On what basic principle are both systems based?

15. Describe at least two ways the oceans can be used as a source of energy.

Thinking Critically

16. The energy transfer shown in **Figure 6-1** on page 152 takes place at the particle level of matter. Thus, the transfer of energy cannot be observed directly. Think of an analogy or model using visible objects that you could use to demonstrate the process of conduction.

17. Explain why the inside of an automobile left sitting in direct sunlight for several hours becomes warm.

18. Design a line of clothing to be used on an Arctic expedition. Describe the articles of clothing and explain why each will keep the wearer warm in extremely cold conditions.

19. The engines of many high-performance cars have four valves per cylinder rather than the usual two-valve arrangement. These engines also have fuel injectors at each cylinder rather than a single carburetor. How do these differences improve the performance of the engine?

Thinking Critically

16. Responses may include energy being passed to a rack of billiard balls when one ball is struck.

17. Radiant energy from the sun strikes materials inside a car and is absorbed, changing to thermal energy. Thermal energy is transferred as heat. The increase in heat causes the temperature inside the car to rise.

18. Dark-colored materials will absorb radiant energy, and layers of fibers will trap air and reduce the loss of body heat. Reflective linings would reflect thermal energy back to the body.

19. Increasing the number of valves increases the amount of fresh air taken into the cylinder during the intake stroke and decreases the amount of exhaust remaining. The fuel injectors provide a more even distribution of fuel. Both help to increase efficiency of combustion and thus increase power.

20. Describe how an automobile air conditioner using Freon gas works. Explain why the engine must be running in order for the air conditioner to work.

Developing Skills

If you need help, refer to the **Skill Handbook.**

21. **Interpreting Data:** Using the *R*-values given in **Table 6-1** on page 157, design an energy-efficient house. Indicate the type and thickness of the different materials to be used to construct the walls and ceilings.
22. **Sequencing:** Order the events that occur in the removal of heat from an object by a refrigerator. Start with the placing of a warm object in the refrigerator and finish with the change in Freon from a gas to a liquid.
23. **Concept Mapping:** Complete the following events chain to show how an active solar heating system works.

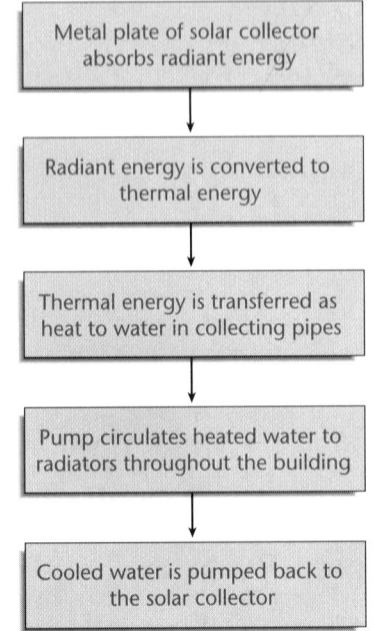

24. **Recognizing Cause and Effect:** Describe the problems that might occur if the following events happened in a fuel-injected gasoline engine. Indicate the engine stroke that will be affected: (a) exhaust valve stuck closed, (b) clogged fuel injector, (c) bad spark plug, (d) intake valve will not close.
25. **Using Variables, Constants, and Controls:** Design an experiment to test the effects of surface area and length of a material on heat conduction. Indicate your hypothesis, variables, and constants. How will you organize and interpret your data?

Performance Assessment

1. **Formulating a Hypothesis:** Review Activity 6-1 on page 160. What do you think would happen if you were to use a funnel to add warm water (colored red) to the bottom of a beaker of cold water? Explain your answer.
2. **Oral Presentation:** Study the four-stroke engine model in Activity 6-2 on page 171. Explain how this model differs from an actual engine.
3. **Making Observations and Inferences:** Identify the building materials used in or on the walls, ceilings, or attic space in your home and school. Which ones are good heat insulators?

Chapter 6 Review

20. As liquid Freon moves through the air conditioner, it evaporates, removing thermal energy from the interior of the car. This thermal energy is then released to the warmer air outside the car. This process requires work to be done by the car's engine.

Developing Skills

21. **Interpreting Data** The combinations of suggested materials and/or their thicknesses should give total *R*-values of at least 19 for the walls and 30 for the roof.
22. **Sequencing** Liquid Freon evaporates, removing thermal energy from the interior of the refrigerator; compressing the gas raises the temperature of the gas; heat flows from the gas to the exterior air, causing the Freon gas to condense to a liquid.
23. **Concept Mapping** See student page.
24. **Recognizing Cause and Effect** (a) Exhaust gases will not be removed during the exhaust stroke, which will limit the amount of fuel-air mixture entering during the intake stroke. (b) The amount of fuel being mixed with air during the intake stroke will be reduced, thus reducing the effectiveness of the power stroke. (c) The fuel-air mixture will not be properly ignited, thus reducing the power stroke. (d) The fuel-air mixture that enters during the intake stroke will be pushed out of the stuck valve during the compression stroke, thus reducing the power stroke.
25. **Using Variables, Constants, and Controls** A sample hypothesis is that increasing surface area increases thermal conduc-

---

tion. Variables would include changing the surface area. Constants would include the type of material tested. Data analysis may include a line graph.

Performance Assessment

1. The warmer, less dense water should rise to the top. The water will mix as it reaches the same temperature. Use the Performance Task Assessment List for Formulating a Hypothesis in **PASC,** p. 21. **P**
2. An actual engine uses a combustible material to force the piston down, which converts potential to kinetic energy. The exhaust in the activity is chemically different from the air-gas mixture in a real engine. Use the Performance Task Assessment List for Oral Presentation in **PASC,** p. 71. **P**
3. Table **6-1** on p. 157 can be used to determine which materials are the better insulators. Use the Performance Task Assessment List for Making Observations and Inferences in **PASC,** p. 17. **P**

Section	Objectives/Standards	Activities/Features
Chapter Opener		**Explore Activity:** How can a machine make it easier to move an object? p. 179
7-1 **Why We Use Machines** (½ session)*	1. **Explain** how machines make work easier. 2. **Calculate** mechanical advantage. **National Content Standards:** (5-8) UCP1-UCP3, UCP5, A1, A2, B2, C1, C5, E1; (9-12) UCP1-UCP3, UCP5, A1, A2, B4, E1	**Connect to Life Science,** p. 182 **Using Math:** Calculating Mechanical Advantage, p. 182 **Using Math,** p. 183 **Skill Builder:** Recognizing Cause and Effect, p. 183 **Activity 7-1:** Access for Everyone, pp. 184-185
7-2 **The Simple Machines** (2½ sessions, 1½ blocks)*	1. **Describe** the six types of simple machines. 2. **Calculate** the ideal mechanical advantage for different types of simple machines. **National Content Standards:** (5-8) UCP1-UCP3, UCP5, A2, C1; (9-12) UCP1-UCP3, UCP5, A2	**Using Math:** Mechanical Advantage of a Lever, p. 187 **MiniLAB:** How do pulleys make work easier? p. 190 **Using Math:** Calculating *IMA* of a Wheel and Axle, p. 192 **Using Computers,** p. 194 **Skill Builder:** Making and Using Tables, p. 194 **Activity 7-2:** Levers, p. 195
7-3 **Science and Society** (1 session)*	1. **Explain** what the science of bionics involves. 2. **Contrast** two methods of using electrical signals to trigger motion of a limb or other bodily process. **National Content Standards:** (5-8) UCP1-UCP3, UCP5, A2, C5, E2, F1, F5, G1, G3; (9-12) UCP1-UCP3, UCP5, A2, E2, F1, F6, G1, G3	**Explore the Technology,** p. 197 **Science Journal,** p. 197
7-4 **Using Machines** (2 sessions, 1 block)*	1. **Recognize** the simple machines that make up a compound machine. 2. **Calculate** the efficiency of a machine. 3. **Describe** the relationship among work, power, and time. **National Content Standards:** (5-8) UCP1-UCP3, UCP5, A2, E2, F1-F3, F5, G1, G3; (9-12) UCP1-UCP3, UCP5, A2, E2, F1, F3-F6, G1, G3	**Problem Solving:** Frivolous Machines, p. 199 **Connect to Chemistry,** p. 200 **Using Math:** Calculating Efficiency, p. 200 **MiniLAB:** Can you measure the power of a toy car? p. 201 **Using Technology:** Nanomachines, p. 202 **Using Math:** Calculating Power, p. 203 **Science Journal,** p. 203 **Skill Builder:** Measuring in SI, p. 203 **Science and History:** Building the Pyramids, p. 204

* A complete Planning Guide that includes block scheduling is provided on pages 32T-35T.

Activity Materials

Explore	Activities	MiniLABs
page 179 2 or 3 books, meterstick	pages 184-185 poster-sized paper, ruler or meterstick, pencils, markers page 195 20 cm × 28 cm (8 1/2" × 11") sheet of paper, 3 coins (quarter, nickel, dime), balance, metric ruler	page 190 1-m length of rope, two 1-m lengths of PVC pipe or broom handles page 201 wind-up toy car, inclined plane made with a board and books, watch or clock with second hand, spring scale, meterstick

Need Materials? Call Science Kit (1-800-828-7777).

Chapter 7 Machines—Making Work Easier

Teacher Classroom Resources

Reproducible Masters	Transparencies	Teaching Resources
Study Guide, p. 30 Reinforcement, p. 30 Enrichment, p. 30 Activity Worksheets, pp. 38-39	Section Focus Transparency 26, The Mechanics of Rowing	Glencoe Physical Science Interactive Videodisc Physical Science CD-ROM Spanish Resources English/Spanish Audiocassettes Cooperative Learning Resource Guide Lab Partner Lab and Safety Skills Lesson Plans
Study Guide, p. 31 Reinforcement, p. 31 Enrichment, p. 31 Activity Worksheets, pp. 40-41, 42 Lab Manual 14, Balanced Levers Multicultural Connections, p. 17 Science Integration Activity 7, On the Rocks Concept Mapping, pp. 19-20 Critical Thinking/Problem Solving, p. 13	Section Focus Transparency 27, An Automobile Steering System Science Integration Transparency 7, Levers in Your Body Teaching Transparency 13, Classes of Levers	**Assessment Resources** Chapter Review, pp. 17-18 Assessment, pp. 36-39 Performance Assessment in the Science Classroom (PASC) MindJogger Videoquiz Alternate Assessment in the Science Classroom Performance Assessment Chapter Review Software Computer Test Bank
Study Guide, p. 32 Reinforcement, p. 32 Enrichment, p. 32 Lab Manual 15, Pulleys Science and Society Integration, p. 11	Section Focus Transparency 28, Driving with a Disability Teaching Transparency 14, A Compound Machine	
Study Guide, p. 33 Reinforcement, p. 33 Enrichment, p. 33 Activity Worksheets, p. 43 Lab Manual 16, The Bicycle: A Well- Engineered Machine Cross-Curricular Integration, p. 11	Section Focus Transparency 29, The Power of Language	**Key to Teaching Strategies** The following designations will help you decide which activities are appropriate for your students.

Key to Teaching Strategies

The following designations will help you decide which activities are appropriate for your students.

L1 Level 1 activities should be within the ability range of all students, including those with learning difficulties.

L2 Level 2 activities should be within the ability range of the average to above-average student.

L3 Level 3 activities are designed for the ability range of above-average students.

LEP LEP activities should be within the ability range of Limited English Proficiency students.

LS These activities are designed to address different learning styles.

COOP LEARN Cooperative Learning activities are designed for small group work.

P These strategies represent student products that can be placed into a best-work portfolio.

GLENCOE TECHNOLOGY

The following multimedia resources are available from Glencoe.

**Science and Technology Videodisc
Series (STVS)**
Physics
 Bipedal Robot
Human Biology
 Tricycle for the Handicapped

National Geographic Society Series
STV: Human Body Volume 2

**Glencoe Physical Science
Interactive Videodisc**
Machines and Forces

Physical Science CD-ROM

Teacher Classroom Resources

This is a representation of key blackline masters available in the Teacher Classroom Resources.

Teaching Aids

Section Focus Transparencies

26 SECTION FOCUS TRANSPARENCY

THE MECHANICS OF ROWING

L1

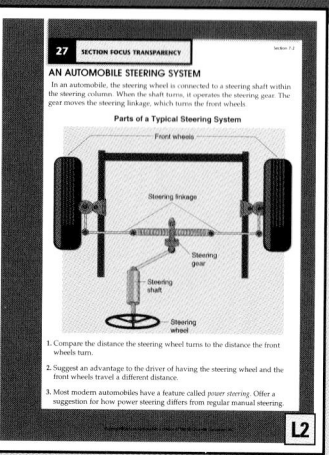

27 SECTION FOCUS TRANSPARENCY

AN AUTOMOBILE STEERING SYSTEM

L2

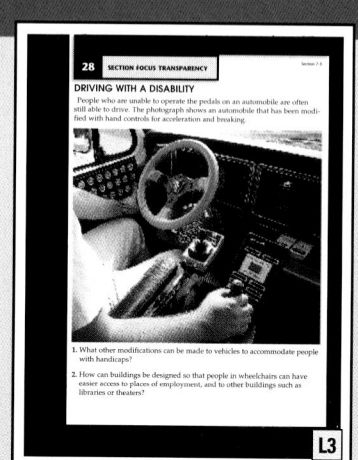

28 SECTION FOCUS TRANSPARENCY

DRIVING WITH A DISABILITY

L3

Science Integration Transparencies

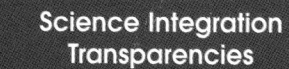
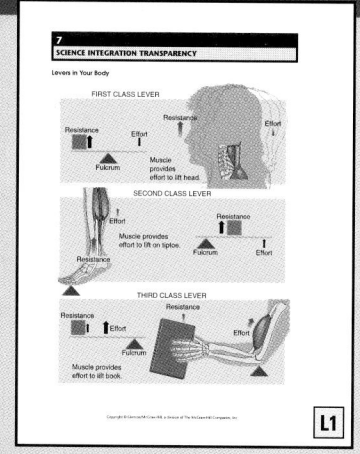

7 SCIENCE INTEGRATION TRANSPARENCY

Levers in Your Body

L1

Teaching Transparencies

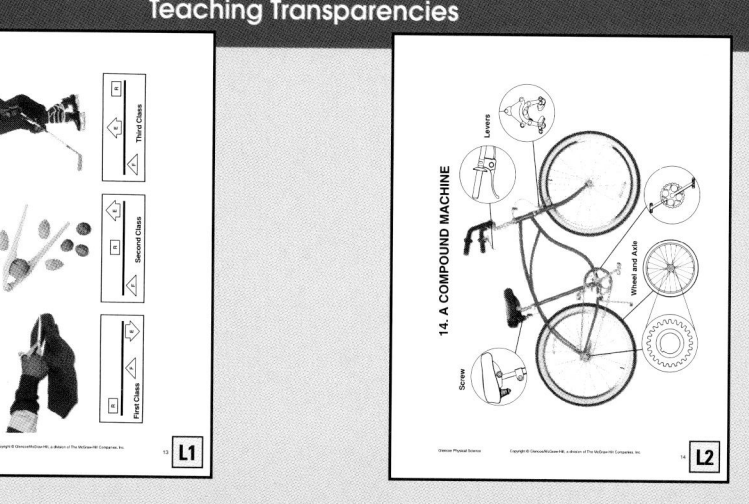

13. CLASSES OF LEVERS

L1

14. A COMPOUND MACHINE

L2

Meeting Different Ability Levels

Study Guide

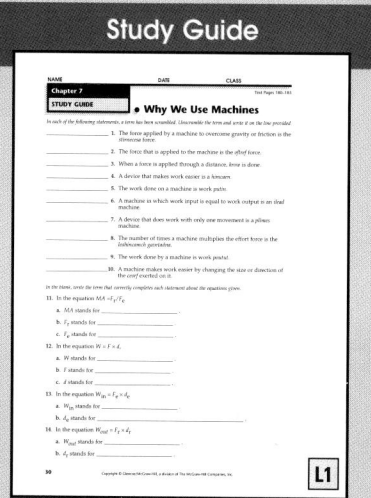

Chapter 7
STUDY GUIDE
Why We Use Machines

L1

Reinforcement

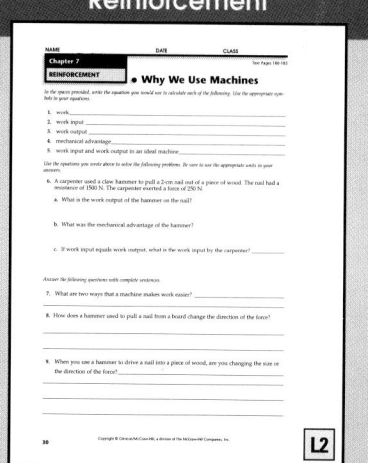

Chapter 7
REINFORCEMENT
Why We Use Machines

L2

Enrichment Worksheets

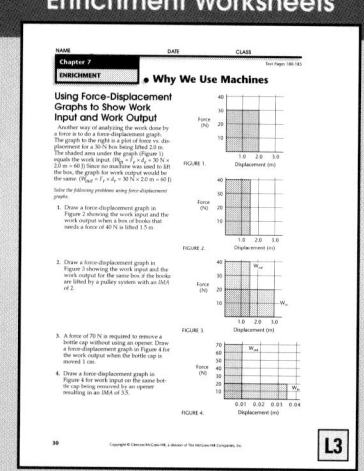

Chapter 7
ENRICHMENT
Why We Use Machines

Using Force-Displacement Graphs to Show Work Input and Work Output

L3

Chapter 7 Machines—Making Work Easier

Hands-On Activities

Science Integration Activity

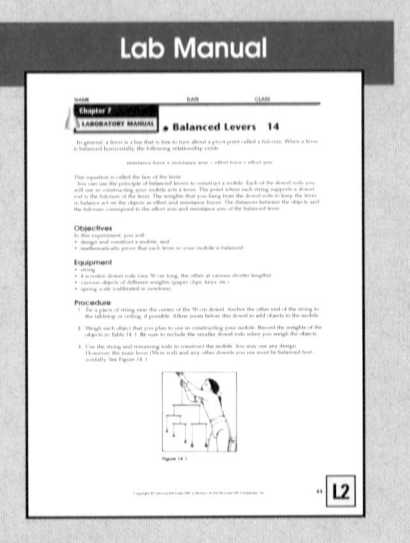

Lab Manual

Assessment

Performance Assessment

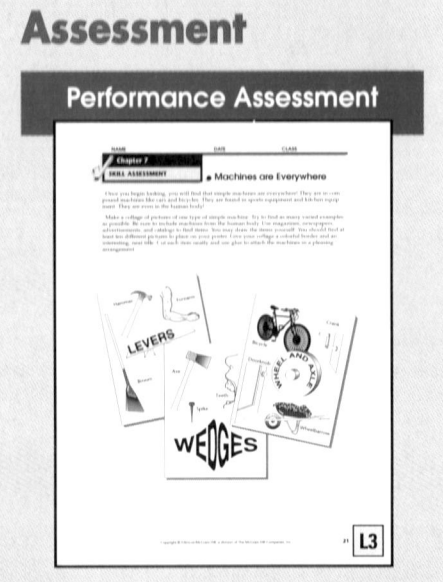

Enrichment and Application

Critical Thinking/ Problem Solving

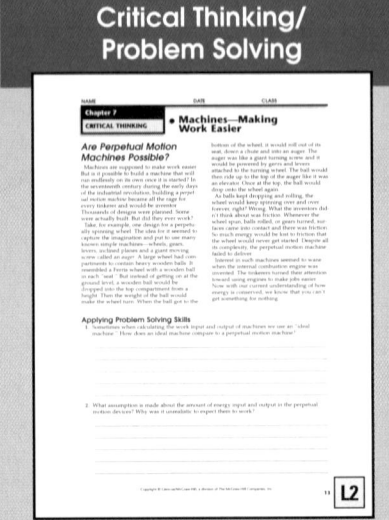

Cross-Curricular Integration

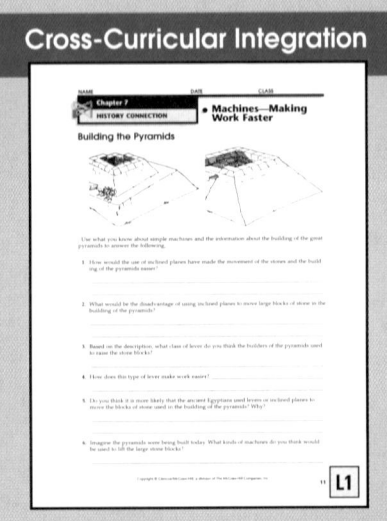

Science and Society Integration

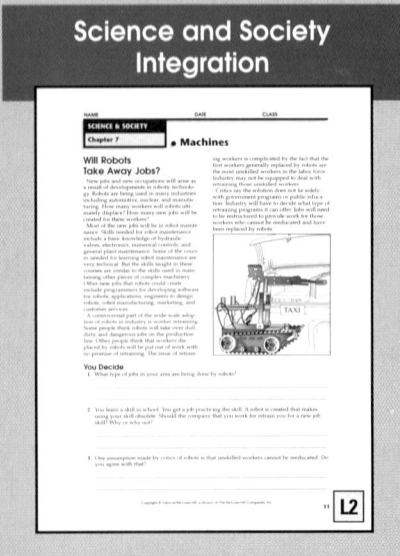

Multicultural Connections

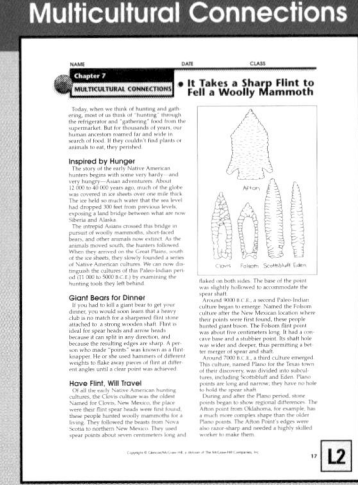

Concept Mapping

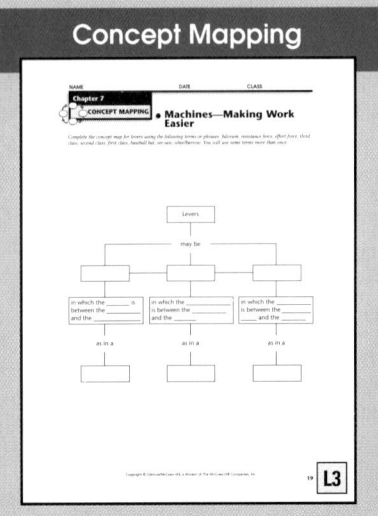

Machines—Making Work Easier

CHAPTER OVERVIEW

Section 7-1 Simple machines are introduced and characterized in this section. The concept of mechanical advantage is developed.

Section 7-2 This section classifies simple machines into six types and discusses how they can be conceptualized into two broad categories: levers and inclined planes.

Section 7-3 Science and Society This section contains a report on the developing science of bionics. Two methods of electrical stimulus use and the new technique of brain-to-computer interfaces are presented and described.

Section 7-4 Compound machines are discussed using a bicycle as an example. The concepts of efficiency and power of a machine are introduced and developed.

Chapter Vocabulary

machine	pulley
simple machine	wheel and
effort force	axle
resistance force	inclined
ideal machine	plane
mechanical	screw
advantage	wedge
lever	bionics
fulcrum	compound
effort arm	machine
resistance	efficiency
arm	power

Theme Connection

Systems and Interactions A simple machine is a system that has work done on it. It, in turn, does work on an object or another system. Emphasize the input-output nature of simple and compound machines throughout the chapter.

Previewing the Chapter

178

Learning Styles Look for the following logo for strategies that emphasize different learning modalities. **LS**

Kinesthetic	Explore, p. 179; Activity, p. 187; Enrichment, p. 192; Close, p. 197
Visual-Spatial	Visual Learning, pp. 182, 186, 187, 188, 189, 190, 192, 193; Reteach, pp. 182, 193, 196; MiniLAB, pp. 190, 201; Demonstration, p. 191
Interpersonal	Activity, pp. 188, 191; Reteach, p. 202; Science and History, p. 204
Logical-Mathematical	Activity 7-1, pp. 184;-185; Using an Analogy, p. 188; Activity 7-2, p. 195; Science and History, p. 204
Linguistic	Across the Curriculum, p. 187; Enrichment, p. 192; Science Journal, p. 197; Science and History, p. 204

Chapter 7

Machines— Making Work Easier

People use machines to make jobs easier. What comes to mind when you think of machines? You might list computers, washing machines, and electric drills as examples. Many machines are simple and do not need electricity to operate. What kinds of machines are in a bicycle?

EXPLORE ACTIVITY

How can a machine make it easier to move an object?

1. Try to lift a stack of two or three books off a table using only your fingertips.
2. Then place the stack of books on the end (at the 5-cm mark) of a meterstick so the entire width of the bottom book is resting on the stick.
3. Push up on the meterstick at the 40-cm mark to lift the books.
4. Repeat the experiment twice, lifting at the 70-cm mark and then at the 100-cm mark (the opposite end).

Observe: Which time did you need the most force to lift the books? The least? In your Science Journal, describe which setup gives you the greatest lifting advantage and try to explain why.

Previewing Science Skills

▶ In the Skill Builders, you will recognize cause and effect, make and use tables, and map concepts.

▶ In the Activities, you will analyze, communicate, measure in SI, and experiment.

▶ In the MiniLABs, you will observe and measure in SI.

179

Section Background

• A machine can be thought of as a system that transfers mechanical energy. A simple machine transfers energy in one movement.

• A machine can multiply forces, distances, or speeds at which the forces are delivered; because of the conservation of energy, it cannot do more work than the energy put into it.

• The mechanical advantage of a machine indicates the factor by which it multiplies an effort force.

Preplanning

Prepare for Activity 7-1 by obtaining poster-sized paper, metersticks, and colored markers for each lab group.

1 Motivate

Bellringer

Before presenting the lesson, display **Section Focus Transparency 26** on the overhead projector. Assign the accompanying **Focus Activity** worksheet. [L1] [LEP]

Tying to Previous Knowledge

Have students recall from Section 5-1 the scientific meaning of *work*. Point out that in this section they will use this definition to learn how machines make accomplishing work easier.

7•1 Why We Use Machines

Science Words

machine
simple machine
effort force
resistance force
ideal machine
mechanical advantage

Objectives

• Explain how machines make work easier.
• Calculate mechanical advantage.

What are machines?

Have you used any machines today? You probably know that a bicycle is a machine. Pencil sharpeners and can openers are also machines. If you have turned a doorknob or twisted off a bottle cap, you have used a machine. A **machine** is a device that makes work easier.

Keeping It Simple

Some machines are powered by engines or electric motors; others are people-powered. Some machines are complex; others are simple. A **simple machine** is a device that does work with only one movement. There are six types of simple machines, examples of which are shown in **Figure 7-1.** You'll learn more about each type in a later section of this chapter.

Figure 7-1

Familiar examples of simple machines can be found in the following places.

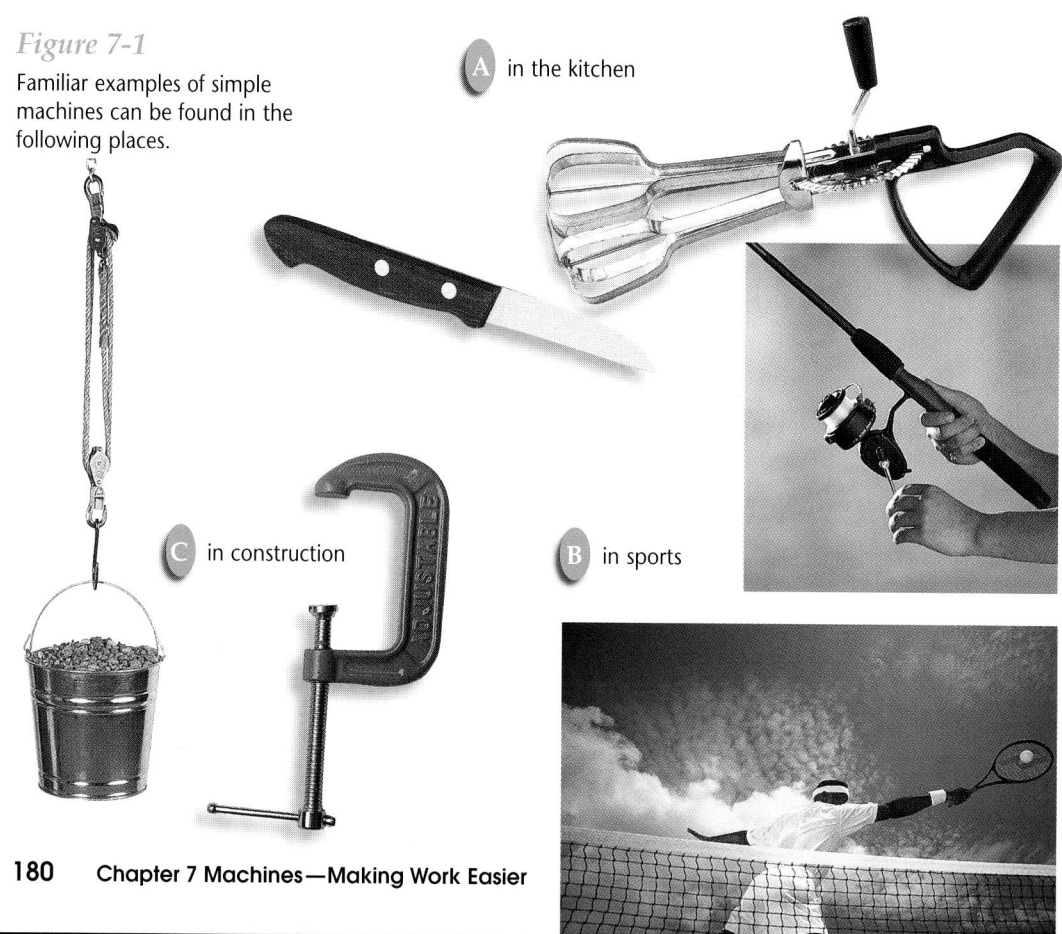

A in the kitchen

C in construction

B in sports

180   Chapter 7 Machines—Making Work Easier

Program Resources

Reproducible Masters
Study Guide, p. 30 [L1]
Reinforcement, p. 30 [L1]
Enrichment, p. 30 [L3]
Activity Worksheets, pp. 38-39 [L1]

 Transparencies
Section Focus Transparency 26 [L1]

Advantages of Simple Machines

Suppose you wanted to pry the lid off a wooden crate with a crowbar. You'd slip the end of the crowbar blade under the edge of the crate lid and push down on the handle. You would do work on the crowbar, and the crowbar would do work on the lid.

Machines make work easier by changing the force you exert in size, direction, or both. **Figure 7-2** shows how the crowbar changes the size and direction of your force as you attempt to lift the lid.

Overcoming Gravity and Friction

When you use a simple machine, you are trying to move something that resists being moved. For example, when you use a crowbar to move a large rock, you are working against gravity—the weight of the rock. When you use a crowbar to remove a lid, you are working against friction—the friction between the nails in the lid and the crate.

Applying Force and Doing Work

Two forces are involved when a machine is used to do work. The force applied *to* the machine is called the **effort force (F_e)**. The force applied *by* the machine to overcome resistance is called the **resistance force (F_r)**. In the crate lid example, you apply the effort force to the crowbar handle. The resistance force is the force the crowbar applies to the lid.

There are also two kinds of work to be considered when a machine is used—the work done *on* the machine and the work done *by* the machine. The work done on the machine is called work input (W_{in}); the work done by the machine is called work output (W_{out}). Recall that work is the product of force and distance: $W = F \times d$. Work input is the product of the effort force and the distance that force is exerted: $W_{in} = F_e \times d_e$. Work output is the product of the resistance force and the distance that force moves: $W_{out} = F_r \times d_r$.

Remember that energy is always conserved. So, you can never get more work out of a machine than you put into it. In other words, W_{out} can never be greater than W_{in}. In fact, whenever a machine is used, some energy is changed to heat due to friction. So, W_{out} is always smaller than W_{in}.

Although a perfect machine has never been built, it helps to imagine a frictionless machine in which no energy is converted to heat. Such an **ideal machine** is one in which work input equals work output. For an ideal machine,

$$W_{in} = W_{out}$$
$$F_e \times d_e = F_r \times d_r$$

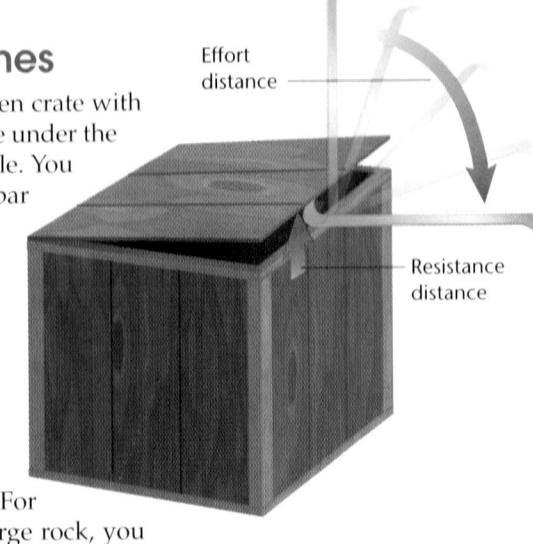

Effort distance

Resistance distance

Figure 7-2

The crowbar is used to multiply the effort force and open the crate. However, to gain force, the effort must push through a greater distance than the resistance distance.

2 Teach

Discussion

Have students list various simple tasks done at home or at school and discuss how different tools or devices help make doing the tasks easier.

Use **Study Guide,** p. 30, as you teach this lesson.

Revealing Preconceptions

Some students may think that machines make work easier by reducing the amount of work that has to be done. Point out that machines actually require more work than do the tasks themselves. Machines allow us to multiply our force, speed, or distance, but not our energy.

GLENCOE TECHNOLOGY

Videodisc

Glencoe Science Interactive Videodisc

Side 1, Lesson 2
Simple Machines and Mechanical Advantage

10630-13318
Mechanical Advantage Comparisons

13320-13399

CD-ROM

Physical Science CD-ROM

Have students perform the interactive exploration for Chapter 7 to reinforce important chapter concepts and thinking processes.

Answer The effort force is exerted by the muscles in the breast of the bird, and the resistance force is that of the wing pushing up on the bird's body.

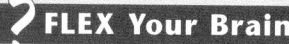

Practice Problem Answer

$$MA = \frac{F_r}{F_e}$$

$$F_e = \frac{F_r}{MA} = \frac{2000 \text{ N}}{10} = 200 \text{ N}$$

3 Assess

Check for Understanding

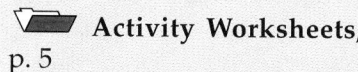
? FLEX Your Brain

Use the Flex Your Brain activity to have students explore MECHANICAL ADVANTAGE.

📁 **Activity Worksheets,** p. 5

Reteach

IS Visual Spatial Set up the demonstration as shown below.

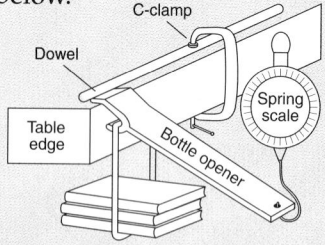

C-clamp
Dowel
Spring scale
Table edge
Bottle opener

Have a volunteer measure the force delivered to the opener by carefully raising it as shown with the spring scale. Remove the books and weigh them. Sketch the demonstration on the chalkboard. **LEP**

Birds fly using their wings to move them through the air. *Identify* the resistance and effort forces for this simple machine by making a labeled drawing.

In most cases, a machine multiplies the force applied to it—F_r is greater than F_e. So, in order for W_{in} to equal W_{out}, the effort force must travel farther than the resistance force—d_e must be greater than d_r.

Mechanical Advantage

Think again about the crate lid example. The distance you move the crowbar handle (d_e) is greater than the distance the crowbar moves the lid (d_r). So, the end of the crowbar must exert more force on the lid (F_r) than you exert on the handle (F_e). The machine multiplies your effort, but you must move the handle a greater distance.

The number of times a machine multiplies the effort force is the **mechanical advantage (MA)** of the machine. To calculate mechanical advantage, you divide the resistance force by the effort force.

$$MA = \frac{\text{resistance force}}{\text{effort force}} = \frac{F_r}{F_e}$$

Work the problems below to see how mechanical advantage is related to effort force and resistance force.

Some machines don't multiply force. They simply change the direction of the effort force. For example, when you pull

USING MATH

Calculating Mechanical Advantage

Example Problem:

A worker applies an effort force of 20 N to pry open a window that has a resistance force of 500 N. What is the mechanical advantage of the crowbar?

Problem-Solving Steps:

1. What is known?
   resistance force, F_r = 500 N
   effort force, F_e = 20 N
2. What is unknown?
   mechanical advantage, *MA*
3. Choose the equation.
   $$MA = \frac{F_r}{F_e}$$

4. **Solution:** $MA = \dfrac{500 \text{ N}}{20 \text{ N}} = 25$

Practice Problem

Find the effort force needed to lift a 2000-N rock, using a jack with a mechanical advantage of 10.

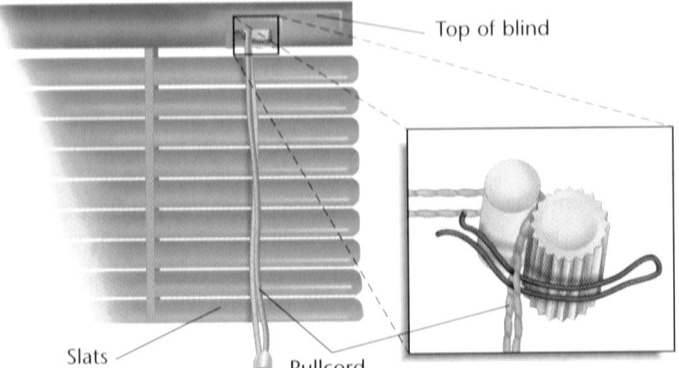

Top of blind

Slats

Pullcord

Figure 7-3
Miniblinds are a familiar example of a simple machine that changes the direction of a force.

down on the cord of window blinds, such as those in **Figure 7-3**, the blinds go up. Only the direction of the force changes; the effort force and resistance force are equal, so the mechanical advantage is 1.

Other machines, such as third-class levers, have mechanical advantages that are less than 1. Such machines are used to increase the distance an object moves or the speed at which it moves.

Section Wrap-up

Review

1. Explain how simple machines can make work easier without violating the law of conservation of energy.

2. A carpenter uses a claw hammer to pull a nail from a board. The nail has a resistance of 2500 N. The carpenter applies an effort force of 125 N. What is the mechanical advantage of the hammer?

3. **Think Critically:** Give an example of a simple machine you've used recently. How did you apply effort force? How did the machine apply resistance force?

Skill Builder
Recognizing Cause and Effect
When you operate a machine, it's often easy to observe cause and effect. For example, when you turn a doorknob, the latch in the door moves. Give five examples of machines and describe one cause-and-effect pair in the action of each machine. If you need help, refer to Recognizing Cause and Effect in the **Skill Handbook.**

USING MATH

Suppose you want to use a simple machine to lift a 6000-N log from a fallen tree. What effort force will you need if your machine has a mechanical advantage of 25? Of 15? Of 1? Show your calculations.

7-1 Why We Use Machines   **183**

Activity 7-1

PREPARATION

Purpose

LS **Logical-Mathematical** Design ways to use simple machines to make a building handicap accessible. **L1** **COOP LEARN**

Process Skills

describing, modeling, hypothesizing, communicating, sequencing, observing and inferring, comparing and contrasting, recognizing cause and effect, predicting, designing

Time

1-2 class periods for preparation, planning, and implementing

Materials

See student page.

Possible Hypotheses

Students may hypothesize the use of inclined planes (ramps) for wheelchair access up stairs, pulleys may help to open doors, a gear assembly might reduce the distance the string must be pulled in order to open the door, etc.

📁 **Activity Worksheets,** pp. 5, 38-39

PLAN THE EXPERIMENT

Possible Procedures

1. Students might start with a drawing of a nonaccessible building; they might modify each hurdle they can identify.
2. They may enlarge each doorway or room to be modified, then show before and after drawings.

Activity 7-1

Design Your Own Experiment
Access for Everyone

If you were in a wheelchair or unable to climb stairs, would you be able to get into your home or school and move around? It is now law that public buildings must be accessible to people who are wheelchair-bound. What are some features you have seen that make this possible? In this experiment, you will use simple machines to design a house, school, or other public building that is wheelchair-accessible.

PREPARATION

Problem

How can you use simple machines to make a building accessible to people in wheelchairs?

Form a Hypothesis

Brainstorm ideas with your group and generate a hypothesis statement about how you could solve the problem stated above.

Objectives

- Apply simple machines in designing a building plan.
- Explain the purpose of each simple machine used in the plan, including how it provides a mechanical advantage.

Possible Materials

- poster-sized paper
- ruler or meterstick
- pencils
- markers

184

PLAN THE EXPERIMENT

1. Agree on the type of building your group wishes to design, and sketch a rough diagram of it in pencil on one side of your paper.
2. Determine at least three features that will help make your building accessible and easy to use for people in wheelchairs. At least two must relate directly to simple machines.
3. Preview the details about these simple machines in Section 7-2.
4. On your rough diagram, sketch how and where your features will be used.

Check the Plan
1. Before you draw your final plan, make sure your group has reached agreement on the features and design.

2. Why would your building be difficult to access without these special features? Think about a non-accessible building as your control.
3. Explain how you can get the greatest advantage out of each simple machine feature. For example, if you are using an inclined plane for a ramp, would a long, gradual ramp or a short, steep ramp be more effective? Why?
4. *Make sure your teacher approves your plan and that you have included any changes suggested in the plan.*

DO THE EXPERIMENT

1. Draw your final plan as clearly as possible on large paper or poster board.
2. Be sure to label your features. Discuss what function the simple machines perform in each case. Prepare to present your plan to the class.

Analyze and Apply
1. Briefly **communicate** your plan to the class and **analyze** the designs of other groups.

Which features are most commonly used?
2. Identify at least two simple machines found in a wheelchair and **discuss** where they are found.
3. You designed a new wheelchair-accessible building. **Infer** what methods might be feasible to make an old building wheelchair-accessible.

Go Further

Organize and conduct a survey of handicap accessibility of buildings in your neighborhood. What things should you check for? You might investigate four categories of buildings: homes, schools, private buildings, and other public buildings.

7-1 Why We Use Machines   **185**

DO THE EXPERIMENT

Expected Outcome
Students will design a building that is completely handicap accessible.

Analyze and Apply
1. Ramps for wheelchairs and pulleys for elevators will probably be the most commonly used features.
2. Wheel and axle—wheels of the chair and the axle they're rotating around; lever—wheelchair brake
3. Where possible, install ramps at all stairways, electric eyes for automatic door openers or buttons to push to open doors, widen doorways to accommodate chair width, install elevator where ramps are impractical, etc.

Go Further

List may include: check for doorway width, stairs with ramp installments, accessibility from parking lot or sidewalk, rest room accessibility, etc.

Assessment

Portfolio Have students put final touches on their drawings, such as colored labels, brief descriptions near each simple machine, etc. Include this design in their portfolios. Use the Performance Task Assessment List for Display in **PASC**, p. 63. **P**

Prepare

Section Background

- The six types of simple machines—lever, pulley, wheel and axle, inclined plane, wedge, and screw—fall into two broad categories: levers and inclined planes. Pulleys and wheels and axles are leverlike machines. Wedges and screws are grouped with inclined planes.

- The mechanical advantage of every simple machine depends upon its design and is calculated by comparing its F_r to its F_e.

Preplanning

- To prepare for the Mini-LAB, obtain and cut pieces of PVC pipe.
- Collect materials for Activity 7-2.

1 Motivate

Bellringer

Before presenting the lesson, display **Section Focus Transparency 27** on the overhead projector. Assign the accompanying **Focus Activity** worksheet. L1 LEP

Tying to Previous Knowledge

Have students recall from the last section the definition of *mechanical advantage*. Point out that in this section they will learn how the design of a simple machine affects its mechanical advantage.

Visual Learning

Figure 7-4 **What acts as the fulcrum?** *The edge of the wheel acts as the fulcrum.* LEP LS

186

Science Words

lever
fulcrum
effort arm
resistance arm
pulley
wheel and axle
inclined plane
screw
wedge

Objectives

- Describe the six types of simple machines.
- Calculate the ideal mechanical advantage for different types of simple machines.

Levers

If you've ever ridden a seesaw, pried the cap from a bottle of soda pop, or swung a tennis racket, you have used a lever. A **lever** is a bar that is free to pivot, or turn, about a fixed point. The fixed point of a lever is called the **fulcrum.** The part of the lever on which the effort force is applied is called the **effort arm.** The part of the lever that exerts the resistance force is called the **resistance arm.**

Suppose you are using a tire tool to pry the tire from a bicycle wheel before you change the tire. You can see in **Figure 7-4** that the rim of the wheel acts as the fulcrum. You push down on the effort arm of the tire tool. The tool pivots about the fulcrum, and the resistance arm exerts a force on the tire, lifting it upward.

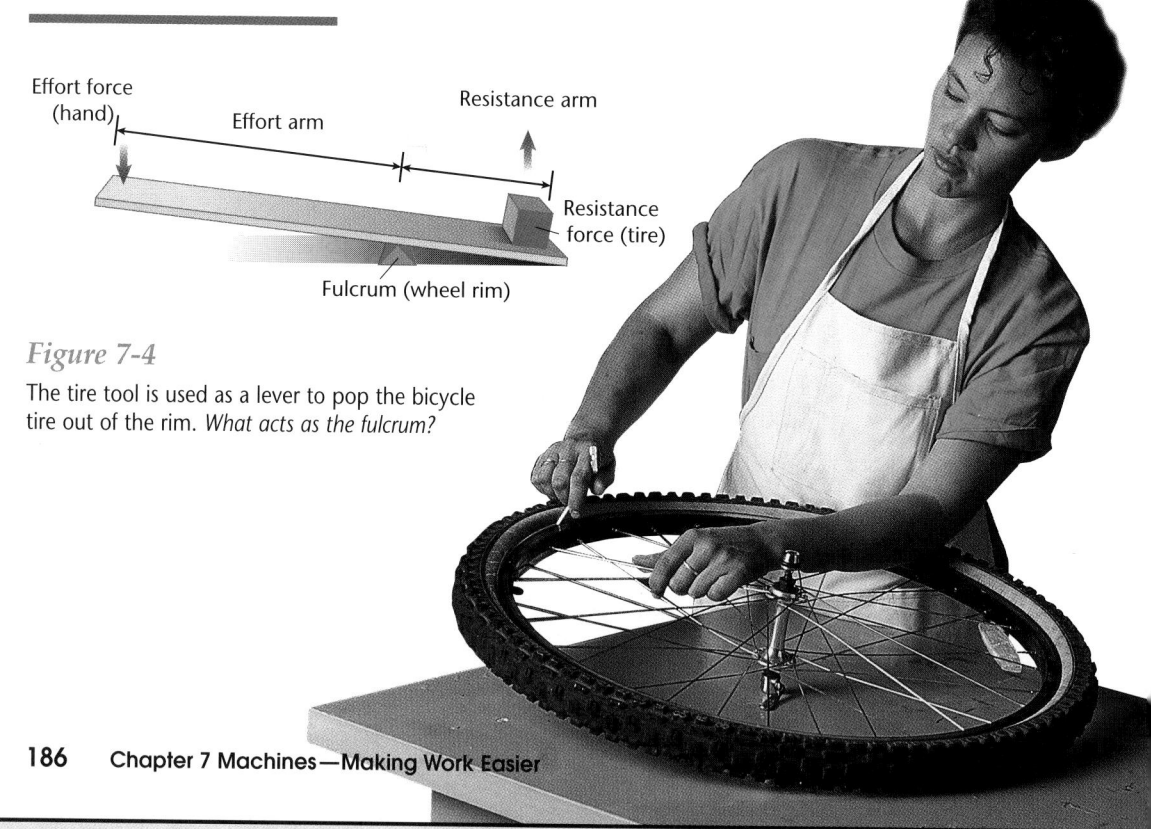

Figure 7-4

The tire tool is used as a lever to pop the bicycle tire out of the rim. *What acts as the fulcrum?*

186 Chapter 7 Machines—Making Work Easier

Program Resources

Reproducible Masters
Study Guide, p. 31 L1
Reinforcement, p. 31 L1
Enrichment, p. 31 L3
Activity Worksheets, pp. 40-41, 42
Science Integration Activities, pp. 13-14
Multicultural Connections, pp. 17-18
Concept Mapping, pp. 19-20

Lab Manual 14
Critical Thinking/Problem Solving, p. 13

Transparencies
Section Focus Transparency 27 L1
Teaching Transparency 13 L1
Science Integration Transparency 7

Finding the Ideal Mechanical Advantage

The type of lever shown in **Figure 7-4** makes work easier by multiplying your effort force and changing the direction of your force. You've learned that the mechanical advantage of any machine can be calculated by dividing the resistance force by the effort force. You can also use the lengths of the arms of a lever to find the *ideal mechanical advantage* of the lever. The length of the effort arm is the distance from the fulcrum to the point where effort force is applied. The length of the resistance arm is the distance from the fulcrum to the point where the resistance force is applied. The following equation, which assumes no friction, can be used to find the ideal mechanical advantage (IMA) of any lever.

$$IMA = \frac{\text{length of effort arm}}{\text{length of resistance arm}} = \frac{L_e}{L_r}$$

Practice calculating ideal mechanical advantage by working the following problems.

USING MATH

Ideal Mechanical Advantage of a Lever

Example Problem:

A worker uses an iron bar to raise a manhole cover weighing 65 N. The effort arm of the lever is 60 cm long. The resistance arm is 10 cm long. What is the ideal mechanical advantage of the bar?

Problem-Solving Steps:

1. What is known? effort arm, L_e = 60 cm
   resistance arm, L_r = 10 cm
2. What is unknown? ideal mechanical advantage, *IMA*
3. Choose the equation. $IMA = \dfrac{L_e}{L_r}$

4. **Solution:** $IMA = \dfrac{60 \text{ cm}}{10 \text{ cm}} = 6.0$

Practice Problems

1. You use a crowbar 140 cm long as a lever to lift a large rock. The rock is 20 cm from the fulcrum. What is the *IMA* of the lever?
2. An oar used to row a boat has the handle 50 cm from the fulcrum and the blade 125 cm from the fulcrum. What is the ideal mechanical advantage of the oar?

---

2 Teach

Activity

LS **Kinesthetic** Provide students with many opportunities to manipulate levers. Allow ample time for them to locate the fulcrum and identify the effort arm and resistance arm in each lever. **LEP**

Discussion

Have students discuss how various kitchen utensils, such as knives, spoons, spatulas, whisks, and hand mixers, fit the definition of a simple machine. Ask students to find similarities among the functions and shapes of the utensils.

Practice Problem Answers

1. $IMA = \dfrac{L_e}{L_r} = \dfrac{140 \text{ cm}}{20 \text{ cm}} = 7$
2. $IMA = \dfrac{L_e}{L_r} = \dfrac{50 \text{ cm}}{125 \text{ cm}} = 0.4$

Videodisc

STV: Human Body, Vol. 2

Knee joint

48683

Pelvis

48888

Scapula

48884

Shoulder joint (art)

48678

Across the Curriculum

Language Arts Have students look up the word *lever* in a dictionary. Students should find that it comes from the Latin root *levare* (to lift). Ask a volunteer to explain how the meaning of the word reflects its root. **L1** **LS**

It isn't necessary to know the exact location of F_e, F_r, and the fulcrum to classify a lever. All one needs to know is what is in between. In a first-class lever, the fulcrum is located somewhere between the effort force and the resistance force. In a second-class lever, F_r is in between, and in a third-class lever, F_e is in between.

Visual Learning

Figure 7-5C What is the purpose of a third-class lever? *It increases speed.* **LEP** **LS**

Activity

LS **Interpersonal** Assign one of the six simple machines to each team of 3-4 students. Each group will have two class periods to prepare a presentation about their machine. They should define their machine, identify its important parts, and give a demonstration of its use. Encourage creativity, such as using a knife to cut some brownies to share. **L1**

📁 Use these teaching resources as you teach this lesson.

Multicultural Connections, pp. 17-18

Science Integration Activities, pp. 13-14

Teaching Transparency 13

Science Integration Transparency 7

Using an Analogy

LS **Logical-Mathematical** Assign problem-solving teams to identify bones that act as levers shown in cardboard skeletons displayed at Halloween. From the functioning of each bone, have the team determine if it is a first-, second-, or third-class lever. **L1** **LEP** **COOP LEARN**

Figure 7-5

The classes of levers differ in the positions of the effort force, the resistance force, and the fulcrum. Their purposes also differ.

Types of Levers

There are three different types, or classes, of levers. These classes are based on the positions of the effort force, resistance force, and fulcrum. **Figure 7-5** shows the three classes of levers.

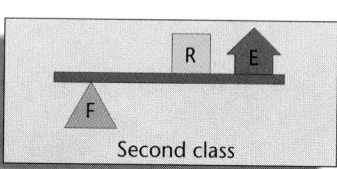

A First Class: These levers, with the fulcrum located between the effort and resistance forces, are usually used to multiply force. If a large effort force is used on the shorter arm of the lever, it can also be used to multiply distance.

B Second Class: The resistance is located between the effort force and the fulcrum in second-class levers. These levers always multiply force.

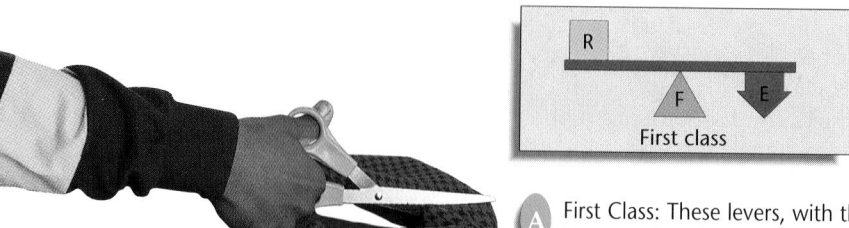

C Third Class: The effort force is located between the resistance force and the fulcrum in third-class levers. The effort arm is always shorter than the resistance arm, so it cannot multiply force and its *MA* is always less than one. *What is the purpose of a third-class lever?*

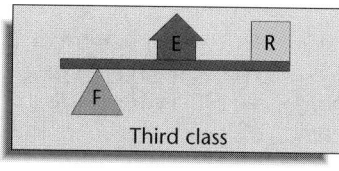

188 Chapter 7 Machines—Making Work Easier

Cultural Diversity

The Great Pyramids It is possible that the great pyramids of Egypt were constructed using inclined planes to move the stones into place. One of them, the pyramid of Cheops, covers 5.2 hectares and contains more than 2.25×10^6 blocks that average 2×10^3 kg each. Most modern cranes would not be able to lift stones that heavy, so it is easy to understand why the Egyptians would need to find an alternative to lifting the stones.

That alternative was the inclined plane, which allowed them to move the heavy blocks with smaller forces exerted over a longer period of time. For example, let's assume you want to move an object to a height of 10 m, but you cannot lift it. However, you can push it up a plane inclined at 10°. This means you must push the object about 60 m to get it 10 m off the ground.

Levers with a Human Touch

Try holding a book or other object at your side with your arm straight down and palm facing forward. Keeping your upper arm at your side, bend your elbow and raise the book to chest height. Weightlifters call this lift a curl. What is supplying the effort force? Think about the mechanics of such a simple lifting task. You have many natural levers in your body composed of bone and muscle systems that give your body its form and ability to function.

Locating Levers

Although all lever classes are found in your body, the majority are third-class levers such as using your arm in the curl. Recall that third-class levers cannot multiply force. What advantage do they provide? As in a hockey stick or fishing rod, third-class levers provide a greater range of distance and speed of motion than the effort force itself can provide. Your biceps muscles used in the curl have a limited length, but they can supply more effort force to increase the range and speed of motion of the resistance force (your forearm). In this case, your elbow serves as the fulcrum. You would also use your arm as a third-class lever to pitch a baseball.

Can you find any first- or second-class levers in your body? Try sitting in an upright position and completely relaxing your neck. Your head tips forward because your neck muscles support the effort force to hold up the resistance weight of your head, acting as a first-class lever. Your foot becomes a second-class lever when you stand on your toes. Examine **Figure 7-6** to find out how you can identify the effort and resistance forces in these levers. What other levers can you identify in your body?

INTEGRATION
Life Science

Figure 7-6

Your body's structural system of muscles and bones contains natural examples of first-, second-, and third-class levers.

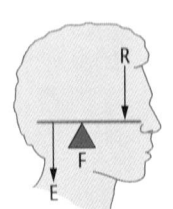

A First-class lever

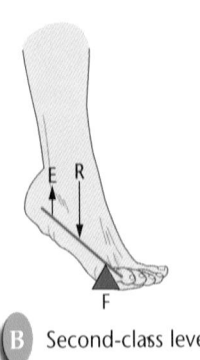

B Second-class lever

C Third-class lever

INTEGRATION
Life Science

There are many bones contained within the skeletal structure of the body. Obtain a diagram of the human skeletal system from a life science book and have students identify as many bones as they can find.

Visual Learning

Figure 7-6 Have students act out each photo in the figure and identify the effort and resistance forces in each photo. F_e—neck holding head upright, F_r—head slumping forward, F_e—calf and foot muscles straightening leg and ankle, F_r—toes bending at foot; F_e—biceps muscle bending arm, F_r—forearm. **LEP** **LS**

NATIONAL GEOGRAPHIC SOCIETY

Videodisc
STV: Human Body Vol. 2
Ankle and foot (art)

48680
Elbow joint (art)

48679
Foot

48887

MiniLAB

Purpose

LS **Visual-Spatial** To observe and compare the *MA* of different pulley systems. **L1**

COOP LEARN

Materials

two 1-m lengths of PVC pipe (or broom handles) per lab group; rope

Teaching Strategies

Troubleshooting Students should pull the rope smoothly and continuously.

Analysis

1. With the rope wrapped around the pipe, the pipe is the fulcrum between the F_e—length being pulled and the F_r—fixed end. Wrapping the rope again would increase the *MA*.

2. With more than a single wrap, force is multiplied, not energy.

✔ Assessment

Portfolio Have students collect pictures and drawings of various pulley systems (i.e., flagpoles, cranes), and make a display indicating whether they are single or multiple systems and indicating which systems have the greater *MA*s. Use the Performance Task Assessment List for Making Observations and Inferences in **PASC**, p. 17. **P**

Activity Worksheets, pp. 5, 42

MiniLAB

How do pulleys make work easier?

Procedure

1. Working in groups of three, tie one end of a rope to a piece of PVC pipe or a broomstick handle. Wrap the other end around another piece of pipe once (see Figure A below).

2. Two of the group members should each hold one pipe horizontally in their hands, with the pipes about 30 cm apart. The third person should grasp the free end of the rope and slowly try to pull the pipes together while the others try to hold the pipes firmly in place.

3. Wrap the rope again around the first pipe (B). What happens now when the third person tries to pull the pipes together?

4. Wrap the rope one or two more times across the pipes (C) and repeat the experiment.

Analysis

1. How is this similar to a pulley system? What increases the mechanical advantage of these pipe pulleys?

2. Does this machine multiply the energy of the person pulling? Why or why not?

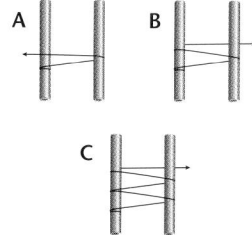

Pulling with Pulleys

Have you ever seen someone raise a flag on a flagpole? A pulley is used to help get the flag to the top of the pole. A **pulley** is a grooved wheel with a rope or a chain running along the groove.

A pulley works something like a first-class lever. Instead of a bar, a pulley has a rope. The axle of the pulley acts like the fulcrum. The two sides of the pulley are the effort arm and the resistance arm.

Positioning Pulleys

Pulleys can be fixed or movable. A *fixed* pulley is attached to something that doesn't move, such as a ceiling, wall, or tree. A fixed pulley, such as the one used at the top of a flagpole, can change the direction of an effort force. When you pull down on the effort arm with the rope, the pulley raises the object attached to the resistance arm. The *IMA* of a single fixed pulley is 1. Thus, a single fixed pulley does not multiply the effort force.

A *movable* pulley is attached to a construction crane, as shown in **Figure 7-7.** The difference between a fixed pulley and a movable pulley is shown in **Figure 7-8.** Unlike the fixed pulley, a movable pulley does multiply the effort force. Because a movable pulley multiplies force, its MA is greater than 1. In fact, the *IMA* of a single movable pulley is 2. This means that, assuming no friction, an effort force of 1 N will lift a weight (resistance) of 2 N. In order to conserve energy, the effort distance must be twice as large as the resistance distance.

IMA = 1
Single Fixed Pulley

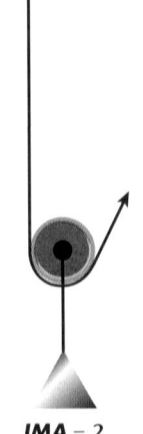

IMA = 2
Single Movable Pulley

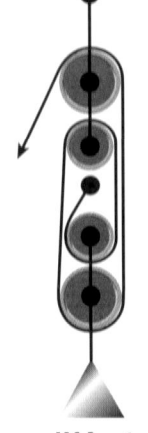

IMA = 4
Block and Tackle

The Block and Tackle

Fixed and movable pulleys can be combined to make a system of pulleys called a block and tackle. Depending on the number of pulleys used, a block and tackle can have a large mechanical advantage. The ideal mechanical advantage of any ideal pulley or pulley system is equal to the number of ropes that support the resistance weight. As **Figure 7-8** shows, the only rope that does not support the resistance weight is the rope leading from the pulley system down to the effort force.

Wheel and Axle

Look closely at a doorknob or the handle of a water faucet. Do you recognize these as simple machines? Do they seem to make work easier? If you don't think so, remove the knob or handle from its narrow shaft. Now, try opening the door or turning on the water by rotating that shaft with your fingers. After a few minutes, you'll appreciate the fact that the knob and handle, together with their shafts, are machines. They do make work easier.

Doorknobs, Faucets, and Ice-Cream Makers

Doorknobs, faucet handles, and ice-cream makers are examples of a wheel and axle. A **wheel and axle** is a simple machine consisting of two wheels of different sizes that rotate together. An effort force is usually applied to the larger wheel. The smaller wheel, called the axle, exerts the resistance force. In many cases, the larger wheel doesn't look like a typical circular wheel. It may be a crank handle, like that of the ice-cream freezer on the next page, or a faucet, as in **Figure 7-9**. But it always travels in a circle.

Figure 7-8

The ideal mechanical advantage of each type of pulley can be found by counting every rope supplying an upward support force on the resistance.

Figure 7-9

A water faucet is an example of a wheel and axle. Without the faucet handle (wheel), the shaft of the faucet (axle) is difficult to turn.

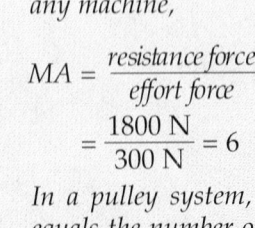

191

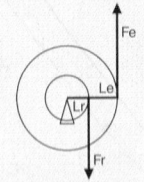
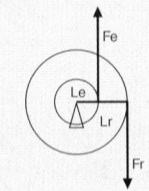

Visual Learning

Figure 7-11 Have students refer to the two sets of gears shown in **Figure 7-11.** Ask them to identify the set in which the gears will move in the same direction (drive chain) and the set in which they will move in opposite directions (mesh teeth). **LEP**
🅛🅢

Teacher F.Y.I.

In mesh teeth gears, the *IMA* is equal to the number of teeth on the effort gear, N_e, divided by the number of teeth on the resistance gear, N_r. In mesh gears, the size, number, and spacing of teeth must be precise for the gears to operate smoothly.

Enrichment

🅛🅢 **Kinesthetic** Have students research various types of gears, such as *helical, bevel, planetary,* and *worm;* display them to the class as working models and discuss their uses. **L3**

🅛🅢 **Linguistic** Have interested students research how simple machines were used in early civilizations such as Babylonian, Egyptian, Chinese, and Incan. **L2**

🅛🅢 **Kinesthetic** Have interested students research and build an Archimedes' screw. **L3**

Figure 7-10

The length of the crank handle represents the radius of the wheel. The axle radius is represented by the turning paddle inside the ice-cream maker.

It might help to think of a wheel and axle as being a lever attached to a shaft. The radius of the wheel is the effort arm, and the radius of the axle is the resistance arm. The center of the axle is the fulcrum. As with the lever, the ideal mechanical advantage of a wheel and axle can be calculated by dividing the radius of the wheel (effort arm) by the radius of the axle (resistance arm).

$$IMA = \frac{\text{radius of wheel}}{\text{radius of axle}} = \frac{r_w}{r_a}$$

USING MATH

Calculating *IMA* of a Wheel and Axle

Example Problem:

In the ice-cream freezer shown in **Figure 7-10**, the wheel has a radius of 20 cm. The axle has a radius of 15 cm. What is the ideal mechanical advantage of the wheel and axle?

Problem-Solving Steps:

1. What is known?   radius of wheel, r_w = 20 cm
                      radius of axle, r_a = 15 cm
2. What is unknown?   ideal mechanical advantage, *IMA*
3. Choose the equation. $IMA = \dfrac{r_w}{r_a}$
4. **Solution:** $IMA = \dfrac{20 \text{ cm}}{15 \text{ cm}} = 1.3$

Practice Problem

An automobile steering wheel with a diameter of 48 cm is used to turn the steering column with a radius of 4 cm. What is the *IMA* of this wheel and axle?

192   Chapter 7 Machines—Making Work Easier

Across the Curriculum

Mathematics Supply the students with silhouettes of several common wheels and axles, such as screwdrivers, house keys, and beaters from an electric mixer; or ask them to trace similar items at home and bring the tracings to class. Have them cut out the silhouettes or tracings, fold each along the center of the wheel and axle, unfold the paper, and measure the approximate radius of the wheel and the radius of the axle for each item. Allow the students to use calculators to determine the *IMA* of each item and relate this value to the task that the item is used for. **L1** 🅛🅢

Visual Learning

Figure 7-11 What causes the rear wheel of the bicycle to turn? *The force of the rider's foot on the pedal is transferred by the chain and gears, causing the rear wheel to turn.* **LEP** **LS**

Figure 7-11

Gears are a form of wheel and axle. *What causes the rear wheel of the bicycle to turn?*

Gearing Up

Gears are modified wheel-and-axle machines. A gear is a wheel with teeth along its circumference. **Figure 7-11** shows two ways gears work. Effort is exerted on one of the gears, causing the other gear to turn. In most gear systems, the larger gear is the effort gear. The ideal mechanical advantage of a pair of gears is found by dividing the radius of the effort gear by the radius of the resistance gear. This is known as the gear ratio of the pair.

Inclined Plane

Suppose you had to move a heavy box from the ground up onto a porch. Would you rather lift the box straight up or slide it up a ramp like the one in **Figure 7-12**? The ramp would make your job easier. A ramp is a type of **inclined plane,** which is a sloping surface used to raise objects.

The amount of work done on the box is the same whether you lift it straight up or slide it up the ramp. But remember that work has two parts—force and distance. When you lift the box straight up, the distance is small, but the force is large. Using the ramp, you cover more distance, but you exert less force.

You can calculate ideal mechanical advantage of an inclined plane using distances.

$$IMA = \frac{\text{effort distance}}{\text{resistance distance}} = \frac{\text{length of slope}}{\text{height of slope}} = \frac{l}{h}$$

Figure 7-12

The ramp is an example of an inclined plane.

7-2 The Simple Machines **193**

3 Assess

Check for Understanding

? FLEX Your Brain

Use the Flex Your Brain activity to have students explore PULLEYS.

📁 **Activity Worksheets,** p. 5

Reteach

LS **Visual-Spatial** From one color of construction paper, cut three arrows, about 10 cm wide and 10, 20, and 30 cm long. Make a similar set in another color. Using a wide marker, label each arrow in the first set F_e. Label the second set F_r. Collect three or four examples of each type of simple machine. Explain that the arrows represent different amounts of effort and resistance force. Ask volunteers to choose a machine and use an arrow from each pile that represents the relative size of F_e and F_r. Showing the locations and directions of F_e and F_r, have the class agree upon the labels and classify each machine. **LEP**

Extension

📁 For students who have mastered this section, use the **Reinforcement** and **Enrichment** masters.

4 Close

Section Wrap-up

Review

1. Student responses will vary. Accept all reasonable responses.

2. Students should describe how pulleys and wheel and axle machines are modified levers. Wedges and screws are modified inclined planes.

3. **Think Critically** Responses may vary but could include the fact that friction makes it possible for people to walk up a ramp and keeps people from sliding down the ramp.

Using Computers

Student responses will vary. Accept all reasonable responses.

Figure 7-13

The blade of the knife and the threads of the screw are special types of inclined planes.

The Screw

The screw and the wedge, **Figure 7-13**, are examples of inclined planes that move. A **screw** is an inclined plane wrapped in a spiral around a cylindrical post. If you look closely at a screw, you'll see that the threads form a tiny ramp that runs from its tip to near its top. As you turn the screw, the threads seem to pull the screw into the wood. The wood seems to slide up the inclined plane. Actually, the plane slides through the wood.

The Wedge

A **wedge** is an inclined plane with one or two sloping sides. Chisels, knives, and ax blades are examples. A typical inclined plane stays in one place while materials move along its surface. A wedge is a moving inclined plane. The material remains in one place while the wedge moves through it.

Perhaps you've noticed that the six types of simple machines are all variations of two basic machines—the lever and the inclined plane. As you go about your daily activities, look for examples of each type of simple machine. See if you can tell how each makes work easier.

Section Wrap-up

Review

1. Give one example of each kind of simple machine. Use examples different from the ones in the text.

2. Explain why the six kinds of simple machines are really variations on just two basic machines.

3. **Think Critically:** When would the friction of an inclined plane be useful?

Skill Builder
Making and Using Tables
Organize information about the six kinds of simple machines into a table. Include the type of machine, an example of each type, and a brief description of how it works. You may include other information if you wish. If you need help, refer to Making and Using Tables in the **Skill Handbook.**

> ### Using Computers
>
> **Word Processor** Using a word processor, write a separate paragraph about each class of lever. Describe at least two examples of each class, identifying the effort force, resistance force, and fulcrum for each. Use **Figure 7-5** to help you.

 **Skill Builder**
Tables should reflect information included in the chapter about each type of machine.

✓ Assessment

Performance Bring out the same simple machine you used in the Skill Builder Assessment for Section 7-1. Have students use their tables to help them identify the type of machine and the effort and resistance forces. Use the Performance Task Assessment List for Analyzing Data in **PASC,** p. 27. **P**

Activity 7-2

Levers

Did you ever play on a seesaw when you were younger? Wasn't it much easier to balance your friend on the other end if both of you weighed about the same? If your friend was lighter, you probably moved toward the fulcrum to balance the seesaw. How did this help? Moving toward the fulcrum shortened the effort arm. In this activity, you will balance a lever to determine the mass of a coin.

Problem
How can a lever measure mass?

Materials
- sheet of paper, 20 cm × 28 cm (8½" × 11")
- coins, 3 (quarter, dime, nickel)
- balance
- metric ruler

Procedure
1. Make a lever by folding the paper into a strip 3 cm wide by 28 cm long.
2. Mark a line 2 cm from one end of the paper strip. Label this line *Resistance.*
3. Slide the other end of the paper strip over the edge of a table until the strip begins to teeter on the edge. Mark a line across the paper at the table edge and label this line *Effort.*
4. Measure the mass of the paper to the nearest 0.1 g. Write this mass on the *Effort* line.
5. Center a dime on the *Resistance* line.
6. Locate the fulcrum by sliding the paper strip until it begins to teeter on the edge. Mark the balance line. Label it *Fulcrum #1.*
7. Measure the lengths of the resistance and effort arms to the nearest 0.1 cm.

8. Calculate the *IMA* of the lever.
9. Multiply the *IMA* times the mass of the lever to find the approximate mass of the coin.
10. Repeat steps 5 through 9 with the nickel and then with the quarter. Mark the fulcrum line #2 for the nickel and #3 for the quarter.

Analyze
1. In this activity, is the effective total length of the lever a constant or a variable?
2. Are the lengths of the resistance and effort arms constants or variables?
3. What provides the effort force?
4. What does it mean if the *IMA* is less than 1.0?

Conclude and Apply
5. Is it necessary to have the resistance line 2.0 cm from the end of the paper? Explain.
6. The calculations are done as if the entire weight of the paper is located at what point?
7. **Infer** why mass units can be used in place of force units in this kind of problem.

7-2 The Simple Machines   195

Answers to Questions
1. The lever's total length is a constant.
2. The lengths of effort and resistance arms are variables.
3. Gravity provides the effort force.
4. The force to move the lever is more than the weight that is moved.
5. No, the resistance line can be placed at any convenient location.
6. The calculations assume the entire weight of the lever is located at the center of gravity of the paper.

7. The forces on both ends of the lever are caused by gravity acting with the same ratio of force to mass.

> **✔ Assessment**

Oral Evaluate student understanding of the lever design by having them explain to the class how they carried through the calculations. Use the Performance Task Assessment List for Oral Presentation in **PASC**, p. 71. **P**

Activity 7-2

Purpose
LS **Logical-Mathematical** Calculate the mechanical advantage of levers to balance unequal forces. **L1**

Process Skills
measuring, using numbers, interpreting data, classifying, observing and inferring, comparing and contrasting, recognizing cause and effect, forming operational definitions, interpreting scientific illustrations

Time
40 minutes

📁 **Activity Worksheets,** pp. 5, 40-41

Teaching Strategies
- This activity applies the idea that a rigid object can act as though its entire mass is located at its center of gravity. One end of the lever is located at the center of gravity of the paper, and the other end is located near one end of the paper. The coin is placed with its center of gravity right on the lever's end mark.
- Have students draw the effective lever and fulcrum on the paper between the effort and resistance lines. Point out that the rest of the paper is not part of the lever but provides effort force and support.
- Typically, *MA* is associated with force or mass, and *IMA* with length. However, for purposes of this activity and ease of calculation, assume *MA = IMA*.
- The following equations are needed for procedure steps 8 and 9.
  Step 8 $IMA = \dfrac{L_e \text{ (paper)}}{L_r \text{ (coin)}}$
  Step 9 $MA = \dfrac{F_r \text{ (coin)}}{F_e \text{ (paper)}}$
  Mass and force are proportional, so $m_r \text{ (coin)} = m_e \text{ (paper)} \times MA$

195

1 Motivate

Bellringer

Before presenting the lesson, display **Section Focus Transparency 28** on the overhead projector. Assign the accompanying **Focus Activity** worksheet. [L1] [LEP]

2 Teach

Enrichment

Invite someone from your local rehabilitation center or hospital to demonstrate various prosthetic devices. A video or movie might be equally informative.

3 Assess

Check for Understanding

? FLEX Your Brain

Use the Flex Your Brain activity to have students explore BIONICS.

📁 **Activity Worksheets,** p. 5

Reteach

[LS] **Visual-Spatial** Show photos and diagrams of different prostheses and joints; discuss how they connect and how they work. Contrast older "passive" devices with newer ones such as those described in this section. [L1]

Extension

📁 For students who have mastered this section, use the **Reinforcement** and **Enrichment** masters.

196

7•3 Mending with Machines

Science Words

bionics

Objectives

- Explain what the science of bionics involves.
- Contrast two methods of using electrical signals to trigger motion of a limb or other bodily process.

The Human Body as a Machine

You have learned that some parts of the body work like machines. Some, such as your arms and legs, act as simple machines and are controlled by nerve impulses from your brain. What happens when a part of the human body no longer functions correctly or has to be removed?

Bionics—Making New Parts

Some organs, such as kidneys, eyes, and hearts, can be transplanted from another person. But many times, an appropriate technique or donor does not exist. Some of the most common replacement procedures involve artificial parts, such as a hip or knee joint. **Bionics** is the science of designing artificial replacements for parts of the human body. Artificial replacements are also called prostheses. Specialized prostheses made of new shock-absorbing material can even make it possible for a leg amputee to ski competitively, as in **Figure 7-14**, or complete a lengthy triathlon.

Making Electrical Connections

In the 1700s, a scientist named Alessandro Volta observed that muscles can be affected by small electric shocks. This observation is used in a technique called functional neuromuscular stimulation (FNS). These FNS systems have allowed people with paralyzed legs to walk again and maintain muscle tone. To cause a walking motion, FNS systems use a control unit to cause the proper amount of electric current to be emitted from small electrodes on the thighs of patients whose leg muscles are inactive due to spinal injury.

Figure 7-14

A special prosthesis helps this skier perform well in competition.

Program Resources

📁 **Reproducible Masters**
Study Guide, p. 32 [L1]
Reinforcement, p. 32 [L1]
Enrichment, p. 32 [L3]
Science and Society Integration, p. 11
Lab Manual 15

🖥 **Transparencies**
Section Focus Transparency 28 [L1]
Teaching Transparency 14 [L1]

Currently, rehabilitation researchers are working on developing an arm prosthesis as shown in **Figure 7-15,** which will be capable of helping an amputee feel sensations such as pressure and texture. To feel these sensations, the brain must receive appropriate neural impulses from the limb. Sensors and a power source capable of generating artificial neural signals would be placed in the prosthesis, and the sensors would be connected to the remaining healthy portion of the natural arm's nerve.

Just Think About It

An exciting new technology may hold possibilities for helping people regain some use of a disfunctional organ or limb. Scientists are experimenting with brain-to-computer interfaces. Electrodes are used to stimulate or collect signals from neurons in the brain. This type of neural prosthesis may allow people to control motors, computers, and even artificial limbs with natural electrical impulses originating in their brains. Some research even suggests that a tiny camera could be linked to these electrodes to provide sight to a person without directly using the eyes. How else might this technology be useful?

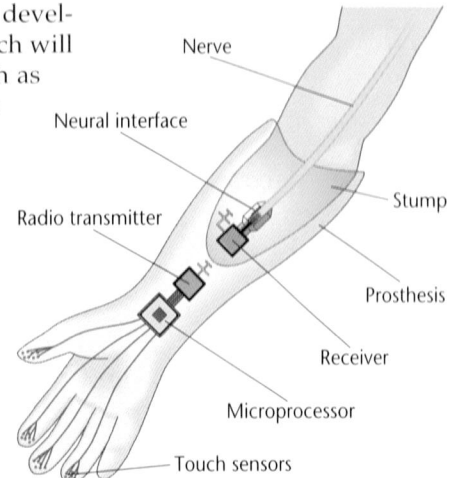

Figure 7-15

Prostheses like this one may soon help amputees experience sensations such as pressure and partially regain their sense of touch.

Section Wrap-up

Review

1. What is the purpose of the study of bionics? Explain at least two helpful inventions that have resulted from bionics research.

2. FNS systems and brain-to-computer interface systems both make use of electrical signals to move muscles. In what ways do these systems differ?

Explore the Technology

Imagine that you are a doctor working with a patient who has recently had his arm amputated below the elbow. If you were trying to advise the patient about what options are available for regaining some practical use of the arm, what would you point out as the strong points of the techniques discussed above? What are the drawbacks of each technique? Write up your arguments in the form of a report.

Science Journal

Knee joint replacements are made of modern materials. In your Science Journal, make a diagram of a replacement joint and label the material each part is made of. Write a paragraph explaining the use of each material.

SCIENCE & SOCIETY

4 Close

LS **Kinesthetic** To stress how important sensory capabilities are in prosthetic devices, have students use pliers, tongs, or similar implements to perform simple tasks such as tying a shoe, blowing their noses, combing their hair, buttoning a coat, opening a door, etc. **LEP**

Section Wrap-up

Review

1. The purpose is to develop artificial devices to replace limbs and organs. FNS systems stimulate paralyzed muscles and help people walk again. Limbs from shock absorbing materials allow people to use the limb more actively.

2. FNS directly stimulates the muscle via a control unit. A brain-to-computer system interacts directly with the neuron signals from the brain and may result in the person being able to control the prosthesis on his or her own.

Explore the Technology

Strong points—limb may be able to move by muscle stimulation, brain-to-computer could allow voluntary movement; Drawbacks—limb may not have sensory stimulation, brain-to-computer is not developed yet.

Science Journal

LS **Linguistic** Journal entries should reflect the new materials being used and how they interact with living tissue. **P**

Section 7•4

Prepare

Section Background

How effectively a machine transfers energy is measured by its efficiency. The rate at which it transfers energy is measured by its power.

Preplanning

You will need a wind-up toy car for each group conducting the MiniLAB.

1 Motivate

Bellringer

 Before presenting the lesson, display **Section Focus Transparency 29** on the overhead projector. Assign the accompanying **Focus Activity** worksheet. L1 LEP

Tying to Previous Knowledge

Have students recall what a simple machine does. In this section they will see how combining simple machines also makes work easier.

GLENCOE TECHNOLOGY

 **Videodisc**

Glencoe Physical Science Interactive Videodisc
Side 1, Lesson 2
Introduction to Bicycles

13404-15729
Gears

19494-20799

Science Words

compound machine
efficiency
power

Objectives

- Recognize the simple machines that make up a compound machine.
- Calculate the efficiency of a machine.
- Describe the relationship among work, power, and time.

Figure 7-16

A racing bike uses wheels and axles in the form of gears to produce an *MA* of less than 1. Force is sacrificed to gain speed.

Compound Machines

Many machines you use, such as a lawn mower or a pencil sharpener, are made up of several simple machines. A combination of two or more simple machines is a **compound machine.** Even a tool as simple as an ax is a compound machine made up of a wedge and a lever.

Often, the simple machines that make up a compound machine are concealed. However, in one familiar compound machine—the bicycle—many of the simple machines are visible and easily recognized.

Pedal Power

Look at the bicycle in **Figure 7-16.** The pedal mechanism is a wheel-and-axle system made up of two wheels attached to the same axle. Each pedal moves in a circle, like a wheel. The axle is a gear. The effort force exerted on the pedals by the rider turns this pedal gear. The bicycle chain transfers the force to a smaller gear attached to the rear wheel. This wheel gear is actually the effort wheel of the rear wheel-and-axle system. As the wheel gear turns, it causes the rear wheel to turn, and the wheel exerts the resistance force on the road. Because the wheel gear is smaller than the pedal gear that drives it, the rear wheel turns faster than the pedals do.

Program Resources

 Reproducible Masters
Study Guide, p. 33 L1
Reinforcement, p. 33 L1
Enrichment, p. 33 L3
Activity Worksheets, p. 43
Cross-Curricular Integration, p. 11
Lab Manual 16

Transparencies
Section Focus Transparency 29 L1

Finding the Mechanical Advantage of a Bicycle

The overall mechanical advantage of a bicycle is the ratio of the resistance force exerted by the tires on the road to the effort force exerted by the rider on the pedals. A one-speed bicycle has only two gears. Its gear ratio is fixed, so its mechanical advantage cannot be changed. This makes it a one-speed bike. A ten-speed bike has two pedal gears and five wheel gears. By shifting gears, the rider can change gear ratios, which changes the mechanical advantage and helps the rider maintain a steadier rate of pedaling over different terrains.

A bicycle has many other simple machines. Some of these machines are shown in **Figure 7-16.** See if you can tell what each does. Can you think of any others?

Problem Solving

Frivolous Machines

Creating machines for various tasks has long been a favorite pastime of inventors and even cartoonists. In the mid-1900s, Rube Goldberg published syndicated cartoons that often featured bizarre devices or machines to tackle simple, everyday tasks. These sketches became popular, so cumbersome, elaborate, or ridiculous machines are frequently called "Rube Goldberg devices."

These creative and wacky devices are often designed from a combination of simple machines made of ordinary materials. This sketch suggests a device to turn off a ringing alarm clock. It begins with the person in bed nudging the cat off his feet and onto a lever to drink some milk. Can you figure out how the device depicted in this cartoon works?

Solve the Problem:

1. What simple machines can you identify in this cartoon?
2. Describe, step by step, what happens as this device functions. What step do you think might be the most likely to malfunction? Why?

Think Critically:

Design your own Rube Goldberg device to blow out a candle, shut off an alarm, open a window, or do some other ordinary task. Be sure your device uses at least four distinct steps. What simple machines are used in your device?

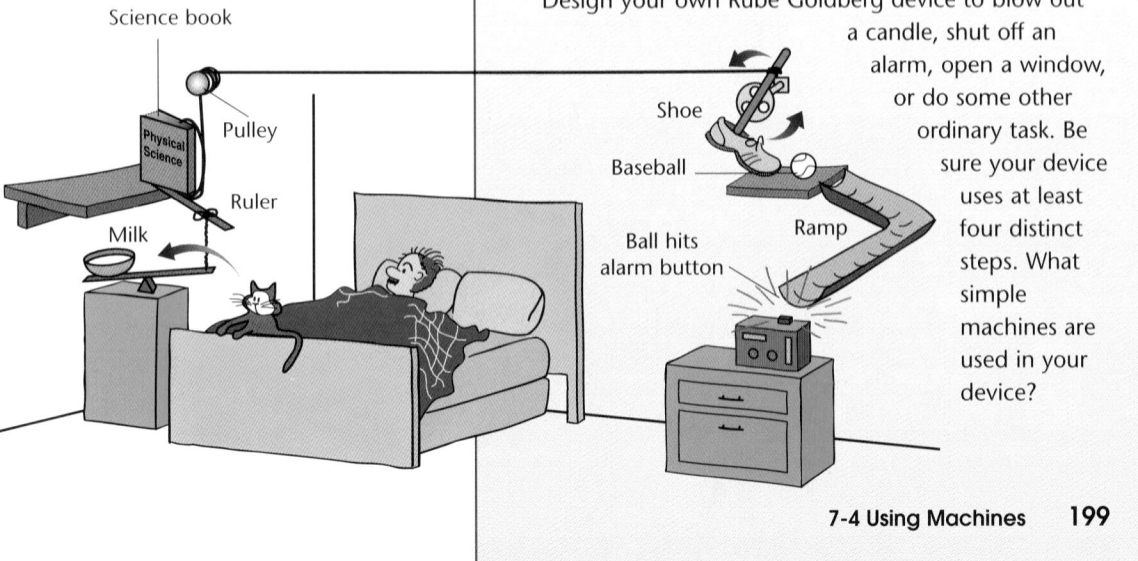

Science book
Pulley
Physical Science
Ruler
Milk
Shoe
Baseball
Ball hits alarm button
Ramp

7-4 Using Machines **199**

Concept Development

The following diagram illustrates energy transfers in a compound machine.

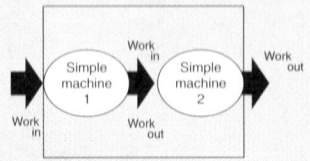

Compound machine

Teacher F.Y.I.

Talk about a complex machine! The longest bicycle is about 20 m long and weighs about 1100 kg. With this vehicle, you could take 34 of your friends along for a ride.

Problem Solving

Solve the Problem

1. first-class lever, fixed pulley, wheel and axle, inclined plane
2. Foot nudges cat awake, cat pulls down lever to slide bowl of milk within reach, pulls down lever with science book on end, science book flips off and falls, pulling string connected to pulley, pulley operates lever with shoe, pulls lever to make shoe kick baseball down the inclined plane. Ball falls on alarm clock and makes it stop ringing. The step most likely to malfunction is the cat. If the cat is not interested in the milk, the whole system won't work.

Think Critically

Student designs will vary. They should include four steps and use at least four simple machines (one for each step).

199

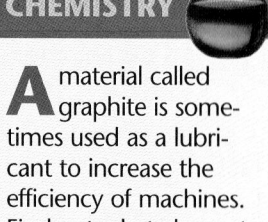
Efficiency

You learned earlier that some of the energy put into a machine is lost as thermal energy produced as a result of friction. The work put out by a machine is always less than the work put into the machine.

Efficiency is a measure of how much of the work put into a machine is changed to useful work put out by a machine. The higher the efficiency of a machine, the greater the amount of work input that is changed to useful work output. Efficiency is calculated by dividing work output by work input and is usually expressed as a percentage.

$$\text{efficiency} = \frac{W_{out}}{W_{in}} \times 100\% = \frac{F_r \times d_r}{F_e \times d_e} \times 100\%$$

Why must the efficiency of a machine always be less than 100 percent? Can you think of ways to increase the efficiency of a machine?

Many machines can be made more efficient by reducing friction. This is usually done by adding a lubricant, such as oil or grease, to surfaces that rub together, as shown in **Figure 7-17**. After a time, dirt will build up on the grease or oil, and the lubricant will lose its effectiveness. The dirty lubricant should be wiped off and replaced with clean grease or oil. What other ways can increase the efficiency of a machine?

USING MATH

Calculating Efficiency

Example Problem:
A sofa weighing 1500 N must be placed in a truck bed 1.0 m off the ground. A worker uses a force of 500 N to push the sofa up an inclined plane that has a slope length of 4.0 m. What is the efficiency of the inclined plane?

Problem-Solving Steps:

1. What is known?   resistance force, $F_r = 1500$ N
                      effort force, $F_e = 500$ N
                      resistance distance, $d_r = 1.0$ m
                      effort distance, $d_e = 4.0$ m
2. What is unknown?   efficiency
3. Choose the equation. $\text{efficiency} = \dfrac{F_r \times d_r}{F_e \times d_e} \times 100\%$

4. **Solution:** $\text{efficiency} = \dfrac{1500 \text{ N} \times 1.0 \text{ m}}{500 \text{ N} \times 4.0 \text{ m}} \times 100\% = 75\%$

 Calculator Hint: Combine all math into one calculator operation. Enter: **1500 × 1.0 ÷ 500 ÷ 4.0 × 100 =** and your calculator will read **75**.

Power

Suppose you and a friend are pushing boxes of books up a ramp to load them into a truck. The boxes weigh the same, but your friend is able to push a box a little faster than you can. Your friend moves a box up the ramp in 30 seconds. It takes you 45 seconds. Do you both do the same amount of work on the books? Yes. This is true because the boxes weighed the same and were moved the same distance. The only difference is in the time it takes you and your friend to do the work.

Figure 7-17

Oiling machines reduces friction and makes them operate more efficiently. Dirty oil should be replaced periodically with clean oil.

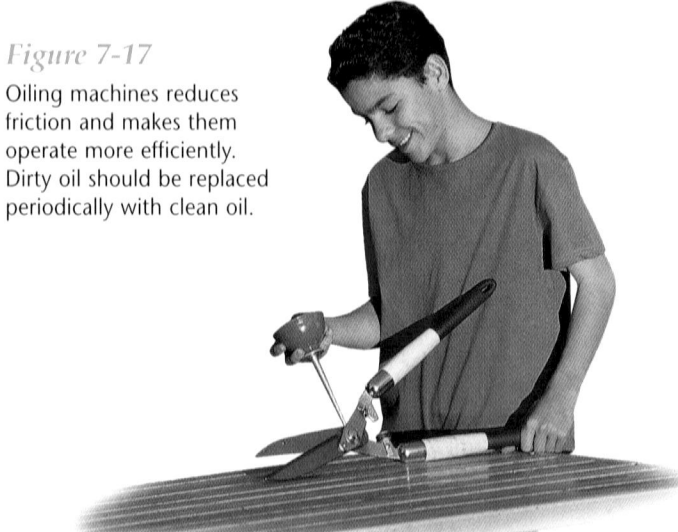

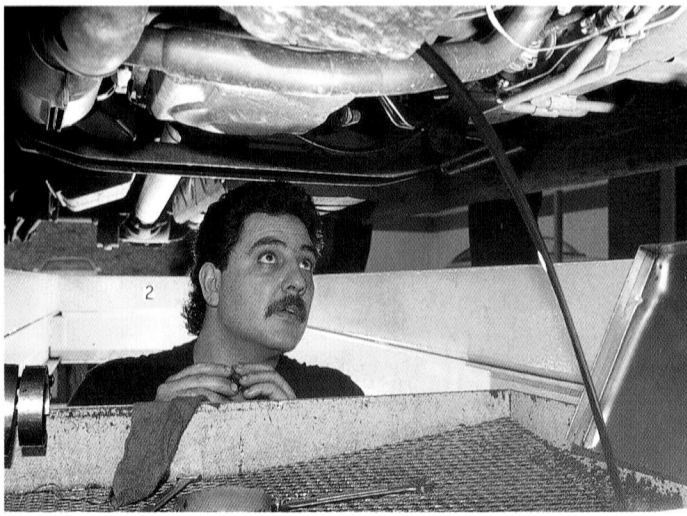

MiniLAB

Can you measure the power of a toy car?

Procedure
1. Wind up a toy car and place it at the bottom of an inclined plane.
2. Experiment to adjust the angle of the inclined plane (using books to stack under a board) so the car will reach the top at the slowest speed possible.
3. Measure the time in seconds for the car to travel to the top of the plane.
4. Measure the weight of the car in newtons using a spring scale.
5. Measure the height of the inclined plane in meters. Be sure to measure from the floor or the tabletop straight up to the top of the inclined plane.

Analysis
1. Use weight and height (as force and distance) to calculate the work done in joules.
2. What is the power of your toy car? Divide your work in joules by the time in seconds to get power in watts.
3. What could you do to increase the power of the car as it travels up the same inclined plane?

Inclusion Strategies

Gifted Have interested students compile a listing of power ratings and efficiency ratings for various appliances. Encourage them to investigate whether the two are related. **L3** **P**

Discussion

To introduce the concept of power, pose the following questions. Suppose two students are asked to unpack identical cartons of books. One student completes the job in 10 minutes, whereas the other takes 20 minutes. Which student worked harder? Which student did more work? Develop the idea of power, the rate of work, by asking students what they mean when they say someone worked harder at a task.

MiniLAB

Purpose

LS **Visual-Spatial** To measure the power of a toy car. **L1**
LEP **COOP LEARN**

Materials
inclined plane board, stopwatch, meterstick, spring scale, and wind-up car

Teaching Strategies
Any toy that will travel up an incline can be used. Encourage students to bring their own.

Analysis
1. $W = F \times d$
2. $P = W/t$
3. increase the car's speed

Assessment

Performance Have students calculate the difference in power as the angle of the ramp is changed, then as the speed of the car is changed. Use the Performance Task Assessment List for Using Math in Science in **PASC**, p. 29. **P**

📁 **Activity Worksheets,** pp. 5, 43

Concept Development

Power is the rate at which energy is converted. For example, a 100-W lightbulb converts 100 J of electrical energy into 100 J of radiant energy and heat during 1 second. Machines and appliances can be compared by their power ratings.

3 Assess

Check for Understanding

?FLEX Your Brain

Use the Flex Your Brain activity to have students explore EFFICIENCY.

Activity Worksheets, p. 5

Reteach

IS Interpersonal Have problem-solving teams determine a method for calculating the amount of power a student generates while climbing a flight of stairs. **COOP LEARN**

USING TECHNOLOGY

Nanomachines

Can you imagine tiny machines one-thousandth the diameter of a human hair? A new branch of science—called nanotechnology—is aimed at producing chemical clusters of less than 50 atoms, electronic components, and even machines or robots by constructing them molecule by molecule. These will be measured in nanometers, a unit representing one-billionth of a meter.

Scientists have already succeeded at building switches out of single atoms and constructing clusters of small numbers of atoms or molecules. This is made possible by the scanning tunneling microscope, which allows individual atoms to be manipulated. This technology is being used in efforts to develop nanomachines.

Nanomachines may be engineered that could be injected into the body to clean out arteries or clean up hazardous materials in the environment. The potential benefits could drastically change industry.

Think Critically:

What are some of the problems currently encountered by mechanical engineers that would likely be helped by constructing machines at the micro or nano level?

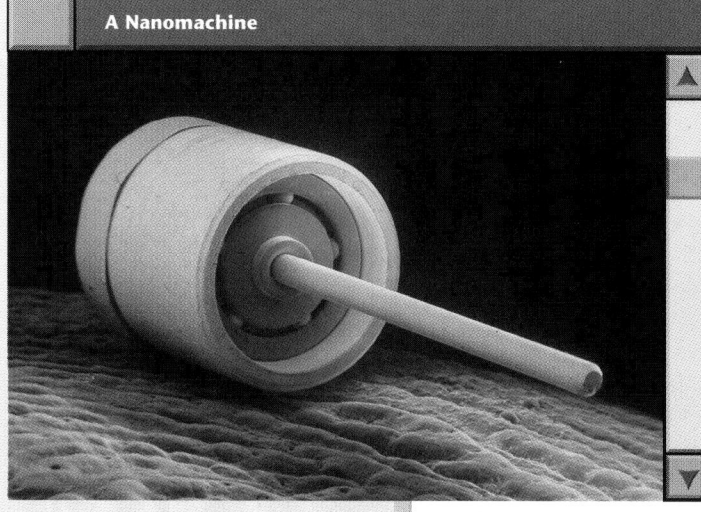

A Nanomachine

202   Chapter 7 Machines—Making Work Easier

Your friend has more power than you have. **Power** is the rate at which work is done. In other words, power is a measure of the amount of work done in a certain amount of time. To calculate power, divide the work done by the time required to do the work.

$$\text{power} = \frac{\text{work}}{\text{time}}$$

$$P = \frac{W}{t}$$

Power is measured in watts, named for James Watt, who worked on the steam engine. A watt (W) is one joule per second. A watt is pretty small—about equal to the power used to raise a glass of water from your knees to your mouth in one second. Because the watt is such a small unit, large amounts of power often are expressed in kilowatts. One kilowatt (kW) equals 1000 watts.

The power ratings for appliances are listed in watts. Power ratings can be used to compare the rate at which machines do work. Where the same amount of work

---

is done at different rates, different amounts of power are required. Study the Example Problem below to see how work and power are related.

USING MATH

Calculating Power

Example Problem:

A figure skater lifts his partner, who weighs 450 N, 1.0 m in 3.0 s. How much power is required?

Problem-Solving Steps:

1. What is known?   force, $F = 450$ N
                              distance, $d = 1.0$ m
                              time, $t = 3.0$ s

2. What is unknown?   power, P

3. Choose the equation. $P = \dfrac{W}{t} = \dfrac{F \times d}{t}$

4. **Solution:**
$$P = \frac{450 \text{ N} \times 1.0 \text{ m}}{3.0 \text{ s}} = 150 \text{ W}$$

Section Wrap-up

Review

1. Give an example of a compound machine. What are the simple machines that make it up?

2. How are power, work, and time related?

3. **Think Critically:** You buy a secondhand lawn mower with an efficiency of 30 percent. By repairing and lubricating it, you increase its efficiency to 40 percent. How does this affect the work put out by the lawn mower for a given amount of work put in?

Skill Builder
Measuring in SI

A 500-N passenger is inside a 24 500-N elevator that rises 30 m in exactly 1 minute. How much power is needed for the elevator's trip? If you need help, refer to Measuring in SI in the **Skill Handbook.**

Science Journal

A pair of scissors is a compound machine. In your Science Journal, draw a diagram of a pair of scissors and label the simple machines you can identify. Explain the purpose of each simple machine in this device.

Skill Builder

$$P = \frac{F \times d}{t} = \frac{25\,000 \text{ N} \times 30 \text{ m}}{60 \text{ s}}$$
$$= 12\,500 \text{ W} = 12.5 \text{ kW}$$

Assessment

Oral Tell students that the efficiency of the elevator is 55% and ask them how this affects the power. *Power is based on W_{out} and would not change.* Use the Performance Task Assessment List for Making Observations and Inferences in **PASC**, p. 17. **P**

Extension

For students who have mastered this section, use the **Reinforcement** and **Enrichment** masters.

4 Close

•MINI•QUIZ•

Use the Mini Quiz to check students' recall of chapter content.

1. A(n) _____ is a combination of two or more simple machines. *compound machine*

2. _____ is a measure of how much of the work put into a machine is changed to useful work put out by a machine. *Efficiency*

Section Wrap-up

Review

1. Student responses will vary. Accept all reasonable responses.

2. By definition, power is a measure of the amount of work done in a given amount of time.

3. **Think Critically** For a given amount of work put into the mower, the work put out by the mower will increase.

Science Journal

Scissors contain wedges and levers. The levers transfer and change the direction of force, and the wedges, or blades, use the force to cut the material.

Sources

- "Herodotus's Theory of How the Pyramids Were Built Gets a Lift," *Discover*, June, 1987.
- *Pyramid*, David Macaulay, Houghton Mifflin Co., 1975.

Teaching Strategies

- Ask students to brainstorm ways each simple machine might be used to raise something too heavy to lift by muscle power alone.

Interpersonal Have students work in teams to design a compound machine for lifting a heavy stone. Have them try to design a machine that uses one of each simple machine. L2 **COOP LEARN**

Linguistic Explain to the students that the Great Pyramids of Giza, the highest of which was 481 feet (147 m) were the tallest structures built until the 19th century. Have students research to discover some early buildings that rose higher than the pyramids, and what machines and techniques made those buildings possible. L2

Linguistic Have students research what hand tools were used in constructing the pyramids. Have them classify the tools in terms of simple machines. L2

Logical-Mathematical Have students make a list of hand tools used in modern construction. Have them classify the tools in terms of simple machines. L1

Building the Pyramids

The ancient Egyptians had to rely on simple machines for building the great pyramids. Somehow they managed to raise, and precisely place, huge building stones weighing as much as 50 tons. How they did it is still largely a mystery.

For many years, the accepted theory has been based on the use of the inclined plane in the form of ramps made of mud and stone.

Beginning with the Lever

Some historians have begun to believe that this theory is wrong. These historians believe that the primary machine was not the inclined plane, but the lever. According to this theory, instead of ramps, there were relatively narrow sets of long stone stairs built perpendicular to each face of the pyramid. Workers would use levers to raise each stone slowly from one level to the next.

Because they couldn't raise the gigantic stones all at once, they did it a little at a time. They'd lever one end up a small distance and slide a piece of wood underneath. Then they would do the same to the other end. Then they would raise the first side a little farther and place a second block of wood on top of the first, and so on until they could slide the stone onto the next level and start again.

This theory was first proposed by the Greek historian Herodotus in the fifth century B.C. While it's finding new popularity among some scholars, it remains controversial.

The truth is, we may never know for certain how the pyramids were built. But these giant monuments do prove that hard work can be accomplished with simple machines.

*inter*NET
CONNECTION

Visit the Chapter 7 Internet Connection at Glencoe Online Science, **www.glencoe.com/ sec/science/physical**, for a link to more information about building the pyramids.

204

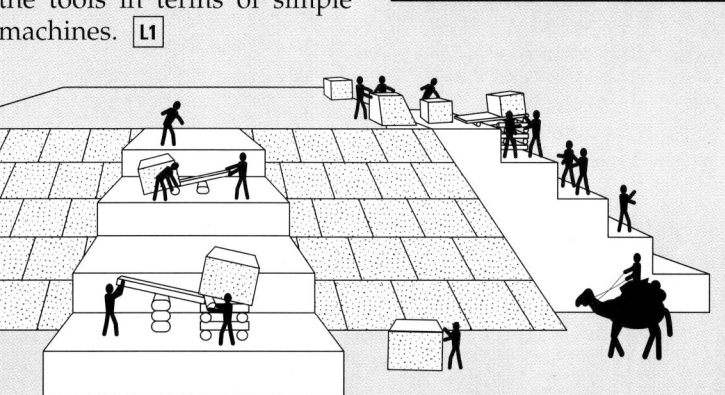

Summary

7-1: Why We Use Machines

1. A machine makes work easier by changing the size of the force applied to it, the direction of that force, or both.
2. The number of times a machine multiplies the force applied to it is the mechanical advantage of the machine.

7-2: The Simple Machines

1. A lever is a bar that is free to pivot about a fixed point called a fulcrum. A pulley is a grooved wheel with a rope running along the groove. A wheel and axle consists of two different-sized wheels that rotate together. An inclined plane is a sloping surface used to raise objects. The screw and the wedge are special types of inclined planes.
2. Each type of simple machine has a specific equation for calculating its ideal mechanical advantage. Each equation is related to the distance the effort force moves divided by the distance the resistance force moves.

7-3: Science and Society: Mending with Machines

1. The science of bionics is the creation of artificial replacements for parts of the body that are not functioning properly.
2. Electronic stimulation can be used to move paralyzed muscles or provide sensations for prosthetic limbs.

7-4: Using Machines

1. A combination of two or more simple machines is called a compound machine.
2. The efficiency of any machine is a ratio of the useful work put out by the machine to the work put into the machine.

3. Power is a measure of the amount of work done in a certain amount of time. The SI unit of power is the watt.

Key Science Words

a. bionics
b. compound machine
c. efficiency
d. effort arm
e. effort force
f. fulcrum
g. ideal machine
h. inclined plane
i. lever
j. machine
k. mechanical advantage
l. power
m. pulley
n. resistance arm
o. resistance force
p. screw
q. simple machine
r. wedge
s. wheel and axle

Reviewing Vocabulary

Match each phrase with the correct term from the list of Key Science Words.

1. any device that accomplishes work with only one movement
2. force applied by a machine
3. a device in which work output equals work input
4. a bar that pivots about a fixed point
5. two different-sized wheels that rotate together
6. a combination of simple machines
7. the work output of a machine divided by the work input
8. the rate at which work is done
9. the fixed point of a lever
10. an inclined plane wrapped around a cylinder

Summary

Have students read the summary statements to review the major concepts of the chapter.

Reviewing Vocabulary

1. q	**6.** b
2. o	**7.** c
3. g	**8.** l
4. i	**9.** f
5. s	**10.** p

✓ Assessment

Portfolio Encourage students to place in their portfolios one or two items of what they consider to be their best work. Examples include:

- Activity 7-2 calculations and answers, p. 195
- Science Journal answer and diagrams, p. 197
- Inclusion Strategy research, p. 201 **P**

Performance Assessment Additional performance assessments may be found in **Performance Assessment** and **Science Integration Activities.** Performance Task Assessment Lists and rubrics for evaluating these activities can be found in Glencoe's **Performance Assessment in the Science Classroom.**

GLENCOE TECHNOLOGY

⊙ Videodisc

Glencoe Physical Science
Interactive Videodisc
Use the videodisc lesson *Machines and Forces* to review aspects of simple and compound machines as they are applied to a bicycle race.

▭ MindJogger Videoquiz

Chapter 7 Have students work in groups as they play the Videoquiz game to review key chapter concepts.

Checking Concepts

1. b	6. b
2. b	7. a
3. a	8. b
4. c	9. b
5. d	10. a

Understanding Concepts

11. Simple machines can make work easier by (a) multiplying the effort force, as with a first-class lever; (b) changing the direction of the applied force, as with a single, fixed pulley; (c) increasing the distance through which the resistance force moves, as with a third-class lever.

12. Most machines increase the amount of work done by increasing the amount of friction involved in doing the work. However, the applied force is smaller.

13. A fixed pulley changes the direction of the effort force; it does not multiply that force. A fixed pulley is similar to a first-class lever. A single, movable pulley moves with the resistance. It multiplies the force by 2. Such a pulley is similar to a second-class lever.

14. The *MA* of an inclined plane can be increased by increasing its length (or decreasing its height). Sharpening a knife has the effect of increasing the length of the inclined plane.

15. Oil lubricates engine parts, thereby reducing friction and making the engine more efficient.

Checking Concepts

Choose the word or phrase that completes the sentence or answers the question.

1. There are _____ types of simple machines.
   a. three   b. six   c. eight   d. ten
2. Which of these cannot be done by a machine?
   a. multiply force
   b. multiply energy
   c. change direction of a force
   d. work
3. In an ideal machine, the work input _____ work output.
   a. is equal to   c. is less than
   b. is greater than   d. is independent of
4. The _____ of a machine is the number of times it multiplies the effort force.
   a. efficiency   c. mechanical advantage
   b. power   d. resistance
5. To raise a resistance 4 m, the effort rope of a single fixed pulley must move _____.
   a. 2 m   b. 8 m   c. 1 m   d. 4 m
6. The ideal mechanical advantage of a pulley system in which five ropes support an object is _____.
   a. 2.5   b. 5   c. 10   d. 25
7. In a wheel and axle, the resistance force is usually exerted by the _____.
   a. axle   c. gear ratio
   b. larger wheel   d. pedals
8. The *IMA* of an inclined plane 8 m long and 2 m high is _____.
   a. 2   b. 4   c. 16   d. 8
9. As the efficiency of a machine increases, the _____ of the machine increases.
   a. work input   c. friction
   b. work output   d. *IMA*
10. The *IMA* of an inclined plane can be increased by _____.
    a. increasing the length
    b. increasing the height
    c. decreasing the length
    d. making its surface smoother

Understanding Concepts

Answer the following questions in your Science Journal using complete sentences.

11. Describe and give examples of three ways that a simple machine can make work easier.
12. If machines make work easier, explain why most machines actually increase the amount of work you do in accomplishing a task.
13. Distinguish between a single fixed pulley and a single movable pulley, and describe the advantages of using each. Compare each type of pulley to its corresponding class of lever.
14. Explain how sharpening a knife changes its *MA*.
15. The efficiency of an automobile is usually expressed in terms of gas mileage. How can changing the engine oil increase the gas mileage of the automobile?

Thinking Critically

16. A cyclist applies a force of 250 N to the pedals of a bicycle. If the rear wheel applies a force of 200 N to the road surface, what is the *MA* of the bicycle?
17. An adult and a small child get on a seesaw that has a movable fulcrum. When the fulcrum is in the middle, the child can't lift the adult. How should the fulcrum be moved so that the two can seesaw? Explain.
18. You have two screwdrivers. One is long with a thin handle, and the other is short with a fat handle. Which would you use to drive a screw into a board? Explain your choice.
19. Using a ramp 4 m long, workers apply an effort force of 1250 N to move a 2000-N crate onto a platform 2 m high. What is the efficiency of the ramp?
20. How much power does a person weighing 500 N need to climb a 3-m ladder in 5 seconds?

Thinking Critically

16. $MA = \dfrac{F_r}{F_e} = \dfrac{200\ N}{250\ N} = 0.8$

17. The fulcrum should be moved toward the adult, which increases the length of the effort arm of the child's lever. Thus, the effort force exerted by the child is multiplied.

18. The screwdriver is being used as a wheel and axle, with the handle acting as the wheel. Thus, the screwdriver with the fat handle will have a greater MA. The length is irrelevant.

19. efficiency = W_{out}/W_{in}
$= \dfrac{2000\ N \times 2\ m}{1250\ N \times 4\ m} \times 100\% = 80\%$

20. $P = \dfrac{W}{t} = \dfrac{F \times d}{t}$
$= \dfrac{500\ N \times 3\ m}{5\ s}$
$= 300$ watts or 0.3 kW

Developing Skills

If you need help, refer to the **Skill Handbook.**

21. **Sequencing:** Make an ordered list to show how some of the simple machines in a bicycle work to make the bicycle move. Start with the feet applying force to the pedals.

22. **Making and Using Graphs:** An ideal lever has a fixed effort arm length of 40 cm. Calculate the effort force needed to raise a 10-N object with these resistance arm lengths: 80 cm, 40 cm, 20 cm, and 10 cm. Construct a line graph relating these resistance arm lengths to effort force. Describe the relationship between these two variables.

23. **Interpreting Scientific Illustrations:** Study the diagram of a garden hoe and answer the questions.
    a. What type of machine does the diagram represent? (Be specific.)
    b. What does the longer, red arrow represent?
    c. What does point A represent?
    d. What is represented by segment AC? Segment AB?
    e. What is the *IMA* of this machine?
    f. How does this machine make work easier?

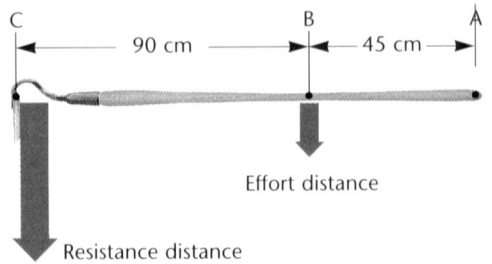

24. **Concept Mapping:** Complete the concept map for simple machines using the following terms: *inclined plane, lever, lever types, pulley, screw, wedge,* and *wheel and axle.*

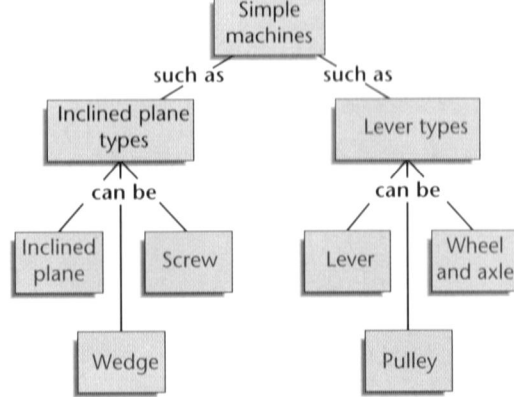

Performance Assessment

1. **Measuring in SI:** The SI unit of power, the watt, is a derived unit. Using the definition of *power,* develop the formula for calculating power and determine the fundamental SI units for expressing power.

2. **Designing an Experiment:** You measured the power of a toy car in the MiniLAB on page 201. Design a procedure to determine your own power as you walk up an incline. Conduct your experiment.

3. **Invention:** Design a human-powered machine of some kind. Describe the simple machines used in your design, and tell what each of these machines does.

Chapter 7 Review

Developing Skills

21. **Sequencing**
    (1) Feet apply force to pedals.
    (2) Pedals move in circle around axle.
    (3) Force is transferred from pedals to wheels through gears and chain.
    (4) Wheels exert resistance force on road surface.

22. **Making and Using Graphs**

$$IMA = \frac{L_e}{L_r} \quad F_e = \frac{F_r}{MA}$$

$\frac{40}{80} = 0.5$	$\frac{10}{0.5} = 20$ N
$\frac{40}{40} = 1$	$\frac{10}{1} = 10$ N
$\frac{40}{20} = 2$	$\frac{10}{2} = 5$ N
$\frac{40}{10} = 4$	$\frac{10}{4} = 2.5$ N

The effort force and the length of the resistance arm are directly related.

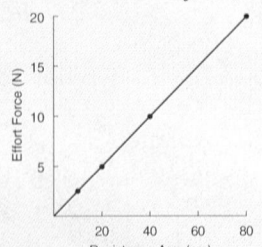

23. **Interpreting Scientific Illustrations**
    a. third-class lever
    b. the distance the resistance moves
    c. fulcrum
    d. AC = resistance arm; AB = effort arm
    e. $IMA = \frac{L_e}{L_r} = \frac{45 \text{ cm}}{135 \text{ cm}} = 0.33$
    f. The machine increases the distance the resistance moves for a given movement of the effort.

24. **Concept Mapping** See student page.

Performance Assessment

1. Power is the rate at which work is done, or

$P = W/t.$ Power is expressed in watts. Using $F \times d$ to represent work, the formula for power becomes $P = (F \times d)/t.$ So, a watt is the name for the derived unit Newton × meter per second. Use the Performance Task Assessment List for Using Math in Science in **PASC,** p. 29. **P**

2. Have a partner time you as you walk up a ramp at maximum speed. Determine your weight in Newtons and measure the vertical height of the

incline in meters. Use $P = (F \times d)/t$ to determine your power in watts. Use the Performance Task Assessment List for Designing an Experiment in **PASC,** p. 23. **P**

3. Student diagrams and designs will vary. Accept any reasonable design. Be sure they identify and describe all simple machines used. Use the Performance Task Assessment List for Invention in **PASC,** p. 45. **P**

Objectives

IS Linguistic Students will analyze the motion and forces involved in a sport or physical activity of their choice. This process will make use of concepts studied in Unit 2, such as measuring average speed and recognizing Newton's laws of motion. L1

Summary

Students will begin this project by researching the history of their sport. They will design and conduct an experiment to determine the average speed of a moving person or object in this sport. Students will describe examples of work being done and recognize how Newton's laws of motion are illustrated in this activity. They will share their sports analysis in a presentation, making use of actual or videotaped demonstrations.

Time Required

This project is designed to be worked on throughout Chapters 3-5 in Unit 2. One day is required to set up groups, explain the tasks, and select topics. Allow groups to meet in several segments of time during this unit to plan and check the progress of their project. One day is required for group presentations at the end of Chapter 5.

Preparation

Establish cooperative groups of three students each. Have each group assign specific roles to encourage participation by all group members.

UNIT PROJECT 2

Physics In Sports

In some sports, such as swimming and cross-country running, the winner of a competition is determined by who can move at the fastest average speed. Other sports, such as basketball, figure skating, and ultimate Frisbee, depend on accuracy, technique, and teamwork. You have learned that forces are used to control motion by changing the speed or the direction of an object's motion. These same forces are used to accelerate equipment or human bodies in sports.

Analyze a Sport

Select a sport or physical activity that your group would like to study. Choose a sport you can actually play or observe. Use direct observations and measurements to explore the dynamics, or the motion and forces, involved in this sport. Suggestions for sports are listed below.

• soccer	• basketball	• tennis
• swimming	• biking	• softball
• volleyball	• gymnastics	• baseball
• handball	• running	• table tennis
• wrestling	• football	• weightlifting

For the sport of your choice, research the following information.

1. What is the history of your sport? When and by whom was it invented? How much has it changed since it was invented?

2. Describe the objects in motion in your sport. Do people move around a lot? Do they move in straight lines or do they change directions? Is a ball or other object the focal point of the game?

3. What two things must be measured to determine the average speed of an object? Design a way to measure the average speed of a moving person or object in this sport, such as the speed of a served tennis ball. Try it! Describe how this speed is controlled by the athlete.

4. Review the definition of *work*. Try to identify and describe at least two examples of an action in which work is done in your sport. Remember, the force and the distance moved must be in the same direction. What information would you need to calculate the amount of work done?

5. Find at least one example of each of Newton's laws of motion in your sport. Describe each example as clearly as possible.

Using Your Research

After you have investigated the questions, design a presentation for the class about the physics of your sport. Your presentation should answer the questions you investigate. Use actual demonstrations of the sport as much as possible to help make your explanations clear. You might consider videotaping some motions involved in your sport if the sport itself cannot be easily demonstrated in the classroom. If possible, dress as a person would dress to participate in the sport.

Go Further

Each and every movement in each and every sport contains a great deal of physics. Make a poster that includes at least one diagram or picture of a specific action. Show an analysis of all forces and motion existing in this action by pointing to specific parts of the body or equipment that are in use. Consider the role friction plays in your sport. In what circumstances is friction desirable? Undesirable? Using your understanding of the physics involved, describe several specific things that an athlete could use to improve performance in this sport.

Unit 2 Project   209

Go Further

Lead a discussion on the role of friction in sports, including air resistance and drag. Friction is necessary for people to walk and run. For example, it is undesirable for a racing swimmer to wear a baggy suit, but a runner will choose shoes that have good traction, and thus increased friction. In order to improve performance in a sport, athletes might want to know how to reduce drag, that larger forces will cause greater accelerations, and that objects moving fast have more momentum, for example.

Teaching Strategies

• To illustrate the concepts without leading the students toward a specific sport or activity, show an animated cartoon that contains a lot of activity to introduce this project. Ask students to identify the kinds of motion and forces involved in a particular action, such as running or striking an object. Have students describe how they would determine the average speed of a person or object moving as they do in the cartoon. Have students identify situations in the cartoon where Newton's laws are not obeyed, and compare this situation to a real-life situation.

• Suggest that, in measuring the average speed of an object, students may wish to make several measures of the same event to assure consistency. Have them describe the sources of error in any measurements.

• You may wish to make up an assessment sheet for grading the presentations. Share this sheet with your students so they know exactly what they need to do to be successful in this project.

References

Doherty, Paul. "That's the Way the Ball Bounces." *The Exploratorium Quarterly*, Fall 1991.

Holmes, Brian. "Spit and Dimples." *The Exploratorium Quarterly*, Fall 1991.

Miller, Mary K. "Racing the Dolphins." *The Exploratorium Quarterly*, Fall 1991.

The Nature of Matter

In Unit 3, students are introduced to the physical and chemical classification of matter. The unit begins by describing the characteristics of the states of matter, continues by differentiating between physical and chemical properties, introduces the periodic table as a means of organizing elements, and relates chemical properties to atomic structure.

CONTENTS

Unit Contents

210

Science at Home

A Model Idea Have students monitor television, magazines, and newspapers to find the uses of models. Have them list the circumstances and how the model was used to convey information. For example, an advertisement may show a floor plan of a model home. This shows the basic features but does not require that every home is exactly like the model.
L1

The Nature of Matter

What's Happening Here?

The interior of a chambered nautilus shell shows how nature uses a pattern to create a useful and beautiful design. You make use of patterns when sorting and organizing things. When sorting coins, you don't have to look at every one, because their properties enable you to feel the differences among them. Atoms also have properties that allow you to predict what they will do. Organizing atoms and the matter that they make up may not be as simple as sorting coins, but the patterns and properties discussed in the next four chapters will help you sort out many of the things you see in nature.

Science Journal

For many years scientists have believed in the existence of atoms. Only in the last few years, however, have their distinctive patterns become visible. Write in your Science Journal about how the scanning tunneling microscope helps scientists to see atoms. Include a drawing if possible.

211

Theme Connection

Scale and Structure A major theme of Unit 3 is the structure of matter and how that structure is related to the atomic elements of which it is composed.

Cultural Diversity

It's Elemental Encourage students to select an element from the periodic table and use reference books to find out the nationality of the person who discovered the element. **L2**

INTRODUCING THE UNIT

What's Happening Here?

Ask students to tell you what's happening in the photos. Point out that in this unit, they will be studying the structure of matter and how its structure determines its characteristics and behavior.

Background

The chambered nautilus is a member of Phylum Mollusca, Class Cephalopoda and is a relative of the squid. Nautiluses are predatory and highly mobile due to their ability to expel water from a ventrally located funnel. Movement resembles a sort of jet propulsion. Tentacles grasp and hold prey. Nautiluses have highly developed nervous systems, with eyes similar in form and function to vertebrate eyes. The gas-filled chambers enable the animal to remain buoyant in spite of the heavy shell.

Previewing the Chapters

Ask students the following: **How would you classify ice? Why?** *Solid. It is "hard."* **How would you classify water? Why?** *Liquid. It "pours."* **Discuss the characteristics you would use to classify butter and fog.** *Butter keeps its shape and melts at a certain temperature. It must be a solid. Fog has characteristics of air, so it must be a gas. However, it feels wet, so it must contain liquid, too.*

Tying to Previous Knowledge

Have students identify changes that occur when one bakes a cake. Ask them to discuss what clues (visual, tactile, olfactory) they use to determine if a change has happened. Ask them to list other processes that indicate chemical changes.

Chapter Organizer

Section	Objectives/Standards	Activities/Features
Chapter Opener		Explore Activity: Observe what happens during a change of state. p. 213
8-1 **Matter & Temperature** (1 session, ½ block)*	1. **Describe** the four states of matter. 2. **Use** the kinetic theory of matter to explain the characteristics of solids, liquids, and gases. 3. **Explain** the thermal expansion of matter. **National Content Standards:** (5-8) UCP1, UCP2, UCP5, A2, B1; (9-12) UCP1, UCP2, UCP5, A2, B2	Using Math, p. 215 Connect to Life Science, p. 218 Problem Solving: Mind over Matter, p. 219 Using Math, p. 220 Skill Builder: Making and Using Tables, p. 220 Activity 8-1: Properties of Liquids, p. 221
8-2 **Science and Society** (1 session, ½ block)*	1. **Describe** how people use and pollute water. 2. **Discuss** how people can save water and stop pollution. **National Content Standards:** (5-8) UCP2, F1-F5; (9-12) UCP2, F1-F6	MiniLAB: Can dissolved chemicals be filtered out of a solution? p. 223 Explore the Issue, p. 223
8-3 Changes in State (2 sessions, 1 block)*	1. **Interpret** state changes in terms of the kinetic theory of matter. 2. **Account** for the energy of the heats of fusion and vaporization in state changes. **National Content Standards:** (5-8) UCP1-UCP3, A2, B1; (9-12) UCP1-UCP3, A2, B2	MiniLAB: Can you detect energy changes during evaporation? p. 225 Science Journal, p. 227 Skill Builder: Sequencing, p. 227
8-4 **Behavior of Gases** (3 sessions, 1½ blocks)*	1. **Explain** how a gas exerts pressure on its container. 2. **State** and **explain** how the pressure of a container of gas is affected when the volume is changed. 3. **Explain** the relationship between the temperature and volume of a gas. **National Content Standards:** (5-8) UCP2, UCP3, UCP5, A1, A2, B1, D1, E1, E2, F5, G1, G3; (9-12) UCP2, UCP3, UCP5, A1, A2, B2, E1, E2, F6, G1, G3	Using Technology: Gaseous Giants, p. 230 Using Computers, p. 231 Skill Builder: Hypothesizing, p. 231 Activity 8-2: More Than Just Hot Air, pp. 232-233
8-5 **Uses of Fluids** (2 sessions)*	1. **State** Archimedes' principle and **predict** whether an object will sink or float in water. 2. **State** Pascal's principle and **describe** the operation of a machine that uses Pascal's principle. 3. **State** Bernoulli's principle and describe a way that Bernoulli's principle is applied. **National Content Standards:** (5-8) UCP2, UCP3, UCP5, A2, G1, G3; (9-12) UCP2, UCP3, UCP5, A2, G1, G3	MiniLAB: How does applied pressure affect different areas of a fluid? p. 235 Connect to Physics, p. 236 Using Math, p. 239 Skill Builder: Measuring in SI, p. 239 People and Science: Debra Moore, Glass Artist, p. 240

Activity Materials

Explore	Activities	MiniLABs
page 213 a small scented candle, matches	page 221 dropper, food coloring, wooden stick, paper cup, graduated cylinder, 4% solution of powdered borax in water, 4% solution of polyvinyl alcohol (PVA) in water pages 232-233 string, marking pen, meterstick, ice, tongs, heat-proof glove, beaker of boiling water, thermometer, 2 balloons	page 223 50 mL water, litmus paper, 50 mL vinegar, filter paper, two 150-mL beakers page 225 dropper, rubbing alcohol, clock or watch page 235 dropper, 2-L soft-drink bottle with cap

* A complete Planning Guide that includes block scheduling is provided on pages 32T-35T.

Chapter 8 Solids, Liquids, and Gases

Teacher Classroom Resources

Reproducible Masters	Transparencies	Teaching Resources
Study Guide, p. 34 Reinforcement, p. 34 Enrichment, p. 34 Activity Worksheets, pp. 44-45 Lab Manual 17, Density of a Liquid Lab Manual 18, Densities of Solutions Multicultural Connections, p. 19	Section Focus Transparency 30, Ice Sculptures Teaching Transparency 15, States of Matter	Glencoe Physical Science Interactive Videodisc Physical Science CD-ROM Spanish Resources English/Spanish Audiocassettes Cooperative Learning Resource Guide Lab Partner Lab and Safety Skills Lesson Plans
Study Guide, p. 35 Reinforcement, p. 35 Enrichment, p. 35 Activity Worksheets, p. 48 Science Integration Activity 8, ...And Not a Drop to Drink Science and Society Integration, p. 12 Cross-Curricular Integration, p. 12 Technology Integration, pp. 11-12	Section Focus Transparency 31, Water Pollution	**Assessment Resources** Chapter Review, pp. 19-20 Assessment, pp. 53-56 Performance Assessment in the Science Classroom (PASC) MindJogger Videoquiz Alternate Assessment in the Science Classroom Performance Assessment Chapter Review Software Computer Test Bank
Study Guide, p. 36 Reinforcement, p. 36 Enrichment, p. 36 Activity Worksheets, p. 49 Concept Mapping, pp. 21-22	Section Focus Transparency 32, Heating Curve for Water	
Study Guide, p. 37 Reinforcement, p. 37 Enrichment, p. 37 Activity Worksheets, pp. 46-47 Lab Manual 19, The Behavior of Gases	Section Focus Transparency 33, Tire Pressure Science Integration Transparency 8, The Birth of a Cloud	**Key to Teaching Strategies** The following designations will help you decide which activities are appropriate for your students.
Study Guide, p. 38 Reinforcement, p. 38 Enrichment, p. 38 Activity Worksheets, p. 50 Critical Thinking/Problem Solving, p. 14	Section Focus Transparency 34, Supertankers Teaching Transparency 16, Archimedes' Principle	**L1** Level 1 activities should be within the ability range of all students, including those with learning difficulties. **L2** Level 2 activities should be within the ability range of the average to above-average student. **L3** Level 3 activities are designed for the ability range of above-average students. **LEP** LEP activities should be within the ability range of Limited English Proficiency students. **LS** These activities are designed to address different learning styles. **COOP LEARN** Cooperative Learning activities are designed for small group work. **P** These strategies represent student products that can be placed into a best-work portfolio.

GLENCOE TECHNOLOGY

The following multimedia resources are available from Glencoe.

Glencoe Physical Science Interactive Videodisc
Behavior of Gases

Physical Science CD-ROM

National Geographic Society Series
STV: Water
Newton's Apple: Physical Sciences

Teacher Classroom Resources

This is a representation of key blackline masters available in the Teacher Classroom Resources.

Teaching Aids

Section Focus Transparencies

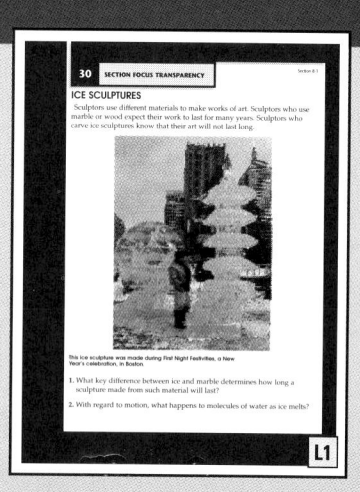

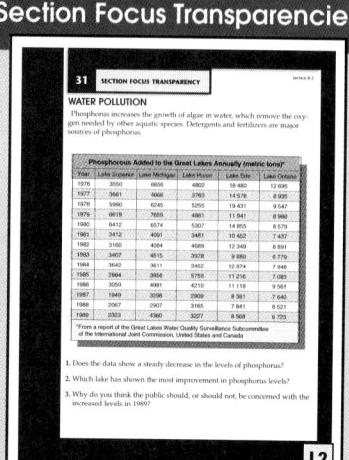

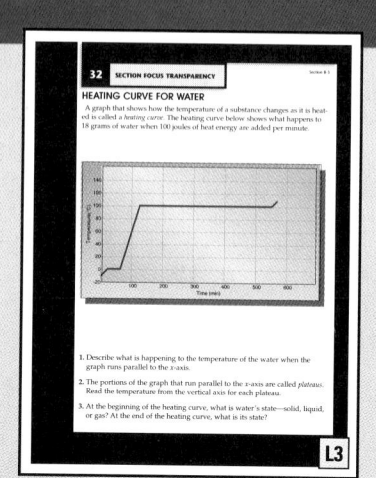

Science Integration Transparencies

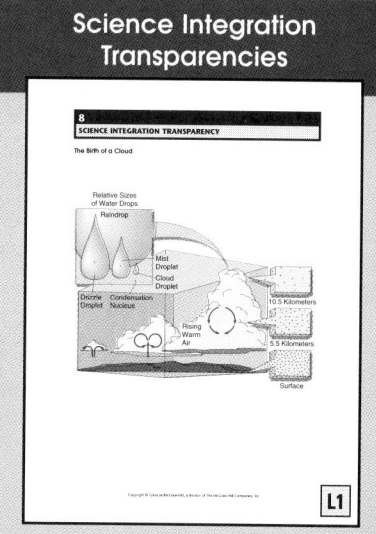

Teaching Transparencies

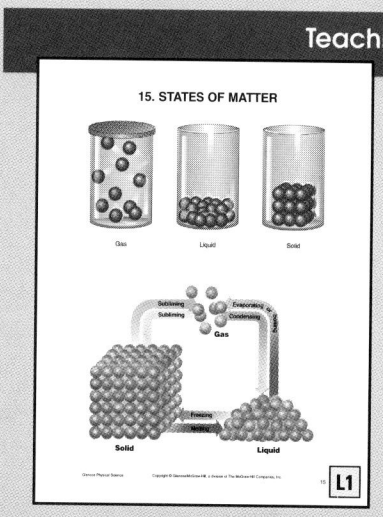

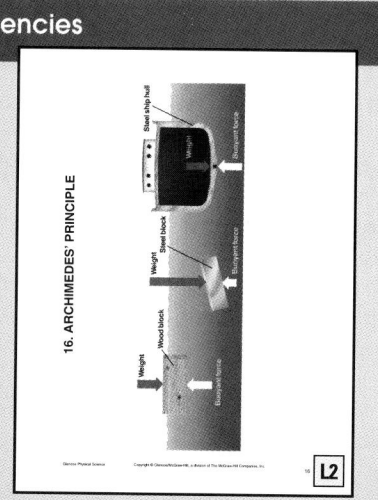

Meeting Different Ability Levels

Study Guide

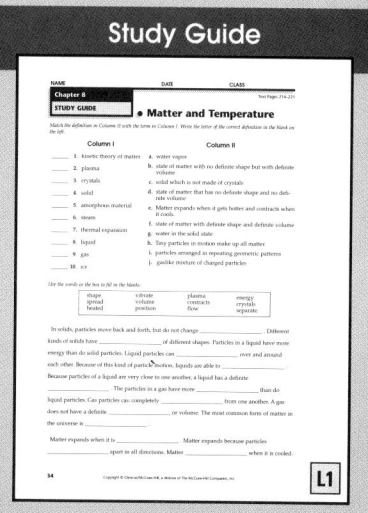

Reinforcement

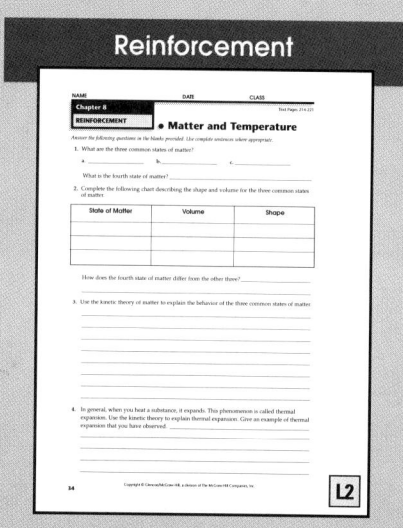

Enrichment Worksheets

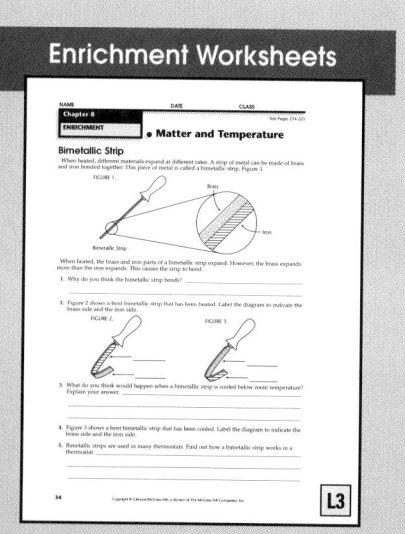

212C

Chapter 8 Solids, Liquids, and Gases

Hands-On Activities

Science Integration Activity

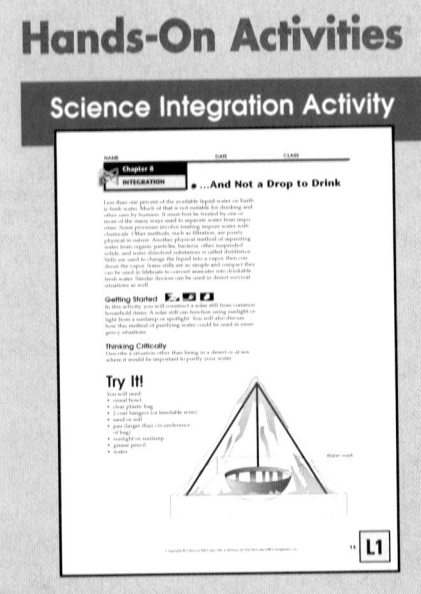

Lab Manual

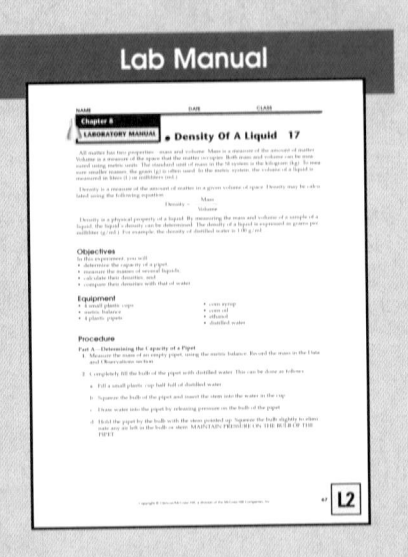

Assessment

Performance Assessment

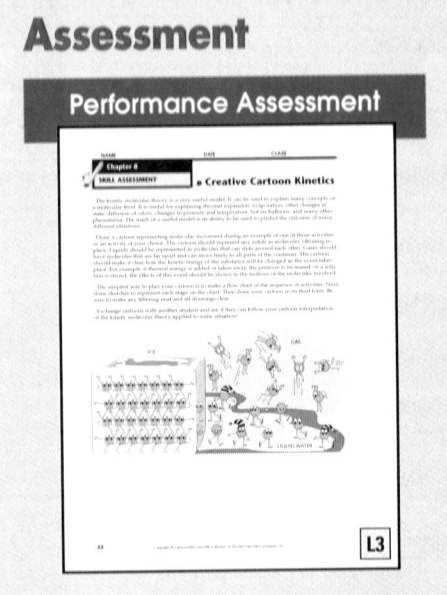

Enrichment and Application

Critical Thinking/ Problem Solving

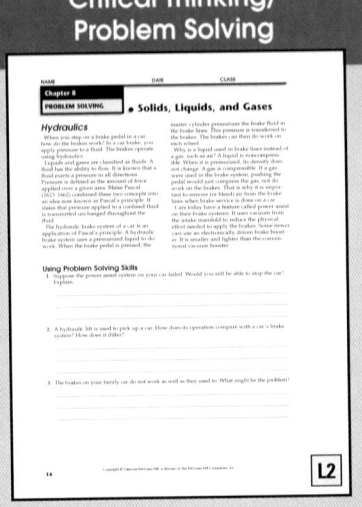

Cross-Curricular Integration

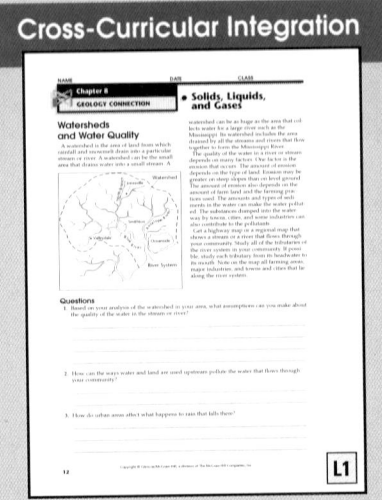

Science and Society Integration

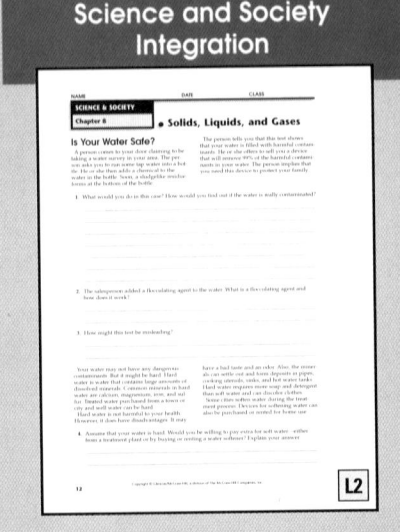

Technology Integration

Multicultural Connections

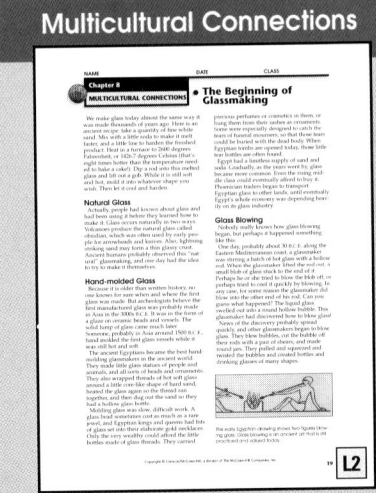

Concept Mapping

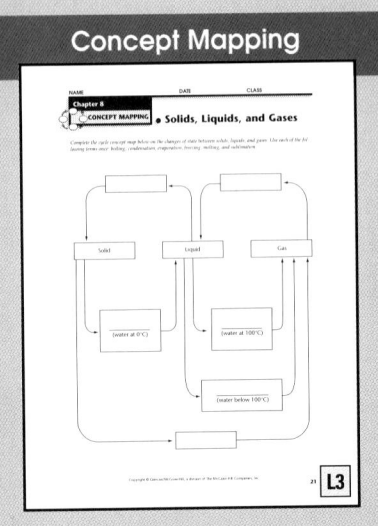

Chapter 8

Solids, Liquids, and Gases

CHAPTER OVERVIEW

Section 8-1 The kinetic theory of matter explains why solids, liquids, gases, and plasmas behave differently.

Section 8-2 Science and Society Students learn how they use, pollute, and regulate fresh water.

Section 8-3 The kinetic theory is applied to explain how thermal energy affects changes of state.

Section 8-4 After *pressure* is defined, the relationships among volume, pressure, and temperature in gases are explored.

Section 8-5 Characteristics of fluids are discussed by exploring the discoveries of Archimedes, Pascal, and Bernoulli.

Chapter Vocabulary

states of matter	pressure
	pascal
evaporation	crystal
Boyle's law	plasma
Charles's law	Venturi effect
buoyant force	condensation
polluted water	
thermal expansion	
Archimedes' principle	
heat of fusion	
Pascal's principle	
thermal pollution	
Bernoulli's principle	
heat of vaporization	
kinetic theory of matter	

Theme Connection

Energy The energy content, or temperature, of particles plays a major role in determining their state of matter. The kinetic theory of matter presents a way for students to visualize particles in motion and to form a mental model of the states of matter.

Previewing the Chapter

212

Learning Styles	Look for the following logo for strategies that emphasize different learning modalities. **LS**
Kinesthetic	MiniLAB, p. 225; Integrating the Sciences; p. 238, People and Science, p. 240
Visual-Spatial	Explore, p. 213; Demonstration, pp. 215, 217, 224, 227, 228, 229, 234, 236; Content Background, p. 216; Activity 8-1, p. 221; MiniLAB, p. 222; Visual Learning, pp. 215, 217, 225, 226, 227, 229, 235, 238
Interpersonal	Activity, p. 216; Reteach, p. 226; MiniLAB, p. 235
Logical-Mathematical	Activity, p. 223; Activity 8-2, pp. 232-233
Linguistic	Across the Curriculum, p. 216; Reteach, pp. 219, 237

Chapter 8

Solids, Liquids, and Gases

You have likely had the experience of eating a rapidly melting ice cream cone. When changing from solid to liquid, all substances absorb energy. The large photo shows liquid ingredients being poured into an ice-cream maker. How does energy play a role in this process? What is the purpose of the ice? You will find the answers to these questions in this chapter as you study the states of matter and ways to change from one state to another.

EXPLORE ACTIVITY

Observe what happens during a change of state.

1. Obtain a small scented candle and place it on a flat, non-flammable surface. In your Science Journal, list the properties of the candle that allow you to classify it as a solid.
2. Under your teacher's direction, light the candle. List observations about changes the candle undergoes. **CAUTION:** *Keep hair and clothing away from the open flame.*
3. As the candle burns, what observations can you make about the three states of matter—solid, liquid, and gas?

Observe: In your Science Journal, describe what role energy played in bringing about the changes of state in the matter making up the candle.

Previewing Science Skills

▶ In the Skill Builders, you will **make and use a table, map concepts, hypothesize,** and **measure in SI.**

▶ In the Activities, you will **observe, organize** and **collect data, infer, communicate, measure,** and **classify.**

▶ In the MiniLABs, you will **observe** and **hypothesize.**

213

Prepare

Section Background

All true solids are crystalline. Their crystal shape reflects the original arrangement of atoms and molecules.

Particles of liquids have little, if any, more spacing than those of solids. For water, liquid particles are actually closer together than those of the solid. Thus, ice is less dense and floats in liquid water. Because it is made of charged particles, plasma is greatly affected by electric and magnetic fields.

Preplanning

For Activity 8-1, make enough 4% sodium borate and polyvinyl alcohol solutions for your lab groups.

1 Motivate

Bellringer

Before presenting the lesson, display **Section Focus Transparency 30** on the overhead projector. Assign the accompanying **Focus Activity** worksheet. L1 LEP

Tying to Previous Knowledge

Ask your students if they have ever heard of fuel-line freeze-up or engine vapor lock. With vapor lock, fuel vaporizes at a hot spot, causing a fuel line to partially fill with gas. A car's fuel pump is designed to pump a liquid, not a gas. In summer, the gasoline can vaporize in the fuel line. In winter, water from condensation in the gas tank can freeze and block the fuel line. Discuss with students how temperature affects the state of matter.

214

Science Words

> states of matter
> kinetic theory of matter
> crystal
> plasma
> thermal expansion

Objectives

- Describe the four states of matter.
- Use the kinetic theory of matter to explain the characteristics of solids, liquids, and gases.
- Explain the thermal expansion of matter.

Figure 8-1

Water can exist in nature as a solid, a liquid, or a gas. *What states do you see in this photo?*

8•1 Matter and Temperature

States of Matter

After swimming on a hot day, Eli was having a refreshing glass of ice water. As he rested by the pool, the water on his arm evaporated and the ice cubes melted in his glass. In how many different forms does water exist in this example? Ice, liquid water, and water vapor are examples of the three most familiar states of matter—solid, liquid, and gas. **Figure 8-1** shows water as it can be found in nature.

All matter takes up space and has mass, yet matter can exist in different states. There are four **states of matter**—solid, liquid, gas, and plasma. The state of a sample of matter depends on its temperature. For example, water ordinarily exists as ice at low temperatures and as liquid water at moderate temperatures. At higher temperatures, water changes to the gas state as water vapor. At still higher temperatures, the matter in water becomes plasma. Each state has characteristics that are used to identify it, as you'll see.

Solids

A metal spoon, a cube of sugar, and a piece of cement are classified as solids. Every solid has a definite shape and a definite volume. For example, a metal spoon stays spoon-shaped whether it's in your hand or in a glass of water. And, because no ordinary amount of force can squeeze the spoon into a smaller space, it has a definite volume.

Program Resources

 Reproducible Masters
Study Guide, p. 34 L1
Reinforcement, p. 34 L1
Enrichment, p. 34 L3
Lab Manual 17, 18
Activity Worksheets, pp. 44, 45 L1
Multicultural Connections, pp. 19, 20

Transparencies
Section Focus Transparency 30 L1
Teaching Transparency 15

Figure 8-2

Although the particles in a solid, such as this crystal of table salt, vibrate, they do not move out of position. *What other kitchen solid can you think of that has crystals you can see?*

Chlorine

Sodium

What accounts for the characteristics of solids? Tiny particles in constant motion make up all matter. This idea is called the **kinetic theory of matter.** The particles in solid matter are held close together by forces between them. This is why a solid can't be squeezed into a smaller space. The particles can vibrate close to their neighbors, but they lack enough energy to move out of position. Thus, they lack enough energy to move over or around each other. This explains why a solid holds its shape.

Crystalline Solids

In most solids, the particles are arranged in repeating geometric patterns. These arrangements are **crystals.** Different kinds of solids have crystals of different shapes. In the magnified view of **Figure 8-2,** you can see that crystals of table salt are little cubes. A snowflake is a crystal of water that has the shape of a hexagon.

Noncrystalline Solids

Some materials, such as glass, many plastics, and some kinds of wax, appear to be solids but are not made of crystals. They are often called amorphous solids. The word *amorphous* means "having no form." Many scientists think some of these noncrystalline materials should be classified as thick liquids.

USING MATH

A hexagon is a six-sided figure that occurs in ice crystals and honeycombs. It is an efficient shape that *tessellates* a surface; that is, it covers the surface with no overlapping or gaps. Name another shape that tessellates a surface. What shape that you see in the salt crystals will also tessellate a surface?

8-1 Matter and Temperature   215

2 Teach

Demonstration

LS **Visual-Spatial** If possible, obtain some moth crystals to use in teaching this section. As you discuss solids, have a student place some moth crystals in a test tube. Stopper the tube and pass it around the class so students can observe the crystalline form. **LEP**

LS **Visual-Spatial** Have students observe the regular shape of salt crystals under a microscope or with a hand lens. **LEP**

Visual Learning

Figure 8-1 What states do you see in this photo? *solid, liquid, and gas*

Figure 8-2 What other kitchen solid can you think of that has crystals you can see? *sugar* **LEP** **LS**

USING MATH

Rectangles and equilateral triangles tessellate. A square is in salt crystals.

📁 Use **Multicultural Connections,** pp. 19, 20 as you teach this lesson.

🔖 Use **Teaching Transparency 15** as you teach this lesson.

Inclusion Strategies

Gifted Have each team prepare a written statement explaining how refineries solve the problems of fuel-line freeze-up and vapor lock. They will find that refineries change gasoline blends for different seasons and different geographical locations. **L3** **COOP LEARN**

GLENCOE TECHNOLOGY

 CD-ROM

Physical Science CD-ROM
Have students perform the interactive exploration for Chapter 8 to reinforce chapter concepts and thinking processes.

Liquids

If you don't eat it quickly enough, a solid scoop of ice cream will turn into a liquid, taking the same shape as your bowl or cup. A liquid flows and takes the shape of its container. However, like solids, liquids can't normally be squeezed to a smaller volume. If you push down on a liter of water with a moderate amount of force, its volume remains a liter.

More Motion

Just as the kinetic theory explains the properties of solids, it also explains the properties of liquids. Because a liquid can't be squeezed, its particles must also be close together, like those of a solid. However, they have enough kinetic energy to move over and around each other. This movement of particles lets a liquid flow and take the shape of its container. Thus, the orange juice poured into a glass in **Figure 8-3** will take the shape of the glass.

Because its particles are held close together, almost as close as those of a solid, liquid matter does have a definite volume. If you pour 1 L of orange juice into a 2-L bottle, it will not spread out to fill the bottle. Likewise, you couldn't force the liter of juice into a half-liter container. The two containers in **Figure 8-4** contain the same volume of liquid.

Figure 8-3

The particles in a liquid are close together, but they have enough energy to move over and around one another.

Science Journal

In your Science Journal, explain how the motion of particles changes when matter changes from one state to another.

Figure 8-4

Although its volume does not change, the shape of a liquid depends on the shape of its container.

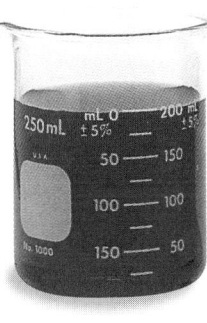

216   Chapter 8 Solids, Liquids, and Gases

216

Gases

You may have pumped air into a basketball, tire, or balloon and noticed that the air takes the shape of the object. Gases are "springy"—they expand or contract to fill the space available to them and can be squeezed into a smaller space. A gas has neither a definite shape nor a definite volume.

According to the kinetic theory of matter, the particles of a gas have enough energy to separate completely from one another. Therefore, the particles are free to move in all directions until they have spread evenly throughout their container. **Figure 8-5** shows a test for a gas that has filled a room as its particles spread. Because the particles of a gas are not close together, they can also be squeezed into a smaller space. When you pump up a bicycle tire, you are forcing more and more air particles into the same space.

Figure 8-6 explains the relationship between particles and energy in a solid, in a liquid, and in a gas.

Figure 8-5
This test disk is used to detect released carbon monoxide in all parts of a building. *Why would a detector be effective even if it were not right next to the gas leak?*

Figure 8-6

The energy of particles is different for each state of matter.

Gas	Liquid	Solid
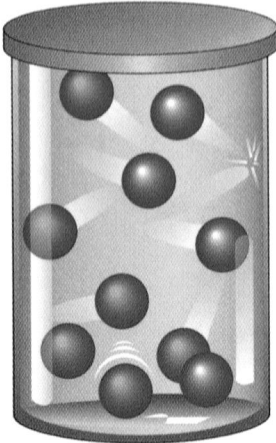		

A In gas samples, the particles have enough energy to overcome attractive forces that would hold them together.

B The particles composing a liquid don't have enough energy to overcome all attractive forces, but they do have enough energy to move around each other.

C Solids are made up of particles that do not have enough energy to move from one place to another. *How does that affect the shape of a solid?*

Demonstration

LS **Visual-Spatial** On the overhead projector, in a clear plastic dish, place five marbles. Agitate the box to simulate gas molecules in motion. Increase agitation to simulate the effect of increasing the temperature. Point out the change in collisions. Relate student observations to the properties and particle model of a gas. **LEP**

Enrichment

Cryogenics is the study of matter at low temperatures. Ask a student to research how a solid's properties change near absolute zero. **L3**

Teacher F.Y.I

At room temperature, oxygen gas molecules have an average speed of 1700 km/h.

Visual Learning

Figure 8-5 Why would a detector be effective even if it were not right next to the gas leak? *because the particles of a gas spread to fill a room*

Figure 8-6 How does that affect the shape of a solid? *It does not change.* **LEP** **LS**

GLENCOE TECHNOLOGY

 Videodisc

Glencoe Physical Science
Interactive Videodisc
Side 1, Lesson 3
Gases Expand to Fill Container

23416

23418-23508

Figure 8-7

These photos show examples of matter in the plasma state.

Plasma

So far, you've learned about the three familiar states of matter. But none of these is the most common state of matter in the universe. For example, 99 percent of the mass of our solar system is contained in the sun. The most common form of matter in the universe is the type found in stars like the sun and the nebula shown in **Figure 8-7.** Such matter is called plasma.

Plasma is a gaslike mixture of positively and negatively charged particles. You know that particles of matter move faster as the matter is heated to higher temperatures. The faster they move, the greater the force with which they bump into other particles, or collide. As matter is heated to very high temperatures, the particles begin to collide violently. As a result, the particles break up into the smaller particles they are made of. These particles are electrically charged.

Besides light from the sun, you can observe the effects of plasma in your home or school. When a fluorescent light is switched on, electricity causes particles of mercury gas inside the tube to form plasma.

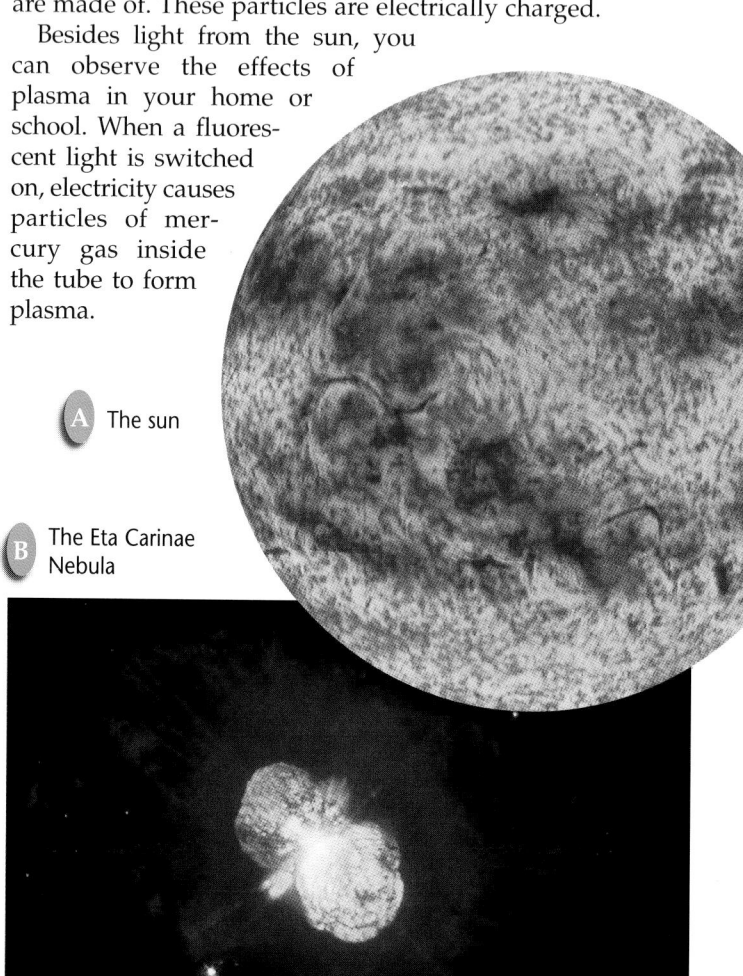

A The sun

B The Eta Carinae Nebula

Thermal Expansion

You have learned how the kinetic theory accounts for characteristics of different states of matter you see and touch every day. The kinetic theory also explains other things you may have observed. For example, have you ever noticed the strips of metal that run across the floors and up the walls in long hallways of concrete and steel buildings? Maybe you've seen these strips in your school. These strips usually cover gaps in the building structure called expansion joints. Expansion joints allow the building to expand in hot weather and shrink in cold weather without cracking the concrete. As you drive onto or off a bridge, you will usually pass over a large steel expansion joint.

The Heat and Motion Connection

Almost all matter expands as it gets hotter and contracts when it cools. This characteristic of matter is called **thermal expansion.** You can compare thermal expansion to a crowd of people. When the people are quiet and still, they are able to stand close

Problem Solving

Mind over Matter

Martin, Rita, and L.J. were working on fixing up an old car and they needed some nuts and bolts to make their repairs. L.J. found a jar of assorted nuts and bolts in the garage, but they had trouble removing the metal lid. After struggling for a few minutes, Rita suggested that running hot water over the metal lid might make it easier to open.

Solve the Problem:

1. Think about the effect of energy on the motion of particles. How would you describe the effect of heat energy on the metal?

2. After the hot-water treatment, the jar easily opened. Why did this work better than just forcing the lid off the jar?

Think Critically:

Sealing containers is important in preserving freshness and preventing spoilage. Explain two reasons why a food container that was closed when it was warm might become tight as it cooled.

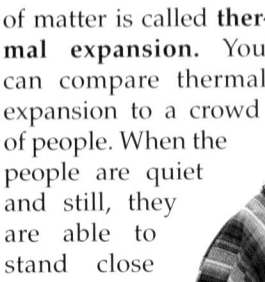

8-1 Matter and Temperature   **219**

Use Lab 18 in the **Lab Manual** as you teach the lesson.

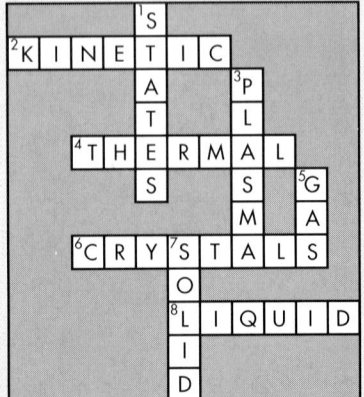
219

4 Close

Review

1. Both solids and liquids have a definite volume because their particles are closely packed. Solids have a definite shape, but liquids take the shape of the container. The particles of solids are held in place by attractive forces. The particles of liquids have enough energy to tumble over and around one another.

2. The material is a gas because it expanded to occupy all the space available in the container.

3. The particles of copper move more slowly as it cools. Attractive forces then pull them closer together.

4. **Think Critically** The particles in solids and liquids are close together compared to those in gases.

USING MATH

Liquid water is 323 times more dense than water vapor. In the liquid, the water molecules are much closer together than in the vapor.

Skill Builder

Possible Solution

State	Properties	Particle description	Examples
Solid	Definite shape and volume	Closely packed; do not easily change position	ice, sugar
Liquid	Definite volume; takes shape of container	Closely packed; able to move past one another	milk, mercury in thermometer
Gas	Occupies shape and volume of container	Spread apart; free to move in all directions	oxygen, steam
Plasma	Occupies shape and volume of container	Gaslike mix of negatively and positively charged particles	mercury vapor in fluorescent tube, sun and stars

220

together. As the people become restless, they jostle one another and the crowd spreads out.

How does the kinetic theory of matter explain thermal expansion? In a solid, forces between the particles hold them together. As the solid is heated, these particles move faster and faster and vibrate against each other with more force. As a result, the particles spread apart slightly in all directions, and the solid expands. This explains why expansion joints are needed in buildings and on bridges. The same effect also occurs in liquids and gases. A common example of this is shown in the thermometers in **Figure 8-8.**

As the liquid in a bulb is heated, the energy causes more movement of the particles. This energy is transferred to other parts of the liquid. The liquid in the narrow part has to expand only slightly to show a large change on the temperature scale.

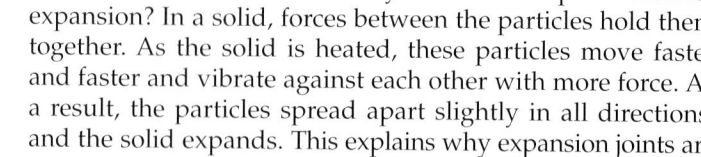

Figure 8-8

An important example of thermal expansion is found when liquid that is heated, even slightly, expands and moves up through a narrow tube in a thermometer.

Section Wrap-up

Review

1. Compare the characteristics of solids and liquids.

2. You pour 500 mL of a green material into a 1-L flask and stopper it. The material completely fills the flask. What state of matter is the material? Explain how you know.

3. In terms of particle motion, explain why copper shrinks when it cools.

4. **Think Critically:** Solids and liquids are considered to be "condensed" states of matter. Explain.

Skill Builder
Making and Using Tables
Make a table to classify several materials as a solid, liquid, gas, or plasma. Include columns for properties and a description of particles for each state. If you need help, refer to Making and Using Tables in the **Skill Handbook.**

USING MATH

The density of a sample of water vapor is 0.0031 g/mL. The density of a second sample of liquid water is 1.0 g/mL. How many times more dense is the liquid than the vapor? How does this relate to the spacing of the water molecules?

Assessment

Oral After students have made their tables, select objects from the classroom and ask students to describe the particles and their movements within the selected objects. Use the Performance Task Assessment List for Data Table in **PASC,** p. 37. **P**

Activity 8-1

Properties of Liquids

Why is a soft drink considered a liquid? Why is a hamburger considered a solid? Each state of matter has its own characteristics that allow us to classify it as that state.

Problem

How can the properties of a material be used to classify it as a solid, a liquid, or a gas?

Materials

- dropper
- food coloring
- wooden stick
- graduated cylinder
- 4% solution of powdered borax in water
- 4% solution of polyvinyl alcohol (PVA) in water
- paper cup
- goggles
- apron

Procedure

1. Copy the data table and use it to record your observations.
2. Using a graduated cylinder, measure 30 mL of PVA solution into a paper cup. Add 2 drops of food coloring.
3. Using a dropper, add about 3 mL of borax solution to the PVA in the cup, and begin to stir vigorously with a wooden stick.
4. After it has been stirred for 2 minutes, what is the consistency of the material?
5. Transfer the material to your hand. **CAUTION:** *Do not taste or eat the material, and be sure to wash your hands after the activity.* Rate the ease with which the material flows.
6. Form the material into a ball and then place it in the cup to test its ability to take the shape of its container.
7. Compare the volume of the new material with the volume of the original material before stirring.

Analyze

1. Is the new material most like a gas, a liquid, or a solid?
2. What other materials have you seen that have similar properties to this one?
3. Considering the flow of the new material, how would you rate the strength of the attraction among its particles?

Conclude and Apply

4. Using the kinetic theory of matter, describe the closeness of the particles of matter in the new material.
5. How can properties of this material be used to **classify** it?

Data and Observations

Sample Data

Property	Observation	Interpretation
Ability to flow	Yes, but slowly (high viscosity)	Molecules hold to each other, but not as much as solids
Shape change	Does change shape	Molecules do allow movement
Volume change	No change	Molecules are not totally free to fill up any volume

they are not attracted strongly enough to mold the material into a definite shape.

4. The particles are close together because the material cannot be pressed into a smaller space.

5. It exhibits the properties of flow, indefinite shape, and constant volume. Therefore, it is classified as a liquid.

✔ Assessment

Portfolio Students should make a table with the following three headings: GAS, LIQUID, SOLID. Under each, describe the general properties and include three examples from home or school. Use the Performance Task Assessment List for Data Table in PASC, p. 37. **P**

Activity 8-1

Purpose

LS **Visual-Spatial** Observe the properties of a material and classify it according to state.
L1 **LEP** **COOP LEARN**

Process Skills

measuring, classifying, formulating models, formulating operational definitions, observing and inferring, making and using tables, comparing and contrasting, recognizing cause and effect

Time

35 minutes

📁 **Activity Worksheets,** pp. 5, 44, 45

Teaching Strategies

Troubleshooting In order to properly mix the borax and PVA, students must stir the solutions quickly and continuously. Do not allow students to taste the resulting gel or take it out of the classroom. Spills may be cleaned up with water.

- The new material is a polymer that is much like products sold in toy stores under names such as Slime. Encourage students to make comparisons to commercial products.
- To prepare 4% borax solution, dissolve 12 g of sodium borate in 288 mL of hot water with constant stirring. Cool before using.
- To prepare 4% PVA solution, add 40 g of 98% hydrolyzed PVA to 960 mL of H_2O. Heat to 80°C with constant stirring. Cool before using.

Answers to Questions

1. a liquid
2. Slime, gelatin dessert, jellies, mucus
3. The particles are fairly strongly attracted. They remain a certain volume and do not fill the container as a gas would. However, unlike a solid,

221

1 Motivate

Bellringer

Before presenting the lesson, display **Section Focus Transparency 31** on the overhead projector. Assign the accompanying **Focus Activity** worksheet. L1 LEP

2 Teach

MiniLAB

Purpose
LS **Visual-Spatial** Students will observe and infer that filtering does not remove dissolved materials. L1

Materials
litmus paper, vinegar, graduated cylinder, filter paper

Analysis
1. No, the paper will appear clean.
2. No, the evidence is the litmus test.
3. Evaluate answers for plausibility.

✓Assessment

Performance Have students filter salt water, evaporate the filtrate, and see the salt residue. Use the Performance Task Assessment List for Making Observations and Inferences in **PASC**, p. 17. P

3 Assess

Check for Understanding

Have students answer Section Wrap-up questions.

ISSUE:
8•2 Fresh Water—Will there be enough?

Science Words
polluted water
thermal pollution

Objectives
- Describe how people use and pollute water.
- Discuss how people can save water and stop pollution.

Fresh Water

For living things, like yourself, the most important liquid on Earth is fresh water. It is not as abundant as you might think. Fresh water, which is water that is not salty, makes up only 0.75 percent of the water available on Earth in the liquid state.

How much water do you think you use in a day? Take a guess. Did you say 1700 gallons? For the average person in the United States, that's the correct answer! Each person uses about 200 gallons a day for cooking, bathing, toilet use, and heating and cooling homes. Add to this the 750 gallons per person a day used to produce materials and energy, and another 750 gallons per person a day used to water crops. **Table 8-1** lists only a few ways you use water.

Water Pollution

Not only do humans use huge amounts of fresh water, humans also pollute their natural supplies of this liquid. **Polluted water** refers to water that contains such high levels of unwanted materials that it is unacceptable for drinking or other specific purposes. Sometimes, the pollution is visible, as in **Figure 8-9**.

Water that runs off land after rain falls and snow melts may wash animal and plant wastes into our water supplies. Human activities in cities may contaminate rivers with raw sewage. Fertilizers, pesticides, and herbicides used on farms

Table 8-1

Ways Water Is Used			
Estimated Gallons of Water Used to:			
Make 1 gallon of milk	5	Produce 1 gallon of gasoline	10
Brush teeth with water running	1-2	Do one load of laundry	32-59
Flush a toilet	5-7	Burn a 100-watt lightbulb for	
Shower for 1 minute	3-5	10 hours	80
Make enough steel for a table knife	6	Wash a car at home with	
Wash car at self-service car wash	5-10	a hose	up to 150

Program Resources

 Reproducible Masters
Study Guide, p. 35 L1
Reinforcement, p. 35 L1
Enrichment, p. 35 L3
Cross-Curricular Integration, p. 12 L3
Science and Society Integration, p. 12 L3
Activity Worksheets, p. 48 L1
Science Integration Activity, pp. 15, 16

Transparencies
Section Focus Transparency 31 L1

may enter the groundwater. Toxic chemicals from home use may end up in our water supply if they are not disposed of properly.

Another form of water pollution can occur when electrical generating plants or industries release large amounts of heated water into rivers. The excess heat in the water is called **thermal pollution.** If water temperature is changed too much, some organisms that live in the water will die.

Figure 8-9

Many sources are possible for this pollution.

2 Points of View

▷ Regulating Pollution

Scientists are trying to develop better ways to contain and dispose of industrial and farming by-products to reduce water pollution. Congress also passes laws governing the release of and the levels of various materials in water supplies.

▷ Cost of Regulation

New methods for stricter control of industrial and farming by-products disposal will raise costs of these processes. Improving water-treatment facilities will also increase our costs for fresh water.

Section Wrap-up

Review

1. What are two main reasons that there could be a shortage of clean water?
2. List five sources of water pollution.

*inter***NET** **CONNECTION** Visit the Chapter 8 Internet Connection at Glencoe Online Science, **www.glencoe.com/sec/science/physical,** for a link to more information about water and water pollution.

MiniLAB

Can dissolved chemicals be filtered out of a solution?

Procedure
1. Test 50 mL of water with litmus paper.
2. Add 50 mL of vinegar to the water.
3. Test with litmus paper.
4. Pour solution through filter paper, and retest with litmus paper.

Analysis
1. Was there any visible residue on the filter?
2. Did the filter remove the acid?
3. What other method could you try to make the water pure?

SCIENCE & SOCIETY

Community Connection

Water Treatment Contact the local water-treatment plant. Invite the chief water-treatment chemist to visit the class and discuss the ways that water is treated prior to use. If the visit is not possible, ask about a videotaped version of a field trip of the plant to show to the class.

Reteach

Have students discuss whether there is a water supply or pollution problem in your area.

Extension

For students who have mastered this section, use the **Reinforcement** and **Enrichment** masters.

Videodisc
STV: Water
Water Quality
Unit 1
Water Quality

00392-17833

4 Close

Activity

LS **Logical-Mathematical** Use Table 8-1 to have students estimate their daily water usage.

Section Wrap-up

Review

1. wasteful use and pollution of water
2. Answers may include toxic chemicals, farm chemicals, radioactive materials, raw sewage, and excess heat.

Prepare

Section Background

Substances that have weak intermolecular forces evaporate easily and melt or boil at relatively low temperatures. Before a substance reaches its melting or boiling point, added energy increases the motion (kinetic energy) of the particles, and the temperature increases.

Preplanning

The MiniLAB requires rubbing alcohol and droppers.

1 Motivate

Bellringer

Before presenting the lesson, display **Section Focus Transparency 32** on the overhead projector. Assign the accompanying **Focus Activity** worksheet. **L1** **LEP**

Demonstration

Visual-Spatial Place a few crystals of iodine in a stoppered test tube. A pale violet color will be visible. **CAUTION:** *Iodine vapors are toxic.* Ask students to suggest a mechanism for this phenomenon based on the particle models of the states of matter.

Tying to Previous Knowledge

- Ask students to recall how cold they felt when they got out of a pool when a strong breeze was blowing.
- Ask your class if anyone has seen dry ice subliming. Ask them to describe it. If possible, obtain a piece to observe. **CAUTION:** *Use gloves when handling dry ice. Never place dry ice in a closed container.*

224

8•3 Changes in State

Science Words

evaporation
condensation
heat of fusion
heat of vaporization

Objectives

- Interpret state changes in terms of the kinetic theory of matter.
- Account for the energy of the heats of fusion and vaporization in state changes.

Identifying Changes in State

If you've ever seen ice cream melt before you could eat it, you have seen matter change state. Solid ice crystals in the ice cream melt when they change from the solid state to the liquid state. When melting, a solid changes into a liquid. You put water in the freezer to make ice cubes. When freezing, matter changes from the liquid state to the solid state.

When you boil water, you observe another change of state, called vaporization. When boiling, you add heat to a liquid until it reaches a temperature at which it changes to bubbles of gas below its surface. Many liquids don't need to boil to change to a gas. During **evaporation,** a liquid changes to a gas gradually at temperatures below the boiling point. When you come out of a pool into warm air, water on your skin soon evaporates. You'll see later how this drying helps cool you. In **Figure 8-10,** evaporation helped dry, or solidify, the concrete.

Figure 8-10

Freshly poured concrete includes liquid water spread into a thin layer over a large surface area. As the water particles gain energy, they can overcome attractions to other particles and evaporate, leaving behind a dry, solid surface.

224 Chapter 8 Solids, Liquids, and Gases

Program Resources

Reproducible Masters
Study Guide, p. 36 **L1**
Reinforcement, p. 36 **L1**
Enrichment, p. 36 **L3**
Activity Worksheets, p. 49 **L1**
Concept Mapping, pp. 21, 22

 **Transparencies**
Section Focus Transparency 32 **L1**

Figure 8-11

Condensation appears on many surfaces.

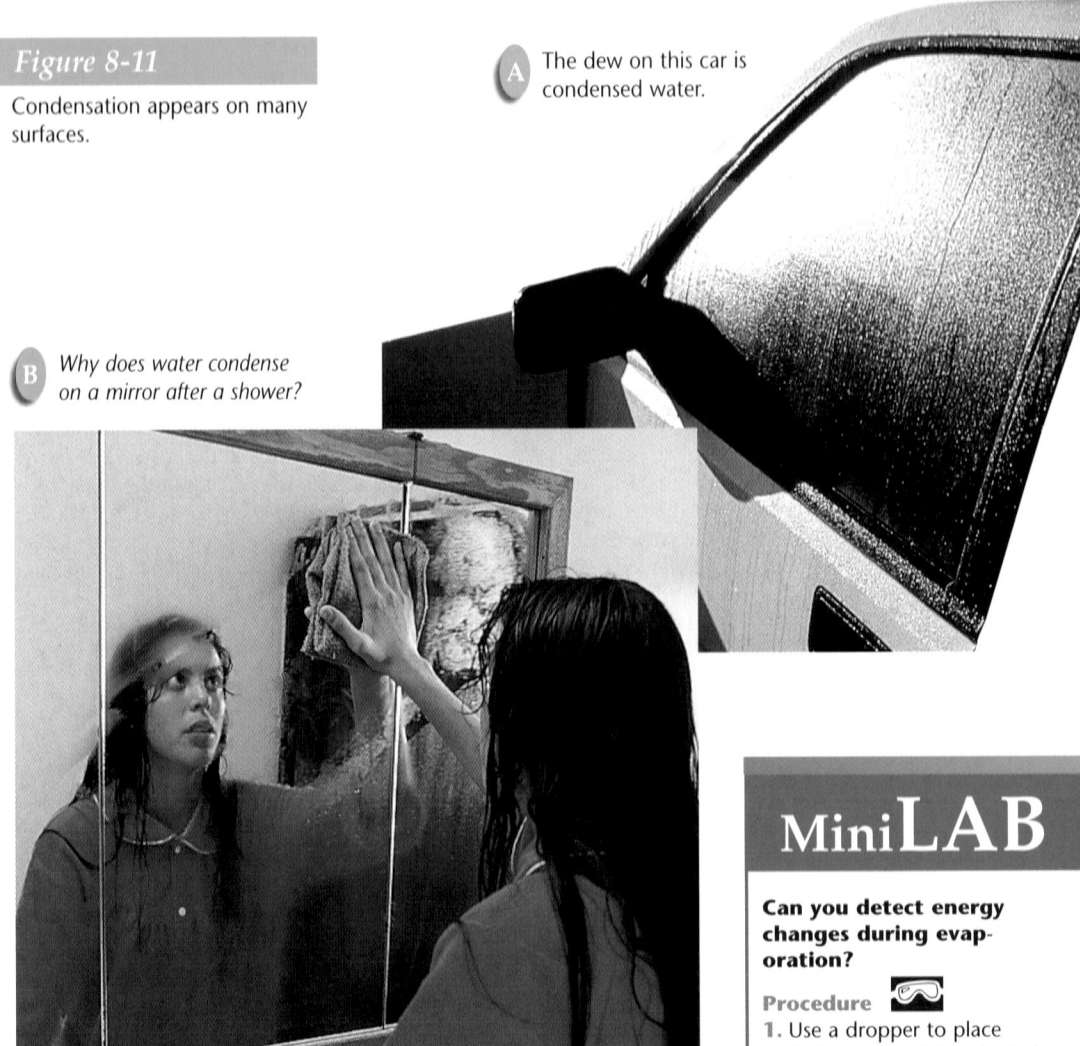

A The dew on this car is condensed water.

B Why does water condense on a mirror after a shower?

Have you noticed that ice cubes seem to shrink when they've been in the freezer for a long time? This shrinkage happens because of another change of state called sublimation. In sublimation, a solid changes directly to a gas without going through the liquid state.

You see another change of state when your glass of ice-cold soft drink "sweats." The drops of water on your glass appear when gaseous water condenses on the cold surface.

Condensation takes place when a gas changes to a liquid. Generally, a gas will condense when cooled to or below its boiling point. Two examples of this are seen in **Figure 8-11.**

MiniLAB

Can you detect energy changes during evaporation?

Procedure 🥽

1. Use a dropper to place 5 drops of rubbing alcohol on the back of your hand.
2. Wait for 2 minutes.

Analysis

1. During the 2 minutes, what sensation involving energy did you notice?
2. What changes in appearance of the rubbing alcohol did you notice?
3. In your Science Journal, explain how you know that evaporation requires an input of energy.
4. From where did that energy input come in this example?

8-3 Changes in State   **225**

MiniLAB

Purpose

LS **Kinesthetic** Students will observe that energy is needed to change a liquid into a gas. **L1** **LEP**

Materials

dropper, rubbing alcohol

Teaching Strategies

- Emphasize that an input of energy is required for continued evaporation.
- Discuss the function of perspiration.
- In a short time, the alcohol will evaporate. Students will feel a cooling sensation.

Analysis

1. Hands felt cooler.
2. Evaporation was observed.
3. All state changes require a change in energy. Also, evidence is the coolness of the hand.
4. Energy came mostly from the skin.

✓ Assessment

Oral Have students explain how the panting of a dog regulates its temperature. Use the Performance Task Assessment List for Making Observations and Inferences in **PASC**, p. 17. **P**

📁 **Activity Worksheets,** pp. 5, 49

Visual Learning

Figure 8-11 **Why does water condense on a mirror after a shower?** *The cold glass cools the water vapor to below its boiling point.* **LEP** **LS**

Heat and State Changes

The kinetic theory of matter explains changes of state. The amount of energy needed to change a material from the solid state to the liquid state is the **heat of fusion.** For water, the heat of fusion is 334 kJ/kg. The amount of energy needed to change a material from a liquid to a gas is the **heat of vaporization.** For water, this value is 2260 kJ/kg. **Figure 8-12** details the energy changes involved in state changes.

Figure 8-12

In this example, watch how temperature, energy supplied, and the motion of particles are related.

A As a sample of ice is warmed and the temperature is measured every 30 seconds, the particles of solid water absorb the applied heat energy. *What effect does this energy have on the motion of the particles?*

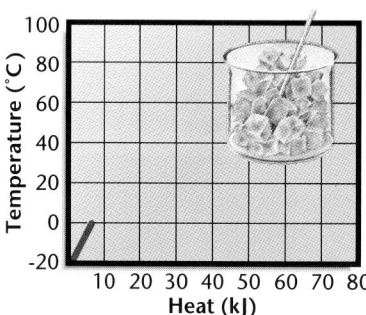

B As heat energy continues to be supplied, you may notice that the solid water now begins to appear as a liquid, but the temperature is not rising. To change a solid to a liquid, many forces must be overcome. For each kilogram of ice you want to melt, 334 kJ must be supplied.

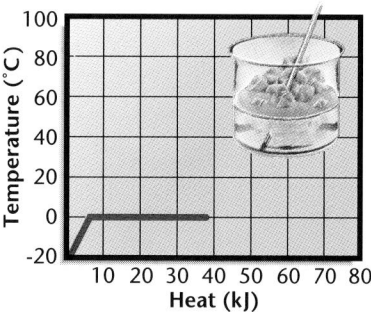

C Once the work of melting is accomplished, the applied heat energy causes the water particles making up the liquid to move faster. As you can see on the graph, the temperature begins to rise.

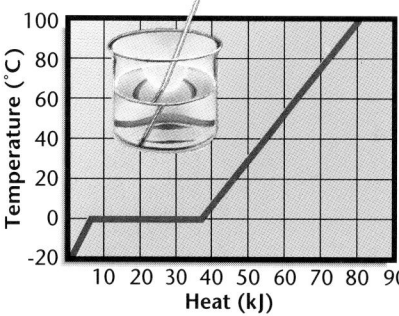

D Once the boiling point has been reached, the liquid becomes a gas, and the temperature again remains constant. At the heat of vaporization, the forces of attraction between particles are totally overcome. This requires 2260 kJ of energy. *Why do you think this value is so much higher than the heat of fusion?*

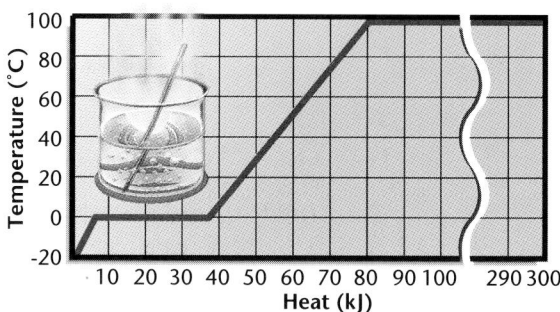

226   Chapter 8 Solids, Liquids, and Gases

Look at **Figure 8-13.** You can also use the kinetic theory to explain how water evaporates from your skin and how it cools you. When a liquid evaporates, the particles do not all have the same kinetic energy. Many fast-moving particles break away from the liquid and become a gas. As these fast-moving water particles leave, the average kinetic energy of the remaining particles becomes less. As a result, the temperature of the remaining water goes down. Because the water is now cooler than your skin, it takes heat from your skin and cools you.

Figure 8-13

The water spread over your skin after swimming absorbs energy from your body and the air and evaporates. *How does the temperature of the air affect the speed of evaporation?*

Section Wrap-up

Review

1. Name and describe the state changes in which solids and liquids become gases.

2. Use the kinetic theory to explain melting.

3. What happens to the energy put into a liquid during boiling?

4. **Think Critically:** Steam must lose its heat of vaporization in order to condense. How does this fact help explain why steam can cause more severe skin burns than liquid water at its boiling point?

 ### Skill Builder
Sequencing
Sequence the processes that occur when ice is heated until it becomes steam and then the steam is cooled until it is ice again. If you need help, refer to Sequencing in the **Skill Handbook.**

Science Journal

A glass of water and a puddle of water, both containing the same volume of water at the same temperature, are left to evaporate. Which do you think will evaporate sooner? In your Science Journal, write and explain your answer.

 Skill Builder
melting, boiling, condensation, freezing

 **Assessment**

Oral Have students tell you the values of heat of fusion or heat of vaporization that go with each change of state. **P**

4 Close

Demonstration

LS **Visual-Spatial** In novelty stores, one can purchase a "drinking bird." If a student has one, have him or her bring it to class and explain how it works. The water evaporates from the head to cool and condense the vapors of the volatile liquid inside.

Section Wrap-up

Review

1. Boiling: Gas bubbles form within the liquid at the boiling point. Sublimation: A solid changes directly to a gas. Evaporation: Particles of a liquid gain enough energy to escape from the surface.

2. Particles absorb enough energy to overcome some attractive forces and become free enough to tumble over one another.

3. The energy is used in overcoming the attractive forces that hold the liquid particles close to one another.

4. **Think Critically** If steam touches the skin, steam will condense and give up its heat of vaporization there. Liquid water at its boiling temperature does not have heat of vaporization to lose.

Science Journal

The spill has a greater surface area so its rate of evaporation is faster.

Prepare

Section Background

The gas laws treat gases as ideal. In ideal gases, each molecule has no volume, and there is no attraction between molecules.

Preplanning

Ice will be needed for Activity 8-2. Two balloons will be needed per lab team.

1 Motivate

Bellringer

 Before presenting the lesson, display **Section Focus Transparency 33** on the overhead projector. Assign the accompanying **Focus Activity** worksheet. L1 LEP

Demonstration

LS **Visual-Spatial** Fill a small glass to overflowing with water. Hold a 3" × 5" card on top of the glass and invert it. Ask your students what will happen if you take your hand away. Remove your hand. Air pressure will hold the card on the inverted glass. LEP

8•4 Behavior of Gases

Science Words

- pressure
- pascal
- Boyle's law
- Charles's law

Objectives

- Explain how a gas exerts pressure on its container.
- State and explain how the pressure of a container of gas is affected when the volume is changed.
- Explain the relationship between the temperature and volume of a gas.

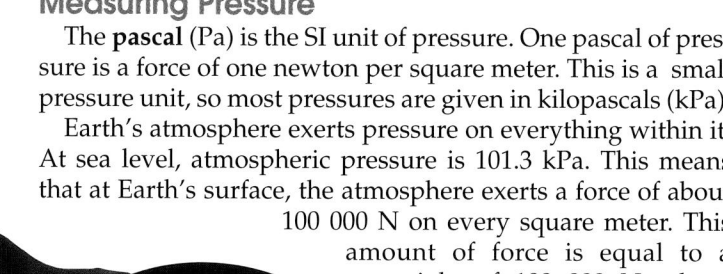

Pressure

Every time you feel the wind on your face, you observe the behavior of a gas—rather, the mixture of gases that is Earth's air, or atmosphere. Even when the wind is calm, the air exerts a force called pressure.

What causes the pressure of a gas? Particles of matter are small—many billions of particles of air fill an inflated toy balloon. When riding on a bike, you're riding on pockets of colliding air particles inside your tires. You have learned that the particles of air, like those in all gases, are constantly moving. They're free to fly about and collide with anything in their way. The collisions with the inside walls keep the balloons or tires inflated and cause the force that you feel when you squeeze them. The motion of the particles in air is illustrated in **Figure 8-14.**

The total amount of force exerted by a gas depends on the size of its container. **Pressure** is the amount of force exerted per unit of area.

$$P = \frac{F}{A}$$

Measuring Pressure

The **pascal** (Pa) is the SI unit of pressure. One pascal of pressure is a force of one newton per square meter. This is a small pressure unit, so most pressures are given in kilopascals (kPa).

Earth's atmosphere exerts pressure on everything within it. At sea level, atmospheric pressure is 101.3 kPa. This means that at Earth's surface, the atmosphere exerts a force of about 100 000 N on every square meter. This amount of force is equal to a weight of 100 000 N—about the weight of a large truck.

Figure 8-14

The force of particles of air in constant motion colliding with the inside walls of the tire keeps a tire inflated.

228

Program Resources

 Reproducible Masters
Study Guide, p. 37 L1
Reinforcement, p. 37 L1
Enrichment, p. 37 L3
Activity Worksheets, pp. 46, 47 L1
Lab Manual 19

🔦 **Transparencies**
Section Focus Transparency 33 L1
Science Integration Transparency 8

Our Atmosphere—A Sea of Air

INTEGRATION
Earth Science

You can see right through it, and most of the time you can't even feel it. But the air we breathe and live in contains many gas particles that cause the air pressure you have just been reading about. The atmosphere closest to Earth's surface, up to approximately 16 to 17 km, is called the troposphere. In the troposphere, you come into direct contact with nitrogen, oxygen, argon, carbon dioxide, and water vapor. These particles are moving incredibly fast, approaching 1610 km per hour. At these speeds, one particle may collide with another, or you, as often as every billionth of a second. Millions of fast-moving, colliding particles create air pressure in every square meter of the troposphere.

Higher Altitudes Mean Less Pressure

Above the troposphere, there are fewer gas particles and fewer collisions, which means less pressure. The stratosphere continues to about 50 km above Earth's surface. At higher distances, the mesosphere and thermosphere continue. If the air pressure is measured at 5 km above Earth, the pressure will have dropped from 101 kPa at Earth's surface to approximately 54 kPa. At 50 km, the pressure may be only 0.15 kPa. An example of the effect of altitude on pressure is shown in **Figure 8-15**.

Boyle's Law

Suppose you have some gas in a sealed flexible container, such as a balloon. You can squeeze or stretch the container without changing the amount of gas trapped inside.

The pressure of a gas depends on how often its particles strike the walls of the container. If you squeeze some gas into a smaller space, its particles will strike the walls more often, giving an increased pressure. This behavior explains why when you squeeze a balloon into a smaller space, it causes the balloon to push back with more force. The reverse happens, too. If you give the gas particles more space, they will hit the walls less often and the gas pressure will be reduced. Robert Boyle (1627-1691), a British scientist, described this property of gases. According to **Boyle's law,** if you decrease the volume of a container of gas, the pressure of the gas will increase, provided the temperature does not change. Increasing the volume causes pressure to drop. As you'll see, it is important that the temperature remains constant.

Figure 8-15

At lower altitudes (bottom), air pressure increases. The air trapped inside the bottle during a high-altitude biking trip (top) exerts less pressure than the air back home. *Why does the bottle have less volume in the bottom photo?*

8-4 Behavior of Gases   **229**

 Use **Science Integration Transparency 8** as you teach this lesson.

Science Journal

Mnemonic Devices Have the students write a plan to help them remember that Charles's law relates temperature and volume and Boyle's law relates pressure and volume. *Plans may include relating the letter C to both Charles and Celsius, reflecting temperature. Or, when someone is under pressure, his or her emotions may be near the boiling point, relating pressure and Boyle's law.*

---

2 Teach

Demonstration

Visual-Spatial Put 20 mL of water into an empty aluminum soft-drink can, and place it on a hot plate to boil. After the can has filled with steam, grasp it with tongs or a hot pad and plunge the can inverted into a beaker of ice water. The sudden drop of pressure inside the can will crush the can. **LEP**

INTEGRATION
Earth Science

Although the troposphere is important, the stratosphere should not be neglected. This is where most of the protective ozone layer is found. Ozone absorbs most of the harmful ultraviolet rays striking Earth from the sun.

Visual Learning

Figure 8-15 Why does the bottle have less volume in the bottom photo? *The pressure of the air inside is less than the pressure outside.* **LEP**

GLENCOE TECHNOLOGY

Videodisc

Glencoe Physical Science Interactive Videodisc
Side 1, Lesson 3
Boyle's Law

24161

24163-24258

24260-24491

3 Assess

Check for Understanding

? Flex Your Brain

Use the Flex Your Brain activity to have students explore BEHAVIOR OF GASES.

**Activity Worksheets,** p. 5

Reteach

Have students seal an air sample in a plastic bag in the freezer overnight. Remove the bag, and with the bag sealed, warm it with a hair dryer. Have students explain their observations.

Extension

For students who have mastered this section, use the **Reinforcement** and **Enrichment** masters.

Use Lab 19 in the **Lab Manual** as you teach this lesson.

USING TECHNOLOGY

Gaseous Giants

Imagine a balloon so immense that, when fully inflated, it will hold a volume of gas equivalent to 168 Goodyear blimps! Scientists are sending such balloons into the stratosphere, the layer of the atmosphere that is above 99 percent of Earth's air. These giant balloons are filled with helium gas, which gives them their lift. One of these balloons may hoist a load higher than a cruising jumbo jet, but lower than an orbiting satellite in space. The initial volume of helium required to lift the load might be the size of a small house. In the stratosphere, the volume may reach the size of 283 of the houses!

Scientists use the balloons for research. For example, a load of sensing equipment may gather data on how ozone gas is formed in the upper atmosphere. Ozone helps protect life by absorbing a dangerous form of energy given off by the sun. Because the sky is clear in the stratosphere, another load may be a telescope for studying the energy given off by the sun and other stars. Such studies help scientists understand the makeup and behavior of matter.

Think Critically:

Using the kinetic theory and Boyle's law, explain why the balloon gets so much larger as it rises through the atmosphere.

Research balloon

Charles's Law

If you've seen a hot-air balloon being inflated, you know gases expand when heated. Jacques Charles (1742-1823) was a French scientist who studied gases. According to **Charles's law,** the volume of a gas increases with increasing temperature, provided the pressure does not change. As with Boyle's law, the reverse is true, also. The volume of a gas shrinks with decreasing temperature. **Figure 8-16** demonstrates Charles's law.

Using his law, Charles was able to calculate the temperature at which a gas would have a volume of zero. The kinetic theory would say that this is the temperature at which all particle motion of matter should stop. Charles found this temperature to be –273°C, or 0 K, also called absolute zero. In reality, gases cannot be cooled to zero volume. Instead, they condense to liquids when cooled below their boiling points.

Integrating the Sciences

Earth Science Bring an aneroid barometer to class. Have the students record the pressure each day and the outside weather conditions for several days. Ask them to search the data for correlations. They may observe that on days of low pressure, it tends to be cloudy with precipitation, while on days of high pressure, it may be clear. L1

Figure 8-16

Hot-air balloonists use Charles's law with every launch. As the temperature of the trapped air is increased, the volume of the balloon also increases.

You can explain Charles's law by using the kinetic theory of matter. As a gas is heated, its particles move faster and faster and its temperature increases. Because the gas particles move faster, they begin to strike the walls of their container more often and with more force. If the walls are free to move, the gas pushes the walls out and expands.

Section Wrap-up

Review

1. When you bounce a basketball on the floor, the air pressure inside the ball increases for a moment. Explain why this increase occurs.

2. Why does a closed, empty, 2-L, plastic soft-drink bottle cave in when placed in a freezer?

3. **Think Critically:** Labels on cylinders of compressed gases state the highest temperature to which the cylinder may be exposed. Give a reason for this warning.

Skill Builder
Hypothesizing
A bottle of ammonia begins to leak. An hour later, you can smell ammonia almost everywhere, especially near the bottle. State a hypothesis to explain your observations. If you need help, refer to Hypothesizing in the **Skill Handbook.**

Using Computers

Spreadsheet Once you have completed Activity 8-2 on pages 232 and 233, use the temperature and volume in a spreadsheet. Generate a graph from the spreadsheet data.

8-4 Behavior of Gases **231**

Skill Builder
The bottle of ammonia gives off ammonia gas. Gases are made of particles that spread farther and farther apart as they move from a bottle into a room. Thus, the ammonia is more concentrated near the bottle and less concentrated farther away.

Assessment

Portfolio After students have stated their hypotheses, have each of them explain an experiment that would show a way to test the hypothesis. Use the Performance Task Assessment List for Designing an Experiment in **PASC,** p. 23. P

•MINI•QUIZ•

Use the Mini Quiz to check students' recall of chapter content.

1. Force per unit area is a measure of _____ . *pressure*

2. Average sea-level air pressure is _____ kPa. *101.3*

3. With temperature held constant, when the volume of a gas increases, the pressure will _____ . *decrease*

4. With pressure held constant, as a gas is heated, the volume will _____ . *increase*

Section Wrap-up

Review

1. The ball momentarily has smaller volume, resulting in a momentary increase in pressure.

2. The air particles in the bottle slow down as the temperature drops. Thus, they collide less with the bottle. The pressure inside the bottle decreases, but the atmospheric pressure remains the same, causing the bottle to decrease in volume.

3. **Think Critically** With increasing temperatures, gas particles move faster, striking the walls of their container more often. The pressure builds up and the cylinder may explode.

Using Computers

Check to see that students have generated a spreadsheet and graph from their lab data. The graph should approximate a straight line.

Activity 8-2

PREPARATION

Purpose

LS **Logical-Mathematical** Design and carry out an experiment to show the relationship between temperature and volume of a gas. **L1** **COOP LEARN**

Process Skills

observing and inferring, recognizing cause and effect, measuring, interpreting data, hypothesizing, making and using graphs, making and using tables, separating and controlling variables, communicating, designing an experiment

Time

40 minutes

Materials

See Student List.

Safety Precautions

Caution students to wear goggles. Do not heat balloons directly with a hot plate or flame or immerse balloons in hot water. When working with heat or steam, use tongs and heatproof gloves.

Possible Hypotheses

If the temperature of a gas sample is increased or decreased, then its volume will also increase or decrease.

📁 **Activity Worksheets,** pp. 5, 46, 47

PLAN THE EXPERIMENT

Possible Procedures

1. Inflate the two balloons to the same size and tie them closed.

2. Use string to determine the circumference of each balloon. Find the volume of each and record the room temperature. Mark dots on the balloon so that you can always put the string around the

---

Design Your Own Experiment

More Than Just Hot Air

The rapidly moving, small particles in a sample of a gas, such as air, respond easily to changes in temperature. In this activity, you will investigate the nature of that response. Does the volume expand or contract when the temperature of a gas is raised? Can you find a regular pattern when you examine several measurements?

PREPARATION

Problem

How does changing the temperature of a gas sample change its volume?

Form a Hypothesis

State your hypothesis about how the volume of a gas changes when its temperature changes.

Objectives

• Observe the changes in the volume of a gas as the temperature of the gas is changed.

• Design an experiment that tests the effects of changing temperature on the volume of a gas in a balloon.

Possible Materials

• string
• marking pen
• meterstick
• bucket of ice
• tongs
• heat-proof gloves
• beaker of boiling water
• thermometer
• safety goggles
• 2 medium, round balloons
• hot plate

Safety Precautions

Wear goggles and use heat-proof gloves or tongs to handle the balloons near heated air or steam.

---

same place on the balloon when other measurements are needed.

3. Use one balloon to place the trapped air in a colder temperature. Determine both the colder temperature and the resulting volume.

4. Use the other balloon to determine the change in the volume of the air at a hotter temperature.

Teaching Strategies

Be sure to place a dot on the inflated balloon so that the string will be measured around the same place on the balloon when the temperature is changed.

You may want to assist students with finding the volume of the balloon; $V = C^3/59$. This formula is derived from the formulas $V = (4/3) \pi r^3$ and $r = C/(2\pi)$.

PLAN THE EXPERIMENT

1. As a group, agree upon and write out your hypothesis statement.
2. As a group, list the steps that you need to take to test your hypothesis. Be specific, describing exactly how you will treat the balloon with various temperatures and how long you will use each treatment.
3. Make a list of the materials that you will need to complete your experiment.
4. Before you begin, make a table to record your data in your Science Journal.

Check the Plan

1. Read over your entire experiment to make sure that all steps are in logical order. Identify how and when you will check temperatures and changes in the size of the balloon.

2. Identify any constants and the variables of the experiment.
3. Have you allowed for a control? How will the control be treated?
4. Do you have to run any tests more than one time?
5. Will the data be summarized in a graph?
6. *Make sure your teacher approves your plan and that you have included any changes in the plan.*

DO THE EXPERIMENT

1. Carry out the approved experiment as planned.
2. While the experiment is going on, write down any observations that you make and complete the data table in your Science Journal.

Analyze and Apply

1. **Compare** your volume and temperature results with those of other groups.

2. How did the balloon's volume change when temperature changed?
3. Use a line **graph** to display your results and those of other groups.
4. From the graph, what can you **conclude** about the relationship between temperature and gas volume?

Go Further

If you extended your graph line to "zero volume," what temperature value would you find?

8-4 Behavior of Gases   233

DO THE EXPERIMENT

Expected Outcome

Temp.	Circumfer-ence	Volume
24°C	36 cm	780 cm^3
1°C	33 cm	609 cm^3
98°C	42 cm	1256 cm^3

Analyze and Apply

1. Student results should be consistent with those of other groups.
2. The volume of the gas increased as the temperature was raised. The volume decreased as the temperature was decreased.
3. Check graphs for straight line fit.
4. There is a direct relationship between temperature and volume of a gas. As temperature increases, so does the volume of gas.

Go Further

The graph suggests a direct relationship between the temperature and volume of a gas. Extending to zero volume results in a predicted temperature of approximately −273°C. Ask the students to give possible reasons if their data does not produce a temperature of −273°C.

✓ Assessment

Oral Ask students to explain the "Do not incinerate; contents under pressure" caution that is found on many aerosol cans. Use the Performance Task Assessment List for Making Observations and Inferences in **PASC**, p. 17. **P**

Prepare

Section Background

An object's density determines whether it will float. Bernoulli's principle applies to both liquids and gases.

Preplanning

The MiniLAB requires 2-L plastic bottles and droppers.

1 Motivate

Bellringer

 Before presenting the lesson, display **Section Focus Transparency 34** on the overhead projector. Assign the accompanying **Focus Activity** worksheet. L1 LEP

Demonstration

 **Visual-Spatial** Place an egg in a beaker of tap water. It sinks. Place one in a saturated salt solution. It floats. Discuss swimming in a pool and in the ocean as it relates to buoyancy. LEP

NATIONAL GEOGRAPHIC SOCIETY

 **Videodisc**

Newton's Apple: Physical Science
Buoyancy
Chapter 1, Side A
Buoyancy principle

4355-5410
Why huge ships float

8175-9712

Science Words

buoyant force
Archimedes' principle
Pascal's principle
Bernoulli's principle
Venturi effect

Objectives

- State Archimedes' principle and predict whether an object will sink or float in water.
- State Pascal's principle and describe the operation of a machine that uses Pascal's principle.
- State Bernoulli's principle and describe a way that Bernoulli's principle is applied.

Archimedes' Principle

Have you ever relaxed by floating quietly on your back in a swimming pool? You seem weightless as the water supports you. If you climb slowly out of the pool, you feel as if you gain weight. The farther out you climb, the more you have to use your muscles to support yourself. When you were in the pool, you experienced buoyancy. Buoyancy is the ability of a fluid—a liquid or a gas—to exert an upward force on an object immersed in it. This force is called **buoyant force.**

The amount of buoyant force determines whether an object will sink or float in a fluid. If the buoyant force is less than the object's weight, the object will sink. If the buoyant force equals the object's weight, as in **Figure 8-17**, the object floats. Sometimes, the buoyant force on an object is greater than its weight. This force is what seems to pull a helium-filled balloon upward in the air. When the balloon is released, the unbalanced buoyant force causes the balloon to accelerate upward.

Figure 8-17

There are different forces acting on a person lying on the ground beside a pool and in water.

Program Resources

 Reproducible Masters
Study Guide, p. 38 L1
Reinforcement, p. 38 L1
Enrichment, p. 38 L3
Critical Thinking/Problem Solving, p. 14 L1
Activity Worksheets, p. 50 L1
Technology Integration, pp. 11, 12

Transparencies
Teaching Transparency 16
Section Focus Transparency 34

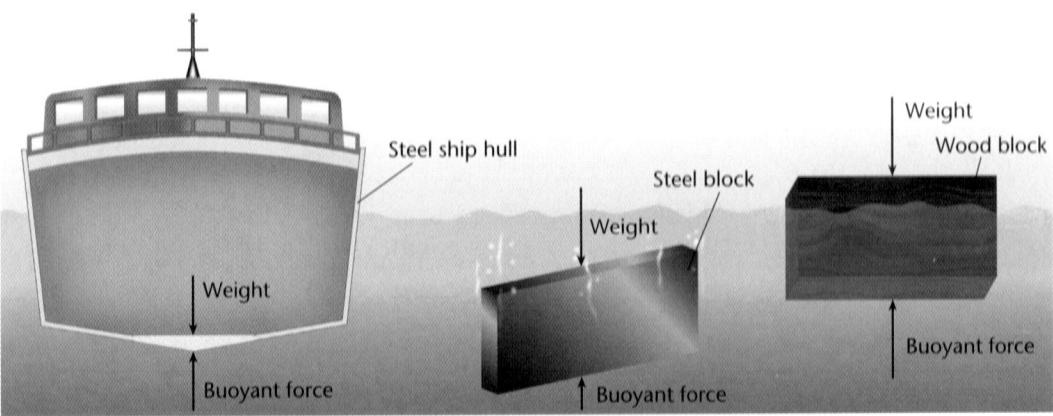

Figure 8-18

Notice that the weight and buoyant forces balance for floating objects. *Which of these objects have balanced forces?*

How do ships float?

Archimedes, a Greek mathematician who lived in the third century B.C., made a discovery about buoyancy. According to **Archimedes' principle,** the buoyant force on an object in a fluid is equal to the weight of the fluid displaced by the object. If you place a block of pine wood in water, it will push water out of the way as it begins to sink—but only until the weight of the water it displaces equals the block's weight. The block floats at this level, as shown in **Figure 8-18.**

Suppose you drop a solid steel block the same size as the wood block into water. When the steel block is placed into the water, the steel block begins to push aside water as it sinks. Buoyant force begins to push up on the block. However, the density of steel is much greater than that of wood. As a result, the buoyant force never becomes great enough to equal the weight of the steel block, and it sinks to the bottom.

How, then, does a steel ship float? Suppose you formed the steel block into a large, hollowed-out bowl shape. As this shape sinks into water, it displaces much more water than the solid block. Soon, it displaces enough water to equal the weight of the steel, and it floats.

Pascal's Principle

If you dive underwater, you can feel the pressure of the water all around you. You live at the bottom of Earth's atmosphere, which also is a fluid that exerts a pressure.

MiniLAB

How does applied pressure affect different areas of a fluid?

Procedure
1. Draw water into a small medicine dropper and place it into a 2-L, clear, soft-drink bottle that is filled to within 1 cm of the top.
2. Draw in or expel water from the dropper until it just barely floats.
3. Place the soft-drink bottle lid tightly back on the bottle.

Analysis
1. How does adding water inside the dropper affect its overall density?
2. When you squeeze the sides of the soft-drink bottle, what effect do you notice on the water level inside the dropper?
3. In your Science Journal, describe how Pascal's principle explains the movement of water inside the contained dropper.

8-5 Uses of Fluids   **235**

 Use **Teaching Transparency 16** as you teach this lesson.

235

Demonstration

Discuss with your class how each of the following demonstrations makes Archimedes' principle visible. Use the principle to explain why the various objects sink or float.

LS Visual-Spatial Place an unopened aluminum can of diet cola and one of non-diet cola in a water-filled aquarium. The can of sugar-free cola is less dense and will float.

LS Visual-Spatial Place an ice cube in a beaker of water. It floats. Place an ice cube in a beaker of rubbing alcohol. It sinks.

CONNECT TO
PHYSICS

See student page for answer.

📁 Use these program resources as you teach this lesson.
Critical Thinking/Problem Solving, p. 14
Technology Integration, pp. 11, 12

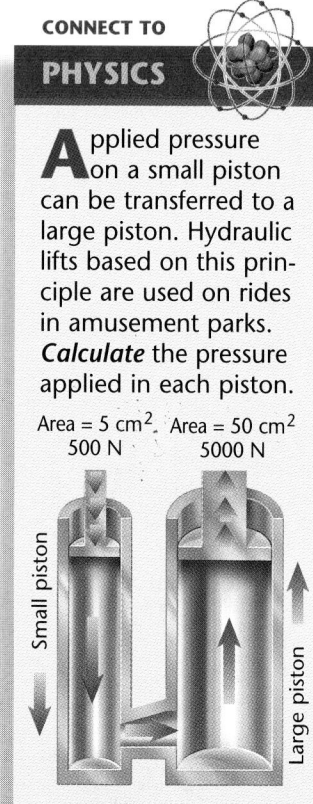

CONNECT TO
PHYSICS

Applied pressure on a small piston can be transferred to a large piston. Hydraulic lifts based on this principle are used on rides in amusement parks. *Calculate* the pressure applied in each piston.

Area = 5 cm²   Area = 50 cm²
500 N   5000 N

Small piston

Large piston

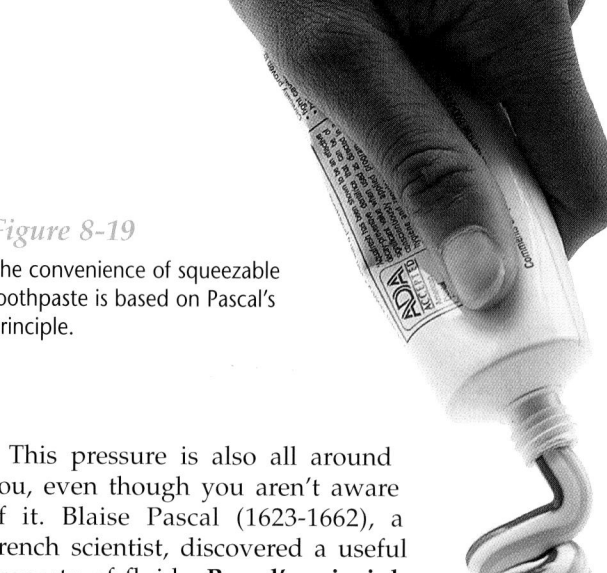

Figure 8-19

The convenience of squeezable toothpaste is based on Pascal's principle.

This pressure is also all around you, even though you aren't aware of it. Blaise Pascal (1623-1662), a French scientist, discovered a useful property of fluids. **Pascal's principle** states that pressure applied to a fluid is transmitted unchanged throughout the fluid. For example, when you squeeze one end of a balloon, the balloon pops out on the other end. When you squeeze one end of a toothpaste tube, as in **Figure 8-19,** toothpaste emerges from the other end.

Applying the Principle

Hydraulic machines that move heavy loads use Pascal's principle. Maybe you've seen a car raised using a hydraulic lift. Look at the hydraulic machine to the left to see how a machine of this type works. A small cylinder and a large cylinder are connected by a pipe. The cross section of the small cylinder has an area of 5 cm². The cross section of the large cylinder is 50 cm². Each cylinder is filled with a hydraulic fluid, usually oil, and has a piston that rests on the oil's surface.

Suppose you apply 500 N of force to the small piston. The pressure on the small piston is expressed in this equation.

$$P = \frac{F}{A} = \frac{500 \text{ N}}{5 \text{ cm}^2} = \frac{100 \text{ N}}{\text{cm}^2}$$

Pascal's principle says that this pressure is transferred unchanged throughout the liquid. Therefore, the large piston will also have a pressure of 100 N/cm² applied to it. But the area of the large piston is 50 cm². So, the total force on the large piston is 100 N/cm² × 50 cm² = 5000 N. With this hydraulic machine, you could use your weight to lift something ten times as heavy as you are.

Cultural Diversity

Kayaks and Boomerangs Many cultures have understood the principles discussed in this chapter, even if they did not scientifically name them. Boomerangs make use of Bernoulli's principle. The boomerang is like a curved wing, and its differences in thickness allow it to fly. Several cultures, including some Native Americans, used non-returning boomerangs to hunt birds and small animals. The most famous boomer-angs are those developed by societies in eastern and western Australia. These form either a deep, even curve or a straight-sided angle with the ends of the arms twisted in opposite directions. These shapes affect the spin of the flight so that the boomerang completes a circle approximately 45 m wide, then several small circles as it returns to the thrower.

continued on p. 237

Bernoulli's Principle

It took humans thousands of years to learn to do what birds do naturally—fly, glide, and soar. Obviously, it was no simple task to build a machine that could lift itself off the ground and fly with people aboard. The ability of an airplane to rise into the air is an example of another property of fluids stated in Bernoulli's principle. Daniel Bernoulli (1700-1782) was a Swiss scientist who studied the properties of moving fluids such as water and air. He published his discovery in 1738. According to **Bernoulli's principle,** as the velocity of a fluid increases, the pressure exerted by the fluid decreases.

Figure 8-20 shows Bernoulli's principle at work. Another way to demonstrate Bernoulli's principle is to blow across the surface of a sheet of paper. The paper will rise. The velocity of the air you blew over the top surface of the paper is greater than that of the quiet air below it. As a result, the downward air pressure above the paper decreases. The higher air pressure below the paper exerts a net force that pushes the paper upward.

> **Figure 8-20**
>
> Blowing between two empty aluminum cans can cause them to be forced together.

A The moving air between the cans exerts less pressure than the air on either side of each can at that moment.

B As a result, the cans move together because of the increased pressure of the air around the cans.

8-5 Uses of Fluids   **237**

3 Assess

Check for Understanding

? Flex Your Brain

Use the Flex Your Brain activity to have students explore PROPERTIES OF FLUIDS.

📁 **Activity Worksheets,** p. 5

Reteach

LS Linguistic Have students test several objects that sink or float in water. Encourage students to explain why each object sinks or floats. Have students use density, volume, and mass in their explanations. **LEP**

Extension

📁 For students who have mastered this section, use the **Reinforcement** and **Enrichment** masters.

Cultural Diversity

continued from p. 236

Kayaks make use of buoyancy principles. These canoes, still used by the Inuit, also known as the Eskimo, skim the water and are ideal for hunting because they are virtually noiseless. The Inuit traditionally used kayaks for hunting seals in the sea and caribou swimming in lakes and rivers. Seal or other animal skins are shrunk over a wooden frame, and the canoe is propelled by a double-bladed paddle. The canoe is completely enclosed except for a center cockpit. The air inside the enclosed shell gives the vessel buoyancy.

4 Close

•MINI•QUIZ•

Use the Mini Quiz to check students' recall of chapter content.

1. The cause of buoyant force was first explained by _____ . *Archimedes*
2. The discovery that pressure is exerted in all directions throughout a fluid was made by _____ . *Pascal*
3. Hydraulic jacks that move heavy loads make use of _____ principle. *Pascal's*

Now, look at the curvature of the airplane wing in **Figure 8-21.** As the plane moves forward, the air passing over the wing must travel farther than air passing below it. To take the same time to get to the rear of the wing, air must travel faster over the top of the wing than below it. Thus, the pressure above the wing is less than the pressure below it. The result is a net upward force on the wing, lifting the airplane in flight.

Figure 8-21

The design of airplane wings is an application of Bernoulli's principle. Note the curvature of the upper portion. *Which side has more pressure exerted on it?*

Did you know that baseball pitchers use science? If the pitcher puts a spin on the ball, air moves faster along the side of the ball that is spinning away from the direction of the throw. Look at **Figure 8-22.** The ball's spin results in a net force that pushes the ball toward the low-pressure direction. The force gives the ball a curve.

Windy Cities

Fluids flow faster when forced to flow through narrow spaces. As a result of this speed increase, the pressure of the fluid drops. This reduction in pressure in these spaces is a special case of Bernoulli's principle called the **Venturi effect.**

Integrating the Sciences

Earth Science Have a student form a sloping streambed out of wet sand. Have the width of the streambed vary. Notice that water speeds up as it passes through the narrow area (Venturi effect). Have students notice the rate of erosion of the banks. LS

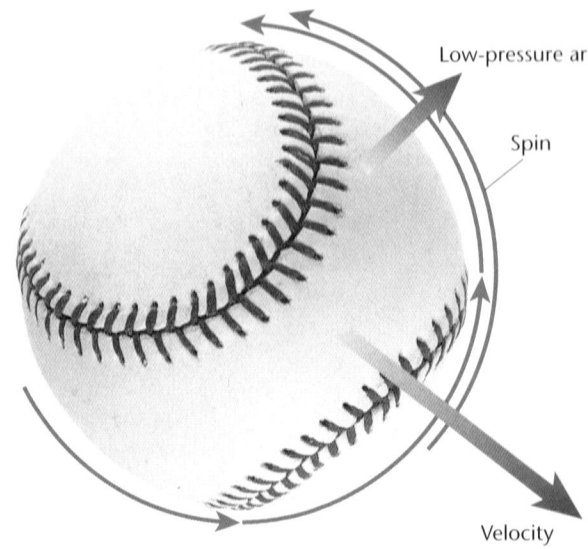

Low-pressure area

Spin

Velocity

Figure 8-22

Air moves faster along the side of a spinning ball that is moving away from the direction of the throw.

A dramatic demonstration of the Venturi effect has occurred in cities where the wind is forced to blow between rows of skyscrapers. The reduced air pressure outside of buildings during strong winds has caused windows to be forced out by the higher air pressure inside the buildings.

Section Wrap-up

Review

1. How is it possible that a boat made of concrete can float?

2. State Bernoulli's principle and tell how it explains the lift on an airplane.

3. **Think Critically:** If you fill up a balloon with air, tie it off, and release it, it will fall to the floor. Why does it fall instead of float?

Skill Builder
Measuring in SI

The density of water is 1.0 g/cm³. How many kilograms of water does a submerged 120-cm³ block displace? One kilogram weighs 9.8 N. What is the buoyant force on the block? If you need help, refer to Chapter 2 and to Measuring in SI in the **Skill Handbook.**

USING MATH

If you wanted to lift an object weighing 20 000 N, how much force would you have to apply to the small piston on page 236?

Section Wrap-up

Review

1. Concrete shaped into a hollowed-out boat displaces enough water to equal its weight.

2. As the velocity of a fluid increases, the pressure exerted by the fluid decreases. Air is a fluid. An airplane's wing is shaped such that air flows at a greater velocity over its upper surface. The difference in pressure above and below the wing creates a net upward force that helps lift the plane.

3. **Think Critically** The air in the balloon is compressed compared to normal air and thus has a greater density. The weight of the denser air along with the weight of the rubber itself is greater than the buoyant force of the surrounding air. Therefore, the balloon sinks.

USING MATH

The machine in the Connect to Physics on p. 236 multiplies force by ten. Therefore, a force of 2000 N is applied to the small piston.

Skill Builder

A submerged block of 120 cm³ displaces 120 cm³ of water. The mass of water displaced is, therefore, 120 cm³ × 1.0 g/cm³ = 120 g. Converting to kilograms, 120 g ÷ 1000 g/kg = 0.12 kg. The buoyant force on the block is then 0.12 kg × 9.8 N/kg = 1.2 N.

Assessment

Performance Have students repeat the Skill Builder using, instead of water, salt water with a density of 1.3 g/cm³. By what percentage would the buoyant force be increased? *27% increase to 1.5 N* Use the Performance Task Assessment List for Using Math in Science in **PASC**, p. 29. **P**

Background

From the field of glass blowing come many colorful terms. Here are a few: *doghouse* – a place where a batch of glass is fed into a furnace, *gaffer* – the person in charge of a group of glass blowers, *punty* – a solid iron rod used to carry small amounts of glass.

The earliest hollow glass vessels date from 1500 B.C. and were made by winding hot glass around a core of clay, letting the glass cool, then removing the clay. The art of glass blowing began around 50 B.C. in the Roman Empire.

Teaching Strategies

Kinesthetic Invite students to collect materials that would otherwise be thrown away or recycled, and use the materials to create artwork. They may wish to arrange a show of their art for the school. L1

Career Connection

Career Path Artists usually take art classes if they are available in their high schools. Another possible place to find art instruction is a community recreation center. After high school, many artists enroll in art schools, such as Seattle's Pratt Art Center or the Pilchuck Glass School, both of which Ms. Moore attended.

For More Information

Students could investigate artistic careers by contacting an art school. For example, they could request a course catalog from the California College of Arts and Crafts, Broadway and College Avenue, Oakland, CA 94618.

240

People and Science

DEBRA MOORE, *Glass Artist*

On the Job

Q **Besides transforming materials from solid to liquid and back to solid, what scientific principles are involved in working with glass?**

A When I spin glass as I heat it, centrifugal force creates a round shape. Also, as I tell my students, gravity is the glassblower's best friend. If you want your molten glass in a stretched-out shape, you just hold it up until it's as long as you want.

Q **Do you teach glassblowing to young people?**

A I'm currently working with a group of 13- to 18-year-olds. They are turning old bottles they collect in their community into a beautiful chandelier. The young people are divided into teams of three. One team takes the solid blue water bottles we find, uses high heat to open them up, and then creates a fluted shape on the rim of the cups while the glass is in the liquid state. The glass is liquefied by breaking up one bottle and melting it on a long, flat shovel in the glory hole. The glory hole is a large, round oven heated to about 1000°F. Meanwhile, another team shapes glass leaves and flowers for decorations. Young people find the glassblowing studio an exciting place.

Personal Insights

Q **What fascinates you about glass?**

A I love the way glass looks in its glowing, molten form. It's like honey. Then, when it's blown and cooled, it has the appearance of a frozen liquid. I also love the colors glass comes in. Even the color names can inspire me—smoky topaz, opal lilac, reddish aurora.

Career Connection

Glass now can be made to have different specifications for different purposes. Chemists and engineers develop the processes used to make different kinds of glass. Investigate a few of the different types of glass and make a chart comparing and contrasting them. Research the following careers to find out how a knowledge of glassmaking is important.

- **Ceramic Engineer**
- **Industrial Designer**

Summary

8-1: Matter and Temperature
1. There are four states of matter: solid, liquid, gas, and plasma.
2. According to the kinetic theory, all matter is made of constantly moving particles.
3. Most matter expands when heated and contracts when cooled.

8-2: Science and Society: Fresh Water—Will there be enough?
1. Fresh water on Earth is scarce, and this water is often wasted or polluted.
2. There are many ways in which people can save water, and laws can be passed to regulate pollution.

8-3: Changes in State
1. Changes of state can be interpreted in terms of the kinetic theory of matter.
2. The energy of the heat of fusion and vaporization overcomes attractive forces between particles of matter.

8-4: Behavior of Gases
1. Gas pressure is caused by moving particles colliding with the inside walls of its container.
2. Boyle's law states that the volume of a gas decreases when the pressure increases, at constant temperature.
3. Charles's law states that the volume of a gas increases when the temperature increases, at constant pressure.

8-5: Uses of Fluids
1. Archimedes' principle states that the buoyant force on an object in a fluid is equal to the weight of the fluid displaced.

2. Pascal's principle states that pressure applied to a fluid is transmitted unchanged throughout the fluid.
3. Bernoulli's principle states that the pressure exerted by a fluid decreases as its velocity increases.

Key Science Words

a. Archimedes' principle
b. Bernoulli's principle
c. Boyle's law
d. buoyant force
e. Charles's law
f. condensation
g. crystal
h. evaporation
i. heat of fusion
j. heat of vaporization
k. kinetic theory of matter
l. pascal
m. Pascal's principle
n. plasma
o. polluted water
p. pressure
q. states of matter
r. thermal expansion
s. thermal pollution
t. Venturi effect

Reviewing Vocabulary

Match each phrase with the correct term from the list of Key Science Words.

1. The volume of a gas is reduced when the temperature is decreased.
2. water that exceeds government limits for impurities
3. liquid changes to gas below the boiling point
4. amount of force exerted per unit of area

Summary

Have students read the summary statements to review the major concepts of the chapter.

Reviewing Vocabulary

1. e	6. l
2. o	7. g
3. h	8. n
4. p	9. k
5. a	10. j

✓ Assessment

Portfolio Encourage students to place in their portfolios one or two items of what they consider to be their best work. Examples include:
- Across the Curriculum interpretation, p. 216
- Flex Your Brain, p. 226
- MiniLAB analysis, p. 235 **P**

Performance Additional performance assessments may be found in **Performance Assessment** and **Science Integration Activities**. Performance Task Assessment Lists and rubrics for evaluating these activities can be found in Glencoe's **Performance Assessment in the Science Classroom**.

GLENCOE TECHNOLOGY

 Videodisc

Glencoe Physical Science
Interactive Videodisc
Use the segments *Introduction to Hot Air Balloons, Hot Air Balloon Navigation, Warm Area (Parking Lot),* and *Cool Area (Lake)* from videodisc Lesson 3, *Behavior of Gases* to review the major concepts presented in the chapter.

MindJogger Videoquiz

Chapter 8 Have students work in groups as they play the Videoquiz game to review key chapter concepts.

Checking Concepts

1. a	6. a
2. d	7. a
3. c	8. b
4. b	9. d
5. c	10. c

Understanding Concepts

11. Some water is suitable for watering plants, gardens, and lawns. Some used wash water is suitable for presoaking clothes yet to be washed. Other answers are possible.

12. Because gases can be squeezed into smaller spaces, more gas can be added to a room, no matter how much is already present.

13. The hot tea transfers energy to particles in the glass and the ice. The tea cools because its particles now have a lower average kinetic energy. The glass warms as its particles absorb energy from the tea. The ice absorbs energy from the tea. Ice will melt after its temperature is raised to 0°C.

14. Water vapor in the air is cooled by the glass. The particles lose energy and condense.

15. The deeper water exerts a much greater buoyant force on your body, reducing the downward force of your feet on the rocks.

Thinking Critically

16. According to Charles's law, as the temperature of a gas changes, so does the volume. As the volume changes, so does the pressure, according to Boyle's law. Point out that bleeding hot tires to lower the pressure is not advisable.

---

5. The buoyant force on an object in a fluid equals the weight of the displaced fluid.
6. the SI unit of pressure
7. particles are arranged in regular patterns
8. a gaslike mixture of charged particles
9. Matter is made of tiny, moving particles.
10. energy needed for a liquid to boil

Checking Concepts

Choose the word or phrase that completes the sentence.

1. The temperature at which all particle motion of matter would stop is _____.
   a. absolute zero   c. 0°C
   b. its melting point   d. 273°C
2. The state of matter that has a definite volume and a definite shape is _____.
   a. gas   c. plasma
   b. liquid   d. solid
3. The most common state of matter is

   _____.
   a. gas   c. plasma
   b. liquid   d. solid
4. Most pressure is measured in _____.
   a. grams   c. newtons
   b. kilopascals   d. kilograms
5. Pascal's principle is the basis for _____.
   a. aerodynamics   c. hydraulics
   b. buoyancy   d. changes of state
6. Bernoulli's principle explains why _____.
   a. airplanes fly   c. pistons work
   b. boats float   d. ice melts
7. Particles separate completely from each other in a(n) _____.
   a. gas   c. solid
   b. liquid   d. amorphous material
8. The state of the matter in the sun and other stars is primarily _____.
   a. amorphous   c. liquid
   b. plasma   d. gas
9. In general, as a solid is heated, it _____.
   a. becomes a gas   c. contracts
   b. condenses   d. expands

---

10. A material's heat of fusion gives the amount of energy needed to _____.
    a. condense a gas   c. melt a solid
    b. boil a liquid   d. evaporate a liquid

Understanding Concepts

Answer the following questions in your Science Journal using complete sentences.

11. How might wastewater in the home be recycled instead of being poured into the sewer system?
12. Why would it be incorrect to say, "This room is full of air"?
13. What energy changes occur when hot tea is poured over ice?
14. Use the kinetic theory to explain why liquid water forms on the outside of a glass of cold lemonade.
15. Why might rocks in a creek hurt your feet more than the same rocks would in deeper water?

Thinking Critically

16. Use Charles's and Boyle's laws to explain why you should check your tire pressure when the temperature changes.
17. Explain how food might get freezer burn. How might you prevent it?
18. Alcohol evaporates more quickly than does water. What can you tell about the forces between the alcohol particles?

---

17. Ice in the food sublimes, drying out the food. This can be prevented by wrapping the food in airtight packaging such as foil or by using heavy plastic containers.

18. The forces between the alcohol particles are probably weaker than those between the water particles.

19. According to Charles's law, as temperature increases, gases tend to expand. However, because the can is not free to

expand, pressure builds up and the can may explode.

20. If the ice is just below its normal melting point, pressure, such as that applied by tires, will melt it and form a slippery layer of liquid water on the ice. At lower temperatures, the pressure of the tires may not be enough to melt the ice.

19. Why do aerosol cans have a "do not incinerate" warning such as the one in the photo on page 242?

20. Applying pressure lowers the melting point of ice. Why might an icy road at –1°C be more dangerous than an icy road at –10°C?

Developing Skills

If you need help, refer to the **Skill Handbook.**

21. **Observing and Inferring:** Infer the air pressure effect on a small car when a tractor-trailer passes it.

22. **Making and Using Graphs:** A group of students heated ice until it melted and then turned to steam. They measured the temperature each minute and graphed the results. Their graph is provided below. In terms of the energy involved, explain what is happening at each letter (a, b, c, d) in the graph.

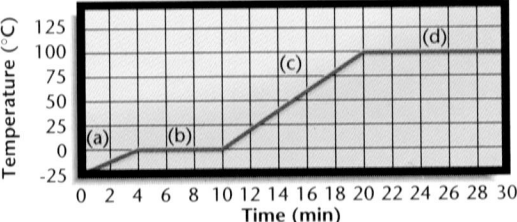

23. **Concept Mapping:** Use a cycle map to show the changes in particles as cool water boils, changes to steam, and then changes back to cool water.

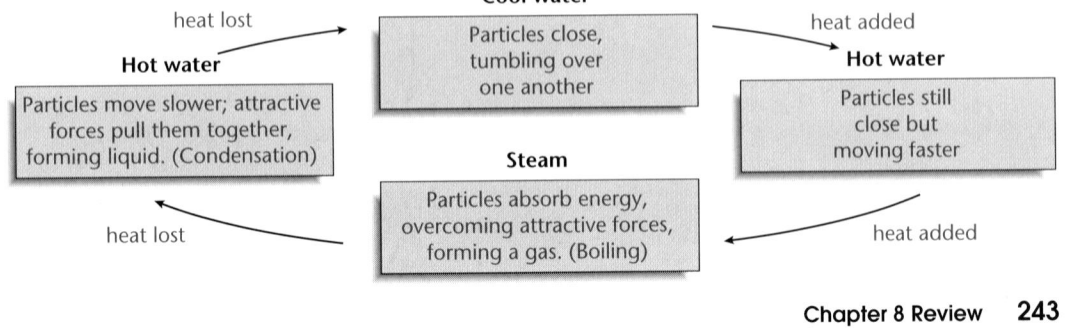

24. **Interpreting Data:** As elevation increases, boiling point decreases. List each of the following locations as *at sea level, above sea level,* or *below sea level.* (Boiling point of water in °C is given.)
Death Valley (100.3), Denver (94), Madison (99), Mt. Everest (76.5), Mt. McKinley (79), New York City (100), Salt Lake City (95.6)

25. **Recognizing Cause and Effect:** List possible effects for each of the following causes.
   a. Perspiration evaporates.
   b. Pressure on a balloon decreases.
   c. Your buoyant force equals your weight.

Performance Assessment

1. **Graph from Data:** Using your data from Activity 8-2, convert the temperature readings from Celsius to Kelvin and replot the data. Is the volume-temperature relationship the same?

2. **Report:** Research the effects of pressure changes on the human body and write a report. Include in your report any precautions that must be taken in dealing with pressure changes in space and when deep-sea diving.

3. **Display:** Research crystal growing and the conditions needed to grow a perfect crystal. Grow crystals of several different materials and display them.

Chapter 8 Review

Developing Skills

21. **Observing and Inferring** When the car and truck are side by side, the space between them is narrow. The air passing through this space must travel faster. This lowers the air pressure, and the higher pressure on the other side of the car pushes the car toward the truck.

22. **Making and Using Graphs**
   a. Ice is warming to its melting point.
   b. Ice is absorbing energy. Ice melts.
   c. Liquid water warms.
   d. Liquid absorbs energy and boils.

23. **Concept Mapping** See student page.

24. **Interpreting Data**
   Above: Denver, Madison, Mt. Everest, Mt. McKinley, Salt Lake City
   At: New York City
   Below: Death Valley

25. **Cause and Effect** Student answers may vary. Sample answers are given.
   a. Skin feels cooler.
   b. Volume increases.
   c. You float.

Performance Assessment

1. Check conversions; yes, the relationship is the same. Use the Performance Task Assessment List for Graph from Data in **PASC,** p. 39. **P**

2. Look for references to Boyle's and Charles's laws. Use the Performance Task Assessment List for Lab Report in **PASC,** p. 47. **P**

3. Ask students to identify the shapes of their crystals. Use the Performance Task Assessment List for Display in **PASC,** p. 63. **P**

Assessment Resources

📁 Reproducible Masters
Chapter Review, pp. 19, 20
Assessment, pp. 53-56
Performance Assessment, p. 22

Glencoe Technology
🔘 **Chapter Review Software**
🔘 **Computer Test Bank**
📼 **MindJogger Videoquiz**

Chapter Organizer

Section	Objectives/Standards	Activities/Features
Chapter Opener		Explore Activity: Making Concrete, p. 245
9-1 **Composition of Matter** (2½ sessions, 1½ blocks)*	1. **Distinguish** between substances and mixtures. 2. **Compare** and **contrast** solutions, colloids, and suspensions. National Content Standards: (5-8) UCP1, UCP2, A2; (9-12) UCP1, UCP2, A2	MiniLAB: How can mixtures be separated? p. 248 Using Math, p. 248 Connect to Life Science, p. 250 Problem Solving: Cooking a Colloid, p. 249 Science Journal, p. 250 Skill Builder: Comparing and Contrasting, p. 250 Activity 9-1: Elements, Compounds, and Mixtures, p. 251
9-2 **Science and Society** (1 session, ½ block)*	1. **Identify** two colloids related to air and water pollution. National Content Standards: (5-8) UCP1, UCP2, E2, F1-F3, F5; (9-12) UCP1, UCP2, E2, F1, F3-F6	Explore the Technology, p. 253
9-3 **Describing Matter** (2½ sessions, 1½ blocks)*	1. **Give examples** of physical properties. 2. **Distinguish** between physical and chemical changes. 3. **Distinguish** between chemical and physical properties. 4. **State** and **explain** the law of conservation of mass. National Content Standards: (5-8) UCP1-UCP3, UCP5, A1, A2, B1, E1, E2, F5; (9-12) UCP1-UCP3, UCP5, A1, A2, B2, D2, E1, E2, F6	Using Math, p. 254 Connect to Earth Science, p. 256 Using Technology: Aerogels, p. 257 Science Journal, p. 259 MiniLAB: What are some indications of changes? p. 260 Using Math, p. 261 Skill Builder: Observing and Inferring, p. 261 Activity 9-2: Checking Out Chemical Changes, pp. 262-263 Science and Art: Copper Art, p. 264 Science Journal, p. 264

* A complete Planning Guide that includes block scheduling is provided on pages 32T-35T.

Activity Materials

Explore	Activities	MiniLABs
page 245 crushed calcium carbonate, fine sand, clean small pebbles, water, plastic cup, goggles	page 251 plastic freezer bag containing tagged items: copper foil, small package of salt, piece of solder, aluminum foil, chalk (calcium carbonate) or baking soda (sodium hydrogen carbonate), piece of granite, sugar water in vial pages 262-263 safety goggles, baking soda, small evaporating dish, hand lens, dilute hydrochloric acid, 10-cm³ graduated cylinder, electric hot plate, apron, heat-proof glove	page 248 soil; clay; sand; gravel; pebbles; clear, plastic gallon jar; water page 260 250-mL beaker, water, potassium permanganate, sodium hydrogen sulfite, stirring rod

Need Materials? Call Science Kit (1-800-828-7777).

Teacher Classroom Resources

Reproducible Masters	Transparencies	Teaching Resources
Study Guide, p. 39 **Reinforcement,** p. 39 **Enrichment,** p. 39 **Activity Worksheets,** pp. 51-52, 55 **Lab Manual 20,** Chromatography **Multicultural Connections,** p. 21 **Science Integration Activity 9,** Eggs Up to You **Concept Mapping,** pp. 23-24	**Section Focus Transparency 35,** Types of Glass **Teaching Transparency 17,** Classification of Matter	**Glencoe Physical Science Interactive Videodisc Physical Science CD-ROM Spanish Resources English/Spanish Audiocassettes Cooperative Learning Resource Guide Lab Partner Lab and Safety Skills Lesson Plans**
Study Guide, p. 40 **Reinforcement,** p. 40 **Enrichment,** p. 40	**Section Focus Transparency 36,** Dialysis of Blood	**Assessment Resources**
Study Guide, p. 41 **Reinforcement,** p. 41 **Enrichment,** p. 41 **Activity Worksheets,** pp. 53-54, 56 **Lab Manual 21,** Properties of Matter **Cross-Curricular Integration,** p. 13 **Science and Society Integration,** p. 13 **Critical Thinking/Problem Solving,** p. 15	**Section Focus Transparency 37,** Making Pottery **Science Integration Transparency 9,** Physical and Chemical Weathering **Teaching Transparency 18,** Conservation of Mass	**Chapter Review,** pp. 21-22 **Assessment,** pp. 57-60 **Performance Assessment in the Science Classroom (PASC) MindJogger Videoquiz Alternate Assessment in the Science Classroom Performance Assessment Chapter Review Software Computer Test Bank**

Key to Teaching Strategies

The following designations will help you decide which activities are appropriate for your students.

[L1] Level 1 activities should be within the ability range of all students, including those with learning difficulties.

[L2] Level 2 activities should be within the ability range of the average to above-average student.

[L3] Level 3 activities are designed for the ability range of above-average students.

[LEP] LEP activities should be within the ability range of Limited English Proficiency students.

[LS] These activities are designed to address different learning styles.

[COOP LEARN] Cooperative Learning activities are designed for small group work.

[P] These strategies represent student products that can be placed into a best-work portfolio.

GLENCOE TECHNOLOGY

The following multimedia resources are available from Glencoe.

Science and Technology Videodisc Series (STVS)
Chemistry
Sandblasting with Dry Ice
Treating Acid Lakes

Glencoe Physical Science Interactive Videodisc
Carbohydrates and Hydrocarbons
Chemical Detectives

Physical Science CD-ROM

National Geographic Society Series
STV: Water

Teacher Classroom Resources

This is a representation of key blackline masters available in the Teacher Classroom Resources.

Teaching Aids

Section Focus Transparencies

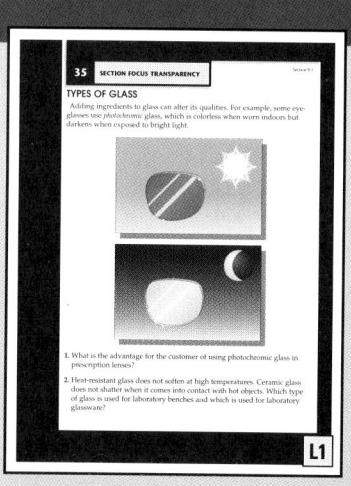

35 SECTION FOCUS TRANSPARENCY

TYPES OF GLASS

L1

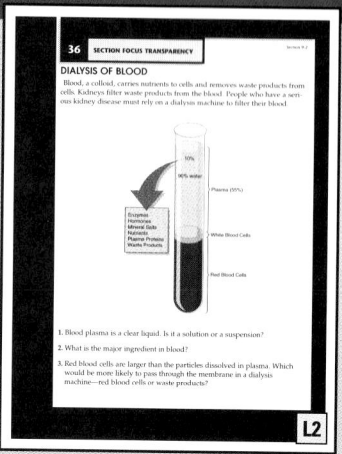

36 SECTION FOCUS TRANSPARENCY

DIALYSIS OF BLOOD

L2

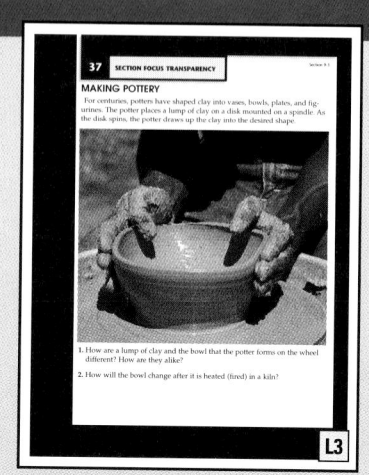

37 SECTION FOCUS TRANSPARENCY

MAKING POTTERY

L3

Science Integration Transparencies

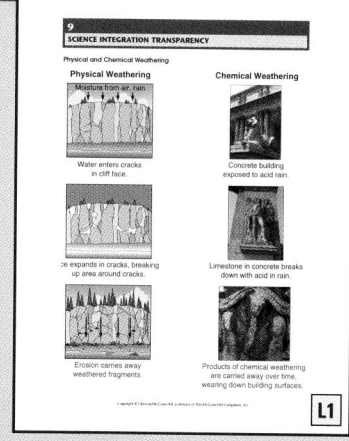

9 SCIENCE INTEGRATION TRANSPARENCY

Physical and Chemical Weathering

L1

Teaching Transparencies

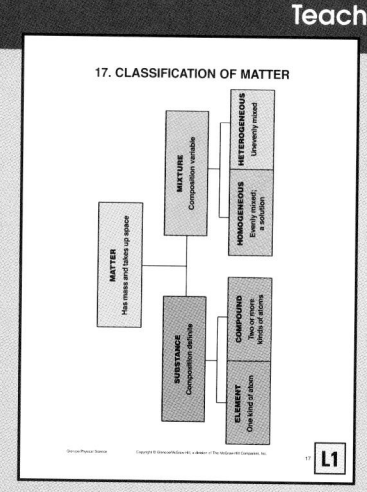

17. CLASSIFICATION OF MATTER

L1

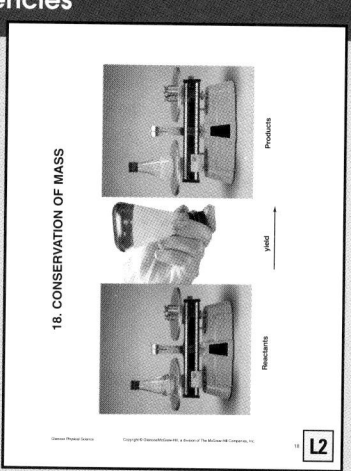

18. CONSERVATION OF MASS

L2

Meeting Different Ability Levels

Study Guide

Chapter 9 STUDY GUIDE • Composition of Matter

L1

Reinforcement

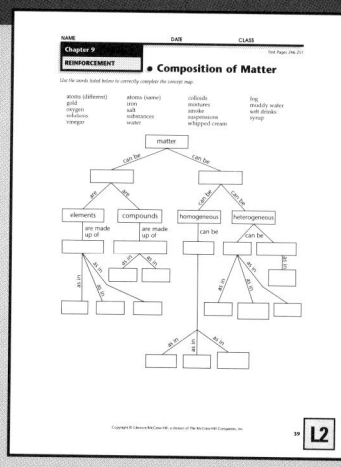

Chapter 9 REINFORCEMENT • Composition of Matter

L2

Enrichment Worksheets

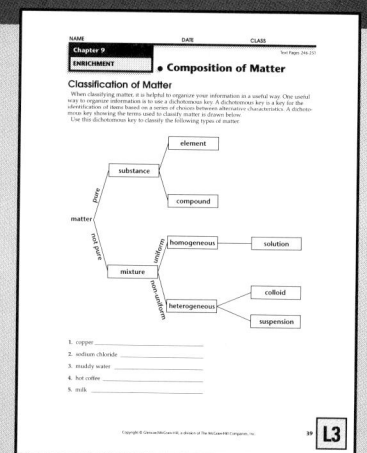

Chapter 9 ENRICHMENT • Composition of Matter

Classification of Matter

L3

244C

Hands-On Activities

Science Integration Activity
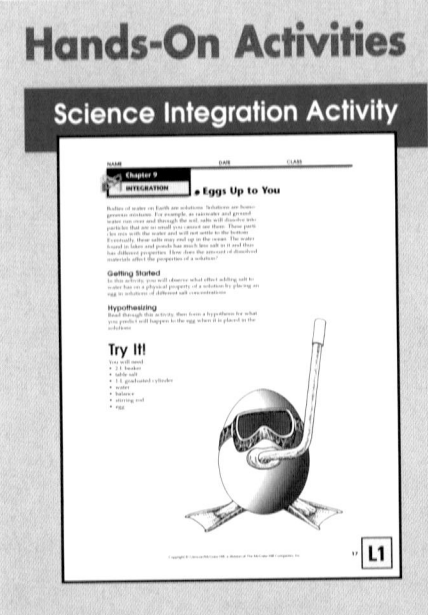

Lab Manual
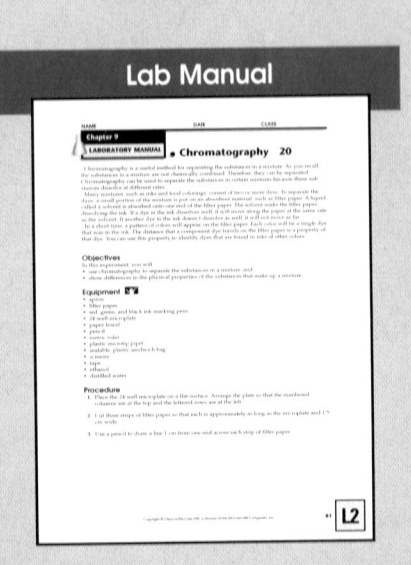

Assessment

Performance Assessment
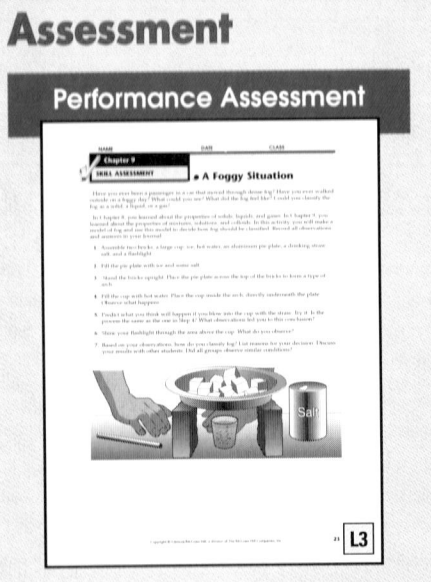

Enrichment and Application

Critical Thinking/Problem Solving
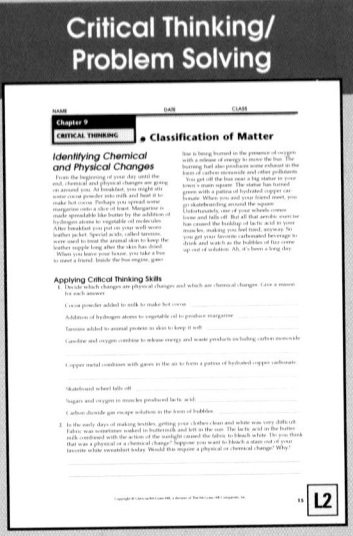

Cross-Curricular Integration
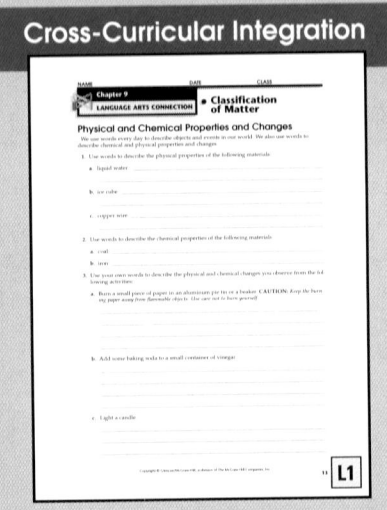

Science and Society Integration
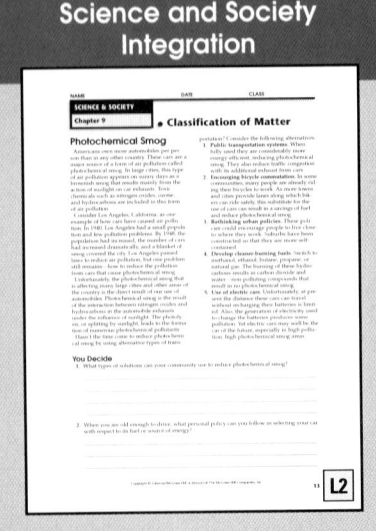

Multicultural Connections
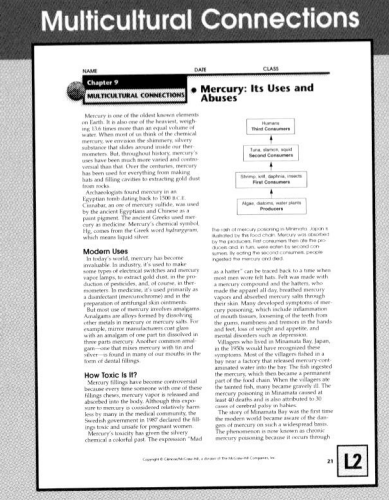

Concept Mapping
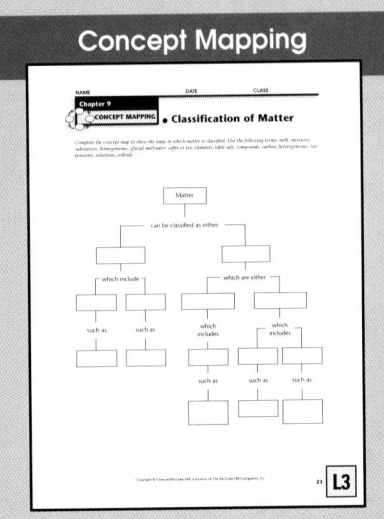

Classification of Matter

CHAPTER OVERVIEW

Section 9-1 This section introduces the student to the vocabulary used by scientists when classifying matter. Also, it distinguishes between true substances and various kinds of mixtures.

Section 9-2 Science and Society Students learn how colloids can be used to purify water. However, some colloids can be an air pollution problem.

Section 9-3 Physical and chemical properties and physical and chemical changes are described and related to the students' daily living in this section.

Chapter Vocabulary

element	compound
substance	solution
colloid	suspension
heterogeneous mixture	
homogeneous mixture	
Tyndall effect	
coagulation	
physical property	
physical change	
chemical change	
chemical property	
law of conservation of mass	

Theme Connection

Stability and Change In Chapter 9, the student learns to classify matter by observing the patterns that occur when it undergoes physical and chemical change to become more stable.

Previewing the Chapter

244

Learning Styles

Look for the following logo for strategies that emphasize different learning modalities. **LS**

Visual-Spatial	Making a Model, p. 247; Reteach, pp. 249, 253; Visual Learning, pp. 247, 248, 253, 255, 256, 258, 260; Revealing Preconceptions, p. 256; Demonstration, pp. 247, 256, 258; MiniLAB, p. 260
Interpersonal	Explore, p. 245; Activity 9-1, p. 251; Activity 9-2, pp. 262-263; Reteach, p. 259
Logical-Mathematical	Enrichment, p. 257; Science & Art, p. 264
Kinesthetic	MiniLAB, p. 248
Linguistic	Science & Art, p. 264

Chapter 9

Classification of Matter

Did you ever watch a construction crew pour concrete for a building? What is concrete? Actually, you can't classify fresh concrete as only one of the states of matter. Rather, it is a mixture of solids—portland cement, sand, rocks—and a liquid, water. Do the following activity to make some classroom concrete and observe its properties. As you study the chapter, you'll find out more about this and other matter around you.

EXPLORE ACTIVITY

Making Concrete

1. Measure 10 g crushed calcium carbonate; 20 g fine sand; 30 g clean, small pebbles; and 5 g water into a plastic cup.
2. Stir the materials together thoroughly.
3. Allow the combined materials to dry in the cup overnight.

Observe: In your Science Journal, describe any changes you noticed in the stirred materials after they dried. Discuss whether your product still would be concrete if you changed the amounts of the materials.

Previewing Science Skills

▶ In the Skill Builders, you will compare and contrast and observe and infer.

▶ In the Activities, you will classify, make and use tables, observe, and hypothesize.

▶ In the MiniLABs, you will observe, infer, and compare.

245

Assessment Planner

Portfolio
Refer to page 265 for suggested items that students might select for their portfolios.

Performance Assessment
See page 265 for additional Performance Assessment options.
Skill Builders, pp. 250, 261
MiniLABS, pp. 248, 260
Activities 9-1, p. 251; 9-2, pp. 262-263

Content Assessment
Section Wrap-ups, pp. 250, 253, 261
Chapter Review, pp. 265-267
Mini Quizzes, pp. 250, 261

Group Assessment
Opportunities for group assessment occur with Cooperative Learning Strategies and Flex Your Brain Activities.

245

Prepare

Section Background

- In science, the word *substance* is limited to elements and compounds. More than 9 million substances are known to chemists. A system of classification is necessary.

- A heterogeneous mixture is one that is composed of more than one phase. A phase is any region with a uniform set of properties. Colloids are composed of two phases—the dispersed phase and the continuous phase.

Preplanning

To prepare for Activity 9-1, label seven items and place them in a plastic bag for each activity group.

1 Motivate

Bellringer

Before presenting the lesson, display **Section Focus Transparency 35** on the overhead projector. Assign the accompanying **Focus Activity** worksheet. [L1] [LEP]

Tying to Previous Knowledge

Ask a student to describe the phases present in an ice cream soda. The ice cream, flavored syrup, soda water, whipped cream, and cherry will probably be mentioned. Compounds, heterogeneous mixtures, homogeneous mixtures, and colloids are present in the ice cream soda.

9•1 Composition of Matter

Science Words

element
compound
substance
heterogeneous mixture
homogeneous mixture
solution
colloid
Tyndall effect
suspension

Objectives

- Distinguish between substances and mixtures.
- Compare and contrast solutions, colloids, and suspensions.

Substances

You can easily tell whether a line is drawn in ink or pencil. The lines look different because they are made of different materials. Look at **Figure 9-1.** Notice that a pencil is made of several kinds of materials. You could classify the materials of the pencil according to the four states of matter. Another way to classify materials is by the units they are made of.

Elements

The units that make up all matter are called atoms. If all the atoms in a sample of matter have the same identity, that kind of matter is an **element.** The carbon used in a pencil point contains only carbon atoms. Carbon is an element. The copper in a penny is an example of another element. In a pure copper sample, all the atoms have the same identity. Altogether, there are 111 recognized elements. The names of most of the elements are in a table on pages 732-733.

Compounds

Materials called **compounds** are made from atoms of two or more elements that are combined. The ratio of the different atoms in a compound is always the same. For example, the elements hydrogen and oxygen can combine to form the compound water. The atoms of elements in water are present in the ratio of two hydrogen atoms to one oxygen atom.

When was the last time you ate a compound whose elements

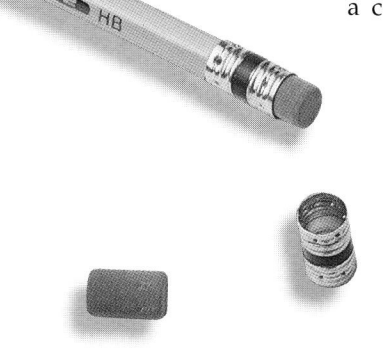

Figure 9-1

A pencil is made up of several different materials. *Which of these materials can you identify as elements?*

Program Resources

 Reproducible Masters
Study Guide, p. 39 [L1]
Reinforcement, p. 39 [L1]
Enrichment, p. 39 [L3]
Activity Worksheets, pp. 51-52, 55
Science Integration Activities, pp. 17-18
Lab Manual 20
Multicultural Connections, pp. 21-22
Concept Mapping, p. 23

Transparencies
Section Focus Transparency 35 [L1]
Teaching Transparency 17

Figure 9-2
Sugar is a compound of carbon, oxygen, and hydrogen.

are a black solid and two invisible gases? One compound that fits this description is sugar. You can recognize sugar by its white crystals and sweet taste. But the elements that form sugar—carbon, hydrogen, and oxygen—are neither white nor sweet, as shown in **Figure 9-2.** Like sugar, compounds usually have a different appearance from the elements that make them up.

Oxygen, carbon, water, sugar, baking soda, and salt are examples of materials classified as substances. A **substance** is either an element or a compound. Elements and compounds cannot be reduced to more basic components by physical processes.

Mixtures

When you have a sore throat, do you gargle with salt water? Salt water is classified as a mixture. A mixture such as salt water is a material made up of two or more substances that can be separated by physical means.

Unlike compounds, mixtures do not always contain the same amounts of the different substances that make them up. You may be wearing clothing made of permanent-press fabric. This fabric is a mixture of fibers of two materials—polyester and cotton. The fabric may contain varying amounts of polyester and cotton, as shown by the labels in **Figure 9-3.** Fabric with more polyester is more resistant to wrinkling than is fabric with less polyester.

Figure 9-3

Permanent-press fabrics are mixtures that have variable composition.

9-1 Composition of Matter **247**

---

2 Teach

Making a Model

LS **Visual-Spatial** It is important that students recognize the difference between combining and mixing substances. You can model the difference between mixing and combining by comparing wooden balls or marbles in a shallow container with modeling clay balls in a similar container. As you shake the container, the clay balls will adhere (combine), while the other balls only mix together. To emphasize the point, the clay spheres should be further stuck together so that their individual natures are lost. **LEP**

Revealing Preconceptions

The word *substance* is popularly used to refer to any kind of matter. Make sure students realize that its scientific usage is limited to either an element or a compound.

Demonstration

LS **Visual-Spatial** Obtain five small, clear glass bottles. In the first, place some copper metal and label it. In the second, place some sulfur and label it. Leave the third empty, and label it *oxygen*. In the fourth bottle, place a little copper and sulfur. Leave plenty of space. The fifth bottle should contain copper(II) sulfate, $CuSO_4$. Use these bottles to demonstrate elements, compounds, and mixtures. Aid students in forming mental models of chemically combined compounds versus mixtures.

Use **Teaching Transparency 17** as you teach this lesson.

Visual Learning

Figure 9-1 **Which of these materials can you identify as elements?** *Pencil lead is carbon; the metal part may be an elemental metal.* **LEP**

LS

247

Purpose

IS **Kinesthetic** Students will observe and infer that heterogeneous mixtures have identifiable components that do not lose their identities while in the mixture. **L1** **LEP**

Materials

transparent gallon jar, soil, clay, sand, gravel, small pebbles, water

Teaching Strategies

Have students write their predictions of the order of settling before they make the mixture.

Troubleshooting Depending on the type of soil and clay available, be prepared to delay observations about settling until the next day.

Analysis

1. gravel, pebbles, sand, soil, clay

2. The materials settle out greatest density first to least density last.

3. Filtering with several filters of varying pore size would allow most components to be separated.

Assessment

Performance Have students use their answers for question 3 to physically separate the components in the mixture. Use the Performance Task Assessment List for Assessing a Whole Experiment and Planning the Next Experiment in **PASC**, p. 33. **P**

Activity Worksheets, pp. 5, 55

USING MATH

Answer 25 g

MiniLAB

How can mixtures be separated?

Find out about the nature of a heterogeneous mixture.

Procedure

1. Put equal amounts of soil, clay, sand, gravel, and pebbles in a clear plastic gallon jar. Add water until the jar is almost full.
2. Stir or shake the mixture thoroughly. Predict the order in which the materials will settle.
3. Observe what happens and compare your observations to your predictions.

Analysis

1. In what order did the materials settle?
2. Explain why the materials settled in the order they did.
3. How could the individual materials in such a mixture be separated?

USING MATH

The concentration of a solution can be expressed using percentages. For example, a 5 percent solution of sodium hydroxide contains 5 g of NaOH in each 100 g of solution. Calculate the number of grams of NaOH in 250 g of a 10 percent solution.

A mixture in which different materials can be easily distinguished is called a **heterogeneous mixture.** Permanent-press fabrics are heterogeneous mixtures, and you can detect the different materials by sight or with a microscope. Granite, concrete, pizza, and dry soup mixes are examples of other heterogeneous mixtures.

Solutions

The salt water you gargle with looks like water and tastes salty. Like a polyester-cotton fabric, salt water is in some ways similar to the substances it contains. But you can't see the particles in salt water even with a microscope. A material, such as salt water, in which two or more substances are uniformly spread out is a **homogeneous mixture.**

For example, rubbing alcohol is a common disinfectant. It appears clear, even though it is made up of particles of alcohol in water. A **solution** is another name for a homogeneous mixture. Particles in solutions are so small that they cannot be seen even with a microscope. The particles have diameters of about 0.000 000 001 m (1 nm). These particles will never settle to the bottom of their container. Solutions remain constantly and uniformly mixed. Substances and mixtures are summarized in **Figure 9-4.**

Figure 9-4

Every sample of matter is an element, a compound, or a mixture. *Which of these types of matter are substances?*

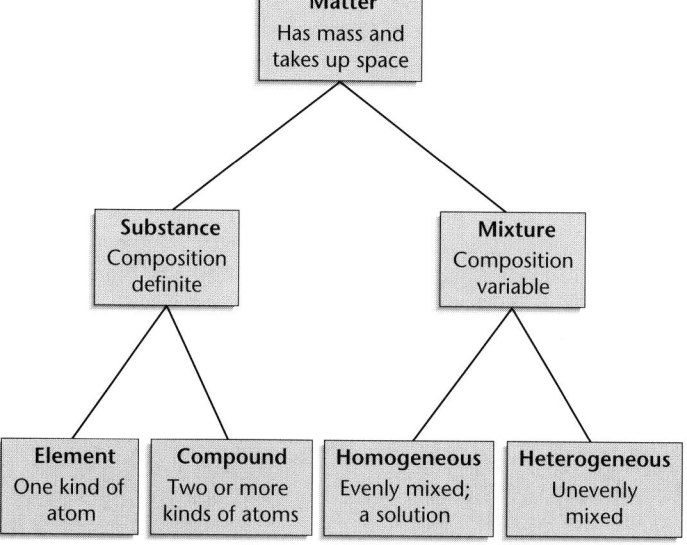

Visual Learning

Figure 9-4 Which of these types of matter are substances? *element, compound* **LEP** **IS**

Use these program resources as you teach this lesson.
Study Guide, p. 39
Lab Manual 20

Figure 9-5

Unlike the particles in a solution, the particles in a colloid are large enough to scatter light.

 A A beam of light goes straight through a solution.

B Gelatin is a colloid that may seem to be clear until you shine a light on it. Then you see that its particles scatter light.

Colloids and Suspensions

When you drink a glass of whole or low-fat milk, you are drinking a mixture of water, fats, proteins, and other substances. Milk is a colloid. A **colloid** is a heterogeneous mixture that, like a solution, never settles. One way to tell a colloid from a solution is shown in **Figure 9-5.** Milk appears white because its particles scatter light. The scattering of light by particles in a mixture is called the **Tyndall effect.** You can see the Tyndall effect in all colloids.

Problem Solving

Cooking a Colloid

The term *colloid* comes from the Greek word for glue. Gelatin, which is used to make some types of glue, was one of the first colloids studied. Gelatin consists of large strands of small units that are twisted and held together by weak attractions. Heat from boiling water is sufficient to break these attractions and uncoil the strands. When the strands are cooled, the attractions re-form, but the strands are tangled into new positions with water trapped in the spaces between the strands. This creates a semisolid colloid.

Presweetened gelatin desserts have a sweetener, such as sugar or aspartame, added to the dry gelatin. This sweetener helps disperse the dry gelatin in the hot water to which it is added. But unsweetened gelatin is first added to cold water, and then the water is heated. Why is unsweetened gelatin added to cold water instead of hot water?

Think Critically:

1. What role does boiling water play in the formation of the colloid?
2. The strands of most of the collagen in gelatin must be separated and dispersed in order to avoid clumps in the final product. How does cold water assist this process?
3. Could presweetened gelatin desserts be added to cold water and then heated? Explain.

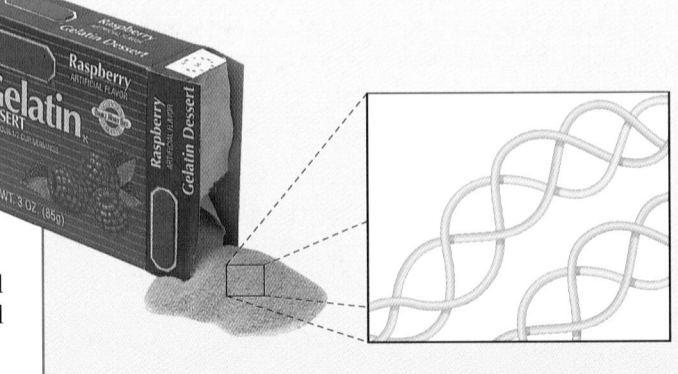

9-1 Composition of Matter **249**

 Use these program resources as you teach this lesson.
Science Integration Activities, pp. 17-18
Multicultural Connections, pp. 21-22
Concept Mapping, p. 23

Problem Solving

Think Critically

1. The heat is used to overcome the weak attractions holding strands of protein together.
2. Cold water allows the individual strands to be dispersed so that when they later come back together, they will trap water. This allows a semisolid to form.
3. Yes, but because the cold water is meant to help disperse the dried, unsweetened gelatin, it is not necessary to do this step if the sweetener is dispersing the gelatin.

3 Assess

Check for Understanding

 **FLEX Your Brain**

Use the Flex Your Brain activity to have students explore COLLOIDS.

📁 **Activity Worksheets,** p. 5

Reteach

LS **Visual-Spatial** Fill a large jar with water. Using a large metal spoon, dissolve some sugar in the water. Follow this with some food coloring, some sand, and a dropperful of milk. Allow everything to stand. Have a student describe the contents of the jar at each step using terms from this section. Note: Don't forget the metal spoon. Stainless steel is an alloy. **LEP**

Extension

📁 For students who have mastered this section, use the **Reinforcement** and **Enrichment** masters.

Table 9-1

Comparing Solutions, Colloids, and Suspensions			
Description	Solutions	Colloids	Suspensions
Settle upon standing	No	No	Yes
Can be separated using filter paper	No	No	Yes
Sizes of particles	0.1-1 nm	1-100 nm	Greater than 100 nm
Scatter light	No	Yes	Yes

Answer Homogenized milk is made by passing fat globules through a uniform sieve that forms the fat into more easily dispersed particles.

4 Close

•MINI•QUIZ•

Use the Mini Quiz to check students' recall of chapter content.

1. **If all the atoms of a sample of matter are alike, the matter is a(n) _____ .** *element*

2. **Substances made from a chemical combination of two or more elements are called _____ .** *compounds*

3. **A mixture that is not the same throughout is called a(n) _____ mixture.** *heterogeneous*

Section Wrap-up

Review

1. The hydrogen and oxygen are separate, uncombined elements. The water vapor is a compound made from these elements.

2. A substance is either an element or a compound. It has a specific composition. A mixture contains two or more substances. The composition of a mixture can vary.

3. **Think Critically** The juice must contain materials that have settled out and must be re-suspended.

Science Journal

Some examples could include milk, colloid; tap water, solution; orange juice, suspension.

Most milk containers carry the label *Homogenized.* Find out what part of the milk is homogenized and **describe** the process in your Science Journal.

Some mixtures are neither solutions nor colloids. If you fill a glass with pond water, you may notice that the water is slightly muddy. If you let it stand long enough, the silt will fall to the bottom of the glass and the water will clear. River deltas are examples of what happens when muddy water slows down and suspended soil particles settle out. Muddy water is a suspension. A **suspension** is a heterogeneous mixture containing a liquid in which visible particles settle. Another example of a suspension is glacial meltwater, which has a milky appearance. Particles carried by the meltwater settle out, forming glacial deposits such as alluvial fans.

Table 9-1 summarizes how the different types of mixtures vary in characteristics. You can use the information in this table to classify different kinds of mixtures.

Section Wrap-up

Review

1. How is a container of hydrogen gas and oxygen gas different from a container of water vapor?

2. Distinguish between a substance and a mixture.

3. **Think Critically:** Why do the words "Shake well before using" on a bottle of fruit juice indicate that the juice is a suspension?

Skill Builder
Comparing and Contrasting
In terms of suspensions and colloids, compare and contrast a glass of milk and a glass of grapefruit juice. If you need help, refer to Comparing and Contrasting in the **Skill Handbook.**

Science Journal

In your Science Journal, make a list of the liquids you consumed yesterday. Classify each as a solution, a colloid, or a suspension.

Skill Builder
They are alike in that both have small particles suspended in a liquid. The particles of the juice will settle out. Thus, the juice is a suspension. The milk does not settle and is therefore a colloid.

✔ Assessment

Portfolio Display some liquid mixtures and have students classify the liquids. Use the Performance Task Assessment List for Making Observations and Inferences in **PASC,** p. 17. **P**

Activity 9-1

Elements, Compounds, and Mixtures

Elements, compounds, and mixtures all contain atoms. In elements, the atoms all have the same identity. In compounds, two or more elements have been combined in a fixed ratio. In a mixture, the ratio of substances present can vary.

Problem
What are some differences among elements, compounds, and mixtures?

Materials
- plastic freezer bag containing the following labeled items
- copper foil
- small package of salt
- piece of solder
- aluminum foil
- chalk (calcium carbonate) or baking soda (sodium hydrogen carbonate)
- piece of granite
- sugar water in a vial

Procedure
1. Copy the data table into your Science Journal and use it to record your observations.
2. Obtain a prepared bag of numbered objects.
3. Use the data table to identify each object and classify it as either an element, a compound, a heterogeneous mixture, or a homogeneous mixture. The names of all the elements appear in the periodic table on pages 732-733. Any of the

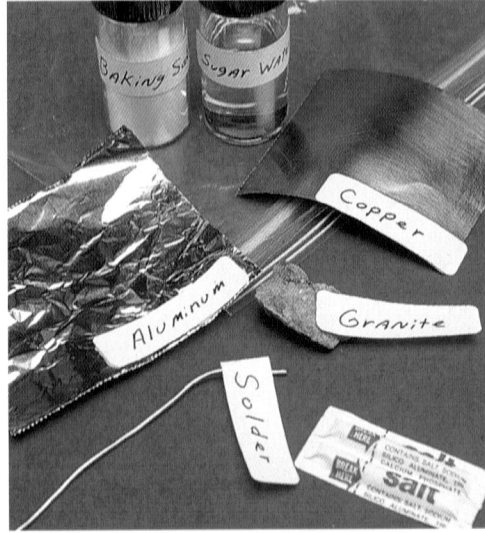

objects that are compounds have been named as examples in Section 9-1.

Analyze
1. If you know the name of a substance, how can you find out whether it is an element?
2. How is a compound different from a mixture?
3. Were the mixtures you identified homogeneous or heterogeneous?

Conclude and Apply
4. Examine the contents of your refrigerator at home. **Classify** what you find as elements, compounds, or mixtures.
5. If they are described, all materials can be classified. **Compare and contrast** elements, compounds, and mixtures.

Data and Observations
Sample Data

Object	Identity	Classification
1	copper	element
2	salt	compound
3	solder	homogeneous mixture
4	aluminum	element
5	baking soda	compound
6	granite	heterogeneous mixture
7	sugar/water	homogeneous mixture

9-1 Composition of Matter   **251**

Purpose

IS **Interpersonal** Classify materials based on their appearance and chemical makeup.
L1 **LEP** **COOP LEARN**

Process Skills
observing, classifying

Time
25 minutes

📁 **Activity Worksheets,** pp. 5, 51-52

Teaching Strategies
Alternate Materials Any equivalent assortment that students can identify may be substituted. Pictures of materials may be substituted for actual objects.
Troubleshooting Make sure vials are tightly sealed.

- Introduce formulas for the compounds or have students look them up.
- If you have not done so already, this would be a good time to hang a large periodic table on the wall.

✓ **Assessment**

Oral Have students classify examples of materials in the classroom as elements, compounds, or mixtures. Use the Performance Task Assessment List for Making Observations and Inferences in **PASC,** p. 17. **P**

Answers to Questions
1. See whether it appears on the periodic table.
2. A compound is a substance that consists of chemically combined elements. A mixture consists of various substances that are not chemically combined.
3. Solder and sugar water were homogeneous. Granite was heterogeneous.
4. Answers will vary but may include: milk, heterogeneous mixture; water, compound; and tea, homogeneous

mixture. Students will find few, if any, elements.
5. Elements are homogeneous substances having a specific composition consisting of only one kind of atom. Compounds are also homogeneous and have a specific composition, but they consist of two or more kinds of atoms that are chemically combined. Mixtures do not have a specific composition and may consist of varying amounts of two or more substances.

Prepare

Section Background

As the colloid aluminum hydroxide, $Al(OH)_3$, forms in water, it can trap smaller particles within its overall gel-like structure.

1 Motivate

Bellringer

 Before presenting the lesson, display **Section Focus Transparency 36** on the overhead projector. Assign the accompanying **Focus Activity** worksheet. L1 LEP

Tying to Previous Knowledge

In Chapter 8, students learned about the importance of freshwater supplies. By purifying water sources with colloids, we can utilize more water for drinking purposes.

2 Teach

NATIONAL GEOGRAPHIC SOCIETY

 **Videodisc**
STV: Water
Water Quality
Unit 1
Natural Filters

09425-10612
Hydrologic Cycle

10619-13542

TECHNOLOGY:
9•2 The Colloid Connection

Science Words

coagulation

Objectives

• Identify two colloids related to air and water pollution.

A Water-Purifying Colloid

Most of us just turn a tap for a drink of fresh water. Did you realize that a useful colloid may help make this possible?

In most municipal water-treatment plants, water is first taken from a river and quickly filtered. Then it is allowed to settle to remove some of the clays and other suspended materials. However, many fine particles still remain in the water, and they need to be removed. Water sanitation engineers have developed techniques that use colloids to remove these fine particles.

Figure 9-6 shows how such a colloid is formed and used. The newly formed colloid has a fluffy, gel-like appearance. It is capable of trapping microorganisms and many of the finely dispersed particles that escaped filtering because some of the particles stick to the gel.

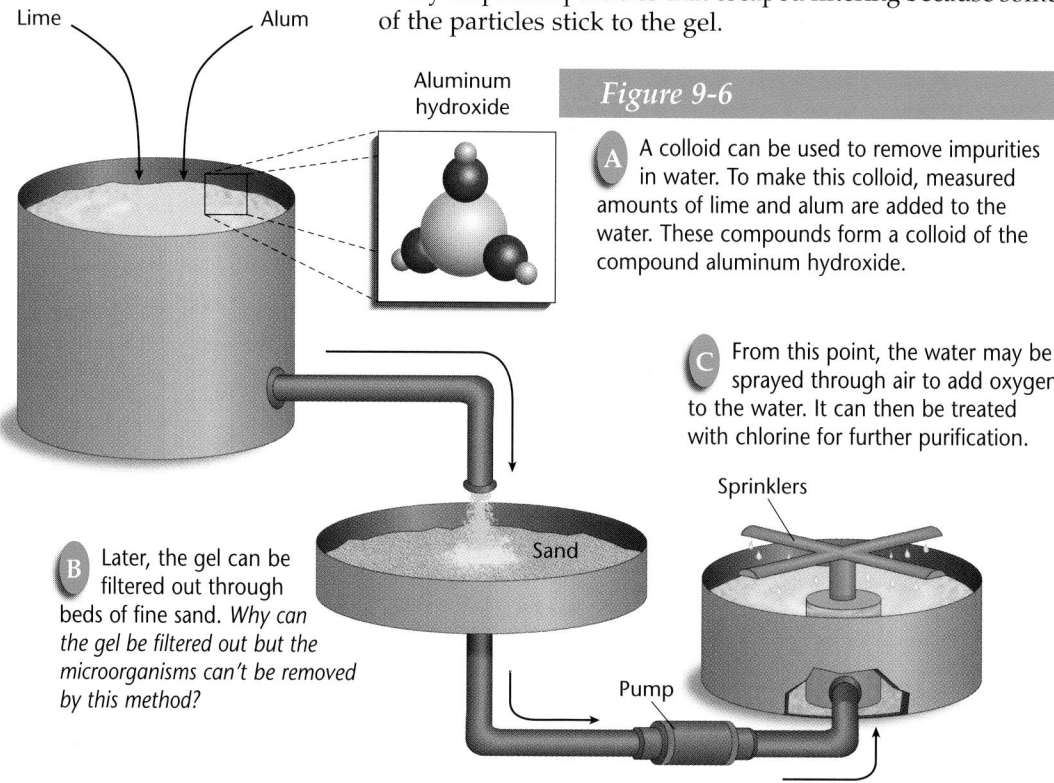

Figure 9-6

A A colloid can be used to remove impurities in water. To make this colloid, measured amounts of lime and alum are added to the water. These compounds form a colloid of the compound aluminum hydroxide.

C From this point, the water may be sprayed through air to add oxygen to the water. It can then be treated with chlorine for further purification.

B Later, the gel can be filtered out through beds of fine sand. *Why can the gel be filtered out but the microorganisms can't be removed by this method?*

Program Resources

Reproducible Masters
Study Guide, p. 40 L1
Reinforcement, p. 40 L1
Enrichment, p. 40 L3

Transparencies
Section Focus Transparency 36 L1

Breaking up a Colloid

Colloids are not always helpful. Some colloids may cause pollution rather than ridding the environment of pollutants. For example, a familiar sight around some factories is large smokestacks with columns of dense smoke billowing skyward. In addition to materials that can't be seen, smoke may contain significant amounts of carbon and uncombusted coal fragments. These particles make up soot, which is suspended in the air. To avoid polluting the air, this colloid must be broken up before it is released into the atmosphere. The colloid structure is destroyed in a process called **coagulation.** In coagulation, small, suspended particles are attracted to each other. They clump together, forming a particle that is too large to remain suspended.

Engineers have designed a way to remove most of the soot in smokestack exhaust in industrial locations, as shown in **Figure 9-7.** The particles can be collected later for further disposal.

Figure 9-7

The soot colloid is vented into a chamber where it is subjected to a high voltage. At this point, the soot particles become negatively charged. They attract uncharged soot particles, and these larger particles are then attracted to a positive part of the chamber, where they accumulate. If soot is effectively removed, the products from a smokestack appear much cleaner.

Section Wrap-up

Review

1. Even though water is settled and filtered, it still is often treated with a colloid before it is used for drinking. Explain what advantage colloid treatment has for purifying water.

2. How does smoke fit the definition of a colloid?

Explore the Technology

Heating a colloid often causes the velocities of the particles to increase to a level at which high-impact collisions cause them to coagulate. In your Science Journal, explain why it is necessary to use some other technique when coagulating soot in smoke.

SCIENCE & SOCIETY

3 Assess

Check for Understanding

? FLEX Your Brain

Use the Flex Your Brain activity to have students explore COAGULATION.

📁 **Activity Worksheets,** p. 5

Reteach

LS Visual-Spatial Demonstrate the formation of a colloid. **CAUTION:** *Wear goggles.* Dissolve $Ca(OH)_2$ in water, then add alum. The aluminum hydroxide gel will soon form. **LEP**

Extension

📁 For students who have mastered this section, use the **Reinforcement** and **Enrichment** masters.

4 Close

Section Wrap-up

Review

1. Colloid treatment may trap materials that are so small that they pass through filters.

2. Smoke contains carbon fragments that do not settle out and can reflect light.

Explore the Technology

Because soot particles from the wastes in smokestacks are already heated, they are too dispersed to have the particles effectively collide with enough consistency and force to charge the particles.

Visual Learning

Figure 9-6 Have students examine **Figure 9-6** to identify as many physical and chemical changes as they can in the process of water treatment. *chemical—reaction of lime and alum; chlorine reacting with bacteria; physical—filtering, evaporating, dissolving* **Why can the gel be filtered out but the microorganisms can't be removed by this method?** *The gel particles are larger.* **LEP LS**

Prepare

Section Background

Mass can be changed into energy and energy can be changed into mass. However, the law of conservation of mass, as stated here, holds true for everyday physical and chemical changes.

Preplanning

For Activity 9-2, you will need to prepare dilute $4M$ hydrochloric acid.

1 Motivate

Bellringer

Before presenting the lesson, display **Section Focus Transparency 37** on the overhead projector. Assign the accompanying **Focus Activity** worksheet. L1 LEP

Tying to Previous Knowledge

A common remedy for a sore throat is to gargle with salt water. Dissolving salt, NaCl, in water is a physical change. The action of the salt on the bacteria is also physical in that it kills the cells by dehydrating them. Salt has been used as a preservative for meat for centuries.

USING MATH

Answer 3 cm

Student Text Questions

Does all matter have physical properties? *Yes. All matter can be described by its physical properties.* **What physical property of the nail is measured with a balance?** *its mass*

9•3 Describing Matter

Science Words

physical property
physical change
chemical change
chemical property
law of conservation of mass

Objectives

- Give examples of physical properties.
- Distinguish between physical and chemical changes.
- Distinguish between chemical and physical properties.
- State and explain the law of conservation of mass.

USING MATH

Is the length of a nail most likely to be 3 mm, 3 cm, or 3 km?

Figure 9-8

Each of the items pictured has its own properties. *What are the physical properties of the CD cases?*

Physical Properties

You can bend an empty aluminum can, but you can't bend a piece of chalk. Chalk doesn't bend—it breaks. Brittleness is a characteristic that describes chalk. Its color and shape also describe the chalk. Any such characteristic of a material that you can observe without changing the substances that make up the material is a **physical property.** Examples of physical properties you have learned about are color, shape, size, density, melting point, and boiling point. You can describe matter using physical properties. Look at **Figure 9-8.** Does all matter have physical properties?

Some physical properties describe the appearance of an object. For example, you might describe an iron nail as a pointy-ended cylinder made of a dull, gray-colored solid. By describing the shape, color, and state of the nail, you have listed several of its physical properties. Some physical properties can be measured. For instance, you could use a metric ruler to measure one property of the nail—its length. What physical property of the nail is measured with a balance?

254   Chapter 9 Classification of Matter

Program Resources

 Reproducible Masters
Study Guide, p. 41 L1
Reinforcement, p. 41 L1
Enrichment, p. 41 L3
Activity Worksheets, pp. 53-54, 56 L1
Critical Thinking/Problem Solving, p. 15
Science and Society Integration, p. 13
Lab Manual 21
Cross-Curricular Integration, p. 13

Transparencies
Teaching Transparency 18
Section Focus Transparency 37 L1
Science Integration Transparency 9

If you had a soft drink in a cup, you could measure its volume and temperature and describe its odor. Each of these characteristics is a physical property of the soft drink. Some physical properties describe the behavior of a material or a substance. As you may know, all objects made of iron are attracted by a magnet. Attraction by a magnet is a property of the substance iron. Every substance has physical properties that distinguish it from other substances.

Identification by Properties

Do you pick out the grapes in a fruit salad and eat them first, last, or maybe not at all? If you do, you are using physical properties to identify the grapes and separate them from the other fruits in the mixture. **Figure 9-9** shows a mixture of pebbles and sand. You can identify the pebbles and grains of sand by differences in color, shape, and size. By sifting the mixture, you can quickly separate the pebbles from the grains of sand because they are different sizes.

Now look at the mixture of iron filings and sand shown in **Figure 9-10B.** It would be impossible to separate this mixture with a sieve because the filings and grains of sand are the same size. A more efficient way is to pass a magnet through the mixture. When you pass a magnet through the mixture, the magnet attracts the iron filings and pulls them from the sand. In this way, the difference in a physical property, such as attraction to a magnet, can be used to separate substances in a mixture. A practical example of using the physical property of magnetism is shown in **Figure 9-10A.**

Figure 9-9

This mixture of pebbles and sand can be separated by physical means, such as using different sizes of filters.

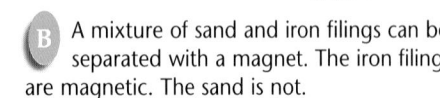
Figure 9-10

Iron can be separated from other materials because it has the physical property of magnetism.

A The iron in these materials to be recycled can be separated by a magnet.

B A mixture of sand and iron filings can be separated with a magnet. The iron filings are magnetic. The sand is not.

9-3 Describing Matter **255**

Visual Learning

Figure 9-8 **What are the physical properties of the CD cases?** *rigid, smooth, flat, etc.* LEP
LS

Figure 9-11

When iron melts and then recools, physical changes occur. *Does the identity of the iron change?*

Physical Changes

If you break a piece of chalk, its original size and shape change. You have caused a change in some of its physical properties. But you have not changed the identity of the substance that makes up the chalk.

The changes in state that you studied in Chapter 8 are all examples of physical changes. When a substance freezes, boils, evaporates, sublimes, or condenses, it undergoes physical changes such as those shown in **Figure 9-11.** Energy changes occur during these changes in state, but the kind of substance—the identity of the element or compound—does not change.

The Identity Remains the Same

As shown in the picture above, iron will change states if it absorbs or loses enough energy. In each state, it will have physical properties that identify it as the substance iron. A change in size, shape, or state of matter is called a **physical change.** Sometimes, a color change indicates a physical change. Physical changes do not change the identities of the substances in a material.

Just as physical properties can be used to separate mixtures, so can physical changes. For example, if you let a cup of salt water stand for a week, you'll find that the water has

256 Chapter 9 Classification of Matter

evaporated, leaving salt crystals inside the cup. The process of evaporating water from salty seawater is used to produce drinking water.

Chemical Changes

From observations of what happens around you, you know that changes do occur in which substances change their identities. Fireworks explode, matches burn, eggs rot, and bikes and car bodies rust. What do changes in these materials have in common?

Burned toast, burned soup, and burned steak all smell burned. The smell is different from the smell of bread, soup, or steak. The odor is a clue that a new substance has been produced. A change of one substance in a material to a different substance is a **chemical change.** Many signs can tell you when a chemical change has taken place. For example, the foaming of an antacid tablet in a glass of water and the smell in the air after

Aerogel

USING TECHNOLOGY

Aerogels

Imagine a block of gelatin dessert in which all of the liquid has been replaced with air. It might look like a frozen cloud and would be called an aerogel.

Light passing through the tiny pores and microscopic framework of an aerogel is bent, giving the aerogel a bluish color against a dark background and a yellowish color in the light. The framework is so weakly connected that the aerogel is a poor conductor of heat and an excellent insulator.

The insulating property of aerogels has caught the interest of industry. Aerogels can be used to replace the current foam insulation used in refrigerators. A one-inch thick aerogel in a double-pane window insulates as well as 30 layers of glass. They block sound so well that the Navy may use them to reduce noise levels in submarines. They are ideal for cameras that focus by response to sound waves.

A black, pure-carbon aerogel has also been made that is an excellent conductor of electricity. Using this aerogel, it would be possible to create a car part that would hold enough electrical charge to accelerate an electric vehicle from a standstill without draining the battery.

GRAPHING CALCULATOR

 Assume a pane of glass has an R-value of 1. Use your calculator to determine the R-value of a layer of aerogel 4.00 cm thick.

9-3 Describing Matter   **257**

USING TECHNOLOGY

For more information on aerogels, see "The Light Stuff" by Andrew Chaikin, *Popular Science,* Feb. 1993, pp. 72-74 and 100, and "A Brighter Future for Silicon Aerocrystals" by R. Lipkin, *Science News,* March 12, 1994.

 Aerogel 1 in. (2.54 cm) thick has an R-value of 30 (30 panes × 1 R/pane). Therefore, 4.00/2.54 × 30, which can be done as a chain function on a graphing calculator, equals an R-value of about 47.

Enrichment

IS **Logical-Mathematical** Some of the chemical changes that occur in the body absorb energy, and others release energy. Have students gather information about the Calorie content of some common foods. Then have them research activities to find out how many Calories are used. Students should compute the energy value of a typical meal and then determine how long they would have to perform various activities in order to use up that energy. Be sure students include basal metabolic rates. [L2]

GLENCOE TECHNOLOGY

 **Videodisc**

STVS: Chemistry
Disc 2, Side 2
Treating Acid Lakes (Ch. 2)

Use these program resources as you teach this lesson.

Teaching Transparency 18
Science Integration Transparency 9

Across the Curriculum

History Nitrates are a primary ingredient in explosives and gunpowder. Before World War I, Germany had to import nitrates from South America. Because the British navy could have cut off this supply, Germans looked for alternate sources of nitrates. Fritz Haber discovered a way to make ammonia from nitrogen and hydrogen. Friedrich Wilhelm Ostwald later discovered a method to oxidize the ammonia to form nitrate. The motivation for these scientists' research was initially to enable Germany to make explosives, but later they were used to make inexpensive fertilizer available to Europe's worn-out farmland. Interested students could read about Haber's and Ostwald's lives and work.

a thunderstorm indicate that new substances have been produced. In some chemical changes, a rapid production of energy, such as the light and sound of an exploding firecracker, is a clue.

A Change in Identity

When iron is exposed to the oxygen and water in the air, the iron and oxygen slowly form a new substance, rust. When hydrogen gas is burned in a rocket engine, the elements hydrogen and oxygen combine to form water. Burning and rusting are chemical changes because different substances are produced.

INTEGRATION
Earth Science

Figure 9-12

Calcium carbonate is found in cave formations and on Earth's surface in formations such as the White Cliffs of Dover, England.

Weathering: Chemical or Physical Change?

The effects of nature's forces on Earth's surface provide many opportunities to observe dramatic changes. Steep canyon walls, shifting sand dunes, and unusual limestone formations are easily observed at various global locations. Would you classify the changes that caused these formations as physical changes or chemical changes? Geologists, using the same criteria that you have learned in this chapter, would classify some weathering changes as physical and some as chemical.

Large rocks can split when water seeps into small cracks, freezes, and expands. However, the smaller pieces of newly exposed rock still have the same properties as the original sample. This change is a physical change.

Limestone, shown in **Figure 9-12,** may be washed away by rain and moving rivers, which is another physical change. Sometimes, however, the water is acidic. If this is the case, the changes may produce new products. Solid calcium carbonate, a compound found in limestone, does not easily dissolve in water. But when calcium carbonate reacts with an acid, it changes into a new substance, calcium hydrogen carbonate, that can dissolve in water. This change in limestone would be classed as a chemical change because the identity of the calcium carbonate changes. The mineral feldspar is also susceptible to acid attack.

Integrating the Sciences

Earth Science Feldspar is a component in granite. It can be represented as $KAlSi_3O_8$. When it reacts with an acid, it is changed to a water-soluble clay and other components. When this happens, even granite can begin to erode as the feldspar no longer provides strong support. A demonstration can be used of some simple acid tests to identify limestone and feldspar materials in rock samples.

Visual Learning

Figure 9-13 How do flammability and combustibility differ? *Flammable materials start to burn at a lower temperature.* **LEP**

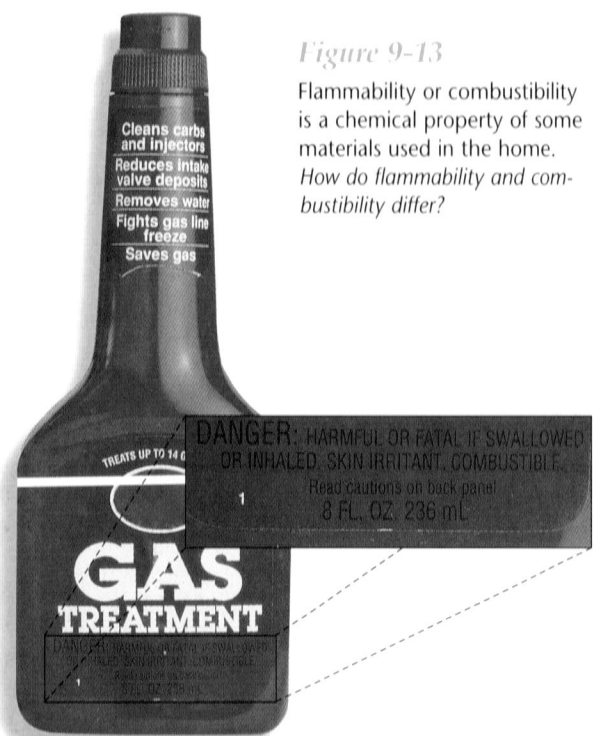

Figure 9-13

Flammability or combustibility is a chemical property of some materials used in the home. *How do flammability and combustibility differ?*

Chemical Properties

Look at **Figure 9-13.** You have probably seen these warnings on cans of paint thinners and lighter fluids for charcoal grills. The warnings indicate that these liquids burn quickly. The tendency of a substance to burn is an example of a chemical property. A **chemical property** is a characteristic of a substance that indicates whether it can undergo a certain chemical change. Many substances are flammable or combustible. Knowing which materials contain substances that have this chemical property allows you to use them safely.

If you look around a drugstore, you might notice that many medicines are stored in dark bottles. These medicines contain compounds with a similar chemical property. Chemical changes will take place in the compounds if they are exposed to light. What physical property do these bottles have in common?

Even though there are thousands of substances and billions of mixtures, they do share a few common physical and chemical properties. You can use these properties to study matter further.

Science Journal

Over the past couple of days, you have observed or been part of many chemical and physical changes. In your Science Journal, describe five chemical changes and five physical changes that you have noticed in the past two days.

9-3 Describing Matter 259

MiniLAB

What are some indications of changes?

Discover some clues that indicate that a physical or a chemical change has occurred.

Procedure

1. Add water to a 250-mL beaker until it is half-full.
2. Add a crystal of potassium permanganate to the water and observe what happens.
3. Add 1 g of sodium hydrogen sulfite to the solution and stir until the solution becomes colorless.

Analysis

1. Is dissolving a chemical or a physical change?
2. What evidence of a chemical change did you see?

The Conservation of Mass

Wood is combustible, or burnable, which is a chemical property. Suppose you burn a large log on a campfire until nothing is left but a small pile of ashes. During the burning, smoke, heat, and light are given off. It's easy to see that a chemical change occurs. At first, you might also think that matter was lost during this change because the pile of ashes looks much smaller than the log. In fact, if you could measure both the mass of the log and the mass of the ashes, the mass of the ashes would be less than that of the log. But suppose that during the burning, you could collect all the oxygen in the air that was combined with the log during the burning. And suppose you could also collect all the smoke and gases that escape from the burning log and measure their masses, too. Then you would find that there is no loss of mass during the burning, as shown in **Figure 9-14.**

Not only is there no loss of mass during burning, there is no loss or gain of mass during any chemical change. In other words, matter is neither created nor destroyed during a chemical change. This statement is known as the **law of conservation of mass.** According to this law, the mass of all substances present before a chemical change equals the mass of all the substances remaining after the change. How does **Figure 9-15** illustrate the law of conservation of mass?

Figure 9-14

Although you can't see the gases used or formed when a log burns, there is no loss or gain of mass. *What gas is necessary for burning to take place?*

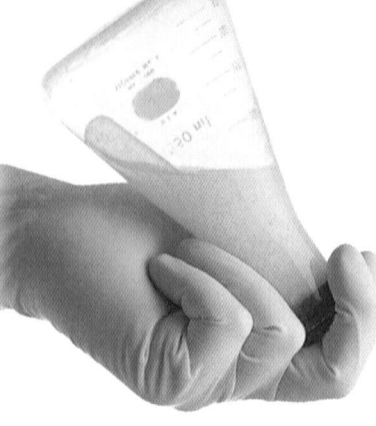
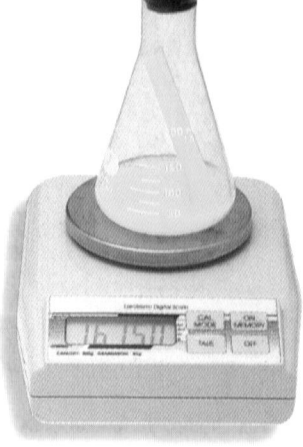

Figure 9-15

A chemical change between two compounds demonstrates the law of conservation of mass.

Why do substances have physical and chemical properties and undergo physical and chemical changes? These properties and changes are the result of what atoms are present and how they are arranged. The composition and arrangement are different for each substance.

Section Wrap-up

Review

1. In terms of substances, explain why evaporation of water is a physical change and not a chemical change.

2. Give an example of a chemical change that occurs when you prepare a meal.

3. Why is being flammable a chemical property rather than a physical property?

4. **Think Critically:** The law of conservation of mass applies to physical changes as well as to chemical changes. How might you demonstrate this law for melting ice?

Skill Builder

Observing and Inferring

Observe a burning candle. What evidence do you have that there are chemical and physical changes in the candle as it burns? If you need help, refer to Observing and Inferring in the **Skill Handbook.**

USING MATH

Two chemicals with a combined mass of 25.48 g react in a flask that has a mass of 142.05 g. A gas is produced that totally escapes into a flask that has an empty mass of 141.65 g. After the reaction, the first flask and its contents have a mass of 167.16 g. Calculate the total mass of the second flask and gas.

9-3 Describing Matter   **261**

Skill Builder

Evidence of chemical changes includes the production of heat and light, the smell of something burning, the darkening of the wick, and the production of smoke (soot) and water vapor. Evidence of physical changes includes the melting and hardening (solidification) of the wax; the change in the shape, mass, and length of the candle; and the shortening of its wick.

✔Assessment

Oral Display several similar, but slightly different, candles. Have a student describe a secretly selected candle using only its physical properties. The other students then try to identify the candle. Use the Performance Task Assessment List for Analyzing the Data in **PASC,** p. 27. **P**

Student Text Question

How does Figure 9-15 demonstrate the law of conservation of mass? *Even though a change takes place, no mass is lost or gained.*

4 Close

•MINI•QUIZ•

Use the Mini Quiz to check students' recall of chapter content.

1. **Which kind of property indicates whether a substance can undergo a certain chemical change?** *chemical*

2. **Which law states that matter can be neither created nor destroyed during a chemical change?** *law of conservation of mass*

3. **The mass of substances remaining after a chemical change _____ the mass of substances present before the change.** *equals*

Section Wrap-up

Review

1. The makeup of the water units is unchanged.

2. Answers may vary but could include toasting bread, rising and browning of pancakes on a grill, or searing of meat in a pan. Processes involving only melting or mixing should not be included.

3. Burning is a chemical change because substances change to new substances.

4. **Think Critically** Measure the mass of a closed container of ice. Allow the ice to melt and measure the mass again. There should be no change in mass.

USING MATH

Answer 142.02 g

Activity 9-2

PREPARATION

Purpose
LS Interpersonal Design and carry out an experiment to show evidence of a chemical change. **L1 COOP LEARN**

Process Skills
observing and inferring, recognizing cause and effect, interpreting data, and making and using tables

Time
one class period

Materials
Prepare dilute (4M) HCl by slowly adding 330 cm^3 of stock concentrated (12M) HCl to 660 cm^3 of distilled water in a well-ventilated area. Allow the solution to cool before use.

Safety Precautions
Caution students to avoid direct contact with HCl, immediately flushing any affected area with water, and to avoid inhaling HCl fumes. Do not use containers that are not heat resistant for the evaporating dishes.

Possible Hypotheses
Students may hypothesize that if mixing together baking soda and hydrochloric acid makes a chemical change, then a new substance will be detected.

📁 **Activity Worksheets,** pp. 5, 53-54

PLAN THE EXPERIMENT

Possible Procedures
Place a small amount of baking soda in an evaporating dish. Add 2 mL of dilute hydrochloric acid. Observe. Allow residue to dry. Use a hand lens to observe residue for comparison to unreacted baking soda. Add 1 mL more acid to the residue to compare the reaction to the original baking soda reaction.

262

Activity 9-2

Design Your Own Experiment

Checking Out Chemical Changes

Mixing materials together does not always produce a chemical change. You must find evidence of a new substance with new properties being produced before you can conclude that a chemical change has taken place. Try this activity and use your observation skills to make careful deductions about changes.

PREPARATION

Problem
What evidence indicates a chemical change within a mixture?

Form a Hypothesis
Think about what happens when small pieces of limestone are mixed with sand. Now think what happens when limestone is mixed with an acid. Based on these thoughts, form a hypothesis about the relationship between mixing substances together and chemical changes.

Objectives
- Observe the results of adding dilute hydrochloric acid to baking soda.
- Infer that the production of new substances indicates a chemical change.
- Design an experiment that allows you to compare the activity of baking soda and the activity of the resulting product.

Possible Materials
- goggles
- baking soda
- small evaporating dish
- hand lens
- dilute hydrochloric acid, HCl
- 10-mL graduated cylinder
- electric hot plate
- apron

Safety Precautions
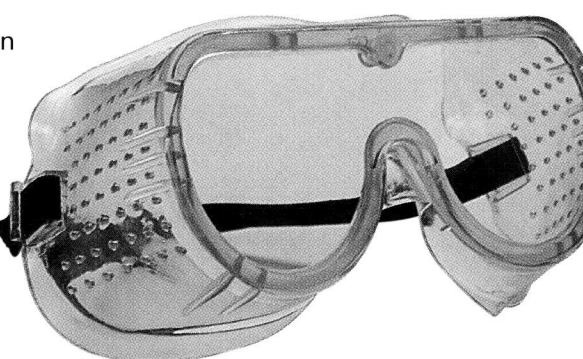
Wear safety goggles during any laboratory work. Use care when pouring the dilute acid. Quickly wash any spills with water, and notify your teacher.

PLAN THE EXPERIMENT

1. As a group, agree upon and write out the hypothesis statement.
2. To test your hypothesis, devise a plan to compare two different mixtures. The first mixture consists of 3 mL of hydrochloric acid and 0.5 g of baking soda. The second mixture is 3 mL of hydrochloric acid and the solid product of the first mixture. Describe exactly what you will do at each step.
3. Make a list of the materials that you will need to complete your experiment.
4. Design a data and observations table in your Science Journal so that it is ready to use as your group observes what happens.

Check the Plan

1. Read over your entire experiment to make sure that all steps are in logical order.

2. Identify any constants and the variables of the experiment.
3. Should you run any test more than one time?
4. How will observations be summarized?
5. *Make sure your teacher approves your entire plan before you begin and that you have included any changes in the plan.*

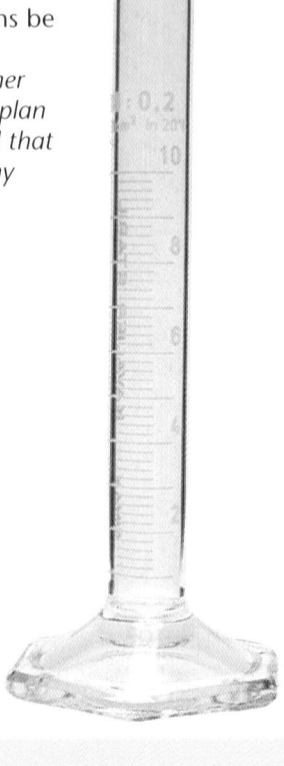

DO THE EXPERIMENT

1. Carry out the experiment as approved.
2. While the experiment is going on, write down any observations that you make and complete the data and observations table in your Science Journal.

Analyze and Apply

1. **Compare** your results with those of other groups.
2. What different properties of any new substances did you **observe** after adding hydrochloric acid to the baking soda?

Go Further

If you had used vinegar, which contains acetic acid, as the acid, do you think your results would be different? Explain.

Theme Connection

Stability and Change The basic chemical identity of a substance remains unchanged, or stable, during a physical change. However, the basic identity changes during a chemical change. Emphasize this theme by discussing the importance of both terms in describing their results.

 **Assessment**

Performance To further assess students' understanding of chemical changes, ask them to write a paragraph in their Student Journals summarizing their observations. Use the Performance Task Assessment List for Science Journal in **PASC**, p. 103. **P**

Teaching Strategies

- If time or supplies are limited, you may measure out the baking soda into watch glasses before starting class.
- The solution should be dried at a slow enough rate to avoid splattering. Turn off the hot plate as the solution approaches dryness, and let residual heat finish the drying process.
- The chemical reaction involved is $NaHCO_3 + HCl \rightarrow NaCl + CO_2 + H_2O$.

DO THE EXPERIMENT

Expected Outcome
Baking soda will react with the acid to produce carbon dioxide bubbles and a white residue (visible after drying). The white residue will have a slightly different appearance than baking soda. The white residue, sodium chloride, will have no further reaction with the acid. These two observations indicate that a new substance, sodium chloride, has been formed.

Analyze and Apply

1. Bubbles formed after HCl was added to baking soda. The dried product did not seem to react with HCl.
2. New products were produced. The dried material had new physical and chemical properties. These were evidence of a chemical change.

Go Further

The physical reaction observed using vinegar will be similar to that using HCl. More vinegar must be used to get the same results because it is a weaker acid. The white residue produced will be sodium acetate.

Science & ART

Teaching Strategies

- Discuss how the patina preserves the sculpture.

- **Logical-Mathematical** Have students go over the characteristics of copper and copper alloys described here and classify each as a physical or chemical property. L1

- Have students locate a copper or copper alloy sculpture in the area. Have them describe the work and the patina. L1

- Have students describe why they might want a certain color of patina—or no patina at all—on a sculpture. Ask them to take into account subject matter and the setting.

- **Linguistic** Have students research to find the most common forms of copper alloy (brass and bronze) and the difference between them. (Traditionally, brass is an alloy of copper and zinc while bronze is copper and tin. Some modern classifications depend upon the percentage of materials used.) L2

Sources

Living Materials: A Sculptor's Handbook, Oliver Andrews, University of California Press, 1983.

Encyclopaedia Britannica, Micropedia, "Copper."

Metal Design & Technique, Wilhelm Braun-Feldweg, Van Nordstron Reinhold Co., 1975.

Copper Art

If you wanted a small copper statue for your room, you'd probably be thinking about something the color of a penny. But if you were a sculptor planning a copper statue for a park, you might be thinking about something green.

Just as iron rusts when it combines with oxygen in the air, copper combines with other substances in the atmosphere to form a green coating of copper carbonates, copper sulfates, and copper chlorides. Artists call this coating a patina. They may decide to make a sculpture out of copper because they want the look of the patina rather than the look of the pure copper.

Patinas do more than merely look nice. They also protect the sculpture. Once the patina forms, air can no longer get to the metal underneath. Therefore, the layer of oxidized copper stops further corrosion.

Working with Copper

There are other reasons some artists like working with copper. It's a soft metal and easy to mold, but it becomes stiffer when hammered or bent. This is known as work-hardening.

Copper also has some disadvantages. It can be too soft for some projects. It also conducts heat so well that it can be difficult to weld. The heat from the welding torch can quickly spread throughout the copper. Then the metal may melt all over—not just around the weld.

To get slightly different properties, artists may decide to work with copper alloys instead of pure copper. An alloy is a homogeneous mixture of a metal with different metals or other elements. Different copper alloys have different degrees of hardness, are more suitable for welding, and develop patinas of various colors.

interNET
CONNECTION

Visit the Chapter 9 Internet Connection at Glencoe Online Science, **www.glencoe.com/ sec/science/physical**, for a link to more information about the Statue of Liberty.

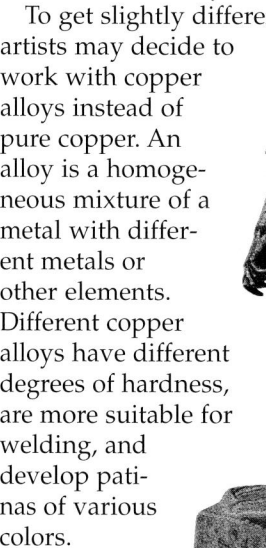

Classics

- Statue of Liberty

Sculptures pictured in these books:

- *200 Years of American Sculpture,* Armstrong et al. David R. Godine & Whitney Museum of American Art, 1976.

- *Masterpieces of Western Sculpture,* Howard Hibbard, Harper & Row, 1977.

- *The Great Bronze Age of China,* En Fong, Ed., Metropolitan Museum of Art, Alfred A. Knopf, Inc., 1980.

- *Western Sculpture: Definitions of Man,* Ruth Butler, New York Graphic Society, Little, Brown & Co., 1975.

- *The History of World Sculpture,* Germain Bazin, Lamplight Publishing, Inc., 1968.

Summary

9-1: Composition of Matter

1. Elements and compounds are substances; a mixture is composed of two or more substances.
2. A solution is a homogeneous mixture. Colloids and suspensions are two kinds of heterogeneous mixtures.

9-2: Science and Society: The Colloid Connection

1. Lime and alum form an aluminum hydroxide colloid that is used to help remove finely suspended particles from water supplies.
2. Exhaust from some industries may be a soot-containing colloid, which can be coagulated by using electricity in the smokestacks.

9-3: Describing Matter

1. Physical properties are characteristics of materials that you can observe without changing the identities of the substances themselves.
2. In physical changes, the identities of substances in materials do not change. In chemical changes, substances in materials change to different substances.
3. Physical properties can be observed without changing the identities of substances; chemical properties indicate chemical changes substances can undergo.
4. The law of conservation of mass states that during any chemical change, matter is neither created nor destroyed.

Key Science Words

a. chemical change
b. chemical property
c. coagulation
d. colloid
e. compound
f. element
g. heterogeneous mixture
h. homogeneous mixture
i. law of conservation of mass
j. physical change
k. physical property
l. solution
m. substance
n. suspension
o. Tyndall effect

Reviewing Vocabulary

Match each phrase with the correct term from the list of Key Science Words.

1. process used to reduce a colloidal form of air pollution
2. mixture of parts that can be easily distinguished
3. all atoms in a sample have the same identity
4. change in size, shape, or state
5. change of substance to a different substance
6. combined atoms of two or more elements
7. either an element or a compound
8. mixture that scatters light and never settles
9. an indication of whether a chemical change can occur in a substance
10. In a chemical change, matter is neither created nor destroyed.

Summary

Have students read the summary statements to review the major concepts of the chapter.

Reviewing Vocabulary

1. c	**6.** e
2. g	**7.** m
3. f	**8.** d
4. j	**9.** b
5. a	**10.** i

✓ Assessment

Portfolio Encourage students to place in their portfolios one or two items of what they consider to be their best work. Examples include:

- Activity 9-1 Assessment, p. 251
- MiniLAB Assessment, p. 260
- Skill Builder Assessment, p. 261 **P**

Performance Additional performance assessments may be found in **Performance Assessment** and **Science Integration Activities.** Performance Task Assessment Lists and rubrics for evaluating these activities can be found in Glencoe's **Performance Assessment in the Science Classroom.**

GLENCOE TECHNOLOGY

 Videodisc

Glencoe Physical Science
Interactive Videodisc

Use the videodisc lesson *Chemical Detectives* to compare and contrast chemical and physical changes and properties.

MindJogger Videoquiz

Chapter 9 Have students work in groups as they play the Videoquiz game to review key chapter concepts.

Checking Concepts

1. b	**6.** c
2. a	**7.** d
3. b	**8.** c
4. a	**9.** b
5. d	**10.** b

Understanding Concepts

11. Answers might include color, mass, smell, taste, temperature, state, volume, texture, density, and shape.

12. Table salt has properties different from those of sodium or chlorine alone.

13. Many materials such as paints and glues are colloids.

14. The mass of the rusty nail equals the mass of the original nail plus the mass of the oxygen that reacted with it.

15. Many substances are dissolved in ocean water. There are also many small particles suspended in the water.

Thinking Critically

16. Coat the iron to prevent oxygen from coming into contact with the iron.

17. Use a magnet.

18. Brass is a homogeneous mixture.

19. As water flows rapidly, particles are kept suspended. At the delta, the water slows down, allowing particles to settle out.

20. Some medications are suspensions. Particles need to be re-suspended evenly before the medication is used.

Developing Skills

21. Recognizing Cause and Effect Causes include dust, ash, unburned materials, and spray products. Effects include health problems, restric-

Checking Concepts

Choose the word or phrase that completes the sentence or answers the question.

1. A copper wire will bend. This is an example of _____.
   a. a chemical property c. conservation
   b. a physical property d. an element

2. Which of the following is not an element?
   a. water   c. oxygen
   b. carbon   d. hydrogen

3. An example of a chemical change is

   _____.
   a. boiling   c. evaporation
   b. burning   d. melting

4. Gelatin is an example of a _____.
   a. colloid   c. substance
   b. solution   d. suspension

5. A sunbeam is an example of _____.
   a. an element   c. a suspension
   b. a solution   d. the Tyndall effect

6. One way to destroy a colloid is called

   _____.
   a. combustion   c. coagulation
   b. filtration   d. technology

7. The red color of a rose is a _____.
   a. chemical change   c. physical change
   b. chemical property d. physical property

8. The process of evaporating water from seawater for drinking is a _____.
   a. chemical change   c. physical change
   b. chemical property d. physical property

9. Which warning label indicates a chemical property of the material being labeled?
   a. "Fragile"   c. "Handle with Care"
   b. "Flammable"   d. "Shake Well"

10. Which of the following is a substance?
   a. colloid   c. mixture
   b. element   d. solution

Understanding Concepts

Answer the following questions in your Science Journal using complete sentences.

11. Describe a carton of milk using its physical properties.

12. The soft metal sodium and the greenish gas chlorine combine to form table salt, sodium chloride. How do you know table salt is a compound?

13. The word *colloid* means "gluelike." Why was *colloid* chosen to name certain mixtures?

14. Use a nail rusting in air to explain the law of conservation of mass.

15. Mai says that ocean water is a solution. Ed says that ocean water is a suspension. Are they both correct? Explain.

Thinking Critically

16. Rust is formed from oxygen and iron. How might you keep an iron pipe from rusting?

17. By mistake, scrap iron was put in with the glass at a recycling center. How might you separate the mixture?

18. Not all solutions are liquid. Why is a metal alloy, such as brass, considered a solution?

19. Use what you know about suspensions to explain why deltas form at the mouths of large rivers.

20. Why do many medications, such as the one in the photo, have instructions to shake well before using?

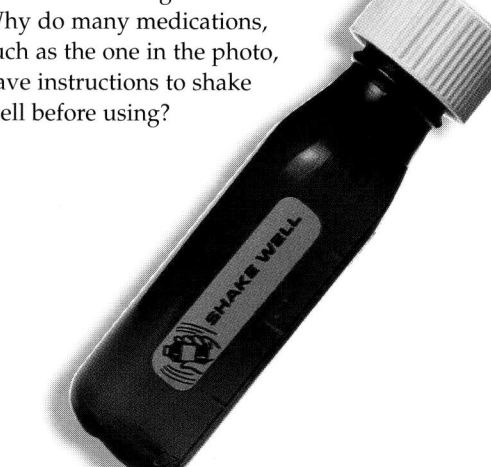

tions on industry and traffic, and a drop in tourism.

22. Making and Using Tables See reduced student page.

23. Using Variables, Constants, and Controls The variable was temperature. Constants include the volume.

24. Interpreting Data The pond water could have been a suspension. As the experiment progressed, settling out was taking place. Thus, each succeeding filter paper received more sediment.

Developing Skills

If you need help, refer to the **Skill Handbook.**

21. Recognizing Cause and Effect: List at least two causes and two effects of industrial colloids.

22. Making and Using Tables: Different colloids may involve different states. For example, gelatin is formed from solid particles in a liquid. Complete the following table, using these common colloids: smoke, marshmallow, fog, paint.

Types of Colloids	
Colloid	**Example**
Gas in solid	marshmallow
Solid in liquid	paint
Solid in gas	smoke
Liquid in gas	fog

23. Using Variables, Constants, and Controls: Marcos took a 100-cm^3 sample of a suspension, shook it well, and poured equal amounts into four different test tubes. He placed one test tube in a rack, one in very hot water, one in warm water, and the fourth in ice water. He then observed the time it took for each suspension to settle. What was the variable in the experiment? What was one constant?

24. Interpreting Data: Starting with a 25-cm^3 sample of pond water, Hannah poured 5 cm^3 through a piece of filter paper. She repeated this with four more pieces of filter paper. She dried each piece of filter paper and measured the mass of the sediment. Why did the last sample have a higher mass than did the first sample?

25. Concept Mapping: Make a network tree to show types of liquid mixtures. Include these terms: *homogeneous mixtures, heterogeneous mixtures, solutions, colloids,* and *suspensions.*

Performance Assessment

1. Designing an Experiment: Assume that some sugar was put into some rice by mistake. Design an experiment to separate the mixture. In your Science Journal, list your hypothesis and your experimental steps. Perform the experiment and summarize the results.

2. Making and Using a Classification System: Design a system for classifying different heterogeneous mixtures. State what the system is based on, and give several examples of using the system.

3. Investigating the Issue Controversy: Research the air pollution problem in Los Angeles. What causes the problem? What health and other problems are caused by the pollution? Could solutions such as those used to remove soot from smoke in industrial smokestacks be used to clear up the Los Angeles pollution? Why or why not? Present your results in an oral presentation.

Chapter 9 Review   **267**

Chapter 9 Review

25. Concept Mapping Mixtures include homogeneous mixtures, which are solutions, and heterogeneous mixtures, which can be colloids or suspensions.

Performance Assessment

1. Hypothesis: If the sugar can dissolve in water, then it can be separated by filtration from the rice. Experimental steps: Add enough water to dissolve the sugar. Pour the mixture through filter paper and rinse. Then, evaporate the solution to reclaim the separated sugar. Results: Rice remains in the filter paper, while the dissolved sugar passes through. Use the Performance Task Assessment List for Designing an Experiment in **PASC,** p. 23.

2. An example: classify based on noticeable phases. Heterogeneous system with liquid and gas components—carbonated water. Heterogeneous system with solid and liquid—tomato sauce. Heterogeneous system with solid and gas—catalytic converters in cars. Use the Performance Task Assessment List for Making and Using a Classification System in **PASC,** p. 49.

3. Weather conditions trap exhaust in the air. Air pollution can cause health and visibility problems. The scrubber systems would be difficult to install on all automobiles. Use the Performance Task Assessment List for Investigating an Issue Controversy in **PASC,** p. 65.

Assessment Resources

 **Reproducible Masters**

Chapter Review, pp. 21-22
Assessment, pp. 57-60
Performance Assessment, p. 23

Glencoe Technology

🔘 **Chapter Review Software**
🔘 **Computer Test Bank**
📼 **MindJogger Videoquiz**

Chapter Organizer

Section	Objectives/Standards	Activities/Features
Chapter Opener		**Explore Activity:** Make a model of an atom using pieces of hardware. p. 269
10-1 **Structure of the Atom** (2 sessions, 1 block)*	**1. List** the names and symbols of common elements. **2. Describe** the present model of the atom. **3. Describe** how electrons are arranged in an atom. **National Content Standards: (5-8) UCP1, UCP2, UCP5, A2, G1-G3; (9-12) UCP1, UCP2, UCP5, A2, B1, G1-G3**	**Using Math,** p. 271 **Connect to Physics,** p. 274 **Using Math,** p. 274 **Skill Builder:** Concept Mapping, p. 274 **Activity 10-1:** Models of Atomic Structure, p. 275
10-2 **Science and Society** (1 session, ½ block)*	**1. Identify** quarks as particles of matter that make up protons and neutrons. **2. Explain** how particle accelerators are used to study particles within atoms. **National Content Standards: (5-8) UCP1, UCP2, E2, F5; (9-12) UCP1, UCP2, B1, E2, F6**	**MiniLAB:** How can indirect evidence be used to identify known objects? p. 277 **Explore the Technology,** p. 277
10-3 **Masses of Atoms** (2 sessions, 1 block)*	**1. Compute** the atomic mass and mass number of an atom. **2. Identify** and **describe** isotopes of common elements. **3. Interpret** the average atomic mass of an element. **National Content Standards: (5-8) UCP1-UCP3, A1, D2, E1; (9-12) UCP1-UCP3, A1, B1, E1**	**Problem Solving:** An Unstable Isotope: Providing Medical Help, p. 280 **Connect to Life Science,** p. 281 **Using Computers,** p. 281 **Skill Builder:** Comparing and Contrasting, p. 281 **Activity 10-2:** Candy-Covered Isotopes, pp. 282-283
10-4 **The Periodic Table** (3 sessions, 2 blocks)*	**1. Describe** the periodic table of elements and use it to find information about an element. **2. Distinguish** between a group and a period. **3. Use** the periodic table to classify an element as a metal, nonmetal, or metalloid. **National Content Standards: (5-8) UCP1, UCP2, UCP5, A2, B1, E2, G1-G3; (9-12) UCP1, UCP2, UCP5, A2, B1, B2, E2, G1-G3**	**MiniLAB:** What are the advantages of organizing chemical elements in a periodic table? p. 288 **Using Technology:** Seeing Atoms, p. 290 **Science Journal,** p. 291 **Skill Builder:** Making and Using Graphs, p. 291 **People and Science:** Dr. Brenna Flaugher, Particle Physicist, p. 292

Activity Materials

Explore	Activities	MiniLABs
page 269 small ball of clay about 5 cm in diameter, small nails, screws, paper clips, ball bearings, toothpicks	page 275 magnetic board about 20 cm × 27 cm, one 0.5-cm piece and 20 1-cm pieces of rubber magnetic tape, circles of white paper 4 cm wide, circles of red paper 1 cm wide, marker pages 282-283 1 container each of red chocolate-center candy, red peanut-center candy, green chocolate-center candy, and green peanut-center candy	page 277 shallow pan, flour or sand, 4 objects page 288 variety of pens and pencils

Need Materials? Call Science Kit (1-800-828-7777).   * A complete Planning Guide that includes block scheduling is provided on pages 32T-35T.

Teacher Classroom Resources

Reproducible Masters	Transparencies	Teaching Resources
Study Guide, p. 42 Reinforcement, p. 42 Enrichment, p. 42 Activity Worksheets, pp. 57-58 Cross-Curricular Integration, p. 16	Section Focus Transparency 38, Element of the Day	Glencoe Physical Science Interactive Videodisc Physical Science CD-ROM Spanish Resources English/Spanish Audiocassettes Cooperative Learning Resource Guide Lab Partner Lab and Safety Skills Lesson Plans
Study Guide, p. 43 Reinforcement, p. 43 Enrichment, p. 43 Activity Worksheets, p. 61	Section Focus Transparency 39, Follow the Tracks (of subatomic particles)	**Assessment Resources**
Study Guide, p. 44 Reinforcement, p. 44 Enrichment, p. 44 Activity Worksheets, pp. 59-60 Science Integration Activity 10, Bean Counters and Isotopes Concept Mapping, pp. 25-26 Critical Thinking/Problem Solving, p. 16	Section Focus Transparency 40, How Old? Ask Carbon	Chapter Review, pp. 23-24 Assessment, pp. 61-64 Performance Assessment in the Science Classroom (PASC) MindJogger Videoquiz Alternate Assessment in the Science Classroom Performance Assessment Chapter Review Software Computer Test Bank
Study Guide, p. 45 Reinforcement, p. 45 Enrichment, p. 45 Activity Worksheets, p. 62 Lab Manual 22, Chemical Activity Multicultural Connections, p. 23 Science and Society Integration, p. 14 Cross-Curricular Integration, pp. 14-15	Section Focus Transparency 41, Family Resemblance Science Integration Transparency 10, Creating the Atom Teaching Transparency 19, The Periodic Table Teaching Transparency 20, The Periodic Table—Blank	

Key to Teaching Strategies

The following designations will help you decide which activities are appropriate for your students.

L1 Level 1 activities should be within the ability range of all students, including those with learning difficulties.

L2 Level 2 activities should be within the ability range of the average to above-average student.

L3 Level 3 activities are designed for the ability range of above-average students.

LEP LEP activities should be within the ability range of Limited English Proficiency students.

LS These activities are designed to address different learning styles.

COOP LEARN Cooperative Learning activities are designed for small group work.

P These strategies represent student products that can be placed into a best-work portfolio.

GLENCOE TECHNOLOGY

The following multimedia resources are available from Glencoe.

Glencoe Physical Science Interactive Videodisc
Periodicity

Physical Science CD-ROM

The Infinite Voyage Series
Unseen Worlds

Teacher Classroom Resources

This is a representation of key blackline masters available in the Teacher Classroom Resources.

Teaching Aids

Section Focus Transparencies

38 SECTION FOCUS TRANSPARENCY

ELEMENT OF THE DAY

L1

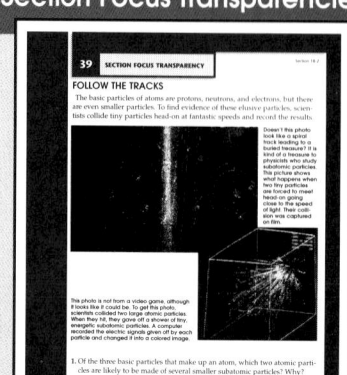

39 SECTION FOCUS TRANSPARENCY

FOLLOW THE TRACKS

L2

40 SECTION FOCUS TRANSPARENCY

HOW OLD? ASK CARBON

L3

Science Integration Transparencies

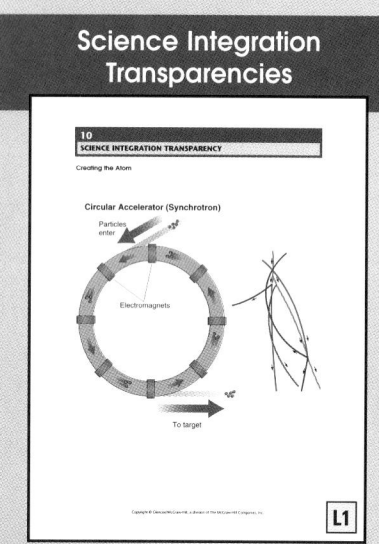

10 SCIENCE INTEGRATION TRANSPARENCY

Creating the Atom

L1

Teaching Transparencies

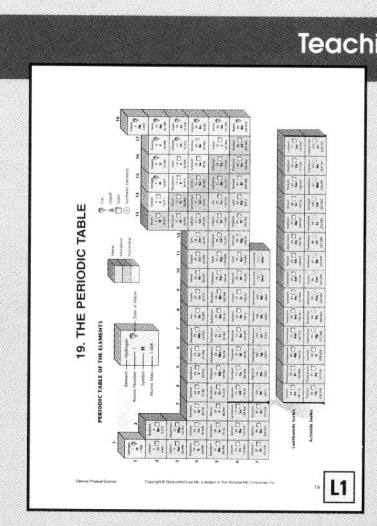

19. THE PERIODIC TABLE

L1

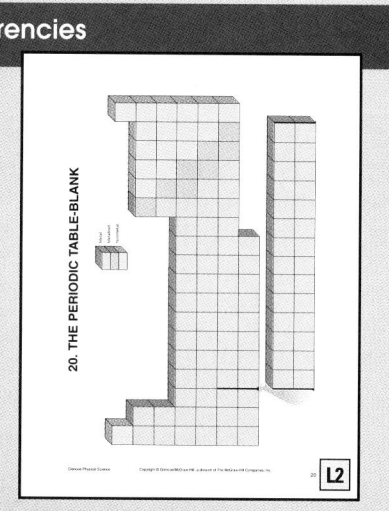

20. THE PERIODIC TABLE-BLANK

L2

Meeting Different Ability Levels

Study Guide

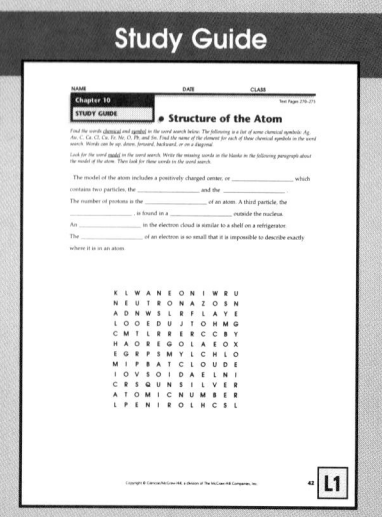

Chapter 10 STUDY GUIDE ● **Structure of the Atom**

L1

Reinforcement

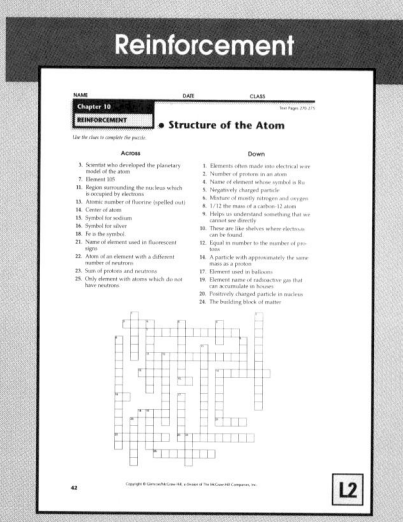

Chapter 10 REINFORCEMENT ● **Structure of the Atom**

L2

Enrichment Worksheets

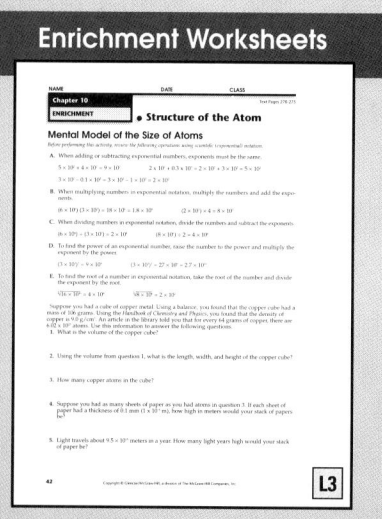

Chapter 10 ENRICHMENT ● **Structure of the Atom**

Mental Model of the Size of Atoms

L3

Chapter 10 Atomic Structure and the Periodic Table

Hands-On Activities

Science Integration Activity

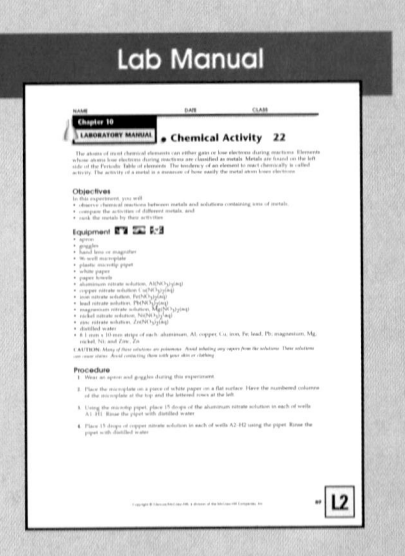

Chapter 10
INTEGRATION • Bean Counters and Isotopes

L1

Lab Manual

Chapter 10
LABORATORY MANUAL • Chemical Activity 22

L2

Assessment

Performance Assessment

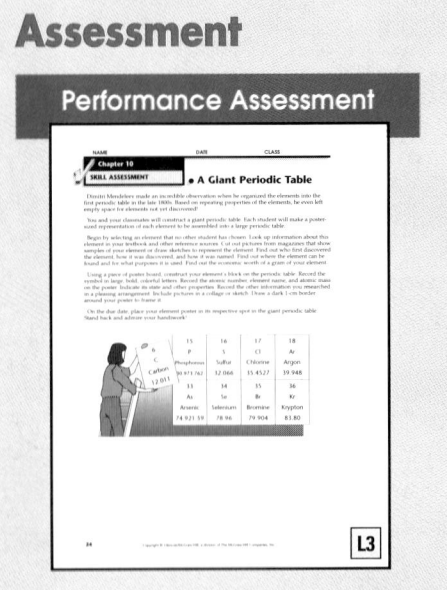

Chapter 10
SKILL ASSESSMENT • A Giant Periodic Table

L3

Enrichment and Application

Critical Thinking/Problem Solving

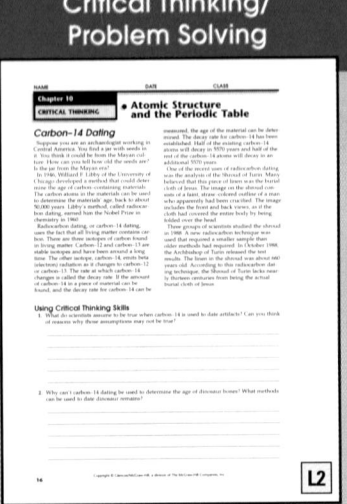

Chapter 10
CRITICAL THINKING • Atomic Structure and the Periodic Table

Carbon–14 Dating

L2

Cross-Curricular Integration

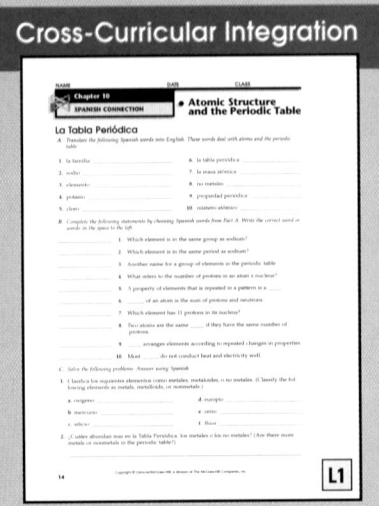

Chapter 10
SPANISH CONNECTION • Atomic Structure and the Periodic Table

La Tabla Periódica

L1

Science and Society Integration

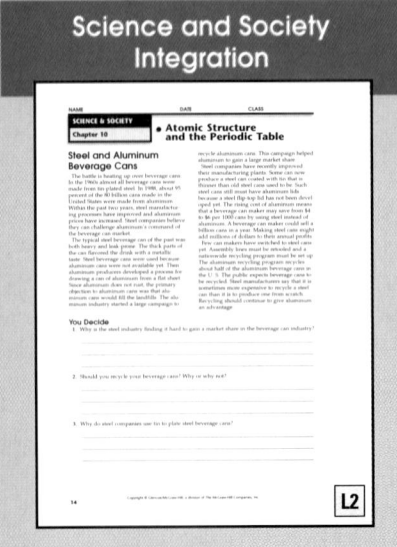

SCIENCE & SOCIETY
Chapter 10 • Atomic Structure and the Periodic Table

Steel and Aluminum Beverage Cans

L2

Multicultural Connections

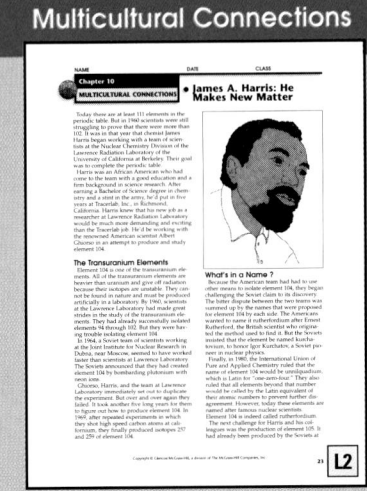

Chapter 10
MULTICULTURAL CONNECTIONS • James A. Harris: He Makes New Matter

L2

Concept Mapping

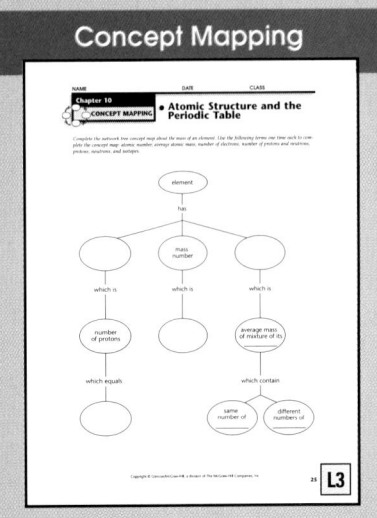

Chapter 10
CONCEPT MAPPING • Atomic Structure and the Periodic Table

L3

Atomic Structure and the Periodic Table

CHAPTER OVERVIEW

Section 10-1 This section introduces the general structure of the atom. The makeup of the nucleus and electron cloud are described.

Section 10-2 Science and Society The student learns that protons and neutrons are made of smaller parts called quarks.

Section 10-3 In this section, the student learns that atoms of different elements have different masses. The term *isotope* is introduced, and the concept of average atomic mass is developed.

Section 10-4 The student learns how the periodic table organizes elements into groups and periods related to atomic structure. Students are introduced to dot diagrams of atoms.

Chapter Vocabulary

chemical	isotope
symbol	average
nucleus	atomic mass
electron	periodic table
proton	group
neutron	dot diagram
atomic	period
number	metal
electron cloud	nonmetal
quark	metalloid
mass number	

Theme Connection

Scale and Structure The theme is developed through a presentation of atomic structure and the periodic table. The structure of an atom is related to its position on the periodic table.

Previewing the Chapter

268

Learning Styles

Look for the following logo for strategies that emphasize different learning modalities. **LS**

Kinesthetic	Explore, p. 269; Activity 10-1, p. 275; MiniLAB, p. 276; People and Science, p. 292
Visual-Spatial	Demonstration, p. 272; Reteach, p. 277
Interpersonal	Activity 10-1, p. 275; Activity 10-2, pp. 282-283; Reteach, p. 289
Intrapersonal	MiniLAB, p. 288
Logical-Mathematical	Explore, p. 269; Reteach, pp. 273, 280; Across the Curriculum, p. 283; Activity, p. 286
Linguistic	Motivate Activity, p. 271; Enrichment, pp. 287, 288; Revealing Preconceptions, p. 279

Chapter 10

Atomic Structure and the Periodic Table

The launch and alignment of the Hubble space telescope shown here has allowed us to have unusually clear views of objects in distant space. The collection of stars shown in this photo could represent a model for the structure of atoms. In the pages that follow, you will learn about scientists' models of atoms.

EXPLORE ACTIVITY

Make a model of an atom using pieces of hardware.

1. Count the number of different pieces of hardware your group was given.
2. Bury these pieces in a ball of modeling clay.
3. Trade balls of clay with another group.
4. Without pulling apart the clay, use toothpicks to find out how many pieces of each kind of hardware are in the ball. Keep track in your Science Journal, and sketch the shapes as they develop.

Observe: How many of each kind of hardware did you determine? Were you right? Now you have experienced using indirect evidence, as scientists do, to infer the identity and quantity of the hidden parts of a model.

Previewing Science Skills

▶ In the Skill Builders, you will map concepts, make and use a table, and make and use a graph.

▶ In the Activities, you will build models, infer, compare and contrast, and interpret data.

▶ In the MiniLABs, you will observe and infer.

269

EXPLORE ACTIVITY

Purpose
LS **Kinesthetic** Use the Explore Activity to introduce atomic structure and the problems that faced scientists who tried to determine atomic structure. **L1** **LEP**

Preparation
Collect small nuts, bolts, and washers. Determine how much clay will be needed for each group.

Materials
assorted small nuts, bolts, and washers; toothpicks; modeling clay

Teaching Strategies
LS **Logical-Mathematical** Have students prepare a detailed map of the interior of the clay ball. Use a broken-off toothpick in a permanent location on the ball as a point of reference. Record the position of each probe and the result. An *O* means the probe hit nothing; an *X* indicates that the probe touched a piece of hardware.

Observe
Answers will vary, but students will find objects as they probe the clay.

Assessment

Process In their Science Journals, have students suggest how scientists might solve the problems of probing something as small as atoms. Use the Performance Task Assessment List for Science Journal in **PASC**, p. 103. **P**

269

Section 10•1

Prepare

Section Background

- The International Union of Pure and Applied Chemistry (IUPAC) has adopted a three-letter symbol for elements beyond 109. The symbols represent the names, which themselves are made up of Latin and Greek prefixes for the corresponding atomic numbers.
- Ca. 400 B.C., Democritus proposed that all matter was composed of small particles he called *atomos*. It wasn't until 1808 that Dalton restated this as a part of the atomic theory of matter.
- Atomic numbers were first assigned to elements as a result of the work of Henry Moseley.
- In 1913, Niels Bohr described the planetary model of the atom. Schrödinger later used quantum mechanics to describe electron clouds.

Preplanning

- Try to have a wall-size copy of the periodic table available to refer to during this and following chapters.
- To prepare for Activity 10-1, cut and glue the magnetic tape to the backs of the paper circles. Allow the glue to dry overnight. As an alternative, obtain magnetic tape that has one peel-and-stick side.

Student Text Question

Which symbols in Table 10-1 might come from Latin?
Cu, Au, Fe, Hg, K, Ag, Na

Science Words

- chemical symbol
- nucleus
- electron
- proton
- neutron
- atomic number
- electron cloud

Objectives

- List the names and symbols of common elements.
- Describe the present model of the atom.
- Describe how electrons are arranged in an atom.

Chemical Symbols

Do the letters C, Al, Ne, and Ag mean anything to you? Each letter or pair of letters is a **chemical symbol**, which is an abbreviated way to write the name of an element. The black material on a burned match is carbon—C. You may wrap food in foil made of aluminum—Al. Have you noticed the bright glow of electrical signs? Many are filled with neon—Ne. You often use coins that contain copper—Cu.

Chemical symbols consist of one capital letter or a capital letter plus one or two small letters. For some elements, the symbol is the first letter of the element's name. For other elements, the symbol is the first letter of the name plus another letter from its name. Some symbols, such as Ag, are derived from Latin. *Argentum* is Latin for "silver." Which symbols in **Table 10-1** might come from Latin?

Table 10-1

Symbols of Some Elements			
Element	**Symbol**	**Element**	**Symbol**
Aluminum	Al	Iodine	I
Calcium	Ca	Iron	Fe
Carbon	C	Magnesium	Mg
Chlorine	Cl	Mercury	Hg
Copper	Cu	Nitrogen	N
Fluorine	F	Oxygen	O
Gold	Au	Potassium	K
Helium	He	Silver	Ag
Hydrogen	H	Sodium	Na

Matter and Atoms

The idea of atoms began more than 2400 years ago with Greek thinkers, who defined atoms as the smallest parts of matter. Atoms consist of a positively charged center, or **nucleus**, surrounded by negatively charged particles called **electrons.** The two major kinds of particles in the nucleus are

Program Resources

Reproducible Masters
Study Guide, p. 42 [L1]
Reinforcement, p. 42 [L1]
Enrichment, p. 42 [L3]
Cross-Curricular Integration, p. 16
Activity Worksheets, pp. 57, 58 [L1]

Transparencies
Section Focus Transparency 38 [L1]

Table 10-2

Comparison of Particles in an Atom			
Particle	Relative Mass	Charge	Location in the Atom
Proton	1	1+	Part of nucleus
Neutron	1	none	Part of nucleus
Electron	0	1−	Moves around nucleus

protons and **neutrons.** Together, these are referred to as nucleons. The nucleus contains most of the mass of the atom. The mass of a proton is about the same as that of a neutron. The mass of an electron is about 1/2000 the mass of a proton. The electron's mass is so small that it is considered negligible when finding the mass of an atom.

A neutron is neutral, which means it has no charge. A proton has a positive charge. As a result, the net charge on the nucleus is positive. The amount of positive charge on a proton is equal to the amount of negative charge on an electron. **Table 10-2** summarizes this information. Although atoms are composed of smaller particles, scientists still consider atoms to be the basic building blocks of matter.

Counting in Atoms

The **atomic number** of an atom is the number of protons in its nucleus. Every atom of the same element has the same number of protons. For example, every carbon atom has six protons. Therefore, it has the atomic number six. Atoms of different elements have different numbers of protons. For example, every carbon atom has six protons, but every oxygen atom has eight protons.

The number of electrons in a neutral atom is equal to the number of protons. A neutral carbon atom, then, has six electrons because it has six protons. Because the numbers of positively charged protons and negatively charged electrons in an atom are equal, the atom as a whole is electrically neutral.

Models of the Atom

As scientists continued to study matter and atoms, they tried to form a mental picture or model of what an atom might look like. A model helps us understand something we cannot see directly, usually because it is too large or too small. A good model of the atom must explain all the information about matter and atoms. As more information was collected, scientists changed their models. Therefore, the model of the atom we use today is the result of the work of many scientists.

USING MATH

You can use the integers +6 and −6 to represent the charge of 6 protons and 6 electrons in a carbon atom. The addition problem $+6 + (−6) = 0$ shows that the charge is neutral. Write an addition problem that shows that an oxygen atom with 8 protons and 8 electrons is neutral.

10-1 Structure of the Atom   271

1 Motivate

Bellringer

Before presenting the lesson, display **Section Focus Transparency 38** on the overhead projector. Assign the accompanying **Focus Activity** worksheet. [L1] [LEP]

Activity

[LS] **Linguistic** Obtain a periodic table that is in another language from your school's foreign language teacher. Explain that the chemical symbols are the same for all languages.

2 Teach

Content Background

By 1700, thirteen elements had been identified in their pure form. When Mendeleev proposed his periodic table in 1869, chemists knew of 26 elements. In 1908, Moseley used X rays to determine the atomic numbers of 81 elements. Today, 111 elements have been discovered or synthesized.

Revealing Preconceptions

Students often do not realize the importance of capital and lowercase letters in chemical symbols. For example, *CO* and *Co* do not have the same meaning. Point out that carbon monoxide is not the same as an atom of cobalt.

Use **Cross Curricular Integration**, p. 16 as you teach this lesson.

USING MATH

$+8 + (−8) = 0$

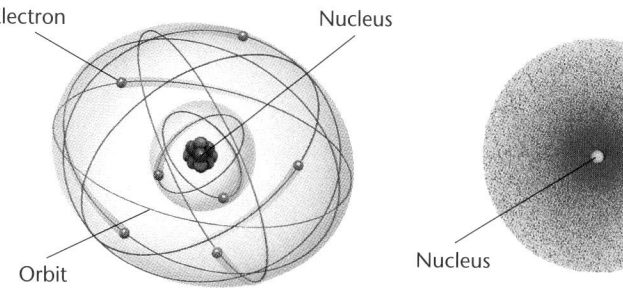

Electron   Nucleus

Orbit

Bohr model

Nucleus

Electron cloud model

Figure 10-1

The left illustration is an early model of an atom with defined paths for electrons. The right illustration is a later model showing a region where electrons are likely to be. *Why is the later model more accurate?*

One of the early models of an atom looked like the one on the left in **Figure 10-1.** It was developed by the Danish scientist Niels Bohr in 1913. Bohr pictured the atom as having a central nucleus with electrons moving around it in well-defined paths, or orbits.

The Atom: A Cloudy View

In 1926, scientists developed a better model of the atom. In this model, the electrons moved about in a region called an **electron cloud** as in **Figure 10-1.** This cloud surrounds the nucleus of the atom. It describes the region where an electron is likely to be at any time. The diameter of the nucleus is about 1/100 000 the diameter of the electron cloud. Look at **Figure 10-2.** Suppose you built a model of an atom with an electron cloud as wide as a football field. The atom's nucleus would be about the thickness of the wire in a paper clip!

Because an electron's mass is so small, it is impossible for you—or anyone—to describe exactly where it is as it moves in the atom. All anyone can give is its probable location. Describing an electron's location around a nucleus is like trying to describe your location in your science class at any given moment. Your most probable location during the class is at your desk. Other possible, though less probable, locations are at the pencil sharpener and at the teacher's desk.

Figure 10-2

In this example, the paper clip wire represents the width of a nucleus. The outer seats of the stadium represent the outer electrons. *What makes up most of the atom?*

Across the Curriculum

Design and Engineering Both engineers and scientists often use models to predict the behavior of actual objects. Design flaws can be corrected because a model behaves just like the real thing. Sometimes, computer models are used instead of physical models. Our model of the atom should also allow us to explain how the real atom behaves.

A Probability Model

You may have heard the expression, "you can't be everywhere at once." The multiple-exposure photo in **Figure 10-3** shows what it might be like if you could. The electron cloud model of the atom is based on the same idea. Scientists make calculations of the electron's most probable locations around the nucleus. If each location were marked with a dot, the closer spacing of the dots would indicate the most probable area for an electron. This is called an electron cloud because the dots give a cloudlike appearance when taken altogether.

Figure 10-3

If you could combine photos taken of a person in a room, you would not know exactly where that person was, but you could predict the most likely places. *In what area of this classroom would you most likely find the student?* Our best understanding of electron locations is similar to this.

Energy Levels and Electrons

Figure 10-4 illustrates another way to look at the placement of electrons. The electrons in an atom make up the electron cloud. Within the electron cloud, electrons are at various distances from the nucleus. Electrons closest to the nucleus have low energy. Electrons farther away from the nucleus

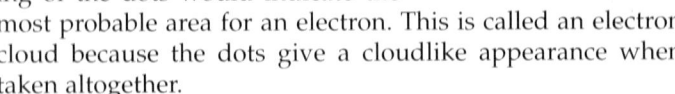
Figure 10-4

Sulfur's atomic number is 16. It has 16 protons and 16 electrons. The electron arrangement of sulfur is shown here.

A The first level is filled with only two electrons.

B The second energy level has one spherical area and three pairs of lobes. This level can accommodate eight electrons. *How many of the electrons in sulfur remain?*

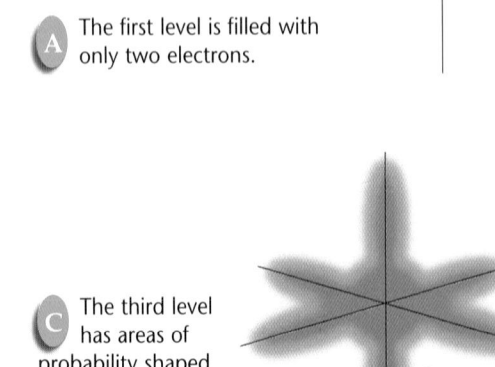

C The third level has areas of probability shaped like the second, only larger.

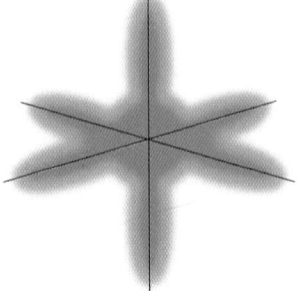

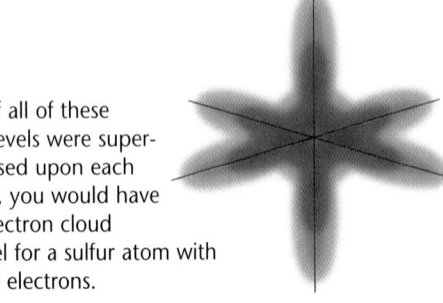

D If all of these levels were superimposed upon each other, you would have an electron cloud model for a sulfur atom with its 16 electrons.

10-1 Structure of the Atom   **273**

CONNECT TO

PHYSICS

See page 274.
Answer electron

3 Assess

? FLEX Your Brain

Use the Flex Your Brain activity to have students explore ATOMIC STRUCTURE.
📁**Activity Worksheets,** p. 5

Reteach

LS **Logical-Mathematical** Have each student complete a duplicate of **Table 10-2** that has one item missing from each of the three lines of information.

Extension

📁 For students who have mastered this section, use the **Reinforcement** and **Enrichment** masters.

4 Close

•MINI•QUIZ•

Use the Mini Quiz to check students' recall of chapter content.

1. **An abbreviated way to write an element's name is to use a(n)** _____ . *chemical symbol*

2. **An atom's nucleus contains** _____ . *protons and neutrons*

3. **An atom's atomic number is the number of** _____ **in its nucleus.** *protons*

Section Wrap-up

Review

1. C, Al, H, O, Na

2. proton: positive, nucleus; neutron: no charge, nucleus; electron: negative, surrounding nucleus

3. the region where electrons are likely to be in an atom

4. **Think Critically** The visual blur formed by the moving blades can be a model for an electron cloud. However, the blades move in a flat plane, whereas electrons have a spherical distribution.

USING MATH

Level	Electrons	Orbitals
1	2 ÷ 2	1
2	8 ÷ 2	4
3	18 ÷ 2	9
4	32 ÷ 2	16

CONNECT TO

PHYSICS

When a glass rod is rubbed with silk, the rod becomes positively charged. *Infer* what type of particle in the atoms in the rod has been removed?

---

have higher energy. You can represent the energy differences of the electrons by picturing the atom as having energy levels.

Energy levels are somewhat like shelves of a refrigerator door. A carton on the lowest shelf represents an electron in the lowest energy level. The difference in spacing of the shelves could indicate the differences in the amount of energy the electrons in that level can have. Shelves of a refrigerator door can usually hold the same number of cartons. Unlike refrigerator shelves, each energy level of an atom has a different maximum number of electrons it can hold. The lowest energy level can hold just two electrons. The second energy level can hold eight electrons, and the third energy level, a maximum of 18 electrons.

Table 10-3

Electrons in Energy Levels	
Energy Level in Atom	**Maximum Number of Electrons**
1	2
2	8
3	18
4	32

Section Wrap-up

Review

1. Write the chemical symbols for the elements carbon, aluminum, hydrogen, oxygen, and sodium.

2. List the names, charges, and locations of three kinds of particles that make up an atom.

3. What does an electron cloud represent?

4. **Think Critically:** How might an electric fan that is turned on be a model of an atom? How is the fan unlike an atom?

Skill Builder

Concept Mapping

Make a concept map for the parts of an atom. Include the following terms: *electron cloud, nucleus, electrons, protons,* and *neutrons.* Also provide the charge of each part. If you need help, refer to Concept Mapping in the **Skill Handbook.**

USING MATH

Scientists often refer to the probability areas for electrons as orbitals. Each orbital, regardless of shape or level, can hold a maximum of two electrons. Using **Table 10-3,** determine how many electron orbitals would exist in each of the first four energy levels.

Skill Builder

Atom

part / part

Electron Cloud / Nucleus

contains / contains / contains

Electrons / Protons / Neutrons

charge / charge / charge

Negative / Positive / Neutral

✓ Assessment

Performance Select several elements. Tell students how many protons, neutrons, and electrons are in an atom of each. Have students model a nucleus of an atom of each element. Use the Performance Task Assessment List for Model in **PASC,** p. 51. **P**

P

Activity 10-1

Models of Atomic Structure

Building a model can help you understand how a complex system operates. Something as simple as a paper and magnet model can give you information about the relationship between the parts of an atom.

Problem
How can you build a model to help you understand how the main parts of an atom are related?

Materials
- magnetic board about 20 cm × 27 cm
- one 0.5-cm piece and 20 1-cm pieces of rubber magnetic tape
- circles of white paper 4 cm wide
- circles of red paper 1 cm wide
- marker

Procedure
1. Choose an element with an atomic number of 1 through 20. Determine the number of each kind of particle needed to make up an atom of that element.
2. Use a marker to write the number of protons and neutrons on a paper circle. This represents the nucleus of the atom.
3. Use a 0.5-cm magnetic strip to attach the model nucleus to one side of a magnetic board.
4. Attach the red paper circles to the 1-cm magnetic strips. Arrange these model electrons in energy levels around the nucleus. Use as many of these electrons as you need for your element.
5. Remove either the model nucleus or the model electrons from the magnetic board and ask classmates to identify the element.
6. Repeat steps 1 through 5 for another element.

Analyze
1. In a neutral atom, which particles will always be present in equal numbers?
2. Hypothesize what you think would happen to the charge of an atom if one of the electrons were removed from the atom.
3. Except for hydrogen, how many first-level electrons did each selected atom contain?

Conclude and Apply
4. How is this **model** of an atom similar to an actual atom?
5. Name two differences, other than size, between your model and an actual atom.
6. **Predict** what happens to the atom if one proton is removed from the nucleus and one electron is also removed.

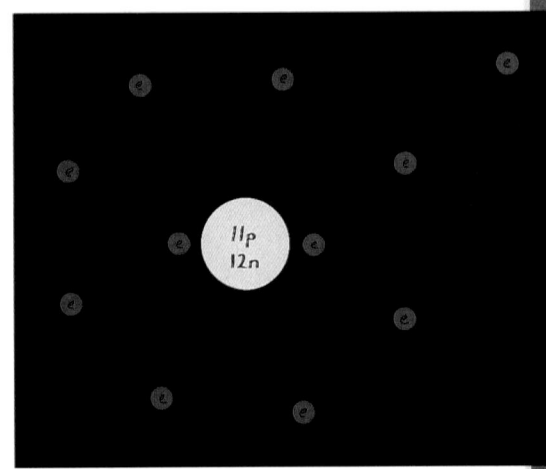

10-1 Structure of the Atom   **275**

---

equal, the protons and neutrons are localized in the nucleus, and electrons are distributed in energy levels outside the nucleus.
5. Answers will vary. Possible answers include the facts that the model's electrons are not moving, the model is two dimensional, and the electrons would be distributed much farther from the nucleus if the model were to scale.
6. The atom becomes a different element with an atomic number one less than that of the original atom.

✔ Assessment

Process Have each group make one example of an atom, and display each as a quiz for the class to identify the represented atom. Use the Performance Task Assessment List for Group Work in **PASC**, p. 97. **P**

Activity 10-1

Purpose

LS **Kinesthetic** Formulate models of atomic structure and identify atoms by their structures. **L1** **LEP**

Process Skills
formulating models, inferring, classifying, forming operational definitions, comparing and contrasting, predicting

Time
40 minutes

📁 **Activity Worksheets,** pp. 5, 57, 58

Teaching Strategies
Alternate Materials Painted steel rectangular snack trays may work well as magnetic boards. Use the bottom if there is a distracting design. Another possibility is to use carpet squares and the stiff half of Velcro strips.

Troubleshooting Make sure the atom models are correctly constructed before having the class identify them. Going beyond element 20 requires a more detailed understanding of the distribution of electrons in energy levels.

- Peel-and-stick magnetic plastic tape is sold in most craft and hobby shops.

LS **Interpersonal** Have one student or pair construct the atom's nucleus and present it to the other two students, who must then decide on the electron arrangement.
COOP LEARN

Answers to Questions
1. protons and electrons
2. The number of protons would be one greater than the number of electrons. Therefore, the atom would have a positive charge.
3. two
4. Answers will vary but should reflect the ideas that the numbers of protons and electrons are

275

1 Motivate

Bellringer

Before presenting the lesson, display **Section Focus Transparency 39** on the overhead projector. Assign the accompanying **Focus Activity** worksheet. L1 LEP

2 Teach

MiniLAB

Purpose
Kinesthetic A model for identifying subatomic particles will be studied. L1
LEP COOP LEARN

Materials
aluminum pie pans; flour or sand; various common objects

Analysis
1. Prominent markings make more specific tracks.
2. It accentuates them. Using more or less energy in particle accelerators can release or not release quarks.

Assessment

Oral Have students explain to each other specific clues that they looked for in identifying objects. Use the Performance Task Assessment List for Making Observations and Inferences in **PASC**, p. 17. P

Activity Worksheets, pp. 5, 61

Science Words

quark

Objectives

- Identify quarks as particles of matter that make up protons and neutrons.
- Explain how particle accelerators are used to study particles within atoms.

TECHNOLOGY:
10•2 Smaller Particles of Matter

Looking for Quarks

You know that protons, electrons, and neutrons make up atoms. Are these particles made up of even smaller particles? Scientists hypothesize that electrons are not made of any smaller particles. Protons and neutrons, however, are made up of smaller particles called **quarks.** At present, scientists have confirmed the existence of six uniquely different quarks. A particular arrangement of three quarks held together with strong nuclear force produces a proton. Another arrangement of three quarks produces a neutron. A breakdown of an atom and its parts is shown in **Figure 10-5.**

Studying Collisions

Fermilab, in Batavia, Illinois, houses a machine that can generate the tremendous forces that are required to study quarks. This machine, the Tevatron, shown in **Figure 10-6,** is approximately 6.4 km in circumference. Electric and magnetic fields are used to accelerate, focus, then collide fast-moving protons. Studying the collisions reveals information about the inner structure of the atom.

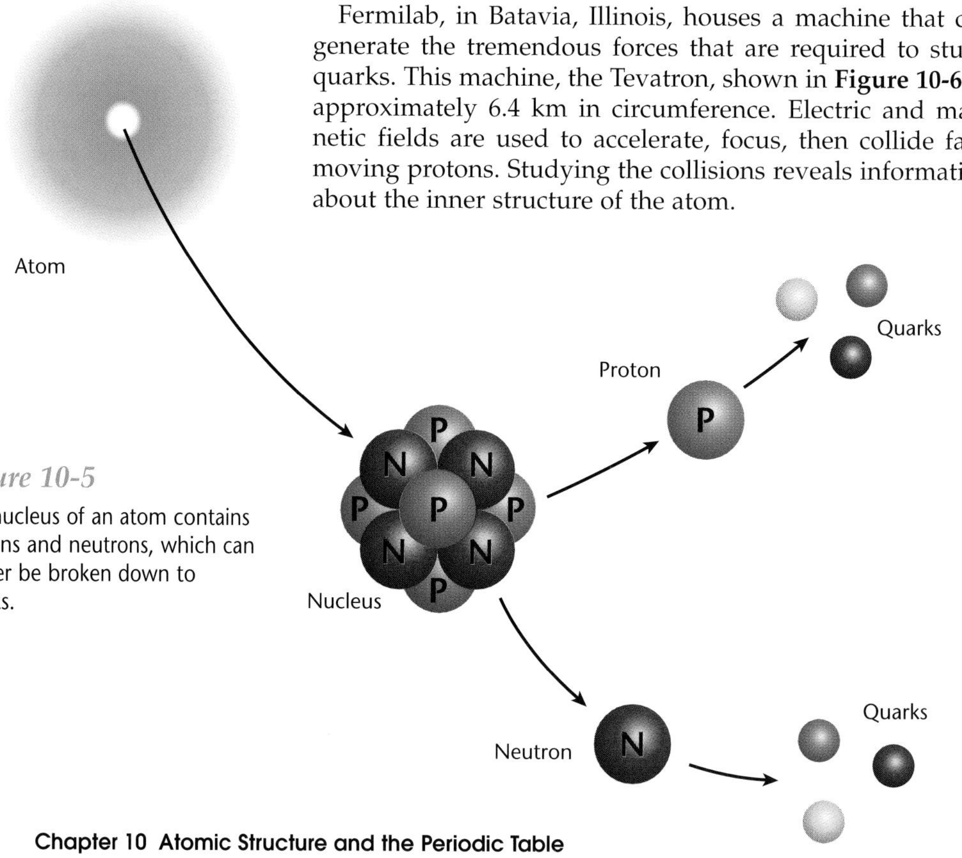

Figure 10-5

The nucleus of an atom contains protons and neutrons, which can further be broken down to quarks.

Atom

Nucleus

Proton

Quarks

Neutron

Quarks

Program Resources

Reproducible Masters
Study Guide, p. 43 L1
Reinforcement, p. 43 L1
Enrichment, p. 43 L3
Activity Worksheets, p. 61

Transparencies
Section Focus Transparency 39 L1

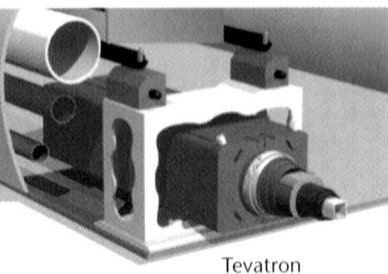

Tevatron

The Last Quark?

Finding the sixth quark took a team of nearly 450 scientists several years. The collisions of protons with oppositely charged particles at high energies were necessary to produce the sixth quark—typically referred to as the "top" quark.

Just as police investigators can reconstruct traffic accidents from clues at the scene, trained scientists are able to examine tracks left after proton collisions. The tracks of the sixth quark were hard to detect. Only about one-billionth of a percent of the proton collisions yielded evidence of a sixth quark.

Section Wrap-up

Review

1. What are quarks?
2. How does the Tevatron help scientists study the atom?

inter NET CONNECTION The Fermilab site on the World Wide Web contains detailed information on the Tevatron and the experiments done to find the top quark. Visit the Chapter 10 Internet Connection at Glencoe Online Science, **www.glencoe.com/sec/science,** for a link to more information about experiments at Fermilab.

MiniLAB

How can indirect evidence be used to identify known objects?

Procedure

1. In your two-member team, have one person make an imprint or track of two objects in a pan of shallow flour or sand.
2. The other partner must now identify which objects were used to make the tracks.
3. Smooth the surface, trade roles, and repeat the process.

Analysis

1. What made some objects easy to identify?
2. How does the force involved in making the imprints affect the observed tracks? How does the same reasoning apply to detecting quarks?

SCIENCE & SOCIETY

10-2 Smaller Particles of Matter 277

Prepare

Section Background

- For a long time, chemists used an atomic mass scale based on oxygen-16. In 1961, it was agreed to use carbon-12 as the standard for atomic mass.
- Be certain students understand that isotopes of the same element are alike chemically but are different in mass.
- The mass of 1 u is approximately 1.67×10^{-24} g.

Preplanning

To prepare for Activity 10-2, you will need to obtain the candy-coated peanuts and candy-coated chocolates. Sort out the needed number of different colors and place them in plastic bags.

1 Motivate

Bellringer

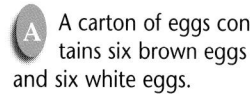 Before presenting the lesson, display **Section Focus Transparency 40** on the overhead projector. Assign the accompanying **Focus Activity** worksheet. L1 LEP

Tying to Previous Knowledge

Students have previously studied mass in different contexts. Refresh their memories by asking them to relate their concepts of mass. Make sure they are not confusing mass with volume.

Visual Learning

Figure 10-7 What number do they have in common? *12*
LEP LS

10•3 Masses of Atoms

Science Words

mass number
isotope
average atomic mass

Objectives

- Compute the atomic mass and mass number of an atom.
- Identify and describe isotopes of common elements.
- Interpret the average atomic mass of an element.

Atomic Mass

If asked to estimate the height of your school building, you would not give an answer in miles. Realizing the scale of the building, you would more likely think about the height in feet. When thinking about the small masses of atoms, scientists found that even grams were not small enough. The unit of measurement of those particles is the atomic mass unit (u). In fact, the mass of a proton or a neutron is almost equal to 1 u. This is not a coincidence—the unit was defined that way. The atomic mass unit is defined as one-twelfth the mass of a carbon atom containing six protons and six neutrons. Remember that the mass of a carbon atom is in its nucleus because the atom's six electrons have a negligible mass. Therefore, each of the 12 particles in the nucleus must have a mass nearly equal to one-twelfth the mass of the carbon atom. Thus, a proton or a neutron has a mass of about 1 u. To help you understand this concept, look at **Figure 10-7**.

Figure 10-7

You can compare a carton of eggs to a carbon atom. *What number do they have in common?*

A A carton of eggs contains six brown eggs and six white eggs.

B You could define a mass unit as one-twelfth the mass of the entire dozen.

C The mass of each egg, brown or white, would be equal to your defined unit.

278   Chapter 10 Atomic Structure and the Periodic Table

Program Resources

Reproducible Masters
Study Guide, p. 44 L1
Reinforcement, p. 44 L1
Enrichment, p. 44 L3
Critical Thinking/Problem Solving, p. 16
Science Integration Activity, pp. 19, 20
Activity Worksheets, pp. 59, 60 L1
Concept Mapping, pp. 25, 26

Transparencies
Section Focus Transparency 40 L1

Mass Number

The **mass number** of an atom is the sum of the number of protons and the number of neutrons in the nucleus of an atom. As you can see in **Table 10-4**, the mass number of an atom is almost equal to the mass of its most common form, expressed in atomic mass units.

If you know the mass number and the atomic number of an atom, you can then calculate the number of neutrons. The number of neutrons is equal to the atomic number subtracted from the mass number.

number of neutrons = mass number − atomic number

Isotopes

Not all the atoms of an element have the same number of neutrons. Atoms of the same element that have different numbers of neutrons are called **isotopes**.

B
Boron Isotopes

Mass number   11
# of protons   -5
# of neutrons   6

Mass number   10
# of protons   -5
# of neutrons   5

Suppose you have a sample of the element boron like the powder shown in **Figure 10-8**. Naturally occurring atoms of boron have mass numbers of 10 or 11. How many neutrons are there in a boron atom? It depends upon the boron to which you are referring. Obtain the number of protons from **Table 10-4** and use the formula above to calculate that a boron isotope may contain five or six neutrons. Use the example above to help you.

Figure 10-8

Boron has two naturally occurring isotopes.

Table 10-4

Mass Numbers of Some Atoms					
Element	Symbol	Protons	Neutrons	Mass Number	Average Atomic Mass*
Boron	B	5	6	11	10.81
Carbon	C	6	6	12	12.01
Oxygen	O	8	8	16	16.00
Sodium	Na	11	12	23	22.99
Copper	Cu	29	34	63	63.55
* to two decimal places					

10-3 Masses of Atoms   279

Use **Concept Mapping**, pp. 25, 26 as you teach this lesson.

279

CONNECT TO
LIFE SCIENCE

Answer Carbon-14 is radio-active and thus decays over time. The amount of carbon-14, compared to the amount of carbon-12, present in the remains of an organism may be used to determine the age of the remains.

📁 Use these program resources as you teach this lesson.
Critical Thinking/Problem Solving, p. 16
Science Integration Activity, pp. 19, 20

3 Assess

Check for Understanding

❓ FLEX Your Brain

Use the Flex Your Brain activity to have students explore ATOMIC MASS.

📁 **Activity Worksheets**, p. 5

Reteach

LS **Logical-Mathematical** What is the weighted average for a class of 24 students on a quiz if five students earn a grade of 100 and 19 students earn a grade of 80?
$5/24$ (100%) + $19/24$ (80%) = 84.2%

Extension

📁 For students who have mastered this section, use the **Reinforcement** and **Enrichment** masters.

Problem Solving

An Unstable Isotope: Providing Medical Help

A common treatment for cancer tumors is intense, high-energy radiation. Where can doctors find such focused high energy? Understanding the makeup of some isotopes helps solve this problem.

The clustering of protons and neutrons in a nucleus of some isotopes does not produce a stable, low-energy situation. On some occasions, the number of protons and neutrons in a nucleus produces a high-energy, unstable nucleus. In this situation, the nucleus spontaneously changes to become more stable. To do this, the nucleus must give off excess energy in the form of radiation.

A particular application of this situation is found in the treatments of thyroid gland tumors. In order to function properly, the thyroid gland must absorb iodine.

Solve the Problem:
1. If an unstable isotope of iodine were carefully administered to a thyroid tumor patient, why might it cure the person?
2. Why is it important to use iodine isotopes in this particular situation instead of carbon isotopes?

Think Critically:
Iodine typically can be found as one of two isotopes: iodine-131 and iodine-130. How many neutrons are present in each? In this case, the isotope that spontaneously gives off radiation is the one with the most neutrons. Which one is the unstable isotope?

Identifying Isotopes

Figure 10-9 shows models of the two isotopes of boron. Because the numbers of neutrons in the isotopes are different, the mass numbers are also different. You use the name of the element followed by the mass number of the isotope to identify each isotope: boron-10 and boron-11. Because most elements have more than one isotope, each element is given an average atomic mass. The **average atomic mass** of an element is the average mass of the mixture of its isotopes.

For example, four out of five atoms of boron are boron-11, and one out of five is boron-10. Thus, in an average sample of five atoms of boron, four atoms are likely to have a mass of 11 u. One atom will have a mass of 10 u. If you measured the mass of the five atoms, you would find it to be 54 u. The average atomic mass of the boron mixture is 54 u divided by 5, or 10.8 u. That is, the average atomic mass of the element boron is 10.8 u. Note that

Community Connection

Nuclear Medicine Contact the nearest hospital to see whether they have a nuclear medicine department. Ask if they could send a representative to your classroom to explain some of their procedures. If this is not possible, have students write to the nuclear medicine department of a hospital to ask for brochures about using isotopes in medicine.

Figure 10-9
The two isotopes of boron have different numbers of neutrons and different mass numbers. *How many neutrons are in each isotope?*

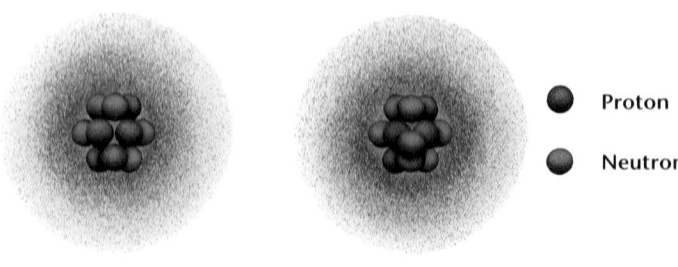

● Proton

● Neutron

Living organisms on Earth contain carbon. Carbon-12 makes up 99 percent of this carbon. Carbon-13 and carbon-14 make up the other one percent. Which isotopes are archaeologists most interested in when they determine the age of carbon-containing remains? *Explain* your answer in your Science Journal.

Student Text Question

How many neutrons does each hydrogen isotope have? *hydrogen-1: none; hydrogen-2: one; hydrogen-3: two*

Visual Learning

Figure 10-9 How many neutrons are in each isotope? *five and six* LEP LS

4 Close

Activity

Have students determine the average student mass of the class after each student records his or her weight on a folded piece of paper.

the average atomic mass of boron, given in **Table 10-4,** is close to the mass of its most abundant isotope, boron-11.

For another example, hydrogen has three isotopes—hydrogen-1, hydrogen-2, and hydrogen-3. Each has one proton. How many neutrons does each have? The most abundant isotope is hydrogen-1, and the element's average atomic mass is 1.008. The average atomic mass of each element can be found in the periodic table of elements.

Section Wrap-up

Review

1. A chlorine atom has 17 protons and 18 neutrons. What is its mass number? What is its atomic number?

2. How are the isotopes of an element alike and how are they different?

3. **Think Critically:** Chlorine is used to treat most city water systems. The atomic number of chlorine is 17. The two naturally occurring isotopes of chlorine are chlorine-35 and chlorine-37. The average atomic mass of chlorine is 35.45 u. Why does this indicate that most chlorine atoms contain 18 neutrons?

Skill Builder
Comparing and Contrasting
How does the average atomic mass relate to the mass number of an atom? If you need help, refer to Comparing and Contrasting in the **Skill Handbook.**

Using Computers

Spreadsheet Use a spreadsheet to construct a table organizing information about the atomic numbers; atomic mass numbers; and the number of protons, neutrons, and electrons in atoms of oxygen-16 and oxygen-17.

Section Wrap-up

Review

1. 35, 17

2. They have the same number of protons (atomic number) but different numbers of neutrons, resulting in a different mass number.

3. **Think Critically** An atom of chlorine has 17 protons. Because the average atomic mass is close to 35, most atoms of chlorine would have 18 neutrons, giving a mass number of 35.

Using Computers

Isotope	oxygen-16	oxygen-17
Atomic #	8	8
Protons	8	8
Electrons	8	8
Neutrons	8	9
Mass #	16	17

Skill Builder
The mass number is the total number of protons and neutrons in an atom. Each isotope of an element has a different mass number, which is always an integer. The average atomic mass also takes into account the protons and neutrons; however, it is an average of all the isotopes, so each element has only one average atomic mass.

Assessment

Process Have the students write an equation to find the average atomic mass using the following variables.

x = 1st isotope mass   A = % of 1st isotope
y = 2nd isotope mass   B = % of 2nd isotope
z = 3rd isotope mass   C = % of 3rd isotope

Use the Performance Task Assessment List for Using Math in Science in **PASC,** p. 29. P

Activity
10-2

PREPARATION

Purpose

IS **Interpersonal** Design and carry out an experiment to show how average atomic mass is found. **L1** **LEP** **COOP LEARN**

Process Skills

comparing and contrasting, communicating, classifying, making and using tables, forming operational definitions, forming hypotheses, designing experiments, making models

Time

40 minutes

Materials

Have candy samples sorted and placed in sealed plastic bags. If students are not permitted to open the bags, samples may be used by more than one class.
Alternate Materials Several different types of candy may be used. Other colors may be substituted as long as each set is all the same color.

Safety Precautions

Caution students to keep the bags closed and to not eat the candy.

Possible Hypotheses

Candy samples of different masses can be used to model the calculations for determining the average mass of an element.

📁 **Activity Worksheets,** pp. 5, 59, 60.

Design Your Own Experiment

Candy-Covered Isotopes

The extremely small size of atoms makes it impossible to count them or determine their individual atomic masses using direct means. An instrument called a mass spectrometer allows for such determinations. As demonstrated earlier in this section, average atomic masses depend on the number and masses of the isotopes of an element.

In this activity, find out how you can use colored candy to represent the atoms in an average mass calculation.

PREPARATION

Problem

How can you calculate the average mass of a mixture of objects?

Form a Hypothesis

Based on the calculations and information in this section, state a hypothesis about samples of isotopes regarding number of objects, their individual masses, and average masses.

Objectives

- Using two types of coated candy, make a model to represent the average atomic mass calculations of isotopes.

- Calculate the average mass of candy samples.

Possible Materials

- container of red chocolate-center candy
- container of red peanut-center candy
- container of green chocolate-center candy
- container of green peanut-center candy

Safety Precautions 🚫

Do not eat candy samples used in this activity.

Data and Observations

Sample Data

	Peanut	Chocolate	Average
	candy × mass	candy × mass	$\dfrac{\text{total mass}}{\text{total candies}}$
Red	$4 \times 2 = 8$	$2 \times 1 = 2$	$\dfrac{8 + 2}{4 + 2} = 1.7$
Green	$3 \times 4 = 12$	$3 \times 3 = 9$	$\dfrac{12 + 9}{3 + 3} = 3.5$

282 Chapter 10 Atomic Structure and the Periodic Table

PLAN THE EXPERIMENT

1. Within your group, for purposes of this activity, make a basic assumption that a red peanut candy is twice as massive as the red chocolate candy. Use the ratio of four times more massive for the green peanut candies and three times more massive for green chocolate candies.

2. Decide on a plan to test your hypothesis. Be sure to design a data table in your Science Journal to use with your plans.

Check the Plan

1. If you assume that the mass of one red chocolate candy is 1 and that the mass of the red peanut candy is 2, can you now calculate the total mass of any mixture of red candy?

2. Using your total mixture, can you now calculate the mass, in terms of red candy units, of the mixture?

3. Using the number of candies, design a plan to calculate the average mass of red candy.

4. Use your plan and other assumptions about green candy to deduce the average mass of green candy.

5. *Make sure your teacher approves your plan and that you have indicated any changes suggested in the plan.*

DO THE EXPERIMENT

1. When you count candy pieces, make sure to record the number of each type you use.

2. Show all of your calculations and label your answers in an orderly data table.

Analyze and Apply

1. **Compare** your results of the average mass of red candy with those of your classmates. Do the same for the average mass of green candy.

2. Did you find that regardless of the number of candies used in your groups, the average masses for each type of candy were the same? **Explain** why or why not.

10-3 Masses of Atoms **283**

Go Further

If a bag of red peanut candies contains eight pieces, how many red chocolate pieces would you have to add to the same bag to produce an average mass equal to your answer in this activity?

Across the Curriculum

Math What is the average atomic mass of hafnium if, out of every 100 atoms, five have mass 176 u, 19 have mass 177 u, 27 have mass 178 u, 14 have mass 179 u, and 35 have mass 180 u? *179 u* **LS**

 **Assessment**

Process Assign the class a calculation based on the weighted average concept. Compute the average atomic mass of silver if 51.83% of the atoms in nature have a mass of 106.905 u and 48.17% have a mass of 108.905 u. **Answer:** 107.868 Use the Performance Task Assessment List for Using Math in Science in **PASC**, p. 29. **P**

PLAN THE EXPERIMENT

Possible Procedures

- Assume that a red chocolate candy has a mass of one. Then assume that 1/the mass of a red peanut-centered candy is two times that. Now, take the weighted average of a mixed sample of red candy.
- Do the same for green candy but use a different peanut-to-chocolate ratio, such as 4 units for peanut to 3 for chocolate.

Teaching Strategies
Tying to Previous Knowledge Review with your students the process used to find weighted averages. Use examples that are familiar to students, such as finding average grades if numbers of each letter grade are known.

DO THE EXPERIMENT

Expected Outcome
See the sample data on the chart.

Analyze and Apply

1. Average should be the same if the same number of each type of candy is used.

2. Average masses were not the same because different units and different numbers of each type of candy were used.

Go Further

The new bag of red candy should contain four pieces of chocolate-centered candy. The 2:1 ratio of peanut to chocolate should be maintained for all samples to produce the same average mass.

Prepare

Section Background

- Dimitri Ivanovich Mendeleev was born in Siberia, the last child of a large family. He graduated from college in St. Petersburg at the top of his class and then studied in France and Germany. In 1866, he became professor of chemistry at the University of St. Petersburg.

- Mendeleev predicted the existence of six undiscovered elements—scandium, gallium, germanium, technetium, rhenium, and polonium.

- Mendeleev's first three predictions, made in 1871, were confirmed by 1885.

Preplanning

- Collect assorted pens and pencils for the MiniLAB on page 288.

- If you have not already obtained one, you should have a display-size periodic table available.

1 Motivate

Bellringer

Before presenting the lesson, display **Section Focus Transparency 41** on the overhead projector. Assign the accompanying **Focus Activity** worksheet. L1 LEP

Tying to Previous Knowledge

The books in the library are classified so that a student can quickly find a book about a particular subject. The elements are classified and arranged on the periodic table. The periodic table contains a library of useful information for the scientist.

Science Words

- periodic table
- group
- dot diagram
- period
- metal
- nonmetal
- metalloid

Objectives

- Describe the periodic table of elements and use it to find information about an element.
- Distinguish between a group and a period.
- Use the periodic table to classify an element as a metal, nonmetal, or metalloid.

Figure 10-10

Among his contributions to science, Mendeleev is best known for his periodic chart of the elements. *Why do you think there are question marks in some places on his chart?*

10•4 The Periodic Table

Structure of the Periodic Table

On clear evenings, you can see the phases of the moon. Each month, the moon seems to grow larger, then smaller, in a repeating pattern. This type of change is periodic. *Periodic* means "repeated in a pattern." Look at a calendar. The days of the week are periodic because they repeat themselves every seven days. The calendar is a periodic table of days. You use calendars to organize your activities.

In the late 1800s, Dimitri Mendeleev, a Russian chemist, searched for a way to organize the elements. When he arranged all the elements known at that time in order of increasing atomic masses, he discovered that there was a pattern. Chemical properties found in lighter elements could be shown to repeat in heavier elements. Because the pattern repeated, it could be considered periodic. Today, we call this arrangement a **periodic table** of elements. An early periodic table is shown in **Figure 10-10**.

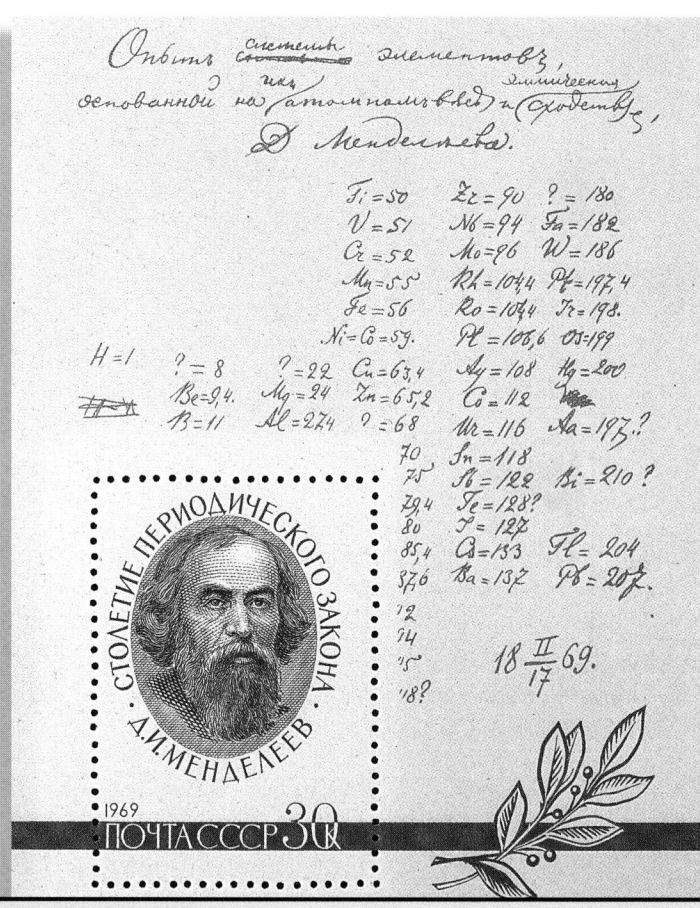
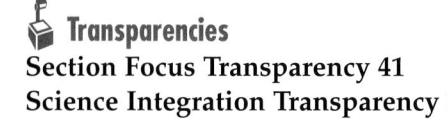

Program Resources

Reproducible Masters
Study Guide, p. 45 L1
Reinforcement, p. 45 L1
Enrichment, p. 45 L3
Activity Worksheets, p. 62 L1
Lab Manual 22
Cross-Curricular Integration, pp. 14, 15
Science and Society Integration, p. 14
Multicultural Connections, pp. 23, 24

Transparencies
Section Focus Transparency 41
Science Integration Transparency 10
Teaching Transparencies 19, 20

Mendeleev's periodic table had blank spaces. He looked at the properties and atomic masses of the elements surrounding these blank spaces. From this information, he predicted the properties and the mass numbers of new elements that had not yet been discovered. Sometimes, you make predictions like this. Suppose someone crossed out a date on a calendar, as shown. You could predict the missing date by looking at the surrounding dates. Mendeleev's predictions proved to be quite accurate. Scientists later discovered elements, such as germanium, having the properties that he had predicted (Table 10-5).

Counting Protons: An Improved Table

Although Mendeleev's arrangement of elements was successful, it did need some changes. On Mendeleev's table, the atomic mass gradually increased from left to right in each row. If you look at the modern periodic table as shown on pages 286 and 287, you will see several examples, such as cobalt and nickel, where the mass decreases from left to right. However, you may notice that the atomic number always increases from left to right. The work of Henry G.J. Moseley, a young English scientist, in 1913 led to the arrangement of elements based on their properties and atomic numbers instead of an arrangement based on atomic masses.

Each box in the periodic table contains information about the elements that you studied earlier in this chapter. Look at **Figure 10-11**. This box represents the element boron. The atomic number, chemical symbol, name, and average atomic mass are included in this box. The boxes for all of the elements are arranged in order of their atomic numbers.

Figure 10-11

The periodic table shows the symbol, name, atomic number, and atomic mass of each element.

Table 10-5

"Ekasilicon" (Germanium)		
Properties	**Predicted**	**Actual**
Atomic mass	72	72.6
Density	5.5 g/cm^3	5.35 g/cm^3
Color	dark gray	gray-white
Effect of water	none	none
Effect of acid	slight	HCl: no effect
Effect of base	slight	KOH: no effect

Cultural Diversity

Dr. Wu and Parity One of the foundations of physics was the law of atomic parity. Dr. Chien-Shiung Wu was able to disprove this basic law experimentally.

Dr. Wu's colleagues in the study were awarded the Nobel prize in 1957 for their theoretical contributions to the project, but Dr. Wu was overlooked. Dr. Wu, born near Shanghai, China, immigrated to the United States in 1936. At the age of 27, she served briefly as a physics instructor at Princeton—at a time when Princeton would not accept women as students. Dr. Wu was the first woman to receive the Comstock Prize from the National Academy of Sciences and the Research Corporation Award. In 1976, she received the U.S. Medal of Science. Today, Dr. Wu is a highly honored scientist.

2 Teach

Visual Learning

Figure 10-10 Why do you think there are question marks in some places on his chart? *Mendeleev knew that elements belonged there, but they had not yet been discovered.* **LEP** **LS**

Demonstration

Using a battery-powered conductivity tester, touch the instrument's probes to samples of elements. Your high school physics teacher may have a conductivity tester that you can borrow. Students can begin to classify elements as conductors and nonconductors. Use this activity to lead into a discussion about the need to classify elements.

Inquiry Question

The noble gases (Group 18) are *chemically inert* under most reaction conditions. What does *chemically inert* mean? What feature of outer electron structure do all noble gases share? *The elements usually do not react with other elements. With the exception of helium, all have eight outer-level electrons.*

Content Background

In the time of Mendeleev, energy levels and sublevels had not been discovered. He didn't even have a modern concept of what an atom was. His predictions, based solely on physical and chemical properties, were a remarkable achievement.

Concept Background

- The elements have an increasing nonmetallic character as you read from left to right across the table. Along the stair-step line are metalloids, which have properties of both metals and nonmetals.
- The noble gases are technically nonmetals in physical character, but they are extremely unreactive and, therefore, form a separate group with their own distinct properties.
- Hydrogen, by its electron arrangement, is part of Group 1. However, because it has only one electron energy level that can hold only two electrons, it has its own unique set of properties.

GLENCOE TECHNOLOGY

 CD-ROM

Physical Science CD-ROM

Have students perform the interactive exploration for Chapter 10 to reinforce important chapter concepts and thinking processes.

Activity

LS Logical-Mathematical List the elements in the first six periods that would be out of order if the table were arranged by atomic mass. *The elements Ar, K, Co, Ni, Te, and I would be out of order.*

Table 10-6

PERIODIC TABLE OF THE ELEMENTS
Based on Carbon 12 = 12.0000

Element — Hydrogen
Atomic Number — 1 — State of Matter
Symbol — H
Atomic Mass — 1.008

	1	2	3	4	5	6	7	8	9
1	Hydrogen 1 **H** 1.008								
2	Lithium 3 **Li** 6.941	Beryllium 4 **Be** 9.012							
3	Sodium 11 **Na** 22.990	Magnesium 12 **Mg** 24.305							
4	Potassium 19 **K** 39.098	Calcium 20 **Ca** 40.078	Scandium 21 **Sc** 44.956	Titanium 22 **Ti** 47.88	Vanadium 23 **V** 50.942	Chromium 24 **Cr** 51.996	Manganese 25 **Mn** 54.938	Iron 26 **Fe** 55.847	Cobalt 27 **Co** 58.933
5	Rubidium 37 **Rb** 85.468	Strontium 38 **Sr** 87.62	Yttrium 39 **Y** 88.906	Zirconium 40 **Zr** 91.224	Niobium 41 **Nb** 92.906	Molybdenum 42 **Mo** 95.94	Technetium 43 **Tc** 97.907	Ruthenium 44 **Ru** 101.07	Rhodium 45 **Rh** 102.906
6	Cesium 55 **Cs** 132.905	Barium 56 **Ba** 137.327	Lanthanum 57 **La** 138.906	Hafnium 72 **Hf** 178.49	Tantalum 73 **Ta** 180.948	Tungsten 74 **W** 183.85	Rhenium 75 **Re** 186.207	Osmium 76 **Os** 190.2	Iridium 77 **Ir** 192.22
7	Francium 87 **Fr** 223.020	Radium 88 **Ra** 226.025	Actinium 89 **Ac** 227.028	Rutherfordium 104 **Rf** (261)	Dubnium 105 **Db** (262)	Seaborgium 106 **Sg** (263)	Bohrium 107 **Bh** (262)	Hassium 108 **Hs** (265)	Meitnerium 109 **Mt** (266)

Lanthanide Series	Cerium 58 **Ce** 140.115	Praseodymium 59 **Pr** 140.908	Neodymium 60 **Nd** 144.24	Promethium 61 **Pm** 144.913	Samarium 62 **Sm** 150.36	Europium 63 **Eu** 151.965
Actinide Series	Thorium 90 **Th** 232.038	Protactinium 91 **Pa** 231.036	Uranium 92 **U** 238.029	Neptunium 93 **Np** 237.048	Plutonium 94 **Pu** 244.064	Americium 95 **Am** 243.061

286   Chapter 10 Atomic Structure and the Periodic Table

Metal
Metalloid
Nonmetal

🎈 Gas
💧 Liquid
⬜ Solid
⊙ Synthetic Elements

18

Helium
2 🎈
He
4.003

13	**14**	**15**	**16**	**17**	
Boron 5 ⬜ **B** 10.811	Carbon 6 ⬜ **C** 12.011	Nitrogen 7 🎈 **N** 14.007	Oxygen 8 🎈 **O** 15.999	Fluorine 9 🎈 **F** 18.998	Neon 10 🎈 **Ne** 20.180

10   **11**   **12**

| Aluminum 13 ⬜ **Al** 26.982 | Silicon 14 ⬜ **Si** 28.086 | Phosphorus 15 ⬜ **P** 30.974 | Sulfur 16 ⬜ **S** 32.066 | Chlorine 17 🎈 **Cl** 35.453 | Argon 18 🎈 **Ar** 39.948 |

10	**11**	**12**						
Nickel 28 ⬜ **Ni** 58.693	Copper 29 ⬜ **Cu** 63.546	Zinc 30 ⬜ **Zn** 65.39	Gallium 31 ⬜ **Ga** 69.723	Germanium 32 ⬜ **Ge** 72.61	Arsenic 33 ⬜ **As** 74.922	Selenium 34 ⬜ **Se** 78.96	Bromine 35 💧 **Br** 79.904	Krypton 36 🎈 **Kr** 83.80
Palladium 46 ⬜ **Pd** 106.42	Silver 47 ⬜ **Ag** 107.868	Cadmium 48 ⬜ **Cd** 112.411	Indium 49 ⬜ **In** 114.82	Tin 50 ⬜ **Sn** 118.710	Antimony 51 ⬜ **Sb** 121.757	Tellurium 52 ⬜ **Te** 127.60	Iodine 53 ⬜ **I** 126.904	Xenon 54 🎈 **Xe** 131.290
Platinum 78 ⬜ **Pt** 195.08	Gold 79 ⬜ **Au** 196.967	Mercury 80 💧 **Hg** 200.59	Thallium 81 ⬜ **Tl** 204.383	Lead 82 ⬜ **Pb** 207.2	Bismuth 83 ⬜ **Bi** 208.980	Polonium 84 ⬜ **Po** 208.982	Astatine 85 ⬜ **At** 209.987	Radon 86 🎈 **Rn** 222.018
(unnamed) 110 ⊙ **Uun**	(unnamed) 111 ⊙ **Uuu**	(unnamed) 112 ⊙ **Uub**						

Gadolinium 64 ⬜ **Gd** 157.25	Terbium 65 ⬜ **Tb** 158.925	Dysprosium 66 ⬜ **Dy** 162.50	Holmium 67 ⬜ **Ho** 164.930	Erbium 68 ⬜ **Er** 167.26	Thulium 69 ⬜ **Tm** 168.934	Ytterbium 70 ⬜ **Yb** 173.04	Lutetium 71 ⬜ **Lu** 174.967
Curium 96 ⊙ **Cm** 247.070	Berkelium 97 ⊙ **Bk** 247.070	Californium 98 ⊙ **Cf** 251.080	Einsteinium 99 ⊙ **Es** 252.083	Fermium 100 ⊙ **Fm** 257.095	Mendelevium 101 ⊙ **Md** 258.099	Nobelium 102 ⊙ **No** 259.101	Lawrencium 103 ⊙ **Lr** 260.105

10-4 The Periodic Table   287

📁 Use **Cross-Curricular Integration,** p. 14 as you teach this lesson.

🔨 Use **Teaching Transparencies 19, 20** as you teach this lesson.

• MINI • QUIZ •

Use the Mini Quiz to check students' recall of chapter content.

1. The arrangement of elements according to repeated changes in properties is called the _____. *periodic table*
2. _____ used his table to make predictions about undiscovered elements. *Mendeleev*
3. The elements on the periodic table are arranged in order of increasing atomic _____. *number*
4. Vertical columns on the periodic table are called _____. *groups or families*
5. The number of outermost electrons determines the _____ of an element. *chemical properties*

MiniLAB

Purpose
⟦LS⟧ Intrapersonal Students will develop an analogy between organizing a sample of pens and pencils and metallic and nonmetallic elements. ⟦L1⟧ **LEP**

Materials
assorted pencils and pens, large table surface

Teaching Strategies
Discuss with students the reasons for their arrangements. Make sure they have examined the properties of objects as they are arranged within columns.

Analysis
1. Students should be able to see similarities within columns.
2. Answers will vary, but should include comparing and contrasting properties of the new pen or pencil to the properties of groupings on the student-designed pen/pencil chart.

✔ Assessment

Performance Have students properly place another pen or pencil in their arrangement. Note the placement based on properties. Use the Performance Task Assessment List for Model in **PASC,** p. 51. **P**

📁 **Activity Worksheets,** pp. 5, 62

Enrichment

⟦LS⟧ Linguistic Lothar Meyer, a German, published a table similar to Mendeleev's at about the same time. Have students find out why he is not credited with the invention. Mendeleev used his own table to predict six undiscovered elements. ⟦L3⟧

MiniLAB

What are the advantages of organizing chemical elements in a periodic table?

Procedure
1. Gather pens and pencils from everyone in class.
2. Group the pens separately from the pencils.
3. Decide on specific properties—such as color, mass, or size—and arrange the writing utensils into a table.

Analysis
1. In the periodic table of the elements, organized columns and rows allow scientists to study properties of elements within the same column. How does your chart allow the same practice?
2. If someone brought a new pen or pencil to class, how would you decide where to place it on your table? Write your answer in your Science Journal.

Figure 10-13

The elements copper, silver, and gold are all found in Group 11 on the periodic table. *What similar properties do they possess?*

Visual Learning

Figure 10-12 **What similar properties do they have?** *Answers may vary but may include that they bounce, are used in sports, etc.*

Figure 10-13 **What similar properties do they possess?** *shiny metal, good conductors*
LEP ⟦LS⟧

Figure 10-12

These objects can be classified as sports balls. *What similar properties do they have?*

Groups of Elements

We often sort or group objects because of the properties they have in common. **Figure 10-12** shows an example. This is also done in the periodic table. The vertical columns in the periodic table are called **groups,** or families. The groups are numbered 1 through 18. Elements in each group have similar properties. For example, **Figure 10-13** shows objects made of three elements in Group 11: copper—Cu, silver—Ag, and gold—Au. These elements have similar properties. Each is a shiny metal and a good conductor of electricity and heat.

Atoms of different elements have different numbers of electrons. However, atoms of different elements may have the same number of electrons in their outer energy levels. It is the number of electrons in the outer energy level that determines the chemical properties of the element. Different elements with the same number of electrons in their outer energy level have similar chemical properties. These outer electrons are so important that a special way to represent them has been developed. A **dot diagram** uses the symbol of the element and dots to represent the electrons in the outer energy level.

Family Traits

The dot diagrams of the atoms of the elements in Group 17, called the halogens, are shown in **Figure 10-14.** They all have seven electrons in their outer energy levels. One similar property of the halogens is the ability to

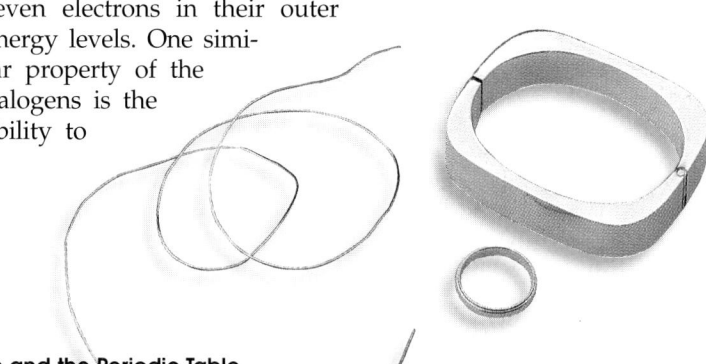

📁 Use these program resources as you teach this lesson.
Lab Manual 22
Cross-Curricular Integration, p.15

Figure 10-14

In each of these dot diagrams, the seven dots represent seven outer electrons. *What group do these elements belong to?*

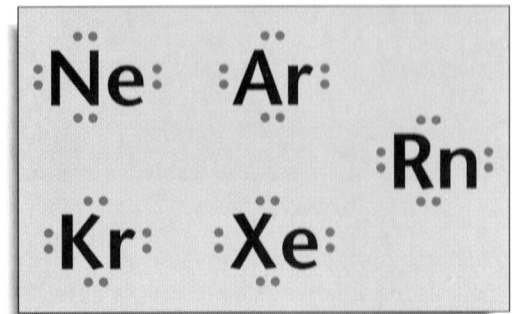

form compounds with elements in Group 1. The elements in Group 18 are known as noble gases. Noble gases do not usually form compounds. We say they are stable, or unreactive. The atoms of all the noble gases except helium have outer energy levels that contain eight electrons, as shown in **Figure 10-15.** You will learn more about the significance of electron arrangement of elements in later chapters.

Identifying a New Element

You have now learned that each element can be specifically identified by its atomic number—the number of protons in the nucleus of an atom. The number of neutrons may vary, as with isotopes. The number of electrons also may vary if some outer ones are removed or added. However, if the number of protons in an atom changes, then the atom has a new identity.

Recipe for an Element

As you will find in later chapters, chemical elements that you commonly encounter only change their number of protons through nuclear reactions. Were there ever more elements than we can now identify? Many scientists believe that the stable elements that we now have may at one time have been part of heavier elements that split apart. If you wanted to test part of that theory, you would have to bring existing elements together with a lot of energy to form superheavy atoms. For example, element 103, lawrencium, can be formed in a particle accelerator by colliding boron with an isotope of californium. How many total protons are found in californium and boron?

A New Member of the Family

Scientists in Berkeley, California; Dubna, Russia; and Darmstatdt, Germany, have been colliding particles to attempt to form elements 110 and 111. The German team has succeeded in showing evidence of creating isotopes of element 110. Although the synthesized element fell apart in

Figure 10-15

In each of these dot diagrams, the eight dots represent eight outer electrons. *What group do these elements belong to?*

 INTEGRATION
Physics

Use **Multicultural Connections,** pp. 23, 24 as you teach this lesson.

Use **Science Integration Transparency 10** as you teach this lesson.

Inclusion Strategies

Learning Disabled Obtain samples of several elements, such as tin, sulfur, copper, carbon, iron, aluminum, lead, zinc, gold, and silver. Let students examine and record the characteristics of each element. They should look for such properties as color, hardness, texture, brittleness, and the like. They can play a game in which one student lists characteristics and others try to identify the element.

 **INTEGRATION**
Physics

Discovering new elements is a practice in the field of physics. The process involves accelerating protons and other particles and colliding them.

3 Assess

Check for Understanding

? FLEX Your Brain

Use the Flex Your Brain activity to have students explore the PERIODIC TABLE.

Activity Worksheets, p. 5

Reteach

LS **Interpersonal** Using an appropriate strategy such as Expert Teams, organize eight class groups. Have them devise dot diagrams for the first 18 elements in the periodic table. Each group should be assigned one of the major groups—1, 2, and 13 through 18. Once the correct diagrams are established, have the groups enter their diagrams in a large table on the chalkboard or on roll paper. Point out the electron similarity in each group and the repeating pattern in the periods. **LEP**

Extension

For students who have mastered this section, use the **Reinforcement** and **Enrichment** masters.

USING TECHNOLOGY

For more information on imaging atoms, read "Seeing Atoms" by James Trefil, *Discover,* June 1990, pp. 55-60.

Think Critically

The radii given are in a ratio of approximately 1:5. Size comparisons could include a golf ball and a softball.

GLENCOE TECHNOLOGY

 Videodisc

The Infinite Voyage: Unseen Worlds

Chapter 10

The Scanning Tunneling Microscope: Observing Atomic Particles

4 Close

•MINI•QUIZ•

Use the Mini Quiz to check students' recall of chapter content.

1. **The halogens are found in which group?** *17*

2. **The noble gases form which group?** *18*

3. **Horizontal rows on the periodic table are called _____ .** *periods*

4. **Elements next to the stair-step line are called _____ .** *metalloids*

5. **Groups 3 through 12 are called the _____ metals.** *transition*

USING TECHNOLOGY

Seeing Atoms

For more than 50 years, scientists have been able to see extremely tiny things with electron microscopes. But it was less than 20 years ago that scientists first saw atoms with scanning probe microscopes. These microscopes are the result of a new approach to how we see objects. Instead of shining light or a beam of electrons on an object, these new microscopes drag a probe across the surface of an object. The position of the probe is then changed, and it is dragged across the surface again. This process, called scanning, is repeated many times to build an image of the peaks and valleys on the surface of the object. Scanning is like going into a dark room and using your hand as a probe to feel a chair to determine its shape, instead of shining a light on it.

The key development in the invention of the scanning probe microscope was a system that moves the probe in precise steps smaller than the width of an atom. In addition to seeing atoms, these probes enable scientists to move single atoms.

Think Critically:

Atoms range in radius from 0.053 nm to 0.27 nm. What everyday objects, like two kinds of balls, would represent this range in size?

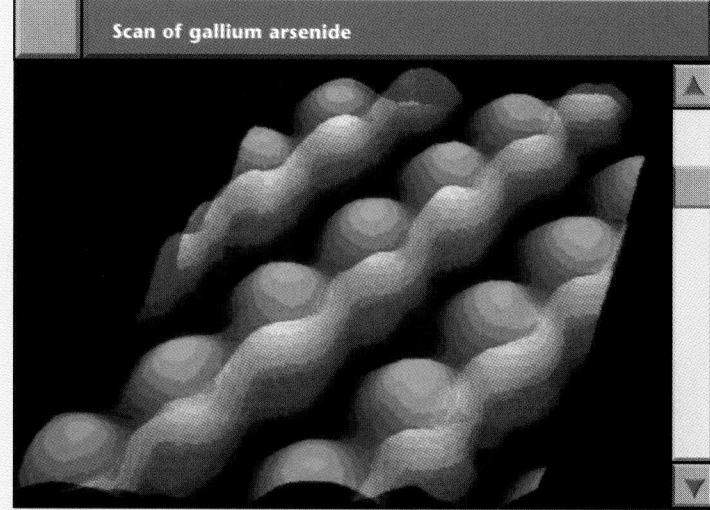

Scan of gallium arsenide

290   Chapter 10 Atomic Structure and the Periodic Table

Use **Science and Society Integration,** p. 14 as you teach this lesson.

---

approximately 270 microseconds, there was enough evidence of its presence to claim production of the heaviest known element. The German physicists used lead and nickel atoms to add the newest member to the family of the chemical elements. Look on the periodic table on pages 286-287 to see where element 110 now resides.

Periods of Elements

The horizontal rows of elements in the periodic table are called **periods.** Notice the stair-step line on the right side of the periodic table. All the elements to the left of this line, except hydrogen, are metals. Iron, zinc, and copper are examples of metals. Most **metals** have the common properties of existing as solids at room temperature and being shiny and good conductors of heat and electricity. **Figure 10-16** shows the metal gallium and its use in a semiconductor.

Those elements to the right of the stair-step line on the periodic table are classified as **nonmetals.** Oxygen, nitrogen, and carbon are examples of nonmetals. At room temperature, most nonmetals are gases and some are brittle solids. Most nonmetals do not conduct heat and electricity well.

The elements next to the stair-step line are **metalloids** because they have properties of both metals and nonmetals. Boron and silicon are examples of metalloids.

Elements in Groups 3 through 12 are called the transition elements. They are metals but have properties not found in elements of other groups. Copper and iron are examples of common transition elements.

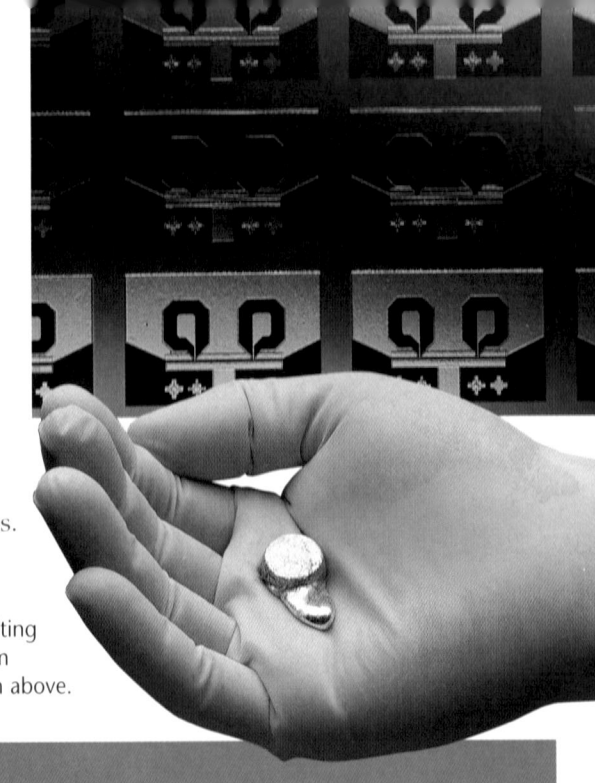

Figure 10-16

Gallium is a metal that has a low melting point. It is used in the form of gallium arsenide in the semiconductor shown above.

Section Wrap-up

Review

1. Use the periodic table to find the name, atomic number, and average atomic mass of the following elements: N, Ca, Kr, and W.

2. Give the period and group in which each of these elements is found: nitrogen, sodium, iodine, and mercury.

3. Write the name of each of these elements and classify it as a metal, a nonmetal, or a metalloid: K, Si, Ba, and S.

4. **Think Critically:** Both the Mendeleev and Mosely periodic charts have gaps for as-then-undiscovered elements. Why do you think the chart used by Mosely could be more accurate at predicting where new elements would be placed?

Skill Builder

Making and Using Graphs
Construct a circle graph showing the elements classified as metals, metalloids, and nonmetals. If you need help, refer to Making and Using Graphs in the **Skill Handbook.**

Science Journal

From the periodic table on pages 286-287, choose a synthetic element and write a brief biography of the element in your Science Journal. Include information about the element's name, location of the synthesis research, and the people responsible.

Section Wrap-up

Review

1. N, nitrogen, 7, 14.01; Ca, calcium, 20, 40.08; Kr, krypton, 36, 83.80; W, tungsten, 74, 183.85. Note: These atomic masses are rounded to two decimal places.

2. nitrogen, period 2, Group 15; sodium, period 3, Group 1; iodine, period 5, Group 17; mercury, period 6, Group 12

3. K, potassium, metal; Si, silicon, metalloid; Ba, barium, metal; S, sulfur, nonmetal

4. **Think Critically** By using an arrangement based on sequential number of protons, you could know how many elements were missing in the gaps.

Science Journal

Answer Answers will vary. Encyclopedias and *The Handbook of Chemistry and Physics* are excellent sources of information.

Skill Builder

The graph should indicate that 77% of the elements are metals, 8% are metalloids, and 15% are nonmetals.

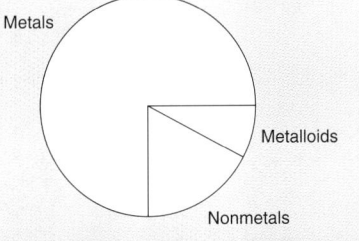

Assessment

Performance If a sensitive balance is available, have students determine the mass of their circle graphs. After then cutting the graph into parts, have students mass each piece and determine what percentage of the whole each slice represents. Use the Performance Task Assessment List for Using Math in Science in **PASC,** p. 29. **P**

People and Science

Background

- Physicists have now discovered all six quarks predicted by the Standard Model: up and down, top and bottom, strange and charm. Up and down quarks, plus clouds of gluons, form protons and neutrons. The other four quarks were present when the universe was formed and are now produced by scientists in particle collisions.

- The Tevatron accelerator contains a 5000-metric-ton detector that encircles the accelerator at the point where particle collisions occur. It contains computers that detect the paths of particles produced during the collisions.

Teaching Strategies

LS **Kinesthetic** Use the game of pool as an analogy in explaining to students how the mass influences the path taken by each particle created during a high-energy collision. Have students imagine that the cluster of balls set up at the beginning of the game is a single proton split apart by a high-energy collision. Then have them imagine what might happen if the setup included not only the plastic balls used for playing pool, but also bowling balls, Ping-Pong balls, golf balls, and other balls of different sizes and masses.

Career Connection

- Physicists design collider experiments and equipment. They work in the detector control room, and instruct the technicians who repair and operate the collider. A Ph.D. is usually required.

- Engineers design and build the collider and its parts. A graduate degree is usually required.

292

DR. BRENNA FLAUGHER, *Particle Physicist*

On the Job

Q Dr. Flaugher, is particle physics the same as atomic physics?

A Atomic physicists study the atom as a whole. Particle physicists work on a smaller scale, inside the atomic nucleus. They are interested in the particles and forces that make up protons and neutrons.

Q You are on a team of physicists who found the sixth quark. How did you know it was there to find?

A Particle physicists have a theory, called the Standard Model, that explains how particles fit together to form protons and neutrons. According to this model, quarks come in pairs. After the bottom quark was found in 1977, physicists began looking for its partner, which was named *top*. We finally found it in 1995.

Q Why did it take so long to find the top quark?

A It turns out to be much bigger than we thought. When physicists first began looking for the top quark, they were searching for a particle with twice the mass of the bottom quark. The top quark is actually about 35 times more massive than its partner.

Personal Insights

Q Why did you choose a career in particle physics?

A It's a real frontier. This is a very exciting field—we are doing things that have never been done before. The collider is like a powerful microscope that lets us look deeper and deeper into the workings of nature. The particles we study are very, very small, but they can help us understand very big things, like why matter has mass, why gravity works the way it does, what happens inside stars, and how the universe was formed.

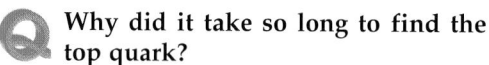

Career Connection

Physicists, electronics engineers, mechanical engineers, civil engineers, computer specialists, science writers, and repair technicians all worked together to discover the top quark. Talk with a career counselor at a local high school or college to find out what kinds of education and training these careers require.

- **Physics Teacher**   • **Nuclear Technician**

- Computer specialists design and build computer equipment and develop software. A graduate degree in computer science is needed.

- Science writers develop materials to inform the public about the scientific activities going on in the lab and to help educate students and teachers.

- Technicians repair and operate collider equipment. Education requirements include a high school diploma and on-the-job training.

For More Information

- Fermilab's Public Affairs Office distributes information about the laboratory and its research and educational activities.

Summary

10-1: Structure of the Atom

1. A chemical symbol is a shorthand way of writing the name of an element.
2. An atom consists of a nucleus made of protons and neutrons surrounded by an electron cloud.
3. The electrons in an atom are arranged in several energy levels, each of which is able to hold a certain number of electrons.

10-2: Science and Society: Smaller Particles of Matter

1. Quarks are particles of matter that make up protons and neutrons.
2. Protons can be broken into quarks by having them collide while traveling near the speed of light.

10-3: Masses of Atoms

1. The number of neutrons in an atom can be computed by subtracting the atomic number from the mass number.
2. The isotopes of an element are atoms of that same element that have different numbers of neutrons.
3. The average atomic mass of an element is the average mass of the mixture of its isotopes.

10-4: The Periodic Table

1. The periodic table of elements is an arrangement of elements according to repeated changes in properties.
2. In the periodic table, the 111 elements are arranged in 18 vertical columns, or groups, and seven horizontal rows, or periods.

3. Metals are found at the left of the periodic table, nonmetals at the right, and metalloids along the line that separates the metals from the nonmetals.

Key Science Words

a. atomic number
b. average atomic mass
c. chemical symbol
d. dot diagram
e. electron
f. electron cloud
g. group
h. isotope
i. mass number
j. metal
k. metalloid
l. neutron
m. nonmetal
n. nucleus
o. period
p. periodic table
q. proton
r. quark

Reviewing Vocabulary

Match each phrase with the correct term from the list of Key Science Words.

1. an abbreviated way to write the name of an element
2. an atomic particle with no charge
3. any of two or more atoms of the same element with a different number of neutrons
4. average mass of the mixture of isotopes
5. positively charged center of an atom
6. particle that makes up protons and neutrons
7. the region formed by electrons
8. the number of protons in the nucleus
9. total number of protons and neutrons in the nucleus
10. a horizontal row in the periodic table

Chapter 10 Review   293

Summary

Have students read the summary statements to review the major concepts of the chapter.

Reviewing Vocabulary

1. c	6. r
2. l	7. f
3. h	8. a
4. b	9. i
5. n	10. o

✓ Assessment

Portfolio Encourage students to place in their portfolios one or two items of what they consider to be their best work. Examples include:

- Enrichment research, p. 287
- Activity 10-1 observations and answers, p. 275
- Skill Builder concept map, p. 274 **P**

Performance Additional performance assessments may be found in **Performance Assessment** and **Science Integration Activities.** Performance Task Assessment Lists and rubrics for evaluating these activities can be found in Glencoe's **Performance Assessment in the Science Classroom.**

GLENCOE TECHNOLOGY

● Videodisc

Glencoe Physical Science

Interactive Videodisc

Use the videodisc lesson *Periodicity* to review the basis for the arrangements of elements in the periodic table.

▣ MindJogger Videoquiz

Chapter 10 Have students work in groups as they play the Videoquiz game to review key chapter concepts.

Chapter 10 Review

Checking Concepts

1. d   6. c
2. c   7. d
3. b   8. a
4. b   9. d
5. a   10. b

Understanding Concepts

11. The tendency of the moving electron to escape is balanced by the electrical attraction of the positive nucleus for the negative electron.

12. Because the names of several elements may start with the same letter, a second or third letter may be needed to distinguish them.

13. Silver atoms have an atomic number of 47 and therefore contain 47 protons.

    Mass of atoms in sample

    $= 52 (60 + 47) + 48 (62 + 47)$

    $= 52 \times 107 + 48 \times 109$

    $= 5564 + 5232$

    $= 10\ 796$

    Average atomic mass

    $= \dfrac{10\ 796}{52 + 48} = 107.96$

14. Protons are thought to be composed of smaller particles called quarks. However, electrons cannot be broken down into smaller particles.

15. They are called nucleons because they are the particles that make up the nucleus.

Thinking Critically

16. Possible answers are tides, time of day, seasons, and moon phases.

17. They have relatively high atomic masses.

18. the same group

19. They are planets. Neptunium is also named for a planet.

294

Checking Concepts

Choose the word or phrase that completes the sentence or answers the question.

1. The state of matter of most of the elements to the left of the stair-step line in the periodic table is _____.
   a. gas   c. plasma
   b. liquid   d. solid

2. If a pattern repeats itself, it is _____.
   a. isotopic   c. periodic
   b. metallic   d. transition

3. _____ is an element that would have similar properties to those of neon.
   a. Aluminum   c. Arsenic
   b. Argon   d. Silver

4. Boron is a _____.
   a. metal   c. noble gas
   b. metalloid   d. nonmetal

5. The element potassium is a _____.
   a. metal   c. nonmetal
   b. metalloid   d. transition element

6. The element bromine is a _____.
   a. metal   c. nonmetal
   b. metalloid   d. transition element

7. The halogens are those elements in Group _____.
   a. 1   c. 15
   b. 11   d. 17

8. In its group, nitrogen is the only element that is a _____.
   a. gas   c. metal
   b. metalloid   d. liquid

9. _____ is a shiny element that conducts electricity and heat well.
   a. Chlorine   c. Hydrogen
   b. Sulfur   d. Magnesium

10. The atomic number of Re is 75. The atomic mass of one of its isotopes is 186. How many neutrons are in an atom of this isotope?
    a. 75   c. 186
    b. 111   d. 261

Understanding Concepts

Answer the following questions in your Science Journal using complete sentences.

11. Why do electrons keep moving around the nucleus and not away from the atom?

12. Why do some chemical symbols have one letter and some have two or three letters?

13. A silver sample contains 52 atoms, each having 60 neutrons, and 48 atoms, each having 62 neutrons. What is the sample's average atomic mass?

14. According to currently accepted ideas, how do protons and electrons differ in structure?

15. Why are protons and neutrons also known as nucleons?

Thinking Critically

16. We know that properties of elements are periodic. List at least two other things in nature that are periodic.

17. Lead and mercury are two pollutants in the environment. From information about them in the periodic table, why are they called *heavy metals*?

18. Ge and Si are used in making semiconductors. Are these two elements in the same group or the same period?

19. U and Pu are named for objects in nature. What are these objects, and what other element is named after a similar object?

20. Ca is used by the body to make bones and teeth. Radioactive Sr is in nuclear waste. These two elements are in the same group. Yet one is safe for people and the other is hazardous. Why is this Sr hazardous to people?

20. Calcium and strontium are in the same group and therefore have similar properties. Strontium, like calcium, is easily absorbed into the body.

Developing Skills

21. **Making and Using Tables** metal, bismuth; metalloid, antimony or arsenic; nonmetal, nitrogen or phosphorus

22. **Comparing and Contrasting** The chemical properties of chlorine and bromine would be similar because they are in the same group. At room temperature, bromine is a liquid, whereas chlorine is a gas. Bromine atoms have a greater mass than do chlorine atoms. It is also reasonable to assume that a bromine atom is larger than a chlorine atom because the bromine atom has one more level of electrons.

23. **Interpreting Data** 37 protons, 37 electrons, 48 neutrons

Developing Skills

If you need help, refer to the **Skill Handbook**.

21. **Making and Using Tables:** Use the periodic table to list a metal, a metalloid, and a nonmetal each with five outer-level electrons.
22. **Comparing and Contrasting:** From the information found in the periodic table and reference books, compare and contrast the properties of chlorine and bromine.
23. **Interpreting Data:** If scientists have determined that a neutral atom of rubidium has an atomic number of 37 and a mass number of 85, how many protons, neutrons, and electrons does the atom have?
24. **Concept Mapping:** As a star dies, it becomes more dense. Its temperature rises to a point where He nuclei are combined with other nuclei. When this happens, the atomic numbers of the other nuclei are increased by 2 because each gains the two protons contained in the He nucleus. For example, Cr fused with He becomes Fe. Complete the concept map below showing the first four steps in He fusion.

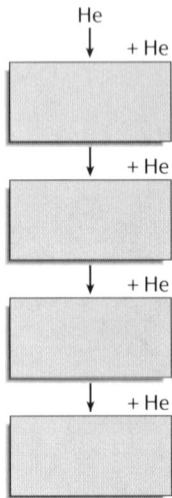

25. **Sequencing:** What changes in the periodic table would occur in periods 1 through 4 if the elements were arranged according to increasing average atomic mass instead of atomic number? How do we know that arrangement by atomic number is correct?

Performance Assessment

1. **Asking Questions:** Research the attempts made by Johann Döbereiner and John Newlands to classify the elements. Research the background and work of these scientists and write your findings in the form of an interview.
2. **Display:** Make a display of samples or pictures of several elements, such as the application of neon in the photograph below. List the name, symbol, atomic number, average atomic mass, and several uses for each element.

3. **Research and Report:** Research and report on models of the atom from the time of the ancient Greeks until the present.

24. **Concept Mapping** Be, C, O, Ne
25. **Sequencing** Argon and potassium would be reversed, as would nickel and cobalt and tellurium and iodine. The resulting arrangement would cause the elements to appear in groups with other elements that do not share their properties.

Performance Assessment

1. The interviews should refer to Döbereiner's triads and Newland's law of octaves. Use the Performance Task Assessment List for Asking Questions in **PASC**, p. 19. **P**
2. Check to see that student samples are truly of that element. Their uses should include some little-known uses they would have had to research. Use the Performance Task Assessment List for Display in **PASC**, p. 63. **P**
3. Students should include at least three models of the atom over time. Use the Performance Task Assessment List for Lab Report in **PASC**, p. 47. **P**

Assessment Resources

Reproducible Masters
Chapter Review, pp. 23, 24
Assessment, pp. 61-64
Performance Assessment, p. 24

Glencoe Technology
- Chapter Review Software
- Computer Test Bank
- MindJogger Videoquiz

Chapter Organizer

Section	Objectives/Standards	Activities/Features
Chapter Opener		Explore Activity: Find evidence of the formation of a chemical bond. p. 297
11-1 **Why Atoms Combine** (2 sessions, 1 block)*	1. **Describe** how a compound differs from the elements that compose it. 2. **Explain** what a chemical formula represents. 3. **State** a reason why chemical bonding occurs. **National Content Standards: (5-8) UCP1-UCP3, UCP5, A2; (9-12) UCP1-UCP3, UCP5, A2, B2**	Using Math, p. 300 Using Math, p. 301 Science Journal, p. 302 Skill Builder: Making and Using Tables, p. 302 Activity 11-1: Eggshell Electrons, p. 303
11-2 **Kinds of Chemical Bonds** (3 sessions, 2 blocks)*	1. **Describe** ionic bonds and covalent bonds. 2. **Identify** the particles produced by ionic bonding and by covalent bonding. 3. **Distinguish** between a nonpolar covalent bond and a polar covalent bond. **National Content Standards: (5-8) UCP1-UCP3, UCP5, A1, A2, C1, C3, E1, E2, F1-F3, F5, G1; (9-12) UCP1-UCP3, UCP5, A1, A2, B2, C1, C5, E1, E2, F1-F6, G1**	Connect to Earth Science, p. 305 MiniLAB: What type of molecule is water? p. 306 Using Technology: Bond Busters, p. 307 Using Math, p. 308 Skill Builder: Concept Mapping, p. 308 People and Science: Dr. Alvaro Garza, Physician, p. 3(Activity 11-2: Become a Bond Breaker, pp. 310-311
11-3 **Science and Society** (1 session, ½ block)*	1. **Describe** the dangers posed by hazardous compounds in the home. 2. **Demonstrate** a knowledge of safer alternative compounds to use. **National Content Standards: (5-8) UCP2, UCP5, F1-F5; (9-12) UCP2, UCP5, F1, F3-F6**	Explore the Issue, p. 313
11-4 **Formulas and Names of Compounds** (2 sessions, 1 block)*	1. **Explain** how to determine oxidation numbers. 2. **Write** formulas for compounds from their names. 3. **Name** compounds from their formulas. 4. **Describe** hydrates and their formulas. **National Content Standards: (5-8) UCP1, UCP2, UCP5, A2; (9-12) UCP1, UCP2, UCP5, A2**	Using Math: Writing Formulas for Binary Compounds p. 316 Using Math: Naming Some Binary Compounds, p. 31 Using Math: Writing Formulas with Polyatomic Ions, p. Problem Solving: The Packet of Mystery Crystals, p. 3 Using Math, p. 320 Skill Builder: Using Variables, Constants, and Controls, p. 320

* A complete Planning Guide that includes block scheduling is provided on pages 32T-35T.

Activity Materials

Explore	Activities	MiniLAB
page 297 2 identical test tubes, steel wool, beaker, water	page 303 modified egg carton, marbles pages 310-311 small samples of crushed ice, table salt, and sugar; wire test-tube holder; 3 test tubes; safety goggles; laboratory burner; timer; oven mitt; apron	page 306 water faucet, balloon, small piece of wool or fur

Need Materials? Call Science Kit (1-800-828-7777).

Teacher Classroom Resources

Reproducible Masters	Transparencies	Teaching Resources

Study Guide, p. 46
Reinforcement, p. 46
Enrichment, p. 46
Activity Worksheets, pp. 63-64
Lab Manual 23, The Six Solutions
  Problem
Science and Society Integration, p. 15
	Section Focus Transparency 42, Puzzle
	  Pieces
	Teaching Transparency 21, Chemical
	  Bonding
		Glencoe Physical Science
		  Interactive Videodisc
		Physical Science CD-ROM
		Spanish Resources
		English/Spanish Audiocassettes
		Cooperative Learning Resource
		  Guide
		Lab Partner
		Lab and Safety Skills
		Lesson Plans

Study Guide, p. 47
Reinforcement, p. 47
Enrichment, p. 47
Activity Worksheets, pp. 65-66, 67
Lab Manual 24, Chemical Bonds
Lab Manual 25, Preparation of Carbon
  Dioxide
	Section Focus Transparency 43, Salt
	  Crystals
	Science Integration Transparency 11,
	  Nerve Impulses and Chemistry
		Assessment Resources
		
		Chapter Review, pp. 25-26
		Assessment, pp. 65-68
		Performance Assessment in the
		  Science Classroom (PASC)
		MindJogger Videoquiz
		Alternate Assessment in the
		  Science Classroom
		Performance Assessment
		Chapter Review Software
		Computer Test Bank

Study Guide, p. 48
Reinforcement, p. 48
Enrichment, p. 48
Science Integration Activity 11, Static
  Cling's the Thing
Concept Mapping, pp. 27-28
Critical Thinking/Problem Solving, p. 17
	Section Focus Transparency 44,
	  Pesticides	

Study Guide, p. 49
Reinforcement, p. 49
Enrichment, p. 49
Multicultural Connections, p. 25
Cross-Curricular Integration, p. 17
	Section Focus Transparency 45,
	  Naming Compounds
	Teaching Transparency 22, Oxidation
	  Numbers of Elements
		Key to Teaching Strategies
		
		The following designations will help you decide which activities are appropriate for your students.
		
		L1 Level 1 activities should be within the ability range of all students, including those with learning difficulties.
		
		L2 Level 2 activities should be within the ability range of the average to above-average student.
		
		L3 Level 3 activities are designed for the ability range of above-average students.
		
		LEP LEP activities should be within the ability range of Limited English Proficiency students.
		
		LS These activities are designed to address different learning styles.
		
		COOP LEARN Cooperative Learning activities are designed for small group work.
		
		P These strategies represent student products that can be placed into a best-work portfolio.

GLENCOE TECHNOLOGY

The following multimedia resources are available from Glencoe.

Science and Technology Videodisc Series (STVS)
Chemistry
  Images of Atoms
  Super Grit
  Dealing with Hazardous Materials

The Infinite Voyage Series
Miracles by Design
Unseen Worlds

Glencoe Physical Science Interactive Videodisc
Periodicity

Physical Science CD-ROM

National Geographic Society Series
STV: Human Body Volume 2

Teacher Classroom Resources

This is a representation of key blackline masters available in the Teacher Classroom Resources.

Teaching Aids

Section Focus Transparencies

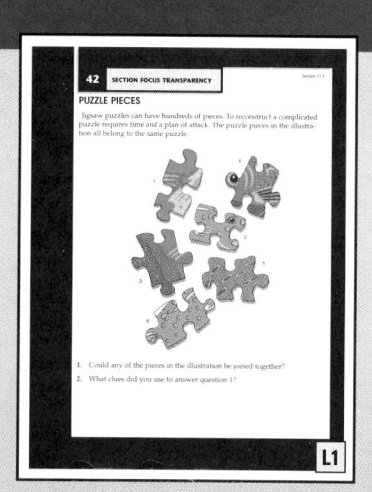

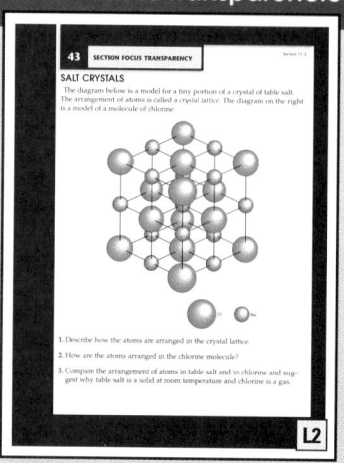

Science Integration Transparencies

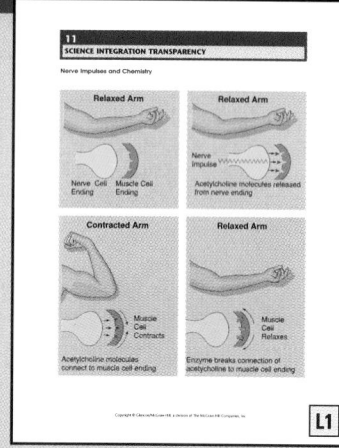

Teaching Transparencies

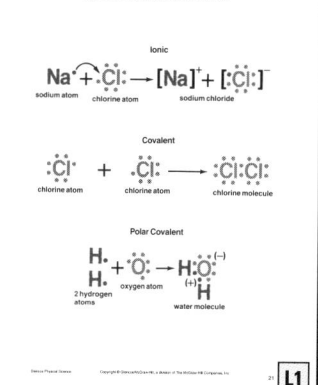

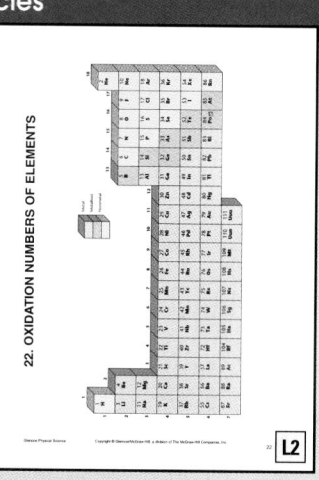

Meeting Different Ability Levels

Study Guide

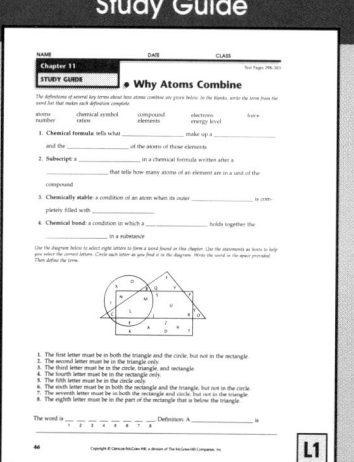

Reinforcement

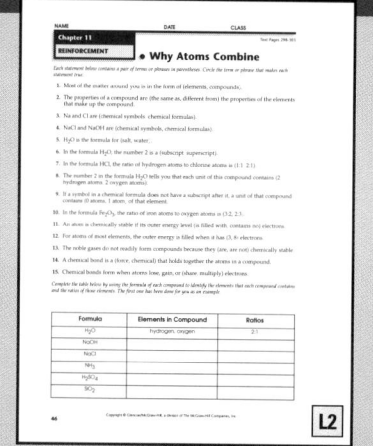

Enrichment Worksheets

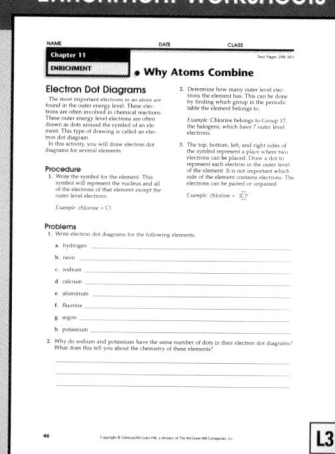

Chapter 11 Chemical Bonds

Hands-On Activities

Science Integration Activity

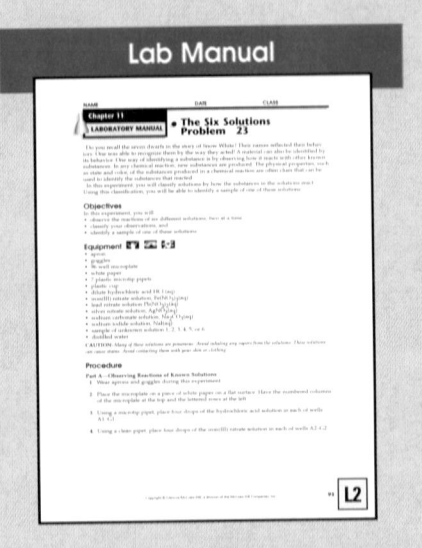

Lab Manual

Assessment

Performance Assessment

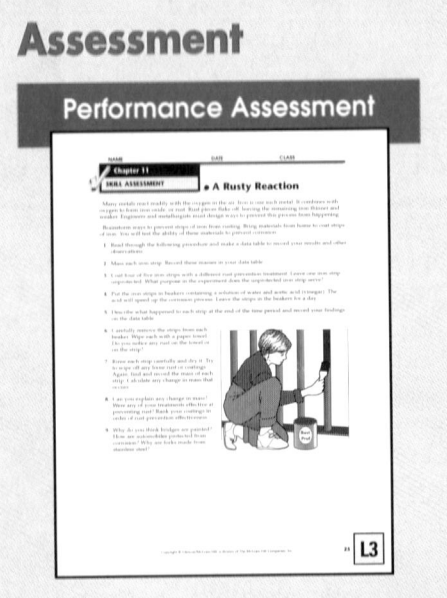

Enrichment and Application

Critical Thinking/ Problem Solving

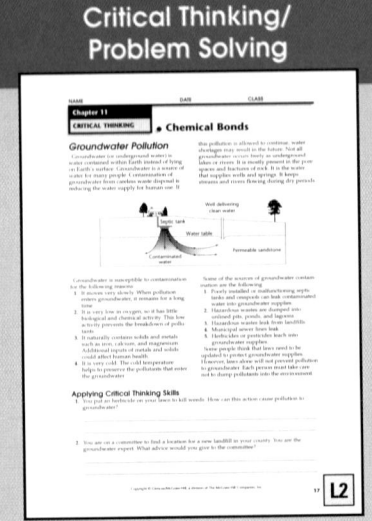

Cross-Curricular Integration

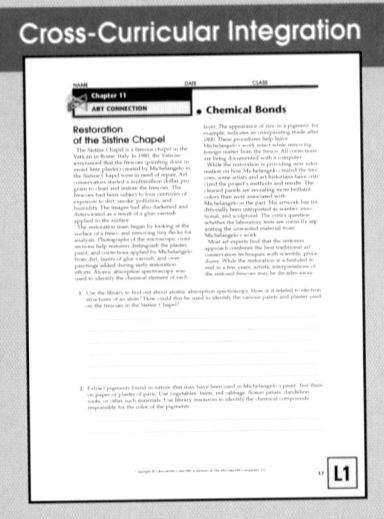

Science and Society Integration

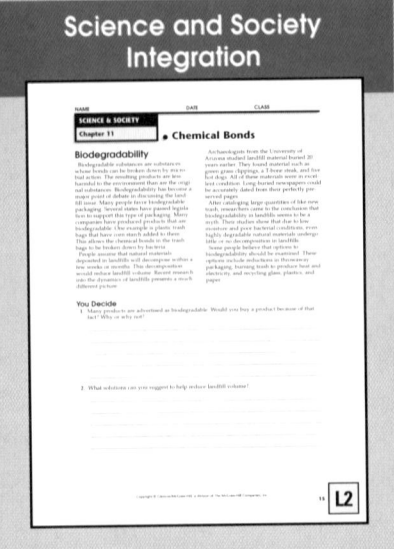

Multicultural Connections

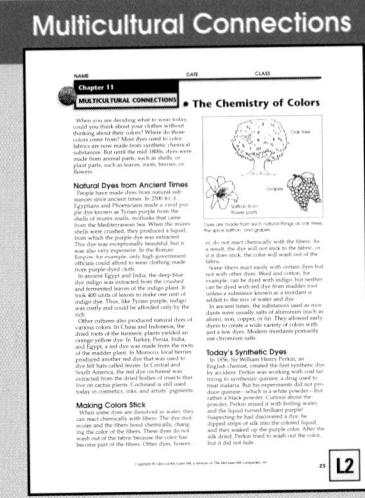

Concept Mapping

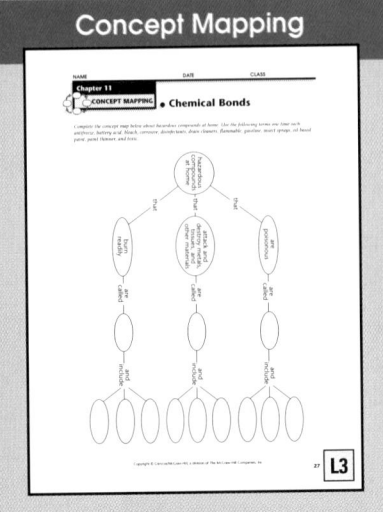

Chapter 11

Chemical Bonds

CHAPTER OVERVIEW

Section 11-1 Atoms react to form chemically stable substances that are held together by chemical bonds and are represented by chemical formulas.

Section 11-2 Ionic, polar covalent, and covalent bonds are conceptualized in this section, as are the resulting polar and nonpolar molecules.

Section 11-3 Science and Society Students learn that care must be exercised when using many chemicals found in the home.

Section 11-4 Oxidation numbers are used to introduce formula writing and naming for both binary and polyatomic compounds. This section also introduces students to hydrates.

Chapter Vocabulary

chemical formula	nonpolar molecule
chemically stable	toxic
chemical bond	corrosive
ion	oxidation number
ionic bond	binary compound
covalent bond	polyatomic ion
polar molecule	hydrate

Theme Connection

Stability and Change The relationship between an atom's electron structure and its stability is discussed. A study of chemical bonding further develops the theme. Chemical properties such as corrosiveness, flammability, and toxicity are related to stability.

Previewing the Chapter

296

Learning Styles Look for the following logo for strategies that emphasize different learning modalities. **LS**

Kinesthetic	Demonstration, p. 305
Visual-Spatial	Explore, p. 297; Demonstration, pp. 299, 307, 315, 317; Visual Learning, pp. 299, 301, 302, 306; People and Science, p. 309
Interpersonal	Activity 11-2, pp. 310-311; Reteach, p. 319
Logical-Mathematical	Across the Curriculum, p. 299; Activity 11-1, p. 303; Activity, p. 316; MiniLAB, p. 306

Chapter
11

Chemical Bonds

Whenever you see a brilliant display of fireworks, like the one over the Tower Bridge in London, you observe a chemical change in which chemical bonds break and others form. Forming a chemical bond is usually an energy-releasing process, as is shown by the energy given off when fireworks explode. Evidence of a chemical change provides evidence of a chemical bond. The materials in fireworks combine, or bond, with oxygen from the air to make new substances. Can you think of other changes that involve combining with oxygen?

EXPLORE ACTIVITY

Find evidence of the formation of a chemical bond.

1. Obtain two identical test tubes.
2. Insert some steel wool into the bottom of one test tube.
3. Invert both tubes and place them into a beaker of water.
4. Adjust the water level in both to be approximately 1 cm into the openings.

Observe: After a few days, what differences do you notice that indicate a chemical change? Summarize your observations in your Science Journal.

Previewing Science Skills

▶ In the Skill Builders, you will **make and use a table; map concepts; and use variables, constants, and controls.**

▶ In the Activities, you will **observe, compare, classify,** and **hypothesize.**

▶ In the MiniLAB, you will **observe and infer.**

297

Prepare

Section Background

- An atom with a full outer energy level is chemically stable, or unreactive. For most atoms, this condition means having eight outer electrons. For this reason, this principle is called the octet rule. Note, however, that if the first energy level is the outer level, only two electrons are needed for the level to be full.
- Most metals have three or fewer outer electrons and tend to lose electrons to form positive ions.
- Most nonmetals have five or more outer electrons and tend to gain electrons to form negative ions.

Preplanning

To prepare for Activity 11-1, prepare modified egg cartons and package marbles for each laboratory team.

1 Motivate

Bellringer

Before presenting the lesson, display **Section Focus Transparency 42** on the overhead projector. Assign the accompanying **Focus Activity** worksheet. L1 LEP

Tying to Previous Knowledge

Ask students to think of how the pieces of a jigsaw puzzle interlock to form a complete picture. In a similar way, atoms can bond together to form a compound. The atoms combine in a definite ratio and in a definite order in a way similar to the way the puzzle pieces combine.

Science Words

chemical formula
chemically stable
chemical bond

Objectives

- Describe how a compound differs from the elements that compose it.
- Explain what a chemical formula represents.
- State a reason why chemical bonding occurs.

Compounds

Most of the matter around you is in the form of compounds or mixtures of compounds. The water you drink, the carbon dioxide you exhale, and the salt contained in some foods are examples of compounds.

Some of the matter around you is in the form of uncombined elements, such as iron and oxygen. But, like many other pairs of elements, iron and oxygen tend to unite chemically to form a compound when the conditions are right. You know how iron exposed to oxygen and water forms rust, a compound. Rusting is a chemical change because a new substance, as shown in **Figure 11-1,** is produced.

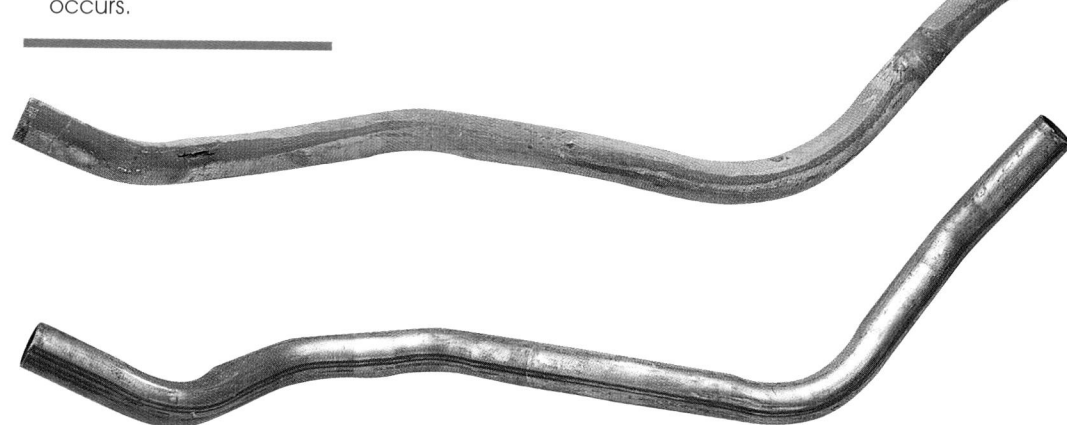

Figure 11-1

When iron and oxygen combine, a new substance, rust, forms. The properties of rust differ from those of iron and oxygen.

Compounds have properties unlike those of their elements. Table salt, for example, is made up of the elements sodium and chlorine. Sodium is a shiny, soft, silvery metal that reacts violently with water. Chlorine is a greenish-yellow gas that can kill an animal that inhales a few deep breaths of the gas. These elements combine to form table salt, or sodium chloride, as shown in **Figure 11-2.**

298 Chapter 11 Chemical Bonds

Program Resources

 Reproducible Masters
Study Guide, p. 46 L1
Reinforcement, p. 46 L1
Enrichment, p. 46 L3
Science and Society Integration, p. 15
Activity Worksheets, pp. 63-64 L1
Lab Manual 23

Transparencies
Section Focus Transparency 42 L1
Teaching Transparency 21

$$Na^{\cdot} - \cdot \longrightarrow [Na]^+$$
$$:\overset{\cdot\cdot}{Cl}\cdot + \cdot \longrightarrow [:\overset{\cdot\cdot}{Cl}:]^-$$

Formulas

Do you recall that the chemical symbols Na and Cl represent sodium and chlorine? When written as NaCl, the symbols make up a formula, or chemical shorthand, for the compound sodium chloride. Another formula you may recognize is H_2O, for water. The formula is a combination of the symbols H and O and the subscript number 2. *Subscript* means "written below." A subscript number written after a symbol tells how many atoms of that element are in a unit of the compound. If a symbol has no subscript, the unit contains only one atom of that element.

Look at the formula for iron(III) oxide: Fe_2O_3. The symbol for iron, Fe, is followed by the subscript 2, and the symbol for oxygen, O, is followed by the subscript 3. Other examples of formulas are shown in **Figure 11-3**.

Figure 11-2

Sodium, a soft, silvery metal, combines with chlorine, a greenish-yellow gas, to form sodium chloride, a white crystalline solid. *How are the properties of table salt different from those of sodium and chlorine?*

Figure 11-3

These models show the ratios and the arrangements of atoms in water and carbon dioxide.

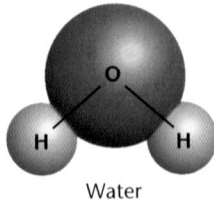

Water

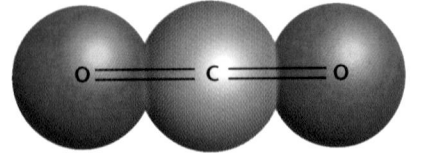

Carbon dioxide

A The formula H_2O shows there are two atoms of hydrogen for one atom of oxygen in one unit of the compound water. Put another way, the ratio of hydrogen atoms to oxygen atoms in water is 2 to 1.

B For one atom of carbon, there are two atoms of oxygen in carbon dioxide.

Content Background

• Some element names are based on German, Greek, or Latin roots. Some are based on names of geographic locations. Some are named to honor scientists.

• The driving force that causes a chemical reaction to go to completion is the atom's tendency to become stable.

Demonstration

LS **Visual-Spatial** Two elements combining in a dramatic way can be shown by reacting 3 g of zinc dust with 2 g of iodine crystals. Place them in a test tube. Shake to mix, then clamp the tube securely upright. Add 1 mL water to start the reaction. **CAUTION:** *The reaction gives off heat and a toxic vapor. Perform only outdoors or in a fume hood.* Point out that the elements become more stable by combining. Tell students that they will learn to name and write the formula for the product of this reaction, zinc iodide.

GLENCOE TECHNOLOGY

 Videodisc

STVS: Chemistry

Disc 2, Side 1

Images of Atoms (Ch. 4)

Super Grit (Ch. 9)

Visual Learning

Figure 11-2 How are the properties of table salt different from those of sodium and chlorine? *Sodium is a soft, greyish metal. Chlorine is a yellow-green gas. Table salt is a white solid.* **LEP** **LS**

Across the Curriculum

Geology The names that geologists use for gemstones and other crystals are not the same names used by chemists. A quartz crystal is silicon dioxide. Fool's gold, iron pyrite, is an iron sulfide, and fluorite is calcium fluoride. Have students research common names, chemical names, and chemical formulas for other gems and minerals. **L1** **LS**

Table 11-1

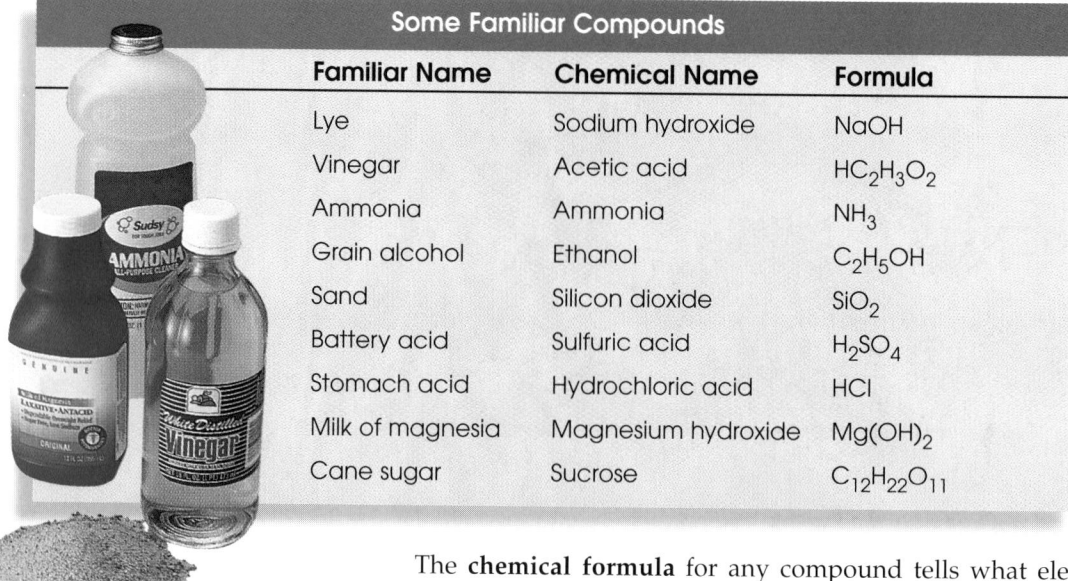

Some Familiar Compounds		
Familiar Name	**Chemical Name**	**Formula**
Lye	Sodium hydroxide	NaOH
Vinegar	Acetic acid	$HC_2H_3O_2$
Ammonia	Ammonia	NH_3
Grain alcohol	Ethanol	C_2H_5OH
Sand	Silicon dioxide	SiO_2
Battery acid	Sulfuric acid	H_2SO_4
Stomach acid	Hydrochloric acid	HCl
Milk of magnesia	Magnesium hydroxide	$Mg(OH)_2$
Cane sugar	Sucrose	$C_{12}H_{22}O_{11}$

The **chemical formula** for any compound tells what elements it contains and the ratio of the atoms of those elements. What elements are in each compound listed in **Table 11-1**? What is the ratio of atoms in each compound?

Chemically Stable Atoms

What causes elements to form compounds? Look at the periodic table on pages 732 and 733. It lists 111 elements, most of which can, and often do, combine with other elements. But the six noble gases in Group 18 seldom combine with other elements. Why do the noble gases so infrequently form compounds? The reason is that the arrangement of electrons in their atoms makes them chemically stable, or resistant to change.

Outer Levels: Getting Their Fill

What electron arrangements make atoms stable? An atom is **chemically stable** if its outer energy level is completely filled with electrons. For the atoms of most elements, the outer energy level is filled when it contains eight electrons. Atoms of the noble gases neon, argon, krypton, xenon, and radon all contain eight electrons in the outermost energy level. For atoms of helium, the outermost energy level is filled when it has two electrons.

Figure 11-4 shows dot diagrams of some of the noble gases. Remember from Chapter 10 that a dot diagram of an atom shows the number of electrons in its outer energy level. Notice that each of these elements has eight outer electrons.

USING MATH

The formula H_2O represents one molecule of water. To represent two molecules of water, use a coefficient as you would in algebra—$2H_2O$. The formula for carbon dioxide is CO_2. How would you represent three molecules of CO_2?

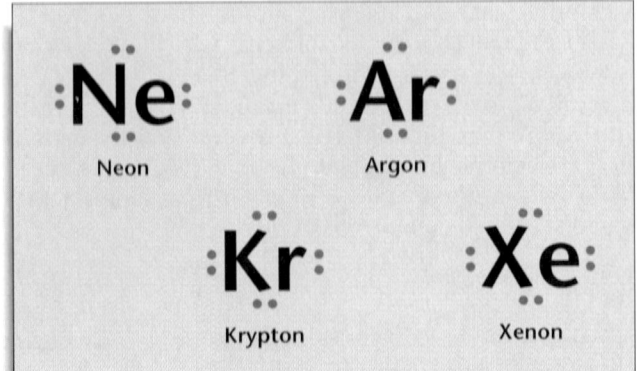

Figure 11-4

Dot diagrams of noble gases show that they all have a stable, filled outer level.

Because each noble gas has an outer level that is filled with electrons, each of these elements is chemically stable. Thus, the noble gases don't form compounds naturally. A few compounds of xenon, krypton, and radon have been prepared in the laboratory under special conditions. Each atom of all the elements *except* the noble gases has from one to seven electrons in its outer energy level. These atoms tend to lose, gain, or share electrons with other atoms. When gaining, losing, or sharing electrons, attractions form between atoms, and compounds form. In one of these ways, an atom may have its outer energy level filled, resulting in an outer energy level similar to that of a noble gas.

Losing, gaining, and sharing electrons are the means by which atoms become stable and form chemical bonds. A **chemical bond** is a force that holds together the atoms in a substance. As sodium and chlorine become bonded, each sodium atom loses one electron and each chlorine atom gains one. As a result, each atom in sodium chloride, NaCl, has eight electrons in its newly formed outer energy level. **Figure 11-5** illustrates this process.

USING MATH

You can use integers to represent positive and negative charges of atoms. If a neutral atom gains an electron, the new charge is $0 + (1-) = 1-$. The atom has a net charge of $1-$. What equation shows a neutral atom losing an electron? What is the net charge?

Figure 11-5

When sodium combines with chlorine, each sodium atom loses an electron and each chlorine atom gains an electron. *Why is the compound sodium chloride stable?*

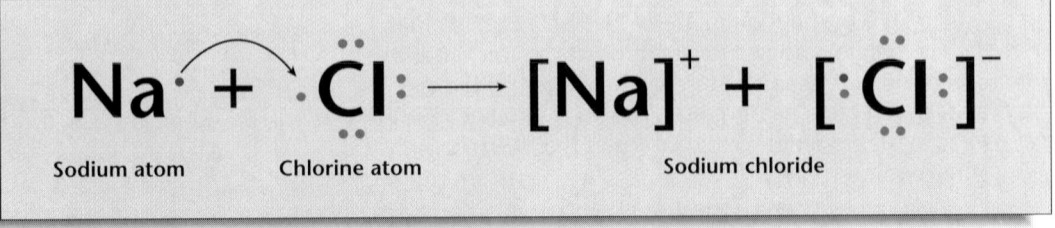

Na + Cl ⟶ [Na]⁺ + [Cl]⁻

Sodium atom　　Chlorine atom　　　　Sodium chloride

Visual Learning

Figure 11-5 Why is the compound sodium chloride stable? *Both Na and Cl now have eight electrons in the outer energy level.* **LEP**

LS

Assessment

Portfolio Use the Flex Your Brain activity to reinforce critical-thinking and problem-solving skills. Use the Performance Task Assessment List for Asking Questions in **PASC,** p. 19. **P**

USING MATH

Answer $0 - (-1) = +1; +1$

GLENCOE TECHNOLOGY

 Videodisc

Glencoe Physical Science Interactive Videodisc

Side 1, Lesson 4

Electron Structure

37983-38622

38625-39566

39568-39685

Use **Teaching Transparency 21** as you teach this lesson.

3 Assess

Check for Understanding

?FLEX Your Brain

Use the Flex Your Brain activity to have students explore CHEMICAL FORMULAS.

 **Activity Worksheets,** p. 5

Reteach

Ask students why they think each compound listed in **Table 11-1** on page 300 is a stable compound.

Extension

For students who have mastered this section, use the **Reinforcement** and **Enrichment** masters.

301

Visual Learning

Figure 11-6 How is the bond in Cl₂ different from the bond in NaCl? *It is formed by sharing electrons.* **LEP** **LS**

4 Close

•MINI•QUIZ•

1. What is used in a formula to tell the number of atoms of an element in a compound? *subscript*

2. When the outer energy level of an atom is filled with electrons, the atom is chemically _____ . *stable*

Section Wrap-up

Review

1. The properties of the elements are replaced by the properties of the compound they form.

2. (a) CaF_2, (b) Al_2S_3

3. The compounds are chemically more stable than the elements.

4. **Think Critically** sodium, carbon, and oxygen, in the ratio of 2 to 1 to 3

Science Journal

In a chemical bond, a force holds atoms together in a substance. The force may result from the sharing or loss or gain of electrons. Atoms generally fill their outer energy levels when they bond, becoming more stable.

In chlorine gas, two chlorine atoms share electrons, as shown in **Figure 11-6.** In its formula, Cl_2, the subscript 2 shows that two atoms are bonded together.

As you have read, compounds make up most of the matter around you. Why is this so? The answer is that the arrangements of electrons in most atoms make them chemically unstable. By sharing or transferring electrons, atoms achieve more stable arrangements of electrons.

Figure 11-6

This dot diagram shows the formation of chlorine gas, Cl_2, from two atoms of chlorine. *How is the bond in Cl_2 different from the bond in NaCl?*

Chlorine atom   Chlorine atom   Chlorine molecule

Section Wrap-up

Review

1. What happens to the properties of elements when atoms form compounds?

2. Write formulas for (a) a compound with one calcium atom and two fluorine atoms and (b) a compound with two aluminum atoms and three sulfur atoms.

3. Why do most elements tend to form compounds?

4. **Think Critically:** The label on a box of washing soda states that it contains Na_2CO_3. Name the elements in this compound. In what ratio are they present?

Skill Builder
Making and Using Tables
The compounds in **Table 11-1** on page 300 that contain carbon are classified as organic, and the others are classified as inorganic. Reorganize the contents of the table using these groups. If you need help, refer to Making and Using Tables in the **Skill Handbook.**

Science Journal

A chemical bond is not a thing that can be touched. Using electron arrangements, energy, and stability, write in your Science Journal a paragraph describing a chemical bond.

 Skill Builder

Type of Compound	Familiar Name	Chemical Name	Formula
Organic	vinegar	acetic acid	$HC_2H_3O_2$
	grain alcohol	ethanol	C_2H_5OH
	cane sugar	sucrose	$C_{12}H_{22}O_{11}$
Inorganic	lye	sodium hydroxide	NaOH
	ammonia	ammonia	NH_3
	sand	silicon dioxide	SiO_2
	battery acid	sulfuric acid	H_2SO_4
	stomach acid	hydrochloric acid	HCl
	milk of magnesia	magnesium hydroxide	$Mg(OH)_2$

Activity 11-1

Eggshell Electrons

You know how important electrons are in chemical bonding and how this bonding is based on the arrangement of electrons. Show how electrons control bonding by forming some model bonds.

Problem
How can a visible model show how energy levels fill when atoms combine?

Materials
- modified egg carton
- marbles

Procedure
1. Obtain a modified egg carton and marbles from your teacher. The carton represents the first and second energy levels of an atom. The marbles represent electrons.
2. Place one marble in each depression of the carton. Start with the pair of depressions that represents the first energy level of the atom.
3. Place the remaining marbles in receptacles representing the second energy level. In which column would your element appear on the periodic chart?
4. Compare your model with those of your classmates. Find one or more other cartons that, when combined with yours, will make it possible for each of the two cartons to have eight marbles in its second energy level.
5. In your Science Journal, make a list of the combinations you were able to make with your classmates' models.

Analyze
1. Generally, do elements in metal groups on the periodic table have more or fewer electrons in their outer energy levels than do nonmetals?
2. The combinations you found could represent chemical formulas. Why did some formulas require more than one atom of an element?

Conclude and Apply
3. Would your **model** be more likely to represent a metal or a nonmetal atom? Explain.
4. **Predict** what group of elements your **model** would be in if you received eight marbles. Explain.

11-1 Why Atoms Combine   **303**

Purpose
IS **Logical-Mathematical** Students will infer chemical formulas by making models of outer electron levels. **L1** **LEP** **COOP LEARN**

Process Skills
classifying, communicating, comparing and contrasting, forming operational definitions, interpreting scientific illustrations, making models

Time
30 minutes

📁 **Activity Worksheets,** pp. 5, 63-64

Teaching Strategies
Alternate Materials Substitutes for marbles could include buttons, candy, or beans.
Troubleshooting Remind students that these examples do not include the transition metals. These egg-carton atoms can represent the elements in the first two rows of the periodic table.

- Using the egg carton, block off two of the egg receptacles, and review with students the capacity of first and second energy levels.
- Give different numbers of marbles to different students so that all elements from lithium through fluorine are represented.
- To answer the question in Procedure step 3, direct students to count the marbles in the second energy level. Have them count over the same number of columns in Period 2.
- Monitor students so that they discover simple matchups such as LiF and then progress to compounds such as BeF_2, Li_2O, BeO, and Be_3N_2. Remind them that atoms can achieve stability by losing, gaining, or sharing electrons.

Answers to Questions
1. Metals have fewer outer-level electrons than do nonmetals.
2. Two atoms combined do not provide enough electrons for each atom to have eight in its outer level.
3. Answers will vary. Usually, fewer than four outer electrons indicates a metal. Four or more usually indicates a nonmetal.
4. Eight marbles indicate a Group 16 element. These atoms have six outer electrons.

✔ Assessment
Content Have students check their combinations of atoms to make sure full energy levels are evident. Use the Performance Task Assessment List for Model in **PASC**, p. 51. **P**

303

Prepare

Section Background

An atom's electronegativity is the tendency of the atom to attract a pair of electrons in a bond. An ionic bond results from electron transfer between two atoms that differ greatly in electronegativity. A covalent bond results from the sharing of a pair of electrons between two atoms that are close in electronegativity. In a polar covalent bond between two atoms, there is unequal sharing of the electron pair due to a moderate difference in electronegativity.

Preplanning

- For Activity 11-2, you will need sugar, salt, and finely crushed ice.
- For the MiniLAB, assemble balloons and pieces of wool or fur.

1 Motivate

Bellringer

Before presenting the lesson, display **Section Focus Transparency 43** on the overhead projector. Assign the accompanying **Focus Activity** worksheet. L1 LEP

Tying to Previous Knowledge

Recall that elements with three or fewer electrons are metals. Metals tend to lose electrons to form positive ions. Nonmetals generally have four or more outer electrons. Nonmetals tend to gain electrons to form negative ions.

11•2 Kinds of Chemical Bonds

Science Words

ion
ionic bond
covalent bond
polar molecule
nonpolar molecule

Objectives

- Describe ionic bonds and covalent bonds.
- Identify the particles produced by ionic bonding and by covalent bonding.
- Distinguish between a nonpolar covalent bond and a polar covalent bond.

Ions and Ionic Bonds

The brilliant flash of a camera and a bright white explosion in a fireworks display both show what happens when magnesium metal combines with oxygen, as shown in **Figure 11-7.** Recall how the atoms of these elements bond chemically when mixed. An atom of magnesium has two electrons in its outer level, as shown below. When that atom loses two electrons from its third level, the second level becomes a complete outer level because it contains eight electrons. But the atom now contains 12 protons, with a total positive charge of 12+, and ten electrons, with a total charge of 10−. Because $(12+) + (10−) = 2+$, the atom now has a net charge of 2+. Atoms that have a charge are called **ions.** Magnesium with a 2+ charge is called a magnesium ion. The charge would be shown as a superscript adjacent to the element's symbol, Mg^{2+}, to indicate its charge. *Superscript* means "written above."

An oxygen atom has six electrons in its outer energy level. When the atom gains two electrons, its outermost energy level contains eight electrons. The atom now has eight protons, with a total charge of 8+, and ten electrons, with a total charge of 10−. Because $(8+) + (10−) = 2−$, the atom has a net negative charge of 2−. In other words, it is now an oxygen ion, which you write with a superscript minus sign, $O^{2−}$, to indicate its charge. However, the compound

Figure 11-7

When burned, magnesium combines with oxygen.

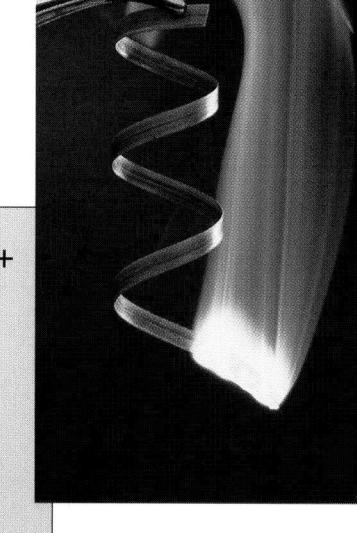

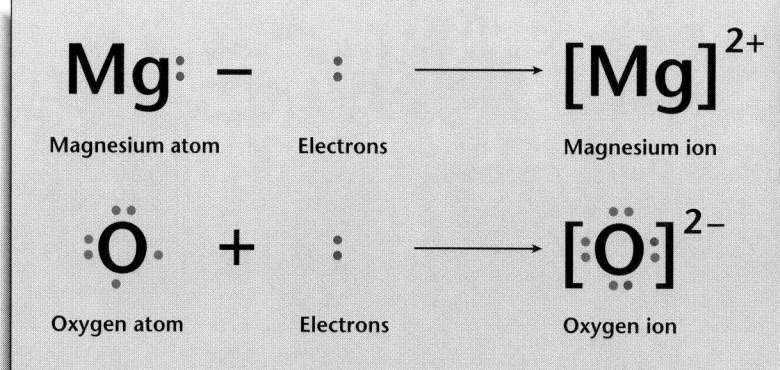

| Magnesium atom | Electrons | Magnesium ion |

| Oxygen atom | Electrons | Oxygen ion |

Program Resources

Reproducible Masters

Study Guide, p. 47 L1
Reinforcement, p. 47 L1
Enrichment, p. 47 L3
Activity Worksheets, pp. 65-67 L1
Lab Manual 24 and **25**

Transparencies

Section Focus Transparency 43
Science Integration Transparency 11

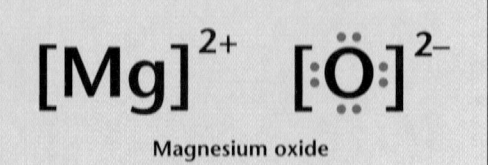
Magnesium oxide

Figure 11-8
This dot diagram for magnesium oxide, MgO, shows that both ions are stable.

MgO as a whole is neutral because the total net positive charge equals the total net negative charge. The dot diagram for MgO is shown in **Figure 11-8**.

Opposites Attract

An ion is an atom that is either positively or negatively charged. Compounds made up of ions are ionic compounds, and the bonds that hold them together are ionic bonds. An **ionic bond** is the force of attraction between the opposite charges of the ions in an ionic compound.

Figure 11-9 shows another example of ionic bonding—the formation of magnesium chloride, $MgCl_2$. When magnesium reacts with chlorine, a magnesium atom loses two electrons and becomes a positively charged ion, Mg^{2+}. Notice that you write 2+ on the symbol for magnesium to indicate the ion's net charge. At the same time, each of the two chlorine atoms gains one electron and becomes a negatively charged chloride ion, Cl^-. The compound as a whole is neutral. Why? Because the sum of the net charges on all three ions is zero.

CONNECT TO
EARTH SCIENCE

Farmers must sometimes add lime, calcium oxide, to their soil. What is the formula of calcium oxide? *Predict* whether calcium oxide is likely to be an ionic compound.

Figure 11-9
Magnesium chloride can be extracted from seawater and used as a source of magnesium metal.

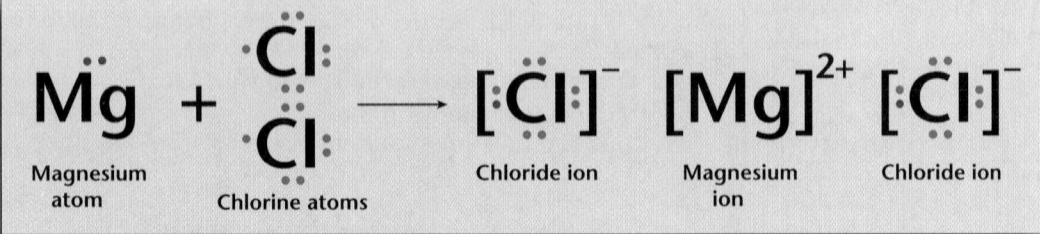

Magnesium atom + Chlorine atoms → Chloride ion · Magnesium ion · Chloride ion

Nervous Ions

Ions are not just something you read about in a science book. Every movement you make—from the slight eye movements needed to read this sentence to the larger movement of turning the page—depends on ions and their relationship to the actions of muscles. How do muscles receive the signal to move? You may be surprised to find out that muscle movements depend on the movement of ions into and out of nerve cells. Nerve cells are surrounded by sodium, potassium, and chloride ions.

INTEGRATION
Life Science

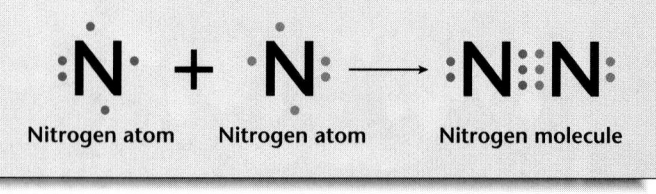

Using an Analogy

A covalent bond can be thought of as a spring that allows the bonded atoms to stretch and bend.

MiniLAB

Purpose

LS **Logical-Mathematical** Determine the type of bonding in water molecules. **L1**

Materials

balloon, scrap of wool or fur

Teaching Strategies

• Have students repeat the activity using a plastic comb that has just combed clean, dry hair instead of using the balloon.

• Be sure students hold the balloon near the stream of water but not touching it.

Analysis

1. Rubbing the balloon gave it a charge. Holding it near the water stream attracts the oppositely charged end of the water molecules.

2. A water molecule is polar covalently bonded.

✔ Assessment

Oral Students don't know whether the charge on the balloon is positive or negative. Ask them to explain why the results would be the same, no matter what the charge is. *A negative charge attracts the positive end of the water molecule, and a positive charge attracts the negative end.* Use the Performance Task Assessment List for Making Observations and Inferences in **PASC**, p. 17. **P**

📁 **Activity Worksheets**, pp. 5, 67

Figure 11-10

Dot diagrams for the formation of nitrogen gas, N_2, from two atoms of nitrogen show that nitrogen atoms share six electrons. *How many covalent bonds are present in N_2?*

MiniLAB

What type of molecule is water?

Find out what type of bonding is present in a water molecule.

Procedure

1. Turn on the faucet to produce a thin stream of water.
2. Rub an inflated balloon with wool or fur.
3. Bring the balloon near the stream of water, and describe your observations.

Analysis

1. Explain your observations.
2. In terms of the observations, explain what type of bonding is in a water molecule.

Figure 11-11

A dot diagram and model for a molecule of hydrogen chloride show the sharing of electrons.

As these ions move in and out of the nerve fibers, through the cell membrane, they set up an electrical pulse that results in a signal moving along a nerve. For example, many sodium ions, which are positively charged, are outside of a normal nerve cell membrane, while negative ions stay on the inside, which makes a slight charge build up in that area of the nerve.

The typical speed for signals moving along nerve cells is between 40 and 60 m/s. As you think about gaining and losing electrons to make ions, remember how important they are to many functions of your body.

Molecules and Covalent Bonds

Most atoms become more chemically stable by sharing electrons, rather than by losing or gaining electrons. In **Figure 11-6** on page 302, you saw how two chlorine atoms share electrons. Notice that the chlorine particle, Cl_2, is not charged—it is neutral. The chlorine is not in the form of ions, but rather molecules. Neutral particles formed as a result of electron sharing are called molecules.

The attraction that forms between atoms when they share electrons is known as a **covalent bond.** In the case of chlorine, a covalent bond is formed between two atoms of the same element. Covalent bonds also form between atoms of nitrogen, the gas that makes up most of the air. As you can see in **Figure 11-10,** two atoms in a molecule of nitrogen share six electrons, forming three covalent bonds between the atoms.

Polar and Nonpolar Molecules

Atoms in molecules do not always share electrons equally. An example is a molecule of hydrogen chloride, HCl. A water solution of HCl is hydrochloric acid, which is used in laboratories, to clean metal, and in your stomach to digest food.

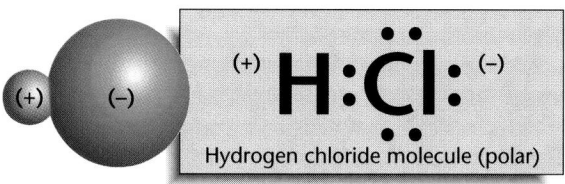

Hydrogen chloride molecule (polar)

Visual Learning

Figure 11-10 How many covalent bonds are present in N_2? *3*

Figure 11-11 The parentheses are used to show that the charges in polar molecules are *partial* charges, and the forces between polar molecules are much weaker than the forces between ions. **LEP** **LS**

Chlorine atoms have a stronger attraction for electrons than hydrogen atoms do. As a result, the electrons shared in hydrogen chloride will spend more time near the chlorine atom than near the hydrogen atom. Look at **Figure 11-11.** This unequal sharing of electrons gives each chlorine atom a slight negative charge and each hydrogen atom a slight positive charge. This type of molecule is called polar. *Polar* means "having opposite ends." A **polar molecule** is one that has a positive end and a negative end. Water is another example of a compound with polar molecules, as shown in **Figure 11-12** on page 308, as is sugar.

Some molecules, such as those of nitrogen, are nonpolar because the two nitrogen atoms share their electrons equally. A **nonpolar molecule** is one that does not have oppositely charged ends. Look again at the diagram of a nitrogen molecule in **Figure 11-10.**

USING TECHNOLOGY

Bond Busters

Human activities generate toxic compounds in huge amounts. One way to get rid of toxic compounds is to change them chemically into other harmless compounds; that is, their chemical bonds must be broken. Then the atoms could bond into different molecules of less harmful substances.

From oil spills on water to toxic solvents at hazardous waste sites, scientists are using microbes as bond busters. The microbes eat the compounds, which change chemically as they are digested. To create a strain of pollutant-eating microbes, scientists first add a pollutant to a population of microbes. Most of the microbes die, leaving only the hardiest ones to reproduce. Scientists then seed polluted areas with these microbes and add fertilizer and oxygen to promote their growth. The microbes break down the toxic compounds and then, with the food source gone, the microbes die. At least 1000 different strains of bacteria and fungi are now helping to get rid of pollutants. The biggest advantage of the microbial approach is that instead of relocating the pollutants, the microbes destroy them.

Think Critically:

What would be the disadvantages of using microbes compared to removing or burning contaminated soil?

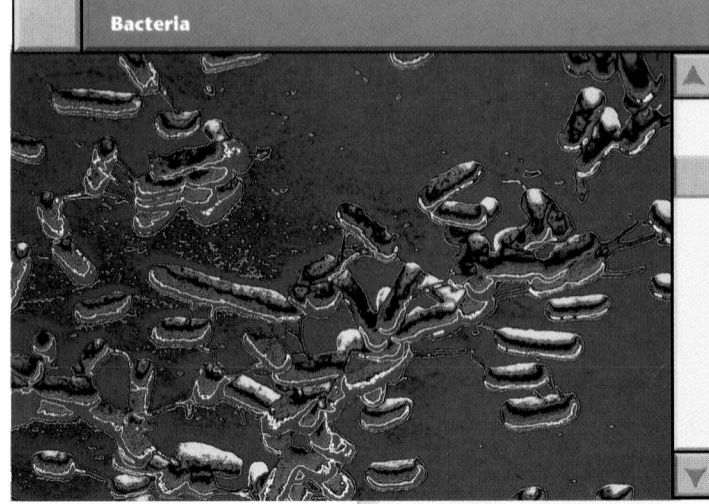

Bacteria

Cultural Diversity

Ocher Ocher is an iron oxide (Fe_2O_3) that occurs naturally. It is formed by the weathering of rocks and has been used since prehistoric times. Ocher can be used for treatment of wounds, to tan hides, and as a pigment for art and body decoration. It was used prehistorically to produce the rock paintings at Lascaux and other sites and continues to be used for artwork today. It comes in a variety of colors, depending on the percentage of iron oxide present. Hematite ore, one source of ocher that can be ground to produce a bright red ore, appears to have been mined in Swaziland, Africa, as early as 44 000 years ago. Other iron oxide ores produce different colors: black (magnetite), yellow (limonite), and brown (siderite and pyrite).

4 Close

Use the Mini Quiz to check students' recall of chapter content.

1. **Atoms that have charges are called _____ .** *ions*

2. **What do we call the force of attraction between oppositely charged ions?** *ionic bond*

3. **How do most atoms bond to become chemically stable?** *by sharing electrons*

4. **Atoms that share electrons are held together by at least one _____ .** *covalent bond*

Section Wrap-up

Review

1. In ionic bonds, ions are held together by opposite charges. In covalent bonds, atoms share electrons. Both are attractive forces.

2. a. ionic bonds—positive and negative ions; b. polar covalent bonds—polar molecules (Note: If the molecule is symmetrical, it may not be polar even though it has polar bonds.); c. nonpolar covalent bonds—nonpolar molecules

3. **Think Critically** Ionic bond: KCl, $CaCl_2$; covalent bond: CaS, K_2S; answers are based on whether transfer of or sharing of electrons results in eight outer electrons.

USING MATH

Each aluminum atom loses three electrons. Each oxygen atom gains two electrons.

$$2(3+) + 3(2-) = 6 - 6 = 0$$

Figure 11-12

The combination of hydrogen atoms and an oxygen atom forms a polar molecule.

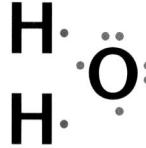

 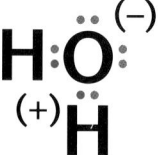

A As hydrogen atoms approach an oxygen atom, their electrons are attracted to the oxygen atom.

B Although the pull of oxygen is strong, hydrogen does not lose its electrons. It shares them with the oxygen atom. Because the electrons are more attracted to the oxygen atom than to the hydrogen atoms, the sharing is unequal.

C Unequal sharing of electrons produces an area of negative charge and an area of positive charge on the molecule. Because the electrons spend more time around the oxygen atom, this atom has a slightly negative charge. The hydrogen atoms have a slightly positive charge.

You have read about two main ways chemical bonding takes place, producing two kinds of compounds—ionic and covalent. These groups of compounds have contrasting properties. Ionic compounds are usually formed by bonding between a metal and a nonmetal. Bonds between nonmetal atoms are covalent. Most ionic compounds are crystalline solids with high melting points. Many covalent compounds are liquids or gases at room temperature. In later chapters, you will learn more about compounds of each type.

Section Wrap-up

Review

1. Compare ionic and covalent bonds.

2. What type of particle is formed by the following bonds: (a) ionic, (b) polar covalent, (c) nonpolar covalent?

3. **Think Critically:** From the following list of symbols, choose two elements that are likely to form an ionic bond. Select two elements that would likely form a covalent bond. Explain your choices.

   Cl, Ne, S, Ca, K

Skill Builder
Concept Mapping

Using the following terms, make a network tree concept map of chemical bonding: *ionic, covalent, ions, positive ions, negative ions, molecules, polar, nonpolar*. If you need help, refer to Concept Mapping in the **Skill Handbook.**

USING MATH

Aluminum oxide, Al_2O_3, can be produced during space shuttle launches. Show that the sum of the positive and negative charges in a unit of Al_2O_3 equals zero.

Skill Builder

Assessment

Performance Assign a compound to each student, and ask him or her to follow it through the concept map, identifying each component. Use the Performance Task Assessment List for Concept Map in **PASC,** p. 89. **P**

People and Science

DR. ALVARO GARZA, *Physician*

On the Job

Q What are the sources of lead poisoning?

A It results from someone ingesting or inhaling something that contains lead. Years ago, gasoline had a high lead content. Now the source is primarily lead-based paints in older homes.

Q How does lead poisoning affect young children?

A At high levels, it can be deadly. Even at lower levels, it may affect behavioral and brain development.

Q What treatments are available to children with high levels of lead in their blood?

A At the extreme, a child may need hospitalization and chelation therapy. That's the injection of a substance that chelates, or forms bonds, to the lead so that it can be excreted through the kidneys. It is most important to stress preventive measures, such as frequent washing of children's hands and toys. Nutrition is important, too, especially adequate iron and calcium in the intestine to decrease the body's absorption of lead.

Personal Insights

Q Childhood lead poisoning occurs mostly in low-income inner-city neighborhoods. How do you feel about this?

A Environmental and social justice is very important to me. I often work with community-based organizations. It's vital to educate neighborhood activists so they can be leaders in health education in their own communities.

Q When you give talks at schools, how do you hope students perceive you?

A I'd like them to think of me as an activist, someone who works for a better world and healthier lives for kids.

Career Connection

Working in the public health sector involves educating people to take care of their own families' health. Use the information given about childhood lead poisoning to make a poster that could be displayed in your community.

- **Public health worker**
- **Pharmacist**

11-2 Kinds of Chemical Bonds   **309**

Activity 11-2

Purpose
IS Interpersonal Students will design and carry out an experiment to show how melting properties of substances relate to the type of bonding found in the substance. **L1**
COOP LEARN

Process Skills
observing, inferring, comparing, making and using tables, forming a hypothesis, designing an experiment, interpreting data, measuring in SI

Time
one class period

Materials
See student page.

Safety Precautions
Caution students to be careful around the hot burner and the heated materials.

Possible Hypotheses
Students may hypothesize that the more easily melted a substance is, the less attraction the particles have for each other.

📁 **Activity Worksheets,** pp. 5, 65-66

PLAN THE EXPERIMENT

Possible Procedures
One possible procedure is for students to use the same amount of each substance, beginning with the ice because it will begin to melt first. One lab partner can watch for the first sign of melting while the other student times the experiment. Students might also note the length of time between the first observable melting and total melting.

Activity 11-2

Become a Bond Breaker

The basic structural units of ionic compounds are ions. For covalent substances, molecules make up the basic units. By using controlled heat to melt substances, you can test various compounds to rate the attractive forces between their basic units. Would a substance that is difficult to melt have strong forces or weak forces holding its basic units to each other?

PREPARATION

Problem
How do the attractive forces between ions compare to the attractive forces between molecules?

Form a Hypothesis
Based on what you know about ions and molecules, state a hypothesis about which would generally have stronger attractions between their structural units.

Objectives
- Observe the effect of heat on melting points of selected substances.
- Design an experiment that allows you to make some inferences that relate ease of melting and forces of attraction between particles of a substance.

Possible Materials
- small samples of crushed ice, table salt, sugar
- wire test-tube holder
- test tubes
- goggles
- laboratory burner
- timer

Safety Precautions

Keep a safe distance from the open flame of the lab burner. Wear proper eye protection. Do not continue heating beyond 5 minutes.

PLAN THE EXPERIMENT

1. As a group, agree upon and write a hypothesis statement.
2. As a group, write a detailed list of steps needed to test your hypothesis.
3. Make a list of materials that you will need to complete your experiment.
4. Design a data table in your Science Journal to record your observations.

Check the Plan

1. As you heat materials in a test tube, what variables are you controlling?
2. How are you timing the heating of substances?
3. Will you run any tests more than one time?
4. *Make sure your teacher approves your plan and that you have included any changes suggested in the plan.*

DO THE EXPERIMENT

1. Carry out the experiment as planned.
2. While you are observing the heating of each substance, think about the movements of the particles. Which particles are held together by ionic bonds? Which are made up of covalent molecules?

3. Be sure to write down how long it takes to melt the tested substances.

Analyze and Apply

1. **Compare** your results with those of other groups.
2. **Classify** your tested substances as more likely ionic or covalent.
3. Which substances are generally more difficult to melt?
4. Sugar is known as a polar covalent compound. **Infer** how polarity affects melting point.

Go Further

Because you tested a limited number of samples, you can't make conclusions about all ionic and covalent substances. But if you had to make a prediction, which do you think would be easier to melt, KBr or candle wax? Candle wax is made up of many of the same types of atoms that make up sugar.

11-2 Kinds of Chemical Bonds   311

✔ **Assessment**

Portfolio Have students construct a graph that reports their observations of the three substances during the heating process. Their observations should contain references to time intervals. Use the Performance Task Assessment List for Data Table in **PASC**, p. 37. **P**

Teaching Strategies
Troubleshooting To avoid premature melting, test tubes should be clean, dry, and at room temperature.

- Discuss with students the importance of constants. Measuring differing amounts of the three substances and holding the test tubes in different parts of the flame may affect results.
- Review the chemical formulas of the substances $C_{12}H_{22}O_{11}$, H_2O, and NaCl.
- Assist students in adjusting the air-gas mix in their burners so that the flames are blue, not yellow.

DO THE EXPERIMENT

Expected Outcome
Ice will melt first, then sugar. Salt will not melt due to its high melting point (801°F).

Analyze and Apply

1. Most results should be the same, within a few seconds.
2. Ice—polar covalent
   Sugar—covalent
   Salt—ionic
3. ionic substances, because the attractions between the particles are strongest
4. Polar covalent bonding means that the bonds are shared unequally, giving the molecule some ionic characteristics. Because it is more ionic, it is more difficult to melt than purely covalent compounds.

Go Further

Potassium bromide is ionic and thus is more difficult to melt.

311

Prepare

Section Background

One quart of oil can contaminate 1 million liters of drinking water because it forms a very thin film on the surface of the water.

1 Motivate

Bellringer

 Before presenting the lesson, display **Section Focus Transparency 44** on the overhead projector. Assign the accompanying **Focus Activity** worksheet. L1 LEP

Brainstorming

Have students brainstorm a list of household chemicals they think are corrosive, toxic, or flammable.

2 Teach

GLENCOE TECHNOLOGY

 **Videodisc**

The Infinite Voyage: Miracles by Design
Chapter 9
"Smart Materials" and Their Construction

STVS: Chemistry
Dealing with Hazardous Materials (Ch. 20)

ISSUE:
11•3 Chemical Risks in the Home

Science Words

toxic
corrosive

Objectives

• Describe the dangers posed by hazardous compounds in the home.
• Demonstrate a knowledge of safer alternative compounds to use.

Chemicals—Not Just in the Lab

Professional chemists regularly work with chemicals that may be hazardous. They have learned to follow safety precautions and minimize risks. Do you realize that you, too, often work around hazardous chemicals? A hazardous chemical is one that can affect the health and safety of people or harm the environment. If you read the labels on household materials, you will often see strong health warnings.

Classifying the Problems

Hazardous chemicals usually fit one of three categories. **Toxic** compounds are poisonous. The compounds that help disinfectants work and repel insects are often toxic. The term **corrosive** comes from Latin roots meaning "to tear up." This type of compound is useful for unclogging drains and cleaning ovens and can be found in car batteries. However, corrosives cause problems when they come in contact with human skin or some metals. The third overall class of hazardous compounds is made up of compounds that burn easily. They are said to be flammable. Gasoline, paint thinners, and some aerosols are common examples of flammable materials.

The type of bonding in a compound helps determine how hazardous a material is. For example, the bonding in acetone, found in some paint removers, is covalent. Many covalently bonded compounds readily evaporate, which makes them easily inhaled and probably flammable. Ionic compounds are not likely to evaporate. Drain cleaners often contain NaOH, an ionically bonded compound.

Is there any way to control what hazardous compounds are used and how they are used? What can you do to minimize risk to yourself, your family, and the environment?

Table 11-2

Tips to Prevent Pollution from Hazardous Materials

1. Use safer alternatives.
2. Recycle oil, antifreeze, and used batteries.
3. If you must buy a hazardous product, buy only the amount you need.
4. Store hazardous materials in their original containers, away from children.
5. Keep hazardous materials that easily evaporate in tightly closed containers.
6. Share your knowledge of the problem of hazardous household compounds with others.

Program Resources

 **Reproducible Masters**
Study Guide, p. 48 L1
Reinforcement, p. 48 L1
Enrichment, p. 48 L3
Science Integration Activities, pp. 21-22
Concept Mapping, p. 27
Critical Thinking/Problem Solving, p. 17

Transparencies
Section Focus Transparency 44

2 Points of View

Laws to Regulate Sale, Use, and Disposal of Hazardous Household Materials

Some people think that laws should be enacted that control hazardous materials. They believe that without any restrictions, it is inevitable that potentially hazardous materials will be released to harm the environment. Laws have been enacted that require an additional disposal fee for proper handling of discarded oil from automobiles, old batteries, and tires. These added fees help pay for recycling and proper disposal of some hazardous products. By paying small fees now for proper handling and recycling efforts, we can avoid more costly water and environmental cleanup costs and possibly even increased health costs. Do you think the fees are worth the cost?

Individual Responsibility

Some consumers resent being told how they must deal with potentially hazardous materials. How can an individual handle such materials in a responsible manner?

One solution is to use less harmful compounds as substitutes for potentially hazardous materials. Other possible solutions are shown in **Table 11-2.**

Figure 11-13

Warning labels on these home chemicals advise consumers about how to use the products safely.

Section Wrap-up

Review

1. What are the three main categories of hazardous household compounds?

2. Freon from air-conditioning systems is almost certainly responsible for some destruction of Earth's protective ozone layer. Improper handling of Freon can expose it to air, where it evaporates easily. Do you think Freon is more likely a covalently bonded or ionically bonded compound?

 Visit the Chapter 11 Internet Connection at Glencoe Online Science, **www.glencoe.com/sec/science/physical,** for more information about household hazardous wastes.

SCIENCE & SOCIETY

---

3 Assess

Check for Understanding

? FLEX·Your Brain

Use the Flex Your Brain activity to have students explore HAZARDOUS MATERIALS.

Activity Worksheets, p. 5

Reteach

Have teams of students survey their homes for hazardous chemicals.

Extension

For students who have mastered this section, use the **Reinforcement** and **Enrichment** masters.

4 Close

Have pairs of students demonstrate some safe alternatives and compare their effectiveness.

Section Wrap-up

Review

1. corrosive, toxic, flammable
2. Freon is more likely a covalently bonded compound.

Prepare

Section Background

- As a general rule, the oxidation numbers of metals can be predicted from their outer electron configurations, as shown in **Figure 11-15**. Most transition metals have variable oxidation numbers.
- The oxidation numbers of nonmetals can be predicted from the outer electron configuration. Those with seven outer-level electrons (one less than eight) have an oxidation number of 1– and so on.
- The noble gases have an oxidation number of zero because they have eight outer electrons.
- Many compounds have water molecules as a part of the crystal. The compounds are referred to as hydrates.

1 Motivate

Bellringer

Before presenting the lesson, display **Section Focus Transparency 45** on the overhead projector. Assign the accompanying **Focus Activity** worksheet. L1 LEP

Tying to Previous Knowledge

Ask students to write on the chalkboard the chemical formulas for as many compounds as they remember. Use the formulas to introduce oxidation numbers, formula writing, and naming.

Science Words

oxidation number
binary compound
polyatomic ion
hydrate

Objectives

- Explain how to determine oxidation numbers.
- Write formulas for compounds from their names.
- Name compounds from their formulas.
- Describe hydrates and their formulas.

11•4 Formulas and Names of Compounds

Oxidation Numbers

The people in **Figure 11-14** seem to have little in common. If the alchemist knew the composition of silver tarnish, how might he write its formula? The modern chemist does know its composition; it is Ag_2S. When you get to the end of this section, you, too, will know how to write formulas of compounds.

Assigning Oxidation Numbers

You can figure out formulas with the help of oxidation numbers. What are these numbers? An **oxidation number** is a positive or negative number assigned to an element to show its combining ability in a compound. In other words, an oxidation number indicates how many electrons an atom has gained, lost, or shared when bonding with other atoms. For example, when sodium forms an ion, it loses an electron and

Figure 11-14

The medieval alchemist and the modern chemist are shown at work investigating matter. Notice how each would write symbols for the elements silver and sulfur. Silver tarnish is silver sulfide, a compound of silver and sulfur.

Program Resources

Reproducible Masters
Study Guide, p. 49 L1
Reinforcement, p. 49 L1
Enrichment, p. 49 L3
Cross-Curricular Integration, p. 17
Multicultural Connections, p. 25

Transparencies
Section Focus Transparency 45
Teaching Transparency 22

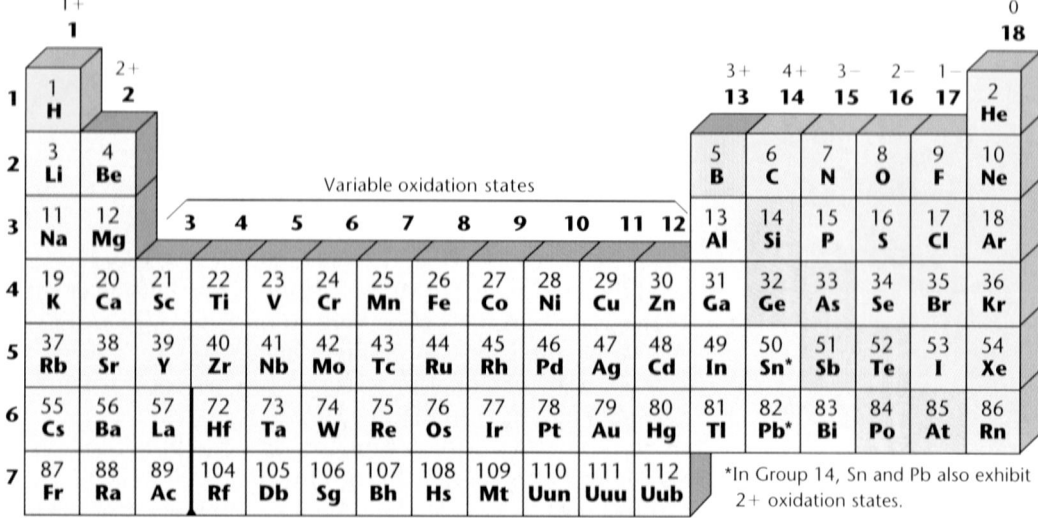

The lanthanide and actinide elements are not shown in this table.

has a charge of 1+. Therefore, the oxidation number of sodium is 1+. When chlorine forms an ion, it gains an electron and has a charge of 1−. Therefore, the oxidation number of chlorine is 1−.

Binary Compounds

Oxidation numbers are often a periodic property of elements. The numbers with signs printed on the periodic table shown in **Figure 11-15** are the oxidation numbers for these elements in many of their binary compounds. *Bi-* means "two," and a **binary compound** is one that is composed of two elements. Sodium chloride is an example of a binary compound.

Like sodium, each metal in Group 1 loses its one outer electron in bonding, so each of these elements has an oxidation number of 1+. Remember, losing electrons produces a positive oxidation number in an element. Each metal in Group 2 loses both its outer electrons in bonding, so each has an oxidation number of 2+.

Like chlorine, each of the nonmetals in Group 17 gains one electron in bonding, so each of these elements has an oxidation number of 1−. Remember, gaining electrons produces a negative oxidation number in an element. Each nonmetal in Group 16 typically gains two electrons in bonding, so each has an oxidation number of 2−.

Some elements have more than one oxidation number. Copper, as you can see in **Figure 11-16**, can be either Cu^+ or

Figure 11-15

This part of the periodic table shows oxidation numbers for elements in Groups 1, 2, 13, 14, 15, 16, 17, and 18.

Figure 11-16

Here are some elements that have variable oxidation numbers. Not all possible oxidation numbers are shown.

Copper(I)	Cu^+
Copper(II)	Cu^{2+}
Iron(II)	Fe^{2+}
Iron(III)	Fe^{3+}
Chromium(II)	Cr^{2+}
Chromium(III)	Cr^{3+}
Lead(II)	Pb^{2+}
Lead(IV)	Pb^{4+}

11-4 Formulas and Names of Compounds   **315**

2 Teach

Demonstration

LS **Visual-Spatial** Exhibit 4 g of powdered sulfur and 7 g of iron filings. Point out the yellow color of sulfur, and use a magnet to demonstrate the magnetic properties of iron. Put on safety goggles. Mix the sulfur and iron on a piece of paper and pour into a test tube. Behind a shield, heat the tube using a burner until the contents glow red. Then, plunge it into a beaker of water to break the test tube. The students will notice that the product is not yellow or magnetic.

$$Fe + S \rightarrow FeS$$

Use the demonstration to introduce oxidation numbers, formula writing, and naming. **LEP**

Revealing Preconceptions

Explain that oxidation numbers are used as an aid in understanding atomic structure and in writing formulas correctly. Formula writing is not just a reasoning exercise but represents a way of predicting the composition of actual chemical substances formed in real chemical processes.

Activity

Give each student a blank copy of the periodic table. Have students write the oxidation numbers of Groups 1, 2, 13, 14, 15, 16, 17, and 18 at the top of each column. **L1**

 Use these program resources as you teach this lesson.
Cross-Curricular Integration, p. 17
Multicultural Connections, p. 25

Use **Teaching Transparency 22** as you teach this lesson.

Concept Development

To help students write compound formulas correctly, remind them of the following.

- The oxidation number of hydrogen in most compounds is 1+.
- The oxidation number of oxygen in most compounds is 2–.
- The sum of the oxidation numbers of all the atoms in a compound formula must equal zero.
- In compounds, the elements of Groups 1 and 2 and aluminum have positive oxidation numbers of 1+, 2+, and 3+, respectively.

Inquiry Questions

- **Determine the oxidation number of chlorine in the following compounds: HClO, HClO$_2$, HClO$_3$, HClO$_4$, HCl, and Cl$_2$.** *1+, 3+, 5+, 7+, 1–, 0*

- **Write the name for each of the following binary compounds.**
  a. KBr *potassium bromide*
  b. CaO *calcium oxide*
  c. CuI *copper(I) iodide*
  d. Al$_2$S$_3$ *aluminum sulfide*

Activity

Logical-Mathematical Have groups of students make up plausible compound names while the other groups make up plausible formulas. Have one group write names for the other group's formulas and vice versa. L1 **COOP LEARN**

Cu^{2+}. In the name of a compound of copper, the Roman numeral equals the oxidation number of copper in that compound. Thus, the oxidation number of copper in copper(II) oxide is 2+. Iron may be Fe^{2+} or Fe^{3+}. Thus, the oxidation number of iron in iron(III) oxide is 3+.

Formulas for Binary Compounds

Once you know how to find the oxidation numbers of elements in binary compounds, you can write formulas by using the following rules: (1) Write first the symbol of the element that has the positive oxidation number. Hydrogen and all metals have positive oxidation numbers. (2) Then write the symbol of the element with the negative oxidation number. (3) Add subscripts so that the sum of the oxidation numbers of all the atoms in the formula is zero. Study the following example of how to write the formula of a specific compound.

USING MATH

Writing Formulas for Binary Compounds

Example Problem:
What is the formula of a compound composed of only sulfur and aluminum?

Problem-Solving Steps:
1. Write the symbol of the positive element followed by the symbol of the negative element.   Al   S
2. Look up oxidation numbers for each element. Write oxidation numbers above the symbols.
   3+   2–
   Al   S
3. Determine how many atoms of each element you need so that the sum of the oxidation numbers is zero.
Hint: The least common multiple of 2 and 3 is 6.
   3+   2–
   Al$_2$ S$_3$
   $2(3+) + 3(2-) = (6+) + (6-) = 0$
4. **Solution:** Write the subscripts.
   Final Formula: Al$_2$S$_3$

Practice Problems
1. Write the formula for the binary compound of calcium and oxygen.
Strategy Hint: Be sure to use the smallest possible subscript numbers to make the sum zero.
2. Write the formula of the binary compound of copper(II) and sulfur.
Strategy Hint: The Roman numeral II tells you the oxidation number of copper.

Across the Curriculum

Geology Geologists use oxidation numbers to infer what conditions were present in past geologic periods. The presence of iron(III) compounds indicates shoreline deposition where plenty of oxygen was available. Iron(II) compounds indicate the sediment was deposited in deeper water where oxygen was lacking.

Table 11-3

Elements in Binary Compounds

Element	*-ide* Naming	Element	*-ide* Naming
Chlorine	Chloride	Oxygen	Oxide
Fluorine	Fluoride	Phosphorus	Phosphide
Nitrogen	Nitride	Sulfur	Sulfide

Naming Binary Compounds

You can name a binary compound from its formula by using these rules: (1) Write the name of the first element. (2) Write the root of the name of the second element. (3) Add the ending *-ide* to the root. **Table 11-3** lists several elements and their *-ide* counterparts. For example, $CaCl_2$ is named calcium chloride.

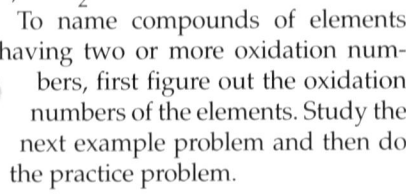

To name compounds of elements having two or more oxidation numbers, first figure out the oxidation numbers of the elements. Study the next example problem and then do the practice problem.

USING MATH

Naming Some Binary Compounds

Example Problem:

What is the name of CrO?

Problem-Solving Steps:

1. Write the name of the positive element. (ANS: chromium)
2. If this element has more than one oxidation number, use the oxidation number of the negative element to figure out the oxidation number of the positive element. Write this number as a Roman numeral after the name of the element.
   $? + (2-) = 0; ? = 2+$
   Cr O
   chromium(II)
3. Add the root of the name of the second element, followed by -ide. (ANS: oxide)
4. **Solution:** chromium(II) oxide

Practice Problem

Name the following compounds.
   Li_2S, MgF_2, FeO, CuCl

Strategy Hint: For names of elements with more than one oxidation number, remember to include the Roman numeral. For names of nonmetals in binary compounds, use **Table 11-3**.

Demonstration

Visual-Spatial To show that like charges repel and opposites attract to form chemical compounds, place two 25-cm-long pieces of cellophane tape on a plastic surface side by side. Pull both up at the same time. They will then have like charges. Bring the two pieces toward each other and watch them repel. Then put one piece down on the plastic surface. Place a second piece on top of it. Pull them off the plastic surface, and then separate them. They will have opposite charges, and, as you bring them close together, they will attract. LEP

Content Background

Emphasize that a Roman numeral indicates the oxidation number of an ion whose oxidation number varies. It is as important as the name of the element.

Practice Problem Answers

Note: Have students check **Figure 11-16** on page 315 for elements with variable oxidation numbers. Many more elements than the ones in the list are variable. However, those listed are the only variable elements used in this text unless specifically stated in a problem.

1. lithium sulfide
2. magnesium fluoride
3. iron(II) oxide
4. copper(I) chloride

Use **Study Guide**, p. 49, as you teach this lesson.

Content Background

- Remind students that polyatomic ions act as a single group. Thus, they should be treated much like a single element when writing formulas.

- Students will probably want to use parentheses in formulas when they are not needed. Remind them that parentheses are to be used only around polyatomic ions when more than one is needed to make the formula electrically neutral.

Student Text Questions

- **What four elements does NaHCO$_3$ contain?** *sodium, hydrogen, carbon, and oxygen*

- **What is the name of Sr(OH)$_2$?** *strontium hydroxide*

- **How does cobalt chloride react to the presence of water vapor?** *by changing color as it forms a hydrate*

Practice Problem Answers

1. 1+ 2–
   Na SO$_4$
   Therefore, Na$_2$SO$_4$ gives a sum of zero for charges.
2. 2+ 1–
   Mg ClO$_3$
   Therefore, the formula is Mg(ClO$_3$)$_2$.

3 Assess

Check for Understanding

? FLEX Your Brain

Use the Flex Your Brain activity to have students explore HYDRATES.

Activity Worksheets, p. 5

Table 11-4

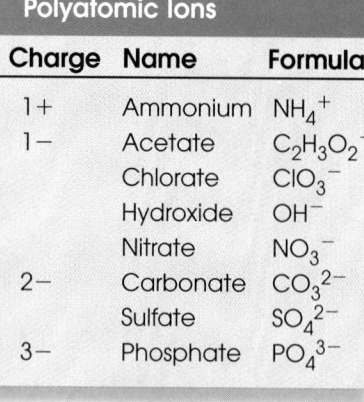

Polyatomic Ions		
Charge	Name	Formula
1+	Ammonium	NH$_4^+$
1–	Acetate	C$_2$H$_3$O$_2^-$
	Chlorate	ClO$_3^-$
	Hydroxide	OH$^-$
	Nitrate	NO$_3^-$
2–	Carbonate	CO$_3^{2-}$
	Sulfate	SO$_4^{2-}$
3–	Phosphate	PO$_4^{3-}$

Not all compounds are binary. Have you ever used baking soda in cooking, as a medicine, or to brush your teeth? Baking soda, which has the formula NaHCO$_3$, is an example of a compound that is not binary. What four elements does it contain? Some compounds, including baking soda, are composed of more than two elements because they contain polyatomic ions. The prefix *poly-* means "many," so *polyatomic* means "having many atoms." A **polyatomic ion** is a positively or negatively charged, covalently bonded group of atoms. So, the compound as a whole contains three or more elements.

Table 11-4 lists several polyatomic ions. To name a compound that contains one of these ions, first write the name of the positive element. For a compound of the ammonium ion, NH$_4^+$, write ammonium first. Then, use **Table 11-4** to find the name of the polyatomic ion. For example, K$_2$SO$_4$ is potassium sulfate. What is the name of Sr(OH)$_2$?

To write formulas for compounds containing polyatomic ions, follow the rules for writing formulas for binary compounds, with one addition. Write parentheses around the group representing the polyatomic ion when more than one of that ion is needed.

USING MATH

Writing Formulas with Polyatomic Ions

Example Problem:
What is the formula for calcium nitrate?

Problem-Solving Steps:
1. Write symbols and oxidation numbers for calcium and the nitrate ion.
   2+   1–
   Ca   NO$_3$
2. Write in subscripts so that the sum of the oxidation numbers is zero. Enclose the NO$_3$ in parentheses.
   2+   1–
   Ca   (NO$_3$)$_2$
3. **Solution:** Final formula: Ca(NO$_3$)$_2$

Practice Problems
1. What is the formula for sodium sulfate?
Strategy Hint: When only one polyatomic ion is needed in a formula, do not enclose the ion in parentheses.
2. What is the formula for magnesium chlorate?
Strategy Hint: Because the subscript 3 in ClO$_3^-$ is part of the ion, it should not be changed when written as part of a formula.

318   Chapter 11 Chemical Bonds

Across the Curriculum

Geology Calcium sulfate dihydrate is plaster of paris. It is also called gypsum. When seawater evaporated from shallow pools, beds of gypsum were formed. One of these gypsum beds is the Paris Basin in France.

Hydrates

Some ionic compounds may have water molecules as part of their structure. These compounds are called hydrates. A **hydrate** is a compound that has water chemically attached to its ions. *Hydrate* comes from a word that means "water." For example, when a water solution of cobalt chloride evaporates, pink crystals that contain six water molecules for each unit of cobalt chloride are formed. The formula for this compound is $CoCl_2 \cdot 6H_2O$ and its name is cobalt chloride hexahydrate. *Hexa-* means "six." You can remove water from these crystals by heating them. The resulting blue compound is called *anhydrous,* which means "without water." The blue paper in **Figure 11-17** on page 320 has been soaked in cobalt chloride solution and heated.

Like many anhydrous compounds, cobalt chloride gains water molecules easily. You may have seen weather predictors made from blue paper that turns pink in humid air. The paper contains cobalt chloride. How does it react to the presence of water vapor?

Problem Solving

The Packet of Mystery Crystals

When a new videocassette recorder (VCR) was delivered to Peter's home, he was eager to start using it to tape programs. He asked for and got the job of unpacking the VCR. When he lifted the equipment from the carton, a small flat packet fell out. A label on the packet read "Contains silica gel. Do not eat." What is silica gel, and why was it in the VCR carton?

Solve the Problem:

1. Use a reference book to find out what silica gel is.
2. Explain what silica gel would do if water molecules were in the air.
3. Why was the packet of silica gel placed in the carton with the VCR?

Think Critically:

1. The chemical formula for silica gel is SiO_2. How does this formula differ from the formula of what is present in the packet after it has been in moist air for several days?

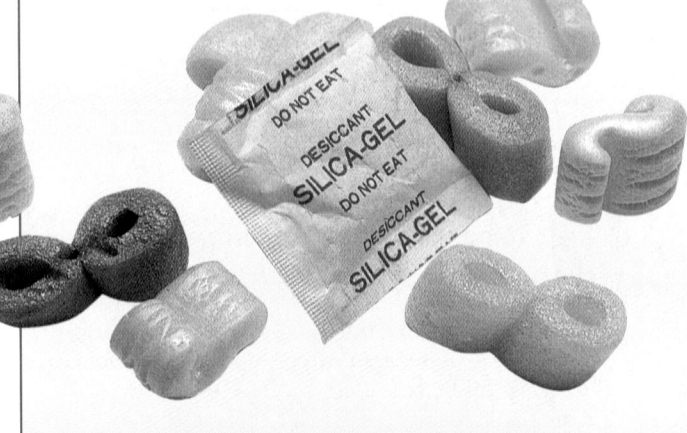

Problem Solving

Solve the Problem

1. Silica gel is a form of SiO_2 that is porous and is able to absorb large amounts of water. The gel is dried, and the dry form is what is used to absorb water.
2. The silica gel would absorb the water and form a hydrate.
3. It was packed with the VCR to absorb water molecules, thus helping to prevent corrosion of the metal parts of the VCR.

Think Critically

Water molecules are now bonded to the SiO_2. The formula is $SiO_2 \bullet H_2O$.

Reteach

IS **Interpersonal** Organize students into teams. Have one-half of each team prepare flash cards showing positive ions and their names. The other half should prepare a similar set of flash cards showing negative ions. Teams should get back together and draw one card from each pack. Have them determine the formula and name of each of the resulting compounds. **COOP LEARN**

Extension

For students who have mastered this section, use the **Reinforcement** and **Enrichment** masters.

How would you write the formula for this powder? $CaSO_4$

4 Close

1. **An element's combining ability in a compound is designated by its _____ .** *oxidation number*

2. **A compound formed from two elements is called a(n) _____ .** *binary compound*

3. **Atoms that lose one electron take on a charge of _____ .** *1+*

4. **How do you show the oxidation number of a metal having more than one possible oxidation number when naming a compound?** *Use a Roman numeral.*

5. **In a correct formula for a compound, the oxidation numbers of all atoms add up to _____ .** *zero*

Section Wrap-up

Review

1. sodium iodide, iron(III) bromide, potassium sulfate, ammonium bromide

2. (a) Li_2S, (b) $Ca(C_2H_3O_2)_2$, (c) BaO

3. (a) $CoCl_2$, (b) $CaSO_4 \cdot 2H_2O$

4. **Think Critically** KNO_3

USING MATH

$$PO_4 = P + 4(2-) = 3-$$
$$P + 8- = 3-$$
$$P = 5$$

Just Add Water

Have you ever made a mold or cast with plaster of paris? If so, you have made a hydrate. When you mix plaster of paris with water, it absorbs water and changes into a hydrated form of calcium sulfate known as gypsum.

Water in a hydrate has lost its properties because it is chemically attached. Not all crystals are hydrates. The only way to detect the presence of water in crystals is to heat the solid material and see whether it gives off steam and changes to powder. For example, if you heat hardened plaster of paris, it will give off steam, crumble, and become powdery. How would you write the formula for this powder?

You have learned how to write formulas of binary ionic compounds and of compounds containing polyatomic ions. Using oxidation numbers to write formulas, you can predict the ratio in which atoms of elements may combine to form compounds. You have also seen how hydrates have water molecules as part of their structures and formulas. As you study the chapters that follow, you will see many uses of formulas.

Figure 11-17

The blue cobalt chloride on these paper strips is anhydrous. A hydrate forms when water is added to anhydrous cobalt chloride.

Section Wrap-up

Review

1. Name the following: NaI, $FeBr_3$, K_2SO_4, NH_4Br.

2. Write formulas for (a) lithium sulfide, (b) calcium acetate, and (c) barium oxide.

3. Write formulas for the following: (a) the anhydrous form of $CoCl_2 \cdot 6H_2O$ and (b) calcium sulfate dihydrate.

4. **Think Critically:** Plant food may list potassium nitrate as one ingredient. What is the formula for this compound?

USING MATH

The charge on the phosphate ion, found in some detergents, is $3-$. Using **Table 11-4**, find the oxidation number of P.

Skill Builder
Using Variables, Constants, and Controls

Design an experiment to distinguish between crystals that are hydrates and those that are not. Include crystals of iron(II) chloride, copper(II) nitrate, and crystals of sucrose. If you need help, refer to Using Variables, Constants, and Controls in the **Skill Handbook.**

Skill Builder

Heat crystals of each substance and observe whether or not they give off steam and turn powdery. Add water to the powder. If the original types of crystals form, then the substance was a hydrate. Crystals of iron(II) chloride and copper(II) nitrate are both hydrates, whereas sucrose is not. When sucrose is heated, it decomposes and gives off water vapor, leaving the black carbon as a residue that will not dissolve in water or form sugar crystals. **P**

Assessment

Content Have students explain why hydrates are heated in their experiments. Use the Performance Task Assessment List for Making Observations and Inferences in **PASC**, p. 17. **P**

Chapter 11 Review

Summary

Have students read the summary statements to review the major concepts of the chapter.

Reviewing Vocabulary

1. m	**6.** f
2. a	**7.** g
3. l	**8.** d
4. h	**9.** k
5. i	**10.** c

✓ Assessment

Portfolio Encourage students to place in their portfolios one or two items of what they consider to be their best work. Examples include:
• MiniLAB Assessment, p. 306
• Reteach, p. 307
• Skill Builder experiment design, p. 320 P

Performance Additional performance assessments may be found in **Performance Assessment** and **Science Integration Activities.** Performance Task Assessment Lists and rubrics for evaluating these activities can be found in Glencoe's **Performance Assessment in the Science Classroom.**

Summary

11-1: Why Atoms Combine

1. The properties of compounds are generally different from those of the elements they contain.
2. A chemical formula for a compound indicates the composition of a unit of the compound.
3. Chemical bonding occurs because atoms of most elements become more stable by gaining, losing, or sharing electrons.

11-2: Kinds of Chemical Bonds

1. Ionic bonds between atoms are formed by the attraction between ions. Covalent bonds are formed by the sharing of electrons.
2. Ionic bonding produces charged particles called ions. Covalent bonding produces units called molecules.
3. The unequal sharing of electrons produces compounds that contain polar bonds, and the equal sharing of electrons produces nonpolar compounds.

11-3: Science and Society: Chemical Risks in the Home

1. Compounds that are toxic, corrosive, or flammable are hazardous.
2. People can protect their health and the environment from hazardous compounds in the home.

11-4: Formulas and Names of Compounds

1. An oxidation number indicates how many electrons an atom has gained, lost, or shared when bonding with other atoms.
2. In the formula of an ionic compound, the element or ion with the positive oxidation number is written first, followed by the one with the negative oxidation number.
3. The name of a binary compound is derived from the names of the two elements that compose the compound.
4. A hydrate is a compound that has water chemically attached to its ions and written into its formula.

Key Science Words

a. binary compound	**h.** ion
b. chemical bond	**i.** ionic bond
c. chemical formula	**j.** nonpolar molecule
d. chemically stable	**k.** oxidation number
e. corrosive	**l.** polar molecule
f. covalent bond	**m.** polyatomic ion
g. hydrate	**n.** toxic

Reviewing Vocabulary

Match each phrase with the correct term from the list of Key Science Words.

1. a charged group of atoms
2. a compound composed of two elements
3. a molecule with opposite charges on each end
4. a positively or negatively charged atom
5. a chemical bond between oppositely charged ions
6. chemical bond formed from shared electrons
7. crystalline substance that contains water
8. outer energy level is filled with electrons
9. shows an element's combining ability
10. tells which elements are in a compound and their ratios

Chapter 11 Review 321

GLENCOE TECHNOLOGY

 Videodisc

Glencoe Physical Science

Interactive Videodisc

Use the videodisc segment *Alkali Metals* from lesson 4, *Periodicity,* to reinforce that the chemical properties of an element depend upon the number of electrons in the outermost energy level.

MindJogger Videoquiz

Chapter 11 Have students work in groups as they play the Videoquiz game to review key chapter concepts.

Checking Concepts

1. b	**6.** d
2. c	**7.** b
3. c	**8.** b
4. c	**9.** a
5. a	**10.** d

Understanding Concepts

11. The molecule is composed of one atom of carbon and two atoms of oxygen.

12. Atoms can lose, gain, or share electrons to become more stable.

13. Chromium is a transition element and has two different oxidation numbers.

14. The hydroxide ion, OH⁻, in $Mg(OH)_2$ is a polyatomic ion that acts as a unit when it combines with the magnesium ion.

15. XZ, X_2Z_3, X_2Z_5, X_3Z_5

Thinking Critically

16. $[:\ddot{C}l:]^-$ $[Mg]^{2+}$ $[:\ddot{C}l:]^-$

17. $NaHCO_3$ (baking soda); $HC_2H_3O_2$ (vinegar)

18. Tl_2CO_3

19. calcium phosphate

20. $(NH_4)_2SO_4$

Checking Concepts

Choose the word or phrase that completes the sentence or answers the question.

1. The elements that are least likely to react with other elements are the _____.
   a. metals   c. nonmetals
   b. noble gases   d. transition elements

2. The oxidation number of Fe in Fe_2S_3 is _____.
   a. 1^+ c. 3^+
   b. 2^+ d. 4^+

3. The name of CuO is _____.
   a. copper oxide   c. copper(II) oxide
   b. copper(I) oxide   d. copper(III) oxide

4. The formula for copper(II) chlorate is _____.
   a. $CuClO_3$ c. $Cu(ClO_3)_2$
   b. $CuCl$ d. $CuCl_2$

5. Which of the following is a nonpolar molecule?
   a. N_2 c. $NaCl$
   b. H_2O d. HCl

6. The number of electrons in the outer energy level of Group 17 elements is _____.
   a. 1   c. 17
   b. 2   d. 7

7. An example of a binary compound is _____.
   a. O_2 c. H_2SO_4
   b. NaF d. $Cu(NO_3)_2$

8. An example of an anhydrous compound is _____.
   a. H_2O c. $CuSO_4 \cdot 5H_2O$
   b. $CaSO_4$ d. $CaSO_4 \cdot 2H_2O$

9. An atom that has gained an electron is a _____.
   a. negative ion   c. polar molecule
   b. positive ion   d. nonpolar molecule

10. An example of a covalent compound is _____.
   a. sodium chloride   c. calcium chloride
   b. calcium fluoride   d. water

Understanding Concepts

Answer the following questions in your Science Journal using complete sentences.

11. What does the formula CO_2 tell you about a molecule of carbon dioxide?

12. By what three ways can atoms become chemically stable?

13. How can chromium form two different compounds with oxygen?

14. Why is the formula for milk of magnesia written $Mg(OH)_2$ instead of MgO_2H_2?

15. What compounds can be formed from element X, with oxidation numbers 3+ and 5+, and element Z, with oxidation numbers 2− and 3−? Write their formulas.

Thinking Critically

16. Anhydrous magnesium chloride is used to fireproof wood. Draw a dot diagram of magnesium chloride.

17. Baking soda, which is sodium hydrogen carbonate, and vinegar, which contains hydrogen acetate, can be used as household cleaners. Write the chemical formulas for these two compounds.

18. Artificial diamonds are made using thallium carbonate. If thallium has an oxidation number of 1+, what is the formula for the compound?

19. The formula for a compound that composes kidney stones is $Ca_3(PO_4)_2$. What is the chemical name of this compound?

20. Ammonium sulfate is used as a fertilizer. What is its chemical formula?

Developing Skills

If you need help, refer to the **Skill Handbook.**

21. **Comparing and Contrasting:** Compare and contrast polar and nonpolar molecules.
22. **Interpreting Scientific Illustrations:** Write the name and formula for the compound illustrated below.

23. **Hypothesizing:** Several uses of HCl were given to you in Section 11-2. HF is another acid, and it is used to etch glass. If HCl is hydrochloric acid, what is the name of HF?
24. **Observing and Inferring:** Ammonia gas and water react to form household ammonia, which contains NH_4^+ and OH^- ions. If the formula for water is H_2O, what is the formula for ammonia gas?
25. **Concept Mapping:** In photosynthesis, green plants that are in sunlight convert carbon dioxide and water to glucose, $C_6H_{12}O_6$, and oxygen, O_2. In respiration, glucose and oxygen react to produce carbon dioxide and water and release energy. In the map below, write in the formulas of the molecules and the names of the processes.

Performance Assessment

1. **Formulating a Hypothesis:** Calcium chloride is sometimes used in winter to help melt ice. Lard is mostly made from carbon and hydrogen and may be used in cooking. Using the same techniques that you used in Activity 11-2 on pages 310-311, what would you hypothesize about the bonding types and melting points of these two materials?
2. **Designing an Experiment:** Cobalt chloride hydrate changes color as the water part of the molecule is attached or removed. Devise a way that this could be used to determine the humidity in your classroom.
3. **Model:** One common form of phosphorus, white phosphorus, has the formula P_4 and is formed by four covalently bonded phosphorus atoms. Make a model of this molecule, showing that all four atoms are now chemically stable.

Assessment Resources

Reproducible Masters
Chapter Review, pp. 25, 26
Assessment, pp. 65-68
Performance Assessment, p. 25

Glencoe Technology
- Chapter Review Software
- Computer Test Bank
- MindJogger Videoquiz

Chapter 11 Review

Developing Skills

21. **Comparing and Contrasting** Both molecules have covalent bonds. Polar molecules always have polar bonds and a slight charge on each end. Nonpolar molecules have no charged ends.
22. **Interpreting Scientific Illustrations** hydrogen sulfide, H_2S
23. **Hypothesizing** hydrofluoric acid
24. **Observing and Inferring** NH_3
25. **Concept Mapping** See student page.

Performance Assessment

1. Calcium chloride is an ionic compound with a high melting point. Lard has covalent bonds between carbon atoms and between carbon and hydrogen atoms. It has a low melting point. Use the Performance Task Assessment List for Formulating a Hypothesis in **PASC,** p. 21. **P**
2. A dilute $CoCl_2 \cdot 6H_2O$ solution could be absorbed onto some white paper. On dry days, the water would tend to go into the air, resulting in a blue color. On more humid days, the water would be reabsorbed, turning the color pink. Use the Performance Task Assessment List for Designing an Experiment in **PASC,** p. 23. **P**
3. Models should show that electrons are shared by the four P atoms in such a way that each atom has eight electrons in its outer energy level. Use the Performance Task Assessment List for Model in **PASC,** p. 51. **P**

Summary

The selection of two scien-tists starts students on a pro-ject that will lead them to an understanding and apprecia-tion of the modern view of atomic structure. Students will place their selected scien-tists in the appropriate time period and explain their sci-entific and social environ-ment. They will then connect the work of those scientists to any related beliefs in more modern times. By delivering a class report with props and costuming, students will bring more interest to some of their findings.

Time Required

This project would be most effective if introduced in the latter part of Unit 3. Students will need one class period to examine the list of prospec-tive scientists for their choices. Time outside of class will be needed for research. Another class period may be set aside for rehearsal of the presentations, and one class period is needed for the ac-tual presentation.

Preparation

Be aware of various re-sources available for student use. Advise any community or school librarian that the project is being done, and en-courage them to gather possi-ble resources.

324

UNIT PROJECT 3

A View of the Atom

Throughout history, we have records of people who won-dered what made up matter. Although it is difficult for many people to visualize, all matter is made up of small pieces of material called atoms. If you asked several peo-ple to describe or diagram the general appearance of an atom, you would get many different responses. In fact, some scientists would also have varying opin-ions. How do we describe some-thing we cannot see? What do you think an atom looks like?

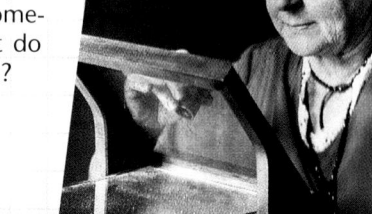
Annie Jump Cannon

Atomic Sleuth

Our current view of the atom is the result of historical investigations and modern experimental results. From the following list, work with other class members and choose two scientists to research. It would be best if all of the following scientists could be researched by the class.

Neils Bohr

- Democritus
- John Dalton
- J.J. Thomson
- Ernst Rutherford
- Marie Curie
- Annie Jump Cannon
- Neils Bohr
- Louis de Broglie
- Werner Heisenberg
- Murray Gell-Mann
- Lise Meitner

Werner Heisenberg

Procedure

For each of your chosen scientists, answer the following questions.

1. In what country or countries did the scientist live and work?

2. When was the scientist born? If no longer living, when did he or she die?

3. What area or areas of science did the scientist study?

4. Summarize the scientist's contributions to atomic structure. How did his or her findings help our modern view of the atom?

5. What are some important or interesting facts about the scientist?

Using Your Research

With the information that you now have, make a presentation to your class. You could dress in costume to give the report. You could work with other students to make a panel of Distinguished Scientists, each student presenting and debating his or her contributions to our view of the atom. Another alternative would be to present the information as if it were a television special. Whether you use one of these suggestions or another type of presentation, be creative.

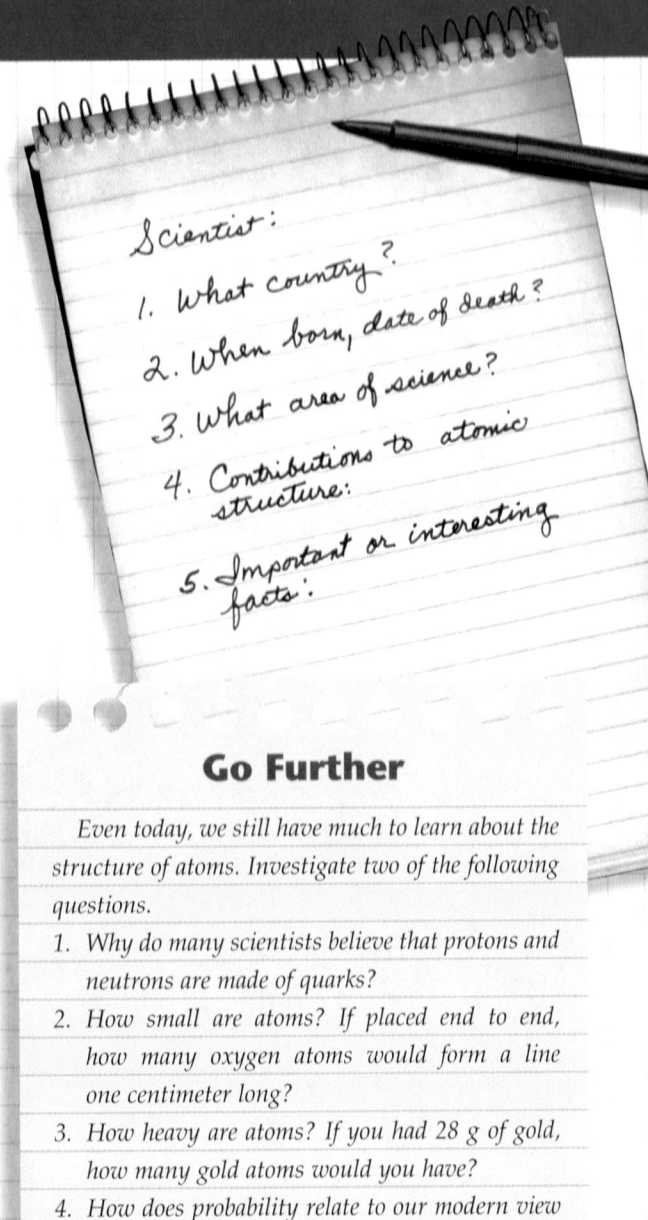

Scientist:
1. What country?
2. When born, date of death?
3. What area of science?
4. Contributions to atomic structure:
5. Important or interesting facts:

Go Further

Even today, we still have much to learn about the structure of atoms. Investigate two of the following questions.

1. Why do many scientists believe that protons and neutrons are made of quarks?

2. How small are atoms? If placed end to end, how many oxygen atoms would form a line one centimeter long?

3. How heavy are atoms? If you had 28 g of gold, how many gold atoms would you have?

4. How does probability relate to our modern view of the atom?

5. How has the scanning tunneling microscope helped us to understand the structure of atoms?

Unit 3 Project 325

Teaching Strategies

- Before beginning this project, ask students how they could gather information about the contents of a container that they could not open. Remind students that, in a related way, this process is what scientists had to do to find out about atomic structure.

- Help students place their selected scientist in a historical setting. Ask students what political events were taking place during the scientist's lifetime, what the style of dress was, what laboratory facilities looked like, and other leading questions.

- Presentations will be more effective if they are done in chronological order.

Go Further

1. Quark theory has received support from observations of energy requirements and particle bombardment patterns of protons.

2. Assuming the diameter of an atom to be 1×10^{-10}, approximately 10^8 atoms would be required.

3. Students learned in Chapter 10 that the masses of atoms are in atomic mass units (u). If $1\ u = 1.66 \times 10^{-24}$ g, and the mass of one atom of gold is 197 u, then

$$\frac{197\ u}{atom} \times \frac{1.66 \times 10^{-24}\ g}{1\ u} =$$

$$\frac{3.27 \times 10^{-22}\ g}{atom} \text{ and}$$

$$28\ g \times \frac{1\ atom}{3.27 \times 10^{-22}\ g} =$$

8.6×10^{22} atoms.

4. Our present atomic view does not deal with exact positions of electrons. Rather, approximations and probabilities are how we best describe electron movement.

5. The scanning tunneling microscope has not led us to understand the interior structure of atoms, but it has been used to observe individual atoms and arrangements of atoms.

References

Asimov, Isaac. *Atom: Journey Across the Subatomic Cosmos.* New York, NY: Plume-Truman Talley Books, 1992.

Salzberg, Hugh W. *From Caveman to Chemist.* Washington, DC: American Chemical Society, 1991.

Snow, C.P. *The Physicists.* Boston: Little, Brown and Co., 1981.

Wolf, Fred Alan. *Taking the Quantum Leap.* New York, NY: HarperCollins, 1989.

Kinds of Substances

In Unit 4, students build on the broad classification of substances, elements, and compounds that they studied in Chapter 9. The common properties and uses of elements categorized as metals, nonmetals, mixed groups, and synthetic elements are discussed and related to their positions in the periodic table. The common structures and characteristics of organic and biological compounds are then explored. Next, students are introduced to materials science with a discussion of alloys, ceramics, plastics, and composites.

CONTENTS

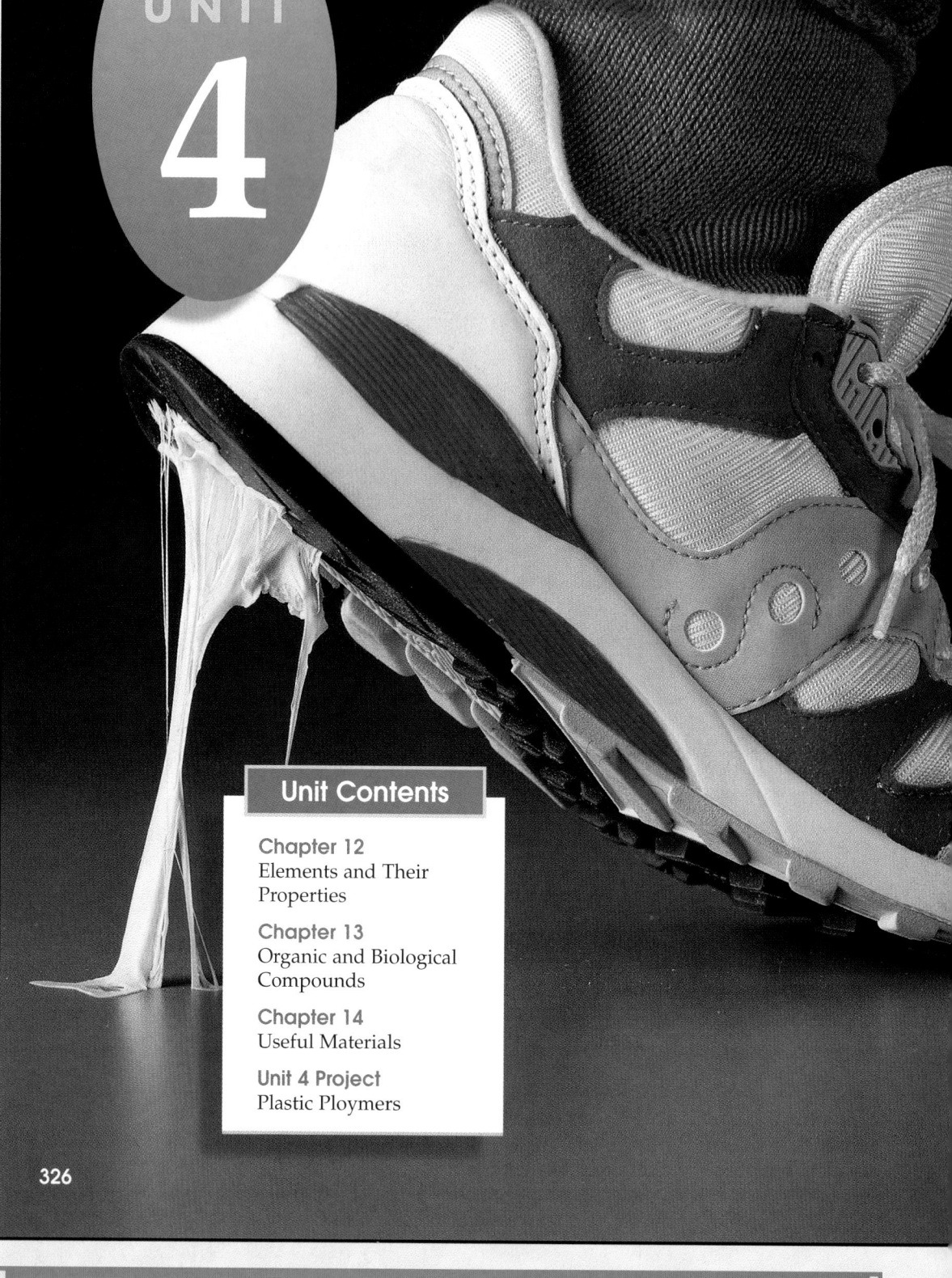

UNIT 4

Unit Contents

326

Science at Home

Community Recycling Assign groups of students to determine community disposal and/or recycling methods for metals, paper, and plastic. Have students describe major differences between the substances and explain why proper handling of each is important. [L1] **COOP LEARN**

Does the nose really know? Have students list various products found around the home that include imitation or artificial flavorings or scents. Have students research what these artificial scents and flavorings are and how they are made. [L2]

Kinds of Substances

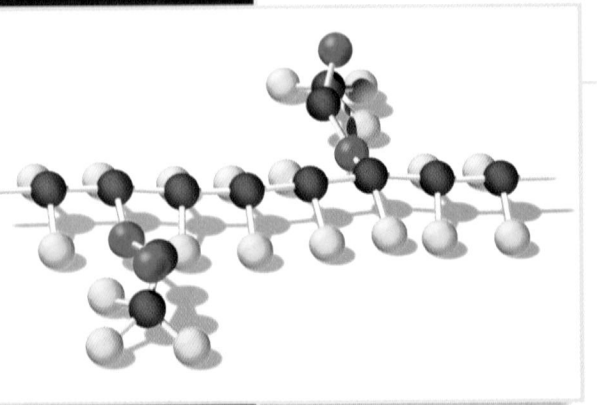

What's Happening Here?

Even after you think *you're* done with a piece of chewing gum, look again—it still has plenty of stretch and give for the shoe of the next passerby. Chewing gum gets its stretch from its structure—a special arrangement of atoms called a polymer. A model of the chewing gum polymer is shown above. In this unit you'll learn about polymers and many other kinds of substances—metal, nonmetal, organic, biological—and the special properties of each. So the next time you get rid of your gum, remember its properties and throw it in the trash.

Science Journal

There are many materials besides gum and tennis shoes that have the characteristics of stretchiness and elasticity. In your Science Journal, write a report on other products and materials with these characteristics and explain why these characteristics are important.

327

Chapter Organizer

Section	Objectives/Standards	Activities/Features
Chapter Opener		**Explore Activity:** Some clues are in color. p. 329
12-1 Metals (2 sessions, 1 block)*	1. **Describe** the properties of a typical metal. 2. **Identify** the alkali and alkaline earth metals. 3. **Differentiate** among three groups of transition elements. **National Content Standards:** (5-8) UCP1-UCP3, UCP5, A2, B1, D1, E2, F1, F5; (9-12) UCP1-UCP3, UCP5, A2, B2, E2, F1, F6	**MiniLAB:** How can metallic bonding explain the flexibility of metals? p. 331 **Using Technology:** Metals with Memory, p. 335 **Using Math,** p. 337 **Science Journal,** p. 337 **Skill Builder:** Interpreting Scientific Illustrations, p. 337
12-2 Science and Society (1 session, ½ block)*	1. **Distinguish** among elements classified as lanthanides, actinides, & transuranium elements. 2. **Determine** the uses of transuranium elements. 3. **Compare** the pros and cons of making new elements. **National Content Standards:** (5-8) UCP1-UCP3, UCP5, A2, B1, E2, F5; (9-12) UCP1-UCP3, UCP5, A2, B2, E2, F6	**Explore the Technology,** p. 339
12-3 Nonmetals (2 sessions, 1 block)*	1. **Recognize** hydrogen as a nonmetal. 2. **Compare** and **contrast** properties of the halogens. 3. **Describe** properties and uses of the noble gases. **National Content Standards:** (5-8) UCP1-UCP3, UCP5, B1; (9-12) UCP1-UCP3, UCP5, B2	**MiniLAB:** Can you detect chlorine in your drinking water? p. 342 **Connect to Earth Science,** p. 342 **Using Computers,** p. 344 **Skill Builder:** Interpreting Data, p. 344 **Activity 12-1:** What type is it? p. 345
12-4 Mixed Groups (2 sessions, 1 block)*	1. **Distinguish** among metals, nonmetals, and metalloids in Groups 13 through 16 of the periodic table. 2. **Describe** the nature of allotropes. 3. **Recognize** the significance of differences in crystal structure in carbon. **National Content Standards:** (5-8) UCP1-UCP3, UCP5, A1, A2, B1, E1, G1, G3; (9-12) UCP1-UCP3, UCP5, A1, A2, B2, E1, G1, G3	**Problem Solving:** Waiting to Be Discovered, p. 348 **Using Math,** p. 348 **Science Journal,** p. 351 **Skill Builder:** Concept Mapping, p. 351 **Activity 12-2:** Slippery Carbon, pp. 352-353 **Science and Art:** Pottery and Clay, p. 354

Activity Materials

Explore	Activities	MiniLABs
page 329 tongs, clean metal paper clip, laboratory burner, copper(II) sulfate solution, strontium chloride solution, sodium chloride solution, safety goggles, apron	page 345 samples of carbon, magnesium, aluminum, sulfur, and tin; dishes; conductivity tester; spatula, small hammer pages 352-353 thin spaghetti, toothpicks, small gumdrops, thin polystyrene sheets, flat cardboard, scissors, safety goggles	page 331 tongs, 2 pieces of metal wire, laboratory burner, cold water, beaker, safety goggles, apron page 342 3 identical test tubes, 2 mL chlorine standard solution, 2 mL distilled water, 2 mL drinking water, silver nitrate solution, safety goggles, apron

Need Materials? Call Science Kit (1-800-828-7777).   * A complete Planning Guide that includes block scheduling is provided on pages 32T-35T.

328A

Chapter 12 Elements and Their Properties

Teacher Classroom Resources

Reproducible Masters	Transparencies	Teaching Resources
Study Guide, p. 50 Reinforcement, p. 50 Enrichment, p. 50 Activity Worksheets, p. 72 Multicultural Connections, p. 27 Science Integration Activity 12, Metal or Not? Cross-Curricular Integration, p. 18 Critical Thinking/Problem Solving, p. 18	Section Focus Transparency 46, Light Metals Teaching Transparency 23, Metallic Bonding	Glencoe Physical Science Interactive Videodisc Physical Science CD-ROM Spanish Resources English/Spanish Audiocassettes Cooperative Learning Resource Guide Lab Partner Lab and Safety Skills Lesson Plans
Study Guide, p. 51 Reinforcement, p. 51 Enrichment, p. 51 Science and Society Integration, p. 16	Section Focus Transparency 47, Naming New Elements	**Assessment Resources** Chapter Review, pp. 27-28 Assessment, pp. 80-83 Performance Assessment in the Science Classroom (PASC) MindJogger Videoquiz Alternate Assessment in the Science Classroom Performance Assessment Chapter Review Software Computer Test Bank
Study Guide, p. 52 Reinforcement, p. 52 Enrichment, p. 52 Activity Worksheets, pp. 68-69, 73 Lab Manual 26, Preparation of Hydrogen	Section Focus Transparency 48, Swimming Pools	
Study Guide, p. 53 Reinforcement, p. 53 Enrichment, p. 53 Activity Worksheets, pp. 70-71 Lab Manual 27, Preparation of Oxygen Concept Mapping, pp. 29-30	Section Focus Transparency 49, Dietary Allowances of Minerals Science Integration Transparency 12, Claiming Earth's Aluminum Teaching Transparency 24, Allotropes of Carbon	**Key to Teaching Strategies** The following designations will help you decide which activities are appropriate for your students.

Key to Teaching Strategies

The following designations will help you decide which activities are appropriate for your students.

L1 Level 1 activities should be within the ability range of all students, including those with learning difficulties.

L2 Level 2 activities should be within the ability range of the average to above-average student.

L3 Level 3 activities are designed for the ability range of above-average students.

LEP LEP activities should be within the ability range of Limited English Proficiency students.

LS These activities are designed to address different learning styles.

COOP LEARN Cooperative Learning activities are designed for small group work.

P These strategies represent student products that can be placed into a best-work portfolio.

GLENCOE TECHNOLOGY

The following multimedia resources are available from Glencoe.

Science and Technology Videodisc Series (STVS)
Chemistry
  Images of Atoms
Human Biology
  Measuring Calcium Deficiency

The Infinite Voyage Series
Miracles by Design

Glencoe Physical Science Interactive Videodisc
Periodicity

Physical Science CD-ROM

National Geographic Society Series
STV: Atmosphere

Teacher Classroom Resources

This is a representation of key blackline masters available in the Teacher Classroom Resources.

Teaching Aids

Section Focus Transparencies

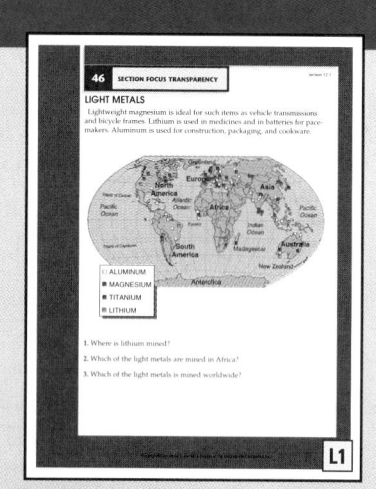

46 SECTION FOCUS TRANSPARENCY
LIGHT METALS

L1

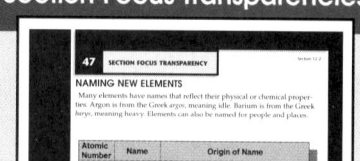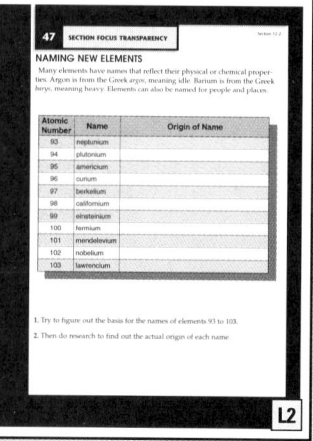

47 SECTION FOCUS TRANSPARENCY
NAMING NEW ELEMENTS

L2

48 SECTION FOCUS TRANSPARENCY
SWIMMING POOLS

L3

Science Integration Transparencies

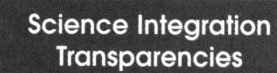

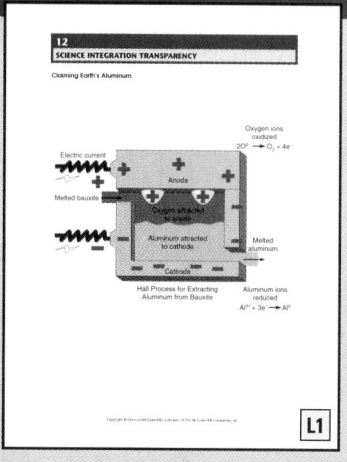

12 SCIENCE INTEGRATION TRANSPARENCY
Claiming Earth's Aluminum

L1

Teaching Transparencies

23. METALLIC BONDING

L1

24. ALLOTROPES OF CARBON

Diamond

Graphite

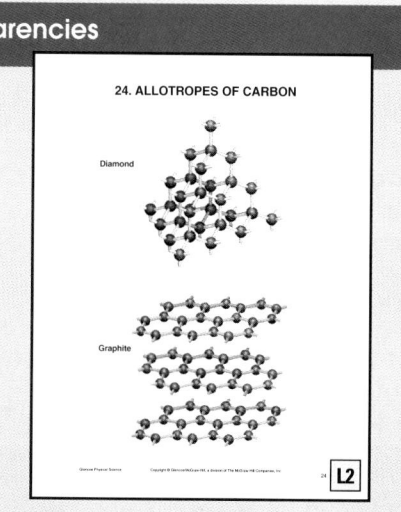

L2

Meeting Different Ability Levels

Study Guide

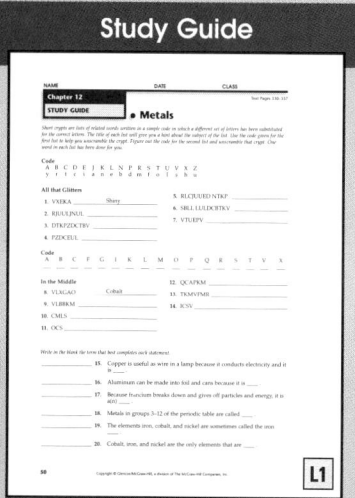

Chapter 12
STUDY GUIDE • Metals

L1

Reinforcement

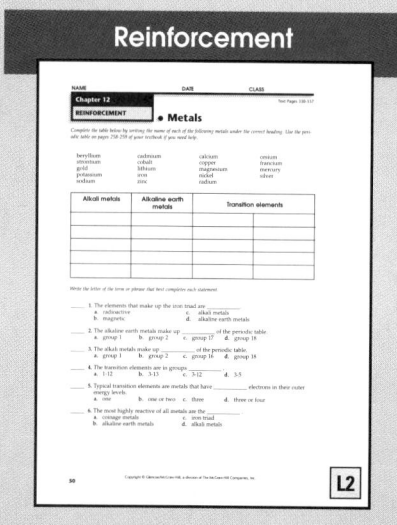

Chapter 12
REINFORCEMENT • Metals

L2

Enrichment Worksheets

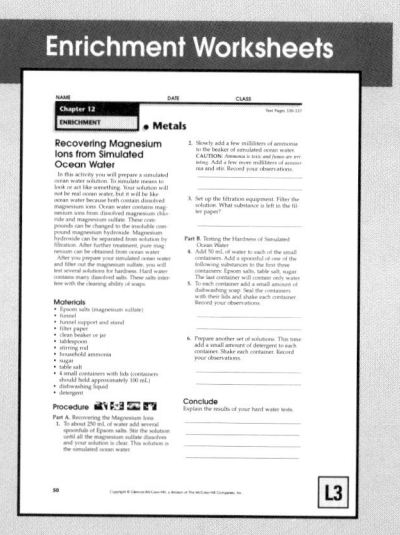

Chapter 12
ENRICHMENT • Metals

Recovering Magnesium Ions from Simulated Ocean Water

L3

Chapter 12 Elements and Their Properties

Hands-On Activities

Science Integration Activity

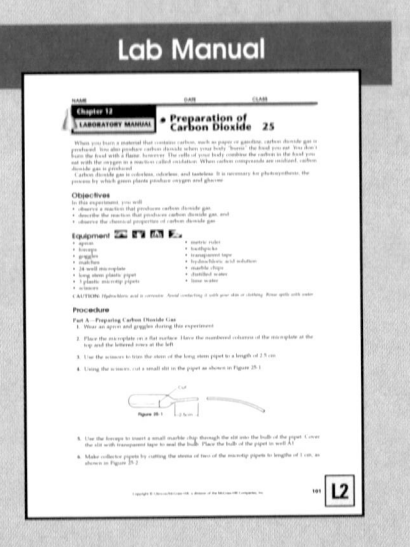

Lab Manual

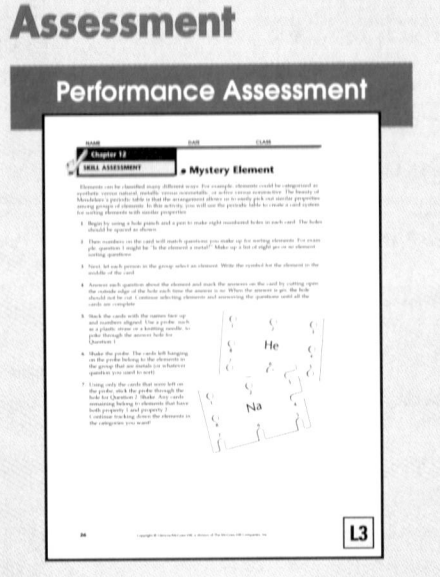

Assessment

Performance Assessment

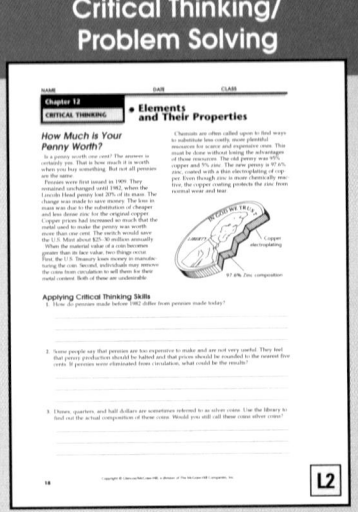

Enrichment and Application

Critical Thinking/Problem Solving

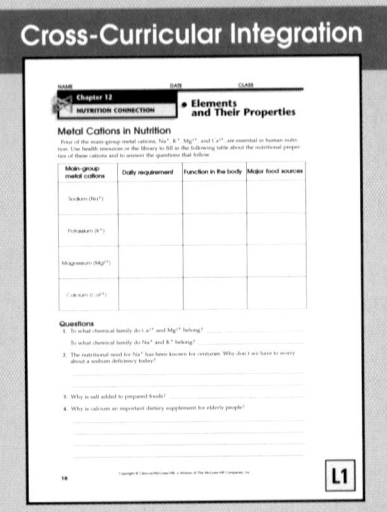

Cross-Curricular Integration

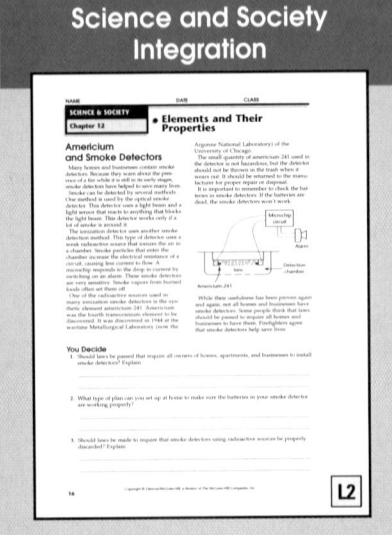

Science and Society Integration

Multicultural Connections

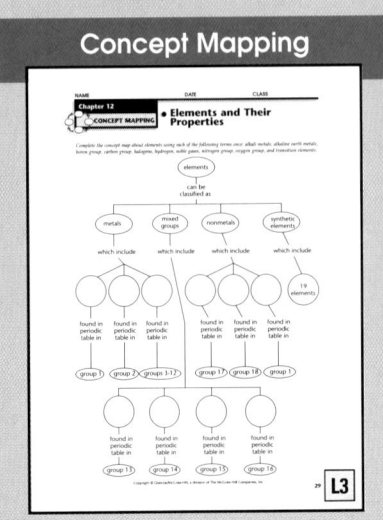

Concept Mapping

Elements and Their Properties

CHAPTER OVERVIEW

Section 12-1 The properties of metals are presented and explained by a study of metallic bonding. The alkali metal and alkaline earth metal families, as well as transition elements, are studied in detail.

Section 12-2 Science and Society The student is introduced to the synthetic elements.

Section 12-3 The properties of hydrogen, the halogens, and noble gases are developed. The properties of nonmetals are contrasted with those of metals.

Section 12-4 Within some groups on the periodic table there are nonmetals, metalloids, and metals. These groups are studied.

Chapter Vocabulary

malleable
ductile
metallic
  bonding
radioactive
  element
transition
  element

transuranium
  element
diatomic
  molecule
sublimation
semiconductor
allotrope

Theme Connection

Scale and Structure This theme is developed through a presentation of metallic and nonmetallic properties of elements and their positions on the periodic table based on their atomic structures.

Al Si P S
.982 28.086 30.974 32.066

llium Germanium ium
31 32.
Ga Ge
.723 72.61

dium Tin ium
49 50 52
In Sn Sb Te
4.82 118.710 121.757 127.60

allium Lead Bismuth Polonium
81 82 83 84
Tl Pb Bi Po

Previewing the Chapter

328

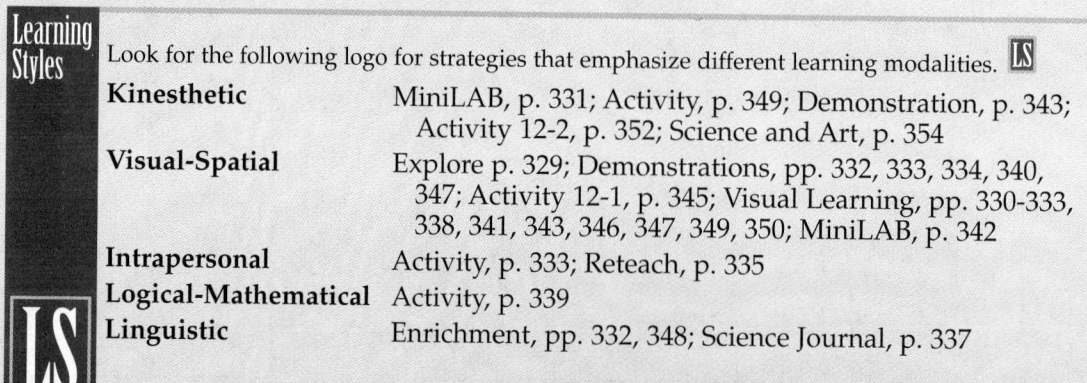

Learning Styles Look for the following logo for strategies that emphasize different learning modalities. **LS**

Kinesthetic MiniLAB, p. 331; Activity, p. 349; Demonstration, p. 343; Activity 12-2, p. 352; Science and Art, p. 354

Visual-Spatial Explore p. 329; Demonstrations, pp. 332, 333, 334, 340, 347; Activity 12-1, p. 345; Visual Learning, pp. 330-333, 338, 341, 343, 346, 347, 349, 350; MiniLAB, p. 342

Intrapersonal Activity, p. 333; Reteach, p. 335

Logical-Mathematical Activity, p. 339

Linguistic Enrichment, pp. 332, 348; Science Journal, p. 337

LS

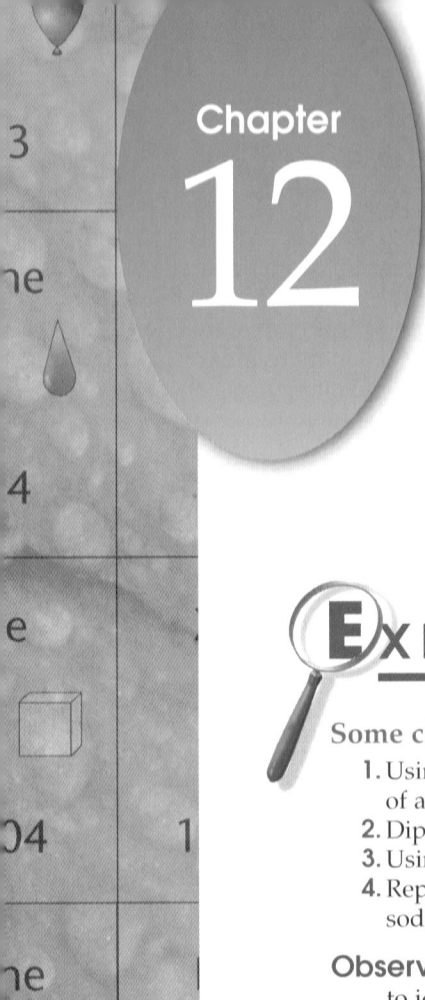

Chapter
12

Elements and Their Properties

This beautiful diamond is made up of atoms of the element carbon. Most matter can be described by the elements of which it is made. Each element has distinct properties. Elements arranged together on the periodic table have similar properties. Knowing the properties of elements allows us to use them in a variety of practical ways. This chapter will describe groups of elements and help you to learn how they are related.

EXPLORE ACTIVITY

Some clues are in color.

1. Using tongs, carefully hold a clean paper clip in the hottest part of a lab burner flame until no additional color is observed.
2. Dip the hot paper clip into a solution of copper(II) sulfate.
3. Using the tongs, repeat step 1, observing any color change.
4. Repeat these steps using solutions of strontium chloride and sodium chloride.

Observe: How can observation of color be used to identify elements in solutions? Based on your observations, devise a plan in your Science Journal to determine whether chlorine or strontium was giving the color you observed in a particular test.

Previewing Science Skills

► In the Skill Builders, you will interpret scientific illustrations, make and use a graph, and map concepts.

► In the Activities, you will observe, compare, predict, infer, and build a model.

► In the MiniLABs, you will observe and draw conclusions

329

Prepare

Section Background

As the atomic numbers of the alkali metals increase, the outer electrons are farther from the nucleus and are held less tightly. This results in more chemically active atoms.

Preplanning

Obtain thin pieces of wire for the MiniLAB.

1 Motivate

Bellringer

Before presenting the lesson, display **Section Focus Transparency 46** on the overhead projector. Assign the accompanying **Focus Activity** worksheet. **L1** **LEP**

Demonstration

Have a student wearing leather gloves attempt to tear an aluminum can in two. It will twist but not tear. Use a stiff piece of wire to scratch a line in the protective plastic film on the inside of the can. Put 125 mL of water into the can. Then add 100 mL concentrated HCl. **CAUTION:** *HCl is corrosive and gives off harmful fumes.* Place can inside a large beaker. After a few minutes, rinse the can. You can now easily tear the treated can. **LEP**

2 Teach

Visual Learning

Figure 12-1 What is this property of metals called? *malleability* **LEP** **IS**

12•1 Metals

Science Words

malleable
ductile
metallic bonding
radioactive element
transition element

Objectives

• Describe the properties of a typical metal.
• Identify the alkali and alkaline earth metals.
• Differentiate among three groups of transition elements.

Figure 12-1

Gold can be hammered into thin sheets. The sheets, called gold leaf, can be applied to many types of surfaces. *What is this property of metals called?*

Properties of Metals

Have you ever seen very old jewelry or statues made of gold and copper? These were two of the first metals used thousands of years ago. The use of silver and tin soon followed. Then came iron. But aluminum—the metal used in thousands of soft-drink cans—wasn't discovered until a little more than 100 years ago.

Gold, copper, silver, tin, iron, and aluminum are typical metals. What do these and other metals have in common? Most metals are hard, shiny solids. Metals are also good conductors of both heat and electricity. These properties make metals suitable for uses ranging from kitchen pots and pans to wires for electric appliances. Because metals reflect light well, they are also used in mirrors. Metals are **malleable,** which means they can be hammered or rolled into sheets. The gold leaf shown in **Figure 12-1** is used for elaborate decoration. Metals are also **ductile,** which means they can be drawn into wires like the ones shown in **Figure 12-2.**

The atoms of metals generally have from one to three electrons in their outer energy levels. Metals tend to give up electrons easily. Remember from Chapter 11 what happens when metals combine with nonmetals? The atoms of the metals tend to lose electrons to the atoms of nonmetals, and form ionic bonds. The second type of bond you studied is the covalent bond, which generally forms between atoms of nonmetals.

330   Chapter 12 Elements and Their Properties

Program Resources

Reproducible Masters
Study Guide, p. 50 **L1**
Reinforcement, p. 50 **L1**
Enrichment, p. 50 **L3**
Cross-Curricular Integration, p. 18
Activity Worksheets, p. 72 **L1**
Science Integration Activity, pp. 23, 24
Critical Thinking/Problem Solving, p. 18
Multicultural Connections, pp. 27, 28

Transparencies
Section Focus Transparency 46
Teaching Transparency 23

Figure 12-2

Metals can be drawn into wires. This property is called ductility. The gauge of a wire is related to its thickness. A large number indicates a thinner wire.

Bonding in Metals

A third type of bonding, neither ionic nor covalent, occurs among the atoms in a metal. In **metallic bonding,** positively charged metallic ions are surrounded by a "sea of electrons." Outer-level electrons are not held tightly to their particular nucleus. Rather, the electrons move freely among many positively charged ions. The electrons form a cloud around the ions of the metal as shown in **Figure 12-3.**

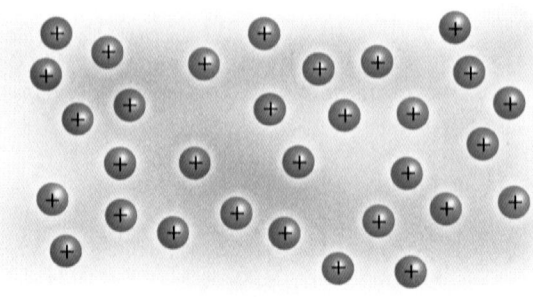

Figure 12-3

In metallic bonding, the moving outer electrons, depicted by the cloud, cover the rest of the atoms no matter where the electrons are in the metal. *How does this differ from covalent bonding?*

Metallic bonding explains many of the properties of metals. For example, when a metal is hammered into a sheet or drawn into a wire, it does not break because the ions are in layers that slide past one another. And because the outer-level electrons are weakly held, metals are good conductors of electricity.

Look at the periodic table on pages 286-287. How many of the elements in the table are classified as metals? Except for hydrogen, all the elements in Groups 1 through 12 are metals. You will learn about metals in some of these groups and others throughout this chapter.

12-1 Metals   **331**

MiniLAB

How can metallic bonding explain the flexibility of metals?

Procedure

1. Using tongs, carefully hold two pieces of metal wire in a lab burner flame until they are glowing red-hot.
2. Quickly drop one of the hot wires into a beaker of cold water. Allow it to cool.
3. Allow the other hot wire to cool slowly.
4. Hold the ends of each piece of cooled wire and bend them.

Analysis

1. What do you observe about the flexibility of the two cooled wires?
2. How can metallic bonding be used to help explain your observations?

MiniLAB

Purpose

L$ **Kinesthetic** Students will observe that metallic bonding allows flexibility while maintaining strength. **L1**

Materials

thin iron wire (hairpins); tongs; burner; cold water

Teaching Strategies

Safety Precaution Warn students to be careful not to drop red-hot wires.

• Place the cold water near the flame so that the wire is hot when immersed.

Analysis

1. The slowly cooled wire retains its flexibility while the quickly cooled wire becomes brittle.
2. Layers of atoms can move past one another. When heated, the structure becomes unstable. Quick cooling "freezes" the atoms in other positions. Slow cooling allows the atoms to resume normal positions.

Assessment

Content Check student responses to determine if they have clearly shown the concept of the "sea of electrons" as a way to explain their results. Use the Performance Task Assessment List for Writing in Science in **PASC,** p. 87. **P**

Activity Worksheets, pp. 5, 72

Use **Teaching Transparency 23** as you teach this lesson.

Across the Curriculum

Medicine Sodium and potassium ions are essential to the proper functioning of the human nervous system. The nerve cell axons do not conduct electricity. The sodium or potassium ions move through the cell membrane, changing the potential. This change moves down the axon like a wave at a rate of 30-50 m/s.

Visual Learning

Figure 12-3 How does this differ from covalent bonding? *In covalent bonding atoms share specific electrons, while in metallic bonding the electrons move freely.* **LEP** **L$**

The Alkali Metals

The elements in Group 1 of the periodic table are the alkali metals. Like other metals, Group 1 metals are shiny, malleable, and ductile. They are good conductors of both heat and electricity. The alkali metals are highly reactive metals. **Figure 12-4** shows how reactive they are with water. As a result of their reactivity, why would you expect to find the alkali metals in nature only in the form of compounds?

Each atom of an alkali metal has one electron in its outer energy level. This electron is given up when an alkali metal combines with another atom. The result is a positively charged ion in a compound such as sodium chloride, NaCl, or potassium bromide, KBr.

Alkali metals and their compounds have many uses. You and other living things need potassium and sodium compounds—such as table salt, NaCl—to stay healthy. Doctors use lithium compounds to treat bipolar disorder. The operation of some photocells depends upon rubidium or cesium compounds.

Francium, the last element in Group 1, is extremely rare and also radioactive. The nucleus of a **radioactive element** breaks down and gives off particles and energy. You will study radioactive elements later.

1

Lithium
3
Li
6.941

Sodium
11
Na
22.990

Potassium
19
K
39.098

Rubidium
37
Rb
85.468

Cesium
55
Cs
132.905

Francium
87
Fr
223.020

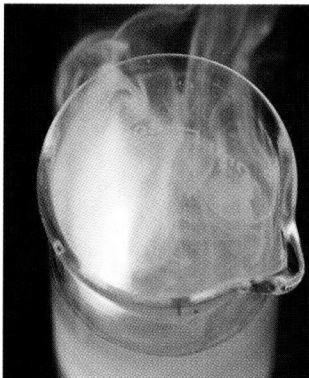
Sodium (Na)

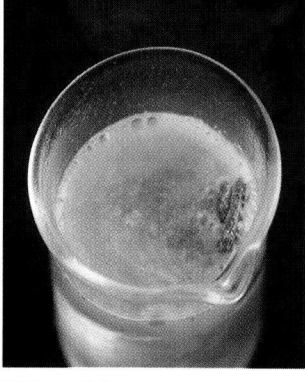
Lithium (Li)

Potassium (K)

Figure 12-4

Alkali metals are very reactive. Li, Na, K, and Cs react violently in water. *Why wouldn't the reaction of francium be shown in the photo?*

332 Chapter 12 Elements and Their Properties

Integrating the Sciences

Life Science When doctors prescribe a low salt (sodium) diet, they often suggest KCl as a salt substitute. **How can the two elements' biochemistries be similar?** *Both Na and K have one outer electron and react in a similar manner with other elements.*

Life Science The potassium ion helps regulate heartbeat and other nerve transmission. **What are some sources of potassium in your diet?** *Bananas, peaches, and orange juice may be mentioned.*

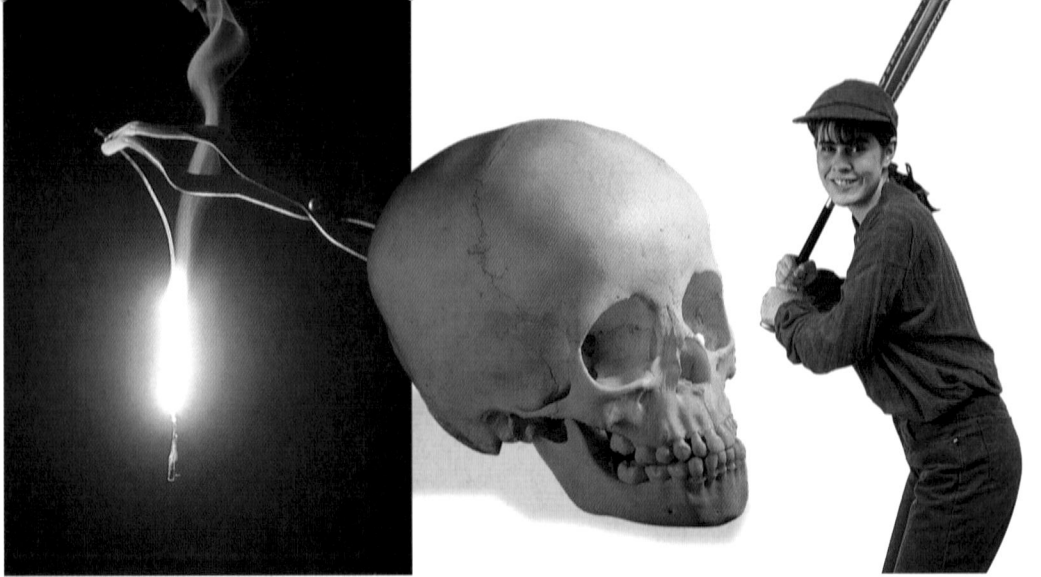

The Alkaline Earth Metals

The alkaline earth metals make up Group 2 of the periodic table. Like the alkali metals, these metals are shiny, malleable, ductile, and so reactive that they are not found free in nature. Each atom of an alkaline earth metal has two electrons in its outer energy level. These electrons are given up when an alkaline earth metal combines with another element. The result is a positively charged ion in a compound such as calcium fluoride, CaF_2.

Bats, Batteries, and Beryls

Emeralds and aquamarines are gemstone forms of beryl, a mineral that contains beryllium. Do you watch fireworks on the Fourth of July? The brilliant red in fireworks is produced by compounds of strontium.

Magnesium can be made into a fine wire that burns so brightly that it is used in some photographic flashbulbs. Magnesium metal is also used in fireworks to produce brilliant white. Magnesium's lightness and strength account for its use in cars, planes, and spacecraft. Magnesium is also used to make such things as household ladders and baseball and softball bats, such as the one shown in **Figure 12-5**. Most life on Earth depends upon chlorophyll, a magnesium compound that enables plants to make food.

Calcium is seldom used as a free metal, but its compounds are needed for life. Calcium phosphate in your bones helps make them strong. Marble and limestone are calcium carbonate.

Figure 12-5

Alkaline earth metals are found in these objects. *Which metals are represented here?*

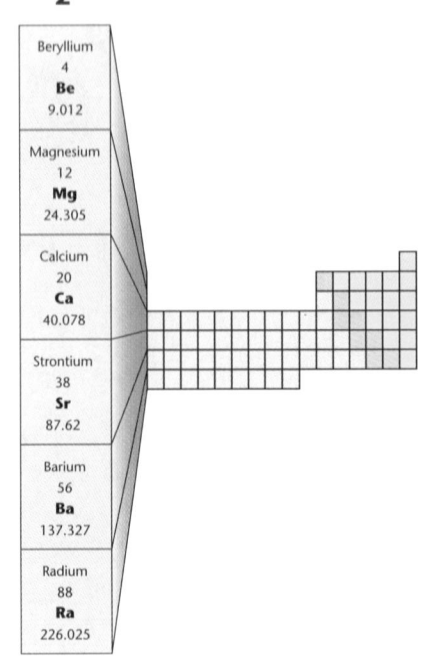

2
Beryllium 4 **Be** 9.012
Magnesium 12 **Mg** 24.305
Calcium 20 **Ca** 40.078
Strontium 38 **Sr** 87.62
Barium 56 **Ba** 137.327
Radium 88 **Ra** 226.025

12-1 Metals   **333**

Use **Cross-Curricular Integration,** p. 18 as you teach this lesson.

Some barium compounds are used to diagnose digestive disorders. The patient swallows a barium compound, which can absorb X rays. As the barium compound goes through the digestive tract, a doctor can study the X rays. Radium, the last element in Group 2, is radioactive. It was once used to treat cancers. Today, other radioactive substances are replacing radium in cancer therapy.

Transition Elements

Figure 12-6

The colors of these gems are due to compounds of transition elements.

An iron nail, a copper wire, and a silver dime are examples of objects made from transition elements. The **transition elements** are those elements in Groups 3 through 12 of the periodic table. Typical transition elements are metals and have one or two electrons in the outer energy level. These metals are less active than those in Groups 1 and 2.

The gems in **Figure 12-6** contain brightly colored compounds of transition elements. Brilliant cadmium yellow and cobalt blue paint pigments are made from compounds of transition elements. But cadmium and cobalt paints are so toxic that their use is now limited.

Iron, Cobalt, and Nickel

Iron, cobalt, and nickel form a unique cluster of transition elements. The first elements in Groups 8, 9, and 10, respectively, are known as the iron triad. The elements of the triad are the only ones known to create a magnetic field.

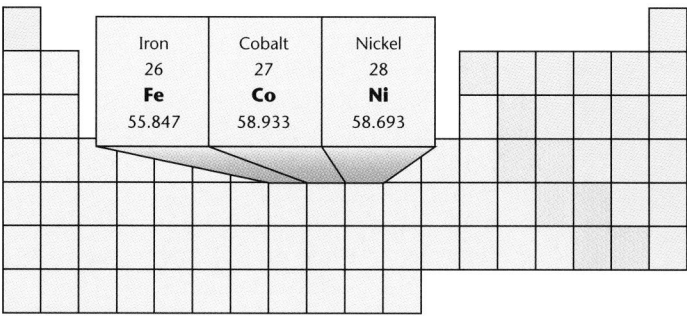

Iron	Cobalt	Nickel
26	27	28
Fe	**Co**	**Ni**
55.847	58.933	58.693

Cultural Diversity

That'll be two cowries and a whale's tooth... Coinage metals got their name from their use as money, but many things have been used as money. The Native Americans' wampum consisted of beads made from shells. Cowrie shells in India, whales' teeth in Fiji, and stone disks in Yap (a Pacific Island) have all served in monetary systems. The use of metals for exchange appeared around 7000 B.C. in the Middle East. Bronze was used first. Gold and silver began to be used because of their value. True coins, of standard weight and value, were probably first produced by the Lydians of Anatolia (modern-day Turkey) in 640 B.C. They used a naturally occurring alloy of gold and silver called electrum. Eventually, the purity and weight of gold and silver in coins was certified by the government.

Iron is second only to aluminum among the metals in abundance in Earth's crust. As the main component of steel, iron is the most widely used of all metals. Some steels also contain cobalt. Nickel is added to other metals to give them strength. Nickel is also often used to give a shiny, protective coating to other metals.

Copper, Silver, and Gold

Can you name the main metals in the coins in **Figure 12-7** on page 336? They are copper, silver, and gold—the three elements in Group 11. They are so unreactive that they can be found as elements in nature. For centuries, these metals have been widely used as coins. For this reason, they are known as the coinage metals.

Copper is often used in electric wiring because of its superior ability to conduct electricity and

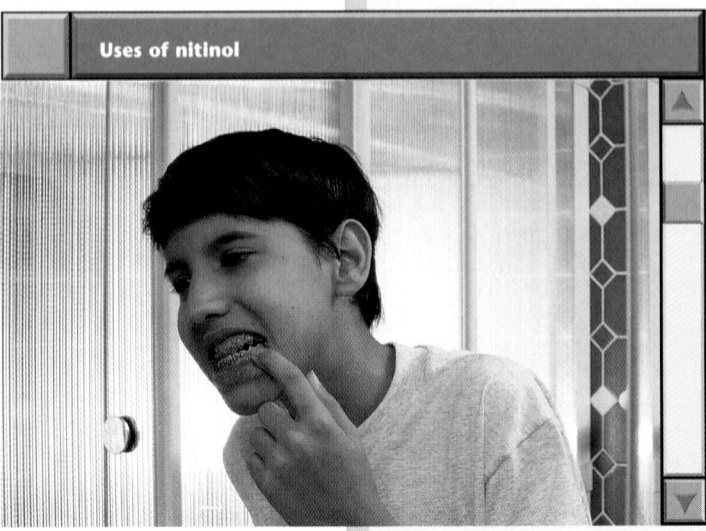
Uses of nitinol

Metals with Memory

What would a radio antenna in space have in common with your smile? They both benefit from nitinol. Nitinol is a metal alloy with a memory.

Nitinol is the name given to the metal mixture, or alloy, made from the transition elements nickel and titanium. The alloy, which is made into a preset shape, can later be deformed into a new shape.

Practical Uses

So how is this used on teeth? The wire connecting brackets in braces is nitinol. If the teeth are not properly aligned, the wire will be deformed from its preset shape. Soon, the wire will attempt to return to its original shape, putting pressure on the bracketed teeth and moving them.

Engineers also made use of nitinol when a large radio antenna was compactly folded during launch into space. Once in orbit, the metal with a memory absorbed heat and began to unfold into its original shape. This property of nitinol is also useful in eyeglass frames.

Think Critically:

Elements that can form shape-memory alloys, SMAs, are found in Groups 3 through 12 on the periodic table. List other properties of these elements in your Science Journal.

Have a local orthodontist visit the class with a sample of nitinol to show the class.

Think Critically
The transition elements are chemically less active than the metals in Groups 1 and 2. The transition elements form many colored compounds and have a variety of oxidation states.

GLENCOE TECHNOLOGY

 Videodisc

The Infinite Voyage Series: Miracles by Design
Chapter 2
Improving Steel: Examining Chemical Makeup

3 Assess

Check for Understanding

? FLEX Your Brain

Use the Flex Your Brain activity to have students explore PROPERTIES OF ELEMENTS. **P**

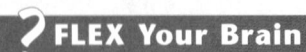 **Activity Worksheets,** p. 5

Reteach

LS **Intrapersonal** Have each student make one flash card listing an element, its group (or transition element if it is one) and its properties. Use the cards to provide review and reinforcement of the entire section. **LEP**

Extension

For students who have mastered this section, use the **Reinforcement** and **Enrichment** masters.

Use these program resources as you teach this lesson.
Multicultural Connections, pp. 27, 28
Critical Thinking/Problem Solving, p. 18

INTEGRATION
Earth Science

Point out to students how mining for metals differs from using other resources. For example, with proper management, food resources can be grown each year. Even synthetic materials may be made from reformulated chemicals. Metals, however, must be taken from Earth's crust. As more metals are mined, our supply dwindles. Recycling and finding metal substitutes will continue to become more important.

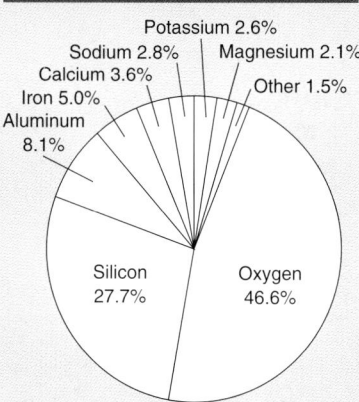
4 Close

•MINI•QUIZ•

Use the Mini Quiz to check students' recall of chapter content.

1. **Metals that can be hammered or rolled into thin sheets are** _____ . *malleable*

2. **When bonding with a nonmetal, metal atoms tend to** _____ **electrons.** *lose*

3. **The** _____ **metals are highly reactive metals.** *alkali*

4. **What part of the radioactive element emits particles or energy?** *nucleus*

5. **Group 2 of the periodic table is called the** _____ **metals.** *alkaline earth*

Figure 12-7
Some Group 11 elements are used to make coins.

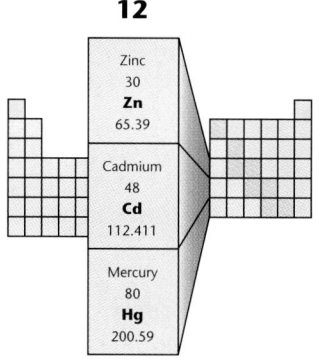
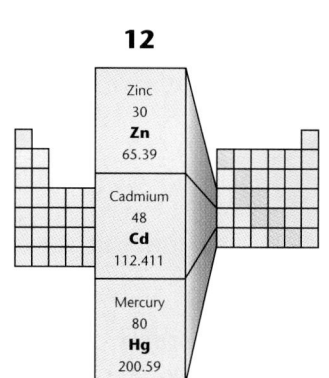

its relatively low cost. Can you imagine a world without photographs and movies? Because silver iodide and silver bromide break down when exposed to light, these compounds are used to make photographic film and paper. Much silver is also used in jewelry. The yellow color, relative softness, and rarity of gold account for its use in jewelry.

Zinc, Cadmium, and Mercury

Zinc, cadmium, and mercury make up Group 12 of the periodic table. Zinc combines with oxygen in the air to form a thin protective coating of zinc oxide on the surface of the metal. Zinc is often used to coat, or plate, other metals, such as iron. Cadmium is also used in plating and in rechargeable batteries.

Mercury is a silvery, liquid metal used in thermometers, thermostats, switches, and batteries. Mercury is poisonous, and mercury compounds can accumulate in the body. People have died of mercury poisoning that resulted from eating fish from mercury-contaminated water.

Look at the periodic table on pages 286-287. Which elements do you think of as being typical metals? Transition elements are the most familiar because they occur in nature as elements. Group 1 and Group 2 metals are found in nature only in compounds.

INTEGRATION
Earth Science

Metals in the Crust

When we examine pictures of metals and their compounds, we are really seeing only half the story. Where are metals found? How do we obtain them? Earth's crust contains many compounds and a few examples of uncombined metals such as gold and copper. Metals must be dug, or mined, from Earth's hardened outer layer.

Due to varying conditions in different areas, some metals are deposited more in one place than in another. For example, most of the world's platinum is found in South Africa. Large amounts of cobalt can be found in Morocco, Zaire, and Canada. Chromium is important to the United States, but we must import most of it from South Africa, the Philippines, and Turkey.

Ores: Minerals and Mixtures

Metals in Earth's crust that are in a combined form are found in ores. Typically, an ore consists of a metal compound, or mineral, within a mixture of clay or rock. Lead metal is usually found combined with sulfur in the form of galena, or lead sulfide. After an ore is mined and brought from Earth's surface, minerals must be separated from the rock. Then the mineral is often converted to another chemical form. Galena is converted to lead oxide. This step involves heat and is called roasting. Finally, the metal is refined into a pure form. It may later be alloyed with other metals. Removing the waste rock can be very expensive. If the cost of removing the waste rock gets higher than the value of the desired material, the mineral will no longer be classified as an ore.

USING MATH

Several elements are present in Earth's crust. The approximate percentage of each element is as follows:
Oxygen, 46.6
Silicon, 27.7
Aluminum, 8.1
Iron, 5.0
Calcium, 3.6
Sodium, 2.8
Potassium, 2.6
Magnesium, 2.1
Other elements include titanium, hydrogen, phosphorus, and manganese, which account for 1.5 percent. Make a circle graph of these data.

Section Wrap-up

Review

1. You are given a piece of the element palladium. How would you test it to see if it is a metal?

2. On the periodic table, how does the arrangement of the iron triad differ from the arrangements of the coinage metals and of the zinc group?

3. **Think Critically:** If X stands for a metal, how can you tell from the formulas XCl and XCl_2 which compound contains an alkali metal and which contains an alkaline earth metal?

Skill Builder
Interpreting Scientific Illustrations
Draw dot diagrams to show the similarity among chlorides of three alkali metals: lithium chloride, sodium chloride, and potassium chloride. If you need help, refer to Interpreting Scientific Illustrations in the **Skill Handbook.**

Science Journal

We encounter examples of metals nearly everywhere. In your Science Journal, write a paragraph describing how a specific metallic element has an effect in your life. Which group is that metal in on the periodic table?

Section Wrap-up

Review

1. Find out if it is hard and shiny, is a good conductor of heat and electricity, and is malleable and ductile.

2. Iron, cobalt, and nickel are successive metals in the same period. Copper, silver, and gold are all in Group 11. Zinc, cadmium, and mercury are all in Group 12.

3. **Think Critically** Chlorine bonds by gaining or sharing one electron. Alkali metals can lose one electron, so they would form the compound XCl. Alkaline earth metals have two outer electrons, so they would form compounds XCl_2.

Science Journal

Linguistic Accept all reasonable answers. The book *The Periodic Table* by Primo Levi contains essays by the Italian author/chemist on this theme.

Skill Builder

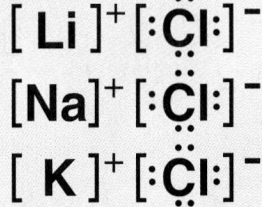

$$[\text{Li}]^+ [:\ddot{\text{C}}\text{l}:]^-$$

$$[\text{Na}]^+ [:\ddot{\text{C}}\text{l}:]^-$$

$$[\text{K}]^+ [:\ddot{\text{C}}\text{l}:]^-$$

Assessment

Oral Ask students to list at least two ways that the three drawn structures are alike. All the Group 1 ions are 1+; all formulas show a 1:1 ratio between ions. Use the Performance Task Assessment List for Making Observations and Inferences in **PASC,** p. 17. **P**

Prepare

Section Background

Transuranium elements have been produced using other elements to bombard target elements.

1 Motivate

Bellringer

Before presenting the lesson, display **Section Focus Transparency 47** on the overhead projector. Assign the accompanying **Focus Activity** worksheet. `L1` `LEP`

Tying to Previous Knowledge

Ask the class if they have ever heard of Merlin in King Arthur's court. He was an alchemist. He tried to change common metals into gold. Today this can be accomplished in a nuclear reactor.

2 Teach

Concept Development

Bring a portable smoke detector to class, open it and show the class the case where the americium is.

Teacher F.Y.I.

The International Union of Pure and Applied Chemists (IUPAC) is responsible for settling discovery and naming disputes for elements.

Visual Learning

Figure 12-8 How would plutonium be classified? *actinide* `LEP` `IS`

TECHNOLOGY:
12•2 New Elements, New Properties

Science Words

transuranium element

Objectives

- Distinguish among elements classified as lanthanides, actinides, and transuranium elements.
- Determine the uses of transuranium elements.
- Compare the pros and cons of making new elements.

Figure 12-8

These fuel rods contain plutonium, element 84. *How would plutonium be classified?*

Elements Beyond 92

What if you made something that always fell apart? You might think you were not successful. Yet nuclear scientists are learning to do just that. By combining existing elements with fast-moving particles, they have been successful at creating elements not typically found on Earth. Except for one, the short-lived elements have more than 92 protons. All of these synthetic elements are unstable and fall apart quickly. Elements having more than 92 protons, the atomic number of uranium, are called **transuranium elements.**

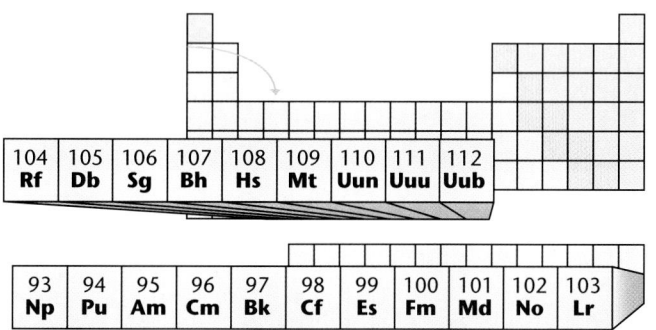

104 Rf	105 Db	106 Sg	107 Bh	108 Hs	109 Mt	110 Uun	111 Uuu	112 Uub

93 Np	94 Pu	95 Am	96 Cm	97 Bk	98 Cf	99 Es	100 Fm	101 Md	102 No	103 Lr

Additions to the Family

If you look at the periodic table on pages 286-287, you will see breaks in periods 6 and 7. The first break includes a series of 14 elements with atomic numbers of 58-71. The elements in this series are known as the lanthanides. The second break includes elements with atomic numbers ranging from 90-103. These elements are known as actinides. Ten synthetic elements are found in the actinide series. Where are the others placed on the periodic table? Technetium is element number 43. The newest synthetic elements are placed in period 7.

Synthesis: A Smashing Success

If atoms and other particles are smashed together with sufficient force, they will fuse to form a new identity. For example, neptunium, element 93, can be made by bombarding

---

Program Resources

 Reproducible Masters
Study Guide, p. 51 `L1`
Reinforcement, p. 51 `L1`
Enrichment, p. 51 `L3`
Science and Society Integration, p. 16

Transparencies
Section Focus Transparency 47 `L1`

uranium with neutrons. Atoms of neptunium do not stay together for long. Half of the synthesized atoms disintegrate in about two days. This may not sound useful, but when neptunium atoms fall apart, they form plutonium. This highly toxic element has been used in control rods of nuclear reactors, such as those shown in **Figure 12-8,** and in bombs. Plutonium can also be changed to americium, element 95. This transuranium element is used in some home smoke detectors such as the one in **Figure 12-9.**

Figure 12-9

Synthesis of elements has resulted in life-saving smoke detectors.

58 Ce	59 Pr	60 Nd	61 Pm	62 Sm	63 Eu	64 Gd	65 Tb	66 Dy	67 Ho	68 Er	69 Tm	70 Yb	71 Lu
90 Th	91 Pa	92 U	93 Np	94 Pu	95 Am	96 Cm	97 Bk	98 Cf	99 Es	100 Fm	101 Md	102 No	103 Lr

Why make elements?

By studying how the synthesized elements form and fall apart, we can gain an understanding of the forces holding other elements together. When these atoms fall apart, they are said to be radioactive. Technetium's radioactivity makes it ideal for many medical applications. At this time, many of the synthetic elements, once made, last only small fractions of seconds and can be made only in small amounts. However, further advances in technology may bring production up to larger amounts. The as-yet-undiscovered applications could prove to be more valuable than their costs.

Section Wrap-up

Review

1. How is an element classified as a transuranium element?
2. Where are the actinides located on the periodic table?

*inter***NET** **CONNECTION** Some people are concerned about the use of radioactive elements such as americium in smoke detectors and other products. Visit the Chapter 12 Internet Connection at Glencoe Online Science, **www.glencoe.com/sec/science/physical,** for a link to more information about radioactive elements.

SCIENCE & SOCIETY

Check for Understanding

?FLEX Your Brain

Use the Flex Your Brain activity to have students explore SYNTHETIC ELEMENTS.

🗂 **Activity Worksheets,** p. 5

Reteach

Have your students use a blank copy of the periodic table to record the symbols, names, and atomic numbers of the 19 synthetic elements.

Extension

🗂 For students who have mastered this section, use the **Reinforcement** and **Enrichment** masters.

4 Close

Activity

LS **Logical-Mathematical** Have students use a periodic chart to predict the atomic number of the next noble gas. *118* L1

Section Wrap-up

Review

1. It has more than 92 protons.
2. They are located in the last row on the periodic table.

Prepare

Section Background

• Atoms of most nonmetals have five or more outer electrons, and tend to gain electrons to complete their outer energy levels.

• Hydrogen has unique properties and is usually considered as a group by itself.

• As the atomic numbers of the halogens increase, the atoms become larger, the outer electrons are farther from the nucleus, and the atoms become less active.

• The noble gases were considered to be inert until 1962 when xenon difluoride was produced.

Preplanning

• To prepare for Activity 12-1 you will need to assemble conductivity testers.

• Prepare the standard chloride and silver nitrate solutions for the MiniLAB.

1 Motivate

Bellringer

Before presenting the lesson, display **Section Focus Transparency 48** on the overhead projector. Assign the accompanying **Focus Activity** worksheet. **L1** **LEP**

Demonstration

LS **Visual-Spatial** Use a piece of roll sulfur to visually reinforce the properties of nonmetals. Show the class the dull surface before you cover it with a cloth and, using a hammer, break off a piece. Students will see that it is brittle and powdery, not malleable. **LEP**

Science Words

diatomic molecule
sublimation

Objectives

• Recognize hydrogen as a nonmetal.
• Compare and contrast properties of the halogens.
• Describe properties and uses of the noble gases.

12•3 Nonmetals

Properties of Nonmetals

Figure 12-10 shows that you're mostly made of oxygen, carbon, hydrogen, and nitrogen. Calcium, a metal, and other elements make up the remaining four percent of your body's weight. Phosphorus, sulfur, and chlorine are among other elements found in your body. These elements are classified as nonmetals.

Look at the periodic table on pages 286 and 287. How many elements are nonmetals? Notice that most nonmetals are gases at room temperature. Several nonmetals are solids, and one nonmetal is a liquid.

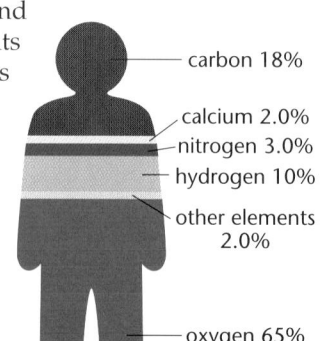

carbon 18%
calcium 2.0%
nitrogen 3.0%
hydrogen 10%
other elements 2.0%
oxygen 65%

Figure 12-10

Humans are made up of mostly nonmetals.

Attracting Electrons

In contrast to metals, solid nonmetals are dull. Because they are brittle and powdery, they are neither malleable nor ductile. The electrons in most nonmetals are tightly attracted and are restricted to one atom. So, as a group, nonmetals are poor conductors of heat and electricity.

Most nonmetals form both ionic and covalent compounds. Examples of these two kinds of compounds are shown in **Figure 12-11.** When nonmetals gain electrons from metals, the nonmetals become negative ions in ionic compounds. An example of such an ionic compound is potassium iodide, KI,

Program Resources

 Reproducible Masters

Study Guide, p. 52 **L1**
Reinforcement, p. 52 **L1**
Enrichment, p. 52 **L3**
Lab Manual 26
Activity Worksheets, pp. 68, 69, 73 **L1**

Transparencies

Section Focus Transparency 48 **L1**

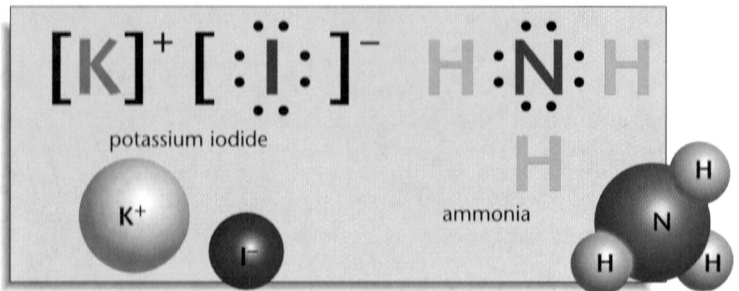
potassium iodide

ammonia

which is often added to table salt. KI is formed from the nonmetal iodine and the metal potassium. On the other hand, when bonded with other nonmetals, atoms of nonmetals usually share electrons and form covalent compounds. An example is ammonia, NH_3, the gas you can smell when you open a bottle of household ammonia.

The noble gases, Group 18, make up the only group of elements that are all nonmetals. Group 17 elements, except for astatine, are also nonmetals. Several other nonmetals, found in Groups 13 through 16, will be discussed later. Except for hydrogen, all of the nonmetals are located on the right side of the periodic table.

Hydrogen

Do you know that approximately 90 percent of all the atoms in the universe are hydrogen? Most hydrogen on Earth is found in the compound water. When water is broken down into its elements, hydrogen forms as a gas made up of diatomic molecules. A **diatomic molecule** consists of two atoms of the same element. Thus, the formula for hydrogen gas is H_2, shown in **Figure 12-12.**

Small, But Active

Hydrogen is highly reactive. A hydrogen atom has a single electron, which the atom shares when it combines with other nonmetals. Hydrogen burns in oxygen to form water, H_2O. In forming water, hydrogen shares electrons with oxygen. Hydrogen also shares electrons with chlorine to produce hydrogen chloride, HCl.

Hydrogen may gain an electron when it combines with alkali and alkaline earth metals. The compounds formed are hydrides, such as sodium hydride, NaH.

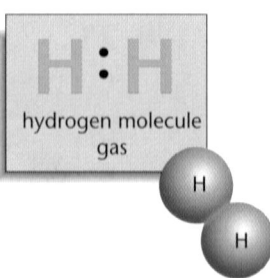
hydrogen molecule gas

Figure 12-12

Hydrogen gas, which consists of diatomic molecules, combines with oxygen in an oxyhydrogen torch.

341

 Use Lab 26 in the **Lab Manual** as you teach this lesson.

2 Teach

Demonstration

A small amount of hydrogen gas can be generated by placing a piece of mossy zinc in a large test tube. Add 4 mL of 6*M* sulfuric acid and stopper the test tube with a 1-hole stopper fitted with a 90° glass bend. **CAUTION:** *The acid is corrosive.* Use an inverted test tube to collect the gas by the downward displacement of air. A burning wood splint can be used to show that the gas is flammable. **CAUTION:** *Use an explosion shield. Never produce more than a small test tube of hydrogen gas.*

Concept Development

Tell students that hydrogen is being investigated as a nonpolluting fuel for autos, buses, and planes. The flammable gas can be used to heat homes with no danger of carbon monoxide poisoning.

Revealing Preconceptions

Ask students what gas is used to fill balloons that are lighter than air. Some think that hydrogen rather than helium is used. Show a picture of the burning Hindenburg to help students remember the flammability of hydrogen.

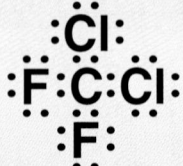

MiniLAB

Can you detect chlorine in your drinking water?

Procedure

1. In separate test tubes, obtain 2 mL of chloride standard solution, distilled water, and drinking water.
2. Carefully add five drops of silver nitrate solution to each and stir. **CAUTION:** *Avoid contact with the silver nitrate solution.*

Analysis

1. Which solution will definitely show a presence of chloride? How did this result compare to the result with distilled water?
2. Which result most resembled your drinking water?

CONNECT TO

EARTH SCIENCE

Compounds called chlorofluorocarbons are used in refrigeration systems. If released, these compounds destroy ozone in the atmosphere. *Draw* the electron dot diagram for CF_2Cl_2 in your Science Journal.

Figure 12-13

The halogens bromine and iodine are used in the bulbs of halogen lamps.

17

Fluorine	9
	F
	18.998
Chlorine	17
	Cl
	35.453
Bromine	35
	Br
	79.904
Iodine	53
	I
	126.904
Astatine	85
	At
	209.987

The Halogens

Look how bright the high-tech lamp is in **Figure 12-13.** Its light and that of the headlights of some cars are supplied by halogen lightbulbs. These bulbs contain small amounts of bromine or iodine. These elements, as well as fluorine, chlorine, and astatine, are called halogens and are found in Group 17.

Because an atom of a halogen has seven electrons in its outer energy level, only one electron is needed to complete this energy level. If a halogen gains an electron from a metal, an ionic compound, called a salt, is formed. The word *halogen* means "salt former." In the gaseous state, the halogens form very reactive diatomic covalent molecules and can be identified by their distinctive colors, as shown in **Figure 12-14.**

Fluorine is the most chemically active of all the elements. Hydrofluoric acid, a mixture of hydrogen fluoride and water, is used to etch glass and to frost the inner surfaces of lightbulbs. Other fluorides are added to toothpastes and to city water systems to prevent tooth decay. Does your community add fluorides to its water?

Figure 12-14

These are the gaseous halogens chlorine, bromine, and iodine.

Uses of Halogens

The odor near swimming pools is the odor of chlorine. Chlorine compounds are used to disinfect water. Chlorine, the most abundant halogen, is obtained from seawater. Household and industrial bleaches used to whiten flour and paper also contain chlorine compounds.

Bromine, the only nonmetal that is liquid at room temperature, is extracted from compounds in seawater. Other bromine compounds are used as dyes in cosmetics, such as the lipsticks shown in **Figure 12-15.**

Iodine, a shiny gray solid at room temperature, is obtained from seawater. When heated, iodine changes directly to a purple vapor. The process of a solid changing directly to a vapor without forming a liquid is called **sublimation** as shown in **Figure 12-16.** Recall that sublimation accounts for ice cubes shrinking in freezers. Iodine is essential in your diet for the production of the hormone thyroxin and to prevent goiter, an enlarging of the thyroid gland in the neck.

Astatine, the last member of Group 17, is radioactive and rare. But it has many properties similar to those of the other halogens.

Figure 12-15

Dyes in some cosmetics contain bromine compounds. *What unique physical property does bromine have?*

Figure 12-16

Ice and dust form the body of a comet. The ice undergoes sublimation, forming the gases contained in the tail of the comet. This particular comet is Hale-Bopp, which was visible from Earth for several months in 1997.

The Noble Gases

Why are the noble gases called noble? It was known that these gases did not naturally form compounds. Thus, they were thought of as the nobility of elements because nobles did not mix with common folk. However, in the early 1960s, scientists were able to prepare some compounds of noble gases.

12-3 Nonmetals   **343**

4 Close

•MINI•QUIZ•

Use the Mini Quiz to check students' recall of chapter content.

1. **In what state are most nonmetals at room temperature?** *gases*
2. **When hydrogen burns in air, what product forms?** *water*
3. **Group 17 elements are also called _____ .** *halogens*
4. **Which element is the most chemically active nonmetal?** *fluorine*
5. **Which nonmetal is a liquid at room temperature?** *bromine*

Section Wrap-up

Review

1. by sharing its electron or by gaining an electron to form a hydride
2. H^+; H; He; Cl^-
3. They are chemically stable. Therefore they can be used to maintain a chemically inert environment around reactive materials.
4. **Think Critically** Hydrofluoric acid is very reactive and will react with glass. It must be stored in bottles made of more inert material.

Using Computers

Note: No metalloids (B-At) are included in the count. All synthetic elements are counted as solid metals.

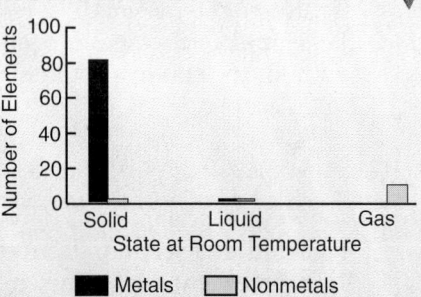

Figure 12-17

Noble gases are used to produce laser light shows.

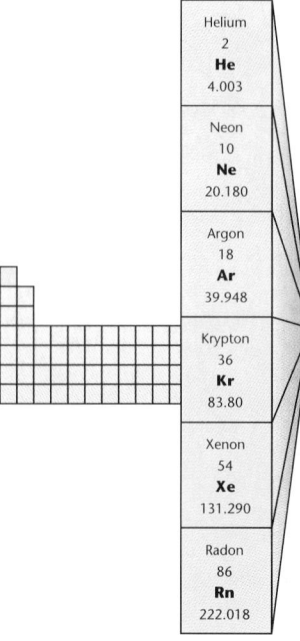

As you recall, each element in Group 18 is stable because its outer energy level is full. The stability of the noble gases plays an important role in their uses, such as those shown in **Figure 12-17.**

Both the halogens and the noble gases illustrate that each element in a group has some similar properties but also has unique properties and uses.

Section Wrap-up

Review

1. What are two ways in which hydrogen combines with other elements?

2. Rank the following nonmetals from lowest number of electrons in the outer level to highest: Cl^-, H^+, He, H.

3. What property of noble gases makes them useful?

4. **Think Critically:** Why must hydrofluoric acid always be stored in plastic bottles?

Skill Builder

Interpreting Data

Within the following compounds, identify the nonmetal and list its oxidation number: MgO, NaH, $AlBr_3$, FeS. If you need help, refer to Interpreting Data in your **Skill Handbook.**

> **Using Computers**
>
> **Graphing** Prepare a bar graph comparing nonmetals and metals as solids, liquids, and gases at room temperature.

Skill Builder

MgO:	O nonmetal, 2−
NaH:	H nonmetal, 1−
$AlBr_3$:	Br nonmetal, 1−
FeS:	S nonmetal, 2−

✓ Assessment

Oral Ask students to explain why nonmetals typically have negative charges when combined with metals. The electron-attracting ability of nonmetals is usually better than that of metals due to their strong nuclei with small atomic size. Use the Performance Task Assessment List for Making Observations and Inferences in **PASC,** p. 17. **P**

Activity 12-1

What type is it?

Suppose you want an element for a certain use. You may be able to use a metal but not a nonmetal. Or maybe a nonmetal better meets your needs. In this activity, you will test several metals and nonmetals and compare their properties.

Problem
How can you use properties to tell metals from nonmetals?

Materials
- samples of carbon, magnesium, aluminum, sulfur, and tin
- dishes for the samples
- conductivity tester
- spatula
- small hammer

Procedure
1. Prepare a table in your Science Journal like the one shown.
2. Observe and record the appearance of each element sample. Include its physical state, color, and whether it is shiny or dull.
3. Remove a small sample of one of the elements. Place it on a hard surface chosen by your teacher. Gently tap the sample with a hammer. The sample is malleable if it flattens when tapped and brittle if it shatters. Record your results.

4. Repeat step 3 for each sample.
5. Test the conductivity of each element by touching the electrodes of the conductivity tester to a sample, as shown in the photo. If the bulb lights, the element conducts electricity. Record your results.

Analyze
1. Which elements showed properties of metals?
2. Which elements showed properties of nonmetals?
3. Which elements, if any, showed properties of both metals and nonmetals?

Conclude and Apply
4. Locate each element you used on the periodic table. **Compare** your results with what you would expect from the location of each element.
5. Locate palladium, Pd, on the periodic table. Use the results of the activity to **predict** some of the properties of palladium.
6. **Infer** why some elements may show both properties of metals and properties of nonmetals.

Data and Observations
Sample Data

Element	Appearance	Malleable or brittle	Electrical conductivity
carbon	solid, black, dull	brittle	no
magnesium	solid, gray, shiny	malleable	yes
aluminum	solid, gray, shiny	malleable	yes
sulfur	solid, yellow, dull	brittle	no
tin	solid, gray, shiny	malleable	yes

12-3 Nonmetals   345

Answers to Questions
1. magnesium, aluminum, tin, maybe carbon
2. carbon, sulfur
3. Depending on the sample, carbon may have conducted a current as well as shown nonmetallic properties.
4. Elements on the left side of the table would show more metallic properties; those further to the right, nonmetallic properties.
5. Predictions may include shiny, malleable, and conductor.

6. They are in an area that shows transition from metallic to nonmetallic.

Assessment

Performance Give each student a periodic table. Based on their activity results, ask students to draw an arrow on the table to show the direction of increasing metallic properties across the table. Use the Performance Task Assessment List for Making Observations and Inferences in **PASC**, p. 17. **P**

Activity 12-1

Purpose
LS **Visual-Spatial** Students will observe properties of metals and nonmetals and use their observations to classify certain elements. **L1** **LEP**

Process Skills
observing, interpreting data, comparing and contrasting, classifying

Time
30 minutes

Safety Precautions
Have students wear goggles. Have them use caution when using the hammer.

Activity Worksheets, pp. 5, 68, 69

Teaching Strategies
Alternate Materials As long as there is an assortment of metals and nonmetals, other elements can be used. The conductivity tester shown can be easily assembled from a 9-V battery, a 9-V battery clip, an LED lamp assembly with a resistor, 3 small wire nuts, and 2 lengths of solid wire, 20-gauge. Other conductivity testers can be used.
Troubleshooting Test the conductivity tester before class to make sure that it works. To observe the nonmetallic properties of carbon, use amorphous carbon, such as charcoal, not graphite, which is shiny and will conduct a current.

- You may want to set up stations for each test. Stations require fewer materials and can be better supervised.
- As a demonstration to show another property of metals, place a sample of each element in its own test tube. Add 5 mL of 1 *M* HCl to each. Students will observe the metals reacting with the acid to produce hydrogen gas. The HCl can be prepared by adding 5 mL concentrated HCl to 55 mL of water.

345

Prepare

Section Background
• Of the elements in Group 13, aluminum has the greatest number of practical uses.
• The elements of Group 14 generally react by sharing electrons.

Preplanning
Assemble the materials needed for Activity 12-2. The polystyrene foam sheets can be obtained from a hardware store or a vinyl siding company. Students can bring spaghetti and gum drops from home.

1 Motivate

Bellringer
Before presenting the lesson, display **Section Focus Transparency 49** on the overhead projector. Assign the accompanying **Focus Activity** worksheet. L1 LEP

Tying to Previous Knowledge
Help students to recall information about Groups 13-16 presented in Chapter 10 when they learned about the periodic table. Remind students about the number of outer electrons in each group and how these electrons relate to the element's chemical behavior.

Visual Learning
Figure 12-18 What metals from Group 13 are used to produce semiconductors? *gallium, indium, and thallium* LEP LS

Science Words
semiconductor
allotrope

Objectives
• Distinguish among metals, nonmetals, and metalloids in Groups 13 through 16 of the periodic table.
• Describe the nature of allotropes.
• Recognize the significance of differences in crystal structure in carbon.

Locating the Mixed Groups
Can an element be both a metal and a nonmetal? In a sense, some elements are. They are the metalloids, and they have both metallic and nonmetallic properties. A metalloid may conduct electricity better than many nonmetals but not as well as some metals. In the periodic table, the metalloids are the elements located along the stair-step line. The mixed groups—13, 14, 15, and 16—contain metals, nonmetals, and metalloids.

The Boron Group
Boron, a metalloid, is the first element in Group 13. If you look around your home, you may find two compounds of boron. One of these is boric acid, a mild antiseptic. The other is borax, which is used in laundry products to soften water. Less familiar are the boranes, which are compounds of boron used in fuels for rockets and jet airplanes.

Aluminum, a metal, is also in Group 13. Aluminum is the most abundant metal in Earth's crust. Aluminum is used in soft-drink cans, foil wrap, and cooking pans. You may also see aluminum on the sides of buildings. And because aluminum is both strong and light, it is used to build airplanes.

When you operate electronic equipment, such as the calculator shown in **Figure 12-18**, you may be using one of the other metals in Group 13. These metals are gallium, indium, and thallium, used to produce semiconductors. **Semiconductors** conduct an electric current under certain conditions. You will learn more about semiconductors in Chapter 23.

Figure 12-18
Calculators contain semiconductors. *What metals from Group 13 are used to produce semiconductors?*

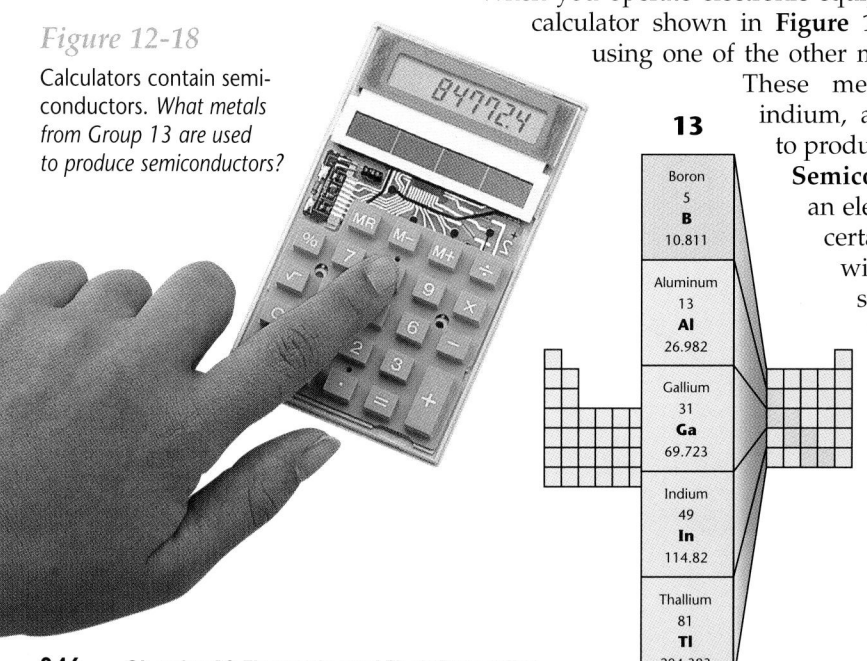

13

| Boron |
| 5 |
| **B** |
| 10.811 |

| Aluminum |
| 13 |
| **Al** |
| 26.982 |

| Gallium |
| 31 |
| **Ga** |
| 69.723 |

| Indium |
| 49 |
| **In** |
| 114.82 |

| Thallium |
| 81 |
| **Tl** |
| 204.383 |

Program Resources

📁 **Reproducible Masters**
Study Guide, p. 53 L1
Reinforcement, p. 53 L1
Enrichment, p. 53 L3
Concept Mapping, pp. 29, 30
Activity Worksheets, pp. 70, 71
Lab Manual 27

Transparencies
Section Focus Transparency 49 L1
Teaching Transparency 24 L1
Science Integration Transparency 12

Figure 12-19

Two allotropes of carbon are graphite (A) and diamond (B). *What geometric shapes make up each allotrope?*

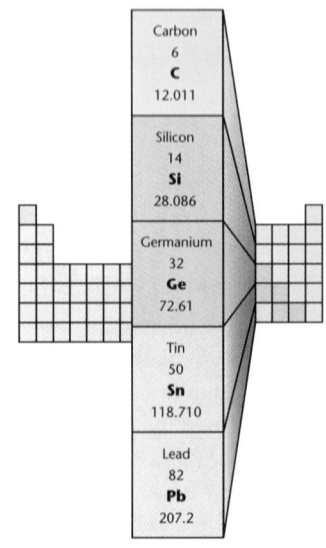

The Carbon Group

Each element in Group 14, the carbon family, has four electrons in its outer energy level. But this is where much of the similarity ends. Carbon is a nonmetal; silicon and germanium are metalloids; tin and lead are metals.

What do the diamond in a diamond ring and the graphite in your pencil have in common? It may surprise you to learn that they are both carbon. How can this be? Diamond and graphite are examples of allotropes. **Allotropes** are different forms of the same element having different molecular structures. Look at **Figure 12-19.** Graphite is a black powder that consists of hexagonal layers of carbon atoms. In the hexagons, each carbon atom is bonded to three other carbon atoms. The fourth electron of each atom is bonded weakly to the layer next to it. This structure allows the layers to slide easily past one another, making graphite an excellent lubricant.

14

Carbon
6
C
12.011

Silicon
14
Si
28.086

Germanium
32
Ge
72.61

Tin
50
Sn
118.710

Lead
82
Pb
207.2

12-4 Mixed Groups   **347**

Problem Solving

Waiting to Be Discovered

In sports, the "Rookie of the Year" award typically designates a player who had an outstanding first year and shows great potential for continued achievement. Most of the selections live up to that expectation, but not all. At the beginning of the 1990s, *Science* magazine selected Buckminsterfullerene as the "Molecule of the Year." This allotrope of carbon was isolated by R.E. Smalley when some carbon materials were subjected to laser bursts. Later, the carbon condensed into unusual spherical elements. The tiny spheres were later shown to be made of 60 carbon atoms arranged in an open cage that could resemble a hollow soccer ball.

Many new uses of this third allotrope were thought possible. Interesting medical applications were proposed to attach special disease-attacking components to the outside of the molecule. Another idea was to insert radioactive atoms inside the cage. They could then be injected inside a patient to arrive at a specific site. Another problem the new molecule could solve is the separation of mixtures. Buckminsterfullerene molecules, also called Buckyballs, can be packed into containers and used as separating sieves.

Solve the Problem:

1. Use a tall cylinder of glass marbles to represent Buckminsterfullerene molecules, and a mixture of sand or salt and small BB pellets to represent a mixture of chemical compounds.
2. Pour the sand and BB mixture over the marbles.

Think Critically:

1. How did the packing and spacing of the spherical marbles enhance the separation of your mixture?
2. How could Buckyballs be used to separate compounds?
3. Can you think of other potential uses for this molecule?

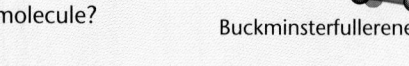

Buckminsterfullerene

348   Chapter 12 Elements and Their Properties

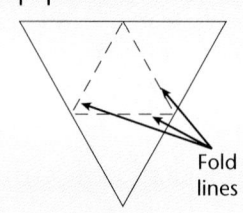
Tough Tetrahedrons

A diamond is clear and extremely hard. In a diamond, each carbon atom is bonded to four other carbon atoms at the vertices, or corner points, of a tetrahedron. In turn, many tetrahedrons join together to form a giant molecule in which the atoms are held tightly in a strong crystal structure. This structure accounts for the hardness of diamond.

Carbon occurs as an element in coal and in compounds in oil, natural gas, and foods. Carbon in these materials may combine with oxygen to produce carbon dioxide, CO_2. In the presence of sunlight, plants utilize CO_2 to make food. In Chapter 13, you will study other carbon compounds—many essential to life.

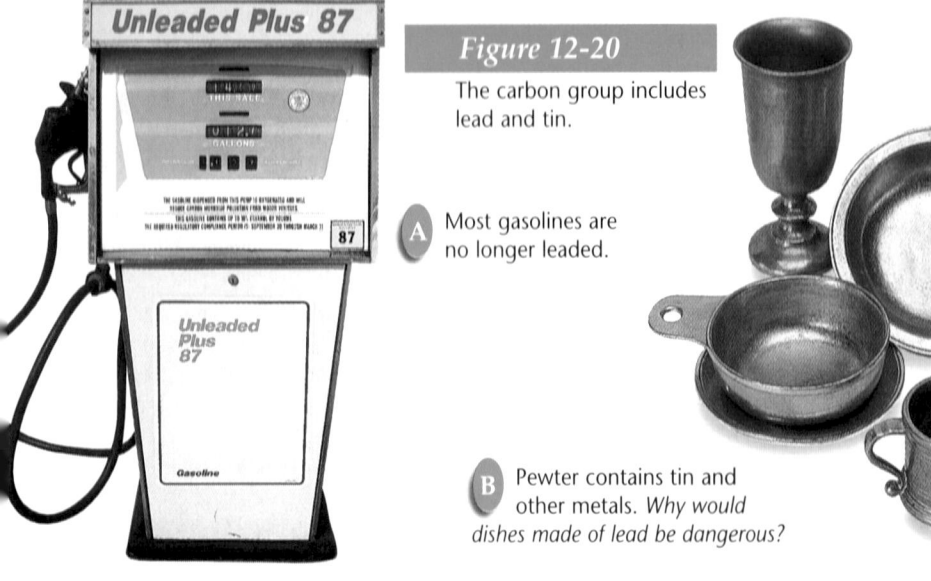

Figure 12-20
The carbon group includes lead and tin.

A Most gasolines are no longer leaded.

B Pewter contains tin and other metals. *Why would dishes made of lead be dangerous?*

Silicon is second only to oxygen in abundance in Earth's crust. Most silicon is found in sand, silicon dioxide, SiO_2, and in almost all rocks and soil. The crystal structure of silicon dioxide is similar to the tetrahedrons in diamond. Silicon occurs as two allotropes. One of these is a hard, gray substance, and the other is a brown powder.

Both silicon and germanium, the other metalloid in the carbon group, are used in making semiconductors, which you'll learn about in Chapter 23. Tin and lead are typical metals. Tin is used to coat other metals to prevent corrosion. Tin is also combined with other metals to produce bronze and pewter. Lead was once used widely in paints and antiknock gasoline. However, lead is poisonous, so it has been replaced in these materials.

The Nitrogen Group

The nitrogen family makes up Group 15. Each element has five electrons in its outer energy level. These elements tend to share electrons and to form covalent compounds with other elements.

Each breath you take is about 80 percent gaseous nitrogen, in the form of diatomic molecules, N_2. If you look again at **Figure 12-10** on page 340, you'll see that nitrogen is the element fourth in abundance in your body. Yet you and other animals and plants can't use nitrogen as N_2. The nitrogen must be combined into compounds, such as nitrates—compounds that contain the nitrate ion, NO_3^-. Much nitrogen is used to make nitrates and ammonia, NH_3, both of which are used in fertilizers.

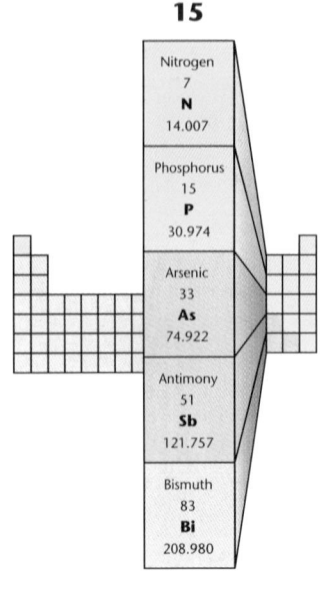

15

Nitrogen	7
N	14.007

Phosphorus	15
P	30.974

Arsenic	33
As	74.922

Antimony	51
Sb	121.757

Bismuth	83
Bi	208.980

12-4 Mixed Groups 349

Activity

LS **Kinesthetic** Duplicate four equilateral triangles for each student. Have the student cut out each triangle, leaving a tab on each edge. Tape or glue the triangles together to form a tetrahedron. Stack the tetrahedrons together in a clear plastic sweater storage box. The interlocking network of tetrahedrons represents the rigid crystal structure of a diamond. **L1** **LEP**

Visual Learning

Figure 12-20 **Why would dishes made of lead be dangerous?** *Lead is poisonous.* **LEP** **LS**

Content Background

White phosphorus is very reactive. It is a component in the chemical weapon napalm. Its structure is a tetrahedron with the formula P_4. Red phosphorus is less reactive and is thought to have a chain structure. Black phosphorus is not common but does exist.

NATIONAL GEOGRAPHIC SOCIETY

Videodisc
STV: Atmosphere
What is the Atmosphere?
Unit 1
Atmospheric Gases

03109-07851
Gas composition of atmosphere; artwork

38663
Oxygen and ozone molecules; artwork

38684

Across the Curriculum

Political Science Nitrogen compounds include fertilizers that help feed the world's population as well as drugs to help treat disease. They also include trinitrotoluene, TNT, and compounds used in poisonous gases. Chemicals are neither good nor bad. People decide how chemicals are to be used. Have a discussion on whether scientists should work on development of explosives and poisons.

3 Assess

Check for Understanding

? FLEX Your Brain

Use the Flex Your Brain activity to have students explore ALLOTROPES.

📁 **Activity Worksheets,** p. 5

Reteach

Analogy Remind students that allotropes are different forms of the same element. Many winter coats are sold as a three-in-one coat. A lining zips out to serve as a jacket. The sleeves can also be zipped off leaving a vest. All three are forms of the same coat. **LEP**

Extension

📁 For students who have mastered this section, use the **Reinforcement** and **Enrichment** masters.

4 Close

•MINI•QUIZ•

Use the Mini Quiz to check students' recall of chapter content.

1. **Substances that conduct an electric current under certain conditions are called** _____ . *semiconductors*

2. **Different forms of the same element are called** _____ . *allotropes*

3. **Most semiconductors are made from silicon or** _____ . *germanium*

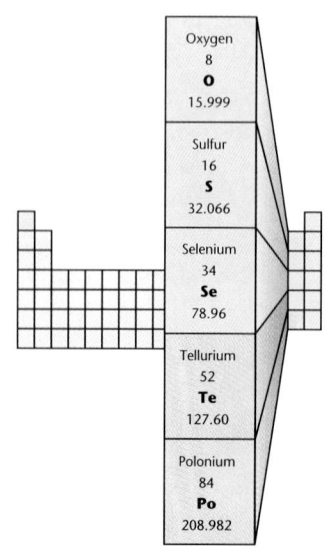

Figure 12-21

Sulfur is typically a bright yellow nonmetal. One allotrope of sulfur has eight-sided crystals. The other allotrope has a needlelike crystal. *What is an eight-sided object called?*

Phosphorus is a nonmetal that has three allotropes. Uses of phosphorus compounds range from water softeners to fertilizers to match heads. Antimony is a metalloid, and bismuth is a metal. Both elements are used with other metals to lower their melting points. It is because of this property that the metal in automatic fire-sprinkler heads contains bismuth.

The Oxygen Group

The oxygen group makes up Group 16 on the periodic table. You can live for only a short time without the nonmetal oxygen, which makes up about 20 percent of air. Oxygen exists in the air as diatomic molecules, O_2. During electrical storms, some oxygen molecules, O_2, change into ozone molecules, O_3. Do you notice that O_2 and O_3 are allotropes?

Nearly all living things on Earth need free oxygen, as O_2, for respiration. Living things also depend on a layer of ozone, O_3, around Earth for protection from some of the sun's radiation, as you will learn in later chapters.

The second element in the oxygen group is sulfur. Sulfur is a nonmetal that can exist in allotropes as different-shaped crystals and as a noncrystalline solid. In **Figure 12-21,** you see a magnified view of two different crystalline allotropes of sulfur. Sulfur combines with metals to form sulfides of such distinctive colors that they are used as pigments in paints.

The nonmetal selenium and two metalloids, tellurium and polonium, are the other Group 16 elements. Selenium is the most common of these. This element is one of several that

350 Chapter 12 Elements and Their Properties

📁 Use these program resources as you teach this lesson.
Concept Mapping, pp. 29, 30
Lab Manual 27

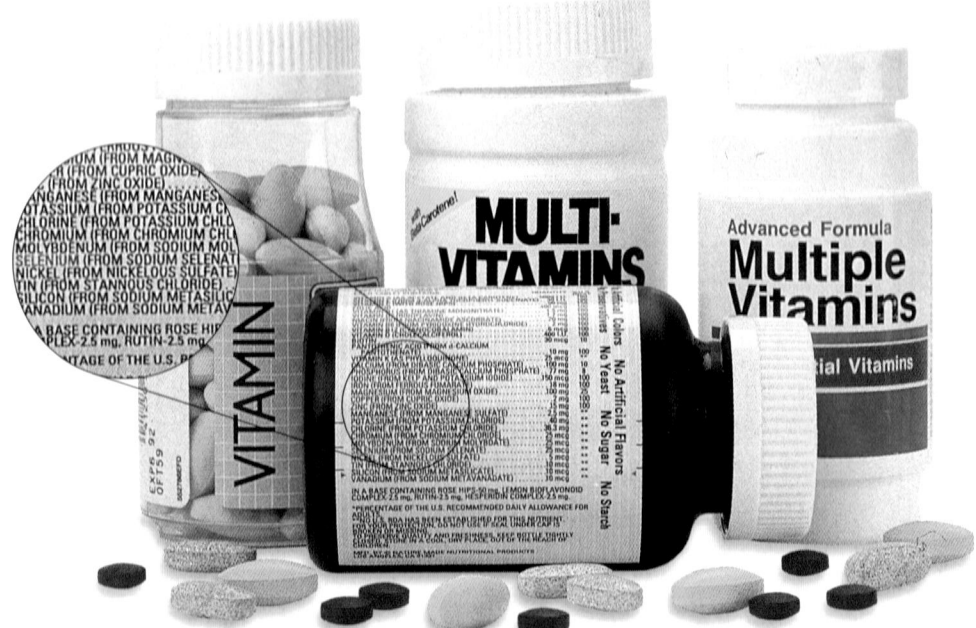

you need in trace amounts in your diet. Many multivitamins contain this nonmetal as one of their ingredients. In **Figure 12-22,** you can read the label and find other elements present in vitamins. Selenium can be toxic if there is too much in your system.

Figure 12-22

Selenium is one of several elements needed by the body in trace amounts.

Section Wrap-up

Review

1. Why are Groups 14 and 15 better representatives of mixed groups than are Group 13 or Group 16?

2. How do the allotropes of sulfur differ?

3. **Think Critically:** Why is graphite a lubricant while a diamond is the hardest gem known?

Skill Builder
Concept Mapping
Make a concept map for allotropes of carbon, using the terms *graphite, diamond, buckminsterfullerene, sphere, hexagon,* and *tetrahedron*. If you need help, refer to Concept Mapping in the **Skill Handbook.**

Science Journal

In your Science Journal, write a paragraph in which you discuss the term *valuable* when applied to an element. Consider rarity, usefulness, and durability.

Section Wrap-up

Review

1. Group 13 contains no elements that are clearly nonmetals. Group 16 contains no elements that are clearly metals.

2. Sulfur can exist as differently shaped crystals or a noncrystalline (amorphous) solid.

3. **Think Critically** In graphite, carbon is arranged in flat sheets of atoms bonded in a hexagonal pattern. Because the sheets can slide over one another, graphite is a good lubricant. A diamond consists of a network of tetrahedrally bonded carbon atoms. Thus the crystal is essentially a giant molecule that results in an extremely hard gem.

Science Journal

"Valuable" depends on several factors. Gold is valuable because it is difficult to find and it has a good resistance to reactions. It is also useful as jewelry. Nitrogen is valuable to life in the production of amino acids. However, nitrogen is plentiful in our atmosphere. Therefore, it is not rare. Student answers will vary—accept all reasonable responses.

✔ Assessment

Performance Have students extend their concept maps to include any uses they know of for the allotropes. Use the Performance Task Assessment List for Concept Map in **PASC,** p. 89. **P**

Skill Builder

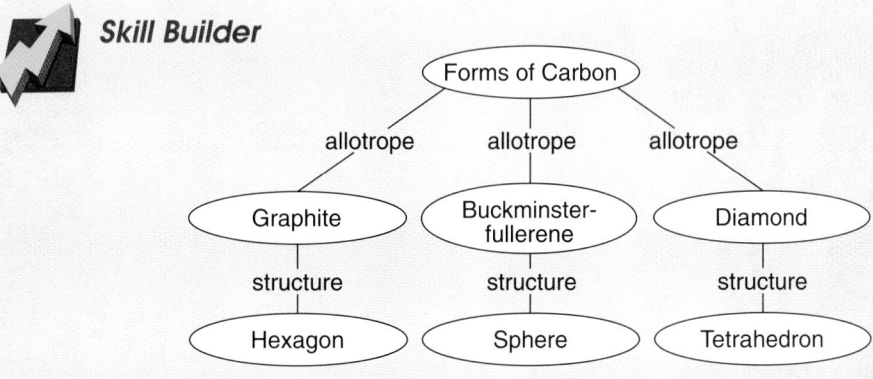

P

Activity 12-2

Activity 12-2

Design Your Own Experiment
Slippery Carbon

Often, a lubricant is needed when two metals touch each other. For example, a sticky lock can sometimes work better with the addition of a small amount of graphite. What gives this allotrope of carbon the slippery property of a lubricant?

PREPARATION

Purpose
IS **Kinesthetic** Students will design and carry out an experiment to show a working model of the layered graphite structure. They will determine the cause and effect relationship between bonding and properties. **L1** **LEP** **COOP LEARN**

Process Skills
making and using models, inferring, communicating, classifying, making and using tables, recognizing cause and effect, forming a hypothesis, designing an experiment, interpreting scientific illustrations

Time
30 minutes

Safety Precautions
Advise students to use care in working with scissors, especially if they use them to make holes in the polystyrene or cardboard.

Possible Hypotheses
If the bonding structure of graphite can be accurately modeled, then its lubricating ability can be demonstrated.

📁 **Activity Worksheets,** pp. 5, 70, 71

PLAN THE EXPERIMENT

Possible Procedures
1. Use scissors to cut hexagon shapes out of the thin polystyrene or cardboard sheets.
2. Carefully punch small holes near each corner of the sheet. Insert a gumdrop into the six holes.
3. Draw lines between the gumdrops on the hexagon to represent bonds between carbon atoms.

PREPARATION

Problem
Why do certain arrangements of atoms in a material cause the material to feel slippery?

Form a Hypothesis
Based on your understanding about how carbon atoms bond, form a hypothesis about the relationship of graphite's molecular structure to its physical properties.

Objectives
- Make a model that will demonstrate the molecular structure of graphite.
- Compare and contrast the strength of types of bonding in graphite.
- Infer the relationship between bonding and physical properties.

Possible Materials
- thin spaghetti
- toothpicks
- small gumdrops

- thin polystyrene sheets
- flat cardboard
- scissors

Safety Precautions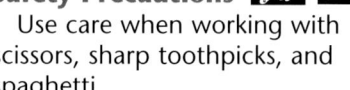
Use care when working with scissors, sharp toothpicks, and spaghetti.

4. Now, use the thin spaghetti to attach hexagons to each other by inserting both ends through the gumdrops.
5. Next, gently push on your connected hexagons to see what happens to your structure.

PLAN THE EXPERIMENT

1. As a group, agree upon a logical hypothesis statement.
2. As a group, sequence and list the steps you need to test your hypothesis. Be specific, describing exactly what you will do at each step to make a model of the types of bonding present in graphite.
3. Remember that graphite consists of rings of six carbons in a flat hexagon. These rings are interconnected. In addition, the flat rings in one layer are weakly attached to other flat layers.

4. Make a list of possible materials you plan to use.

Check the Plan
1. Read over the entire experiment to make sure that all steps are in logical order.
2. Will your model be constructed with materials that show both weak and strong attractions?
3. *Make sure your teacher approves your plan and that you have included any changes suggested.*

DO THE EXPERIMENT

1. Have you selected materials to use in your model that demonstrate both weak and strong attractions?
2. Once your model has been constructed, write down any observations that you make and include a sketch for your Science Journal.

Analyze and Apply
1. **Compare** your model with designs and results of other groups.

2. How does your model illustrate two types of attractions found in the graphite structure?
3. How does the bonding in graphite explain graphite's lubricating properties? Write your answer in your Science Journal.

Go Further

Oils are used on bike chains, engine parts, and other areas where metals touch. Find out how these lubricants work.

12-4 Mixed Groups   353

Teaching Strategies
Troubleshooting Have students compare lines made by pencils of various hardnesses and relate the results to the activity.

DO THE EXPERIMENT

Expected Outcome
- Student structures should show a model that represents six carbon atoms bonded together in flat layers that do not break under pressure. These layers may be represented by a hexagonal piece of polystyrene or cardboard. Two flat layers should be connected by easily broken objects, such as spaghetti, that represent the weak bonds in graphite.
- When pressure is applied to graphite, the weak bonds between layers break. This breaking allows one layer of hexagonal structures to move over another layer, causing a lubricating effect.

Analyze and Apply
1. All should show that some bonds between carbon atoms in graphite are easy to break.
2. C atoms in the same plane, on the flat sheets, are bonded more strongly than the bonds represented by spaghetti between the sheets of cardboard.
3. The weaker bonds between the flat hexagons allow flakes of graphite to slide over each other.

Go Further

Oils provide a smooth contact between metals. This smooth contact reduces friction between surfaces.

353

Sources

- Cosentino, Peter. *The Encyclopedia of Pottery Techniques.* Running Press Book Publishers: Philadelphia, 1990.
- Chavarria, Joaquim. *The Big Book of Ceramics.* Watson-Guptill Publications: New York, 1994.
- Camusso, Lorenzo, and Sandro Bortone, eds. *Ceramics of the World—From 4000 B.C. to the Present.* Harry N. Abrams, Inc., Publishers: New York, 1991.

Background

Clays used in pottery can be classified into six major groups. Ball clays are very plastic clays that become sticky when wet and vitrify at about 1300°C. White earthenware clays have less than 1 percent iron oxide and are used in the production of tableware. Refractory clays are very pure, contain essentially no iron, and vitrify at about 1600-1750°C. Bentonite is an oily volcanic clay that increases in volume by 10 to 15 times when it comes in contact with water. Stoneware clays change color after firing and can be gray to tannish-yellow, to brown. Red earthenware clays are very plastic, contain much iron oxide, and are easy to fire.

Teaching Strategies

Kinesthetic Obtain some pottery clay from a craft store and allow students to make simple pieces of art. If a kiln is available, fire pieces. If you don't have access to such an oven, allow the pieces to dry in the sun. L1

Troubleshooting If the clay is dry and cracks upon handling, mix it with some moister clay or add sufficient amounts of water to increase its plasticity. Also, make sure the clay is thoroughly kneaded to rid the material of air bubbles

Science & ART

Pottery and Clay

In this chapter, you found out that the element hydrogen makes up nearly 90 percent of the universe. However, 82 percent of Earth is made up of oxygen, silicon, and aluminum. The clays formed by these three elements have been used by people for nearly 7000 years.

Pottery clay is approximately 40 percent aluminum oxide, 46 percent silicon oxide, and about 14 percent water. The presence or absence of impurities in clay creates the wide variety of pottery from delicate, porcelain Chinese teacups to Hopi earthenware pots.

Stains and Glazes

Prehistoric potters did little to change or enhance the colors of their works of art. Glazes, or glassy coatings, appeared about 2700-2100 B.C., when Persian and Egyptian potters made pastes of quartz and sand mixed with either copper or manganese. Copper glazes left the ceramics turquoise, and manganese glazes made them violet or purple.

Today, many artists use metal oxides to enhance or even alter the basic color of the clay. Wedgwood jasperware gets its blue color from an oxide of cobalt. Iron oxides, often present in clay, provide the earthy red and brown tones common among Native American pottery.

Firing

Firing is the application of heat to a piece of pottery. In oxidized firing, a kiln, or oven, rich in oxygen allows the combination of metals within the clays and the glazes through combustion. This results in clean, unspotted pottery. Reduction firing—firing with no oxygen present—takes place in a kiln fueled by natural gas, wood, or oil, and leaves a characteristic speckled look.

Different colors can be obtained from the same glazes. Reduction firing of copper oxide glazes will produce a deep, blood-red color. When the same glazes undergo oxidized firing, various shades of green are produced.

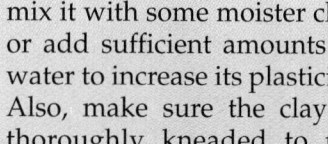

Science Journal

Design a simple piece of pottery and describe in your Science Journal the methods and glazes you would use to produce your finished look.

and to evenly distribute any moisture present in the clay.

- Invite a potter to class to demonstrate using a potter's wheel.
- Review the results of standard flame tests with students. Have them use this information to identify the metals present in a variety of metal oxide glazes. The powdered glazes can be purchased commercially or you might simply show students pictures of different colored pottery and have them identify the glazes used.

Summary

12-1: Metals

1. A typical metal is a hard, shiny solid that—due to metallic bonding—is malleable, ductile, and a good conductor.
2. Groups 1 and 2 are the alkali and alkaline earth metals, which have some similar and some contrasting properties.
3. The iron triad; the coinage metals; and the zinc, cadmium, and mercury group are among the transition elements, which make up Groups 3-12 on the periodic table.

12-2: Science and Society: New Elements, New Properties

1. The lanthanides and actinides have atomic numbers 58-71 and 90-103, respectively, whereas transuranium elements have atomic numbers greater than 92.
2. The making of synthetic elements with no immediate practical use is controversial.

12-3: Nonmetals

1. As a typical nonmetal, hydrogen is a gas that forms compounds by sharing electrons with other nonmetals and also with metals.
2. All the halogens, Group 17, have seven outer electrons and form both covalent and ionic compounds, but each halogen has some properties unlike the others.
3. The noble gases, Group 18, are elements whose properties and uses are related to their chemical stability.

12-4: Mixed Groups

1. Groups 13 through 16 of the periodic table include metals, nonmetals, and metalloids.
2. Allotropes are forms of the same element having different molecular structures.

3. The properties of three forms of carbon—graphite, diamond, and buckminster-fullerene—depend upon the differences in their crystal structures.

Key Science Words

a. allotrope
b. diatomic molecule
c. ductile
d. malleable
e. metallic bonding
f. radioactive element
g. semiconductor
h. sublimation
i. transition element
j. transuranium element

Reviewing Vocabulary

Match each phrase with the correct term from the list of Key Science Words.

1. can be drawn out into a wire
2. after uranium on the periodic table
3. can be hammered into a thin sheet
4. process of changing directly from a solid to a gas
5. will conduct an electric current under certain conditions
6. different structural forms of the same element
7. composed of two atoms
8. breaks down and gives off particles, radiation, and energy
9. in Groups 3-12 on the periodic table
10. Positively charged ions are surrounded by freely moving electrons.

Chapter 12 Review   355

Chapter 12

Review

Summary

Have students read the summary statements to review the major concepts of the chapter.

Reviewing Vocabulary

1. c	6. a
2. j	7. b
3. d	8. f
4. h	9. i
5. g	10. e

✓Assessment

Portfolio Encourage students to place in their portfolios one or two items of what they consider to be their best work. Examples include:

- Flex Your Brain, p. 335
- Activity 12-1 observations and answers, p. 345
- Skill Builder concept map, p. 351 P

Performance Additional performance assessments may be found in **Performance Assessment** and **Science Integration Activities**. Performance Task Assessment Lists and rubrics for evaluating these activities can be found in Glencoe's **Performance Assessment in the Science Classroom.**

GLENCOE TECHNOLOGY

Videodisc

Glencoe Physical Science
Interactive Videodisc
Use the videodisc lesson *Periodicity* to review the principles in this chapter.

MindJogger Videoquiz
Chapter 12 Have students work in groups as they play the Videoquiz game to review key chapter concepts.

Checking Concepts

1. b	**6.** c
2. a	**7.** d
3. d	**8.** d
4. c	**9.** a
5. a	**10.** d

Understanding Concepts

11. francium, because it is at the bottom of Group 1 and at the left of the table

12. oxygen, silicon, aluminum, and iron, respectively

13. Atoms of gaseous elements other than the noble gases have unstable electron structures. These atoms acquire stable structures by sharing electrons in diatomic molecules. The atoms of a noble gas, on the other hand, are stable and thus do not share electrons to form diatomic molecules. Instead, they occur as single atoms.

14. Elements are grouped on the periodic table according to their atomic numbers and electron arrangements. Hydrogen is placed first because its atomic number is 1. It is placed above the alkali metals because, like those elements, it has one electron in its outer energy level. Note, however, that hydrogen's properties differ from those of Group 1 metals in several significant ways.

15. Zinc, cadmium, and nickel are three metals used to coat other metals to form a protective layer that resists corrosion.

Thinking Critically

16. Mercury remains a liquid throughout the ordinary temperature range and expands as temperature increases. Mercury is poisonous.

Checking Concepts

Choose the word or phrase that completes the sentence or answers the question.

1. When magnesium and fluorine react, what type of bond is formed?
   a. metallic c. covalent
   b. ionic d. diatomic
2. What type of bond is found in a piece of pure gold?
   a. metallic c. covalent
   b. ionic d. diatomic
3. Because electrons move freely in metals, metals are _____.
   a. brittle c. dull
   b. hard d. conductors
4. The _____ make up the most reactive group of all metals.
   a. iron triad c. alkali metals
   b. coinage metals d. alkaline earth metals
5. The most reactive of all nonmetals is

   _____.
   a. fluorine c. hydrogen
   b. uranium d. oxygen
6. The element _____ is always found in nature combined with other elements.
   a. copper c. magnesium
   b. gold d. silver
7. The least magnetic of these metals is

   _____.
   a. cobalt c. nickel
   b. iron d. titanium
8. When neptunium atoms fall apart, they form _____.
   a. ytterbium c. americium
   b. promethium d. plutonium
9. An example of a radioactive element is

   _____.
   a. astatine c. chlorine
   b. bromine d. fluorine
10. The only group that is completely nonmetallic is Group _____.
    a. 1 c. 17
    b. 2 d. 18

Understanding Concepts

Answer the following questions in your Science Journal using complete sentences.

11. Reading from top to bottom on the periodic table, metallic properties of elements increase; reading from left to right, metallic properties decrease. Which element is the most metallic of all? Explain your answer.
12. The most abundant elements in Earth's crust are a nonmetal, a metalloid, and two metals. List the four elements.
13. Why do oxygen and nitrogen occur in the air as diatomic molecules, but argon, neon, krypton, and xenon occur as single atoms?
14. Explain why hydrogen, a nonmetal, is on the metal side of the periodic table.
15. Name three metals used to coat other metals. Why is one metal used to coat another?

Thinking Critically

16. Why was mercury used in clinical thermometers, and why is it no longer used for that purpose?
17. The density of hydrogen is so low that it can be used to fill balloons to make them lighter than air. Why is helium used more frequently?
18. Why is aluminum used instead of steel in building airplanes?
19. Why are silver compounds used in photography? Name two nonmetals extracted from seawater that are also part of these compounds.
20. Like selenium, chromium is poisonous but is needed in trace amounts in your diet. How does this information apply to the safe use of vitamin-mineral pills?

---

17. Hydrogen burns readily in air. Helium is a noble gas and therefore is stable.
18. Aluminum is strong but lightweight. Therefore, it is more fuel efficient than is steel.
19. Some silver compounds change chemically when exposed to light. Bromine and iodine, used to make silver bromide and silver iodide, are obtained from brine.

20. Do not take more than the recommended dosage. Exceeding the trace amounts needed may produce toxic reactions.

Developing Skills

21. **Making and Using Tables** lanthanides: europium, cerium; actinides: californium, nobelium

Developing Skills

If you need help, refer to the **Skill Handbook.**

21. **Making and Using Tables:** Use the periodic table to classify each of the following as a lanthanide or actinide: californium, europium, cerium, nobelium.

22. **Comparing and Contrasting:** Aluminum is close to carbon on the periodic table. Explain why aluminum is a metal and carbon is not.

23. **Observing and Inferring:** You are shown two samples of phosphorus. One is white and burns if exposed to air. The other is red and burns if lit. Infer why the properties of two samples of the same element differ.

24. **Recognizing Cause and Effect:** Plants need nitrogen compounds. Nitrogen fixing changes free nitrogen into nitrates. Lightning and legumes are both nitrogen fixing. What are the cause and effect of nitrogen fixing in this example?

25. **Concept Mapping:** Complete the concept map below for some common metals. You may use symbols for the elements.

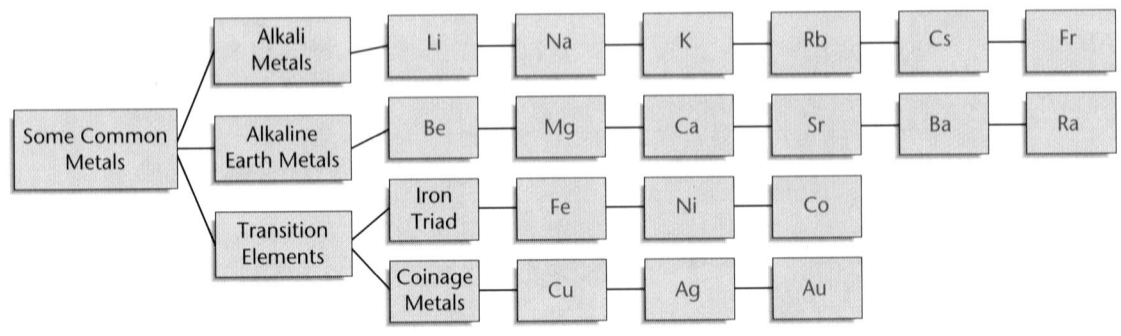

Performance Assessment

1. **Analyzing the Data:** The model you made in Activity 12-2 on pages 352-353 helps explain the lubricating ability of graphite. Use that same model to explain how the graphite in a pencil is able to leave a mark on a piece of paper.

2. **Investigating an Issue Controversy:** Research the pros and cons of using nuclear energy to produce electricity. Prepare a report that includes data as well as your informed opinion on the subject. If several class members do research, conduct a debate on the issue.

3. **Newspaper Article:** Research the source, composition, and properties of asbestos. What properties made it suitable for use in construction in the past? How did these same properties cause asbestos to become a health hazard, and what is being done now to eliminate the hazard? Write a newspaper article that shares your findings with the public.

Assessment Resources

Reproducible Masters
Chapter Review, pp. 27, 28
Assessment, pp. 80-83
Performance Assessment, p. 26

Glencoe Technology
◉ **Chapter Review Software**
◉ **Computer Test Bank**
▭ **MindJogger Videoquiz**

---

22. **Comparing and Contrasting** Aluminum has metallic properties such as being malleable and ductile, and it transfers electrons when it combines with nonmetals. Carbon has none of these properties.

23. **Observing and Inferring** They are allotropes of phosphorus, that is, forms of the element with different properties because they have different structures.

24. **Recognizing Cause and Effect** Cause: lightning and legumes; Effect: changing atmospheric nitrogen into nitrogen compounds

25. **Concept Mapping** See SE annotations.

Performance Assessment

1. A pencil mark is made on paper when the bonds between the hexagonal layers are broken apart and carbon is left on the paper. Use the Performance Task Assessment List for Analyzing the Data in **PASC,** p. 27. [P]

2. Pros could include the lack of air pollution and no contribution to global warming. Cons could include the environmental damage done by mining uranium and the risk of radiation leakage. Use the Performance Task Assessment List for Investigating an Issue Controversy in **PASC,** p. 65. [P]

3. Asbestos is obtained from the chrysotile variety of the serpentine group of minerals by mining. It is nonflammable, but has been found to be dangerous because of fiber inhalation. Use the Performance Task Assessment List for Newspaper Article in **PASC,** p. 69. [P]

357

Chapter Organizer

Section	Objectives/Standards	Activities/Features
Chapter Opener		**Explore Activity:** What element forms from the breakdown of bread and paper? p. 359
13-1 Simple Organic Compounds (2 sessions, 1 block)*	1. **Describe** structures of organic compounds and explain why carbon forms so many compounds. 2. **Distinguish** between saturated and unsaturated hydrocarbons. 3. **Identify** isomers of organic compounds. **National Content Standards:** (5-8) UCP1-UCP3, UCP5, A2; (9-12) UCP1-UCP3, UCP5, A2, B2	**Using Math,** p. 361 **MiniLAB:** How many structures can be shown for octane? p. 363 **Connect to Earth Science,** p. 362 **Problem Solving:** The Case of the Overripe Tomatoes, p. 364 **Using Math,** p. 365 **Skill Builder:** Making and Using Graphs, p. 365
13-2 Other Organic Compounds (2 sessions, 1/2 block)*	1. **Describe** characteristics of aromatic compounds. 2. **Classify** groups of organic compounds as substituted hydrocarbons. **National Content Standards:** (5-8) UCP1, UCP2, UCP5, C1; (9-12) UCP1, UCP2, UCP5, B2	**MiniLAB:** Are esters aromatic compounds? p. 367 **Science Journal,** p. 369 **Skill Builder:** Interpreting Scientific Illustrations, p. 369
13-3 Science and Society (1 session, 1/2 block)*	1. **Describe** the role of biomass and biogas in increasing our energy supply. 2. **Analyze** the positive and negative aspects of raising crops to be harvested for their energy value. **National Content Standards:** (5-8) UCP2, UCP5, B3, F1-F5; (9-12) UCP2, UCP5, F1-F6	**Explore the Issue,** p. 371 **Activity 13-1:** Alcohols and Organic Acids, p. 372
13-4 Biological Compounds (2 sessions, 1/2 block)*	1. **Describe** the formation of polymers and discuss their importance as biological compounds. 2. **Compare** and **contrast** proteins, nucleic acids, carbohydrates, and lipids. **National Content Standards:** (5-8) UCP1-UCP3, UCP5, A1, A2, C1-C5, E1, E2, F5, G1; (9-12) UCP1-UCP3, UCP5, A1, A2, B2, C1, C2, C4, C5, E1, E2, F6, G1	**Using Technology:** DNA Fingerprinting, p. 377 **Connect to Life Science,** p. 378 **Using Math,** p. 379 **Skill Builder:** Comparing and Contrasting, p. 379 **Activity 13-2:** Mixing Many Monomers, pp. 380-381 **People and Science:** Norma Martinez, Food Scientist, p. 382

* A complete Planning Guide that includes block scheduling is provided on pages 32T-35T.

Activity Materials

Explore	Activities	MiniLABs
page 359 small piece of bread, 2 test tubes, laboratory burner, test-tube tongs, heat-proof glove, small piece of paper, safety goggles	**page 372** test tube and stopper, 0.01M potassium permanganate solution, 6M sodium hydroxide solution, ethanol, safety goggles, apron, graduated cylinder, dropper **pages 380-381** note cards of 2 different colors, paper clips, fine-line markers	**page 363** soft gumdrops, raisins, toothpicks **page 367** hot plate, beaker, water, thermometer, medium test tube, salicylic acid, methyl alcohol, dropper, concentrated sulfuric acid, safety goggles, watch or clock, heat-proof glove

Need Materials? Call Science Kit (1-800-828-7777).

Teacher Classroom Resources

Reproducible Masters	Transparencies	Teaching Resources
Study Guide, p. 54 Reinforcement, p. 54 Enrichment, p. 54 Activity Worksheets, p. 78 Lab Manual 28, The Breakdown of Starch	Section Focus Transparency 50, Tastes Good, But Is It Organic? Teaching Transparency 25, Hydrocarbon Structure	Glencoe Physical Science Interactive Videodisc Physical Science CD-ROM Spanish Resources English/Spanish Audiocassettes Cooperative Learning Resource Guide Lab Partner Lab and Safety Skills Lesson Plans
Study Guide, p. 55 Reinforcement, p. 55 Enrichment, p. 55 Activity Worksheets, p. 79 Technology Integration, pp. 13-14	Section Focus Transparency 51, All-Weather Compounds	**Assessment Resources** Chapter Review, pp. 29-30 Assessment, pp. 84-87 Performance Assessment in the Science Classroom (PASC) MindJogger Videoquiz
Study Guide, p. 56 Reinforcement, p. 56 Enrichment, p. 56 Science Integration Activity 13, Old King Coal Critical Thinking/Problem Solving, p. 19	Section Focus Transparency 52, Help from Kelp	Alternate Assessment in the Science Classroom Performance Assessment Chapter Review Software Computer Test Bank
Study Guide, p. 57; Reinforcement, p. 57 Enrichment, p. 57; Activity Worksheets, pp. 74-75, 76-77 Lab Manual 29, Testing for a Vitamin Multicultural Connections, p. 29 Concept Mapping, pp. 31-32 Cross-Curricular Integration, p. 19 Science and Society Integration, p. 17	Section Focus Transparency 53, Organic Compounds on the Trail Science Integration Transparency 13, Marching Ants and Pheromones Teaching Transparency 26, Biological Polymers	

GLENCOE TECHNOLOGY

The following multimedia resources are available from Glencoe.

Science and Technology Videodisc Series (STVS)
Chemistry
  Oil from Wood
Plants & Simple Organisms
  Plant Chemical Repels Cockroaches
Animals
  Fire Ants

Ecology
  Pheromone Rope
  Energy-Integrated Farm

The Infinite Voyage Series
Insects: The Ruling Class
The Geometry of Life
The Champion Within
A Taste of Health

National Geographic Society Series
GTV: Planetary Manager

Glencoe Physical Science Interactive Videodisc
Carbohydrates and Hydrocarbons

Physical Science CD-ROM

The Secret of Life
What's in Stetter's Pond: The Basics of Life

Key to Teaching Strategies

The following designations will help you decide which activities are appropriate for your students.

L1 Level 1 activities should be within the ability range of all students, including those with learning difficulties.

L2 Level 2 activities should be within the ability range of the average to above-average student.

L3 Level 3 activities are designed for the ability range of above-average students.

LEP LEP activities should be within the ability range of Limited English Proficiency students.

LS These activities are designed to address different learning styles.

COOP LEARN Cooperative Learning activities are designed for small group work.

P These strategies represent student products that can be placed into a best-work portfolio.

Teacher Classroom Resources

This is a representation of key blackline masters available in the Teacher Classroom Resources.

Teaching Aids

Section Focus Transparencies

TASTES GOOD, BUT IS IT ORGANIC?

L1

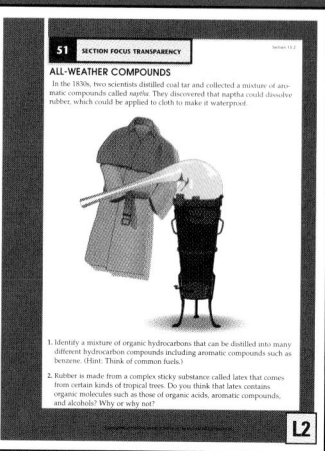

ALL-WEATHER COMPOUNDS

L2

HELP FROM KELP

L3

Science Integration Transparencies

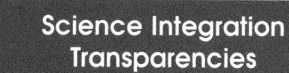
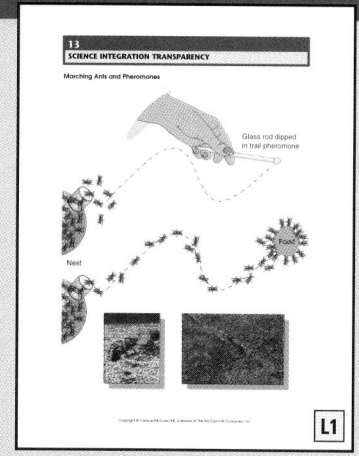

L1

Teaching Transparencies

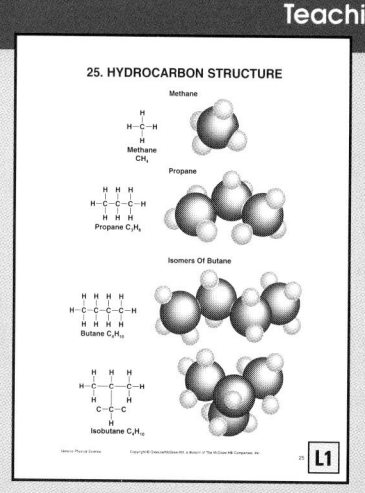

25. HYDROCARBON STRUCTURE

L1

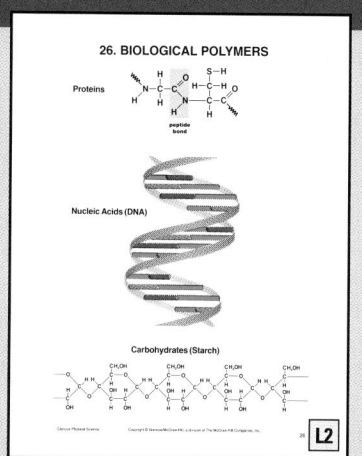

26. BIOLOGICAL POLYMERS

L2

Meeting Different Ability Levels

Study Guide

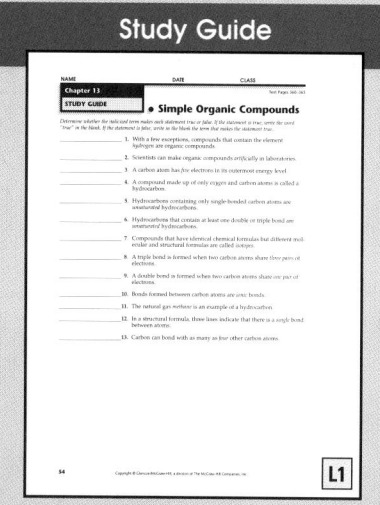

L1

Reinforcement

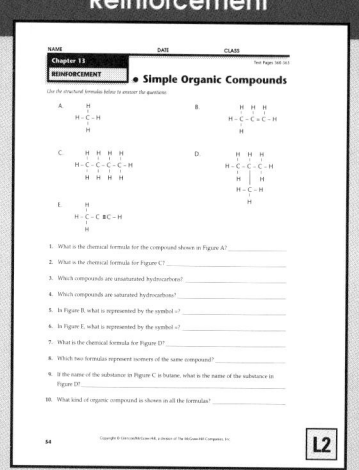

L2

Enrichment Worksheets

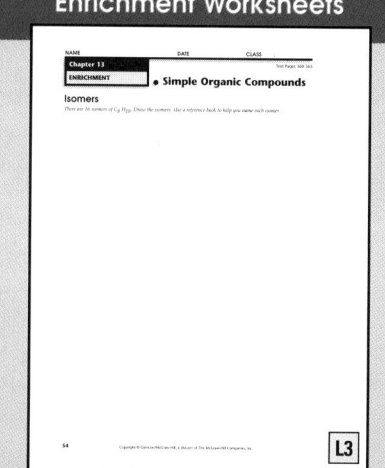

L3

Chapter 13 Organic and Biological Compounds

Hands-On Activities

Science Integration Activity

Lab Manual

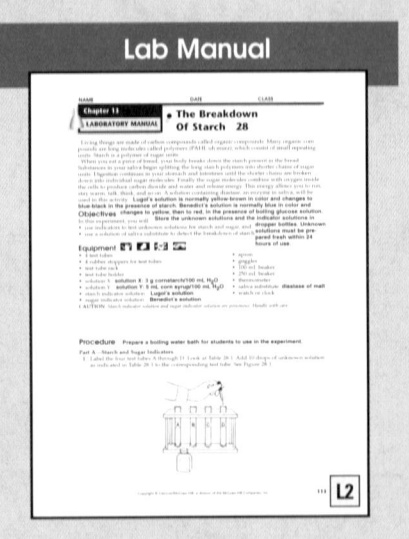

Assessment

Performance Assessment

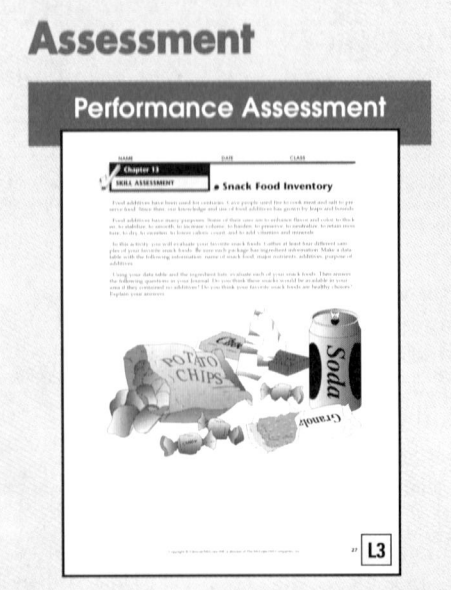

Enrichment and Application

Critical Thinking/ Problem Solving

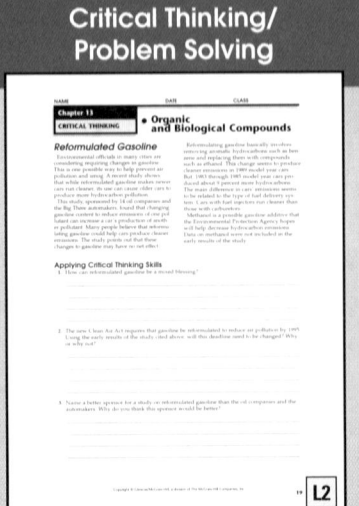

Cross-Curricular Integration

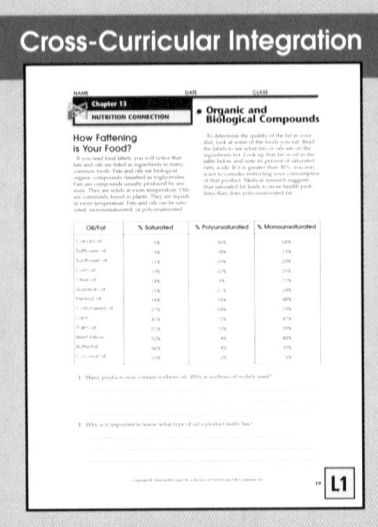

Science and Society Integration

Technology Integration

Multicultural Connections

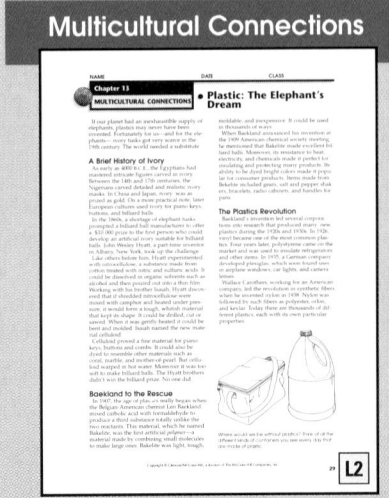

Concept Mapping

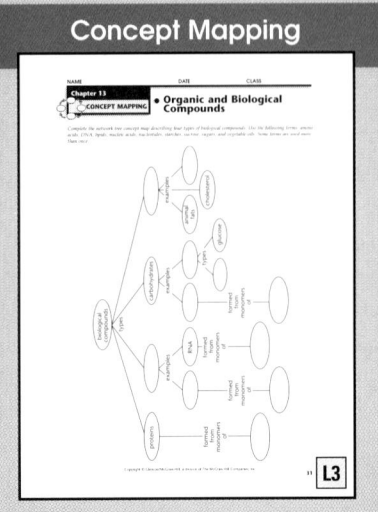

Organic and Biological Compounds

CHAPTER OVERVIEW

Section 13-1 The chemistry of carbon as it forms different molecular shapes is presented in this section. The student is introduced to saturated and unsaturated hydrocarbons as well as isomers.

Section 13-2 Aromatic compounds and substituted hydrocarbons are described in this section.

Section 13-3 Science and Society Biomass conversion as an alternate energy source is described in detail.

Section 13-4 This section introduces biochemistry and describes the structure of the main classes of biochemicals: proteins, fats, and carbohydrates. The concept of polymerization is presented to unify the section.

Chapter Vocabulary

organic compound	biomass
hydrocarbon	biogas
saturated hydrocarbon	energy farming
isomer	gasohol
unsaturated hydrocarbon	polymer
aromatic compound	protein
substituted hydrocarbon	nucleic acid
alcohol	carbohydrate
	lipid

Theme Connection

Scale and Structure Scale and structure as a theme is developed through a presentation of organic compounds and related biopolymers. The relationships that exist between the structure of organic molecules and their properties are described and developed.

Previewing the Chapter

Section 13-1 Simple Organic Compounds
▶ Organic Compounds
▶ Hydrocarbons
▶ Single Bonds
▶ Multiple Bonds

Section 13-2 Other Organic Compounds
▶ Aromatic Compounds
▶ Substituted Hydrocarbons

Section 13-3 Science and Society Issue: Growing Energy on the Farm

Section 13-4 Biological Compounds
▶ Proteins
▶ Nucleic Acids
▶ Carbohydrates
▶ Lipids

358

Learning Styles Look for the following logo for strategies that emphasize different learning modalities. **LS**

Kinesthetic	Across the Curriculum, p. 367; Reteach, p. 368; Activity, p. 375; Demonstration, p. 376; Activity 13-2, pp. 380-381
Visual-Spatial	Visual Learning, pp. 361, 362, 374, 377; MiniLAB, p. 363; Demonstration, pp. 364, 374, 376
Interpersonal	Explore, p. 359; MiniLAB, p. 367
Logical-Mathematical	Reteach, pp. 364, 379; Activity 13-1, p. 372
Linguistic	Across the Curriculum, pp. 371, 375; Science Journal, p. 374

Chapter 13

Organic and Biological Compounds

Many of the objects in the photo have one main element in common. The food, the plastic bags and containers, the wool sweater, and even the person contain many molecules based on the element carbon. Why does carbon form so many compounds? Why are the compounds so different? In this chapter, find out about this unique and important element.

EXPLORE ACTIVITY

What element forms from the breakdown of bread and paper?

1. Place a small piece of bread in a test tube.
2. Using a test-tube holder, hold the tube over the flame of a laboratory burner until you observe changes in the bread.
3. Using a clean test tube and a small amount of paper instead of bread, repeat the procedure.

Observe: In your Science Journal, compare the changes that occurred in the two test tubes.

Previewing Science Skills

▶ In the Skill Builders, you will make and use a graph, interpret scientific illustrations, and compare and contrast.

▶ In the Activities, you will observe, predict, and make a model.

▶ In the MiniLABs, you will make a model, observe, predict, and compare and contrast.

359

Prepare

Section Background

- Organic compounds can be classified as aromatic and aliphatic. Aromatic compounds contain one or more benzene rings. All other hydrocarbons are classified as aliphatic.
- An atom or group of atoms that is attached to a main chain is called a branch or substituent.
- A hydrocarbon in which all carbon-carbon bonds are single bonds is called an alkane. The general formula for alkanes is C_nH_{2n+2}.
- An alkene contains a double bond. The general formula for an alkene is C_nH_{2n}.
- An alkyne contains a triple bond. The general formula for an alkyne is C_nH_{2n-2}.

Preplanning

To prepare for the MiniLAB on page 363, obtain soft gumdrops, raisins, and toothpicks.

1 Motivate

Bellringer

Before presenting the lesson, display **Section Focus Transparency 50** on the overhead projector. Assign the accompanying **Focus Activity** worksheet. [L1] [LEP]

Tying to Previous Knowledge

There are expensive and inexpensive perfumes. Ask your class what they think the difference is. Some fine perfumes are still made from natural ingredients. Inexpensive perfumes usually use synthetic esters to imitate natural esters (oils).

Science Words

organic compound
hydrocarbon
saturated hydrocarbon
isomers
unsaturated
  hydrocarbon

Objectives

- Describe structures of organic compounds and explain why carbon forms so many compounds.
- Distinguish between saturated and unsaturated hydrocarbons.
- Identify isomers of organic compounds.

Figure 13-1

Many items used every day are made from organic compounds.

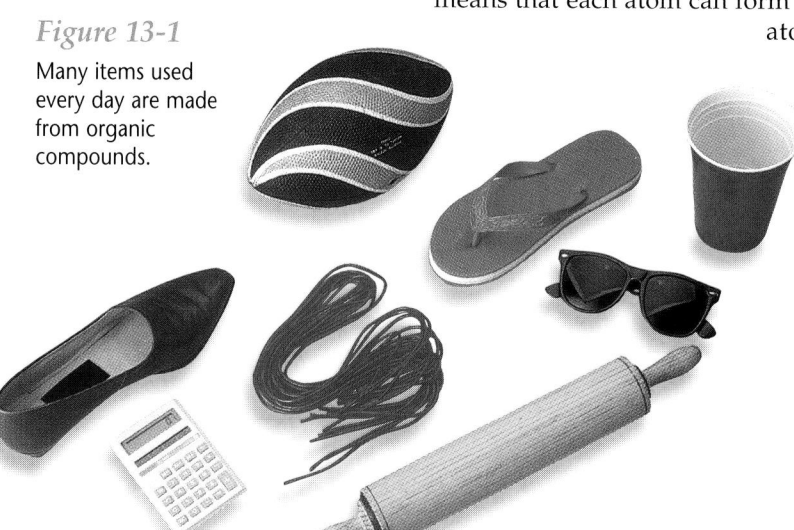

13•1 Simple Organic Compounds

Organic Compounds

What do cassette tapes and CDs, you, and your athletic shoes have in common? Each is made up mostly of substances called organic compounds. Most compounds that contain the element carbon are **organic compounds.** There are several exceptions, such as carbon monoxide, carbon dioxide, and carbonates. Overall, more than 90 percent of all compounds are organic compounds.

You probably recognize that the word *organic* is similar to the word *organism*. At one time, scientists assumed that only living organisms could produce organic compounds. However, by 1830, scientists were making organic compounds artificially in laboratories. But scientists didn't change the name of these carbon-containing compounds. Today, most of the millions of different organic compounds that exist are synthetic, and the manufacturing of organic compounds is one of the world's largest industries.

Bonding—Carbon Connections

You may wonder why there are so many organic compounds. There are several reasons, and they are related to the element carbon. First, a carbon atom has four electrons in its outer energy level. Recall that this electron arrangement means that each atom can form four covalent bonds with atoms of carbon or with other elements. Four is a large number of bonds compared to the number of bonds that atoms of other elements can form. **Figure 13-1** shows the results of some of the ways carbon can bond in an organic compound.

Program Resources

📁 **Reproducible Masters**
Study Guide, p. 54 [L1]
Reinforcement, p. 54 [L1]
Enrichment, p. 54 [L3]
Activity Worksheets, p. 78 [L1]
Lab Manual 28

🔦 **Transparencies**
Section Focus Transparency 50
Teaching Transparency 25

Figure 13-2

Carbon atoms bond to form chains, branched chains, or closed rings.

A Heptane, found in gasoline mixtures, is an example of carbon atoms bonding in a chain. Each dashed line represents a single covalent bond.

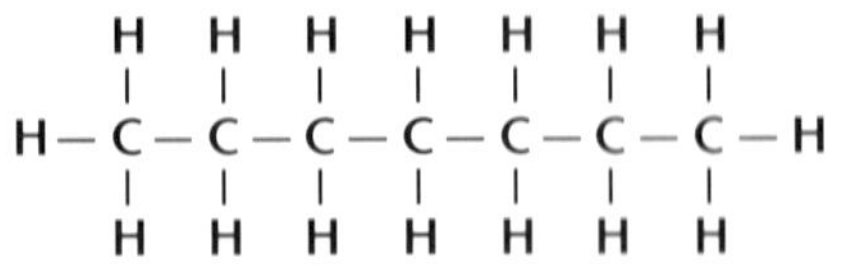

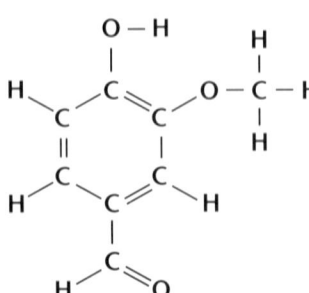

B Isoprene, a molecule that can connect to itself to form natural rubber, has one branch in its carbon chain. In addition, some of the carbon atoms have formed double bonds, shown by the double lines. *How many total bonds do you notice for each carbon atom?*

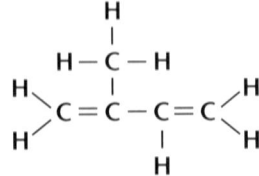

C Vanillin, the compound responsible for the flavor of vanilla beans, shows carbon atoms bonded in a closed ring. Notice that atoms of other elements may be attached to the ring.

Second, many different arrangements of single, double, and triple bonds can exist between carbon atoms. Each arrangement forms a molecule of a different organic compound. A carbon atom can also bond with atoms of many other elements, such as hydrogen and oxygen, as shown in **Figure 13-2**.

Hydrocarbons

Carbon forms an enormous number of compounds with hydrogen alone. A compound made up of only carbon and hydrogen atoms is called a **hydrocarbon.** You can learn a lot about other organic compounds by studying hydrocarbons. Does the furnace, stove, or water heater in your home burn natural gas? Almost all the natural gas used for these purposes is the hydrocarbon methane. The chemical formula of methane is CH_4. There are other ways to represent methane. The structural formula in **Figure 13-3** shows that four hydrogen atoms are bonded to one carbon atom in a methane mol-

Figure 13-3

Natural gas is mostly methane, CH_4.

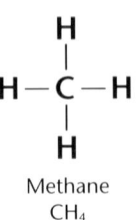

Methane
CH_4

GRAPHING CALCULATOR

The melting point of methane (CH_4) is −183°C, of ethane (C_2H_6) is −172°C, of propane (C_3H_8) is −188°C, and of butane (C_4H_{10}) is −138°C. Use these data to make a scatter plot in which the number of carbon atoms is the independent variable and the melting point is the dependent variable. Use the graph to predict the melting point of octane (C_8H_{18}).

13-1 Simple Organic Compounds **361**

Visual Learning

Figure 13-2 How many total bonds do you notice for each carbon atom? *four* LEP IS

Table 13-1

Some Hydrocarbons		
Name	Chemical Formula	Structural Formula
Methane	CH_4	
Ethane	C_2H_6	
Propane	C_3H_8	
Butane	C_4H_{10}	
Pentane	C_5H_{12}	

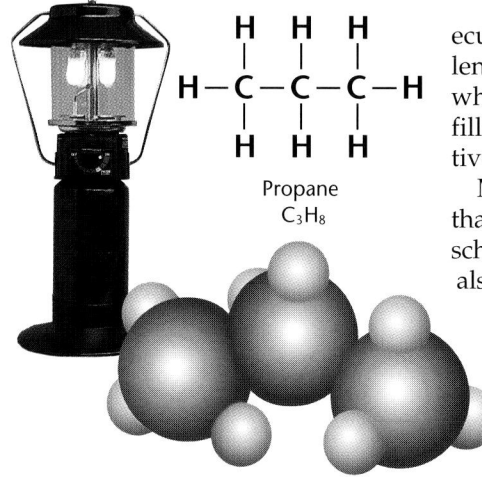

Propane
C_3H_8

Figure 13-4

Bottled gas is mostly propane, C_3H_8.

ecule. Each line between atoms represents a single covalent bond. As you recall, a covalent bond is formed when two atoms share a pair of electrons. The space-filling model in **Figure 13-3** on page 361 shows the relative volumes of the electron clouds in the molecule.

Methane and other hydrocarbons account for more than 90 percent of the energy sources used in homes, schools, industry, and transportation. Hydrocarbons are also important in manufacturing almost all the organic compounds used in products ranging from fertilizers to skateboards.

Some stoves, most outdoor grills, and hot-air balloons burn the bottled gas propane, another hydrocarbon. Its chemical formula is C_3H_8, as shown in **Figure 13-4.**

Single Bonds

In some hydrocarbons, the carbon atoms are joined by single covalent bonds. Hydrocarbons containing only single-bonded carbon atoms are called **saturated hydrocarbons.** The carbon atoms in a molecule of propane are bonded by single covalent bonds. Propane is a saturated hydrocarbon. As you can see, the carbon atoms in propane seem to form a short chain. Propane is a member of a group of saturated hydrocarbons in which the carbon atoms form chains. **Table 13-1** lists

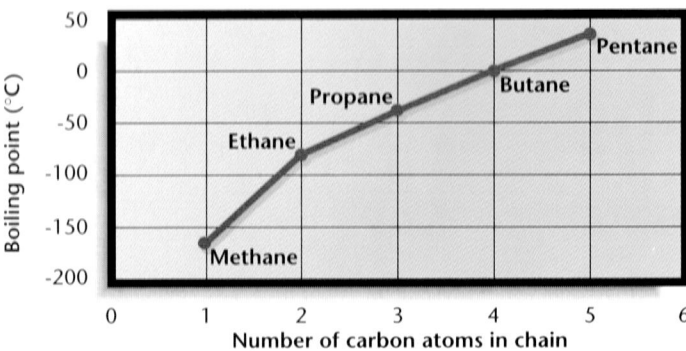

Figure 13-5

Boiling points of hydrocarbons in natural gas are shown here. *What happens to the boiling point as the number of carbon atoms in the chain increases?*

methane, propane, and three other similar hydrocarbons. Notice how each carbon atom appears to be a link in the chain within the molecule. **Figure 13-5** shows a graph of the boiling points of these hydrocarbons.

Isomers

Perhaps you have seen or know about butane, which is a gas sometimes burned in camping stoves and lighters. The chemical formula of butane is C_4H_{10}. Another hydrocarbon also has the same chemical formula as butane. How can this be? The answer lies in the arrangement of the four carbon atoms. Look at **Figure 13-6**. In a molecule of butane, the carbon atoms form

Table 13-2

Properties of Butane Isomers		
	Butane	**Isobutane**
Description	Colorless gas	Colorless gas
Density	0.60 kg/L	0.62 kg/L
Melting point	–138°C	–160°C
Boiling point	–0.5°C	–12°C

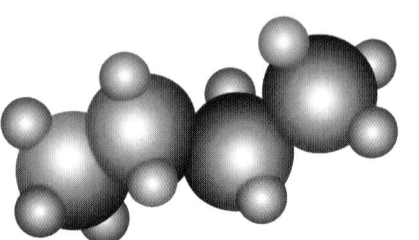

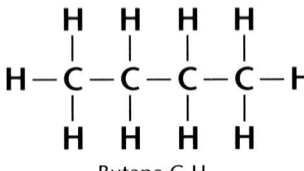

Butane C_4H_{10}

MiniLAB

How many structures can be shown for octane?

Octane, C_8H_{18}, is a hydrocarbon used in fuels. Remembering that each carbon atom can form four bonds and each hydrogen one, make models of the possible structures of octane.

Procedure
1. Use soft gumdrops to represent carbon atoms.
2. Use raisins to represent hydrogen atoms.
3. Use toothpicks for chemical bonds.

Analysis
1. How do you distinguish one structure from another?
2. Do you think the different shapes cause different properties?
3. What was the total of different molecules found in your class?

Figure 13-6

Two isomers of butane are shown.

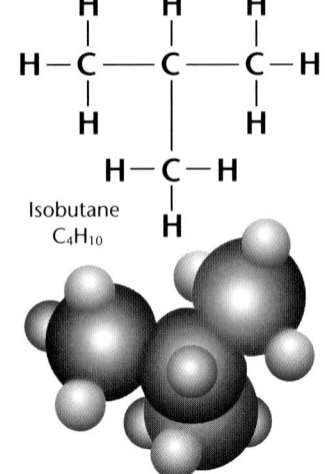

Isobutane
C_4H_{10}

MiniLAB

Purpose

[IS] **Visual-Spatial** Students will model possible structures of the hydrocarbon octane. [L1] [LEP]

Materials

gumdrops of one color, raisins, toothpicks

Teaching Strategies

Troubleshooting Start by having all teams construct models of the linear, unbranched molecule. In this way, everyone will have all carbons and hydrogens in correct relative positions.

• This MiniLAB allows students to discover structural possibilities in hydrocarbons. It could be done as a review after the upcoming discussion about isomers.

📁 **Activity Worksheets,** pp. 5, 78

Analysis

1. by the position and arrangement of the carbon atoms
2. Students may say no because the number of atoms remains the same. However, physical properties such as boiling and freezing points may be different.
3. There can be a maximum of 18 different structures. See structures at the bottom of the page.

✓ Assessment

Content Monitor the structure comparisons so that students do not consider identical structures to be different because they are oriented differently. Use the Performance Task Assessment List for Model in PASC, p. 51. [P]

MiniLAB Results

Octane has 18 isomers. Shown here are the carbon chains for the isomers. Each student model should match one of these structures.

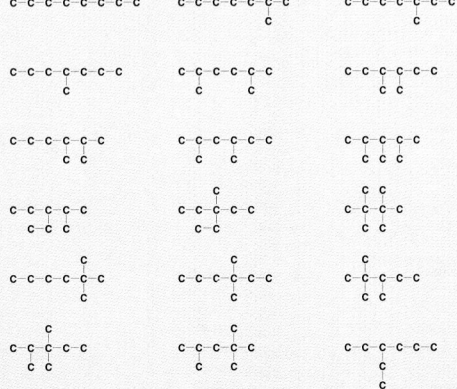

Problem Solving

Think Critically

The avocado was already overripe, so the ripening process continued until it rotted. Ethylene gas given off by the overripe avocado was trapped in the bag, causing the tomatoes to ripen more rapidly than expected.

Demonstration

LS **Visual-Spatial** Bromine bonds to hydrocarbons having double bonds. You can show the presence and reactivity of double bonds by adding a few milliliters of bromine water to cyclohexene in a test tube. Stopper and shake. Add bromine water to cyclohexane as a comparison. **CAUTION:** *Hydrocarbons are flammable and toxic. Bromine is a severe skin irritant.*

3 Assess

Check for Understanding

? FLEX Your Brain

Use the Flex Your Brain activity to have students explore ISOMERS.

📁 **Activity Worksheets,** p. 5

Reteach

LS **Logical-Mathematical** On a handout, draw structural formulas for several hydrocarbons. Have students write the molecular formula for each and decide if it is a saturated or unsaturated hydrocarbon.

Extension

📁 For students who have mastered this section, use the **Reinforcement** and **Enrichment** masters.

364

Problem Solving

The Case of the Overripe Tomatoes

Marc helped out at home by preparing dinner on Mondays and Thursdays. He was also responsible for buying any groceries needed to prepare the meals. One Monday, Marc bought an overripe avocado to use for guacamole that evening and some green tomatoes to use for a salad on Thursday. He thought the tomatoes would be juicy but still firm by Thursday evening. The grocery clerk put the avocado and tomatoes in a brown paper bag.

When Marc arrived at home, his father decided to take the family out to dinner, so he didn't have to prepare dinner that evening.

On Thursday, Marc opened the paper bag. The avocado was rotten, and the tomatoes were overripe—not firm as he had expected. Marc knew that fruits give off ethylene gas as they ripen. But the tomatoes ripened more quickly than Marc expected.

Think Critically:

Why did the avocado rot? What caused the tomatoes to ripen more rapidly than expected?

364

Inclusion Strategies

Learning Disabled Ask students to make models of organic compounds. They can paint polystyrene foam balls various colors for the specific elements (carbon—black, hydrogen—blue, and so forth). Have them combine the atoms to make models of the molecules shown in this chapter. This will afford opportunities for visual, tactile, and kinesthetic learning.

a continuous chain. The carbon chain of isobutane is branched. The arrangement of carbon atoms in each compound changes the shape of the molecule. Isobutane and butane are isomers.

Isomers are compounds that have identical chemical formulas but different molecular structures and shapes. There are thousands of isomers among the hydrocarbons. **Table 13-2** on page 363 summarizes several properties of the two isomers of butane.

Multiple Bonds

Do you like bananas? They are among the many fruits and vegetables that ripen with the help of ethylene. Ethylene is the common name of the hydrocarbon ethene, C_2H_4. Did you ever watch a welder using an acetylene torch? The common name of the hydrocarbon ethyne, C_2H_2, which is fuel for this torch, is acetylene. Compare the structural formulas for ethene and ethyne, which are shown in **Figure 13-7.**

Hydrocarbons that contain at least one double or triple bond between carbon atoms, such as ethene and ethyne, are called **unsaturated hydrocarbons**. Remember that *saturated* hydrocarbons contain only single bonds between carbon atoms. Fats and oils can also be classified as unsaturated or saturated because parts of their molecules are similar to saturated and unsaturated hydrocarbon molecules.

Figure 13-7

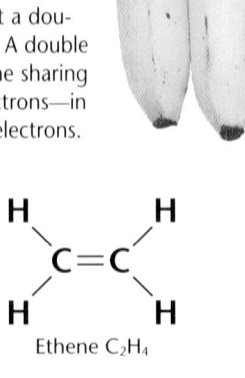

Hydrocarbons may contain double or triple bonds between carbon atoms.

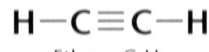

Ethyne C₂H₂

A Ethene gas ripens fruit. The double lines between the carbon atoms in ethene represent a double covalent bond. A double covalent bond is the sharing of two pairs of electrons—in other words, four electrons.

B Ethyne, also called acetylene is used in welding. The three lines in ethyne represent a triple covalent bond.

Ethene C₂H₄

Section Wrap-up

Review

1. Why can carbon form so many organic compounds?

2. Compare and contrast ethane, ethene, and ethyne.

3. How is an unsaturated hydrocarbon different from a saturated hydrocarbon?

4. **Think Critically:** Cyclopropane is a saturated hydrocarbon containing three carbon atoms. In this compound, each carbon atom is bonded to two other carbon atoms. Draw its structural formula. Are cyclopropane and propane isomers? Explain.

Skill Builder
Making and Using Graphs
Make a graph of **Table 13-1.** For each compound, plot the number of carbon atoms on one axis and the number of hydrogen atoms on the other axis. Use the graph to predict the formula of hexane, which has six carbon atoms. If you need help, refer to Making and Using Graphs in the **Skill Handbook.**

USING MATH

Compare the formulas of three saturated hydrocarbons: C_2H_6, C_3H_8, and C_4H_{10}. What relationship do you see between the number of carbon atoms and the number of hydrogen atoms in each? Compare the number of carbon and the number of hydrogen atoms in the unsaturated hydrocarbons C_2H_4, C_3H_6, and C_4H_8.

Skill Builder

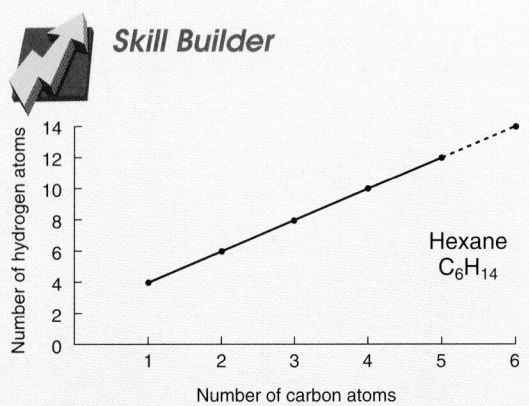

Number of hydrogen atoms (y-axis: 0, 2, 4, 6, 8, 10, 12, 14)
Number of carbon atoms (x-axis: 1, 2, 3, 4, 5, 6)

Hexane
C₆H₁₄

✓ Assessment

Performance Have students use their graphs to predict the formula for decane, which has ten carbon atoms. Use the Performance Task Assessment List for Asking Questions in **PASC,** p. 19. **P**

4 Close

•MINI•QUIZ•

Use the Mini Quiz to check students' recall of chapter content.

1. **Organic compounds contain the element _____ .** *carbon*

2. **How many bonds do carbon atoms form?** *four*

3. **What do you call a compound that is made of only carbon and hydrogen atoms?** *hydrocarbon*

Section Wrap-up

Review

1. Carbon can form four covalent bonds with other atoms, including other carbon atoms. Thus, carbon atoms can form long, branched chains. Also, the bonds can be single, double, or triple.

2. The three compounds are hydrocarbons, each containing two carbon atoms. Ethane is a saturated hydrocarbon; ethene contains a double covalent bond, and ethyne contains a triple covalent bond.

3. A saturated hydrocarbon has only single bonds between carbon atoms. An unsaturated hydrocarbon has one or more double or triple bonds between carbon atoms.

4. **Think Critically** Cyclopropane, C_3H_6, is not an isomer of propane, C_3H_8, because it has a different chemical formula.

USING MATH

Saturated = C_xH_{2x+2}
Unsaturated = C_xH_{2x}

13•2 Other Organic Compounds

Prepare

Section Background

- Aromatic compounds are economically important to many chemical industries including rubber, plastics, fibers, explosives, paint, and petroleum.
- Substituted hydrocarbons are generally more reactive than unsubstituted compounds.
- Alcohols, ethers, aldehydes, ketones, carboxylic acids, and esters are classes of oxygen-containing organic chemicals.

Preplanning

To prepare for the MiniLAB, acquire a hot plate for each group.

1 Motivate

Bellringer

Before presenting the lesson, display **Section Focus Transparency 51** on the overhead projector. Assign the accompanying **Focus Activity** worksheet. L1 LEP

Tying to Previous Knowledge

Have students recall when they have seen explosives such as TNT detonated in movies. Make a model of benzene. Then, in front of the class, remove a hydrogen atom and replace it with a methyl, –CH$_3$, group. Now you have toluene. Replace the two hydrogens on either side of the methyl group with two nitro, –NO$_2$, groups. Remove the hydrogen opposite the methyl and replace it with a third nitro group. You now have a model of trinitrotoluene, TNT.

Science Words

> aromatic compound
> substituted hydrocarbon
> alcohol

Objectives

- Describe characteristics of aromatic compounds.
- Classify groups of organic compounds as substituted hydrocarbons.

Aromatic Compounds

Chewing flavored gum or dissolving a candy mint in your mouth releases pleasant flavors and aromas. Many chemical compounds produce pleasant odors. Other compounds have unpleasant smells. The compound that produces the fragrance of wintergreen is depicted in **Figure 13-8.**

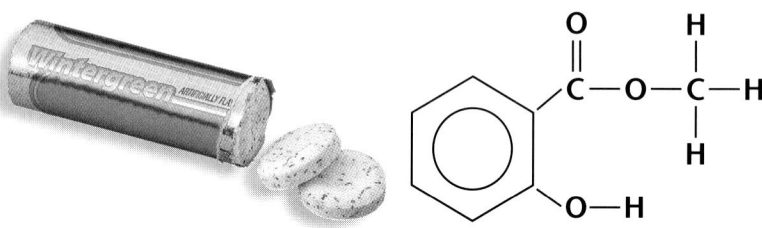

Figure 13-8

Wintergreen is an example of an aromatic compound.

You might guess that aromatic compounds are smelly—and most of them are. An **aromatic compound** is a compound that contains the benzene ring structure. Look at a model of benzene, C$_6$H$_6$, and its structural formula in **Figure 13-9.** As you can see, the benzene molecule has six carbon atoms bonded into a ring. The electrons shown as three double bonds and

Figure 13-9

Benzene, C$_6$H$_6$, can be represented by (A) a space-filling model, (B) a structural formula, or (C) the benzene ring symbol.

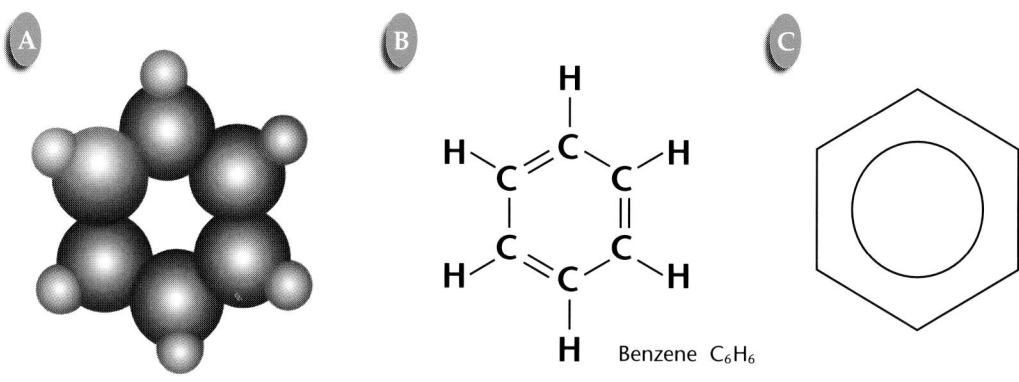

Benzene C$_6$H$_6$

Program Resources

📁 **Reproducible Masters**
Study Guide, p. 55 L1
Reinforcement, p. 55 L1
Enrichment, p. 55 L3
Activity Worksheets, p. 79 L1
Technology Integration, p. 13

📽 **Transparencies**
Section Focus Transparency 51

three single bonds that form the ring are actually shared by all six carbon atoms in the ring. This equal sharing of electrons is represented by the special symbol shown in **Figure 13-9C.** The sharing of these electrons causes the benzene molecule to be very stable. Many compounds contain this stable ring structure. Knowing the structure of benzene can help you to picture hundreds of different aromatic organic compounds.

Can you recognize the odor of moth crystals? One example of an aromatic compound is naphthalene. One type of moth crystal is made of naphthalene, as shown in **Figure 13-10.** Notice that naphthalene is made up of two ring structures.

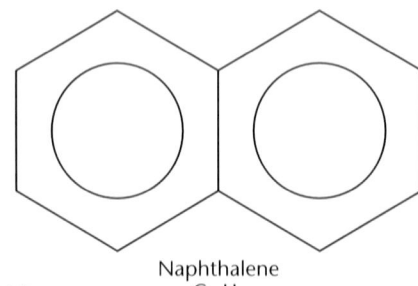

Figure 13-10

Naphthalene
$C_{10}H_8$

Moth crystals are naphthalene.

Substituted Hydrocarbons

Usually, a cheeseburger is a hamburger covered with melted American cheese and served on a bun. However, you can make a cheeseburger with Swiss cheese and serve it on slices of bread. If you ate this cheeseburger, you would notice that the substitutions affect the taste.

Chemists make similar changes to hydrocarbons. These changes produce compounds called substituted hydrocarbons. A **substituted hydrocarbon** has one or more of its hydrogen atoms replaced by atoms of other elements. Examples of substituted hydrocarbons are shown in **Figure 13-11** on page 368.

13-2 Other Organic Compounds   367

Alcohols and Acids

Have you ever used rubbing alcohol to soothe aching muscles after too much exercise? Did you know you were using a substituted hydrocarbon? Alcohols are an important group of organic compounds. An **alcohol** is formed when −OH groups replace one or more hydrogen atoms in a hydrocarbon, as shown in **Figure 13-11B. Table 13-3** lists several alcohols.

Organic acids form when a carboxyl group, −COOH, is substituted for three hydrogen atoms attached to the same carbon atom. Look at **Figure 13-11C.** The structures of ethane, ethanol, and acetic acid are similar. Do you see that acetic acid is a substituted hydrocarbon?

 *Figure 13-11*

Several types of compounds are substituted hydrocarbons.

A A compound used in dry cleaning, tetrachloroethene, C_2Cl_4, is a substituted hydrocarbon. The prefix *tetra-* means "four," and *chloro-* refers to chlorine. C_2Cl_4 is formed when four chlorine atoms replace the hydrogen atoms in ethene, C_2H_4.

Tetrachloroethene
C_2Cl_4

C Recall that vinegar contains acetic acid. Acetic acid is an example of an organic acid. Many fruit juices contain similar organic acids.

Acetic acid CH_3COOH

B Ethanol, C_2H_5OH, is an alcohol produced by the fermentation of sugar in corn and other grains and in many fruits. Alcoholic beverages contain ethanol. As you will learn in the next section, this alcohol is sometimes added to gasoline. Ethanol is also a good cleaning fluid because it can dissolve substances that don't dissolve in water.

Ethanol C_2H_5OH

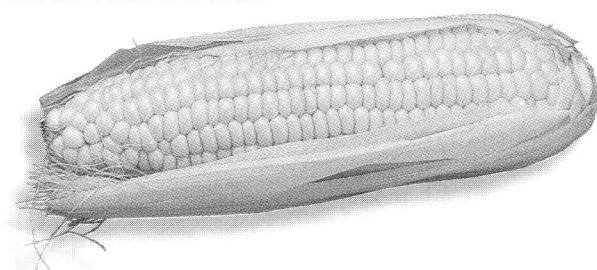

368 Chapter 13 Organic and Biological Compounds

Table 13-3

Common Alcohols

	Methanol	Ethanol	Isopropyl alcohol (rubbing alcohol)	Phenol
Structures	H H–C–OH H	H H H–C–C–OH H H	H OH H H–C–C–C–H H H H	⬡–OH

Uses		Methanol	Ethanol	Isopropyl alcohol	Phenol
Uses	Fuel	✓	✓		
	Cleaner	✓	✓	✓	
	Disinfectant			✓	
	Manufacturing chemicals	✓		✓	✓

You have learned about a wide range of organic compounds and have seen how they affect you individually. In the next section, investigate how organic compounds affect all of society.

Section Wrap-up

Review

1. What do the structures of all aromatic compounds have in common?

2. How is each of the following a substituted hydrocarbon: (a) tetrachloroethene, (b) ethanol, (c) acetic acid?

3. **Think Critically:** Chloroethane, C_2H_5Cl, can be used as a spray-on anesthetic for localized injuries. How does chloroethane fit the definition of a substituted hydrocarbon? Diagram its structure.

Skill Builder
Interpreting Scientific Illustrations
Formic acid, HCOOH, is the simplest organic acid. Draw its structural formula by referring to the structure of acetic acid in **Figure 13-11C.** If you need help, refer to Interpreting Scientific Illustrations in the **Skill Handbook.**

Science Journal

Examine the different ways a benzene molecule can be represented, as shown in **Figure 13-9.** In your Science Journal, write a paragraph that explains how benzene can be represented by a six-sided figure with a circle in it.

Skill Builder

O
‖
H — C — OH

Assessment

Oral Ask students to count and report the number of bonds they made to the carbon atom in the structure. Use the Performance Task Assessment List for Model in **PASC,** p. 51. **P**

4 Close

•MINI•QUIZ•

Use the Mini Quiz to check students' recall of chapter content.

1. **A compound that contains a benzene ring is classified as a(n) _____ compound.** *aromatic*

2. **Why is the benzene molecule stable?** *Electrons are shared by all six carbon atoms.*

3. **Which element has been replaced in a substituted hydrocarbon?** *hydrogen*

Section Wrap-up

Review

1. All aromatic compounds contain the benzene ring.

2. (a) Chlorine atoms have replaced the hydrogen atoms in a hydrocarbon; (b) an –OH group has replaced a hydrogen atom in a hydrocarbon; (c) a –COOH group has replaced a hydrogen atom in a hydrocarbon.

3. **Think Critically** Ethane is C_2H_6. In C_2H_5Cl, a Cl atom has been substituted for an H atom.

H H
| |
H—C—C—Cl
| |
H H

Science Journal

Students should discuss how the electrons in the benzene molecule are shared equally by the six carbon atoms. The circle takes the place of double and single bonds in the diagram.

369

Prepare

Section Background

Synthetic fuels or synfuels are derived from organic sources. Grains, sugars, and wood can produce ethanol and methanol. Tar sands, oil shale, and coal can be used to produce synthetic fuels.

Preplanning

For Activity 13-1, prepare a $0.01M$ solution of potassium permanganate, and a $6.0M$ solution of sodium hydroxide. Allow the base solution to cool before using.

1 Motivate

Bellringer

 Before presenting the lesson, display **Section Focus Transparency 52** on the overhead projector. Assign the accompanying **Focus Activity** worksheet. L1 LEP

2 Teach

GLENCOE TECHNOLOGY

 **Videodisc**
STVS: Ecology
Disc 6, Side 2
Energy-Integrated Farm
(Ch. 20)

Science Words

- biomass
- biogas
- energy farming
- gasohol

Objectives

- Describe the role of biomass and biogas in increasing our energy supply.
- Analyze the positive and negative aspects of raising crops to be harvested for their energy value.

Energy Production: A Global Problem

Transportation, heating, cooling, and making electricity all require energy. In most cases in the United States, fossil fuels—which consist of natural gas, coal, or petroleum—are the primary sources of energy. Fossil fuels are convenient energy sources. Coal, the most abundant energy source in the United States, and other fossil fuels are considered to be nonrenewable because they require millions of years to be replaced. Conservation will help us prolong our supplies of fossil fuels, but we will eventually deplete them.

Energy Farming

Approximately half of the world's population does not have access to fossil fuel energy sources. Instead, they use biomass and biogas as an energy source. **Biomass,** as shown in **Figure 13-12,** is organic material from wood, sugarcane, corn, crop, and animal wastes. Many people burn biomass such as wood and peat on a regular basis, but so far, in some areas, it has been impractical to use as a major source of energy for such things as electricity. Biomass is being burned in the power plant shown in **Figure 13-13.** Brazil has done studies also and is well on the way to having a large wood biomass power plant. If biomass is allowed to rot in the absence of air, bacteria break down the wastes to produce biogas. **Biogas** is a mixture of gases, mostly methane, produced from biomass.

Controlled planting and usage can make these energy sources renewable. Can enough plants be grown and harvested to provide biomass and biogas for realistic energy supplies? Growing plants for this purpose is called **energy farming.**

Figure 13-12

Sources of energy include biomass, such as animal wastes. Here, cow manure is collected to be used as an energy source.

370   Chapter 13 Organic and Biological Compounds

Program Resources

Reproducible Masters
Study Guide, p. 56 L1
Reinforcement, p. 56 L1
Enrichment, p. 56 L3
Activity Worksheets, pp. 74-75 L1
Science Integration Activities, pp. 25-26
Critical Thinking/Problem Solving, p. 19

Transparencies
Section Focus Transparency 52

2 Points of View

Harvesting Energy—The Good News

Burning biomass or biogas to make heat to produce steam to generate electricity may produce less air pollution than the burning of fossil fuels. Fermenting corn to make alcohol to mix with gasoline, a mixture called **gasohol,** will allow current gasoline supplies to last longer. Because most of these techniques make use of existing technology or technology similar to what exists, they can be used readily.

Hidden Problems

While advantages exist, biomass production on a larger scale could result in problems. Using scarce agricultural land and diverting water to energy farms make growing food crops more difficult. Heating, fermenting, and extracting alcohol from plant material may require more energy than it produces. Combustion of biomass may create air pollution problems. For example, a form of bromine has been detected in the smoke from burning trees in forest fires. Bromine is capable of destroying more ozone than chlorine destroys.

The advantages and disadvantages of energy farming should be evaluated. Can solutions be found for the problems associated with energy farming? Will it provide a major source of energy in the future?

Figure 13-13

This biomass power plant provides energy for hundreds of homes in California.

Section Wrap-up

Review

1. What compound can be found in biogas mixtures?
2. What reasons can be given for developing energy farms?

Explore the Issue

Although it is difficult to estimate, some forecasters predict that the world's oil reserves could be down to less than ten percent of the current level within the next 50 years. How could well-managed energy farms provide some worldwide answers to this problem?

SCIENCE & SOCIETY

13-3 Growing Energy on the Farm   371

Across the Curriculum

History When the great buffalo herds thundered across the plains, the pioneers found an unending fuel supply in the form of dried buffalo chips. Students might like to read passages from books such as *Pioneer Women, Voices from the Kansas Frontier* by Joanna L. Stratton. [LS]

---

3 Assess

Check for Understanding

? FLEX Your Brain

Use the Flex Your Brain activity to have students explore BIOMASS.

📁 **Activity Worksheets,** p. 5

Reteach

Have students bring in and report on news articles about alternative energy sources.

Extension

📁 For students who have mastered this section, use the **Reinforcement** and **Enrichment** masters.

4 Close

Discussion

Conduct a discussion and create a chart listing the pros and cons of energy farming.

Section Wrap-up

Review

1. methane and other carbon-based compounds
2. They would provide a replenishable supply of energy. We wouldn't have to rely on fossil fuels.

Explore the Issue

They could provide renewable energy sources to replace oil-based fuels if machines that now use fossil fuels are adapted to effectively use biomass or biogas. **P**

Activity 13-1 Alcohols and Organic Acids

Have you ever wondered how chemists change one substance into another? You have learned that changing the bonding among atoms holds the key to that process.

Activity 13-1

Purpose
LS **Logical-Mathematical** Students will recognize the evidence of the chemical reaction of an organic compound. L1
COOP LEARN

Process Skills
observing, predicting, classifying, recognizing cause and effect, interpreting data

Time
25 minutes

Safety Precautions
Caution students against spilling, skin contact, or inhaling fumes of any chemicals used. If a spill does occur, immediately rinse with water.

📁 **Activity Worksheets,** pp. 5, 74-75

Teaching Strategies
Alternate Materials Use 95% ethanol if you have it. If you use ethanol from a drugstore, increase the amount.

- Show students how to shake a tube by holding it at the lip with fingers of one hand while swinging it against the palm of the other hand.
- Prepare 6.0M NaOH by dissolving 24 g of solid NaOH in 100 mL of distilled water. **CAUTION:** *Sodium hydroxide is caustic, and the solution will become hot. Prepare only in a heat-resistant glass container.*
- Prepare 0.01M KMnO$_4$ by dissolving 0.16 g KMnO$_4$ in 100 mL of distilled water.

Answers to Questions
1.

```
     H   H
     |   |
 H — C — C — OH
     |   |
     H   H
```

2. the hydroxyl group, –OH
3. the carboxylic acid group, –COOH
4. The color changes from purple to green to brown.
5. CH$_3$COOH
6. acetic acid

Problem
How can the bonding in an alcohol change in a way that changes the alcohol into an acid?

Materials
- test tube and stopper
- 1 mL 0.01M potassium permanganate solution
- 1 mL 6M sodium hydroxide solution
- 3 drops of ethanol
- goggles
- apron
- 10-mL graduated cylinder

Procedure
1. Pour 1 mL of 0.01M potassium permanganate solution and 1 mL of 6M sodium hydroxide solution into a test tube. **CAUTION:** *Handle both of these chemicals with care; immediately flush any spill with water.*
2. Add 3 drops of ethanol to the test tube.
3. Stopper the test tube and gently shake it for 1 minute. Observe and record any changes you notice in the tube for the next 5 minutes.

Analyze
1. What is the structural formula for ethanol?
2. What part of a molecule identifies a compound as an alcohol?
3. What part of a molecule identifies a compound as an organic acid?

Conclude and Apply
4. What evidence did you **observe** of a chemical change taking place in the test tube?
5. In the presence of potassium permanganate, an alcohol may undergo a chemical change into an acid. If the alcohol used is ethanol, what would you **predict** to be the formula of the acid produced?
6. The acid from ethanol is found in a common household product—vinegar. What is the acid's common name?

Data and Observations   Sample Data

Changes in Mixture
Within a few seconds, the mixture should change from purple to green. Over the next few minutes, the solution should turn brown.

✔ Assessment

Process If the formulas of an acid and an alcohol were written as C$_3$H$_8$O and C$_3$H$_6$O$_2$, have students explain how to determine which formula is for an acid and which is for an alcohol. *Organic acids all must have at least two oxygen atoms.* Use the Performance Task Assessment List for Analyzing the Data in **PASC**, p. 27. P

Biological Compounds

Proteins

Did you ever loop together strips of paper to make paper chains for decorations? Have you ever strung paper clips together? Both a paper chain and a string of paper clips can represent the structure of polymers. **Polymers** are huge molecules made of many smaller organic molecules that have formed new bonds and linked together. The smaller molecules are called monomers. Monomers within a certain polymer are usually very similar in size and structure. You can picture a monomer as an individual loop of paper or an individual paper clip. **Figure 13-14** shows this model of a polymer. As you will learn in Chapter 14, plastics are polymers. Many of the important biological compounds in your body also are polymers. Among them are the proteins.

Proteins are polymers formed from organic compounds called amino acids. Even though there are millions of different proteins, there are only 20 common amino acids.

Versatile Molecules

Different types of proteins make up many of the tissues in your body, such as muscles and tendons, as well as your hair and fingernails. A protein in your blood called hemoglobin carries oxygen, **Figure 13-14.**

Science Words

- polymer
- protein
- nucleic acid
- carbohydrate
- lipid

Objectives

- Describe the formation of polymers and discuss their importance as biological compounds.
- Compare and contrast proteins, nucleic acids, carbohydrates, and lipids.

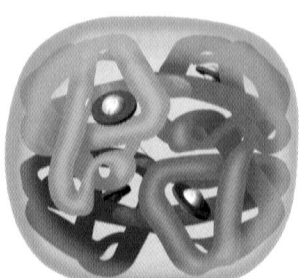

Hemoglobin

Figure 13-14

This paper chain is similar to a polymer because smaller units are linked to form a larger unit. Blood contains a complex protein, which is a polymer called hemoglobin.

Red blood cells

13-4 Biological Compounds 373

Prepare

Section Background

- Approximately one-half of our non-water mass consists of proteins. These form muscle, cartilage, and tendons. One-half of the protein in the human body is used as biological catalysts called enzymes.

- Nucleic acids are polymers of nucleotides. The nucleotide is composed of three parts—a nitrogen base, a sugar, and a phosphate group.

- The carbohydrates glycogen, starch, and cellulose differ simply in the way glucose monomers are linked together. The process of breaking down carbohydrates into CO_2 and H_2O is the chief energy source of most organisms.

Preplanning

To prepare for Activity 13-2, you will need index cards and paper clips. Crayons or colored markers can be used if available.

1 Motivate

Bellringer

Before presenting the lesson, display **Section Focus Transparency 53** on the overhead projector. Assign the accompanying **Focus Activity** worksheet. L1 LEP

Tying to Previous Knowledge

In the party supplies section of toy stores, you can find a spray can of string confetti. This polymer can be shown to students by spraying a string from the can. A 100-g can will produce about 100 m of polymer for about $2.00.

Demonstration

Visual-Spatial Show students a polymer being formed by making nylon. Commercially available solutions of sebacoyl chloride and 1,6-diamino-hexane are available from your science supplier. Layer the two solutions in a small beaker. Using forceps, pick up the center of the film that forms at the interface of the two liquids. Slowly pull the nylon string from the beaker and roll it up on a graduated cylinder that has been wrapped with a paper towel.

Visual-Spatial Place a small amount of acetone or alcohol in the bottom of a 600-mL beaker. Add to the solvent a large volume of the foam "peanuts" used in shipping containers. Allow the acetone to evaporate overnight. A disk of solid polystyrene will remain.

Visual Learning

Figure 13-15 Why is it important to eat foods that contain protein? *Digestion breaks the proteins down into monomers of amino acids, which are then used to make new proteins.* **LEP**

Enrichment

Have students research the structure of DNA and its functions. Have students show the chemical structure of the bases. They can make a three-dimensional model of the double helix and show how the four bases pair. Some students could research the way base sequences code for amino acids in proteins. The students will need to find out how protein is synthesized as a consequence of the DNA code. **L3**

Figure 13-15

Some high-protein foods are shown. *Why is it important to eat foods that contain protein?*

The enzymes that regulate chemical reactions in your body are proteins. In fact, proteins account for 15 percent of your weight. That's about half the weight of all materials other than water in your body. Several high-protein foods are shown in **Figure 13-15.**

Two amino acids are shown in **Figure 13-16A.** The –COOH group is the carboxylic acid group. If you look again at **Figure 13-11C** on page 368, you'll see that the structure of this group appears also in acetic acid. The $-NH_2$ group is called the amine group. Both groups appear in every amino acid.

Peptide bonds form when the amine group of one amino acid combines with the organic acid group of another amino acid. Look at **Figures 13-16B** and **C.**

Your body makes proteins from amino acids. The amino acids come from eating and digesting foods that contain proteins. Digestion breaks the protein polymers into monomers of amino acids, which your body uses to make new proteins.

Figure 13-16

In a protein polymer, peptide bonds link together molecules of amino acids.

B When amino acids react to form a protein, water is formed from the –OH from a carboxylic acid group of one amino acid and a –H from an amine group in the other amino acid.

A Notice that each of these amino acids contains a –COOH group and a –NH₂ group.

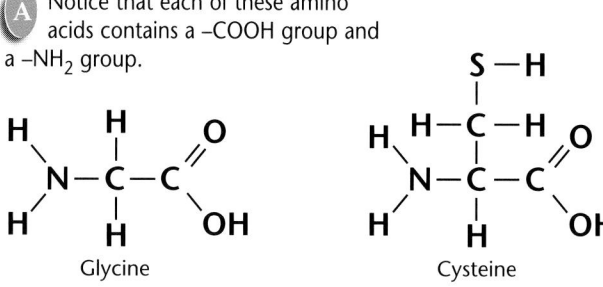

Glycine   Cysteine

C The remaining parts of the molecules form a peptide bond.

374   Chapter 13 Organic and Biological Compounds

Science Journal

Word Scramble Have students write the word *protein* in their journals and make as many words as they can from the seven letters in the word. Explain how this activity is similar to what could be done with seven amino acids being used to make proteins. **L1**

Use these program resources as you teach this lesson.

Teaching Transparency 26
Science Integration Transparency 13

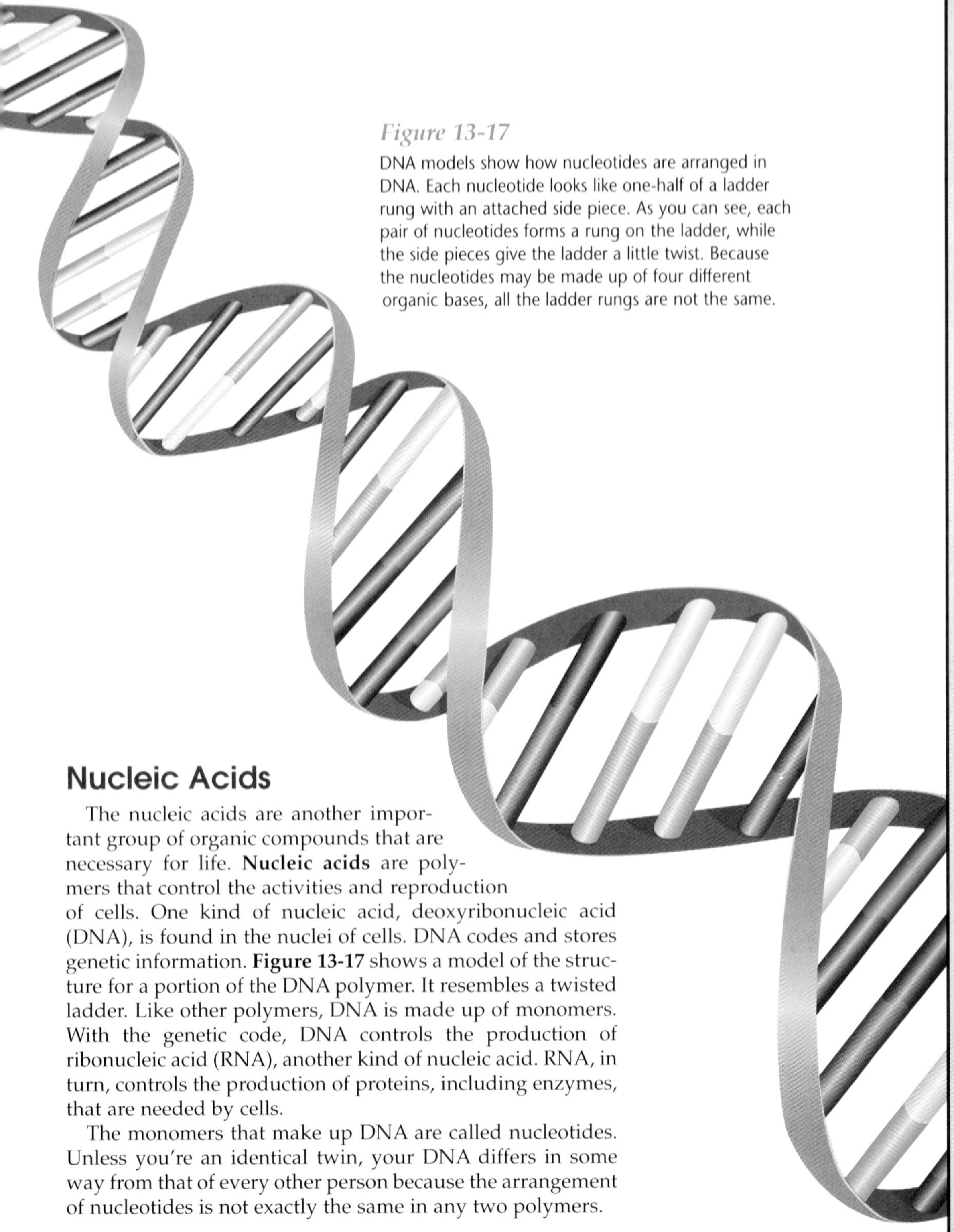

Figure 13-17

DNA models show how nucleotides are arranged in DNA. Each nucleotide looks like one-half of a ladder rung with an attached side piece. As you can see, each pair of nucleotides forms a rung on the ladder, while the side pieces give the ladder a little twist. Because the nucleotides may be made up of four different organic bases, all the ladder rungs are not the same.

Nucleic Acids

The nucleic acids are another important group of organic compounds that are necessary for life. **Nucleic acids** are polymers that control the activities and reproduction of cells. One kind of nucleic acid, deoxyribonucleic acid (DNA), is found in the nuclei of cells. DNA codes and stores genetic information. **Figure 13-17** shows a model of the structure for a portion of the DNA polymer. It resembles a twisted ladder. Like other polymers, DNA is made up of monomers. With the genetic code, DNA controls the production of ribonucleic acid (RNA), another kind of nucleic acid. RNA, in turn, controls the production of proteins, including enzymes, that are needed by cells.

The monomers that make up DNA are called nucleotides. Unless you're an identical twin, your DNA differs in some way from that of every other person because the arrangement of nucleotides is not exactly the same in any two polymers.

13-4 Biological Compounds **375**

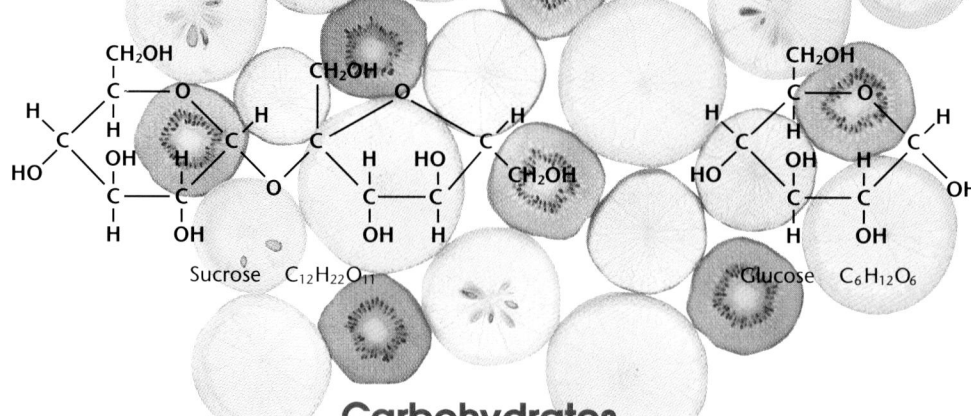

Sucrose $C_{12}H_{22}O_{11}$ Glucose $C_6H_{12}O_6$

Carbohydrates

Figure 13-18

Sucrose and glucose are sugars found in foods. Fruits contain glucose and another simple sugar called fructose. *Why are sugars carbohydrates?*

If you hear the word *carbohydrate,* you may think of a sweet fruit or sugary treat. Do you also think of carbohydrate loading by athletes? Runners, for example, often prepare in advance for a long-distance race by eating, or "loading," carbohydrates in starchy foods such as vegetables and pasta. Starches provide high-energy, long-lasting fuel for the body.

Carbohydrates are organic compounds in which there are twice as many hydrogen atoms as oxygen atoms. One group of carbohydrates is the sugars, as shown in **Figure 13-18.** The sugar glucose is found in your blood and also in many sweet foods such as grapes and honey. The sugar sucrose is common table sugar.

Figure 13-19 shows a part of a starch molecule. Starch is a carbohydrate that is also a polymer. During digestion, sucrose and starch are broken down into smaller molecules of glucose and other similar sugars.

Figure 13-19

Starch is the major component of pasta.

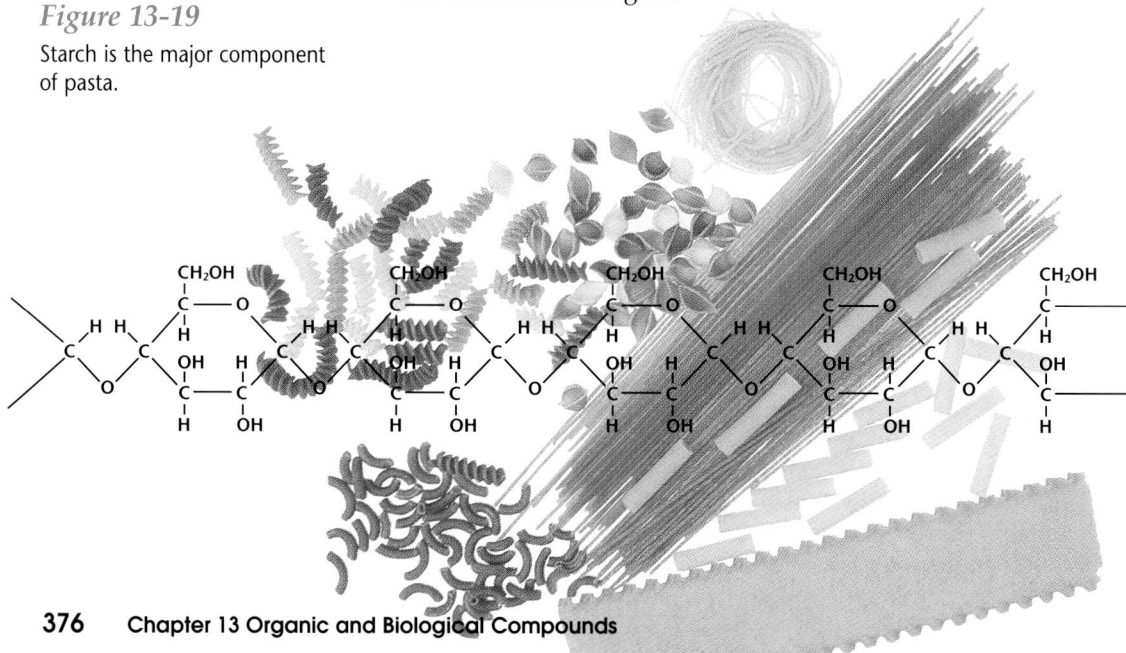

376   Chapter 13 Organic and Biological Compounds

Cultural Diversity

Hold the Cheese! Lactase is an enzyme that allows humans to break down lactose, the sugar in milk. Normally, human infants produce lactase, but lactase production stops in most humans as they age. This causes a condition known as lactose intolerance. If a person lacks lactase, digesting milk products is difficult.

Only 18% of adult Americans of northern European ancestry are lactose intolerant, but 80% of African-American adults and 60% of Mexican-American adults do not produce lactase. In some Asian populations, such as the Thai, 98% of the adults are lactose intolerant.

Scientists suggest that populations that have used dairy farming as an important food resource for thousands of years have adapted by retaining the ability to produce lactase in adulthood.

Figure 13-20

At room temperature, fats are normally solids, and oils are usually liquids.

Lipids

Fats, oils, and related compounds make up a group of organic compounds known as **lipids.** Lipids, examples of which are shown in **Figure 13-20,** include animal fats, such as butter, and vegetable oils, such as corn oil. Lipids contain the same elements as carbohydrates, but in different proportions and arrangements.

USING TECHNOLOGY

DNA Fingerprinting ▼ ▲

Fingerprints! When crime investigators find a print left by the oil of someone's skin, they know that an identifying piece of evidence has been found. Could other complex molecules at a crime scene be just as important? Each of us carries a unique arrangement of several billion nucleotide monomers. Within the DNA of our cells, nucleotides are arranged in groups called genes. These structures make the synthesis of proteins possible.

With the possible exception of identical twins, each person's DNA is unique. Therefore, the fragment break-up pattern can be used to identify someone. From a tissue sample, such as blood, chemists extract DNA, break it up, separate the fragments, obtain a radioactive image that emits X rays, and develop a film picture of the nucleotide pattern.

*inter*NET CONNECTION

What other applications can you think of for DNA fingerprinting? Visit the Chapter 13 Internet Connection at Glencoe Online Science, **www. glencoe.com/sec/ science/physical,** for a link to more information about DNA fingerprinting.

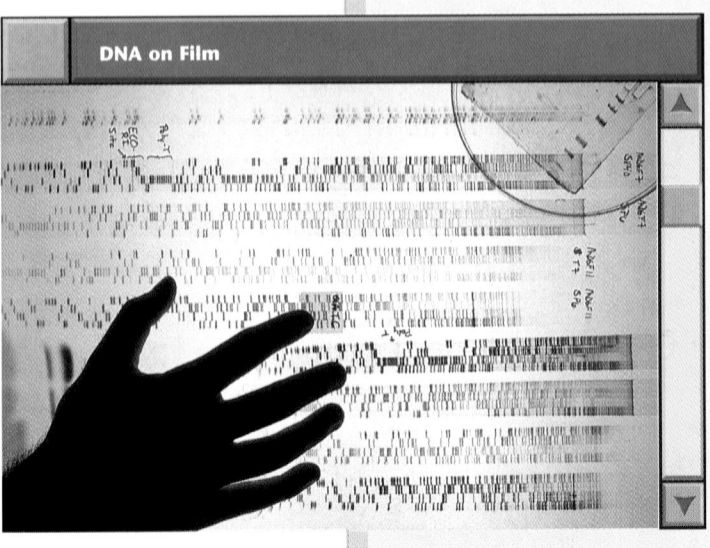

DNA on Film

USING TECHNOLOGY

For more information on DNA fingerprinting, see "Testing the Gene Fit" by L. Hancock, *Newsweek,* January 9, 1995, p. 64, or "DNA on Trial," *Maclean's,* February 6, 1995, pp. 56-63.

GLENCOE TECHNOLOGY

 Videodisc

The Infinite Voyage: The Champion Within

Chapter 6
Glycogen: Fuel for Muscles

Glencoe Physical Science Interactive Videodisc

Side 2, Lesson 5
Introduction to Carbohydrates and Hydrocarbons

13-825
Carbohydrates

827-2605
Source of Energy in Carbohydrates

2608-2828

Visual Learning

Figure 13-18 Why are sugars carbohydrates? *Sugars contain twice as many hydrogen atoms as oxygen atoms.* LEP LS

INTEGRATION
Earth Science

Show students the following videos and discuss pheromones and insects.

GLENCOE TECHNOLOGY

 **Videodisc**

STVS: Ecology
Disc 6, Side 2
Pheromone Rope (Ch. 3)

STVS: Plants and Simple Organisms
Disc 4, Side 2
Plant Chemical Repels Cockroaches (Ch. 14)

The Infinite Voyage: Insects: The Ruling Class
Chapter 8
Prospecting for Healing Medicine from Insects

3 Assess

Check for Understanding

 **FLEX Your Brain**

Use the Flex Your Brain activity to have students explore LIPIDS. **P**

 Activity Worksheets, p. 5

INTEGRATION
Life Science

Figure 13-21
Pheromones enable bees to communicate information such as the location of a food source or the invasion of the hive by an enemy.

Use these program resources as you teach this lesson.
Study Guide, p. 57
Lab Manual 29
Cross-Curricular Integration, p. 19
Science and Society Integration, p. 17
Multicultural Connections, pp. 29-30
Concept Mapping, p. 31

Fats and oils are similar in structure to hydrocarbons, so they can be classified as saturated and unsaturated, according to the types of bonds in their carbon chains. Saturated fats contain only single bonds between carbon atoms, and unsaturated fats contain at least one double bond. Most animal fats contain a great deal of saturated fats. Oils from plants are mainly unsaturated.

Have you heard that eating too much fat can be unhealthy? There is evidence that too much saturated fat and cholesterol in the diet contributes to heart disease and that unsaturated fats may help to prevent heart disease. A balanced diet includes some fats, just as it includes proteins and carbohydrates.

Cholesterol

Another lipid found in meat and fish is cholesterol. Even if you never ate foods containing cholesterol, your body would make its own supply. Cholesterol is used by the body to build cell membranes and is also found in bile, a digestive fluid. Although cholesterol is a lipid, it is not a fat.

Pheromones

In other species of the animal kingdom, cholesterol plays an unusual role. Female ticks combine cholesterol, which they have obtained from blood, with another compound to form an attractant for male ticks. This chemical attractant coats the outside of their bodies and is an example of a pheromone.

The term *pheromone* is used to describe a chemical used in communication among individuals within the same species. Ants from a nest are able to follow an individual food trail by detecting organic compounds left by the first ant. In beehives, as shown in **Figure 13-21**, the queen emits controlling chemicals to other members of the colony. Pheromones have dramatic effects even when present in extremely small amounts.

One way to trap insects for study or pest control is to use the attracting pheromone from a species as bait in traps, as shown in **Figure 13-22**. Chemicals that indicate a warning alarm to a species may be used in a repellent in place of using a more harsh chemical treatment.

Are there human pheromones? Human behavior is complex, and we react in many ways to outside stimuli. Aromas, both natural and those added from fragrances, do have perceived effects. Research is still being done on human pheromones. From insects to larger animals, communication of many important feeding, fighting, mating, and migration activities is regulated by organic molecules, including cholesterol.

Figure 13-22

Traps for certain insect pests, such as the Japanese beetle, use bait consisting of pheromones that attract the pest to the trap.

Section Wrap-up

Review

1. What is a polymer? Why are polymers important organic compounds?

2. Compare and contrast proteins and nucleic acids.

3. **Think Critically:** Is ethanol a carbohydrate? Explain.

Skill Builder
Comparing and Contrasting
In terms of DNA, compare and contrast identical twins and two people who are not identical twins. If you need help, refer to Comparing and Contrasting in the **Skill Handbook.**

USING MATH

By changing just one amino acid in a chain, the function of the entire protein changes. If the letters *G, L, C,* and *I* represent four amino acids, how many different four-amino acid protein chains could be made?

13-4 Biological Compounds   **379**

Skill Builder
Characteristics of identical twins are the same because the order of nucleotides in the DNA ladder is the same. Other people are different because the order of nucleotides in the DNA ladder is different.

 **Assessment**

Portfolio Use the Flex Your Brain activity to reinforce critical-thinking and problem-solving skills by investigating DNA. Use the Performance Task Assessment List for Asking Questions in **PASC,** p. 19. **P**

Reteach
IS **Logical-Mathematical** List two or three healthful menus on the chalkboard. Have students evaluate each item as a contributor of protein, fat, or carbohydrate.

Extension

For students who have mastered this section, use the **Reinforcement** and **Enrichment** masters.

4 Close

•MINI•QUIZ•

Use the Mini Quiz to check students' recall of chapter content.

1. Proteins result from the linking of many monomers called _____ .
   amino acids

2. Fats and oils are _____ .
   lipids

Section Wrap-up

Review

1. It is a large molecule made up of smaller molecules called monomers. Many important biological compounds are polymers.

2. Proteins are formed when amino acids form polymers. They are important in making body tissues and enzymes. Nucleic acids are polymers of nucleotides. They contain a code that controls cell activities.

3. **Think Critically** No. A molecule of ethanol does not have twice as many hydrogen atoms as oxygen atoms.

USING MATH

There are 24 different combinations possible.

$4! = 1 \times 2 \times 3 \times 4 = 24$

Activity 13-2

Activity 13-2

Design Your Own Experiment

Mixing Many Monomers

A common expression about food is, "You are what you eat." This statement is true because your digestive system changes food into useful compounds in your body, such as glucose and amino acids.

PREPARATION

Purpose
LS **Kinesthetic** Students will model the bonding between monomers found in starches and proteins. **L1** **LEP**
COOP LEARN

Process Skills
building models, predicting, communicating, classifying, sequencing, comparing and contrasting, interpreting scientific illustrations, inferring

Time
40 minutes

Materials
Different-colored markers may be used on white index cards if colored cards are not available.

Possible Hypotheses
One hypothesis is the monomers will be able to bond wherever a free –OH or –H exists.

Activity Worksheets, pp. 5, 76-77

PLAN THE EXPERIMENT

Possible Procedures
One possible set of procedures is to draw five or six glucose and amino acid monomers on index cards. Then, by using paper clips to model bonds, link glucose monomer cards to simulate a starch polymer, and link amino acid cards to simulate a protein.

Teaching Strategies
- Be sure students understand the importance of placing the bonding sites on the edges of the cards.
- The structures for the amino acids glycine and cysteine are shown on page 374 of the student text. Other amino acids may be used in addition to those shown on page 374.

PREPARATION

Problem
How can you demonstrate the bonding in the basic components of food?

Form a Hypothesis
From what you know about polymers, state a hypothesis about how models of glucose monomers and amino acid monomers can be shown to bond into starch and protein, respectively.

Objectives
- Make a model of glucose and amino acids that can be used to demonstrate linkage of monomers into biological polymers.
- Observe that certain areas of molecules have structures that can be used for attachments to other molecules.

Possible Materials
- note cards of 2 different colors
- paper clips
- fine-line markers

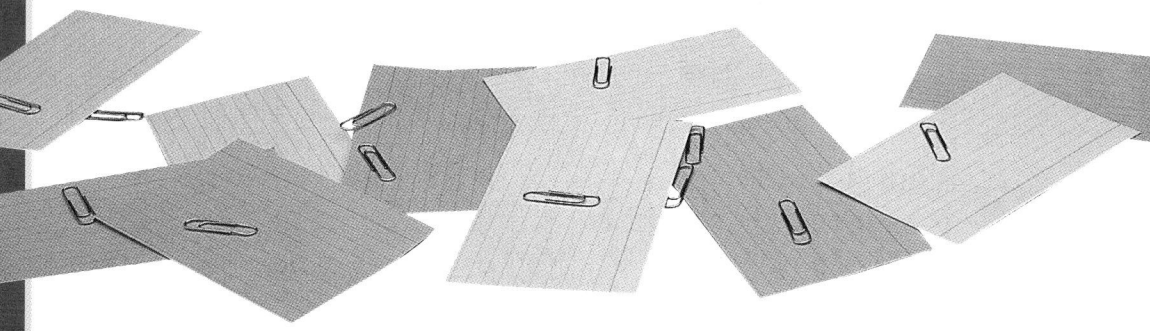

380 Chapter 13 Organic and Biological Compounds

PLAN THE EXPERIMENT

1. As a group, agree upon and write out the hypothesis statement about your model.
2. As a group, list the steps that you need to take to produce your models. Be specific, describing exactly what you will do at each step.
3. Make a list of the materials that you will need to complete your model.

Check the Plan

1. Have you shown the structure of glucose on a note card so that it can hook on either end to another glucose model?
2. What have you used to show linkage between monomers of glucose?

3. Have you found amino acid structures to make a model protein?
4. *Make sure your teacher approves your plan and that you have included any changes suggested in the plan.*

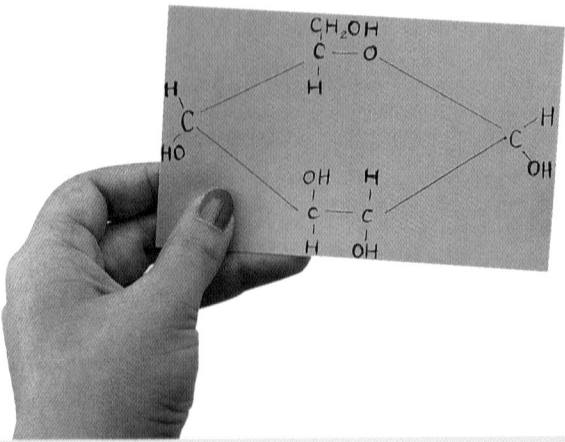

DO THE EXPERIMENT

1. Prepare your monomer models as planned and approved.
2. How many monomers can you hook together?

Analyze and Apply

1. **Compare** your note card models to those of your classmates.
2. **Contrast** the glucose and amino acid monomer models.
3. How can you use the same amino acid models to make a different protein?

Go Further

What small molecule is also produced in the actual formation of biological polymers such as starch and protein?

13-4 Biological Compounds **381**

DO THE EXPERIMENT

Expected Outcome

Students will see that the –OH from one monomer can join with an –H from another monomer, leaving an area on both monomers that can bond to another.

Analyze and Apply

1. Comparison will yield the same models for starch and different models for amino acids.
2. Starch model—monomer units are identical. Protein model—monomers can be from different amino acids.
3. When the sequence of amino acid cards is varied, a new protein can be represented. For example: val-gly-val is different from val-val-gly.

Go Further

water

✓ Assessment

Content To further assess students' understanding of monomers and polymers, ask the following questions: **In this activity, you showed how monomers can make larger molecules. Reversing this process, where would the structure of a protein most likely break apart? What does your body do with amino acid components from digested proteins?** *The peptide bonds between amino acids can be broken to release amino acids. The amino acids that you have eaten as proteins can be reassembled to make needed proteins for your body.* Use the Performance Task Assessment List for Making Observations and Inferences in **PASC**, p. 17. **P**

Background

- The original sports drink was invented in the early 1960s at the University of Florida. The story goes that the school's football team was falling further and further behind in an important game. At halftime, the coach ladled up a concoction created in the school's science lab, and the Gators came roaring back in the second half to win the game.

- The American soft-drink industry grew from the traveling medicine shows of the 1800s. Products incorporating traditional Native American healers' ingredients were hawked by sometimes-unscrupulous salesmen as cures for everything from gout to listlessness. In those days, Native Americans possessed much of the medical knowledge that existed in North America, so traveling medicine shows tried to associate themselves with centuries-old expertise by taking names such as the Kickapoo Indian Medicine Company. *The United States Pharmacopeia* from 1820 listed more than 200 drugs supplied by the natives, including quinine and coca.

Teaching Strategies

Encourage students to look on their kitchen shelves or in a food market, and list some of the places where packaged food is manufactured. If a food manufacturing plant is located near your school, you might arrange a student tour.

People and Science

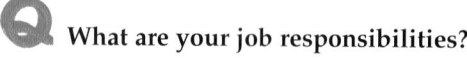 

NORMA MARTINEZ, *Food Scientist*

On the Job

Q What are your job responsibilities?

A I monitor the manufacturing process of a very popular sports drink. I analyze the formulations to make sure the salt, sugar, and water are in the right proportions. That involves measuring such properties as the amount of detectable acidity and the acid-base balance.

Q How do sports drinks differ from soft drinks?

A Sports drinks are formulated so that the electrolytes they contain "hook up" with the electrolytes in the body and speed the absorption of liquid. It's rather like tricking your body into thinking it's the same fluid so your body will absorb the water faster. Sports drinks have been scientifically proven to do that.

Q Did your training require education in food science?

A Yes. I took courses in areas such as microbiology, food engineering, and dairy science, in which we learned to make ice cream. That was my favorite class.

Personal Insights

Q Do you prefer practical application of science over doing research or lab work?

A Yes. I enjoy real-world application of science. The area of food science is constantly changing. Although people will always need to eat and drink, they seek novelty along with nutrition.

Q How does your job affect your outlook on life?

A As a Hispanic in a managerial position, I try to set a positive example for the people who actually work on the line making our product. When I was in high school, I was encouraged by a group that provided support for minorities in engineering. I'd like to help in a similar way.

Career Connection

Write to the consumer information division of the manufacturer of your favorite packaged food. (You can get the address from the label.) Ask for product information and data about nutrition. Prepare a poster display showing what you learned.

- Caterer   - Dietician   - Nutritionist

382   Chapter 13 Organic and Biological Compounds

Career Connection

Career Path Courses necessary for a college degree in food science include food engineering, microbiology, and dairy science.

Summary

13-1: Simple Organic Compounds
1. Carbon is an element with a structure that enables it to form a large number of compounds, known as organic compounds.
2. Saturated hydrocarbons contain only single bonds between carbon atoms, and unsaturated hydrocarbons contain double or triple bonds.
3. Isomers of organic compounds have identical formulas but different molecular shapes.

13-2: Other Organic Compounds
1. Aromatic compounds, many of which have odors, contain the benzene ring structure.
2. A substituted hydrocarbon contains one or more atoms of other elements that have replaced hydrogen atoms.

13-3: Science and Society: Growing Energy on the Farm
1. Biomass is the source of biogas and gasohol, two fuels that can be used to increase our energy supply.
2. Ethanol and gasohol are useful as substitutes for gasoline, but their production may damage the environment.

13-4: Biological Compounds
1. Many important biological compounds are polymers—huge organic molecules made of many smaller units, or monomers.
2. Proteins, nucleic acids, carbohydrates, and lipids are major groups of biological compounds.

Key Science Words

a. alcohol
b. aromatic compound
c. biogas
d. biomass
e. carbohydrate
f. energy farming
g. gasohol
h. hydrocarbon
i. isomers
j. lipid
k. nucleic acid
l. organic compound
m. polymer
n. protein
o. saturated hydrocarbon
p. substituted hydrocarbon
q. unsaturated hydrocarbon

Reviewing Vocabulary

Match each phrase with the correct term from the list of Key Science Words.

1. a combination of ethanol and gasoline
2. a hydrocarbon containing only single bonds
3. all animal and plant material
4. specific type of compound formed when an –OH group replaces one or more hydrogen atoms in a hydrocarbon
5. compounds with identical chemical formulas but different molecular structures
6. a hydrocarbon containing at least one double or triple bond between carbons
7. contains the benzene ring structure
8. a fat, oil, or related compound
9. formed from plant and animal waste
10. growing plants for use as fuel

Chapter 13

Review

Summary

Have students read the summary statements to review the major concepts of the chapter.

Reviewing Vocabulary

1. g	6. q
2. o	7. b
3. d	8. j
4. a	9. c
5. i	10. f

✔ Assessment

Portfolio Encourage students to place in their portfolios one or two items of what they consider to be their best work. Examples include:
- Activity 13-1 observations and answers, p. 372
- Explore the Issue, p. 371
- Flex Your Brain, p. 378 P

Performance Additional performance assessments may be found in **Performance Assessment** and **Science Integration Activities.** Performance Task Assessment Lists and rubrics for evaluating these activities can be found in Glencoe's **Performance Assessment in the Science Classroom.**

GLENCOE TECHNOLOGY

⊙ Videodisc

Glencoe Physical Science

Interactive Videodisc

Use the videodisc segments *Molecular Structures Compared* and *Combustion* in Lesson 5 to review the differences and similarities between carbohydrates and hydrocarbons.

▭ MindJogger Videoquiz

Chapter 13 Have students work in groups as they play the Videoquiz game to review key chapter concepts.

Chapter 13 Review

Checking Concepts

1. b	**6.** a
2. c	**7.** b
3. b	**8.** c
4. a	**9.** d
5. d	**10.** d

Understanding Concepts

11. Saturated fats contain only single bonds between carbon atoms. Unsaturated fats contain one or more double bonds between carbon atoms.

12. Both compounds contain the same number of each kind of atom, but the arrangements of atoms are different.

13. CCl_4

14. Answers will vary. Answers may include use of wood and peat for fuel.

15. four; C, H, N, and O

Thinking Critically

16. Organic compounds make up all living things, or things that have lived. In addition to these, many are made artificially. Also, organic compounds have a large amount of structural variation.

17. In a single bond, two atoms share a pair of electrons. In a double bond, the atoms share two pairs of electrons. In a triple bond, three pairs of electrons are shared. (Student diagrams should show four bonds for each carbon atom.)

$$H-\overset{\overset{\displaystyle H}{|}}{\underset{\underset{\displaystyle H}{|}}{C}}-\overset{\overset{\displaystyle H}{|}}{\underset{\underset{\displaystyle H}{|}}{C}}-\overset{\overset{\displaystyle H}{|}}{C}=\overset{\overset{\displaystyle H}{|}}{C}-\overset{\overset{\displaystyle H}{|}}{\underset{\underset{\displaystyle H}{|}}{C}}-C\equiv C-H$$

Single bond   Double bond   Triple bond

18. Marshes have abundant plant and animal life, and the water provides a medium for the air-free decomposition of dead organisms.

Checking Concepts

Choose the word or phrase that completes the sentence or answers the question.

1. A benzene ring is _____.
- a. rare
- b. stable
- c. unstable
- d. saturated

2. Alcohols and organic acids are both _____ hydrocarbons.
- a. aromatic
- b. saturated
- c. substituted
- d. unsaturated

3. Two examples of _____ are cow manure and a tree.
- a. biogas
- b. biomass
- c. energy farming
- d. hydrocarbons

4. The units of polymers are _____.
- a. monomers
- b. isomers
- c. plastics
- d. carbohydrates

5. Some examples of _____ are enzymes and hemoglobin.
- a. carbohydrates
- b. lipids
- c. nucleic acids
- d. proteins

6. DNA codes and stores _____.
- a. genetic information
- b. nucleic acids
- c. proteins
- d. lipids

7. DNA is made up of _____.
- a. amino acids
- b. nucleotides
- c. polymers
- d. carbohydrates

8. Glucose and fructose, both $C_6H_{12}O_6$, are _____.
- a. amino acids
- b. alcohols
- c. isomers
- d. polymers

9. If a carbohydrate has 16 oxygen atoms, how many hydrogen atoms does it have?
- a. 4
- b. 8
- c. 16
- d. 32

10. Cholesterol is a type of _____.
- a. sugar
- b. starch
- c. protein
- d. lipid

Understanding Concepts

Answer the following questions in your Science Journal using complete sentences.

11. Too much saturated or unsaturated fat in the diet is unhealthful. What is the difference in the composition of saturated fat and unsaturated fat?

12. Rubbing alcohol is isopropyl alcohol. How does this differ from propyl alcohol?

13. Carbon tetrachloride, a former dry cleaning fluid, is formed when all the hydrogen in methane is replaced by chlorine. Write the formula for carbon tetrachloride.

14. Explain how biomass has been used as an energy source throughout history.

15. Luminol, found in light sticks, is an example of an organic compound. How many different elements are bonded to carbon in a molecule of luminol, shown below?

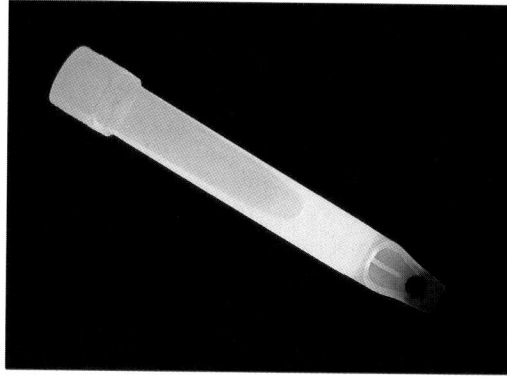

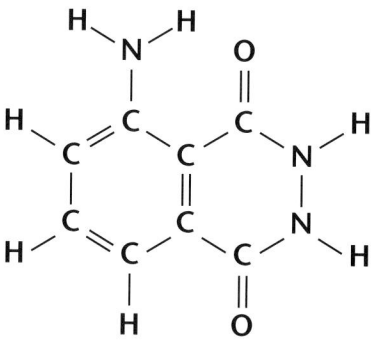

Assessment Resources

Reproducible Masters
Chapter Review, pp. 29-30
Assessment, pp. 84-87
Performance Assessment, p. 27

Glencoe Technology
- Chapter Review Software
- Computer Test Bank
- MindJogger Videoquiz

Thinking Critically

16. Why are more than 90 percent of all compounds organic compounds?
17. Use a diagram to explain single, double, and triple bonds in hydrocarbons. Draw a chain of carbon atoms that shows each type of bond.
18. Explain why a marsh provides an environment that produces biogas.
19. Show how the structure of an amino acid explains the term *amino acid*.
20. Some fats are polyunsaturated. Draw a chain of polyunsaturated carbon atoms.

Developing Skills

If you need help, refer to the **Skill Handbook.**

21. **Making and Using Graphs:** Using the following table, plot the number of carbon atoms on one axis and the boiling point on the other axis on a graph. Use the graph to predict the boiling points of butane, octane, and dodecane ($C_{12}H_{26}$).

Graph Data

Name	Formula	Boiling Point (°C)
Methane	CH_4	−162
Ethane	C_2H_6	−89
Propane	C_3H_8	−42

22. **Interpreting Scientific Illustrations:** Which of the following terms apply to the illustration below? *alcohol, aromatic, carbohydrate, hydrocarbon, lipid, organic compound, polymer, saturated, substituted hydrocarbon, unsaturated*

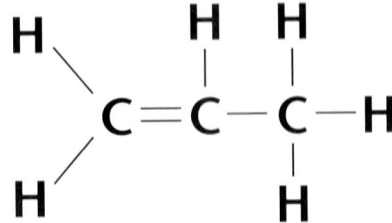

23. **Concept Mapping:** Use a network tree to describe types of fats. Include the terms *saturated fats, unsaturated fats, single bonds,* and *double bonds.*
24. **Recognizing Cause and Effect:** Our society has a definite need for more energy farming. List several causes and several effects of this need.
25. **Hypothesizing:** Sarah decided to go on a weight-reduction diet. She is eating only lettuce and fruit. What do you predict will happen to Sarah as a result?

Performance Assessment

1. **Scientific Drawing:** Research hydrocarbons containing five carbon atoms. Draw diagrams of their structures, name them, and tell their uses.
2. **Conducting a Survey and Graphing the Results:** Survey local gasoline stations to find out whether they sell gasohol or only gasoline. Find out how much of each fuel is sold per month. Display results in a graph.
3. **Oral Presentation:** Research ways crops are used as energy resources. Make an oral presentation of your findings. Suggest how these ways could be implemented locally.

3. Students may choose to add visual aids to illustrate their findings. Use the Performance Task Assessment List for Oral Presentation in **PASC,** p. 71. **P**

---

19. All amino acids contain an –NH_2 (amine) group and a –COOH (carboxylic acid) group.
20. One possible answer is shown.

$$C=C-C=C-C=C$$

Developing Skills

21. **Making and Using Graphs** Answers should be close to −1°C, 126°C, and 216°C, respectively.
22. **Interpreting Scientific Illustrations** hydrocarbon, organic compound, unsaturated
23. **Concept Mapping** Concept maps should show that fats consist of saturated fats, which contain only single bonds, and unsaturated fats, which contain double as well as single bonds.
24. **Recognizing Cause and Effect** Answers will vary. Causes may include need for a renewable resource and conservation of a dwindling supply of fuel. Possible effects include usage of food-producing land, increased fuel supply, and environmental damage.
25. **Hypothesizing** Answers will vary. Answers may include negative effects of protein and lipid deficiencies.

Performance Assessment

1. Student answers will vary. Use the Performance Task Assessment List for Scientific Drawing in **PASC,** p. 55. **P**
2. Answers will differ depending on the gasoline vendor and the type of area. Graphs may be any form (bar, line, pie, etc.). Use the Performance Task Assessment List for Conducting a Survey and Graphing the Results in **PASC,** p. 35. **P**

Chapter Organizer

Section	Objectives/Standards	Activities/Features
Chapter Opener		**Explore Activity:** Ceramic Smiles, p. 387
14-1 **Materials with a Past** (1 session, ½ block)*	1. **Identify** common alloys and ceramics. 2. **Compare** and **contrast** alloys and ceramics. **National Content Standards:** (5-8) UCP1, UCP3, UCP5, A2, B1; (9-12) UCP1, UCP3, UCP5, A2, B2	**Using Math,** p. 389 **Connect to Earth Science,** p. 390 **Problem Solving:** Ceramic Decoys, p. 392 **MiniLAB:** How do properties of ceramics change? p. 393 **Using Math,** p. 394 **Skill Builder:** Concept Mapping, p. 394 **Activity 14-1:** Preparing an Alloy, p. 395
14-2 **New Materials** (1 session, ½ block)*	1. **Compare** and **contrast** plastics and synthetic fibers. 2. **Describe** a composite. **National Content Standards:** (5-8) UCP1-UCP3, UCP5, A1, A2, C1, E1, E2, F5, G1, G3; (9-12) UCP1-UCP3, UCP5, A1, A2, E1, E2, F6, G1, G3	**Using Technology:** A Sticky Lesson, p. 399 **Science Journal,** p. 401 **Skill Builder:** Observing and Inferring, p. 401 **Activity 14-2:** Compose Your Own Composite, pp. 402-403
14-3 **Science and Society** (½ session, ½ block)*	1. **Compare** and **contrast** the advantages of new materials used in sports with older materials. 2. **Identify** chemical elements and materials of technology used in sports. 3. **Determine** whether sports should alter rules about the use of technology. **National Content Standards:** (5-8) UCP2, UCP3, UCP5, A2, F4, F5, G1; (9-12) UCP2, UCP3, UCP5, A2, D2, F6, G1	**People and Science:** Martin Caveza, Industrial Designer, p. 406 **Explore the Issue,** p. 405

* A complete Planning Guide that includes block scheduling is provided on pages 32T-35T.

Activity Materials

Explore	Activities	MiniLAB
page 387 No materials needed; pencil and paper activity.	page 395 copper penny, 30-mesh zinc, hot plate, dilute nitric acid, dilute sodium hydroxide, 2 evaporating dishes, tongs, safety goggles, heat-proof glove, apron pages 402-403 plaster of paris, water, measuring cup, heavy-duty aluminum foil, hammer, paper towels, beaker, paper clips, plastic soda straws, pipe cleaners, crayons, safety goggles	page 393 100 mL flour, 100 mL salt, 100 mL water, microwave-safe bowl (about 400-mL capacity), 5 mL cream of tartar, 5 mL light cooking oil, spoon, microwave oven

Need Materials? Call Science Kit (1-800-828-7777).

Teacher Classroom Resources

Reproducible Masters	Transparencies	Teaching Resources
Study Guide, p. 58 **Reinforcement**, p. 58 **Enrichment**, p. 58 **Activity Worksheets**, pp. 80-81, 84 **Multicultural Connections**, p. 31 **Science and Society Integration** p. 18 **Critical Thinking/Problem Solving**, p. 20	**Section Focus Transparency 54**, Out-of-This-World Materials **Teaching Transparency 27**, Steel Alloys	**Physical Science CD-ROM** **Spanish Resources** **English/Spanish Audiocassettes** **Cooperative Learning Resource Guide** **Lab Partner** **Lab and Safety Skills** **Lesson Plans**
Study Guide, p. 59 **Reinforcement**, p. 59 **Enrichment**, p. 59 **Activity Worksheets**, pp. 82-83 **Science Integration Activity 14**, Package Particulars **Concept Mapping**, pp. 33-34 **Cross-Curricular Integration**, p. 20	**Section Focus Transparency 55**, Improving on Nature **Science Integration Transparency 14**, What Makes a Good Insulator? **Teaching Transparency 28**, Production of Nylon	**Assessment Resources** **Chapter Review**, pp. 31-32 **Assessment**, pp. 88-91 **Performance Assessment in the Science Classroom (PASC)** **MindJogger Videoquiz** **Alternate Assessment in the Science Classroom** **Performance Assessment** **Chapter Review Software** **Computer Test Bank**
Study Guide, p. 60 **Reinforcement**, p. 60 **Enrichment**, p. 60	**Section Focus Transparency 56**, New Composites Really Go Far	

Key to Teaching Strategies

The following designations will help you decide which activities are appropriate for your students.

L1 Level 1 activities should be within the ability range of all students, including those with learning difficulties.

L2 Level 2 activities should be within the ability range of the average to above-average student.

L3 Level 3 activities are designed for the ability range of above-average students.

LEP LEP activities should be within the ability range of Limited English Proficiency students.

LS These activities are designed to address different learning styles.

COOP LEARN Cooperative Learning activities are designed for small group work.

P These strategies represent student products that can be placed into a best-work portfolio.

GLENCOE TECHNOLOGY

The following multimedia resources are available from Glencoe.

Science and Technology Videodisc Series (STVS)
Chemistry
   Composite Materials
Human Biology
   Orthopedic Implants

The Infinite Voyage Series
The Future of the Past
Miracles by Design

National Geographic Society Series
Newton's Apple: Life Sciences

Physical Science CD-ROM

Teacher Classroom Resources

This is a representation of key blackline masters available in the Teacher Classroom Resources.

Teaching Aids

Section Focus Transparencies

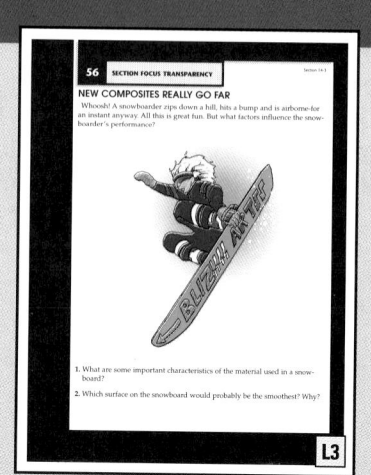

Science Integration Transparencies

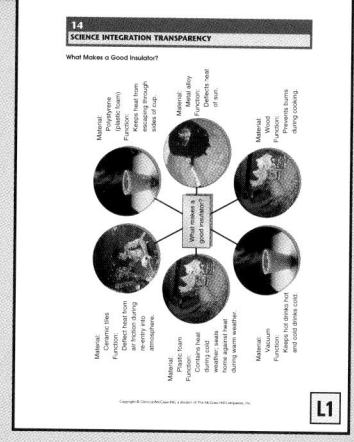

Teaching Transparencies

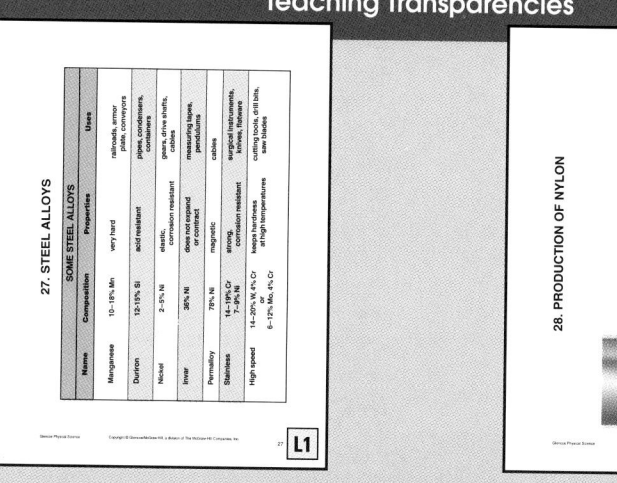

Meeting Different Ability Levels

Study Guide

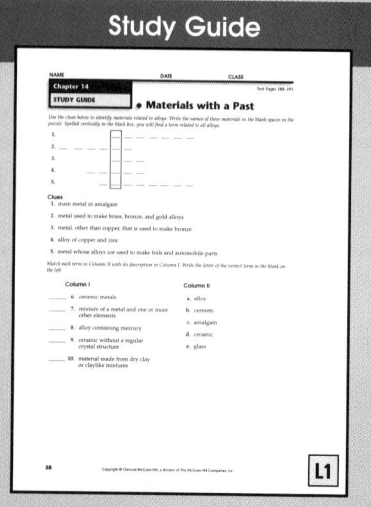

Reinforcement

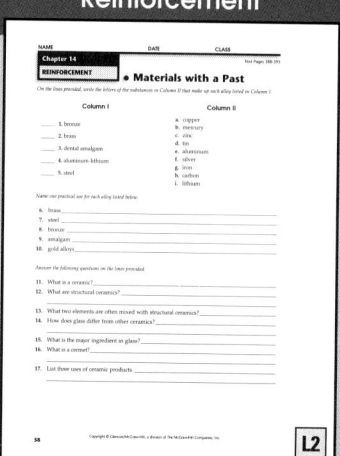

Enrichment Worksheets

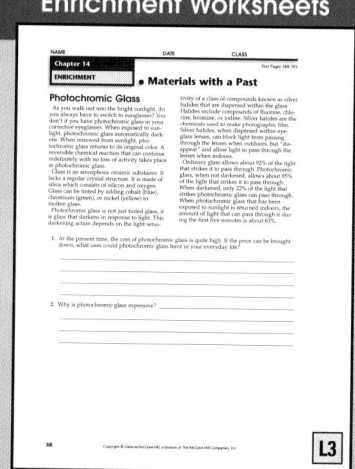

Hands-On Activities

Science Integration Activity

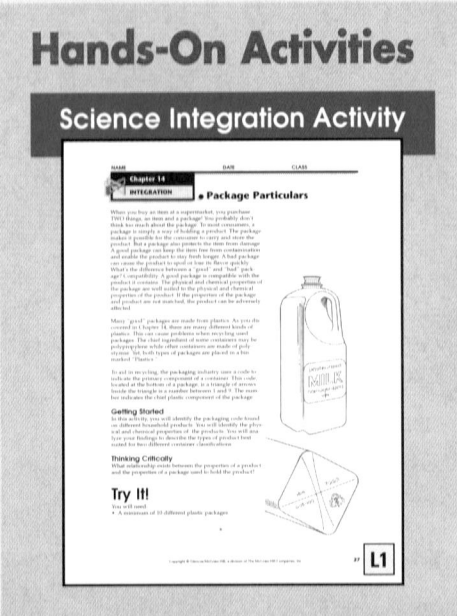

Assessment

Performance Assessment

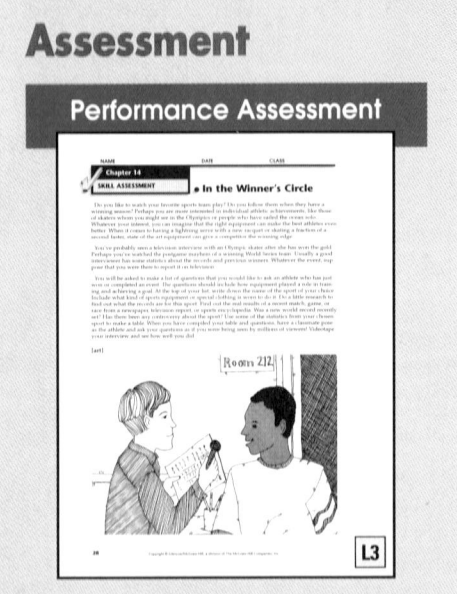

Enrichment and Application

Critical Thinking/ Problem Solving

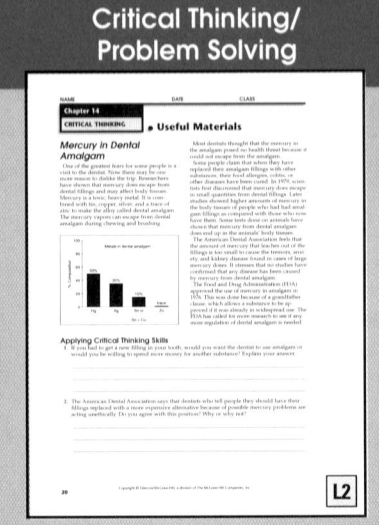

Cross-Curricular Integration

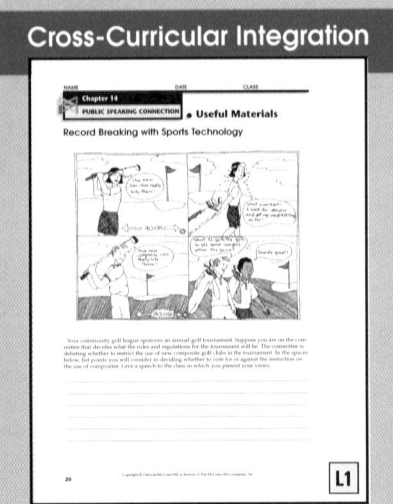

Science and Society Integration

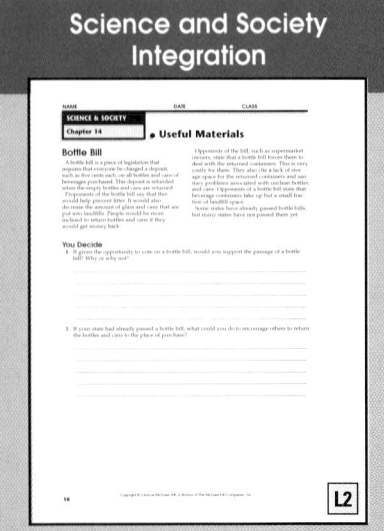

Multicultural Connections

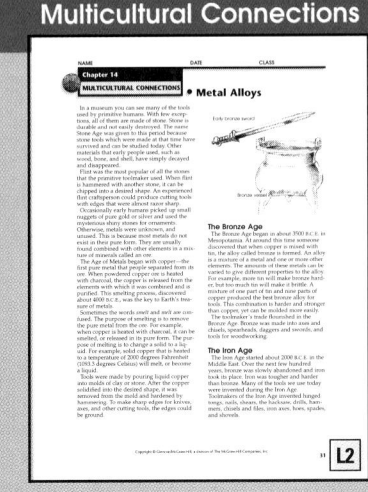

Concept Mapping

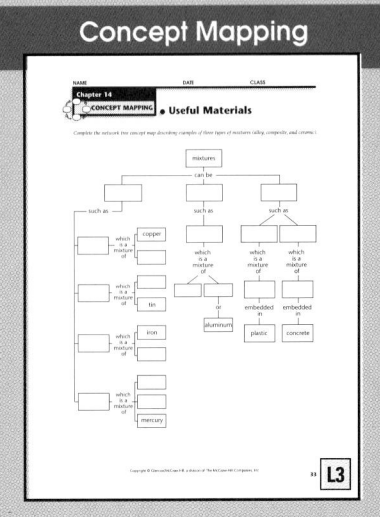

Chapter 14

Useful Materials

CHAPTER OVERVIEW

Section 14-1 The section differentiates between an amalgam and other alloys. The uses and properties of ceramic and glass materials are also presented.

Section 14-2 This section describes the properties and uses of plastics, synthetic fibers, and composite materials.

Section 14-3 Science and Society The effect technology has on sports is the main focus of this section. The question of whether new technology is always useful or might drastically change a sport is presented.

Chapter Vocabulary

alloy	cermet
amalgam	plastic
ceramic	synthetic fiber
glass	composite

Theme Connection

Scale and Structure The student is introduced to the structure of alloys, ceramics, plastics, synthetic fibers, and composite materials. A discussion of the use of new materials in sports causes the student to think about the scale of the effects of new materials on sports and sports equipment.

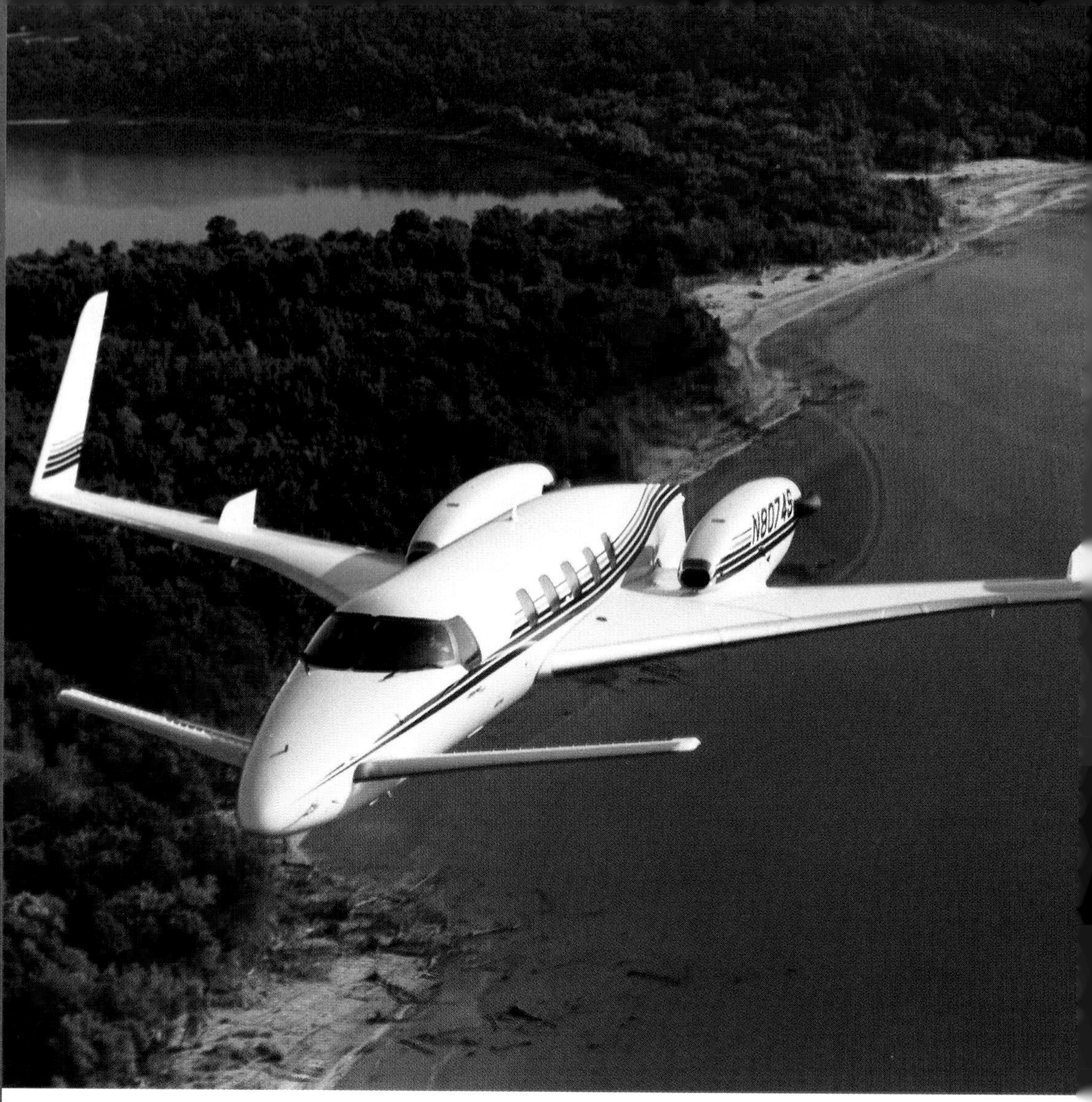

Previewing the Chapter

Learning Styles

Look for the following logo for strategies that emphasize different learning modalities. LS

Kinesthetic	Demonstration, p. 391; Across the Curriculum, p. 397; People and Science, p. 406
Visual-Spatial	Activity 14-1, p. 395; Inclusion Strategies, p. 397; Demonstration, p. 397; Across the Curriculum, p. 399
Interpersonal	Explore, p. 387; MiniLAB, p. 392; Community Connection, p. 398; Activity 14-2, p. 402
Logical-Mathematical	Integrating the Sciences, p. 400
Linguistic	Across the Curriculum, p. 391; Science Journal, pp. 392, 405; Discussion, p. 397; Enrichment, p. 399

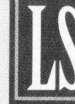

Useful Materials

Many products are redesigned to take advantage of special properties of new materials such as alloys, plastics, and composites. These new materials add strength to a product, cut production time, reduce weight, and protect against corrosion. The airplane shown here is made up of a graphite-carbon composite material that makes it stronger, lighter-weight, and keeps it free from corrosion. In the Explore Activity below, you will investigate another type of useful material.

EXPLORE ACTIVITY

Ceramic Smiles

A ceramic material called dental porcelain is used to restore damaged teeth to their original appearance.
1. Predict the properties porcelain needs to replace tooth enamel.
2. The substances commonly used in dental porcelain are alumina and silica. What chemical elements do you think make up alumina and silica?
3. How are ceramic materials made?

Observe: Survey the members of your class to determine whether they, or members of their families, have made dental visits during which dental porcelain has been used. In your Science Journal, record different reasons people have for having extensive dental work. Analyze these reasons.

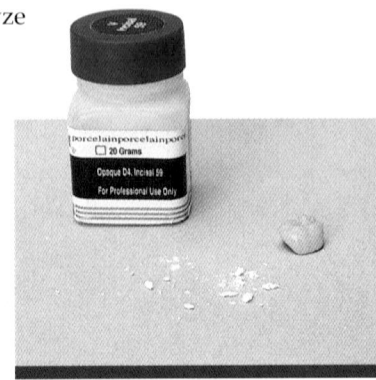

Previewing Science Skills

▶ In the Skill Builders, you will **map concepts, observe,** and **infer.**

▶ In the Activities, you will **observe, infer,** and **predict.**

▶ In the MiniLABs, you will **observe, compare,** and **contrast.**

387

Assessment Planner

Portfolio
Refer to page 407 for suggested items that students might select for their portfolios.

Performance Assessment
See page 407 for additional Performance Assessment options.
Skill Builders, pp. 394, 401
MiniLAB, p. 392
Activities 14-1, p. 395; 14-2, pp. 402, 403

Content Assessment
Section Wrap-ups, pp. 394, 401, 405
Chapter Review, pp. 407-409
Mini Quizzes, p. 394

Group Assessment
Opportunities for group assessment occur with Cooperative Learning Strategies and Flex Your Brain Activities.

EXPLORE ACTIVITY

Purpose
IS **Interpersonal** Use the Explore Activity to introduce students to useful materials.
L1 **COOP LEARN**

Preparation
Ask if a student has a porcelain repair or crown that he or she would show to the class.

Teaching Strategies
• Use teams to find out information about the composition of silica and alumina. **COOP LEARN**
• Students can contact a dentist to determine the composition and properties of the material.
• Ask students to list the properties that dental porcelain should have. It should be water resistant, the same color as the tooth enamel, durable, and able to withstand staining and both hot and cold temperatures.
• Silica is composed of silicon dioxide, and alumina is aluminum oxide. Both materials are hard, light colored, and relatively inexpensive.

Observe
Select two members of the class to conduct the survey and report the results to the class.

✔Assessment

Performance Have groups of students compare their answers to the questions in the Explore Activity. Students can discuss any diversity they notice in their answers. Use the Performance Task Assessment List for Group Work in **PASC**, p. 97. **P**

Prepare

Section Background

- Many common metals are not pure elements, but rather alloys. Some pairs of metals are soluble in each other in all proportions. Homogeneous examples are Cu/Ni, Cu/Au, W/Mo, Pt/Au. Some pairs that do not dissolve completely and therefore form heterogeneous alloys are Al/Si, Pd/Sn, Cu/Sn, and Ag/Cu.

- The solubility of one metal in another is determined mainly by the size of the atoms. Metals with atoms of similar size tend to be soluble in each other. Also, elements whose atoms are much smaller than the other element are usually soluble. For example, hydrogen dissolves in palladium.

Preplanning

- To prepare for the Mini-LAB, obtain flour, salt, cream of tartar, and a light cooking oil.

- To prepare for Activity 14-1, ask students to bring to class bright, shiny pennies. Pennies dated before 1982 are preferred because they are copper and not copper-clad zinc. If bright pennies are not available, dilute the nitric acid. Prepare the sodium hydroxide solution in advance.

1 Motivate

Bellringer

 Before presenting the lesson, display **Section Focus Transparency 54** on the overhead projector. Assign the accompanying **Focus Activity** worksheet. L1 LEP

14•1 Materials with a Past

Science Words

- alloy
- amalgam
- ceramic
- glass
- cermet

Objectives

- Identify common alloys and ceramics.
- Compare and contrast alloys and ceramics.

Alloys

Imagine interstate highways without steel bridges, cities without brick buildings, and rooms without glass windows. It's almost impossible to imagine such things because materials such as steel, brick, and glass are part of your everyday life. As you will see, the characteristics that make these and many more familiar materials useful are rooted in discoveries made thousands of years ago.

When a marching band goes by, you can easily recognize the brass section. The instruments are brilliant in both sound and color. Trumpets, trombones, and tubas are made of an alloy called brass. An **alloy** is a mixture of a metal and one or more other elements. An alloy retains the properties common to metals, but it is not a pure metal. Brass is a mixture of copper and zinc. The brass used in musical instruments and in some hardware is composed of about 80 percent copper and 20 percent zinc. Brass of this quality is a solid solution of the two metals. As you recall from Chapter 9, a solution is a homogeneous mixture. Any brass that contains more than 40 percent zinc is a heterogeneous mixture. This means that the mixture of atoms of the two metals is not completely uniform.

Award-Winning Alloy

Have you ever been awarded a plaque or trophy? Chances are it is made of bronze, an alloy similar to brass. **Figure 14-1** shows several bronze-covered objects. About 5000 years ago, people discovered that a new material could be made by mixing melted copper and tin. The new material—bronze—was

Figure 14-1

These bronze objects show two of the many uses of bronze. *Why is bronze more expensive than brass?*

Program Resources

📁 **Reproducible Masters**
Study Guide, p. 58 L1
Reinforcement, p. 58 L1
Enrichment, p. 58 L3
Science and Society Integration, p. 18
Activity Worksheets, pp. 80, 81, 84 L1
Multicultural Connections, pp. 31, 32
Critical Thinking/Problem Solving, p. 20

🖥 **Transparencies**
Section Focus Transparency 54 L1
Teaching Transparency 27

stronger and more durable than either copper or tin. This first known alloy became so popular and so widely used that a 2000-year span of history is known as the Bronze Age. The Bronze Age ended with the discovery of iron.

Bronze is more expensive than brass because tin is more expensive than zinc. However, bronze is especially useful in plumbing fixtures and in hardware that will be exposed to salt water. Brass is not suitable for such items as boat propellers because the zinc reacts with minerals in salt water, leaving porous copper behind. The tin in bronze does not react with saltwater minerals.

Metals can be combined with other elements in different amounts to produce many different mixtures. **Table 14-1** lists some common alloys, and several are shown in **Figure 14-2**. Thousands of alloys have been created since the time bronze was first made.

Figure 14-2

Alloys are in use all around us.

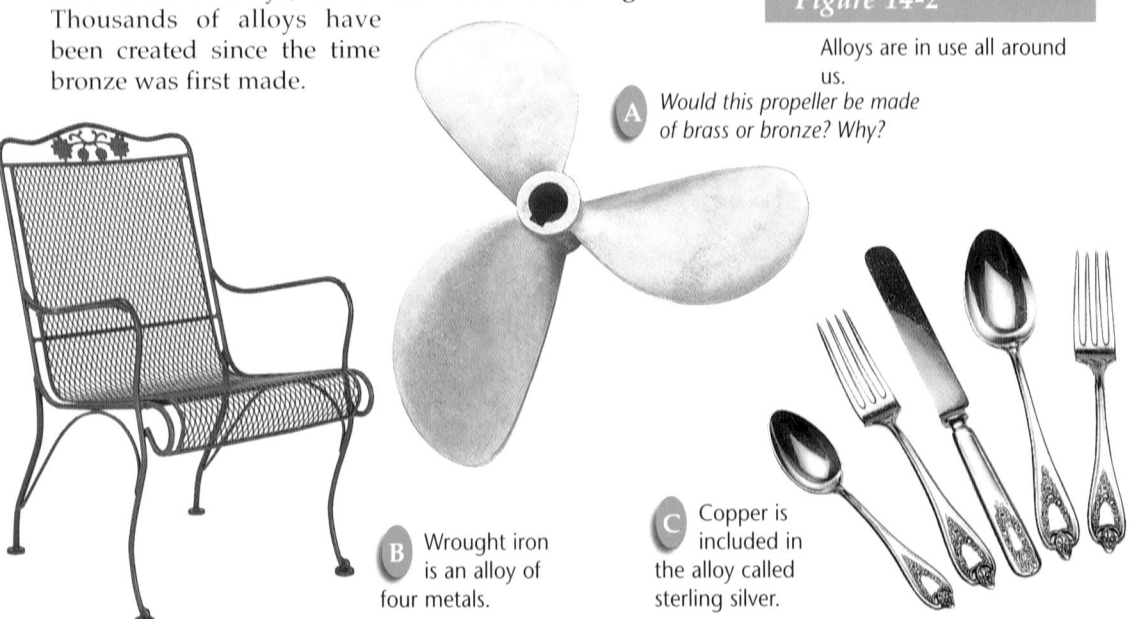

A Would this propeller be made of brass or bronze? Why?

B Wrought iron is an alloy of four metals.

C Copper is included in the alloy called sterling silver.

Table 14-1

Common Alloys		
Name	**Composition**	**Use**
Bronze	copper, tin	jewelry, marine hardware
Brass	copper, zinc	hardware
Sterling silver	silver, copper	tableware
Pewter	tin, copper, antimony	tableware
Solder	lead, tin	plumbing
Wrought iron	iron, lead, copper, magnesium	porch railings, fences

14-1 Materials with a Past   **389**

Cultural Diversity

Lead Isotopes Aslihan Yener, born in Turkey, moved to the United States when she was six months old. In college, she planned to be a chemist. She returned to Turkey and majored in art history, where she became interested in archaeology. She decided to combine her talents and "to apply atomic bomb techniques to archaeology…by analyzing the lead isotopes in mines and metals throughout the Near East." The distinct ratios of lead isotopes to other metals in Bronze Age objects are like a fingerprint, matching the object to the mine the metal came from. Yener has found large Bronze Age industrial parks in the Taurus Mountains. She has also located a semisubterranean city built into the mountainside. Explain how you think determining which mine a metal came from enabled Yener to find Bronze Age industrial parks.

An alloy has properties that are different from and more useful than the properties of the elements in it. Look at any gold jewelry. It appears to be made of pure gold. It is actually made of an alloy. As you know, gold is a bright, expensive metal that is soft and bends easily. Copper, on the other hand, is a somewhat dull, inexpensive metal that is harder than gold. When melted gold and copper are mixed and then allowed to cool, an alloy forms. It has most of the brilliance of gold and most of the sturdiness of copper. By varying the amounts of gold and copper, alloys with properties designed for different purposes are produced. **Figure 14-3** shows several gold alloys.

The alloys shown in **Figure 14-3** contain only two metals. Not all alloys are that simple. Do you or someone you know wear dental braces? Some of the wire used in the braces is an alloy of gold, silver, platinum, palladium, zinc, and copper.

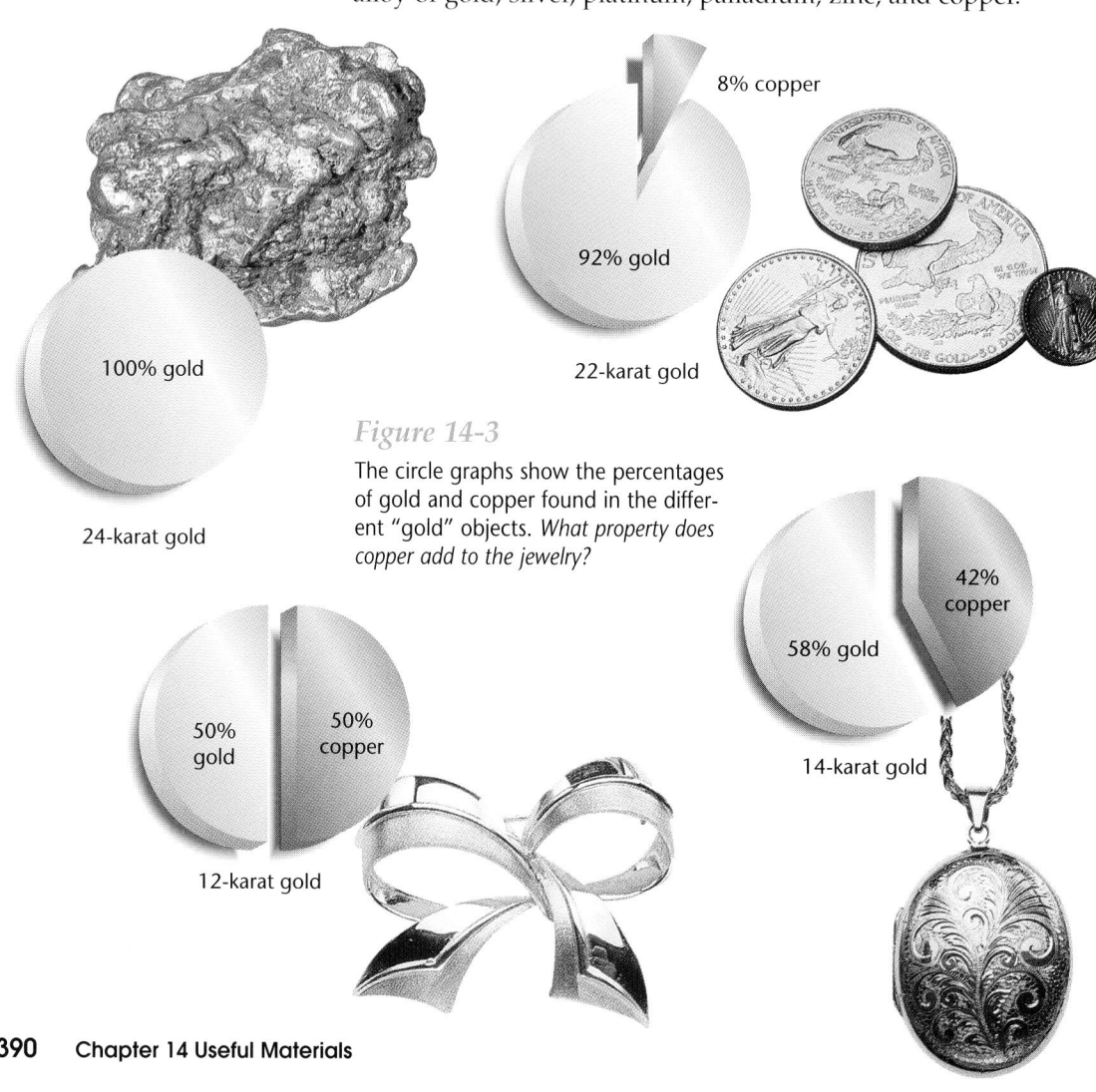

Figure 14-3

The circle graphs show the percentages of gold and copper found in the different "gold" objects. *What property does copper add to the jewelry?*

390 Chapter 14 Useful Materials

Theme Connection

Scale and Structure Adding copper to gold produces an alloy that is more useful for jewelry than is pure gold. The structure of jewelry is enhanced and the items produced are more sturdy. Rings, bracelets, and other jewelry made of pure gold would be too soft. Large pieces of gold jewelry would also be much heavier than that made of a gold-copper alloy.

Some other alloys used by dentists are called amalgams. An **amalgam** is an alloy that contains mercury. Dental amalgams consist of mercury, silver, and zinc. You may have one in your mouth right now. Amalgams are used as fillings for cavities.

Uses of Alloys

Various types of steel make up an important group of alloys. **Figure 14-4** shows a medical use of a steel alloy. Steel production is greater than that of any other alloy. All types of steel are alloys of iron and carbon. You may have seen advertisements for carving knives with blades made of high-carbon steel. The carbon content of this alloy ranges from 1.0 percent to 1.7 percent. Because this steel is hard, the knife blades will be sharp. As you can see from **Table 14-2**, steels of different compositions are made for different uses.

Aluminum alloys are the second largest group of alloys produced. Aluminum alloys are used to manufacture products ranging from foils used for food wraps to automobile bodies.

In the last decade, aluminum-lithium alloys have become important. Because lithium is the lightest metal, these alloys are lightweight, yet exceedingly strong. Unlike other alloys, aluminum-lithium alloys maintain their strength at very high temperatures. In the future, these alloys may be used to make the bodies of aircraft that will travel near the outer limits of Earth's atmosphere.

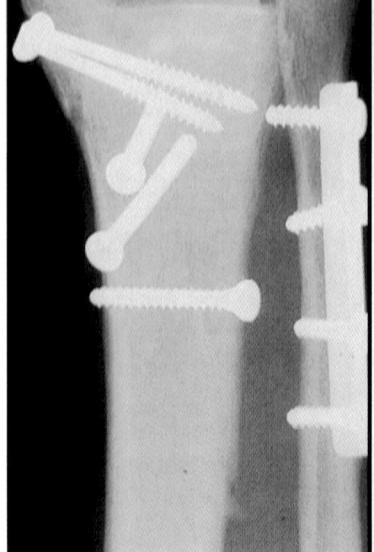

Figure 14-4

Surgical steel can be used to join bones together. *What properties of this type of steel are most important?*

Table 14-2

Some Steel Alloys			
Name	**Composition**	**Properties**	**Uses**
Manganese	10-18% Mn	very hard	railroads, armor plate, conveyors
Duriron	12-15% Si	acid resistant	pipes, condensers, containers
Nickel	2-5% Ni	elastic, corrosion resistant	gears, drive shafts, cables
Invar	36% Ni	does not expand or contract	measuring tapes, pendulums
Permalloy	78% Ni	magnetic	cables
Stainless	14-19% Cr 7-9% Ni	strong, corrosion resistant	surgical instruments, knives, flatware
High speed	14-20% W, 4% Cr or 6-12% Mo, 4% Cr	keeps hardness at high temperatures	cutting tools, drill bits, saw blades

Demonstration

IS **Kinesthetic** An empty aluminum soft-drink can may be used to demonstrate the strength of this lightweight metal. The aluminum skin of an airplane must flex many times without tearing. Have a student wearing leather gloves attempt to tear a soft-drink can into two pieces by gripping the ends and twisting in one direction. The can will usually twist but will not tear.

Visual Learning

Figure 14-4 What properties of this type of steel are most important? *strong, corrosion resistant* **LEP** IS

GLENCOE TECHNOLOGY

 Videodisc
STVS: Human Biology
Disc 7, Side 2
Orthopedic Implants (Ch. 21)

 CD-ROM
Physical Science CD-ROM
Have students perform the interactive exploration for Chapter 14 to reinforce important chapter concepts and thinking processes.

Use these program resources as you teach this lesson.
Critical Thinking/Problem Solving, p. 20
Multicultural Connections, pp. 31, 32

Use **Teaching Transparency 27** as you teach this lesson.

Across the Curriculum

Language Arts Have students write a science fiction story that makes use of a new alloy with unique properties. In the story, have them tell how the alloy is made and what the properties are. An example is the legend surrounding Jim Bowie's knife. It was said to be made from the metal of a meteorite he found. The metal blade was said to be indestructible, and never needed sharpening. **L1** IS

Community Connection

Dental Technology Invite a dentist to visit the class and explain why dental amalgams are preferred over plastic fillings. The dentist could also explain how amalgams are made. Have students ask about the issue of possible health effects of mercury amalgams.

Solve the Problem

1. The outer protein layer of a virus causes our immune system to begin producing disease-fighting compounds.
2. The outer protein layer of a virus could be made to attach to a small ceramic structure that then replaces the infectious virus core.

Think Critically

cancer cells, bacteria, blood cells (Accept any reasonable answer.)

MiniLAB

Purpose

Ⓘ **Interpersonal** Students will observe changing characteristics of ceramics. L1

LEP COOP LEARN

Materials

100 mL each of flour, salt, and water; 5 mL each of cream of tartar and a light cooking oil

Teaching Strategies

Use teams to discover how properties of ceramics change. COOP LEARN

Analysis

1. The material will be a solid that is brittle. It will no longer be wet.
2. Small, rigid structures that bond together form when water is removed.

✓Assessment

Oral Have students explain why properties of ceramics change as they are heated and then cooled. Use the Performance Task Assessment List for Analyzing the Data in **PASC**, p. 27. P

📂 **Activity Worksheets,** pp. 5, 84

Problem Solving

Ceramic Decoys

How does our immune system tell the difference between disease-causing organisms and healthy, necessary cells? You may remember receiving vaccinations against certain diseases. The vaccine probably contained some protein from a virus. That noninfectious part could trigger your body's production of chemicals that would be ready to fight an actual virus invasion.

Finding a noninfectious portion of some viruses has proven difficult. Most viruses consist of an outer protein layer and a core of genetic material. Typically, the outer protein layer of a virus is the part that will begin the immune process without causing a major infection. Can the infectious core of a virus be replaced while retaining the outer protein layer? Researchers are looking to ceramics for a possible answer.

Solve the Problem:

1. Describe the part of a typical virus organism that causes our immune system to begin producing disease-fighting compounds.
2. Ceramic structure can be made strong and small. Ceramics can also be made to attach to some proteins. How could ceramic beads be used to make imitation virus particles?

Think Critically:

If other attachments can be made to small ceramic beads, there may be additional medical applications. What other attachments might there be for these ceramic messengers?

Ceramics

Have you ever watched a potter making a bowl? If you have, you know that the potter spins wet clay on a wheel while shaping it into a bowl by hand. The bowl is then heated in a hot oven, or kiln, to dry it and bring about important chemical changes that harden the material. Pottery like this is made of a material called ceramic. A **ceramic** is a material made from dried clay or claylike mixtures. People began making ceramic containers like the ones in **Figure 14-5** thousands of years ago to store food and carry water.

Figure 14-5

This ancient ceramic pottery is made of a mixture of clay and water.

As with alloys, there are thousands of different ceramics. Is your school building made of bricks? Does it have tiled hallways? Bricks and tiles are examples of structural ceramics. The construction industry uses structural ceramics because of their rigidity and strength. These ceramics contain silicon or aluminum,

---

📝 Science Journal

Ⓘ **Linguistic** Two ways to describe alloys are substitutional and interstitial. Use a dictionary to determine the meaning of these two words. Then, in your Science Journal, write an explanation of how the definitions could be applied to alloys. *Substitutional means*

"replacing something with something else." Interstitial means "fitting in between." Some alloys have atoms from one metal substituted by atoms from another metal. Other alloys have smaller atoms between larger metal atoms.

which can form small, rigid structures that bond together when water is removed. These bonds make the ceramics strong and chemically stable. The porcelain ceramics used on sinks and bathtubs are called whitewares. **Figure 14-6** shows examples of several types of ceramics.

Fragile: Handle with Care

Look at some bricks or bathroom tiles, or at the porcelain enamel around the burner plates of a stove. You'll probably see small breaks in their surfaces. These breaks are evidence of one disadvantage of some ceramics: they are so rigid that they crack. If you've ever had to pick up the pieces of a broken dinner plate, you are well aware of this.

Figure 14-6

Ceramics are used in a variety of ways.

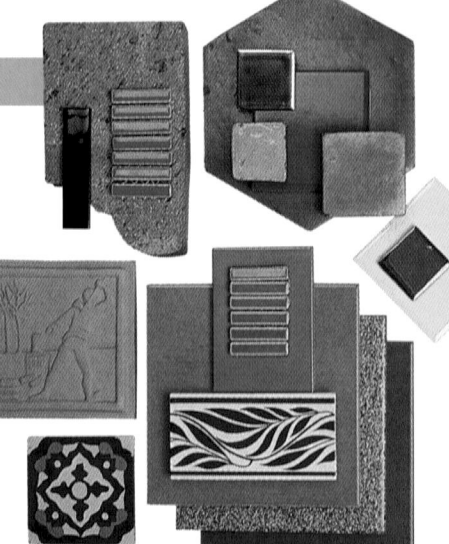

A These structural ceramics are used in construction. *What property of ceramics makes them useful for this?*

B Porcelain enamel is a durable coating for surfaces that are subject to a lot of wear.

C This ceramic piece is an example of the delicate, breakable nature of ceramics.

14-1 Materials with a Past   393

MiniLAB

How do properties of ceramics change?

Procedure

1. Mix 100 mL each of flour, salt, and water in a microwave-safe bowl. Add 5 mL each of cream of tartar and a light cooking oil.
2. Heat for one minute in the microwave.
3. Cool the material and observe its properties.

Analysis

1. Did any physical properties of this material change? If so, explain what differences you notice.
2. In the formation of actual ceramics, high temperatures are used to drive off the water. Explain why the finished product could withstand extreme heat.

Theme Connection

Scale and Structure The theme of scale is involved with a study of small, rigid structures that bond together in a ceramic when water is removed. These rigid structures and the use of structural ceramics in the construction industry reinforce structure as a theme.

Visual Learning

Figure 14-6A What property of ceramics makes them useful for this? *They are strong and chemically stable.*
LEP LS

3 Assess

Check for Understanding

? FLEX Your Brain

Use the Flex Your Brain activity to have students explore ALLOYS.

 Activity Worksheets, p. 5

Reteach

Have students prepare flash cards with examples of each type of material described in the text. For example, on one side of the card is the word *brick,* and on the other side is the word *ceramic.* LEP

Extension

 For students who have mastered this section, use the **Reinforcement** and **Enrichment** masters.

GLENCOE TECHNOLOGY

Videodisc

The Infinite Voyage: Miracles by Design
Chapter 6
New Ceramics: New Uses

Chapter 7
Ceramic Superconductors: Rapid Transportation

4 Close

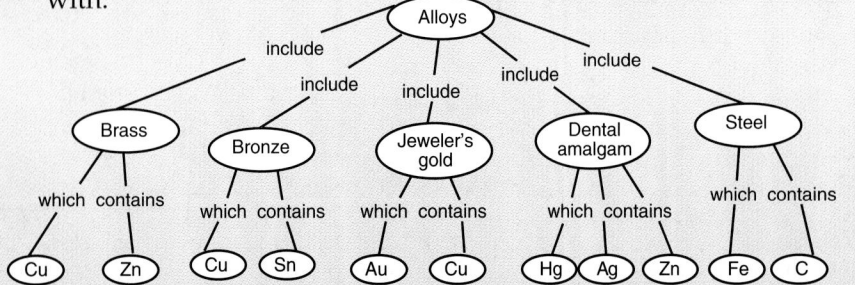

Figure 14-7

Glass can be made to suit different purposes. This dish can be taken directly from the freezer (A) and placed in a hot oven (B). *Why can't all glass products handle the drastic temperature changes shown in these pictures?*

Almost half of the ceramics produced today are classified as glasses. A **glass** is a ceramic without a regular crystal structure. Glasses come in thousands of varieties. You are probably most familiar with the type of glass that you see in windows. This glass contains mostly oxygen and silicon, with smaller amounts of sodium and calcium and a trace of aluminum. The major ingredient that is used to make glass is silicon dioxide, SiO_2—sand.

Elemental Additions

Glasses can be made to have desired properties by changing the kinds and proportions of elements that make them up. Crystal pendants and vases are made from glass that contains lead as well as silicon and oxygen. Food-storage dishes, like the ones in **Figure 14-7** that you can pull from a freezer and place in an oven to heat, are made of glass that contains boron and magnesium or lithium.

Do you have magnets stuck to your refrigerator door? If so, chances are that the magnet itself is a ceramic magnet. These magnets are examples of cermets. **Cermets,** or ceramic-metals, are materials that have properties of both ceramics and alloys. Cermets are tough and heat resistant and are used in gas turbines and rocket motors.

Section Wrap-up

Review

1. What is an alloy?

2. What is a ceramic?

3. **Think Critically:** Why are alloys often preferred to a pure element or metal?

Skill Builder
Concept Mapping

Make a network tree to describe the composition of common alloys using the alloys and elements mentioned in this section. If you need help, refer to Concept Mapping in the **Skill Handbook.**

USING MATH

Use the information from **Figure 14-3** to calculate the grams of actual gold in a 65-g, 14-karat gold necklace.

Activity 14-1

Preparing an Alloy

You have learned that brass is a mixture of copper and zinc. Brass looks very much like gold. In ancient times, people called alchemists attempted to change other elements into gold. They never succeeded, but they did produce some alloys.

Problem
How can you produce an alloy that gives the same appearance as gold?

Materments

Materials
- copper penny
- zinc, 30-mesh
- hot plate
- nitric acid, HNO_3, dilute
- sodium hydroxide solution, NaOH, dilute
- evaporating dishes (2)
- tongs

CAUTION: *Acids and bases can cause burns. Rinse spills immediately. Use nitric acid where there is adequate ventilation.*

Procedure
1. Carefully pour enough dilute HNO_3 into one evaporating dish to fill the dish halfway. Using tongs, grasp the penny and hold it in the acid for about 20 seconds.
2. Still using tongs, remove the penny from the acid and rinse it with cold tap water.

3. Place one teaspoon of 30-mesh zinc in the second evaporating dish. Carefully add dilute NaOH to the dish to a depth of about 2 cm above the zinc.
4. Using tongs, carefully place the penny on top of the zinc. Gently heat the contents of the dish and observe the color of the penny.
5. Set the control of the hot plate on medium high. Using tongs, carefully remove the penny from the dish and rinse it in cold tap water. Dry the penny and place it on the preheated hot plate until the penny has a golden color. **CAUTION:** *Do not touch the hot penny.*

Analyze
1. Observe and describe the appearance of the penny after it was immersed in the nitric acid. Why did its appearance change?
2. In step 4, the penny turned a silver color. Was it actually becoming silver? Was it, at this point, an alloy?

Conclude and Apply
3. Do alloys require specific amounts of ingredients to form?
4. What alloy has this procedure produced?
5. **Infer** why heat is usually necessary for two metals to combine to form an alloy.

Answers to Questions
1. The dark penny became bright. The nitric acid reacted with surface corrosion and removed it.
2. No, the penny was being coated with fresh zinc. It was not an alloy at this point because the metal atoms had not mixed with each other.
3. No. Most alloys are formed from elements that can dissolve in each other in all proportions.
4. brass

5. Heating increases the motion of the metal atoms, causing the components of the alloy to mix.

Assessment

Performance To observe how heat affects the formation of an alloy, perform the procedure without adding heat. Use the Performance Task Assessment List for Analyzing the Data in **PASC**, p. 27. **P**

Purpose
IS **Visual-Spatial** Observe that the properties of an alloy differ from those of the component metals. **L1** **LEP**
COOP LEARN

Process Skills
observing, inferring, communicating, classifying, comparing and contrasting, forming operational definitions

Time
35 minutes

Safety Precautions
- Do not permit students to carry the coin to the sink to rinse it because acid or alkali will drip.
- Acids and bases can cause burns. Rinse spills immediately. Use nitric acid where there is adequate ventilation.

Activity Worksheets, pp. 5, 80, 81

Teaching Strategies
Alternate Materials If you start with clean, shiny pennies (new or cleaned by the teacher), you can have students bypass steps 1 and 2. If you do not have 30-mesh zinc, try granulated zinc and adjust the time according to your results.

Troubleshooting The penny should stay in the zinc long enough to become uniformly coated. Turn the penny or push it into the zinc if needed. Warming speeds the reaction.

- To prepare dilute nitric acid, carefully add 15 mL of concentrated acid to 85 mL of distilled water with constant stirring.
- To prepare 2*M* sodium hydroxide, dissolve 80 g of NaOH in 1000 mL of distilled water. Because the solution will become hot, it should be made well in advance of the activity.

Prepare

Section Background

- Plastics can be made by addition polymerization. In this process, compounds with double bonds add on to each other end-to-end to form the polymer.
- Plastics can also be made by condensation polymerization. To form one of these polymers, one molecule loses a hydrogen atom and the next loses a hydroxyl group.

Preplanning

For Activity 14-2, you will need paper clips and plaster of paris. Pipe cleaners may be used by some students.

1 Motivate

Bellringer

Before presenting the lesson, display **Section Focus Transparency 55** on the overhead projector. Assign the accompanying **Focus Activity** worksheet. L1 LEP

Tying to Previous Knowledge

Students who have camped in a tent will recall that the bottom was probably a plastic called polypropylene. Most plastic wrap used in the kitchen is polyethylene. Vinyl is used to cover notebooks. List on the chalkboard the uses of plastic that students recall.

Visual Learning

Figure 14-8 Can you think of objects you use that are plastic and sustain rough treatment? *Accept all reasonable answers.* LEP IS

Science Words

plastic
synthetic fiber
composite

Objectives

- Compare and contrast plastics and synthetic fibers.
- Describe a composite.

Figure 14-8

The versatility and ruggedness of plastics make them ideal for use in objects that receive a lot of rough handling. *Can you think of objects you use that are plastic and sustain rough treatment?*

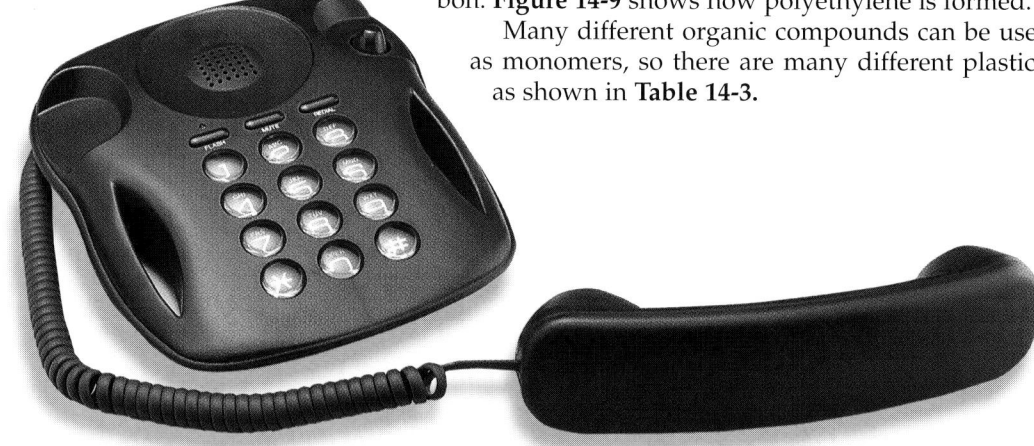

Plastics

Think how often you and your family use the telephone. How often have you accidentally dropped the receiver or knocked a telephone over? Did the phone still work? Phones last because most of their parts are made of durable plastics. A **plastic** is a polymer-based material that can be easily molded into various shapes. **Figure 14-8** illustrates a sturdy plastic telephone. The first synthetic plastic was made only about a century ago. You may be thinking that one hundred years is a long time. However, if you remember that alloys and ceramics have been around for thousands of years, you'll realize that plastics are really modern materials.

Gigantic Molecules

Recall from the last chapter that a polymer is a gigantic molecule formed from thousands of smaller organic molecules, such as hydrocarbons. Molecules that form polymers are called monomers. You are made up of natural polymers, such as proteins and nucleic acids. Polymers that do not form naturally can be manufactured from organic compounds. These polymers are synthetic polymers.

You are probably familiar with the rolls of plastic bags that you find in the produce section of supermarkets. These bags are made of polyethylene, one of the world's most widely produced and used plastics. Among other things, polyethylene is used to make food-storage containers, bottles, and the bands used around beverage six-packs. This synthetic polymer is made of monomers of ethene, an unsaturated hydrocarbon. **Figure 14-9** shows how polyethylene is formed.

Many different organic compounds can be used as monomers, so there are many different plastics, as shown in **Table 14-3**.

Program Resources

Reproducible Masters

Study Guide, p. 59 L1
Reinforcement, p. 59 L1
Enrichment, p. 59 L3
Activity Worksheets, pp. 82, 83 L1
Concept Mapping, pp. 33, 34
Science Integration Activities, pp. 27-28
Cross-Curricular Integration, p. 20

Transparencies

Section Focus Transparency 55 L1
Science Integration Transparency 14
Teaching Transparency 28

Figure 14-9

Polymers are formed through a series of repeating steps.

A An initiating molecule is attracted to the double bond in the monomer.

Initiator

B The double bond in the monomer breaks, causing one electron to be available for bonding to another molecule.

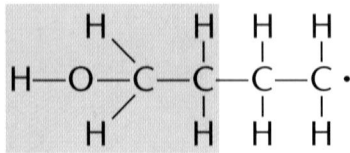
1st monomer

C This monomer is now active and will bond to another monomer, breaking its double bond.

D Eventually, long chains called polymers are formed.

Table 14-3

Common Plastics		
Name	**Polymer Structure**	**Uses**
Polypropylene	$\cdots$C–CH$_2$–C–CH$_2$–C–CH$_2$$\cdots$ with CH$_3$ groups	Rope, protective clothing, textiles, carpet
Polystyrene	$\cdots$CH–CH$_2$–CH–CH$_2$–CH–CH$_2$$\cdots$ with benzene rings	Containers, boats, coolers, insulation, furniture, models
Polyvinyl chloride (PVC)	$\cdots$CH$_2$–CH–CH$_2$–CH–CH$_2$–CH$\cdots$ with Cl groups	Rubber substitute, cable covering, tubing, rainwear, gaskets
Teflon (polytetrafluoroethane)	$\cdots$C–C–C–C–C–C$\cdots$ with F groups	Nonstick cookware surfaces
Saran (polyvinylidene chloride)	$\cdots$C–C–C–C–C–C$\cdots$ with H and Cl groups	Clinging food wraps

Demonstration

LS **Visual-Spatial** Place 12-15 clothes hangers in a sack. Tell students that these hangers represent monomers. Shake the closed sack for 30 seconds. Reach in the sack and slowly pull a hanger out. Many of the monomers will now be attached, forming a polymer. **LEP**

Using an Analogy

Skin is composed mostly of proteins. A protein is a polymer. Skin is tough yet flexible. It is waterproof and provides a barrier against infection. It is also a mixture of polymers. Skin has many of the characteristics of a good plastic.

Discussion

LS **Linguistic** Have a student give a brief report about the Bronze Age. Then ask the class whether they believe this is the Plastic Age. **L2**

GLENCOE TECHNOLOGY

 Videodisc

The Infinite Voyage: Miracles by Design
Chapter 5
Burn Patients and Artificial Skin

Across the Curriculum

Art Have students bring in different small items that are made from plastic. Arrange them using stiff wire and string to form a mobile. Hang the mobile from the ceiling under a sign that reads "New Materials." **LS**

Inclusion Strategies

Gifted Ask students to research the history of the development of plastics. They should find out how plastics have influenced the way we live and the environment. Students can make a mural showing the development of plastics, highlighting historic development changes. Students may choose instead to make a poster showing the new classifications for recycling plastics. **L3** **LS**

Do you know that making plastic soft-drink bottles is similar to blowing up a balloon? Refer to **Figure 14-10** as you read how polyethylene bottles are manufactured. First, a tube of warm polyethylene is placed inside a bottle-shaped mold. Then the mold is closed, sealing the bottom of the tube. Next, compressed air is blown into the polyethylene tube; the tube expands and takes the shape of the mold. The mold is then opened and the bottle is removed.

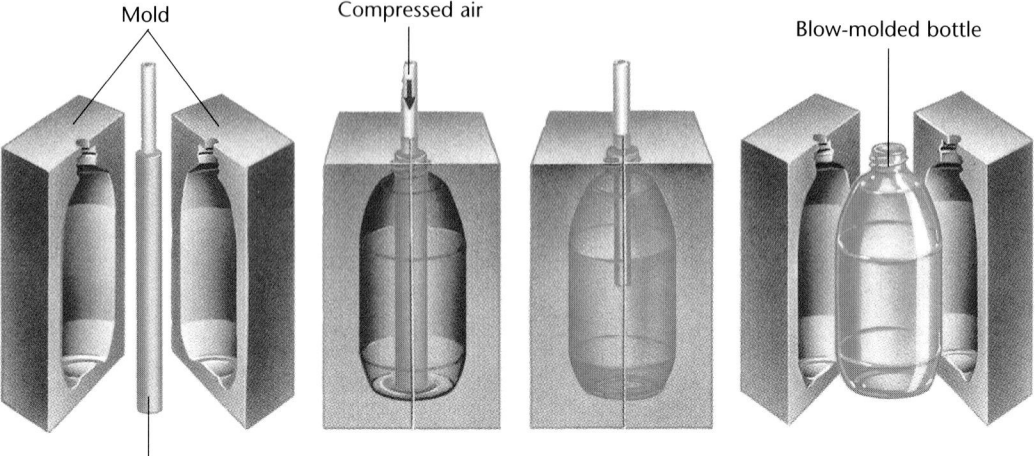

Figure 14-10

The major use of high-density polyethylene (HDPE) is to make blow-molded products. *How does this process illustrate Boyle's law?*

Table 14-4

Common Polymers		
Name	**Structural Formula**	**Uses**
Dacron (a polyester)	$\cdots C-\bigcirc-C-OCH_2CH_2O-C-\bigcirc-C-OCH_2CH_2O\cdots$	Textiles, arterial grafts
Nylon 66	$\cdots C(CH_2)_4-CN(CH_2)_6N-C(CH_2)_4C-N(CH_2)_6N\cdots$	Tire cord, textiles, brush bristles, netting, carpet, athletic turf, sutures
Polyethylene	$\cdots CH_2-CH_2-CH_2-CH_2-CH_2-CH_2\cdots$	Tubing, prosthetic devices, packaging materials, kitchen utensils, paper coating
Orlon (polyacrylonitrile)	(structural formula)	Textiles

398   Chapter 14 Useful Materials

Polymers in Fabrics

The clothing you are wearing may be made of fabrics that have been woven from synthetic fibers. A **synthetic fiber** is a strand of a synthetic polymer. A strand of a synthetic fiber called Kevlar is five times stronger than a similar strand of steel. It is so strong it is used to make bulletproof vests. Synthetic fibers are used to weave both indoor and indoor-outdoor carpeting, upholstery coverings, and other textiles. Today, the use of synthetic fibers is greater than that of natural fibers. **Table 14-4** lists the names and polymer structures of several familiar synthetic fibers. Fabrics with new properties are made by weaving natural fibers, such as cotton and silk, with synthetic fibers. If you look at the label of the permanent-press shirt in **Figure 14-11**, on page 400, you will see that the fabric is a mixture of natural and synthetic fibers.

Model for new polymer

USING TECHNOLOGY

A Sticky Lesson

In the 1940s, a walk in the woods led to the discovery of a popular method of fastening objects. Today, scientists are using the same idea to improve the manufacturing of composite materials.

Instead of merely complaining about the way some burrs from pesky plants hooked to his trousers, Swiss engineer George de Mestral studied the persistent clinging seeds and saw a possible improvement on the traditional zipper. A few years later, he patented his idea for a fastener called Velcro.

When scientists want to experiment with blending two polymers to make a composite, they must find a way for the polymers to remain connected. Compounds known as compatibilizers can be added to bring the two polymers together. Scientists are studying a new type of polymer that may hold materials together in new composites. Like Velcro, this polymer has two important structural features. It has a rigid framework with long, branched parts that stick out in loops and curves.

Think Critically:

Look at the structures of the polymers on pages 397 and 398. Explain in your Science Journal why this new connecting polymer might prove to be rather effective.

Inquiry Question

Why are synthetic fibers necessary in order to clothe the world's population? *Natural fibers rely on agricultural processes. Land that will grow cotton and graze sheep is also needed to produce food. Oil wells don't require much land and can yield the raw materials needed to produce synthetic fibers.*

USING TECHNOLOGY

Think Critically
The loop-and-curve structures present in the polymers enable them to bind together similar to Velcro. As two or more polymers bind in this way, new composites are formed.

Enrichment

LS **Linguistic** Have students investigate the monomers that go into the production of synthetic fibers such as Kodel, Acrilan, Dynel, and Mylar. Kodel is a copolymer of ethylene glycol and terephthalic acid. Acrilan is a polymer of acrylonitrile. Dynel is a copolymer of acrylonitrile and vinyl chloride. Mylar is a copolymer of ethylene glycol and terephthalic acid. **L3**

• Send a student team to interview a fire-prevention officer to determine whether there is a danger in having curtains, drapes, carpeting, and upholstery made of synthetic fibers. Have students ask what gaseous products form when various fibers burn. **L1**
COOP LEARN

Across the Curriculum

Art Have each student bring in a swatch of cloth made from synthetic fibers. Have the class decide on a picture they would like and draw a rough sketch on heavy card stock. "Paint" the picture by cutting and gluing the cloth over the rough sketch. **LEP** **LS**

Theme Connection

Scale and Structure The theme of structure is reinforced with the description of the interwoven structure of Kevlar used in making bullet-proof vests. Scale is supported by the discussion of polymers, gigantic molecules formed from thousands of smaller organic molecules.

Inquiry Question

Silk is made by a caterpillar that eats the leaves of mulberry trees. Compare the cost of a man's silk tie to a polyester one. Why is there a price difference? Why are silk ties still preferred? *The production of silk requires more labor, which is expensive. Silk is also in shorter supply than are synthetic fibers. However, silk usually gives a better appearance.*

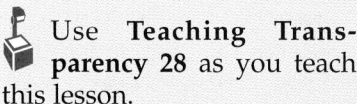

Use **Teaching Transparency 28** as you teach this lesson.

3 Assess

Check for Understanding

? FLEX Your Brain

Use the Flex Your Brain activity to have students explore COMPOSITES.

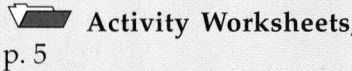 **Activity Worksheets,** p. 5

Reteach

Have students make flash cards with names of modern materials on one side and examples listed in the textbook on the reverse side. **LEP**

Extension

For students who have mastered this section, use the **Reinforcement** and **Enrichment** masters.

Figure 14-11

Wrinkle-resistant materials can be produced by combining synthetic and natural fibers. *What synthetic fibers can you identify in this label?*

One synthetic fiber that is widely used today is nylon. **Figure 14-12** shows how nylon fibers are manufactured. Nylon chips are heated until they melt. The melted nylon is pumped into a high-pressure chamber. It is then forced through tiny openings of a nozzle called a spinneret. As it cools, the nylon forms long strands.

You may wonder where the raw materials—the hydrocarbons and organic compounds—used to produce plastics and synthetic fibers come from. Most of them are found in petroleum—crude oil—and in natural gas. Can you explain why plastics and synthetic fibers are sometimes called petrochemical products? As the production of plastics and synthetic fibers continues to increase, so will the demand for crude oil.

Figure 14-12

Nylon production results in long strands just as silk production does.

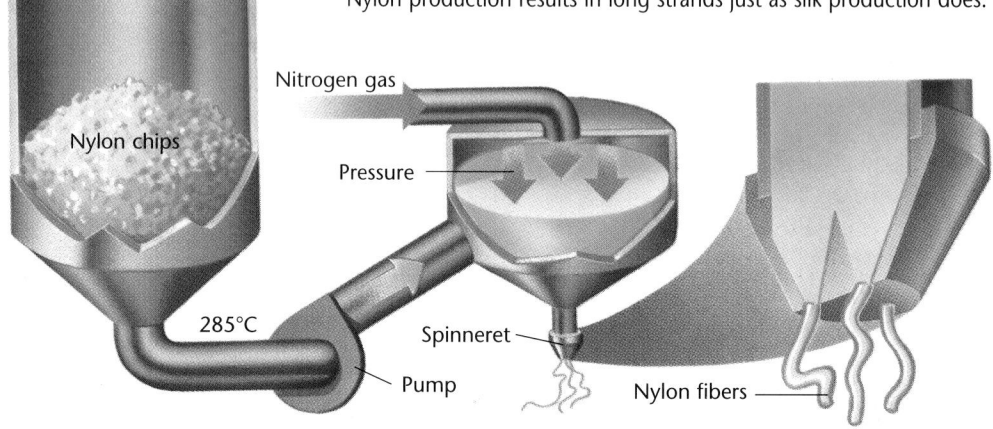

 **INTEGRATION** **Life Science**

A Natural Polymer

Silk is a natural polymer. It is made when silkworms stop eating and begin to make a cocoon in which to live as they change into moths. The worm secretes a protein polymer and gumlike material from its mouth. The protein dries and stiffens as the worm swings its head from side to side, weaving a continuous string of silk. Eventually, the worm seals itself inside of the lightweight but strong cocoon.

Today, silk farmers harvest this natural polymer by pulling off the long continuous fiber to later weave it into fabrics. One thin silk strand from a cocoon may be from 500 to 800 m long. Natural silk is lower in density than rayon and stronger than nylon. It has excellent dying properties and is warmer than cotton.

400 Chapter 14 Useful Materials

Integrating the Sciences

Life Science Have students compare the initial cost and cost of care for fabrics made by life processes and those made by synthetic processes. Then have them conduct a survey of clothing worn by their families. They should list the clothing item, its fiber content, and the cleaning instructions. **L1** **LS**

Community Connection

Car Composition Use the telephone directory to assign a car dealer to each student. With prior approval from the dealer, have the student visit the showroom to determine which cars have composite body panels or plastic exterior parts. Compile the results of the survey. **L1**

When synthetic polymers such as nylon became cheaper to produce and exhibited many of the same properties as silk, they replaced silk in many uses.

Composites

Many high school seniors take college entrance exams. Their score is called a composite score. It is a rating that depends on several subtest scores in different subjects. The word *composite* means "made of two or more parts."

A bridge built of concrete has long steel rods running through it. These rods reinforce the concrete, giving it additional strength and support. Reinforced concrete is an example of a composite. A **composite** is a mixture of two materials, one embedded in the other.

Have you ever heard of cars with fiberglass bodies? These bodies are made of a glass-fiber composite that is a mixture of small threads, or fibers, of glass embedded in a plastic. The structures of the fiberglass reinforce the plastic, making a strong, lightweight composite. A glass-fiber composite, like the one shown in **Figure 14-13,** is an example of a ceramic embedded in a plastic. How would you describe reinforced concrete? Many different composites can be made using various metals, plastics, and ceramics. In the future, you may be driving a car or living in a house that is made almost entirely of composites.

Figure 14-13

This fiberglass piece shows clearly the glass fibers embedded in the plastic. *What advantage is gained by using composites?*

4 Close

Have students discuss the advantages of a metal-ceramic composite automobile engine over one made of an alloy.

Section Wrap-up

Review

1. Compare and contrast plastics and synthetic fibers.

2. Describe a composite.

3. **Think Critically:** Many athletes now play on nylon-based artificial playing surfaces. What properties of plastic make this surface an advantage for the athletes?

Skill Builder

Observing and Inferring

Look at the figures in this section. Describe a way to manufacture a roll of polyethylene bags. If you need help, refer to Observing and Inferring in the **Skill Handbook.**

Science Journal

In the future, many objects will be made of plastic that are not today. In your Science Journal, write a paragraph about one of those objects, including new properties it might have.

Skill Builder

The following is a description of one method of manufacture. A continuous sheet of polyethylene is rolled into a continuous tube. The tube is inflated by compressed air to form a long cylinder. One end of the cylinder is wrapped around a roller that begins to pull and roll up the cylinder. At certain intervals, the rolled polyethylene is pinched shut to form the bottom of an individual bag. **P**

Assessment

Portfolio Have students diagram their production description. Use the Performance Task Assessment List for Scientific Drawing in **PASC,** p. 55. **P**

Section Wrap-up

Review

1. Plastics and synthetic fibers are both made from synthetic polymers. Plastics are formed by making polymers into large sheets or molding them into different forms. Synthetic fibers are formed by drawing synthetic polymers into long strands.

2. A composite is made from two materials, one embedded in the other.

3. **Think Critically** Plastics are waterproof and somewhat flexible.

Science Journal

Answers will vary. Allow all ideas that are supported by scientific explanations.

Purpose

IS **Interpersonal** Design and carry out an experiment to show how a composite material will have properties of the utilized materials. **L1**
COOP LEARN

Process Skills

interpreting data, observing, measuring, predicting, communicating, classifying, making and using tables, forming a hypothesis, designing an experiment, separating and controlling variables, making models

Time

two class periods separated by a day for drying

Materials

common items, such as straws, pipe cleaners, paper clips, crayons, string, and popsicle sticks

Safety Precautions

Caution students to have all means of comparing products approved by you. Any striking of materials by a hammer should be done when the material is wrapped in a towel and any person nearby is wearing goggles.

Possible Hypotheses

- Embedding long, thin materials such as pipe cleaners or paper clips will add strength to the composite.
- Embedding soda straws will make the composite lighter but also weaker.

📁 **Activity Worksheets,** pp. 5, 82, 83

Activity 14-2

Design Your Own Experiment
Compose Your Own Composite

When designing composites, engineers carefully consider the properties of the individual materials that they want to blend. The resulting composite utilizes the most desirable properties of each component to make new materials.

PREPARATION

Problem

How do various embedded materials give new properties to plaster of paris?

Form a Hypothesis

Based on the properties of the embedding materials and the brittleness of plaster of paris, form a hypothesis about the properties of the new composite.

Objectives

- Observe the effects of blending various materials on changing the properties of plaster of paris.
- Infer the relationship between the characteristics of individual components and the final properties of a new composite.

Possible Materials

- plaster of paris
- water
- measuring cup
- heavy-duty aluminum foil
- hammer
- paper towels
- beaker
- miscellaneous possible embedding materials; i.e., paper clips, plastic soda straws, pipe cleaners, crayons, etc.

Safety Precautions

Be sure to use protective eyeware.

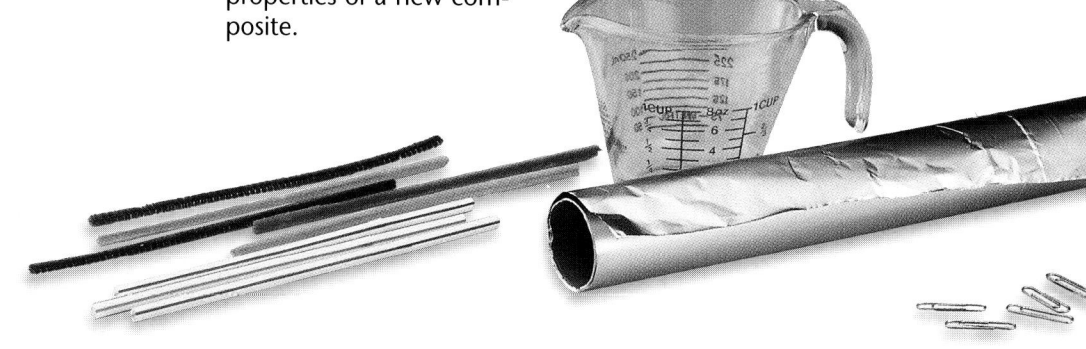

Inclusion Strategies

Visually Impaired Have teams divide the activity tasks into measuring, mixing, and constructing. A reader can give instructions and monitor each task. Students with impaired vision can mix the materials and help with construction tasks.
COOP LEARN

PLAN THE EXPERIMENT

1. As a group, agree upon and write out the hypothesis statement.
2. As a group, list the exact activity you need for each step in testing the hypothesis.
3. Decide what type of material you want to embed in the plaster of paris. Think about the properties of the embedding material and its possible effect on the resulting composite.
4. Decide how you will compare the properties of the composite with those of the original plaster of paris.

Check the Plan

1. Make sure that your plan makes sense as you have stated it.

2. Have you included the directions for making your aluminum foil compartments and for making the plaster of paris?
3. Will you need to repeat any of your steps?
4. How will you report your findings?
5. *Make sure that your teacher approves all embedding components and comparative tests.*

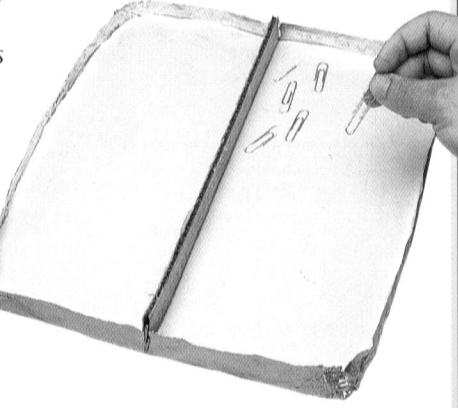

DO THE EXPERIMENT

1. Carry out the experiment, being sure to record in your Science Journal the placement of the embedding materials.
2. Write down any new properties your composite exhibits.

Analyze and Apply

1. **Compare** the materials and results other groups had with their new composites.

2. How do you think the density of the composite **compares** to the density of plaster of paris?
3. **Hypothesize** what would happen if you doubled the amount of embedding material while holding the amount of plaster of paris constant.

Go Further

Think about a few familiar concrete structures you have seen. In what way is your composite similar to these concrete structures?

14-2 New Materials **403**

✔Assessment

Performance Have students embed overlapping small sheets of metal screen (from an old window screen) in plaster of paris and observe any change in properties. Use the Performance Task Assessment List for Making Observations and Inferences in **PASC**, p. 17. **P**

Go Further

Like reinforced concrete, plaster of paris is poured around another strengthening material.

PLAN THE EXPERIMENT

Possible Procedures

Screening may cause the composite to resist crumbling. Try to get students to relate the desired property to a logical choice of embedding material.

Teaching Strategies

Tying to Previous Knowledge Help students recall that a composite is a mixture of two materials, one embedded in the other.

Troubleshooting If possible, try the plaster mix ahead of time so that you can tell students how long to wait before pouring it. The plaster should become thick enough to support the reinforcing materials but not so firm that it will not adhere.

• For best results, observe the 24-hour drying time.

DO THE EXPERIMENT

Expected Outcome

Most results will reflect changes in composite properties, depending upon the material embedded in the plaster of paris.

Analyze and Apply

1. Student results will vary depending on what was embedded.
2. The density of the composite will increase if the embedded material has a greater density than does plaster of paris. If the embedded material is light or causes air to be trapped, its presence will decrease density.
3. Doubling the amount of material may be beneficial if it does not prevent the plaster of paris from adhering to the material.

403

Prepare

Section Background

Graphite embedded in a plastic is a composite that is used in golf clubs, tennis rackets, and fishing rods. It is also being used to replace aluminum panels in aircraft skins. It is lighter, stronger, and can be made to absorb radar.

1 Motivate

Bellringer

Before presenting the lesson, display **Section Focus Transparency 56** on the overhead projector. Assign the accompanying **Focus Activity** worksheet. L1 LEP

Tying to Previous Knowledge

Have students interview members of the athletic department to review possible connections of new school athletic records to new materials in use for sports equipment and sportswear.

2 Teach

Visual Learning

Figure 14-15 How might the properties of a new tennis racket or golf club compare with the older wooden models? *New equipment is stronger, which may improve the speed of a serve or power of a swing.* LEP LS

Objectives

- Compare and contrast the advantages of new materials used in sports with older materials.
- Identify chemical elements and materials of technology used in sports.
- Determine whether sports should alter rules about the use of technology.

Figure 14-14

The use of composite materials has benefits for many sports.

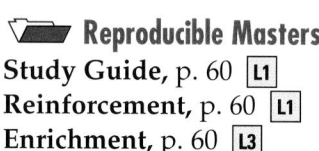

ISSUE:
14•3 Record Breaking with Sports Technology

Records: Made to be broken?

Can you think of any sports record in your school, or are you aware of a recent professional accomplishment in sports? It seems that when each record or "best-ever" achievement is set, someone quickly makes plans to break it. Many top competitors have come to realize that record performances can be helped by technology.

Making Every Second Count

In the Olympic sport of speed skating, races are won and records are set by less than a tenth of a second. How can technology give skaters that tiny time advantage? Composite materials such as steel and carbon for the blade and other carbon composites for the shoe cut the overall mass and air drag on a skate. Silicon-based polymers in the skating suit shown in **Figure 14-14** can help direct the wind current around a fast-moving skater for a slight air pressure advantage. In other winter sports, new plastic materials are used to stabilize skis during jumping and to aid bobsled racers.

Carbon and boron fibers are used to build lightweight, streamlined auto racing-car bodies. Reducing the weight of an Indianapolis racing car could increase the gas mileage, alter the friction, and result in a faster race time.

Performance Advantages

Not every sports record or win depends on time. Tennis and golf are examples. One of the premier tennis events is the annual Wimbledon grass court tournament. How has technology affected this traditional contest? In tennis, an accurate and  fast serve gives a player a tremendous advantage. Strong rackets, such as the one in **Figure 14-15,** made from carbon and boron composites, are now being used to improve the speeds of some serves. These improvements give the receiving player little time to react. New boron-composite golf clubs are also giving golfers more powerful swings than ever.

404   Chapter 14 Useful Materials

Program Resources

Reproducible Masters
Study Guide, p. 60 L1
Reinforcement, p. 60 L1
Enrichment, p. 60 L3

Transparencies
Section Focus Transparency 56 L1

2 Points of View

If It Works, Use It

Many improvements in sports performance have been brought about by using scientific analysis to change training methods, athlete nutrition, and coaching techniques. Why not use technology when designing the equipment? A running shoe with new lightweight, but supportive materials will still require a talented runner to win the race.

Winning Isn't Everything

Setting a record or winning the game is not the only reason we enjoy sports. Many fans and participants enjoy the strategy and efforts in a contest based on the present restrictions. Changes in some equipment could cause drastic changes in the basic aspects of the game. For example, would baseball sluggers have such an advantage with special composite bats that home runs would become as common as singles?

Figure 14-15

These athletes are using materials made with new technology. *How might the properties of a new tennis racket or golf club compare with the older wooden models?*

Section Wrap-up

Review

1. What are two chemical elements that are finding new uses in the world of sports?

2. What advantages do composite-based materials offer players?

interNET CONNECTION The properties of each composite are dependent on how the material is made. Visit the Chapter 14 Internet Connection at Glencoe Online Science, **www.glencoe.com/sec/science/physical,** for a link to more information about carbon-carbon composites.

SCIENCE & SOCIETY

Science Journal

LS Linguistic Have students research how athletic shoes have changed from the 1960s through the 1990s. Have them write in their Science Journals a summary of changes and how these changes may have affected sports records. **L2**

---

3 Assess

Check for Understanding

Call on one student to summarize the lesson to another student as if that student had missed class.

Reteach

Ask students to consider whether a record in sports is truly a record if new materials, not available to everyone and not recognized as legitimate, are used in setting the record.

Extension

For students who have mastered this section, use the **Reinforcement** and **Enrichment** masters.

4 Close

Have the students take on the role of a sports record holder from the past and tell why they would or would not want new materials used for sports equipment.

Section Wrap-up

Review

1. Carbon, silicon, and boron may all be listed.
2. improved speed of servers in tennis, more powerful golf swings, new lightweight supportive materials in running shoes, among other answers

405

Background

- Toy companies test their products with their most important customers—children. They might arrange to bring a prototype of a new toy to a daycare center for informal testing. In a more formal testing situation, children of the appropriate ages will be brought into a room with the toys to be tested. Company personnel will observe the children at play through a one-way window, taking notes on how the children use the toys. This kind of market research might show that a toy demands more physical dexterity than anticipated, or else that kids find the toy babyish. Then modifications are made before the toy is put on the market.

- In the hands of artists, plastic building blocks have recreated national monuments such as Mount Rushmore (1.5 million blocks), and the Statue of Liberty (1.4 million blocks). Both are on view at Legoland in Denmark. If laid end to end, one year's production of Lego blocks would circle Earth eleven times.

- During a severe storm in January of 1992, a cargo vessel was making its way across the Pacific Ocean. A huge steel container fell overboard spilling 29,000 plastic bath toys into the waves. The inadvertent launching of brightly colored ducks, frogs, and turtles was an oceanographer's dream-come-true. Ten months later, some of the creatures were sighted on the shores of southeastern Alaska. The arrival of the toys provided a valuable record of the northwest Pacific's winds and currents.

People and Science

MARTIN CAVEZA, *Industrial Designer*

On the Job

Q How do the designers at your toy company come up with new product ideas?

A We regularly hold brainstorming sessions in which the major guideline is "There's no such thing as a bad idea." Even a really silly idea can snowball into something great. Other times, we might go to the people in the chemistry lab and say, "What have you guys found that might be fun for us to look at?" Some classic toys have had their starts in this somewhat accidental fashion.

Q What determines the kind of plastic you use in a particular toy?

A For the durability most of our toys demand, we use a super-high-impact styrene. For an even more impact-resistant, scratch-resistant version, we use a plastic called ABS, acrylic butyl styrene. However, some toys need bendable parts. For those we use a polypropylene, which has a property we call a living hinge. Think of the lid on a plastic box, for example. Its hinge must be flexible enough to bend back and forth many times without breaking.

Personal Insights

Q How did you get into the business of designing toys?

A After beginning in engineering and intending to be an architect, I found out about the field of industrial design, designing products. Industrial designers design everything from cars, to blenders, to drills, to almost anything the consumer touches. Then I was a winner in a design contest held by a toy company. My entry was a clear plastic space vehicle with rubber-band power. The company liked my concept and offered me a job.

Career Connection

The manufacturers of toys are interested in feedback from their customers. Think of a toy that you have enjoyed, or even one that you found disappointing. Write a letter to the manufacturer. Outline what you consider to be the best (or worst) features of the toy. Investigate the following related careers.

- **Architect**
- **Manufacturing Engineering**

Teaching Strategies

KS **Kinesthetic** Collect some everyday objects, such as rubber bands, small rubber balls, paper clips, and pieces of scrap wood. Challenge students to use the materials to create a toy. **L1**

Career Connection

- Martin Caveza feels that industrial design is a combination of engineering and art, because industrial designers need both to understand how something works and to be able to communicate ideas through drawing. Many universities offer degrees in industrial design through their college of art.

Summary

14-1: Materials with a Past

1. People have been making and using alloys and ceramics for thousands of years. Some common alloys include bronze, brass, amalgams, and various alloys of iron. Some common ceramics include structural ceramics, such as brick and tile, and various kinds of glass.

2. An alloy is a mixture of a metal with one or more other elements. Alloys exhibit metallic properties. Ceramics are composed of clay or claylike mixtures. Except for cermets—ceramic metals—ceramics generally have nonmetallic properties.

14-2: New Materials

1. Plastics and synthetic fibers are materials made from synthetic polymers. Plastics can be produced in many forms, ranging from thin films to thick slabs or blocks. Synthetic fibers are produced in thin strands that can be woven into fabrics.

2. A composite is a mixture of two materials, one embedded in the other. Reinforced concrete is an example of a composite.

14-3: Science and Society: Record Breaking with Sports Technology

1. New composite materials used in sporting events allow top-level athletes to challenge records. Examples in speed events such as auto racing, bobsledding, and skating point out the advantages of strong, lightweight materials.

2. Some people feel that either equipment updates should not be permitted, or some game rules must be changed to preserve traditions and skills in games such as tennis and baseball.

Key Science Words

a. alloy
b. amalgam
c. ceramic
d. cermet
e. composite
f. glass
g. plastic
h. synthetic fiber

Reviewing Vocabulary

Match each phrase with the correct term from the list of Key Science Words.

1. a strand of a synthetic polymer
2. ceramic with no regular crystal structure
3. mixture consisting of a metal and one or more other elements
4. made from dried clay or claylike mixtures
5. material made from polymers; can be easily molded
6. mixture of one material embedded in another
7. has properties of ceramics and alloys
8. alloy containing mercury

Chapter 14

Review

Summary

Have students read the summary statements to review the major concepts of the chapter.

Reviewing Vocabulary

1. h	**5.** g
2. f	**6.** e
3. a	**7.** d
4. c	**8.** b

✓ Assessment

Portfolio Encourage students to place in their portfolios one or two items of what they consider to be their best work. Examples include:

- Activity 14-1 observations and answers, p. 395
- Skill Builder description and Assessment diagram, p. 401
- Explore the Issue responses, p. 405 **P**

Performance Additional performance assessments may be found in **Performance Assessment** and **Science Integration Activities.** Performance Task Assessment Lists and rubrics for evaluating these activities can be found in Glencoe's **Performance Assessment in the Science Classroom.**

GLENCOE TECHNOLOGY

▭ MindJogger Videoquiz

Chapter 14 Have students work in groups as they play the Videoquiz game to review key chapter concepts.

Checking Concepts

1. d	**6.** b
2. a	**7.** a
3. a	**8.** d
4. c	**9.** b
5. d	**10.** b

Understanding Concepts

11. The primary raw material is silicon dioxide, sand. It is abundant and inexpensive, and the production process is relatively simple.

12. Decreasing the mass of objects makes them lighter, reducing the pull of gravity and air drag on sports equipment such as skates.

13. Rigidity is an advantage when you need a strong, chemically stable product. It is a disadvantage because a rigid item will break under stress.

14. Copper can be found in its elemental form in nature. Iron is found only in a combined state and must be separated chemically to be usable.

15. Steel is an alloy of a metal with a nonmetal, carbon. The other alloys are made from two or more metals.

Thinking Critically

16. Strength and lightness

17. Silk is lower in density than rayon and stronger than nylon. It is also warmer than cotton.

18. The 10-karat ring will be less expensive and less easily bent, worn away, or scratched.

19. Composites may give athletes advantages that make sports less challenging and change the game.

20. Some synthetic fibers will not decompose and may be found in the environment thousands of years later.

408

Checking Concepts

Choose the word or phrase that completes the sentence.

1. The production of _____ is greater than that of any other alloy.
a. amalgam   c. bronze
b. brass   d. steel

2. Steel is an alloy of iron and _____.
a. carbon   c. tin
b. mercury   d. zinc

3. Structural ceramics contain silicon or _____.
a. aluminum   c. copper
b. carbon   d. lithium

4. A source of the materials used to make plastics is _____.
a. clay   c. petroleum
b. ore   d. synthetic fiber

5. _____ is an element used in fibers to build lightweight car bodies.
a. Sodium   c. Lithium
b. Chlorine   d. Boron

6. Brass and bronze both contain _____.
a. mercury   c. tin
b. copper   d. zinc

7. _____ alloys are lightweight and strong.
a. Aluminum-lithium   c. Copper-tin
b. Copper-zinc   d. Iron-carbon

8. _____ is a natural protein polymer.
a. Nylon   c. Teflon
b. Polyethylene   d. Silk

9. Clay is used to make _____.
a. alloys   c. ores
b. ceramics   d. plastics

10. _____ is not a recent use of technology that has improved speed-skating performance.
a. New fiber in suits
b. Colder ice
c. Composite blades
d. Lighter shoes

Understanding Concepts

Answer the following questions in your Science Journal using complete sentences.

11. Explain why most glass products are relatively inexpensive.

12. One reason to use new materials in sports technology is to decrease the mass of objects. Explain how this can be an advantage in a sport.

13. Explain why the rigidity of ceramics can be an advantage as well as a disadvantage.

14. Iron bonds more readily with other elements than does copper. Why do you think copper was discovered so much earlier than iron was?

15. In what way is steel different from such alloys as brass, bronze, and most aluminum alloys?

Thinking Critically

16. Describe two benefits carbon and boron fibers give to race cars.

17. List three ways silk is better than other natural or synthetic fibers.

18. A lower karat gold has less gold in it than a higher karat gold. Why might you prefer a ring that is 10-karat gold over a ring that is 20-karat gold?

19. Explain why the use of composites can be considered as a problem in the world of sports.

20. A synthetic fiber might be preferred over a natural fiber for use outdoors because it will not rot. Why might this property become a negative feature in the environment?

Developing Skills

21. Comparing and Contrasting Both alloys and ceramics are mixtures designed to use the most favorable properties of their components. Alloys are usually mixtures of two or more metals. Ceramics are mixtures of claylike materials.

22. Measuring in SI 143 g

Chapter 14 Review

Developing Skills

If you need help, refer to the **Skill Handbook.**

21. **Comparing and Contrasting:** Compare and contrast alloys and ceramics.

22. **Measuring in SI:** A bronze trophy has a mass of 952 g. If the bronze is 85 percent copper, how many grams of tin are contained in the trophy?

23. **Interpreting Scientific Illustrations:** Look at the polymer below. Draw the monomer upon which the polymer is based.

$$\cdots -\underset{\underset{H}{|}}{\overset{\overset{H}{|}}{C}} - \underset{\underset{CN}{|}}{\overset{\overset{H}{|}}{C}} - \underset{\underset{H}{|}}{\overset{\overset{H}{|}}{C}} - \underset{\underset{CN}{|}}{\overset{\overset{H}{|}}{C}} - \underset{\underset{H}{|}}{\overset{\overset{H}{|}}{C}} - \underset{\underset{CN}{|}}{\overset{\overset{H}{|}}{C}} - \cdots$$

24. **Recognizing Cause and Effect:** A student performing Activity 14-1 finished with step 5, but did not see any change. What are some possible causes of this effect?

25. **Concept Mapping:** Draw a network tree classifying matter, moving from the most general term to the most specific. Use the terms *compounds, elements, heterogeneous, heterogeneous mixtures, homogeneous, homogeneous mixtures, materials, solutions, substances.* Check (✓) the most specific term that describes an alloy and underline the most specific term that describes a composite.

Performance Assessment

1. **Data Table:** Research gold, its alloys, and karats. Find out several uses for gold and its alloys and what karat gold is used for each. Show your findings in a table.

2. **Poster:** Make a poster of the floor plan of your school or an area of the school. On the floor plan, list locations where ceramics are used and the types of ceramics used there.

3. **Project:** Research the production of silk fabric and compare it with production of synthetic fabrics such as nylon. Display samples of each.

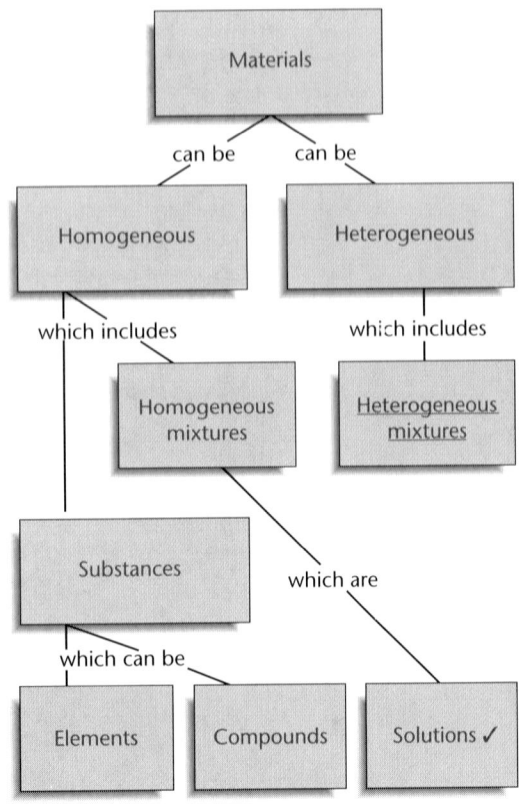

---

23. **Interpreting Scientific Illustrations** The monomer is the repeating pattern based on the first two linked carbons.

24. **Recognizing Cause and Effect** The penny may not have been in the acid long enough. The penny may not have been in the zinc and NaOH long enough. The penny may not have been warmed long enough.

25. **Concept Mapping** See annotated student page.

Performance Assessment

1. Data tables should list gold, its alloys, and their uses. The karat of gold used in each alloy and materials made from the alloys should be indicated. Use the Performance Task Assessment List for Data Table in **PASC,** p. 37. P

2. Answers will vary. Posters should reflect general layout of the school. Use the Performance Task Assessment List for Poster in **PASC,** p. 73. P

3. Students' research should reflect the role of the silkworm in silk production compared to the machinery used to produce nylon. Use the Performance Task Assessment List for Science Portfolio in **PASC,** p. 105. P

Assessment Resources

Reproducible Masters

Chapter Review, pp. 31, 32
Assessment, pp. 88-91
Performance Assessment, p. 28

Glencoe Technology

⊙ **Chapter Review Software**
⊙ **Computer Test Bank**
▭ **MindJogger Videoquiz**

Objectives

LS **Kinesthetic** By using the plastic container code system, students will classify plastics, experiment, and communicate their findings. **L1** **COOP LEARN**

Summary

This project directs students to apply their knowledge of carbon chemistry as they look at the practical uses of plastics and study the properties of an easily produced polymer that they make. Students will use the plastic container code to classify and identify the various plastic containers that they may encounter. Part of the project is to increase the awareness of plastic use and to possibly get involved in plastic recycling efforts.

Time Required

This project should not be started until students begin Chapter 13. Students will need one to two class periods to find the structures of the polymers named. They will need one class period to involve other students in a plastics awareness week. One class period will be needed to synthesize the polymer. If class time is not easily available, the entire project can be done out of class.

Preparation

- Have chemistry textbooks or other references available for use by the students.
- If the polymer preparation is to be done in class, have a saturated borax solution, white school glue, paper cups, and paper towels available.

References

Mark, Herman F. *Giant Molecules.* New York, NY: Time, Inc., Life Science Library, 1966.

Saunders, K.J. *Organic Polymer Chemistry.* London: Chapman and Hall, 1988.

UNIT PROJECT 4

Plastic Polymers

As you got ready for school today, you may have poured the milk for your breakfast cereal from a plastic milk jug. You probably brushed your teeth using a toothbrush that has plastic bristles and a plastic handle. As you left your home, you may have grabbed your plastic-coated binder on your way out the door. Think of all the objects around you that are made of plastic. The chemicals that compose plastics typically share one common property. They can hook together to form long molecules called polymers. Polymers, including plastics, require a lot of energy to produce, and after they are formed, they can be difficult to break down. Because of these energy and disposal problems, people are now finding ways to reuse plastics by saving and recycling them.

Recycling: As Easy as 1, 2, 3

Many industries have begun to imprint a code number on their plastic products. This number refers to a code that allows consumers to sort their plastic products based on chemical makeup. For example, if you looked on the bottom of a plastic milk container, you may likely see the number 2 in a triangle formed from three arrows, perhaps followed by the letters *HDPE*. By examining several types of containers and using the classifying codes, you will learn more about polymer chemistry and possibly help improve our environment.

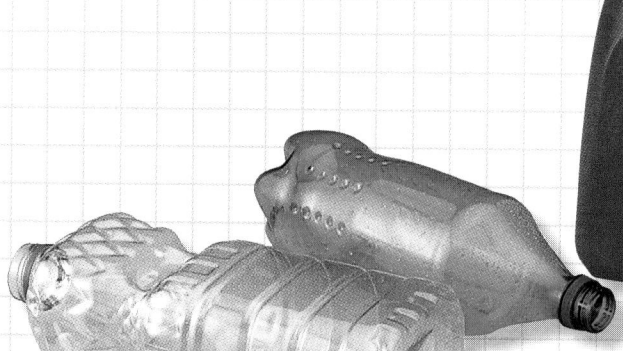

Purpose

Use Section 14-2 of your textbook and other references to determine the general formula for the materials named in the plastic code table. Then, use the following directions to explore plastics and to make a sample polymer.

Procedure

1. Find examples of plastic products that are marked examples of each number.

2. Prepare a display for the class that shows each number and examples of plastic products that illustrate that number.

3. Make a list of all the objects in the classroom that are made from plastics.

4. Prepare your own polymer with common white school glue and a saturated solution of borax and water. Squeeze some of the glue into a cup containing the saturated borax solution. As the product forms, use your hands to pinch it together and form it into a ball. Dry the formed ball with a paper towel.

Using Your Research

Start a plastics awareness week in your class. Have students bring clean plastics from home and sort them into boxes pre-marked with the recycling number. At the end of a week decide what type is the most common.

Contact your local landfill to determine the availability of plastic recycling operations.

Use your polymer glue ball to illustrate some of the properties of polymers. Does the glue ball bounce? Does it stretch? What caused the glue to go from a liquid to a flexible solid? Many everyday types of polymers also have a flexible nature. Some, however are more rigid.

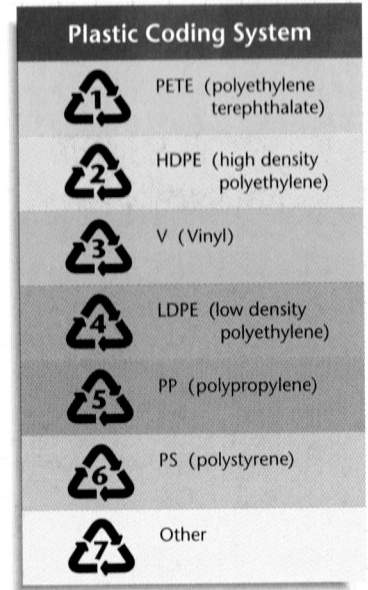

Plastic Coding System

Code	Symbol
1	PETE (polyethylene terephthalate)
2	HDPE (high density polyethylene)
3	V (Vinyl)
4	LDPE (low density polyethylene)
5	PP (polypropylene)
6	PS (polystyrene)
7	Other

Go Further

The use of plastic in many of our everyday objects is a relatively new development. Ask parents or older family members or friends to describe objects that are now made of plastics but were formerly made of other materials.

Contact a local recycling center to ask about how recycled plastics are used.

Unit 4 Project   **411**

Teaching Strategies

- Plastics make up one class of polymers that are mid-range on an extensibility scale. This scale measures how much a material can be stretched and then returned to its original dimensions. Elastomers are highest on this scale, while fibers are the least extensible.

- The plastics' identification numbers, abbreviations, general formulas, and common uses are summarized in the table below.

- If your community has an active recycling center, ask a representative to visit the class to discuss the importance and present status of plastic recycling.

- Polyethylene may be branched or linear in its structure. When it is branched, a more rigid, higher-density polymer is formed.

- In the polymer that the students make, a cross-linked molecule is responsible for making bonds between polymer chains. The borate attaches some amino acids in the casein protein of the glue to other amino acids.

Go Further

The current address and phone number for the Plastics Institute of America is 277 Fairfield Road, Suite 100, Fairfield, NJ 07004-1932. Recycled plastics are currently being used for carpets, textiles, and clothing (PETE); packaging for motor oil, household detergents, and bleach and plastic lumber (HDPE); and garbage bags (all types of retail purchase bags).

Code Number	Symbol	General Formula	Use	Code Number	Symbol	General Formula	Use
1	PETE	$-O-\overset{\text{H}}{\underset{\text{H}}{C}}-\overset{\text{H}}{\underset{\text{H}}{C}}-O-\overset{\text{O}}{C}-\bigcirc-\overset{\text{O}}{C}-$	Soft-drink bottles	4	LDPE	$-\overset{\text{H}}{\underset{\text{H}}{C}}-\overset{\text{H}}{\underset{\text{H}}{C}}-\overset{\text{H}}{\underset{\text{H}}{C}}-\overset{\text{H}}{\underset{\text{H}}{C}}-$	Cottage cheese containers
2	HDPE	$-\overset{\text{H}}{\underset{\text{H}}{C}}-\overset{\text{H}}{\underset{\text{H}}{C}}-\overset{\text{H}}{\underset{\text{H}}{C}}-\overset{\text{H}}{\underset{\text{CH}_2-}}{C}-$ (branches)	Plastic milk bottles	5	PP	$-\overset{\text{H}}{\underset{\text{H}}{C}}-\overset{\text{H}}{\underset{\text{CH}_3}}{C}-\overset{\text{H}}{\underset{\text{H}}{C}}-\overset{\text{H}}{\underset{\text{CH}_3}}{C}-$	Shampoo bottles
3	V	$-\overset{\text{H}}{\underset{\text{H}}{C}}-\overset{\text{Cl}}{\underset{\text{Cl}}{C}}-\overset{\text{H}}{\underset{\text{H}}{C}}-\overset{\text{Cl}}{\underset{\text{Cl}}{C}}-$	Garden hoses, binders	6	PS	$-\overset{\text{H}}{\underset{\text{H}}{C}}-\overset{\text{H}}{\underset{\bigcirc}{C}}-\overset{\text{H}}{\underset{\text{H}}{C}}-\overset{\text{H}}{\underset{\bigcirc}{C}}-$	Foam cups and plates

UNIT 5

Interactions of Matter

In Unit 5, students are introduced to broad categories of physical and chemical changes in matter. Common properties of solutions and their behaviors are discussed. Chemical changes in matter are categorized as chemical reactions that can be represented by chemical equations. Four broad classes of chemical reactions are discussed, as well as the role of energy in determining chemical reactions. The unit concludes with a discussion of acids and bases.

CONTENTS

Unit Contents

412

Science at Home

Safety Labeling Have groups of students begin collecting labels from empty containers of household materials that contain acids and bases and that have safety instructions for emergency antidotal procedures. Have students prepare a household alert bulletin board to describe dangers, proper uses, and first-aid procedures. **L1 COOP LEARN**

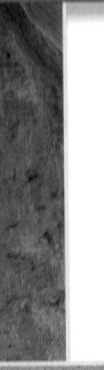

Interactions of Matter

What's Happening Here?

Some chemical reactions may take years before an effect is noticed. Acids in air and water that eventually form caves from what was once solid rock can also damage buildings and statues that are exposed to them. But other chemical reactions can be over in an instant. The chemical changes that enable you to see the images on these pages take place in fractions of a second. What do all these chemical reactions have in common? They all depend on atoms in action.

Science Journal

Stalactites and stalagmites make the underground world of caves breath-takingly beautiful, but where did they come from? Read about cave formation and write a story or poem that describes cave formation.

413

413

Chapter Organizer

Section	Objectives/Standards	Activities/Features
Chapter Opener		Explore Activity: Make a model filtration system. p. 415
15-1 How Solutions Form (2 sessions, 1 block)*	1. **Classify** solutions into three types, and **identify** their solutes and solvents. 2. **Explain** the dissolving process. 3. **Describe** the factors that affect the rates at which solids and gases dissolve in liquids. National Content Standards: (5-8) UCP1-UCP3, UCP5, A2; (9-12) UCP1-UCP3, UCP5, A2	Using Math, pp. 417, 418 MiniLAB: How does solute surface area affect dissolving? p. 419 Problem Solving: Solubility and Surface Area, p. 420 Science Journal, p. 421 Skill Builder: Comparing and Contrasting, p. 421
15-2 Science and Society (1 session, ½ block)*	1. **Compare** and **contrast** the effects of organic solvents. 2. **Determine** whether government agencies should regulate all usage of known organic solvents. National Content Standards: (5-8) UCP2, UCP5, F1-F5, G1; (9-12) UCP2, UCP5, F1, F4-F6, G1	Explore the Issue, p. 423
15-3 Solubility and Concentration (2 sessions, 1½ blocks)*	1. **Discuss** how solubility varies among different solutes and for the same solute at different temperatures. 2. **Demonstrate** an understanding of solution concentrations. 3. **Compare** and **contrast** a saturated, an unsaturated, and a supersaturated solution. National Content Standards: (5-8) UCP1-UCP4, A1, A2, E1, E2, F1, F5; (9-12) UCP1-UCP4, A1, A2, E1, E2, F1, F6	Connect to Earth Science, p. 425 Using Math: Calculating the Mass of Solute Per Liter of Solution, p. 426 Using Technology: IV Solution, p. 428 Using Computers, p. 429 Skill Builder: Making and Using Graphs, p. 429 Activity 15-1: Saturation Situation, pp. 430-431
15-4 Particles in Solution (1 session, ½ block)*	1. **Compare** and **contrast** the behavior of polar and nonpolar substances in forming solutions. 2. **Relate** the processes of dissociation and ionization to solutions that conduct electricity. 3. **Explain** how the addition of solutes to solvents affects the freezing and boiling points of solutions. National Content Standards: (5-8) UCP1-UCP3, UCP5, C1, C5; (9-12) UCP1-UCP3, UCP5	Science Journal, p. 434 Skill Builder: Concept Mapping, p. 434 Science and Literature: Native American Ocean Lore, p. 435 Science Journal, p. 435 Activity 15-2: Boiling Points of Solutions, p. 436

Activity Materials

Explore	Activities	MiniLAB
page 415 sharp pencil, large foam cup, fine gravel, clean sand, powdered charcoal, clear container that will hold at least 100 mL, dirty water	pages 430-431 ice, distilled water, large test tubes, thermometer inserted into a 2-hole stopper, table sugar, copper wire stirrer, test-tube holder, 10-mL graduated cylinder, electric hot plate, heat-proof glove page 436 400 mL distilled water, thermometer, ring stand, 72 g table salt, laboratory burner, 250-mL beaker	page 419 sugar cubes, mortar and pestle, two 100-mL beakers, water, graduated cylinder, watch or clock with second hand

Need Materials? Call Science Kit (1-800-828-7777).   * A complete Planning Guide that includes block scheduling is provided on pages 32T-35T.

Teacher Classroom Resources

Reproducible Masters	Transparencies	Teaching Resources
Study Guide, p. 61 **Reinforcement**, p. 61 **Enrichment**, p. 61 **Activity Worksheets**, p. 89 **Lab Manual 30**, Solutions **Concept Mapping**, pp. 35-36 **Science and Society Integration**, p. 19	**Section Focus Transparency 57**, Solutions Are Everywhere **Teaching Transparency 29**, The Solution Process	**Glencoe Physical Science Interactive Videodisc** **Physical Science CD-ROM** **Spanish Resources** **English/Spanish Audiocassettes** **Cooperative Learning Resource Guide** **Lab Partner** **Lab and Safety Skills** **Lesson Plans**
Study Guide, p. 62 **Reinforcement**, p. 62 **Enrichment**, p. 62 **Lab Manual 31**, Solubility	**Section Focus Transparency 58**, Organic Solvents	**Assessment Resources** **Chapter Review**, pp. 33-34 **Assessment**, pp. 101-104 **Performance Assessment in the Science Classroom (PASC)** **MindJogger Videoquiz** **Alternate Assessment in the Science Classroom** **Performance Assessment** **Chapter Review Software** **Computer Test Bank**
Study Guide, p. 63 **Reinforcement**, p. 63 **Enrichment**, p. 63 **Activity Worksheets**, pp. 85-86 **Science Integration Activity 15**, A Solution Worth Its Salt **Multicultural Connections**, p. 33 **Critical Thinking/Problem Solving**, p. 21	**Section Focus Transparency 59**, Making Maple Syrup	
Study Guide, p. 64 **Reinforcement**, p. 64 **Enrichment**, p. 64 **Activity Worksheets**, pp. 87-88 **Cross-Curricular Integration**, p. 21	**Section Focus Transparency 60**, Ice Isn't Nice **Science Integration Transparency 15**, Solutions in Life **Teaching Transparency 30**, Dissociation and Ionization	

Key to Teaching Strategies

The following designations will help you decide which activities are appropriate for your students.

L1 Level 1 activities should be within the ability range of all students, including those with learning difficulties.

L2 Level 2 activities should be within the ability range of the average to above-average student.

L3 Level 3 activities are designed for the ability range of above-average students.

LEP LEP activities should be within the ability range of Limited English Proficiency students.

LS These activities are designed to address different learning styles.

COOP LEARN Cooperative Learning activities are designed for small group work.

P These strategies represent student products that can be placed into a best-work portfolio.

GLENCOE TECHNOLOGY

The following multimedia resources are available from Glencoe.

Science and Technology Videodisc Series (STVS)
Ecology
Evaluating Artificial Reefs
Aquaculture

National Geographic Society Series
STV: Biodiversity
STV: Atmosphere

Glencoe Physical Science Interactive Videodisc
Behavior of Gases

Physical Science CD-ROM

Teacher Classroom Resources

This is a representation of key blackline masters available in the Teacher Classroom Resources.

Teaching Aids

Section Focus Transparencies

SOLUTIONS ARE EVERYWHERE

L1

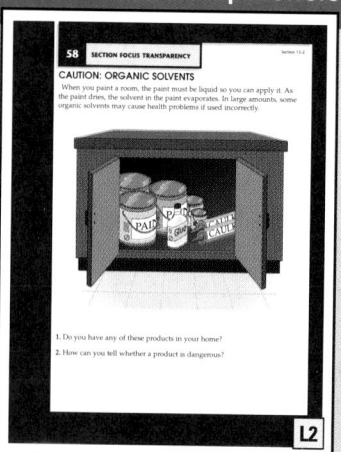

CAUTION: ORGANIC SOLVENTS

L2

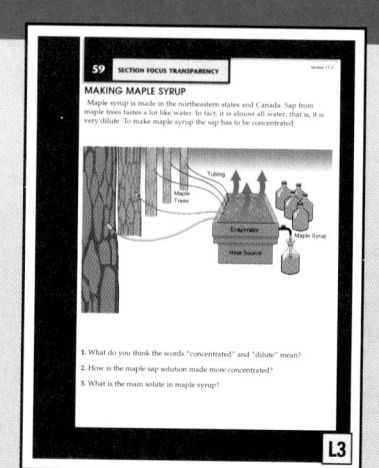

MAKING MAPLE SYRUP

L3

Science Integration Transparencies

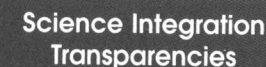 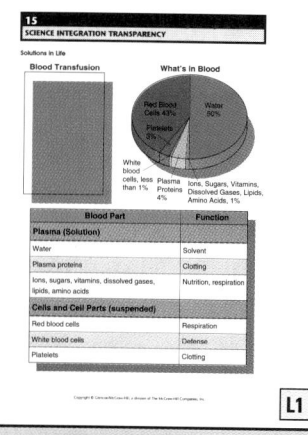

L1

Teaching Transparencies

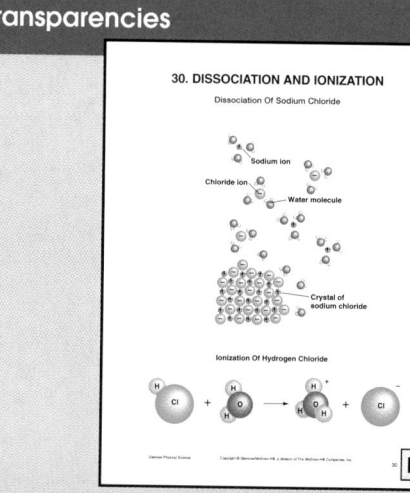

29. THE SOLUTION PROCESS

L1

30. DISSOCIATION AND IONIZATION

L2

Meeting Different Ability Levels

Study Guide

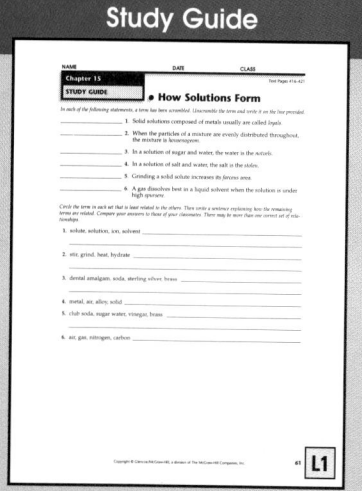

L1

Reinforcement

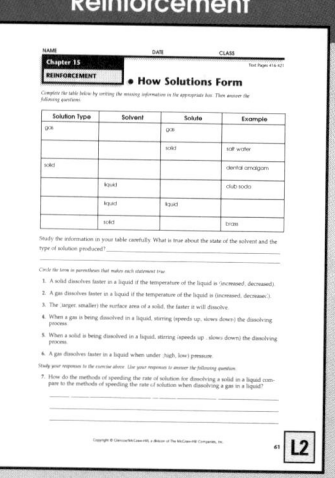

L2

Enrichment Worksheets

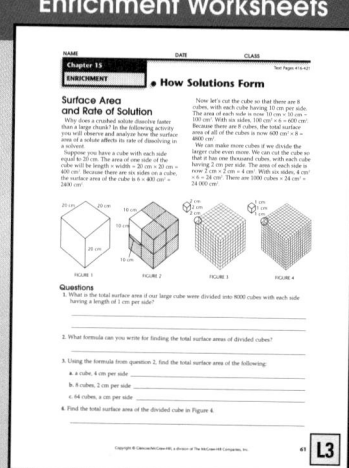

L3

Hands-On Activities

Science Integration Activity

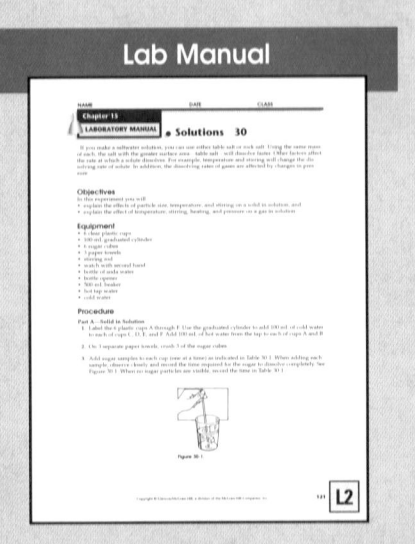

L1

Lab Manual

L2

Assessment

Performance Assessment

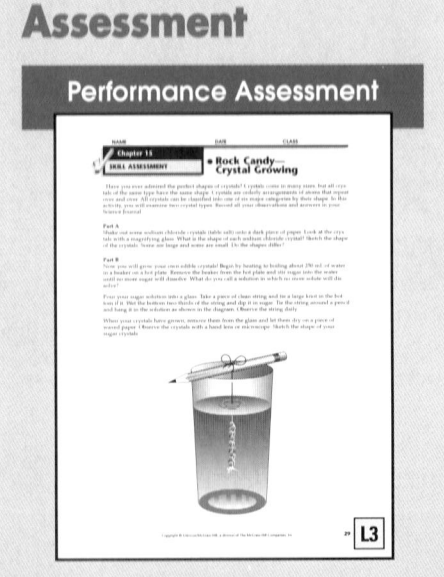

L3

Enrichment and Application

Critical Thinking/ Problem Solving

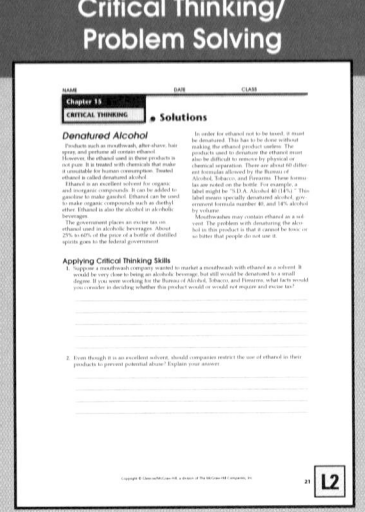

L2

Cross-Curricular Integration

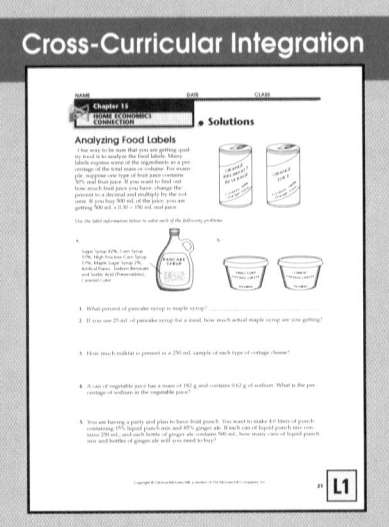

L1

Science and Society Integration

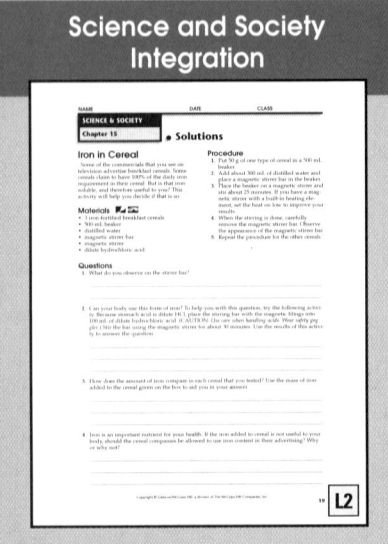

L2

Multicultural Connections

L2

Concept Mapping

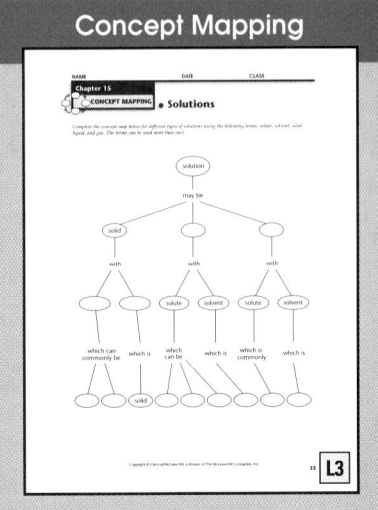

L3

Chapter 15

Solutions

CHAPTER OVERVIEW

Section 15-1 The different types of solutions and the dissolving process are presented, as are factors that affect the rate of dissolving.

Section 15-2 Science and Society Students are asked to discover that some materials in their homes may be hazardous.

Section 15-3 Solubility is defined and developed. Ways used to describe solution concentration are presented.

Section 15-4 The effect of solute and solvent polarity on solubility is presented. Electrolytes are introduced and students learn that solutes have an effect on a solvent's boiling and freezing points.

Chapter Vocabulary

solute	supersaturated solution
solvent	dissociation
organic solvent	ionization
solubility	electrolyte
saturated solution	nonelectrolyte
unsaturated solution	

Theme Connection

Systems and Interactions The student is introduced to systems and interactions by seeing the interaction of solute and solvent. The dissolving process enables the student to see the overall system involved with this interaction to form solutions.

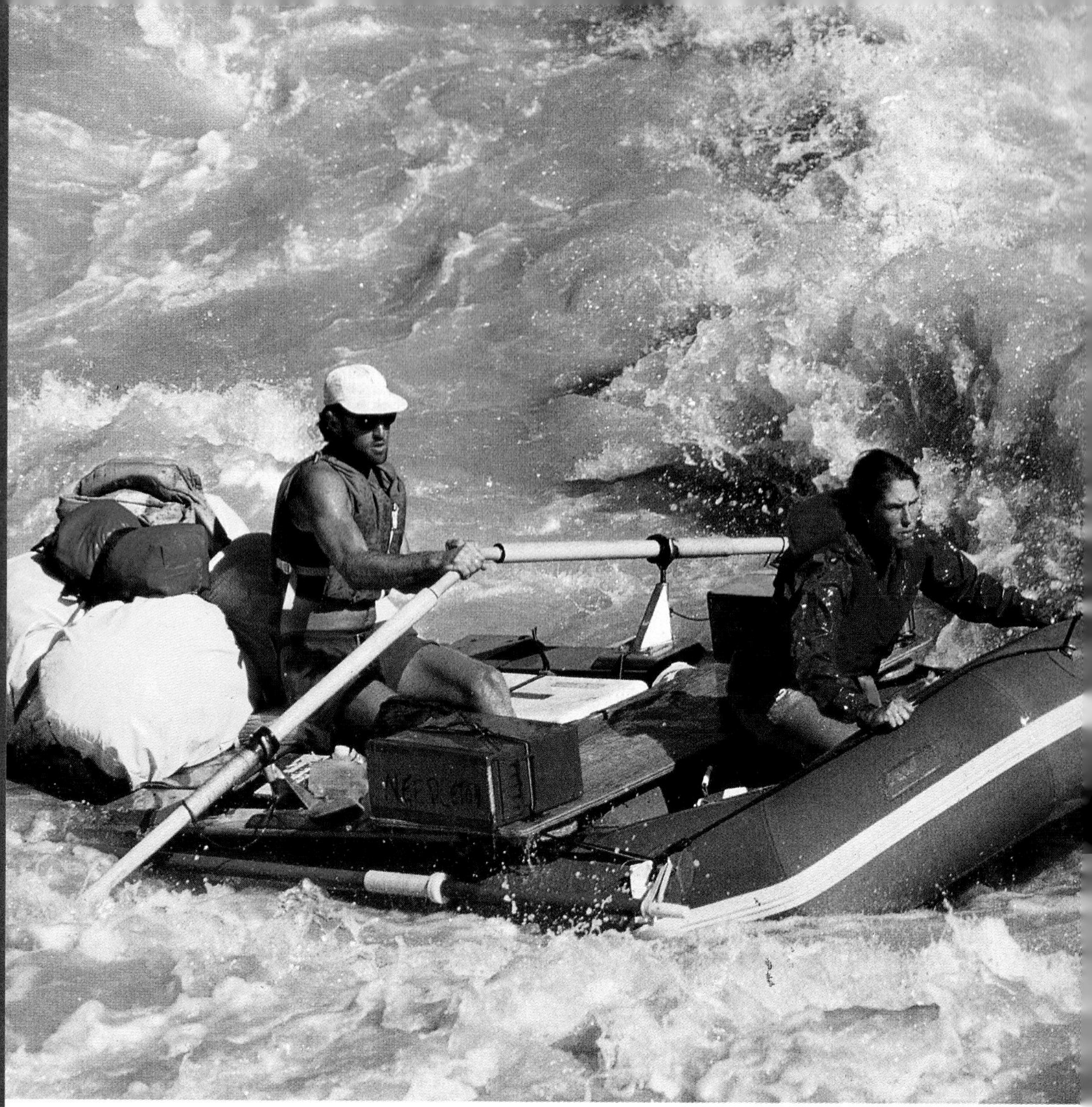

Previewing the Chapter

Section 15-1 How Solutions Form
- ▶ Types of Solutions
- ▶ Dissolving

Section 15-2 Science and Society Issue: Regulating Organic Solvents

Section 15-3 Solubility and Concentration
- ▶ Solubility
- ▶ Concentration
- ▶ Limits of Solubility

Section 15-4 Particles in Solution
- ▶ The Nature of Solvents and Solutes
- ▶ Particles in Solution

414

Learning Styles LS

Look for the following logo for strategies that emphasize different learning modalities. LS

Kinesthetic	Across the Curriculum, p. 426
Visual-Spatial	Explore, p. 415; Demonstration, pp. 417, 418, 426, 433; Across the Curriculum, p. 426
Interpersonal	MiniLAB, p. 419; Activity, p. 425; Activity 15-2, p. 436
Intrapersonal	Reteach, p. 423
Logical-Mathematical	Reteach, p. 420; Using Math, p. 426; Using Computers, p. 429; Activity 15-1, p. 430; Science Journal, p. 434
Linguistic	Science Journal, pp. 421, 423; Integrating the Sciences, p. 433

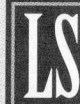

Chapter
15

Solutions

Do you know that people in the United States consume about 1.5 trillion liters of water daily from sources such as the Colorado River, shown in the photo? Almost all of this drinking water has been filtered before it reaches you. Filtering water removes some contaminants, but not all. In the following activity, find out about contaminants that can be removed by a simple filter. Then, in the chapter that follows, learn why not everything can be filtered from water.

EXPLORE ACTIVITY

Make a model filtration system.
1. Use a sharp pencil point to poke ten holes in the bottom of a foam cup.
2. Put 1 cm of fine gravel in the cup, add 3 cm of clean sand, then 1 cm of powdered charcoal, and finally, another layer of fine gravel.
3. Put a clear container under the filter, and slowly pour 100 mL of dirty water through the model filtration system.

Observe: In your Science Journal, compare the filtered water and the original water.

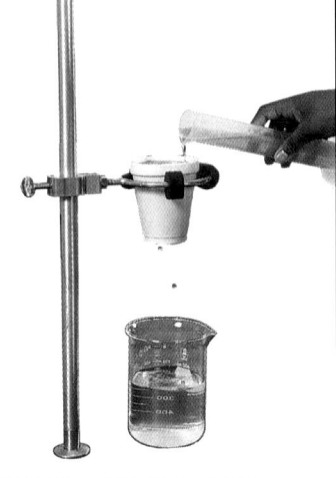

Previewing Science Skills

▶ In the Skill Builders, you will **compare and contrast, make and use a graph,** and **map concepts.**

▶ In the Activities, you will **measure, graph, calculate,** and **infer.**

▶ In the MiniLABS, you will **observe** and **infer.**

415

Section 15•1

Prepare

Section Background

- Without stirring, the dissolving process slows as the solution reaches saturation around each piece of solute. Eventually, the number of particles leaving the surface of the solute equals the number of particles returning to the surface. Stirring brings a fresh supply of solvent near the solute.

- Most solids have positive enthalpies of solution, and are more soluble in hot water than in cold. Gases and some solids have negative enthalpies of solution and are more soluble in cold than in hot water.

- The mass of a gas that will dissolve in a liquid at a given temperature varies directly with the partial pressure of that gas.

Preplanning

To prepare for the MiniLAB, obtain 4 sugar cubes per laboratory team.

1 Motivate

Bellringer

Before presenting the lesson, display **Section Focus Transparency 57** on the overhead projector. Assign the accompanying **Focus Activity** worksheet. L1 LEP

Tying to Previous Knowledge

Have students recall various solvents used around the house. Water is the most common, but paint thinner, alcohol, spot removers, and nail polish removers are also frequently used.

Science Words

solute
solvent

Objectives

- Classify solutions into three types, and identify their solutes and solvents.
- Explain the dissolving process.
- Describe the factors that affect the rates at which solids and gases dissolve in liquids.

Figure 15-1

There are three classes of solutions: gaseous solutions, liquid solutions, and solid solutions. The type of solution is based on the state of the final mixture.

Types of Solutions

When you hear the word *solution*, you may think of a liquid mixture, such as salt water or soda. Both of these liquids are examples of solutions. But did you know that the air you breathe is also a solution? So is the sterling silver used to make fine jewelry. Solutions, then, can be gaseous or solid as well as liquid, as the examples show in **Figure 15-1.** What makes all these mixtures solutions? You may recall from Chapter 9 that a solution is a homogeneous mixture, in which the particles of the mixing substances are evenly distributed throughout.

A A gaseous solution is a mixture of two or more gases. A gaseous solution is used in scuba tanks.

B Solid solutions form when a solid dissolves in a solid, such as when tin dissolves in copper to form bronze.

416   Chapter 15 Solutions

Program Resources

 Reproducible Masters
Study Guide, p. 61 L1
Reinforcement, p. 61 L1
Enrichment, p. 61 L3
Science and Society Integration, p. 19
Activity Worksheets, p. 89 L1
Concept Mapping, p. 35
Lab Manual 30

Transparencies
Section Focus Transparency 57
Teaching Transparency 29

Solutes and Solvents

To describe a solution, you may say that one substance is dissolved in another. The substance being dissolved is the **solute.** The substance that dissolves the solute is the **solvent.** When a solid dissolves in a liquid, the solid is the solute and the liquid is the solvent. Thus, in salt water, sodium chloride is the solute, and water is the solvent. In a liquid-gas solution, the gas is the solute. In soda, carbon dioxide is the solute and water is the solvent.

Generally, the substance present in the largest amount is considered to be the solvent. Air is 78 percent nitrogen, 21 percent oxygen, and one percent argon. Thus, nitrogen is the solvent. In the alloy sterling silver, made of 92.5 percent silver and 7.5 percent copper, silver is the solvent and copper is the solute.

USING MATH

Make a circle graph that shows the composition of air.

Dissolving

Think about lemonade for a minute. A good way to make lemonade is to prepare a solution of sugar in water before you add the lemon juice and ice. By what process do a solid and a liquid, such as sugar and water, form a solution?

The Process

The dissolving of a solid in a liquid occurs at the surface of the solid. For water solutions, keep in mind two things you have learned about water. Like the particles of any liquid, water molecules are constantly moving. A water molecule is polar, which means it has a positive area and a negative area. Molecules of sugar are also polar. Study **Figure 15-2** on page 418 to see how a sugar crystal dissolves in water.

C A liquid solution results when a gas, liquid, or solid is dissolved in a liquid. A fish uses oxygen that is dissolved in water. Vinegar is a solution of two liquids—water and acetic acid. Many solids, such as salt, dissolve easily in liquids.

15-1 How Solutions Form **417**

Community Connection

Road Maintenance If you live in an area that uses salt and other materials for snow and ice removal on roads, have students investigate what materials are used and why. Have them relate these reasons for usage to the chapter concepts. L2

2 Teach

Demonstration

LS **Visual-Spatial** Place 20 g each of calcium acetate and potassium nitrate in separate, labeled beakers that contain 50 mL of water each. Stir, and observe that all the calcium acetate dissolves, but only part of the potassium nitrate dissolves. Then heat both beakers on a hot plate and stir occasionally. More potassium nitrate will dissolve, whereas the calcium acetate will come out of solution. LEP

LS **Visual-Spatial** To demonstrate the dissolving process, partially fill a petri dish with water and place it on the overhead projector. Place a crystal of potassium permanganate in the dish. Have students note the purple streamers as the water is stirred gently. LEP

USING MATH

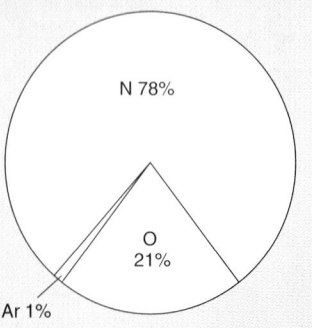

Use these program resources as you teach this lesson.
Study Guide, p. 61
Science and Society Integration, p. 19
Concept Mapping, p. 35
Lab Manual 30

Use **Teaching Transparency 29** as you teach this lesson.

417

Figure 15-2
The dissolving of sugar in water can be thought of as a three-step process.

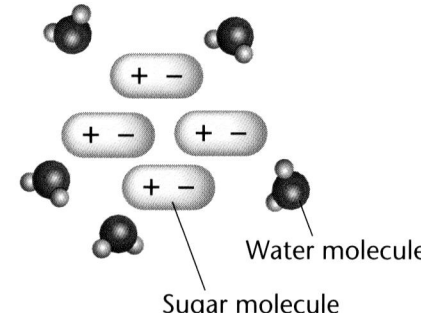

A The moving water molecules cluster around the sugar molecules. The negative ends of the water molecules are attracted to the positive ends of the sugar molecules.

Water molecule
Sugar molecule

B The water molecules pull the sugar molecules into solution.

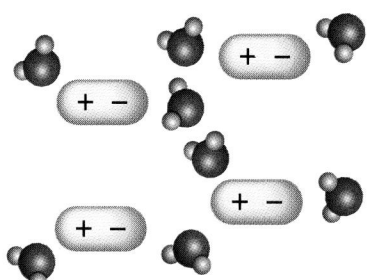

C The moving water molecules spread the sugar molecules out equally throughout the solution.

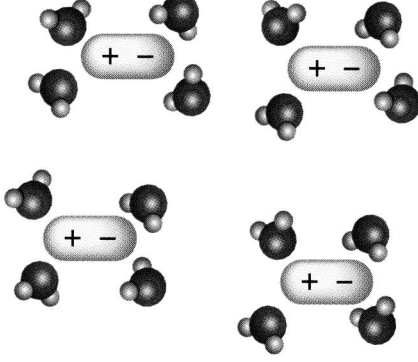

 **U**SING **M**ATH

Suppose the length of each edge of a cube is 1 cm. Find the surface area of the cube. Now imagine cutting the cube in half vertically. What is the total new surface area of the two shapes?

The process described in **Figure 15-2** repeats itself as layer after layer of sugar molecules moves away from the crystal, until all the molecules are evenly spread out. The same three steps occur when any polar or ionic compound dissolves in a polar liquid. A similar, but more difficult, process also occurs when a gas dissolves in a liquid.

You know from your study of fluids in Chapter 8 that particles of liquids and gases move much more freely than do the particles of solids. When gases dissolve in gases, or when liquids dissolve in liquids, this movement results in uniform distribution. Alloys are made by first melting the individual components and then mixing them together, which causes the particles in an alloy to be evenly spread out.

418 Chapter 15 Solutions

Inclusion Strategies

Learning Disabled Students can observe substances in a filtered liquid. They should stir table salt into 50 mL of water until no more salt will dissolve, then add 5 mL of powdered chalk and stir. Have them pour the liquid through filter paper quickly. They should observe the substance on the filter paper after it has dried. Have them evaporate the liquid and observe the crystals that remain.

Rate of Dissolving

Think again about making a sugar solution for lemonade. When you add the sugar to the water, you stir it. Stirring a solution speeds up dissolving because it brings more fresh solvent into contact with more solute. The fresh solvent attracts the particles in the solute, causing the solid solute to move into solution faster.

A second way to speed the dissolving of a solid in a liquid is to grind large crystals into smaller ones. Suppose you had to use a 5-g crystal of rock candy, which is made of sugar, to sweeten your lemonade. If you put the crystal into a glassful of water, it might take 2 minutes to dissolve, even with stirring. However, 5 g of rock candy ground into a powder would dissolve in the same amount of water in a few seconds with stirring.

Why does powdering a crystal cause it to dissolve faster? Breaking the crystal into smaller pieces greatly increases its surface area, as you can see in **Figure 15-3.** Because dissolving takes place at the surface of the solid, increasing the surface area allows more solvent to come into contact with more solid solute. Thus, the speed of the solution process increases.

Other Methods

A third way to increase the rate at which most solids dissolve is to increase the temperature of a solvent. Think again of making a sugar solution for lemonade. You can make the sugar dissolve faster by putting it in hot water instead of cold water. Increasing the temperature of a solvent speeds up the movement of its particles. This increase causes more solvent particles to bump into the solute. As a result, solute particles break loose faster.

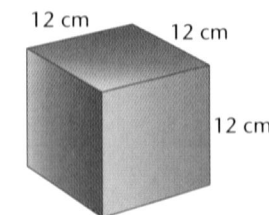

Total surface area 864 cm²

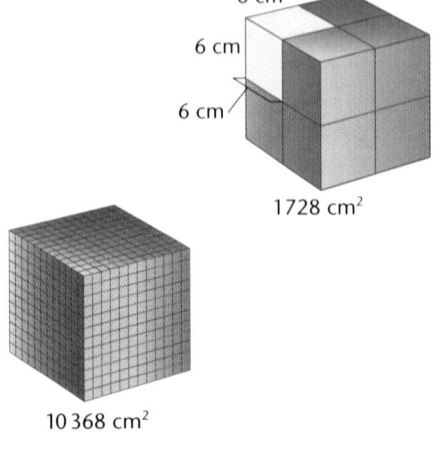

6 cm
6 cm
6 cm
1728 cm²

Figure 15-3

As a crystal is broken down into smaller pieces, the total surface area of the crystal increases.

Total surface area 6 cm²
1 cm   1 cm
1 cm

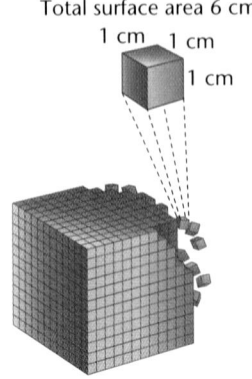

10 368 cm²

10 368 cm²

15-1 How Solutions Form   **419**

Mini**LAB**

How does solute surface area affect dissolving?

Find out what happens when you increase the surface area of sugar.

Procedure

1. Grind up two sugar cubes.
2. Place the granules and two other cubes into separate 100-mL beakers with equal amounts of water.

Analysis

1. Compare the times required to dissolve each.
2. What do you conclude about the dissolving rate and surface area?

Mini**LAB**

Purpose

IS **Interpersonal** Students will observe differences in the dissolving rates of crushed and uncrushed sugar cubes. **L1** **LEP**

COOP LEARN

Materials

4 sugar cubes, 2 100-mL beakers, and 2 stirring rods per group

Teaching Strategies

To compare surface areas of cubes versus granules, have students spread granules in one tight layer on a piece of graph paper and outline the area in pencil. Count squares. Next, outline the area of one surface of the cube on the graph paper. Count squares and multiply by 12 (2 cubes × 6 surfaces per cube). If cubes are not square, have students measure and calculate surface area.

Analysis

1. Even though the sugar cubes are somewhat porous, granules should dissolve faster than the cubes.
2. Greater surface area results in quicker dissolving.

◢ **Assessment**

Oral Have students use their MiniLAB results and **Figure 15-3** to explain how the surface area of the two cubes increases when they are ground. Use the Performance Task Assessment List for Analyzing the Data in **PASC,** p. 27. **P**

📁 **Activity Worksheets,** pp. 5, 89

Theme Connection

Systems and Interactions The effects of stirring a solution and grinding the solute into smaller pieces on rate of dissolving reinforces the theme of interaction.

Integrating the Sciences

Health and Recreation Have an athletic trainer talk about the use of hot and cold packs. Striking the pack brings a dry chemical in contact with water. As the chemical dissolves, the temperature will rise if the dissolving process is exothermic, as with calcium chloride; it will go down if the process is endothermic, as with ammonium chloride.

419

Problem Solving

Solve the Problem

1. Answers will vary but should indicate that not all of the powdered sugar dissolves.

2. Students will find that some ingredient (cornstarch) in the powdered sugar forms a suspension in the water instead of completely dissolving.

3. Students may just use the granulated sugar and the sugar cube to determine the relationship between solubility and surface area.

Think Critically

Amounts of sugar and water were controlled. The composition of the sugar was not controlled. To keep it from packing together, powdered sugar usually contains cornstarch.

3 Assess

Check for Understanding

? FLEX Your Brain

Use the Flex Your Brain activity to have students explore RATE OF DISSOLVING. **P**

 Activity Worksheets, p. 5

Reteach

 **Logical-Mathematical** Have students make a list of factors that affect the rate of solution for different types of solutes.

Extension

For students who have mastered this section, use the **Reinforcement** and **Enrichment** masters.

 # Problem Solving

Solubility and Surface Area

If surface area is related to the rate of dissolving of a crystal, then it makes sense that granulated sugar dissolves faster than a sugar cube. But does a third form, powdered sugar, dissolve even faster? To test this, try an experiment in which you stir the same amount of each of the three types of sugar in equal volumes of water at the same temperature.

Solve the Problem:

1. **Describe what happens to the powdered sugar.**
2. **Explain the results.**
3. **How could you revise the experiment to see if surface area is indeed related to solubility?**

Think Critically:

Reading the ingredients label on the powdered sugar package reveals that there's more than just sugar in powdered sugar.

1. What variables were controlled in the experiment described above?
2. What important variable was not controlled?
3. How can you explain why all of the powdered sugar does not dissolve?

420 Chapter 15 Solutions

Gases and Stirring

How do you think these same factors will affect a gas-liquid solution? When you shake or stir an opened bottle of soda, it bubbles up and usually spills over or squirts out. Did you ever wonder why? Stirring or shaking a solution of a gas in a liquid causes the gas to come out of solution faster. This change occurs because as you shake or stir the solution, more gas molecules are exposed to the surface. These molecules escape more freely. Shaking and stirring also increases the temperature of the solution slightly.

Temperature and Pressure

What might you do if you wanted the gas to dissolve faster in the liquid? The answer is simple: cool the liquid solvent and increase the pressure of the gas. In a soda-bottling plant, both of these things are done. The machinery cools the solution and keeps it under pressure.

In **Figure 15-4A,** an unopened bottle of soda has no bubbles. In the sealed bottle, pressure keeps all of the gas dissolved. When the bottle is opened, the pressure of the carbon dioxide is greatly reduced. Then the carbon dioxide comes out of solution quickly. Maybe you have noticed the difference between opening a bottle of cold soda and a bottle of warm soda, **Figure 15-4B.** All gases are more soluble in cooler solvents.

Figure 15-4

A Reduced pressure inside an opened bottle of soda causes the gas to come out of solution.

B When bottles of soda are opened, more gas comes out of solution from the warm bottle than from the cold one.

Section Wrap-up

Review

1. What are the three types of solutions? Give an example of each type.

2. What are three ways to increase the rate of dissolving a solid in a liquid?

3. **Think Critically:** Amalgams, sometimes used in tooth fillings, are alloys of mercury with other metals. Is an amalgam a solution? Explain.

Skill Builder
Comparing and Contrasting

Compare and contrast the effects on the rate of dissolving (1) a solid in a liquid and (2) a gas in a liquid, when (a) the solution is cooled, (b) it is stirred, and (c) the pressure on it is lowered. If you need help, refer to Comparing and Contrasting in the **Skill Handbook.**

Science Journal

Write a paragraph in your Science Journal explaining why many fish aquariums have a device that bubbles air into the water. Explain why it is necessary to constantly bring air into the water.

Skill Builder

Cooling retards the rate of dissolving a solid in a liquid but increases the rate for a gas in a liquid. Stirring increases the rate for a solid and decreases the rate for a gas. Lowering pressure has little effect on a solid in a liquid and decreases the rate for a gas in a liquid.

Assessment

Performance Have students prepare a table that presents their conclusions. Use the Performance Task Assessment List for Data Table in **PASC,** p. 37. **P**

---

4 Close

•MINI•QUIZ•

Use the Mini Quiz to check students' recall of chapter content.

1. **List three different types of solutions with an example of each.** *See Figure 15-1 for possible answers.*

2. **What are two parts that make up a solution?** *solute and solvent*

3. **Polar solvents are able to dissolve which types of solute?** *polar and ionic*

4. **How can the speed of solvent particles be increased?** *increase the temperature or stir*

Section Wrap-up

Review

1. Types are gaseous solutions, liquid solutions, and solid solutions. Examples will vary but may include those in **Figure 15-1.**

2. For most solids, increase temperature, stir or shake, and break solute into small particles to increase surface area.

3. **Think Critically** Yes, because an amalgam is an alloy, and alloys are solutions.

Science Journal

Linguistic Journal entries should reflect that air bubbled through water dissolves oxygen in the water. This dissolved oxygen is necessary for organisms, such as fish, that use it as their oxygen supply. As the fish use dissolved oxygen, more is needed to replenish it.

SCIENCE & SOCIETY
Section 15•2

Prepare

Section Background

Students may be surprised to learn how many common products used in the home every day contain organic solvents. Many home cleaning and painting supplies have warnings on their containers: "Use only in well-ventilated areas. Open windows and use fans while using this product."

1 Motivate

Bellringer

 Before presenting the lesson, display **Section Focus Transparency 58** on the overhead projector. Assign the accompanying **Focus Activity** worksheet. L1 LEP

Tying to Previous Knowledge

Have students recall the last time they used paint thinner to clean paint from brushes. Care should always be taken to clean brushes outside or in well-ventilated areas.

2 Teach

NATIONAL GEOGRAPHIC SOCIETY

 **Videodisc**

STV: Atmosphere
What Is the Atmosphere?
Unit 1
Atmospheric Gases

03109-07851

SCIENCE & SOCIETY

ISSUE:
15•2 Regulating Organic Solvents

Science Words

organic solvent

Objectives

- Compare and contrast the effects of organic solvents.
- Determine whether government agencies should regulate all usage of known organic solvents.

Figure 15-5

Potentially hazardous organic solvents are found in many common materials.

Organic Solvents

Think about home and you probably think about a place that is safe. However, scientists warn that the buildings in which people live may be hazardous to their health because of the number of potentially harmful chemicals found in materials used in construction and remodeling.

Many health problems could be caused by organic solvents present in almost all materials used in building or remodeling. **Organic solvents,** such as those shown in **Figure 15-5,** are found in glues, paints, paint remover, paint and varnish thinners, and in some caulking compounds and carpeting. Organic solvents often give a new house that new smell.

But there are some drawbacks. Organic solvents vaporize easily. They enter the body by being absorbed through the skin or when fumes are inhaled. Many organic solvents are substances that are known to affect the nervous system and to cause growth and development problems. In some instances, a few have been shown to cause cancer.

Those most at risk are professional painters and do-it-yourselfers who work with these substances over long periods of time. What is being done to protect people from these hazardous substances?

Existing Regulations

The United States Occupational Safety and Health Administration (OSHA) has established time limits beyond which a worker cannot be exposed to health-threatening solvents. Another government agency, the Consumer Products Safety Commission (CPSC), requires labels on all corrosive or flammable chemicals. Some labels list harmful ingredients and itemize the dangers to health.

422 Chapter 15 Solutions

Program Resources

📁 **Reproducible Masters**
Study Guide, p. 62 L1
Reinforcement, p. 62 L1
Enrichment, p. 62 L3
Lab Manual 31

🖌 **Transparencies**
Section Focus Transparency 58

2 Points of View

▷ Regulating Products and Professionals

The OSHA regulations have been established to protect employees of large building contractors. Fines may be incurred if workers are found to be exposed to solvents for extended periods of time. The cost of labeling alone has caused some manufacturers to switch to other materials that may not be as effective. Also, large professional contractors ask who will regulate the do-it-yourselfer. Some professionals feel that these people should have to obtain permits from the local building code inspector before using dangerous solvents.

▷ Regulating the Home Remodeler

Home remodelers or do-it-yourselfers balk at the idea that they should obtain permits to use certain products with organic solvents. They claim that their projects are much smaller and that they probably control ventilation better for their own home projects than professional builders working on large new buildings, **Figure 15-6.**

Figure 15-6

Good ventilation is required when working with materials that contain organic solvents.

Section Wrap-up

Review

1. In what types of products are organic solvents found?

2. What problems occur when organic solvent vapors are inhaled or absorbed by the body?

interNET **CONNECTION** What kinds of precautions should be taken when using dangerous organic solvents such as varnishes and stains? Visit the Chapter 15 Internet Connection at Glencoe Online Science, **www.glencoe.com/sec/science/physical,** for a link to more information about organic solvents.

SCIENCE & SOCIETY

3 Assess

Check for Understanding

? FLEX Your Brain

Use the Flex Your Brain activity to have students explore ORGANIC SOLVENTS.

🗂 **Activity Worksheets,** p. 5

Reteach

LS Intrapersonal Have students create a sticker to place on each product that contains an organic solvent, warning others about the solvent content.

Extension

🗂 For students who have mastered this section, use the **Reinforcement** and **Enrichment** masters.

4 Close

Brainstorming

Have students brainstorm and list key concepts from the lesson.

Section Wrap-up

Review

1. glues, paints, paint remover and thinners, and in some caulking compounds and carpeting

2. They affect the nervous system and cause growth and development problems. A few have been shown to cause cancer.

Prepare

Section Background

- If a water solution of two salts is allowed to evaporate, the less soluble one at the temperature of evaporation will crystallize first.

- Supersaturation is possible because solids will not crystallize unless there is a suitable surface, such as a crystal, upon which to start crystallization.

Preplanning

- To prepare for Activity 15-1, you will need to prepare copper wire stirrers.

- For the demonstration on page 428, add 50 to 75 g of sodium acetate trihydrate to a clean 250-mL Erlenmeyer flask. Warm the flask on a hot plate until the crystals melt. Rinse down the sides of the flask with about 8 mL of water. Gently swirl the flask to mix, then cover it and allow to cool.

1 Motivate

Bellringer

Before presenting the lesson, display **Section Focus Transparency 59** on the overhead projector. Assign the accompanying **Focus Activity** worksheet. L1 LEP

Tying to Previous Knowledge

Ask students to check at home to see if there is a very old jar of honey on a shelf. As the water slowly evaporates from the honey, sugar crystals begin to form and cause the honey to have a coarse texture. The solution has become saturated.

Science Words

solubility
saturated solution
unsaturated solution
supersaturated solution

Objectives

- Discuss how solubility varies among different solutes and for the same solute at different temperatures.
- Demonstrate an understanding of solution concentrations.
- Compare and contrast a saturated, an unsaturated, and a supersaturated solution.

Figure 15-7

The amount of sugar that can dissolve in a glass of water is limited.

Solubility

Sugar, as you know, dissolves easily in water. Suppose you want to make super-sweet lemonade. You stir two, three, four—or more—teaspoons of sugar into a cup of water, and it all dissolves. But eventually, as shown in **Figure 15-7,** you add one more teaspoon of sugar and it no longer dissolves.

Change the scene from your kitchen to a laboratory. Suppose now you measure out 1.3 g of lithium carbonate and add it to 100 g of water at 20°C. You observe that all of the lithium carbonate dissolves. But if you add more lithium carbonate to the same solution, none of it dissolves. However, you could add 34 g of potassium chloride to 100 g of water before no more will dissolve. You would have shown that the amount of these two substances that dissolves in 100 g of water at 20°C varies greatly.

Generally, the **solubility** of a substance is expressed as the maximum number of grams of the substance that will dissolve in 100 g of solvent at a certain temperature. **Table 15-1** shows how the solubility of several substances varies at 20°C. For solutes that are gases, the pressure must also be given, for a reason you will learn soon.

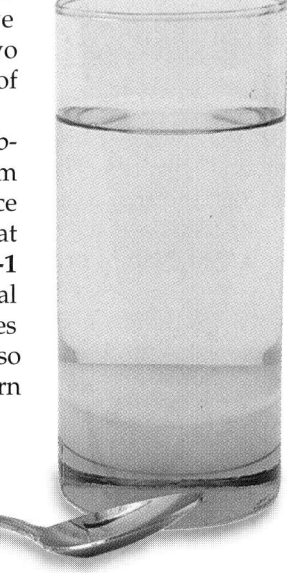

Concentration

Suppose you added one spoonful of lemon juice to a glass of water to make lemonade. Your friend decided to add four spoonfuls of lemon juice to another glass of water. You could say that your glass of lemonade is dilute. Your friend's glass of lemonade is concentrated. It would have more lemon flavor than yours. A concentrated solution is one in which there is a large amount of solute in the solvent. A dilute solution is one in which there is a small amount of solute in the solvent.

424 Chapter 15 Solutions

Table 15-1

Solubility of Substances in Water at 20°C	
Substance	**Solubility in g/100 g of Water**
Solid Substances	
Barium sulfate	0.00025
Lithium carbonate	1.3
Potassium chloride	34.0
Sodium nitrate	87.6
Lithium bromide	166.0
Sucrose (sugar)	203.9
Gaseous Substances*	
Hydrogen	0.00017
Oxygen	0.005
Carbon dioxide	0.16
*when pressure = 1 atmosphere	

CONNECT TO

EARTH SCIENCE

Bronze is stronger and more durable than copper or tin. Bronze is an alloy that can be 90 percent copper and 10 percent tin. **Determine** which metal is the solvent. **Conclude** which metal is the solute.

2 Teach

Activity

LS **Interpersonal** Have students determine the solubility of an unknown substance by repeatedly adding 1-g samples to 100 mL of water at 20°C. When the solution is saturated, ask students to analyze their solubility data and determine the identity of their unknown substance. Use only substances that are safe. **COOP LEARN**

Revealing Preconceptions

Many students believe that all concentrated acids are 100% acid. Have small labeled bottles of concentrated acids on display. Have students read from the label the percentage of acid in each.

CONNECT TO

EARTH SCIENCE

Answer Copper is the solvent; tin is the solute.

Precise Concentrations

Concentrated and *dilute* are not precise terms. However, solution concentrations can be described precisely. One way is to state the percentage by volume of the solute. Do you ever have a fruit juice drink with your lunch or for a snack? The next time you do, read the label to see how much actual juice you are getting. The percentage by volume of the juice in the drink shown in **Figure 15-8** is 100 percent. Commonly, fruit-flavored drinks contain ten percent juice and about 90 percent water. Adding 10 mL of juice to 90 mL of water makes 100 mL of drink. Generally, if two or more liquids are being mixed, the concentration is given in percentage by volume, stated in number of milliliters of solute plus enough solvent to make 100 mL of solution.

You can express concentration of a solid dissolved in a liquid as percentage by mass. This percentage is the mass in grams of solute plus enough solvent to make 100 g of solution. Because 1 mL of water has a mass of 1 g, you would mix 10 g of sodium chloride with 90 mL of water to make a ten percent solution of sodium chloride.

Figure 15-8

The concentrations of fruit juices are often given in percent by volume.

Across the Curriculum

Dietetics People who are trained in food preparation and nutrition know the economy of buying concentrates and diluting them to make the desired solutions. Shipping costs and fuel consumption are reduced when you add the water at the destination.

Community Connection

A Juicy Topic Have students visit a local grocery store or invite the manager of the store to the classroom. Have students question the manager about the differences between ready-mixed juice drinks and juice concentrates located in the frozen-foods section.

You can also express concentration as mass of solute in a liter of solution. When a certain volume of solution is measured out, the number of grams of solute it contains can be calculated easily.

USING MATH

Calculating the Mass of Solute Per Liter of Solution

Example Problem:
A potassium chloride solution contains 88 g of KCl in 1.0 L of solution. How many grams of solute are in 50 mL of solution?

Problem-Solving Steps:
1. Change 1.0 L to mL.
   1.0 L = 1000 mL
2. Determine the number of grams of solute in 1 mL of solution.
   88 g/1000 mL = 0.088 g/mL
3. Use the equation: Mass of solute = concentration of 1 mL solution × volume of solution.
4. **Solution:** Mass = 0.088 g/mL × 50 mL = 4.4 g

Calculator Hint: Steps 2 and 3 can be combined into one calculator operation.
Enter: **88 ÷ 1000 × 50 =** and your calculator will read **4.4.**

Practice Problem
Forty grams of lye (NaOH) were dissolved in enough water to make 1.0 L of solution. How many grams of lye would be in 20 mL of the solution?
Strategy Hint: Divide 40.0 g by 1000 to find the mass of solute in 1 mL of solution.

Limits of Solubility

How much solute can dissolve in a given amount of solvent? That depends on a number of factors. Examine the types of solutions based on the amount of solute dissolved.

Saturated Solutions

If you add 30.0 g of potassium chloride, KCl, to 100 g of water at 0°C, only 28 g will dissolve. You have a saturated solution because no more KCl can dissolve. A **saturated solution** is a solution that has dissolved all the solute it normally can hold at a given temperature. But if you heat the mixture to a higher temperature, more KCl can dissolve. As the temperature of the liquid solvent increases, the amount of solute that can dissolve in it usually increases. **Table 15-2** gives the amounts of a few

Table 15-2

Solubility of Compounds in Water at Various Temperatures

Solubility in g/100 g of Water at the Temperature Indicated

Compound	Formula	0°C	20°C	60°C	100°C
Ammonium chloride	NH_4Cl	29.4	37.2	55.3	77.3
Barium hydroxide	$Ba(OH)_2$	1.67	3.89	20.94	101.4
Copper(II) sulfate	$CuSO_4$	23.1	32.0	61.8	114
Lead(II) chloride	$PbCl_2$	0.67	1.00	1.94	3.20
Potassium bromide	KBr	53.6	65.3	85.5	104
Potassium chloride	KCl	28.0	34.0	45.8	56.3
Potassium nitrate	KNO_3	13.9	31.6	109	245
Sodium acetate	$NaC_2H_3O_2$	36.2	46.4	139	170.15
Sodium chlorate	$NaClO_3$	79.6	95.9	137	204
Sodium chloride	NaCl	35.7	35.9	37.1	39.2
Sodium nitrate	$NaNO_3$	73.0	87.6	122	180
Sucrose (sugar)	$C_{12}H_{22}O_{11}$	179.2	203.9	287.3	487.2

solutes that can dissolve in water at different temperatures, forming saturated solutions.

Another way to picture the effect of higher temperatures on solubility is with a line graph, as shown in **Figure 15-9.** Each line is called a solubility curve for a particular substance. You can use a curve to figure out the amount of solute that will dissolve at any temperature given on the graph. For example, at 47°C, about 82 g of both KBr and KNO_3 form saturated solutions with l00 g of water. How much NaCl will form a saturated solution with 100 g of water at the same temperature?

Unsaturated Solutions

An **unsaturated solution** is any solution that can dissolve more solute at a given temperature. Each time a saturated solution is heated to a higher temperature, it may become unsaturated. The term *unsaturated* isn't precise. If you look at **Table 15-2,** you'll see that at 20°C, 37.2 g of NH_4Cl dissolved in 100 g of water is a saturated solution. However, an unsaturated solution of NH_4Cl could be any amount less than 37.2 g in 100 g of water at 20°C.

Figure 15-9

The solubility of a substance changes with water temperature. *For which substance does solubility increase the least?*

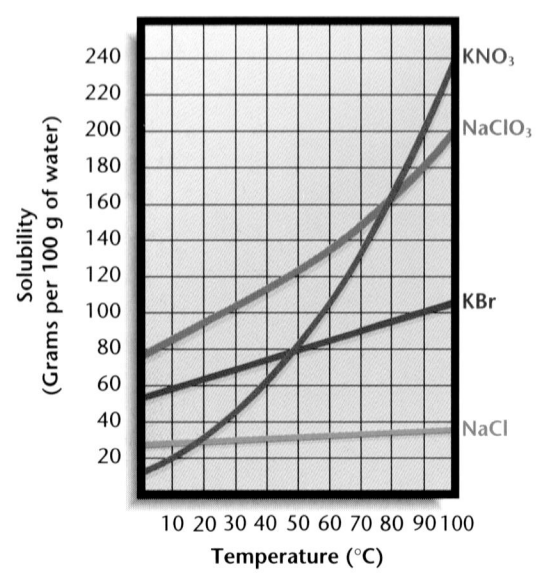

15-3 Solubility and Concentration   **427**

Cultural Diversity

Sandbox Trees Tall sandbox trees are found in the jungles of South and Central America. Their seed capsules were once used as boxes to hold sand for blotting ink. The trees, however, are dangerous. The bark of the tree is poisonous and is covered with short spines. The sap is acidic and will burn skin or cause blindness. The poisonous leaves form a structure at the top of the trees that contains the seed capsules. Hu-

mans and animals can be injured when the seed capsules explode.

When the poisonous sap is thrown into lakes and streams that have been dammed, the sap solution stuns the fish, so that people can gather all they need for food. When they have collected the fish, they remove the dams and water dilutes the sap so that the remaining fish recover completely.

For more information on how and why an IV is used, contact a local health facility. The school nurse could explain its use and demonstrate one setup for students to see.

Think Critically

Dehydration can be a result of loss of fluids. The balance between liquid and tissue is destroyed when fluid level is upset.

3 Assess

Check for Understanding

? FLEX Your Brain

Use the Flex Your Brain activity to have students explore UNSATURATED SOLUTIONS.

 **Activity Worksheets,** p. 5

Reteach

Demonstration Drop one crystal of sodium acetate trihydrate into a previously prepared supersaturated solution of sodium acetate. Have students note changes in appearance and temperature.

Extension

 For students who have mastered this section, use the **Reinforcement** and **Enrichment** masters.

USING TECHNOLOGY

IV Solution

"**S**tart an IV with 0.9 percent normal saline." You've no doubt heard a doctor use this phrase or something similar on a medical television show. What exactly is an IV, and why is it important? An IV is an apparatus used to administer medication or fluids directly into the blood through a vein. An IV is used in a sick or injured person to restore the fluid and electrolyte balance to the body. In a healthy person, blood is in equilibrium; there are enough of the various types of blood cells and plasma in the solution in which cells are transported. Plasma is a solution made mostly of water. It also contains proteins, food substances, enzymes, oxygen, carbon dioxide, and electrolytes.

Out of Balance

When a person is sick or has been injured, the blood may be out of equilibrium. By using an IV, needed fluids, electrolytes, and sugars flow directly into the person's blood, and balance can be restored quickly. If the person has lost a lot of blood, an IV can replace lost blood volume quickly using a saline solution that is similar to plasma.

Think Critically:

Why is it important to replace lost blood and fluids quickly in a sick or injured person?

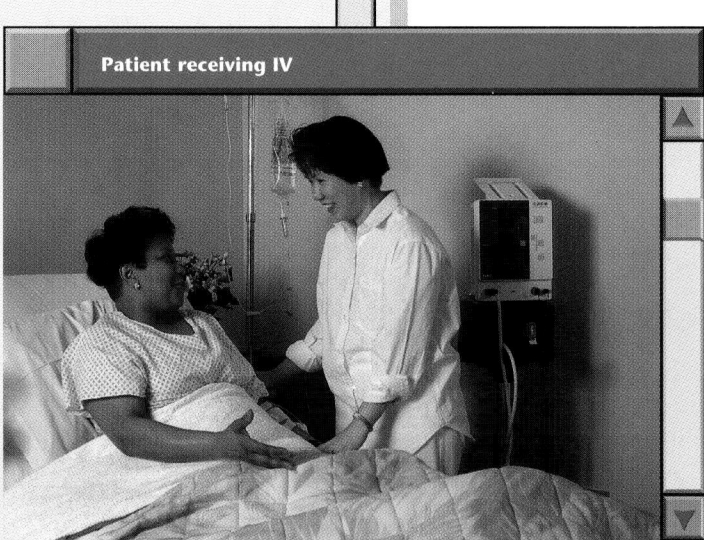

Patient receiving IV

428   Chapter 15 Solutions

Supersaturated Solutions

If you make a saturated solution of potassium nitrate at 100°C and then let it cool to 20°C, part of the solute comes out of solution. At the lower temperature, a saturated solution is formed with less solute. Most other saturated solutions behave in a similar way when cooled. But if you cool a saturated solution of sodium acetate from 100°C to 20°C, with proper precautions, no solute comes out. This solution is supersaturated. A **supersaturated solution** contains more solute than a saturated one has at that temperature. This kind of solution is unstable. When a small crystal of the solute is added to a supersaturated solution, the excess solute quickly crystallizes out, as is shown in **Figure 15-10.**

How can you determine whether a solution is saturated, unsaturated, or supersaturated? Suppose you add a solute crystal

---

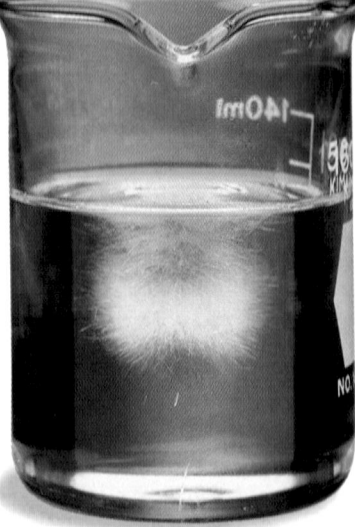

Figure 15-10

A solution of sodium acetate (left) crystallizes out when a crystal of sodium acetate is added (right).

to a solution. If the crystal dissolves, the solution is unsaturated. If the crystal does not dissolve, the solution is saturated. And if excess solute comes out, the solution is supersaturated.

Section Wrap-up

Review

1. Do all solutes dissolve to the same extent in the same solvent? How do you know?

2. Using **Table 15-2** on page 427, state the following: the mass of $NaNO_3$ that would have to be dissolved in 100 g of water to form a saturated, an unsaturated, and a supersaturated solution of $NaNO_3$ at 20°C.

3. **Think Critically:** By volume, orange drink is ten percent orange juice and ten percent corn syrup. A 1500-mL can of the drink costs $0.95. A 1500-mL can of orange juice is $1.49, and 1500 mL of corn syrup is $1.69. How could you make orange drink? Is it cheaper to make or to buy?

Skill Builder
Making and Using Graphs

Using **Table 15-2** on page 427, make a graph showing the solubility curves for $CuSO_4$ and $NaNO_3$. How would you make a saturated water solution of each substance at 80°C? If you need help, refer to Making and Using Graphs in the **Skill Handbook**.

Using Computers

Spreadsheet You work in the lab stockroom, making solutions for teachers to use in their classes. Use **Table 15-2** on page 427 to prepare a spreadsheet showing the number of grams of solute needed to make 10 mL, 50 mL, 100 mL, 500 mL, and 1000 mL of a saturated solution at 20°C of the following compounds: ammonium chloride, lead(II) chloride, sodium chlorate, and sucrose.

Skill Builder

Masses of substances to add to 100 g of water at 80°C to make saturated solutions:

$NaNO_3$ — about 148 g

$CuSO_4$ — about 80 g

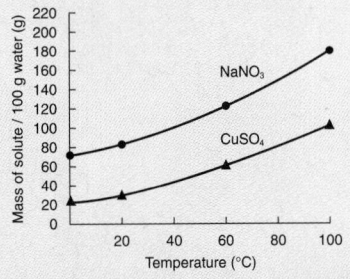

Assessment

Performance To assess the accuracy of the graphs, ask students to determine from the graphs the temperatures at which only half of their reported saturation amounts would be dissolved. *$NaNO_3$—about 0°C; $CuSO_4$—about 40°C.* Use the Performance Task Assessment List for Analyzing the Data in **PASC**, p. 27. **P**

---

4 Close

•MINI•QUIZ•

Use the Mini Quiz to check students' recall of chapter content.

1. The maximum number of grams of a substance that will dissolve in 100 g of solvent at a specific temperature is known as the _____ of the substance. *solubility*

2. A solution that contains all the solute it can hold at a given temperature is _____ . *saturated*

3. A solution that can dissolve more solute at a given temperature is said to be _____ . *unsaturated*

Section Wrap-up

Review

1. No. The solubility data in **Tables 15-1** and **15-2** show wide variation.

2. 87.6 g; less than 87.6 g; more than 87.6 g

3. **Think Critically** The simplest approach is to use all the orange juice and corn syrup plus eight 1500-mL cans of water. This makes 15 000 mL of orange drink for $3.18. Made this way, 1500 mL would cost $0.318 (rounded to $0.32). This is $0.95 − $0.32 = $0.63 cheaper to make than to buy.

Using Computers

LS **Logical-Mathematical** One spreadsheet lists volumes in cells A2-A6 and compound names in cells B1-E1. Empty cells contain amounts of solute. For NH_4Cl: 3.72 g, 18.6 g, 37.2 g, 186 g, 372 g; for $PbCl_2$: 0.10 g, 0.500 g, 1.00 g, 5.00 g, 10.0 g; for $NaClO_3$: 9.59 g, 48.0 g, 95.9 g, 480 g, 959 g; for $C_{12}H_{22}O_{11}$: 20.39 g, 102 g, 203.9 g, 1020 g, 2039 g.

Purpose

LS **Logical-Mathematical** Students will compare the solubility of a solute at different temperatures. **L1** **COOP LEARN**

Process Skills

measuring, using numbers, communicating, making and using graphs, making and using tables, and inferring

Time

one class period

Materials

- For student safety, insert thermometers and stirrers into stoppers ahead of time. To make a stirrer, coil one end of a piece of copper wire several times around a pencil. Extend the wire out of one hole of the stopper, and bend it over the end. Experiment with making the stirrers ahead of time so that they will move freely in the test tube and will not interfere with the use of the thermometer.
- Lubricate thermometers with glycerin and wrap in a towel the part you are holding before inserting the thermometers into the stoppers.

Alternate Materials Salts may be used instead of sugar. Possible salts to use are KBr, AlCl$_3$, or CuSO$_4$.

Safety Precautions

Caution students to use a test-tube holder when handling hot test tubes. Never stir with a thermometer.

Possible Hypotheses

Most student hypotheses will reflect that an increase in temperature will increase the rate of dissolving for sugar.

📁 **Activity Worksheets,** pp. 5, 85-86

430

Activity 15-1

Design Your Own Experiment
Saturation Situation

Two major factors to consider when you are dissolving a solute in water are temperature and the ratio of solute to solvent. What happens to a solution as the temperature changes? Can you learn to predict when a solution will become saturated at a particular temperature?

PREPARATION

Problem
How does solubility change with a change in temperature or amount of solvent?

Form a Hypothesis
Based on observations you previously made, state a hypothesis about how the dissolving rate of table sugar is affected by temperature.

Objectives
- Observe the effects of temperature on the rate at which a solute dissolves.
- Design an experiment that tests the effects of a variable, such as change in temperature or amount of solvent, on the rate at which a solute dissolves.

Possible Materials
- ice
- distilled water

- large test tubes
- thermometer inserted in a two-hole rubber stopper
- table sugar
- copper wire stirrer
- test-tube holder
- 10-mL graduated cylinder
- apron
- safety goggles
- electric hot plate
- oven mitt
- test-tube rack

Safety Precautions

Protect clothing and eyes, and be careful if using a hot plate and heated solutions.

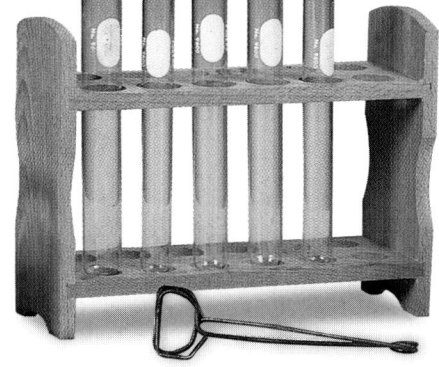

430 Chapter 15 Solutions

Inclusion Strategies

Visually Impaired Place students who have difficulty reading temperatures, massing solute, and recording results in charge of formulating and evaluating the proposed hypothesis.

PLAN THE EXPERIMENT

1. As a group, agree upon and write out the hypothesis statement.
2. As a group, list the steps you need to take to test your hypothesis. Be specific; describe exactly what you will do at each step.
3. Make a list of the materials that you will need to complete your experiment.
4. If you need a data table, design one in your Science Journal so that it is ready to use as your group collects data.

Check the Plan

1. Read over your entire experiment to make sure that all steps are in logical order.
2. Identify any constants and the variables of the experiment.

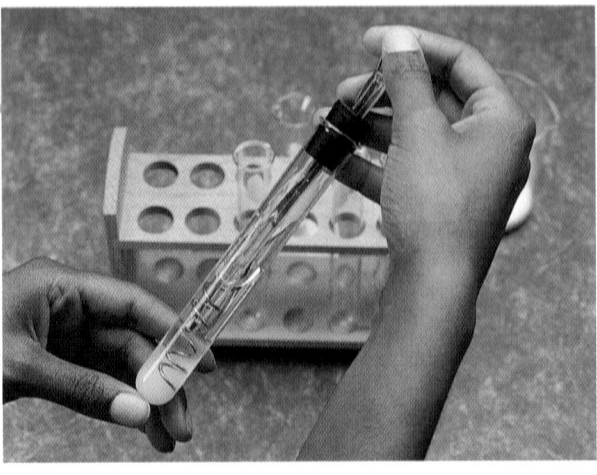

3. Have you allowed for a control? How will the control be treated?
4. Do you have to run any tests more than one time?
5. Will the data be summarized in a graph?
6. *Make sure your teacher approves your plan and that you have included any changes suggested in the plan.*

DO THE EXPERIMENT

1. Carry out the approved experiment as planned.
2. While the experiment is going on, write down any observations that you make, and complete the data table in your Science Journal.

Analyze and Apply

1. **Compare** your results with those of other groups.
2. How did the saturation change as the temperature was changed?

3.  **Graph** your results using a line graph. Place grams of solute per 100 g of water on the *y*-axis and temperature on the *x*-axis.
4. Using your graph, **estimate** the solubility of sugar at 100°C, the boiling point of water.

15-3 Solubility and Concentration   **431**

Go Further

Compare the effects of temperature on dissolving cane sugar versus beet sugar.

PLAN THE EXPERIMENT

Possible Procedures
One possible procedure is to mass the amount of sugar that will form a saturated solution in a measured amount of water at a specific temperature. This procedure is then repeated at different temperatures and data are recorded.

Teaching Strategies
Tying to Previous Knowledge Most students have dissolved materials such as sugar or salt in hot liquids and in cold liquids and are aware that most materials dissolve better in hot liquids.

DO THE EXPERIMENT

Expected Outcome
Most results will reflect an increase in solubility with an increase in temperature.

Analyze and Apply

1. Students' results should be consistent with those of other groups.
2. As temperature increased, the amount of solute needed for saturation increased.

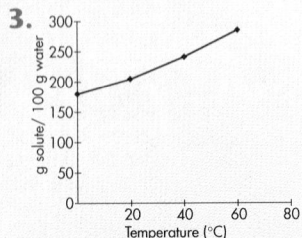
3.

4.  487 g sugar/100 g water

Go Further

Both sugars are of basically the same composition. Students will find few, if any, differences. **P**

Assessment

Oral Have students explain how the relationship between solubility and temperature shown in this activity can be used in making candy. Use the Performance Task Assessment List for Making Observations and Inferences in **PASC**, p. 17. **P**

Prepare

Section Background

In 1887, Arrhenius explained why acids and bases conduct an electric current. These substances ionize in water to produce ions.

Preplanning

To prepare for Activity 15-2, instruct students in the use of a laboratory thermometer.

1 Motivate

Bellringer

Before presenting the lesson, display **Section Focus Transparency 60** on the overhead projector. Assign the accompanying **Focus Activity** worksheet. L1 LEP

Tying to Previous Knowledge

Ask students why it is unwise to take a bath while an electric appliance is plugged in nearby. *Most tap water contains enough dissolved salts to make it conductive.*

2 Teach

Use **Science Integration Transparency 15** as you teach this lesson.

Science Words

dissociation
ionization
electrolyte
nonelectrolyte

Objectives

* Compare and contrast the behavior of polar and nonpolar substances in forming solutions.
* Relate the processes of dissociation and ionization to solutions that conduct electricity.
* Explain how the addition of solutes to solvents affects the freezing and boiling points of solutions.

Figure 15-12

Ethanol, C_2H_5OH, has a polar –OH group at one end and a nonpolar –CH_3 group at the other end.

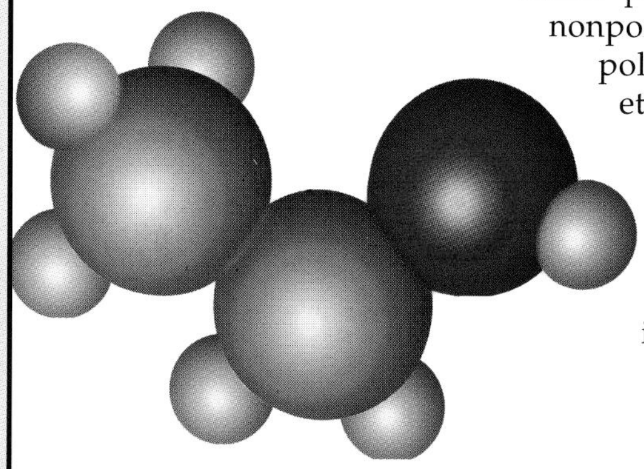

432 Chapter 15 Solutions

The Nature of Solvents and Solutes

Why do you have to shake an oil-and-vinegar salad dressing, such as shown in **Figure 15-11**, immediately before you use it? Oil and vinegar do not form a solution. Oil does not dissolve in vinegar, which is approximately 95 percent water.

What dissolves what?

Why does a substance dissolve in one solvent but not in another? Solvents with nonpolar molecules dissolve solutes with nonpolar molecules. A solvent such as hexane dissolves grease because both the solvent and the solute have nonpolar molecules. Similarly, polar solvents dissolve polar solutes. Water is a polar solvent, and it dissolves sucrose, a polar solute. Polar solvents also dissolve ionic solutes, such as sodium chloride. A polar solvent won't normally dissolve a nonpolar solute. That's why oil, which is nonpolar, won't dissolve in vinegar, which is polar. In general, when predicting which solutes will dissolve in which solvents, remember the phrase, "like dissolves like."

Some substances form solutions with both polar and nonpolar solutes because their molecules have a polar and a nonpolar end. Ethanol, shown in **Figure 15-12**, is such a molecule. The polar end dissolves polar substances, and the nonpolar end dissolves nonpolar substances. Thus, ethanol dissolves iodine, which is nonpolar, and water, which is polar. How important are polar and nonpolar solvents and solutes in nature?

Figure 15-11
Vinegar and oil don't form a solution.

Program Resources

Reproducible Masters
Study Guide, p. 64 L1
Reinforcement, p. 64 L1
Enrichment, p. 64 L3
Cross-Curricular Integration, p. 21
Activity Worksheets, pp. 87-88 L1

Transparencies
Section Focus Transparency 60
Science Integration Transparency 15
Teaching Transparency 30

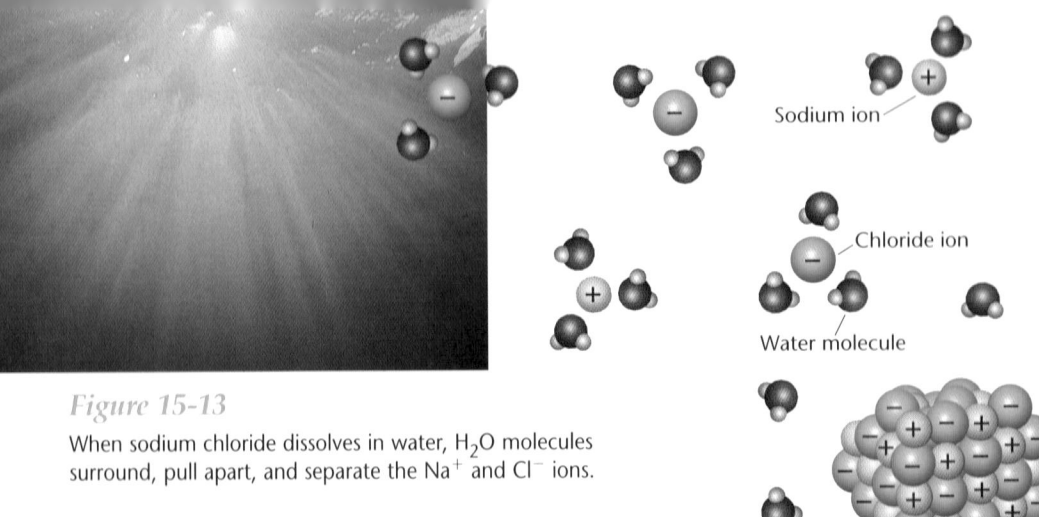

Sodium ion

Chloride ion

Water molecule

Figure 15-13

When sodium chloride dissolves in water, H_2O molecules surround, pull apart, and separate the Na^+ and Cl^- ions.

Particles in Solution

Ions in solution determine the fate of a coral reef, but they also may affect other situations. Why should you never use a hair dryer when standing on a wet floor? You take such precautions because you are an excellent conductor of electricity. Your body is about 70 percent water, with many ionic compounds dissolved in it. Water solutions of ionic compounds and of some polar compounds conduct electricity.

Breaking Up

As an ionic solid dissolves in water, the positive and negative ions separate from one another. This separation process is called **dissociation.** For example, sodium chloride dissociates when it dissolves in water, as shown in **Figure 15-13.** When certain polar substances dissolve in water, the water pulls their molecules apart. This process is called **ionization.** Hydrogen chloride ionizes in water, as shown in **Figure 15-14** below. A substance that separates into ions or forms ions in a water solution is called an **electrolyte.** Thus, both NaCl and HCl are examples of electrolytes.

Pure water is a **nonelectrolyte,** which means it does not conduct electricity. If a water solution is a conductor, the solute must be an electrolyte.

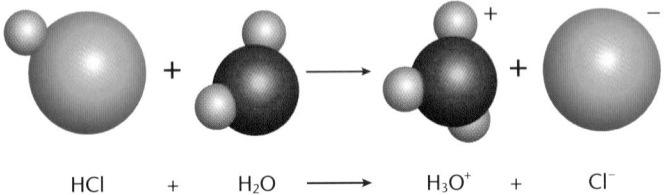

HCl   + H_2O ⟶ H_3O^+ + Cl^-

Figure 15-14

When hydrogen chloride dissolves in water, H_2O molecules surround and pull apart HCl molecules.

15-4 Particles in Solution   **433**

INTEGRATION
Life Science

Ionic solutes have more effect on freezing and boiling points than do covalent solutes because units of ionic compounds dissociate into more than one particle. Thus, glucose in the legs of caribou would not have as much effect as would salt, which forms both Na$^+$ and Cl$^-$ ions.

4 Close

•MINI•QUIZ•

Use the Mini Quiz to check students' recall of chapter content.

1. **When an ionic solid dissolves in water, the separation process is called _____ .** *dissociation*

2. **What happens to the freezing point of a solvent when a solute is added?** *It is lowered.*

Section Wrap-up

Review

1. **a.** Both are polar. **b.** Water is polar but oil is not. **c.** Both are nonpolar.

2. **a.** ions; **b.** molecules

3. **Think Critically** As long as the temperature does not drop lower than the new, lower freezing point of the solution, the water will remain a liquid.

Science Journal

Logical-Mathematical Journal entries should reflect that eggs contain much less water by percentage than does the human body. The solute in fluids contained in insects is more concentrated than it is in the human body. **L3**

Effects of Solute Particles

Antifreeze that you may add to water in a car radiator makes freezing more difficult. Adding a solute to a solvent lowers the freezing point of the solvent. How much the freezing point goes down depends upon how many solute particles you add.

How does antifreeze work? As a substance freezes, its particles organize themselves into an orderly pattern. The solute particles interfere with the formation of this orderly pattern. This prevents the solvent from freezing at its normal freezing point. To overcome this interference, a lower temperature is required to freeze the solvent. The same principle applies to certain animals that live in extremely cold climates. Caribou, for example, contain substances in the lower section of their legs that prevent their freezing in subzero temperatures. The caribou can stand for long periods of time in snow and ice with no harm to their legs.

You may also know that antifreeze raises the boiling point of the water in a car radiator. What causes this effect? The amount that the boiling point is raised depends upon the number of solute molecules present. Solute particles interfere with the evaporation of solvent particles. Thus, more energy is needed to allow the solvent particles to evaporate, and the solution boils at a higher temperature.

INTEGRATION
Life Science

Section Wrap-up

Review

1. Explain why (a) water dissolves sugar, (b) water does not dissolve oil, and (c) benzene does dissolve oil.

2. What kinds of solute particles are present in water solutions of (a) electrolytes and (b) nonelectrolytes?

3. **Think Critically:** In cold weather, people often put salt on ice that forms on sidewalks and driveways. The salt helps melt the ice, forming a saltwater solution. Explain why this solution may not refreeze.

Skill Builder
Concept Mapping

Draw a concept map to show the relationship among the following terms: *electrolytes, nonelectrolytes, dissociation, ionization, ionic compounds, certain polar compounds, other polar compounds.* If you need help, refer to Concept Mapping in the **Skill Handbook.**

Science Journal

Many insect eggs can survive extremely cold temperatures. What can you conclude about the fluids in these eggs? Summarize your conclusions in your Science Journal.

Skill Builder

Compounds that dissolve in water
- conduct electricity → Electrolytes
  - type → Ionic compounds → process → Dissociation
  - type → Certain polar compounds → process → Ionization
- do not conduct electricity → Nonelectrolytes
  - type → Other polar compounds

✔ Assessment

Oral Have students use their concept maps to explain why water is a nonelectrolyte. Use the Performance Task Assessment List for Concept Map in **PASC,** p. 89. **P**

Native American Ocean Lore

In this chapter, you've seen that various types of solutions play important and useful roles in the everyday lives of people. However, no solution is more important to living things on Earth than the ocean. The ocean is the world's largest solution, made up of dissolved sodium chloride and many other salts. Three-fourths of Earth's surface is water, and about 97 percent of this total is ocean water. Thus, it's not surprising that the oceans are home to millions of different living things, and that countless other species depend on oceans indirectly.

Historically, oceans have also occupied a special place in human life. For many indigenous peoples who live by or near shore, the importance of oceans is often deeply woven into the fabric of the culture. Consider this folktale entitled "Octopus and Raven" from the Nootka people, a Native American culture from the Pacific Northwest. This story, about a raven who asks a silly question of an octopus looking for food, demonstrates the Nootka culture's knowledge about oceans and ocean life. In this excerpt, Raven is having an argument with Octopus as the tide is coming in.

Science Journal

At your local library, find a children's story, myth, or folktale to read. In your Science Journal, write a summary of the story. Are there any images or characters in the story that might give you clues about the type of environment or animal life in the native country?

> . . . *The water was now over their waists. Raven saw that it would soon be even deeper. "Octopus, can you please let me go?" But Octopus did not let go. "Raven," she said, "I was digging clams!"*
>
> *Again Raven begged, but the water continued to get deeper and Octopus held tight. The water came up to their necks and then it was over their heads.*
>
> *Up on the beach, above the tide line, the old people watched.*
>
> *"Octopus can hold her breath longer than Raven," one person said.*
>
> *They watched and finally, after a very long time, Raven could hold his breath no longer and he drowned. Octopus let go and Raven floated up to the surface.*

As you can see from this excerpt, the Nootka people show knowledge of ocean tides, as well as about some organisms that live there. The appearance of ocean creatures and information about tides in a children's story illustrates just how important the ocean is in Nootka culture.

15-4 Particles in Solution **435**

- Caduto, Michael J., and Joseph Bruchac. *Keepers of the Earth: Native American Stories and Environmental Activities for Children.* Fulcrum Publishing: Golden, CO, 1988.
- Shannon, George. *The Oryx Multicultural Folktale Series: A Knock at the Door.* Oryx Press: Phoenix, AZ, 1992.

Source

Caduto, Michael J., and Joseph Bruchac. *Keepers of the Animals.* Fulcrum Publishing: Golden, CO, 1991.

Background

The oceans have had a great cultural influence on the many Native North American peoples who once lived by or near the oceans and thrived on their rich resources. This fact is particularly evident when one reads the many Native American myths, legends, and children's stories devoted to ocean life and processes.

Teaching Strategies

- Point out the types and amounts of dissolved salts in the ocean. Provide students with data about the differences in salinity of ocean waters around the world. Discuss the significance of salinity differences to life in the ocean.

- Have students discuss the composition of ocean water and consider the reasons why oceans are capable of supporting a great diversity of life.

- Provide students with a copy of any Native American tale involving oceans. Have students read the story, then have them write Science Journal entries summarizing the story. Have students highlight the parts of the story that demonstrate knowledge about oceans or the organisms that live there.

Other Works

- Barton, Bob, and David Booth. *Stories in the Classroom: Storytelling, Reading Aloud and Role Playing with Children.* Heinemann: Portsmouth, NH, 1990.

Boiling Points of Solutions

Purpose

IS Interpersonal Students will interpret data and draw conclusions regarding the effect of a solute on the boiling point of water. **L1 COOP LEARN**

Process Skills

measuring in SI, communicating, interpreting data, graphing

Time

35 minutes

Safety Precautions

Caution students not to let the thermometer touch the sides or bottom of the beaker.

📁 **Activity Worksheets,** pp. 5, 87-88

Teaching Strategies

Alternate Materials Other soluble ionic compounds, such as calcium chloride, will also raise the boiling point.

Troubleshooting Be sure students have thermometer bulbs immersed when reading temperatures and that thermometers read clearly in a range above 100°C. The boiling temperature of water may be less than 100°C because atmospheric pressure is usually less than 1 atmosphere.

• Have one student in a group measure out the additional NaCl needed while others handle heating. Students should work on graphs in groups.

• The following is a graph using sample data. Students should obtain similar graphs.

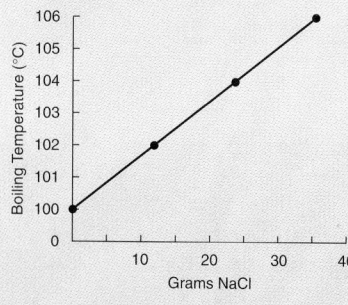

Adding small amounts of salt to water being boiled to cook pasta and adding antifreeze to a car's radiator have a common result. In each case, the boiling point of the water is now higher than that of pure water.

Problem

How much can the boiling point of a solution be changed?

Materials

• 400 mL distilled water
• thermometer
• ring stand
• 72 g table salt, NaCl
• hot plate
• 250-mL beaker

Procedure

1. Copy the data table and use it to record your observations.
2. Bring 100 mL of distilled water in a 250-mL beaker to a gentle boil. Record the temperature.
3. Dissolve 12 g of NaCl in 100 mL of distilled water. Bring this solution to a gentle boil and record its boiling point. **CAUTION:** *Always keep the thermometer away from the flame.*

4. Repeat step 3, using 24 g of NaCl.
5. Repeat step 3 again, using 36 g of NaCl.
6. Plot your results on a graph that shows boiling point on the *y*-axis and grams of NaCl on the *x*-axis.

Analyze

1. What difference is there between the boiling point of a pure solvent and that of a solution?
2. Instead of doubling the amount of NaCl in step 4, what would have been the effect of doubling the amount of water?
3. What would be the result of using tap water instead of distilled water?

Conclude and Apply

4. In cooking, why would adding salt to water cause some foods to cook faster?
5. If you continued to add more salt, would you **predict** that your graph would continue in the same pattern or level off? Explain your prediction.

Data and Observations Sample Data

Grams of NaCl Solute	Boiling Point (°C)
0	~100
12	~102
24	~104
36	~106

Answers to Questions

1. The boiling point of the solution is higher.
2. The boiling point would have gone down but would still be higher than the boiling point of pure water.
3. Most tap water contains dissolved salts, so its boiling point would be above 100°C.
4. The saltwater solution would boil at a slightly higher temperature.

5. The graph would continue until saturation was reached. Then it would level off because no more salt would dissolve.

 **Assessment**

Process Have students use their graphs to predict the boiling point of a solution that contains 18 g of NaCl. *about 103°C* Use the Performance Task Assessment List for Analyzing the Data in **PASC,** p. 27. **P**

Summary

15-1: How Solutions Form

1. Solutions are classified into three types according to their final state: gaseous, liquid, and solid.
2. Dissolving occurs when constantly moving solvent molecules attract particles of solute and surround them with solvent.
3. Stirring, surface area, temperature, and pressure affect the rate of dissolving.

15-2: Science and Society: Regulating Organic Solvents

1. Many materials used for building or furnishing homes contain potentially harmful chemicals.
2. OSHA and the CPSC protect people by controlling use of these materials.

15-3: Solubility and Concentration

1. Solubility of substances varies among different solutes and for the same solute at different temperatures.
2. Two precise ways to express solution concentrations are (1) percentage by volume and (2) percentage by mass.
3. A solution may be saturated, unsaturated, or supersaturated.

15-4: Particles in Solution

1. Usually, among polar and nonpolar solvents and solutes, like dissolves like.
2. Ionic compounds dissociate when dissolved in water, and some polar compounds ionize when dissolved in water.
3. Adding a solute to a solvent lowers its freezing point and raises its boiling point.

Key Science Words

a. dissociation
b. electrolyte
c. ionization
d. nonelectrolyte
e. organic solvent
f. saturated solution
g. solubility
h. solute
i. solvent
j. supersaturated solution
k. unsaturated solution

Reviewing Vocabulary

Match each phrase with the correct term from the list of Key Science Words.

1. a solution that can dissolve more solute at a given temperature
2. the substance being dissolved
3. has more solute than a saturated solution has at that temperature
4. exists as ions in a water solution
5. molecules are pulled apart into ions in solution
6. separation of negative and positive ions
7. the most solute, in grams, that will dissolve in 100 g of solvent at a certain temperature
8. a solution that has dissolved all the solute it can hold at a given temperature
9. the substance that dissolves a solute
10. liquid often found in building materials

Summary

Have students read the summary statements to review the major concepts of the chapter.

Reviewing Vocabulary

1. k	6. a
2. h	7. g
3. j	8. f
4. b	9. i
5. c	10. e

✓ Assessment

Portfolio Encourage students to place in their portfolios one or two items of what they consider to be their best work. Examples include:

- Flex Your Brain, p. 420
- Science Journal, p. 423
- Activity 15-1 results and answers, pp. 430-431 **P**

Performance Additional performance assessments may be found in **Performance Assessment** and **Science Integration Activities.** Performance Task Assessment Lists and rubrics for evaluating these activities can be found in Glencoe's **Performance Assessment in the Science Classroom.**

Chapter 15 Review **437**

GLENCOE TECHNOLOGY

 Videodisc

Glencoe Physical Science

Interactive Videodisc

Use the videodisc segment *Gases Exert Pressure* from Chapter 1, to review the principles of gases in solution.

MindJogger Videoquiz

Chapter 15 Have students work in groups as they play the Videoquiz game to review key chapter concepts.

Chapter 15 Review

Checking Concepts

1. d	**6.** b
2. a	**7.** b
3. b	**8.** a
4. d	**9.** b
5. a	**10.** c

Understanding Concepts

11. Organic solvents vaporize easily. They can enter the body by absorption or as fumes. Some are known to affect the nervous system and cause growth and development problems.

12. The negative ends of polar water molecules attract the positive copper ions at the surface of the solid and pull them into solution. The positive ends of water molecules attract the negative sulfate ions at the surface of the solid and pull them into solution.

13. There is no set amount of solute that makes a solution concentrated or dilute.

14. One possibility is to add 25 mL of apple juice to 75 mL of water.

15. Solutions may be prepared using solvents other than water. Water also could be the solute in a solution.

Thinking Critically

16. Tetrachloroethene and grease are both composed of nonpolar molecules. Water is polar and is not attracted to nonpolar grease molecules.

17. The salt dissociates into ions when it dissolves. The presence of these ions lowers the freezing point of water.

18. The salted water boils at a slightly higher temperature than unsalted water because of the presence of

Checking Concepts

Choose the word or phrase that completes the sentence or answers the question.

1. Which of the following is not a solution?
   a. a glass of soda
   b. air in a SCUBA tank
   c. bronze (an alloy)
   d. muddy water
2. Solutions may not be _____.
   a. colloidal   c. liquid
   b. gaseous   d. solid
3. When iodine is dissolved in alcohol, the alcohol is the _____.
   a. alloy   c. solution
   b. solvent   d. solute
4. If a bronze alloy is 85 percent copper and 15 percent tin, tin is the _____.
   a. alloy   c. solution
   b. solvent   d. solute
5. Forty-nine mL of water at room temperature have a mass of _____.
   a. 49 g   c. 100 g
   b. 51 g   d. 4900 g
6. A polar solvent will dissolve _____.
   a. any solute   c. a nonpolar solute
   b. a polar solute   d. no solute
7. If a water solution conducts electricity, the solute must be a(n) _____.
   a. gas   c. liquid
   b. electrolyte   d. nonelectrolyte
8. In a water solution, an ionic compound undergoes _____.
   a. dissociation   c. ionization
   b. electrolysis   d. no change
9. A gas becomes more soluble in a liquid when you increase _____.
   a. particle size   c. stirring
   b. pressure   d. temperature
10. Solute may come out of a solution that is _____.
    a. unsaturated   c. supersaturated
    b. saturated   d. miscible

Understanding Concepts

Answer the following questions in your Science Journal using complete sentences.

11. How do the health effects of an organic solvent differ from another solvent such as water?
12. Explain what happens when copper(II) sulfate, $CuSO_4$, an ionic compound, dissolves in water.
13. Why are *concentrated* and *dilute* not exact terms?
14. Explain how you would make a 25 percent solution of apple juice.
15. Why is the statement "Water is the solvent in a solution" not always true?

Thinking Critically

16. Explain why tetrachloroethene, a dry-cleaning fluid, will dissolve grease and water will not.
17. Explain why salt will melt ice on a sidewalk.
18. Why might potatoes cook more quickly in salt water than in unsalted water?
19. Explain why an unrefrigerated glass of soda will go flat more quickly in hot weather.
20. Why might a tropical lake have more minerals dissolved in it than a lake in Minnesota?

dissolved ions. At the higher temperature, the potatoes cook more quickly.

19. The higher room temperature in hot weather causes decreased solubility of carbon dioxide gas in water. The gas comes out of solution more rapidly.

20. A tropical lake is likely to be warmer. Most solutes are more soluble in water at higher temperatures.

Developing Skills

21. Comparing and Contrasting Both processes result in ions in solution. Dissociation involves the separation of ions in an ionic compound. Ionization involves the pulling apart of polar molecules into ions.

Developing Skills

If you need help, refer to the **Skill Handbook.**

21. **Comparing and Contrasting:** Compare and contrast the processes of ionization and dissociation.

22. **Hypothesizing:** You are stocking a pond in a tropical climate and a pond in a temperate climate with fish. Both ponds contain about the same amount of water. Based on the amount of oxygen dissolved in the water, to which pond would you be able to add more fish? Explain why.

23. **Interpreting Data:** The label on a bottle of rubbing alcohol might read "70 percent isopropanol by volume." Assuming the rest of the solution is water, what does this label tell you about the contents and preparation of the solution in the bottle?

24. **Measuring in SI:** 153 g of potassium nitrate have been dissolved in enough water to make 1.00 L of solution. You use a graduated cylinder to measure 80.0 mL of solution. What mass of potassium nitrate is in the 80.0-mL sample?

25. **Making and Using Tables:** Using the data in **Table 15-2** on page 427, fill in the following table. Use the terms *saturated, unsaturated,* and *supersaturated* to describe the type of solution.

Performance Assessment

1. **Poster:** Visit a SCUBA shop. Find out about the composition of the gas used in diving tanks. In terms of solution, report your findings on a poster.

2. **Making and Using a Classification System:** Based upon information about composition of solutions in **Figure 15-1**, classify 20 different solutions in a grocery store. Explain your classification system to the class.

3. **Display:** Research crystal growing using saturated and supersaturated solutions. Grow several crystals and display them. Accompany the display with a report explaining what you did and your results.

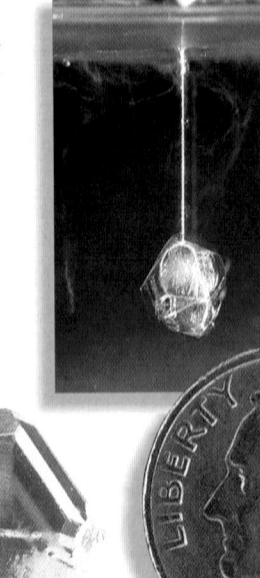

Limits of Solubility

Compound	Type of Solution	g Dissolved in 100 g Water at 20°C
Ba(OH)$_2$	unsaturated	2.96
CuSO$_4$	saturated	32.0
KCl	supersaturated	45.8
KNO$_3$	saturated	31.6
NaClO$_3$	unsaturated	79.6

22. **Hypothesizing** In theory, more fish could live in the temperate climate. At the colder temperature, more oxygen would dissolve in the water.

23. **Interpreting Data** For every 100 mL of the rubbing alcohol, 70 mL are alcohol, and 30 mL are water.

24. **Measuring in SI**
153 g KNO$_3$/1000 mL solution × 80.0 mL solution = 12.2 g KNO$_3$

25. **Making and Using Tables** unsaturated, saturated, supersaturated, saturated, unsaturated

Performance Assessment

1. Most diving is done using compressed air, which is 79 percent N$_2$ and 21 percent O$_2$. Specialized mixtures can have O$_2$ content up to 36 percent. Use the Performance Task Assessment List for Poster in **PASC,** p. 73. **P**

2. Answers will vary, but the system must be explained using science concepts. Use the Performance Task Assessment List for Making and Using a Classification System in **PASC,** p. 49. **P**

3. Answers will vary, but reports should explain the procedure followed and the results obtained. Use the Performance Task Assessment List for Display in **PASC,** p. 63. **P**

Assessment Resources

Reproducible Masters
Chapter Review, pp. 33, 34
Assessment, pp. 101-104
Performance Assessment, p. 29

Glencoe Technology
Chapter Review Software
Computer Test Bank
MindJogger Videoquiz

Chapter Organizer

Section	Objectives/Standards	Activities/Features
Chapter Opener		**Explore Activity:** Find out why copper surfaces often turn green. p. 441
16-1 Chemical Changes in Matter (1 session, ½ block)*	1. **Identify** reactants and products in a chemical reaction. 2. **Explain** how a chemical reaction satisfies the law of conservation of mass. 3. **Interpret** chemical equations. **National Content Standards:** (5-8) UCP1-UCP3, UCP5, A2, G1, G3; (9-12) UCP1-UCP3, UCP5, A2, B3, G1, G3	**Connect to Earth Science,** p. 442 **MiniLAB:** Does mass change in a reaction? p. 443 **Problem Solving:** Metals and the Atmosphere, p. 444 **Using Math,** p. 445 **Skill Builder:** Recognizing Cause and Effect, p. 445
16-2 Science and Society (½ session, ½ block)*	1. **Explain** the effect of CFCs on ozone in the atmosphere. 2. **Compare** possible alternatives to CFCs. **National Content Standards:** (5-8) UCP2, UCP3, UCP5, D1, F1-F5; (9-12) UCP2, UCP3, UCP5, B3, D1, F1-F6	**Science Journal,** p. 447 **Explore the Issue,** p. 447
16-3 Chemical Equations (2½ sessions, 1½ blocks)*	1. **Explain** what is meant by a balanced chemical equation. 2. **Demonstrate** how to write balanced chemical equations. **National Content Standards:** (5-8) UCP1-UCP3; (9-12) UCP1-UCP3, B3	**Using Computers,** p. 450 **Skill Builder:** Observing and Inferring, p. 450
16-4 Types of Chemical Reactions (2 sessions, ½ block)*	1. **Describe** four types of chemical reactions using their generalized formulas. 2. **Classify** various chemical reactions by type. **National Content Standards:** (5-8) UCP1-UCP3, UCP5, A1, E1; (9-12) UCP1-UCP3, UCP5, A1, B3, E1	**Connect to Life Science,** p. 451 **MiniLAB:** Must the sums of reactant and product coefficients be equal? p. 452 **Using Math,** p. 453 **Skill Builder:** Hypothesizing, p. 453 **Activity 16-1:** Metal Activity, pp. 454-455
16-5 Energy and Chemical Reactions (1 session, ½ block)*	1. **Differentiate** between an exothermic reaction and an endothermic reaction. 2. **Describe** the effects of catalysts and inhibitors on the speed of chemical reactions. **National Content Standards:** (5-8) UCP1-UCP3, UCP5, A2, B3, C1, C3, C5, E2, F1, F2, F5; (9-12) UCP1-UCP3, UCP5, A2, B3, E2, F1, F4-F6	**Using Technology:** Catalytic Converters, p. 457 **Science Journal,** p. 458 **Skill Builder:** Concept Mapping, p. 458 **Activity 16-2:** Catalyzed Reaction, p. 459 **Science and Art:** Fresco Painting, p. 460

Activity Materials

Explore	Activities	MiniLABs
page 441 table salt, tarnished pennies, vinegar, water, plastic wrap, saucer, safety goggles	pages 454-455 identical test tubes, test-tube rack, dilute hydrochloric acid, 10-mL graduated cylinder, tongs, wooden splint, small pieces of metals, safety goggles, apron page 459 3 medium-sized test tubes, test-tube stand, 15 mL hydrogen peroxide, graduated cylinder, small plastic spoon, sand, manganese dioxide, wooden splint, hot plate, beaker of hot water, safety goggles, apron	page 443 NaOH solution, dropper bottle, dropper, $FeCl_3$ solution, small plug of soap, balance page 452 index cards

Need Materials? Call Science Kit (1-800-828-7777).   * A complete Planning Guide that includes block scheduling is provided on pages 32T-35T.

Teacher Classroom Resources

Reproducible Masters	Transparencies	Teaching Resources
Study Guide, p. 65 Reinforcement, p. 65 Enrichment, p. 65 Activity Worksheets, p. 94	Section Focus Transparency 61, Chemical Changes	Glencoe Physical Science Interactive Videodisc Physical Science CD-ROM Spanish Resources English/Spanish Audiocassettes Cooperative Learning Resource Guide Lab Partner Lab and Safety Skills Lesson Plans
Study Guide, p. 66 Reinforcement, p. 66 Enrichment, p. 66 Concept Mapping, pp. 37-38	Section Focus Transparency 62, Avoiding Ultraviolet	
Study Guide, p. 67 Reinforcement, p. 67 Enrichment, p. 67 Lab Manual 32, Conservation of Mass Cross-Curricular Integration, p. 22	Section Focus Transparency 63, Balancing a Buck Teaching Transparency 31, Chemical Equations	**Assessment Resources** Chapter Review, pp. 35-36 Assessment, pp. 105-108 Performance Assessment in the Science Classroom (PASC) MindJogger Videoquiz
Study Guide, p 68; Reinforcement, p. 68 Enrichment, p. 68 Activity Worksheets, pp. 90-91, 95 Lab Manual 33, Reaction Rates and Temperature Science Integration Activity 16, Vanishing Buildings Critical Thinking/Problem Solving, p. 22	Section Focus Transparency 64, Building Blocks Teaching Transparency 32, Types of Chemical Reactions	Alternate Assessment in the Science Classroom Performance Assessment Chapter Review Software Computer Test Bank
Study Guide, p. 69 Reinforcement, p. 69 Enrichment, p. 69 Activity Worksheets, pp. 92-93 Lab Manual 34, Chemical Reactions Multicultural Connections, p. 35 Science and Society Integration, p. 20	Section Focus Transparency 65, Hot and Cold Packs Science Integration Transparency 16, Enzymes in Body Chemistry	

Key to Teaching Strategies

The following designations will help you decide which activities are appropriate for your students.

L1 Level 1 activities should be within the ability range of all students, including those with learning difficulties.

L2 Level 2 activities should be within the ability range of the average to above-average student.

L3 Level 3 activities are designed for the ability range of above-average students.

LEP LEP activities should be within the ability range of Limited English Proficiency students.

LS These activities are designed to address different learning styles.

COOP LEARN Cooperative Learning activities are designed for small group work.

P These strategies represent student products that can be placed into a best-work portfolio.

GLENCOE TECHNOLOGY

The following multimedia resources are available from Glencoe.

The Infinite Voyage Series
The Future of the Past
Secrets from a Frozen World

National Geographic Society Series
GTV: Planetary Manager

Glencoe Physical Science Interactive Videodisc
Periodicity
Chemical Detectives

Physical Science CD-ROM

Teacher Classroom Resources

This is a representation of key blackline masters available in the Teacher Classroom Resources.

Teaching Aids

Section Focus Transparencies

Science Integration Transparencies

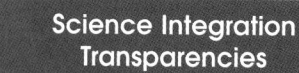

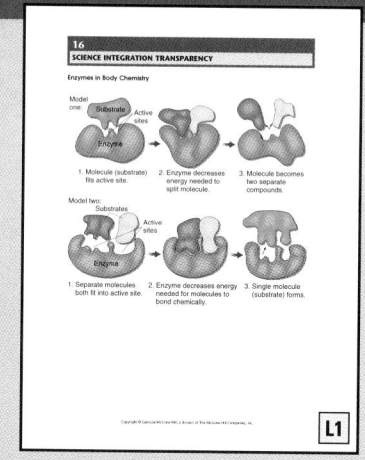

Teaching Transparencies

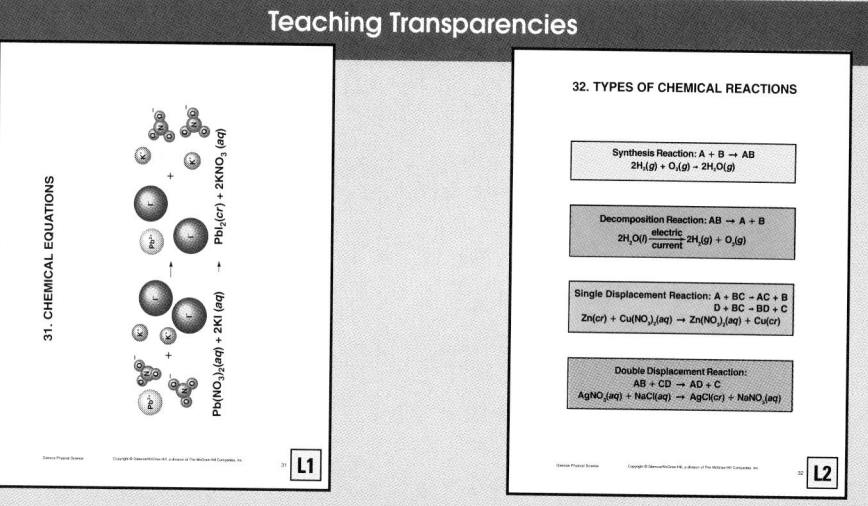

Meeting Different Ability Levels

Study Guide

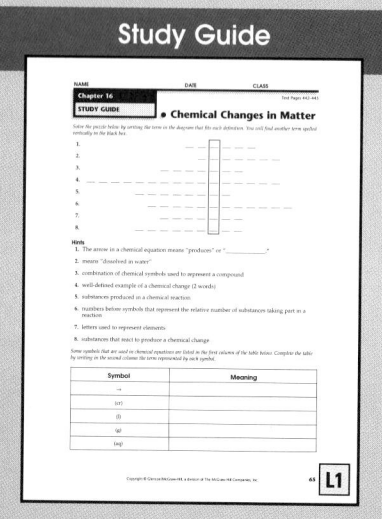

Reinforcement

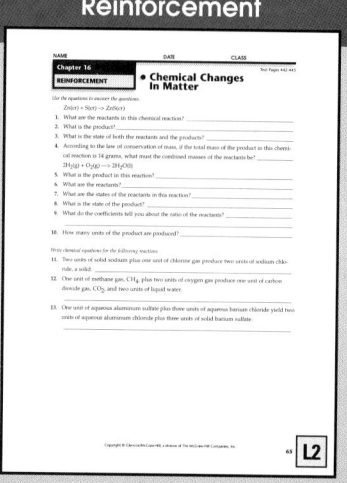

Enrichment Worksheets

Hands-On Activities

Science Integration Activity

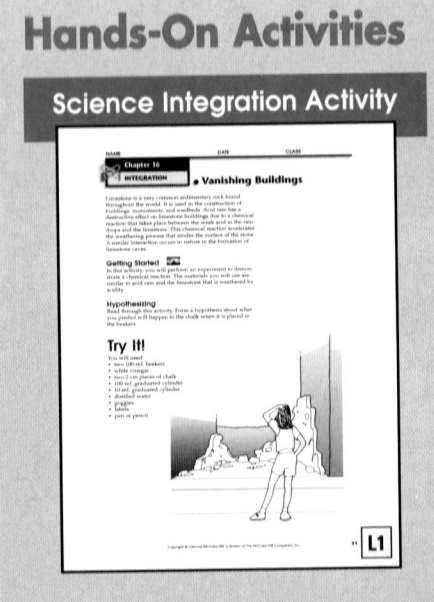

Vanishing Buildings

L1

Lab Manual

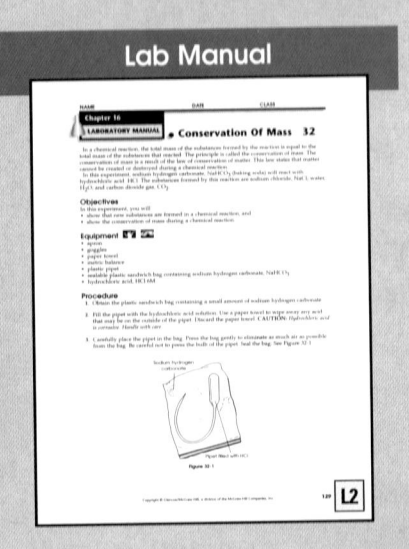

Conservation Of Mass 32

L2

Assessment

Performance Assessment

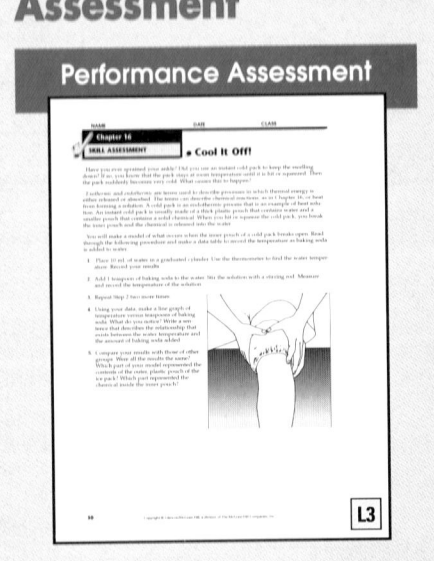

Cool It Off!

L3

Enrichment and Application

Critical Thinking/ Problem Solving

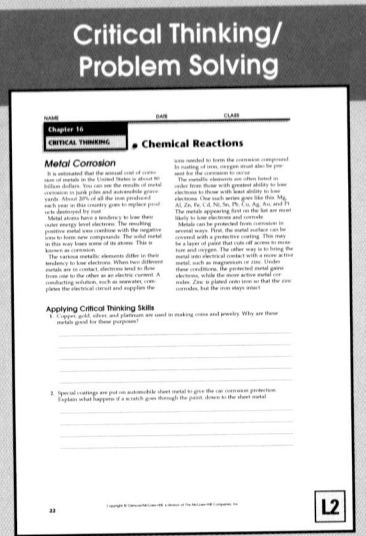

Chemical Reactions

Metal Corrosion

L2

Cross-Curricular Integration

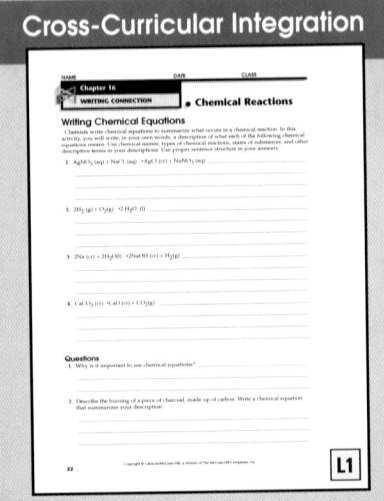

Chemical Reactions

Writing Chemical Equations

L1

Science and Society Integration

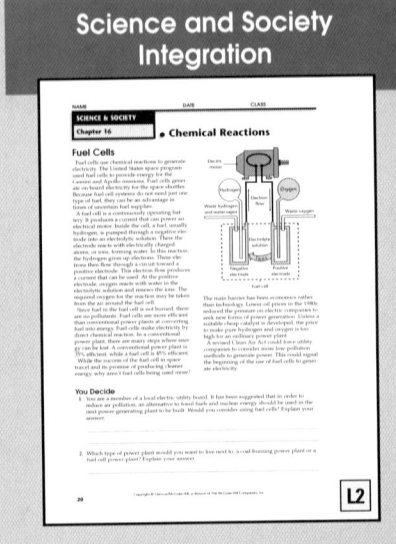

Chemical Reactions

Fuel Cells

L2

Multicultural Connections

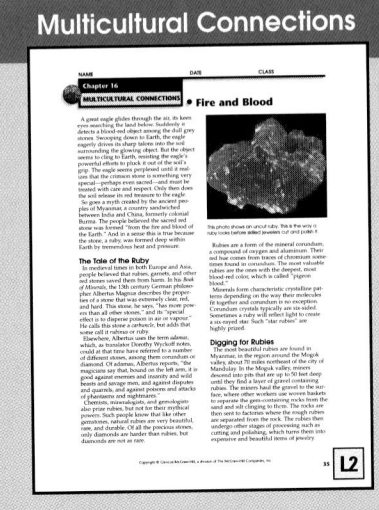

Fire and Blood

L2

Concept Mapping

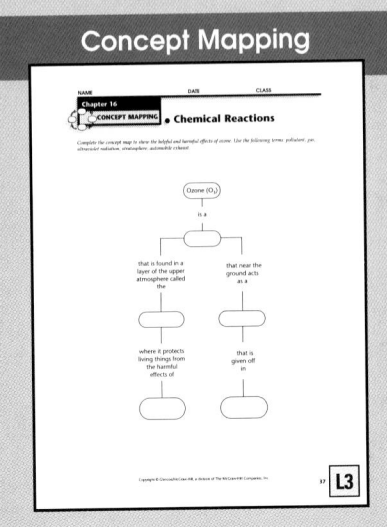

Chemical Reactions

L3

Chemical Reactions

CHAPTER OVERVIEW

Section 16-1 The chemical equation is introduced as a way of representing chemical reactions in a way that demonstrates the law of conservation of mass.

Section 16-2 Science and Society This section describes how the ozone layer is affected by CFCs.

Section 16-3 How to balance a chemical equation is presented in this section.

Section 16-4 Four common types of chemical reactions are presented. This section demonstrates how chemists are able to predict products for a chemical reaction.

Section 16-5 Energy as a product or reactant in a chemical change is discussed. Catalysts and inhibitors are introduced.

Chapter Vocabulary

chemical reaction
reactant
product
coefficient
chlorofluorocarbon (CFC)
balanced chemical equation
synthesis reaction
decomposition reaction
single-displacement reaction
double-displacement reaction
precipitate
exothermic reaction
catalyst
inhibitor
endothermic reaction

Theme Connection

Stability and Change Change as a theme of the textbook is developed through a presentation of how chemical changes are classified and represented by balanced chemical equations.

440

Previewing the Chapter

440

Learning Styles

Look for the following logo for strategies that emphasize different learning modalities. **LS**

Kinesthetic Activity, p. 448; Reteach, p. 449

Visual-Spatial Explore, p. 441; Demonstration, pp. 444, 446; Reteach, p. 447; Activity 16-2, p. 459

Interpersonal MiniLAB, p. 443; Activity 16-1, pp. 454-455

Logical-Mathematical Inclusion Strategies, p. 449; MiniLAB, p. 452; Reteach, p. 453

Linguistic Across the Curriculum, p. 452; Community Connection, p. 457; Science & Art, p. 460

Chapter 16

Chemical Reactions

You know that metals can be made into a variety of shapes. The Statue of Liberty, given to the United States by France, is made primarily of copper. However, the green color of the statue is different from the color of a typical copper penny in your pocket. What causes copper items to lose their shiny surfaces and turn green?

EXPLORE ACTIVITY

Find out why copper surfaces often turn green.

1. Sprinkle some table salt on the tops of two dark pennies.
2. Carefully place five or six drops of vinegar on each of the pennies.
3. After 20 minutes, rinse the pennies with water.
4. Wrap one of the pennies tightly in plastic wrap.
5. Place the other penny in water in a saucer so that the penny is not completely covered.
6. Let both pennies remain undisturbed for one day.

Observe: Record your observations in your Science Journal. From your observations, what can you suggest as the cause of the color changes in the pennies?

Previewing Science Skills

▶ In the Skill Builders, you will recognize **cause and effect, observe and infer, hypothesize,** and **map concepts.**

▶ In the Activities, you will **observe, infer,** and **classify.**

▶ In the MiniLABs, you will **observe and infer.**

441

EXPLORE ACTIVITY

Purpose
Visual-Spatial Introduce students to chemical reactions and how to describe them. **L1** **LEP** **COOP LEARN**

Preparation
Have students bring in pennies from home.

Materials
two darkened pennies, table salt, vinegar, medicine dropper, paper towel, plastic wrap, one dish per group

Teaching Strategies
Tell students that for copper to turn green, it must oxidize and then be exposed to water and sulfur compounds in air that has a relative humidity of at least 75 percent. This process is very slow.

Observe
Vinegar and salt solution removed the dark patina. The sealed penny remained shiny. The penny exposed to the air began to darken again. The Cu reacted with oxygen and water.

✔ Assessment

Performance Give students small pieces of different metals to moisten and leave outside. Have students discuss possible solutions to the building industry's oxidation problem. Use the Performance Task Assessment List for Carrying Out a Strategy and Collecting Data in **PASC**, p. 25. **P**

Assessment Planner

Portfolio
Refer to page 461 for suggested items that students might select for their portfolios.

Performance Assessment
See page 461 for additional Performance Assessment options.
Skill Builders, pp. 445, 450, 453, 458
MiniLABS, pp. 443, 452
Activities 16-1, pp. 454-455, 16-2, p. 459

Content Assessment
Section Wrap-ups, pp. 445, 447, 450, 453, 458
Chapter Review, pp. 461-463
Mini Quizzes, pp. 445, 450, 453, 458

Group Assessment
Opportunities for group assessment occur with Cooperative Learning Strategies and Flex Your Brain Activities.

Section 16•1

Prepare

Section Background

The symbols used to indicate the state of matter can vary in scientific literature. The (*s*) for solid sometimes is replaced with (*cr*) for crystalline solid. If the material is not crystalline, it is not a true solid, and the symbol (*amor*) for amorphous is used.

Preplanning

For the MiniLAB, to make the needed solutions, dissolve 4 g NaOH in 100 mL H_2O and dissolve 8 g $FeCl_3$ in 100 mL H_2O. **CAUTION:** *These substances are corrosive.*

1 Motivate

Bellringer

Before presenting the lesson, display **Section Focus Transparency 61** on the overhead projector. Assign the accompanying **Focus Activity** worksheet. L1 LEP

Tying to Previous Knowledge

Remind students of the indications that a chemical change has occurred: color change, evolution of gas, heat, light, formation of a precipitate, production of electric current.

2 Teach

Student Text Question

What are the reactants and products in the photosynthesis reaction described in the first paragraph? *reactants—carbon dioxide and water; products—sugar and oxygen*

Science Words

chemical reaction
reactant
product
coefficient

Objectives

- Identify reactants and products in a chemical reaction.
- Explain how a chemical reaction satisfies the law of conservation of mass.
- Interpret chemical equations.

CONNECT TO

EARTH SCIENCE

Some darkening of the Statue of Liberty occurs over time. *Infer* what environmental factor may cause this problem to increase.

16•1 Chemical Changes in Matter

Describing Chemical Reactions

In Chapter 9, you learned about chemical changes. These changes are taking place all around you and even inside your body. One of the most important chemical changes on Earth is photosynthesis. During photosynthesis, plants use sunlight, carbon dioxide, and water to make sugar and oxygen. Another important chemical change takes place in cells in your body during respiration. This change combines sugar and oxygen to produce energy, carbon dioxide, and water.

A **chemical reaction** is a well-defined example of a chemical change. In a chemical reaction, one or more substances are changed to new substances. The substances that are about to react are called **reactants.** The new substances produced are called **products.** This relationship can be written as follows.

$$\text{reactants} \xrightarrow{\text{produce}} \text{products}$$

What are the reactants and products in the photosynthesis reaction described in the first paragraph?

Conservation of Mass

In the early 1700s, scientists wondered what happened to the masses of reactants and products in chemical reactions. The French chemist Antoine Lavoisier, shown in **Figure 16-1,** performed many experiments to find out. In one experiment, Lavoisier placed a carefully measured amount of solid mercury(II) oxide into a sealed flask. When he heated this flask, oxygen gas and liquid mercury were produced.

Figure 16-1

Antoine Lavoisier used precise balances to perform experiments that led to the law of conservation of mass.

Program Resources

📁 **Reproducible Masters**
Study Guide, p. 65 L1
Reinforcement, p. 65 L1
Enrichment, p. 65 L3
Activity Worksheets, p. 94 L1

📽 **Transparencies**
Section Focus Transparency 61 L1

Lavoisier found that the total mass of the oxygen gas and the mercury produced was equal to the mass of the mercury(II) oxide that he started with.

mercury(II) oxide produces oxygen plus mercury
$$10.0 \text{ g} = 0.7 \text{ g} + 9.3 \text{ g}$$

This and other experiments allowed Lavoisier to state the law of conservation of mass: in a chemical reaction, matter is not created or destroyed but is conserved. In Chapter 9, you learned how this law applies to chemical changes. For a chemical reaction in which all of the reactants change to products, this law means that the starting mass of the reactants equals the final mass of the products. The law of conservation of mass must be satisfied when describing a chemical reaction.

Writing Equations

If you wanted to write a description of the chemical reaction shown in **Figure 16-2**, it might look like this.

solid lead(II) nitrate, dissolved in water, plus
solid potassium iodide, dissolved in water, produces
solid lead(II) iodide plus potassium nitrate,
dissolved in water

This series of words is rather long and cumbersome. However, all of the information in this expression is important. The same is true of descriptions of most chemical reactions—many words are needed to state all the important information. As a result, scientists have developed a shorthand method to describe chemical reactions. A chemical equation is an expression that describes a chemical reaction using chemical formulas and other symbols. In Chapter 11, you learned

Figure 16-2

When solutions of lead(II) nitrate and potassium iodide are mixed, a bright yellow solid forms. The yellow solid, lead(II) iodide, settles to the bottom of the beaker. The liquid in the beaker is a potassium nitrate solution.

16-1 Chemical Changes in Matter   **443**

MiniLAB

Does mass change in a reaction?

Procedure
1. Add NaOH solution to a dropper bottle until the bottle is approximately half full. Fill the dropper with a solution of $FeCl_3$. Stop up the dropper with a soap plug. Screw the dropper onto the bottle. Find and record the mass of the bottle, dropper, and their contents.
2. Gently squeeze the dropper so that the soap plug is forced out of the dropper.
3. Again, determine the mass of the bottle, dropper, and products.

Analysis
1. What changes did you observe to indicate that a chemical reaction has taken place?
2. How did the mass before the reaction compare to the mass after the reaction?
3. What scientific law does this experiment demonstrate?

Demonstration

LS Visual-Spatial To easily illustrate the law of conservation of mass, determine the mass of a flashbulb on the balance. Fire the flashbulb. Show that there has been no change in mass. **LEP**

Problem Solving

Think Critically

- corrosion of iron—reactants: iron and oxygen; product: hydrated iron(III) oxide
- The coating was not removed because it protects the copper from further corrosion.

3 Assess

Check for Understanding

? FLEX Your Brain

Use the Flex Your Brain activity to have students explore COEFFICIENTS.

📁 **Activity Worksheets,** p. 5

Reteach

On the chalkboard, write the equation for a reaction of vinegar and baking soda. Show the reaction and discuss the evidence that a chemical change is taking place. Point out that if the gas were collected, you could show that mass was conserved. **LEP**

$NaHCO_3(cr) + CH_3COOH(aq) \rightarrow$
$NaCH_3COO(aq) + H_2O(l) + CO_2(g)$

Extension

📁 For students who have mastered this section, use the **Reinforcement** and **Enrichment** masters.

Problem Solving

Metals and the Atmosphere

When metals are exposed to air, they often corrode. Rusting is one type of corrosion. Rust is hydrated iron(III) oxide.

Aluminum also reacts with oxygen in the air to form aluminum oxide. Unlike rust, which crumbles and exposes more iron to the air, aluminum oxide adheres to the aluminum surface on which it forms, protecting the aluminum underneath.

Copper is another metal that corrodes when exposed to air. When copper corrodes, a green coating called basic copper carbonate is formed. You may have seen this type of corrosion on the Statue of Liberty.

Think Critically:

Identify the reactants and products in the corrosion of iron. When the Statue of Liberty was restored, the green coating was not removed. Why?

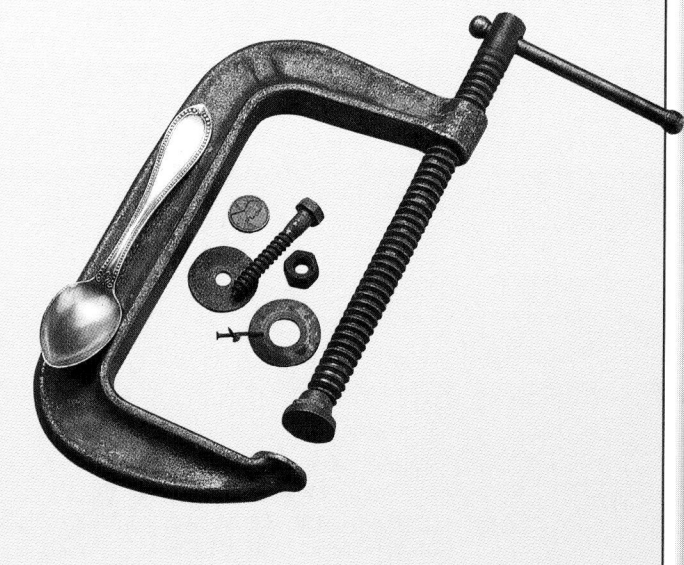

444 Chapter 16 Chemical Reactions

📁 Use **Study Guide,** p. 65, as you teach this lesson.

how to use chemical symbols and formulas. Some of the other symbols used in chemical equations are listed in **Table 16-1.**

What would the chemical equation for the reaction in **Figure 16-2** look like?

$Pb(NO_3)_2(aq) + 2KI(aq) \rightarrow$
$\qquad PbI_2(cr) + 2KNO_3(aq)$

The symbols to the right of the formulas are (cr), for *crystalline solid*, and (aq) for *aqueous*, which means "dissolved in water."

Coefficients—Unit Managers

What do the numbers to the left of the formulas for reactants and products mean? Remember that the law of conservation of

Table 16-1

Symbols Used in Chemical Equations	
Symbol	**Meaning**
$\longrightarrow$	produces or forms
+	plus
(cr)	crystalline solid
(l)	liquid
(g)	gas
(aq)	aqueous, a solid is dissolved in water
heat $\longrightarrow$	the reactants are heated
light $\longrightarrow$	the reactants are exposed to light
elec. $\longrightarrow$	an electric current is applied to the reactants

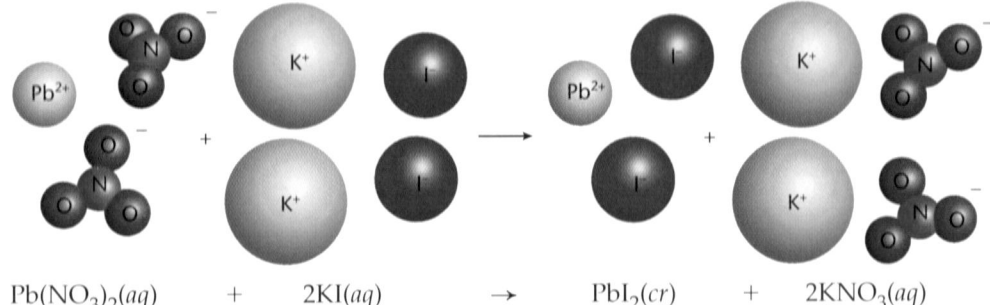

$$Pb(NO_3)_2(aq) \quad + \quad 2KI(aq) \quad \rightarrow \quad PbI_2(cr) \quad + \quad 2KNO_3(aq)$$

mass states that matter is neither created nor destroyed during chemical reactions. Atoms are rearranged but never lost or destroyed. These numbers, called **coefficients,** represent the number of units of each substance taking part in a reaction. For example, in the above reaction, one unit of $Pb(NO_3)_2$ reacts with two units of KI to produce one unit of PbI_2 and two units of KNO_3. **Figure 16-3** will help you visualize this reaction.

You don't always have to analyze the reaction mixture to find out what the coefficients in a chemical reaction are. In the next section, you will find out how to choose coefficients for a chemical equation.

Figure 16-3

Each coefficient in the equation represents the number of units of each type in this reaction.

Section Wrap-up

Review

1. Identify the reactants and the products in the following chemical equation.

   $$2B(cr) + 3I_2(g) \rightarrow 2BI_3(cr)$$

2. What is the name and state of matter of each substance in the following reaction?

   $$4Al(cr) + 3O_2(g) \rightarrow 2Al_2O_3(cr)$$

3. **Think Critically:** Identify the reactants and the products in the body cell reaction described on page 442.

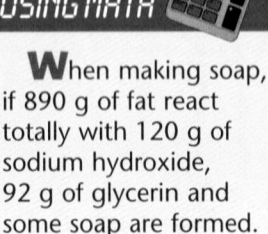

USING MATH

When making soap, if 890 g of fat react totally with 120 g of sodium hydroxide, 92 g of glycerin and some soap are formed. Calculate the mass of soap formed to satisfy the law of conservation of mass.

Skill Builder

Recognizing Cause and Effect
Lavoisier heated mercury(II) oxide in a sealed flask. Explain the effect on Lavoisier's conclusions about the law of conservation of mass if he had used an open flask. If you need help, refer to Recognizing Cause and Effect in the **Skill Handbook.**

GLENCOE TECHNOLOGY

 Videodisc
Glencoe Physical Science Interactive Videodisc
Side 2, Lesson 6
Introduction : Discovering the Cave

8593-12380

 CD-ROM
Physical Science CD-ROM
Have students perform the interactive exploration for Chapter 16 to reinforce important chapter concepts and thinking processes.

4 Close

•MINI•QUIZ•

1. A chemical reaction begins with substances called _____ . *reactants*

2. The new substances produced in a chemical reaction are called _____ . *products*

3. If you begin with 12.5 g of reactants that completely react, what mass of products can you expect to form? *12.5 g*

Section Wrap-up

Review

1. reactants—B and I_2; product—BI_3

2. Al—aluminum, solid; O_2—oxygen, gas; Al_2O_3—aluminum oxide, solid

3. **Think Critically**
   Reactants: O_2, $C_6H_{12}O_6$
   Products: CO_2, H_2O

USING MATH

890 g + 120 g = 92 g + x
1010 g = 92 g + x
918 g = x

Skill Builder

Any gas produced could escape, making the apparent mass of the products less than the mass of the reactants.

P

✔Assessment

Oral Have students explain the importance of a closed system when collecting and measuring reactants and products. Use the Performance Task Assessment List for Making Observations and Inferences in **PASC,** p. 17.

P

Prepare

Section Background

Some ultraviolet energy has a devastating effect on the DNA molecules that control the function of living cells of many microorganisms.

1 Motivate

Bellringer

 Before presenting the lesson, display **Section Focus Transparency 62** on the overhead projector. Assign the accompanying **Focus Activity** worksheet. L1 LEP

Demonstration

LS **Visual-Spatial** In a darkened room, use an ultraviolet lamp to show how some chemicals glow when exposed to ultraviolet rays. Boxes of detergent are often printed with fluorescent inks. Discuss how ultraviolet rays differ from light. LEP

2 TEACH

GLENCOE TECHNOLOGY

 **Videodisc**

The Infinite Voyage Series: Secrets from a Frozen World

Chapter 7

Solar Ultraviolet Light and the Diminishing Ozone Layer

Science Words

chlorofluorocarbon (CFC)

Objectives

- Explain the effect of CFCs on ozone in the atmosphere.
- Compare possible alternatives to CFCs.

The Ozone Layer: Protecting Earth

You may have noticed labels on many sunglasses claiming 100 percent UV protection. *UV* stands for "ultraviolet radiation." Exposure to this type of radiation can lead to many problems for living cells. In some cases, the cells may become cancerous. Other cell mutations associated with UV radiation can mean death to plants and animals.

How are people protected from this potentially harmful radiation? A gas called ozone, O_3, found in Earth's upper atmosphere, helps keep some harmful UV radiation from reaching Earth's surface, shown in **Figure 16-4.** Ozone molecules are formed from three oxygen atoms. Ozone isn't always helpful. Near the ground, ozone is a pollutant found in automobile exhaust. It can be a major irritant in areas with large amounts of air pollution.

So, what's the problem?

With ozone in the stratosphere giving us this protection, why is there any concern? Satellite photos, such as the one shown in **Figure 16-5,** allow scientists to measure the amount of ozone in a layer around Earth. Unmistakable decreases in the level of ozone, particularly around the poles, are now being detected. This is often referred to as the "hole in the ozone layer." Less ozone in the stratosphere means more UV radiation can reach Earth.

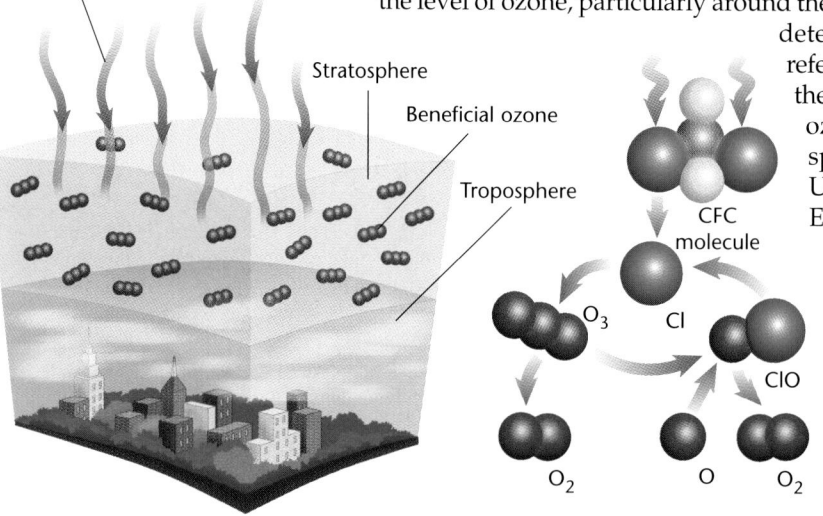

Figure 16-4

Ozone in the stratosphere absorbs ultraviolet rays, breaking O_3 into O_2 and O. O_2 and O react, forming more ozone.

Program Resources

 Reproducible Masters
Study Guide, p. 66 L1
Reinforcement, p. 66 L1
Enrichment, p. 66 L3
Concept Mapping, p. 37

Transparencies
Section Focus Transparency 62 L1

2 Points of View

A Global Plan

Most scientists now believe that a group of compounds called **chlorofluorocarbons (CFCs)** migrate through Earth's atmosphere and decompose. This decomposition releases chlorine atoms that destroy ozone. CFCs are used in air-conditioning systems and in making some types of polymer foams. If CFC use were banned and substitute compounds found, more protective ozone could be conserved.

Don't Overreact

The decrease of the ozone levels in the atmosphere does appear to be related to chlorine concentrations in the stratosphere. But mounting evidence suggests that human activity is not the only source of chlorine. Scientists have found approximately 2000 chlorine and related compounds that are released into the air from natural phenomena—plants, ocean life, fungi, forest fires, and volcanoes. Blaming CFCs for all the ozone damage will cause economic problems as we try to switch millions of cooling systems to some other coolant.

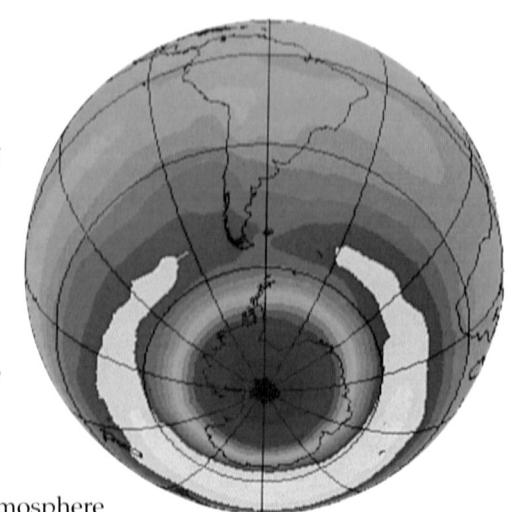

Figure 16-5

The colors in this satellite photo indicate how much ozone is present in the atmosphere. Yellow indicates greater amounts of ozone. Purple shows the least amount.

Section Wrap-up

Review

1. Why is the amount of ozone in Earth's atmosphere important to us?

2. The following equations show chlorine reacting with ozone and what happens to the ClO formed.

$$Cl(g) + O_3(g) \rightarrow ClO(g) + O_2(g)$$
$$ClO(g) + O(g) \rightarrow Cl(g) + O_2(g)$$

Taking these two reactions together, what can you conclude about the potential problem of controlling the amount of Cl in the atmosphere?

Explore the Issue

Would you be willing to pay more for air-conditioning services in order to regulate and recycle CFCs to help keep them out of the environment? Explain.

Science Journal

Many environmental groups are calling for CFC recycling programs to prevent CFCs from entering the atmosphere. In your Science Journal, write a paragraph summarizing your thoughts about whether recycling will solve the problem.

SCIENCE & SOCIETY

3 Assess

Check for Understanding

? FLEX Your Brain

Use the Flex Your Brain activity to have students explore OZONE. **P**

📁 **Activity Worksheets,** p. 5

Reteach

LS **Visual-Spatial** Have students draw molecules to represent the reaction in which ozone is destroyed.

Extension

📁 For students who have mastered this section, use the **Reinforcement** and **Enrichment** masters.

4 Close

Illustrate the connection among ozone, pollution, and high temperatures by bringing in copies of summer weather reports from big cities. They should include ozone levels.

Section Wrap-up

Review

1. It shields living things on Earth from harmful UV radiation from the sun.

2. Once ClO is formed, it can react again with O(g), creating Cl(g). This gas is available again to destroy more ozone, and the cycle continues.

Explore the Issue

Student opinions will differ between those who see the current problem as serious and those who see the need for more research.

Cultural Diversity

The sky is falling, or is it? Punta Arenas, Chile, is a town of 115 000 that happens to be located under an ozone hole for part of the year. In Punta Arenas, reproductive patterns have been changing in local plants and animals. Plants are turning yellow, burned by the high levels of ultraviolet radiation. The vision of some rabbits has become poor.

Prepare

Section Background

Coefficients are used to balance an equation; subscripts are never changed. To change a subscript is to change the identity of a substance.

1 Motivate

Bellringer

Before presenting the lesson, display **Section Focus Transparency 63** on the overhead projector. Assign the accompanying **Focus Activity** worksheet. L1 LEP

Tying to Previous Knowledge

Give students a simple addition problem followed by a simple multiplication problem. Inform the class that if they got both problems correct, they have the math skills needed to balance equations.

2 Teach

Activity

Kinesthetic To model the reaction on page 448, give each student four red, four white, and two black gumdrops. The white ones represent silver atoms; the red ones represent hydrogen atoms; the black ones represent sulfur atoms. Have students draw an arrow on a sheet of paper. Ask them to place their models on the paper to show the balanced equation that appears at the top of page 449. L1

Science Words

balanced chemical equation

Objectives

* Explain what is meant by a balanced chemical equation.
* Demonstrate how to write balanced chemical equations.

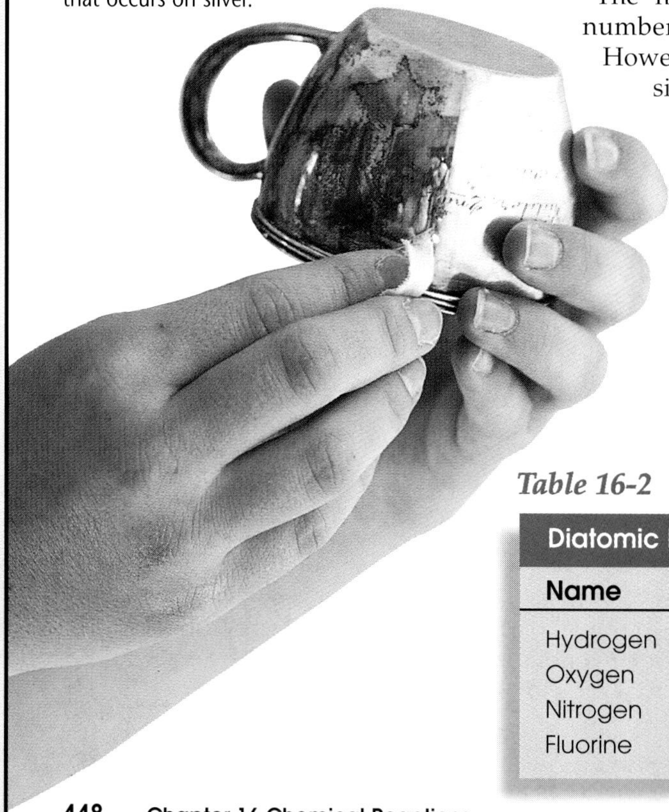

Figure 16-6

Tarnish is a type of corrosion that occurs on silver.

16•3 Chemical Equations

Checking for Balance

As you read in Chapter 11, silver tarnish, shown in **Figure 16-6**, is silver sulfide, Ag_2S. It forms when sulfur-containing compounds in the air or food react with silver. Suppose you write this chemical equation for tarnishing.

$$Ag(cr) + H_2S(g) \rightarrow Ag_2S(cr) + H_2(g)$$

Now, examine the equation. Remember that matter is never created or destroyed in a chemical reaction. Notice that one silver atom appears in the reactants, $Ag + H_2S$. However, two silver atoms appear in the products, $Ag_2S + H_2$. As you know, one silver atom can't just become two. The equation must be balanced so it shows a true picture of what takes place in the reaction. A **balanced chemical equation** has the same number of atoms of each element on both sides of the equation. To find out if this equation is balanced, make a chart like that shown in **Table 16-3**.

The number of hydrogen atoms and the number of sulfur atoms are balanced. However, two silver atoms are on the right side of the equation and only one is on the left side. This equation isn't balanced. To balance an equation, never change subscripts of a correct formula. Instead, place whole number coefficients to the left of the formulas of the reactants and products so that equal numbers of silver atoms are on both sides of the equation. If no number is written, the coefficient is 1.

Table 16-2

Diatomic Molecules

Name	Formula	Name	Formula
Hydrogen	H_2	Chlorine	Cl_2
Oxygen	O_2	Bromine	Br_2
Nitrogen	N_2	Iodine	I_2
Fluorine	F_2		

448 Chapter 16 Chemical Reactions

Program Resources

Reproducible Masters
Study Guide, p. 67 L1
Reinforcement, p. 67 L1
Enrichment, p. 67 L3
Cross-Curricular Integration, p. 22
Lab Manual 32

Transparencies
Section Focus Transparency 63 L1
Teaching Transparency 31

Choosing Coefficients

How do you find out which coefficients to use to balance an equation? This decision is usually a trial-and-error process. With practice, the process becomes simple to perform.

In the chemical equation for tarnishing, the sulfur atoms and the hydrogen atoms are already balanced. Putting a coefficient before formulas that contain these atoms is not necessary at this time. So, look at the formulas containing silver atoms: Ag and Ag_2S. Two atoms of silver are on the right side and only one is on the left side. If you put a coefficient of 2 before Ag, the equation is balanced, as shown in **Table 16-4**.

$$2Ag(cr) + H_2S(g) \rightarrow Ag_2S(cr) + H_2(g)$$

Writing Balanced Chemical Equations

When a piece of magnesium ribbon burns in a flask of oxygen, a white powder called magnesium oxide is formed. To write a balanced chemical equation for this and most other reactions, follow these four steps.

Step 1 Describe the reaction in words, putting the reactants on the left side and the products on the right side.

magnesium plus oxygen produces magnesium oxide

Step 2 Write a chemical equation for the reaction using formulas and symbols. Review how to write formulas for compounds in Section 11-4. The formulas for elements are generally just their symbols. However, as you read in Chapter 12, some elements ordinarily exist as diatomic molecules, as shown in **Table 16-2**. Oxygen is a diatomic molecule.

$$Mg(cr) + O_2(g) \rightarrow MgO(cr)$$

Step 3 Check the equation for atom balance. Set up a chart similar to **Table 16-5** to help you. The magnesium atoms are balanced, but the oxygen atoms are not. Therefore, this equation isn't balanced.

Step 4 Choose coefficients that balance the equation. Remember, never change subscripts of a correct formula to balance an equation. Try putting a coefficient of 2 before MgO to balance the oxygen.

$$Mg(cr) + O_2(g) \rightarrow 2MgO(cr)$$

Now, there are two Mg atoms on the right side and only one is on the left side. So, a coefficient of 2 is needed before Mg also.

$$2Mg(cr) + O_2(g) \rightarrow 2MgO(cr)$$

Table 16-6 indicates that the equation is now balanced.

Table 16-3

Atoms in Unbalanced Equation

Kind of Atom	Number of Atoms $Ag + H_2S \rightarrow Ag_2S + H_2$	
Ag	1	2
H	2	2
S	1	1

Table 16-4

Atoms in Balanced Equation

Kind of Atom	Number of Atoms $2Ag + H_2S \rightarrow Ag_2S + H_2$	
Ag	2	2
H	2	2
S	1	1

Table 16-5

Atoms in Unbalanced Equation

Kind of Atom	Number of Atoms $Mg + O_2 \rightarrow MgO$	
Mg	1	1
O	2	1

Table 16-6

Atoms in Balanced Equation

Kind of Atom	Number of Atoms $2Mg + O_2 \rightarrow 2MgO$	
Mg	2	2
O	2	2

Concept Development

Here are some helpful hints to give students to help them learn to balance equations.

1. Treat a polyatomic ion as a unit if it is not changed during a reaction.

2. If an element appears in more than one compound on the same side of the equation, leave that element until last to balance.

3. If there is an even number of atoms of an element on one side of the equation and an odd number on the other, place a coefficient of 2 in front of the compound containing an odd number of atoms.

4. Representing balanced chemical equations can best be done using models. If you do not have commercial kits, substitute gumdrops and toothpicks.

3 Assess

Check for Understanding

? FLEX Your Brain

Use the Flex Your Brain activity to have students explore BALANCED EQUATIONS.

📁 **Activity Worksheets,** p. 5

Reteach

LS **Kinesthetic** Have students use gumdrops and toothpicks or magnet-backed posterboard "atoms" to build models of the molecules in equations to be balanced. **LEP**

Extension

📁 For students who have mastered this section, use the **Reinforcement** and **Enrichment** masters.

Inclusion Strategies

Gifted Challenge students to balance the following equations. The equations are given here in balanced form. Rewrite them, leaving out the coefficients. **L3** **LS**

1. $2KNO_3 \rightarrow 2KNO_2 + O_2$
2. $2C_2H_6 + 7O_2 \rightarrow 4CO_2 + 6H_2O$
3. $2Na + 2H_2O \rightarrow 2NaOH + H_2$
4. $3O_2 + CS_2 \rightarrow CO_2 + 2SO_2$
5. $2KClO_3 \rightarrow 2KCl + 3O_2$

4 Close

•MINI•QUIZ•

Use the Mini Quiz to check students' recall of chapter content.

1. A _____ chemical equation has the same number and kinds of atoms on each side. *balanced*

2. The number placed in front of a chemical formula to balance an equation is called a(n) _____. *coefficient*

3. Balance the equation
$NaBr + Cl_2 \rightarrow NaCl + Br_2$.
$2NaBr + Cl_2 \rightarrow 2NaCl + Br_2$

4. Balance the equation
$Al(OH)_3 + CO_2 \rightarrow Al(HCO_3)_3$.
$Al(OH)_3 + 3CO_2 \rightarrow Al(HCO_3)_3$

Section Wrap-up

Review

1. to accurately show what takes place in a reaction and to show that mass is conserved

2. (a) $2Cu(cr) + S(cr) \rightarrow Cu_2S(cr)$
   (b) $2Na(cr) + 2H_2O(l) \rightarrow 2NaOH(aq) + H_2(g)$

3. **Think Critically** The subscripts are counted in the balancing procedure as well.

Using Computers

$CaCO_3 + 2HCl \rightarrow CO_2 + CaCl_2 + H_2O$

Table 16-7

Atoms in AgCl Formation

Kind of Atoms	Number of Atoms $AgNO_3 + NaCl \rightarrow AgCl + NaNO_3$			
Ag	1		1	
N	1			1
O	3			3
Na		1		1
Cl		1	1	

Work through the following example.

When a silver nitrate solution is mixed with a sodium chloride solution, insoluble silver chloride and sodium nitrate solution form. Write a balanced equation for this reaction. Refer to **Table 16-7** as needed.

Step 1 Describe the reaction in words.
aqueous silver nitrate plus aqueous sodium chloride produces solid silver chloride plus aqueous sodium nitrate

Step 2 Write the chemical equation.
$AgNO_3(aq) + NaCl(aq) \rightarrow AgCl(cr) + NaNO_3(aq)$

Step 3 Check the equation for balance.
The equation already has equal numbers of each element on both sides. This equation is balanced.

Step 4 Choose coefficients.
This equation is balanced, so no coefficients other than 1 are needed.

Section Wrap-up

Review

1. What are two reasons for balancing equations for chemical reactions?

2. Write balanced chemical equations for the following reactions: (a) copper metal plus sulfur produces solid copper(I) sulfide, (b) sodium metal plus water produces aqueous sodium hydroxide plus hydrogen gas.

3. **Think Critically:** Explain why the sum of the coefficients on the reactant side of a balanced equation does not have to equal the sum of the coefficients on the product side of the equation.

Skill Builder
Observing and Inferring
Hard water contains ions of calcium, magnesium, and/or iron. When Na_2CO_3 is added to hard water, water is softened because these ions are removed as a white solid. For Mg, this reaction may occur.

$MgCl_2(aq) + Na_2CO_3(aq) \rightarrow MgCO_3(cr) + 2NaCl(aq)$

Write a balanced equation showing how water containing $Ca(NO_3)_2$ could be softened. If you need help, refer to Observing and Inferring in the **Skill Handbook.**

Using Computers
Spreadsheet Use a spreadsheet to produce your own balance table for the reaction of $CaCO_3$ with HCl to form CO_2, $CaCl_2$, and water.

---

450 Chapter 16 Chemical Reactions

Skill Builder
$Ca(NO_3)_2(aq) + Na_2CO_3(aq) \rightarrow 2NaNO_3(aq) + CaCO_3(cr)$
The $CaCO_3$ precipitates as a white solid.

Assessment

Performance Have students use their balanced chemical equation and the periodic table on pages 732-733 to make a poster that illustrates the law of conservation of mass. Use the Performance Task Assessment List for Poster in **PASC,** p. 73. **P**

Types of Chemical Reactions

Classifying Chemical Reactions

There are hundreds of kinds of chemical reactions. Rather than try to memorize them, it is easier to group reactions by their similarities. You can learn a great deal about a reaction by comparing it to others in its group.

One system of classification is based upon the way the atoms rearrange themselves in a chemical reaction. Most reactions are either synthesis, decomposition, single-displacement, or double-displacement reactions.

Synthesis Reactions

One of the easiest reaction types to recognize is a synthesis reaction. In a **synthesis reaction,** two or more substances combine to form another substance. The generalized formula for this reaction type is as follows.

$$A + B \rightarrow AB$$

The reaction in which hydrogen burns in oxygen to form water is an example of a synthesis reaction.

$$2H_2(g) + O_2(g) \rightarrow 2H_2O(g)$$

This reaction between hydrogen and oxygen occurs between two elements. One type of rocket fuel is hydrogen, which burns explosively in oxygen when the rocket is fired, as shown in **Figure 16-7.**

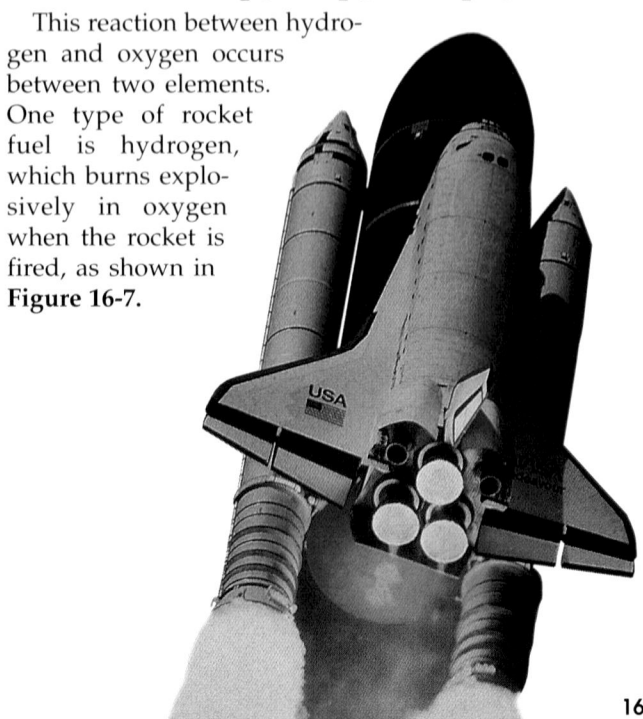

Science Words

- synthesis reaction
- decomposition reaction
- single-displacement reaction
- double-displacement reaction
- precipitate

Objectives

- Describe four types of chemical reactions using their generalized formulas.
- Classify various chemical reactions by type.

Figure 16-7

The synthesis reaction between hydrogen and oxygen fuels the main stage of the space shuttle.

CONNECT TO

LIFE SCIENCE

Substances in both plants and animals react to form proteins. *Research* the compounds that are the basic units in proteins.

16-4 Types of Chemical Reactions   **451**

Program Resources

 Reproducible Masters
Study Guide, p. 68 L1
Reinforcement, p. 68 L1
Enrichment, p. 68 L3
Activity Worksheets, pp. 90-91, 95 L1
Science Integration Activities, pp. 31-32
Critical Thinking/Problem Solving, p. 22
Lab Manual 33

Transparencies
Section Focus Transparency 64 L1
Teaching Transparency 32

Section 16•4

Prepare

Section Background

Chemists are able to predict the products of decomposition reactions because the reactions often follow known patterns. For example, some metal carbonates produce metal oxides and carbon dioxide gas. Some metal chlorates yield metal chlorides and oxygen gas. Some metal hydroxides decompose into metal oxides and water.

Preplanning

- To prepare for Activity 16-1, safely dilute hydrochloric acid by adding concentrated acid to water and prepare the small pieces of metal.
- Gather markers and cards for the MiniLAB.

1 Motivate

Bellringer

Before presenting the lesson, display **Section Focus Transparency 64** on the overhead projector. Assign the accompanying **Focus Activity** worksheet. L1 LEP

Tying to Previous Knowledge

Ask students how they would describe the chemical reactions that occur when a car's engine burns gasoline.

CONNECT TO

LIFE SCIENCE

Answer Student research results should reflect information about amino acids.

2 Teach

MiniLAB

Purpose

LS **Logical-Mathematical**
Demonstrate why chemical equations must be balanced. L1

Materials
cards, markers

Teaching Strategies
Other analogies can be used, such as ingredients to make one cake.

Analysis

1. 2G + 2F + C = T
2. Example: (1) There are reactants (individual players) and a product (the whole team).
   (2) There are the same number of atoms (players) on each side of the equation.
3. The addition of the extra players would cause the equation to become unbalanced. There would be too many on the reactant side. If you change the subscripts, you change the formula of the compounds. Here, you would change the "formula" for the team make-up.

✓Assessment

Portfolio Ask students to explain why books, grouped into types, making up a library is not a good analogy for balancing equations. *Numbers of types and books within types will vary.* Use the Performance Task Assessment List for Making Observations and Inferences in **PASC**, p. 17. P

📁 **Activity Worksheets,** pp. 5, 95

MiniLAB

Must the sums of reactant and product coefficients be equal?

A basketball team is made up of individuals acting as a unit. A chemical compound also consists of individual atoms in one unit.

Procedure

1. Obtain a card marked with one of five basketball team positions.
2. Group together with others in the class to form one basketball team. Assume that a team needs two guards, two forwards, and one center.

Analysis

1. Write the formation of the basketball team in the form of an equation. Use coefficients in front of each type of player needed to make one team.
2. What are two ways this equation is like a chemical equation?
3. Suppose, after the formation of as many complete teams as possible, there were one guard and two centers remaining. Why couldn't these players be added to existing teams? How does this compare to restrictions on the use of subscripts when balancing equations?

Figure 16-8
Copper in a wire replaces silver in silver nitrate, forming a blue solution of a copper nitrate.

Across the Curriculum

Language Arts Have students look up the word *decompose* in a dictionary. Have them write a paragraph relating decomposition reactions to a possible solution to the problem of solid waste disposal caused by the closing of landfills. LS

Decomposition Reactions

The opposite of a synthesis reaction is a decomposition reaction. The equation below illustrates this relationship.

In a **decomposition reaction,** one substance breaks down, or decomposes, into two or more simpler substances. The generalized formula for this type of reaction is as follows.

$$AB \rightarrow A + B$$

Most decomposition reactions require the use of heat, light, or electricity. The decomposition of water by an electric current produces hydrogen and oxygen.

$$2H_2O(l) \xrightarrow{\text{elec.}} 2H_2(g) + O_2(g)$$

Displacement Reactions

A **single-displacement reaction** occurs when one element replaces another in a compound. There are two generalized equations for this type of reaction.

In the first case, A replaces B as follows.

$$A + BC \rightarrow AC + B$$

In the second case, D replaces C as follows.

$$D + BC \rightarrow BD + C$$

In **Figure 16-8,** a copper wire is put into a solution of silver nitrate. Because copper is a more active metal than silver, it

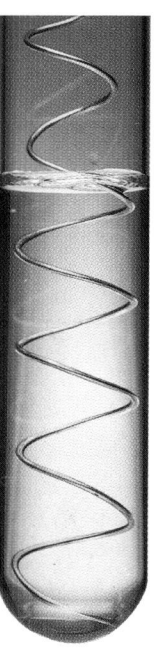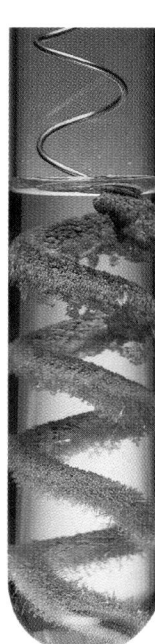

replaces the silver, forming a blue copper(II) nitrate solution. The silver forms as an insoluble solid.

Table 16-8 lists metals according to how active they are. A metal will replace any metal beneath it in the table. Notice that copper, silver, and gold are near the bottom of the list. Their inactivity is why these elements often occur in Earth as veins of relatively pure element. Most other metals are more active and occur as compounds.

A **double-displacement reaction** takes place if a precipitate, water, or a gas forms when two ionic compounds in solution are combined. A **precipitate** is an insoluble compound formed during this type of reaction. In a double-displacement reaction, the positive ion of one compound replaces the positive ion of the other compound to form two new compounds. The generalized formula for this type of reaction is as follows.

$$AB + CD \rightarrow AD + CB$$

The reaction of silver nitrate with sodium chloride is an example of this type of reaction. A precipitate—silver chloride—is formed. The chemical equation is as follows.

$$AgNO_3(aq) + NaCl(aq) \rightarrow AgCl(cr) + NaNO_3(aq)$$

You have seen just a few examples of chemical reactions classified into types. Many more reactions of each type occur around you.

Table 16-8

Activity Series of Metals	
Lithium	**Most active**
Potassium	
Barium	
Calcium	
Sodium	
Magnesium	
Aluminum	
Zinc	
Iron	
Nickel	
Tin	
Lead	
(Hydrogen)	
Copper	
Mercury	
Silver	
Gold	**Least active**

Section Wrap-up

Review

1. Classify the following reactions by type.
   (a) $2KClO_3(cr) \rightarrow 2KCl(cr) + 3O_2(g)$
   (b) $CaBr_2(aq) + Na_2CO_3(aq) \rightarrow CaCO_3(cr) + 2NaBr(aq)$
   (c) $Zn(cr) + S(cr) \rightarrow ZnS(cr)$
   (d) $Ba(cr) + FeBr_2(aq) \rightarrow BaBr_2(aq) + Fe(cr)$

2. The copper bottoms of some cooking pans turn black after being used. The copper reacts with oxygen, forming black copper(II) oxide. Write a balanced chemical equation for this reaction.

3. **Think Critically:** One method used to obtain gold from its ore involves forming a gold compound, then recovering the gold by using zinc. Balance the following equation and identify the reaction type.
   $$Au(CN)_2^-(aq) + Zn(cr) \rightarrow Au(cr) + Zn(CN)_4^{2-}$$

USING MATH

Most balanced equations use the smallest coefficients possible. Rewrite the following equation, calculating the smallest coefficients that still indicate a balanced equation.

$$9Fe(cr) + 12H_2O(g) \rightarrow$$
$$3Fe_3O_4(cr) + 12H_2(g)$$

Skill Builder
Hypothesizing

Group 1 metals replace hydrogen in water in reactions that are often violent. A sample equation for one reaction is as follows.

$$2K(cr) + 2HOH(l) \rightarrow 2KOH(aq) + H_2(g).$$

Use **Table 16-8** to help you hypothesize why these metals are often stored in kerosene. If you need help, refer to Hypothesizing in the **Skill Handbook**.

16-4 Types of Chemical Reactions   **453**

Skill Builder

Evidence that these metals are very active is their position near the top of the activity series. Because they are very active, Group 1 metals will react with water vapor and oxygen in the air. Therefore, they must not come in contact with air.

✓ Assessment

Oral Have students explain why it might be useful to write water as HOH in a chemical equation. Use the Performance Task Assessment List for Making Observations and Inferences in **PASC**, p. 17. **P**

3 Assess

Check for Understanding

?FLEX Your Brain

Use the Flex Your Brain activity to have students explore SYNTHESIS.

📁 **Activity Worksheets,** p. 5

Reteach

LS Logical-Mathematical Provide additional equations to balance and classify.

Extension

📁 For students who have mastered this section, use the **Reinforcement** and **Enrichment** masters.

4 Close

•MINI•QUIZ•

1. **What type of reaction occurs when two substances combine to form one?** *synthesis*

2. **What type of reaction occurs when one substance breaks down into two?** *decomposition*

Section Wrap-up

Review

1. (a) decomposition; (b) double displacement; (c) synthesis; (d) single displacement
2. $2Cu(cr) + O_2(g) \rightarrow 2CuO(cr)$
3. **Think Critically**
   $2Au(CN)_2^-(aq) + Zn(cr) \rightarrow 2Au(cr) + Zn(CN)_4^{2-}(aq)$; single displacement

USING MATH

All coefficients are divisible by 3. $3Fe + 4H_2O \rightarrow Fe_3O_4 + 4H_2$

Purpose

IS **Interpersonal** Students will design and carry out an experiment to show how to analyze and compare reaction results. **L1** **COOP LEARN**

Process Skills

observing, measuring, comparing, communicating, making and using tables, designing an experiment, interpreting data, inferring

Time

45 minutes; to shorten time, the hydrochloric acid could be premeasured and poured into test tubes for students.

Materials

Dilute (0.1M) HCl may be prepared by adding 9 mL of concentrated HCl to 1 L of distilled water. **CAUTION:** *Avoid fumes of and direct contact with HCl.* Assemble small, identified samples of magnesium, zinc, copper, lead, aluminum, tin, and silver for testing.

Safety Precautions

Caution students to be careful around HCl. Avoid skin contact and fumes. Immediately wash any affected area with water. Wear goggles and aprons at all times. Do not allow students to test for the presence of hydrogen.

Possible Hypotheses

Possible hypotheses may include that heavier metals are more reactive, lighter metals are more reactive, hydrogen gas may form from the reaction of HCl with the metals, or a precipitate may form.

📁 **Activity Worksheets,** pp. 5, 90-91

Activity 16-1

Design Your Own Experiment
Metal Activity

How can you compare the chemical activity of two substances? One test that can be applied is to allow each substance to react with a third substance. Then the results can be compared and ranked.

PREPARATION

Problem

How can the chemical activity of metals be compared?

Form a Hypothesis

Using a dilute acid for comparison, state a hypothesis concerning a way to rank the metals.

Objectives

- Observe the chemical reactions between metals and a dilute acid.
- Design an experiment that compares the degree of chemical reactions among selected metals.

Possible Materials

- test tubes, all the same size
- test-tube rack
- dilute hydrochloric acid, HCl
- 10-mL graduated cylinder
- tongs
- wooden splint
- small pieces of metals

Safety Precautions

Handle hydrochloric acid with care. Report any spills immediately. Wear safety goggles at all times during the experiment. Use only pieces of metal that are less than the size of a pea.

PLAN THE EXPERIMENT

1. As a group, agree upon and write out the hypothesis statement.
2. As a group, list your procedure steps in a logical sequence.
3. Make a list of the materials that you will need to complete your experiment.
4. In your Science Journal, design a data table to make comparisons of your tested metals.

Check the Plan

1. After testing the small pieces of metals, how will you compare your test results?
2. What will be held constant? What will you allow to vary?
3. Should you run any tests more than once?
4. *Make sure your teacher approves of your plan and that you have included any changes suggested in the plan.*

DO THE EXPERIMENT

1. Carry out the experiment as planned.
2. Testing the gas production of a metal and acid reaction will require igniting the gas. Your teacher will carry out this part of the experiment.
3. In the data table in your Science Journal, write down your observations of the chemical reaction between the metals and the acid.

Analyze and Apply

1. **Compare** the results of each metal's reaction with the acid.
2. **Compare** your results with those of other class members.
3. **Sequence** all the metals tested in the order of most reactive first, down to least reactive.

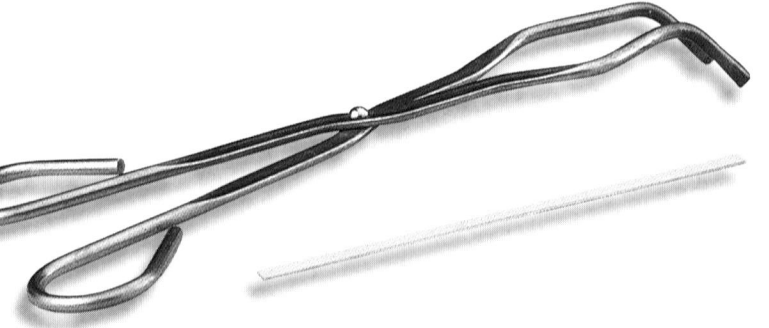

Go Further

Write balanced chemical equations for all of the chemical reactions that took place during the experiment.

16-4 Types of Chemical Reactions   **455**

Possible Procedures

One possible procedure is to use the same amount of acid in each test tube, record the beginning weight of each sample, drop samples in tubes at approximately the same time, observe and record any chemical reactions (bubbling, precipitate, color change), continue until reaction stops (noting which ends first, etc.), have teacher test for presence of H_2 gas, weigh any remaining sample, and rank reactivity of metals.

Teaching Strategies

- Possible metals to use are Mg, Zn, Cu, Pb, Al, Sn, and Ag.
- Collect gas by inverting a larger test tube over the sample tube. The H_2 gas will rise up and collect in the larger tube.

Troubleshooting If the zinc is old, there may be no immediate reaction visible. Warming the tube between hands or in warm water may help.

Troubleshooting If an oxide forms, students may be comparing the reactivity of the oxide, not the metal, with the acid. The metal could be cleaned with sandpaper immediately before adding the sample to the acid.

DO THE EXPERIMENT

Expected Outcome

H_2 gas will form from the reaction of HCl and Mg, Zn, Sn, Al, and Pb (barely).

Analyze and Apply

1. Have students compare their lists with Table 16-8.
2. Ranking will differ due to accuracy of observation and testing for gas.
3. For the elements listed: Mg, Al, Zn, Sn, Pb, Cu, Ag.

✔ Assessment

Performance Use the following two equations to complete and balance the third equation. The symbols A and B represent metals.

$$A(cr) + 2HCl(aq) \rightarrow ACl_2(aq) + H_2(g)$$
$$B(cr) + HCl(aq) \rightarrow \text{no reaction}$$
$$A(cr) + BCl(aq) \rightarrow ?$$

The first two equations show that A is more active than B, so A would replace B in a compound. Also, in a compound, A would have an oxidation number of 2+, and B would have an oxidation number of 1+.

$$A(cr) + 2BCl(aq) \rightarrow ACl_2(aq) + 2B(cr)$$

Use the Performance Task Assessment List for Analyzing the Data in **PASC**, p. 27. **P**

Go Further

Chemical equations should reflect that any metal more active than hydrogen would displace hydrogen from the acid. Metals less active than hydrogen would not react with the acid.

455

Section Background

Exothermic reactions generally take place spontaneously. Endothermic reactions are generally not spontaneous.

Preplanning

For Activity 16-2, obtain sand and manganese dioxide.

1 Motivate

Bellringer

 Before presenting the lesson, display **Section Focus Transparency 65** on the overhead projector. Assign the accompanying **Focus Activity** worksheet. L1 LEP

Discussion

Ask your students to describe how they feel when they have a fever. Ask them why the body gets hot. Guide the students' thinking toward the making and breaking of chemical bonds with the associated release of energy.

2 Teach

For more information on catalytic converters, read "They Forgot More Chemistry Than I Ever Knew" by Dennis Simanaitis, *Road and Track,* May 1985, page 154.

Think Critically

Rhodium and platinum are catalysts and are therefore not consumed.

Science Words

exothermic reaction
catalyst
inhibitor
endothermic reaction

Objectives

- Differentiate between an exothermic reaction and an endothermic reaction.
- Describe the effects of catalysts and inhibitors on the speed of chemical reactions.

INTEGRATION
Life Science

16•5 Energy and Chemical Reactions

Energy Changes in Chemical Reactions

Did you ever watch the explosion of dynamite? An explosion results from a rapid chemical reaction. In all chemical reactions, energy is either released or absorbed. This energy can take many forms. It might be in the form of heat, light, sound, or even electricity.

When most chemical reactions take place, some chemical bonds in the reactants must be broken. To break chemical bonds, energy must be provided. In order for products to be produced, new bonds must be formed. Bond formation releases energy.

Exothermic Reactions

In many chemical reactions, less energy is required to break original bonds than is released when new bonds form. In these reactions, called **exothermic reactions,** some form of energy is given off by the reaction. In many instances, the heat given off may cause the reaction mixture to feel hot, as shown in **Figure 16-9.**

The burning of wood and the explosion of dynamite are exothermic reactions. They are exothermic reactions because they give off energy as the reaction proceeds. Rusting is also exothermic, but the reaction proceeds so slowly that it is difficult to detect any temperature change.

Catalysts and Inhibitors

Some reactions are too slow to be useful. To speed up these reactions, you can add a catalyst. A **catalyst** is a substance that

456

Figure 16-9

The burning of wood is an example of an exothermic reaction.

Program Resources

 Reproducible Masters
Study Guide, p. 69 L1
Reinforcement, p. 69 L1
Enrichment, p. 69 L3
Science and Society Integration, p. 20
Activity Worksheets, pp. 92-93 L1
Multicultural Connections, pp. 35-36
Lab Manual 34

Transparencies
Section Focus Transparency 65 L1
Science Integration Transparency 16

speeds up a chemical reaction without itself being permanently changed. When you add a catalyst to a reaction, you end up with the same amount of the catalyst that you started with.

Certain proteins known as enzymes act as catalysts in living organisms. Enzymes allow reactions to occur at faster rates and lower temperatures than would otherwise be possible. Without these enzymes, life as we know it would not be possible.

At times, it is worthwhile to prevent certain reactions from occurring. Substances called **inhibitors** are used to combine with one of the reactants. The reactant is tied up and cannot undergo the original reaction. The food preservatives BHT and BHA are inhibitors. They prevent spoilage of certain foods.

Using catalysts and inhibitors are just a few of the ways scientists control chemical reactions.

USING TECHNOLOGY

Catalytic Converters

The exhaust systems of cars today are equipped with catalytic converters that use rhodium and platinum as the catalysts. These catalysts speed up reactions of three types of toxic compounds given off by cars. The rhodium and platinum don't enter into a chemical reaction, but the rough surface of the catalyst provides a high-surface-area place for reactions to occur. As pollutants adhere to the surface of the catalyst, chemical bonds in the compounds are weakened, so another reactant can more easily break the bond. Catalytic converters change unburned hydrocarbons, carbon monoxide, and nitrogen oxides into harmless compounds.

Two types of reactions are catalyzed. In one type of reaction, oxygen is added. Adding oxygen converts carbon monoxide to carbon dioxide, and unburned hydrocarbons to carbon dioxide and water. In the other type of reaction, oxygen is removed. Removing oxygen converts nitrogen oxides to nitrogen.

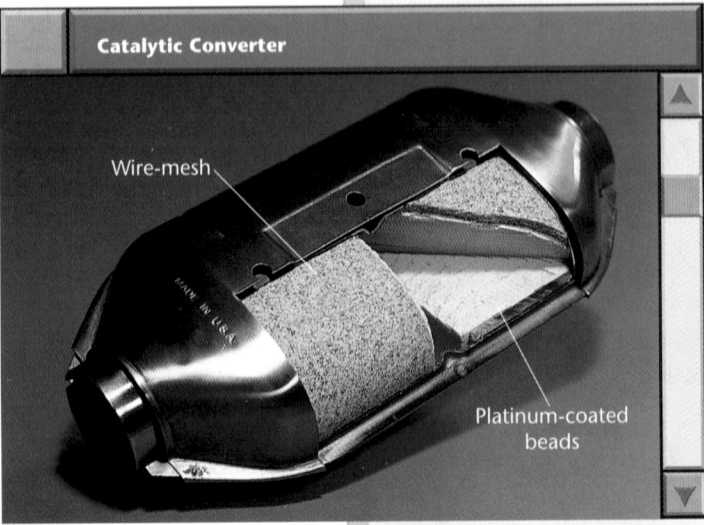
Catalytic Converter

Wire-mesh

Platinum-coated beads

Think Critically:

Why is it not necessary to add more rhodium and platinum to a catalytic converter as these substances are used?

16-5 Energy and Chemical Reactions   **457**

NATIONAL GEOGRAPHIC SOCIETY

 Videodisc

GTV: Planetary Manager
Energy

47916

47917

 **INTEGRATION Life Science**

Enzymes are especially important in digestion. At different points in the digestive process, different enzymes are added to assist in the breakdown of proteins, carbohydrates, and lipids into simpler molecules.

3 Assess

Check for Understanding

? FLEX Your Brain

Use the Flex Your Brain activity to have students explore EXOTHERMIC REACTIONS.

 Activity Worksheets, p. 5

Reteach

Discuss how homes are heated, cars are fueled, and cooking is done. Have students identify where exothermic chemical reactions play a role.

Extension

For students who have mastered this section, use the **Reinforcement** and **Enrichment** masters.

Community Connection

Cleaner Cars Have students research the effect catalytic converters have had on auto emissions. Compare exhaust composition before and after converters were installed. Results could be presented on a poster for display in the classroom. ⌊2⌋ ⌊LS⌋

4 Close

•MINI•QUIZ•

1. If energy is given off, a reaction is said to be _____ . *exothermic*

2. How does a catalyst affect the rate of a chemical reaction? *increases it*

3. A catalyst in a living organism is called a(n) _____ . *enzyme*

4. How does an inhibitor affect the rate of a chemical reaction? *decreases it*

Section Wrap-up

Review

1. An exothermic reaction releases more energy than it absorbs. An endothermic reaction absorbs more energy than it releases.

2. The MnO_2 would not be used up in the reaction.

3. **Think Critically** You would choose an exothermic reaction so that the heat given off could be used to warm cold hands.

Science Journal

Money invested is like the energy needed to start a chemical reaction. Money earned is like energy given off in an exothermic reaction. Money lost is like energy absorbed in an endothermic reaction.

Figure 16-10

The refining of ores to produce useful metals frequently involves an endothermic reaction.

Endothermic Reactions

Sometimes, more energy is required to break bonds than to form new ones in a chemical reaction. In these reactions, called **endothermic reactions,** energy must be provided for the reaction to take place.

When some endothermic reactions proceed, so much heat is absorbed that their containers feel cold to the touch. In other endothermic reactions, so little heat is absorbed that you would need a thermometer to determine that a temperature change has occurred.

An endothermic reaction is frequently used to obtain a metal from its ore. For example, aluminum metal is obtained by passing an electric current through molten aluminum ore, as shown in **Figure 16-10.**

$$2Al_2O_3 \xrightarrow{\text{elec.}} 4Al + 3O_2$$

In this case, electrical energy, not thermal energy, provides the energy needed to keep the reaction going.

Section Wrap-up

Review

1. What is the difference between exothermic and endothermic reactions?

2. Oxygen is produced when potassium chlorate is heated. What is true about manganese dioxide if it is a catalyst for this reaction?

3. **Think Critically:** Suppose you wanted to develop a product for warming people's hands at outdoor winter events. Would you choose an exothermic reaction or an endothermic reaction to use in your device? Why?

Skill Builder
Concept Mapping

Construct a concept map to show the relationship between energy and bond formation and bond breakage. If you need help, refer to Concept Mapping in the **Skill Handbook.**

Science Journal

In your Science Journal, discuss how the profit or loss in an investment or business deal is like the loss or gain of energy in a chemical reaction.

Skill Builder

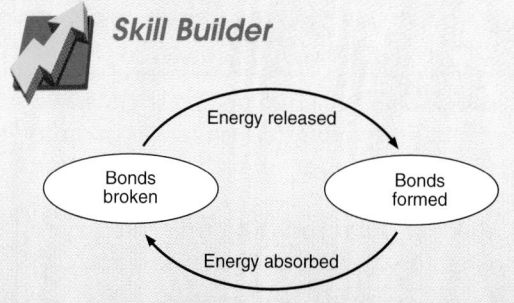

Assessment

Oral Have students explain what is happening in terms of bonding if a reaction in a solution in a beaker makes the beaker feel cold. Use the Performance Task Assessment List for Making Observations and Inferences in **PASC,** p. 17. **P**

Catalyzed Reaction

A balanced chemical equation provides much information about a chemical reaction, but it doesn't tell you about the rate of the reaction. Temperature, concentration, and physical state of reactants and products may all affect the rate of reaction.

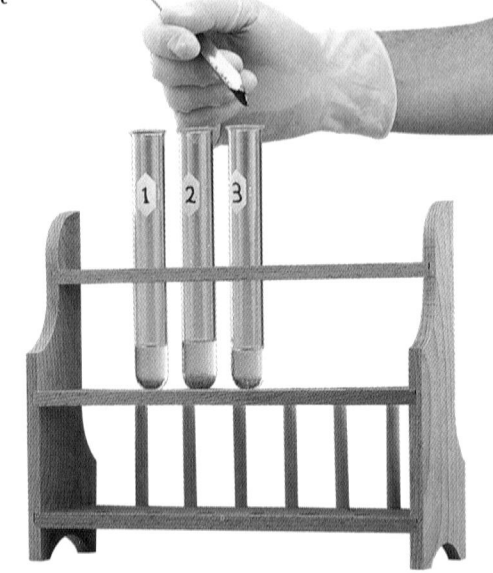

Problem
How does the presence of a catalyst affect the rate of a chemical reaction?

Materials
- 3 medium-sized test tubes
- test-tube stand
- 15 mL 3% hydrogen peroxide, H_2O_2
- graduated cylinder
- small plastic spoon
- sand
- manganese dioxide, MnO_2
- wooden splint
- hot plate
- beaker of hot water

  CAUTION: *Hydrogen peroxide can irritate skin and eyes. Wear goggles and an apron.*

Procedure
1. Set three test tubes in a test-tube stand. Pour 5 mL of hydrogen peroxide into each tube.
2. Place about 1/4 spoonful of sand in tube 2 and the same amount of manganese dioxide in tube 3.
3. In the presence of some substances that act as catalysts, H_2O_2 decomposes rapidly, producing oxygen gas, O_2. A glowing splint placed in O_2 will relight. Test the gas produced in any of the tubes as follows. Light a wooden splint, blow out the flame, and insert the glowing splint into the tube.
4. Place all three tubes in a beaker of hot water. Heat on a hot plate until all the remaining H_2O_2 is driven away and no liquid is left in the tubes.

Analyze
1. What changes did you observe when the solids were added to the tubes?
2. In which tube was oxygen produced rapidly, and how do you know?
3. Which substance, sand or manganese dioxide, caused the rapid production of gas from the hydrogen peroxide?
4. How did you identify the gas produced?
5. What remained in each tube after the hydrogen peroxide was driven away?

Conclude and Apply
6. What are two characteristics common to catalysts? Which substance in this activity has both characteristics?
7. The word *catalyst* has uses beyond chemistry. If a person who is added to a basketball team acts as a catalyst, **infer** what effect that person is likely to have on the team.

---

5. tube with H_2O_2 alone—nothing; tube with H_2O2 and sand—sand; tube with H_2O_2 and MnO_2—MnO_2
6. A catalyst speeds up a chemical reaction. A catalyst is not permanently changed by a reaction; MnO_2.
7. The person energizes the team and causes the team to be more active than it would normally be.

✓ Assessment

Portfolio Have students define the term *control*. Then, using examples from the activity, have them write out the role of the control in helping make conclusions in an experiment. Use the Performance Task Assessment List for Writing in Science in **PASC**, p. 87. **P**

Purpose
IS **Visual-Spatial** Students will operationally define a catalyst and observe its action.
L1 **COOP LEARN**

Process Skills
observing, inferring, forming operational definitions, classifying, recognizing cause and effect

Time
30 minutes

📁 **Activity Worksheets,** pp. 5, 92-93

Teaching Strategies
Troubleshooting Students should have the glowing splint ready when the MnO_2 is placed in the peroxide solution. Demonstrate the technique for students. Hydrogen peroxide must be freshly opened.

- Blood or small pieces of raw liver contain an enzyme that catalyzes the breakdown of H_2O_2. Save some blood from a beef roast and demonstrate its catalytic activity to students by substituting it for MnO_2.
- Repeat the activity as a demonstration by using larger test tubes and larger quantities.
- Ordinary 3 percent H_2O_2 available from drugstores should work well if it is fresh.

Answers to Questions
1. tube with H_2O_2 and sand—no change; tube with H_2O_2 and MnO_2—rapid bubbling
2. Oxygen was produced rapidly in the tube containing H_2O_2 and MnO_2. Bubbles appeared, and the glowing splint test for oxygen was positive.
3. MnO_2
4. The glowing splint burst into flame.

Science & ART

Source

El pan nuestro by Diego Rivera, Mexico City

Background

Frescoes were painted in ancient Greece and Italy. Also, ancient Buddhist frescoes have been discovered in China, and late Byzantine frescoes exist in Moscow. However, fresco painting reached its height during late medieval times and the Italian Renaissance (1300s-1500s). In the 1920s and 1930s, Diego Rivera helped inspire a brief revival of fresco painting in Mexico and the United States.

Teaching Strategies

LS Linguistic Ask volunteers to share the descriptions of the frescoes that they wrote about in their Science Journals. Have them determine which sections of their painting that they think they could get done each day. More artistic students could be encouraged to draw each section.

• Ask students to think of and list other techniques for which artists may depend upon natural chemical reactions. Have them discuss how scientific researchers might contribute to developing better equipment and materials for artists.

interNET CONNECTION

Visit the Chapter 16 Internet Connection at Glencoe Online Science, **www.glencoe.com/sec/ science/physical,** for a link to more information about frescoes.

Fresco Painting

A fresco is a kind of painting made on a plaster wall. The word *fresco* is Italian for "fresh." Why would a painting be called fresh? Possibly because only as much plaster as can be painted in a day may be applied each day. Fresco painting makes use of chemical reactions that cause color pigments to crystallize within the plaster as it dries.

The Process

To paint a fresco, a full-scale drawing showing exactly what will be painted that day is prepared. A section of the wall is covered with plaster made of sand and lime, which is calcium hydroxide. The drawing is traced onto the plaster wall, which is then painted. The paint itself is made from natural pigments diluted with water. As the plaster dries, carbon dioxide in the air turns the lime to calcium carbonate, crystallizing the pigments. The color then becomes part of the plaster itself. Any plaster that has not been painted by the end of the day must be cut away. New, wet plaster must be applied the next day.

The lime in plaster bleaches many pigments. Only pigments that resist the action of lime can be used. These pigments are usually earth colors.

Frescoes have been painted for hundreds and hundreds of years. The fresco shown here was painted by Mexican artist Diego Rivera. He is well-known for the magnificent frescoes he painted of Mexican history and politics. Notice the natural golden earth tones Rivera used.

Classics

Michelangelo's frescoes on the walls of the Sistine Chapel in the Vatican, Italy

Other Works

Twentieth-century frescoes include Diego Rivera's *Great Tenochtitlan*, 1929-1945 (National Palace, Mexico City) and *Agrarian Leader Zapata*, 1931 (Museum of Modern Art, New York City), as well as *An Epic of American Civilization* by Jose Clemente Orozco, 1934 (Dartmouth College).

GLENCOE TECHNOLOGY

 Videodisc

The Infinite Voyage Series: The Future of the Past

Chapter 1

Preserving Frescoes in Florence

Summary

16-1: Chemical Changes in Matter

1. In a chemical reaction, the reactants are changed into the products, which are different substances.
2. According to the law of conservation of mass, the mass of the reactants in a chemical reaction equals the mass of the products.
3. A chemical equation is a shorthand way of describing a chemical reaction using symbols, coefficients, and formulas.

16-2: Science and Society: Chemical Reactions—Up in the Air

1. The ozone layer helps protect Earth by absorbing ultraviolet radiation from the sun.
2. In the upper atmosphere, CFCs release atoms of chlorine that destroy ozone molecules.
3. Replacing CFCs and regulating their use may slow down the loss of ozone.

16-3: Chemical Equations

1. A balanced chemical equation has the same number of atoms of each element on both sides of the equation.
2. The final step in the process of balancing a chemical equation is the choice of the correct coefficients.

16-4: Types of Chemical Reactions

1. Many specific chemical reactions can be classified as one of four reaction types—synthesis, decomposition, single-displacement, and double-displacement.

16-5: Energy and Chemical Reactions

1. In an exothermic reaction, energy is released; in an endothermic reaction, energy is absorbed.

2. A catalyst increases the speed of a chemical reaction; an inhibitor prevents a chemical reaction.

Key Science Words

a. balanced chemical equation
b. catalyst
c. chemical reaction
d. chlorofluoro-carbon (CFC)
e. coefficient
f. decomposition reaction
g. double-displacement reaction

h. endothermic reaction
i. exothermic reaction
j. inhibitor
k. precipitate
l. product
m. reactant
n. single-displacement reaction
o. synthesis reaction

Reviewing Vocabulary

Match each phrase with the correct term from the list of Key Science Words.

1. energy is given off
2. has same number of atoms on both sides
3. energy is absorbed
4. an element replaces another in a compound
5. prevents a reaction
6. substance breaks down into simpler substances
7. two ionic compounds react to form a precipitate, water, or a gas
8. two or more substances combine
9. compound that contains chlorine, fluorine, and carbon
10. well-defined example of chemical change

Chapter 16 Review   461

Summary

Have students read the summary statements to review the major concepts of the chapter.

Reviewing Vocabulary

1. i	6. f
2. a	7. g
3. h	8. o
4. n	9. d
5. j	10. c

✓ Assessment

Portfolio Encourage students to place in their portfolios one or two items of what they consider to be their best work. Examples include:

- Skill Builder explanation, p. 445
- Flex Your Brain, p. 447
- Activity 16-1 observations and answers, pp. 454-455 **P**

Performance Additional performance assessments may be found in **Performance Assessment** and **Science Integration Activities**. Performance Task Assessment Lists and rubrics for evaluating these activities can be found in Glencoe's **Performance Assessment in the Science Classroom**.

GLENCOE TECHNOLOGY

⊙ Videodisc

Glencoe Physical Science

Interactive Videodisc

Use the videodisc segments *Alkali Metals, Electron Structure, Stability and Reactivity,* and *Noble Gases* from Lesson 4, *Periodicity,* to review related chapter concepts.

▭ MindJogger Videoquiz

Chapter 16 Have students work in groups as they play the Videoquiz game to review key chapter concepts.

Checking Concepts

1. d	6. a
2. b	7. c
3. a	8. d
4. a	9. b
5. c	10. c

Understanding Concepts

11. 76 g Cr_2O_3 + 27 g Al = 51 g Al_2O_3 + X g Cr
mass of Cr = 52 g

12. $C_3H_8(g) + 5O_2(g) \rightarrow 3CO_2(g) + 4H_2O(g)$

13. decomposition of ClO and CFCs

14. $2NO(g) + 5H_2(g) \rightarrow 2NH_3(g) + 2H_2O(g)$

15. exothermic

Thinking Critically

16. Equations will vary, but for all equations, the number of atoms of each element in the reactants must equal the number of atoms of each element in the products.

17. Zinc will replace copper in $Cu(NO_3)_2$ because zinc is the more active metal. Copper metal will form. No reaction will occur in the other case.

18. $PbSO_4(cr)$

19. Changing subscripts changes the identity of a substance.

20. $2CuO(cr) \rightarrow 2Cu(cr) + O_2(g)$

Developing Skills

21. **Concept Mapping** See reduced student page.

22. **Recognizing Cause and Effect** The simple sugars glucose and fructose are formed. The acid is a catalyst.

23. **Observing and Inferring** The symbol *cr* would be used. A precipitate must be formed.

24. **Interpreting Data** 46 g Na + X g O_2 = 62 g Na_2O mass of O_2 = 16 g

Checking Concepts

Choose the word or phrase that completes the sentence.

1. An example of a chemical reaction is _____.
   a. bending
   b. evaporation
   c. melting
   d. photosynthesis

2. Lavoisier's experiments gave examples of the law of _____.
   a. chemical reaction
   b. conservation of mass
   c. coefficients
   d. gravity

3. An element that is more _____ will replace another element in a compound.
   a. active
   b. catalytic
   c. inhibiting
   d. soluble

4. In the expression $4Ca(NO_3)_2$, the 4 is a _____.
   a. coefficient
   b. formula
   c. subscript
   d. symbol

5. BHA is an example of a(n) _____.
   a. catalyst
   b. formula
   c. inhibitor
   d. CFC

6. If a substance is dissolved in water, _____ follows its formula in an equation.
   a. (aq)   b. (cr)   c. (g)   d. (l)

7. In the tarnishing of silver, Ag_2S is a(n) _____.
   a. catalyst
   b. inhibitor
   c. product
   d. reactant

8. In the burning of hydrogen, O_2 is a(n) _____.
   a. catalyst
   b. inhibitor
   c. product
   d. reactant

9. _____ is an allotrope of oxygen, O_2, that absorbs UV radiation.
   a. H_2 b. O_3 c. Cl   d. Cl_2

10. If a substance is a solid, _____ follows its formula in an equation.
   a. (l)   b. (g)   c. (cr)   d. (aq)

Understanding Concepts

Answer the following questions in your Science Journal using complete sentences.

11. Chromium is produced by reacting its oxide with aluminum. If 76 g of Cr_2O_3 and 27 g of Al react to form 51 g of Al_2O_3, how many grams of Cr are formed?

12. Propane, $C_3H_8(g)$, burns in oxygen to form carbon dioxide and water vapor. Write a balanced equation for burning propane.

13. $Cl(g) + O_3(g) \rightarrow ClO(g) + O_2(g)$ plays a role in ozone destruction. What is the source of the monatomic chlorine, Cl, in this reaction?

14. In one reaction of a catalytic converter, nitrogen(II) oxide reacts with hydrogen to form ammonia and water vapor. Write a balanced equation for this reaction.

15. If lye, NaOH, is put in water, the solution gets hot. What kind of energy process is this?

Thinking Critically

16. Write a balanced chemical equation and use it to explain the law of conservation of mass.

17. If Zn is placed in a solution of $Cu(NO_3)_2$, and Cu is placed in a $Zn(NO_3)_2$ solution, what reaction takes place? Explain.

18. $PbCl_2$ and Li_2SO_4 react to form LiCl and what other substance?

19. Why should subscripts not be changed to balance an equation?

20. Write a balanced equation for the decomposition of copper(II) oxide.

25. **Classifying** Sample outline:
   I. Chemical reactions
      A. Synthesis
         1. Two or more substances react to form another substance.
         2. Example: $2H_2(g) + O_2(g) \rightarrow 2H_2O(g)$
      B. Decomposition
         1. A substance breaks down into other substances.
         2. Example: $2HI(g) \rightarrow H_2(g) + I_2(g)$
      C. Single displacement
         1. One element replaces another in a compound.
         2. Example: $Zn(cr) + 2AgNO_3(aq) \rightarrow 2Ag(cr) + Zn(NO_3)_2(aq)$
      D. Double displacement
         1. Takes place when two ionic compounds combine and water, a gas, or a precipitate is formed.
         2. Example: $NaCl(aq) + AgNO_3(aq) \rightarrow AgCl(cr) + NaNO_3(aq)$

Developing Skills

If you need help, refer to the **Skill Handbook.**

21. **Concept Mapping:** The arrow in a chemical equation tells the reaction direction. Some reactions are reversible because they don't go in only one direction. Sometimes, the bond formed is weak, and a product breaks apart as it's formed. Double arrows are used in the equation to indicate this reaction. Fill in the cycle map, using the words *product(s)* and *reactant(s)*. In the blanks in the center, fill in the formulas for the substances appearing in the reversible reaction.

$$H_2(g) + I_2(g) \rightleftarrows 2HI(g)$$

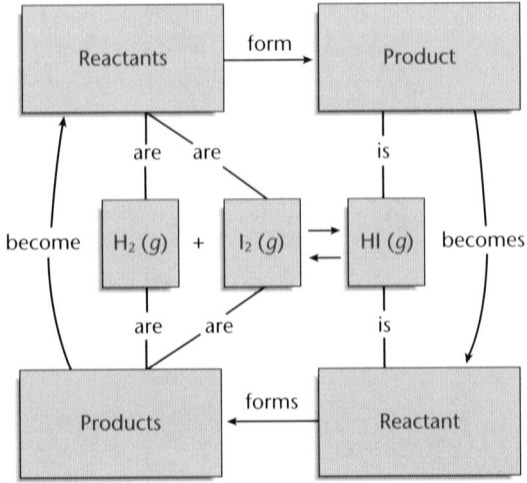

22. **Recognizing Cause and Effect:** Sucrose, table sugar, is a disaccharide. This means that sucrose is composed of two simple sugars chemically bonded together. Sucrose can be separated by digestion or by heating it in an aqueous sulfuric acid solution. Find out what products are formed by breaking up sucrose. What role does the acid play?

23. **Observing and Inferring:** What belongs in the parentheses in the following double-displacement reaction?

$$BaCl_2(aq) + K_2CO_3(aq) \rightarrow$$
$$BaCO_3(\ ) + 2KCl(aq)$$

24. **Interpreting Data:** When 46 g of sodium were exposed to dry air, 62 g of sodium oxide formed over a period of time. How many grams of oxygen from the air were used?

25. **Classifying:** Make an outline with the general heading "Chemical Reactions." Include the four types of reactions, with a description and example of each.

Performance Assessment

1. **Asking Questions:** Investigate the label on a package of cold cuts. Research as many of the substances added as possible. Is anything added as an inhibitor? Report on your findings.

2. **Oral Presentation:** Research common chemical reactions that occur safely in your home. Demonstrate one of these reactions. Tell what type of reaction it is and how it is used. Include in your report a balanced equation for the reaction.

3. **Model:** Use toothpicks and Styrofoam balls of different sizes or colors to balance each of the following chemical reactions and demonstrate the law of conservation of mass.
   a. $Cl_2O(g) + H_2O(l) \rightarrow HClO(aq)$
   b. $Fe_2O_3(cr) + CO(g) \rightarrow Fe(cr) + CO_2(g)$
   c. $H_2(g) + N_2(g) \rightarrow NH_3(g)$
   d. $ZnO(cr) + HCl(aq) \rightarrow ZnCl_2(aq) + H_2O(l)$

Chapter 16 Review

Performance Assessment

1. Students may find dextrose, hydrolyzed sodium caseinate, sodium phosphates, sodium erythorbate, sodium nitrite, etc. (bologna ingredients). Use the Performance Task Assessment List for Asking Questions in **PASC,** p. 19. **P**

2. Possible reactions: lighting a match, using H_2O_2 on a cut. Use the Performance Task Assessment List for Oral Presentation in **PASC,** p. 71. **P**

3. For each of the following balanced equations, the law of conservation of mass is shown by there being an equal number of each type of atom on both sides of each equation. Use the Performance Task Assessment list for Model in **PASC,** p. 51 **P**
   a. $Cl_2O + H_2O \rightarrow 2HClO$
   b. $2Fe_2O_3 + 6CO \rightarrow 4Fe + 6CO_2$
   c. $3H_2 + N_2 \rightarrow 2NH_3$
   d. $ZnO + 2HCl \rightarrow ZnCl_2 + H_2O$

Assessment Resources

📁 **Reproducible Masters**
Chapter Review, pp. 35-36
Assessment, pp. 105-108
Performance Assessment, p. 30

Glencoe Technology
💿 **Chapter Review Software**
💿 **Computer Test Bank**
📼 **MindJogger Videoquiz**

Chapter Organizer

Section	Objectives/Standards	Activities/Features
Chapter Opener		**Explore Activity:** Explore how marble reacts with some rainwater. p. 465
17-1 Acids and Bases (3 sessions, 2 blocks)*	1. **Define** *acid* and *base*. 2. **Describe** the characteristic properties of acids and bases. 3. **List** the names, formulas, and uses of some common acids and bases. 4. **Relate** the processes of ionization and dissociation to the formation of acids and bases. **National Content Standards:** (5-8) UCP1-UCP3, UCP5, A2, B1, C1, E2, F5; (9-12) UCP1-UCP3, UCP5, A2, B2, E2, F6	**Using Technology:** Sulfuric Acid, p. 469 **MiniLAB:** Are acids and bases conductors of electricity? p. 470 **Connect to Life Science,** p. 473 **Using Computers,** p. 473 **Skill Builder:** Observing and Inferring, p. 473
17-2 Strength of Acids and Bases (1 session, ½ block)*	1. **Explain** what determines the strength of an acid or a base. 2. **Differentiate** between strength and concentration. 3. **Define** *pH*. 4. **Describe** the relationship between pH and the strength of an acid or a base. **National Content Standards:** (5-8) UCP1-UCP3, UCP5, C1, C3; (9-12) UCP1-UCP3, UCP5	**Connect to Life Science,** p. 475 **Science Journal,** p. 476 **Skill Builder:** Concept Mapping, p. 476 **Activity 17-1:** Strong and Weak Acids, p. 477
17-3 Science and Society (1 session, ½ block)*	1. **Use** pH units to define *acid rain*. 2. **Describe** the factors contributing to the formation of acid rain. 3. **Discuss** the effects of acid rain, and **evaluate** methods of controlling this problem. **National Content Standards:** (5-8) UCP2, UCP3, UCP5, F1-F5; (9-12) UCP2, UCP3, UCP5, B3, F1, F3-F6	**Explore the Issue,** p. 479
17-4 Acids, Bases, and Salts (3 sessions, 1½ blocks)*	1. **Describe** a neutralization reaction. 2. **Explain** what a salt is and how salts form. 3. **Differentiate** between soaps and detergents. 4. **Explain** how esters are made and what they are used for. **National Content Standards:** (5-8) UCP2, UCP3, UCP5, A1, C1, C3, E1, G1; (9-12) UCP2, UCP3, UCP5, A1, B3, E1, F1, G1	**Problem Solving:** An Upsetting Situation, p. 481 **MiniLAB:** How can fruit juices be tested for ascorbic acid, vitamin C? p. 484 **Science Journal,** p. 485 **Skill Builder:** Interpreting Data, p. 485 **Activity 17-2:** Be a Soda Scientist, pp. 486-487 **People and Science:** Dennis Abrahamson, Paper Mill Technical Director, p. 488

Activity Materials

Explore	Activities	MiniLABs
page 465 5.0 g marble chips, beaker, 50.0 mL soda water, filter paper, paper towels, balance	page 477 typing paper, thick cardboard, 2 test tubes, 2 droppers, graduated cylinder, dilute hydrochloric acid, dilute acetic acid, pH paper, timer with second hand pages 486-487 2 different colorless soft drinks, 2 test tubes, graduated cylinder, 2 droppers, 1% phenolphthalein indicator solution, dilute sodium hydroxide solution	page 470 conductivity tester, glass of orange juice, ammonia-based household cleaner page 484 10 mL each of different fruit juices, cornstarch, pipet or dropper, iodine solution, stirrer

Need Materials? Call Science Kit (1-800-828-7777). * A complete Planning Guide that includes block scheduling is provided on pages 32T-35T.

Teacher Classroom Resources

Reproducible Masters	Transparencies	Teaching Resources
Study Guide, p. 70 **Reinforcement,** p. 70 **Enrichment,** p. 70 **Activity Worksheets,** p. 100 **Multicultural Connections,** p. 37 **Concept Mapping,** pp. 39-40	**Section Focus Transparency 66,** In a Pickle **Teaching Transparency 33,** Acids in Solutions	**Physical Science CD-ROM** **Spanish Resources** **English/Spanish Audiocassettes** **Cooperative Learning Resource Guide** **Lab Partner** **Lab and Safety Skills** **Lesson Plans**
Study Guide, p 71 **Reinforcement,** p. 71 **Enrichment,** p. 71 **Activity Worksheets,** pp. 96-97 **Science Integration Activity 17,** Jumping Curdles **Lab Manual 35,** Acids, Bases, and Indicators	**Section Focus Transparency 67,** DANGER: Acids **Teaching Transparency 34,** The pH Scale	**Assessment Resources** **Chapter Review,** pp. 37-38 **Assessment,** pp. 109-112 **Performance Assessment in the Science Classroom (PASC)** **MindJogger Videoquiz**
Study Guide, p. 72 **Reinforcement,** p. 72 **Enrichment,** p. 72 **Lab Manual 36,** Acid Rain **Science and Society Integration,** p. 21 **Critical Thinking/Problem Solving,** p. 23	**Section Focus Transparency 68,** Acid Rain and Snow **Science Integration Transparency 17,** Acid Rain in North America	**Alternate Assessment in the Science Classroom** **Performance Assessment** **Chapter Review Software** **Computer Test Bank**
Study Guide, p. 73 **Reinforcement,** p. 73 **Enrichment,** p. 73 **Activity Worksheets,** pp. 98-99, 101 **Cross-Curricular Integration,** p. 23 **Technology Integration,** pp. 15-16	**Section Focus Transparency 69,** Wash Your Hands	

Key to Teaching Strategies

The following designations will help you decide which activities are appropriate for your students.

L1 Level 1 activities should be within the ability range of all students, including those with learning difficulties.

L2 Level 2 activities should be within the ability range of the average to above-average student.

L3 Level 3 activities are designed for the ability range of above-average students.

LEP LEP activities should be within the ability range of Limited English Proficiency students.

LS These activities are designed to address different learning styles.

COOP LEARN Cooperative Learning activities are designed for small group work.

P These strategies represent student products that can be placed into a best-work portfolio.

GLENCOE TECHNOLOGY

The following multimedia resources are available from Glencoe.

Science and Technology Videodisc Series (STVS)
Ecology
   Acid Rain and Plants
   Fish and Acid Rain

National Geographic Society Series
STV: Human Body Volume 1
GTV: Planetary Manager

Physical Science CD-ROM

Teacher Classroom Resources

This is a representation of key blackline masters available in the Teacher Classroom Resources.

Teaching Aids

Section Focus Transparencies

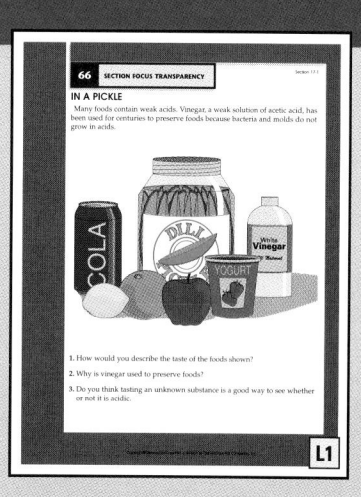

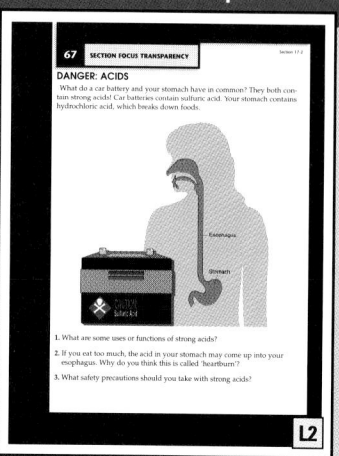

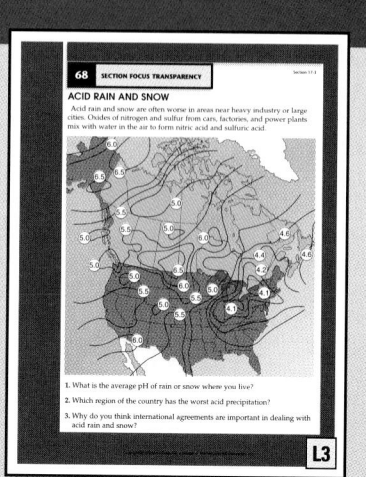

Science Integration Transparencies

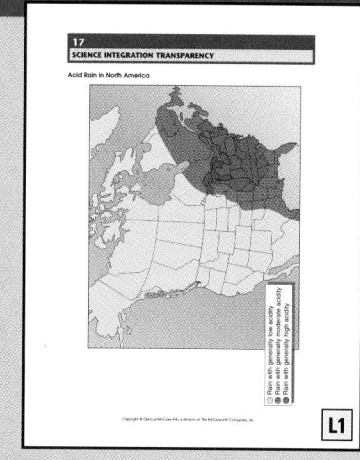

Teaching Transparencies

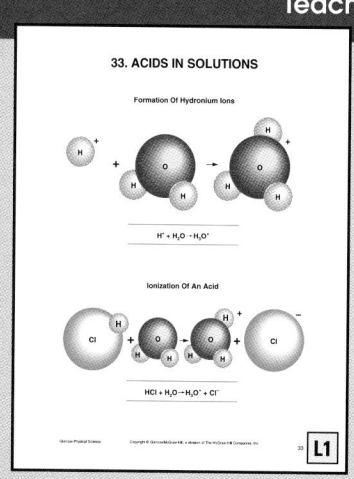

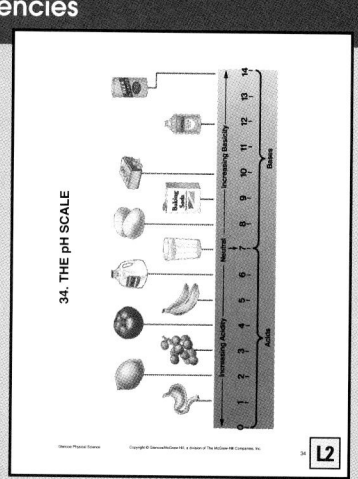

Meeting Different Ability Levels

Study Guide

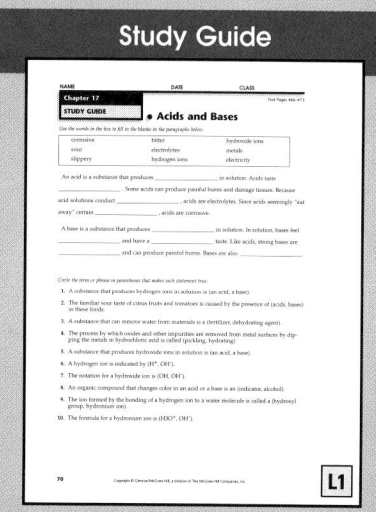

Reinforcement

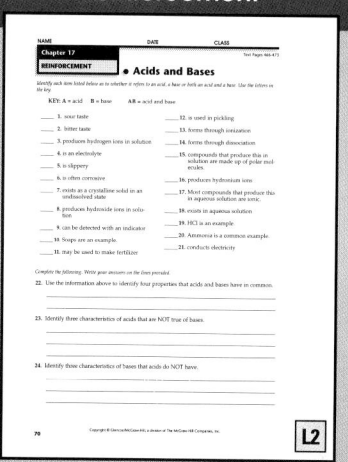

Enrichment Worksheets

Hands-On Activities

Science Integration Activity

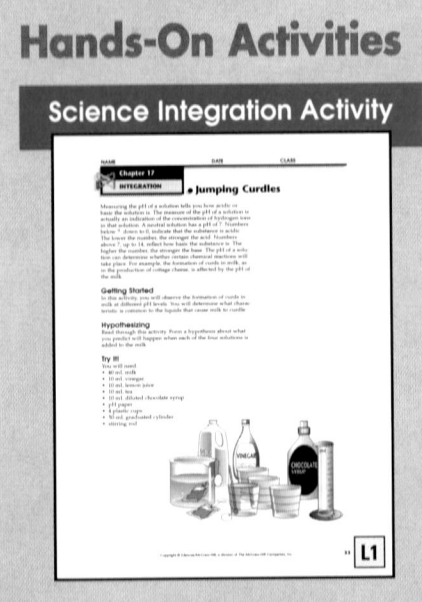

L1

Lab Manual

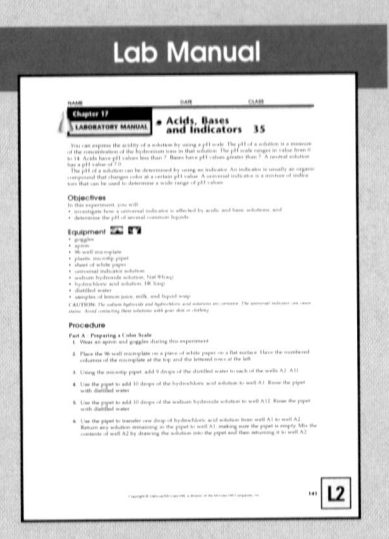

L2

Assessment

Performance Assessment

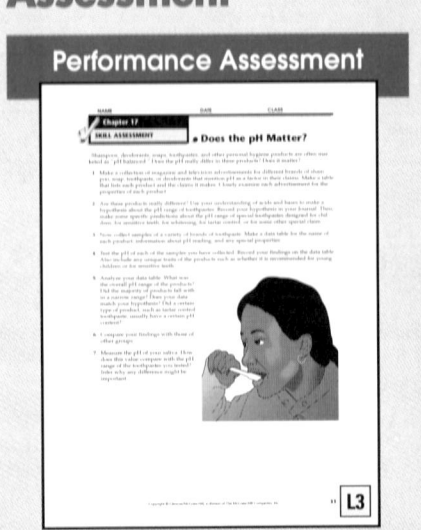

L3

Enrichment and Application

Critical Thinking/ Problem Solving

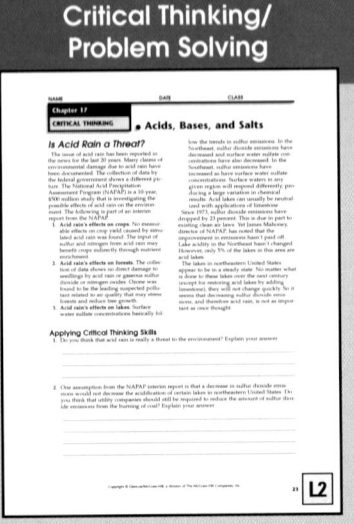

L2

Cross-Curricular Integration

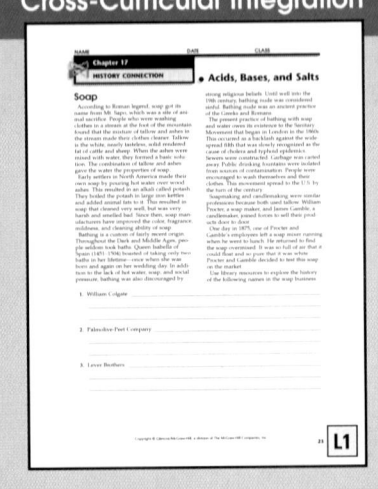

L1

Science and Society Integration

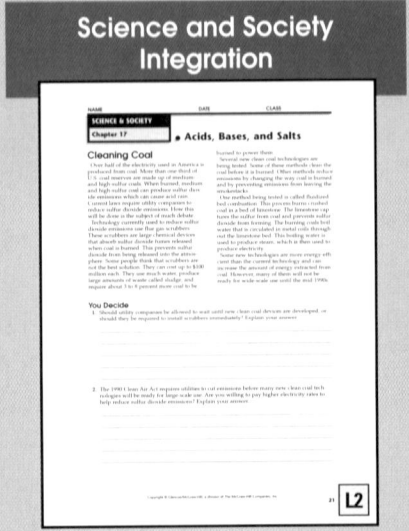

L2

Technology Integration

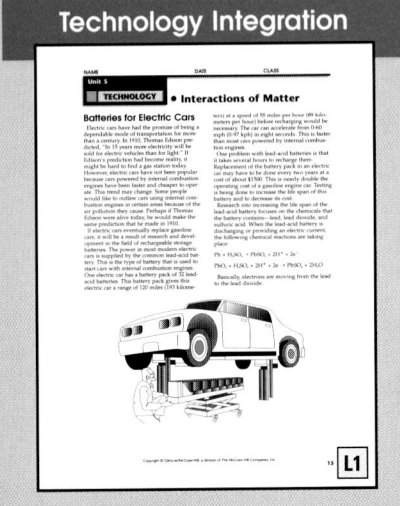

L1

Multicultural Connections

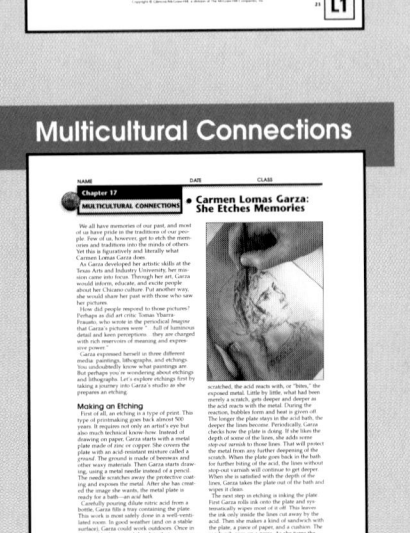

L2

Concept Mapping

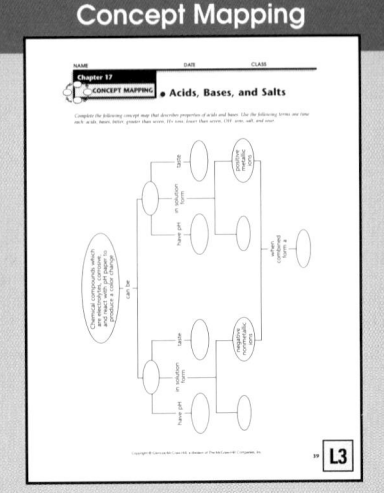

L3

Acids, Bases, and Salts

CHAPTER OVERVIEW

Section 17-1 Acids and bases are defined in this section. Their properties are described and explained according to the theories of ionization and dissociation.

Section 17-2 This section explains what determines acid or base strength. The concept of pH and how it relates to the strength of an acidic or basic solution is developed.

Section 17-3 Science and Society Probable causes and effects of acid rain are presented. Two points of view concerning possible solutions and the cost of this environmental problem are presented.

Section 17-4 This section describes how a salt is formed in a neutralization reaction. Soaps, detergents, organic acids, and esters are introduced.

Chapter Vocabulary

acid	weak base
indicator	pH
dehydrating agent	acid rain
	plankton
pickling	neutralization
base	salt
hydronium ion	titration
	soap
strong acid	saponification
weak acid	detergent
strong base	ester

Theme Connection

Systems and Interactions The student is introduced to systems and interactions as a theme in this chapter through a presentation of the Arrhenius system of acids and bases. The interactions that produce a salt and other acid/base interactions are also developed in the chapter.

Previewing the Chapter

464

Learning Styles Look for the following logo for strategies that emphasize different learning modalities. **LS**

Kinesthetic	Demonstration, p. 471
Visual-Spatial	Demonstration, pp. 466, 467, 471, 474, 481, 482, 483; Reteach, p. 472
Interpersonal	Explore, p. 465; MiniLAB, pp. 470, 484; Activity 17-1, p. 477; Activity 17-2, p. 486
Intrapersonal	Integrating the Sciences, p. 468
Linguistic	Across the Curriculum, p. 469; Science Journal, pp. 469, 475, 476, 481; Close, p. 479
Logical-Mathematical	People and Science, p. 488

Chapter 17

Acids, Bases, and Salts

Imagine the tremendous amount of work that must have gone into this impressive structure. Unfortunately, repair work is needed. What do you think is causing the deterioration of the marble of this ancient structure?

Durable marble structures can resist rain and many weathering factors, but not all. Marble is made mostly of calcium carbonate, $CaCO_3$, also called calcite. In the following activity, you will use a solution that is similar to polluted rainwater. In the chapter that follows, you will find out what kinds of solutions can cause these changes.

EXPLORE ACTIVITY

Explore how marble reacts with some rainwater.

1. Mass approximately 5.0 g of marble chips and place them in a beaker.
2. Add 50.0 mL of soda water, which represents acid rain.
3. After several minutes, stir the mixture and allow it to settle.
4. When the mixture stops reacting, pour the mixture through filter paper.
5. Dry the trapped marble chip residue and determine its mass.

Observe: What was the change in mass of the marble chips? In your Science Journal, write your conclusions about the effect that acid rain may have on marble.

Previewing Science Skills

▶ In the Skill Builders, you will **observe** and **infer**, **map concepts**, and **interpret data**.

▶ In the Activities, you will **use models, hypothesize, measure, observe,** and **infer**.

▶ In the MiniLABs, you will **observe** and **infer**.

465

EXPLORE ACTIVITY

Purpose

IS Interpersonal Use the Explore Activity to introduce students to acids and bases.
L1 **LEP** **COOP LEARN**

Preparation

Mass samples of marble chips ahead of time.

Materials

soda water, marble chips, balances, filter paper, funnels, heat lamp or portable hair dryer, beaker, graduated cylinder

Teaching Strategies

• Several portable hair dryers will quickly dry the calcium carbonate chips so they can be re-massed during the same laboratory period.

Observe

Conclusions may vary but should indicate that a reaction occurred between the marble and soda water. The reaction released material from the marble, thus causing a loss of mass.

Assessment

Oral Have students of one group explain the cause or causes of what they observed in this activity to students in another group. Use the Performance Task Assessment List for Group Work in **PASC**, p. 97. **P**

Assessment Planner

Portfolio
Refer to page 489 for suggested items that students might select for their portfolios.

Performance Assessment
See page 489 for additional Performance Assessment options.
Skill Builders, pp. 473, 476, 485
MiniLABS, pp. 470, 484
Activities 17-1, p. 477; 17-2, pp. 486, 487

Content Assessment
Section Wrap-ups, pp. 473, 476, 479, 485
Chapter Review, pp. 489-491
Mini Quizzes, pp. 469, 473, 485

Group Assessment
Opportunities for group assessment occur with Cooperative Learning Strategies and Flex Your Brain Activities.

Prepare

Section Background

This text limits the discussion of acids and bases to the Arrhenius theory. The theories proposed by Brønsted, Lowry, and Lewis build on the Arrhenius theory and extend the definitions of acids and bases.

Preplanning

To prepare for the MiniLAB on page 470, obtain a conductivity tester, orange juice, and ammonia cleaner.

1 Motivate

Bellringer

Before presenting the lesson, display **Section Focus Transparency 66** on the overhead projector. Assign the accompanying **Focus Activity** worksheet. L1 LEP

Demonstration

IS **Visual-Spatial** Prepare two solutions. The acid solution is made by adding 0.6 mL of 18M sulfuric acid to 99.4 mL of water. Add 10 mL of this solution to 90 mL of water to prepare a 0.01M acid solution. Separately prepare a base solution by adding 0.32 g of Ba(OH)$_2$•8H$_2$O to 100 mL of water. This gives a 0.01M solution. Half fill the beaker of a conductivity apparatus with the acid solution. Place the electrodes in the beaker. While stirring the solution, add the 0.01M Ba(OH)$_2$ solution dropwise. Have students observe both the salt formation and the decreasing conductivity.

Science Words

- acid
- indicator
- dehydrating agent
- pickling
- base
- hydronium ion

Objectives

- Define *acid* and *base*.
- Describe the characteristic properties of acids and bases.
- List the names, formulas, and uses of some common acids and bases.
- Relate the processes of ionization and dissociation to the formation of acids and bases.

Figure 17-1

These fruits contain citric acid. *What other foods have a sour taste that may be due to the presence of acids?*

17•1 Acids and Bases

Acids

What comes to mind when you hear the word *acid*? Do you think about something that has a sour taste? Do you think of a substance that can burn your skin, or even burn a hole through a piece of metal? Although many acids are not strong enough to burn through metal, or burn skin, these characteristics are, indeed, properties of some acids.

Acids make up an important group of compounds that contain hydrogen. When an acid dissolves in water, some of the hydrogen is released as hydrogen ions, H$^+$. Thus, an **acid** is a substance that produces hydrogen ions, H$^+$, in solution. It is the ability to produce H$^+$ ions that gives acids their characteristic properties.

Properties of Acids

Acids taste sour. The familiar sour taste of the foods shown in **Figure 17-1** is due to the presence of acids. *However, taste should **NEVER** be used to test for the presence of acids.* Some acids can produce painful burns and damage to tissues.

Acids are electrolytes. Because they always contain some ions, acid solutions conduct electricity.

Acids are corrosive. Some acids react strongly with certain metals, seeming to eat away the metals as metallic compounds and hydrogen gas are produced.

Acids react with certain compounds, called indicators, to produce predictable changes in color. An **indicator** is an organic compound that changes color in an acid or a base. **Figure 17-2** shows this interesting property. Some indicators are soaked into paper strips, such as litmus paper. Litmus paper is red in the presence of an acid and blue in the presence of a base.

Program Resources

Reproducible Masters
Study Guide, p. 70 L1
Reinforcement, p. 70 L1
Enrichment, p. 70 L3
Activity Worksheets, p. 100 L1
Concept Mapping, pp. 39, 40
Multicultural Connections, pp. 37, 38

Transparencies
Section Focus Transparency 66
Teaching Transparency 33

Common Acids

Many familiar items that you use every day contain acids. Acids are components of many foods. The gastric juice in your stomach contains an acid that helps to break down food during digestion. Car batteries, some flashlight batteries, and some household cleaners also contain acids. **Table 17-1** gives the names and formulas of a few common acids and tells where they are found.

In addition to these acids that you come in contact with every day, certain acids are vital to industry. The four most important industrial acids are sulfuric acid, phosphoric acid, nitric acid, and hydrochloric acid.

Figure 17-2

Litmus paper is a common acid-base indicator. *Which of these solutions is an acid?*

Sulfuric Acid

Sulfuric acid, H_2SO_4, is the most widely used chemical in the world. The Using Technology feature on page 469 discusses the uses of sulfuric acid in detail. Because of its use in automobile storage batteries, sulfuric acid is sometimes called battery acid. However, most sulfuric acid produced is used in the manufacture of fertilizer.

Table 17-1

Some Common Acids		
Name	**Formula**	**Where Found**
Acetic acid	CH_3COOH	Vinegar
Acetylsalicylic acid	$HOOC-C_6H_4-OOCCH_3$	Aspirin
Ascorbic acid (vitamin C)	$H_2C_6H_6O_6$	Citrus fruits, tomatoes, vegetables
Boric acid	H_3BO_3	Eyewash solutions
Carbonic acid	H_2CO_3	Carbonated drinks
Hydrochloric acid	HCl	Gastric juice in stomach
Nitric acid	HNO_3	Making fertilizers, explosives (TNT)
Phosphoric acid	H_3PO_4	Making detergents, fertilizers
Sulfuric acid	H_2SO_4	Car batteries, making fertilizers

Integrating the Sciences

Geology Earth scientists use dilute hydrochloric acid to test different rocks for the presence of carbonate. A piece of limestone will react and give off carbon dioxide gas when a few drops of dilute HCl are added.

---

2 Teach

Demonstration

 **Visual-Spatial** Acids are electrolytes. Materials to make a lemon or potato clock are available from commercial science suppliers. Cu/Zn electrodes connected to a digital clock are placed 1 cm apart in a lemon, and the clock operates. A voltmeter can be used to show that the electrodes in a lemon do produce voltage.

Revealing Preconceptions

Students have the misconception that all acids are strong. Bring in some common items found at home that contain acids. You can include vitamin C (ascorbic acid), eyewash (boric acid), drink mixes (citric acid), and soft drinks (phosphoric acid). Students should know that most acids are weak acids.

Visual Learning

Figure 17-1 What other foods have a sour taste that may be due to the presence of acids? *Answers will vary; possible answers include tomatoes, pickles and olives.*

Figure 17-2 Which of these solutions is an acid? *the one with the red litmus paper* **LEP**

 Use **Multicultural Connections,** pp. 37, 38 as you teach this lesson.

GLENCOE TECHNOLOGY

CD-ROM

Physical Science CD-ROM Have students perform the interactive exploration for Chapter 17 to reinforce important chapter concepts and thinking processes.

Figure 17-3

Sulfuric acid reacts with the hydrogen and oxygen in sugar, removing them as water and leaving only carbon.

Sulfuric acid is also an important dehydrating agent. A **dehydrating agent** is a substance that can remove water from materials. As **Figure 17-3** illustrates, organic compounds can be dehydrated by removing hydrogen and oxygen in the form of water from the compounds, leaving only carbon. When sulfuric acid comes in contact with human tissue, this dehydrating action produces painful burns.

Sugar · Sulfuric acid · Carbon

Phosphoric Acid

Phosphoric acid, H_3PO_4, is another acid important to industry worldwide. Eighty percent of phosphoric acid produced is used to make fertilizers.

Dilute phosphoric acid has a slightly sour but pleasant taste and is used in soft drinks. It is also used to make phosphates, which may be added to detergents to enhance their cleaning power. Unfortunately, phosphates cause pollution problems in lakes and streams.

Nitric Acid

Nitric acid, HNO_3, is best known for its use in making explosives. However, as with sulfuric and phosphoric acids, most nitric acid is used in the manufacture of fertilizers. Concentrated nitric acid is a colorless liquid. When exposed to light, the acid slowly changes to yellow. This change in

Theme Connection

Systems and Interactions Sulfuric acid used as a dehydrating agent reinforces the interaction of acids with other materials, in this case removing water. Properties common to all acids are presented, reinforcing the concept of looking at acids as a system.

Integrating the Sciences

Geology Have a student contact the local soil and water conservation office or county extension agent to find out how soil is tested to determine its acidity or alkalinity, what the test costs, and what soils are like in your area. Even if you live in a large city, these offices and services may be available. **LS**

color occurs because some of the nitric acid decomposes into nitrogen dioxide, NO_2, which is a brownish gas. Nitric acid can cause serious burns on human skin. It can even cause burns if it is inhaled or if the vapor comes into contact with skin.

Hydrochloric Acid

Hydrogen chloride, HCl, is a colorless gas. When this gas is dissolved in water, hydrochloric acid forms. Both HCl fumes and the ions that form in solution can harm human tissue.

Hydrochloric acid, when used in industry to clean surfaces of materials, is commonly called muriatic acid. Large quantities of the acid are used by the steel industry for pickling. **Pickling** is a process in which oxides and other impurities are removed from metal surfaces by dipping the metals in hydrochloric acid.

Several methods can be used to make phosphoric, nitric, and

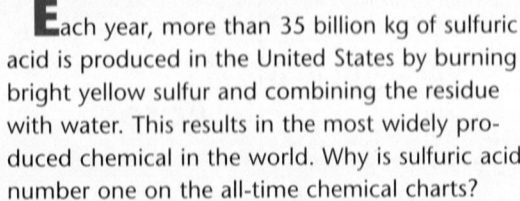

USING TECHNOLOGY

Sulfuric Acid

Each year, more than 35 billion kg of sulfuric acid is produced in the United States by burning bright yellow sulfur and combining the residue with water. This results in the most widely produced chemical in the world. Why is sulfuric acid number one on the all-time chemical charts?

Have you eaten cereal or bread lately? Grain foods come from crops that need phosphates in order to grow. Often, the phosphate in soil is not in a usable form. Sulfuric acid (H_2SO_4) is used to break down phosphate rock into an inexpensive fertilizer.

$$Ca_3(PO_4)_2(cr) + 3H_2SO_4(aq) \rightarrow$$
$$3CaSO_4(cr) + 2H_3PO_4(aq)$$

Sulfuric acid is also used to make clothing, dyes, and synthetic materials like rayon. In addition, it is used in car batteries and to process paper. The widespread use of sulfuric acid reveals many applications of the science of acid reactions.

Think Critically:

The amount of sulfuric acid used by a country has been proposed as a measure of its level of industrialization. In your Science Journal, explain why this might be an accurate measure.

Uses of Sulfuric Acid

Iron and steel 2%
Petroleum 2%
Rayon and film 3%
Paints and other pigments 6%
Other industries 7%
Chemicals 19%
Fertilizer 61%

Across the Curriculum

Language Arts The old name for sulfuric acid was oil of vitriol. Have students use a dictionary to look up *vitriol* and *vitriolic*. Ask them to use *vitriolic* in a sentence. L1
LS

USING TECHNOLOGY

Demonstrate the reaction shown in **Figure 17-3.** Explain how sulfuric acid can be a dehydrating agent. Ask students to explain what property of sulfuric acid is being used when the acid is in an automobile storage battery.

Think Critically

For one answer, students could refer to increased use of storage batteries in countries with higher levels of industrialization. Also, students may refer to production of greater amounts of synthetic materials requiring dyes.

•MINI•QUIZ•

Use the Mini Quiz to check students' recall of chapter content.

1. **What ions are produced by an acid that give it many of its properties?** *hydrogen ions*

2. **List four properties that are common to acids.** *Acids taste sour, conduct electricity, cause indicators to change color, and are corrosive.*

3. **A substance that can remove water from another substance is called a(n) _____ agent.** *dehydrating*

NATIONAL GEOGRAPHIC SOCIETY

 Videodisc
GTV: Planetary Manager
Industry

47891

Purpose

LS **Interpersonal** Students will observe whether acids and/or bases conduct an electric current. **L1** **LEP**
COOP LEARN

Materials

conductivity tester, orange juice, cleaner containing ammonia, small jars or beakers

Teaching Strategies

Safety Precautions Tell students not to taste any of the substances being tested or breathe ammonia fumes.

- Both the acid solution and the base solution will conduct electricity.
- Do not allow students to use 110-volt conductivity meters. Use only battery-powered apparatuses.

Analysis

1. ascorbic acid (vitamin C) and citric acid; typically, acids produce H⁺ ions.
2. ammonia; bases produce OH⁻ ions in solution.

✔ Assessment

Performance Have students test other juices and other bases, such as milk of magnesia, Mg(OH)$_2$. Use the Performance Task Assessment List for Carrying Out a Strategy and Collecting Data in **PASC**, p. 25. **P**

📁 **Activity Worksheets,** pp. 5, 100.

Visual Learning

Figure 17-4 What property of bases is evident in soaps? *bitter taste, slippery* **LEP** **LS**

MiniLAB

Are acids and bases conductors of electricity?

Find out whether acids and bases conduct an electric current.

Procedure 🧪 🥽 🧤

1. Carefully place the electrodes of a conductivity tester into a glass of orange juice. **CAUTION:** *Do not taste any substance being tested in this lab.*
2. Perform the same test with a solution of ammonia-based household cleaner. **CAUTION:** *Avoid direct contact with these fumes.*

Analysis

1. Identify an acid present in orange juice. What important ion is found in acids?
2. Identify a base present in the cleaner. What important ion is found in bases?

Figure 17-4

Many household materials contain bases. *What property of bases is evident in soaps?*

Inclusion Strategies

Hearing Impaired Show students a picture of a child with shampoo on his or her hair and one with shampoo in his or her eyes. Label the photograph, "Why So Irritable?" *Shampoo may contain a base.* Have the students make a display of other household products that may irritate because of bases.

---

hydrochloric acids. But one general method can be used to make all three acids. This method involves reacting concentrated sulfuric acid with a compound containing the negative ion of the desired acid.

$$Ca_3(PO_4)_2(cr) + 3H_2SO_4(aq) \rightarrow 2H_3PO_4(aq) + 3CaSO_4(cr)$$

$$NaNO_3(cr) + H_2SO_4(aq) \rightarrow HNO_3(g) + NaHSO_4(aq)$$

$$NaCl(cr) + H_2SO_4(aq) \rightarrow HCl(g) + NaHSO_4(aq)$$

Bases

Like acids, bases are an important group of chemical compounds. Although acids and bases seem to have some features in common, the two groups are surprisingly different.

A **base** is a substance that produces hydroxide ions, OH⁻, in solution. Does this OH combination look familiar? You may recall from Chapter 13 that an alcohol has a hydroxyl group, –OH, as part of its molecule. This hydroxyl group is not an ion and should not be confused with the OH⁻ ions that are produced by all bases. Because alcohols do not produce OH⁻ ions, alcohols are not classified as bases.

The hydroxide ions present in all basic solutions are negative ions, whereas the hydrogen ions present in acidic solutions are positive ions. This difference in charge accounts for some of the differences between acids and bases.

Properties of Bases

In the pure, undissolved state, most bases are crystalline solids. In solution, bases feel slippery and have a bitter taste. Like acids, strong bases are corrosive, and contact with the skin may result in severe burns and tissue damage. *Therefore, taste and touch should **NEVER** be used to test for the presence of a base.* Basic solutions are electrolytes. This property is due to the presence of ions in the solutions. Finally, bases react with indicators to produce predictable changes in color.

Science Journal

Alcohols Even though alcohols have the hydroxyl group, –OH, as part of their molecule, they are not bases. Have students write a paragraph in their Science Journal explaining why alcohols are not classified as bases. **L1** **LS**

Table 17-2

Common Bases and Their Uses		
Name	**Formula**	**Uses**
Aluminum hydroxide	$Al(OH)_3$	Deodorant, antacid
Calcium hydroxide	$Ca(OH)_2$	Leather production, manufacture of mortar and plaster
Magnesium hydroxide	$Mg(OH)_2$	Laxative, antacid
Sodium hydroxide	$NaOH$	Drain cleaner, soap making
Ammonia	NH_3	Household cleaners, fertilizer, production of rayon and nylon

Common Bases

Many household products contain bases. **Figure 17-4** shows several of these products. Most soaps, for example, contain bases, which explains their bitter taste and slippery feel. **Table 17-2** contains the names and formulas of some common bases and tells how each base is used.

The most widely used base is ammonia, NH_3. Pure ammonia is a colorless gas with a distinctive, irritating odor. It can be extremely irritating to nasal passages and lungs if inhaled. When ammonia gas dissolves in water, some of its molecules react with water molecules, forming ammonium ions (NH_4^+) and hydroxide ions.

$$NH_3(g) + H_2O(l) \rightarrow NH_4^+(aq) + OH^-(aq)$$

The solution containing ammonium ions and hydroxide ions is called ammonia water. Dilute ammonia water is an excellent household cleaning agent. Ammonia is also used as a fertilizer, as shown in **Figure 17-5**.

Metallic Hydroxides

Another widely used base is calcium hydroxide, $Ca(OH)_2$, commonly called caustic lime. Calcium hydroxide is used in the production of mortar and plaster and to "sweeten" acidic soils.

Sodium hydroxide, $NaOH$, is a strong, corrosive base. This compound, commonly called lye, dissolves readily in water in an extremely exothermic process. Sodium hydroxide reacts with organic materials such as oils and fats to produce soap and glycerine. This property also enables it to be an effective oven cleaner and drain cleaner.

Figure 17-5

Ammonia is an important base in many products.

17-1 Acids and Bases **471**

3 Assess

Check for Understanding

? FLEX Your Brain

Use the Flex Your Brain activity to have students explore ACIDS and BASES. P

Activity Worksheets, p. 5

Reteach

LS **Visual-Spatial** In a fume hood, add a small piece of copper wire to 2 mL of 6*M* nitric acid in a test tube fitted with a one-hole rubber stopper and delivery tube. Collect the dense gas by upward displacement of air in a small bottle that contains 8 mL of water that has universal indicator added. Close the bottle and shake the water as you observe the color change. Ask students to explain what has happened to cause the color change. You can relate this to acid rain, which is presented in Section 17-3. LEP

Extension

For students who have mastered this section, use the **Reinforcement** and **Enrichment** masters.

Solutions of Acids and Bases

You have learned that substances such as HCl, HNO₃, and H₂SO₄ are acids because of their ability to produce H⁺ ions. This ionization process is shown in **Figure 17-6,** using HCl as an example. When a polar molecule, such as HCl, is dissolved in water, the negatively charged area of nearby water molecules attracts the positively charged area of the polar molecule. The H⁺ is removed from the polar molecule and a **hydronium ion,** H_3O^+, is formed. Thus, *acid* describes any compound that can be ionized in water to form hydronium ions.

When an organic acid ionizes, the H at the end of the —COOH group combines with water to form the hydronium ion.

Figure 17-6

These equations demonstrate what happens to acids in solution.

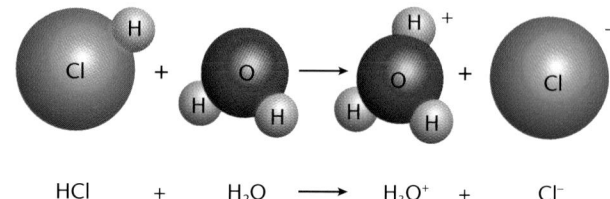

$$HCl \quad + \quad H_2O \longrightarrow H_3O^+ \quad + \quad Cl^-$$

A When hydrogen chloride gas dissolves in water, it ionizes to produce hydronium ions and chloride ions. *What is the name of this common acid?*

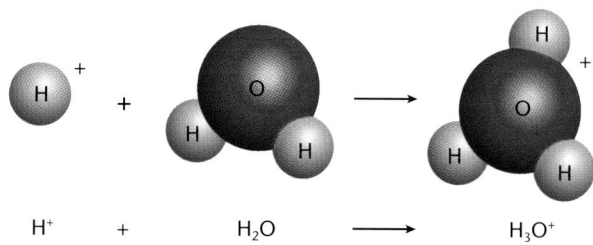

$$H^+ \quad + \quad H_2O \longrightarrow H_3O^+$$

B When an acid dissolves in water, hydrogen ions from the acid combine with water molecules to produce hydronium ions. *What is the overall charge on a hydronium ion?*

Compounds that can form hydroxide ions, OH⁻, in solution are classified as bases. If you look back at **Table 17-2,** you will find that most substances listed contain OH within their formulas. Except for ammonia, inorganic compounds that produce bases in aqueous solution are ionic compounds. That is, they are already made up of ions. As the equation at the top of page 473 shows, when such a compound dissolves, the ions dissociate, or pull apart, and exist as individual ions in

472 Chapter 17 Acids, Bases, and Salts

Inclusion Strategies

Learning Disabled Have students collect and bring to class magazine or newspaper photographs that deal with the properties or effects of acids or bases. Students may collect photographs of toxic waste sites, effects of acid rain, uses of soaps, or others. Other students in the class could produce a collage of the photographs. LEP

solution. These substances release OH⁻ ions in water. If a solution contains more OH⁻ ions than H_3O^+ ions, it is referred to as a basic solution.

$$NaOH(cr) \xrightarrow{H_2O} Na^+(aq) + OH^-(aq)$$

Ammonia is a polar compound. In solution, ionization takes place when the ammonia molecule attracts a hydrogen atom from a water molecule, as shown in **Figure 17-7**.

The next time you use any of the products described in this section, think about the properties the product has and why it has them.

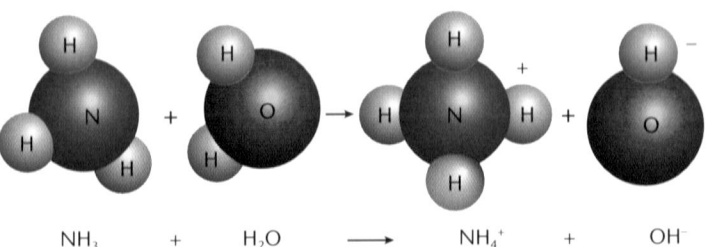

$$NH_3 \quad + \quad H_2O \quad \longrightarrow \quad NH_4^+ \quad + \quad OH^-$$

Figure 17-7

Even though ammonia is not a hydroxide, it reacts with water to produce some hydroxide ions; therefore, it is a base.

Section Wrap-up

Review

1. Why are acids and bases electrolytes?

2. Name three important acids and three important bases and describe some uses of each.

3. **Think Critically:** When you are stung by an ant, the insect injects formic acid, HCOOH, into your body. What kind of acid is formic acid—organic or inorganic? How do you know? Write an equation showing how formic acid ionizes.

Skill Builder
Observing and Inferring
You have observed that alcohols are not classified as bases, even though they have OH in their formulas. What do you infer about the bonding between OH and carbon in alcohols? If you need help, refer to Observing and Inferring in the **Skill Handbook.**

Using Computers

Spreadsheet Make a spreadsheet that compares acids and bases.

17-1 Acids and Bases   **473**

Skill Builder
The OH group in an alcohol is not an ion. It is covalently bonded to carbon atoms in alcohols.

✓Assessment

Process Have students produce a drawing of the structure of one of the alcohols learned earlier. Have them designate the covalent bond in their drawings. Use the Performance Task Assessment List for Scientific Drawing in **PASC,** p. 55. **P**

---

4 Close

•MINI•QUIZ•

Use the Mini Quiz to check students' recall of chapter content.

1. **What substance produces hydroxide ions in solution?** *a base*

2. **Give the name and formula for the ions that form when ammonia gas reacts with water.** *ammonium, NH_4^+, and hydroxide, OH^-, ions*

3. **When an ionic compound is pulled apart by the solvent, the process is called ____ .** *dissociation*

Section Wrap-up

Review

1. Acids and bases are electrolytes because they produce ions in solution and conduct electric currents.

2. Responses will vary but should include the acids and bases discussed in this section.

3. **Think Critically** The carboxyl, –COOH, group indicates that formic acid is an organic acid.
$HCOOH + H_2O \rightarrow HCOO^- + H_3O^+$

Using Computers

Be sure students compare acids and bases in a spreadsheet format.

Prepare

Section Background

• The binary acids HCl, HBr, and HI are strong; all other binary acids are weak.

• Ternary acids contain oxygen. Some are weak and some are strong.

• Polyprotic acids have more than one ionizable hydrogen.

• Hydroxides of Group 1 and 2 metals (except beryllium and magnesium) are strong bases. All others are weak.

Preplanning

To prepare for Activity 17-1, have dilute HCl and CH_3COOH on hand.

1 Motivate

Bellringer

Before presenting the lesson, display **Section Focus Transparency 67** on the overhead projector. Assign the accompanying **Focus Activity** worksheet. L1 LEP

Demonstration

IS **Visual-Spatial** Place two petri dishes, each containing a small piece of magnesium ribbon, on an overhead projector. Place several drops of $1M$ HCl on one piece and a similar amount of vinegar on the other piece. Students will see that the action of a strong acid contrasts greatly with that of a weak acid. LEP

Visual Learning

Figure 17-8 Which beaker contains the stronger acid? *the one with hydrochloric acid* LEP IS

17•2 Strength of Acids and Bases

Science Words

strong acid
weak acid
strong base
weak base
pH

Objectives

• Explain what determines the strength of an acid or a base.
• Differentiate between strength and concentration.
• Define *pH.*
• Describe the relationship between pH and the strength of an acid or a base.

Figure 17-8

This simple conductivity test indicates the presence of ions in solution. *Which beaker contains the stronger acid?*

Strong and Weak Acids and Bases

You have learned that all acids have certain properties in common. Yet, some acids are safe enough to swallow or put in your eyes, whereas other acids destroy human tissue and corrode metal. Obviously, all acids are not alike. Some acids are stronger than others. The same is true of bases. **Table 17-3** lists some strong and weak acids and bases.

The strength of an acid or base depends on how completely a compound is pulled apart to form ions when dissolved in water. An acid that ionizes almost completely in solution is a **strong acid.** Conversely, an acid that only partly ionizes in solution is a **weak acid. Figure 17-8** shows how the amount of ions in solution can be compared using a measure of conductivity.

Ionization Evidence

$$HCl(g) + H_2O(l) \rightarrow H_3O^+(aq) + Cl^-(aq)$$

$$CH_3COOH(l) + H_2O(l) \rightleftarrows H_3O^+(aq) + CH_3COO^-(aq)$$

The first equation, with only a forward arrow, indicates that ions are formed when HCl is dissolved in water. All of the HCl ionizes. The second equation illustrates a characteristic of weak acids. Only some of the CH_3COOH ionizes; thus the reaction is incomplete, as indicated by the arrows pointing in opposite directions. In an acetic acid solution, mostly CH_3COOH is present. There are only a few ions in solution. These equations represent the general ionization properties of strong and weak acids.

Remember that many bases are ionic compounds that dissociate to produce individual ions when they dissolve. A **strong base** dissociates completely in solution. The following equations show the dissociation of sodium hydroxide, a strong base, and the ionization of ammonia, a weak base.

$$NaOH(cr) \rightarrow Na^+(aq) + OH^-(aq)$$

$$NH_3(aq) + H_2O(l) \rightleftarrows NH_4^+(aq) + OH^-(aq)$$

Whereas all of the NaOH has broken up into ions, most of the ammonia remains in the form of NH_3. Because ammonia produces few ions, ammonia is a **weak base.**

Program Resources

 Reproducible Masters
Study Guide, p. 71 L1
Reinforcement, p. 71 L1
Enrichment, p. 71 L3
Activity Worksheets, pp. 96, 97 L1
Science Integration Activity, pp. 33, 34
Lab Manual 35

Transparencies
Section Focus Transparency 67 L1
Teaching Transparency 34

Table 17-3

Strengths of Some Acids and Bases

Acid	Strength	Base
Hydrochloric, HCl Sulfuric, H_2SO_4 Nitric, HNO_3	Strong	Sodium hydroxide, NaOH Potassium hydroxide, KOH
Acetic, CH_3COOH Carbonic, H_2CO_3 Boric, H_3BO_3	Weak	Ammonia, NH_3 Aluminum hydroxide, $Al(OH)_3$ Iron(III) hydroxide, $Fe(OH)_3$

Strength and Concentration

Often, the *strength* of an acid or base is confused with its *concentration*. As you learned in Chapter 15, *dilute* and *concentrated* are terms used to indicate the concentration of a solution. The concentration of an acid or base refers to the amount of acid or base dissolved in solution. It is possible to have dilute concentrations of strong acids or bases and concentrated solutions of weak acids or bases. *Strong* or *weak* refers to the ease with which an acid or base forms ions in solution.

pH of a Solution

If you have a pool or keep tropical fish, you know that the pH of the water must be controlled. **pH** is a measure of the concentration of hydronium ions in a solution. To determine pH, a scale typically ranging from 0 to 14 has been devised, as shown in **Figure 17-9.**

As the scale shows, solutions with a pH lower than 7 are acidic. The lower the value, the more acidic the solution. Solutions with pH greater than 7 are basic, and the higher the pH, the more basic the solution. A solution with a pH of exactly 7 is neutral—neither acidic nor basic. Pure water has a pH of 7.

A pH meter can be used to determine the pH of a solution. This meter is operated by immersing the electrodes in the solution to be tested and reading the dial. Small, battery-

CONNECT TO
LIFE SCIENCE

The aromas of many fish are caused by compounds called amines. If the pH of a dilute solution of an amine is 11.4, *infer* whether the amine is an acid or a base.

Figure 17-9

The pH scale helps us classify solutions as acidic or basic.

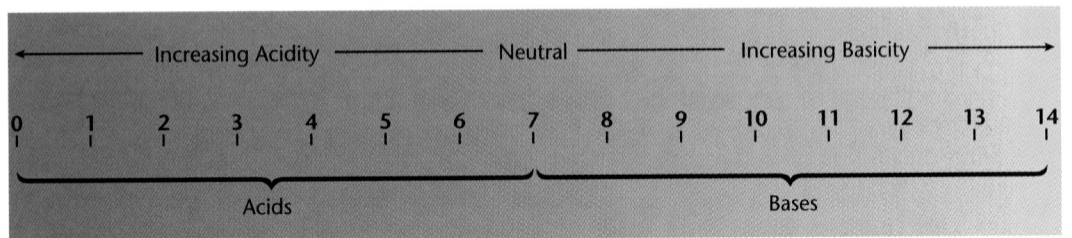

Increasing Acidity — Neutral — Increasing Basicity

0 1 2 3 4 5 6 7 8 9 10 11 12 13 14

Acids   Bases

Science Journal

Testing pH Obtain a sample of water from a nearby stream, lake, or reservoir; have students test its pH and write a paragraph in their Science Journals about what may have caused the water to develop the measured pH. **L1** **LS**

2 Teach

CONNECT TO
LIFE SCIENCE

Answer a base

INTEGRATION
Life Science

(See page 476.) Human blood is buffered by the hydrogen carbonate ion. This ion can react with excess acid or base in the blood to help maintain the pH at 7.4. The reaction with excess hydronium ions from acid is $HCO_3^- + H_3O^+ \rightarrow H_2CO_3 + H_2O$. The reaction with excess hydroxide ions from base is $HCO_3^- + OH^- \rightarrow H_2O + CO_3^{2-}$.

Visual Learning

Figure 17-10 (See p. 476) **What would you conclude about the pH of the rainwater being tested in this photo?** *pH = 6* **LEP** **LS**

3 Assess

Check for Understanding

⑦ FLEX Your Brain

Use the Flex Your Brain activity to have students explore pH OF SOLUTIONS.

 **Activity Worksheets,** p. 5

Reteach

Have students look at food labels and record two items that contain weak acids.

Extension

For students who have mastered this section, use the **Reinforcement** and **Enrichment** masters.

475

4 Close

Have students complete the following statement. **Hydronium ion is to acid as _____ is to base.** *hydroxide ion*

Section Wrap-up

Review

1. The strength of an acid depends on the relative ease with which the hydrogen ion is given up. For bases, it depends on the hydroxide ions.

2. A strong acid can be made dilute by adding a large amount of water.

3. **Think Critically** slightly acidic; moderately acidic; strongly basic; moderately basic; neutral.

Science Journal

Linguistic A strong acid has a lower pH because of the completeness of its ionization, not because of its concentration.

Skill Builder

Concept maps will vary. A sample is given here.

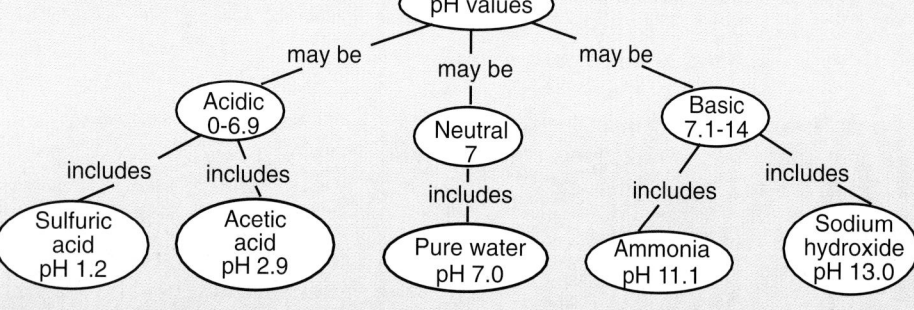

Figure 17-10

By moistening this treated paper and comparing the resulting color to the chart, the pH of the solution can be approximated. *What would you conclude about the pH of the rainwater being tested in this photo?*

operated pH meters with digital readouts make measuring the pH of materials convenient.

If a pH meter is not available, a universal indicator or pH paper can be used. Both of these undergo a color change in the presence of H_3O^+ ions and OH^- ions in solution. The final color of the solution or the pH paper is matched with colors in a chart to find the pH, as shown in **Figure 17-10.**

INTEGRATION
Life Science

pH of Your Blood

Your blood circulates throughout your body carrying oxygen, removing carbon dioxide, and absorbing nutrients from food that you have eaten. In order to carry out its many functions properly, the pH of blood cannot be allowed to change significantly. Why is the pH of blood affected so little by most foods? Some compounds in blood, called buffers, enable small amounts of acids or bases to be absorbed without harmful effects. Buffers are acids, bases, and salts that react with additional acids or bases to minimize their effects. Buffers help keep your blood close to a nearly constant pH of 7.4.

Section Wrap-up

Review

1. What determines the strength of an acid? A base?

2. How can you make a dilute solution of a strong acid? Include an example as part of your explanation.

3. **Think Critically:** What does the pH of each of the solutions indicate about the solution? Rainwater, 5.8; soda water, 3.0; drain cleaner, 14.0; seawater, 8.9; pure water, 7.0.

Skill Builder

Concept Mapping
Make a concept map of pH values. Start with three boxes labeled *acidic, neutral,* and *basic* and indicate the pH range of each box. Below each box, give some examples of solutions that belong in each pH range. If you need help, refer to Concept Mapping in the **Skill Handbook.**

Science Journal

In your Science Journal, explain why a weak acid in solution has a higher pH than a strong acid of the same concentration.

Assessment

Process List several common materials and their pH values on the chalkboard. Have students classify each material as an acidic, a basic, or a neutral solution. Use the Performance Task Assessment List for Analyzing the Data in **PASC,** p. 27.

Activity 17-1

Strong and Weak Acids

When you read about the behavior of things as small as molecules, it is difficult to visualize what causes the behavior.

Problem
How can a model be made that represents the relative ionization abilities of weak and strong acids?

Materials
- typing paper (22 cm × 28 cm) marked with rectangles
- thick cardboard (22 cm × 28 cm) marked with rectangles
- scissors
- test tubes (2)
- droppers (2)
- graduated cylinder
- dilute hydrochloric acid (HCl)
- dilute acetic acid (CH₃COOH)
- pH paper
- timer, with second hand

Procedure
1. Prepare a data table like the one shown.
2. Mark your paper and cardboard with equal numbers of rectangles. Mark every other rectangle with a plus sign (+) to represent hydronium ions. Then mark the blank rectangles with a minus sign (−) to represent negative ions.
3. Cut the typing paper into as many rectangles as possible in 10 seconds.
4. Repeat step 3 with the piece of cardboard.
5. Obtain 2 mL each of dilute HCl and dilute CH₃COOH in test tubes. **CAUTION:** *Handle these acids with care. Avoid direct contact. Report any spills immediately.*

6. Using a different dropper for each acid, add 2 drops of each to separate strips of pH paper. Determine their respective pH values and record them in the data table.

Analyze
1. Which was easier to separate into ions, the typing paper or the cardboard?
2. Which acid separated into ions most easily? How do you know?

Conclude and Apply
3. Which acid is stronger, HCl or CH₃COOH? How do you know?
4. Which represents the stronger acid, the typing paper or the cardboard? Explain.
5. **Infer** which has the stronger bonds between fibers, the typing paper or the cardboard. Explain.
6. **Predict** which acids have stronger bonds within their molecules, strong acids or weak acids. Explain.

Data and Observations   Sample Data

_8-12_ ions removed from typing paper	_1_ pH of HCl solution
_4-6_ ions removed from cardboard	_2_ pH of acetic acid solution

17-2 Strength of Acids and Bases   **477**

but HCl was more ionized, as shown by its lower pH.

4. The typing paper represented the stronger acid. It separated more readily.
5. The cardboard was more strongly bonded because it was more difficult to separate.
6. Weak acids are more strongly bonded. The attractions of water molecules do not cause them to break into ions as readily as do strong acids.

✔ Assessment

Process Think about how you represented strong and weak acids above. If an acid is described as weak, would its negative ions have a strong or a weak attraction for any hydronium ions in solution? Use the Performance Task Assessment List for Making Observations and Inferences in **PASC**, p. 17.
P

Activity 17-1

Purpose
IS **Interpersonal** Students will distinguish between weak and strong acids by comparing ease of ionization. **L1**
LEP **COOP LEARN**

Process Skills
using models, hypothesizing, classifying, sequencing, comparing and contrasting, recognizing cause and effect, forming operational definitions, interpreting scientific illustrations

Time
20 minutes

Safety Precautions
Students must handle acids with care and avoid contact with the skin or clothing.

📁 **Activity Worksheets,** pp. 5, 96, 97

Teaching Strategies
Troubleshooting Caution students to cut at a steady pace and not to rush.

- To emphasize the differences in the separation of ions, make a chalkboard data table and have each team record its results.
- To save time, have the rectangles marked on the paper and cardboard before class.
- To prepare 0.1*M* acetic acid, add 5.9 mL of 17*M*, glacial acetic acid to 1.0 L of distilled water.
- To prepare 0.1*M* hydrochloric acid, add 8.4 mL of 12*M* HCl to 1.0 L of distilled water.

Answers to Questions
1. The ions on the typing paper were easier to separate.
2. The HCl separated more easily. It had a lower pH, indicating a greater concentration of H₃O⁺ ions.
3. The HCl is the stronger acid. The acids were of the same concentration,

477

ISSUE:
17•3 Acid Rain

Prepare

Section Background
The term *acid rain* was first used in 1872 by Angus Smith in England.

1 Motivate

Bellringer

Before presenting the lesson, display **Section Focus Transparency 68** on the overhead projector. Assign the accompanying **Focus Activity** worksheet. L1 LEP

Tying to Previous Knowledge
Ask students where they have observed the effects of acid rain.

2 Teach

GLENCOE TECHNOLOGY

Videodisc
STVS: Ecology
Disc 6, Side 2
Acid Rain and Plants (Ch. 10)

Fish and Acid Rain (Ch. 11)

Science Words
acid rain
plankton

Objectives
- Use pH units to define *acid rain.*
- Describe the factors contributing to the formation of acid rain.
- Discuss the effects of acid rain, and evaluate methods of controlling this problem.

Figure 17-11
Many industrialized countries use coal as their main source of energy. As coal burns, so does the sulfur that it contains.

A The sulfur dioxide oxidizes and combines with moisture in the air to form sulfuric acid droplets. In addition, nitrogen oxide from automobile exhaust combines with moisture in the air to make dilute solutions of nitric acid.

How can rain be an acid?
Popping up umbrellas, grabbing raincoats, and running for shelter are all common occurrences when a rainstorm approaches. We all seek protection from an uncomfortable drenching. However, rain may be causing more problems for us than getting damp. Unpolluted rain typically has a pH value of 5.6, which is slightly acidic, but not harmful. However, what happens if the pH drops below that level?

If rain in your location has a pH of 4.6, just one unit lower than normal rain, that may not sound like a problem. However, the math used in the pH scale reveals that 4.6 is ten times more acidic than 5.6. Any rain that is below 5.6 is considered **acid rain.** Although acid rain has been documented since the 1870s near large cities in England, only since the 1970s have people in North America become concerned about the causes. **Figure 17-11** illustrates the process that makes acid rain.

Come in Out of the Rain
What could be some of the effects of rain that is more than 100 times more acidic than normal rain? As seen in the chapter opening photograph, acid rain can dissolve marble buildings and statues. Some of the effects are not as easy to notice. Among the first organisms to suffer are plankton. **Plankton** are aquatic plants and animals that form the base of the food chain for small fish. As plankton die, the fish that feed on them eventually feel the effects. If soil near a polluted area becomes more acidic, nutrients and metals may dissolve away. These become unavailable for proper plant growth. These effects are now being noticed in some forest areas.

478 Chapter 17 Acids, Bases, and Salts

Program Resources

Reproducible Masters
Study Guide, p. 72 L1
Reinforcement, p. 72 L1
Enrichment, p. 72 L3
Science and Society Integration, p. 21
Lab Manual 36
Critical Thinking/Problem Solving, p. 23

Transparencies
Section Focus Transparency 68 L1
Science Integration Transparency 17

2 Points of View

No Easy Solutions

Preventing sulfur oxides and nitrogen oxides from reaching the atmosphere is expensive. Chemical attachments, called scrubbers, on smokestacks reduce the amount of these gases. However, power companies would pass the cost to consumers. Using nuclear fuel also would result in less sulfur oxides. However, do we want a nuclear waste problem? Exhaust from automobiles contributes to the formation of acid rain. Are we willing to give up the convenience of personal automobiles to lower acid rain levels?

Pay Now or Pay Later

Reversing the effects of acid rain is already costing nations tens of millions of dollars. Developing cleaner fuel technology must be a research priority in order to save soils, lakes, and crops for the future. Air pollutants released in one country drift to distant environments. International agreements about emissions would help reduce the overall level of pollutants that later contribute to acid rain.

C The acid rain then falls and begins to deteriorate buildings and pollute water.

B This sequence of reactions shows one process by which acid rain forms.

$$S(cr) + O_2(g) \longrightarrow SO_2(g)$$
from burning coal containing sulfur

$$2SO_2(g) + O_2(g) \longrightarrow 2SO_3(g)$$
converting to sulfur trioxide

$$SO_3(g) + H_2O(l) \longrightarrow H_2SO_4(aq)$$
dissolving sulfur trioxide to form sulfuric acid

Section Wrap-up

Review

1. How low must the pH level of rain drop before most scientists classify it as acid rain?
2. What are the steps in burning coal that can lead to the formation of acid rain?

You have read about some of the effects of acid rain. How else do you think acid rain affects the environment? What can be done? Visit the Chapter 17 Internet Connection at Glencoe Online Science, **www.glencoe.com/sec/science/physical,** for a link to more information about acid rain.

SCIENCE & SOCIETY

Community Connection

Fossil Fuels Invite operators of local power plants that use fossil fuels to visit class and describe the methods used at their plants to control harmful emissions.

Prepare

Section Background

- When an alcohol reacts with either an organic acid or an organic acid anhydride, an ester is formed.
- An ester can be split into an alcohol and carboxylic acid by the addition of water, called hydrolysis.
- If a metallic base is used instead of water, the metal salt of the carboxylic acid is obtained, not the acid. This process is called saponification. Soap is a metal salt of a fatty acid. The natural fat or oil is an ester.

Preplanning

- To prepare for Activity 17-2, purchase clear carbonated beverages. Prepare a 0.20M NaOH solution.
- To prepare for the Mini-LAB, obtain fruit juices, cornstarch, and an iodine solution.

1 Motivate

Bellringer

Before presenting the lesson, display **Section Focus Transparency 69** on the overhead projector. Assign the accompanying **Focus Activity** worksheet. L1 LEP

Tying to Previous Knowledge

Ask students whether they have ever taken something to relieve stomach discomfort. Stomach remedies contain a base to neutralize the hydrochloric acid.

17•4 Acids, Bases, and Salts

Science Words
neutralization
salt
titration
soap
saponification
detergent
ester

Objectives
- Describe a neutralization reaction.
- Explain what a salt is and how salts form.
- Differentiate between soaps and detergents.
- Explain how esters are made and what they are used for.

Neutralization

You have probably seen television commercials for antacids that describe how effectively these products neutralize excess stomach acid. Would the pH of such a product be higher or lower than 7? If you answered higher, you are correct. Only a base can neutralize an acid.

Neutralization is a chemical reaction between an acid and a base. During a neutralization reaction, hydronium ions from the acid combine with hydroxide ions from the base to produce water.

$$H_3O^+(aq) + OH^-(aq) \rightarrow 2H_2O(l)$$

As the reactive hydronium and hydroxide ions combine to form water, the acidic and basic properties of the reactants are canceled, or neutralized.

The equation above accounts for only half of the ions present in the solution. What happens to the remaining ions? They react to form a salt. A **salt** is a compound formed when the negative ions from an acid combine with the positive ions from a base.

Neutralization reactions are ionic. The following equations show what happens to all of the ions during a neutralization reaction.

$$H_2O(l) + HCl(aq) + NaOH(aq) \rightarrow$$

$$H_3O^+(aq) + Cl^-(aq) + Na^+(aq) + OH^-(aq)$$

Water Formation: $H_3O^+(aq) + OH^-(aq) \rightarrow 2H_2O(l)$

Salt Formation: $Na^+(aq) + Cl^-(aq) \rightarrow NaCl(cr)$*

*Note: The formation of crystalline NaCl occurs only if the water is removed by boiling or evaporation.

Salts

Many substances that you come in contact with every day are salts. Of course, the most familiar salt is sodium chloride—common table salt. **Table 17-4** on page 482 contains information about some common salts. As the table shows, most salts are composed of a metal and a nonmetal other than oxygen, or a metal and a polyatomic ion. Ammonium salts contain the polyatomic ammonium ion, NH_4^+, rather than a

---

Program Resources

 **Reproducible Masters**
Study Guide, p. 73 L1
Reinforcement, p. 73 L1
Enrichment, p. 73 L3
Cross-Curricular Integration, p. 23
Activity Worksheets, pp. 5, 98, 99, 101 L1
Technology Integration, pp. 15, 16

Transparencies
Section Focus Transparency 69 L1

metal. Salts also form when acids react with metals. Hydrogen gas is released during such reactions.

Acid + Metal →
        Salt + Hydrogen

$$H_2SO_4(aq) + Zn(cr) →$$
$$ZnSO_4(aq) + H_2(g)$$

$$6HCl(aq) + 2Fe(cr) →$$
$$2FeCl_3(aq) + 3H_2(g)$$

Reactions between metals and acids are single-displacement reactions in which the metal displaces hydrogen from the acid.

Titration

It is sometimes necessary to find the concentration of an acidic or basic solution. This can be accomplished by titration. **Titration** is the process in which a solution of known concentration is used to determine the concentration of another solution. **Figure 17-12** on page 482 shows a titration experiment.

In a titration, a solution of known concentration, called the standard solution, is added to a solution of unknown concentration to which an indicator has been added. If the solution of

Stomach Acid		Common Antacid Bases	
		Ca(OH)$_2$	
HCl	+	Al(OH)$_3$	⟶ NEUTRALIZATION
		NaHCO$_3$	

Problem Solving

An Upsetting Situation

Most of us have, at some time, experienced an upset stomach. Often, the cause is the presence of excess acid within our stomach. For digestive purposes, our stomachs contain dilute hydrochloric acid. A doctor might recommend an antacid treatment for an upset stomach. What type of compound would be "anti acid"?

You have learned that neutralization reactions change acids and bases into salts. The ingredients of antacids typically contain small amounts of $Ca(OH)_2$, $Al(OH)_3$, or $NaHCO_3$, which are bases. While an excess acid condition lowers the pH of your stomach contents, these compounds raise the pH.

Solve the Problem:

1. **What compounds would be produced from a reaction of HCl and Ca(OH)$_2$?**
2. **Why is it important for there to be some acid in your stomach?**

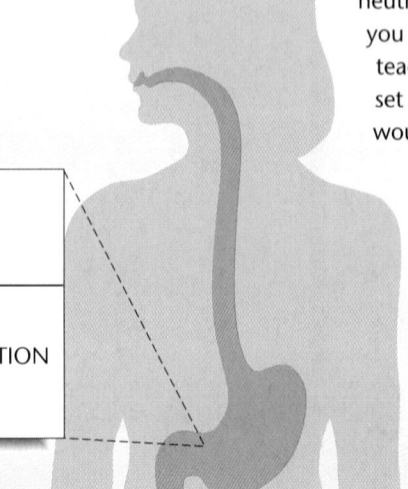

Think Critically:

If you wanted to know which antacid could neutralize the most acid, you could ask your teacher to help you set up a test. How would you know when the acid in a sample was neutralized? What could you do to make your comparison a fair test?

2 Teach

Demonstration

 **Visual-Spatial** Dissolve a common stomach acid remedy in warm water. Add a few drops of universal indicator. Add drops of dilute 0.1M HCl while stirring. Use this demonstration to introduce the idea of neutralization.

Problem Solving

Solve the Problem
1. H$_2$O, CaCl$_2$
2. to digest food

Think Critically
Use bromothymol blue in the acid. When it turns a pale green, the acid is neutralized. The most effective antacid would be the one that changes the color of the acid solution using the smallest amount of antacid.

 Videodisc
STV: Human Body Volume 1
Digestive System
Unit 2, Side 2
Stomach

13799-17725

Theme Connection

Systems and Interactions The fact that a salt and water are produced when an acid reacts with a base reinforces the theme of interactions. The fact that similar products are produced in similar reactions reinforces the theme of systems.

Science Journal

Neutralization Reactions Have students write a brief paragraph in their Science Journals that explains the fact that neutralization reactions are ionic. Chemical equations may be used to show examples of what they are writing about. LS

When an acid reacts with a base, a salt and water are the products. Although the term *neutralization* is applied to the reaction, the resulting salt solution isn't always neutral. It is neutral when the acid and base are both strong or both weak. When a strong acid neutralizes a weak base, the salt solution has an acidic pH. When a strong base neutralizes a weak acid, the salt solution has a basic pH.

Demonstration

Visual-Spatial Show students that equal volumes of strong acid and strong base of the same concentration produce a neutral solution. Use an overhead projector and a small beaker. Place 50 drops of $0.1M$ HCl and a drop of universal indicator in the beaker. Then slowly add, with swirling, 50 drops of $0.1M$ NaOH. Have students match the final color with the chart that comes with universal indicator.

Teacher F.Y.I.

Alkaloids are naturally occurring bases that contain nitrogen. Some alkaloids are listed below.

Quinine—found in cinchona bark

Morphine—narcotic found in opium poppy flower

Caffeine—found in coffee beans and tea leaves

Nicotine—found in tobacco leaves

Mescaline—found in peyote cactus

Visual Learning

Figure 17-12 How would you calculate how much standard solution you had added? *subtract the beginning reading on the buret from the end reading* **LEP** **LS**

Table 17-4

Some Common Salts			
Name	**Formula**	**Common Name**	**Uses**
Sodium chloride	NaCl	Salt	Food preparation; manufacture of chemicals
Sodium hydrogen carbonate	$NaHCO_3$	Sodium hydrogen carbonate (baking soda)	Food preparation
Calcium carbonate	$CaCO_3$	Calcite (chalk)	Manufacture of paint and rubber tires
Potassium nitrate	KNO_3	Saltpeter	In fertilizers; manufacture of explosives
Potassium carbonate	K_2CO_3	Potash	Manufacture of soap and glass
Sodium phosphate	Na_3PO_4	TSP	In detergents
Ammonium chloride	NH_4Cl	Sal ammoniac	In dry cells

unknown concentration is a base, a standard acid solution is used. If the unknown is an acid, a standard base solution is used.

The Endpoint Has a Color

As an example, assume that you need to find the concentration of an acid solution. First, you would add a few drops of phenolphthalein indicator to a carefully measured amount of this solution. Phenolphthalein is colorless in an acid, but is pink in a base.

A standard base solution is added to this acid. At some point, one drop of the base just begins to turn the acid solution pink. This is known as the endpoint of the titration. The volume of the base used to neutralize the known volume of acid is determined, and the concentration of the acid can be calculated.

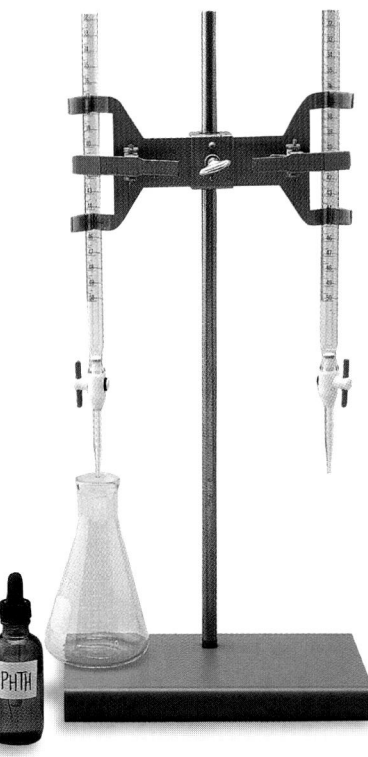

Figure 17-12

This apparatus is used for titration experiments. *How would you calculate how much standard solution you had added?*

Community Connection

Making Medicines Invite a chemist from a local pharmacy to class. Ask him or her to demonstrate procedures that are followed in dealing with acids and bases involved with preparation of medicines.

Use these program resources as you teach this lesson.
Cross-Curricular Integration, p. 23
Technology Integration, pp. 15, 16

Soaps and Detergents

Soaps are organic salts. Soaps are made by reacting fats or oils with sodium hydroxide or potassium hydroxide. One formula for soap made using sodium hydroxide is $C_{17}H_{35}COONa$. The structural formula for this soap is provided below.

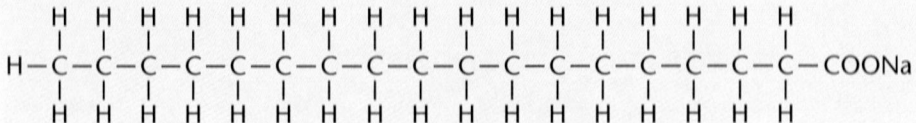

A soap made using potassium hydroxide might have the formula $C_{15}H_{31}COOK$. Below is the structural formula for this soap.

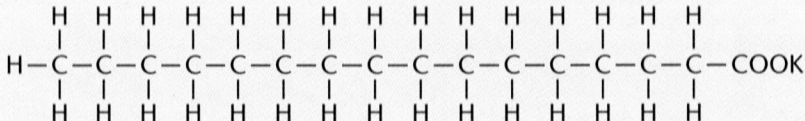

Sodium hydroxide produces solid soaps. Potassium hydroxide makes liquid soaps. Another product of this reaction is glycerin, a softening agent in hand creams.

The process of making soap is called **saponification. Figure 17-13** shows how soap works to clean your hands.

Detergents are organic salts having structures similar to those of soaps. The most common detergent in use today has the structural formula shown on page 484.

Figure 17-13

Soap helps the grease and oil on your hands mix with water so they can be carried away when you rinse your hands.

A Soaps have a long, nonpolar hydrocarbon structure like a tail, and an ionic head.

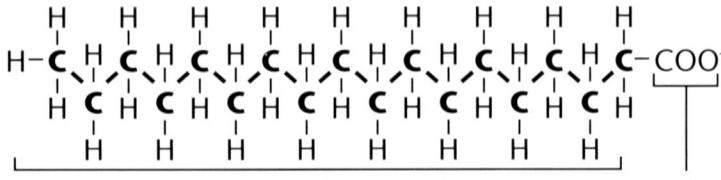

water molecules

oil molecules

nonpolar hydrocarbon tail

ionic head

B Grease and oil on your skin are nonpolar hydrocarbons that do not easily mix with polar water.

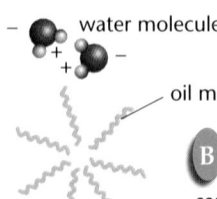

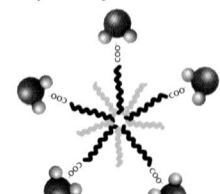

C The nonpolar tail of soap molecules can easily mix with the nonpolar grease and oil molecules, while the ionic head interacts with water.

17-4 Acids, Bases, and Salts   **483**

Cultural Diversity

Laundry List Soap has been known for more than 2000 years. In 600 B.C., soap was so valued that the Phoenicians used it for barter with other groups such as the Gauls. The Phoenicians made soap by mixing goat tallow (melted fat) and wood ashes. Ancient Mediterranean peoples and the Celts of Britain also developed soaps. In fact, the English word *soap* came from the Celtic word *saipo.*

Inquiry Questions

- **What is the purpose of performing a titration?** *to determine the concentration of an unknown solution*
- **What is an endpoint in a titration experiment?** *point at which just enough acid has been added to react with the base to produce color change in an indicator*

Demonstration

LS **Visual-Spatial** Show students how soap is made by placing 25 g of solid vegetable shortening in a 250-mL beaker. Add 10 mL of ethanol and 5 mL of 6*M* NaOH (6 g NaOH in 25 mL water). Heat the mixture using a hot plate and stir for 15 minutes. **CAUTION:** *Ethanol is flammable.* A small piece of colored crayon can be added to make colored soap. Cool the mixture in an ice-water bath. Add 25 mL of water and 25 mL of saturated NaCl solution. The soap will appear as curds. Collect the soap by filtering it through cheesecloth and pressing it into an evaporating dish to mold it. Allow the soap to dry and harden for several days.

LS **Visual-Spatial** To show an effect of soap, pour 3.5% butterfat milk 0.5 cm deep in a petri dish. Place a drop each of four different food colors in four different areas of the milk. Dip a toothpick into liquid dish detergent and then touch the surface of the milk in the center of the petri dish. Hold the toothpick in the milk, but don't touch the bottom of the dish. The detergent affects the fat in the milk to produce a colorful mixing.

LEP

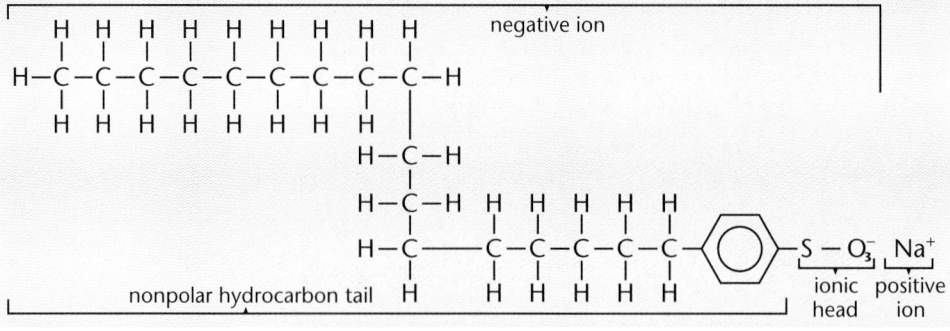

negative ion

nonpolar hydrocarbon tail

ionic head

positive ion

Like soap, the negative ions of detergents have a nonpolar hydrocarbon tail and an ionic head. Detergents enhance the cleaning action of water.

When soap is used in hard water, it forms soap scum. Hard water contains ions of calcium, magnesium, and iron that react with the soap to form a precipitate. Detergents do not form this scum. Therefore, most laundry products are detergents, not soaps.

Organic Acids and Esters

An **ester** is an organic compound formed by the reaction of an organic acid with an alcohol. The reaction requires a compound that will catalyze the formation of a molecule of water from the organic acid and alcohol. Concentrated sulfuric acid, a powerful dehydrating agent, is often present for the organic reaction to take place. **Figure 17-14** shows the reaction of acetic acid and methyl alcohol to produce an ester—methyl acetate—and water.

Esters are responsible for the many wonderful odors and flavors of flowers, fruits, and other foods. Sometimes, esters are added to gelatin desserts or candy to give the characteristic flavors of strawberry, banana, or apple.

acetic acid   methanol

Figure 17-14
This structural equation shows the formation of an ester.

methyl acetate
(an ester)

+ H_2O
water

MiniLAB

MiniLAB

How can fruit juices be tested for ascorbic acid, vitamin C?

Procedure

1. Measure 10 mL of each juice sample.
2. Add a pea-sized sample of cornstarch to each sample.
3. Using a pipet or dropper, carefully add 1 drop of iodine solution to each sample and stir. Add one more drop and stir. **CAUTION:** *Do not allow any contact with iodine.*

Analysis

1. Iodine reacts with both cornstarch and ascorbic acid. Iodine will react first with ascorbic acid. When iodine and cornstarch react, a dark color appears. How can you tell which of your samples contains the most ascorbic acid?
2. What common arrangement of carbon, oxygen, and hydrogen atoms would you expect to find in ascorbic acid?

Integrating the Sciences

Life Science Esters can be prepared by placing 6 mL of each reactant into a test tube with 5 drops of concentrated sulfuric acid. **CAUTION:** *Sulfuric acid is corrosive, and alcohols are flammable.* Heat the test tube in a warm-water bath. Pineapple smell can be created by reacting butyric acid with ethanol, apples by reacting acetic acid with ethanol, and wintergreen by reacting salicylic acid and methanol.

Synthetic fibers known as polyesters are made from an organic acid that has two –COOH groups and an alcohol that has two –OH groups. Because the two compounds form long chains with the ester linkage, a polymer results. This is shown in **Figure 17-15.**

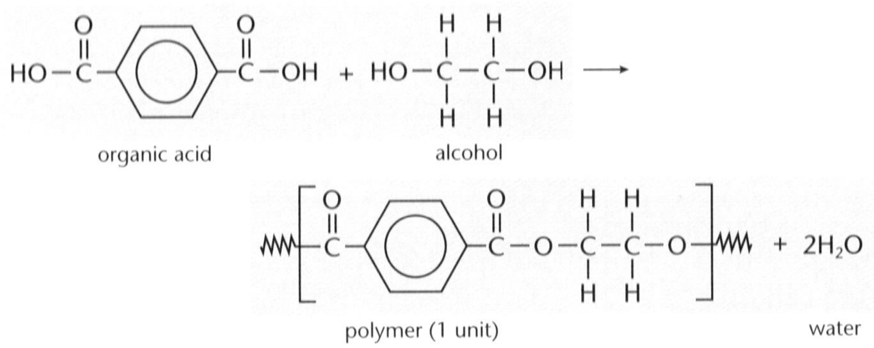

organic acid

alcohol

polymer (1 unit)

water

Figure 17-15

Polymers of this type are important to the textile industry today. Check a label on a piece of clothing you are wearing. *What percent of the garment is polyester?*

Section Wrap-up

Review

1. What is a neutralization reaction? What are the products of such reactions?

2. When doing laundry, what advantage does using a detergent have over using a soap?

3. **Think Critically:** Give the names and formulas of the salt that will form in these neutralizations:

   a. hydrochloric acid and calcium hydroxide

   b. nitric acid and potassium hydroxide

   c. carbonic acid and aluminum hydroxide

Skill Builder
Interpreting Data
Three salts—calcium sulfate, sodium chloride, and potassium nitrate—were obtained in reactions of the acid-base pairs shown here. Match each salt with the acid-base pair that produced it. If you need help, refer to Interpreting Data in the **Skill Handbook.**

   HCl + NaOH   HNO₃ + KOH   H₂SO₄ + Ca(OH)₂

Science Journal

In political and labor disputes, a mediator is often called in to bring opposing sides together. In your Science Journal, explain how soap could be thought of as a mediator for cleaning grease.

17-4 Acids, Bases, and Salts   **485**

Skill Builder
calcium sulfate from H_2SO_4 + $Ca(OH)_2$; sodium chloride from HCl + NaOH; potassium nitrate from KOH + HNO_3

Assessment

Performance Ask students to predict the formula of the salt formed when H_2SO_4 and $Al(OH)_3$ react. $Al_2(SO_4)_3$ Use the Performance Task Assessment List for Analyzing the Data in **PASC**, p. 27. **P**

Extension

For students who have mastered this section, use the **Reinforcement** and **Enrichment** masters.

Visual Learning

Figure 17-15 What percent of the garment is polyester? *Answers will vary.* **LEP** **LS**

4 Close

•MINI•QUIZ•

Use the Mini Quiz to check students' recall of chapter content.

1. **When an acid reacts with metal, what is produced?** *a salt and hydrogen*

2. **A fatty acid reacts with an inorganic base to produce _____.** *soap*

3. **What forms when an organic acid reacts with an alcohol?** *an ester*

Section Wrap-up

Review

1. Neutralization is a reaction between an acid and a base; the products are a salt and water.

2. Detergents do not produce soap scum.

3. **Think Critically a.** calcium chloride, $CaCl_2$; **b.** potassium nitrate, KNO_3; **c.** aluminum carbonate, $Al_2(CO_3)_3$

Science Journal

Entries should reflect how soap's polar and nonpolar ends enable unlike substances, such as water and grease, to be brought together.

Activity 17-2

Purpose

IS Interpersonal Students will design and carry out an experiment to show that a neutralization reaction can be used to find the level of acidity in a solution. **COOP LEARN** **L1**

Process Skills

observing, comparing, measuring, inferring, communicating, making and using tables, recognizing cause and effect, forming a hypothesis, designing an experiment, separating and controlling variables, interpreting data, measuring in SI

Time

45 minutes

Materials

To prepare 0.2*M* NaOH solution, dissolve 8.0 g of NaOH pellets in enough water to make 1.0 L of solution. **CAUTION:** *Sodium hydroxide is caustic. Do not touch the pellets or the solution.* Be sure the carbonated beverages are all freshly opened so that escape of CO_2 does not affect the acidity.

Safety Precautions

Caution students to avoid all contact with the NaOH solution. Do not allow students to drink the beverages.

Possible Hypotheses

• A neutralization reaction can be used as a method of comparing acid levels in solutions.

• The NaOH solution will neutralize the acidity of soft drinks. The number of drops of NaOH solution needed for neutralization is an indicator of acidity level in the soft drinks.

📂 **Activity Worksheets,** pp. 5, 98, 99

Activity 17-2

Design Your Own Experiment
Be a Soda Scientist

Carbonated soft drinks contain carbonic acid and sometimes phosphoric acid. You have learned that bases can neutralize acids. Using a proper indicator and a base solution, how could you compare the acidity levels in soft drinks?

PREPARATION

Problem

How can the acidity level of soft drinks be effectively compared?

Form a Hypothesis

Based on observations about acids and bases, develop a hypothesis about how neutralization reactions can be used to rank the acidity of soft drinks.

Objectives

• Observe evidence of a neutralization reaction using an indicator.
• Compare the acidity levels in soft drinks.
• Design an experiment that uses the independent variable of acid content of soft drinks, and the dependent variable of amount of base added to the soft drinks, to determine the acidity of the drinks.

Possible Materials

• 2 different colorless soft drinks
• 2 test tubes
• graduated cylinder
• 2 droppers
• 1% phenolphthalein indicator solution
• dilute NaOH solution

Safety Precautions

Sodium hydroxide is caustic. Wear safety eye protection and avoid any skin contact with the solution.

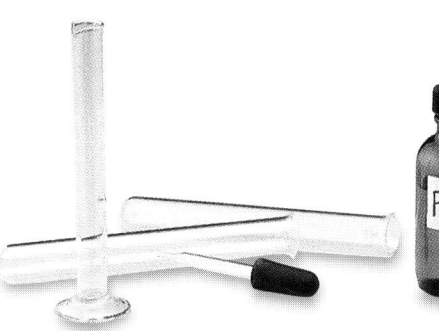

Inclusion Strategies

Learning Disabled Learning-disabled students could carry out their own titrations of vinegar, lemon juice, and orange juice to compare the acidity with that of carbonated beverages. Learning-disabled students might like to perform the suggested extension of this activity in which they compare the acidity of different brands and types of beverages.

PLAN THE EXPERIMENT

1. As a group, agree upon and write out the hypothesis statement.
2. As a group, in a logical manner, list the specific steps that you will use to test your hypothesis.
3. Make a list of all the materials that you will need to test your hypothesis.
4. Design a data table in your Science Journal that will allow you to record the amount of NaOH required to neutralize each soda sample.

Check the Plan

1. Did you control the amount of soda to be tested in each trial?
2. Is it important to repeat each trial?

3. Will you be able to test only colorless solutions with this procedure?
4. *Make sure your teacher approves your plan and that you have included any suggested changes.*

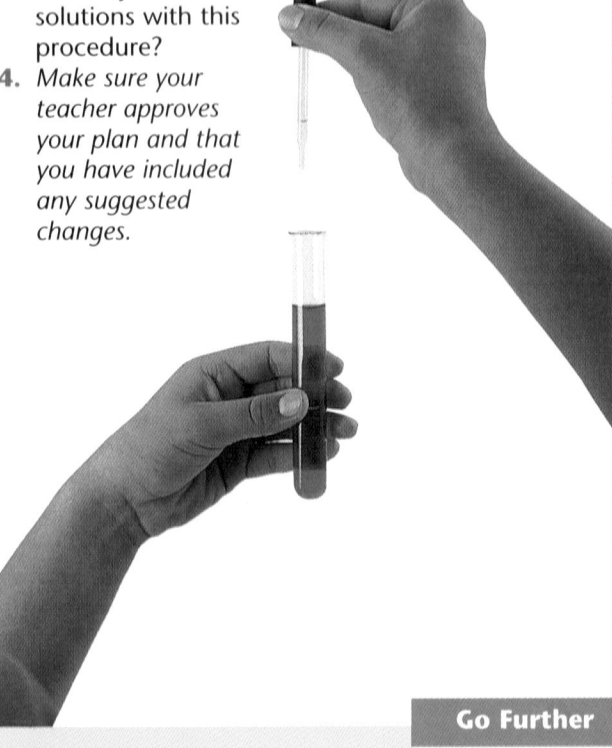

DO THE EXPERIMENT

1. What color change does phenolphthalein undergo in a solution that changes from an acid pH to a basic pH?
2. While the experiment is going on, write down any observations that you make and complete the data table in your Science Journal.

Analyze and Apply

1. **Compare** your soda acid rankings with other class groups.
2. Can your rankings be compared if other groups used a different amount of soda from your group?

Go Further

Dissolving CO_2 in water causes the formation of carbonic acid in soda. At warmer temperatures, less CO_2 dissolves. How would comparing the acidity of two sodas stored at different temperatures affect the validity of the comparison?

17-4 Acids, Bases, and Salts   **487**

Go Further

If the two soft drinks were produced at the same temperature and sealed, storing them at different temperatures should not drastically affect the comparison, if they are tested at the same temperature. However, if the two soft drinks were opened and tested at different temperatures, the levels of carbonic acid would differ.

 **Assessment**

Performance Ask one student from each group to demonstrate the procedures his or her group used to compare the soft drinks. Use the Performance Task Assessment List for Designing an Experiment in **PASC,** p. 23. **P**

PLAN THE EXPERIMENT

Possible Procedures
- Be sure students mix the contents well after each drop of NaOH is added.
- The pink color can best be seen if the test tubes are held against a piece of white paper.
- For most carbonated beverages, approximately 5 mL of beverage requires 15-25 drops of NaOH solution.

Teaching Strategies
Troubleshooting Be sure to try the activity ahead of time to make sure that the two beverages chosen have a measurable difference in acidity.

DO THE EXPERIMENT

Expected Outcome
Results will show that the acidity level of soft drinks affects the number of drops of NaOH solution needed to neutralize the soft drinks.

1. PHTH changes to pink from colorless.
2. Answers will vary but should include the change in color and how many drops of NaOH solution are needed to produce the change.

Analyze and Apply
1. Answers should be similar between groups.
2. Yes. Their rankings should be the same. The number of NaOH drops needed may vary, but the relationship would be the same.

Background

- Here's how Mr. Abrahamson describes the transformation of wood chips into wood pulp. "A ton of chips are put into a digester, which is a sort of pressure cooker about 50 feet high and 20 feet in diameter. Sodium hydroxide and sodium sulfide, the cooking 'liquor,' are added. Then the digester is capped off, and steam is injected to achieve a pressure of about 150 pounds per square inch. Two to three hours of cooking dissolves the lignin, which is a sort of glue that holds together the cellulose fibers of wood. When the batch of pulp is suddenly blown out of the digester, the release of pressure literally explodes the chips into fibers. That original ton of wood chips is reduced to about one-half ton of wood pulp. The solid residue is ash and a black, tarry mass that are burned in a furnace to generate steam for making another batch of wood pulp."

- The basics of papermaking haven't changed since the first known papermaker, a Chinese man named T'sai Lun, made paper out of pieces of hemp rope and old fishing nets. Both his handmade paper and a giant roll of paper that streams from a modern paper machine are produced by pulping plant fibers and then removing the water.

- In an experiment, paper made from kenaf was recently used to print several U.S. newspapers. This paper performed satisfactorily on the presses. Its strength and brightness were higher than necessary, and the opacity and print-through were below normal. Scientists think that

488

People and Science

DENNIS ABRAHAMSON,
Paper Mill Technical Director

On the Job

Q What is your job at the paper mill?

A I'm responsible for environmental compliance, which includes testing for pH, conductivity, and total dissolved solids in the wastewater produced in papermaking.

Q How does pH affect paper?

A Neutral or slightly alkaline paper can last for hundreds of years. Traditionally, paper is manufactured under acid conditions, with a pH of 4.5 to 5.0. It's the residual acid in the sheet that eventually causes it to break down. Of course, many kinds of paper, like newsprint, don't need to last a long time.

Q How has paper manufacturing changed during your career?

A Today, most paper is made from recycled materials because of state and federal mandates. The only nonrecycled paper is for uses that require very white paper, such as art books, and may have a brightness reflectance of up to 96. Newly fallen snow probably registers at 100. Paper mills make white paper that is anywhere from 78 to 96 on the scale.

Personal Insights

Q Are there new materials for paper?

A I've seen some samples made from kenaf, a type of hibiscus plant. Our company also consults with artists who make handmade artistic and ornate paper. The Japanese make paper out of rice.

Q How did you get interested in the paper business?

A My college major was forest science, and we took a field trip to a paper mill. I guess I was impressed by the immensity of the equipment. Our company's paper machine is comparatively small, at about 10 feet wide and 300-400 feet long, but it turns out over 100 tons of paper each day.

Career Connection

Contact your state's department of natural resources and ask about careers having to do with paper production or environmental management. Make a poster detailing the career options and the education requirements for each career. Start with:

- **Environmental Engineer**
- **Forest Scientist**

kenaf shows promise as newsprint pulp and could be especially useful in countries that lack adequate trees.

Teaching Strategies

LS **Logical-Mathematical** Encourage students to test the pH of several kinds of paper found in the classroom. They might also create a chart to rank what they consider the ideal longevity of the paper used for encyclopedias, historic documents, textbooks, magazines, advertising flyers, and newspapers. L1

Career Connection

Career Path Management positions in the paper industry generally require a college degree in paper science. High school and college courses in math, physics, chemistry, and other branches of science are required.

For More Information

Students could ask a librarian how to contact the public information office of a large paper-manufacturing plant.

Summary

17-1: Acids and Bases

1. An acid is a substance that produces hydrogen ions, H^+, in solution. A base produces hydroxide ions, OH^-, in solution.
2. Properties of acids and bases are due, in part, to the presence of the H^+ and OH^- ions.
3. Common acids include hydrochloric acid, sulfuric acid, nitric acid, and phosphoric acid. Common bases include sodium hydroxide, calcium hydroxide, and ammonia.
4. Acidic solutions form when certain polar compounds ionize as they dissolve in water. Except for ammonia, basic solutions form when certain ionic compounds dissociate upon dissolving in water.

17-2: Strength of Acids and Bases

1. The strength of an acid or base is determined by how easily it forms ions when in solution.
2. Strength and concentration are not the same thing. Concentration involves the relative amounts of solvent and solute in a solution.
3. pH is a measure of the hydronium ion concentration of a solution.
4. For acidic solutions of equal concentration, the stronger the acid, the lower its pH; for basic solutions of equal concentration, the stronger the base, the higher the pH.

17-3: Science and Society: Acid Rain

1. Acid rain is produced by substances in the air reacting with rainwater to make it acidic. Acid rain production is increased by the release of certain oxides into the air when fossil fuels are burned.
2. Acid rain is harmful to plants and animals and increases the rate of weathering.

17-4: Acids, Bases, and Salts

1. In a neutralization reaction, the H_3O^+ ions from an acid react with the OH^- ions from a base to produce water molecules. The products of neutralization are a salt plus water.
2. Salts form by negative ions from an acid combining with the positive ions from a base.
3. Soaps and detergents are organic salts. Unlike soaps, detergents do not react with compounds in hard water to form scum deposits.
4. Esters are organic compounds formed by the reaction of an organic acid and an alcohol.

Key Science Words

a. acid
b. acid rain
c. base
d. dehydrating agent
e. detergent
f. ester
g. hydronium ion
h. indicator
i. neutralization
j. pH
k. pickling
l. plankton
m. salt
n. saponification
o. soap
p. strong acid
q. strong base
r. titration
s. weak acid
t. weak base

Summary

Have students read the summary statements to review the major concepts of the chapter.

Reviewing Vocabulary

1. g	**6.** m
2. i	**7.** k
3. p	**8.** j
4. t	**9.** a
5. d	**10.** n

✓Assessment

Portfolio Encourage students to place in their portfolios one or two items of what they consider to be their best work. Examples include:

- Flex Your Brain, p. 472
- Activity 17-1 results and answers, p. 477
- Explore the Issue responses, p. 479. **P**

Performance Additional performance assessments may be found in **Performance Assessment** and **Science Integration Activities.** Performance Task Assessment Lists and rubrics for evaluating these activities can be found in Glencoe's **Performance Assessment in the Science Classroom.**

GLENCOE TECHNOLOGY

📼 MindJogger Videoquiz

Chapter 17 Have students work in groups as they play the Videoquiz game to review key chapter concepts.

Checking Concepts

1. a   6. a
2. a   7. c
3. b   8. b
4. d   9. d
5. a   10. c

Understanding Concepts

11. HCl ionizes, forming H$^+$ and Cl$^-$ ions.
12. In alcohol, OH forms the hydroxyl group, which is not ionized. In a base, OH forms the hydroxide ion, OH$^-$.
13. In water, ammonia molecules react with water molecules, producing ammonium ions and hydroxide ions.
14. A concentrated acid contains more solute and less water than does a dilute acid. The terms have nothing to do with how much an acid dissociates in water, which is the measure of its strength.
15. Ashes were used as the source of hydroxide ions. These ashes were mixed with animal fat to make soap.

Thinking Critically

16. Acids that you eat are weak. Strong acids can cause severe burns.
17. The pH of the concentrated solution would be lower (more acidic) than that of the dilute solution. In the dilute solution, the hydronium ions are less concentrated than in the concentrated solution.
18. hydrogen gas and iron(II) sulfate
19. calcium hydroxide, Ca(OH)$_2$, and carbonic acid, H$_2$CO$_3$
20. You would use a detergent because a detergent will not react with minerals in the water to form a scum (precipitate).

490

Reviewing Vocabulary

Match each phrase with the correct term from the list of Key Science Words.

1. a hydrogen ion bonded to a water molecule
2. a reaction between an acid and a base
3. acid that ionizes completely in solution
4. base that partly dissociates in solution
5. can remove water from materials
6. formed by negative ions from an acid combining with positive ions from a base
7. process that removes impurities from metal surfaces
8. measures the concentration of H$_3$O$^+$ ions
9. any solution that has a pH less than 7
10. the process of making soap

Checking Concepts

Choose the word or phrase that completes the sentence.

1. Solutions of equal concentrations of HCl and CH$_3$COOH _____.
   a. will not have the same pH
   b. will react the same with metals
   c. will make the same salts
   d. will have the same amount of ionization
2. Sulfuric acid is also known as _____.
   a. battery acid   c. stomach acid
   b. citric acid   d. vinegar
3. Most sulfuric acid is used to produce _____.
   a. batteries   c. petroleum products
   b. fertilizer   d. plastics
4. _____ is used to make phosphoric, nitric, and hydrochloric acids.
   a. Dehydration   c. Pickling
   b. A base   d. Sulfuric acid
5. The most widely used base is _____.
   a. ammonia   c. lye
   b. caustic lime   d. milk of magnesia
6. Carrots have a pH of 5.0, so carrots are _____.
   a. acidic   c. neutral
   b. basic   d. an indicator

490   Chapter 17 Acids, Bases, and Salts

7. Pure water has a pH of _____.
   a. 0   b. 5.2   c. 7   d. 14
8. Certain materials can act as indicators because they change _____.
   a. acidity   c. concentration
   b. color   d. taste
9. KBr is an example of a(n) _____.
   a. acid   c. indicator
   b. base   d. salt
10. You might use a solution of _____ to titrate an oxalic acid solution.
   a. HBr   c. NaOH
   b. Ca(NO$_3$)$_2$ d. NH$_4$Cl

Understanding Concepts

Answer the following questions in your Science Journal using complete sentences.

11. When hydrogen chloride, HCl, is dissolved in water to form hydrochloric acid, what happens to the HCl?
12. Explain how the hydroxide ion differs from the –OH group in an alcohol.
13. Why is ammonia considered a base, even though it contains no hydroxide ions?
14. Explain why a concentrated acid is not necessarily a strong acid.
15. Ashes from a wood fire are basic. Explain how early settlers used this fact when making soap.

Thinking Critically

16. Explain why you should never use taste to test for an acid, even though acetic, citric, and dilute phosphoric acids are in things you eat and drink.
17. How would the pH of a dilute solution of HCl compare with the pH of a concentrated solution of the same acid? Explain.
18. Acid rain containing sulfuric acid, H$_2$SO$_4$, will react with iron objects. What products will result from this reaction?

Developing Skills

21. **Making and Using Tables** Tables may vary, but should resemble the following.

	Physical Properties	**Chemical Properties**
Acid	sour taste, electrolyte	corrosive, reacts with indicators, forms H$_3$O$^+$ in solution
Base	bitter taste, feels slippery, electrolyte	corrosive, reacts with indicators, forms OH$^-$ in solution

19. Chalk, $CaCO_3$, is a salt. What acid and what base react to form this salt?
20. Suppose you have hard water in your home. Would you use a soap or a detergent for washing your clothes and dishes? Explain your answer.

Developing Skills

If you need help, refer to the **Skill Handbook.**

21. **Making and Using Tables:** Make a table that lists the chemical and physical properties of acids and bases.
22. **Recognizing Cause and Effect:** Complete the table below by describing the cause and effect of each change listed in the left-hand column.
23. **Comparing and Contrasting:** Compare and contrast the reactions that would be produced by pouring sulfuric acid on a piece of paper and burning a marshmallow.
24. **Observing and Inferring:** You have equal amounts of three colorless liquids: A, B, and C. You add several drops of phenolphthalein to each liquid. A and B remain colorless, but C turns pink. Next, you add some of liquid C to liquid A and the pink color disappears. You add the rest of C to liquid B and the mixture remains pink. What can you infer about each of these liquids? Which liquid probably has a pH of 7?

25. **Interpreting Data:** A soil test indicates that the pH of the soil in a field is 4.8. To neutralize the soil, would you add a substance containing H_3PO_4 or one containing $Ca(OH)_2$? Explain your answer.

Performance Assessment

1. **Display:** Make a display and write a report about common household acids and bases, how they are used, and cautions to be taken.
2. **Lab Report:** Do research to find out how acids are used in making fertilizer. Name several acids and tell why each is needed and how their proportions may vary from fertilizer to fertilizer. Conduct an investigation by growing seedlings using different fertilizers. Describe your results in a report.
3. **Analyzing the Data:** In Activity 17-2 on pages 486 and 487, you neutralized an unknown acid solution with a basic solution. If you found that 1 mL of the NaOH would neutralize 5 mL of the carbonated beverage, which is more concentrated, the acid or the base? Explain in your Science Journal.

Sample Data

Chemical Changes

Process	Cause	Effect
Saponification	NaOH or KOH and fat react	Soap and glycerine form
Esterification	Organic acid and alcohol react	Ester and water formed
Acid rain	Pollutants from burning fossil fuels react with water in air	Destruction of life, soil, metal, minerals, buildings
Neutralization	Acid and base react	Salt and water formed
Corrosion	Metal reacts with a corrosive agent	Metal forms a compound and loses strength

22. **Recognizing Cause and Effect** See annotated student page.
23. **Comparing and Contrasting** Both processes involve dehydration of an organic material, removing the water and leaving only carbon. Thus, both final products are dark in appearance. They differ in that acid is used as a dehydrating agent on the paper, and heat drives off the water in the marshmallow.
24. **Observing and Inferring** Liquid C is a base because it turns phenolphthalein pink. Liquids A and B are not bases. Liquid A is an acid because it turns the pink base solution colorless. Liquid B is probably not acid and is most likely to have a pH of 7.
25. **Interpreting Data** You would use $Ca(OH)_2$. The soil has a pH of 4.8, indicating that it is acidic. To neutralize this acidic soil, a base should be added.

Performance Assessment

1. Answers will vary. Be sure reports include how the materials are used and what cautions should be taken. Use the Performance Task Assessment List for Display in **PASC**, p. 63. **P**
2. Be sure students use up-to-date research and follow accepted scientific procedures while conducting investigations. Use the Performance Task Assessment List for Lab Report in **PASC**, p. 47. **P**
3. The basic NaOH solution is more concentrated. Use the Performance Task Assessment List for Analyzing the Data in **PASC**, p. 27. **P**

Assessment Resources

📁 **Reproducible Masters**

Chapter Review, pp. 37, 38
Assessment, pp. 109–112
Performance Assessment, p. 31

Glencoe Technology

⊙ **Chapter Review Software**
⊙ **Computer Test Bank**
▭ **MindJogger Videoquiz**

Objectives

LS **Visual-Spatial** Students will gather evidence that catalysts have a special role in changing the rate of a reaction. They will control variables as they test for the presence of a catalyst and make a model of a generalized enzyme reaction. **L1** **COOP LEARN**

Summary

Enzymes are made of proteins. In the body, enzymes act as biological catalysts, with each enzyme catalyzing a specific reaction. Using common food items, students will develop an investigation that shows the role the catalyst protease plays in breaking down protein. After comparing the results to a control, students will infer what has taken place. Then, students will construct a model that proposes a lock-and-key mechanism for a generalized, enzyme-catalyzed reaction.

Time Required

This project would best be introduced when catalysts are discussed in Chapter 16. If this project is to be done in class, students will require part of one class period to prepare the gelatin dessert. An additional class period will be needed to test the gelatin with the other materials. Students will need some time to research other catalyzed reactions. One class period will be required to model the enzyme reaction.

Preparation

- Obtain dry gelatin mix and materials to be tested.
- Arrange with the school cafeteria or home arts department for refrigerator access.
- Provide poster board, scissors, and markers for students to use in preparing the enzyme model.

UNIT PROJECT 5

Enzymes: Your Body's Catalysts

As you read this, think of the last meal that you ate. The meal may have included some foods that are similar in makeup, such as bread and potatoes. It may have included some foods that are quite different, such as meat and vegetables. How is your body able to break down such different, complex food materials into simple, useful products that your body can use? The chemicals responsible for this relatively quick process are produced by your body and are called enzymes. Enzymes are part of a broader classification of chemicals called catalysts.

A Catalyzed Reaction

Enzymes are typically named after a specific reaction and given an *-ase* ending. For example, *protease* refers to enzymes that break down proteins. Proteases are found in some fresh fruits. Meat-tenderizing products also contain proteases.

Purpose
In this project, investigate the ways in which catalysts change reaction rates. You will find out more about the importance of these interesting chemicals that bring about change without being permanently changed themselves.

Procedure

1. Using the printed instructions on the box, prepare a gelatin dessert and divide it into five samples.

2. Place a piece of freshly cut kiwi fruit on the first sample, freshly cut pineapple on the second, canned pineapple on the third, some meat tenderizer on the fourth, and use the other sample as a control.

3. After one hour, examine the gelatin to note any changes near the fruits and tenderizer areas.

4. As the protein in the gelatin breaks down, water from the gelatin is released. Make a chart showing the degree of protein breakdown in each sample.

5. Which of your samples showed the most protein breakdown in one hour? What could cause canned and fresh pineapple to have different results? What are the general breakdown products of proteins?

Using Your Research

Do other fruits contain protease enzymes? Design a testing procedure to determine the relative levels of protease in fruits supplied by classmates.

Go Further

Just as opening a lock is most effectively done with a specific key, enzymes are able to effectively change the rate of specific reactions. In fact, the fit between a lock and key is often used as a model to explain enzyme activity.

Model Molecule

You can illustrate an enzyme model by cutting the shapes shown below from a piece of poster board.

A and B represent two molecules that will react when the two dotted areas are close. That particular positioning takes place more quickly when the enzyme molecule is present. The enzyme enables A and B to fit together. Once A and B are attached to the enzyme, the dots are aligned and A attaches to B. The enzyme then releases and is free to attach to other A and B molecules.

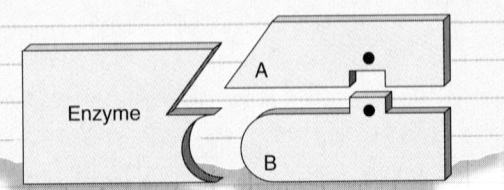

Unit 5 Project **493**

Go Further

The lock-and-key model can be constructed in various ways. It should be used as the group explains how the activity of a catalyst may depend on forming some attachments that more effectively align the reactants to allow them to bond more easily.

Reference

Metzler, David E. *Biochemistry.* New York, NY: Academic Press Inc., 1977.

Teaching Strategies

- The enzyme protease, which is found in some fruits, speeds up the breakdown of proteins. Water is trapped when the proteins bond together during the gel formation. Evidence for the breakdown of the protein structure is the release of the trapped water.

- The structure and bonding within an enzyme can be changed by heating, exposure to acids and bases, or oxidation. Any of these factors could be responsible for changing the structure of the protease in canned pineapple. Without its structure intact, an enzyme will lose its catalytic activity.

- A general reaction that involves a catalyst can be expressed as

$$X + Y \rightarrow XY$$
$$XY + Z \rightarrow XZ + Y,$$

where substance Y is the catalyst because it is regenerated. A reaction of this type can be seen in the ozone-depleting reaction shown on page 446, where chlorine atoms act to speed the breakdown of ozone.

- Other enzymes that could be studied include catalase, which will break down hydrogen peroxide into hydrogen and oxygen; amylase, the enzyme in saliva that catalyzes the conversion of starch to glucose; and papain, an enzyme found in some meat tenderizers that also breaks down proteins.

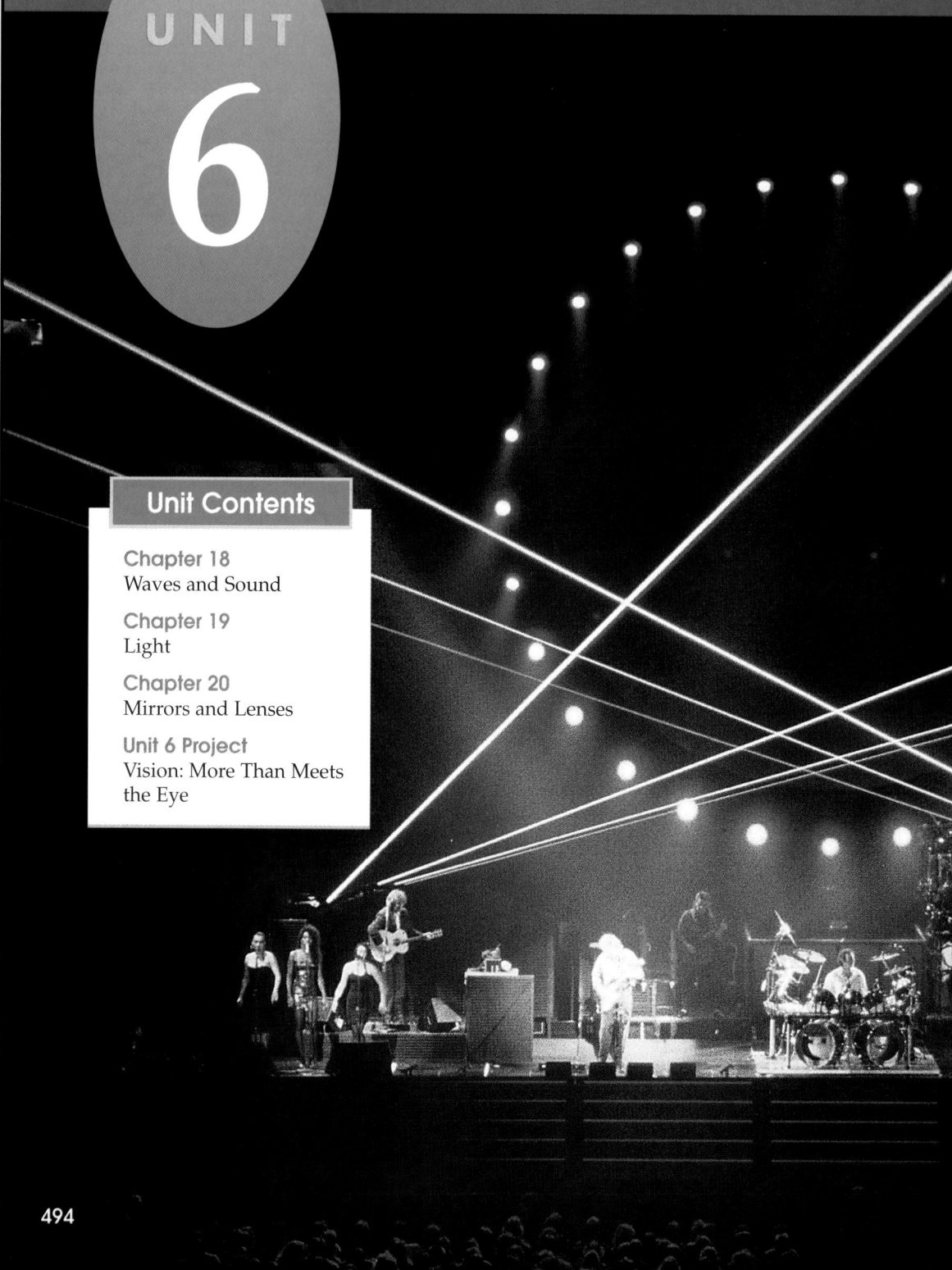

UNIT 6

Waves, Light, and Sound

In Unit 6, students are introduced to wave phenomena. The properties and behavior of transverse mechanical waves are discussed and related to those of longitudinal waves. The various properties of sound are explained by its wave characteristics. Light is then introduced to demonstrate the wave characteristics of electromagnetic waves. The unit closes by having students explore optics and optical instruments.

CONTENTS

UNIT 6

Unit Contents

494

Science at Home

A Colorful World Have students list the titles of songs that they have heard that contain colors. Have them conjecture on how color is used to portray or affect a mood.
L1

Waves, Light, and Sound

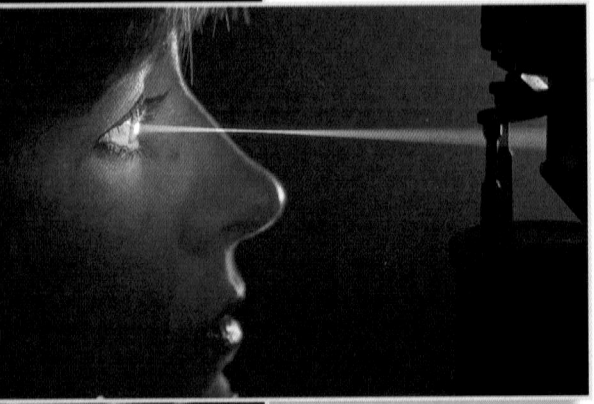

What's Happening Here?

Perhaps you have been to a concert or play that used light for unusual effects. The combination of special sound and lighting is often used to enhance entertainment. In fact, most human communication makes use of light or sound in some way. How does a spoken comment actually reach another person? What happens to light when it shines through a prism? Light and sound both travel in the form of waves. In addition to communication and entertainment, waves are used for many other purposes. How do you think light is being used in the small photo above?

Science Journal

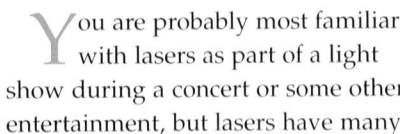

You are probably most familiar with lasers as part of a light show during a concert or some other entertainment, but lasers have many other uses. Research the different uses for lasers and write a report in your Science Journal.

495

Chapter Organizer

Section	Objectives/Standards	Activities/Features
Chapter Opener		**Explore Activity:** Making Music with a Ruler, p. 497
18-1 **Characteristics of Waves** (2 sessions, 1 block)*	1. **Sketch** a transverse wave and identify its characteristics. 2. **Discuss** the relationship between the frequency and wavelength in a transverse wave. 3. Using the relationship among wavelength, frequency, and velocity, **find** one variable when two are given. **National Content Standards:** (5-8) UCP1-UCP3, UCP5, A1, A2, B3, E1; (9-12) UCP1-UCP3, UCP5, A1, A2, B6, D1, E1	**Connect to Earth Science,** p. 499 **MiniLAB:** How do different transverse waves compare? p. 500 **Using Math:** Calculating the Velocity of a Wave; Calculating the Frequency of a Wave, p. 501 **Activity 18-1:** Making Waves, pp. 502-503 **Using Math,** p. 504 **Skill Builder:** Comparing and Contrasting, p. 504
18-2 **The Nature of Sound** (1 session, 1 block)*	1. **Describe** the transmission of sound through a medium. 2. **Recognize** the relationships between intensity and loudness and frequency and pitch. 3. **Illustrate** the Doppler effect with a practical example. **National Content Standards:** (5-8) UCP1-UCP3, UCP5, A2, C1, E2, F5 (9-12) UCP1-UCP3, UCP5, A2, B6, E2, F6	**MiniLAB:** How is sound different when it travels through solids? p. 506 **Connect to Physics,** p. 508 **Using Technology:** Searching with Sound, p. 509 **Science Journal,** p. 510 **Using Computers,** p. 512 **Skill Builder:** Concept Mapping, p. 512 **Activity 18-2:** Frequency of Sound Waves, p. 513
18-3 **Science and Society** (1 session, ½ block)*	1. **Explain** how sound waves can be used to create images of organs inside the body. 2. **Describe** some of the uses of ultrasound technology in medicine. **National Content Standards:** (5-8) UCP2-UCP5, A2, E2, F1, F5; (9-12) UCP2, UCP3, UCP5, A2, B6, E2, F1, F6	**Science Journal,** p. 515 **Explore the Technology,** p. 515
18-4 **Music to Your Ears** (1 session, 1 block)*	1. **Distinguish** between music and noise. 2. **Describe** why different instruments produce sounds of different quality. 3. **Explain** two types of wave interference. **National Content Standards:** (5-8) UCP1-UCP3, UCP5; (9-12) UCP1-UCP3, UCP5, B6	**Problem Solving:** The Mysterious Bouncing Bridge, p. 518 **MiniLAB:** How can a hearing loss change the sounds you hear? p. 520 **Science Journal,** p. 521 **Skill Builder:** Recognizing Cause and Effect, p. 521 **Science and Literature:** The Sound of Poetry

Activity Materials

Explore	Activities	MiniLABs
page 497 thin ruler	pages 502-503 coiled spiral spring, meterstick, stopwatch, safety goggles page 513 plastic pipe, rubber band, metric ruler	page 500 metric ruler, paper page 506 metal object such as a hanger or spoon, piece of string page 520 radio, heavy cloth or foam pads

Need Materials? Call Science Kit (1-800-828-7777).   * A complete Planning Guide that includes block scheduling is provided on pages 32T-35T.

Teacher Classroom Resources

Reproducible Masters	Transparencies	Teaching Resources
Study Guide, p. 74 **Reinforcement**, p. 74 **Enrichment**, p. 74 **Activity Worksheets**, pp. 102-103, 106 **Lab Manual 37**, Velocity of a Wave **Science Integration Activity 18**, Surf's Up **Science and Society Integration**, p. 22 **Concept Mapping**, pp. 41-42	**Section Focus Transparency 70**, People Waves **Teaching Transparency 35**, Wave Motion	**Glencoe Physical Science Interactive Videodisc** **Physical Science CD-ROM** **Spanish Resources** **English/Spanish Audiocassettes** **Cooperative Learning Resource Guide** **Lab Partner** **Lab and Safety Skills** **Lesson Plans**
Study Guide, p. 75 **Reinforcement**, p. 75 **Enrichment**, p. 75 **Activity Worksheets**, pp. 104-105, 107 **Lab Manual 38**, Sound Waves and Pitch **Critical Thinking/Problem Solving**, p. 24	**Section Focus Transparency 71**, Out of Range **Science Integration Transparency 18**, Sound Waves and Hearing **Teaching Transparency 36**, Decibel Scale	**Assessment Resources** **Chapter Review**, pp. 39-40 **Assessment**, pp. 122-125 **Performance Assessment in the Science Classroom (PASC)** **MindJogger Videoquiz** **Alternate Assessment in the Science Classroom** **Performance Assessment** **Chapter Review Software** **Computer Test Bank**
Study Guide, p. 76 **Reinforcement**, p. 76 **Enrichment**, p. 76	**Section Focus Transparency 72**, Seeing with Sound	
Study Guide, p. 77 **Reinforcement**, p. 77 **Enrichment**, p. 77 **Activity Worksheets**, p. 108 **Multicultural Connections**, p. 39 **Cross-Curricular Integration**, p. 24	**Section Focus Transparency 73**, Music...or Noise?	

Key to Teaching Strategies

The following designations will help you decide which activities are appropriate for your students.

L1 Level 1 activities should be within the ability range of all students, including those with learning difficulties.

L2 Level 2 activities should be within the ability range of the average to above-average student.

L3 Level 3 activities are designed for the ability range of above-average students.

LEP LEP activities should be within the ability range of Limited English Proficiency students.

LS These activities are designed to address different learning styles.

COOP LEARN Cooperative Learning activities are designed for small group work.

P These strategies represent student products that can be placed into a best-work portfolio.

GLENCOE TECHNOLOGY

The following multimedia resources are available from Glencoe.

Science and Technology Videodisc Series (STVS)
Animals
  Songbird Study
  How Bats Hear
Human Biology
  Ear Implants
  Hearing by Touch

Glencoe Physical Science Interactive Videodisc
Waves and Sound

Physical Science CD-ROM

National Geographic Society Series
Newton's Apple: Physical Sciences

Teacher Classroom Resources

This is a representation of key blackline masters available in the Teacher Classroom Resources.

Teaching Aids

Section Focus Transparencies

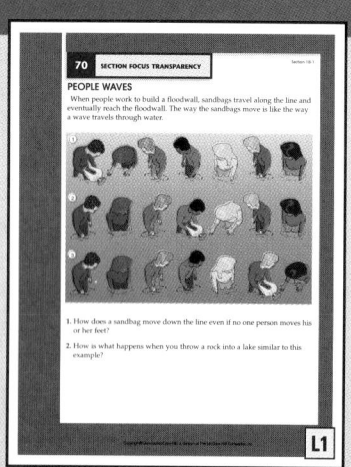

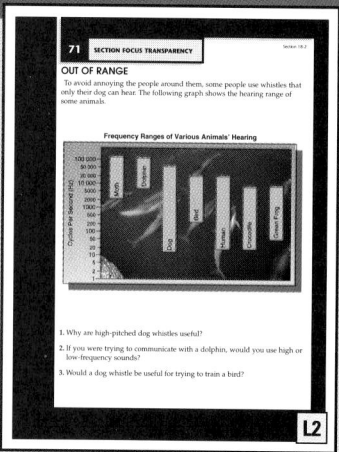

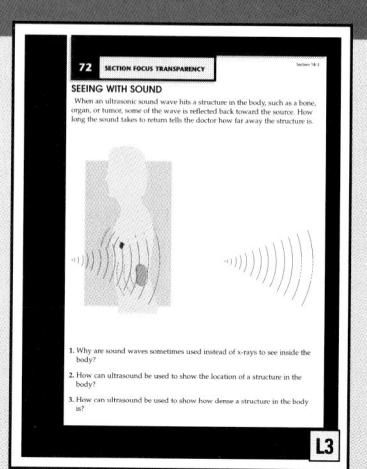

Science Integration Transparencies

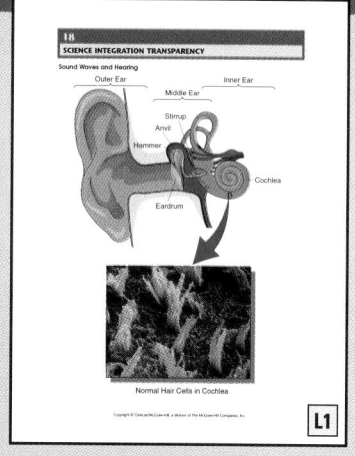

Teaching Transparencies

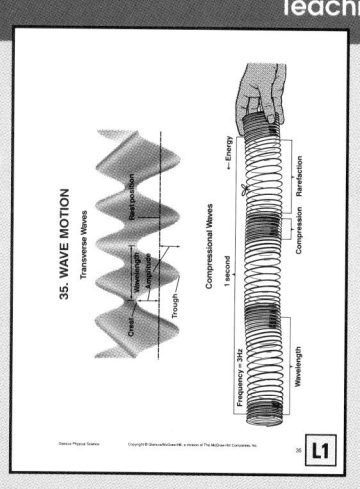

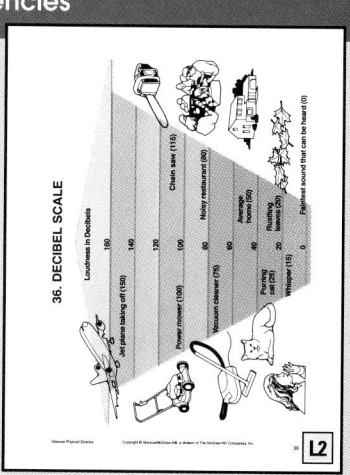

Meeting Different Ability Levels

Study Guide

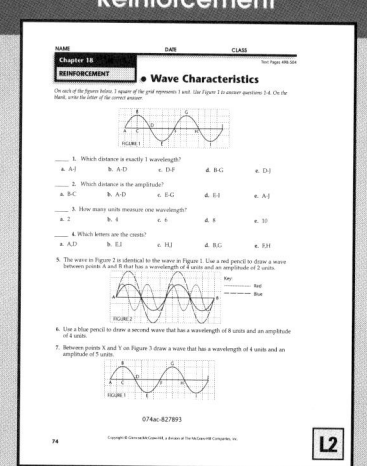

Reinforcement

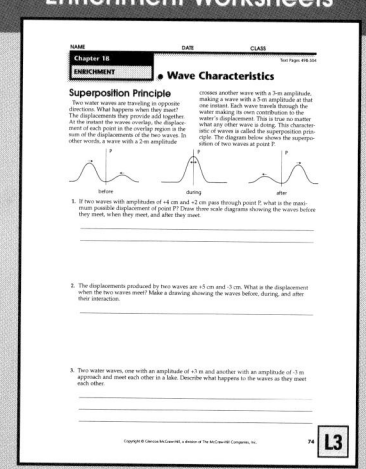

Enrichment Worksheets

Hands-On Activities

Science Integration Activity

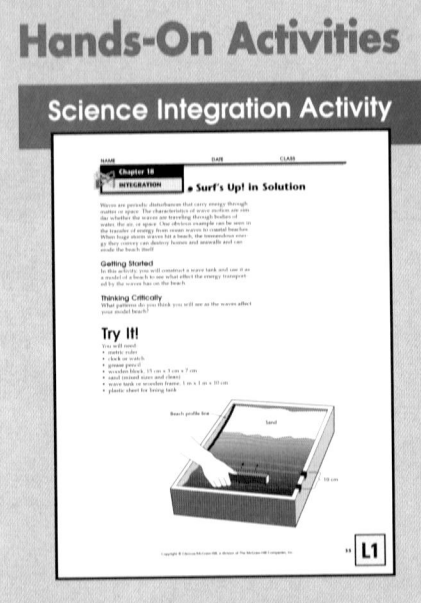

Lab Manual

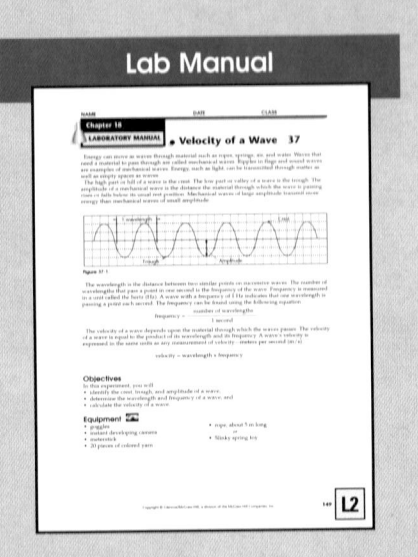

Assessment

Performance Assessment

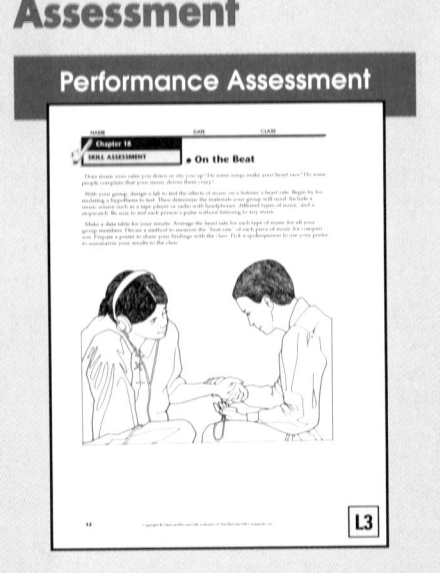

Enrichment and Application

Critical Thinking/Problem Solving

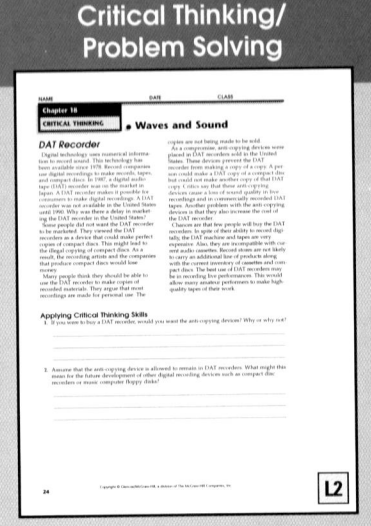

Cross-Curricular Integration

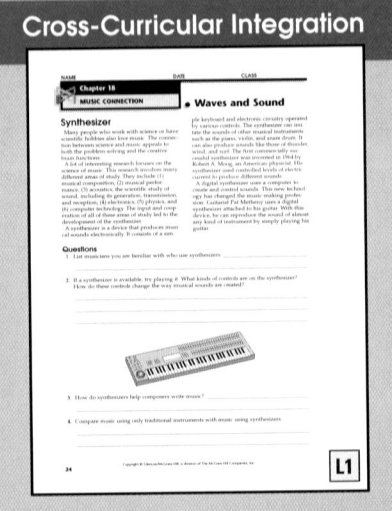

Science and Society Integration

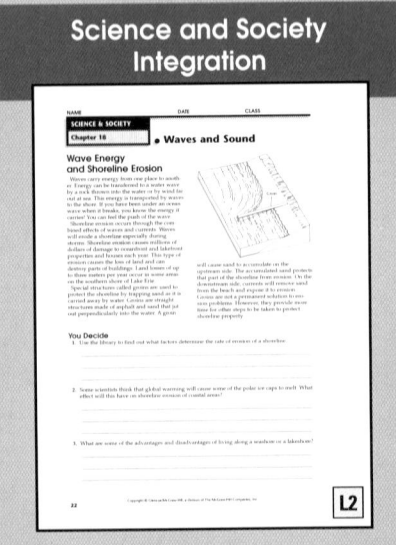

Multicultural Connections

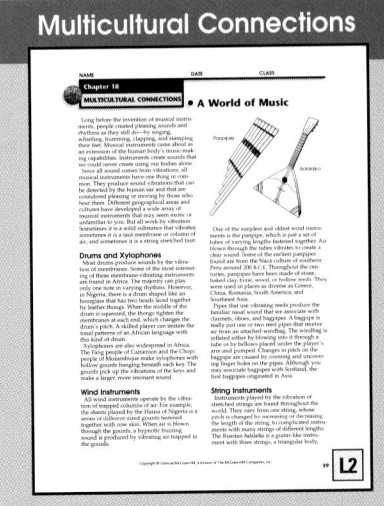

Concept Mapping

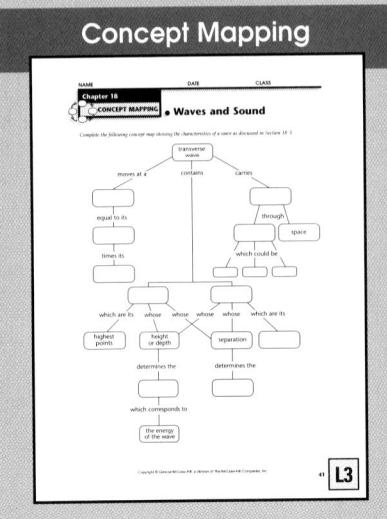

Chapter 18

Waves and Sound

CHAPTER OVERVIEW

Section 18-1 This section introduces transverse waves. The parts of a wave are identified, and the relationship among frequency, wavelength, and velocity is discussed.

Section 18-2 Compressional waves are explained. The differences between frequency and pitch and intensity and loudness are examined. A discussion of the Doppler effect concludes this section.

Section 18-3 Science and Society The use of ultrasonic technology in the medical field and its ability to view soft tissues and fetal development is explained. The section ends by introducing the use of ultrasound instead of surgery to remove kidney stones.

Section 18-4 This section contrasts music and noise. Musical quality and interference are illustrated.

Chapter Vocabulary

- wave
- medium
- transverse wave
- crest
- trough
- wavelength
- amplitude
- frequency
- compressional wave
- pitch
- intensity
- loudness
- ultrasound technology
- music
- noise
- resonance
- quality
- interference
- reverberation
- acoustics

Theme Connection

Energy Energy is one of the main themes of this chapter, as waves are rhythmic disturbances that carry energy. Emphasize how the amplification of waves increases the energy. Use examples to show both kinds of waves carrying energy.

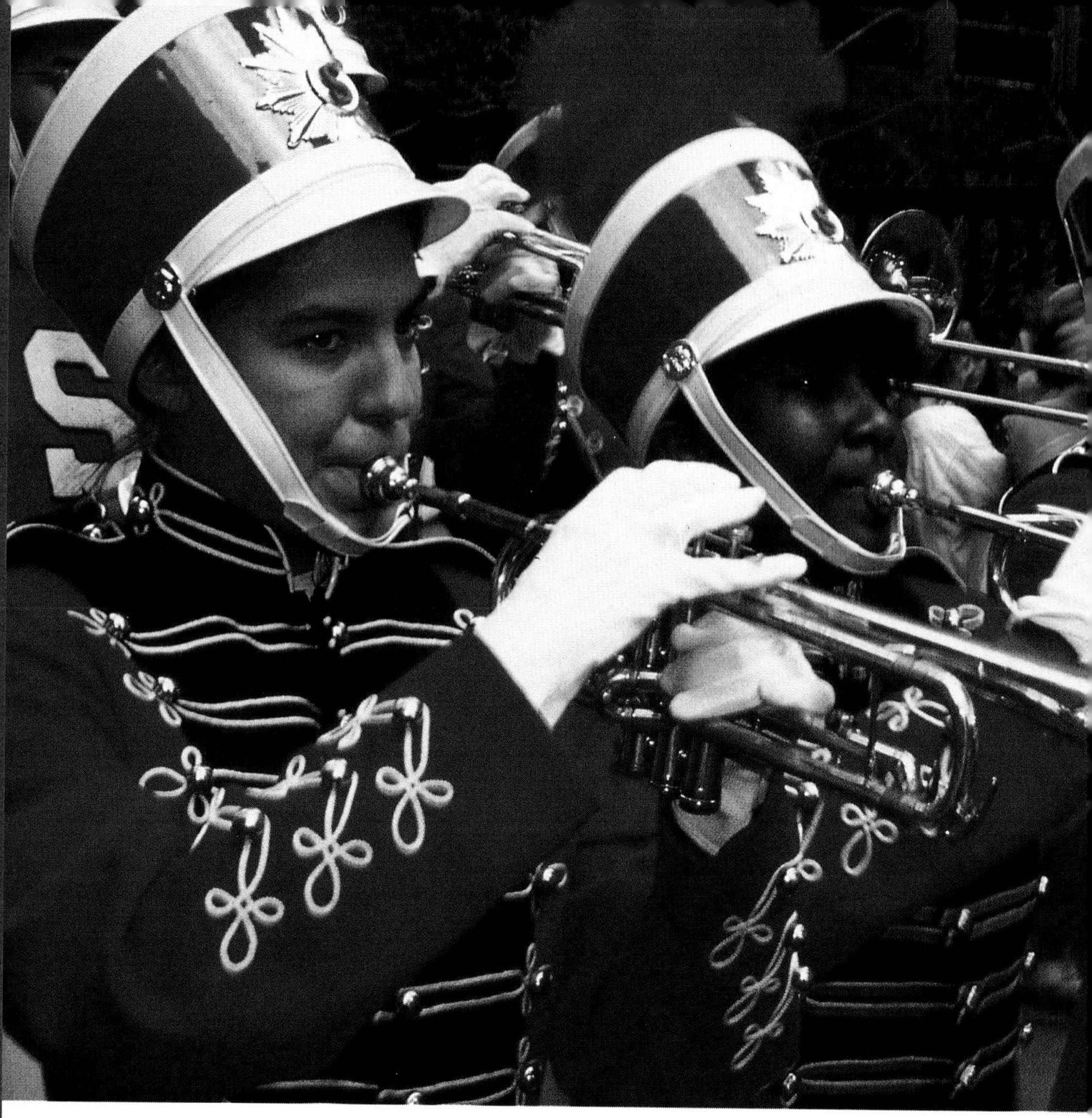

Previewing the Chapter

Section 18-1 Characteristics of Waves
- ▶ What are waves?
- ▶ Wave Frequency
- ▶ Wave Velocity

Section 18-2 The Nature of Sound
- ▶ What causes sound?
- ▶ Moving Through Media
- ▶ Human Hearing
- ▶ Frequency and Pitch
- ▶ Intensity and Loudness
- ▶ The Doppler Effect

Section 18-3 Science and Society Technology: Using Sound Advice in Medicine

Section 18-4 Music to Your Ears
- ▶ What is music?
- ▶ Musical Sounds
- ▶ Interference
- ▶ Acoustics

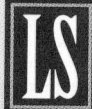

Learning Styles

Look for the following logo for strategies that emphasize different learning modalities. LS

Kinesthetic	Reteach, p. 501; Activity, p. 508
Visual-Spatial	Demonstration, pp. 499, 517; Across the Curriculum, p. 509; Discussion, p. 517
Interpersonal	Activity 18-1, p. 502; Reteach, p. 519
Logical-Mathematical	MiniLAB, p. 500; Across the Curriculum, p. 508; Science Journal, p. 510; Activity 18-2, p. 513
Linguistic	Science Journal, p. 518; Science & Literature, p. 522
Auditory-Musical	Explore, p. 497; MiniLAB, pp. 506, 520; Demonstration, p. 511; Across the Curriculum, p. 519

Chapter 18

Waves and Sound

Music can be made in a variety of different ways: plucking the strings on a guitar, blowing air into a flute or oboe, beating the membrane on a drum, or even pushing air through your own vocal chords. All of these are ways to vibrate air, creating sound waves.

EXPLORE ACTIVITY

Making Music with a Ruler

1. Hold one end of a thin metal, plastic, or wood ruler firmly down on the edge of a desk, allowing the free end to extend beyond the desk.
2. Gently pull up on and release the free end of the ruler. What do you see and hear?
3. Move the ruler to allow more or less length to hang over the edge and repeat the experiment several times. Do you hear different sounds?

Observe: How does the length of ruler extending beyond the table affect the sound you hear? In your Science Journal, write a paragraph about how the ruler could be moved to play a musical scale. Try tuning the ruler to a piano and writing a few measures of music. Then let someone play the tune using the ruler.

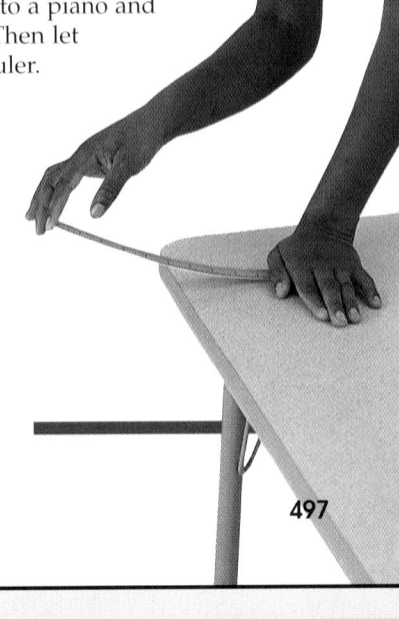

Previewing Science Skills

► In the Skill Builders, you will **map concepts, recognize cause and effect,** and **compare and contrast.**

► In the Activities, you will **observe, measure,** and **define operationally.**

► In the MiniLABs, you will **observe, interpret,** and **compare.**

497

Prepare

Section Background

- Although students sometimes name water waves as examples of transverse waves, these waves are actually a distinct kind of wave called a surface wave. In addition to moving up and down, the water molecules also move in circles.

- Earthquakes under the ocean floor can cause giant tidal waves called tsunamis. They move at speeds up to 800 km/h in deep waters, and wave height can exceed that of a 10-story building.

- One hertz is one cycle per second.

Preplanning

- Collect metric rulers for the MiniLAB.

- Obtain a class set of small Slinkys for Activity 18-1.

1 Motivate

Bellringer

Before presenting the lesson, display **Section Focus Transparency 70** on the overhead projector. Assign the accompanying **Focus Activity** worksheet. L1 LEP

Tying to Previous Knowledge

Review the idea of energy introduced in Chapter 5. Ask students if they ever made waves in the bathtub. Ask what you have to do to make a wave. *You must move your body to put energy into the water.*

18•1 Characteristics of Waves

Science Words

wave
medium
transverse wave
crest
trough
wavelength
amplitude
frequency

Objectives

- Sketch a transverse wave and identify its characteristics.
- Discuss the relationship between the frequency and wavelength in a transverse wave.
- Using the relationship among wavelength, frequency, and velocity, find one variable when two are given.

What are waves?

What examples come to mind when you think of waves? Do water waves, microwaves, sound waves, and radio waves have anything in common with each other? These and all other types of waves transfer energy from one place to another.

Look at **Figure 18-1.** Water waves are probably the easiest type of wave to visualize. If you've been in a boat on a lake, you know that approaching waves bump against the boat but do not carry the boat along with them as they pass. The boat mostly moves up and down as the waves pass by. Only energy carried by the waves moves forward.

Waves are rhythmic disturbances that carry energy through matter or space. Water waves transfer energy through the water. Earthquakes transfer energy in powerful waves that travel through Earth. Both types of waves travel through a **medium,** a material through which a wave transfers energy. This medium may be a solid, a liquid, a gas, or a combination of these. Radio waves and light waves, however, are types of waves that can travel without a medium.

Figure 18-1

Waves transfer energy. The energy of a water wave does work on anything in its path. *What instances have you seen that indicated that waves have done work?*

498 Chapter 18 Waves and Sound

Program Resources

 **Reproducible Masters**
Study Guide, p. 74 L1
Reinforcement, p. 74 L1
Enrichment, p. 74 L3
Science and Society Integration, p. 22
Activity Worksheets, pp. 102-103, 106 L1
Science Integration Activities, pp. 35-36
Lab Manual 37
Concept Mapping, p. 41

Transparencies
Section Focus Transparency 70 L1
Teaching Transparency 35 L1

Transverse Waves

Two types of waves carry energy. These are transverse and compressional waves. In one type, a **transverse wave,** the medium moves at right angles to the direction the wave travels. **Figure 18-2** shows how you can make transverse waves by shaking the end of a rope up and down while your friend holds one end. Compressional waves will be described in the next section.

Measuring Transverse Waves

The transverse wave in **Figure 18-2** can be described by its characteristics. When you snap the rope up and down, you may notice that high points and low points form. The highest points of a wave are called the **crests,** and the lowest points are called the **troughs.** Waves are measured by their wavelength. **Wavelength** is the distance between a point on one wave and the identical point on the next wave, such as from crest to crest or trough to trough. The wavelength between two crests is labeled on **Figure 18-2.**

Ocean or lake waves can be described by how high they appear above the normal water level. **Amplitude** is the distance from the crest (or trough) of a wave to the rest position of the medium, as shown in **Figure 18-2.** The amplitude corresponds to the amount of energy carried by the wave. Waves that carry great amounts of energy have large heights or amplitudes, and waves that carry less energy have smaller amplitudes.

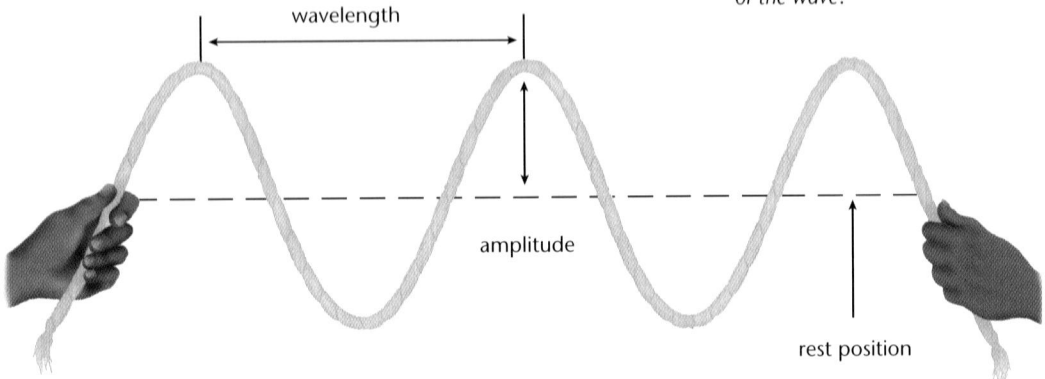

wavelength

amplitude

rest position

Wave Frequency

Do you know the frequency of your favorite radio station? When you tune your radio to a station, you are actually looking for waves of a certain frequency. The **frequency** of a wave

Earthquakes and volcanic activity under the ocean floor can cause giant tidal waves. *Research* what these waves are called and why they are so dangerous.

Figure 18-2

The parts of a transverse wave are shown on the wave in this rope. *How could you measure the wavelength on other parts of the wave?*

---

2 Teach

Demonstration

Visual-Spatial Place a table tennis ball in a wide pan of water about halfway between the center and the edge of the pan. When the water is still, drop a rock into the pan. The wave moves, but the ball has no horizontal motion. Compare the motion of the wave to that of the energy. **LEP**

CONNECT TO
EARTH SCIENCE

Answer These waves are called tsunamis and carry extremely large amounts of energy. As the water becomes shallower near shore, the wave height can exceed that of a 20-story building.

GLENCOE TECHNOLOGY

 Videodisc
Glencoe Physical Science Interactive Videodisc
Side 2, Lesson 7
Waves Defined

20742-23097
Transverse Waves

23099-24005
Longitudinal vs. Transverse Waves

24007-25402

 CD-ROM

Physical Science CD-ROM
Have students perform the interactive exploration for Chapter 18 to reinforce important chapter concepts and thinking processes.

Visual Learning

Figure 18-1 What instances have you seen that indicated that waves have done work? *Students may mention eroded beaches or wrecked boats.*

Figure 18-2 How could you measure the wavelength on other parts of the wave? *You can measure wavelength from any part of one wave to the same part of the next wave (crest to crest, trough to trough, and so on).*
LEP

MiniLAB

Purpose

LM Logical-Mathematical Compare and contrast the wavelength and frequencies of transverse waves. **L1**

Materials

paper, pencil, metric ruler

Teaching Strategies

Lined paper may help students draw wavelengths of uniform height.

Analysis

1. Some waves will be nearly identical; others will differ in wavelength, amplitude, or both.
2. $f = v/\lambda$;
   $f = $ (20 cm/sec)/(distance between 2 crests)
3. Highest $f = $ many crests passing one point, lowest $f = $ few crests passing one point; the lower-frequency waves will have longer wavelengths than the high-frequency waves.

Assessment

Performance Have students visualize and describe how the amplitude and frequency of water waves change as they approach the shore from the open ocean. A diagram can be used to illustrate their answers. *Both amplitude and frequency increase as the waves approach shore.* Use the Performance Task Assessment List for Making Observations and Inferences in **PASC**, p. 17. **P**

📁 **Activity Worksheets,** pp. 5, 106

Teacher F.Y.I.

Lightning flashes move at speeds up to 140 000 km/s and last about 0.003 s, with temperatures of about 30 000°C.

Figure 18-3

How do the waves shown on this drawing compare? Wave **A** has a longer wavelength and a lower frequency than wave **B**. The speed of the wave is the same for both waves because wave speed is determined by the medium.

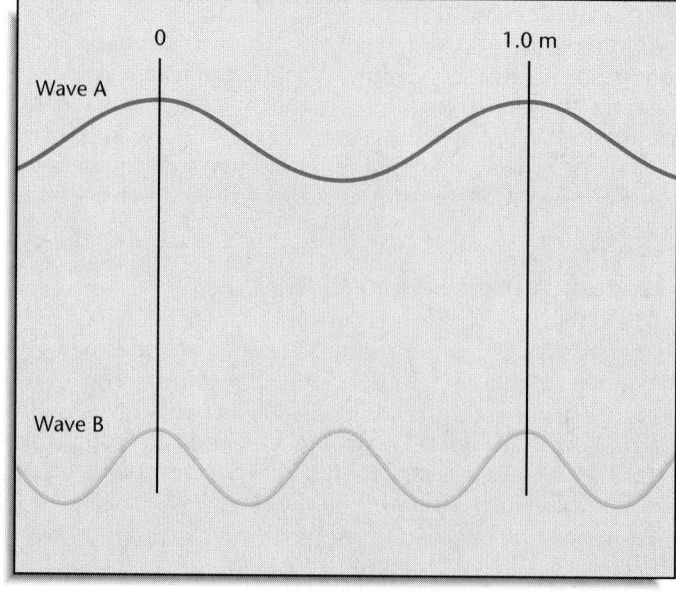

MiniLAB

How do different transverse waves compare?

Procedure

1. On your paper, draw a transverse wave with at least three complete wavelengths.
2. Label one wavelength, the amplitude, a crest, and a trough on your wave. Use a ruler to measure the wavelength and the amplitude to the nearest tenth of a centimeter.
3. In groups of four, compare your wave sketches.
4. In your group, rank your waves from longest to shortest wavelength and then from largest to smallest amplitude.

Analysis

1. How were your waves similar? How did they differ?
2. Assume all of your waves traveled at the same speed of 20 cm/s. Count the number of complete waves in a 20-cm stretch of your wave to determine its frequency.
3. Identify the waves in your group with the highest and lowest frequency.

---

is the number of wave crests that pass one place each second. Frequency is expressed in hertz (Hz). One hertz is the same as one wave per second.

How do you increase the frequency of a wave? To do this with a rope, simply move the rope up and down faster to create more crests per second. Because the speed of a wave in a given medium is constant and does not depend on the wavelength or frequency, as the frequency increases, the wavelength becomes shorter. In other words, as the frequency increases, the wavelength decreases. Using a rope, you can demonstrate this principle, as shown in **Figure 18-3.**

Wave Velocity

Sometimes you may want to know how fast a wave is traveling. For example, earthquakes below the ocean can produce giant waves. You will want to know how soon such a wave will reach you and whether you need to seek shelter. Wave velocity, v, describes how fast the wave moves forward.

Wave velocity can be determined by multiplying the wavelength and frequency as shown below. Wavelength is represented by the Greek letter *lambda*, λ. If you know any two variables in the following equation, you can find the remaining unknown variable.

$$velocity = wavelength \times frequency$$

$$v = \lambda \times f$$

Across the Curriculum

Math Students have learned that as frequency decreases, the wavelength increases if the speed of the wave is constant. Ask them what the relationship between wavelength and frequency is called. *inverse proportion*

The following examples show how you can use this equation to solve for the unknown variable. Rewrite the equation to make the desired quantity the unknown.

USING MATH

Calculating the Velocity of a Wave

Example Problem:

A wave is generated in a wave pool at a water amusement park. The wavelength is 3.2 m. The frequency of the wave is 0.60 Hz. What is the velocity of the wave?

1. What is known?   wavelength, λ = 3.2 m
frequency, f = 0.60 Hz

Strategy Hint: Another way to express hertz is 1/second, therefore, m × 1/s = m/s.

2. What is unknown?   velocity, v
3. Choose the equation. $v = \lambda \times f$
Solution: 3.2 m × 0.60 Hz = 1.92 m/s

Practice Problems

1. A wave moving along a rope has a wavelength of 1.2 m and a frequency of 4.5 Hz. How fast is the wave traveling along the rope?
2. An ocean wave has a length of 10.0 m. Two waves pass a fixed point every 1.0 s. What is the speed of the wave?
Strategy Hint: Make sure all units correspond.

Calculating the Frequency of a Wave

Example Problem:

Earthquakes can produce three types of waves. One of these is a transverse wave called an S wave. A typical S wave travels at 5000 m/s. Its wavelength is about 417 m. What is its frequency?

1. What is known?   velocity, v = 5000 m/s
wavelength, λ = 417 m

Strategy Hint: Remember, Hz = 1/s, so m/s ÷ m = 1/s = 1 Hz.

2. What is unknown?   frequency, f
3. Choose the equation. $v = \lambda \times f$ so $f = v/\lambda$
Solution: f = (5000 m/s)/(417 m) = 12 Hz

Practice Problem

A tuning fork produces a sound wave with a wavelength of 0.20 m and a velocity of 25.6 m/s. What is the frequency of the tuning fork?
Strategy Hint: What formula do you use to find frequency?

Calculating the Velocity of a Wave

1. wavelength, λ = 1.2 m
frequency, f = 4.5 Hz
$v = \lambda \times f$ = 1.2 m × 4.5 Hz
= 5.4 m/s

2. wavelength, λ = 10.0 m
frequency, f = 2.0 Hz
$v = \lambda \times f$ = 10.0 m × 2.0 Hz
= 20.0 m/s

Calculating the Frequency of a Wave

velocity, v = 25.6 m/s
wavelength, λ = 0.20 m
$f = v/\lambda$ = (25.6 m/s)/(0.20 m)
= 128 Hz **P**

3 Assess

Check for Understanding

? FLEX Your Brain

Use the Flex Your Brain activity to have students explore WAVE CHARACTERISTICS.

📁 **Activity Worksheets,** p. 5

Reteach

LS Kinesthetic Give students a piece of string about 60 cm long. Instruct them to form the string into the shape of a transverse wave. Count the number of waves and measure the average wavelength. Now change the wave so it has a greater amplitude; then change it again so it has shorter wavelengths. **What happens to the frequency?** *It increases with shorter wavelengths; no change with change in amplitude.* **LEP**

Extension

📁 For students who have mastered this section, use the **Reinforcement** and **Enrichment** masters.

501

PREPARATION

Purpose
IS **Interpersonal** Use Slinkys and rope or string to investigate some of the properties and behaviors of transverse waves. **L1** **LEP** **COOP LEARN**

Process Skills
observing, inferring, hypothesizing, communicating, making and using tables, recognizing cause and effect, forming operational definitions, interpreting data, making models

Time
one class period

Materials
See student page.

Possible Hypotheses
Students may hypothesize that the frequency and amplitude can be changed, or that the spring will stop moving at a barrier or bounce back (send the waveform back toward the source).

📁 **Activity Worksheets,** pp. 5, 102-103

PLAN THE EXPERIMENT

Possible Procedures
A set of procedures might be: Have one partner hold one end still; place a meterstick at the end near the wave maker; have the wave maker create a transverse pulse by horizontally shaking the spring approximately 30 cm; shake the spring vertically the same distance and compare. Try to change the wavelength and amplitude by shaking the spring larger or smaller distances; record the distances and all observations. Note what happens to the waveform when it reaches the end.

Design Your Own Experiment
Making Waves

All waves carry energy from one place to another. How can you make waves in a spiral spring? What happens to these waves when they hit a solid barrier? In this activity, you will create waves in a spiral spring and observe the characteristics of the waves.

PREPARATION

Problem
Which wave quantities can you influence when making transverse waves in a spiral spring? How does a transverse wave in a spring behave when it hits an immovable barrier?

Form a Hypothesis
State a hypothesis concerning the expected behavior of the spring in the situations described above.

Objectives
- Design an experiment to test the behavior of transverse waves in a spiral spring.

- Observe what happens to a transverse wave when it hits a fixed boundary.

Possible Materials
- coiled, spiral spring
- meterstick
- stopwatch

Safety Precautions
Use caution with possible sharp ends of the springs. Do not twist or stretch the spiral springs.

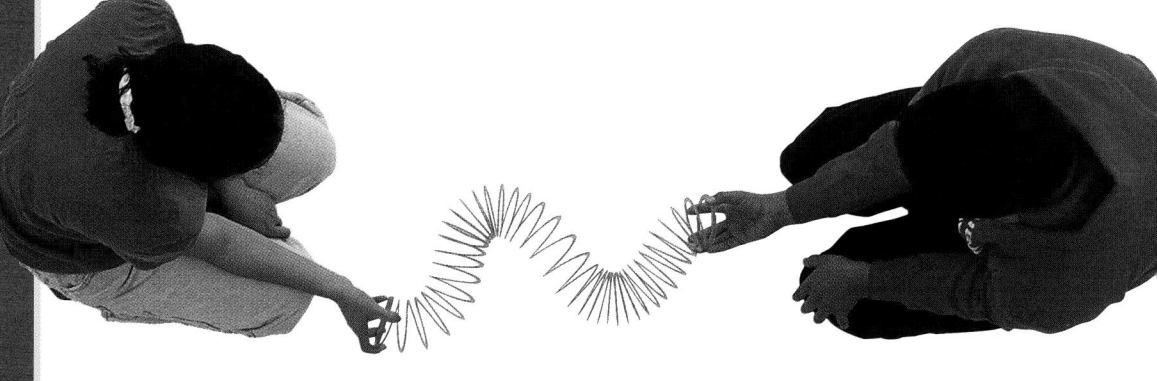

PLAN THE EXPERIMENT

1. As a group, agree upon and write out your hypothesis statements.
2. List the steps you will take to test your hypothesis. Be specific in your directions for each step.
3. Collect all materials you will need. Design a data table in your Science Journal if needed.

Check the Plan

1. Two people in your group should be seated on a smooth floor with the spring stretched between them. Decide who will be the wave maker. The other person will hold his or her end as still as possible.

2. The wave maker should start by quickly shaking the end of the spring horizontally about 30 cm to create a single transverse wave pulse. How else could you make a transverse wave?
3. How will you attempt to change the amplitude of the wave?
4. How will you shake one end to experiment with the relationship between wave frequency and wavelength?
5. *Make sure your teacher approves your plan and that you have included any changes suggested in the plan.*

DO THE EXPERIMENT

1. Carry out the experiment as planned.
2. During the experiment, write down any observations that you make in your Science Journal.

Analyze and Apply

1. How did you change the amplitude of the wave? Did the amplitude stay the same as the pulse traveled along the spring? Explain why or why not.
2. **Predict** what happens when the wave hits the fixed boundary.
3. **Infer** how you can make waves of greater frequency. **Compare** frequency and wavelength.

18-1 Characteristics of Waves   **503**

Go Further

Can you make a different kind of wave by pushing the spring straight forward and back, rather than moving it side to side or up and down? Does this kind of wave seem to have an amplitude or a wavelength? What happens when you push it to compress it with more force and speed? You will learn more about this type of wave in the next section.

Assessment

Performance Have students sketch a series of diagrams to show what happens to a single transverse wave as it travels along the spring, hits a fixed barrier, and returns to its point of origin. Use the Performance Task Assessment List for Scientific Drawing in **PASC**, p. 55. **P**

Teaching Strategies

Troubleshooting Students might tape the meterstick to the floor to avoid unwanted movement.

- Remind the students not to overstretch the coiled springs. The small coiled springs sold in toy stores work very well. To avoid having tangled coils, it is wise to have a box or sack for each spring.
- You also might have groups attempt the activity using lightly pulled ropes instead of coiled springs, but ropes do not hold the energy of the wave as well.

DO THE EXPERIMENT

Expected Outcome

Moving the Slinky a greater distance will create a larger wave amplitude; shorter distances will create smaller amplitudes; the waveform disappears sooner with vertical movement.

Analyze and Apply

1. Answers may include shaking over a longer or shorter distance. No, friction from contact with the floor caused some energy to dissipate and the wave amplitude decreased.
2. The wave is reflected back with smaller amplitude in an inverted position.
3. Shake the spring faster. As frequency increases, wavelength decreases.

Go Further

Yes, a different kind of wave can be made. It has an amplitude and wavelength. It is a compressional wave; the wave has a larger amplitude.

4 Close

Section Wrap-up

Review

1.

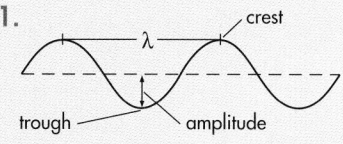

2. The wavelength decreases as the frequency increases, and the wavelength increases as the frequency decreases.
3. $v = \lambda \times f$, so $\lambda = v/f$
   $= (4.0 \text{ m/s}) \div (3.5 \text{ Hz})$
   $= 1.14$ m
4. **Think Critically** Pendulums oscillate back and forth between high and low points, as do transverse waves.

USING MATH

Because both travel at the same speed, the lowest-frequency station, 101.9 MHz, sends out waves with longer wavelengths.

Figure 18-4

Earth experiences hundreds of earthquakes in a typical day, but most are too small to be easily noticed. An earthquake measuring 7.1 on the Richter scale shook northern California on October 17, 1989, collapsing this highway in Oakland. Every increase of 1 on the Richter scale represents an increase by a factor of 10 in amplitude.

Section Wrap-up

Review

1. Sketch a transverse wave and label a crest, a trough, a wavelength, and the amplitude.
2. What is the relationship between the frequency and wavelength of a water wave?
3. A wave travels at a velocity of 4.0 m/s and has a frequency of 3.5 Hz. What is the wavelength?
4. **Think Critically:** You learned in Chapter 5 that gravity causes a pendulum to oscillate. How is its motion similar to that of a transverse wave?

Skill Builder
Comparing and Contrasting
Use **Figure 18-2** and information from this section to compare the frequency, amplitude, and wavelength of a wave. Which of these measurements depends on energy? Which is measured in meters? Which depends on the number of waves? If you need help, refer to Comparing and Contrasting in the **Skill Handbook.**

USING MATH

FM radio stations use frequencies in the megahertz (MHz) range. Your favorite radio station is at 104.1 MHz and your friend prefers a station at 101.9 MHz. Calculate whose station uses longer wavelengths. Explain your answer.

Skill Builder
Amplitude depends on energy; wavelength would be measured in meters; and frequency depends on the number of waves.

✓ Assessment

Performance Have students use a Slinky to determine which of the three variables compared in this Skill Builder cannot be directly controlled when shaking the Slinky to create a transverse wave. *Wavelength can only be controlled indirectly by changing the frequency. The speed is the same no matter what the frequency is.* Use the Performance Task Assessment List for Model in **PASC,** p. 51. **P**

The Nature of Sound

What causes sound?

Think of all the sounds you've heard since you awoke this morning. Did you hear a blaring alarm, honking horns, human voices, or lockers slamming? Your ears allow you to recognize these different sounds, but do you know what they all have in common? These sounds are all produced by the vibrations of objects. For example, your voice is produced by the vibrations of your own vocal cords. The energy produced by these vibrations is carried to your friends' ears by sound waves.

Compressional Waves

The waves discussed in the last section were described as transverse waves because the matter moved at right angles to the direction the wave was traveling. You could produce this type of wave in a rope or spring by moving one end from side to side or up and down. Sound waves carry energy by a different type of wave motion. You can model sound waves with a coil spring. If you hold one end of a spring, squeeze the coils together, and then release the coils while holding the end of the spring, you will produce a compressional wave. Matter vibrates in the same direction as the wave travels in a **compressional wave. Figure 18-5** shows how a compressional wave made with a spring should look.

Science Words
compressional wave
pitch
intensity
loudness

Objectives
- Describe the transmission of sound through a medium.
- Recognize the relationships between intensity and loudness and frequency and pitch.
- Illustrate the Doppler effect with a practical example.

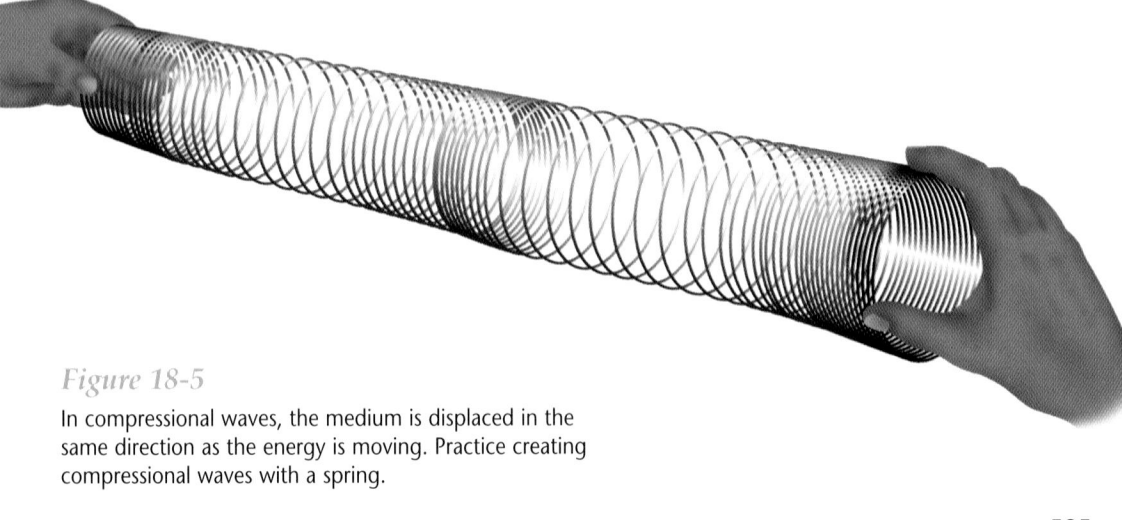

Figure 18-5

In compressional waves, the medium is displaced in the same direction as the energy is moving. Practice creating compressional waves with a spring.

Program Resources

Reproducible Masters
Study Guide, p. 75 [L1]
Reinforcement, p. 75 [L1]
Enrichment, p. 75 [L3]
Activity Worksheets, pp. 104-105, 107 [L1]
Lab Manual 38
Critical Thinking/Problem Solving, p. 24

Transparencies
Section Focus Transparency 71 [L1]
Science Integration Transparency 18
Teaching Transparency 36

Prepare

Section Background
- The speed of sound in air is called Mach 1. A common reference value for Mach 1 at a 12-km altitude is 1056 km/h. Altitude and air temperature affect this value.
- The speed of sound waves depends on the temperature and elasticity of the matter in the medium. Atoms are usually close together in solids. As a result, solids transmit sound faster than air.

Preplanning
- Collect metal objects for the MiniLAB.
- Get plastic pipe for Activity 18-2.

1 Motivate

Bellringer
 Before presenting the lesson, display **Section Focus Transparency 71** on the overhead projector. Assign the accompanying **Focus Activity** worksheet. [L1] [LEP]

Tying to Previous Knowledge
Ask students what they have to do to sing a tune. (They must make their voices go higher and lower.) Explain that in this lesson they will find out what *higher* and *lower* really mean.

Visual Learning
Figure 18-5 Push part of the spring together, then release it or stretch out part of the spring and release it. [LEP] [IS]

505

Use **Teaching Transparency 36** as you teach this lesson.

MiniLAB

Purpose

Auditory-Musical Observe the change in sound as it travels through different media. `L1` `LEP`

Materials

metal object (i.e., spoon, wire coat hanger, etc.), string

Teaching Strategies

- Use about 1.5 m of string per student.
- Try a variety of metal objects—ring stand rods make excellent bell-like sounds. Wire coat hangers make gonglike sounds. Curtain rods, metal tools, dinner knives, and forks could be tested. Oven racks give impressive results.

Analysis

1. Comparisons will vary from student to student. Low-energy, low-frequency sounds are not easily heard through air but travel well through the string, giving the objects sounds similar to those of large bells.

Assessment

Content Ask students why the sound travels better through the string than the air. *The string molecules are closer together than the air molecules.* Use the Performance Task Assessment List for Making Observations and Inferences in **PASC**, p. 17. `P`

Activity Worksheets, pp. 5, 107

MiniLAB

How is sound different when it travels through solids?

Procedure

1. Tie a metal object, such as a wire hanger or a spoon, to the center of a piece of string.
2. Wrap each of the two ends of the string around one finger on each hand.
3. Gently place the fingers holding the string in your ears. Let the object swing until it bumps against the edge of a chair or table and listen to the sound.
4. Listen to the sound made by the collisions when your fingers are not in your ears.

Analysis

1. Compare the sounds you hear when your fingers are and are not in your ears. Do sounds travel better through air or through the string?

Describing Compressional Waves

Notice in **Figure 18-5** that as the wave moves, some of the coils are squeezed together, just as you squeezed the ones on the end of the spring. This crowded area is called a *compression*. The compressed area then expands, spreading the coils apart and creating a less dense area. This less dense area of the wave is called a *rarefaction*. Does the whole spring move forward? Tie a piece of string onto one of the coils and observe its motion. The string moves back and forth with the coils. Therefore, the matter in the medium does not move forward with the wave. Instead, the wave carries only the energy forward.

Recall that transverse waves have wavelengths, frequencies, amplitudes, and velocities. Compressional waves also have these characteristics. A wavelength in a compressional wave is made of one compression and one rarefaction, as shown in **Figure 18-6.** Notice that one wavelength is the distance between two compressions or two rarefactions of the same wave. The frequency is the number of compressions that pass a place each second. If you could repeatedly squeeze and release the end of the spring two times each second, you would produce a wave with a frequency of 2 Hz. The amount of compression is like the amplitude of the transverse wave, and it depends on the energy content of the wave. Think about what you would do to increase the amount of compression in the coil spring. You would have to squeeze harder, and you would be putting more energy into the wave.

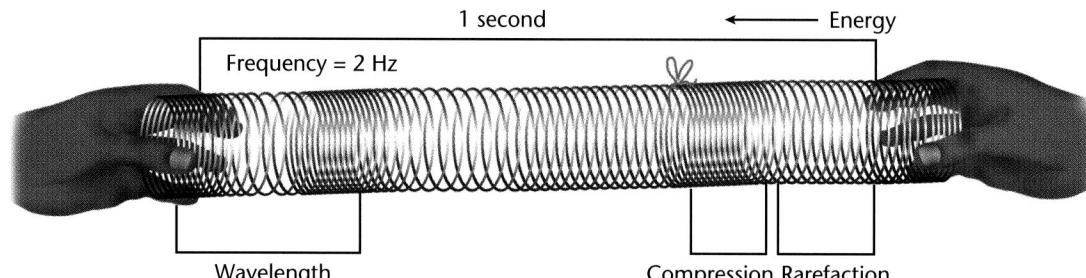

Figure 18-6

The spring vibrates back and forth, but the energy of this compressional wave moves forward. *How could you measure the wavelength of a compressional wave?*

Moving Through Media

How does sound travel when you have a conversation with a friend? The vibrations generated by your vocal cords produce compressional waves that travel through the air to your friend. This process is similar to what you saw when you made compressional waves on a spring. Your voice causes compressions and rarefactions among the particles in the air.

Comparing Media: Solids, Liquids, and Gases

The speed of sound waves depends on two things—the medium through which the waves travel and the temperature of the medium. Air is the most common medium through which you hear sound waves, such as the sounds of the porpoise in **Figure 18-7**. Sound waves, however, can be transmitted through any type of matter. Liquids and solids are even better conductors of sound than air is because the individual particles in a liquid or solid have greater influence on each other than the particles in air, making the transmission of energy easier. Can sound be transmitted if there is no matter to form a medium? Astronauts on the moon would find it impossible to talk to each other without the aid of modern electronic communication systems. Because the moon has no atmosphere, there is no air to compress and expand.

The Speed of Sound

The temperature of the medium is also an important factor in determining the speed of sound waves. As the temperature of a substance increases, the molecules move faster and, therefore, collide more frequently. This increase in molecular collisions transfers more energy in a shorter amount of time. This allows the sound waves to be transmitted faster. Sound travels through air at 344 m/s if the temperature is 20°C, but at only 332 m/s when the temperature is 0°C.

Have you seen fireworks explode in the sky before you heard the boom? The speed of sound is much slower than the speed of light. Have you ever tried to guess how far away a lightning bolt is by counting the time interval between when you see a lightning flash and when you hear the thunder? You see the light before you hear the sound because light waves travel through air about 1 million times faster than sound waves do.

Human Hearing

Think of the last conversation you had with a friend. How were you able to understand what your friend was saying? It is amazing that compressional waves caused by speech, music, or other sounds around us can be turned into meaningful sounds by our sense of hearing. Making sense of these waves involves three stages: the gathering and amplification of the compressional waves by the ear mechanism, the conversion of these waves into nerve impulses, and the decoding of these signals in the brain.

Figure 18-7

Porpoises communicate with each other vocally. *Does the sound of this porpoise's voice travel slower in air or water?*

INTEGRATION
Life Science

18-2 The Nature of Sound   507

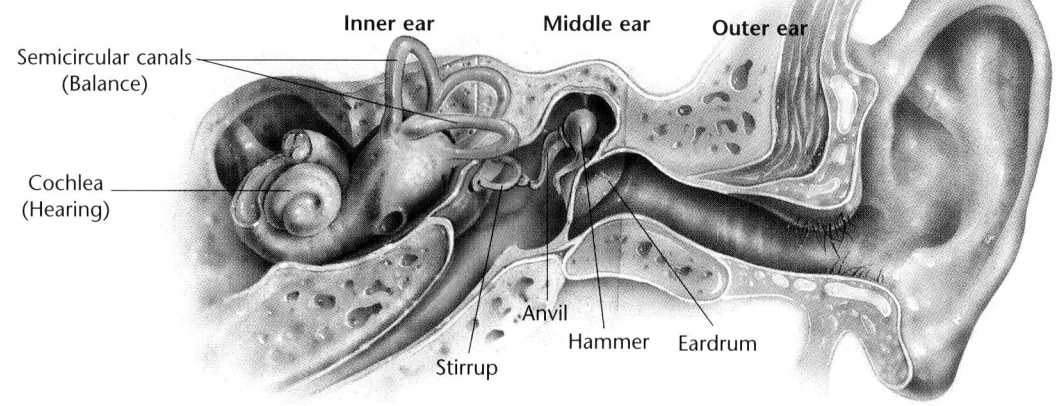

Inner ear Middle ear Outer ear
Semicircular canals (Balance)
Cochlea (Hearing)
Anvil
Hammer Eardrum
Stirrup

Figure 18-8

The human ear is made up of the outer ear, the middle ear, and the inner ear. *Which parts vibrate and multiply the force and pressure of the sound waves?*

CONNECT TO
PHYSICS

You have just formed a new company, Ultrasonics Unlimited. **Design** an advertisement for a product that uses ultrasonic energy.

The Ear as a Sound Detector

Sound detectors, such as microphones, collect and convert kinetic energy from the moving particles in sound waves into another form of energy, usually electrical energy. Your ear is a versatile sound detector. It is sensitive to a range of intensities and frequencies. The human ear, seen in **Figure 18-8,** has three sections: the outer ear, the middle ear, and the inner ear.

Although you probably think of your ear as the fleshy, visible outer part, this part contributes little to your hearing abilities. It helps direct sound waves into the ear canal, which is a little narrower than your index finger and is 2-3 cm long. It leads to the eardrum, which is a tough membrane about 0.1 mm thick that is vibrated by incoming sound waves. The eardrum vibrates three tiny bones in the middle ear: the hammer, the anvil, and the stirrup. These bones, which are full adult size at birth, make a lever system that multiplies the force and pressure of the sound wave.

They transmit the sound vibrations to the oval window in the inner ear. The inner ear contains the liquid-filled, spiral-shaped cochlea. Tiny hair cells in the cochlea are vibrated, causing nerve impulses to be sent to the brain through the auditory nerve. Hearing damage inflicted by sudden or ongoing loud sounds is generally the result of destruction of these tiny, hair-like cells. Do you like to listen to your music loud? Remember, intense sounds can result in permanent hearing damage.

Frequency and Pitch

If you have ever taken a music class, you are probably familiar with the scale "do re mi fa so la ti do." As you sing this scale, your voice starts low and becomes higher with each note. You hear a change in pitch. **Pitch** is the highness or lowness of a sound. The pitch you hear depends on the frequency of the sound waves. The higher the frequency is, the higher the pitch is, and the lower the frequency is, the lower the pitch is. A healthy human ear can hear sound frequencies from about 20 Hz to 20 000 Hz. As people age, they often have some trouble hearing high frequencies.

Ultrasonic and Infrasonic Waves

Most people can't hear sound frequencies above 20 000 Hz, which are called ultrasonic waves. Bats, however, can detect frequencies as high as 100 000 Hz. Ultrasonic waves are used in sonar as well as in medical diagnosis and treatment. Sonar, or sound navigation ranging, is a method of using sound waves to estimate the size, shape, and depth of underwater objects. Infrasonic, or subsonic, waves have frequencies below 20 Hz. These are produced by sources such as heavy machinery and thunder. Although you probably can't hear them, you may have sensed these sound waves as a disturbing rumble inside of your body.

USING TECHNOLOGY

Searching with Sound

More than 130 years ago, a ship named the *Central America* disappeared in a hurricane off the coast of South Carolina. In its hold lay three tons of newly minted gold coins and bars worth more than $560 million in the current market. How would you find such a vessel?

The wrecked ship was located under 2800 m of water using advanced sonar technology. In sonar techniques, a sound pulse is emitted toward the bottom of the ocean. The sound travels through the water and is reflected when it encounters solid barriers. When this time interval and the speed of sound in water are both known, the distance to the barrier or object can be calculated. Ultrasound frequencies are typically used because they provide better resolution of small details. Sonar can locate an object as small as a book at a depth of 7000 m!

In addition to locating sunken ships, sonar technology has made it possible to map geologic features on the bottom of the ocean, to locate reefs, and to monitor the location of submarines. Sonar is commonly used in commercial fishing to locate schools of fish.

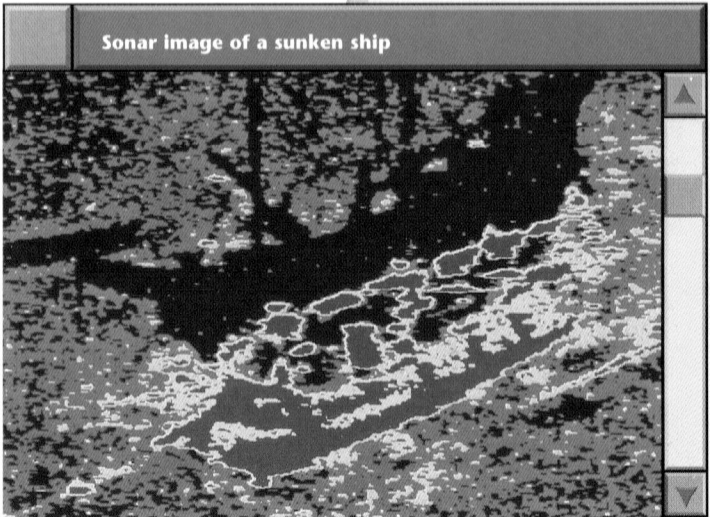
Sonar image of a sunken ship

Think Critically:

How do you think sonar technology might be useful in locating deposits of oil and mineral resources?

USING TECHNOLOGY

For more information on the use of low-frequency sonar to find underwater treasures, see "Deep Quest" by Abe Dane, *Popular Mechanics*, Jan. 1990, pp. 56-59.

Think Critically

The speed of reflection differs depending upon the medium encountered. Knowing the speed for reflected oil or other features would allow explorers to locate them.

Teacher F.Y.I.

The frequencies of the human voice range from about 250 to 2000 Hz in a normal conversation.

 Use these program resources as you teach this lesson.
Science Integration Transparency 18
Teaching Transparency 36

Use **Critical Thinking/Problem Solving**, p. 24, as you teach this lesson.

GLENCOE TECHNOLOGY

 **Videodisc**
Glencoe Physical Science Interactive Videodisc
Side 2, Lesson 7
Echolocation

26624-28086

Across the Curriculum

Geology Prior to WWII, the topography of the ocean floor was basically unknown. Echo-sounding equipment was present on American naval ships during WWII. Commander (and geologist) Harry Hess kept it operating while patrolling the Atlantic Ocean. Not only was a map of the Atlantic Ocean floor created, but data was also collected to support the theory of sea-floor spreading. Have students compare the detail in pre- and post-WWII maps of the Atlantic Ocean floor and discuss the improvements. Advanced students might explore how this proved the theory of sea-floor spreading. L3 LS

Science Journal

Intensity and Loudness

Have you ever been told to turn down your CD player? If so, you probably adjusted the volume. The music still had the same notes, so the frequencies didn't change, but the amplitude of each sound wave was reduced. The **intensity** of a sound wave depends on the amount of energy in each wave. This, in turn, corresponds to the wave's amplitude. Intensity of a sound wave increases as its amplitude increases.

Sensing Loudness

Loudness is the human perception of sound intensity. The higher the intensity and amplitude, the louder the sound. People vary in sensitivity to different frequencies. What seems loud to one person may not seem loud to you. The intensity level of a sound is measured in units called decibels, abbreviated *dB*. On this scale, the faintest sound that can be heard by most humans is 0 dB. Sounds with intensity levels above 120 dB may cause pain and permanent hearing loss.

Table 18-1

Decibel Scale

Loudness in Decibels

Description	dB
	160
Jet plane taking off	150
	140
Pain threshold	120
Chain saw	115
Power mower	100
Noisy restaurant	80
Vacuum cleaner	75
	60
Average home	50
	40
Purring cat	25
Rustling leaves	20
Whisper	15
Faintest sounds that can be heard	0

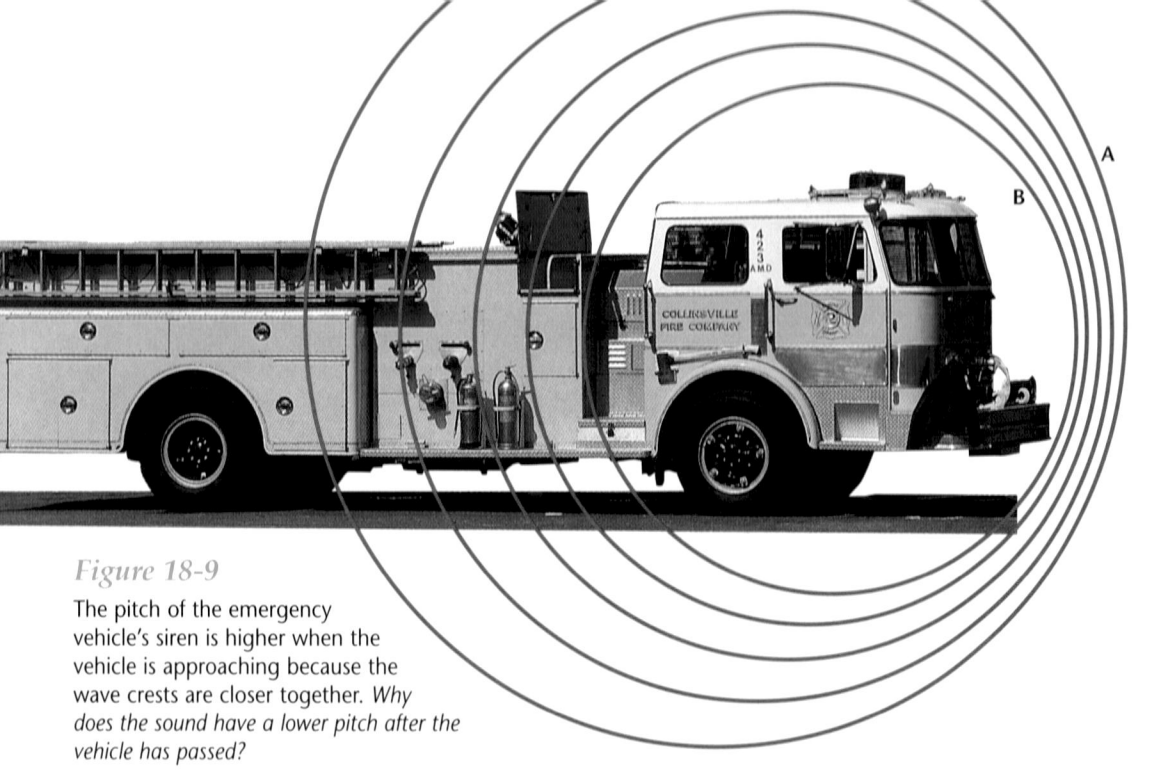

Figure 18-9

The pitch of the emergency vehicle's siren is higher when the vehicle is approaching because the wave crests are closer together. *Why does the sound have a lower pitch after the vehicle has passed?*

Sounds at this intensity level occur during some rock concerts. **Table 18-1** shows some familiar sounds and their intensity levels in decibels.

The Doppler Effect

Imagine the sound you hear when an emergency vehicle with its siren on rapidly approaches and then passes you. As the truck is moving toward you, the pitch of the siren sounds higher than it would if the truck were not moving. Each wave crest forms an expanding circle around the spot where it started. See crests **A** and **B** in **Figure 18-9.** By the time crest **B** left the siren, the truck has moved forward causing crest **B** to be closer to crest **A** in front of the truck. The result is a higher frequency and pitch in front of the moving truck. This change in wave frequency due to a moving wave source is called the Doppler effect. You can also see from **Figure 18-9** that the frequency is lower in the area behind the moving truck.

What would you expect would happen if you were moving past a stationary sound source? Suppose you were riding a school bus and passed by a building with a ringing alarm bell. The pitch would sound higher as you approached the building and lower as you rode away from it. The Doppler

18-2 The Nature of Sound **511**

511

4 Close

•MINI•QUIZ•

Use the Mini Quiz to check students' recall of chapter content.

1. **A change in the pitch of a sound is caused by a change in _____ .** *frequency*

2. **The amount of energy in a sound wave determines its _____ .** *intensity, amplitude, loudness*

3. **Can sound travel through empty space?** *No, it must have a medium through which it can travel.*

4. **The change in wave frequency due to a moving wave source is the _____ .** *Doppler effect*

Section Wrap-up

Review

1. compressional waves; in the same direction the waves move, by colliding particles

2. No; there is no matter in space; compressional waves don't transfer energy in a vacuum.

3. Intensity, amplitude, and loudness change.

4. **Think Critically** The bat is flying toward the prey; the prey is flying toward the bat; or both are flying toward each other.

Using Computers

Outlines should include: transverse waves move perpendicular to the medium, compressional waves move parallel to the medium. Both can be measured for wavelength, amplitude, frequency, and velocity.

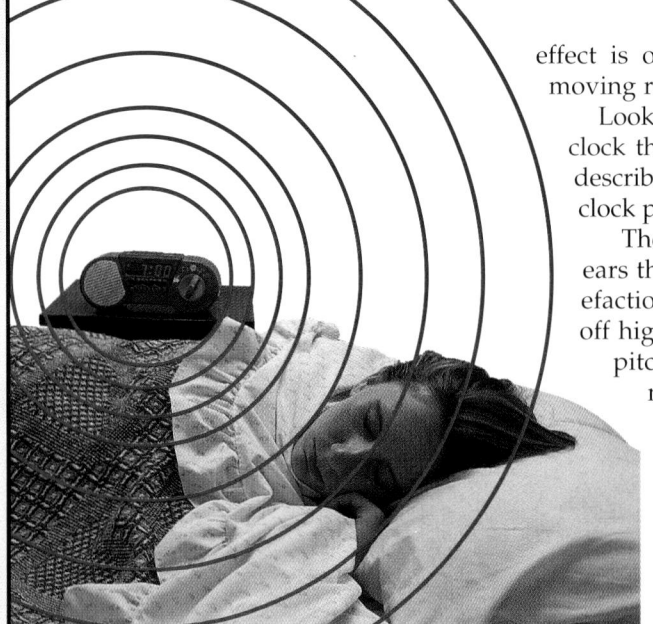

effect is observed when the source of sound is moving relative to the observer.

Look at **Figure 18-10.** Think again of the alarm clock that woke you up this morning. Can you describe how the rhythmic vibrations of the clock produced compressional waves in the air?

The sound waves carried energy to your ears through a series of compressions and rarefactions of air molecules. If your clock gives off high-frequency vibrations, you hear a high pitch. The more intense the waves, or the more energy they carry, the louder your clock sounds.

Figure 18-10

When the alarm clock starts to ring in the morning, it makes compressional waves in the air that are interpreted by your ears as sound.

Section Wrap-up

Review

1. What type of waves are sound waves, and how do they transfer energy?

2. While watching a "Star Trek" rerun, you hear the roaring approach of another ship in space. Is this possible? Explain.

3. While on your way to school, you turn up the volume on the car radio. Which of the following quantities change as a result: velocity of sound, intensity, pitch, amplitude, frequency, wavelength, loudness?

4. **Think Critically:** A bat in a dark cave sends out a high-frequency sound wave and detects an increase in frequency after the sound reflects off the prey. Describe the possible motions of the bat and the prey.

Skill Builder
Concept Mapping

Prepare a concept map that shows the series of events that occur to produce sound. Include the terms *rarefaction, vibration,* and *compression.* If you need help, refer to Concept Mapping in the **Skill Handbook.**

Using Computers

Word Processors Use a word processor to make an outline showing the characteristics of transverse and compressional waves.

Skill Builder

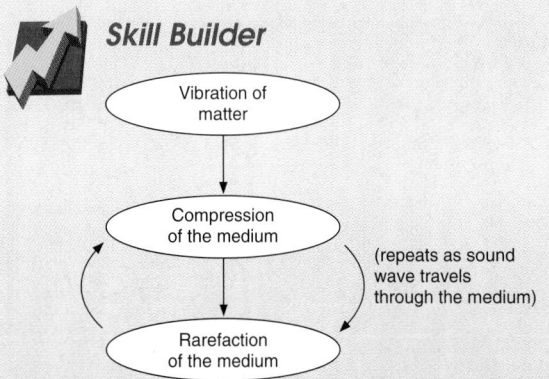

Vibration of matter

↓

Compression of the medium

↓

(repeats as sound wave travels through the medium)

Rarefaction of the medium

✓ Assessment

Oral Ask students to examine their concept maps. **How would this process differ for a sound traveling through a solid versus one traveling through air?** *The process would be the same, but sound would travel faster through a solid.* Use the Performance Task Assessment List for Concept Map in **PASC,** p. 89.
P

Frequency of Sound Waves

Sounds originate with a vibration. Sometimes this vibration also causes nearby objects to vibrate, or resonate, at the same frequency. Many musical instruments make use of a column of air that vibrates at a certain frequency. How is pitch controlled in instruments such as trumpets? By closing and opening specific valves, the length of the air column, the wavelength, and the frequency are changed.

Problem
Can you find the wavelength and frequency of the sound made in an open-ended pipe?

Materials
- rubber band
- plastic pipe (open at both ends)
- metric ruler

Procedure
1. Measure the length of the pipe and record it on the data table.
2. Stretch a rubber band across one open end of the pipe and hold it firmly in place as shown. **CAUTION:** *Be careful not to release your grip on the ends of the rubber band.*
3. Hold the rubber band close to your ear and pluck it. Listen for a *double* note.
4. Slowly relax the tightness of the rubber band. Listen for one part of the double note to change and the other part to remain the same.
5. Continue to adjust the tightness until you hear only one note.
6. Exchange pipes with another group and repeat the experiment.

Analyze
1. The wavelength you obtained in step 5 is twice the length of the pipe. **Calculate** the wavelength.
2. Assume the velocity of sound to be 34 400 cm/s. Use the equation frequency = velocity/wavelength to **calculate** the frequency of the note.
3. What were the wavelength and frequency of the sound waves in the second pipe?
4. Use your graphing calculator to graph $Y = 34\ 400/X$, where Y is frequency and X is wavelength. Use your graph to find the frequency of sound waves in a tube 2 meters long.

Conclude and Apply
5. How does the length of a pipe **compare** with the frequency and wavelength of the sound it can make?
6. **List** musical instruments that use lengths of pipe to produce musical notes.

Data and Observations
Sound Frequencies Produced by Open Pipes   Sample Data

Length of pipe	Length of wave	Frequency of sound
0.2 m	0.4 m	855 Hz
0.5 m	1.0 m	342 Hz

18-2 The Nature of Sound   **513**

2. Example:
   34 400 cm/s/40 cm = 860 Hz
3. Answers will vary. Wavelength will increase and frequency decrease as the pipes become longer.
4. The trace function can be used to determine that 86 Hz is the Y value that corresponds to an X value of 400 cm.
5. The longer the pipe, the longer the wavelength and the lower its frequency.
6. All horns and woodwinds as well as the human voice use a vibrating air column. A xylophone uses open pipes.

✔ Assessment

Performance Have students predict how six pipes of different lengths should be arranged from highest to lowest pitch. Then, test their predictions. Use the Performance Task Assessment List for Formulating a Hypothesis in **PASC**, p. 21. **P**

Purpose
LS **Logical-Mathematical** Students will compare wavelength and frequency with the size of an object producing sound. **L1** **COOP LEARN**

Process Skills
observing, measuring, using numbers, defining operationally, sequencing, comparing and contrasting, making models

Time
30 minutes

📁 **Activity Worksheets,** pp. 5, 104-105

Teaching Strategies
Alternate Materials Any kind of hard-walled pipe will work. Rolled paper will absorb sound without resonating.
Troubleshooting Students can hear the resonant frequency of the pipe by tapping either open end with a pencil.

- Use PVC pipe sold in home supply or hardware stores. For best results, use thin-walled pipe of 2.5 cm diameter or more. Pipe can easily be cut with a hacksaw. Cut into various lengths from 20 cm to 1 m. Mark each pipe with an identifying number.
- The pipe produces its fundamental frequency regardless of the frequency of the rubber band. But overtones of the fundamental can be heard as slightly louder sounds when the rubber band hits those frequencies.
- If a pitch pipe, tuning fork, or musical instrument is available, have students try to match the sound of the pipe to a note of known pitch and frequency.

Answers to Questions
1. longest wavelength = 2 × pipe length
   (Example) 2 × 20 cm = 40 cm

SCIENCE & SOCIETY
Section 18•3

Prepare

Section Background

Ultrasound techniques are in use in many fields including oceanography, dentistry, and medicine. Medically, it allows doctors to see soft tissues without costly and invasive surgical procedures. It is practically standard procedure in obstetrics.

1 Motivate

Bellringer

Before presenting the lesson, display **Section Focus Transparency 72** on the overhead projector. Assign the accompanying **Focus Activity** worksheet. [L1] [LEP]

2 Teach

Visual Learning

Figure 18-12 What are some of the advantages of ultrasound testing? *Ultrasound is noninvasive and not as damaging as X rays.* [LEP] [LS]

3 Assess

Check for Understanding

? FLEX Your Brain

Use the Flex Your Brain activity to have students explore ULTRASOUND.

Activity Worksheets, p. 5

TECHNOLOGY:
18•3 Using Sound Advice in Medicine

Science Words

ultrasonic technology

Objectives

- Explain how sound waves can be used to create images of organs inside the body.
- Describe some of the uses of ultrasound technology in medicine.

Figure 18-11

A technician applies an ultrasonic probe to the surface of a patient's body. The sound waves bounce off of structures inside the body and are converted to images on a monitor.

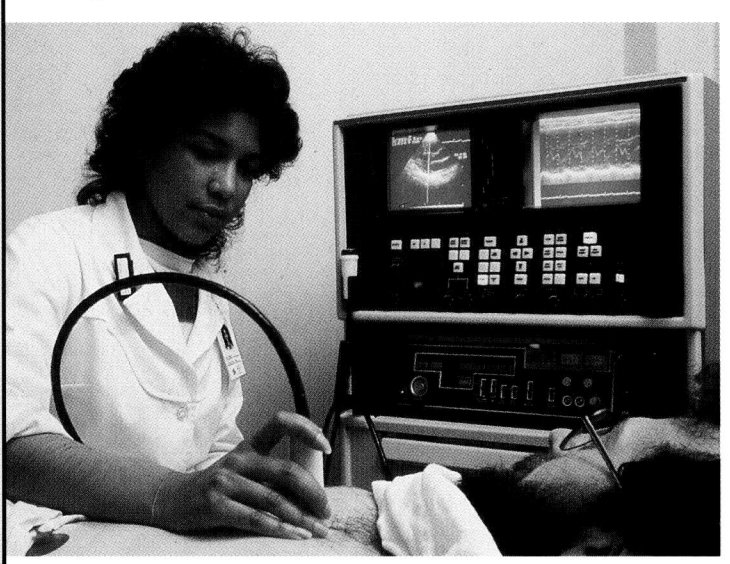

Ultrasonic Technology

Can you think of ways a doctor could check the function of organs or other tissue inside your body? You might think of stethoscopes, X rays, or even surgical methods, but did you think of using sound waves to create images? In the last section, you learned how sonar techniques can help oceanographers in exploring ocean depths. Sound can also be used to aid medical professionals in "seeing" inside your body without surgery or using the ionizing radiation found in X rays.

Ultrasonic technology is used for a wide variety of purposes. Recall that *ultrasonic* refers to sound waves of frequencies greater than 20 000 Hz, above which most humans cannot hear. Jewelers use ultrasonic sound to remove stubborn deposits from jewelry, and chemists sometimes use ultrasonic baths to clean glassware. Medical professionals use high-frequency sound waves to interact with tissue inside the body, **Figure 18-11.**

Seeing with Sound

How can we see with sound? Medical ultrasound technologists use a special probe to send high-frequency sound waves into a specific part of the body. These sound waves reflect off of the targeted organs or tissues and are used to produce electrical signals. A computer program converts these electrical signals into video images. Trained physicians can use these images to detect potential problems without invasive surgery.

Probably the best-known use of ultrasound in medicine is to monitor the development of a fetus in a woman's uterus. This procedure produces a sonogram. An image of a fetus from a sonogram is shown in **Figure 18-12.** Ultrasound is commonly

---

Program Resources

📁 Reproducible Masters
Study Guide, p. 76 [L1]
Reinforcement, p. 76 [L1]
Enrichment, p. 76 [L3]

 **Transparencies**
Section Focus Transparency 72 [L1]

used to examine a variety of abdominal organs, including the liver, pancreas, gallbladder, and spleen. Physicians can also use ultrasound Doppler shifts with cardiac patients to gather information about blood flow.

Preventing Surgery with Ultrasound

Sometimes small, hard deposits of calcium oxalate form in the kidneys. Surgical removal of kidney stones used to be the primary treatment option, but ultrasonic treatments create vibrations that can often break up the stones without surgery. The smaller pieces can then be naturally passed with urine. A similar treatment is available for gallstones. Patients successfully receiving ultrasound treatment make a quicker recovery than those who required surgical treatment.

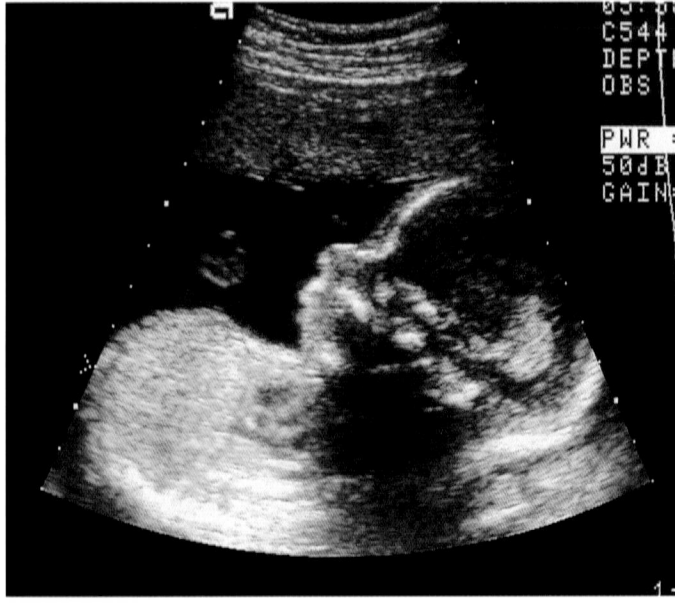

Figure 18-12

A sonogram of a developing fetus helps doctors check on the progress of the pregnancy. *What are some advantages of ultrasound testing?*

Section Wrap-up

Review

1. Describe at least three uses of ultrasonic technology in medicine.
2. How can images of an organ in your body be created from sound waves?

 Visit the Chapter 18 Internet Connection at Glencoe Online Science, **www.glencoe.com/sec/science/physical,** for a link to more information about ultrasonography.

Science Journal

Ultrasound gives doctors and parents a first view of a baby. In your Science Journal, list some things you think might be learned from a sonogram and write a short paragraph about how parents might feel on first seeing their baby.

SCIENCE & SOCIETY

18-3 Using Sound Advice in Medicine   515

Science Journal

Answers will vary. Accept all reasonable ideas. Information might include sex of child and multiple births.

Prepare

Section Background

• When disturbed, an object made of elastic materials vibrates at a distinct set of frequencies, called its natural frequency. The natural frequency is determined by the elasticity and shape of the object. At this frequency, the minimum energy is required to begin the vibrations.

• Some objects can resonate much like waves. Consider the resonance and collapse of the Tacoma Narrows Bridge in Washington State.

Preplanning

• Obtain a radio for the Mini-LAB.

• Find out which students in the class play musical instruments. Arrange for a "musical instrument day" so they can bring their instruments to class to demonstrate.

1 Motivate

Bellringer

Before presenting the lesson, display **Section Focus Transparency 73** on the overhead projector. Assign the accompanying **Focus Activity** worksheet. L1 LEP

Tying to Previous Knowledge

Students have taken music at some point in their education. Ask what kinds of things the choir teacher looks for when directing a choir. Answers may include pitch and a specific rhythm. This should introduce music as related to physics.

516

18•4 Music to Your Ears

Science Words

music
noise
resonance
quality
interference
reverberation
acoustics

Objectives

• Distinguish between music and noise.
• Describe why different instruments produce sounds of different quality.
• Explain two types of wave interference.

Figure 18-13

Music and pushing a chair across a floor both generate sound waves. *How do the waveforms representing music and noise differ?*

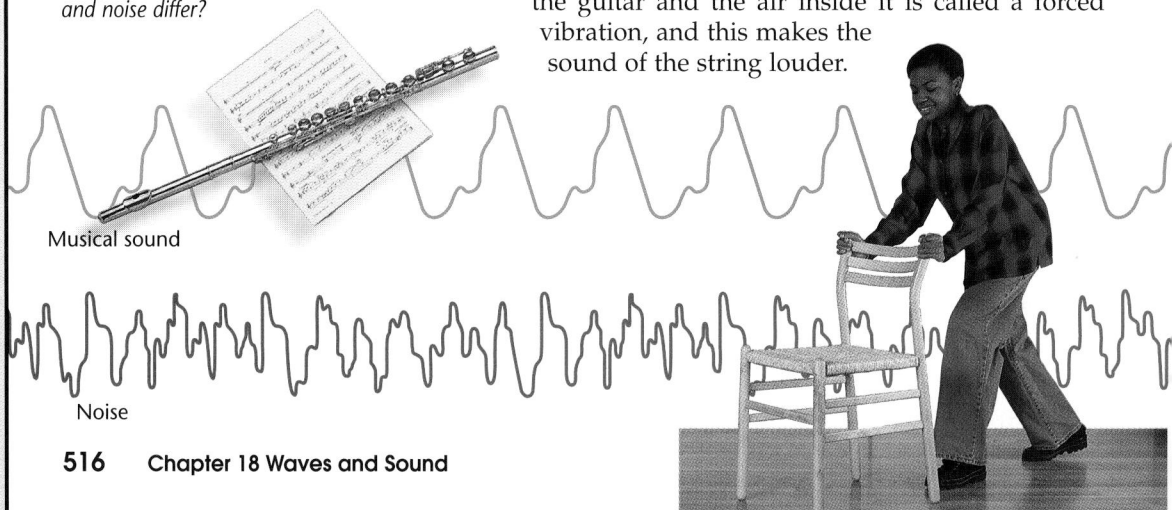

Musical sound

Noise

What is music?

Has anyone ever commented that the music you were listening to sounded like a jumble of noise? Both music and noise are caused by vibrations, but there are some important differences. You can easily make a noise by just speaking a word or tapping a pencil on a desk, but it takes some deliberate actions to create music. Of course, you also may be able to create music with your voice or your pencil if you try. **Music** is created using specific pitches and sound quality and by following a regular pattern. The most common kind of sound is **noise,** which has no set pattern and no definite pitch. If all the frequencies of noise are present in equal amplitude, the result is called white noise. White noise has been found to have a relaxing effect and is sometimes used by dentists or to help people sleep. **Figure 18-13** shows a comparison of noise and music patterns.

Natural Frequencies

A stringed musical instrument, such as a guitar, generates a sound when you pluck a string. Because the ends of the string are fastened, the waves created reflect back and forth between the ends. This causes the string to vibrate at its natural frequency. Most objects have a natural frequency of vibration.

What kind of sound would be produced if you held a guitar string tightly between your hands while a friend plucked it? It would be much quieter than if the string were fixed on a guitar because the guitar frame and the air inside the instrument absorb energy from the vibrating string. The vibration of the guitar and the air inside it is called a forced vibration, and this makes the sound of the string louder.

Program Resources

📁 **Reproducible Masters**
Study Guide, p. 77 L1
Reinforcement, p. 77 L1
Enrichment, p. 77 L3
Activity Worksheets, p. 108 L1
Cross-Curricular Integration, p. 24
Multicultural Connections, pp. 39-40

 **Transparencies**
Section Focus Transparency 73 L1

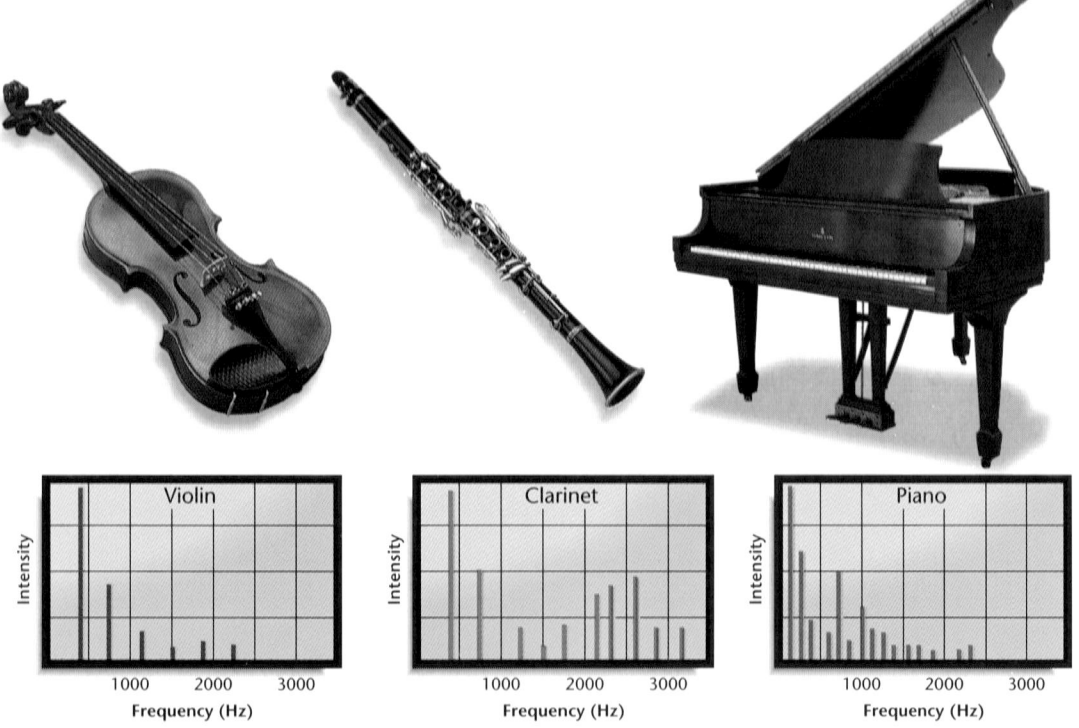

If the sound that reaches an object is at the same frequency as the natural frequency of the object, the object will begin to vibrate at this frequency. This type of vibration is called **resonance.**

Musical Sounds

If you were to play a note of the same pitch and loudness on a flute and on a piano, the sound wouldn't be the same. These instruments have a different quality of sound. This quality does not refer to how good or bad the instrument sounds. Sound **quality** describes the differences among sounds of the same pitch and loudness. All sounds are produced by vibrations of matter, but most objects vibrate at more than one frequency. Distinct sounds from musical instruments are produced by different combinations of these wave frequencies. The patterns of frequencies of three different instruments are shown in **Figure 18-14.**

Overtones Cause Unique Quality

Imagine producing a tone by plucking a guitar string. The tone produced when the string vibrates along its entire length is called the fundamental frequency. At the same time, each half-length of the string can vibrate on its own.

Figure 18-14

A violin, clarinet, and piano have a characteristic pattern of frequencies. *Which instrument has more higher-pitched sounds for a given note?*

18-4 Music to Your Ears   517

Figure 18-15 **How is its sound different from a piano?** *Piano strings are of different lengths, so their fundamental frequencies and overtones will be different. A piano string is hit with a rubber hammer; guitar strings are strummed or plucked.*

Figure 18-16 **How could beats be used to tune orchestra instruments?** *All instruments could be tuned to a specific instrument until beats are no longer heard.* **LEP** **IS**

Problem Solving

The wind set the bridge swaying and, because it was connected at both ends, the wave form was reflected back and forth as it vibrated. Eventually, the natural frequency was reached.

Solve the Problem

1. vibrations that are the same as the natural frequency of the object
2. wind

Think Critically

They build bridges in segments so that it is impossible for the entire span to vibrate at the same frequency.

Problem Solving

The Mysterious Bouncing Bridge

You have learned that resonance is important in amplifying sounds made in musical instruments. Resonance was also blamed in the destruction of this bridge on November 7, 1940, shortly after it was completed.

The Tacoma Narrows Bridge, in the state of Washington, was a large suspension bridge constructed largely of steel and concrete. In a storm with gusty winds, the bridge began to oscillate in a wavelike pattern. Over several hours, the amplitude of the waving bridge increased and it finally collapsed. What do you think caused the bridge to vibrate at its natural frequency?

Solve the Problem:
1. What is resonance?
2. What might have produced the force that caused the bridge to vibrate on this day?

Think Critically:
How do you think engineers could prevent something like this from happening again?

518 Chapter 18 Waves and Sound

This produces the first overtone. Its frequency is twice the fundamental frequency. Overtones have frequencies that are multiples of the fundamental frequency. The intensity and number of overtones vary with each instrument to form a distinct quality of sound. **Figure 18-15** illustrates the fundamental frequency and the first three overtones of a guitar string.

Music is written and played based on a musical scale of eight notes. Each note has a characteristic frequency, and all eight notes span a frequency range called an octave. The highest note in an octave has exactly twice the frequency of the lowest note in that octave.

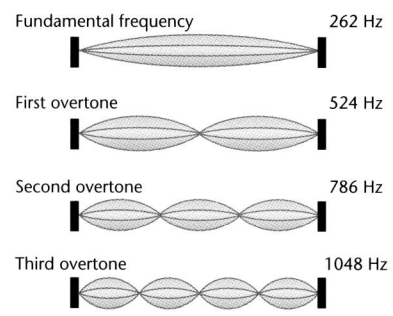

Fundamental frequency   262 Hz

First overtone   524 Hz

Second overtone   786 Hz

Third overtone   1048 Hz

Figure 18-15

A guitar string can vibrate in different ways to produce overtones, giving it a unique sound quality. *How is its sound different from a piano?*

Interference

In a band performance, you may have heard several instruments playing the same notes at the same time. What is the purpose of having more than one instrument creating the same sound? The waves are combining to form a new wave. **Interference** is the ability of two or more waves to combine and form a new wave. Because the musicians are simultaneously playing the same note, their compressions overlap to form a greater compression. As a result, the music sounds much louder. Constructive interference occurs when the compressions of different waves arrive at the same place at the same time.

Figure 18-16

Two or more waves can interfere and add together to produce a new wave.

A In constructive interference, the sound compressions overlap to produce a louder sound.

B In destructive interference, the compressions and rarefactions combine to cancel the sound waves.

C Two slightly different frequencies will result in alternating constructive and destructive interference. This results in a pattern of increases and decreases in sound intensity called beats. *How could beats be used to tune orchestra instruments?*

Sometimes the compression of one wave will arrive with the rarefaction of another wave. They cancel each other, resulting in a decrease in loudness. This is an example of destructive interference.

Destructive Interference

If a tuning fork with a frequency corresponding to A at 440 Hz is used to tune a piano, the piano frequency of A should blend perfectly with the tuning fork. If the piano produces a slightly different frequency from A, the compressions from the fork can't continue to arrive at the same time as the compressions from the piano. The musician will hear variations of sound intensity called beats. The sum of the amplitudes of the waves causes the loudness to regularly rise and fall. Have you ever heard two flutes play the same note when they weren't properly tuned? You could clearly hear the beats until the musicians correctly adjusted their instruments. **Figure 18-16** shows a comparison of the types of sound wave interference.

18-4 Music to Your Ears   **519**

Use these program resources as you teach this lesson.
Study Guide, p. 77
Cross-Curricular Integration, p. 24
Multicultural Connections, pp. 39-40

3 Assess

Check for Understanding

? FLEX Your Brain

Use the Flex Your Brain activity to have students explore INTERFERENCE.

Activity Worksheets, p. 5

Reteach

IS **Interpersonal** Have groups of two students complete a written analysis of one of the illustrations **(Figure 18-13, 18-14,** or **18-16)** in this section. Have each group prepare a written explanation of the concept the diagram is representing, including definitions of terms.

Extension

For students who have mastered this section, use the **Reinforcement** and **Enrichment** masters.

Across the Curriculum

Music Group students in threes and have each group select an instrument. Each group should investigate what vibrates, discover how to produce a change in pitch, and decide to which family of instruments its instrument belongs. Have groups share their findings with the class.

L1 IS

MiniLAB

Purpose

IS **Auditory-Musical** Observe and interpret how hearing loss changes the sounds we hear. **L1** **LEP** **COOP LEARN**

Materials

one radio for not more than six students
Alternate Materials A small radio or small speaker could be enclosed in a pad of cloth.

Teaching Strategies

Work might be best done at home by individuals. Each listener needs two pads of several layers of cloth or small pillows to muffle sound.

Analysis

1. High frequencies are harder to hear, so women's voices are harder to understand.

2. Most affected are consonant sounds.

3. People with hearing losses should be spoken to face-to-face, at a steady, unrushed pace with a slight emphasis on consonant sounds.

✓ Assessment

Performance Have students hypothesize how these results would be affected if the bass were minimized and the treble maximized. Have them test their hypotheses. Use the Performance Task Assessment List for Formulating a Hypothesis in **PASC**, p. 21. **P**

📁 **Activity Worksheets,** pp. 5, 108

MiniLAB

How can a hearing loss change the sounds you hear?

Procedure

1. To simulate a hearing loss, tune a radio to a news or talk station. Turn the volume down to the lowest level you can hear and understand.
2. Turn the bass to maximum and the treble to minimum. If the radio does not have these controls, mask out the higher frequency sounds with heavy pads over your ears.
3. Observe which kinds of sounds are most difficult to hear and which are easiest to hear.

Analysis

1. Are high or low pitches harder to hear? Are men's or women's voices harder to understand?
2. Are vowel or consonant sounds more difficult to hear?
3. How could you help a person with a hearing loss understand what you say?

Acoustics

At a concert or school assembly where a speaker sound system is used, someone usually speaks into the microphone to test the system. Sometimes you hear the sound linger for a couple of seconds. Perhaps you hear echoes of the sound. The sound reaches your ears at different times because it has been reflected off different walls and objects around you. This effect produced by many reflections of sound is called **reverberation.**

Concert halls and theaters like the one shown in **Figure 18-17** are designed by scientists and engineers who specialize in **acoustics,** the study of sound. You will often see carpets

Figure 18-17

This auditorium has good acoustic properties. Its shape was carefully designed by acoustical engineers, and it has both hard and soft surfaces.

and draperies lining the walls of concert halls. Soft, porous materials and certain room shapes can reduce excess reverberation. An anechoic chamber, **Figure 18-18**, is designed to deaden all reverberations. Acoustic scientists also work on understanding human hearing and speaking processes.

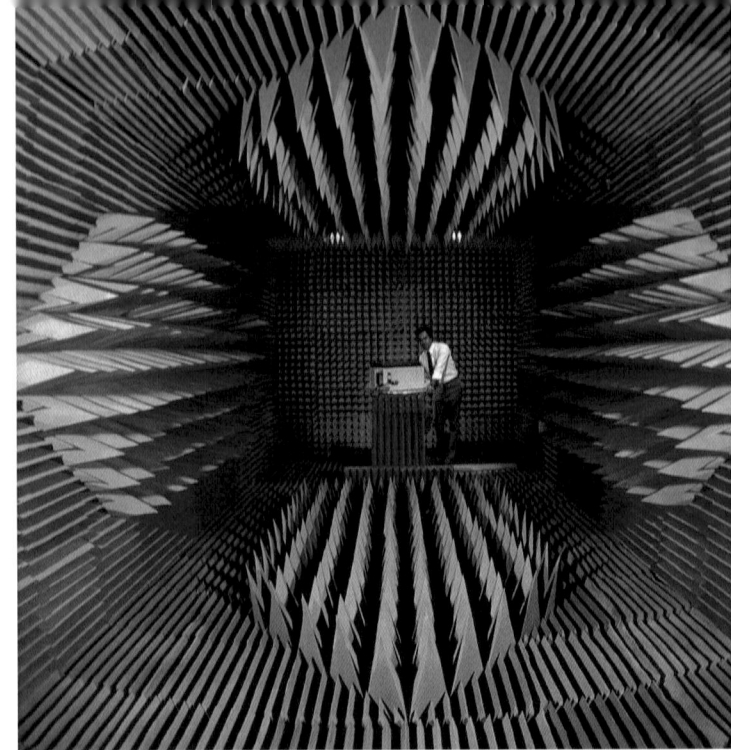

Figure 18-18

An anechoic chamber is designed to eliminate nearly all reflection of sound. Such rooms are used for testing sound equipment.

Section Wrap-up

Review

1. Compare and contrast music and noise.

2. If you were to close your eyes and listen to middle C played both on a flute and on a cello, what musical property would enable you to distinguish one instrument from the other?

3. Explain the difference between constructive and destructive interference.

4. **Think Critically:** Intense, high-frequency sound can actually cause glass to shatter. What might be happening to the glass that causes it to break?

Skill Builder

Recognizing Cause and Effect
What is the effect when sound waves interact by constructive interference? Describe the cause-and-effect sequence for destructive interference. If you need help, refer to Recognizing Cause and Effect in the **Skill Handbook.**

Science Journal

It can be difficult to converse in a crowded room. In your Science Journal, write a paragraph explaining why it might be hard to distinguish the words from a nearby friend in such a situation.

18-4 Music to Your Ears   **521**

Skill Builder

Constructive interference causes an increase in amplitude, intensity, and loudness of the sound heard. Destructive interference is caused when a compression arrives along with a rarefaction, causing a decrease in sound intensity. As a result, less sound is heard.

✓ Assessment

Oral Ask students to explain why there are sometimes spots in auditoriums where little sound is heard. Which type of interference might explain this? Use the Performance Task Assessment List for Making Observations and Inferences in **PASC**, p. 17. **P**

4 Close

•MINI•QUIZ•

Use the Mini Quiz to check students' recall of chapter content.

1. **Two different instruments that play at the same pitch and loudness may have different _____.** *qualities*

2. **What is sound with no set pattern and no definite pitch?** *noise*

3. **_____ occurs when two or more waves combine to form a new wave.** *interference*

Section Wrap-up

Review

1. Music has definite pitches and a rhythmic pattern; noise has irregular patterns and random pitches.

2. Instruments differ in their sound qualities.

3. Compressions overlap, arrive together, and amplify sound in constructive interference; in destructive interference, cancellation occurs when compressions and rarefactions overlap.

4. **Think Critically** If the glass is vibrated by the compressional waves and resonates at the same frequency, it might shatter.

Science Journal

Sounds from many directions can cause complicated interference. The sound waves from your friend will combine with those from many others, making all sounds less clear.

Source

• Laurence Perrine. *Sound and Sense.* New York: Harcourt Brace, 1977.

Biography

American poet Vachel Lindsay (1879-1931) was known for the powerful rhythms of his work. He often read his poems aloud because he believed poetry should be an oral art. For several summers, Lindsay traveled and recited his poems to earn food and shelter.

Background

• *Alliteration* is one type of repetition, using words that begin with the same sound, often a consonant.

• A high proportion of vowels tends to give a poem pleasant, musical tones. The consonants that are formed by building up oral pressure and releasing it suddenly (*p, b, t, d, g, k*) can serve as harsh, interrupting sounds. Incorporating many of these kinds of consonants can make a poem abrupt and difficult to read aloud—which might be the poet's intention.

Teaching Strategies

• Ask volunteers to read aloud the poems they wrote in their Science Journals.

• Invite pairs of students to create a phrase or sentence that is soothing and one that is jarring.

Discuss with students how the use of sound in poetry affects readers who do not hear the poem read aloud.

The Sound of Poetry

The sound of words is an essential part of literature in general and poetry in particular. All sounds start with an object vibrating. These vibrations cause compressional waves in a medium, usually air, that are interpreted by the ear as sound. It is the human brain that gives special meaning to certain sounds. Skilled poets choose words with sounds that help reinforce the meaning and the feeling of their poetry.

Sound Words

Onomatopoeia is the use of real or made-up words that mimic sounds, such as *buzz, hiss, rustle, sip,* and *snap.* See how poet Vachel Lindsay uses onomatopoeia in this section of "The Congo":

> Beat an empty barrel with the handle of a broom
> Hard as they were able,
> Boom, boom, BOOM,
> With the silk umbrella and the handle of a broom
> Boomlay, boomlay, boomlay, BOOM.

The rhythm of the lines combined with the use of onomatopoeia gives life to Lindsay's poem.

Letter Power

Words with long vowels, such as *seamless* and *ooze,* take time to pronounce and slow down the pace of a poem. On the other hand, *-att-* in the middle of a word suggests quick, uncontrolled movement, as in *rattle, shatter, clatter,* and *splatter.* A final *-ck* can suggest movement that stops abruptly, as in *peck, hack,* and *flick.*

The sounds of certain letter combinations can foster visual and tactile images. For example, an initial blend *fl-* often suggests a quickly moving light, as in *flicker, flare,* and *flash.* The sensation of slippery wetness can enter a poem through words beginning with *sl-*, such as *slick, slide*—and *slippery.* The short *i* vowel can add a feeling of smallness, as in *thin, inch, little, chip,* and *sliver.*

While the physical characteristics of sound waves determine their loudness and pitch, carefully chosen combinations of sound help poets share their ideas with readers and listeners.

Science Journal

In your Science Journal, write a poem that uses onomatopoeia and/or the sound of words to reinforce the viewpoint, idea, or feeling you want to share.

522   Chapter 18 Waves and Sound

Linguistic Encourage students to share poems they like. Look for the use of onomatopoeia and other devices based on sound.

• Explore ways that onomatopoeia and the sounds of words and letter combinations might be helpful in prose writing.

Other Works

By Vachel Lindsay: "The Congo," "The Santa Fe Trail."

Summary

18-1: Characteristics of Waves

1. In a transverse wave, the medium moves at right angles to the direction the wave travels.
2. If the speed of a wave remains constant, as the frequency increases, the wavelength decreases and vice versa.
3. The velocity of a wave is equal to its wavelength multiplied by its frequency.

18-2: The Nature of Sound

1. Sound begins as a vibration that is transferred through a medium in a series of compressions and rarefactions.
2. The pitch of a sound becomes higher as the frequency increases. Both amplitude and intensity increase as energy is added to a wave.
3. The Doppler effect is a change of frequency and pitch of a sound as a result of motion.

18-3: Science and Society: Using Sound Advice in Medicine

1. High-frequency sound waves can be used to generate images of tissue inside the body.
2. Ultrasonic technology has many uses, including sonar, cleaning jewelry and glassware, medical diagnosis, and medical treatment.

18-4: Music to Your Ears

1. Music is created using specific pitches, sound quality, and a pattern.
2. Sound quality describes the differences among sounds of the same pitch and loudness.
3. Interference is the ability of two or more waves to combine and form a new wave.

Key Science Words

a. acoustics
b. amplitude
c. compressional wave
d. crest
e. frequency
f. intensity
g. interference
h. loudness
i. medium
j. music
k. noise
l. pitch
m. quality
n. resonance
o. reverberation
p. transverse wave
q. trough
r. ultrasonic technology
s. wave
t. wavelength

Reviewing Vocabulary

Match each phrase with the correct term from the list of Key Science Words.

1. distance between identical points on two waves
2. using high-frequency sound waves
3. the study of sound
4. matter vibrates in the same direction as the wave travels
5. the highness or lowness of a sound
6. material through which a wave travels
7. expressed in hertz
8. the highest points of a transverse wave
9. distance from the rest position of a medium to the trough or crest of a wave
10. human perception of sound intensity

Summary

Have students read the summary statements to review the major concepts of the chapter.

Reviewing Vocabulary

1. t
2. r
3. a
4. c
5. l
6. i
7. e
8. d
9. b
10. h

Assessment

Portfolio Encourage students to place in their portfolios one or two items of what they consider to be their best work. Examples include:
- Practice Problem calculations, p. 501
- MiniLAB answers, p. 506
- Activity 18-2 results and answers, p. 513 **P**

Performance Additional performance assessments may be found in **Performance Assessment** and **Science Integration Activities**. Performance Task Assessment Lists and rubrics for evaluating these activities can be found in Glencoe's **Performance Assessment in the Science Classroom**.

GLENCOE TECHNOLOGY

MindJogger Videoquiz

Chapter 18 Have students work in groups as they play the Videoquiz game to review key chapter concepts.

Checking Concepts

1. b	**6.** d
2. a	**7.** a
3. b	**8.** b
4. d	**9.** a
5. c	**10.** c

Understanding Concepts

11. In a transverse wave, amplitude is the distance between the rest position of the medium and the crest or trough; frequency is the number of crests that pass a point in one second; and wavelength is the distance between two consecutive crests or troughs. In a compressional wave, amplitude is the amount of compression of the medium's particles; frequency is the number of compressions that pass a point in one second; and wavelength is the distance between two consecutive compressions or rarefactions.

12. All three occur when two waves meet. In constructive interference, waves of equal frequency meet crest-to-crest so that their amplitudes add up and loudness increases. In destructive interference, waves of equal frequency meet crest-to-trough so that their amplitudes cancel one another and loudness decreases. Beats are formed when waves of different frequencies meet, resulting in alternating periods of loudness (constructive interference) and softness (destructive interference).

13. A note played on a musical instrument is associated with a fundamental frequency. Whole-number multiples of this frequency produce overtones. Each instrument has a different number of overtones for a

---

Checking Concepts

Choose the word or phrase that completes the sentence.

1. All waves carry _____ forward.
 a. matter
 b. energy
 c. matter and energy
 d. the medium

2. A wave that carries a large amount of energy will always have a _____.
 a. large amplitude
 b. small amplitude
 c. high frequency
 d. short wavelength

3. A sound with a low pitch always has a low _____.
 a. amplitude
 b. frequency
 c. wavelength
 d. wave velocity

4. As _____, sound intensity decreases.
 a. wave velocity decreases
 b. wavelength decreases
 c. quality decreases
 d. amplitude decreases

5. Sounds with the same pitch and loudness traveling in the same medium may differ in _____.
 a. frequency
 b. amplitude
 c. quality
 d. wavelength

6. Sound cannot travel through _____.
 a. solids
 b. liquids
 c. gases
 d. empty space

7. Variations in the loudness of sound that are caused by wave interference are called _____.
 a. beats
 b. standing waves
 c. pitch
 d. forced vibrations

8. _____ is shown when a windowpane vibrates at the same frequency as a thunderclap.
 a. The Doppler effect
 b. Resonance
 c. Reverberation
 d. Destructive interference

9. Wave frequency is measured in _____.
 a. hertz
 b. decibels
 c. meters
 d. meters/second

10. When a sound source moves away from you, the sound's _____.
 a. velocity decreases
 b. loudness increases
 c. pitch decreases
 d. frequency increases

Understanding Concepts

Answer the following questions in your Science Journal using complete sentences.

11. Describe how amplitude, frequency, and wavelength are determined in transverse and compressional waves.

12. What do constructive interference, destructive interference, and the formation of beats have in common? How are they different?

13. Explain how different combinations of wave frequencies make a guitar and a trumpet sound different even when the same note is played.

14. In what ways can ultrasonic technology be used in medicine?

15. Why does the ringing of an alarm clock enclosed in an airtight container become softer as the air is drawn out of the container?

Thinking Critically

16. A wave has a wavelength of 6 m and a wave velocity of 420 m/s. What is its frequency?

17. A bus driver is rounding a curve approaching a railroad crossing. She hears a train's whistle and then hears the whistle's pitch become lower. What assumptions can she make about what she will see when she rounds the curve and looks at the crossing?

18. When a little boy blows a dog whistle, his dog comes, even though the boy can't hear the whistle. Explain why the boy can't hear the whistle, but his dog can.

---

given note. This produces distinct sound qualities that make the instruments sound different from one another.

14. It can detect tumors and other soft-tissue growths, determine fetal development, monitor organ function, crush kidney stones, etc.

15. Compressional waves need a medium to be transferred. As the density of the medium decreases, less energy is carried by the waves.

Thinking Critically

16. Given: *wavelength* = 6 m
 velocity = 420 m/s
 Unknown: *frequency*
 Equation: $v = \lambda \times f$
 Solution: 420 m/s = 6 m $\times f$
 f = (420 m/s)/6 m= 70 Hz

19. An earthquake beneath the middle of the Pacific Ocean produces a tidal wave that hits a remote island. Is the water that hits the island the same water that was above the earthquake? Explain.

Developing Skills

If you need help, refer to the **Skill Handbook.**

20. **Hypothesizing:** Sound travels slower in air at high altitudes than at low altitudes. State a hypothesis to explain this observation.
21. **Observing and Inferring:** Infer the effect of increasing wave velocity on the wavelength of a compressional wave that has a constant frequency.
22. **Concept Mapping:** Design a concept map that shows the characteristics of compressional waves. Include the terms *pitch, compression, medium, rarefaction, speed, loudness, wavelength, frequency,* and *intensity.* Indicate how pitch and the loudness you hear in sound waves relate to these characteristics.
23. **Using Numbers:** Study the speed of sound for various materials at 25°C in the table below. Rank the materials from the fastest to slowest transmitter of sound waves. Approximately how many times faster does sound travel in steel compared to in air?

Sound Transmission

Substance	Speed of Sound 25°C
Air	347 m/s
Brick	3650 m/s
Cork	500 m/s
Water	1498 m/s
Steel	5200 m/s

24. **Making and Using Tables:** You have started a lawn-mowing business during summer vacation. Your family's power lawn mower has a sound level of 100 dB. Using the table below, determine how many hours a day you can safely work mowing lawns. If you want to work longer hours, what can you do to protect your hearing? If your family purchases a new lawn mower with a sound level of 95 dB, how will your business be affected?

Federally Recommended Noise Exposure Limits

Sound Level (decibels)	Time Permitted (hours per day)
90	8
95	4
100	2
105	1
110	0.5

Performance Assessment

1. **Poster:** In Activity 18-1 on pages 502-503, you observed some of the behaviors of transverse waves on a Slinky. Use the Slinky to create constructive and destructive interference by sending waves toward each other. Make a poster showing the spring before, during, and after the waves overlap.
2. **Designing an Experiment:** Try to use an empty cardboard paper-towel roll to repeat the procedure used in Activity 18-2 on page 513 to find the tube's natural frequency. Do you hear as clear a sound? Explain why or why not.
3. **Project:** Using materials you have at home, make a musical instrument. Play your instrument for your classmates and explain how you can change the pitch of your instrument.

17. The engine will have passed the crossing. The drop in the pitch of the whistle indicates that the engine is moving away.
18. The whistle's frequency is outside the range of human hearing.
19. No. Waves transfer energy, not matter. The water that hits the shore was already near the shore.

Developing Skills

20. **Hypothesizing** Air temperature is lower and air is less dense at high altitudes, and the slower-moving particles do not transfer sound as quickly.
21. **Observing and Inferring** If frequency does not change, then wavelength increases with velocity.
22. **Concept Mapping** See concept map below.
23. **Using Numbers** Steel, brick, water, cork, air; 15 times faster.
24. **Making and Using Tables** You can work two hours. You should wear hearing protection. You can work twice as long.

Performance Assessment

1. Student posters should be neatly labeled. Use the Performance Task Assessment List for Poster in **PASC**, p. 73. **P**
2. No, the cardboard serves as a dampening medium and absorbs the sound waves. Use the Performance Task Assessment List for Designing an Experiment in **PASC**, p. 23. **P**
3. Student instruments will vary. Use the Performance Task Assessment List for Model in **PASC**, p. 51. **P**

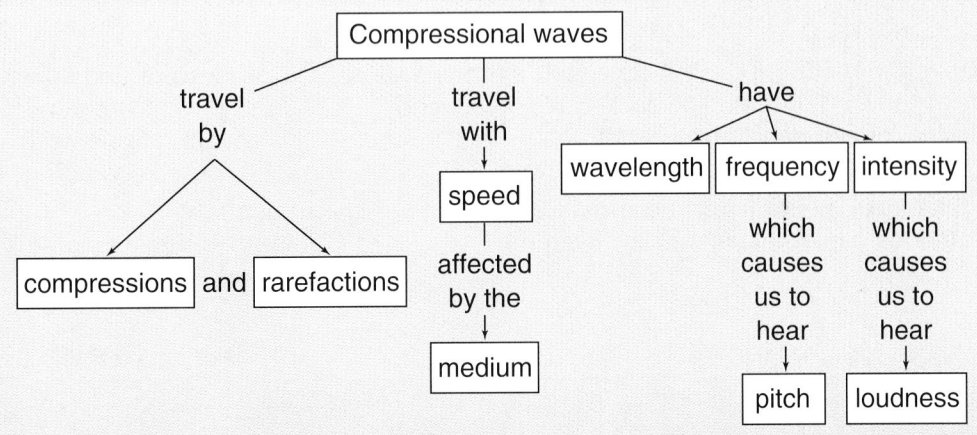

Chapter Organizer

Section	Objectives/Standards	Activities/Features
Chapter Opener		**Explore Activity:** Find out how you can make a rainbow. p. 527
19-1 Electromagnetic Radiation (1 session, 1 block)*	1. **Contrast** electromagnetic waves with other kinds of waves. 2. **Describe** the arrangement of electromagnetic waves on the electromagnetic spectrum. 3. **Explain** at least one application of each type of electromagnetic wave. **National Content Standards:** (5-8) UCP1-UCP3, UCP5, B3, D1, G1, G3; (9-12) UCP1-UCP3, UCP5, B6, D1, G1, G3	**MiniLAB:** Do lightbulbs waste energy? p. 532 **Connect to Chemistry,** p. 533 **Science Journal,** p. 535 **Skill Builder:** Observing and Inferring, p. 535
19-2 Light and Color (1 session, ½ block)*	1. **Describe** the differences among opaque, transparent, and translucent materials. 2. **Explain** how you see color. 3. **Describe** the difference between light color and pigment color. **National Content Standards:** (5-8) UCP1-UCP3, UCP5, A1, A2, E1; (9-12) UCP1-UCP3, UCP5, A1, A2, B6, E1	**MiniLAB:** How do filters affect the colors you see? p. 539 **Problem Solving:** Color in the Sunday Comics, p. 540 **Using Computers,** p. 541 **Skill Builder:** Concept Mapping, p. 541 **Activity 19-1:** Spectrum Inspection, pp. 542-543
19-3 Science and Society (1 session, ½ block)*	1. **Explain** how incandescent and fluorescent bulbs work. 2. **Analyze** the advantages and disadvantages of different light sources. **National Content Standards:** (5-8) UCP2, UCP5, B3, E2, F2, F4, F5; (9-12) UCP2, UCP5, B6, E2, F3, F4, F6	**Connect to Chemistry,** p. 545 **Explore the Technology,** p. 545
19-4 Wave Properties of Light (2 sessions, 1½ blocks)*	1. **State** and **give an example** of the law of reflection. 2. **Explain** how refraction is used to separate white light into the colors of the spectrum. 3. **Describe** how diffraction and interference patterns demonstrate the wave behavior of light. **National Content Standards:** (5-8) UCP1-UCP3, UCP5, A2, E2, F5, G1, G3; (9-12) UCP1-UCP3, UCP5, A2, B6, E2, F6, G1, G3	**MiniLAB:** How does water bend light? p. 548 **Using Technology:** Holograms, p. 549 **Using Math,** p. 550 **Skill Builder:** Observing and Inferring, p. 550 **Activity 19-2:** Make a Light Bender, p. 551 **Science and Art:** The Art of Lighting, p. 552 **Science Journal,** p. 552

Activity Materials

Explore	Activities	MiniLABs
page 527 flashlight, glass prism, sudsy dish water, compact disc	pages 542-543 diffraction grating; power supply with dimmer switch; clear, tubular lightbulb and socket; red, blue, and yellow colored pencils; oven mitt page 551 light source, pencil, clear rectangular container, water, notebook paper, modeling clay, oven mitt	page 532 incandescent and fluorescent lightbulbs of the same wattage, foam cup, plastic wrap, thermometer, watch or clock page 539 red and green colored pencils, red and green plastic filters page 548 penny, short opaque cup, water

Need Materials? Call Science Kit (1-800-828-7777).   * A complete Planning Guide that includes block scheduling is provided on pages 32T-35T.

Teacher Classroom Resources

Reproducible Masters	Transparencies	Teaching Resources
Study Guide, p. 78 **Reinforcement**, p. 78 **Enrichment**, p. 78 **Activity Worksheets**, p. 113 **Lab Manual 39**, Light Intensity **Concept Mapping**, pp. 43-44 **Science and Society Integration**, p. 23	**Section Focus Transparency 74,** Radiation Is All Around Us **Science Integration Transparency 19,** Introducing Roy G. Biv **Teaching Transparency 37,** The Electromagnetic Spectrum	**Physical Science CD-ROM** **Spanish Resources** **English/Spanish Audiocassettes** **Cooperative Learning Resource** Guide **Lab Partner** **Lab and Safety Skills** **Lesson Plans**
Study Guide, p. 79 **Reinforcement**, p. 79 **Enrichment**, p. 79 **Activity Worksheets**, pp. 109-110, 114 **Lab Manual 40**, Producing a Spectrum **Science Integration Activity 19,** Rainbows, Too! **Multicultural Connections**, p. 41 **Cross-Curricular Integration**, p. 25	**Section Focus Transparency 75,** Paint Pigments	
Study Guide, p. 80 **Reinforcement**, p. 80 **Enrichment**, p. 80	**Section Focus Transparency 76,** Counting Kilowatt-Hours	
Study Guide, p. 81 **Reinforcement**, p. 81 **Enrichment**, p. 81 **Activity Worksheets**, pp. 111-112, 115 **Critical Thinking/Problem Solving**, p. 25	**Section Focus Transparency 77,** Rainbow Refraction **Teaching Transparency 38,** Wave Interference	

Assessment Resources

Chapter Review, pp. 41-42
Assessment, pp. 126-129
Performance Assessment in the
Science Classroom (PASC)
MindJogger Videoquiz
Alternate Assessment in the
Science Classroom
Performance Assessment
Chapter Review Software
Computer Test Bank

Key to Teaching Strategies

The following designations will help you
decide which activities are appropriate for
your students.

L1 Level 1 activities should be within the
ability range of all students, including
those with learning difficulties.

L2 Level 2 activities should be within the
ability range of the average to above-average student.

L3 Level 3 activities are designed for the ability range of above-average students.

LEP LEP activities should be within the ability range of Limited English Proficiency
students.

LS These activities are designed to address
different learning styles.

COOP LEARN Cooperative Learning activities are designed for small
group work.

P These strategies represent student products that can be placed into a best-work
portfolio.

GLENCOE TECHNOLOGY

The following multimedia resources are available from Glencoe.

Science and Technology Videodisc
Series (STVS)
Physics
Schlieren Photography
Photo-Acoustic Cell
Laser Eye Surgery
Human Biology
Vision Diagnosis

The Infinite Voyage Series
Unseen Worlds
Crisis in the Atmosphere
The Living Clock
National Geographic Society Series
STV: Plants

Physical Science CD-ROM

Teacher Classroom Resources

This is a representation of key blackline masters available in the Teacher Classroom Resources.

Teaching Aids

Section Focus Transparencies

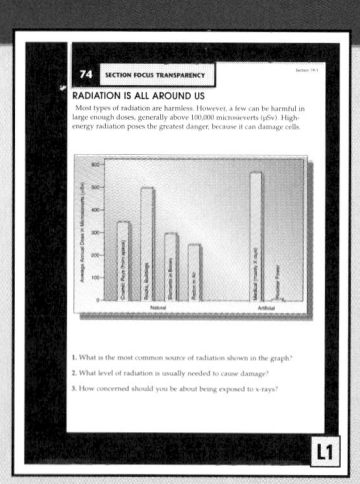

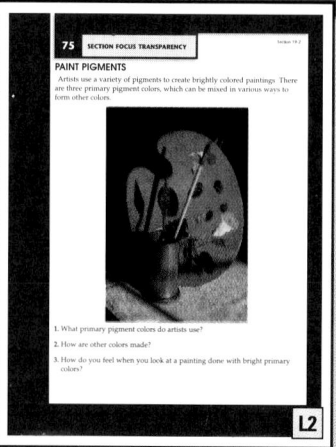

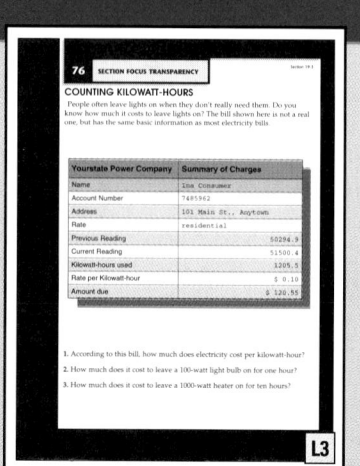

Science Integration Transparencies

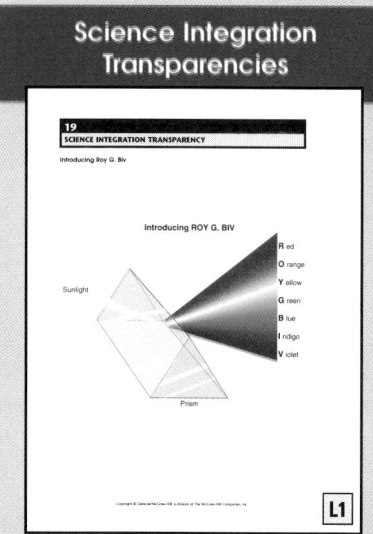

Teaching Transparencies

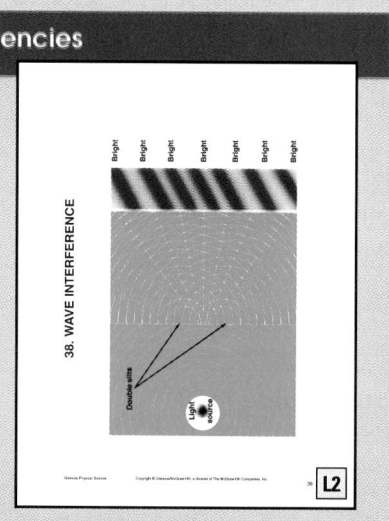

Meeting Different Ability Levels

Study Guide

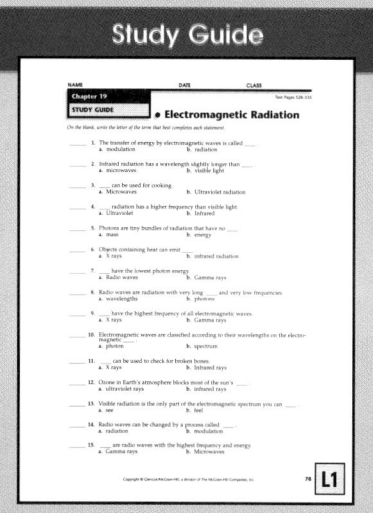

Reinforcement

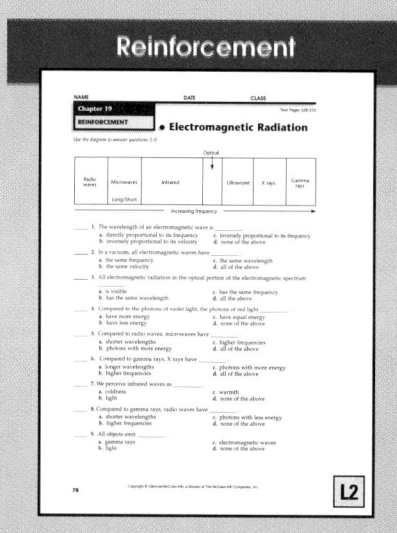

Enrichment Worksheets

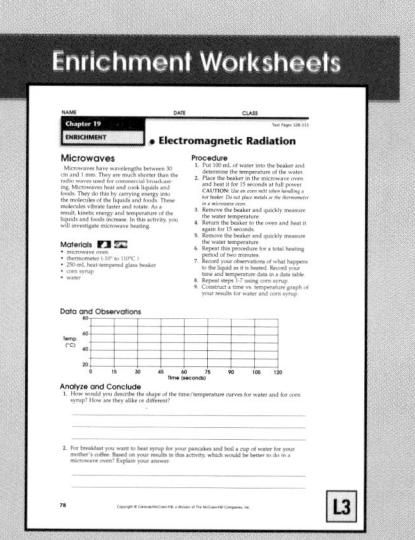

Hands-On Activities

Science Integration Activity

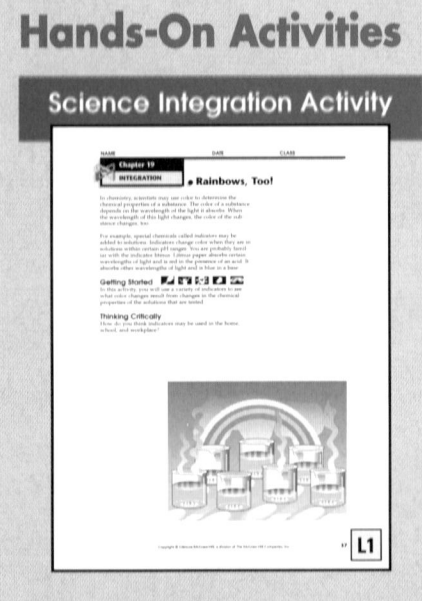

L1

Lab Manual

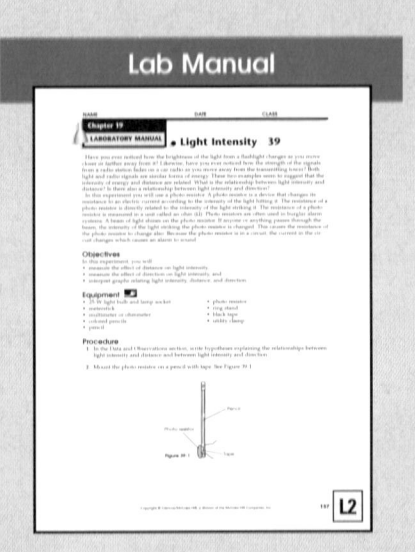

L2

Assessment

Performance Assessment

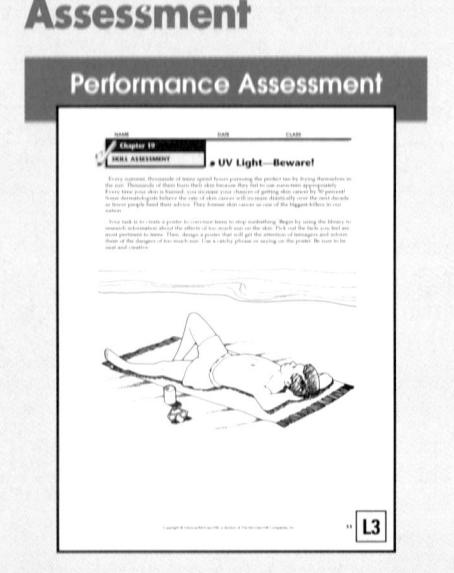

L3

Enrichment and Application

Critical Thinking/ Problem Solving

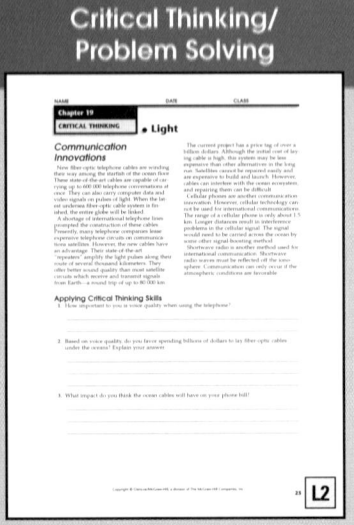

L2

Cross-Curricular Integration

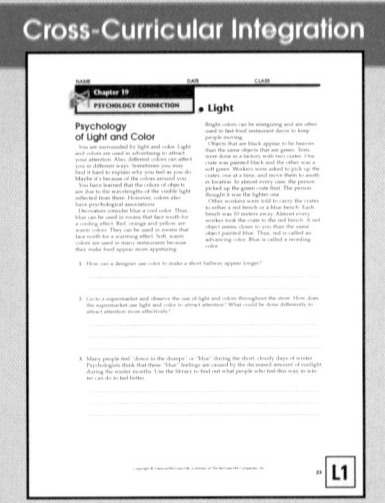

L1

Science and Society Integration

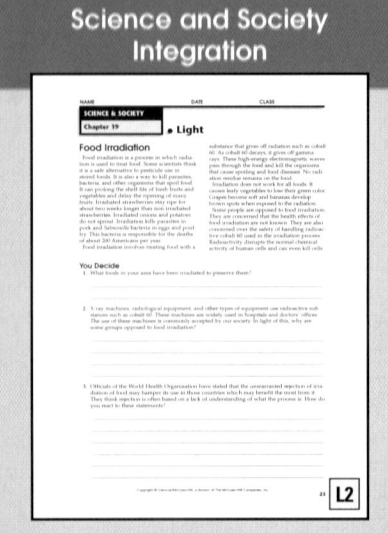

L2

Multicultural Connections

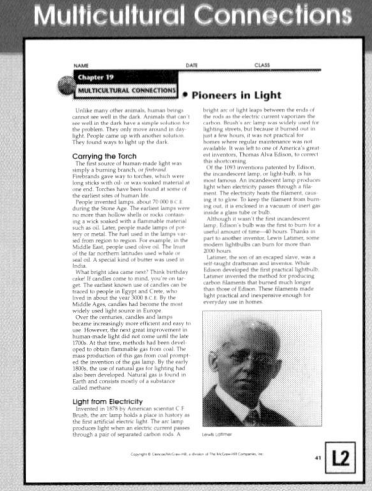

L2

Concept Mapping

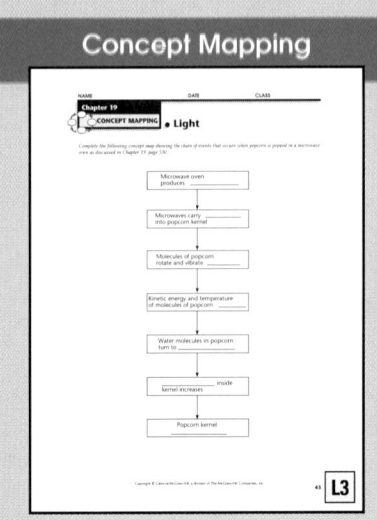

L3

Light

CHAPTER OVERVIEW

Section 19-1 This section introduces the electromagnetic spectrum and discusses each type of radiation.

Section 19-2 The spectrum of light is used to explain colors, and the differences among transparent, translucent, and opaque objects are discussed. Pigments are explained.

Section 19-3 Science and Society The differences between incandescent and fluorescent bulbs are explored. Students are encouraged to make informed decisions as energy-conscious consumers.

Section 19-4 Reflection and refraction are discussed. Diffraction and interference patterns are identified as wave properties of light.

Chapter Vocabulary

electromagnetic radiation	opaque material
photon	transparent material
radio wave	translucent material
modulation	pigment
microwave	incandescent light
infrared radiation	fluorescent light
visible radiation	reflection
ultraviolet radiation	refraction
X ray	diffraction
gamma ray	diffraction grating

Theme Connection

Energy Electromagnetic radiation is a form of energy. Light is one form of electromagnetic radiation. Help students make connections using applications of radiation they are already familiar with. Stress that light and other types of radiation are actually forms of energy.

526

Previewing the Chapter

Section 19-1   Electromagnetic Radiation
► The Electromagnetic Spectrum
► Radio Waves
► Infrared Radiation
► Visible Radiation
► Ultraviolet Radiation
► X Rays and Gamma Rays

Section 19-2   Light and Color
► Light and Matter

► Colors
► Pigments

Section 19-3   Science and Society Technology: Battle of the Bulbs

Section 19-4   Wave Properties of Light
► Reflection
► Refraction
► Diffraction and Interference

526

Learning Styles

Look for the following logo for strategies that emphasize different learning modalities. **LS**

Kinesthetic	MiniLAB, p. 532; Reteach, pp. 540, 549
Visual-Spatial	Explore Activity, p. 527; Demonstration, pp. 530, 533, 537, 539; MiniLAB, pp. 538, 548; Visual Learning, pp. 529, 531, 533, 537, 544, 547; Across the Curriculum, p. 539; Science & Art, p. 552
Interpersonal	Reteach, p. 534; Activity 19-2, p. 551; Science & Art, p. 552
Logical-Mathematical	Activity 19-1, pp. 542-543
Linguistic	Across the Curriculum, p. 548

Chapter

19

Light

Light passing through a prism can produce exciting patterns of color. Imagine what your surroundings would look like now if humans could see only shades of gray, and not distinct colors. The ability to see color depends on the cells in your eyes that are sensitive to different wavelengths of light. What color is the light produced by a flashlight or the sun?

EXPLORE ACTIVITY

Find out how you can make a rainbow.

1. In a darkened room, shine a flashlight through a glass prism and try to project the resulting colors onto a white wall or ceiling.
2. In a darkened room, shine a flashlight over the surface of some water with dishwashing liquid bubbles in it. What do you see?
3. Aim a flashlight at the surface of a compact disc.

Observe: What did you see while aiming the white light of the flashlight at these three different materials? How did your observations differ? In your Science Journal, explain where you think the colors came from.

Previewing Science Skills

▶ In the Skill Builders, you will **observe and infer**, and **map concepts**.

▶ In the Activities, you will **observe, hypothesize, compare,** and **infer**.

▶ In the MiniLABs, you will **observe, interpret,** and **measure**.

527

EXPLORE ACTIVITY

Purpose
LS **Visual-Spatial** Use the Explore Activity to introduce students to properties of light through observations of several phenomena. L1

Preparation
Bring in a compact disc (CD) in advance.

Materials
prism, flashlight, container with water containing dishwashing liquid, compact disc

Teaching Strategies
Darkening the room will make it easier to see the colors from the prism, dishwashing liquid, and compact disc.

Observe
Students should observe that a small color spectrum formed on the ceiling or in the reflections from soap or the compact disc.

✔ Assessment

Portfolio Have students prepare a poster accurately illustrating the light source, how light reached their eye, and color observations in each of the three steps of the Explore Activity. Use the Performance Task Assessment List for Poster in **PASC**, p. 73. P

Assessment Planner

Portfolio
Refer to page 553 for suggested items that students might select for their portfolios.

Performance Assessment
See page 553 for additional Performance Assessment options.
Skill Builders, pp. 535, 541, 550
MiniLABS, pp. 532, 538, 548
Activities 19-1, pp. 542-543; 19-2, p. 551

Content Assessment
Section Wrap-ups, pp. 535, 541, 545, 550
Chapter Review, pp. 553-555
Mini Quizzes, pp. 535, 541, 545, 550

Group Assessment
Opportunities for group assessment occur with Cooperative Learning Strategies and Flex Your Brain Activities.

19•1 Electromagnetic Radiation

Prepare

Section Background

- Scientists now agree that light has both particle and wave nature, but there is a rich scientific history leading to this idea. In the fifth century B.C., Socrates and Plato thought light was made of streams of light emitted by the eye. Newton formed a particle theory of light at the same time Huygens concluded that light was a wave. Einstein's theory of light as particles called photons was published in an explanation of the photoelectric effect in 1905.

- The photoelectric effect is the ejection of electrons from certain photosensitive metals when photons of light are absorbed by electrons in the material. The photon must have enough energy for the electron to escape from the metal. This is evidence for the particle, or quantum, theory of light.

- Electromagnetic radiation is energy emitted from vibrating electric charges in the form of transverse waves. These waves are composed of an electric field and a magnetic field oscillating at right angles to each other.

Preplanning

To prepare for the MiniLAB, obtain incandescent bulbs and fluorescent bulbs of the same wattage, thermometers, polystyrene cups, and some plastic wrap.

Science Words

electromagnetic radiation
photon
radio wave
modulation
microwave
infrared radiation
visible radiation
ultraviolet radiation
X ray
gamma ray

Objectives

- Contrast electromagnetic waves with other kinds of waves.
- Describe the arrangement of electromagnetic waves on the electromagnetic spectrum.
- Explain at least one application of each type of electromagnetic wave.

The Electromagnetic Spectrum

Do you listen to the radio, watch television, play video games, or cook with a microwave oven? These devices all make use of different kinds of electromagnetic waves. Light, radio waves, and microwaves are all examples of electromagnetic waves.

Electromagnetic waves are transverse waves produced by the motion of electrically charged particles. We often call these waves **electromagnetic radiation** because they radiate from the particles. The transfer of energy by electromagnetic waves is called radiation. One way these waves differ from those you learned about in Chapter 18 is that electromagnetic waves do not need a medium to transfer energy. They can travel through a vacuum—empty space—at a speed of 300 000 km/s. The light from the stars in **Figure 19-1** is one type of electromagnetic wave. These waves travel slower when they pass through any type of matter, but they still travel much faster than sound or water waves.

Describing Electromagnetic Radiation

All electromagnetic waves travel at the same speed in a vacuum, but their frequencies and wavelengths may vary. The shorter the wavelength of a wave, the higher its frequency. **Figure 19-2** shows how electromagnetic waves are classified from low-frequency, long-wavelength radio waves to high-frequency, short-wavelength gamma rays according to their wavelengths on the *electromagnetic spectrum.*

Figure 19-1

Objects in outer space emit different types of electromagnetic radiation.

528 Chapter 19 Light

Program Resources

Reproducible Masters
Study Guide, p. 78 [L1]
Reinforcement, p. 78 [L1]
Enrichment, p. 78 [L3]
Science and Society Integration, p. 23
Activity Worksheets, p. 113 [L1]
Lab Manual 39
Concept Mapping, p. 43

Transparencies
Section Focus Transparency 74 [L1]
Teaching Transparency 37
Science Integration Transparency 19

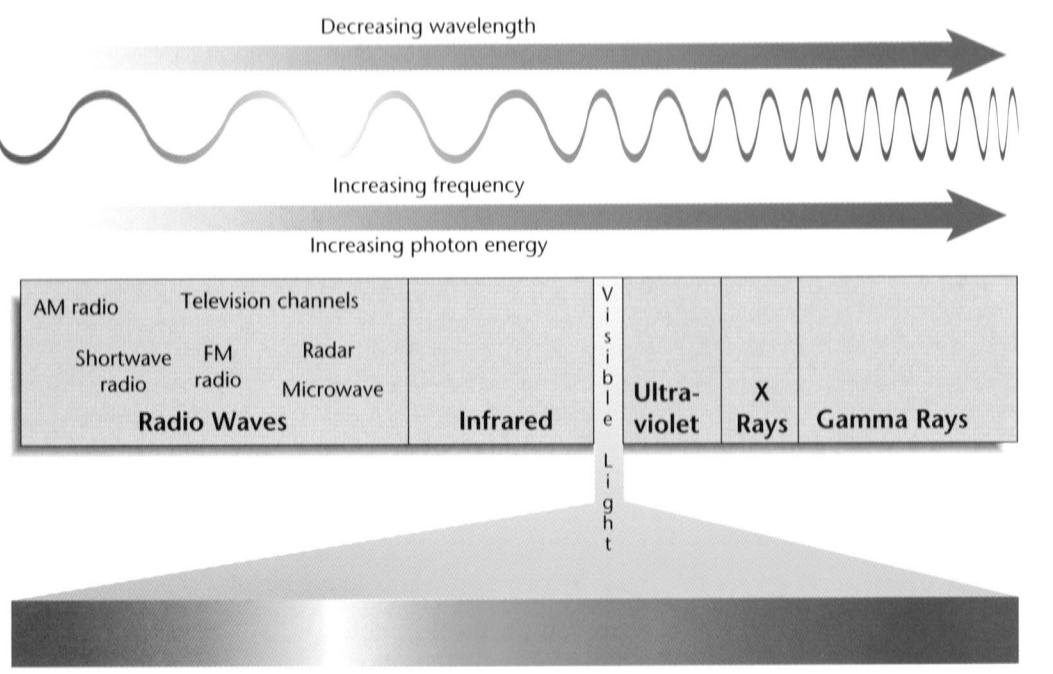

Decreasing wavelength

Increasing frequency

Increasing photon energy

AM radio	Television channels			V i s i b l e L i g h t				
Shortwave radio	FM radio	Radar			Ultra-violet	X Rays	Gamma Rays
		Microwave					
Radio Waves					**Infrared**			

Figure 19-2

All electromagnetic waves are classified by their wavelengths on the electromagnetic spectrum.

Scientists have observed that radiation not only carries energy, but also has momentum. This means that electromagnetic radiation may have particlelike behavior as well as wavelike behavior. In 1905, Albert Einstein hypothesized that light is composed of tiny particles. These tiny, particle-like bundles of radiation are called **photons.** Photons with the highest energy correspond to light with the highest frequency. Very high-energy photons can actually damage matter, such as the cells in your body.

Radio Waves

When the listener in **Figure 19-3** tunes his radio to a station, he is actually adjusting it to respond to radio waves of a certain frequency. Even though radio waves are all around you right now, you can't sense them. **Radio waves** have long wavelengths and low frequencies. Thus, radio waves have the lowest photon energy. Locate radio waves on the electromagnetic spectrum in **Figure 19-2.**

Figure 19-3

You use the tuner to select the station you want to hear. *What characteristic of the radio wave are you selecting?*

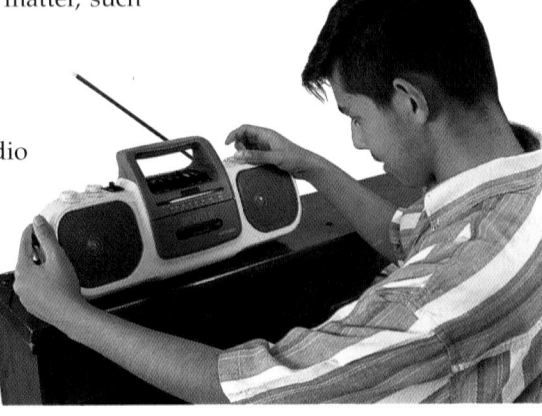

19-1 Electromagnetic Radiation   **529**

In addition to radio communications, radio waves are used in television, cellular telephones, and cordless telephones. How is your voice transmitted by radio waves on a cordless telephone? The sound waves produced by your voice are changed to electrical currents that represent your speech patterns and pitches. These electrical currents are used to vary either the amplitudes or frequencies of radio waves that are transmitted from the handset to the base of the telephone. This process of varying radio waves is called **modulation.** Besides your voice, light images, computer information, and music can also be used to modulate radio waves.

Figure 19-4

Conventional ovens heat food from the outside in. Microwave ovens heat the food molecules internally.

A The popcorn container absorbs very little energy so it remains cool enough to handle.

Cooking with Radio Waves

Radio waves with the highest frequency and energy are called **microwaves.** They are used in communications, but, like the person in **Figure 19-4A,** you are probably most familiar with microwaves for their speedy cooking abilities. Have you ever popped popcorn in a microwave oven? Do you wonder how the kernels pop so quickly? Microwaves transfer energy to the water molecules that are inside the popcorn kernels, causing them to vibrate and rotate faster. As a result, the kinetic energy and temperature of the water molecules increase; the water inside the popcorn kernel turns to steam; and the kernel explodes due to the increased pressure built up by the steam. Containers made of glass, paper, and some plastics are used in microwave ovens because the microwaves pass through them easily and little energy is absorbed by the container. **Figure 19-4B** illustrates the operation of a microwave oven.

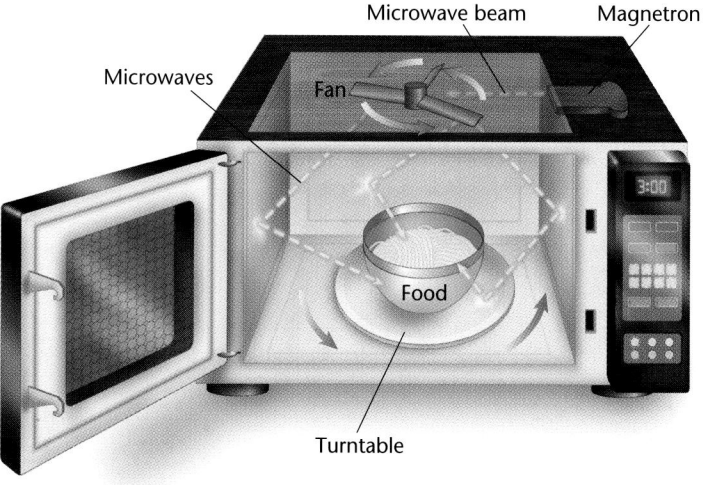

B The spinning fan reflects microwaves onto the food from all directions inside the oven. This heats the food evenly.

Infrared Radiation

Do you like to nap outdoors on a hot summer day? Even with your eyes closed, you feel when clouds move in front of the sun. When the sunlight isn't blocked by the clouds, you feel its warmth on your skin. This warm feeling is caused by the infrared radiation from the sun. **Infrared radiation** has a wavelength slightly longer than visible light, as you can see by its location on the electromagnetic spectrum. Your skin feels warm because it is absorbing some of the infrared radiation from the sun. This causes the molecules in your skin to vibrate more, increasing their kinetic energy as well as your skin's temperature.

How We Use Infrared Radiation

Warm objects, such as your body, give off more infrared radiation than do cool objects. Some parts of your body are warmer than others and, as a result, give off or emit more infrared radiation. Measurement of the body's emission of infrared radiation helps doctors make some medical diagnoses. A thermogram, such as the one shown in **Figure 19-5A**, is produced by measuring the infrared radiation given off by different body parts. Tumors are sometimes detected in a thermogram because they are warmer than the healthy tissue around them.

If you live in a colder climate, an infrared photograph like the one in **Figure 19-5B** can show you where your house or apartment needs insulation. People using infrared-sensitive binoculars can see humans and animals in complete darkness. These binoculars detect infrared waves given off by warm bodies. Some home security systems are designed to detect objects giving off infrared radiation and to respond by activating a light or an alarm.

Figure 19-5

Temperature differences can be detected by devices that are sensitive to infrared radiation. Warmer areas show up as yellow, orange, and red. Cooler areas are green, blue, and black.

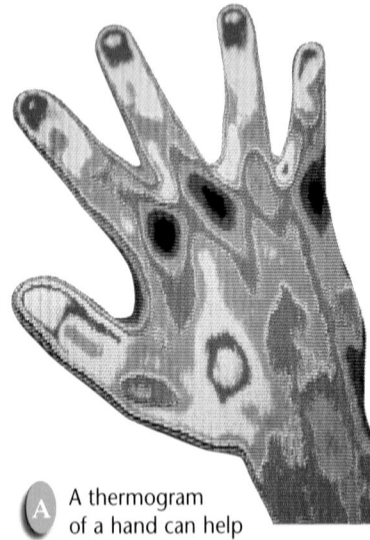

A A thermogram of a hand can help detect circulation problems. *Would the problem area show up as warmer or cooler than the other parts of your hand?*

B Heat loss in your home can be detected by infrared radiation.

531

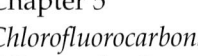
Cultural Diversity

Some Light on the Subject Christiaan Huygens was a Dutch physicist and astronomer who lived at about the same time as Isaac Newton. When he proposed a wave theory for the structure of light that conflicted with Newton's particle theory of light, his skills and insights were questioned. Because Huygens was not English (and was not Newton), his wave theory was ignored for almost a century in England. It took another English physicist, Thomas Young, to propose a wave theory that was accepted. Even so, efforts to fully verify his theory were thwarted in England. The work to overturn Newton's particle theory had to be done in France by Arago and Fresnel.

MiniLAB

Purpose

LS Kinesthetic Students will determine which lightbulb, a fluorescent or incandescent, wastes more energy.

L1 **COOP LEARN**

Materials

fluorescent light and incandescent light of the same wattage, 2 thermometers, 2 polystyrene cups, watch or stopwatch, plastic food wrap

Teaching Strategies

Safety Precautions The bulbs may become hot. Students should not allow the cups or the windows to touch them.

- The heat collector will reach its heat capacity when it absorbs heat as fast as it radiates it. Therefore, the temperature increase is a rough measure of the rate at which heat flows through the window.

Analysis

Students should find that the incandescent lightbulb has more heat flowing from it.

✔ Assessment

Oral Ask students which of the bulbs would give off high levels of infrared radiation. *the incandescent lightbulb* Use the Performance Task Assessment List for Making Observations and Inferences in **PASC**, p. 17. **P**

📁 **Activity Worksheets,** pp. 5, 113

Figure 19-6
You may have seen infrared lamps like these in restaurants, keeping food warm until it is served.

MiniLAB

Do lightbulbs waste energy?

Procedure

1. Obtain an incandescent bulb and a fluorescent bulb of identical wattage.
2. Make a heat collector by covering the top of a foam cup with a plastic food wrap to make a window. Carefully make a small hole (diameter less than the thermometer's) in the side of the cup. Push a thermometer through the hole while supporting the cup with your hand.
3. Measure the temperature of the air inside the cup. Then, hold the window of the tester 1 cm from one of the lights for 2 minutes and measure the temperature.
4. Cool the heat collector and thermometer. Repeat step 3 using the second type of bulb.

Analysis

1. What was the temperature inside the cup for each bulb?
2. Which bulb appears to give off more heat?

Because infrared radiation raises the temperature of matter, it can be used to warm and dry objects. Infrared lamps, like the ones shown in **Figure 19-6,** are used in some restaurants to keep cooked food hot until it is served. Some auto paint shops use infrared radiation to dry car finishes. Can you think of any other uses of this form of radiant energy?

Visible Radiation

Visible radiation, or light, is the only part of the electromagnetic spectrum you can see. Find visible radiation on the electromagnetic spectrum diagram in **Figure 19-2.** Notice that it covers a small range of the spectrum compared to the other types of radiation.

Hot objects, such as the sun, radiate a great deal of electromagnetic energy. Some of this energy is in the range of the electromagnetic spectrum that is visible to humans.

Light can cause chemical reactions to occur. It provides the energy for the process of photosynthesis in which green plants make their own food through a series of chemical reactions. Light also stimulates chemical reactions in your eyes that allow you to see. As you learned in Chapter 16, light can also stimulate decomposition reactions of materials such as hydrogen peroxide.

Ultraviolet Radiation

Why do you wear sunglasses or use sunscreen when you're in strong sunlight? These two products are designed to protect you from large doses of ultraviolet radiation. **Ultraviolet**

Theme Connection

Energy One of the products of doing work is heat. Heat is readily sensed from lightbulbs producing visible light and from infrared sources.

📁 Use these program resources as you teach this lesson.

Study Guide, p. 78
Science and Society Integration, p. 23
Lab Manual 39
Concept Mapping, p. 43

📦 Use these program resources as you teach this lesson.

Teaching Transparency 37
Science Integration Transparency 19

radiation has a higher frequency than visible light, so its photons are more energetic and have greater penetrating power than photons of visible light. Exposure to ultraviolet radiation enables the skin cells to produce vitamin D, which is needed for healthy bones and teeth. Ultraviolet lamps are used in hospitals to kill bacteria and viruses and to sterilize surgical instruments. However, overexposure to ultraviolet radiation kills healthy cells. Prolonged and frequent overexposure can lead to sagging, dry skin and even skin cancer. For this reason, it is important to use a sunscreen regularly if you spend a lot of time outdoors. Sunscreens, such as that being used in **Figure 19-7,** contain chemicals that absorb many of the ultraviolet rays before they penetrate your skin.

Ozone, a gaseous form of oxygen, is found in Earth's upper atmosphere and protects us by blocking most of the sun's ultraviolet radiation, as **Figure 19-8** illustrates. It is produced naturally through photochemical reactions and electric discharge reactions (such as lightning) in the atmosphere.

Figure 19-7

Tanning is the skin's method of trying to protect itself against ultraviolet radiation. Sunscreens provide additional protection.

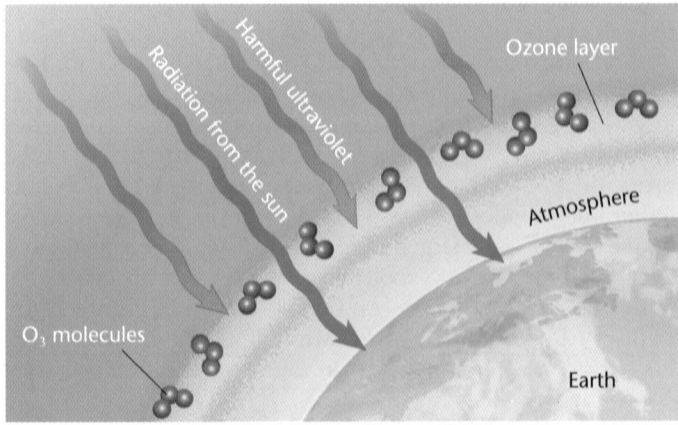

Figure 19-8

A layer of ozone molecules in Earth's atmosphere blocks most of the sun's ultraviolet rays. *Why should we be concerned if this layer gets thinner or develops holes?*

Ozone—Our Protector

Earth's atmosphere has a natural filter, ozone, that blocks most of the sun's ultraviolet rays. Ozone is a form of oxygen where three oxygen atoms combine to make each ozone molecule. A reduction in the amount of ozone in Earth's atmosphere results in an increase in the ultraviolet radiation reaching Earth's surface. Some chemicals break down ozone molecules. Use of some spray-can propellants and refrigerants containing these chemicals may be breaking down the ozone layer, so use of these compounds should be limited.

19-1 Electromagnetic Radiation   **533**

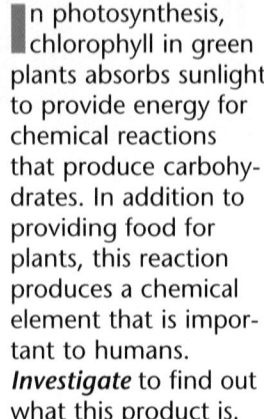

CONNECT TO

CHEMISTRY

In photosynthesis, chlorophyll in green plants absorbs sunlight to provide energy for chemical reactions that produce carbohydrates. In addition to providing food for plants, this reaction produces a chemical element that is important to humans. *Investigate* to find out what this product is.

Student Text Question

See p. 532.

Can you think of any other uses of this form of radiant energy? *Animals use infrared radiation to locate prey; molecules in an unknown material can be identified using infrared radiation.*

Demonstration

[LS] **Visual-Spatial** Show students how a black light can be used to cause certain minerals to fluoresce, or emit light, after absorbing the ultraviolet radiation. You can also turn out the lights and shine the black light on the students' clothes and teeth. Whiteners in clothing and natural tooth enamel fluoresce, as well. [LEP]

CONNECT TO

CHEMISTRY

Answer Oxygen is also produced during photosynthesis.

Inquiry Question

Which would indicate more heat—a blue or a yellow flame? *The blue flame. The color blue indicates a higher frequency, and therefore more photon energy, than the color yellow. Recall that heat is a form of energy.*

Visual Learning

Figure 19-8 Why should we be concerned if this layer gets thinner or develops holes? *More damaging radiation will pass through, increasing the health hazards.* [LEP]

[LS]

Emphasize that light is only a small part of the entire electromagnetic spectrum. We are most familiar with it because we can see it. It differs from other forms of electromagnetic radiation in wavelength and frequency range. In a vacuum, light and all other forms of electromagnetic radiation travel at the same speed, $c = 2.9979 \times 10^8$ m/s.

3 Assess

Check for Understanding

? FLEX Your Brain

Use the Flex Your Brain activity to have students explore WAVE-PARTICLE DUALITY.

📂 **Activity Worksheets,** p. 5

Reteach

LS **Interpersonal** Divide the class into six groups. Present each group with a sign of one of the six types of radiation. Give them about 10 minutes to become experts on their topic, finding out the relative penetrating power and several applications. Have one speaker from each group carry their sign as they line up from the longest to shortest wavelength and make a short presentation. **COOP LEARN**

Extension

📂 For students who have mastered this section, use the **Reinforcement** and **Enrichment** masters.

Figure 19-9

Irradiated food has been treated with ultraviolet light to kill microorganisms.

On a more positive side, because ultraviolet rays can penetrate cells, they can be used to kill microorganisms in food and on hospital equipment. The label in **Figure 19-9** informs the customer that the food has been irradiated. Some minerals, such as fluorite and scheelite, become fluorescent when exposed to ultraviolet light. Fluorescence occurs when a material absorbs ultraviolet radiation and re-emits visible light.

X Rays and Gamma Rays

Your dentist uses X rays to examine the parts of your teeth that can't be seen in a visual inspection. If you ever had a bad fall, you probably had to have an X ray taken to check for broken bones. **X rays** have a shorter wavelength and higher frequency than ultraviolet radiation. X-ray photons carry higher energy and have a greater penetrating power than any of the forms of electromagnetic radiation discussed so far. This higher energy allows X rays to travel through some types of matter, such as your skin and muscles. When X rays hit a more dense material, such as tooth or bone, they are absorbed. An X-ray photograph of a broken bone is shown in **Figure 19-10.** An X-ray photograph is a negative; thus, the bones appear much brighter than the surrounding tissues because they absorb most of the X rays.

As **Figure 19-11** shows, X rays also help us in other ways. They are used in airports to scan passengers' luggage. Personnel can examine luggage and packages without opening them.

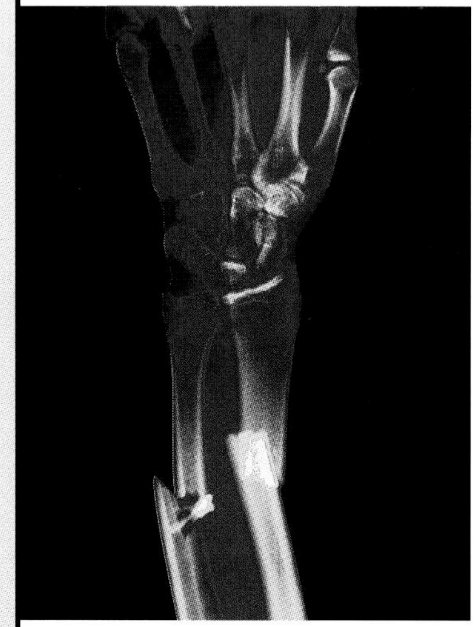

Figure 19-10

X rays can help locate the break in a bone.

Destroying Cells

Gamma rays have the highest frequency and are the most penetrating of all the electromagnetic waves, so they are located at the opposite end of the electromagnetic spectrum from radio waves. Gamma rays are emitted from the nuclei of radioactive atoms. Earth also receives some gamma radiation from space. Concentrated gamma rays are destructive to human cells and can be used to kill cancerous cells. People who undergo gamma radiation therapy for cancer frequently suffer side effects, such as fatigue, nausea, and hair loss, from the therapy because healthy cells are also damaged.

534 Chapter 19 Light

Across the Curriculum

Health Obtain some old X rays from a doctor's or dentist's office and view them. For reinforcement, ask students why only bony tissue is visible. Point out that when having X rays taken, patients usually wear protective lead aprons to cover body parts not being viewed.

Doctors and technicians who give gamma radiation therapy or take X-ray pictures protect themselves from this penetrating radiation by standing behind lead shields. Patients are protected by aprons lined with lead. These shields and aprons absorb the high-energy photons. Excessive exposure to X rays and gamma radiation is harmful and should be avoided.

Think of all the uses of the waves of the electromagnetic spectrum that you read about in this section. Can you think of some uses that weren't mentioned? In the next section, you will learn more about the visible part of the spectrum, light.

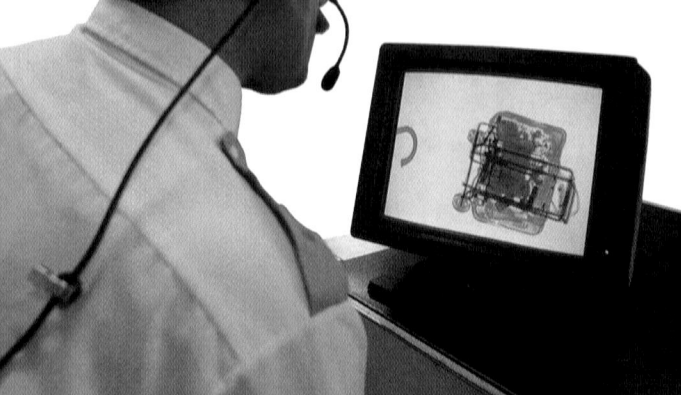

Figure 19-11

X rays assist in providing airport security. An X-ray image shows the contents of the suitcase.

Section Wrap-up

Review

1. Describe at least two ways electromagnetic waves differ from sound waves.

2. Using **Figure 19-2,** rank the six main types of electromagnetic radiation by wavelength, from shortest to longest.

3. **Think Critically:** Describe at least one helpful use for each of the six main types of electromagnetic radiation.

Skill Builder
Observing and Inferring
Observe a colorful object under a regular light-bulb and then under a fluorescent light. Which light has more waves in the red end of the spectrum? If you need more help, refer to Observing and Inferring in the **Skill Handbook.**

Science Journal

After reading this section about radio waves, write a paragraph in your Science Journal explaining whether or not you can hear radio waves. Give evidence for your answer.

19-1 Electromagnetic Radiation   **535**

Skill Builder
Incandescent bulbs make objects appear more red or orange, and part of their heat is due to waves in the infrared part of the spectrum. Fluorescent bulbs make objects appear bluish, so they emit fewer waves at the red end of the spectrum.

 **Assessment**

Oral Ask students to look at their list of uses for each type of electromagnetic radiation. Discuss which type of radiation is least likely to affect or be detected by the human body and why. Use the Performance Task Assessment List for Making Observations and Inferences in **PASC,** p. 17. **P**

4 Close

•MINI•QUIZ•

Use the Mini Quiz to check students' recall of chapter content.

1. **Which type of electromagnetic radiation would you use to identify a fluorescent mineral?** *You use ultraviolet.*

2. **Arrange the following electromagnetic waves in order of decreasing wavelength: radio waves, gamma rays, ultraviolet rays, light, infrared rays.** *radio, infrared, light, ultraviolet, gamma*

3. **Which type of radiation is used in cancer therapy?** *gamma*

Section Wrap-up

Review

1. Electromagnetic waves don't need a medium to transfer energy. They are all transverse waves, and they travel much faster than sound waves.

2. Rank as follows: gamma rays, X rays, ultraviolet rays, visible radiation or light, infrared radiation, and radio waves.

3. **Think Critically** Radio waves—cooking, radar; infrared radiation—medical diagnosis; visible light—enables us to see objects; ultraviolet radiation—germicide, disinfectant; X rays—medical examinations; gamma rays—cancer treatment.

Science Journal

You don't hear radio waves; you hear the sound waves produced by the receiver that collects the radio waves.

Prepare

Section Background

Materials that are transparent to light are not necessarily transparent to other types of radiation. For example, glass is not transparent to ultraviolet radiation. This explains why you do not get a sunburn from light that has passed through a window.

Preplanning

- To prepare for the Mini-LAB, obtain green and red plastic light filters and green and red pens, markers, or crayons.
- To prepare for Activity 19-1, obtain colored pencils (at least red, yellow, and blue); diffraction gratings; power supplies with dimmer switch; and clear, tubular lightbulbs with sockets.

1 Motivate

Bellringer

Before presenting the lesson, display **Section Focus Transparency 75** on the overhead projector. Assign the accompanying **Focus Activity** worksheet. `L1` `LEP`

Tying to Previous Knowledge

Ask students to describe sunlight. They will probably use the phrase *white light*. Point out that if you can see it, then it is part of the visible spectrum. This, in turn, is a part of the electromagnetic spectrum, as they read in the last section.

19•2 Light and Color

Science Words

opaque material
transparent material
translucent material
pigment

Objectives

- Describe the differences among opaque, transparent, and translucent materials.
- Explain how you see color.
- Describe the difference between light color and pigment color.

Figure 19-12

A lighting store is a good place to see materials with different light-transmitting properties.

Light and Matter

Have you ever been asleep in a dark room when someone suddenly opened the curtains and bright light came streaming in? You probably squinted because of the brightness of the light. As your eyes adjusted to the light, you could see all the objects in the room and their colors. But when there was little light in the room, your eyes couldn't distinguish those objects or their colors. What you are able to see depends on the amount of light and the color of the objects as they reflect or absorb light. In order for you to see an object, it must reflect some light.

The type of matter in an object determines the amount of light it absorbs and reflects. For example, heavy window curtains that are completely closed may keep a room dark. The curtains are opaque. **Opaque materials** absorb or reflect all light, and you cannot see objects through them. When the curtains are opened, bright sunlight shines through the glass windowpanes. Most glass windows are transparent. **Transparent materials** allow light to pass through and you can clearly see objects through them. Other materials, such as frosted glass, sheer curtains, or waxed paper, allow some light to pass through, but you cannot clearly see objects through them. These materials are **translucent materials.** The use of these different types of materials results in the variety of lighting fixtures and effects shown in **Figure 19-12.**

Program Resources

📁 **Reproducible Masters**
Study Guide, p. 79 `L1`
Reinforcement, p. 79 `L1`
Enrichment, p. 79 `L3`
Activity Worksheets, pp. 109-110, 114 `L1`
Science Integration Activity, pp. 37-38
Lab Manual 40
Cross-Curricular Integration, p. 25
Multicultural Connections, pp. 41-42

🔦 **Transparencies**
Section Focus Transparency 75 `L1`

Colors

Perhaps you have wondered why your blue jeans look blue or why grass looks green. What makes them reflect color differently from one another? They are both opaque materials, either absorbing or reflecting nearly all of the light that strikes them. But when struck by white light, which is actually a blend of a spectrum of colors, your blue jeans reflect blue light back to your eyes and absorb all of the other colors. Knowing this, explain what happens when white light strikes green grass. White objects appear white because they reflect all colors of light in the visible spectrum. The colors present in white light, arranged in order of decreasing wavelengths, are red, orange, yellow, green, blue, indigo, and violet. The costumes in **Figure 19-13** reflect many different colors. Objects that appear black absorb all colors of light and reflect little or no light back to your eye.

Filtering Light

If you placed a colored, transparent plastic sheet over this white page, what color would the page seem to be? The paper would appear to be the same color as the plastic. The plastic sheet is a filter. A filter is a transparent material that transmits one or more colors of light but absorbs all others. The color of a filter is the same as the color of light it transmits. What happens when you look at colored objects through colored filters? In the white light in **Figure 19-14A,** a green jersey

Figure 19-13

You can see many colorful costumes at parades and celebrations. *Do the costumes reflect or absorb the colors that you see?*

19-2 Light and Color   **537**

Revealing Preconceptions

Ask students if they can name the three primary light colors. They may name the primary pigment colors instead. (Children learn red, yellow, and blue.) From their art education, they may have a working knowledge of color. Emphasize that the primary pigment colors are different from the primary light colors. We will investigate what makes things look green, blue, and red.

Visual Learning

Figure 19-13 **Do the costumes reflect or absorb the colors that you see?** *They reflect the colors you see.* **LEP**
LS

Demonstration

LS **Visual-Spatial** Begin by convincing students that there is more than one color in white light. Aim an overhead projector toward a white screen. Narrow the light path by passing it through a slit, and then aim the light beam through a prism. You should be able to project the spectrum onto the wall. Remember that refraction isn't discussed until later in the chapter. **LEP**

LS **Visual-Spatial** Bring in three white or colorless objects. Make sure one is opaque, one is transparent, and the other is translucent. Have students describe, as specifically as they can, the different ways light behaves when it shines on the objects. **LEP**

MiniLAB

Purpose

IS Visual-Spatial Students will investigate the color of light striking an object and what color the object appears. **L1** **LEP** **COOP LEARN**

Materials

red and green pencils, pens, or crayons; red and green color filters

Teaching Strategies

Troubleshooting This activity is most successful if the filters are reasonably monochromatic in color and the colors of the writing instruments are close to the same shade.

Analysis

1. The red letters should appear visible through the red filter, and the green letters should appear visible through the green filter. With both filters, everything should look black.

2. A multicolored picture will appear shades of red and black through a red filter.

✓ Assessment

Oral Have students use their MiniLAB results to explain how a yellow object will appear through a blue filter. Use the Performance Task Assessment List for Making Observations and Inferences in **PASC**, p. 17.
P

 Activity Worksheets, pp. 5, 114

Student Text Question

Why does it [the jersey] appear this way? *The green jersey appears dark, almost black, when viewed through a red filter. The red filter absorbs green light, so none of the green light reflected by the jersey passes through the filter to your eyes.*

538

Figure 19-14

The green jersey is shown in white light (A). The jersey appears green when viewed through a green filter (B) and black when viewed through a red filter (C).

 INTEGRATION
Life Science

Figure 19-15

White light is produced when the three primary colors of light are mixed.

looks green because it reflects only the green light in the white light striking it. It absorbs light of all other colors. If you look at the jersey through the green filter in **Figure 19-14B**, the jersey still looks green because the filter transmits the reflected green light. **Figure 19-14C** shows how the jersey looks when you look at it through a red filter. Why does it appear this way?

Red, green, and blue are the primary colors of light. They can be mixed in different amounts to produce any color of light. **Figure 19-15** shows how all three primary colors form white light.

Color—Seeing the Light

As you approach a busy intersection, the color of the traffic light changes from green to yellow to red. On the cross street, the color changes from red to green. In this situation, traffic safety depends on your ability to detect immediate color changes. How do you actually see colors?

Light enters your eye and is focused on the retina, which is made up of two types of nerve cells that act as photoreceptors. When these photoreceptors absorb light, chemical reactions convert light energy into nerve impulses that are transmitted to the brain. One type of photoreceptor, called a cone, allows you to distinguish colors and detailed shapes of objects. Cones are most effective in daytime vision. There are three types of cones, each of which absorbs a different range of wavelengths. Red cones absorb mostly red and yellow, green cones absorb mostly yellow and green, and blue cones absorb mostly blue and violet. The second type of photoreceptor, called a rod, is more sensitive to dim light and is useful for night vision.

Interpreting Color

What happens in your eye when you look at a yellow banana? Both your red and green cones respond by absorbing the yellow light, and your brain interprets the combined signal as yellow. Your brain would get the same signal if a mixture of red light and green light reached your eye. Again, your red and green cones would respond, and you would see yellow light because your brain can't perceive the difference between incoming yellow light and yellow light produced by combining red and green light. What happens when you look at a white T-shirt? You see white light when a mixture of all visible wavelengths of light enters your eye because all of your cones are stimulated.

Color Blindness

What would happen to your vision if one or more of your sets of cones did not function properly? This in an inherited condition that affects a small portion of the population. About eight percent of men and one-half percent of women have a problem with one of their sets of color receptors. This condition is sometimes called *color blindness* because it results in an inability to distinguish between certain colors. Most people who are said to be color blind are not truly blind to color, but they have difficulty distinguishing between a few colors, most commonly red and green. Because these two colors are used in traffic signals, drivers and pedestrians must be able to identify them. The person in **Figure 19-16** is taking a color-blindness test.

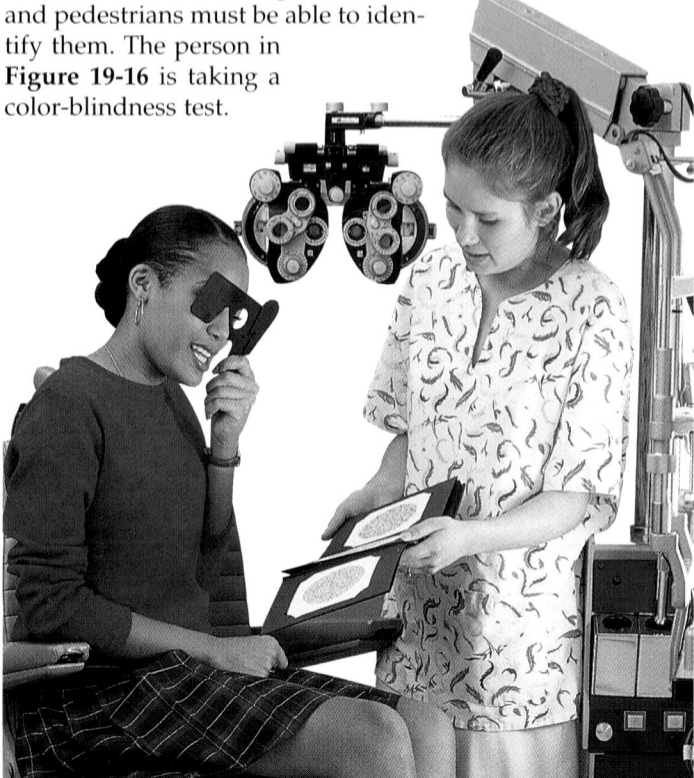

Figure 19-16

Color blindness is a sex-linked condition, more often revealed in males than in females, in which cones for "green" are deficient.

19-2 Light and Color 539

Mini**LAB**

How do filters affect the colors you see?

Procedure

1. Write the word COLOR on your paper, writing the C, L, and R in red and the two Os in green.
2. View the word *COLOR* by looking through a green plastic filter. How does the color of each letter appear?
3. View the word *COLOR* through a red filter.
4. View the word *COLOR* through the red and green filters stacked together.

Analysis

1. What did you observe through the green filter? The red filter? Both filters together? Explain your observations.
2. If you viewed a multicolored picture or article of clothing through a red filter, what colors would you most likely observe? Try it.

Problem Solving

Solve the Problem
A mixture of cyan and yellow produces green.

Think Critically
It is possible. A mixture of all three will produce black. Yellow absorbs blue, cyan absorbs red, and magenta absorbs green. With all color absorbed, black will be seen.

3 Assess

Check for Understanding

 FLEX Your Brain

Use the Flex Your Brain activity to have students explore COLORS.

📁 **Activity Worksheets,** p. 5

Reteach

K **Kinesthetic** Obtain some spot plates and some primary pigment paints from the art department. Have students work in groups and combine small amounts of various combinations of pigments to make other colors. For each color, have them describe which colors are reflected and which are absorbed. Be sure they mix all three primary pigments to produce black. Point out that black pigment absorbs all light and reflects none. **LEP** **COOP LEARN**

Extension

📁 For students who have mastered this section, use the **Reinforcement** and **Enrichment** masters.

 # Problem Solving

Color in the Sunday Comics

Where are you most likely to find color in a daily newspaper? More and more, you find color photographs, but most of the color in newspapers is found in the Sunday comics or in the ads. Are you curious about how color gets into comics?

Use a magnifying glass to examine a color comic strip closely. What do you see? If you look closely enough, you will probably see many small, colored dots. The dots are made up of the three pigment colors—magenta (pinkish-purple), cyan (blue-green), and yellow. Red areas are composed of varying amounts of magenta dots and yellow dots. What color of dots do you think would create a blue sky in a comic?

Combinations of these three pigments are used to create the wide variety of colors you see in the comics. For example, an orange shirt would be made of a mixture of varying amounts of magenta and yellow dots, and a purple book would be a mixture of magenta and cyan dots. What other colors could you easily make?

Solve the Problem:
1. **What colors would make it possible to have green grass in a comic?**
2. **Examine one particular comic strip with a magnifying glass. Make a list of all of the colors you observe and what combinations of colored dots create each color.**

Think Critically:
If you had only magenta, cyan, and yellow paints, could you color a dog's ear black on a painting? Explain why this would or would not be possible.

540   Chapter 19 Light

©1993 Bill Watterson

Pigments

A **pigment** is a colored material that absorbs some colors and reflects others. Pigments are used by artists to make various colors of paints. Paint pigments are usually made of powdered insoluble chemicals such as titanium(IV) oxide, a bright white pigment, and lead(II) chromate, used in painting yellow lines on highways.

You can make any pigment color by mixing different amounts of the three primary pigments—magenta, cyan, and yellow. A primary pigment's color depends on the color of light it reflects. For example, in white light, the yellow pigment appears yellow because it reflects red and green light but absorbs blue light. The color of a mixture of two primary pigments is determined by the primary colors of light that both pigments can reflect.

Look at **Figure 19-17.** Its center appears black because the three blended primary pigments absorb all the primary colors of light. Recall that in **Figure 19-15,** the primary colors of *light* combine to produce white light; they are called additive colors. However, the primary *pigment* colors combine to produce black. Because black results from the absence of reflected light, the primary pigments are called subtractive colors.

Pigmentation is also found in skin. Few people have exactly the same skin color because the amount of pigmentation varies.

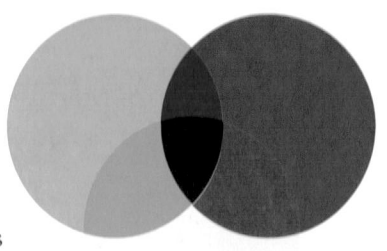

Figure 19-17

The three primary colors of pigment appear black when they are mixed.

Section Wrap-up

Review

1. Contrast opaque, transparent, and translucent materials. Give at least one example of each.

2. If white light shines on a red shirt, what colors are reflected and what colors are absorbed?

3. **Think Critically:** Consider the following parts of your body: the lens of your eye, a fingernail, your skin, and a tooth. Decide whether each of these is opaque, transparent, or translucent. Explain.

Skill Builder
Concept Mapping

Design a concept map to show the chain of events that must happen for you to see a blue object. Work with a partner. If you need help, refer to Concept Mapping in the **Skill Handbook.**

Wavelength and Frequency of Colors		
Color	Wavelength (nm) (in vacuum)	Frequency (10^{14} Hz)
Red	700	4.3
Orange-Yellow	600	5.0
Green-Blue	500	6.0
Violet	400	7.5

Using Computers

Graphing How are the wavelength and frequency of light related to each other? The table below shows the wavelength and frequency that correspond to colors of visible light. Use a computer and a graphing program to make a graph of these data. Put wavelength on the horizontal axis and frequency on the vertical axis. What does the shape of the graph tell you about the relationship between wavelength and frequency?

19-2 Light and Color   **541**

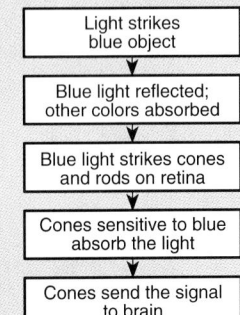

Activity 19-1

PREPARATION

Purpose

Logical-Mathematical Use a diffraction grating to identify the color spectrum of a light source and relate the wavelength of the colors in the spectrum to the temperature of the light source. **L2**

COOP LEARN

Process Skills

observing, interpreting data, inferring, communicating, making and using tables, comparing and contrasting, recognizing cause and effect, forming operational definitions, forming a hypothesis, designing an experiment, using numbers

Time

40 minutes

Materials

• diffraction grating, clear incandescent light with dimmer switch, colored pencils

Alternate Materials Use clear 4-, 15-, 25-, 60-, and 100-watt lightbulbs if a dimmer is not available.

Safety Precautions

When working with electrical power from a 120-VAC wall socket, use caution. Lethal currents are available. Caution students that lightbulbs can become hot.

Possible Hypotheses

Students may hypothesize that the spectra remain the same regardless of the brightness of the light source, the shorter wavelengths (purple and blue) are not present at decreased brightness, or that longer wavelengths (red) are not present at increased brightness.

📁 **Activity Worksheets,** pp. 5, 109-110

Activity 19-1

Design Your Own Experiment
Spectrum Inspection

Have you ever used a dimmer switch to control the brightness of the lighting in a room? Are all of the same wavelengths present at all settings of a dimmer switch, or do they vary with brightness? The following experiment will help you find out.

PREPARATION

Problem
How does the brightness of a lightbulb relate to the wavelengths of light produced by the bulb?

Form a Hypothesis
Which colors of the visible spectrum have the highest frequency? The highest energy? Write a hypothesis describing which colors will be produced by the lightbulb and how they will be affected by reducing its brightness.

Objectives
• Design an experiment that tests the relationship between bulb brightness and the colors of light produced by the bulb.

• Observe the color spectrum produced by a variety of bulbs.

Possible Materials
• diffraction grating
• power supply with variable resistor (dimmer switch)
• clear, tubular lightbulb and socket
• colored pencils (at least red, yellow, and blue)

Safety Precautions
Be sure all electrical cords and connections are intact and that you have a dry working area. Do not touch the bulbs, as they may be hot.

PLAN THE EXPERIMENT

1. As a group, agree upon and write out the hypothesis statement.
2. As shown in the photo at the right, you will look toward the light through the diffraction grating to detect the colors of light emitted by the bulb. The color spectrum will appear to the right and left of the bulb. Given this information, list the specific steps you will need to take to test your hypothesis.
3. Make a list of the needed materials and gather them.

Check the Plan
1. Be sure you have identified any constants and variables in your experiment.
2. Will you first test the bulb at a bright or dim setting? How many settings will you test? (Try at least three.)

3. Make sure each person in your group observes each setting. How will you record each person's observations in an organized way?
4. *Make sure your teacher approves your plan and that you have included any changes suggested in the plan.*

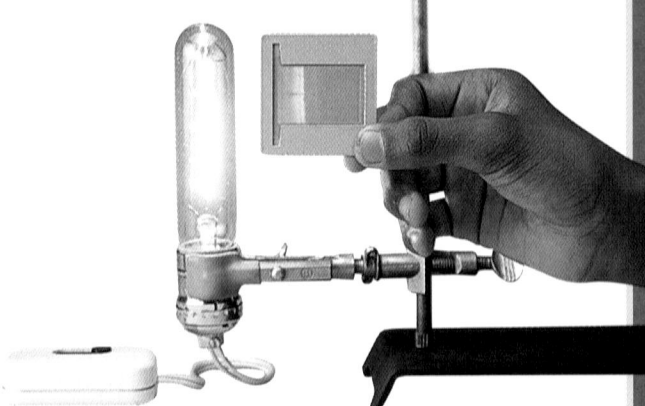

DO THE EXPERIMENT

1. Carry out the approved experiment as planned.
2. While doing your experiment, be sure to write down any observations you make in your Science Journal.

Analyze and Apply
1. **Compare** the spectra from the brightest and dimmest settings you tested.

2. What can you **infer** about the connection between color, frequency, and the brightness of the light source?
3. **Hypothesize** how you might make the spectrum easier to see.

Go Further

Recall your original spectrum from your bulb on the highest setting. Would you expect to find the same spectrum if you used a different kind of bulb, such as a colored lightbulb, neon light, or fluorescent bulb? Try it and see.

19-2 Light and Color   **543**

543

Prepare

Section Background

Standard 60- to 75-watt incandescent bulbs can be replaced by 18-watt compact fluorescent bulbs. They use about 75% less electricity to produce the same amount of light.

1 Motivate

Bellringer

Before presenting the lesson, display **Section Focus Transparency 76** on the overhead projector. Assign the accompanying **Focus Activity** worksheet. L1 LEP

2 Teach

CONNECT TO
CHEMISTRY

Answer Tungsten (atomic number 74) is a hard, heavy, gray, metallic element. It is often added to steel to improve its strength and can also be found in surgical instruments and solar-energy devices.

Visual Learning

Figure 19-18 **What property of phosphors makes them useful in fluorescent bulbs?** *They give off light when they absorb ultraviolet radiation.* LEP LS

Science Words

incandescent light
fluorescent light

Objectives

- Explain how incandescent and fluorescent bulbs work.
- Analyze the advantages and disadvantages of different light sources.

Figure 19-18

Fluorescent lightbulbs do not use filaments. *What property of phosphors makes them useful in fluorescent bulbs?*

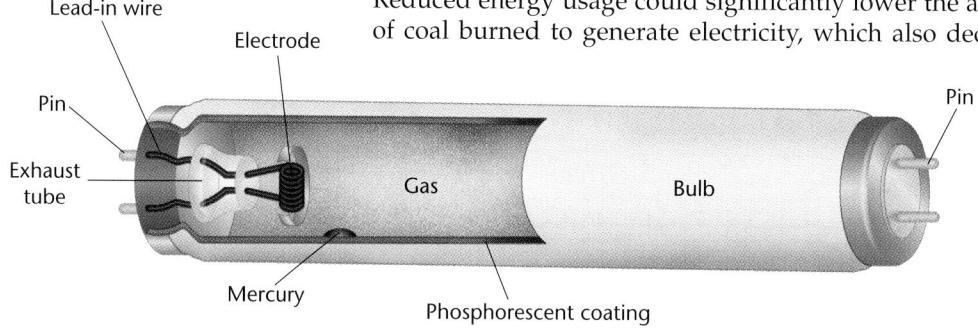

Lead-in wire
Electrode
Pin
Exhaust tube
Gas
Bulb
Pin
Mercury
Phosphorescent coating

544 Chapter 19 Light

Two Kinds of Lighting

Are you using light from a lightbulb to illuminate this page as you read it? Do you know what actually produces this light?

Incandescent Lighting

If you touch a lightbulb after it has been on for a while, it may feel hot. **Incandescent light** is produced by a thin wire called a filament. If you look into an unlit incandescent lightbulb, you will see the filament. It is usually made of the element tungsten. When you turn on the light, electricity flows through the filament and causes it to heat up. When it gets hot, tungsten gives off light. However, more than 80 percent of the energy given off by incandescent bulbs is in the form of heat.

Fluorescent Lighting

Fluorescent lighting is an alternative to incandescent lighting. A fluorescent bulb is filled with gas, typically argon, at a low pressure. As **Figure 19-18** shows, the inner side of the bulb is coated with phosphors, which are fluorescent materials that give off light when they absorb ultraviolet radiation. When the electricity is turned on, electrons collide with the gas molecules in the bulb to make them give off ultraviolet radiation. You can't see the ultraviolet radiation, but the phosphors absorb it and give off visible light.

A **fluorescent light** produces light without excessive loss of thermal energy. These bulbs use as little as one-fifth of the electricity as ordinary incandescent bulbs to produce the same amount of light. Fluorescent bulbs also last much longer than incandescent bulbs. This difference in energy usage will save money during the life of the fluorescent bulb. Reduced energy usage could significantly lower the amount of coal burned to generate electricity, which also decreases

---

Program Resources

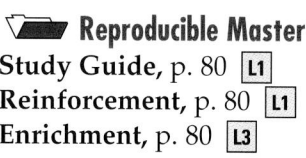 **Reproducible Masters**
Study Guide, p. 80 L1
Reinforcement, p. 80 L1
Enrichment, p. 80 L3

 **Transparencies**
Section Focus Transparency 76

the release of carbon dioxide and other pollutants into Earth's atmosphere.

Some users have observed that fluorescent bulbs can sometimes interfere with their TV remote controls and use of a cable box. Most remote controls transmit pulsed signals of infrared radiation in codes that cause the TVs or VCRs to respond in a desired way. The frequencies at which the fluorescent tubes operate occasionally produce infrared signals that the remote receiver confuses with the remote's infrared signals. Scientists are working to solve this problem.

Lighting enables the student in **Figure 19-19** to study and learn at any time of day. About 20 percent of all electricity consumed in the United States is used for lighting. Advances in developing more efficient lights could save a great deal of energy. What types of lighting are used in your home and school? What products or habits could be changed to save energy and money?

CONNECT TO
CHEMISTRY

Tungsten, used in making the filament in incandescent bulbs, has the highest melting point of all metals. *Research* at least one other property and one other use of tungsten.

Figure 19-19

This student's lamp uses an incandescent lightbulb.

Section Wrap-up

Review

1. Explain how light is produced in an ordinary incandescent bulb.

2. What are the advantages and disadvantages of using a fluorescent bulb instead of an incandescent bulb?

Explore the Technology

You are at the store to purchase a new study lamp. The fluorescent and incandescent lamps cost about the same, but the fluorescent bulb costs about ten times as much as the incandescent bulb. The fluorescent bulb is also guaranteed to last ten times longer than the incandescent bulb. Which would you buy? Explain your reasoning.

SCIENCE & SOCIETY

19-3 Battle of the Bulbs   **545**

Explore the Technology

Note that although the fluorescent bulb is more expensive, it lasts much longer and uses as little as 20% of the energy to produce the same amount of light as an incandescent bulb.

3 Assess

Check for Understanding

? FLEX Your Brain

Use the Flex Your Brain activity to have students explore HALOGEN LIGHTING.

📁 **Activity Worksheets,** p. 5

Reteach

Have students diagram and label the parts of an incandescent light and a fluorescent light, and write captions as if these were diagrams in a book.

Extension

📁 For students who have mastered this section, use the **Reinforcement** and **Enrichment** masters.

4 Close

•MINI•QUIZ•

Use the Mini Quiz to check students' recall of chapter content.

1. **What light is produced by heat?** *incandescent*

2. **What happens when ultraviolet radiation strikes phosphors?** *They fluoresce.*

Section Wrap-up

Review

1. The tungsten filament inside the bulb heats up as electricity passes through it, giving off light.

2. advantages—more energy-efficient, more economical to use; disadvantages—more costly to purchase

Prepare

Section Background

- Light that falls on a rough surface is reflected in many directions. This phenomenon is called diffuse reflection. You see most objects by diffuse reflection of light. Otherwise, you would always see your own reflection in objects.

- The ratio of the speed of light in a vacuum to the speed of light in another material is the index of refraction. The higher this value is, the slower light travels through the other material.

- Christiaan Huygens is usually considered the originator of the wave theory of light, which he published in 1678. Huygens' principle states that each wave front is actually composed of a series of smaller wave fronts, or wavelets. Using this model, diffraction and interference are easier to understand.

Preplanning

- To prepare for the Mini-LAB, obtain short, opaque cups and a supply of pennies.

- Get a light source, clay, and a clear rectangular container for Activity 19-2.

1 Motivate

Bellringer

Before presenting the lesson, display **Section Focus Transparency 77** on the overhead projector. Assign the accompanying **Focus Activity** worksheet. L1 LEP

Science Words

reflection
refraction
diffraction
diffraction grating

Objectives

- State and give an example of the law of reflection.
- Explain how refraction is used to separate white light into the colors of the spectrum.
- Describe how diffraction and interference patterns demonstrate the wave behavior of light.

Figure 19-20

Seeing yourself in a mirror (A) is an illustration of the reflection of light, illustrated in B. *How many times is light reflected when you use a mirror?*

19•4 Wave Properties of Light

Reflection

Just before you left for school this morning, you might have glanced in a mirror one last time to check your appearance. In order for you to see your reflection in the mirror, light had to reflect off you, hit the mirror, and be reflected off the mirror into your eye. **Reflection** occurs when a wave strikes an object and bounces off. **Figure 19-20A** shows an example of reflection.

Reflection occurs with all types of waves: electromagnetic waves, sound waves, and water waves. Look at the light beam striking a mirror in **Figure 19-20B**. The beam striking the mirror is called the incident beam. The beam that bounces off the mirror is called the reflected beam.

Law of Reflection

Notice that in **Figure 19-20B,** a line is drawn perpendicular to the surface of the mirror. This line is called the *normal.* The angle formed by the incident beam and the normal is the angle of incidence, labeled *i.* The angle formed by the reflected beam and the normal is the angle of reflection, labeled *r.* The law of reflection states that *the angle of incidence is equal to the angle of reflection.* Any reflected light, whether it is reflected from a mirror, a piece of foil, or the moon, follows the law of reflection.

If you took a smooth piece of aluminum foil and looked into it, you would see a slightly distorted image of yourself. What if you crumpled up the foil and then unfolded it again? What would you see? The creases in the foil create an uneven surface. You couldn't see an image of yourself because the reflected light is scattered in different directions. Smooth surfaces reflect light in one direction, but rough surfaces scatter light in many directions.

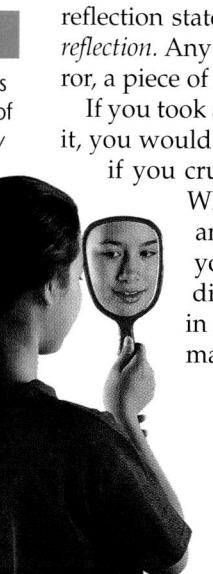

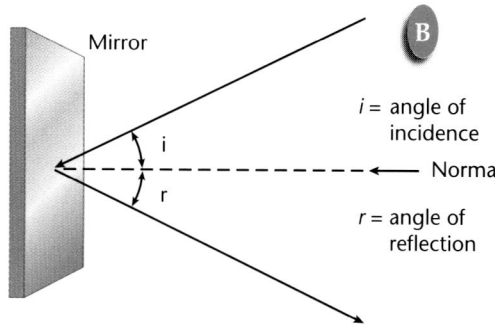

546   Chapter 19 Light

Program Resources

📂 **Reproducible Masters**
Study Guide, p. 81 L1
Reinforcement, p. 81 L1
Enrichment, p. 81 L3
Activity Worksheets, pp. 111-112, 115 L1
Critical Thinking/Problem Solving, p. 25

🖥 **Transparencies**
Section Focus Transparency 77 L1
Teaching Transparency 38

Refraction

Place your finger at an angle halfway into a glass of water and look at it through the side of the glass. Do you notice anything unusual? Your finger appears to be bent or even split into two pieces at the point where it enters the water. This happens because of refraction.

Bending Light

Refraction is the bending of waves caused by a change in their speed when they move from one medium to another. The amount of bending that occurs depends on the speed of light in both materials. The greater the difference between the speeds of light in two media, the more the light is bent as it passes at an angle from one medium to another. The light-refracting properties of certain materials can be especially pleasing to the human eye. When we admire the sparkle in a diamond or other gems, we are seeing refraction in action. Perhaps you have a decorative glass prism hung in a window where you live. When the sun is shining, the prism refracts the sunlight. Have you observed the colorful refraction pattern that the prism projects in the room? **Figure 19-21A** shows an example of refraction.

When light passes into a material that slows it down, **Figure 19-21B** shows that the light is refracted (bent) toward an imaginary line, called the *normal*, drawn perpendicularly through the surface of the medium at the point where the wave strikes the surface. The angle between the normal and the ray entering the medium is called the angle of incidence. The angle between the normal and the ray after it has entered the medium is called the angle of refraction. As **Figure 19-21C** shows, when light passes into a material in which light travels faster, it will be refracted (bent) away from the normal.

The amount of refraction also depends on the wavelength of the light. In the visible spectrum, the wavelengths of light vary from the longer red waves to the shorter violet waves.

Figure 19-21

The handle of the net looks bent in (A) because the light waves are refracted as they change speed when they pass from the water to the air. Light slows down and refracts toward the normal as it passes into a more dense medium (B). Light speeds up and bends away from the normal as it passes into a less dense medium (C).

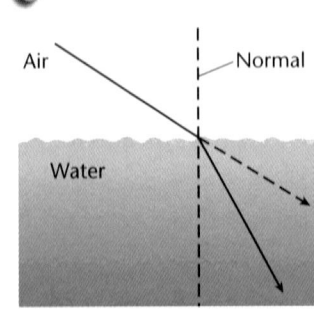

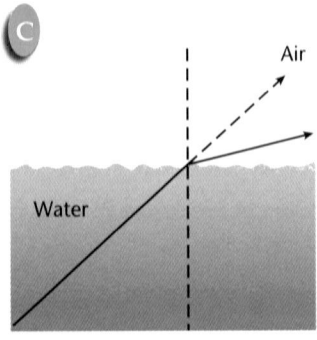

19-4 Wave Properties of Light   **547**

Use these program resources as you teach this lesson.
Study Guide, p. 81
Critical Thinking/Problem Solving, p. 25

Use these program resources as you teach this lesson.
Teaching Transparency 38

Student Text Question

Which color of light would you expect to bend the least? *Because red light has the longest wavelength, it would bend the least.*

MiniLAB

Purpose

LS **Visual-Spatial** Students will investigate how water bends light. **L1** **LEP**

Materials

opaque cup, penny, glass of water

Teaching Strategies

Have students make a prediction of what may happen as the cup is filled with water.

Analysis

1. The penny became visible as water was added because the light from the penny was bent at the surface of the water.

2. Viewed from the side of the cup, a light ray travels diagonally upward from the penny to the surface of the water. At the surface, the ray is refracted away from the normal (more toward the horizontal).

✓ Assessment

Oral Have students use the results of this MiniLAB and describe what a person wearing a mask who is underwater in a swimming pool, looking upward, would see. Use the Performance Task Assessment List for Making Observations and Inferences in **PASC**, p. 17. **P**

📁 **Activity Worksheets,** pp. 5, 115

Figure 19-22

White light is refracted into the colors of the visible spectrum as it passes through a prism.

MiniLAB

How does water bend light?

Procedure

1. Place a penny at the bottom of a short, opaque cup. Set it on a table in front of you.
2. Have a partner slowly slide the cup away from you just until you can no longer see the penny.
3. Without disturbing the penny, have your partner slowly pour water into the cup until you can see the penny.
4. Reverse roles and repeat the experiment.

Analysis

1. What did you observe? Explain how this is possible.
2. Sketch the light path from the penny to your eye after the water was added.

548 Chapter 19 Light

Figure 19-22 shows what happens when white light passes through a prism. The triangular prism refracts the light twice: once when it enters the prism and again when it leaves the prism and reenters the air. Because the shorter wavelengths of light are refracted more than the longer wavelengths, violet light is bent the most. Which color of light would you expect to bend the least? As a result of this varied refraction, the different colors are separated when they emerge from the prism.

Does the light leaving a prism remind you of a rainbow? Like prisms, rain droplets also refract light. The refraction of the different wavelengths can cause white light from the sun to separate into the individual colors of the visible spectrum. Isaac Newton recognized that white light actually includes all of the seven colors of the rainbow. In order of decreasing wavelength, the colors you should see in the rainbow are red, orange, yellow, green, blue, indigo, and violet—the same order as the electromagnetic spectrum.

Diffraction and Interference

One of the sounds of summer is the bells or music played by vendors selling ice cream as they drive their trucks through your neighborhood. Maybe you have heard cars drive by your home with blasting stereos. You can hear these sounds even when buildings separate you from the source. Sound waves can bend around corners to reach you. **Diffraction** is the bending of waves around a barrier. In the 1600s, an Italian physicist, Francesco Grimaldi, observed light and dark areas on the edge of a shadow. If you observe the shadow formed when light passes through an open door, you'll see no clearly defined boundary between light and dark. Instead, you will see a gradual transition. Grimaldi explained this phenomenon by suggesting that light could be bent around the edges of barriers.

Electromagnetic waves, sound waves, and water waves can all be diffracted. Water waves bend around you as you swim. Sandbars form along shorelines when water that

Across the Curriculum

Language Arts Have students write a fictitious, creative short story explaining why the sky is blue. Then have them research and write the factual reason that the sky appears blue. **LS**

carries sand is diffracted by some obstruction in the water. The pattern of the waves changes as the waves are diffracted. Diffraction is important in the transfer of radio waves. Longer wavelengths, such as those used by AM radio stations, are easier to diffract than short FM waves. That is why AM reception is often better than FM reception around tall buildings and natural barriers.

Interference—A Wave Property

Recall from Section 18-4 that sound waves can interfere with each other to form constructive and destructive interference. In the early 1800s, Thomas Young, an English scientist, expanded the diffraction theory Grimaldi had developed. In a famous experiment, he passed light of a uniform wavelength through a narrow slit. The new diffraction pattern of the light waves then passed through a barrier with two slits. You can

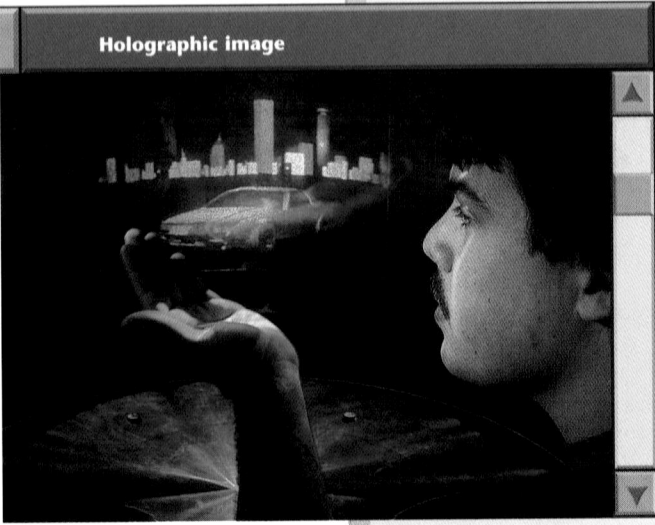
Holographic image

USING TECHNOLOGY

Holograms

A hologram is a vivid, three-dimensional image that appears to float in space. As you walk by a hologram, you see the image from a slightly different angle.

Holograms are produced by illuminating the object you wish to make an image of with light from a laser. Laser light reflected from the object shines on photographic film simultaneously with light from a second laser of the same wavelength. The interference pattern created on the photographic film produces a hologram image when laser light shines through it.

How We Use Holograms

A hologram conveys more information to your eye than a conventional two-dimensional photograph by controlling both the intensity and direction of the light that departs from the image. As a result, these vivid, three-dimensional objects are difficult to copy. For this reason, holographic images are used on credit cards, identification cards, and on the labels of some clothing to help prevent counterfeiting. Scientists are also working on ways to record and display data holographically.

*inter*NET
CONNECTION

Visit the Chapter 19 Internet Connection at Glencoe Online Science, **www.glencoe.com/sec/ science/physical**, for a link to more information about holography.

19-4 Wave Properties of Light   549

USING TECHNOLOGY

A three-dimensional display would be advantageous in situations such as air-traffic control, automobile design, architecture, etc.

3 Assess

Check for Understanding

? FLEX Your Brain

Use the Flex Your Brain activity to have students explore X-RAY DIFFRACTION IN CRYSTALS.

📁 **Activity Worksheets,** p. 5

Reteach

LS Kinesthetic Give each pair of students a cup of water and a round washer. Drop the washer into the cup. Try to put a pen or pencil through the hole in the washer. Have students explain why it is difficult to hit the hole directly. The light reflected from the washer bends away from the normal as it emerges from the water. Because the light ray that reaches your eye is bent, the washer appears higher in the water than it actually is. **LEP**

Extension

📁 For students who have mastered this section, use the **Reinforcement** and **Enrichment** masters.

4 Close

•MINI•QUIZ•

Use the Mini Quiz to check students' recall of chapter content.

1. _____ occurs when **waves bend due to a change in their speed.** *Refraction*

2. **A line drawn perpendicular to the surface of a barrier is the** _____ . *normal*

3. _____ **is the bending of waves around a barrier.** *Diffraction*

4. _____ **is caused when light waves overlap each other.** *Wave interference*

Section Wrap-up

Review

1. The angle of incidence equals the angle of reflection.

2. Toward; light moves more slowly in glass.

3. **Think Critically** Light spreads out as it passes through a diffraction grating; this behavior is characteristic of waves, not particles. Light waves interfere with each other, causing the light and dark bands observed when light passes through a diffraction grating. Particles, like waves, will reflect, but they cannot interfere. Diffraction supports the wave theory of light.

USING MATH

8.3×10^{-5} cm wide slit

Figure 19-23

Diffraction patterns (A) can be produced by compact discs (B).

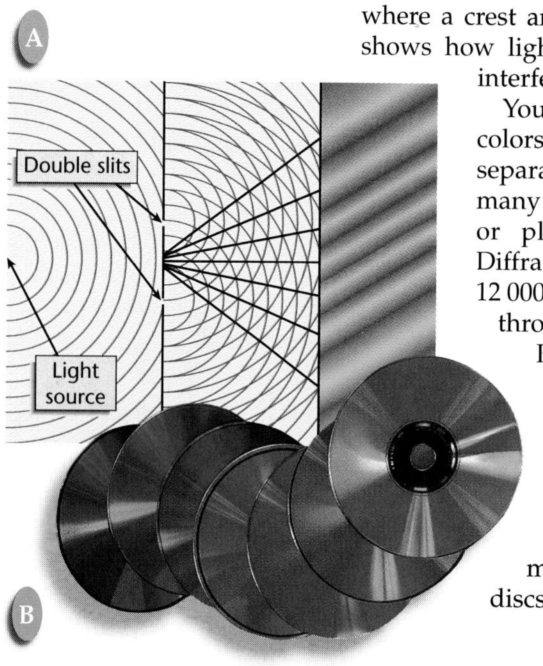

A

Double slits

Light source

B

---

see the results, wave interference, in **Figure 19-23A.** The wave crests are shown by the semicircles. The light from these slits was projected onto a screen. Young observed light bands where two crests or two troughs combined, and dark areas where a crest and a trough crossed paths. This experiment shows how light behaves as a wave. Only waves exhibit interference; particles do not.

You know that white light can be separated into colors by refraction in a prism. Colors can also be separated by diffraction and interference using many slits. A **diffraction grating** is a piece of glass or plastic made up of many parallel slits. Diffraction gratings commonly have as many as 12 000 slits per centimeter. When white light shines through a diffraction grating, the colors separate.

Reflective materials can be ruled with closely spaced grooves to produce diffraction patterns. For example, shiny bumper stickers sometimes diffract light into a wide array of colors. You can also see the dazzling results of simple diffraction gratings by looking at the reflection of light from the microscopic recording pits in the compact discs in **Figure 19-23B.**

Section Wrap-up

Review

1. State the law of reflection.

2. When light moves from air into glass at an angle, is the light refracted toward or away from the normal?

3. **Think Critically:** Scientists say that light has both particle and wave properties. Explain why observations of diffraction and interference patterns conflict with the idea that light consists of particles.

Skill Builder
Observing and Inferring

Imagine you are on the shore of a large lake and see waves moving toward you from the center of the lake. However, the waves pass by a boat dock and move toward you at a slightly different angle afterward. What would you infer is happening? If you need help, refer to Observing and Inferring in the **Skill Handbook.**

USING MATH

If you are using a diffraction grating with 12 000 slits per centimeter, how wide is each slit? Express your answer as a decimal in centimeters.

Skill Builder

The boat dock is acting as a barrier and causing the wave front to bend as it tries to move around it. This changes the direction of the approaching wave.

✓ Assessment

Performance Have students test their inference by building a small model of this situation in a shallow pan. Ask them to sketch what they see as a wave encounters a barrier, such as a wood block. Use the Performance Task Assessment List for Model in **PASC,** p. 51. **P**

Make a Light Bender

Can you recall seeing your reflection in the surface of a lake or other body of water? You can see your image because some of the light that reflects off your face also strikes the water's surface and reflects into your eyes. However, you don't see a clear, color image because much of the light enters the water rather than being reflected.

Problem
How does being underwater affect the viewer's image of an object that is above the water's surface?

Materials 🧤 🔌
- light source
- pencil
- clear rectangular container
- water
- notebook paper
- clay

Procedure
1. Fill the container with water and place it on a sheet of notebook paper.
2. Outline the container on the paper.
3. Stand the pencil on end in the clay and place it on the paper in front of the container as shown in the figure below.
4. Place the light in an upright position near the paper and turn it on.
5. Adjust the position of the paper and the pencil so the pencil's shadow on the paper goes completely through the container.
6. Draw lines on the shadow where it enters and leaves the container. Draw a line on the shadow that reflects from the surface of the container.
7. Move the container. Connect the lines to show the shadow through the container.

Analyze
1. Draw a reference line (normal) at right angles through the outline of the container at the point where the shadow touches its surface. As the shadow enters the container, does it refract toward the reference line or away from it? What happens when the shadow leaves the other side of the container?
2. Alter the experiment by changing a variable—the angle between the reference line and the shadow. What happens to the reflected and refracted shadows?

Conclude and Apply
3. **Compare** the angles of reflection and refraction for light striking the flat surface of the container.
4. **Predict** how the angles of reflection or refraction would change in question 3 if the surface of the container were curved. **Explain** your answer.

19-4 Wave Properties of Light   **551**

---

4. For an individual light ray, the angles would not change. For groups of rays, the images are distorted with curved surfaces.

✔ Assessment

Oral Ask students whether they would expect the same results if a different clear liquid, such as alcohol or mineral oil, were in the box. You may wish to have them try this. Use the Performance Task Assessment List for Formulating a Hypothesis in **PASC**, p. 21. **P**

Purpose
IS **Interpersonal** Predict how light reflects and refracts. **L1**
LEP   **COOP LEARN**

Process Skills
observing, experimenting, predicting

Time
40 minutes

Safety Precautions
If a standard 120-VAC wall light source is used, caution students about the presence of dangerous voltages and shock hazard and remind them that lightbulbs can become hot.

📁 **Activity Worksheets,** pp. 5, 111-112

Teaching Strategies
Alternate Materials A clear baking dish can be substituted for the rectangular container.

Troubleshooting The light source should be taped down so it cannot move during the activity. The pencil must be standing straight up for good results.

Answers to Questions
1. The shadow refracts toward the reference line when it enters the box and refracts away when it leaves.
2. The angles of the refracted and reflected rays increase when the angle of the incoming shadow is increased. Students might also notice that the intensity of the reflected ray increases as the incident angle increases.
3. Light rays *refract* to make smaller angles inside water than outside, but rays always *reflect* with the same angle as the incoming ray.

Background

A lighting designer for a stage play has to understand how light modifies the appearance of actors or objects on display. There are four qualities of light that the designer can control: distribution, intensity, movement, and color. Distribution refers to the direction from which the light reaches the actor, area, or object. Intensity is the level of brightness and can be varied to achieve the proper effect. Movement is guided by definite cues; it often entails moving a narrow beam to follow an actor on the stage. The color of lighting can greatly affect the audience's reaction to and understanding of a play.

Teaching Strategies

Students may never have thought of lighting design as a career. Help them to see its importance to the theater, museums, architecture, interior design, and the beauty of the outdoors at night.

Visual-Spatial Provide lights for students to experiment with. Have them use what they have learned about lighting design. Let them compare downlighting and uplighting. Then have them experiment with illuminating a plant, a painting, or a statue. L1

Interpersonal Have students work in groups to plan outdoor lighting to illuminate their school, a public building, or other outdoor location they are familiar with. Have them draw up a design plan showing where the lights would be placed and the angles at which they would illuminate the area. L1 **COOP LEARN**

The Art of Lighting

Candles burning in jack-o'-lanterns create an eerie mood at a Halloween party. Light has the ability to cause a special psychological effect in a viewer. The kind of effect depends upon the angle at which the light strikes an object. Artists know this and use light accordingly. They adapt light to convey effect.

Stage Lighting

When the stage in a theater is dimly lighted and dark shadows conceal parts of the faces of the actors, the audience gets an uneasy feeling that something bad has happened or is about to happen. They become edgy, fearing that someone is lurking in the shadows to harm the characters they identify with.

Faces can be distorted by the direction of the light, as seen in the photograph of the girl on this page. Light striking from a low angle, such as from a low fire or a lantern, can transform a pleasant face into a frightening one.

On the other hand, bright lights or front lights alone tend to flatten and wash out an actor. Both top lights and side lights create highlights along the edges of the head and body, but cause deep shadows across the head and the front of the body. However, when side and top lights are combined with front lights, proper lighting can be achieved.

Outdoor Lighting

Downlighting is a lighting technique that imitates nature. Just as the morning sun falls on a patio, downlights placed in tall trees provide pleasant light in a garden at night. Uplighting, which is lighting from below, is more dramatic but rarely occurs in nature. That's why it gives a less natural look. Uplighting can be used to emphasize a statue or a large tree, but should be used less often than downlighting. All in all, there's much more to artistic lighting than turning on a switch.

Science Journal

Think about a video in which your favorite singer performs. What do you recall about the lighting? In your Science Journal, describe the lighting effects used for the performance.

Summary

19-1: Electromagnetic Radiation

1. Electromagnetic waves differ from other types of waves because they do not need a medium to transfer energy.
2. Electromagnetic waves are arranged on the electromagnetic spectrum by their wavelength and frequency.
3. Radio waves are used in communications. Infrared radiation indicates the presence of heat. Visible light is only a small part of the electromagnetic spectrum. Ultraviolet radiation causes fluorescent bulbs to glow. X rays aid medical diagnosis. Gamma rays can be used to destroy cancer cells.

19-2: Light and Color

1. You can't see through opaque materials. You can see clearly through transparent materials and hazily through translucent materials.
2. You see color when light is reflected off objects and into your eyes. Inside your eyes, chemical reactions convert light into nerve impulses.
3. The three primary colors of light can be mixed to form all other colors. The colors of pigments are determined by the colors they reflect.

19-3: Science and Society: Battle of the Bulbs

1. Incandescent bulbs produce light by heating a tungsten filament until it glows brightly. Fluorescent bulbs give off light when ultraviolet radiation produced inside the bulb causes a fluorescent coating inside the bulb to glow.

2. Although expensive to purchase, fluorescent bulbs waste less energy, last longer, and cost much less to use than incandescent bulbs.

19-4: Wave Properties of Light

1. The law of reflection states that the angle of incidence is equal to the angle of reflection.
2. White light can be separated into the colors of the visible spectrum because each wavelength refracts at a different angle as it passes into a medium.
3. Diffraction and interference show wave patterns of light by showing that light waves bend around a barrier and can cancel each other.

Key Science Words

a. diffraction
b. diffraction grating
c. electromagnetic radiation
d. fluorescent light
e. gamma ray
f. incandescent light
g. infrared radiation
h. microwave
i. modulation
j. opaque material
k. photon
l. pigment
m. radio wave
n. reflection
o. refraction
p. translucent material
q. transparent material
r. ultraviolet radiation
s. visible radiation
t. X ray

Summary

Have the students read the summary statements to review the major concepts of the chapter.

Reviewing Vocabulary

1. c	**6.** n
2. k	**7.** a
3. g	**8.** b
4. j	**9.** e
5. f	**10.** h

✓ Assessment

Portfolio Encourage students to place in their portfolios one or two items of what they consider to be their best work. Examples include:

- MiniLAB data and conclusions, p. 532
- The answer to Section 19-2 Wrap-up question 1, p. 541. **P**

Performance Additional performance assessments may be found in **Performance Assessment** and **Science Integration Activities.** Performance Task Assessment Lists and rubrics for evaluating these activities can be found in Glencoe's **Performance Assessment in the Science Classroom.**

GLENCOE TECHNOLOGY

 **MindJogger Videoquiz**

Chapter 19 Have students work in groups as they play the Videoquiz game to review key chapter concepts.

Checking Concepts

1. d	**6.** b
2. a	**7.** c
3. a	**8.** d
4. a	**9.** a
5. b	**10.** c

Understanding Concepts

11. Sound waves are converted to electrical signals that correspond to the sound's characteristic patterns. These signals can then be used to modulate, or vary, either the amplitude or frequency of a radio wave. This is done in radio and TV transmissions, cellular phones, and cordless phones.

12. Light color is determined by the wavelength of light transmitted, while pigment color is determined by the color of light reflected.

13. Light reflects off a white wall or a mirror because of the law of reflection. Both the wall and mirror reflect all light frequencies. Light reflected off a wall is scattered because the surface is rough; light reflected off a mirror is not scattered.

14. Rainbows are produced when white light from the sun passes through droplets of water suspended in the atmosphere after a rain. The water droplets act as prisms, *refracting* light.

15. Light can be bent around barriers and diffracted to show interference patterns. This suggests that light behaves as a wave. Light's wavelike behavior was studied by Francesco Grimaldi and Thomas Young.

Thinking Critically

16. Blue light has a higher frequency than red light, so

554

Reviewing Vocabulary

Match each phrase with the correct term from the list of Key Science Words.

1. transverse wave produced by the motion of electrically charged particles
2. bundle of radiation with no mass
3. electromagnetic radiation that is felt as heat
4. object that can't be seen through
5. light produced by a heated filament
6. occurs when a wave strikes an object and then bounces off
7. the bending of waves around a barrier
8. contains many parallel slits that can separate white light into colors
9. highest frequency electromagnetic wave
10. radio wave with the greatest energy

Checking Concepts

Choose the word or phrase that completes the sentence.

1. Electromagnetic waves are different from other types of waves in that they do not _____.
   a. have amplitude
   c. transfer energy
   b. have frequency
   d. need a medium
2. When light enters matter, it ____.
   a. slows down
   b. speeds up
   c. travels at 300 000 km/s
   d. travels at the speed of sound
3. A contact lens is _____.
   a. transparent
   c. opaque
   b. translucent
   d. square
4. Electromagnetic waves with the longest wavelengths are _____.
   a. radio waves
   c. X rays
   b. visible light
   d. gamma rays

5. The process of changing the frequency or amplitude of radio waves in order to encode a signal is _____.
   a. diffraction
   c. reflection
   b. modulation
   d. refraction
6. When food molecules absorb microwaves, they vibrate faster and their _____.
   a. kinetic energy decreases
   b. kinetic energy increases
   c. temperature decreases
   d. temperature remains constant
7. Your body gives off _____.
   a. radio waves
   c. infrared radiation
   b. visible light
   d. ultraviolet radiation
8. Fluorescent bulbs glow when the phosphors inside absorb _____.
   a. microwaves
   c. infrared radiation
   b. gamma rays
   d. ultraviolet radiation
9. X rays are best absorbed by _____.
   a. bone
   c. muscle
   b. hair
   d. skin
10. Objects that partially scatter light that passes through them are called _____.
    a. reflective
    c. translucent
    b. opaque
    d. transparent

Understanding Concepts

Answer the following questions in your Science Journal using complete sentences.

11. Explain how sound waves are converted to radio waves. Give some examples.
12. What is the difference between light color and pigment color?
13. How is the reflection of light off a white wall similar to the reflection of light off a mirror? How is it different?
14. Explain which property of light helps produce a rainbow.
15. Describe the wave model of light and note one of the scientists who studied it.

its particles carry more energy. The hotter an object is, the more energy it gives off.

17. A blue filter transmits the blue light reflected from the shirt, and the shirt appears blue. Because the red and green filters each absorb blue light, the shirt will appear black.

18. You can see only areas whose images are reflected from the mirror into your eyes. Other areas behind you are reflected at angles that prevent the images from entering your eyes.

19. Violet, because it is bent the most by a prism. The greater the difference in the speed of light in two different media, the more it is refracted.

20. Waves with longer wavelengths are diffracted more easily than waves with shorter wavelengths. Sound has a much longer wavelength than light, and is therefore bent more easily around a street corner.

Thinking Critically

16. Heated objects often give off light of a particular color. Explain why an object that glows blue is hotter than one that glows red.

17. How would a blue shirt appear if a blue filter were placed in front of it? A red filter? A green filter?

18. Use the law of reflection to explain why you see only a portion of the area behind you when you look into a mirror.

19. Which color of light changes speed the most when it passes through a prism? Explain.

20. Explain why you can hear a fire engine coming around a street corner before you can see it.

Developing Skills

If you need help, refer to the **Skill Handbook.**

21. **Sequencing:** List the types of radiation in the electromagnetic spectrum in order of decreasing penetrating power. Use **Figure 19-2,** the electromagnetic spectrum, as a reference.

22. **Making and Using Tables:** Construct a table to show the applications of each type of electromagnetic wave. Which type would an electronics engineer be most interested in? A doctor?

23. **Observing and Inferring:** Most mammals, such as dogs and cats, can't see colors. Infer how a cat's eye might be different from your eye.

24. **Concept Mapping:** Use the blank concept map in the column to the right to show the five steps in the production of fluorescent light.

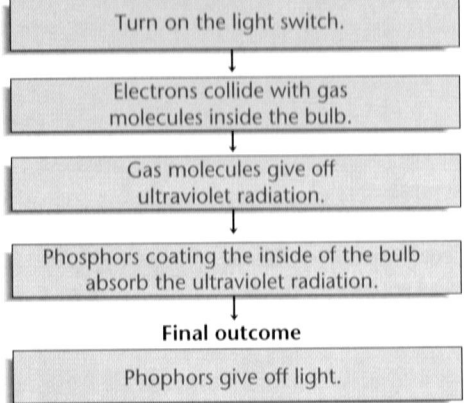

Initiating step

Turn on the light switch.

Electrons collide with gas molecules inside the bulb.

Gas molecules give off ultraviolet radiation.

Phosphors coating the inside of the bulb absorb the ultraviolet radiation.

Final outcome

Phophors give off light.

25. **Using Scientific Illustrations:** Whales make sounds that can be detected for miles. Make a drawing that shows the wave properties that enable the sound to travel so far.

Performance Assessment

1. **Scrapbook:** Make a scrapbook of newspaper and magazine articles describing the dangers of ultraviolet radiation and how this harmful radiation is absorbed by Earth's ozone layer. Write a short summary of each article.

2. **Poster:** Make a poster to show how the three primary pigments are combined to produce common colors such as blue, red, yellow, green, purple, brown, and black.

3. **Making Observations and Inferences:** Repeat Activity 19-2 on page 551. Measure the angle of incidence and the angle of refraction as the shadow enters and leaves the container. Show the angles on the diagram. What do you notice?

3. Both angles in the box are the same, and both angles in the air are the same. The angles of the shadows entering and leaving the pan are the same. Use the Performance Task Assessment List for Making Observations and Inferences in **PASC,** p. 17. **P**

Type of Radiation	Applications
Radio waves	Radio, TV, cordless phones
Infrared radiation	Thermograms to detect tumors
Light	Solar-powered calculators
UV radiation	Fluorescent lights
X rays	Medical diagnoses
Gamma rays	Cancer treatment

Developing Skills

21. **Sequencing** Penetrating power is related to wave energy, which is directly related to wave frequency. The order of decreasing penetrating power is the same as the order of decreasing frequency: gamma rays, X rays, ultraviolet radiation, light, infrared radiation, radio waves.

22. **Making and Using Tables** A possible table is shown below. Electronics engineer—radio waves and light. Doctor—infrared radiation, X rays, and gamma rays.

23. **Observing and Inferring** The best inference would be that a cat's eye does not have the cone nerve cells that a human eye does.

24. **Concept Mapping** Possible steps might be, in order: Turn on the light switch. Electrons collide with gas molecules in the bulb. Gas molecules give off ultraviolet radiation. Phosphors coating the inside of the bulb absorb the ultraviolet radiation. Phosphors give off light.

25. **Using Scientific Illustrations** The drawing should show sound waves moving under water and reflecting off of the ocean bottom, rock formations, and ship bottoms.

Performance Assessment

1. Booklet should contain at least several articles and a summary of each one. Use the Performance Task Assessment List for Booklet or Pamphlet in **PASC,** p. 57. **P**

2. Posters should show the proper combinations of yellow, magenta, and cyan. Use the Performance Task Assessment List for Poster in **PASC,** p. 73. **P**

Chapter Organizer

Section	Objectives/Standards	Activities/Features
Chapter Opener		**Explore Activity:** Find out how familiar surfaces can produce different images. p. 557
20-1 The Optics of Mirrors (1 session, ½ block)*	1. **Explain** how an image is formed in three types of mirrors. 2. **Identify** examples and uses of plane, concave, and convex mirrors. National Content Standards: (5-8) UCP1-UCP3, UCP5; (9-12) UCP1-UCP3, UCP5, B6	**MiniLAB:** What happens when light strikes a concave mirror? p. 559 **Problem Solving:** Looking Back with Car Mirrors, p. 561 **Science Journal,** p. 562 **Skill Builder:** Recognizing Cause and Effect, p. 562 **Activity 20-1:** Reflections of Reflections, p. 563
20-2 The Optics of Lenses (2 sessions, 1 block)*	1. **Describe** the types of images formed with convex and concave lenses. 2. **Cite examples** of how these lenses are used. 3. **Explain** how lenses are used to correct vision. National Content Standards: (5-8) UCP1-UCP3, UCP5, C1; (9-12) UCP1-UCP3, UCP5, B6	**MiniLAB:** Can lenses be made of liquid? p. 566 **Using Computers,** p. 567 **Skill Builder:** Concept Mapping, p. 567
20-3 Optical Instruments (½ session, ½ block)*	1. **Compare** refracting and reflecting telescopes. 2. **Explain** how a camera creates an image. National Content Standards: (5-8) UCP1-UCP3, UCP5; (9-12) UCP1-UCP3, UCP5, B6	**MiniLAB:** How can the location of a lens affect the observed image? p. 569 **Science Journal,** p. 571 **Skill Builder:** Hypothesizing, p. 571
20-4 Science and Society (½ session)*	1. **Describe** the development and repair of the Hubble Space Telescope. 2. **Discuss** the goals and uses of the Hubble Space Telescope. National Content Standards: (5-8) UCP1, UCP2, UCP5, D3, E2, F5, G1; (9-12) UCP1, UCP2, UCP5, B6, D3, D4, E2, F6, G1	**Science Journal,** p. 573 **Explore the Technology,** p. 573
20-5 Applications of Light (1 session, ½ block)*	1. **Describe** polarized light and the uses of polarizing filters. 2. **Explain** how a laser produces coherent light and how it differs from incoherent light. 3. **Apply** the concept of total internal reflection to the uses of optical fibers. National Content Standards: (5-8) UCP1-UCP3, UCP5, A1, A2, E1, E2, F1, F5, G1; (9-12) UCP1-UCP3, UCP5, A1, A2, B6, E1, E2, F1, F6, G1	**Connect to Chemistry,** p. 576 **Using Technology:** The Light Scalpel, p. 578 **Science Journal,** p. 579 **Skill Builder:** Sequencing, p. 579 **Activity 20-2:** What's behind those shades? pp. 580-581 **People and Science:** Jeff Baldwin, Telescope Mirror Maker, p. 582

Activity Materials

Explore	Activities	MiniLABs
page 557 flat mirror, large shiny spoon	page 563 2 plane mirrors, cellophane tape, masking tape, protractor, paper clip pages 580-581 2 polarizing filters, masking tape, light source, objects with flat hard surfaces, heat-proof glove, safety goggles	page 559 large shiny spoon or convex mirror, bright flashlight, metric rulers, index card page 566 plastic wrap, printed text, dropper, penny page 569 test tube with stopper, pencil, index card, ruler

Need Materials? Call Science Kit (1-800-828-7777).   * A complete Planning Guide that includes block scheduling is provided on pages 32T-35T.

Teacher Classroom Resources

Reproducible Masters	Transparencies	Teaching Resources
Study Guide, p. 82 Reinforcement, p. 82 Enrichment, p. 82 Activity Worksheets, pp. 116-117, 120 Lab Manual 41, Reflection of Light Concept Mapping, pp. 45-46	Section Focus Transparency 78, Fun-House Mirrors Teaching Transparency 39, Optics of Concave Mirrors	Physical Science CD-ROM Spanish Resources English/Spanish Audiocassettes Cooperative Learning Resource Guide Lab Partner Lab and Safety Skills Lesson Plans
Study Guide, p. 83 Reinforcement, p. 83 Enrichment, p. 83 Activity Worksheets, p. 121 Lab Manual 42, Magnifying Power Science and Society Integration, p. 24	Section Focus Transparency 79, Magnifying Glasses Science Integration Transparency 20, Lenses and the Human Eye Teaching Transparency 40, Vision Corrections	
Study Guide, p. 84 Reinforcement, p. 84 Enrichment, p. 84 Activity Worksheets, p. 122 Science Integration Activity 20, Nature Film Cross-Curricular Integration, p. 26	Section Focus Transparency 80, Spyglasses	**Assessment Resources** Chapter Review, pp. 43-44 Assessment, pp. 130-133 Performance Assessment in the Science Classroom (PASC) MindJogger Videoquiz Alternate Assessment in the Science Classroom Performance Assessment Chapter Review Software Computer Test Bank
Study Guide, p. 85 Reinforcement, p. 85 Enrichment, p. 85 Critical Thinking/Problem Solving, p. 26	Section Focus Transparency 81, Earth's Atmosphere	
Study Guide, p. 86 Reinforcement, p. 86 Enrichment, p. 86 Activity Worksheets, pp. 118-119 Multicultural Connections, p. 43 Technology Integration, pp. 17-18	Section Focus Transparency 82, Remote-Control Devices	**Key to Teaching Strategies** The following designations will help you decide which activities are appropriate for your students.

GLENCOE TECHNOLOGY

The following multimedia resources are available from Glencoe.

Science and Technology Videodisc Series (STVS)

Physics
 Laser Eye Surgery
 Diseased Cells and Lasers
 Laser Identification of Fibers
 Multi-Mirror Telescope

Earth & Space
 Making Giant Mirrors

Physical Science CD-ROM

National Geographic Series
STV: Solar System
Newton's Apple: Physical Sciences

L1 Level 1 activities should be within the ability range of all students, including those with learning difficulties.

L2 Level 2 activities should be within the ability range of the average to above-average student.

L3 Level 3 activities are designed for the ability range of above-average students.

LEP LEP activities should be within the ability range of Limited English Proficiency students.

LS These activities are designed to address different learning styles.

COOP LEARN Cooperative Learning activities are designed for small group work.

P These strategies represent student products that can be placed into a best-work portfolio.

Teacher Classroom Resources

This is a representation of key blackline masters available in the Teacher Classroom Resources.

Teaching Aids

Section Focus Transparencies

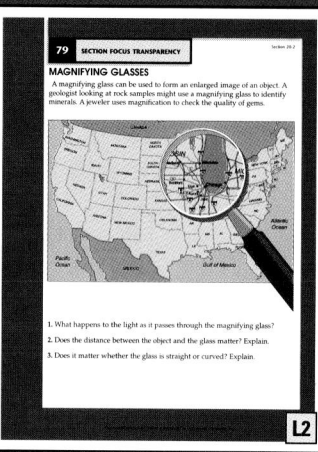

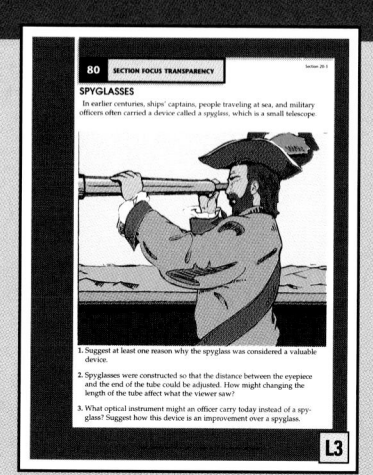

Science Integration Transparencies

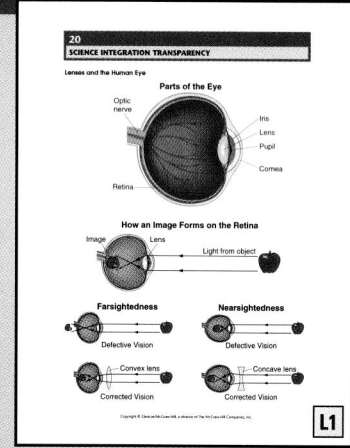

Teaching Transparencies

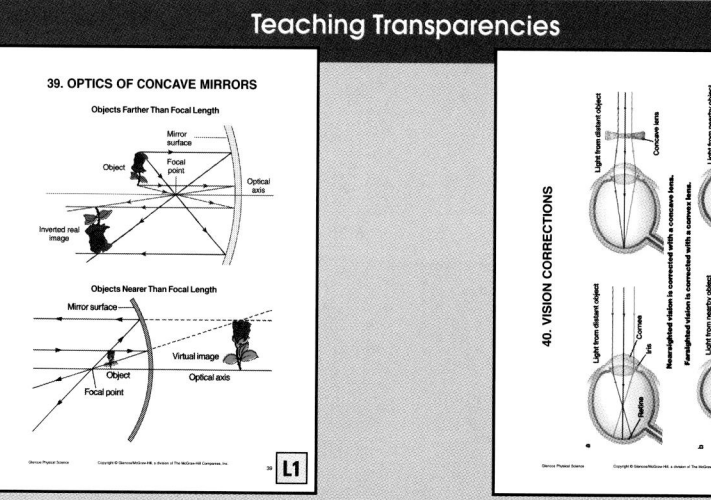

Meeting Different Ability Levels

Study Guide

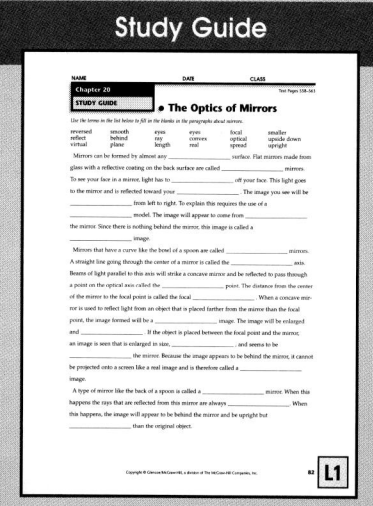

Reinforcement

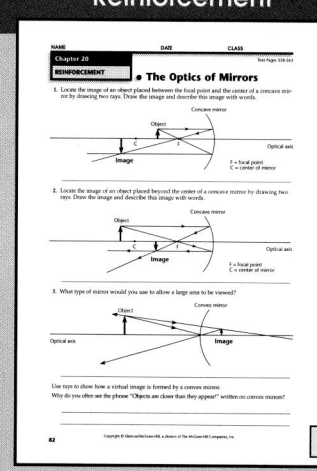

Enrichment Worksheets

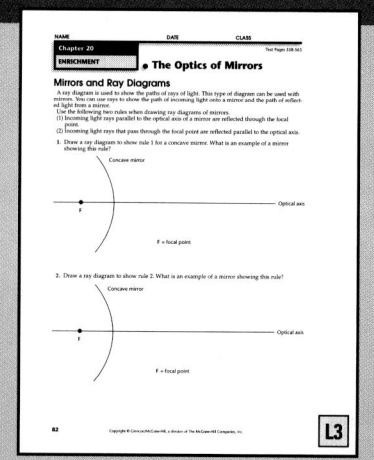

Hands-On Activities

Science Integration Activity

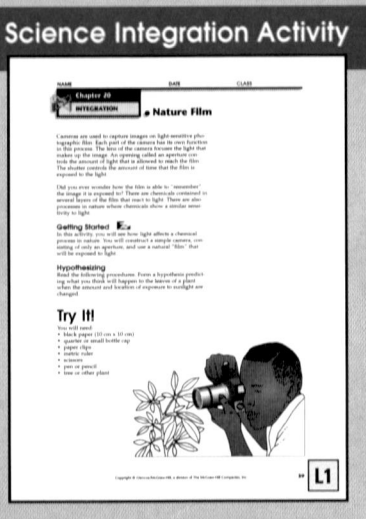

Lab Manual

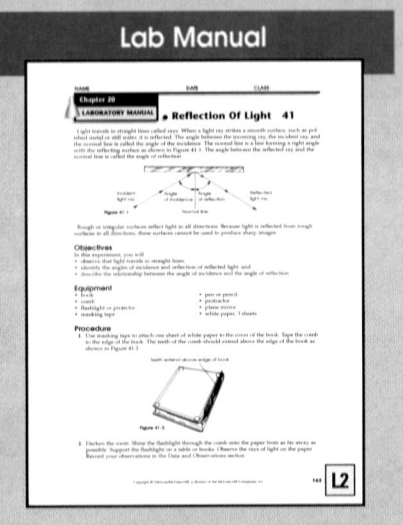

Assessment

Performance Assessment

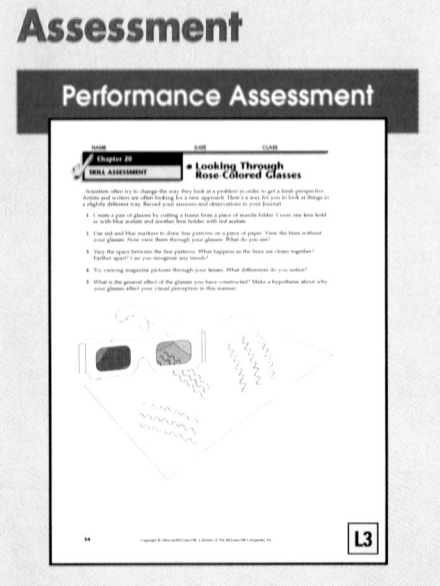

Enrichment and Application

Critical Thinking/Problem Solving

Cross-Curricular Integration

Science and Society Integration

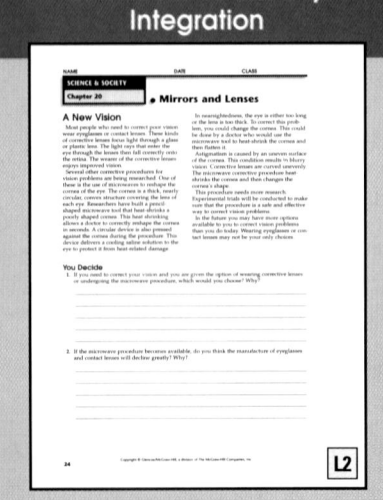

Technology Integration

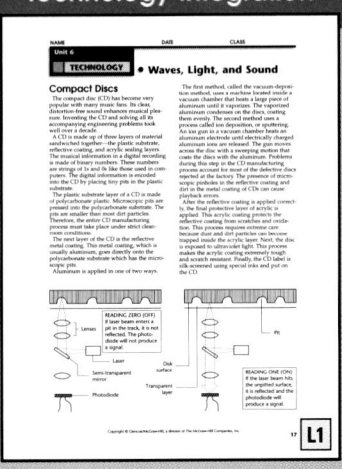

Multicultural Connections

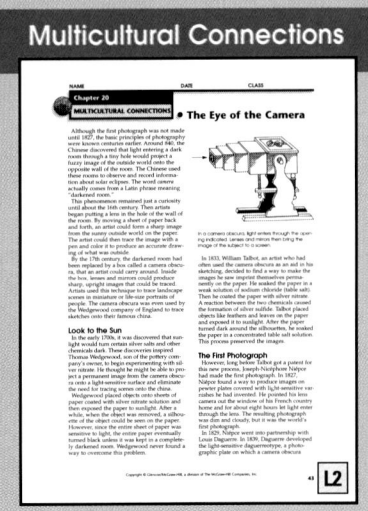

Concept Mapping

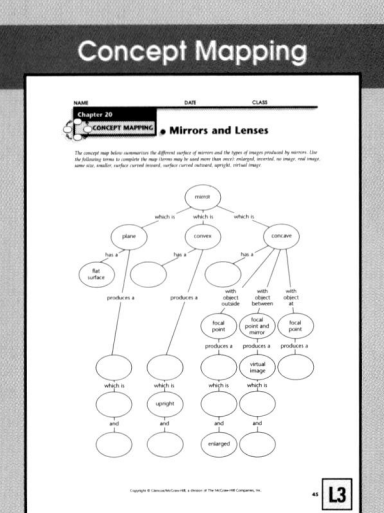

Chapter 20

Mirrors and Lenses

CHAPTER OVERVIEW

Section 20-1 This section introduces the optics of plane, concave, and convex mirrors. Methods of finding virtual and real images from reflections are illustrated.

Section 20-2 Concave and convex lenses are discussed, as well as how they affect vision.

Section 20-3 This section explains the optics of telescopes, microscopes, and cameras.

Section 20-4 **Science and Society** NASA's Hubble Space Telescope project is presented.

Section 20-5 Technological applications of light are discussed. Polarized light, lasers, and fiber optics are highlighted.

Chapter Vocabulary

plane mirror	microscope
virtual image	wide-angle lens
concave mirror	telephoto lens
focal point	polarized light
focal length	laser
real image	coherent light
convex mirror	incoherent light
convex lens	total internal reflection
concave lens	optical fiber
refracting telescope	
reflecting telescope	

Theme Connection

Systems and Interactions This chapter discusses the principles and applications of geometric optics. Systems and interactions are a major theme of this chapter. Light interacts with mirrors and lenses to produce images. Mirrors and lenses allow us to take pictures, view distant stars and planets, and perform surgery using lasers and fiber optics.

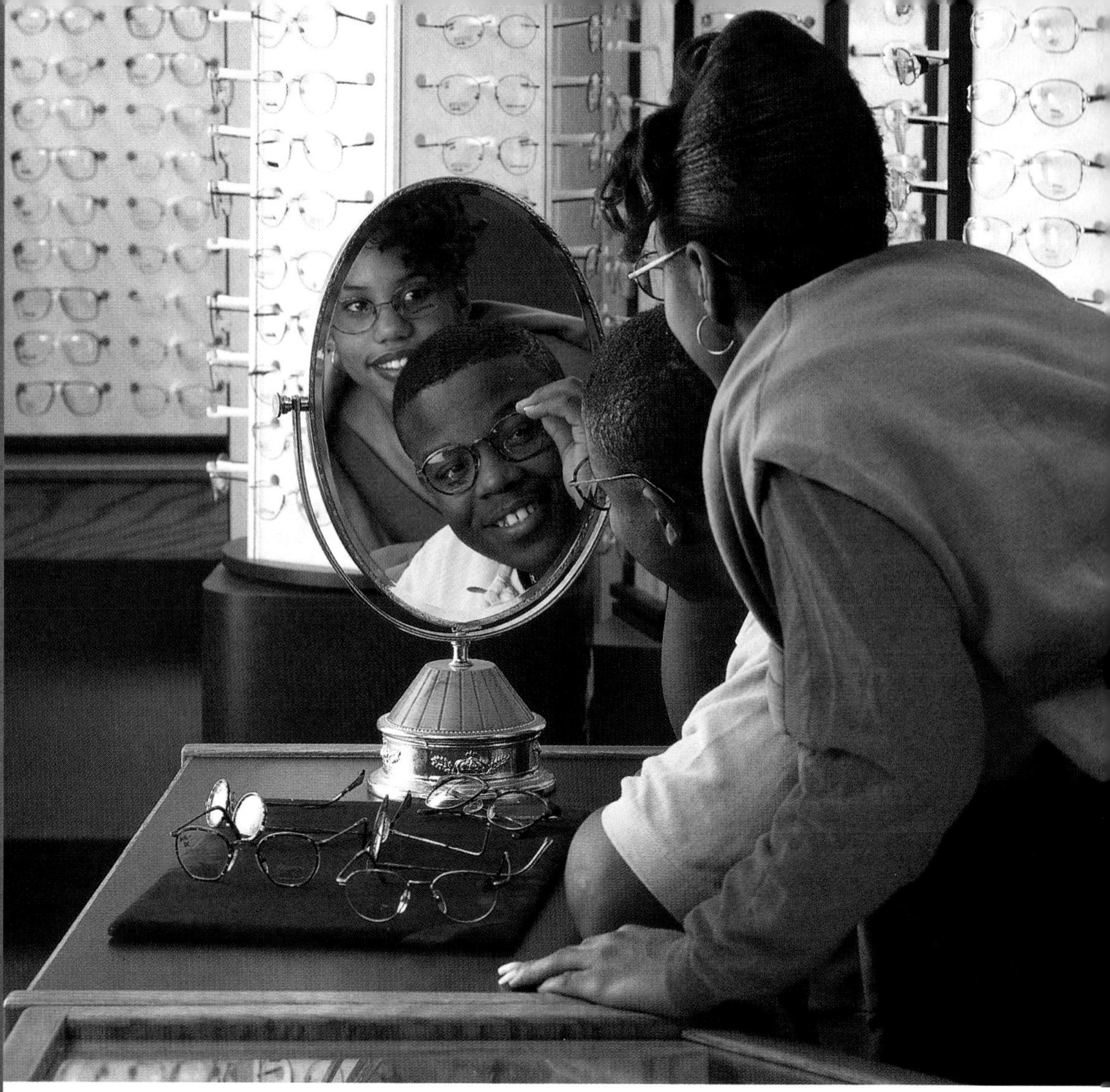

Previewing the Chapter

556

Learning Styles

Look for the following logo for strategies that emphasize different learning modalities. LS

Visual-Spatial	Explore Activity, p. 557; MiniLAB, pp. 559, 569; Activity 20-1, p. 563; Activity, p. 565; Reteach, pp. 567, 570; Across the Curriculum, p. 570; Demonstration, pp. 575, 576; Visual Learning, pp. 560, 564, 565, 575, 577
Interpersonal	MiniLAB, p. 566; Activity 20-2, pp. 580-581
Linguistic	Across the Curriculum, pp. 565, 577; Science Journal, p. 576

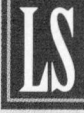

Chapter 20

Mirrors and Lenses

How do mirrors and lenses help the people in this photograph? In the last chapter, you learned about the wave properties of light—reflection, refraction, diffraction, and interference. Which of these are most useful in helping people see images clearly? Mirrors, which reflect light, and lenses, which refract light, are used to change the way you see things. Knowing this, do you still believe everything you see?

EXPLORE ACTIVITY

Find out how familiar surfaces can produce different images.

1. Observe your reflection in a flat mirror. What do you see? Is the reflection the same size as your face? Is it right side up?
2. Look at your image in the back of a large, shiny spoon. Slowly move the spoon close to your face and then far away. How does the image change?
3. Turn the spoon over and look into the inside of the spoon. Slowly move it close to your face and then far away.

Observe: In your Science Journal, compare and contrast the images produced by the three shiny surfaces.

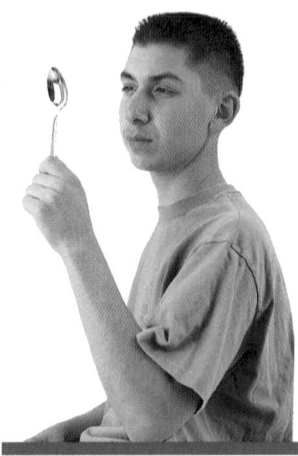

Previewing Science Skills

► In the Skill Builders, you will recognize cause and effect, map concepts, hypothesize, and sequence.

► In the Activities, you will observe, measure, predict, and infer.

► In the MiniLABs, you will hypothesize, infer, and measure.

557

EXPLORE ACTIVITY

Purpose

LS **Visual-Spatial** Use the Explore Activity to introduce students to images produced by reflection. Inform students that they will investigate how the shape of mirrors and lenses affects the kind of image produced. **L1**

Materials

spoons, flat mirrors

Teaching Strategies

Display a large concave and a large convex mirror and have students compare each part of the spoon to the mirrors. Have them view their images in the mirrors and find similarities between their images in the mirrors and those in the spoon.

Observe

Plane mirrors will produce upright images. The inside of spoons will provide an inverted image; students may observe an upright image if extremely close to the spoon. The back of a spoon will produce an upright image. The image size varies with distance to the spoon.

✔ Assessment

Process Have students experiment with both sides of the spoon to determine which shape would be most suitable as a rearview mirror on a car. The back of the spoon, as a convex mirror, allows viewing of a wider angle and produces upright images. Use the Performance Task Assessment List for Making Observations and Inferences in **PASC,** p. 17. **P**

Assessment Planner

Portfolio
Refer to page 583 for suggested items that students might select for their portfolios.

Performance Assessment
See page 583 for additional Performance Assessment options.
Skill Builders, pp. 562, 567, 571, 579
MiniLABS, pp. 559, 566, 569
Activities 20-1, p. 563; 20-2, pp. 580-581

Content Assessment
Section Wrap-ups, pp. 562, 567, 571, 573, 579
Chapter Review, pp. 583-585
Mini Quizzes, pp. 562, 567, 571, 579

Group Assessment
Opportunities for group assessment occur with Cooperative Learning Strategies and Flex Your Brain Activities.

20•1 The Optics of Mirrors

Section Background

Principles of reflection and refraction are consistent with the wave model of light. However, the analysis of images formed by mirrors and lenses is based on the ray model of light, which assumes that light travels in straight-line paths called rays. We assume light travels from objects to our eyes in straight paths.

Preplanning

Gather a few full-sized mirrors, smaller makeup mirrors, light sources, a collection of concave and convex lenses, spoons, index cards, tape, protractor, and paper clips to use in demonstrations and activities.

1 Motivate

Bellringer

Before presenting the lesson, display **Section Focus Transparency 78** on the overhead projector. Assign the accompanying **Focus Activity** worksheet. L1 LEP

Science Words

plane mirror
virtual image
concave mirror
focal point
focal length
real image
convex mirror

Objectives

- Explain how an image is formed in three types of mirrors.
- Identify examples and uses of plane, concave, and convex mirrors.

Figure 20-1

A plane mirror forms an upright, same-size virtual image.

Plane Mirrors

Close your eyes and picture your own face. How do you know what you look like? In what objects have you seen your reflection? The most obvious answer is a mirror, but you probably have also seen your reflection in windows and pieces of aluminum foil. Any smooth object that reflects light to form an image is a mirror.

Virtual Images

You are probably used to seeing your image in a **plane mirror,** one with a flat surface. A plane mirror is a piece of glass with a reflective coating on the front or the back. What do you see when you stand in front of the mirror and look directly into it? Your reflection is upright and appears to be the same size as you. **Figure 20-1** shows how your image is formed by a plane mirror. Light is reflected off you toward the mirror, which reflects the light back to your eyes.

The image appears to be behind the mirror because you perceive the reflected light as coming from somewhere beyond the mirror. However, these light rays do not exist *behind* the mirror; it is opaque. Because the image only *seems* to be behind the plane mirror, it is called a virtual image. A **virtual image** is an image in which no light rays pass through the image. The virtual image formed by a plane mirror is erect and appears as far behind the mirror as the object is in front of it.

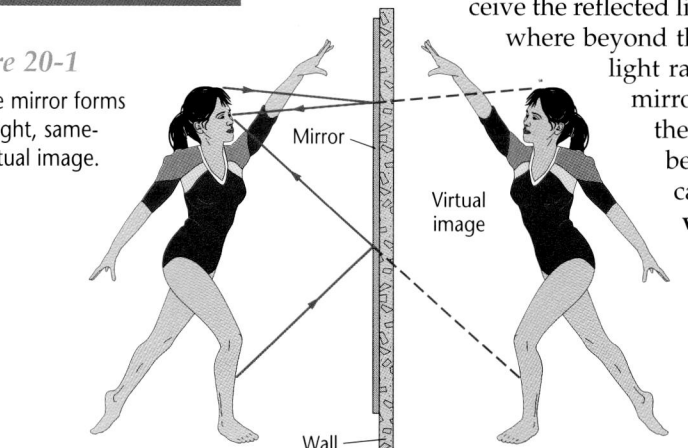

Mirror

Virtual image

Wall

Concave Mirrors

Mirrors are not always flat. If the surface of a mirror is curved inward, like the inside of a spoon, it is a **concave mirror.** The flower in **Figure 20-2A** is being reflected in a concave mirror. Concave mirrors form images differently from the way plane mirrors do. The way an image is formed depends on the position of the object in front of the mirror. **Figure 20-2B** shows one way an image can be formed in a concave mirror.

Program Resources

📁 Reproducible Masters
Study Guide, p. 82 L1
Reinforcement, p. 82 L1
Enrichment, p. 82 L3
Activity Worksheets, pp. 116, 117, 120 L1
Concept Mapping, p. 45
Lab Manual 41

🖲 Transparencies
Section Focus Transparency 78 L1
Teaching Transparency 39

The straight line drawn through the center of the mirror is the optical axis. Light rays parallel to the optical axis are all reflected to pass through one point on the optical axis, called the **focal point**. The distance from the center of the mirror to the focal point is called the **focal length**.

Suppose that the distance from the object to the mirror is a little greater than the focal length, as in **Figure 20-2B**. You can locate the top of the image by following the two red rays. One ray is drawn from the top of the flower through the focal point to the mirror. All rays that pass through the focal point on the way to the mirror are reflected parallel to the optical axis. A second ray is drawn from the top of the flower parallel to the optical axis. This ray is reflected through the focal point. The point where the two red rays meet is the top of the reflected image. Using the same method, blue rays from the bottom of the flower locate the bottom of the image. The image is real, enlarged, and upside-down. This is a **real image** because the rays of light actually meet at the image, so you could hold a screen there and see it. Recall that light does not pass through a virtual image.

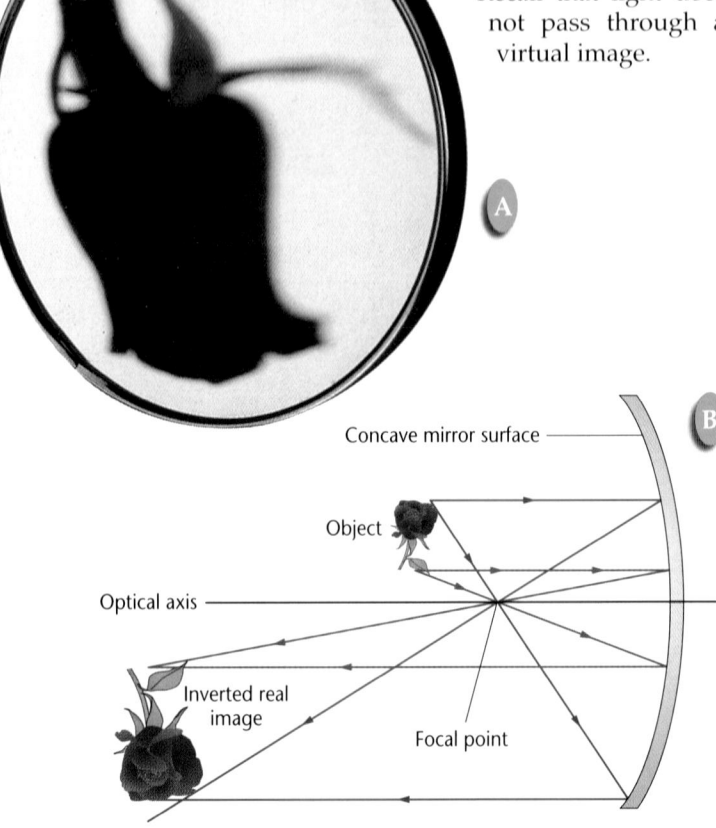

Optical axis

Concave mirror surface

Object

Inverted real image

Focal point

A

B

Figure 20-2

The image in this concave mirror is enlarged and inverted (A). The diagram in (B) shows how the image is created.

Figure 20-3

Some light beams are used to prevent accidents.

B The mirror produces a beam of parallel light rays that do not converge.

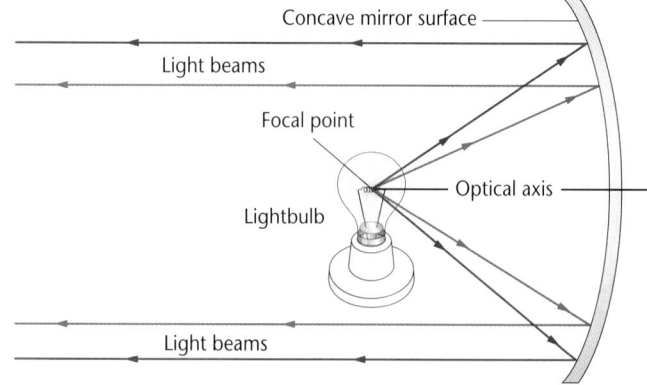

Figure 20-4

The image of the rose is virtual, upright, and enlarged (A) when the rose is placed between the focal point and the surface of a concave mirror (B). *Can you think of places where such a mirror would be helpful?*

What if you placed an object exactly at the focal point of the concave mirror? The lighthouse in **Figure 20-3A** gives us a hint. **Figure 20-3B** shows that if the object is at the focal point, the mirror reflects all light rays parallel to the optical axis. No image can be seen because the rays do not converge. Therefore, a light placed at the focal point is reflected in a beam. Devices such as car headlights, flashlights, and spotlights use this technique to create a concentrated light beam of nearly parallel rays.

Mirrors That Magnify

What if you placed an object between the mirror and the focal point as shown in **Figure 20-4A**? You can find the top of the image using two rays from the top of the flower in **Figure 20-4B**. One ray is drawn as if it comes through the focal point

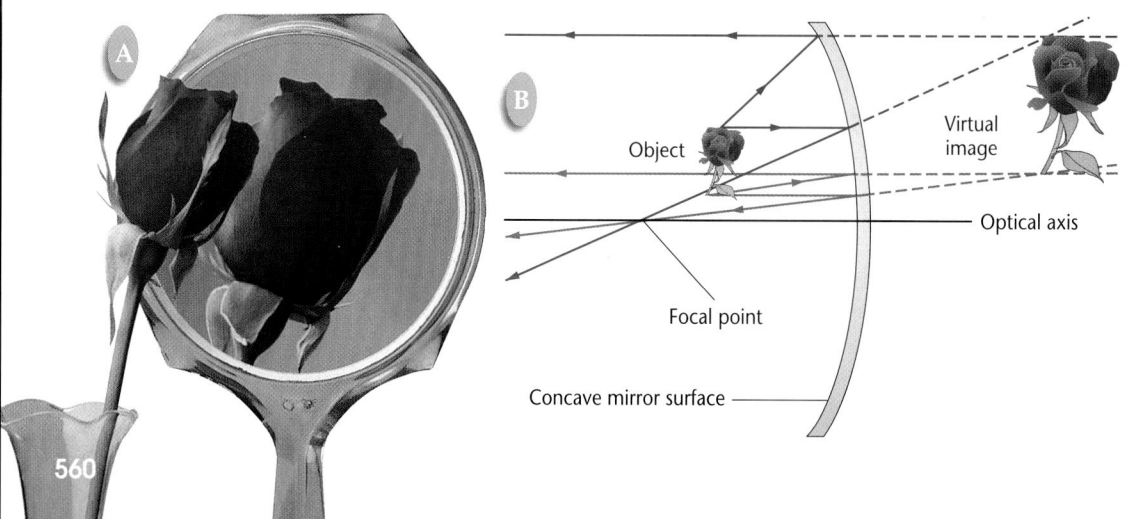

and is reflected parallel to the optical axis. The second ray leaves the flower parallel to the optical axis and reflects through the focal point. These rays never meet to form a real image. Instead, they appear to come from a single point behind the mirror. You can use rays from the bottom of the object to locate the bottom of the upright, enlarged, and virtual image. Hand mirrors that magnify reflections use this technique. The effect can also be seen when the bowl of a shiny spoon is placed close to your face.

Convex Mirrors

Have you noticed a large mirror mounted above the aisles in a store? This type of mirror that curves outward is called a **convex mirror. Figure 20-5** on the next page shows how convex mirrors produce an image. The reflected rays never meet, so the image is always virtual, upright, and smaller than the actual object.

Because convex mirrors spread out the reflected light,

Problem Solving

Looking Back with Car Mirrors

What kinds of mirrors would you find in or on a car? Mirrors that provide a view of the traffic beside or behind a car are necessary for safe, defensive driving. Have you ever noticed any side mirrors on cars and trucks that have a warning stating "OBJECTS IN MIRROR ARE CLOSER THAN THEY APPEAR"? The rearview mirror inside typically does not have these warnings.

If you were to examine the shapes of these mirrors carefully, you would notice that the inside mirror is a plane mirror. However, the mirrors with the warning stated above are convex mirrors because they bulge slightly outward. Can you think of any other places you would expect to find reflective or mirrored surfaces on a car?

Solve the Problem:

1. Describe the images you would see if you looked at your face closely in each type of mirror discussed above.
2. What type of image, real or virtual, is formed in a convex mirror? Explain.

Think Critically:

Explain why convex mirrors are sometimes used as outside rearview mirrors on cars. Why are they stamped with a warning?

20-1 The Optics of Mirrors   **561**

 Problem Solving

Solve the Problem

1. In a plane mirror, your face would appear the same size but reversed. In a convex mirror, your face would appear upright and slightly distorted.
2. A convex mirror diverges light rays, so they never meet. This results in a virtual image.

Think Critically

Convex mirrors give drivers a wider view of the road. They also show an image that is smaller than the actual object. They are stamped with the warning so that drivers do not misjudge the distance of the car behind them.

3 Assess

Check for Understanding

Ask students to draw a ray diagram that represents looking into a makeup or shaving mirror. They should draw the face between the mirror and the focal point. Have them find the image. The image is enlarged, upright, and virtual (behind the mirror). Refer to **Figure 20-4.** They should use rays from the top and bottom of the face.

Reteach

Guide the students through the solution to the problem in the Problem Solving feature.

Extension

For students who have mastered this section, use the **Reinforcement** and **Enrichment** masters.

Cultural Diversity

Arabian Lights One of the greatest early physicists was an Arabian scientist named Abu Ali al-Hassan ibn al-Hagthan, better known in the West as al-Hazen. Born in Basra, he lived from about 965 to 1038, and his book, *The Optical Thesaurus*, was unsurpassed until Johannes Kepler's work 600 years later. Among the contributions he made was the theory that vision is based on light coming from a source and being reflected from an object. Another achievement was the development of the camera obscura, which we know as the pinhole camera.

Section Wrap-up

Review

Figure 20-5

A convex rearview mirror forms a virtual image that is always upright and smaller than the object (A). The mirror diverges reflected rays (B) to form the image.

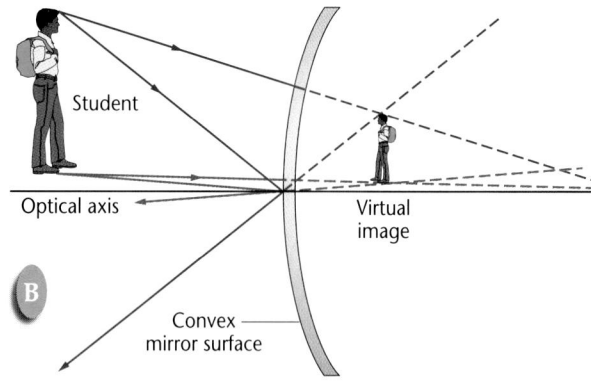

they allow large areas to be viewed. In addition to increasing the field of view in places like grocery stores and factories, convex mirrors can widen the view of traffic that can be seen in rear- or sideview mirrors of automobiles. However, your perception of distance can be distorted; objects are always closer than they appear in a convex mirror. Some mirrors that are on automobiles warn the driver that distances and sizes as seen in the mirror are not realistic.

So far, you have read about three ways that light can be reflected to form images. Review the three types of mirrors and describe the images that each mirror produces. In Section 20-2, you will find out the ways in which light can be refracted to form images.

Section Wrap-up

Review

1. What are convex mirrors used for? Explain.

2. Contrast the differences between the surfaces of plane, concave, and convex mirrors.

3. **Think Critically:** What kind of mirror would you use to focus light entering a telescope? Explain.

Skill Builder

Recognizing Cause and Effect

Suppose you drop a flashlight that has a concave mirror in it. When you turn the flashlight on, you notice that the light is less intense than it was before you dropped it. What may have happened? If you need help, refer to Recognizing Cause and Effect in the **Skill Handbook.**

Science Journal

In your Science Journal, make a list of all the mirrors you might see during an average day. Describe what each is used for and try to identify each one as plane, concave, or convex.

Activity 20-1

Reflections of Reflections

Y ou have probably looked at your face in a mirror sometime today, but have you seen a reflected image of the back of your head? If so, you most likely looked into a mirror held at an angle in front of your face into a mirror directly behind you. You actually saw a reflection of the original reflection of the back of your head.

Problem

How can you create multiple reflections of an object?

Materials

- 2 plane mirrors
- cellophane tape
- masking tape
- protractor
- paper clip

Procedure

1. Lay the two mirrors side by side and tape them together so they will open and close. Label them *R* and *L* as shown in the photo to the right.
2. Place the mirrors on a sheet of paper and, using the protractor, close the mirrors to an angle of 72°. Mark the position of the *R* mirror on the paper.
3. Bend one leg of a paper clip up 90° and place it close to the front of the *R* mirror.
4. Count the number of images of the paper clip you see in the *R* and *L* mirrors. Don't move the clip.
5. Count the images as you slowly open the mirrors to 90° and then to 120°, keeping the paper clip close to the *R* mirror.
6. Make a data table to record the number of images you can see in the *R* and *L* mirrors when they are at 72°, 90°, and 120°.

Analyze

1. The mirror arrangement creates an image of a complete circle divided into

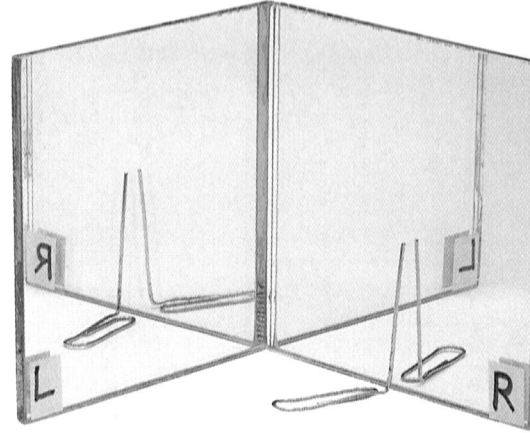

wedges by the mirrors. How many wedges did you observe with the 72°, 90°, and 120° angles?

Conclude and Apply

2. What angle would divide a circle into six wedges? **Hypothesize** how many images would be produced.
3. **Analyze** your results to determine which is the better **predictor** of the number of paper clip images that can be seen—the number of mirror images or the number of wedges.

Data and Observations   Sample Data

Angle of Mirrors	Number of Paper Clip Images	
	R	L
72°	2	2
90°	1	2
120°	1	1

Activity 20-1

Purpose

Visual-Spatial Students will predict and observe image formation by plane mirrors.

L1 **COOP LEARN**

Process Skills

measuring, predicting, inferring, recognizing cause and effect, comparing and contrasting, forming operational definitions

Time

one class period

Activity Worksheets, pp. 5, 116, 117

Teaching Strategies

- Use rectangular mirrors at least 5 cm across. Thin glass is better. Mirror tiles (hardware store) can be cut to size, but sharp edges must be ground or taped.
- Place a protractor on a copy machine and make lab worksheets upon which students can measure the appropriate angles.

Answers to Questions

1. 72°—five wedges
   90°—four wedges
   120°—three wedges
2. The angle producing six segments is 360/6 = 60°. Five images will be produced.
3. The number of mirror images is the number of paper clip images. This number is one less than the number of wedges.

Assessment

Performance Have students face two mirrors angled at 45° to each other. How many images of their faces do they see? What about smaller angles? Larger angles? Have them explain their results. Use the Performance Task Assessment List for Making Observations and Inferences in **PASC**, p. 17.

P

Section 20•2

Prepare

Section Background

Note that in both concave and convex lenses, the greatest bending of light rays occurs near the edges of the lens. Near the edges, the angle between the two glass surfaces is greater, and more total refraction occurs. Light is refracted twice when passing through a lens, once entering and once leaving. At the point in the center of the lens, the glass sides are parallel to each other and the light ray passes straight through.

Preplanning

Obtain plastic wrap, pennies, and medicine droppers for the MiniLAB.

1 Motivate

Bellringer

Before presenting the lesson, display **Section Focus Transparency 79** on the overhead projector. Assign the accompanying **Focus Activity** worksheet. [L1] [LEP]

Tying to Previous Knowledge

Many students wear corrective lenses, either as eyeglasses or contact lenses. What do the eyeglasses do? They change the light path from the object to your eyes by refracting the light before it enters your eyes.

Visual Learning

Figure 20-6 Which lens has the shorter focal length? *the thicker lens* [LEP] [IS]

Science Words

convex lens
concave lens

Objectives

- Describe the types of images formed with convex and concave lenses.
- Cite examples of how these lenses are used.
- Explain how lenses are used to correct vision.

Figure 20-6

A thick convex lens (A) bends light more than a thin convex lens (B). *Which lens has the shorter focal length?*

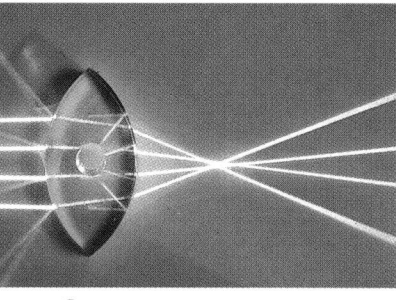
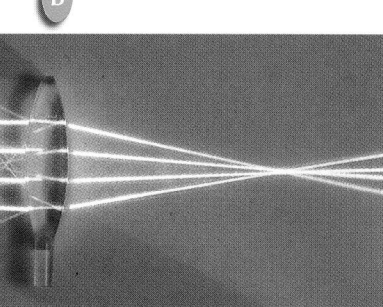

Convex Lenses

Do you wear glasses or contact lenses? If so, you use lenses to improve your ability to see. Like curved mirrors, lenses are described as convex or concave, depending on their shape.

Convex lenses are thicker in the middle than at the edges. Light rays approaching the lens parallel to the optical axis are refracted toward the center of the lens. They converge at the focal point, so they are capable of forming real images that can be projected onto a screen.

The amount of refraction depends on the change in the speed of light as it passes through a material and the shape of the object. As shown in **Figure 20-6**, thick lenses with highly curved surfaces bend light much more than thin ones with less curved surfaces. The focal length of the thick convex lens in the top diagram in **Figure 20-6** is shorter than that of the thin convex lens below it. Convex lenses can produce many kinds of images, both real and virtual, upright, inverted, enlarged, or reduced. The type of image formed depends on the position of the object and the focal length of the lens. The diagrams in **Figure 20-7** show the images that result from three different locations of the object and focal point.

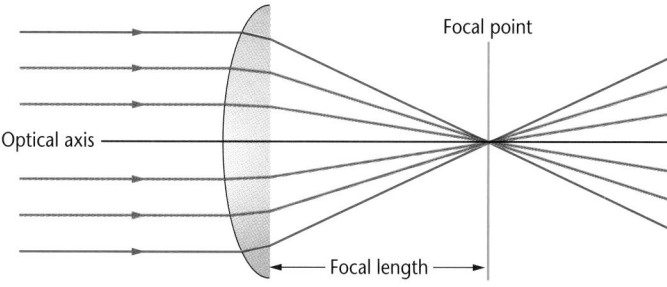
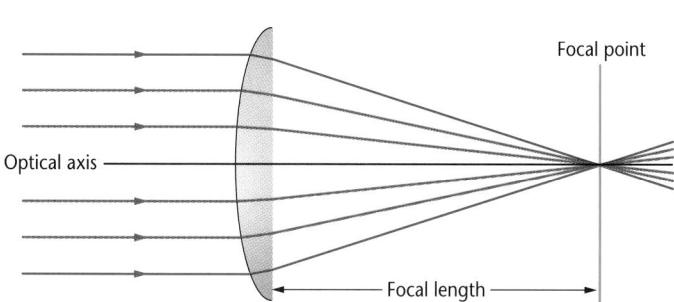

Focal point

Optical axis

Focal length

Focal point

Optical axis

Focal length

Program Resources

Reproducible Masters
Study Guide, p. 83 [L1]
Reinforcement, p. 83 [L1]
Enrichment, p. 83 [L3]
Science and Society Integration, p. 24
Activity Worksheets, p. 121 [L1]
Lab Manual 42

Transparencies
Section Focus Transparency 79 [L1]
Science Integration Transparency 20
Teaching Transparency 40

Figure 20-7

The image formed by a convex lens depends upon the location of the object relative to the focal length of the lens.

A If a person is photographing a faraway object, it's likely that the object is more than two focal lengths from the camera lens. If you follow the light paths, you'll notice that the real image is smaller than the object, and inverted. The lens in your eye forms images in the same way that a camera forms images.

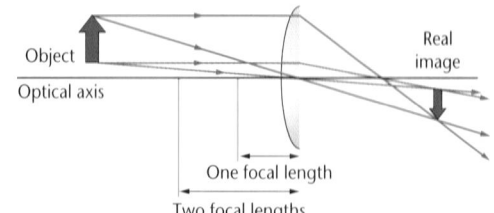

B If an object is between one and two focal lengths from the lens, the real image is inverted and larger than the object. This is the method used to project a movie from a small film to the large screen of a theater or from an overhead projector to a screen in your classroom.

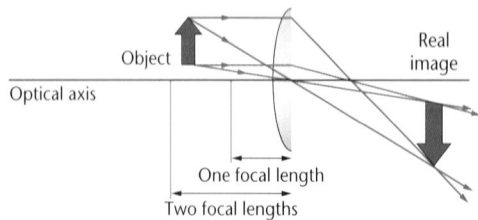

C Have you ever used a magnifying glass to closely examine an object? A magnifying glass is a convex lens, so you must hold it less than one focal length from the object. The light rays can't converge, and an enlarged, upright, and virtual image is formed. Look at the position of the image. Notice that the object seems larger and farther away than it really is.

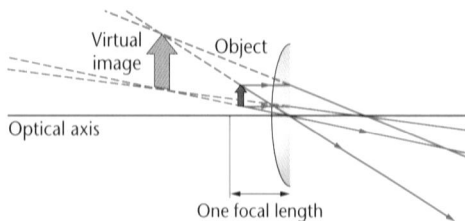

Concave Lenses

Concave lenses are thinner in the middle and thicker at the edges. As is shown in **Figure 20-8,** light passing through a concave lens bends toward the edges. The rays diverge and never form a real image. The image is virtual, upright, and smaller than the actual object. The image formed by a concave lens is similar to the image produced by a convex mirror because they both diverge light to form virtual images.

Concave lenses are usually used in combination with other lenses. They can be used with convex lenses in telescopes and cameras to spread out incoming light and extend the focal length so you can see a clear image of a faraway object. Concave lenses are also used to correct nearsighted vision.

Figure 20-8

Light rays passing through a concave lens diverge and form a virtual image. *Are concave lenses similar to or different from convex mirrors?*

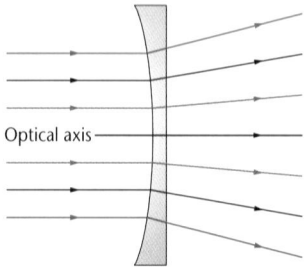

20-2 The Optics of Lenses   **565**

Activity

LS **Visual-Spatial** Distribute some convex lenses of varying focal lengths to your class. Show students how to find the focal length of a lens by focusing the clearest possible image of an overhead light on a white piece of paper (held by a partner) and measuring the distance from the paper to the lens. **L1**

Visual Learning

Figure 20-8 **Are concave lenses similar to or different from convex mirrors?** *Similar. Both diverge light and produce virtual images.* **LEP** **LS**

INTEGRATION
Life Science

See page 566.

The cornea does about two-thirds of the focusing and is responsible for most human vision problems, leaving the flexible lens to conduct the fine focusing. The area between the cornea and the lens is filled with a liquid called the aqueous humor. Glaucoma, a condition of increased pressure in the eye, results when the tube which drains excess amounts of this liquid becomes blocked.

Across the Curriculum

Language Arts Have students write a paper about how convex and concave lenses are manufactured for use in eyeglasses and optical instruments. Eyeglasses were used as early as the thirteenth century, and gems were used even earlier as natural magnifying glasses by the Greeks and Arabs. **LS** **L2**

Integrating the Sciences

Life Science In biology, convergence refers to the tendency of different organisms to have some common characteristics when living in the same conditions. Concave mirrors are said to converge light. **From these uses, can you deduce a general definition of converge? Explain.** *An appropriate definition from these contexts would be "to come together."*

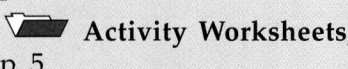

INTEGRATION
Life Science

MiniLAB

Can lenses be made of liquid?

Procedure
1. Cut a 10-cm × 10-cm piece of plastic wrap. Set it on a page of printed text.
2. Place a small water drop on the plastic. Look at the text through the drop. What do you observe?
3. Make your water drop larger and observe the text through it again. Did anything change?
4. Use a medicine dropper to drop water onto the head of a penny. How does the penny look? How many drops of water can you fit on the head of a penny?

Analysis
1. What kind of lens does the drop form?
2. How would your image look if you moved the liquid lens farther from the text you are observing? Try it.

Lenses and Eyesight

What determines how well you can see the words on this page? Your ability to focus on these words depends on the way your eye is designed. Light enters your eye through the transparent covering of your eye, the cornea. The light then passes through an opening called the pupil. The colored part of your eye, the iris, adjusts the pupil size to control how much light can pass through a flexible convex lens behind the pupil. The light then converges to form an inverted image on your retina. The lens in your eye is soft, and flexible muscles in your eye can change its shape. When you look at a distant object, you need a longer focal length, so your eye muscles adjust your lens to a less convex shape. When you focus on a nearby object, the eye muscles increase the curvature of the lens.

Improving Your Vision

If you have healthy vision, you should be able to see objects clearly from a distance of about 25 cm or more. Many people need their vision corrected. To have normal vision, the image of an object should focus on the retina inside your eye. If the image is focused in front of or behind the retina, vision problems will result. **Figure 20-9A** illustrates how concave lenses correct nearsightedness by diverging the light rays before they enter the eye. **Figure 20-9B** shows how farsightedness can be corrected with a convex lens.

Another vision problem is blurry vision from astigmatism, which occurs when the surface of the cornea is curved

Figure 20-9

Some vision problems can easily be corrected with concave (A) and convex (B) lenses. *Examine some eyeglasses. Can you tell which type of lens they use?*

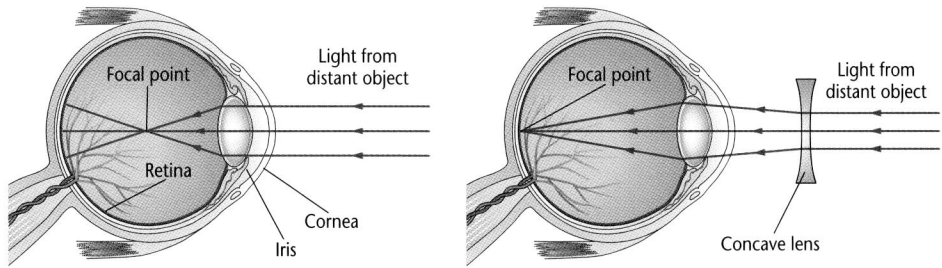

A A nearsighted person has difficulty seeing distant objects clearly. The eyeball is too long or the cornea bulges out, focusing the image in front of the retina.

566 Chapter 20 Mirrors and Lenses

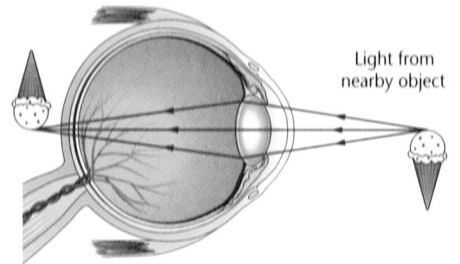

Light from nearby object

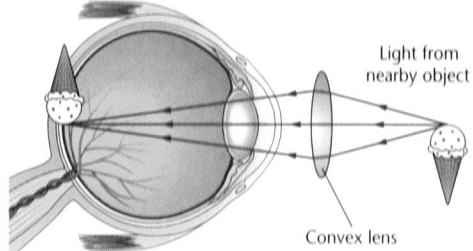

Light from nearby object

Convex lens

unevenly. Corrective lenses for this condition have an uneven curvature, as well.

There are currently several ways to correct poor vision caused by lens problems in the eye. Artificial lenses in the form of eyeglasses or contacts can be worn to refract light before it enters the eye, so the light will focus on the retina. As a result, the wearer will see clear images. Contact lenses are actually worn over the cornea. In some cases, vision can be corrected by using surgical lasers to reshape the cornea.

Can you recall what all lenses have in common? They all refract light that passes through them. Convex lenses refract light toward the center of the lens, and concave lenses refract light away from the center of the lens. As you read the next section, you'll see how these two kinds of lenses are used in cameras, microscopes, and telescopes.

B Farsighted people can see faraway objects, but they can't focus clearly on nearby objects. Their eyeballs are either too short or their corneas are too flat to allow the rays to converge on the retina. As a result, the image is focused behind the retina.

Section Wrap-up

Review

1. When using a slide projector, why are the slides inserted in the projector upside down?

2. What type of lens would you use to examine a tiny spider on your desk?

3. **Think Critically:** If you have difficulty reading the chalkboard from the back row, what is most likely your vision problem? How could it be corrected?

Skill Builder
Concept Mapping
Mirrors and lenses are the simplest optical devices. Design a network tree concept map to show some uses for each shape of mirror and lens. If you need help, refer to Concept Mapping in the **Skill Handbook**.

Using Computers

Spreadsheet Prepare a spreadsheet that organizes the information you have acquired about vision problems. For cases of nearsightedness, farsightedness, and astigmatism, display the symptom, the cause, and the method of correction.

[S] Visual-Spatial Compare eyeglasses that correct nearsightedness and farsightedness. Have students examine near and far objects through each type of lens and explain how each lens corrects vision problems. **LEP**

Extension

For students who have mastered this section, use the **Reinforcement** and **Enrichment** masters.

4 Close

•MINI•QUIZ•

Use the Mini Quiz to check students' recall of chapter content.

1. A(n) _____ lens causes light rays to diverge. *concave*

2. Parallel light rays meet at the focal point after passing through a(n) _____ lens. *convex*

3. What kind of lens is found in your eye? *convex*

Section Wrap-up

Review

1. Convex lenses form enlarged, inverted images when slides are between one and two focal lengths from the lens.

2. a convex lens

3. **Think Critically** Nearsightedness; use a concave lens to diverge the light rays entering your eye.

Using Computers

The spreadsheet can be formatted with the vision problems listed across the top of the page in columns or along the side in three rows.

Skill Builder

OPTICAL DEVICES

Includes → Mirrors
Includes → Lenses

Mirrors — Such as → Plane, Convex, Concave
Plane — Diverges light → Bathroom mirror
Convex — Diverges light → Exterior auto rearview mirror
Concave — Converges light → Reflecting telescope
Concave — Diverges light → Flashlight or beacon

Lenses — Such as → Convex, Concave
Convex — Converges light → Refracting telescope
Convex — Converges light → Slide/Movie projector
Convex — Converges light → Farsighted vision correction
Concave — Diverges light → Nearsighted vision correction
Concave — Diverges light → Complex lenses to alter the focal length

Assessment

Portfolio Have students improve their concept maps and prepare a copy for their portfolios. Use the Performance Task Assessment List for Concept Map in **PASC**, p. 89. **P**

Prepare

Section Background

Galileo did not invent the telescope, but he was among the first to use it in monitoring the heavens. For example, he discovered sunspots, some moons of Jupiter, and the phases of Venus.

Preplanning

- If your school does not have a telescope, try to borrow one for demonstrations. Bring in a microscope and several cameras.
- Obtain test tubes and lids or stoppers for the Mini-LAB.

1 Motivate

Bellringer

Before presenting the lesson, display **Section Focus Transparency 80** on the overhead projector. Assign the accompanying **Focus Activity** worksheet. L1 LEP

Tying to Previous Knowledge

Ask students if they have ever taken a blurry picture with a camera. Have them suggest why this might have happened. The function of a camera will be discussed in this section.

20•3 Optical Instruments

Science Words

refracting telescope
reflecting telescope
microscope
wide-angle lens
telephoto lens

Objectives

- Compare refracting and reflecting telescopes.
- Explain how a camera creates an image.

Telescopes

Lenses and mirrors are important components of optical instruments—devices that are designed to aid the human eye in making observations. In this section, you'll learn about three common optical instruments—telescopes, microscopes, and cameras.

Have you ever looked at the moon through a telescope? With a good telescope, you should be able to see the craters and other features on the moon's surface clearly. Telescopes are designed to collect light from faraway objects. Much of the information we have today about the moon, the planets, our galaxy, and other galaxies has been gathered by viewing these celestial bodies through telescopes.

Around the year 1600, lens makers in Holland constructed a telescope to view distant objects. In 1609, Galileo built and used his own telescope to discover the moons of Jupiter, the phases of Venus, and some details of the Milky Way galaxy. Today, scientists use several kinds of telescopes with many design improvements. **Figure 20-10** shows how a telescope can improve our view of the moon.

Refracting Telescopes

One common telescope is the **refracting telescope.** A simple refracting telescope uses two convex lenses to gather and focus light from distant objects. **Figure 20-11A** is a diagram of a refracting telescope.

Several problems are associated with refracting telescopes. The objective lens must be large to allow enough light in to form a bright image. These heavy glass lenses are hard to make and costly. Their own weight can cause them to sag and distort the image.

Figure 20-10

The moon can be seen in greater detail when viewed through a telescope, as in the lower photo.

568   Chapter 20 Mirrors and Lenses

Program Resources

 Reproducible Masters

Study Guide, p. 84 L1
Reinforcement, p. 84 L1
Enrichment, p. 84 L3
Cross-Curricular Integration, p. 26
Activity Worksheets, p. 122 L1
Science Integration Activity, pp. 39, 40

Transparencies

Section Focus Transparency 80 L1

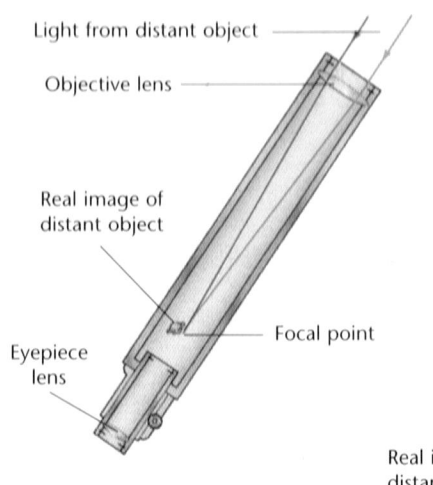

Light from distant object

Objective lens

Real image of distant object

Eyepiece lens

Focal point

Figure 20-11

Refracting and reflecting telescopes collect light to produce images of distant objects.

A Light enters the refracting telescope through a convex lens called the objective lens. The real image formed by this lens is magnified by a second convex lens, called the eyepiece, with a shorter focal length. You see an enlarged, inverted, virtual image of the real image.

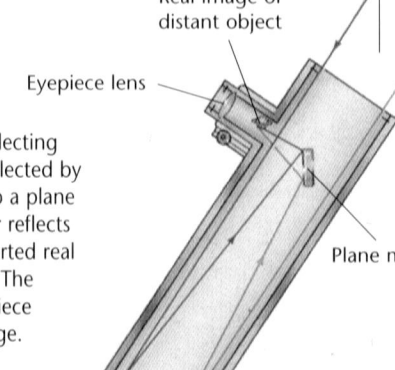

Light from distant object

Real image of distant object

Eyepiece lens

Plane mirror

Concave mirror

B Light enters the reflecting telescope and is reflected by the concave mirror onto a plane mirror. The plane mirror reflects the rays to form an inverted real image in the telescope. The convex lens in the eyepiece then magnifies this image.

Reflecting Telescopes

Because of the problems described on the opposite page, most large telescopes are reflecting telescopes. A **reflecting telescope** uses a concave mirror, a plane mirror, and a convex lens to collect and focus light from distant objects. **Figure 20-11B** is a diagram of a reflecting telescope.

Sometimes, you might want to view distant objects so they appear upright. Imagine trying to watch a baseball game through binoculars if the image were upside down. Binoculars work on the same principle as a refracting telescope, except there are two sets of lenses—one for each eye. A third lens or a pair of reflecting prisms has been added to binoculars to invert the upside-down image so it appears upright. Terrestrial telescopes, such as those used for bird-watching, are also designed to produce an upright image.

MiniLAB

How can the location of a lens affect the observed image?

Procedure
1. Fill a glass test tube with water and seal it with a lid or stopper.
2. Type or print the compound name "SULFUR DIOXIDE" in capital letters on a piece of paper or a note card.
3. Set the test tube horizontally over the words and observe them. What do you notice?
4. Hold the tube 1 cm over the words and observe them again. Record your observations. Repeat, holding the tube at several other heights above the words.

Analysis
1. What were your observations of the words at these distances? Can you explain your observations?
2. Are the images you see at these heights real or virtual?

MiniLAB

Purpose
LS **Visual-Spatial** Students will observe what happens when the distance from a lens to the object being observed is changed. **L1**
LEP **COOP LEARN**

Materials
glass test tube with lid or stopper, water, paper

Teaching Strategies
The words "SULFUR DIOXIDE" should be printed or typed in capital letters smaller than the test tubes used to view them. The words can also be written in two different colors.

Analysis
1. When the tube is sitting on the paper, the words will be slightly magnified. When the tube is held about 1 cm above the words, SULFUR will appear upside down. Note that DIOXIDE appears unchanged because the letters are symmetrical. The whole image is really inverted because the object is beyond the focal length of the lens.
2. The upright, magnified image is virtual. The inverted images are real. (See **Figure 20-7** for explanations.)

✓Assessment

Content Ask students to explain whether a magnifying glass for reading (a convex lens) could produce inverted images of the words. *yes, if the glass is placed far enough from the written words* Use the Performance Task Assessment List for Making Observations and Inferences in **PASC**, p. 17. **P**

📁 **Activity Worksheets,** pp. 5, 122

Integrating the Sciences

Chemistry The reflective coating on many mirrors used in telescopes is made of a valuable metallic element. This same element is found in the light-sensitive compounds used in photographic film. Ask students: **What are the name, chemical symbol, and atomic number of this element?** *silver, Ag, atomic number 47*

Theme Connection

Systems and Interactions This section dealing with optical instruments naturally reinforces the systems and interactions theme. Most optical instruments involve a system of mirrors and/or lenses designed to interact with light in a way that produces images with desired properties.

3 Assess

Check for Understanding

 **FLEX Your Brain**

Use the Flex Your Brain activity to have students explore CAMERA APERTURES.

 Activity Worksheets, p. 5

Reteach

LS Visual-Spatial Point to a somewhat distant object that can be seen separately from its surroundings. Tell students to imagine they are taking two photos of the same object, one with a wide-angle lens and one with a telephoto lens. Have them sketch what they might see with each lens. **Which lens has a longer focal length?** *telephoto lens*

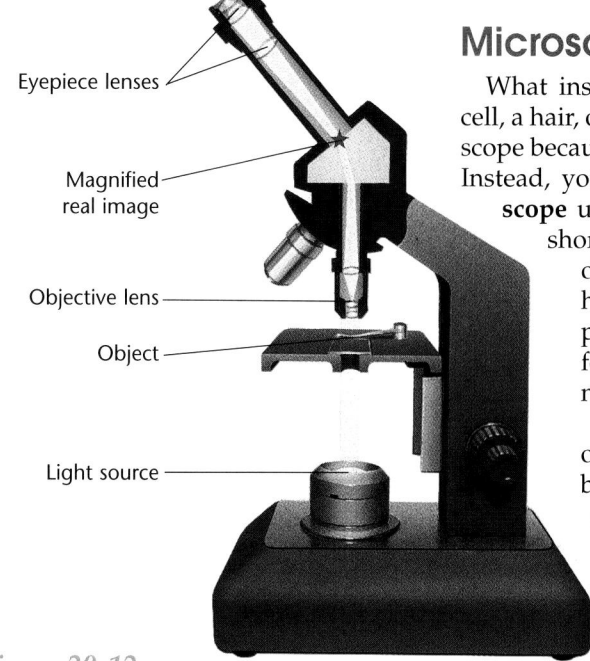

Figure 20-12

A microscope contains two convex lenses so it can magnify small objects.

Figure 20-13

This photo was taken with a wide-angle lens; it includes much of the surroundings. Observe how this photo differs from the one of the same scene on the next page.

Microscopes

What instrument would you use to look at a cell, a hair, or an amoeba? You wouldn't use a telescope because it is used to look at faraway objects. Instead, you would use a microscope. A **microscope** uses two convex lenses with relatively short focal lengths to magnify small, close objects. A microscope, like a telescope, has both an objective lens and an eyepiece lens. However, it is designed differently because the objects viewed are not far away.

Figure 20-12 shows the arrangement of lenses in a microscope. The object to be viewed is placed on a transparent slide and illuminated from below. The light travels through the objective lens and a real, enlarged image is formed. It is enlarged because the object is between one and two focal lengths from the lens. The real image is magnified again by the eyepiece to create a virtual, enlarged image. This results in an image up to several hundred times larger than the actual object.

Cameras

Do you keep a photograph album with your favorite pictures? Have you stopped to think about how a camera transfers an image onto film? A camera gathers light through a lens and projects an image onto light-sensitive film.

When you take a picture with a camera, the light reflected off the subject enters an opening in the camera called the aperture. A shutter on the aperture opens to allow light to enter the camera.

The light passes through the camera lens, which focuses the image on the photographic film. The image is real, inverted, and smaller than the actual object. The size of the image depends upon the focal length of the lens and how close the lens is to the film.

Suppose you and a friend photograph the same object at the same distance. Your picture would look different from that of your friend if the cameras had different lenses. Some lenses have short focal

Across the Curriculum

Art Have students find a book on photography and investigate what variables in a camera can be controlled to gain the desired effects in a photograph. They should explore aperture, shutter speed, wide-angle and telephoto lenses, and flashes. **L1 LS**

lengths that produce a relatively small image of the object but include much of its surroundings. These lenses are called **wide-angle lenses,** and they must be placed close to the film to focus the image with their short focal length. The photo in **Figure 20-13** was taken with a wide-angle lens. **Telephoto lenses** have longer focal lengths and are located farther from the film than wide-angle lenses. Telephoto lenses are easy to recognize because they protrude from the camera to increase the distance between the lens and the film. The image seems enlarged and the object seems closer than it actually is, as shown in **Figure 20-14.** These lenses are preferred when photographing people's faces from a distance.

Telescopes, microscopes, and cameras are examples of instruments that contain mirrors and lenses and help you to make observations. How many other types of optical instruments can you name?

Figure 20-14

This photo was taken with a telephoto lens. It gives a close-up view that omits most of the surroundings.

Section Wrap-up

Review

1. Compare and contrast reflecting and refracting telescopes.

2. If you wanted to photograph a single rose on a rosebush, what kind of lens would you use? Explain why you chose this lens.

3. Think Critically: Which optical instrument—a telescope, a microscope, or a camera—forms images in a way most like your eye? Explain.

Skill Builder
Hypothesizing
You notice that all the objects in a photograph you've taken are blurry. Use your knowledge of lenses and focal lengths to form a hypothesis that could explain why the photo was blurred. If you need help, refer to Hypothesizing in the **Skill Handbook.**

Science Journal

Investigate the latest technologies and designs being used in new telescopes. Write several paragraphs about one of the designs and explain its advantages. Make a drawing that illustrates this design.

Skill Builder
A reasonable hypothesis would be that the lens was too close or too far away from the film. A more basic hypothesis would be that the camera was not held still while taking the picture. **P**

✓ Assessment

Oral Ask students what the photographer would need to do to fix this problem. Many cameras have focusing mechanisms that move the lens closer to or farther from the film. Also, the photographer must hold the camera still while taking pictures. Use the Performance Task Assessment List for Making Observations and Inferences in **PASC,** p. 17. **P**

Prepare

Section Background

- NASA has a teacher resource center that will provide free Hubble Space Telescope materials.
- Have students research the significant error found shortly after launching.

1 Motivate

Bellringer

Before presenting the lesson, display **Section Focus Transparency 81** on the overhead projector. Assign the accompanying **Focus Activity** worksheet. L1 LEP

2 Teach

Making a Model

Draw a line 2.4 m long to show how large the diameter of the primary mirror is. The large size allows the telescope to capture more light so that faint, distant stars can be observed. These stars could not be seen from Earth.

3 Assess

Check for Understanding

? FLEX Your Brain

Use the Flex Your Brain activity to have students explore ATMOSPHERIC DISTORTION.

📁 **Activity Worksheets,** p. 5

TECHNOLOGY:
20•4 The Hubble Space Telescope

Objectives

- Describe the development and repair of the Hubble Space Telescope.
- Discuss the goals and uses of the Hubble Space Telescope.

Seeking a Clear View from Space

Imagine trying to read a sign from the bottom of a swimming pool. The water distorts your view of objects beyond the water's surface. In a similar way, Earth's atmosphere blurs our view of objects in outer space. On April 20, 1990, the National Aeronautics and Space Administration (NASA) used the space shuttle *Discovery* to launch the Hubble Space Telescope into an orbit about 600 km above Earth. The telescope has produced images sharper than powerful telescopes on Earth, and has enabled scientists to detect visible light and infrared and ultraviolet radiation from the planets, stars, and distant galaxies usually blocked by Earth's atmosphere.

Look at the diagram of the Hubble Space Telescope on the opposite page. The solar panels provide electrical power to the system. A 2.4-m primary and a smaller secondary mirror collect and focus light to form an image. Various instruments on the telescope interpret the data and communicate it to scientists on Earth.

Troubleshooting on the Telescope

The Hubble Space Telescope is a complex instrument that at first experienced a major problem. Once in orbit, the telescope sent back blurry images because of a defect in the shape of its primary mirror. In December 1993, a repair mission was launched using the space shuttle. In five space walks, astronauts repaired or replaced several faulty parts and updated the telescope's detectors. As **Figure 20-15** illustrates, this dramatically improved the sharpness of the images sent back to Earth. The diagram in **Figure 20-16** shows the main parts of the telescope.

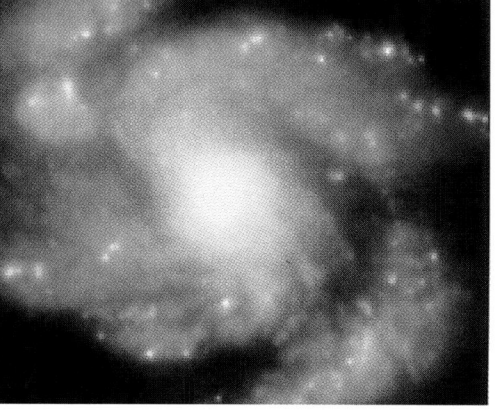

Figure 20-15

An early Hubble photo is shown at the top. The same view is shown at the bottom, after the Hubble's primary mirror was corrected.

Program Resources

 **Reproducible Masters**
Study Guide, p. 85 L1
Reinforcement, p. 85 L1
Enrichment, p. 85 L3
Critical Thinking/Problem Solving, p. 26

 Transparencies
Section Focus Transparency 81 L1

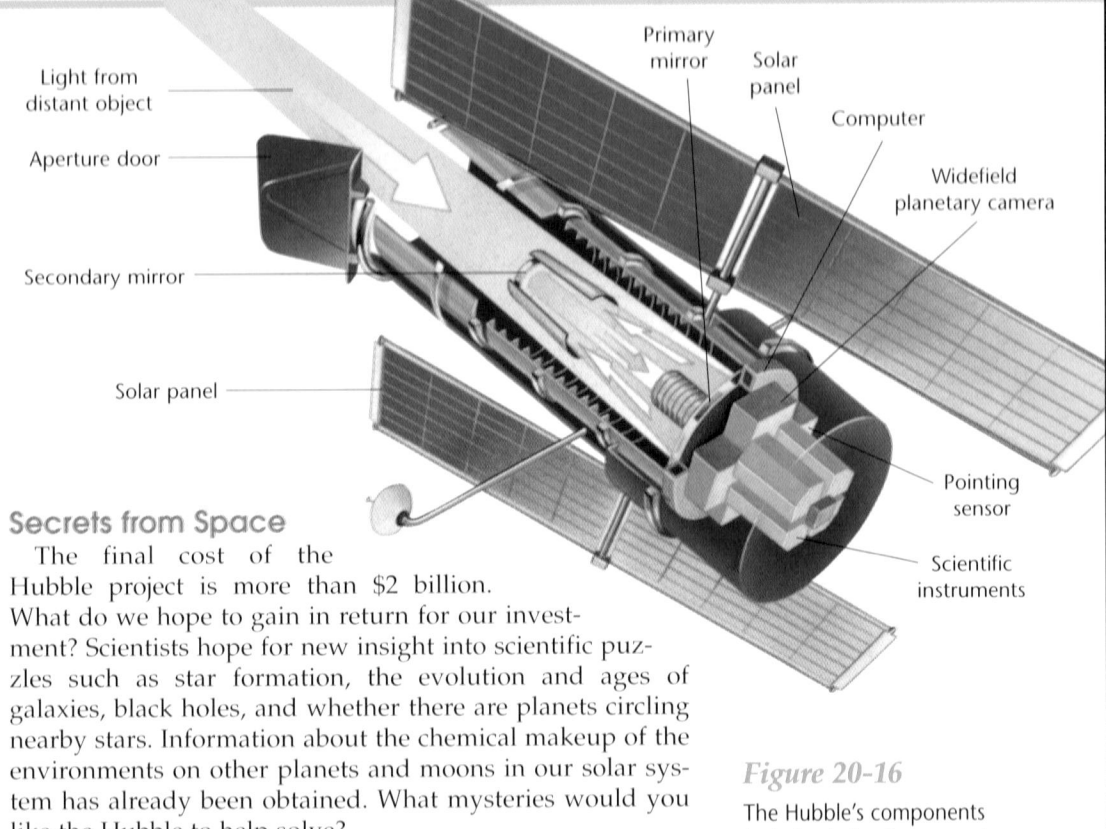

Light from
distant object

Aperture door

Secondary mirror

Solar panel

Primary
mirror

Solar
panel

Computer

Widefield
planetary camera

Pointing
sensor

Scientific
instruments

Secrets from Space

The final cost of the Hubble project is more than $2 billion. What do we hope to gain in return for our investment? Scientists hope for new insight into scientific puzzles such as star formation, the evolution and ages of galaxies, black holes, and whether there are planets circling nearby stars. Information about the chemical makeup of the environments on other planets and moons in our solar system has already been obtained. What mysteries would you like the Hubble to help solve?

Figure 20-16

The Hubble's components include aiming instruments, camera, and power supply.

Section Wrap-up

Review

1. Why can an orbiting telescope form clearer images than one on Earth?

2. What kinds of knowledge do scientists hope to gain from the Hubble Space Telescope?

 **inter**NET **CONNECTION** Study telescopes and their capabilities. Visit the Chapter 20 Internet Connection at Glencoe Online Science, **www.glencoe.com/sec/science/physical,** for a link to more information about telescopes.

Science Journal

Several environmental factors need to be considered when deciding where to build a new Earth telescope. Write down at least two of these factors that you think would be important and explain why.

SCIENCE & SOCIETY

4 Close

After discussing the benefits and drawbacks of operating the Hubble Space Telescope, ask some students to explain their point of view.

Section Wrap-up

Review

1. Telescopes on Earth form images distorted by Earth's nonuniform atmosphere.

2. unsolved questions about the universe—such as how it began and how old it is, the birth of stars, the dynamics of pulsars, quasars, and black holes, or if other planets exist outside our solar system.

Science Journal

Answers can refer to weather, climate, atmospheric pollution, proximity to city lights, ground vibrations, and geologic stability.

Integrating the Sciences

Earth Science The Hubble Space Telescope should be able to collect information for many years. **What source of energy is used to move its parts?** *Solar energy is collected and converted using solar panels.*

Section 20•5

Prepare

Section Background

- Lasers are available in a variety of colors; the wavelength and color of light emitted depend on the materials in the laser. The most common one is probably the helium-neon laser, which explains why so many people imagine lasers as a red beam of light.

- Laser stands for *Light Amplification* by *Stimulated Emission* of *Radiation*. In contrast with the spontaneous release of light during a flame test of an element, the atoms in a laser remain unexcited until struck by a photon of light emitted by another atom.

- In most light sources, the electrons are vibrating in infinitely many directions and light is emitted in every imaginable plane. Use polarization to support the fact that light waves are transverse, not longitudinal like sound.

Preplanning

- If you don't have a laser, arrange to borrow one from a nearby high school, university, or business.

- Obtain polarizing filters, masking tape and light sources for Activity 20-2.

1 Motivate

Bellringer

Before presenting the lesson, display **Section Focus Transparency 82** on the overhead projector. Assign the accompanying **Focus Activity** worksheet. L1 LEP

Science Words

polarized light
laser
coherent light
incoherent light
total internal reflection
optical fiber

Objectives

- Describe polarized light and the uses of polarizing filters.
- Explain how a laser produces coherent light and how it differs from incoherent light.
- Apply the concept of total internal reflection to the uses of optical fibers.

Polarized Light

Have you seen sunglasses with a sticker on them that says *polarized?* What makes them different?

Recall modeling a transverse wave on a rope in Chapter 18. You could make the waves vibrate in any direction—horizontal, vertical, or anywhere in between. Most light sources, such as incandescent lamps and the sun, emit light that vibrates in many directions. If this light passes through a special filter, called a polarizing filter, the light becomes polarized. In **polarized light,** the transverse waves vibrate in only one plane. A polarizing filter contains molecules that act like parallel slits to allow only those light waves vibrating in one direction to pass through. If a second polarizing filter is aligned so its molecules are oriented at right angles to the first filter, no light passes through, as **Figure 20-17** shows.

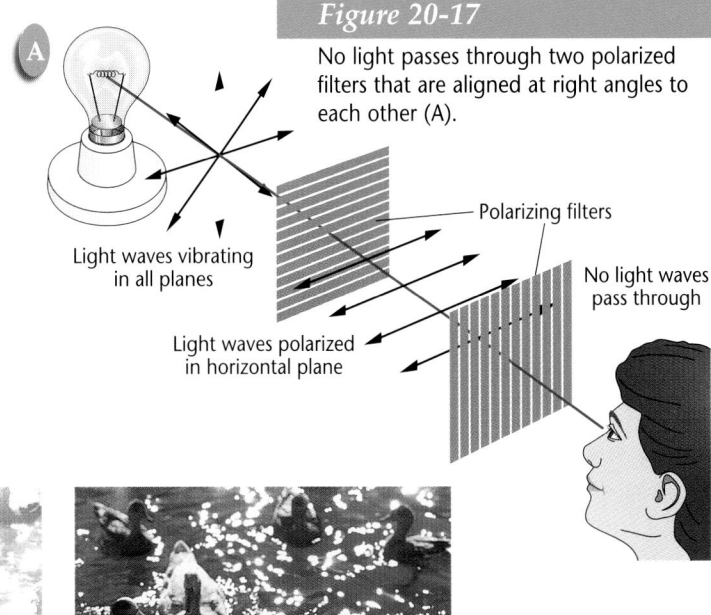

Figure 20-17

No light passes through two polarized filters that are aligned at right angles to each other (A).

Light waves vibrating in all planes

Polarizing filters

No light waves pass through

Light waves polarized in horizontal plane

B Light that is not polarized can produce glare and unwanted reflections.

C Using a polarizing filter on the camera produces a picture in which glare is blocked.

574

Program Resources

Reproducible Masters
Study Guide, p. 86 L1
Reinforcement, p. 86 L1
Enrichment, p. 86 L3
Activity Worksheets, pp. 118, 119 L1
Multicultural Connections, pp. 43, 44
Technology Integration, p. 17

Transparencies
Section Focus Transparency 82 L1

Figure 20-18

More and more, shoppers want to buy sunglasses that are polarized. *Why might this shopper prefer polarized lenses?*

A The *polarized* label means that these sunglasses can do something that other sunglasses can't. They can reduce glare and reflections.

Light reflected from horizontal surfaces, such as a lake or a car hood, is partially polarized horizontally through natural processes. Polarizing sunglasses are made with vertically polarizing filters to block out most of the glare while allowing vertically polarized light through.

The shopper in **Figure 20-18** knows that some sunglasses are polarized to reduce glare and others are not.

B In addition to style, fit, color, and price, this shopper is also examining the polarization of the sunglasses.

Have you ever watched a three-dimensional (3-D) movie? Did you wear special glasses to see the effects? The characters probably seemed to be right in front of you. One kind of 3-D glasses uses polarizing filters. The filter in front of one eye is aligned vertically, and the other filter is aligned horizontally. The movie is shown through two projectors, one with a horizontally polarizing filter and the other with a vertically polarizing filter. As a result, your right eye sees a slightly different picture from what is seen by your left eye. This creates the impression of depth, or three dimensions. Polarizing filters, such as those in **Figure 20-19**, can also be placed over camera lenses to reduce glare in photographs.

Lasers

The narrow beams of light that zip across the stage and through the auditorium during a rock concert are produced by **lasers.** Beams of laser light do not spread out because laser light is coherent. **Coherent light** is electromagnetic energy of only one wavelength that travels with its crests and troughs aligned. The beam does not spread out because all the waves

Figure 20-19

Polarizing filters are often an important part of a photographer's equipment.

20-5 Applications of Light **575**

Use these program resources as you teach this lesson.
Study Guide, p. 86
Multicultural Connections, pp. 43, 44
Technology Integration, p. 17

Tying to Previous Knowledge

Recall from Chapter 19 that light can be modeled as a transverse wave. Ask students to show how they would model a transverse wave (a wave on a rope). This conceptual understanding is important to the three new applications of light in this section.

2 Teach

Visual Learning

Figure 20-17 Use two Polaroid filters to illustrate **Figure 20-17.** Place one filter on the overhead projector and then slowly rotate the second one over it. Ask students to identify when the filters are parallel (when it is light) and when they are perpendicular (when it is dark).

Figure 20-18 **Why might this shopper prefer polarized lenses?** *Because the polarized sunglasses reduce glare and bright reflections.* LEP LS

Demonstration

LS **Visual-Spatial** Use the same setup as for Visual Learning, **Figure 20-17,** but put a piece of stretched plastic wrap or pieces of cut-up storage bags between the filters. Rotate the second filter over it again. You should see an array of continually changing colors and brightness. The chains of molecules in the plastic are rearranged when stretched, and this produces the polarization of light of different wavelengths. LEP

Demonstration

Visual-Spatial Have a laser behind your desk. When you turn it on, tell students they must find the position of the laser beam. They will be disappointed to find only a small dot on the wall. How are laser beams visible in the pictures and laser shows students may have seen? Scatter some chalk dust in the air to make the laser light visible. Discuss how the laser light must be reflected off a particle into your eye for you to see the beam.

Teacher F.Y.I.

The stimulated emissions that produce laser light are similar to the phenomenon that produces flame tests. As a result, the color of laser light depends on the nature of the materials used to produce it.

576

---

travel in the same direction. Light from an ordinary lightbulb is incoherent. **Incoherent light** may contain more than one wavelength, and its electromagnetic waves do not travel in the same direction, causing the beam to spread out.

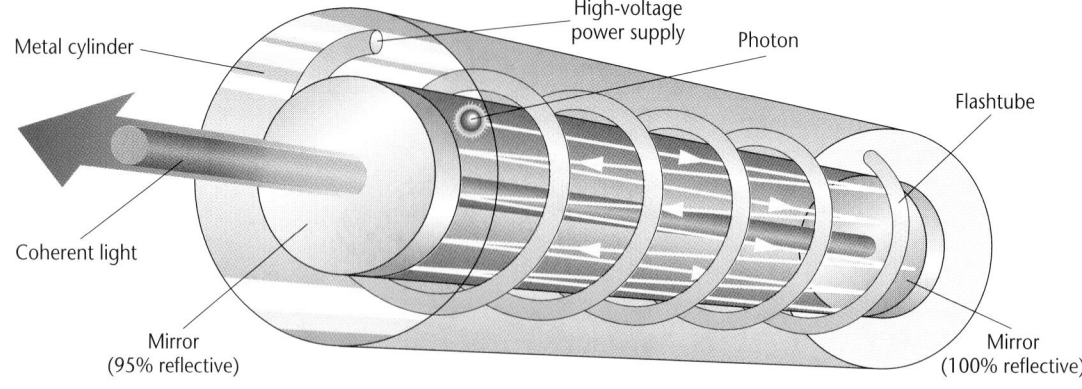

Figure 20-20
A laser produces a coherent beam of visible light of the same wavelength.

CONNECT TO
CHEMISTRY

A particular helium-neon laser contains a mixture of 15 percent He and 85 percent Ne. Where are these gases located on the periodic table? *Analyze* their chemical characteristics. Would you be concerned about their reacting chemically in the laser? Explain.

Creating Laser Light

Photons in a beam of coherent light are identical, travel in the same direction, and can be produced by a laser. A laser's light begins when a photon is spontaneously emitted from an atom. This photon is reflected between two facing mirrors at opposite ends of the laser, one of which is only partially coated to allow some light to pass through. Some emitted photons travel back and forth between the mirrors many times, stimulating other atoms to also emit identical photons. This continual process produces a coherent beam of laser light.

Lasers can be made with many different materials, including gases, liquids, and solids. One of the most common is the helium-neon laser, which produces a beam of red light. A mixture of helium and neon gases sealed in a tube with mirrors at both ends is excited by a flashtube, as shown in **Figure 20-20.** The excited atoms then lose their excess energy by emitting photons.

Using Lasers

Lasers are inefficient. In many types, no more than one percent of the electrical energy that is used by the laser is converted to light energy. Despite this inefficiency, lasers have many unique properties that make them useful. The beam of the laser is narrow and focused. It does not spread out as it travels over long distances.

Laser light has many applications, such as reading the bar codes on packages or maintaining high sound quality from compact discs by not scratching the discs' surfaces. Surgeons

576 Chapter 20 Mirrors and Lenses

Science Journal

New Uses for Lasers Have students write a paragraph describing several situations in which you know lasers are used. **What new uses do you think we might find for lasers in the future?** *Answers may include surgery, surveying, welding, weapons, etc.*

use lasers in place of scalpels to cut cleanly through body tissues. The energy from the laser seals off blood vessels in the incision to reduce bleeding. Lasers are routinely used to remove cataracts and repair the retina in the eye. In industry, powerful lasers are used for cutting and welding materials. Surveyors and builders use lasers for measuring and leveling. To measure the moon's orbit with great accuracy, scientists use laser light reflected from instruments on the moon's surface. Lasers also provide a coherent light source for fiber-optic communications. **Figure 20-21** shows a laser being used in an industrial application.

Figure 20-21

Figure 20-21
There are many practical uses for laser light. *How many can you think of?*

Optical Fibers

Did you ever dangle your legs into the water from the side of a swimming pool, and bending your knees, slowly raise your feet? If so, you saw your feet seem to disappear as you raised them in the water. The disappearance of your feet is an example of total internal reflection. **Total internal reflection** occurs when light striking a surface between two materials reflects totally back into the first material. As you know, to see your feet in the pool, light reflecting from your feet must reach your eyes. As you raise your feet in the pool, the light reflecting from your feet strikes the surface between the water and the air and reflects back into the water. Your feet seem to disappear because light reflecting from your feet never reaches your eyes. Total internal reflection depends on the speed of light in the two materials and the angle at which the light strikes the surface. The speed of light in the first material must be less than it is in the second. Also, the angle at which the light strikes the surface of the first material must be large. The sparkle of the gemstones in **Figure 20-22** is a result of total internal reflection.

Figure 20-22
Total internal reflection causes gems to sparkle.

20-5 Applications of Light   577

3 Assess

Check for Understanding

? FLEX Your Brain

Use the Flex Your Brain activity to have students explore TOTAL INTERNAL REFLECTION.

Activity Worksheets, p. 5

Reteach

To reinforce the idea of coherent light, first have everyone in the class clap their own individual rhythm. It sounds scattered and jumbled. Now have everyone clap on the number as you count "one and two and three and. . . ." The sound is "coherent" or together and more intense.

LEP

Extension

For students who have mastered this section, use the **Reinforcement** and **Enrichment** masters.

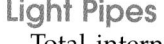

USING TECHNOLOGY

The Light Scalpel

Optical fibers transmitting laser energy deep inside the human body are replacing some surgical procedures. The effect of the laser on human tissue depends on the wavelength and intensity of the laser light and the type of the tissue. By varying the wavelength and intensity of a laser, a particular kind of tissue can be targeted.

Low-power lasers are used to seal soft tissue and to treat ulcers in the stomach, intestine, and colon. However, many other surgical applications require high-power lasers. One of the most exciting applications would be in the treatment of cardiovascular diseases. An optical fiber-conducted laser may be used to deliver pulses of laser light to clear a blocked artery.

Surgical Lighting

Lasers are also used to light the way for other surgical tools. Do you know anyone who has had arthroscopic surgery? In this procedure, tubes containing optical fibers are inserted through small incisions to help doctors see inside the body.

Think Critically:

What problems might arise in using a high-power laser to clear an artery of an obstruction?

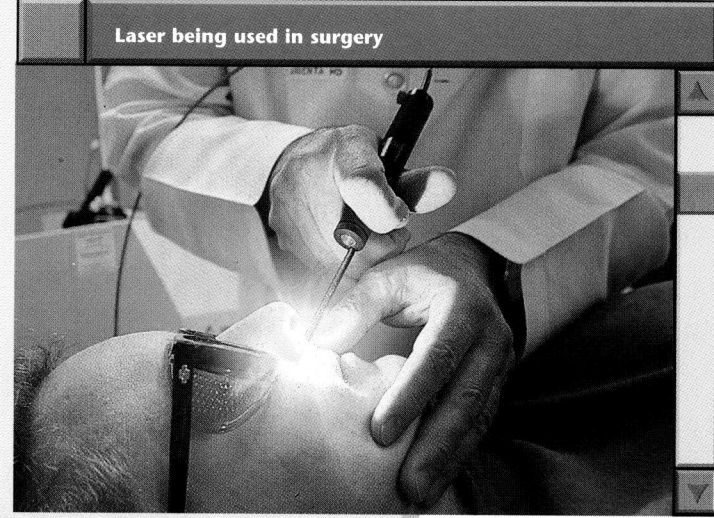

Laser being used in surgery

578   Chapter 20 Mirrors and Lenses

Light Pipes

Total internal reflection makes light transmission in optical fibers possible. **Optical fibers** are transparent glass fibers that can transmit light from one place to another. As shown in **Figure 20-23**, light entering one end of the fiber is continuously reflected from the sides of the fiber until it emerges from the other end. Little light is lost or absorbed in optical fibers.

Optical fibers are most commonly used in communications. Telephone conversations, television programs, and computer information can be coded in modulated light beams. The signals are transmitted by a laser through the optical fibers with far less loss of signal than if similar electric signals were transmitted through copper wires. Because of total internal reflection, signals can't leak from one fiber to another and interfere with other messages. As a result, the signal is transmitted clearly.

Community Connection

Optical Fibers   Invite a physician to visit your class to discuss the diversity of uses of optical fibers in medicine. Have him or her describe how several surgical procedures have been revolutionized by using optical fibers in viewing scopes and cutting devices.

One optical fiber can carry thousands of phone conversations at the same time because the signals can be produced rapidly and travel at high speed through the fiber. **Figure 20-24** shows the size of typical optical fibers.

Optical fibers are also used to explore the inside of the human body. One bundle of fibers transmits light, while the other carries the reflected light back to the doctor. Plants have also been shown to use the same principles used in optical fibers to transport light to cells that use light energy.

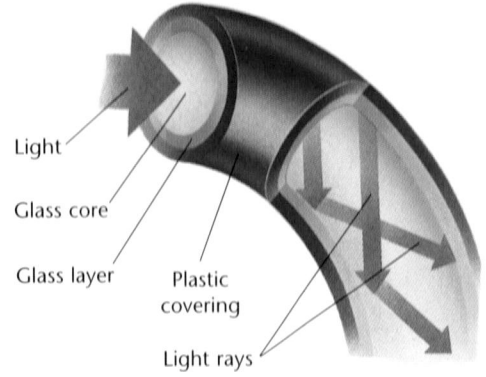

Light

Glass core

Glass layer

Plastic covering

Light rays

Figure 20-23

An optical fiber is designed to reflect light so that it is piped through the fiber without leaving it, except at the ends.

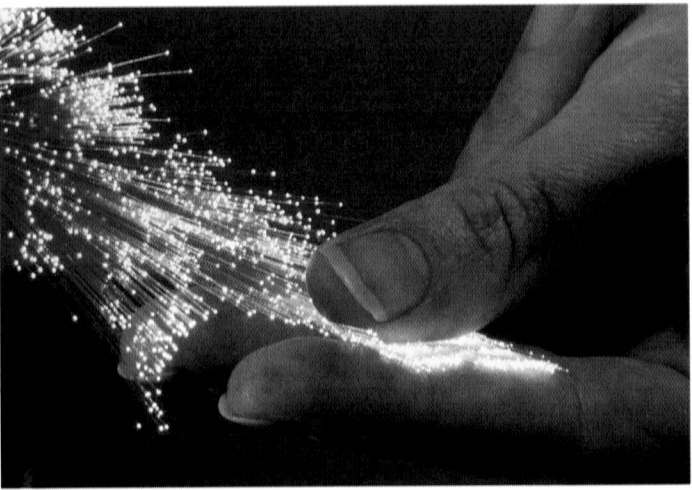

Figure 20-24

Just one of these optical fibers can carry thousands of phone conversations at the same time.

Section Wrap-up

Review

1. What is polarized light?

2. Explain how an optical fiber transmits light.

3. **Think Critically:** Geologists and surveyors often use lasers for aligning equipment, measuring, and mapping. Explain why.

Skill Builder
Sequencing

Sequence the events that occur in a laser in order to produce coherent light. Begin with the emission of a photon from an atom. If you need help, refer to Sequencing in the **Skill Handbook.**

Science Journal

Many people wear polarized sunglasses while they are working. Write a list of jobs or occupations in which wearing polarized sunglasses is helpful and desirable. Explain why.

PREPARATION

Purpose

IS Interpersonal Investigate the effects of polarizing filters and relate these observations to practical applications of polarized light. **COOP LEARN** **L1**

Process Skills

experimenting, observing, inferring, making and using tables, communicating, comparing and contrasting, recognizing cause and effect, forming a hypothesis, separating and controlling variables, interpreting data, using numbers

Time

40 minutes

Materials

Alternate Materials Polarized light filters can be cut into small pieces, but keep the edges in line with the plane of polarization.

Possible Hypotheses

Students may hypothesize that polarizing filters prevent the transmission of some light and that they block glare reflected from horizontal surfaces.

📁 **Activity Worksheets,** pp. 5, 118, 119

PLAN THE EXPERIMENT

Possible Procedures

Students will probably devise a plan that involves testing the light blocking abilities of first one and then two polarizing filters at various orientations. Note that only one filter should be rotated when observing the effect on glare.

Activity 20-2

Design Your Own Experiment
What's behind those shades?

Imagine being near a swimming pool or lake on a bright, sunny day. You would likely be squinting because some sunlight is reflected from the water's surface rather than being absorbed.

PREPARATION

Problem

How do polarizing lenses affect light from the sun and glare?

Form a Hypothesis

In what plane is glare from the ground polarized? How should the polarizing filters in sunglasses be oriented? Write a hypothesis about how polarizing filters affect light and how they affect light reflected off a horizontal surface.

Objectives

- Design an experiment to test the effects of polarizing filters on light from a bulb and on glare from sunlight.
- Relate your observations to some practical uses of these filters.

Possible Materials

- 2 polarizing filters
- masking tape
- light source
- objects with flat, hard surfaces (such as the tabletop)

Safety Precautions 🧤 🥽

Be sure not to touch the light source. Do not look directly at the sun or the lighted bulb at any time, even with a polarizing filter.

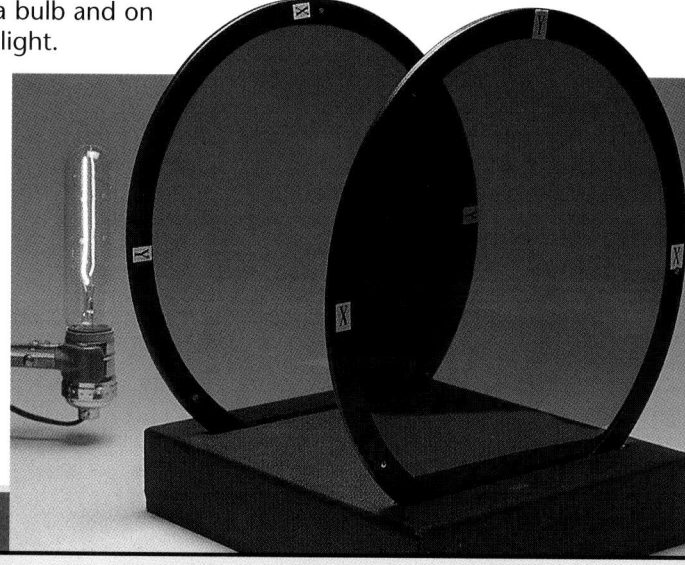

580   Chapter 20 Mirrors and Lenses

Inclusion Strategies

Visually Impaired For students who are unable to see the light change resulting when polarizing filters are placed at right angles to each other, provide a tactile model of the experiment. Use the vertical slits in two oven racks or cookie cooling racks to model how two polarizing filters can be aligned to transmit light vibrating in one plane or placed at right angles to block light.

PLAN THE EXPERIMENT

1. As a group, agree upon and write out the hypothesis statement.
2. Develop a procedure to test your hypothesis. Make a detailed list of the steps you will take. Remember that a filter's orientation affects the blocking or transmitting of light.
3. List the materials you will need and gather them.

Check the Plan

1. Before you start the experiment, label the axes of the polarizing filters. To do this, hold both filters still and look through them. Rotate one filter while holding the other still. When maximum light can be seen, use tape labels to mark the top and bottom edges of both filters X and the side edges Y.
2. How will you find a way to block nearly all light using the two polarizing filters?
3. How will you test the effect of a polarizing filter on glare from both horizontal and vertical surfaces? Be sure to rotate only one filter as you test glare.
4. *Make sure your teacher approves your plan and that you have included any changes suggested in the plan.*

DO THE EXPERIMENT

1. Carry out the approved experiment as planned.
2. Be sure to record any observations that you make during the experiment in your Science Journal.

Analyze and Apply

1. **Estimate** how many degrees you must turn one filter relative to the other to change light passing through from brightest to darkest.
2. When the glare from the horizontal surface nearly disappears, **observe** which edge of the rotated filter is up. Is this also true for glare from the vertical surface?
3. What advantage do polarizing sunglasses have over regular sunglasses? Glare reflected from a horizontal surface is horizontally polarized. **Predict** how polarizing lenses in sunglasses should be oriented to block this glare.

Go Further

Light from the portion of the sky that is 90° from the sun is polarized. Look through the filter at this part of the sky and rotate it to make the sky darken. If there are clouds in the sky, observe their appearance as the filter darkens the sky.

20-5 Applications of Light   581

Teaching Strategies

• To create the glare, use blocks of polished or painted wood or plastic. Lab tables or desks may also work fine. Avoid dull or mirrored surfaces.
• Try to have several pairs of polarizing sunglasses available for students to experiment with.

DO THE EXPERIMENT

Expected Outcome

Students will discover that polarizing filters block nearly all light when the filters are rotated at right angles to each other (Xs of one match Ys of another). They will also discover that glare is polarized.

Analyze and Apply

1. The filter must be turned 90°.
2. In the horizontal and vertical glare questions, if X is the answer to part one, then Y is the answer to part two, and vice versa.
3. Polarizing sunglasses can block glare. Vertically oriented filters would block horizontally polarized glare.

Go Further

Although the sky darkens as one filter is turned, the clouds do not change.

✓ Assessment

Oral Ask students if polarizing sunglasses would work as well if they were turned sideways. *no* Upside down? *yes* Have them explain using observations from this activity. Use the Performance Task Assessment List for Making Observations and Inferences in **PASC**, p. 17. **P**

People and Science

JEFF BALDWIN, *Telescope Mirror Maker*

On the Job

Q How do you create a mirror for a telescope?

A I take a piece of borosilicate glass and use abrasives to grind it to a general curve. Then I smooth and polish it so it's accurate to about one-millionth of an inch. All this work is done by hand. Finally, a very thin layer of aluminum is applied to make the glass reflective, creating a mirror.

Q Is a telescope's mirror as important as its lens?

A Mirrors are more important. A lot of people think that the idea of a telescope is to make things look bigger. However, the main interest of astronomers is making distant objects look brighter so they can be observed closely. The larger a telescope's mirror, the more photons it can reflect from an object and the brighter that object will appear.

Q Is astronomy popular with kids and adults?

A Our astronomy club here in Stockton, California, has members ranging from 11 to 93 years old.

Background

"More things have happened in the last few years than have happened in many astronomers' lifetimes," exclaims an amateur astronomer in California. The year 1995 was full of exciting discoveries in astronomy.

- A new planet outside our solar system orbits a star similar to our sun. Located in the constellation Pegasus, the planet is 42 light-years from Earth.
- Stars are being "born" in enormous pillars of gas within the constellation Serpens.
- The birthplace of comets has been detected in an area of the solar system beyond Pluto.
- GS2000+25, a second black hole in the Milky Way galaxy, has a mass equivalent to five suns.
- The astronomy club mentioned in the interview hosts "Star Parties" at schools and reaches 4000 students yearly.

Teaching Strategies

How powerful is the Hubble Space Telescope? To get an idea, imagine this: Two fireflies are ten feet apart in Washington, DC. Around the globe, in Tokyo, a telescope comparable to the Hubble can detect whether Firefly A or Firefly B is flashing.

Personal Insights

Q Where do you set up your own telescope, and what do you look for?

A Since I live in a light-polluted area, I take my truck and telescope up into the sierra. Right now, I'm studying what are called deep-sky objects—clusters of galaxies that are 4 billion light-years away. Wherever I point my telescope in the sky, I can see galaxies. Each galaxy has about 2 billion stars.

Career Connection

Is there an astronomy club in your area? Find out by inquiring at a museum, a library, a university, or a store that sells telescopes. If you locate a club, invite a member to host a "Star Party" for your class. Then make posters to announce the event. Gather and distribute information about the following careers:

- **Astronomer**
- **Precision Lens Grinder/Polisher**

Career Connection

Career Path The field of astronomy is, well, astronomical. Some specialized areas include stellar, solar, and planetary astronomy. Cosmologists study the origin and structure of the universe, and astrophysicists study physical and chemical changes in the universe. High school and college courses essential for astronomers are mathematics (including calculus), chemistry, and physics.

For More Information

Students could write to:
American Astronomical Society
Education Office
2000 Florida Avenue NW
Suite 30
Washington, DC 20009

Summary

20-1: The Optics of Mirrors

1. Plane mirrors and convex mirrors produce virtual images. Concave mirrors produce real or virtual images.
2. Plane mirrors are used in bathroom mirrors, concave mirrors are used to create flashlight beams and magnify images, and rearview mirrors and shoplifting mirrors are convex.

20-2: The Optics of Lenses

1. Convex lenses form real or virtual images; concave lenses form virtual images.
2. Convex lenses converge light rays and can form images on a screen. Concave lenses diverge light rays and are often used in combination with other lenses.
3. Corrective lenses can be used to focus images on the retina. Farsighted people must wear convex lenses, and nearsighted people must wear concave lenses.

20-3: Optical Instruments

1. A refracting telescope uses convex lenses to magnify distant objects; a reflecting telescope uses concave and plane mirrors and a convex lens to magnify distant objects.
2. Light passing through the lens of a camera is focused on photographic film inside the camera. The image on the film is real, inverted, and smaller than the object being photographed.

20-4: Science and Society: The Hubble Space Telescope

1. By avoiding atmospheric distortion, the Hubble Space Telescope produces sharper images than telescopes on Earth.

In addition, it can detect infrared and ultraviolet radiation.

2. Scientists hope space telescopes will lead to new information about the universe.

20-5: Applications of Light

1. Polarized light consists of transverse waves that vibrate along only one plane. Polarizing filters block light waves that aren't vibrating in the same plane as the filter.
2. A laser produces coherent light by emitting a beam of photons that travel in the same phase and direction. Light that spreads out from its source is incoherent.
3. Optical fibers can pipe light rays because the fibers are made of a material that allows the light rays to reflect totally inside the fibers.

Key Science Words

a. coherent light
b. concave lens
c. concave mirror
d. convex lens
e. convex mirror
f. focal length
g. focal point
h. incoherent light
i. laser
j. microscope
k. optical fiber
l. plane mirror
m. polarized light
n. real image
o. reflecting telescope
p. refracting telescope
q. telephoto lens
r. total internal reflection
s. virtual image
t. wide-angle lens

Summary

Have students read the Summary statements to review the major concepts of the chapter.

Reviewing Vocabulary

See page 584.

1. l	**6.** p
2. s	**7.** j
3. n	**8.** m
4. e	**9.** a
5. d	**10.** k

✔ Assessment

Portfolio Encourage students to place in their portfolios one or two items of what they consider to be their best work. Examples include:

- Skill Builder hypothesis, p. 571
- Activity 20-2 answers and results, p. 581 **P**

Performance Additional performance assessments may be found in **Performance Assessment** and **Science Integration Activities.** Performance Task Assessment Lists and rubrics for evaluating these activities can be found in Glencoe's **Performance Assessment in the Science Classroom.**

GLENCOE TECHNOLOGY

MindJogger Videoquiz

Chapter 20 Have students work in groups as they play the Videoquiz game to review key chapter concepts.

Checking Concepts

1. c	**6.** c
2. b	**7.** b
3. c	**8.** d
4. c	**9.** d
5. b	**10.** c

Understanding Concepts

11. Light rays parallel to the optical axis are reflected through the mirror's focal point. Light rays approaching through the focal point of a concave mirror are reflected parallel to the optical axis.

12. to give us more information and to answer some unsolved questions about the universe by avoiding atmospheric distortion

13. In convex lenses, light is refracted toward the center of the lens and converges behind the lens, hence the name *converging lens.* In contrast, light is refracted toward the edges of concave lenses. Therefore, it diverges away from the lens's focal point. Concave lenses are thus called diverging lenses.

14. Both collect light. Telescopes collect and focus light from faraway objects, while microscopes magnify very small objects that are close at hand. The refracting telescope is built according to the same principles as the microscope.

15. A photon is emitted from an atom and reflected between two mirrors at the ends of the laser. This photon excites other atoms within the laser and they also emit photons. Some of the photons penetrate one of the mirrors and leave the end of the laser in a narrow beam of coherent light.

Reviewing Vocabulary

Match each phrase with the correct term from the list of Key Science Words.

1. mirror with a flat surface
2. image that cannot be projected onto a screen
3. image formed where light rays actually meet
4. curved mirror that diverges reflected light
5. lens that converges light from a distant object at its focal point
6. a telescope that uses two convex lenses
7. an instrument used to study very small objects
8. light waves that vibrate in only one plane
9. light produced by lasers
10. often used to transmit telephone signals

Checking Concepts

Choose the word or phrase that completes the sentence.

1. Images formed by plane mirrors are not _____.
   a. upright   c. enlarged
   b. reversed   d. virtual
2. An object that reflects light and curves inward is called a _____.
   a. plane mirror   c. convex mirror
   b. concave mirror   d. concave lens
3. Mirrors that can magnify a reflection are _____.
   a. convex   c. concave
   b. plane   d. transparent
4. The lightbulb in a headlight, flashlight, or spotlight is placed at the focal point of a _____.
   a. concave lens   c. concave mirror
   b. convex lens   d. convex mirror
5. Lenses form images by _____.
   a. reflecting light   c. diffracting light
   b. refracting light   d. interfering with light

6. A concave lens bends light toward its _____.
   a. optical axis   c. edges
   b. center   d. focal point
7. Farsighted people must wear _____.
   a. flat lenses
   b. convex lenses
   c. concave lenses
   d. unevenly curved lenses
8. Reflecting telescopes don't contain a _____.
   a. plane mirror   c. convex lens
   b. concave mirror   d. concave lens
9. Some sunglasses and 3-D glasses use _____.
   a. concave lenses   c. telephoto lenses
   b. convex lenses   d. polarizing filters
10. Lasers are often used in _____.
    a. cooking food   c. surgery
    b. traffic control   d. headlights

Understanding Concepts

Answer the following questions in your Science Journal using complete sentences.

11. Describe how a concave mirror reflects light rays that approach it parallel to its optical axis. How does such a mirror reflect approaching light rays that pass through its focal point?
12. Why was the Hubble Space Telescope built?
13. Convex lenses are often called converging lenses, while concave lenses are often called diverging lenses. Explain these different names by describing how each type of lens refracts light.
14. Compare and contrast the uses of telescopes and microscopes. Which type of telescope is built most like a microscope?
15. Explain how a laser produces coherent light.

Thinking Critically

16. If the object the audience was looking at were really an image in a concave mirror, the magician could make this image disappear by moving the object to the focal point of the mirror, where no image is formed.

17. It would produce an enlarged, upright, virtual image that can't be projected onto a screen.

18. a convex lens

19. No. The convex mirror would produce a virtual image that cannot be magnified by the eyepiece.

20. A wide-angle lens; wide-angle lenses are easier to use and can be used in many different and more common situations.

Thinking Critically

16. Magicians often make objects disappear by using trick mirrors. How might a magician seem to make an object disappear by using a concave mirror?

17. What would happen if a movie projector's lens were less than one focal length from the film?

18. If you were an optician, what type of lens would you prescribe for a patient who can't focus clearly on close objects?

19. Would a reflecting telescope work properly if its concave mirror were replaced by a convex mirror? Explain.

20. You have enough money to buy only one lens for your camera. What type of lens would be most useful? Explain.

Developing Skills

If you need help, refer to the **Skill Handbook.**

21. **Classifying:** Classify the different types of images formed by plane, concave, and convex mirrors.

22. **Observing and Inferring:** Infer the effects of a hard, rigid eye lens on human vision. Would this make the eye more or less like a simple camera?

23. **Recognizing Cause and Effect:** Distinguish and describe the causes and effects of the following vision problems: nearsightedness, farsightedness, and astigmatism.

24. **Hypothesizing:** Rough, uncut diamonds lack the sparkle of diamonds that have been cut by a gem cutter. Propose a hypothesis to explain this observation.

25. **Concept Mapping:** Make a concept map summarizing characteristics of coherent and incoherent light. Use the following terms to complete the map (terms may be used more than once): *wavelength(s), frequency(ies), color(s), sun, lasers, coherent light, incoherent light.*

Performance Assessment

1. **Writing in Science:** Write a report tracing the development of the telescope from the time of Galileo to the Hubble Space Telescope. Illustrate your report using diagrams or pictures.

2. **Oral Presentation:** Investigate the types of mirrors used in fun houses. Explain how these mirrors are formed and give distorted images. Demonstrate your findings to the class.

3. **Making Observations and Inferences:** Liquid Crystal Displays (LCDs) in digital watches and calculators have polarizing filters. Observe an LCD through a polarizing filter and rotate the filter while you view the display. Why does it matter which way a digital watch's polarizing filter is oriented?

Developing Skills

21. **Classifying** More than one classification method is possible. Accept all reasonable classifications.

22. **Observing and Inferring** A hard, rigid eye lens would have a fixed focal length. This means that the image of an object could be focused on the retina only when the object is a specific distance from the eye. Such an eye would be more like a simple camera, which has a lens of fixed focal length.

23. **Recognizing Cause and Effect** Nearsightedness—cause: an abnormally long eyeball or bulging cornea; effect: light from a distant object is focused in front of the retina.

    Farsightedness—cause: abnormally short eyeball or flattened cornea; effect: light from nearby objects is focused behind the retina. Astigmatism—cause: unevenly curved cornea; effect: all light is poorly focused on the retina.

24. **Hypothesizing** One possible hypothesis might be: Cutting diamonds in specific ways gives them inner surfaces that refracted light can strike at angles large enough to produce total internal reflection, and thus the diamond sparkles.

25. **Concept Mapping** Student maps will vary. Sample map is given at left.

Performance Assessment

1. Accept all reasonable reports. Use the Performance Task Assessment List for Writing in Science in **PASC**, p. 87. P

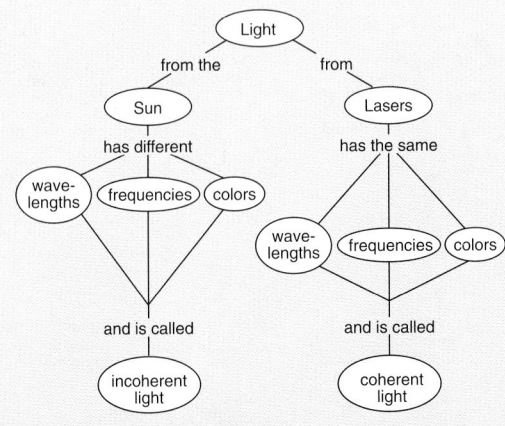

2. Curved mirrors contain both concave and convex mirrors to produce distorted images. Use the Performance Task Assessment List for Oral Presentation in **PASC**, p. 71. P

3. The polarizing filter must be oriented so people wearing polarized glasses can see the time. Use the Performance Task Assessment List for Making Observations and Inferences in **PASC**, p. 17. P

Objectives

LS **Kinesthetic** Students will investigate the cause of eye problems and the effect that these problems have on vision. Students will also study the parts of the eye and relate them to the optics of corrective lenses. L1 **COOP LEARN**

Summary

In Unit 6, students learned about the properties of light and other waves. This project will help students use these principles to understand human vision. After studying the structure of the eye, students will build a convex lens to model the lens in the human eye and find the focal length of this and other lenses. Students will contact a person employed in an eye-care career in the process of researching as many vision or eye health problems as possible. They will share all this information in the form of poster presentations.

Time Required

This project can be done while studying Chapter 20. One class period is required to establish groups and define the task. While discussing convex lenses in class, procedure steps 2 and 3 can be done in about 30 minutes. Allow for group meeting time throughout this chapter. One class period will be required to share and discuss the posters.

Preparation

- For each group, have available two matching watch glasses, a selection of convex and concave lenses from old eyeglasses, grafting wax or waterproof epoxy, a bowl of water, and a ruler.

- You may want to arrange for a vision-care specialist to speak to the class.

Vision: More Than Meets the Eye

How are your eyes able to focus on the words on this page? Your eye contains a lens that causes light that is reflected from objects to converge on the retina. But why do so many people wear corrective eyewear? In Chapter 20, you learned that a number of vision problems, such as farsightedness, nearsightedness, and astigmatism, can result when one or more parts of the eye do not function correctly. Eyeglasses and contact lenses are the most common ways to correct these vision problems, although laser surgery is becoming increasingly popular. Perhaps after researching some common vision problems and their methods of correction, you will see the world of vision through new eyes.

Investigating Vision

Have you ever wondered who invented and wore the first eyeglasses? What kinds of materials are used in making corrective eyewear? Who is responsible for testing and treating health of the eyes? With your group, research information in three different categories: eye and vision problems, lenses and correction of eye problems, and careers related to eye care.

Materials

- a selection of convex and concave lenses
- a ruler
- two watch glasses
- grafting wax
- a bowl of water

Procedure

1. Sketch a diagram of the eye and label each part. Find out the role of each part in human vision.

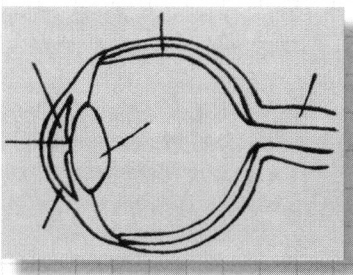

2. Build a convex lens to model the lens in your eye. Do this by placing two watch glasses together under water to trap water inside. Seal them together with grafting wax. How is this model lens similar to and different from the lens in your eye?

3. Find the focal length of several convex lenses, including the one you made. Hold your lens against a piece of white paper across the room from a window in a slightly darkened room. Slowly move the lens away from the paper until you see a clear image of a light source or a brightly lighted outside object on the paper. Measure the distance from the lens to the paper to find the focal length.

4. Identify as many vision or eye health problems as you can. Be sure to investigate near-sightedness, farsightedness, astigmatism, cataracts, glaucoma, and colorblindness. How are each of these problems corrected?

5. What is the difference between an optometrist and an ophthalmologist? Write a list of questions you would like to ask, and interview a person employed in the field of eye care.

Go Further

Think about the advances in diagnosing and treating problems of the eyes that have occurred during your lifetime. What new kinds of contact lenses and eyeglasses are available? What kinds of materials are used in corrective eyewear? What are the advantages of having glass lenses? Plastic lenses? How are lasers currently being used to treat eye problems? Will there be a time when most common eyesight problems are corrected surgically rather than with eyeglasses?

Using Your Research

Design two posters to share your information with other students in the class. One poster should illustrate the human eye, including the parts and their functions. You can also describe on this poster what optometrists and ophthalmologists do. On the second poster, illustrate some of the types of eye problems you learned about. Be sure you can explain how these problems can be corrected. Present your posters to the class and encourage students to identify the problems they or someone they know have been examined or treated for.

Teaching Strategies
- You may wish to divide procedure step 4 among the groups, giving each group a particular eye problem to focus on.
- If lenses are made with watch glasses of different sizes, have groups compare the focal lengths. Also, students can experiment with the effects of filling the lenses with different clear liquids.
- If a model of the eye is available, this model may be used to reinforce the parts of the eye and their functions.

Go Further
Have students survey people younger than 18 and also adults to find out what percent wear corrective lenses. The survey can be expanded to include what type of lens is chosen. Students can investigate the benefits and drawbacks of using daily wear contacts, extended wear contacts, disposable contacts, and glass and plastic eyeglass lenses.

Unit 6 Project   587

References
Ahrens, Kathleen. *Opportunities in Eye Careers.* Chicago, IL: VGM Career Horizons, 1991.

D'Alonzo, Dr. T.L. *Your Eyes!* Clifton Heights, PA: Avanti Publishing, 1991.

Falk, David, Dieter Brill, and David Stork. *Seeing the Light.* New York, NY: John Wiley and Sons, 1986.

Electricity and Energy Resources

In Unit 7, students are first introduced to electricity and magnetism and how they are related. The world of microelectronics is then discussed. Radioactivity is introduced, with emphasis given to its application. The unit ends with a discussion of energy consumption, alternative energy sources, and the role of energy conservation.

CONTENTS

Unit Contents

588

Science at Home

Radioactive Wastes Have groups of students monitor TV and newspapers for coverage of problems concerning the processing and storing of radioactive wastes. [L1]

Energy-Consumption Log Have students keep daily logs of their electrical energy uses. Have them calculate their personal electrical energy consumption by using tables supplied by most utility companies. [L1]

Electricity and Energy Resources

What's Happening Here?

As the sun sets, electric lights start to go on. Most cities generate electricity for lighting by burning fossil fuel, by harnessing the energy in falling water, or by controlling nuclear reactions. The solar car in the small picture runs on electricity generated when light from the sun falls on solar cells attached to the car's surface. But what does it mean to generate electricity? Why is it an advantage to be able to use different energy sources to generate electricity? How do we safely bring electricity to our homes, businesses, and schools to operate our computers and other appliances? In this power-packed unit, you'll investigate some of these questions.

Science Journal

City skylines are beautiful at night with so many electric lights shining in the dark, but electric lights are only a small part of the use of electricity in an office building. In your Science Journal, write a story about what happened to one person in an electrical blackout.

589

INTRODUCING THE UNIT

What's Happening Here?

Ask students to tell you what's happening in the photos. Point out that in this unit, they will learn what electrical energy is; how it is converted to light, heat, and other forms of energy; and how other forms of energy are converted into electrical energy.

Background

About 70% of the electricity produced in the United States is produced by the burning of fossil fuels, 20% from nuclear fission, and 10% from moving water. Most electricity is used for heating, cooling, and operating industrial machinery. In most cases, electricity is produced by generators that convert mechanical energy into electrical energy. However, in solar cells, electricity is produced by silicon wafers that convert radiant energy into electrical energy.

Previewing the Chapters

Ask students the following: **What is the most apparent way that electricity is being used in this picture?** *lighting* **How else do you suppose electricity is being used in these buildings?** *heating, cooling, office cleaning equipment, elevators, computers, movie projectors*

Tying to Previous Knowledge

Have students brainstorm lists of electrical appliances and devices. Then have them categorize the appliance or device as to its major function, such as heating, cooling, lighting, motion, etc. Have them conjecture which category probably accounts for their greatest personal use of electrical energy.

Theme Connection

Energy/Stability and Change Although several themes occur in this unit, these two predominate. Electricity and the production and use of electrical energy are discussed in Chapters 21 and 25. Changes in magnetic materials and in the nuclei of the atoms have multiple applications, discussed in Chapters 22 and 24.

Cultural Diversity

Energy Alternatives Have interested students research how people of various cultures have adapted their lifestyles to the consumption of limited sources of fossil fuels or the exploitation of alternative energy sources. L2

Chapter Organizer

Section	Objectives/Standards	Activities/Features
Chapter Opener		Explore Activity: Can you move matter with electric charges? p. 591
21-1 Electric Charge (1 session, ½ block)*	1. **Describe** the effects of static electricity. 2. **Distinguish** between conductors and insulators. 3. **Recognize** the presence of charge in an electroscope. **National Content Standards:** (5-8) UCP1-UCP5, B2; (9-12) UCP1-UCP5, B4, B6	MiniLAB: How do charged objects interact with each other? p. 593 Problem Solving: Take Charge of Pollution, p. 594 Science Journal, p. 597 Skill Builder: Observing and Inferring, p. 597
21-2 Science and Society (1 session)*	1. **Explain** the occurrence of lightning in terms of induction and static discharge. 2. **Evaluate** the positive and negative aspects of lightning-induced forest fires. **National Content Standards:** (5-8) UCP1-UCP3, UCP5, F1-F5; (9-12) UCP1-UCP3, UCP5, B6, F1, F3-F6	Explore the Issue, p. 599
21-3 Electric Current (1½ sessions, 1 block)*	1. **Describe** how static electricity is different from current electricity. 2. **Explain** how a dry cell is a source of electricity. 3. Conceptually and mathematically **relate** potential difference, resistance, and current. **National Content Standards:** (5-8) UCP1-UCP3, UCP5, A2, B2, E2, F5, G1, G3; (9-12) UCP1-UCP3, UCP5, A2, B4, B6, E2, F6, G1, G3	Connect to Chemistry, p. 602 MiniLAB: Why do some flashlights require more C- or D-cells than others? p. 603 Using Technology: Resistive Heating, p. 604 Using Math: Calculating Current, p. 606 Using Math, p. 606 Skill Builder: Making and Using Tables, p. 606 Activity 21-1: Modeling Ohm's Law, p. 607
21-4 Electrical Circuits (2½ sessions, 1½ blocks)*	1. **Sketch** a series and a parallel circuit, and list applications of each type of circuit. 2. **Recognize** the function of circuit breakers and fuses. **National Content Standards:** (5-8) UCP1-UCP3, UCP5, A1, B2, E1; (9-12) UCP1-UCP3, UCP5, A1, B4, E1	Connect to Earth Science, p. 609 Science Journal, p. 611 Skill Builder: Hypothesizing, p. 611 Activity 21-2: Comparing Electric Circuits, pp. 612-613
21-5 Electrical Power and Energy (1 session, ½ block)*	1. **Explain** and **calculate** electrical power. 2. **Calculate** the amount of electrical energy in kilowatt-hours. **National Content Standards:** (5-8) UCP1-UCP3, UCP5, A2, B3, G1; (9-12) UCP1-UCP3, UCP5, A2, G1	Using Math: Calculating Power, p. 615 MiniLAB: How much power operates an electric toy? p. 616 Using Math: Calculating Electrical Energy, p. 617 Using Computers, p. 617 Skill Builder: Concept Mapping, p. 617 People and Science: Giovannae Anderson, Ph.D., Electrical Engineer, p. 618

Activity Materials

Explore	Activities	MiniLABs
page 591 balloon, salt, pepper, small plate, shredded tissue paper, small (less than 1 cm²) pieces of aluminum foil, piece of wool	page 607 plastic funnel, meterstick, ring stand with ring, two 250-mL beakers, 1 m rubber tubing, stopwatch or clock, water pages 612-613 6-V dry cell, 3 small lights with sockets, aluminum foil, cellophane tape, scissors, paper clips	page 593 roll of cellophane tape; smooth, dry, clean surface page 603 2 lightbulbs, 2 C- or D-cell batteries, wire page 616 electric toy with battery; wire, tape, or rubber bands; ammeter

Need Materials? Call Science Kit (1-800-828-7777).   * A complete Planning Guide that includes block scheduling is provided on pages 32T-35T.

Teacher Classroom Resources

Reproducible Masters	Transparencies	Teaching Resources
Study Guide, p. 87 Reinforcement, p. 87 Enrichment, p. 87 Activity Worksheets, p. 127 Concept Mapping, pp. 47-48	Section Focus Transparency 83, Electron Conductivity	Glencoe Physical Science Interactive Videodisc Physical Science CD-ROM Spanish Resources English/Spanish Audiocassettes Cooperative Learning Resource Guide Lab Partner Lab and Safety Skills Lesson Plans
Study Guide, p. 88 Reinforcement, p. 88 Enrichment, p. 88	Section Focus Transparency 84, Lightning Safety	
Study Guide, p. 89 Reinforcement, p. 89 Enrichment, p. 89 Activity Worksheets, pp. 123-124, 128 Lab Manual 43, Wet Cell Battery Multicultural Connections, p. 45 Science Integration Activity 21, Electrodes and Conductivity Science and Society Integration, p. 25	Section Focus Transparency 85, The Power of Water Science Integration Transparency 21, Where Lightning Comes From Teaching Transparency 41, Series and Parallel Circuits	**Assessment Resources** Chapter Review, pp. 45-46 Assessment, pp. 143-146 Performance Assessment in the Science Classroom (PASC) MindJogger Videoquiz Alternate Assessment in the Science Classroom Performance Assessment Chapter Review Software Computer Test Bank
Study Guide, p. 90 Reinforcement, p. 90 Enrichment, p. 90 Activity Worksheets, pp. 125-126 Lab Manual 44, Simple Circuits Cross-Curricular Integration, p. 27 Critical Thinking/Problem Solving, p. 27	Section Focus Transparency 86, Rear-Window Defoggers Teaching Transparency 42, Household Circuit	
Study Guide, p. 91 Reinforcement, p. 91 Enrichment, p. 91 Activity Worksheets, p. 129	Section Focus Transparency 87, Production of Electricity	

GLENCOE TECHNOLOGY

The following multimedia resources are available from Glencoe.

Science and Technology Videodisc Series (STVS)
Chemistry
  Battery Science

The Infinite Voyage Series
Miracles by Design

Glencoe Physical Science Interactive Videodisc
Electricity and Magnetism

Physical Science CD-ROM

National Geographic Society Series
GTV: Planetary Manager

Key to Teaching Strategies

The following designations will help you decide which activities are appropriate for your students.

L1 Level 1 activities should be within the ability range of all students, including those with learning difficulties.

L2 Level 2 activities should be within the ability range of the average to above-average student.

L3 Level 3 activities are designed for the ability range of above-average students.

LEP LEP activities should be within the ability range of Limited English Proficiency students.

LS These activities are designed to address different learning styles.

COOP LEARN Cooperative Learning activities are designed for small group work.

P These strategies represent student products that can be placed into a best-work portfolio.

Teacher Classroom Resources

This is a representation of key blackline masters available in the Teacher Classroom Resources.

Teaching Aids

Section Focus Transparencies

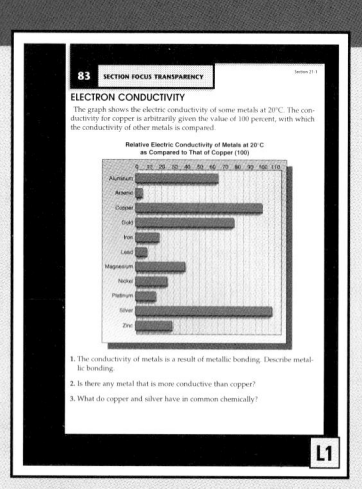

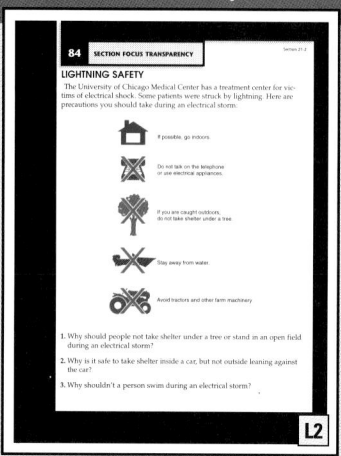

Science Integration Transparencies

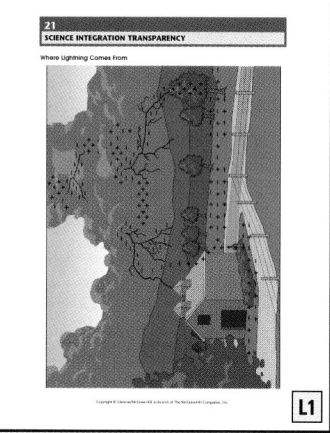

Teaching Transparencies

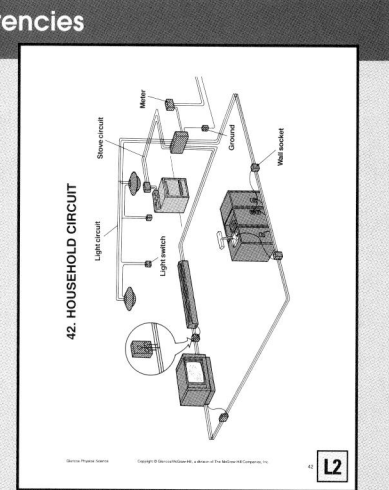

Meeting Different Ability Levels

Study Guide

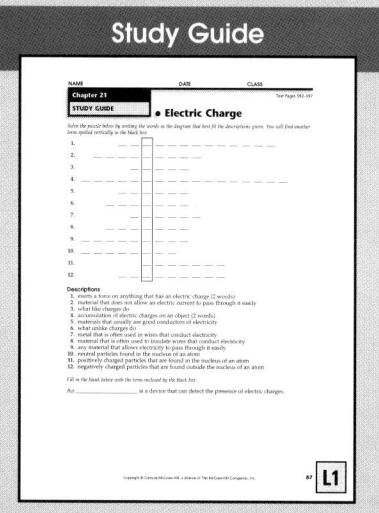

Reinforcement

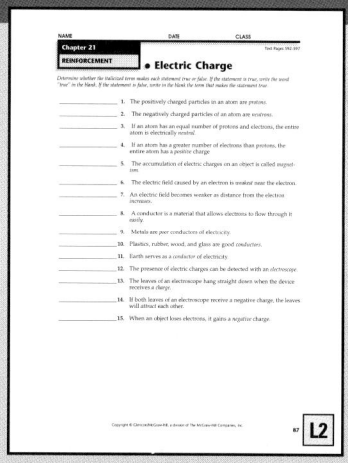

Enrichment Worksheets

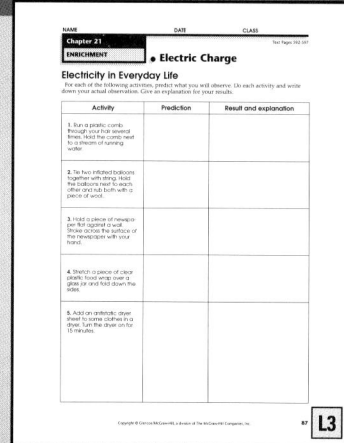

Hands-On Activities

Science Integration Activity

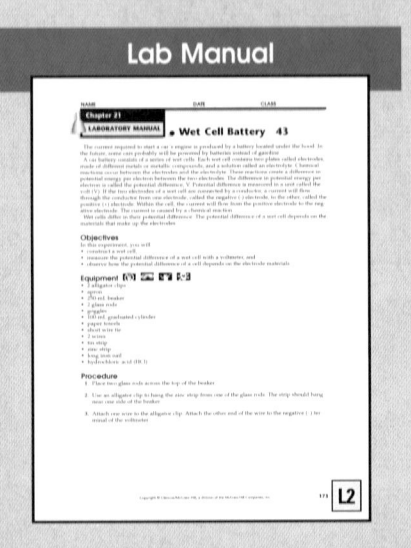

Lab Manual

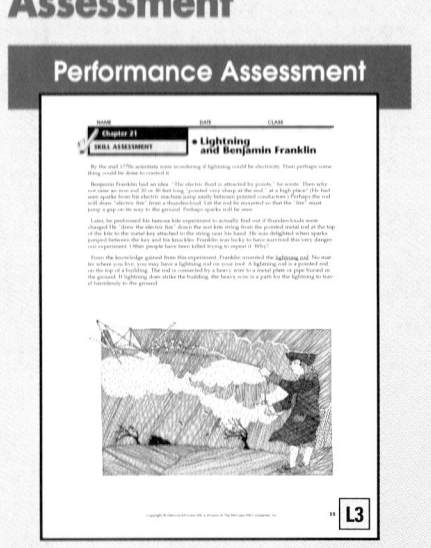

Assessment

Performance Assessment

Enrichment and Application

Critical Thinking/ Problem Solving

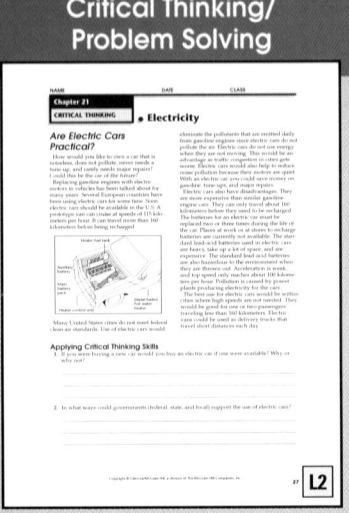

Cross-Curricular Integration

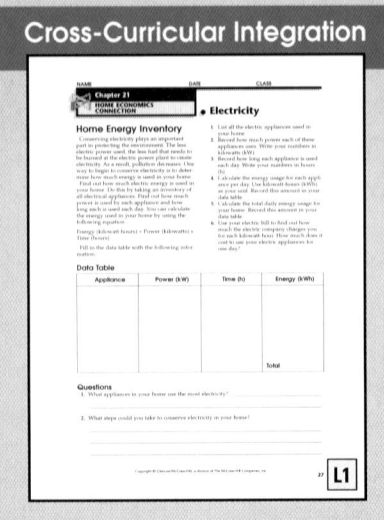

Science and Society Integration

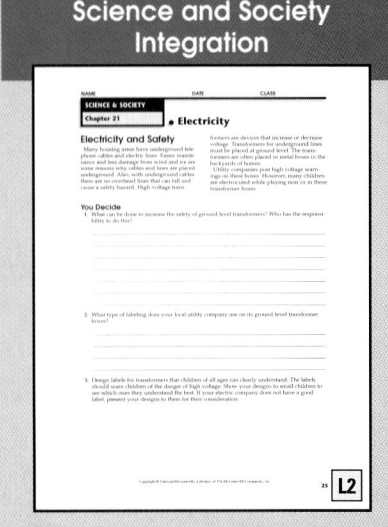

Multicultural Connections

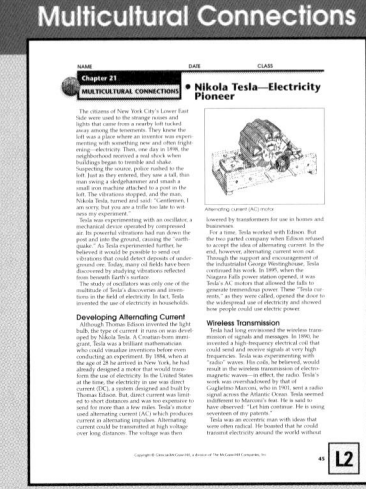

Concept Mapping

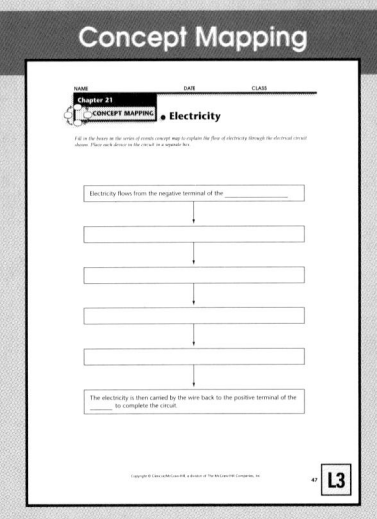

Electricity

CHAPTER OVERVIEW

Section 21-1 This section investigates the causes and effects of static electricity, the behavior of conductors and insulators, and the use of an electroscope.

Section 21-2 Science and Society The science behind lightning is explained and followed by a discussion of lightning-induced forest fires.

Section 21-3 Electric currents are introduced and explained. Resistance, current, and potential difference are related conceptually and mathematically.

Section 21-4 This section illustrates series, parallel, and complex circuits through diagrams and examples.

Section 21-5 Electrical power and energy are distinguished, and calculations of power and energy are included.

Chapter Vocabulary

static	dry cell
electricity	wet cell
electric field	resistance
conductor	Ohm's law
insulator	series circuit
electroscope	parallel
potential	circuit
difference	electrical
circuit	power
current	kilowatt-hour

Theme Connection

Energy This chapter introduces the fundamental principles and applications of electricity. Emphasize conservation of energy. Electricity is a convenient way to deliver energy, but it is often changed to other forms such as heat, light, and mechanical energy as we use it.

Previewing the Chapter

590

Learning Styles

Look for the following logo for strategies that emphasize different learning modalities. **LS**

Kinesthetic	Explore, p. 591; MiniLAB, p. 593
Visual-Spatial	Visual Learning, pp. 592, 596, 598, 601, 602, 604, 609, 610; Demonstration, pp. 604, 608
Interpersonal	Close, p. 599; MiniLAB, p. 616
Intrapersonal	Enrichment, p. 602
Logical-Mathematical	MiniLAB, p. 603; Activity 21-1, p. 607; Activity 21-2, pp. 612-613
Linguistic	People and Science, p. 618

Electricity

How many devices can you locate in this office that make use of electricity? Many people in the world have come to depend on electricity to carry out routine daily activities. Electricity provides us with transportation, the ability to control the temperature of indoor environments, entertainment, and many other conveniences. But what is electricity and where does it come from?

EXPLORE ACTIVITY

Can you move matter with electric charges?

1. Blow up a latex balloon and tie it off.
2. Gather small pieces of a variety of different materials, such as a small plate of salt and pepper flakes, shreds of tissue paper, or a pile of small pieces of aluminum foil (less than 1 cm^2).
3. Briskly rub the balloon with a piece of wool or across your clothing to give it an electric charge.
4. Bring the balloon near the different materials you gathered. Record your observations about the behavior of each material.
5. Touch the balloon to a metal surface and then repeat step 4.

Observe: In your Science Journal, write a paragraph describing how the balloon interacted with each material in step 4. Did the results change after you caused the balloon to lose its charge in step 5?

Previewing Science Skills

▶ In the Skill Builders, you will **observe and infer, hypothesize,** and **map concepts.**

▶ In the Activities, you will **measure in SI, predict,** and **formulate models.**

▶ In the MiniLABs, you will **observe, hypothesize,** and **interpret data.**

591

EXPLORE ACTIVITY

Purpose
LS **Kinesthetic** Use the Explore Activity to introduce the effects of electric charges on various types of matter to students. **L1** **LEP**

Preparation
Obtain at least one balloon for every two students. Small pieces of paper can be used instead of tissue paper.

Materials
balloons, tissue paper, aluminum foil, salt and pepper flakes, wool (optional)

Teaching Strategies
Most students are familiar with the term *static electricity* from prior experiences such as removing clothes from the dryer. Charging of the balloon is the same type of phenomenon. Ask why the balloon does not remain charged after touching it to a metal surface.

Observe
Students should have observed that the balloon attracted salt, pepper, and paper pieces. After touching the balloon to the metal surface, it lost most of its charge, and the material may have fallen off.

✔Assessment

Process Have students classify materials into two groups: those that were attracted to the balloon and those that were not. Have students make a hypothesis about types of materials that should be attracted to the charged balloon. Have them test their hypothesis. Use the Performance Task Assessment List for Formulating a Hypothesis in **PASC**, p. 21. **P**

Assessment Planner

Portfolio
Refer to page 619 for suggested items that students might select for their portfolios.

Performance Assessment
See page 619 for additional Performance Assessment options.
Skill Builders, pp. 597, 606, 611, 617
MiniLABS, pp. 593, 603, 616
Activities 21-1, p. 607, 21-2, pp. 612-613

Content Assessment
Section Wrap-ups, pp. 597, 599, 606, 611, 617
Chapter Review, pp. 619-621
Mini Quizzes, pp. 597, 606, 611, 617

Group Assessment
Opportunities for group assessment occur with Cooperative Learning Strategies and Flex Your Brain Activities.

Section 21•1

Prepare

Section Background

Benjamin Franklin arbitrarily named the two types of electric charge positive and negative.

Preplanning

- Try to obtain access to a Van de Graaff generator and several electroscopes for demonstration purposes.
- Obtain transparent tape for the MiniLAB.

1 Motivate

Bellringer

Before presenting the lesson, display **Section Focus Transparency 83** on the overhead projector. Assign the accompanying **Focus Activity** worksheet. L1 LEP

Tying to Previous Knowledge

Ask students what sometimes happens when they take clothes out of the dryer. They may mention static cling and that they can hear it crackle. Explain that this is caused by the discharge of static electricity, the topic of this section.

Visual Learning

Figure 21-1 How do charged objects interact with each other? *Like charges repel; unlike charges attract.*

Figure 21-2 What happens if it actually touches the sleeve? *Because the sleeve and balloon are both insulators, the charges stay separated and the balloon sticks to the sleeve.* LEP
LS

21•1 Electric Charge

Science Words

static electricity
electric field
conductor
insulator
electroscope

Objectives

- Describe the effects of static electricity.
- Distinguish between conductors and insulators.
- Recognize the presence of charge in an electroscope.

Static Electricity

Have you ever walked across a carpeted floor and gotten a sharp, painful shock by a spark as you reached out to touch something? If you immediately touched the object a second time, you might have felt a very small spark or none at all. What caused this startling and sometimes painful phenomenon?

When your feet rubbed on the carpet, some of the atoms in the carpet were disturbed. Recall from Chapter 10 that atoms contain protons, neutrons, and electrons. Neutrons have no charge, protons are positively charged, and electrons are negatively charged. An atom is electrically neutral if it has an equal number of protons and electrons. Sometimes, electrons are not held tightly in the atom. As you walked on the carpet, some electrons that were loosely held by the atoms were transferred from the carpet to your shoes. As a result, your shoes gained electrons, and they were no longer neutral but instead had a negative charge. The carpet lost electrons, leaving it positively charged. The excess electrons, stored on your body, gave you an overall negative electric charge. This is an example of static electricity. **Static electricity** is the net accumulation of electric charges on an object. Think of other examples of static electricity.

Opposites Attract

Have you noticed how clothes sometimes cling together when removed from the dryer? Electrons rub off some clothes

Figure 21-1
There are just two kinds of electric charge, positive and negative. *How do charged objects interact with each other?*

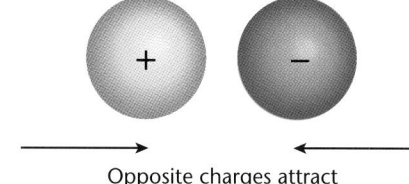
Opposite charges attract

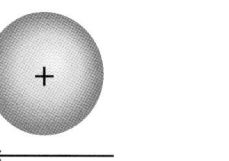 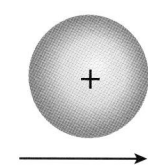
Like charges repel

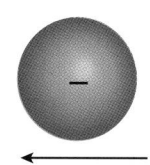 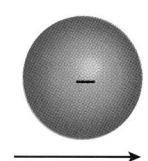
Like charges repel

Program Resources

📁 **Reproducible Masters**
Study Guide, p. 87 L1
Reinforcement, p. 87 L1
Enrichment, p. 87 L3
Activity Worksheets, p. 127 L1
Concept Mapping, p. 47

📽 **Transparencies**
Section Focus Transparency 83 L1

while they are tumbling around inside the dryer. Clothes that gain electrons become negatively charged, whereas those that lose electrons become positively charged. Clothes with opposite charges cling together, but clothes with the same charge repel each other. As shown in **Figure 21-1**, electrically charged objects obey the following rule: opposite charges attract, and like charges repel.

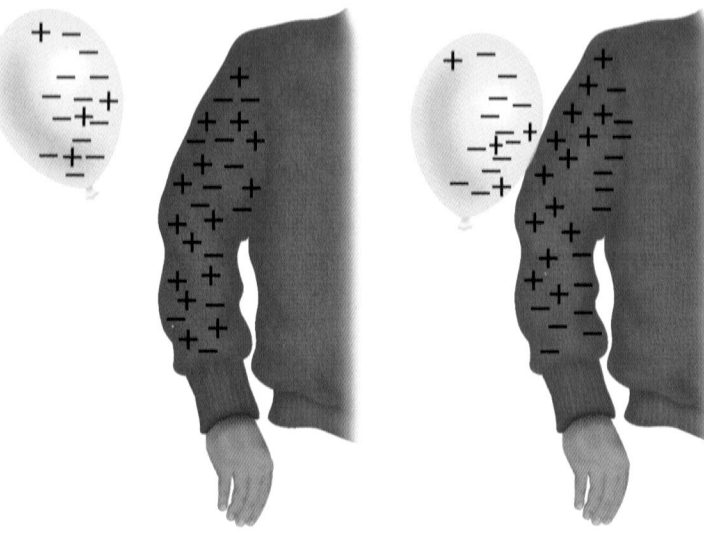

Figure 21-2

The negatively charged balloon induces a positively charged area in the sleeve by repelling its electrons away from the surface. *What happens if it actually touches the sleeve?*

Because of attractive and repulsive forces, charged objects, when brought near a neutral object, can cause electrons to rearrange their positions on the neutral object. Look at **Figure 21-2.** Suppose you charge a balloon by rubbing it with a cloth. If you bring the negatively charged balloon near your sleeve, the extra electrons on the balloon repel the electrons in the sleeve. The electrons near the surface move away from the balloon, leaving a positively charged area on the surface of the sleeve. As a result, the negatively charged balloon attracts the positively charged area of the sleeve. The rearrangement of electrons on a neutral object caused by a nearby charged object is called charging by induction.

Electric Fields

How does an electron exert a force on another charged particle that is some distance away? It does so by setting up

MiniLAB

How do charged objects interact with each other?

Procedure

1. Fold over about 1 cm on the end of a roll of tape to make a handle that is not sticky. Tear off a strip about 10 cm long.
2. Stick the strip on a clean, dry, smooth surface, such as your desktop. Make another identical strip and tape it directly on top of the first.
3. Quickly pull both pieces off the desk together and pull them apart. Then bring the nonsticky sides of both tapes together. What happens?
4. Now stick the two strips of tape side by side onto the smooth surface. Pull them off with a quick motion and bring the nonsticky sides near each other again. What do you observe?

Analysis

1. What happened when you brought the first pair of tapes close together? Were they charged alike or opposite to each other? What may have caused this?
2. What did you observe when you brought the second pair of tapes together? How were they charged? What did you do differently that may have changed the behavior?

Content Background

When discussing static discharges, reinforce the idea that only electrons move from one object to another to cause overall positive and negative charges in the objects.

MiniLAB

Purpose

LS **Kinesthetic** Students will analyze how charged objects interact with each other. **L1** **LEP**

Materials

transparent tape

Teaching Strategies

While cellophane tape often works, the removable transparent tape that can be written on works best.

Analysis

1. The tapes should attract each other. When pulled apart from each other, one became positively charged and one became negatively charged.
2. Both repelled. They were charged alike. Both were removed from the desk, charging them in the same manner.

✓ Assessment

Oral Using the results of their MiniLAB, have students make a generalization about the interaction of charges. Use the Performance Task Assessment List for Making Observations and Inferences in **PASC**, p. 17. **P**

📁 **Activity Worksheets,** pp. 5, 127

Inclusion Strategies

Gifted Have students investigate conductors and insulators. First, they can brainstorm ways to test a list of objects and predict the results. Students should design and make a circuit tester to test the objects. They can make a chart to show which items are conductors and which are insulators. **L3**

Problem Solving

Solve the Problem

1. Opposite charges attract each other. Charged particles in smoke must be attracted to an oppositely charged plate.
2. Sulfur dioxide gas is not charged, so it would not be effective.

Think Critically

1. It can be used to clean particulate matter and smoke from the air.
2. In homes or businesses, the electrostatic air-cleaner plates can be washed with water, washing the runoff down the sewer.

Content Development

- When we categorize substances as conductors, semiconductors, and insulators, we are actually ranking them according to the ease with which electrons move around in them (their conductivity). Ask students to explain, in terms of electrons, the two extreme limits of conductivity; that is, what would zero conductivity mean and what would the maximum possible conductivity mean? *Zero conductivity means that no charge carriers could move. Maximum conductivity means that all charge carriers were in motion.*
- Be consistent when explaining how an object acquires charge by saying it loses or gains electrons, not protons.

▶ Use **Concept Mapping,** p. 47, as you teach this lesson.

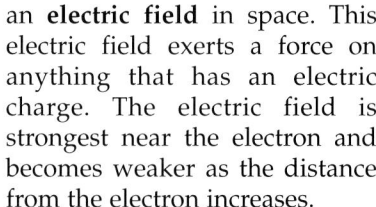

Problem Solving

Take Charge of Pollution

Have you ever noticed a layer of smog in the air over a city? This situation is caused mostly by particulates emitted in combustion processes. Coal-burning power plants, waste incinerators, and other industrial processes contribute to this type of air pollution. How can this pollution be reduced?

A leading method of cleaning up these particulate emissions makes use of static electricity. Electrostatic precipitators are used to trap particles in smoke and ash before they escape into the atmosphere. The ash passes through an electric field to give the particles a charge. They are then trapped as they pass near plates with an opposite electric charge. The collected ash is disposed of in hazardous waste sites or in sanitary landfills.

Solve the Problem:

1. **How do opposite charges interact with each other? Why is such behavior critical to this pollution-control device?**
2. **Would this technique be effective at reducing the emissions of sulfur dioxide gas, which contribute to acid rain formation? Why or why not?**

Think Critically:

Schools, businesses, and even homes also make use of the technology of electrostatic precipitators.

1. Excluding incineration, for what purpose might electrostatic precipitators be used in a shopping mall? Explain.
2. Would there be any kind of waste-disposal problem when using particle precipitators?

an **electric field** in space. This electric field exerts a force on anything that has an electric charge. The electric field is strongest near the electron and becomes weaker as the distance from the electron increases.

Conductors and Insulators

You now know you can build up a negative charge by walking across a carpet. Suppose you then could touch either a wooden door or a metal doorknob. Which one would you prefer to touch? If your choice was the metal doorknob, be prepared to feel a spark.

Carrying Charges

Look at **Figure 21-3** on page 595. The electrons can move through your body to the metal doorknob because both your body and the metal are conductors. Your skin is not as good a conductor as the metal doorknob. A **conductor** is a material that allows electrons to

Student Text Question

Would you expect static discharge to occur if you touched the wooden door instead of the metal doorknob? *No, wood is an insulator.*

Teacher F.Y.I.

The process of charging things by rubbing (electrification by friction) is poorly understood. This branch of science is called triboelectricity. With our increased ability to study surface layers of materials, triboelectricity has attracted new research efforts.

GLENCOE TECHNOLOGY

 Videodisc

The Infinite Voyage Series: Miracles by Design
Chapter 6
New Ceramics: New Uses

Chapter 7
Ceramic Superconductors: Rapid Transportation

Chapter 8
Ceramic Superconductors: Effects on the Computer

 CD-ROM

Physical Science CD-ROM
Have students perform the interactive exploration for Chapter 21 to reinforce important chapter concepts and thinking processes.

move easily through it. Metals such as copper and silver are made of atoms that don't hold their electrons tightly, so electrons move easily through materials made up of these kinds of atoms. For this reason, electric wires are usually made of copper, a good conductor. Silver wire also conducts electricity very well, but silver is much more expensive than copper.

The largest object you are in touch with is Earth. Earth contains a large supply of electrons and functions as a conductor of electricity. It is sometimes desirable to provide a path for the static discharge to reach Earth so that a large static charge doesn't build up on an object and cause damage when it discharges. An object connected to Earth, or the ground, by a good conductor is said to be grounded. Look around you. Do you see anything that might act as a path to the ground? Plumbing fixtures, such as metal faucets, sinks, and pipes, often provide a convenient ground connection.

Insulators—Blocking the Flow

What covers the metal wires in cords attached to telephones and other household appliances? They are usually coated with some type of plastic, an insulating material. An **insulator** is a material that doesn't allow electrons to

Figure 21-3

In the winter, when the dry air indoors is a good insulator, your body builds up a static charge as you walk across a carpeted floor. When you reach for a metal doorknob, the charge flows between your hand and the doorknob and you feel a shock.

21-1 Electric Charge   595

Theme Connection

Energy Ask: **If electric charges cause the attraction between socks in your dryer, do you think this electricity could be used to run your portable tape player? Why or why not?** *No. The movement, or discharge, of electricity happens rapidly and does not continue. It could not be used as a continuous supply of energy.*

Figure 21-4 Which do you think are conductors and which are insulators? *The metals are conductors; the plastic, glass, rubber, and paper are insulators.*

Figure 21-5 Could you tell whether a charged electroscope was positively or negatively charged by looking at it? *No, the leaves will react the same no matter which way the electroscope is charged.* `LEP` `LS`

3 Assess

Check for Understanding

? FLEX Your Brain

Use the Flex Your Brain activity to have students explore static electricity charging by CONDUCTION and INDUCTION.

📁 **Activity Worksheets, p. 5**

Reteach

Carefully demonstrate the function of the electroscope again, using a diagram to show positive and negative areas. Have a student rub a glass rod with silk, so the electrons leave the glass rod and build up on the silk. Touch the electroscope with the glass rod. **Why do the leaves split apart?** *Electrons flow from the leaves into the glass rod. The leaves are positive and repel each other.* `LEP`

Extension

📁 For students who have mastered this section, use the **Reinforcement** and **Enrichment** masters.

Figure 21-4
Pictured here are examples of conducting and insulating materials. *Which do you think are conductors and which are insulators?*

move through it easily. In addition to plastic, wood, rubber, and glass are good insulators. Would you expect static discharge to occur if you touched the wooden door instead of the metal doorknob? **Figure 21-4** shows some conducting and insulating materials.

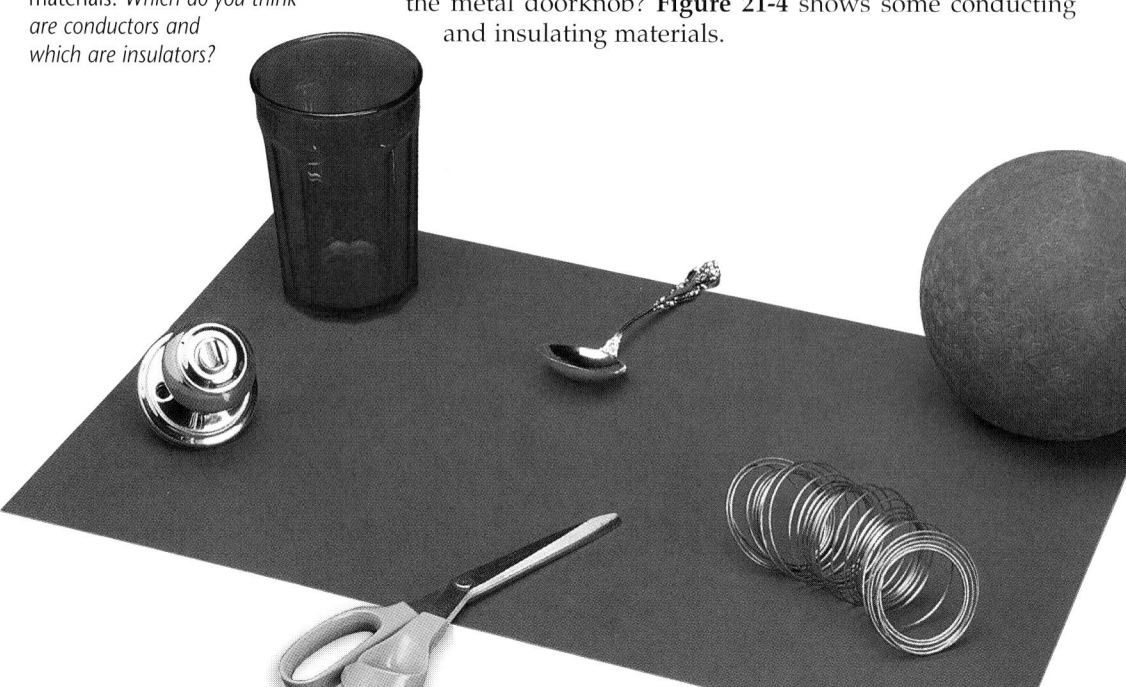

The Electroscope

The presence of electric charges can be detected by an electroscope. An **electroscope** is made of two thin metal leaves attached to a metal rod with a knob at the top. The leaves are allowed to swing freely from the metal rod. When the device is not charged, the leaves hang straight down, as shown in **Figure 21-5A.**

Detecting Positive and Negative Charges

Suppose a negatively charged balloon touches the knob. Because the metal is a good conductor, electrons travel down the rod into the leaves. Both leaves become negatively charged as they gain electrons, **Figure 21-5B.** Now because the leaves have similar charges, they repel each other.

If a glass rod is rubbed with silk, electrons leave the glass rod and build up on the silk. The glass rod becomes positively charged. When the positively charged glass rod is touched to the metal knob, electrons are conducted out of the metal leaves and onto the rod. The leaves repel each other because

596 Chapter 21 Electricity

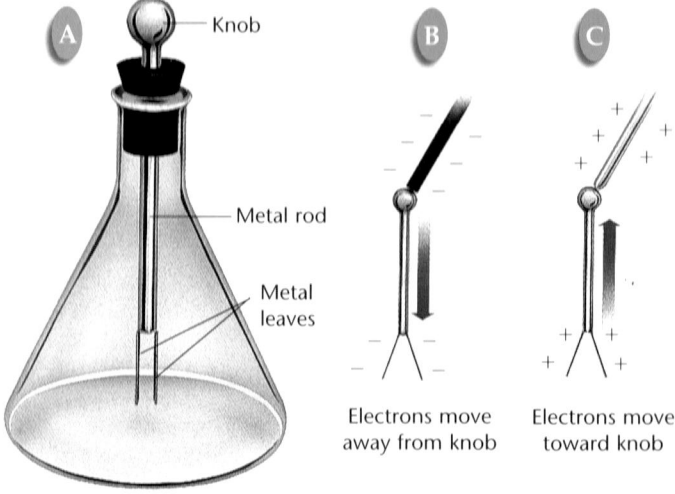

A — Knob

— Metal rod

— Metal leaves

B

Electrons move away from knob

C

+ + +
+ + +
+ + +
+ + +
+ + +

Electrons move toward knob

Figure 21-5

Notice the position of the leaves on the electroscope when they are uncharged (A), negatively charged (B), and positively charged (C). *Could you tell whether a charged electroscope was positively or negatively charged by looking at it?*

each leaf becomes positively charged as it loses electrons, **Figure 21-5C.**

Can you think of any other effects of static electricity you have seen? Can you explain them in terms of like or opposite charges? How do objects become charged, and what happens when they discharge?

Section Wrap-up

Review

1. What is static electricity?

2. Distinguish between electrical conductors and insulators and give an example of each.

3. **Think Critically:** Assume you have already charged an electroscope with a positively charged glass rod. Hypothesize what would happen if you touched the knob again with another positively charged object.

Skill Builder
Observing and Inferring

Suppose that you observe the individual hairs on your arm rise up when a balloon is placed near them. Using the concept of induction and the rules of electricity, what could you infer about the cause of this phenomenon? If you need help, refer to Observing and Inferring in the **Skill Handbook.**

Science Journal

Moist air is more of a conductor than dry air. It is more difficult to observe events related to static electricity, such as clothes clinging or hair standing out, on humid days. More water is in the air on humid days. In your Science Journal, explain how humidity affects static electricity.

21-1 Electric Charge   **597**

Skill Builder

The attraction between the balloon and the hairs on your arm was caused by opposite charges. The negatively charged balloon induced a positive charge on the surface of your arm by repelling electrons.

Assessment

Oral Ask students if the arm has a net charge in this case. Discuss the difference between a net charge and separation of charges. Use the Performance Task Assessment List for Making Observations and Inferences in **PASC,** p. 17. **P**

4 Close

•MINI•QUIZ•

Use the Mini Quiz to check students' recall of chapter content.

1. The buildup of charges on an object is called _____ . *static electricity*

2. Electrons move easily through materials called _____ and do not move easily through materials called _____ . *conductors, insulators*

3. What is the region around the balloon where you notice electric forces acting on other particles called? *electric field*

Section Wrap-up

Review

1. the accumulation of electric charges on an object

2. A conductor is a material that allows electrons to move easily through it, such as copper and silver. An insulator is a material that doesn't allow electrons to pass easily through it, such as plastic, wood, rubber, and glass.

3. **Think Critically** The leaves might spread farther apart.

Science Journal

Humidity reduces static electricity buildup. Charges are more easily able to move from a charged object and combine with an opposite charge when there is water vapor in the air.

Prepare

Section Background

A bolt of lightning is so high in energy and happens so fast that it can have 3 750 000 000 kW of power. The surrounding air temperature can rise by 30 000°C.

1 Motivate

Bellringer

Before presenting the lesson, display **Section Focus Transparency 84** on the overhead projector. Assign the accompanying **Focus Activity** worksheet. L1 LEP

Tying to Previous Knowledge

Ask students how they think most forest fires are started. They will say careless campers, smokers, and perhaps lightning. Ask them to think of a good reason why a park ranger might decide to let a lightning-induced fire burn.

2 Teach

Visual Learning

Figure 21-7 **In what ways might fires help new vegetation grow?** *Fires clean out old growth and give new plants sun. Some seeds need the heat of a fire to germinate.* LEP LS

Objectives

- Explain the occurrence of lightning in terms of induction and static discharge.
- Evaluate the positive and negative aspects of lightning-induced forest fires.

Lightning—A Natural Fire Starter

Have you ever seen lightning strike Earth? Lightning is a large discharge of static electricity. The friction from the movement of water droplets in a cloud can build up areas of positive and negative charges, just as scuffing your feet on a carpet builds opposite charges on you and the carpet. The bottom portion of a cloud develops a negative charge, causing a positive charge to be induced on Earth's surface. As the difference in charge increases, electrons may be attracted to the positively charged ground. A lightning bolt occurs when many billions of electrons are transferred at the same time. The electricity in lightning can have a potential of up to 100 million volts!

Much of the lightning you see doesn't strike Earth's surface. Lightning bolts also occur between the negative area of one cloud and a positive area of another cloud. The electrical energy in a lightning bolt ionizes atoms in the atmosphere and produces great amounts of heat, warming the surround-

ing air to temperatures as high as 30 000°C—several times hotter than the sun's surface! The heat causes air in the bolt's path to expand rapidly, producing compressional waves that we hear as thunder.

The sudden discharge of so much electricity can be dangerous. As seen in **Figure 21-6**, lightning strikes Earth many times each day. It can cause power outages, injury or loss of life, and fires. There is some controversy about whether to put out fires started by lightning in regions unpopulated by humans. Because these fires are natural, they can be viewed as part of nature's cycle of growth and renewal. What do you think?

Figure 21-6

Lightning is beautiful to look at but it can be dangerous to property. Many forest fires are started every year by lightning.

598 Chapter 21 Electricity

Program Resources

Reproducible Masters
Study Guide, p. 88 L1
Reinforcement, p. 88 L1
Enrichment, p. 88 L3

Transparencies
Section Focus Transparency 84

2 Points of View

Letting Nature Take Its Course

Natural fires, like the one in **Figure 21-7**, that are started by lightning serve an important function in many environments. Some people point out that without occasional fires, a thick layer of dead debris can accumulate on a forest floor. This prevents sunlight from reaching the soil and reduces the amount of new growth. The accumulated tinder also increases the likelihood of a fast-moving, destructive, hot fire when one inevitably does occur. Some ecologists point out that few animals actually die in fires. Drought and starvation take a much greater toll on animal populations. In the years following a fire, the variety of plant and animal life in the area typically increases. With the National Park Service supporting this view, laws were passed in the 1970s to allow fires started by lightning to burn naturally unless they threaten people or private property.

Preserve and Protect

Many people were upset in 1988 when one-third of Yellowstone National Park burned. The scenery caused by the seemingly lush growth will not be as spectacular for many years to come. Some argue that the loss of plant and animal life due to fires should be reduced whenever possible. They believe that this destruction of public forests should be prevented by fighting all fires, regardless of their origin.

Figure 21-7

Often, fires creep along the ground and don't destroy large, healthy trees. *In what ways might fires help new vegetation grow?*

Section Wrap-up

Review

1. How are lightning and thunder produced?

2. What is the purpose of a lightning rod?

*inter*NET
CONNECTION Visit the Chapter 21 Internet Connection at Glencoe Online Science, **www.glencoe.com/sec/science/physical,** for a link to more information about lightning and fires.

SCIENCE & SOCIETY

21-2 To Burn or Not 599

Integrating the Sciences

Life Science Animals may be killed or injured by the fire. Many may escape to other areas but put additional pressure on food and habitat as a result. Animals surviving the fire may experience a food shortage. The habitat may be altered, such as by establishment of fields rather than mature forest, as a result of the fire, favoring different species of animals afterward.

Section 21●3

Prepare

Section Background

In 1800, an Italian physicist named Alessandro Volta observed that two metals connected by a conducting liquid produced a continuous transfer of electrons. This phenomenon was different from the rapid static discharge observed by Ben Franklin. This moving electric charge was later called an electric current.

Preplanning

• Acquire the C or D cells for the MiniLAB.
• Collect materials for Activity 21-1.

1 Motivate

Bellringer

🖎 Before presenting the lesson, display **Section Focus Transparency 85** on the overhead projector. Assign the accompanying **Focus Activity** worksheet. [L1] [LEP]

Tying to Previous Knowledge

Ask students if they have ever tried to play a tape in a portable tape player only to find the sound slow and sluggish. They will say the problem was dead batteries, but what is the real problem when the battery "dies"? It stops producing an adequate electric current to operate the tape player. In this section, students will learn what an electric current is and how it is produced.

Science Words
potential difference
circuit
current
dry cell
wet cell
resistance
Ohm's law

Objectives
• Describe how static electricity is different from current electricity.
• Explain how a dry cell is a source of electricity.
• Conceptually and mathematically relate potential difference, resistance, and current.

Flowing Electrons

You learned in Section 21-1 that if you touch a conductor after building up a negative charge in your body, electrons will move from you to the conductor. Could you light a bulb in this manner? Probably not, because the bulb needs a continuous flow of electrons to stay lit.

Potential Difference

Recall from Chapter 5 that thermal energy flows from objects with higher temperatures to objects with lower temperatures. This flow ceases when the temperatures of the objects become the same. Similarly, a negatively charged object has electrons with more potential energy to move and do work than those of an uncharged object. This difference in potential energy causes the electrons to flow from places of higher potential energy to those with lower potential energy. In a static discharge, the potentials quickly become equal and electron flow stops.

The potential energy difference per unit of charge is called the electrical potential. The difference in potential between two different places is the **potential difference.** Potential difference

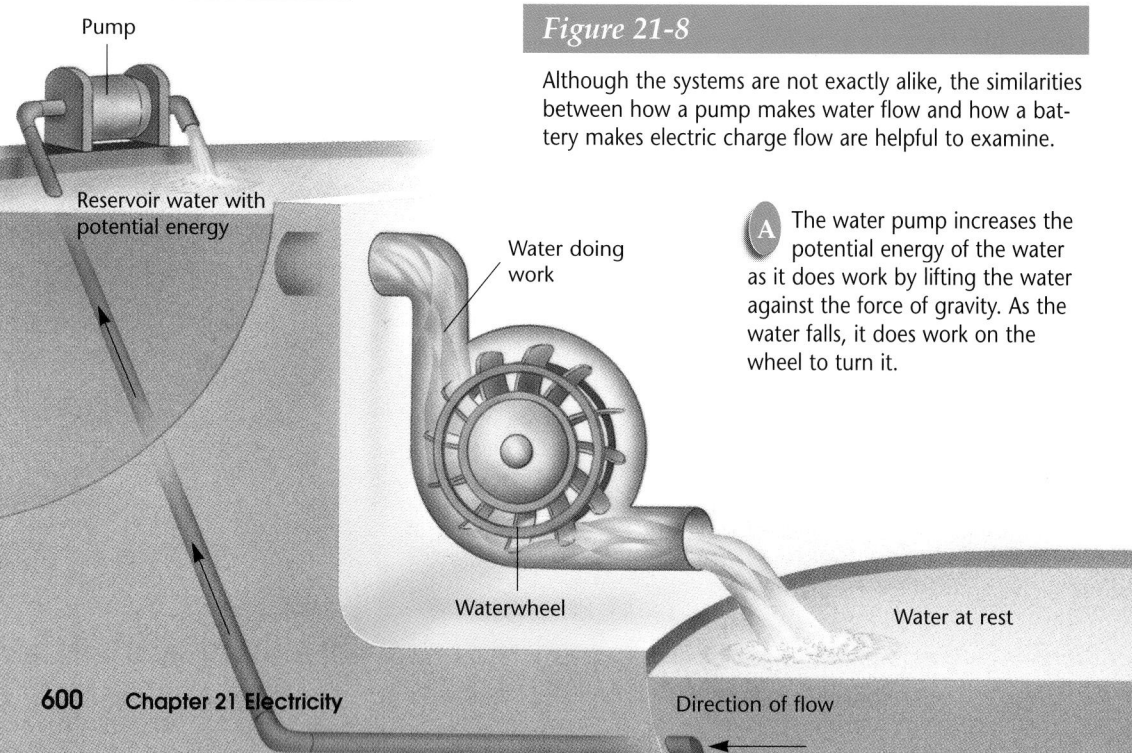

Figure 21-8

Although the systems are not exactly alike, the similarities between how a pump makes water flow and how a battery makes electric charge flow are helpful to examine.

Pump

Reservoir water with potential energy

Water doing work

Ⓐ The water pump increases the potential energy of the water as it does work by lifting the water against the force of gravity. As the water falls, it does work on the wheel to turn it.

Waterwheel

Water at rest

Direction of flow

Program Resources

📂 **Reproducible Masters**
Study Guide, p. 89 [L1]
Reinforcement, p. 89 [L1]
Enrichment, p. 89 [L3]
Activity Worksheets, pp. 123-124, 128 [L1]
Science Integration Activity, pp. 41-42
Multicultural Connections, pp. 45-46
Lab Manual 43
Science and Society Integration, p. 25

🖎 **Transparencies**
Section Focus Transparency 85 [L1]
Science Integration Transparency 21
Teaching Transparency 41

is measured in volts (V). Potential difference, often called voltage, is measured by a voltmeter. The voltage doesn't depend on the number of electrons flowing, but on a comparison of the energy carried by electrons at different points.

Circuit Connections

How can you get electrons to flow through a lamp continuously? You must connect it in an electric circuit. A **circuit** is a closed path through which electrons can flow. Because the lamp is part of a circuit, there is a potential difference across it. If the lamp is turned on, electrons will move through it, causing it to produce light. The electrons will continue to flow in the circuit as long as there is a potential difference and the path of the flowing electrons is unbroken.

The flow of electrons through a wire or any conductor is called **current.** The amount of electric current depends on the number of electrons passing a point in a given time. The current in a circuit is measured in amperes (A). **Figure 21-8** shows how an electrical circuit resembles a water-pumping system.

In order to keep the current moving through a circuit, there must be a device that maintains a potential difference. One common source of potential difference is a battery. Unlike a static discharge from your finger to a doorknob, a battery can light a lamp by maintaining a potential difference in the circuit.

Batteries—A Series of Chemical Cells

Have you ever noticed that a tape or CD begins to slow down in a portable player after using it for several hours? You may also have noticed that a flashlight becomes dim with prolonged use. Perhaps you decided the batteries were dead and you replaced them. Some devices can be plugged into a wall outlet or can use batteries for the energy they need to operate. How do batteries enable these devices to function?

INTEGRATION
Chemistry

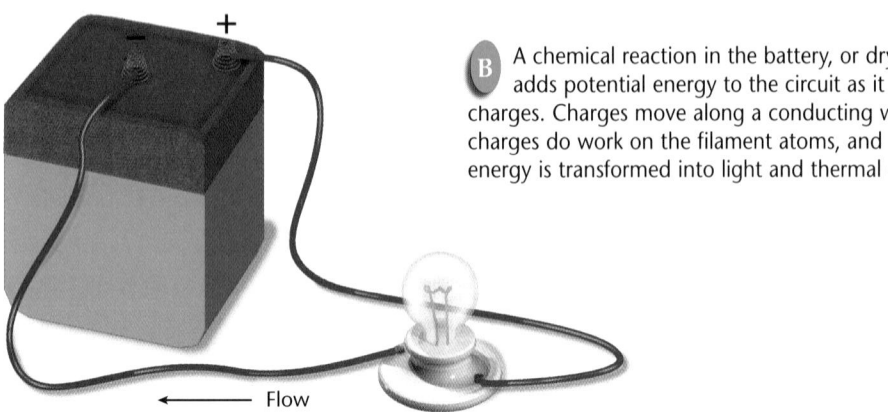

 A chemical reaction in the battery, or dry cell, adds potential energy to the circuit as it separates charges. Charges move along a conducting wire. The charges do work on the filament atoms, and electrical energy is transformed into light and thermal energy.

Flow

21-3 Electric Current   **601**

Demonstration

Place a strip of copper and one of zinc into a glass containing lemon juice, tomato juice, or vinegar. Clip an alligator clip to each electrode and attach these to a sensitive voltmeter (0-1 V DC). A silver electrode may be substituted for the zinc one. Have a student read the potential difference that is established between the two electrodes.

 INTEGRATION
Chemistry

Have students recall how chemical reactions take place between two substances until one is consumed. Remind them of half-cell reactions from chemistry and the resulting energy produced.

Visual Learning

Figure 21-8 Use this figure to explain potential difference and current in a way that is easier for the students to understand. Have them brainstorm some analogies that could represent the same idea. **LEP** **LS**

Use **Multicultural Connections,** pp. 45-46, as you teach this lesson.

Cultural Diversity

Lewis H. Latimer In 1918, the Edison Pioneers was formed to honor some of the people who worked with Thomas Edison as the "creators of the electric industry." One member was the son of a freed slave. Lewis H. Latimer taught himself to be a draftsman when he was a teenager. The office Latimer worked in was near a school where Alexander Graham Bell was developing a device so his deaf, mute wife could hear him. Latimer prepared the drawings that Bell needed to patent the telephone. In 1879, Latimer was hired by the United States Electric Lighting Company, where he invented the first carbon-filament electric lamp and an inexpensive way to produce the filaments. In 1883, he joined Edison, and in 1890 he published the first textbook on electrical lighting—*Incandescent Electric Lighting.*

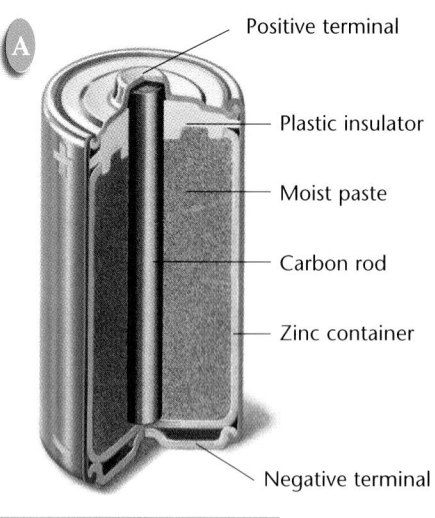

Positive terminal
Plastic insulator
Moist paste
Carbon rod
Zinc container
Negative terminal

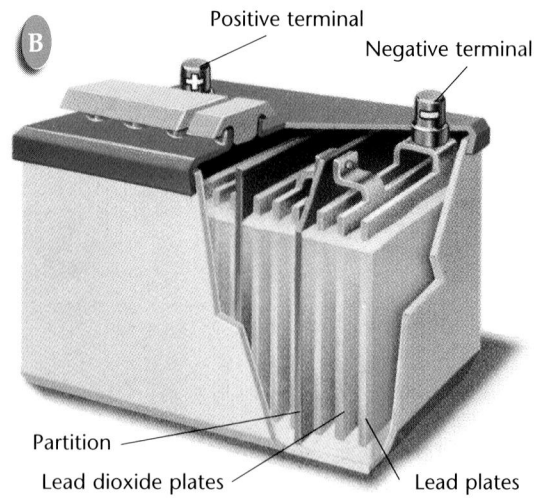

Positive terminal
Negative terminal
Partition
Lead dioxide plates
Lead plates

Figure 21-9

The dry cell (A) can be connected with other dry cells to form a battery. A single dry cell is also called a battery. The car battery (B) is a series of wet cells. *How are these devices similar?*

CONNECT TO
CHEMISTRY

In a car battery, also called a lead storage battery, the following chemical reaction occurs: $Pb + PbO_2 + H_2SO_4 \rightarrow PbSO_4 + H_2O$. Use coefficients to **balance** this equation.

Dry Cell Batteries—The Inside Story

The individual batteries you are most familiar with are dry cells. Look at the dry cell shown in **Figure 21-9A**. Notice that the zinc container of the dry cell surrounds a moist chemical paste with a solid carbon rod suspended in the middle. Can you locate the positive and negative terminals of the dry cell in the diagram? A **dry cell** can act as an electron pump because it has a potential difference between the positive and negative terminals. What causes this potential difference?

When the two terminals of a standard D dry cell are connected in a circuit, such as in a flashlight, a chemical reaction involving zinc and several chemicals in the paste (MnO_2, NH_4Cl, and carbon) occurs. The carbon rod does not take part in the reaction; it serves as a conductor to transfer electrons. Electrons are transferred between some materials in this chemical reaction. As a result, the carbon rod becomes positive, forming the positive (+) terminal of the battery. Electrons accumulate on the zinc, making it the negative (−) terminal of the battery.

The potential difference between these two terminals causes current to flow through a closed circuit, such as when you turn on a portable CD player. You make a battery when you connect two or more cells together in series to produce a higher voltage. Can you think of a device in your home or school that requires more than one dry cell to operate?

Wet Cells—Starting Your Engines

Batteries can also be wet cells. A **wet cell**, **Figure 21-9B**, contains two connected plates made of different metals or metallic compounds in an electrolyte solution. Have you ever examined the battery under the hood of a car? Most car batteries, called lead storage batteries, contain a series of six wet cells made up of lead and lead dioxide plates in a sulfuric acid solution. The chemical reaction in each cell provides a potential difference of about 2 V, giving a total potential of 12 V. As a car is driven, the alternator helps recharge the battery by sending current through the battery in the opposite direction to reverse the chemical reaction.

Plug It In

Most types of household devices are designed to use the potential difference provided by a wall socket. In the United States, the potential difference across the two holes in a wall socket is usually 120 V. The electrical energy is provided by a generator at a power plant. In a later section, you will learn about how this energy is purchased from the electric company.

Resistance

One function of the car battery mentioned earlier is to light various lightbulbs in the car's electric circuits. Do you know what makes a lightbulb glow? Look at the lightbulb in **Figure 21-10.** Part of the circuit through the bulb contains a filament. As a current flows through the filament, electrical energy is converted by the filament into light and heat. The current loses electrical energy as it moves through the filament because the filament resists the flow of electrons, as do most materials.

Resistance is the tendency for a material to oppose the flow of electrons, changing electrical energy into thermal energy and light. With the exception of a few substances that become superconductors at very low temperatures, all conductors have some resistance. The amount of resistance varies with each conductor. Resistance is measured in ohms (Ω).

Figure 21-10

The wire filament of a lightbulb has a high resistance. Electrical energy is converted into light and heat energy.

Why do some flashlights require more C or D cells than others?

Procedure
1. Make a complete circuit by linking two bulbs and one C or D cell battery in a complete loop. Observe the brightness of the bulbs.
2. Disconnect the D cell. Use a wire to connect the positive terminal of one D cell to the negative terminal of the other D cell.
3. Assemble the new circuit by attaching the wire from one bulb to the remaining negative terminal and attaching the wire from the other bulb to the remaining positive terminal. Observe the brightness of the bulbs.

Analysis
1. What is the potential difference of each D cell? Add them together to find the total potential for the circuit you tested in step 3.
2. Assuming that a brighter bulb indicates a greater current, what can you conclude about the relationship between voltage and current?

Purpose

LS **Logical-Mathematical** Students will observe a flashlight lamp operated by one or more cells in a constructed circuit to determine why some flashlights require more than one cell to operate. **L1** **COOP LEARN**

Materials
2 C- or D-cells, 2.7- to 3-volt lamp and socket, 3 pieces of connecting wire

Teaching Strategies
Do not allow students to hook up more than two cells in series per lamp, which will shorten the bulb life or cause it to burn out. Lightbulbs can become hot—caution students to leave the bulbs burning only long enough to make observations.

Analysis
1. 1.5 volts, 3.0 volts
2. As the voltage increases to a lamp, the current flowing through it will increase.

✓Assessment

Performance Have students design and label several possible arrangements showing how to light a lightbulb at different levels of brightness. As time permits, have them demonstrate each method. Use the Performance Task Assessment List for Model in **PASC,** p. 51. **P**

📁 **Activity Worksheets,** pp. 5, 128

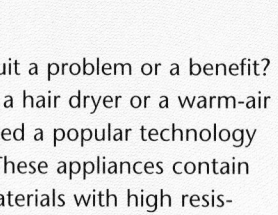

USING TECHNOLOGY

Resistive Heating ▼ ▲

Is resistance in a circuit a problem or a benefit?

If you have ever used a hair dryer or a warm-air hand dryer, you have used a popular technology called *resistive heating*. These appliances contain coils of wire made of materials with high resistances, such as tungsten or nichrome. Some electric energy is converted to heat, along with light, as current passes through these wires. Hot coils are then used to heat the air around them as a fan blows the air toward you.

Resistive heating is also used to warm devices such as clothes irons, electric blankets, toasters, and car window defoggers. Can you think of other appliances in your home or school that use resistive heating?

Think Critically:

Many devices, such as irons and hair dryers, that use resistive heating have several temperature settings. Because the potential difference is the same for all settings, how is the temperature controlled? How do you think the resistance and the current are affected by choosing a hotter setting?

High-resistance wire makes up the heating element.

604   Chapter 21 Electricity

Choosing Resistances for Different Tasks

Copper is an excellent conductor; it has low resistance to the flow of electrons. Copper is used in household wiring because little electrical energy is converted to thermal energy as current passes through the wires. In contrast, tungsten wire glows white-hot as current passes through it. Tungsten's high resistance to current makes it suitable for use as filaments in lightbulbs.

The size of wires also affects their resistance. **Figure 21-11** illustrates how electrons move more efficiently through thick wires

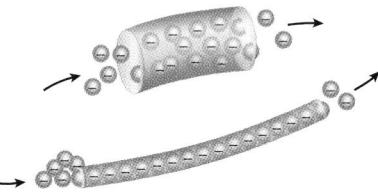

Figure 21-11

The resistance of a short, thick piece of wire is less than the resistance of a long, thin piece of wire.

than thin wires. In wires of the same length and material, thinner wires have greater resistance to electron flow. Likewise, if the diameters of two wires of a material are the same, the longer wire offers a greater resistance. In most conductors, the resistance also increases as the temperature increases.

Figure 21-12

The amount of current flowing through a circuit is related to the amount of resistance in the circuit.

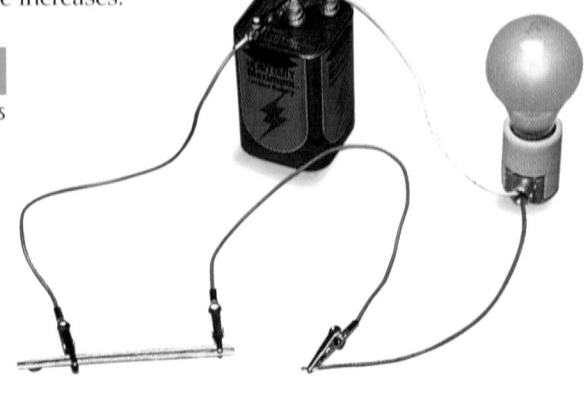

A By changing the length of the graphite rod the current must pass through, the resistance of the circuit can be changed. Recall that longer wires of a given material have higher resistances than shorter wires.

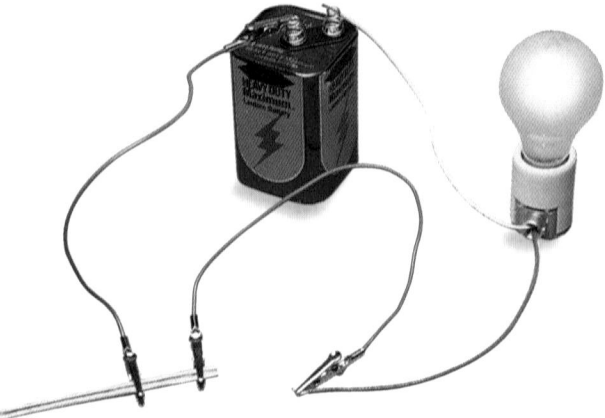

B Notice that the contact to the graphite has been moved so that the circuit includes only a short portion of the graphite. This decreases the total resistance of the circuit, while the voltage remains the same. *How does the brightness of the bulb compare to the brightness of the bulb in the first photo?*

Ohm's Law

When you try to understand the relationship among voltage, current, and resistance, it is helpful to think of the way water behaves in a pipe. If one end of the pipe is higher than the other, there is a difference in the potential energy of the water due to gravity, and pressure at the nozzle increases. This causes a stream, or current, of water to flow. If the height difference increases, the current increases. In a similar way, a greater potential difference in a circuit also causes the electric current to increase. Also, just as the walls and any obstructions in the pipe resist the flow of water, atoms in a wire resist the flow of electricity. As a result, the current in a circuit depends on both the voltage and the resistance, **Figure 21-12.**

George Simon Ohm, a German physicist, found experimentally that the current in a metal conductor is directly

When discussing the relationship between the variables in Ohm's law, expand on the analogy relating water traveling through a pipe to current through a wire. Water is like the current, pressure is like the potential difference, and the size and roughness of the pipe are like the electrical resistance.

3 Assess

Check for Understanding

? FLEX Your Brain

Use the Flex Your Brain activity to have students explore OHM'S LAW.

📁 **Activity Worksheets,** p. 5

Reteach

Students often have trouble rearranging the equation for Ohm's law. Show them the following diagram to go with the expression $V = IR$. When the desired variable is covered, the other two variables are in appropriate mathematical order.

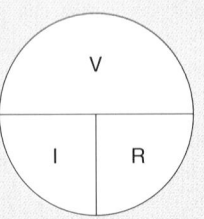

Extension

📁 For students who have mastered this section, use the **Reinforcement** and **Enrichment** masters.

4 Close

•MINI•QUIZ•

1. **What are batteries?** *combinations of dry or wet cells that have a potential difference between positive and negative terminals*

2. **The tendency for a material to oppose the flow of electrons is _____ .** *resistance*

3. **What three quantities are related by Ohm's law?** *voltage, current, resistance*

Section Wrap-up

Review

1. A circuit has continuous current provided by a voltage source. A static discharge happens very rapidly. Both are caused by potential difference.

2. A chemical reaction causes a negative charge on the zinc terminal and a positive charge on the carbon rod. This creates a potential difference and causes a current.

3. **Think Critically**

$$V = IR = (0.3A)(25\ \Omega) = 7.5\ V$$

USING MATH

$I = V/R = 12\ V/60\ \Omega = 0.2\ A$
With two bulbs in series,
$I = 12\ V/120\ \Omega = 0.1\ A$
With three bulbs in series,
$I = 12\ V/180\ \Omega = 0.07\ A$

proportional to the potential difference across its ends and inversely proportional to the resistance. This is expressed mathematically in the equation called **Ohm's law:**

Potential difference = current × resistance

V (volts) = I (amperes) × R (ohms)

USING MATH

Calculating Current

Example Problem:

A lightbulb with a resistance of 160 Ω is plugged into a 120-V outlet. What is the current flowing through the bulb?

Problem-Solving Steps:

1. What is known?   resistance, R = 160 Ω
                    voltage, V = 120 V
2. What is unknown?   current, I
3. Choose the equation. $I = V/R$
4. **Solution:** I = 120 V/160 Ω = 0.75 A

Practice Problem

Find the current flowing through a 20-Ω wire connected to a 12-V battery. What if it were connected to a 6-V battery?

Section Wrap-up

Review

1. How does a current traveling through a circuit differ from the static discharge?

2. Briefly describe how a carbon-zinc dry cell supplies electricity for your tape player.

3. **Think Critically:** Calculate the potential difference across a 25-Ω resistor if a 0.3-A current is flowing through it.

Skill Builder

Making and Using Tables

Suppose you individually connect three copper wires of unequal length to a 1.5-V dry cell and an ammeter. The following currents are obtained: wire 1, 1.2 A; wire 2, 1.4 A; wire 3, 1.1 A. Make a table showing current (given) and resistance (use Ohm's law). If you need help, refer to Making and Using Tables in the **Skill Handbook.**

USING MATH

How is the current in a simple series circuit affected by an increase in the resistance of the circuit? Suppose you placed a 60-Ω bulb in a circuit with a 12-V battery. What would be the current through this circuit? Predict how the current would change if you added one more bulb. Two more bulbs?

Skill Builder

Students should use the equation $R = V/I$ to calculate the resistance needed to complete the table for each of the three wires.

Wire	Current (A)	Resistance (Ω)
1	1.2A	1.3Ω
2	1.4A	1.1Ω
3	1.1A	1.4Ω

✔ Assessment

Performance Have students use the data in their table to predict the relative lengths (longest, medium length, shortest) of the wires. Record this in a fourth column. The length is directly proportional to resistance, so the higher the resistance, the longer the wire. Use the Performance Task Assessment List for Making Observations and Inferences in **PASC,** p. 17. **P**

Activity 21-1

Modeling Ohm's Law

After several hours of use, batteries may go dead. What has changed in the batteries? For the batteries to cause a current in your stereo, there must be a potential difference across the ends of the batteries. In this activity, you will use gravitational potential energy and water to model potential difference and current in an electric circuit.

Problem
How is a water system like an electric circuit?

Materials
- plastic funnel
- meterstick
- ring stand with ring
- 2 beakers (250-mL)
- rubber tubing (1 m)
- stopwatch or clock

Procedure
1. Create a data table like the one below.
2. Assemble the apparatus as shown. Place the funnel as high as possible.
3. Measure the height from the top of the funnel to the outlet end of the rubber tubing, in meters. Record your data on the table.
4. Pour 200 mL of water into the funnel fast enough to keep it full, but not overflowing.
5. Measure the time for 0.10 L of water to flow into the lower beaker, and record it.
6. Repeat steps 3 through 5 at least three more times, lowering the funnel for each trial.

Analyze
1. Gravity causing water to move can be compared to voltage causing electrons to move. Which trial represents a **model** of a circuit with the highest voltage?
2. The rate (L/sec) of flow of water from the tubing can be compared to current. Which trial represents the highest current?
3. If voltage is increased, **predict** what happens to current.

Conclude and Apply
4. According to Ohm's law, what should happen to the current if the voltage stays the same but the resistance is reduced?
5. If a long tube has more resistance, **predict** what should happen to the rate of flow of water if the tube is shorter.

Data and Observations
Sample Data

Trial	Height (m)	Time (s)	Rate (L/s)
1	0.60	10	0.010
2	0.50	12	0.0083
3	0.40	15	0.0067

21-3 Electric Current   **607**

5. A shorter tube offers less resistance, resulting in higher rate of water flow.

✓ Assessment

Oral This activity uses an analogy between electrical potential energy and gravitational potential energy. Ask students to explain the real differences between these two quantities. Use the Performance Task Assessment List for Making Observations and Inferences in **PASC**, p. 17. **P**

Purpose
LS **Logical-Mathematical** Construct and operate a water analogy of an electric circuit.
L1 **LEP** **COOP LEARN**

Process Skills
measuring, predicting, experimenting, making models, sequencing, observing and inferring, comparing and contrasting, recognizing cause and effect

Time
1 class period

📁 **Activity Worksheets,** pp. 5, 123-124

Teaching Strategies
The water should be poured fast enough to keep the funnel full. The timer should measure the flow into the lower beaker. The bottom of the tube should be kept at a constant height for all trials. Students may need help calculating the rate. The actual rate of flow depends upon the diameter of the tubing.

Answers to Questions
1. The highest funnel position represents the greatest voltage. (The most work done on a quantity of water is similar to the most work done on a quantity of electrons.)
2. The greatest amount of water flowing per second occurs when the funnel is at its highest position. The quantity of water flowing per second compares to the quantity of electrons flowing per second (current).
3. A greater voltage in a circuit produces a greater current.
4. According to Ohm's law, if the voltage remains the same, the current will increase as the resistance is decreased.

21•4 Electrical Circuits

Section Background

In studying and teaching circuits, keep in mind that resistance, potential difference, and current are related by Ohm's law, $V = IR$. Normally, R is constant. If you increase the potential, current will increase, and if you decrease potential, current will decrease.

Preplanning

Gather the materials needed for Activity 21-2.

1 Motivate

Bellringer

Before presenting the lesson, display **Section Focus Transparency 86** on the overhead projector. Assign the accompanying **Focus Activity** worksheet. L1 LEP

Demonstration

Visual-Spatial Try to have several small objects with electrical circuits, partially taken apart, in the room for students to examine. You might display a small radio, a flashlight, and/or a telephone.

Tying to Previous Knowledge

Point out that in the last section, students learned about how the resistance, voltage, and current are related. This section will apply those terms to circuits in real devices.

Science Words
series circuit
parallel circuit

Objectives
- Sketch a series and a parallel circuit, and list applications of each type of circuit.
- Recognize the function of circuit breakers and fuses.

Finding Familiar Circuits

Look around you. How many electrical devices, such as lights, alarm clocks, stereos, and televisions, do you see that are plugged into wall outlets? These devices all rely on a source of electrical energy and wires to complete an electrical circuit. Most circuits include a voltage source, a conductor, and one or more devices that use the electricity to do work.

Consider, for example, a circuit that includes an electric hair dryer. The dryer must be plugged into a wall outlet to operate. A generator at a power plant produces a potential difference across the outlet, causing the electrons to move when the circuit is complete. The dryer and the circuit in the house both contain conducting wires to carry current. The hair dryer turns the electricity into thermal energy and mechanical energy to do work. When you unplug the hair dryer or turn off its switch, you are opening the circuit and breaking the path of the current.

Series Circuits—Only One Path

There are several kinds of circuits. One kind of circuit is called a series circuit. Some holiday lights are wired together in a series circuit. In a **series circuit**, the current has only one path it can travel along. Look at the diagram of the series circuit in **Figure 21-13**. If you have ever decorated a window or a tree with a string of lights, you may have had the frustrating experience of trying to find one burned out bulb. How can one faulty bulb cause the whole string to be out? Because the parts of a series circuit are wired one after another, the amount of current is the same through every part. When any part of a series

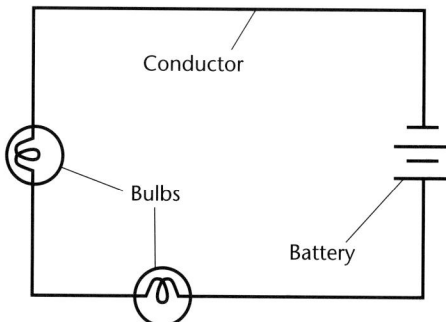

Figure 21-13

A series circuit provides only one path for the current to follow. *What happens to the brightness as more bulbs are added?*

608 Chapter 21 Electricity

Program Resources

Reproducible Masters
Study Guide, p. 90 L1
Reinforcement, p. 90 L1
Enrichment, p. 90 L3
Cross-Curricular Integration, p. 27
Activity Worksheets, pp. 125-126 L1
Lab Manual 44
Critical Thinking/Problem Solving, p. 27

Transparencies
Section Focus Transparency 86 L1
Teaching Transparency 42

Table 21-1

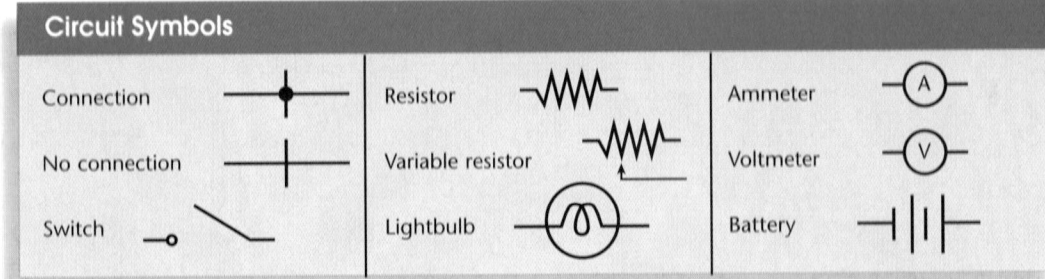

Circuit Symbols

Connection		Resistor		Ammeter	(A)
No connection		Variable resistor		Voltmeter	(V)
Switch		Lightbulb		Battery	

circuit is disconnected, no current flows through the circuit. This is called an open circuit. Does this explain how a broken bulb can ruin a whole string of lights?

Table 21-1 shows the symbols used in diagramming electric circuits. Notice that the switch must be closed for the circuit to be continuous.

Parallel Circuits—Branching Out

What would happen if your home were wired in series and you turned off a light? All other lights and appliances in your home would go out, too. Fortunately, houses are wired in parallel. **Parallel circuits** contain separate branches for current to move through. Look at the parallel circuit in **Figure 21-14**. The current splits up to flow through the different branches. More current flows through the paths of lowest resistance. Because all branches connect the same two points of the circuit, the potential difference is the same in each branch.

CONNECT TO
EARTH SCIENCE

Rivers sometimes form different branches that separate and then rejoin, possibly making an island. Write a paragraph *describing* which kind of circuit this is most like and why.

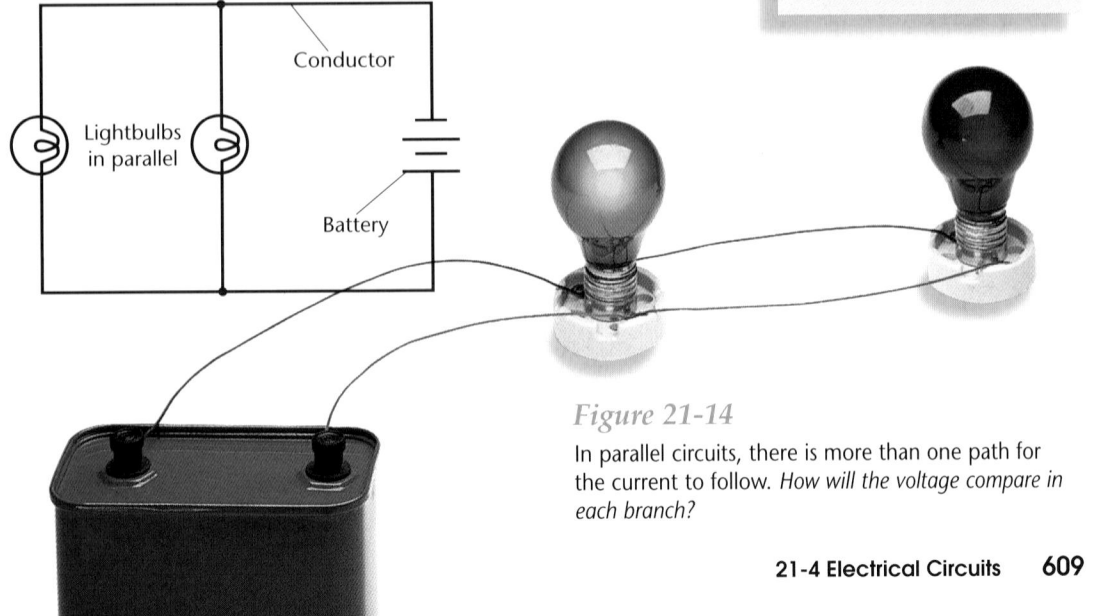

Conductor

Lightbulbs in parallel

Battery

Figure 21-14

In parallel circuits, there is more than one path for the current to follow. *How will the voltage compare in each branch?*

21-4 Electrical Circuits **609**

2 Teach

Visual Learning

Figure 21-13 What happens to the brightness as more bulbs are added? *The brightness decreases.*

Figure 21-14 How will the voltage compare in each branch? *The voltage will be the same.* **LEP** [IS]

Teacher F.Y.I.

In a series circuit, the total resistance is equal to the sum of the individual resistances. $R(total) = R(1) + R(2) + ...$ The potential difference across any resistance is equal to the current times that specific resistance. In a parallel circuit, the total resistance can be calculated by the following equation: $1/R(total) = 1/R(1) + 1/R(2) + ...$ The total resistance is less than any single resistance in a parallel circuit.

Use **Teaching Transparency 42** as you teach this lesson.

CONNECT TO
EARTH SCIENCE

Answer This would be analogous to a parallel circuit. The change in potential energy is the same in both branches, and the branches rejoin with the same current as before the split. **P**

Enrichment

Have students investigate other electrical components commonly found in electrical circuits (see Chapter 23). Resistors, capacitors, and diodes have unique appearances and functions. [L3]

Theme Connection

Energy Emphasize that electrons move in all parts of a circuit simultaneously. Electrons in the circuit outside the battery lose energy as they move through devices in the circuit that convert electrical energy into other forms of energy. Within the battery, electrons gain energy produced by chemical reactions. The battery converts chemical energy to electrical energy.

3 Assess

Check for Understanding

? FLEX Your Brain

Use the Flex Your Brain activity to have students explore CIRCUIT BREAKER CONSTRUCTION. **P**

📂 **Activity Worksheets,** p. 5

Reteach

Demonstration Set up a simple series circuit with a battery and a couple of small lightbulbs. You might want to connect ammeters in two different places to show that the current is the same everywhere in the circuit. Now, disconnect one lamp to show that the entire circuit is broken. Do the same for a parallel circuit, showing that the current is not the same everywhere in a parallel circuit. Disconnect a lamp to show that only that branch is affected. The broken branch no longer carries current, but the current in the other branches remains the same.

Extension

📂 For students who have mastered this section, use the **Reinforcement** and **Enrichment** masters.

Figure 21-15

What type of circuit is most common in household wiring?

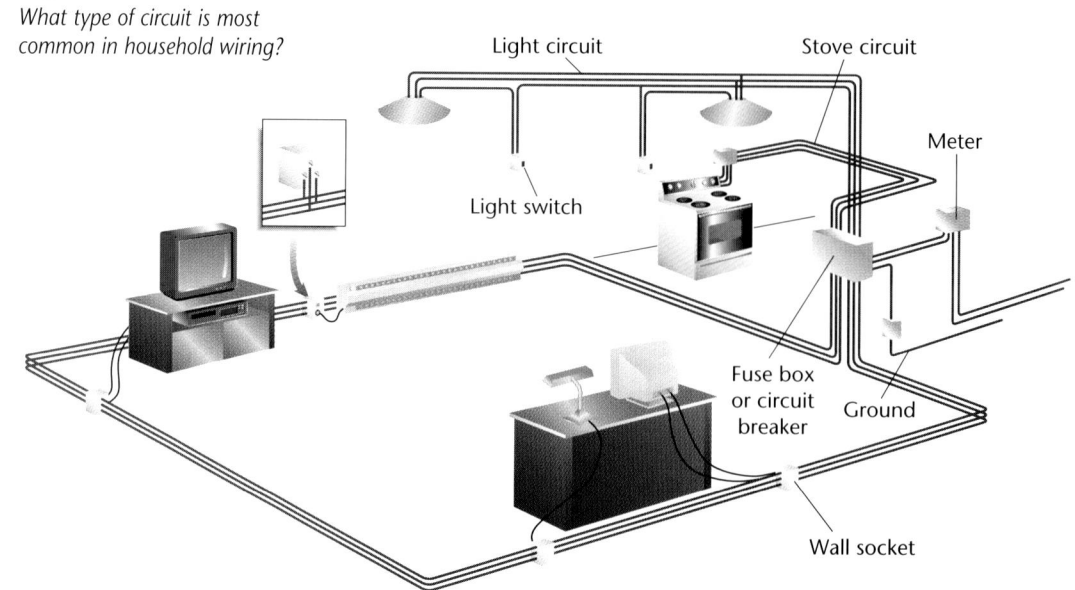

Figure 21-16

The fuses in the fuse box below prevent circuits from overheating.

Parallel circuits have several advantages. When one branch of the circuit is opened, such as when you turn a light off, the current continues to flow through the other branches. Because each branch of a parallel circuit has the same potential difference across it, lights or appliances on parallel branches receive more current than if they were placed in a series arrangement.

Household Circuits

Try to count how many different things in your home require electricity. You don't see the wires because most of them are hidden behind the walls, ceilings, and floors. This wiring is composed mostly of a combination of parallel circuits connected in an organized and logical network. **Figure 21-15** shows how electrical energy enters your house and is distributed. Each branch receives the standard voltage from the electric company, which is 120 V in the United States. The main switch and circuit breaker box serve as a sort of electrical headquarters for your home. Parallel circuits branch out from the breaker box to wall sockets, major appliances, and lights.

Many appliances can draw current from the same circuit, so protection against overheating must be built in to prevent fires. Either a fuse or a circuit breaker is wired between every parallel circuit and the main switch box as a safety device.

Fuses and Circuit Breakers

What does it mean to say somebody has "blown a fuse"? Usually, it refers to somebody losing his or her temper. This expression comes from the function of an electrical fuse, **Figure 21-16,** which contains a small piece of metal that melts if the current becomes too high and heats the circuit wire too much. When it melts, it causes a break in the circuit and prevents more current from flowing through the overloaded circuit. To fix this, you must replace the damaged fuse with a new one rated for the same amount of current as the original fuse was rated.

A circuit breaker, **Figure 21-17,** is another guard against overheating a wire. A circuit breaker contains a piece of metal that bends when it gets hot. The bending causes a switch to open the circuit, preventing the flow of more current. Circuit breakers can usually be reset by flipping the switch. Before you reset a circuit breaker or replace a blown fuse, you should unplug some of the appliances from the overloaded circuit.

Figure 21-17

A newer device to prevent circuits from overheating is a circuit breaker. *Which device, a fuse or a circuit breaker, seems more convenient to have in the home?*

Section Wrap-up

Review

1. Use circuit diagram symbols to draw a series circuit containing a battery, an open switch, a resistor, and a lightbulb.

2. Use symbols to draw a parallel circuit with a battery and two resistors wired in parallel.

3. Compare and contrast fuses and circuit breakers. Which is easier to use? Why?

4. **Think Critically:** Pennies are made of copper and are excellent conductors of heat and electricity. Explain why you should **never** replace a blown fuse with a penny.

Skill Builder
Hypothesizing
You are walking by a sign made of lighted bulbs. One of the lights begins to flicker and goes out. All other bulbs are still lit. Use your knowledge of circuits to form a hypothesis to explain why only one bulb went out instead of the entire sign. If you need help, refer to Hypothesizing in the **Skill Handbook.**

Science Journal

In your Science Journal, draw a circuit diagram of your bedroom or another room. Show all lights or appliances, and consider whether they behave as if wired in series or parallel. Would you like to have two or more devices controlled by the same switch?

Skill Builder
If one bulb went out and the others are still lit, there must be current in all other bulbs. The burned-out bulb must have been connected in parallel, so current stopped in only one branch.

Assessment

Oral Why does an entire lightbulb fail to light if the filament is broken? There is only one path for the current to follow. Breaking the filament is like opening a switch; there is no current in the bulb. Use the Performance Task Assessment List for Making Observations and Inferences in **PASC,** p. 17. **P**

---

4 Close

Section Wrap-up

Review

1.
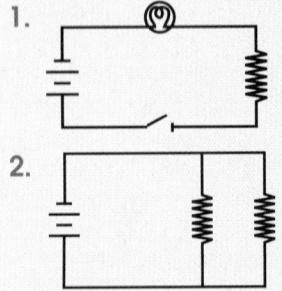

2.

3. Both break overloaded circuits. Fuses melt and must be replaced. Circuit breakers bend to flip a switch that can be easily reset. Circuit breakers are easier to use.

4. **Think Critically** The copper penny would allow too much current in the circuit, and this could cause a fire.

Science Journal
They are wired in parallel. It is not desirable to have one switch control several devices.

Purpose

IS Logical-Mathematical Students will construct, test, and compare working examples of parallel and series circuits.

L1 **COOP LEARN**

Process Skills

observing, communicating, classifying, sequencing, making and using tables, comparing and contrasting, recognizing cause and effect, forming operational definitions, interpreting data, using numbers, experimenting, hypothesizing

Time

1 class period

Materials

Be certain that the light-bulb current requirements closely match the current that will result from the battery voltage so bulbs are not readily burned out. If you are concerned about overheating or just wish to have students use wire, substitute insulated wire for the aluminum foil.

Alternate Materials Economical low-voltage lightbulbs can be obtained by cutting apart a string of minilights for Christmas trees. A short length of wire can be stripped from the ends of the wires.

Safety Precautions

The foil, or wires, can become hot if the current is left turned on. Remind students to hook up the battery only long enough to make their observations. This will conserve battery energy and prolong its lifetime also.

Possible Hypotheses

If a bulb is removed from a series circuit, the lights will all go off. In a parallel circuit, the remaining bulbs will stay lit.

📁 Activity Worksheets, pp. 5, 125-126

Activity 21-2

Design Your Own Experiment

Comparing Electric Circuits

Imagine what a bedroom might be like if it were wired in series. In order for an alarm clock to keep time and wake you in the morning, your lights and anything else that uses electricity would have to be on! Fortunately, most outlets in homes are wired on separate branches of the main circuit. Can you design simple circuits that have specific behaviors and uses?

PREPARATION

Problem

How do the behaviors of series and parallel circuits compare?

Form a Hypothesis

Predict what will happen if one bulb is disconnected when connected in series and in parallel. Also, write a hypothesis predicting in which circuit the lights shine the brightest.

Objectives

• Design and construct series and parallel circuits.
• Compare and contrast the behaviors of series and parallel circuits.

Possible Materials

• dry-cell battery (6 V)
• 3 small lights with sockets
• aluminum foil
• cellophane tape
• scissors
• paper clips

Safety Precautions

Some parts of circuits can become hot. Do not leave the battery connected or the circuit completed for more than a few seconds at a time. *Never* connect the positive and negative terminals of the dry-cell battery directly without including at least one bulb in the circuit.

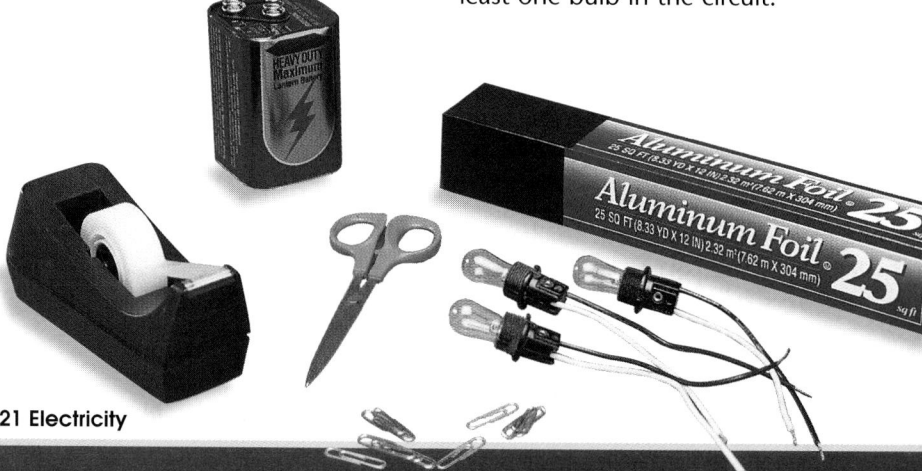

PLAN THE EXPERIMENT

1. As a group, agree upon and write out the hypothesis statement.
2. Work together to determine and write out the steps you will take to test your hypothesis.
3. Make a list of the materials you will need in your experiment and gather them.

Check the Plan

1. How will your circuits be arranged? On a piece of paper, draw a large series circuit of three lights and the dry-cell battery as shown. On the other side, draw another circuit with the three bulbs arranged in parallel.
2. Make conducting wires by taping a 30-cm piece of cellophane tape to a sheet of aluminum foil and folding the foil over twice to cover the tape. Cut these to any length that works in your design.
3. Design several experiments you would like to try with each type of circuit. Be sure you have planned to test your hypothesis.
4. *Make sure your teacher approves your plan and that you have included any changes suggested in the plan.*

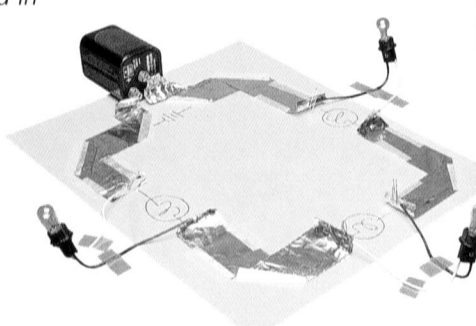

DO THE EXPERIMENT

1. Carry out the experiment. Be sure to leave the circuit on for only a few seconds at a time to avoid overheating.
2. As you do the experiment, record your observations in your Science Journal.

Analyze and Apply

1. **Compare** your results with those of other groups.
2. **Predict** what happens in the series circuit when one end of a bulb is left unconnected.

What happens in the parallel circuit? Explain your observations.

3. **Compare** the brightness of the lights in the different circuits. Explain.
4. **Predict** what happens to the brightness of the series circuit if you complete it with two bulbs instead of three bulbs. How does this demonstrate Ohm's law?

21-4 Electrical Circuits 613

---

4. The bulbs would be brighter because there would be less resistance in the circuit.

Go Further

To control all three lights in a parallel circuit, place the switch between the battery and a wire leading to one side of the lights. To control only one light in a parallel circuit, place the switch along the wire going to one of the lights.

✓ Assessment

Oral Ask students to discuss why houses and buildings are wired in parallel and not in series. In a series circuit, all lights and appliances would have to be on at the same time; if one is off or burned out, no current flows anywhere in the circuit. Use the Performance Task Assessment List for Making Observations and Inferences in **PASC**, p. 17. **P**

Possible Procedures
   Possible circuit designs may include the following: A series circuit should have a single chain or pathway from one light to a second light and then through it to the third light—much like fence posts in a fence row. The battery should be attached to the free wire at each end of the chain. A parallel circuit will have one wire from each of the three lights hooked together and the remaining three wires from the other sides hooked together. These two connection points can be attached to the battery.

Teaching Strategies
   Students may accidentally design a circuit having two lightbulbs in parallel and that combination wired in series with the third light. A combination series-parallel circuit may produce some interesting results, but it will not help students to completely explore the behavior of each of the circuits independently.

DO THE EXPERIMENT

Expected Outcome
   Students should observe that in a series circuit, when any one of the three lights is removed, the remaining lights will go out. In a parallel circuit, when any one of the lights is removed, the others will remain lit.

Analyze and Apply

1. Results should be consistent with those of other groups.
2. In a series circuit, current cannot continue if the path is broken. In a parallel circuit, the current can still move through the other intact branches.
3. Lights shine the brightest in the parallel circuit.

Go Further

Hypothesize where in the parallel circuit you would place a switch to control all three lights. To control only one light? Try it.

Prepare

Section Background

Electricity must often be transmitted many miles to reach people. There is some power loss during this transmission. However, it can be minimized if the electricity is transmitted at high voltages. The voltage in electric power lines may be as high as 765 kV (765 000 V).

Preplanning

Collect battery-operated toys and other equipment for the MiniLAB. Find some sample electric bills, or ask students to bring in some from home.

1 Motivate

Bellringer

Before presenting the lesson, display **Section Focus Transparency 87** on the overhead projector. Assign the accompanying **Focus Activity** worksheet. L1 LEP

Tying to Previous Knowledge

Recall earlier definitions of work and power. Point out that, as a charge moves through a circuit, it does work. Electrical work is done as electrical energy is changed into another form of energy, such as mechanical or thermal energy. The rate at which this occurs is electrical power.

21•5 Electrical Power and Energy

Science Words

electrical power
kilowatt-hour

Objectives

- Explain and calculate electrical power.
- Calculate the amount of electrical energy in kilowatt-hours.

Electrical Power

What do you think of when you hear the word *power?* It has many different meanings. In Chapter 7, you read that power is the rate at which work is done. Electricity can do work for us. Electrical energy is easily converted to other types of energy. For example, the blades of a fan can rotate and cool you as electrical energy is changed into mechanical energy. An iron changes electrical energy into thermal energy. **Electrical power** is the rate at which electrical energy is converted to another form of energy.

The rate at which different appliances use energy varies. Appliances are often advertised with their power rating, which depends on the amount of electrical energy each appliance needs to operate. Electric heating elements in ovens and hair dryers have large power ratings. However, they aren't usually on all the time. Appliances that run a lot of the time, such as refrigerators, usually use more energy. **Table 21-2** shows the power requirements of some appliances, and **Figure 21-18** shows a label with power information.

Table 21-2

Energy Used by Home Appliances			
Appliance	Time of Usage (hours/day)	Power Usage (watts)	Energy Usage (kWh/day)
Hair dryer, blower	0.25	1000	0.25
Microwave oven	0.5	700	0.35
Stereo	2.5	109	0.27
Range (oven)	1	2600	2.60
Refrigerator/freezer (15 cu ft, frostless)	10	615	6.15
Television (color)	3.25	200	0.65
Electric toothbrush	0.08	7	0.0006
100-watt lightbulb	6	100	0.60
40-watt fluorescent lightbulb	1	40	0.04

614 Chapter 21 Electricity

Calculating Power

Electrical power is expressed in watts (W) or kilowatts (kW). The amount of power used by an appliance can be calculated by multiplying the potential difference by the current.

$$power = current \times voltage$$

$$watts = amperes \times volts$$

$$P = I \times V$$

One watt of power is produced when one ampere of current flows through a circuit with a potential difference of one volt. Look again at **Table 21-2**. Which appliance requires the most electrical power to operate? You can tell by looking at the number of watts listed for that appliance under the power usage column. The example problem below shows you how to calculate the electrical power usage for an appliance. This can be done easily as long as you know the values of the current and voltage.

Work the following problems to practice calculating power.

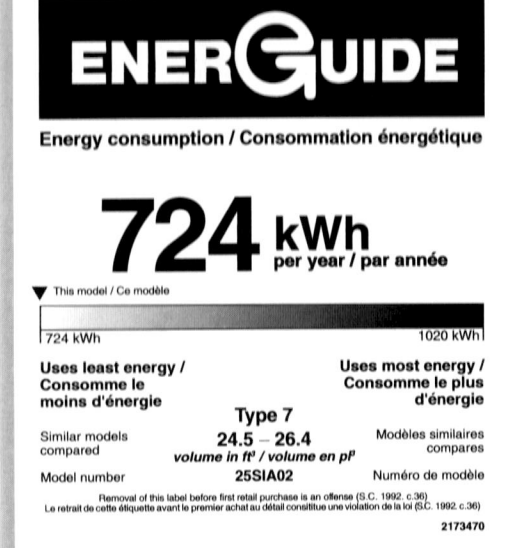

ENERGUIDE

Energy consumption / Consommation énergétique

724 kWh
per year / par année

▼ This model / Ce modèle

724 kWh | 1020 kWh

Uses least energy / Consomme le moins d'énergie		Uses most energy / Consomme le plus d'énergie
	Type 7	
Similar models compared	**24.5 – 26.4** volume in ft³ / volume en pi³	Modèles similaires comparés
Model number	**25SIA02**	Numéro de modèle

Removal of this label before first retail purchase is an offense (S.C. 1992, c.36)
Le retrait de cette étiquette avant le premier achat au détail constitue une violation de la loi (S.C. 1992, c.36)

2173470

Figure 21-18

Most new electrical appliances have labels that show how much power they use.

USING MATH

Calculating Power

Example Problem:

A calculator has a 0.01-A current flowing through it. It operates with a potential difference of 9 V. How much power does it use?

Problem-Solving Steps:

1. What is known?   current, I = 0.01 A
                              potential difference, V = 9 V
2. What is unknown?   power, P
3. Choose the equation. $P = I \times V$
4. **Solution:** P = 0.01 A $\times$ 9 V = 0.09 W

Practice Problems

1. A lamp operates with a current of 0.625 A and a potential difference of 120 V. How much power does the lamp use?
2. A microwave oven uses 1000 W of power. The voltage source is 120 V. What is the current flowing through the microwave?

21-5 Electrical Power and Energy   **615**

2 Teach

Brainstorming

Bring in some advertisements for microwave ovens, hair dryers, and stereos, as well as some lightbulb boxes. Find the rate at which each appliance uses energy in watts. Have students make a list of these ranging from high to low users of electric power. Using the list suggested above, estimate the usage time of each appliance in an average day. Ask which appliances now seem to be the largest users of electrical power.

Practice Problem Answers

1. $P = IV = (0.625$ A$)(120$ V$)$
   $P = 75$ W
2. $P = IV$, so $I = \dfrac{P}{V}$
   $I = \dfrac{1000 \text{ W}}{120 \text{ V}} = 8.3$ A

NATIONAL GEOGRAPHIC SOCIETY

🔘 **Videodisc**

GTV: Planetary Manager

Energy

47575

47576

47922

Industry

47572

MiniLAB

Purpose

IS **Interpersonal** Determine the amount of power an electric toy uses. **L1**

Materials

battery-operated toy; DC milliammeter; 3 alligator clips

Analysis

1. $P = I \times V$
2. The higher the speed of the motor, the greater the power usage.

Assessment

Performance Have students calculate the energy in kWh that the toy would use in one hour. Use the Performance Task Assessment List for Using Math in Science in **PASC**, p. 29. **P**

 **Activity Worksheets,** pp. 5, 129

Practice Problem Answers

$P = 100W \times \dfrac{1\ kW}{1000\ W}$

$= 0.10\ kW$

$E = P \times t$

$\quad (0.10\ kW)(5.5\ h)$

$E = 0.55\ kWh$

---

3 Assess

Check for Understanding

⚡ FLEX Your Brain

Use the Flex Your Brain activity to have students explore POWER COMPANY TRANSMISSION.

 **Activity Worksheets,** p. 5

MiniLAB

How much power operates an electric toy?

Procedure

1. Remove the battery from an electric toy. Fasten wires to the battery with tape or rubber bands.
2. Attach one wire to the proper battery connection in the toy. Attach the other wire to an ammeter.
3. Use a wire to connect the other pole of the ammeter to the other connection in the toy.
4. Turn on the toy and measure the current on the ammeter.
5. Determine the voltage from the battery size. If two cells were used, be sure to add their voltages together to get the total potential.

Analysis

1. Use the voltage and current to calculate the electric power in watts.
2. Would you expect power usage to change if the toy is allowed to run on different speeds? If possible, try it. When is power usage the highest?

Electrical Energy

Why is it important that you not waste electricity? Most electrical energy is produced from natural resources, which are limited in supply. Electrical energy also costs you money. All the electricity you use in your home is measured by a device called an electric meter. You may have noticed that the meter for your home has a wheel that spins quickly when you are using a great deal of electricity and is stopped when no electricity is being used. The amount of electrical energy you use depends on the power required by appliances in your home and how long they are used. For example, you can calculate the amount of energy a refrigerator uses in a day by multiplying the power required by the amount of time it uses that power.

$$energy = power \times time$$

$$kWh = kW \times h$$

$$E = P \times t$$

The unit of electrical energy is the **kilowatt-hour** (kWh). One kilowatt-hour is 1000 watts of power used for one hour. The electric utility company charges you periodically for each kilowatt-hour you use. You can figure your electric bill by multiplying the energy used by the cost per kilowatt-hour. **Table 21-3** shows some sample costs of running electrical appliances.

Table 21-3

Energy Used by Home Appliances

	Appliance		
	Hair Dryer	Stereo	Color Television
Average power in watts	1000	109	200
Hours used daily	0.25	2.5	2.5
Hours used monthly	7.5	75.0	75.0
Monthly watt hours	7500	8175	15 000
kWh used a month	7.5	8.175	15.000
Rate charge	$0.09	$0.09	$0.09
Monthly cost	$0.68	$0.74	$1.35

Using Computers

Spreadsheet headings should include the following: Appliance, Time of Usage (hours/day), Power Usage (watts), Energy Usage (Wh/day). Students may want to convert energy usage to kWh/day.

Use the energy equation to calculate the energy used by the refrigerator in the following example problem.

USING MATH

Calculating Electrical Energy

Example Problem:

A refrigerator is one of the major users of electrical power in your home. If it uses 700 W and runs 10 hours each day, how much energy (in kWh) is used in one day?

Problem-Solving Steps:

1. What is known? power, $P = 700$ W $= 0.7$ kW
   time, $t = 10$ h
2. What is unknown? energy, E
3. Choose the equation. $E = P \times t$
4. **Solution:** $E = 0.7$ kW $\times 10$ h $= 7$ kWh

Practice Problem

A 100-W lightbulb is left on for 5.5 hours. How many kilowatt-hours of energy are used?

Section Wrap-up

Review

1. What is electrical power?

2. A television uses a current of 1.5 A at 120 V. The television is used for 2 hours. Calculate the power used in kW and the energy used in kWh.

3. **Think Critically:** How many kWh of energy would be needed for brushing your teeth with an electric toothbrush daily for the month of May? How much would it cost at $0.09 per kWh?

Skill Builder
Concept Mapping

Prepare a concept map that shows the steps in calculating the energy used in operating an electrical device with known voltage and current for a known amount of time. If you need help, refer to Concept Mapping in the **Skill Handbook.**

> **Using Computers**
>
> **Spreadsheet** On a spreadsheet, list the appliances your family uses daily, the estimated hours per day, and, from **Table 21-2,** the power usage. Multiply the power usage by the hours per day to find the daily energy use for each appliance. Which is responsible for the greatest energy use?

21-5 Electrical Power and Energy **617**

Skill Builder
Possible solution:

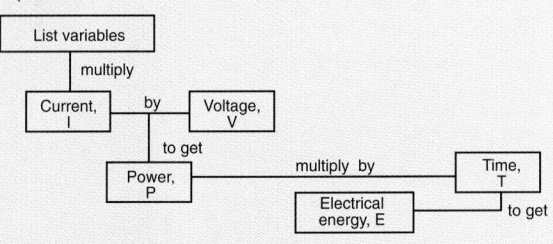

Assessment

Performance Have students design another concept map for determining the current in a device when energy, time, and voltage are known. Use the Performance Task Assessment List for Concept Map in **PASC,** p. 89.

P

---

4 Close

•MINI•QUIZ•

Use the Mini Quiz to check students' recall of chapter content.

1. **What is electrical power?**
   the rate at which electrical energy is changed to another form of energy

2. **The product of electric power used and time is** _____ . *energy*

3. **The unit of electrical energy is the** _____ . *kilowatt-hour*

Section Wrap-up

Review

1. the rate at which electrical energy is changed to another form of energy

2. $P = I \times V = (1.5\ \text{A})(120\ \text{V})$
   $= 180\ \text{W} \times \dfrac{1\ \text{kW}}{1000\text{W}}$
   $= 0.18\ \text{kW}$
   $E = P \times t = (0.18\ \text{kW})(2\ \text{h})$
   $= 0.36\ \text{kWh}$

3. **Think Critically**
   $\dfrac{0.0006\ \text{kWh}}{\text{day}} \times 31\ \text{days} =$
   0.019 kWh
   $0.019\ \text{kWh} \times \dfrac{\$0.09}{\text{kWh}} =$
   \$0.0017

People and Science

Background

Both electricity and electronics involve electrons and electric currents. Electricity usually refers to the conversion of the kinetic energy of electrons into heat and light, as in electric arc furnaces used in industry, or lamps and heaters used in the home. In electronics, alternating current is used to carry information. By carefully controlling the movement of electrons, electronics devices can be used to do a variety of tasks, such as send messages, as in telephones, do calculations, as in computers, or even operate a robot.

Teaching Strategies

- **Visual-Spatial** Demonstrate how electronics devices are shrinking in size and improving in performance by showing students pictures—or actual examples, if available—of vacuum tube devices, such as radios and stereos, and comparing them with more recent devices based on semiconductor technology.

- Have students turn to the periodic table in Chapter 10 and note the location of copper, silicon, and germanium. Point out that both silicon and germanium are metalloids, while copper is a metal. Students may also be interested to know that two elements commonly used to control the conductive properties of semiconductors are gallium and arsenic, which flank germanium on the Table.

- **Linguistic** Review the differences among nonconductors (insulators), semiconductors, conductors, and superconductors by listing the terms on the blackboard. Ask student volunteers to describe the characteristics of each. Record their observations.

618

People and Science

GIOVONNAE ANDERSON, Ph.D., *Electrical Engineer*

Personal Insights

On the Job

Q Dr. Anderson, what kinds of electronics devices do you design?

A I help design things that test the performance of electronics equipment, such as high-speed telephone lines or satellite communications systems. Right now, I'm working on a device that will test the semiconductor chips used in cellular telephones.

Q What is a semiconductor chip?

A It's a tiny, thin piece of metal with very small electrical circuits printed on it. The chip itself is made of a semiconductor, usually silicon. When you hear people talk about a "computer chip," they're talking about a semiconductor chip that runs a computer. Radios, computers, and virtually all electronics are gradually getting smaller and less expensive.

Q What is a typical workday like for you?

A I spend a lot of time thinking about how electric fields will act under certain conditions, and using a computer to model what will happen if I design a circuit a particular way. I also talk to other engineers to find out if we can really make whatever I've come up with. Then technicians build a prototype for me to test. That's when I actually get to sit at a lab bench and play with wires.

Q What advice do you have for students who find physical science interesting but difficult?

A Ask questions. The old saying, "The only dumb question is the one not asked" is really true. People learn in different ways. Often, the person who seems to be struggling is just looking at the information a different way. If you can help that person to understand, he or she will show you something you hadn't noticed before because you've been so busy looking at it your way.

Career Connection

Electrical engineers design and develop electrical equipment, such as household appliances and industrial machinery, and electronics equipment, such as computers and communications equipment. Electronics technicians help build devices designed by engineers. Talk to a career counselor at a local high school or college to find out what kind of education and training these careers require.

- **Electrical Engineer**
- **Electronics Technician**

Career Connection

- Electrical engineers earn at least a bachelor's degree at a four-year college. Or, an engineer might earn a bachelor's degree in physics and then go on to earn a master's degree or Ph.D. in applied engineering.
- **Career Path** Electronics technicians have talent and training in assembling circuits and other electrical components. A college or technical school degree is usually required.

For More Information

Students can contact the following society for more information.

Institute of Electrical and Electronics Engineers (IEEE)
U.S. Activities Office
Pre-College Education Department
1828 L Street, N.W., Suite 1202
Washington, DC 20036-5104

Summary

21-1: Electric Charge

1. Static electricity is the accumulation of electric charges on an object.
2. An electrical conductor allows electrons to move through it easily. An electrical insulator doesn't allow electrons to move through it easily.
3. An electroscope contains two suspended metal leaves in a jar that move apart when charged.

21-2: Science and Society: To Burn or Not

1. Lightning occurs when excess electrons flow between a cloud and the ground.
2. Some lightning-caused forest fires are allowed to burn as part of the natural processes in the forest.

21-3: Electric Current

1. The potential energy stored in an electron decreases as it moves through a circuit.
2. A dry cell creates a potential difference in a circuit, causing the electrons to flow.
3. Ohm's law states that $V = IR$.

21-4: Electrical Circuits

1. Current has only one path in a series circuit. Parallel circuits provide more than one path for current to follow.
2. Circuit breakers and fuses are safety devices.

21-5: Electrical Power and Energy

1. Electrical power is the rate at which electrical energy can be transformed.
2. A kilowatt-hour is 1000 watts of power used for 1 hour.

Key Science Words

a. circuit
b. conductor
c. current
d. dry cell
e. electric field
f. electrical power
g. electroscope
h. insulator
i. kilowatt-hour
j. Ohm's law
k. parallel circuit
l. potential difference
m. resistance
n. series circuit
o. static electricity
p. wet cell

Reviewing Vocabulary

Match each phrase with the correct term from the list of Key Science Words.

1. buildup of electric charges on an object
2. exerts a force on an electric charge
3. a material through which electrons can move easily
4. a device used to detect electric charges
5. closed path through which electrons flow
6. two plates made of different metals or metallic compounds in an electrolyte solution
7. opposes electron flow in a conductor
8. electric current has at least two separate paths
9. the rate at which electrical energy is converted to a different form of energy
10. relates voltage to resistance and current

Summary

Have students read the summary statements to review the major concepts of the chapter.

Reviewing Vocabulary

1. o	**6.** p
2. e	**7.** m
3. b	**8.** k
4. g	**9.** f
5. a	**10.** j

✔ Assessment

Portfolio Encourage students to place in their portfolios one or two items of what they consider to be their best work. Examples include:

- Connect to Chemistry answer, p. 602
- Connect to Earth Science answer, p. 609
- Flex Your Brain, p. 610 **P**

Performance Additional performance assessments may be found in **Performance Assessment** and **Science Integration Activities.** Performance Task Assessment Lists and rubrics for evaluating these activities can be found in Glencoe's **Performance Assessment in the Science Classroom.**

GLENCOE TECHNOLOGY

🔘 Videodisc

Glencoe Physical Science

Interactive Videodisc

Use the videodisc lesson *Electricity and Magnetism* to review the principle of electricity.

📼 MindJogger Videoquiz

Chapter 21 Have students work in groups as they play the Videoquiz game to review key chapter concepts.

Checking Concepts

1. a	**6.** d
2. a	**7.** a
3. d	**8.** c
4. c	**9.** d
5. b	**10.** d

Understanding Concepts

11. Both involve charged particles. In static electricity, these charges accumulate in one place. In current electricity, these charges are continually flowing.

12. In the atoms of conductors, electrons are more free to move. In insulators, electrons are tightly bound and are not free to move.

13. Electrons flow in a circuit because of a difference in electrical potential between two points in the circuit. Electrons flow from the point of highest potential energy (negative electrode) to the point of lowest potential energy (positive electrode). When electrons move through an appliance, they do work on the appliance, and some of their potential energy is converted to some other form of energy.

14. Lightning occurs when opposite charges build up in a cloud and Earth's surface. Lightning is actually a large static discharge.

15. According to Ohm's law, resistance and current are inversely related. As resistance increases, current decreases. As the potential difference increases, current increases.

Checking Concepts

Choose the word or phrase that completes the sentence or answers the question.

1. An object becomes positively charged when it _____.
 a. loses electrons
 b. loses protons
 c. gains electrons
 d. gains neutrons

2. When two negative charges are brought close together, they will _____.
 a. repel
 b. attract
 c. neither attract nor repel
 d. ground

3. As the distance from a charged particle increases, the strength of the electric field _____.
 a. can't be determined
 b. remains the same
 c. increases
 d. decreases

4. An example of a good insulator is _____.
 a. copper
 b. silver
 c. wood
 d. salt water

5. Connecting a charged object to Earth in order to discharge the object into Earth is called _____.
 a. charging
 b. grounding
 c. draining
 d. induction

6. The difference in potential energy per unit charge between two electrodes is measured in _____.
 a. amperes
 b. coulombs
 c. ohms
 d. volts

7. The difference in energy carried by electrons at different points in a circuit will determine the _____.
 a. voltage
 b. resistance
 c. current
 d. power

8. Resistance in an electrical wire causes electrical energy to be converted to _____.
 a. chemical energy
 b. nuclear energy
 c. thermal energy
 d. sound

9. Which of the following wires would tend to have the least amount of electrical resistance?
 a. long
 b. fiberglass
 c. hot
 d. thick

---

10. Electrical energy is measured in _____.
 a. volts
 b. newtons
 c. kilowatts
 d. kilowatt-hours

Understanding Concepts

Answer the following questions in your Science Journal using complete sentences.

11. Compare and contrast static electricity and current electricity.

12. How are the atoms of conductors different from the atoms of insulators?

13. What causes electrons to move through a circuit? What happens to the electrons' potential energy as they pass through an electrical appliance?

14. How does lightning occur?

15. How are resistance and potential difference related to the amount of current flowing through a circuit?

Thinking Critically

16. Lightning rods are grounded conductors located on roofs of buildings. How do they protect buildings from lightning?

17. Explain how an electroscope could be used to detect a negatively charged object.

Thinking Critically

16. Because they are higher than the roof of a building, they are struck first. They discharge the electric charges of a lightning strike into Earth.

17. To detect a negatively charged object, place a positive charge on an electroscope. Observe the electroscope's leaves as you bring the positively charged electroscope close to the object. If the object is negatively charged, the leaves will move closer together.

18. $I = V/R$ or $V = I \times R$
  $V = (1.5 A)(2\ \Omega) = 3\ V$

19. (1) $P = V \times I = (120\ V)(2\ A)$
   $P = 240\ W = 0.240\ kW$
  (2) $E = P \times t$
   $= (0.240 kW)(4\ h)$
   $E = 0.96\ kWh$

18. A toy car has a 1.5-A current, and its internal resistance is 2 ohms. How much voltage does the car require?

19. The current flowing through an appliance connected to a 120-V source is 2 A. How many kilowatt-hours of electrical energy does the appliance use in 4 hours?

20. You are asked to connect a stereo, a television, a VCR, and a lamp in a single, complex circuit. Would you connect these appliances in parallel or in series? How would you prevent an electrical fire? Explain your answers.

Developing Skills

If you need help, refer to the **Skill Handbook.**

21. **Making and Using Graphs:** The resistance in a 1-cm length of copper wire at different temperatures is shown below.

Sample data	
Resistance in Microohms	**Temperature in °C**
2	50
3	200
5	475

Construct a line graph for the above data. Is copper a better conductor on a cold day or a hot day?

22. **Interpreting Data:** Look at the power usage of the appliances in **Table 21-2** and calculate the current each appliance pulls from a 120-V source. Which appliance draws the most current?

23. **Concept Mapping:** Make a network concept map sequencing the events that occur when an electroscope is brought near a positively charged object and a negatively charged object. Be sure to indicate which way electrons flow and the charge and responses of the leaves.

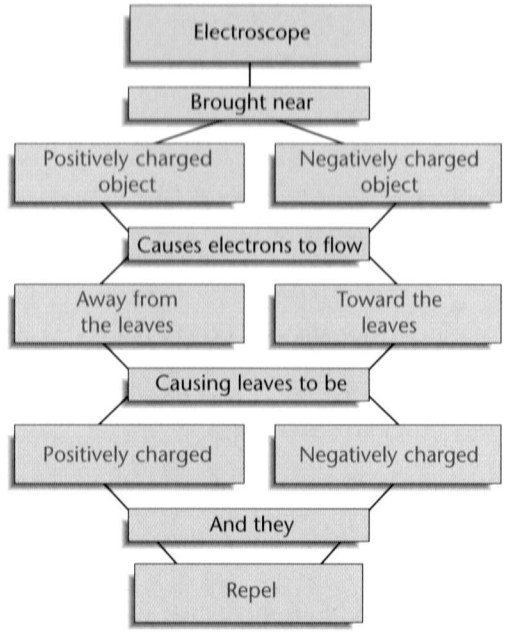

24. **Using Variables, Constants, and Controls:** Design an experiment to test the effect on current and voltage in a circuit when two batteries of equal voltage are connected in series. What is your hypothesis? What are the variables and control?

Performance Assessment

1. **Make a Model:** In Activity 21-1 on page 607, you created a model of Ohm's law with a water system. How would you expect the results of this experiment to differ if you used a tube with a wide diameter?

2. **Design an Experiment:** Review your circuit designs from Activity 21-2 on pages 612-613. How would you design a switch from the materials used in these circuits?

3. **Problem Solving:** You have probably seen warnings about contacting overhead power lines. However, as seen in the photo, birds can safely perch on these power lines. Make a report about how this is possible.

20. The appliances should be connected in parallel so if one appliance goes out, the others still work. Either a fuse or a circuit breaker could be used to protect the circuit from excess current.

Developing Skills

21. **Making and Using Graphs** According to the graph, copper would be a better conductor on a cold day because its resistance decreases with temperature.

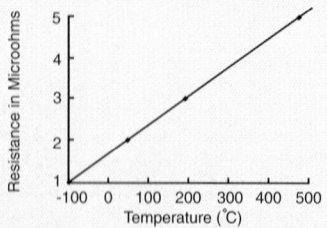

22. **Interpreting Data** The current drawn by each appliance can be solved as follows.

$$I = \frac{P \text{ as given in table}}{120 \text{ V}}$$

The appliance drawing the highest current must have the highest power. The range oven has the highest current.

23. **Concept Mapping** See student page.

24. **Using Variables, Constants, and Controls** Answers will vary somewhat, but the conclusions of the experiment should be that connecting two batteries of equal voltage in series increases the voltage and the current.

Performance Assessment

1. Assuming a constant height, a wider diameter should decrease the resistance and increase water flow. Use the Performance Task Assessment List for Model in **PASC**, p. 51. [P]

2. Answers will vary. Use the Performance Task Assessment List for Designing an Experiment in **PASC**, p. 23. [P]

3. Birds are not grounded. All parts of the bird are at the same high potential. Use the Performance Task Assessment List for Making Observations and Inferences in **PASC**, p. 17. [P]

Chapter Organizer

Section	Objectives/Standards	Activities/Features
Chapter Opener		Explore Activity: Find out where a bar magnet has its greatest attraction for other magnetizable objects. p. 623
22-1 Characteristics of Magnets (2 sessions, 1 block)*	1. **Describe** the properties of magnets. 2. **Define** the region of force around a magnet. 3. **Model** magnetic behavior using domains. National Content Standards: (5-8) UCP1-UCP3, UCP5, D1, D2; (9-12) UCP1-UCP3, UCP5	Problem Solving: Magnet Metal Recycling, p. 626 MiniLAB: What interferes with a magnet? p. 627 Science Journal, p. 627 Skill Builder: Hypothesizing, p. 627
22-2 Uses of Magnetic Fields (2 sessions, 1 block)*	1. **Explain** the magnetic effects of a current in a wire. 2. **Compare** and **contrast** ammeters and voltmeters. 3. **Describe** the function of an electric motor. National Content Standards: (5-8) UCP1-UCP3, UCP5, A2, B2, E2, F1, F5, G1, G3; (9-12) UCP1-UCP3, UCP5, A2, B4, E2, F1, F6, G1, G3	Using Technology: MRI—The Inside Story, p. 629 MiniLAB: How can you observe the relationship between electricity and magnetism? p. 630 Using Computers, p. 631 Skill Builder: Comparing and Contrasting, p. 631 Activity 22-1: Electricity and Magnetism, p. 632
22-3 Producing Electric Current (3 sessions, 1 block)*	1. **Describe** how a generator produces an electric current using electromagnetic induction. 2. **Distinguish** between alternating current and direct current. 3. **Explain** how a transformer can step up or step down the voltage of an alternating current. National Content Standards: (5-8) UCP1-UCP3, UCP5, A1, A2, B2, E1, G1, G3; (9-12) UCP1-UCP3, UCP5, A1, A2, B4, E1, G1, G3	Connect to Earth Science, p. 634 Using Math, p. 637 Skill Builder: Concept Mapping, p. 637 Activity 22-2: Trying Transformers, pp. 638-639
22-4 Science and Society (1 session, ½ block)*	1. **Describe** the characteristics of superconductors. 2. **Consider** various applications of superconductivity. National Content Standards: (5-8) UCP2, UCP4, UCP5, A2, E2, F5, G1, G3; (9-12) UCP2, UCP4, UCP5, A2, E2, F6, G1, G3	Connect to Chemistry, p. 641 Explore the Technology, p. 641 Science and Art: Guitars—Plugged or Unplugged, p. 642 Science Journal, p. 642

Activity Materials

Explore	Activities	MiniLABs
page 623 small metal paper clips, bar magnet	page 632 thin, flexible, insulated wire; cardboard tube; scissors; galvanometer or milliammeter; bar magnet pages 638-639 6-V dry cell battery, low-voltage AC power supply, low-voltage light, 32-gauge insulated wire, large nail, soda straw cut to the length of the nail, scissors, paper clip	page 627 bar magnet; ring stand with clamp; thread; paper clip; pieces of paper, cardboard, plastic wrap, etc.; small mass to anchor thread on the table page 630 insulated wire, lightbulb, 6-V dry cell, magnetic compass

Need Materials? Call Science Kit (1-800-828-7777).   * A complete Planning Guide that includes block scheduling is provided on pages 32T-35T.

Teacher Classroom Resources

Reproducible Masters	Transparencies	Teaching Resources
Study Guide, p. 92 Reinforcement, p. 92 Enrichment, p. 92 Activity Worksheets, p. 134 Lab Manual 45, Magnets Multicultural Connections, p. 47	Section Focus Transparency 88, How'd They Do That?	Glencoe Physical Science Interactive Videodisc Physical Science CD-ROM Spanish Resources English/Spanish Audiocassettes Cooperative Learning Resource Guide Lab Partner Lab and Safety Skills Lesson Plans
Study Guide, p. 93 Reinforcement, p. 93 Enrichment, p. 93 Activity Worksheets, p. 135 Lab Manual 46, Electromagnets Concept Mapping, pp. 49-50 Technology Integration, pp. 19-20	Section Focus Transparency 89, Find the Hidden Motors Teaching Transparency 43, Electric Motor—DC Generator	**Assessment Resources** Chapter Review, pp. 47-48 Assessment, pp. 147-150 Performance Assessment in the Science Classroom (PASC) MindJogger Videoquiz Alternate Assessment in the Science Classroom Performance Assessment Chapter Review Software Computer Test Bank
Study Guide, p. 94 Reinforcement, p. 94 Enrichment, p. 94 Activity Worksheets, pp. 130-133 Science Integration Activity 22, The Electric Lemon Science and Society Integration, p. 26 Critical Thinking/Problem Solving, p. 28	Section Focus Transparency 90, Harnessing Nature's Power Science Integration Transparency 22, Making Usable Electricity Teaching Transparency 44, Transformers	
Study Guide, p. 95 Reinforcement, p. 95 Enrichment, p. 95 Cross-Curricular Integration, p. 28	Section Focus Transparency 91, Superconductors: Cool, *Very* Cool	

GLENCOE TECHNOLOGY

The following multimedia resources are available from Glencoe.

Science and Technology Videodisc Series (STVS)
Human Biology
    Detecting the Body's Magnetic Field

Glencoe Physical Science Interactive Videodisc
Electricity and Magnetism

Physical Science CD-ROM

Key to Teaching Strategies

The following designations will help you decide which activities are appropriate for your students.

L1 Level 1 activities should be within the ability range of all students, including those with learning difficulties.

L2 Level 2 activities should be within the ability range of the average to above-average student.

L3 Level 3 activities are designed for the ability range of above-average students.

LEP LEP activities should be within the ability range of Limited English Proficiency students.

LS These activities are designed to address different learning styles.

COOP LEARN Cooperative Learning activities are designed for small group work.

P These strategies represent student products that can be placed into a best-work portfolio.

Teacher Classroom Resources

This is a representation of key blackline masters available in the Teacher Classroom Resources.

Teaching Aids

Section Focus Transparencies

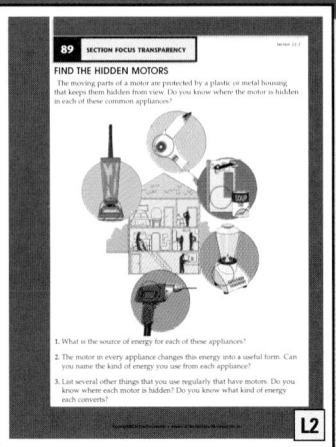

Science Integration Transparencies

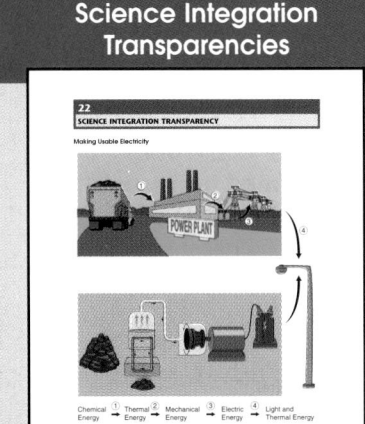

Teaching Transparencies

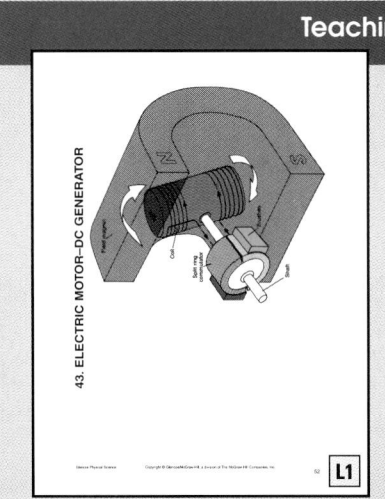

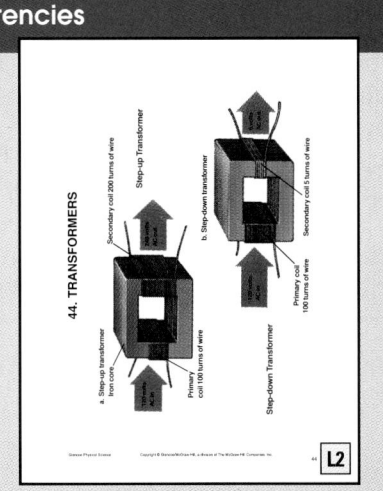

Meeting Different Ability Levels

Study Guide

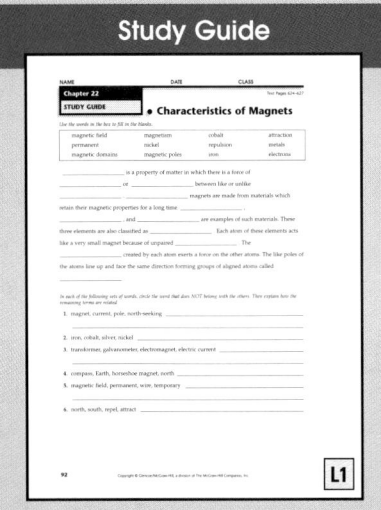

Reinforcement

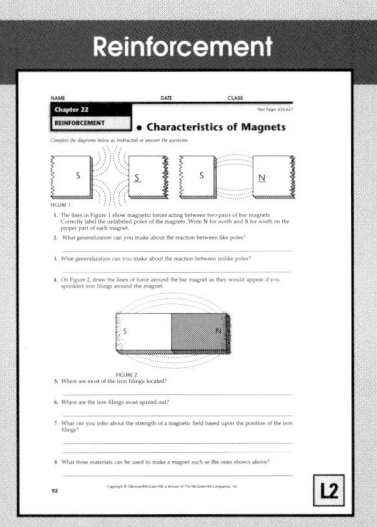

Enrichment Worksheets

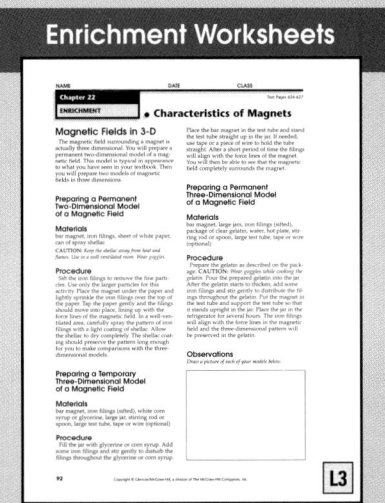

Hands-On Activities

Science Integration Activity

L1

Lab Manual

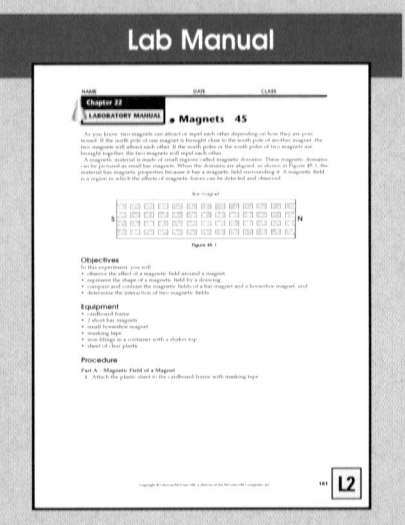

L2

Assessment

Performance Assessment

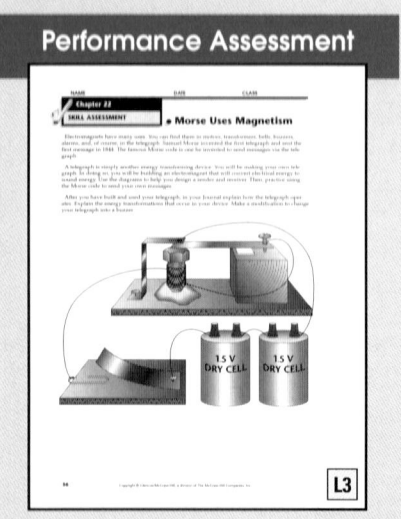

L3

Enrichment and Application

Critical Thinking/ Problem Solving

L2

Cross-Curricular Integration

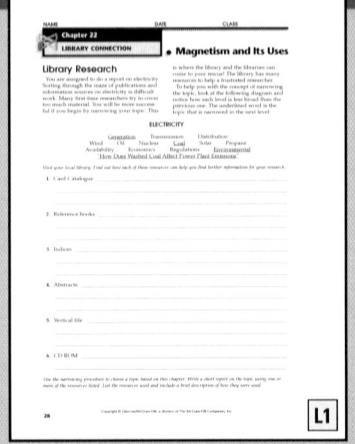

L1

Science and Society Integration

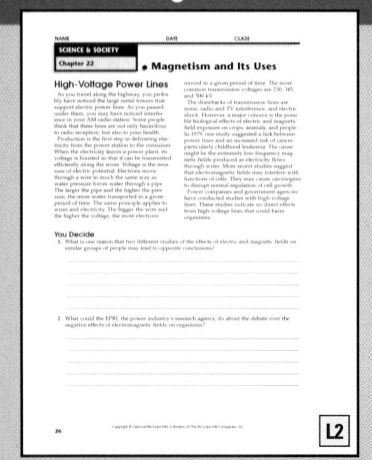

L2

Technology Integration

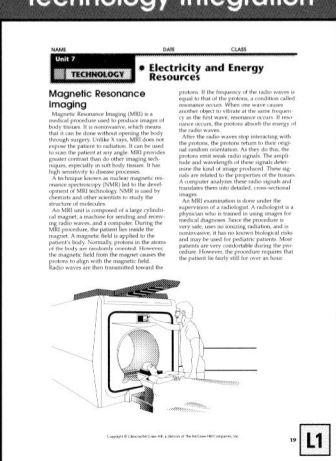

L1

Multicultural Connections

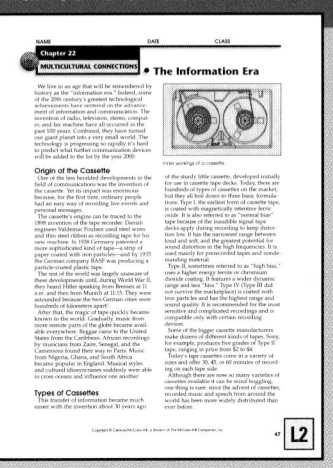

L2

Concept Mapping

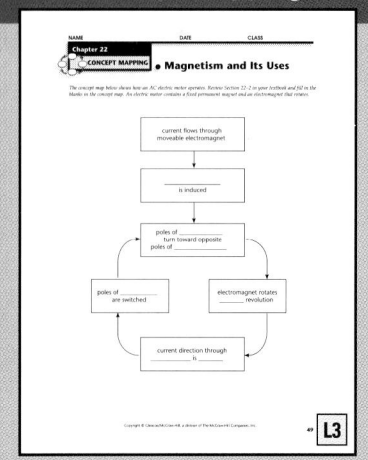

L3

Magnetism and Its Uses

CHAPTER OVERVIEW

Section 22-1 Students have direct experiences with magnets. A model for understanding magnetism is introduced.

Section 22-2 Electrical flow always produces magnetic effects. Electrical meters and electric motors have magnets as important components.

Section 22-3 Electrical flow can be produced (generated) by moving magnets in coils. Changing the direction of motion changes the direction of flow (AC).

Section 22-4 Science and Society Superconductors are synthetic materials that allow electrical flow without wasteful heat. Superconductors may have important uses in the future.

Chapter Vocabulary

magnetism
magnetic pole
magnetic field
magnetic domain
electromagnet
ammeter
voltmeter
electric motor
electromagnetic induction
generator
direct current (DC)
alternating current (AC)
transformer
superconductor

Theme Connection

Systems and Interactions Electrical flow causes magnetic effects, and moving magnets are used to produce electrical energy.

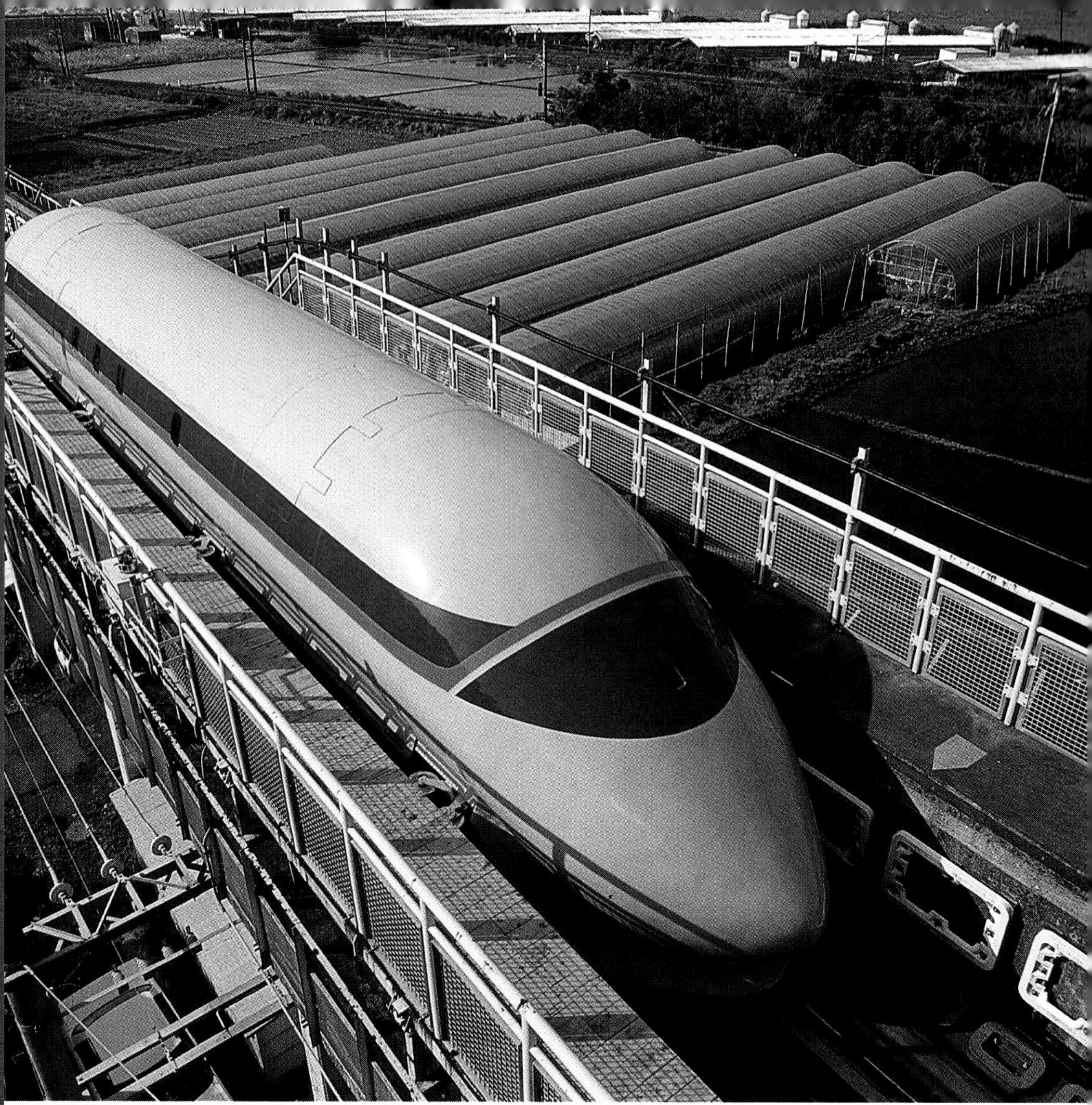

Previewing the Chapter

Section 22-1 Characteristics of Magnets
- ► Magnets
- ► A Giant Bar Magnet
- ► A Model for Magnetism

Section 22-2 Uses of Magnetic Fields
- ► Electromagnets
- ► Meters

- ► Electric Motors—From Electrical to Mechanical Energy

Section 22-3 Producing Electric Current
- ► Generators—From Mechanical to Electrical Energy
- ► Direct and Alternating Currents
- ► Transformers

Section 22-4 Science and Society Technology: Superconductivity

622

Chapter 22

Magnetism and Its Uses

What do a speeding train, stereo speakers, TV screens, and floppy disks for storing computer programs and information all have in common? Although they serve different purposes, the technology that makes each of these devices work involves the use of magnets. You can start learning about magnets by doing the following activity. Then, in the chapter that follows, you can continue your study of magnets and their uses.

EXPLORE ACTIVITY

Find out where a bar magnet has its greatest attraction for other magnetizable objects.

1. Suspend a paper clip from the north end of a bar magnet, while holding the bar magnet still. Add a second paper clip to the end of the first one. Continue adding paper clips to the chain until the magnet will hold no more. Record the number of clips held.
2. Repeat step 1, suspending the paper clips about 2 cm from the end of the magnet, and again, suspending the clips near the center of the magnet.

Observe: In your Science Journal, compare the number of clips suspended from each point. Infer where you think the magnetic field is strongest and explain why.

Previewing Science Skills

▶ In the Skill Builders, you will **hypothesize, compare and contrast,** and **map concepts.**

▶ In the Activities, you will **observe, predict, interpret,** and **formulate models.**

▶ In the MiniLABs, you will **hypothesize, observe,** and **interpret data.**

623

Prepare

Section Background

Demonstration magnets are often made of alloys. ALNICO is a common commercial magnetic material made from *al*uminum, *ni*ckel, *co*balt, and iron.

Preplanning

Round up a selection of magnets of various shapes and strengths. You will need to have at least one bar magnet per lab group for the MiniLAB.

1 Motivate

Bellringer

Before presenting the lesson, display **Section Focus Transparency 88** on the overhead projector. Assign the accompanying **Focus Activity** worksheet. L1 LEP

Tying to Previous Knowledge

Ask students to list the other forces that they have studied so far this year. Ask how magnetism is similar to the other forces. They should at least recall gravitation and the interaction between opposite electrical charges. Note that magnetic forces are related to electrical forces.

Science Words

- magnetism
- magnetic pole
- magnetic field
- magnetic domain

Objectives

- Describe the properties of magnets.
- Define the region of force around a magnet.
- Model magnetic behavior using domains.

Figure 22-1

Magnetite is a mineral with natural magnetic properties. *What kinds of materials are attracted to magnetite?*

624 Chapter 22 Magnetism and Its Uses

Magnets

More than 2000 years ago, the Greeks experimented with a mineral that pulled iron objects toward it. One end of this mineral always pointed north when suspended freely on a string. They described this mineral, shown in **Figure 22-1**, as being magnetic. Today, we know that magnetism is related to electricity. Together, magnetic and electric forces generate electricity and operate electric motors.

Magnetism is a property of matter in which there is a force of attraction or repulsion between unlike or like poles. The magnetic forces are strongest near the ends, or **magnetic poles,** of the magnets. All magnets have two magnetically opposite poles, north (N) and south (S). If a bar magnet is suspended so it turns freely, the north end will point north.

How Magnets Interact with Each Other

When you bring the north ends of two magnets close together, they repel each other. However, the north and south ends will attract. Like magnetic poles repel, and opposite magnetic poles attract. These forces decrease as the distance between the magnets increases.

Only a few materials show strong magnetic properties. Permanent magnets are made from materials such as iron, cobalt, and nickel, which retain their magnetic properties for a long time. Being near or rubbing against a magnet can cause paper clips and nails to become temporary magnets, but they lose their magnetic properties soon after they are separated from the other magnet.

Magnetic Fields

Hold one bar magnet still and move another magnet slowly around it. What happens? Depending on the positions of the magnets, you may feel attractive forces, repulsive forces, or no force at all. As shown in **Figure 22-2,** the **magnetic field** is the region around the magnet where magnetic forces act.

Program Resources

Reproducible Masters
Study Guide, p. 92 L1
Reinforcement, p. 92 L1
Enrichment, p. 92 L3
Activity Worksheets, p. 134 L1
Multicultural Connections, pp. 47-48
Lab Manual 45

Transparencies
Section Focus Transparency 88

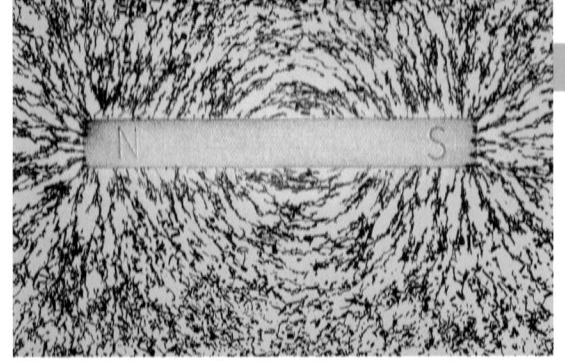

Figure 22-2

You can model the magnetic field by sprinkling iron filings around a bar magnet. The filings line up along the magnetic field lines.

 **A** The magnetic field around this bar magnet can be modeled with iron filings. *Where does the field appear the strongest?*

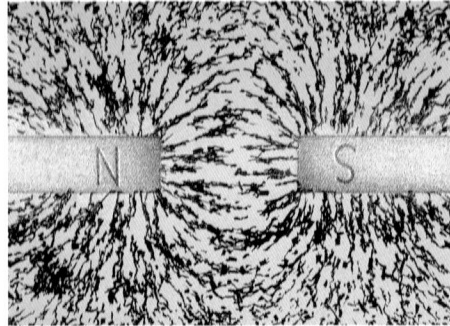

B Like poles repel.

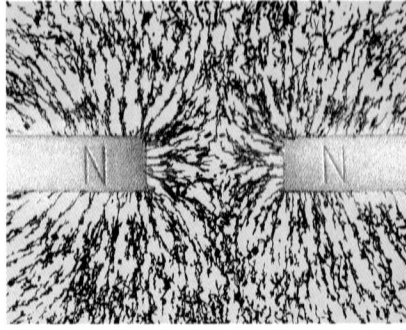

C Unlike poles attract.

A Giant Bar Magnet

Have you ever used a compass to find out which direction you were traveling? A compass contains a magnetic needle that freely rotates in a circle and always points north. This happens because Earth acts like a giant bar magnet surrounded by a magnetic field that extends beyond the atmosphere. The compass needle aligns with Earth's magnetic field lines. Any other direction can be determined on a compass if the northern direction is known.

Does it seem surprising that the north pole of a compass needle is attracted to Earth's north pole? The place we conveniently call Earth's magnetic north pole is actually a south pole. It follows that Earth's actual magnetic north pole is near the geographic south pole, as shown in **Figure 22-3.**

A Model for Magnetism

What happens if you hold an iron nail close to a refrigerator door and let go? It falls to the floor. Many magnets are made of iron, so why doesn't the nail stick to the refrigerator? You can make the nail behave like a magnet. Because you can't see magnetism, a model, or mental picture, will help you understand how.

 **INTEGRATION**
Earth Science

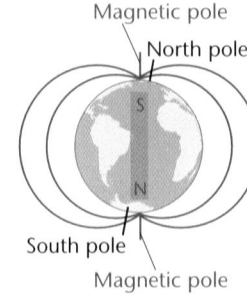

Figure 22-3

The geographic poles are not exactly in the same location as the magnetic poles. What we commonly call the north magnetic pole is actually a south pole.

 **INTEGRATION**
Earth Science

Geologists use devices called seismographs or geophones to measure the motion of Earth's crust. A geophone uses a suspended coil and a permanent magnet. Motion between the two causes an electrical flow which is an indication of the motion of Earth.

---

2 Teach

Demonstration

LS **Visual-Spatial**   Show the areas of intensity around the ends of bar magnets by scattering iron filings around a bar magnet on an overhead projector. Be sure to put an acetate sheet under the filings to avoid messes. By convention, magnetic field lines go from the north to the south pole.

Visual Learning

Figure 22-1 **What kinds of materials are attracted to magnetite?** *Materials made of iron, nickel, or cobalt are attracted to magnetite.*

Figure 22-2 **Where does the field appear the strongest?** *near the poles* **LEP** **LS**

GLENCOE TECHNOLOGY

Videodisc

STVS: Human Biology
Disc 7, Side 1
Detecting the Body's Magnetic Fields (Ch. 18)

CD-ROM

Physical Science CD-ROM
Have students perform the interactive explanation for Chapter 22 to reinforce important chapter concepts and thinking processes.

Problem Solving

Solve the Problem

1. Only the iron and steel will be separated.

2. Most materials do NOT have aligned magnetic domains.

Think Critically

Strong magnets can be used to separate the two types of cans.

Problem Solving

Magnet Metal Recycling

Have you ever seen a junkyard for old cars and trucks by the side of a highway? Most of the 10 million cars discarded by Americans each year go through a recycling process to reclaim some of the more than 600 different materials they contain. Magnets play an important role in auto recycling.

An average U.S. passenger car built in the early 1990s contains more than 700 kg of steel (an alloy of iron and other elements), 180 kg of iron, 100 kg of plastics, about 70 kg each of aluminum and fluids, 60 kg of rubber, 20 kg of copper, and more than 50 kg of other materials. After fluids and reusable parts are removed, the car is fed into a huge shredder. Large magnets are then used to separate some of the valuable materials from the shredded car. Inseparable or less valuable materials, commonly called fluff, are usually incinerated or taken to a landfill.

Solve the Problem:

1. Which materials from the list above would be pulled away from the shredded car by strong magnets?

2. Why do some materials behave magnetically and other materials do not?

Think Critically:

Many recycling centers collect aluminum beverage cans and steel food cans in the same containers. Aluminum and steel cannot be processed together, so why do you think they are collected together if they must later be separated?

626 Chapter 22 Magnetism and Its Uses

In iron, cobalt, nickel, and other magnetic materials, the magnetic field created by each atom exerts force on the other atoms. This causes groups of atoms to align their magnetic poles so all like poles are facing the same direction. These groups of atoms with aligned magnetic poles are called **magnetic domains. Figure 22-4A** shows how the domains are randomly arranged in an ordinary, unmagnetized nail. If a permanent magnet strokes the nail or comes near it, the domains rearrange to orient themselves in the direction of the nearby magnetic field, as in **Figure 22-4B.** The nail now acts as a magnet itself. When the external magnetic field is removed, the magnetic domains in the nail soon return to their random arrangement. For this reason, the nail is a temporary magnet. Even permanent magnets can lose some of their magnetic properties if they are dropped or heated. Their magnetic domains would be jostled out of alignment due to increased motion of the particles.

Can a pole be isolated?

What happens when a magnet is broken in two? Would you expect one piece to be a north pole and one piece to be a south pole? Look again at the domain models in **Figure 22-4.** Because each magnet is actually made of many aligned smaller magnets, even the smallest pieces have both a north and south pole.

Figure 22-4

Groups of atoms are arranged in domains.

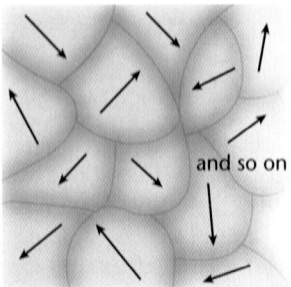
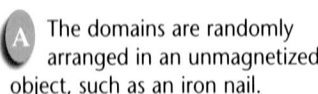

A The domains are randomly arranged in an unmagnetized object, such as an iron nail.

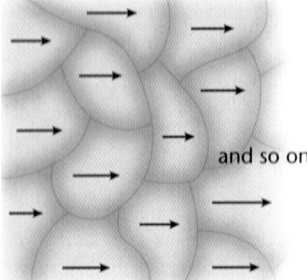

B In a magnetized object, the domains are aligned.

Although the magnetic domain model explains many observations about magnets, some questions are still unanswered. Scientists have, however, found many uses and applications for magnetism. As you read the next section, you will find out more about these applications.

Section Wrap-up

Review

1. Describe what happens when you bring two magnetic poles together.

2. What is a magnetic field, and where is it strongest?

3. **Think Critically:** Explain in terms of magnetic domains what happens when another magnet is brought close to a suspended magnet.

Skill Builder
Hypothesizing

Suppose you allowed your younger brother or sister to play with a strong bar magnet. When you got it back, it was barely magnetic. Write a hypothesis to explain what might have happened to your magnet. How could you fix it? If you need help, refer to Hypothesizing in the **Skill Handbook.**

Science Journal

In your Science Journal, make a list of all of the uses you can think of for magnets. Write a paragraph describing what these magnets seem to have in common.

22-1 Characteristics of Magnets **627**

MiniLAB

What interferes with a magnet?

Procedure
1. Clamp a bar magnet to a ring stand. Tie a thread around one end of a paper clip, and stick the paper clip to one pole of the magnet.
2. Anchor the other end of the thread under a mass on the table. Slowly pull the thread until the paper clip is suspended below the magnet.
3. Without touching the paper clip, slip some paper between the magnet and the paper clip. Does the paper cause the clip to fall?
4. Try other materials.

Analysis
1. What materials interfered with the magnetic field? What materials did not?

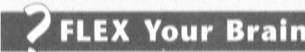

Prepare

Section Background

Interestingly, Oersted's discovery was accidental, as many discoveries are. While doing a demonstration, he observed that a current moving through a wire produced a response in a nearby compass. Oersted was a physics professor.

Preplanning

• Gather the following materials for Activity 22-1: cardboard or plastic tubes, bar (or horseshoe) magnets, 30 m of insulated wire, galvanometers or milliammeters.

• Collect wires, compasses, lamps, and batteries for the MiniLAB.

1 Motivate

Bellringer

Before presenting the lesson, display **Section Focus Transparency 89** on the overhead projector. Assign the accompanying **Focus Activity** worksheet. `L1` `LEP`

Demonstration

Visual-Spatial Begin with Oersted's experiment. Make about 30 loops of wire and connect it to the positive and negative ends of a battery. Set it near a compass. Then reverse the connections—show how the magnetic field reverses.

Science Words

> electromagnet
> ammeter
> voltmeter
> electric motor

Objectives

• Explain the magnetic effects of a current in a wire.
• Compare and contrast ammeters and voltmeters.
• Describe the function of an electric motor.

Figure 22-5

Magnetic fields form around any wire that is conducting current.

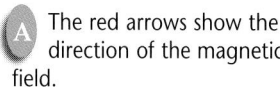
A The red arrows show the direction of the magnetic field.

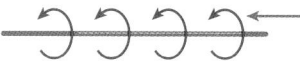
B An electromagnet is made of a coil of wire around a soft iron core.

Electromagnets

In 1820, Hans Christian Oersted, a Danish physics teacher, observed that a current moving through a wire moved the needle on a nearby compass. When the current was reversed, the compass needle was deflected in the opposite direction. These magnetic effects ceased when the current in the wire stopped. Therefore, Oersted hypothesized, the electric current must produce a magnetic field around the wire, the direction of which changes with the direction of the current.

Figure 22-5A shows magnetic field lines around wires with current in them. **Figure 22-5B** shows that adding turns of the wire to make a coil causes more overlapping of the magnetic field lines, and, as a result, the magnetic field grows stronger. When an iron core is inserted into such a coil and a current is passed through the coil, a strong temporary magnet called an **electromagnet** is formed. The iron core becomes a magnet, and its magnetic field is aligned with that of the coil carrying the current. One end of the coil acts as the north pole and the other end as the south pole. The strength of the magnetic field can be increased by adding more turns to the wire coil and by increasing the current passing through the wire.

Music to Your Ears

How does an electrical signal from a CD player become sound you can hear? Electromagnets are used in loudspeakers to change electrical energy to mechanical energy that vibrates the speakers to produce sound. Electromagnets operate doorbells and loudspeakers and lift large metal objects in construction machines. They change electrical energy to mechanical energy to do work and can be turned on and off by controlling the flow of current through the coil. What would happen if the magnetic field in a doorbell were permanent instead of temporary?

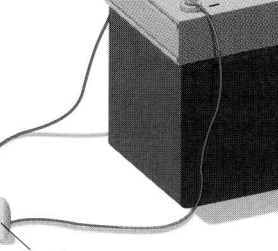
Battery

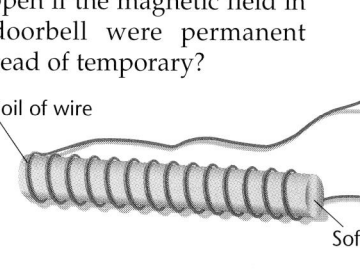
Coil of wire

Soft iron core

Program Resources

Reproducible Masters
Study Guide, p. 93 `L1`
Reinforcement, p. 93 `L1`
Enrichment, p. 93 `L3`
Activity Worksheets, pp. 130-131, 135 `L1`
Concept Mapping, p. 49
Technology Integration, p. 19
Lab Manual 46

Transparencies
Section Focus Transparency 89 `L1`
Teaching Transparency 43 `L1`

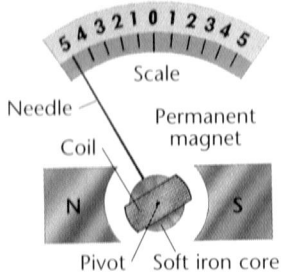

Figure 22-6

A galvanometer uses an electromagnet to detect electric currents. *What causes the needle to move?*

Meters

Because electromagnets are sensitive to electrical currents, they can detect electric current. An instrument used to detect currents is called a galvanometer, shown in **Figure 22-6.** It is made of a coil of wire connected to a circuit and suspended so it can rotate in the magnetic field of a permanent magnet. When current flows through the coil, the magnetic force causes the coil to rotate against a spring. A needle attached to the coil turns with it to provide a reading on a scale.

MRI Image of a Human Brain

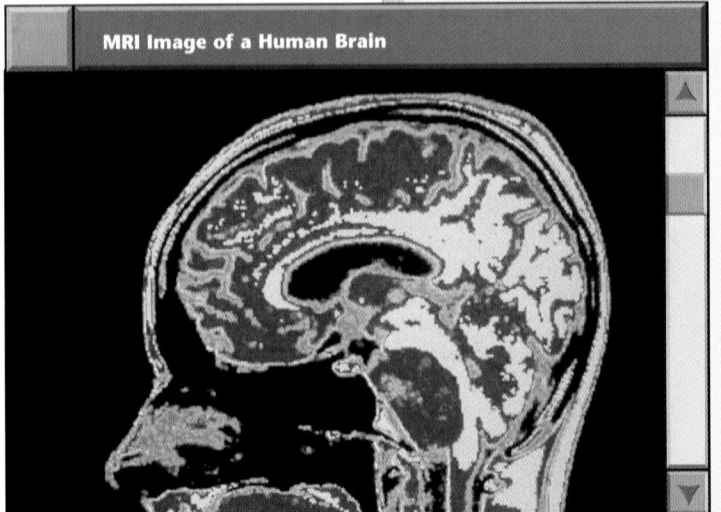

USING TECHNOLOGY

MRI—The Inside Story

Did you know you have billions and billions of nuclei in your body that will respond to an applied magnetic field? Doctors can safely use this property to develop clear, three-dimensional images of overlapping structures that can aid in the identification of unhealthy tissue.

In a technique called magnetic resonance imaging, or MRI, a person is placed in the center of a chamber that contains large coils capable of creating a stable magnetic field. This magnetic field causes the nuclei of certain atoms—most notably hydrogen atoms—to align parallel to the magnetic field. A pulse of radio-frequency electromagnetic waves is sent into the region of the body to be examined. If this frequency can be absorbed by atoms in the tissue, the nuclei respond by flipping to an antiparallel alignment.

This change in magnetic alignment of the nuclei provides information a computer can use to create a three-dimensional image of the desired area of the body. MRI can be used to detect such problems as tumors, skeletal injuries, arthritic swelling, and blocked blood vessels.

inter**NET**
CONNECTION

Visit the Chapter 22 Internet Connection at Glencoe Online Science, **www.glencoe.com/ sec/science/physical,** for a link to more information about MRI technology.

22-2 Uses of Magnetic Fields **629**

Use **Study Guide,** p. 93, as you teach this lesson.

Use **Teaching Transparency 43** as you teach this lesson.

2 Teach

Student Text Question

What would happen if the magnetic field in a doorbell were permanent instead of temporary? *If the magnetic field inside a doorbell were permanent, then the clapper would not vibrate back and forth.*

Visual Learning

Figure 22-6 What causes the needle to move? *When current flows through the coil, the magnetic field causes the coil to rotate and move a needle attached to it.* **LEP**

USING TECHNOLOGY

Students may be familiar with the name *nuclear magnetic resonance* (NMR) rather than the newer magnetic resonance imaging (MRI).

GLENCOE TECHNOLOGY

Videodisc

Glencoe Physical Science Interactive
Side 2, Lesson 8
Motors

35390-35717

35719-36031

36033-36297

MiniLAB

Purpose

LS Interpersonal Students will observe the connection between electricity and magnetism. **L1 COOP LEARN**

Materials

1-m length of wire with both ends stripped, compass, small lamp with socket, and a battery or low-voltage power source for each lab group

Answers to Questions

3. The compass needle will deflect when a current flows in the wire.
4. The needle will remain pointed in the same direction.
5. As the compass is moved farther from the wire, the needle will slowly move back toward Earth's north pole.

Analysis

1. The compass needle is magnetic and will point to the magnetic pole.
2. The electrical flow in the wire produced a magnetic field that caused the needle to deflect.

Assessment

Oral Ask: **Suppose that you reverse the connections to the wire. What would happen to the needle? Why?** *Reversing the direction of flow will reverse the direction of the magnetic field, which will cause the needle to deflect in the opposite direction.* Use the Performance Task Assessment List for Making Observations and Inferences in **PASC**, p. 17. **P**

Activity Worksheets, pp. 5, 135

MiniLAB

How can you observe the relationship between electricity and magnetism?

Procedure

1. Attach an insulated wire to each side of a bulb. Obtain a 6-V dry cell and a compass.
2. Set the compass on top of one of the two connecting wires. Before connecting the dry cell, observe the direction the compass needle is pointing.
3. While watching the compass, complete the circuit. The bulb should light if the circuit is complete. What happened to the compass needle?
4. Slowly rotate the compass, keeping it over the wire. Does the needle change directions or is it still pointing the same way?
5. Gradually move the compass away from the current-carrying wire. What do you notice?
6. While watching the compass, disconnect the circuit. Record your observations.

Analysis

1. What direction does the needle point when the circuit is not connected? What causes it to be oriented in this direction?
2. What can you infer about the relationship between electricity and magnetism from your observations?

Figure 22-7

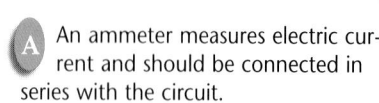

Ammeters and voltmeters are wired into circuits differently.

A An ammeter measures electric current and should be connected in series with the circuit.

B A voltmeter measures potential difference across part of a circuit. *How is the voltmeter connected in this diagram?*

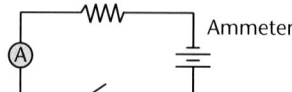 Ammeter

 Voltmeter

A galvanometer can be calibrated to measure current or electrical potential, depending on whether it will be used as an ammeter or a voltmeter. **Ammeters** measure the electrical current passing through a circuit in amperes and should be connected in series with the circuit, **Figure 22-7A.**

Voltmeters measure the potential difference of a circuit in volts. Unlike ammeters, voltmeters should be placed in parallel across a part of a circuit, as in **Figure 22-7B.** The reading in volts depends on the potential difference across the voltmeter.

Electric Motors—From Electrical to Mechanical Energy

Do you ever use an electric fan to keep cool? Your fan uses an **electric motor,** a device that changes electrical energy into mechanical energy, to turn the blades. The turning blades push air toward you so your skin feels cooler.

Like galvanometers, electric motors contain an electromagnet that is free to rotate. It rotates between opposite poles of a permanent, fixed magnet. When a current flows through the movable electromagnet, a magnetic field is induced. This causes enough attraction and repulsion with the permanent magnet to force the coil to turn. But the rotation would stop once the magnetic fields aligned. How could you make the coil turn again? Study **Figure 22-8** to see how an electric motor works.

As you will learn in the next section, current supplied by the potential difference in a wall outlet reverses itself at a regular rate, and many devices that operate on household current do not require a commutator.

630 Chapter 22 Magnetism and Its Uses

Visual Learning

Figure 22-7B **How is the voltmeter connected in this diagram?** *The voltmeter is wired in parallel.* **LEP LS**

Using Computers

Devices may include refrigerators, air conditioners, electric yard tools (trimmers, lawn mowers), table saws, washers, and dryers. Caution students to examine motorized devices only under adult supervision. Students should never try to take apart devices to see how they work.

Figure 22-8

Electric motors use both a permanent magnet and an electromagnet.

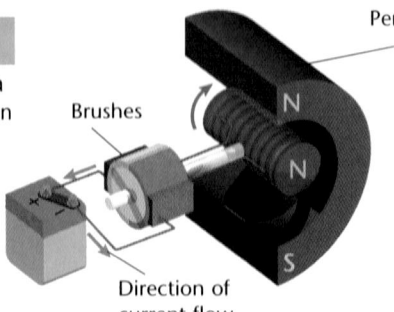

Permanent magnet

Brushes

Direction of current flow

A When the battery causes a current to flow through the coil, a magnetic field is induced. Attraction and repulsion between the coil and the permanent magnet cause the coil to rotate.

B To make the coil continue to spin, the direction of the current through the coil must be reversed after each half revolution. A commutator is a reversing switch that rotates with the electromagnet and alternately contacts the positive and negative terminals of the battery. The reversing current causes the poles of the rotating electromagnet to switch and rotate toward the opposite pole of the permanent magnet.

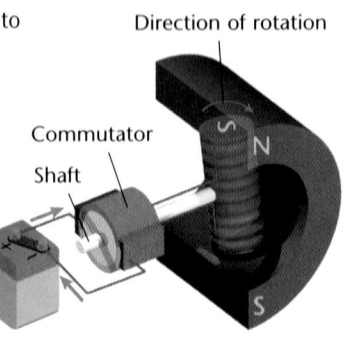

Direction of rotation

Commutator

Shaft

C This process is repeated many times each second.

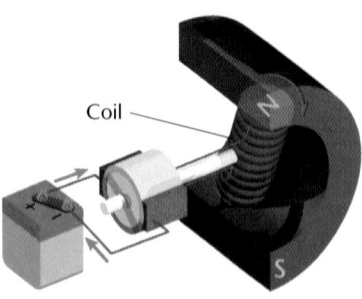

Coil

Section Wrap-up

Review

1. Does a straight wire or a looped wire have a stronger magnetic field when both carry the same amount of current? Explain your answer.

2. Compare and contrast the wiring of a voltmeter and an ammeter.

3. **Think Critically:** Trace the energy conversion from a coal-burning power plant, through an electric motor, and into a CD player.

Skill Builder

Comparing and Contrasting

Compare and contrast ammeters and voltmeters. Discuss what they measure and how they are connected to the rest of the circuit. If you need help, refer to Comparing and Contrasting in the **Skill Handbook.**

> **Using Computers**
>
> **Spreadsheet** Take an inventory of all the devices in your home or school that make use of an electric motor. Organize your inventory using a data base or spreadsheet indicating the name of the device, the place you found it, the voltage source used, and which parts the motor causes to move.

Skill Builder

Ammeters are designed to measure current and should be connected in series. Voltmeters measure potential difference and should be connected in parallel.

Assessment

Process Make a circuit drawing on the overhead of a battery, a lamp, and a switch. Ask students to copy the drawing into their Science Journals and to indicate where the connections should be made to measure the current. Students should indicate that the circuit must be broken and the ammeter should be placed in series with the other components. Point out that the ammeter could be placed at *any* point in the circuit and still give the correct reading. Use the Performance Task Assessment List for Scientific Drawing in **PASC**, p. 55. **P**

Electricity and Magnetism

Can you make a magnet by passing a current through a coil of wire? This discovery led to the development of the electric motor. If you can use electricity to make a magnet, do you think it might be possible to use a magnet to help generate an electric current? Try this activity to find out.

Purpose

IS **Interpersonal** Students will produce electric currents by moving magnets in coils of wire. **L1** **COOP LEARN**

Process Skills

interpreting data, drawing conclusions, communicating, observing and inferring, comparing and contrasting, recognizing cause and effect, making models

Time

45 minutes

Activity Worksheets, pp. 5, 130-131

Teaching Strategies

Alternate Materials Students may use one end of a horse-shoe magnet in place of the bar magnet.

Troubleshooting Be sure that the galvanometer is on the most sensitive scale and that students form a complete circuit with the coil and meter.

Answers to Questions

1. The needle will move when the pole is moving near the coil. The needle moves farther when the magnet is moved inside of the coil.

2. Faster motion produces more deflection of the needle.

3. Moving the magnet *in* produces a deflection in one direction. Moving the magnet *out* reverses the deflection.

4. Fewer coils produce a weaker (smaller) effect.

5. The cardboard tube does not affect the results because it is not magnetic.

Problem

How can a magnet cause electric current?

Materials

- thin, flexible insulated wire
- cardboard tube
- scissors
- galvanometer or milliammeter
- bar magnet

Procedure

1. Leaving about 15 cm for a lead at the end of the wire, make a coil of about 20 turns by wrapping the wire around a cardboard tube. Remove the tube from the coil.

2. Carefully, use the scissors to scrape about 2 cm of insulation off both ends of the wire.

3. Connect the ends of the wire to a galvanometer or milliammeter. Take the reading on your meter.

4. While closely watching the meter, insert one end of the bar magnet into the coil. Then, pull it out of the coil. Repeat this motion and record any observations or measurements you can make.

5. Vary the speed with which you move the magnet. Record your measurements from the meter.

6. Have one partner hold the magnet still while another partner moves the coil over the magnet until it is inside the coil. Record your observations.

7. While watching the meter, move the bar magnet in different directions around the outside of the coil. What do you observe?

Analyze

1. What information does the meter tell you in steps 4-7? Which circumstances that you tested generated the greatest current?

2. How does the speed at which you move the magnet influence the amount of current produced?

3. Does the current you generate by moving the magnet always move in the same direction? What observations help you with this answer?

Conclude and Apply

4. **Predict** what effect using fewer coils would have on the current produced by moving the magnet in and out of the center of the coil. Try it to check your prediction.

5. **Infer** whether a current would have been generated if the cardboard tube were left in the coil. Why or why not? Try it.

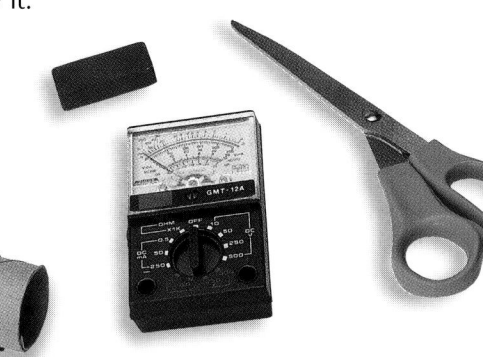

Assessment

Oral Ask students to predict what will happen if the magnet is stationary and the coil is moved back and forth. The results are the same. It doesn't matter which is moving as long as they move compared to each other. Use the Performance Task Assessment List for Making Observations and Inferences in **PASC**, p. 17. **P**

Producing Electric Current 22•3

Generators—From Mechanical to Electrical Energy

Once scientists knew that magnetism could be produced from electric currents, they tried to produce an electric current using magnets. Working independently in 1831, a British scientist, Michael Faraday, and an American scientist, Joseph Henry, found that moving a wire through a magnetic field induced an electric current in that wire. **Electromagnetic induction** is the process by which moving a wire through a magnetic field produces a current. Moving a magnet in and out of a coil of wire also produces a current. This important discovery led to numerous applications.

Where Household Electricity Comes From

What produces the electricity that comes to your home and school? Most of the electricity you use each day was electromagnetically induced in generators. A **generator** produces electric current by rotating a loop of wire in a magnetic field. The wire loop is connected to a source of mechanical energy and placed between the poles of a magnet, as shown in **Figure 22-9**. The design of the generator is much like that of an electric motor, except the wire loop is rotated by external forces. When it rotates, an electric current is produced. This is the opposite of how a motor acts. In a generator, the wire crosses through the magnetic lines of force as it rotates, causing electrons to move along the wire in one direction. After one-half revolution of the wire loop, each side of the coil passes near the opposite pole of the

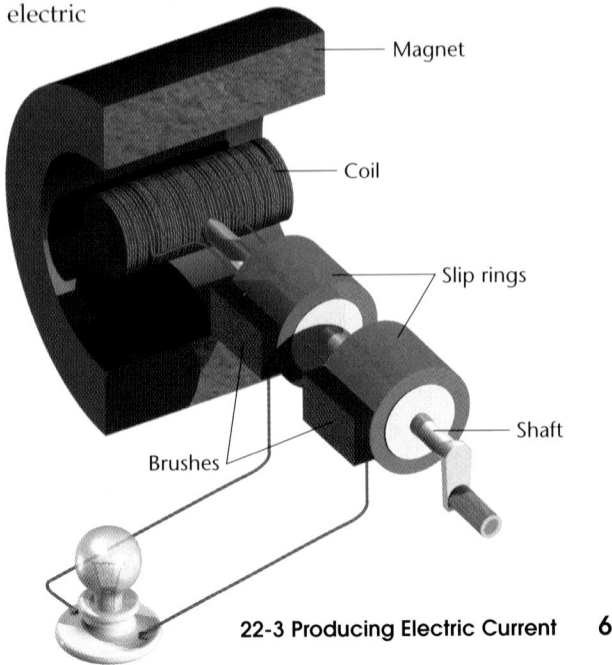

Figure 22-9

Electric current is produced in a generator when a loop of wire is rotated in a magnetic field. *How does a generator differ from a motor?*

Labels: Magnet, Coil, Slip rings, Shaft, Brushes

22-3 Producing Electric Current   **633**

Science Words

electromagnetic induction
generator
direct current (DC)
alternating current (AC)
transformer

Objectives

- Describe how a generator produces an electric current using electromagnetic induction.
- Distinguish between alternating current and direct current.
- Explain how a transformer can step up or step down the voltage of an alternating current.

Section 22•3

Prepare

Section Background

Throughout the world, electrical energy is produced in the same manner. Giant coils of wire are rotated in strong magnetic fields.

Preplanning

For Activity 22-2, each lab group will require two lengths of wire (1.5 m each), drinking straw, minilamp with socket, 6-volt battery, paper clip, large nail, scissors, and a low-voltage AC power supply.

1 Motivate

Bellringer

Before presenting the lesson, display **Section Focus Transparency 90** on the overhead projector. Assign the accompanying **Focus Activity** worksheet. L1 LEP

Tying to Previous Knowledge

Ask students where the energy is generated that they use in their homes. What is the energy that runs the turbines? If the class is unsure, then ask them to ask their parents.

2 Teach

Visual Learning

Figure 22-9 How does a generator differ from a motor? *The motor and generator have exactly the same parts. In a generator, the mechanical energy produces electrical energy. In a motor, electrical energy is changed into mechanical energy.* LEP LS

Program Resources

📁 Reproducible Masters
Study Guide, p. 94 L1
Reinforcement, p. 94 L1
Enrichment, p. 94 L3
Science and Society Integration, p. 26
Activity Worksheets, pp. 132-133 L1
Science Integration Activities, pp. 43-44
Critical Thinking/Problem Solving, p. 28

📦 Transparencies
Section Focus Transparency 90 L1
Science Integration Transparency 22
Teaching Transparency 44 L1

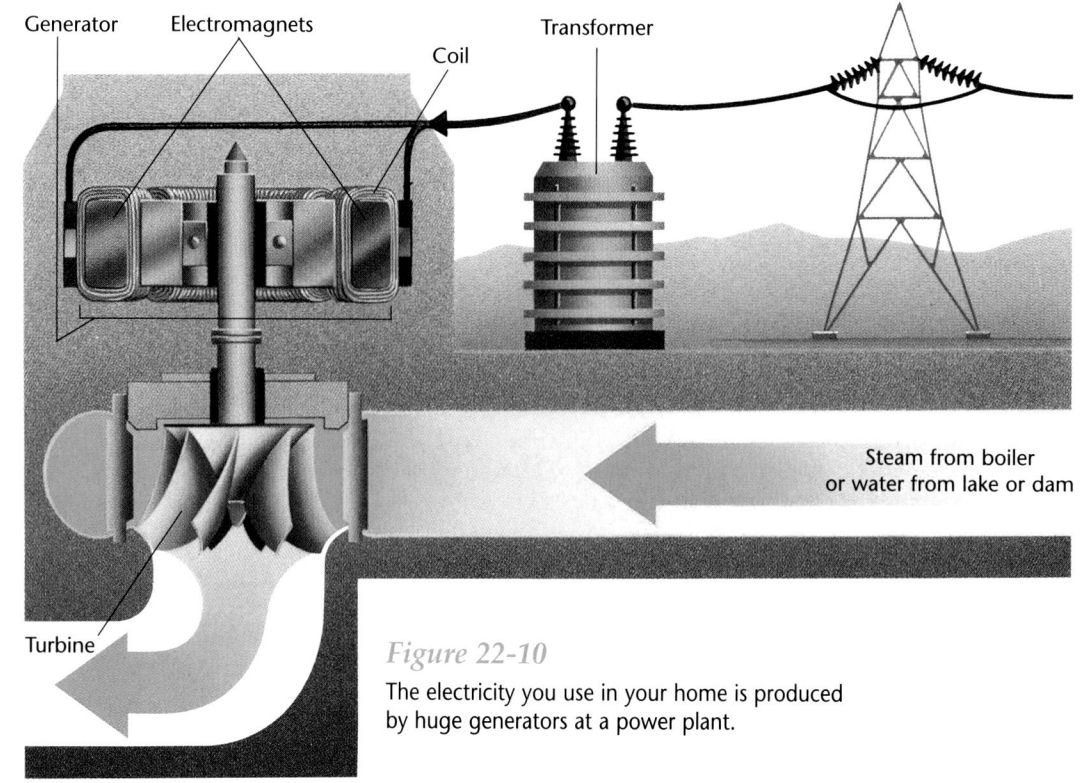

Figure 22-10

The electricity you use in your home is produced by huge generators at a power plant.

CONNECT TO
EARTH SCIENCE

Study the generator diagrammed above. What role does water play in generating electricity? *Infer* what might be some advantages and disadvantages of using water from a dam to push a turbine.

magnet, causing the current to change direction. As a result, the direction of the current changes twice with each revolution. The rotation speed of generators is regulated so the current always changes direction with the same frequency.

Electricity in Your Home

Do you have a generator in your home that supplies all the electricity you need to watch television or wash your clothes? Probably not. You get your electricity from huge generators at a power plant, like the one in **Figure 22-10**. These generators are more complex than the ones discussed here. The electromagnets in these generators are made of many loops of wire wrapped around iron cores. A source of mechanical energy is needed to rotate the loop in the generator. The loop is usually connected to a turbine, which is a large wheel that rotates when pushed by water, wind, or steam. Potential energy released by the burning of fossil fuels or from nuclear reactions can heat water to produce steam. The thermal energy of the steam changes to mechanical energy as it pushes the turbine. The generator then changes this mechanical energy into an electric current that is easily conducted to your home.

634 Chapter 22 Magnetism and Its Uses

Community Connection

Use of Electricity Call your local electric company and ask to have a representative come to speak with your classes about electricity use in your community. Be sure to ask for a speaker who has general knowledge about the local power network. If possible, get your students' questions to the speaker before the day of the visit.

Direct and Alternating Currents

Do you have a portable stereo like the one in **Figure 22-11** that operates either on batteries or on the electric current in your home? Is the electric current from a generator the same as the current produced by a dry cell? Both devices cause the electrons to move through a wire and can operate appliances. However, the currents produced by these electric sources are not the same.

When you use a dry-cell battery to run your stereo, you are using direct current. **Direct current (DC)** flows only in one direction through a wire. When you plug a CD player or any appliance into the wall outlet, you are using alternating current. **Alternating current (AC)** reverses its direction in a regular pattern. In North America, generators produce alternating current at a frequency of 60 cycles per second (60 Hz). Because current in a generator changes direction twice during each rotation of the shaft, 60-Hz alternating current changes direction 120 times each second.

Figure 22-11

Some electrical devices, such as the portable stereo shown, are designed to operate on AC or DC. *Why wouldn't you want to use dry-cell batteries to operate it all of the time?*

Transformers

The alternating current traveling through power lines is at an extremely high voltage. Before alternating current from the power plant can enter your home, its voltage must be decreased. The current must flow through a device called a transformer to decrease the voltage. A **transformer** can increase or decrease the voltage of an alternating current. The operation of a transformer involves principles of both electromagnetism and electromagnetic induction.

22-3 Producing Electric Current   **635**

Enrichment

IS Logical-Mathematical Compare the cost of using dry cells and house current. A typical 9-volt dry cell costs $1.75 and delivers about 0.16 kWh of energy. **What is the cost of energy per kilowatt-hour?** *Cost is $10.94 per kilowatt-hour.* L1

IS Logical-Mathematical Have students ask to see an electric bill for their home. Have them copy the total cost and the number of energy units used (kWh), then find the cost of electricity per kilowatt-hour. *Cost will be less than $0.20 per kilowatt-hour.* L1

Content Background

- Emphasize that transformers are only effective at stepping up or stepping down alternating currents. The current must alternate to change the direction of the magnetic field and induce a current in the secondary coil.

- Because energy is conserved, the power produced in the primary coil of a transformer is equal to that produced in the secondary coil. That is,

$$P_{primary} = P_{secondary}$$
$$(V \times I_{primary}) = (V \times I_{secondary})$$

In a step-up transformer, the voltage of the alternating current produced in the secondary coil is much greater than that supplied to the primary coil. Thus, the magnitude of the alternating current produced in the secondary coil is much less than that supplied to the primary coil.

Use **Science Integration Activities,** pp. 43-44, as you teach this lesson.

  Use **Teaching Transparency 44** as you teach this lesson.

3 Assess

Check for Understanding

636

Figure 22-12

Transformers make it possible to transport current to homes at very high voltage.

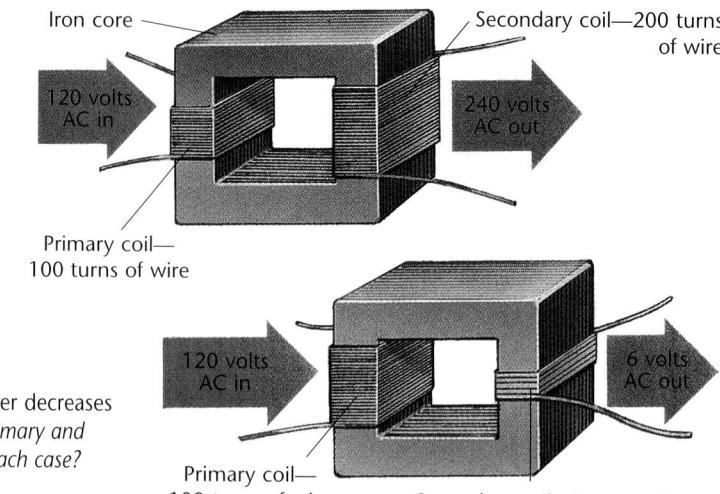

A A step-up transformer increases voltage.

B A step-down transformer decreases voltage. *How do the primary and secondary coils compare in each case?*

Stepping Up and Stepping Down

A simple transformer is made of two coils of wire called the primary and secondary coils. These coils are wrapped around an iron core, as shown in **Figure 22-12.** As an alternating current passes through the primary coil, the iron core becomes an electromagnet. Because the current changes direction many times each second, the magnetic field also changes its direction and induces an alternating current in the secondary coil. If the secondary coil has more turns of wire than the primary coil, then it increases, or steps up, voltage.

A transformer reduces the voltage of the alternating current to 120 volts before it enters your home. A transformer that reduces voltage is called a step-down transformer. **Figure 22-12B** shows how the output voltage of a transformer will be decreased if the number of turns in the secondary coil is less than the number of turns in the primary coil. Suppose the secondary coil of a transformer has half as many turns as the primary coil. An input voltage of 120 V will yield an output voltage of 60 V. Step-down transformers, like the one shown in **Figure 22-13,** allow you to operate devices such as CD players, model trains, and doorbells with 120-V household current.

Transmitting Alternating Current

Power plants commonly produce alternating current because its voltage can be increased or decreased with transformers. To transmit alternating current efficiently over long distances, power plants increase their voltages to high values. In a step-up transformer, **Figure 22-12A,** the output voltage is greater than the input voltage because the secondary coil has more turns than the primary coil. For example, the secondary

---

Theme Connection

Energy On the surface it seems that a step-up transformer gives you free electricity. Review the law of conservation of energy. The same energy is transferred from the primary to the secondary coil, with no gain in energy. In fact, some energy may be lost as low-temperature thermal energy to the iron core of the transformer.

coil of a step-up transformer at a generating plant may have 100 times the number of turns as the primary coil. That means that an input voltage of 2000 V would increase to 200 000 V—high voltage indeed!

Think back over this section. Could you describe how electromagnetic induction, generators, alternating current, and transformers all affect your CD player? See if you can recall the series of steps in which AC current is produced, transported, and delivered to your home in a form that you can safely use.

Figure 22-13

Transformers are found in most residential neighborhoods.

Section Wrap-up

Review

1. How does a generator use electromagnetic induction to produce a current?

2. A transformer in a neon sign contains 20 turns in the primary coil and 80 turns in the secondary coil. Which is greater—the output voltage or the input voltage?

3. Compare and contrast alternating current and direct current.

4. **Think Critically:** Explain why a transformer can't be used to step up the voltage in a direct current.

Skill Builder
Concept Mapping
Prepare an events chain concept map to show how electricity is produced by a generator, as discussed on pages 633 and 634. If you need help, refer to Concept Mapping in the **Skill Handbook**.

USING MATH

How would a transformer convert 120-V AC input current into a 6-V output in a portable CD player? Explain the device and use math to determine the ratio of input coils to output coils.

Activity 22-2

Purpose
IS **Interpersonal** Students will construct a simple transformer and compare the energy-carrying abilities of magnetic fields. **COOP LEARN** **L2**

Process Skills
observing and interpreting data, communicating, sequencing, making and using tables, comparing and contrasting, recognizing cause and effect, forming a hypothesis, making models

Time
30 to 40 minutes

Materials
Each lab group will require 2 lengths of wire (1.5 m each), drinking straw, mini-lamp with socket, 6-volt battery, paper clip, large nail, scissors, and a low-voltage AC power supply.

Safety Precautions
The output voltage of the AC supply should not exceed 7 volts. Remind the students that unauthorized experiments should not be performed in this activity.

Possible Hypotheses
Most students will realize that the output can be increased by increasing the number of turns on the straw. Most students will *not* realize that the transformer will not work with DC.

📁 **Activity Worksheets,** pp. 5, 132-133

Activity 22-2

Design Your Own Experiment
Trying Transformers

How do transformers illustrate the connection between electricity and magnetism? Review the design of transformers in the previous sections. Build and experiment with the transformer described below to find out how current can pass between unconnected coils of wire.

PREPARATION

Problem
How does a current move between the unconnected coils of wire in a transformer?

Form a Hypothesis
Write a hypothesis about how you can build a simple transformer and what kind of current can pass through a transformer.

Objectives
- Design and build a transformer.
- Observe the effectiveness of a transformer at transferring direct and alternating current.

Possible Materials
- dry-cell battery, 6-V
- AC power supply, low-voltage
- lightbulb, low-voltage, with holder
- insulated wire, 32-gauge
- large nail

- soda straw, cut to length of nail
- scissors
- paper clip

Safety Precautions

Make sure the output voltage of the AC supply does not exceed 7 volts. Do not perform any experiments unless your teacher has approved them.

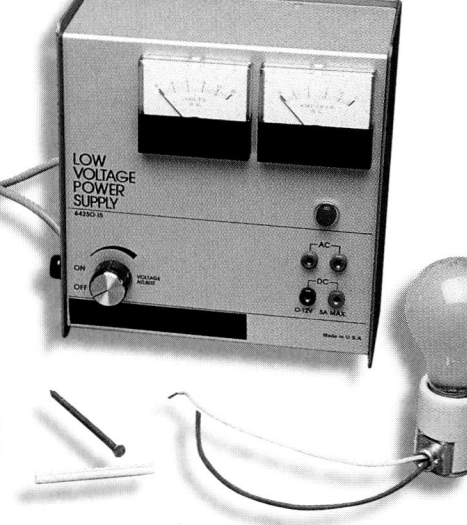

PLAN THE EXPERIMENT

1. In your group, agree upon and write a hypothesis.
2. Determine and write out the steps you will need to take to test your hypothesis.
3. Make a list of the materials you will need and gather them.

Check the Plan

1. How will you construct your transformer? You can make one version by inserting the nail in the straw. Wrap about 300 turns of insulated wire in a tight coil at one end of the straw. Use another piece of wire to make a similar coil at the other end of the straw. Scrape the insulation off the wire ends with scissors to give you electrical contacts.

2. Be sure you place a light in the circuit so you can detect whether the transformer is working and to provide a resistance in the circuit. Where will you place it?
3. How will you test the transformer with both the dry-cell battery and the AC power supply?
4. Where will you make a switch, or a simple contact point, to easily turn the circuit on and off?
5. *Make sure your teacher approves your plan and that you have included any changes suggested in the plan.*

DO THE EXPERIMENT

1. Carry out the experiment as planned.
2. While the experiment is going on, write down any observations you make in your Science Journal.

Analyze and Apply

1. Did both sources of potential difference allow the transformer to function? In your Science Journal, explain your observations. **Compare and contrast** the magnetism produced by the DC current in battery-driven circuits with

the AC current in circuits with AC generators.

2. The two coils of insulated wire in this circuit are not physically in contact with each other. How does this circuit function?
3. **Make a model** of a step-down transformer using the same materials. Where might you find a step-down transformer in your household?

Go Further

Observe the light as you slide the nail out of the straw. What happens? The nail is similar to a standard part in most transformers. Infer what the role of the nail is in this design.

22-3 Producing Electric Current   **639**

Go Further

As you slide the nail out, the magnetic field inside the coil decreases and the light goes out. The nail acts as the iron core in a transformer.

Assessment

Portfolio Have students design electrical systems for remote locations and describe how this electricity could change the way people live. Use the Performance Task Assessment List for Scientific Drawing in **PASC**, p. 55. **P**

PLAN THE EXPERIMENT

Possible Procedures

Students will construct a device similar to the one shown in the photograph. Some groups might try putting a different number of turns on the coils.

Teaching Strategies

Troubleshooting Remind students to leave both ends loose when they make the coils so that they will be able to make connections to the lamp or power supply.

DO THE EXPERIMENT

Expected Outcomes

- The light will be connected to one coil. (Check the Plan, no. 2)
- The power supply (or battery) will be connected to the other coil. (Check the Plan, no. 3)
- The connection to the battery will serve as the switch. (Check the Plan, no. 4)

Analyze and Apply

1. The AC worked fine. The nail became a magnet and the lamp went on. With DC, the nail became a strong magnet, but the lamp did not light.
2. The two coils are connected by the magnetic field in the nail. The flow from the battery causes the nail to become magnetic.
3. A step-down transformer can be made by removing most of the wraps on the lamp coil. You might find a step-down transformer where the current enters the house.

Prepare

Section Background

The levitation of a magnet over a superconductor can be explained by the Meissner effect. The current carried by a superconductor must move on the surface.

1 Motivate

Bellringer

Before presenting the lesson, display **Section Focus Transparency 91** on the overhead projector. Assign the accompanying **Focus Activity** worksheet. L1 LEP

2 Teach

Visual Learning

Figure 22-14 **Can you think of any practical use for magnetic levitation?** *Magnetically levitated trains would not touch the rails, and little energy would be lost to friction.* LEP LS

CONNECT TO

CHEMISTRY

Answer Y, 39; Ba, 56; Cu, 29; O, 8. All are metals except O, a nonmetal.

Science Words

superconductor

Objectives

- Describe the characteristics of superconductors.
- Consider various applications of superconductivity.

Figure 22-14

The magnetic field of the levitating magnet and the current-carrying superconductor repel each other. *Can you think of any practical use for magnetic levitation?*

In Search of a Perfect Conductor

What do you think about the idea of having train systems that would be levitated, or suspended over magnets, instead of running on rails? How would this be possible?

Recall that conducting materials all have some resistance to electron flow. Some of the electricity moving through the conductor is lost as heat due to this resistance. Likewise, as the temperature of a material increases, the resistance of the material increases. Ideally, the most efficient transfer of electricity would occur if conducting materials had no electrical resistance.

Superconductors are materials that have no electrical resistance. In 1911, a Dutch physicist, Heike Kamerlingh Onnes, discovered that some materials lose all electrical resistance when cooled to temperatures near absolute zero (0 K), –273°C. The temperature at which a material becomes superconducting is called the critical temperature.

One way to cool a material to superconducting temperatures is to submerge it in liquid helium. Helium is normally a gas, but it liquefies at 4.2 K. In 1987, J.G. Bednorz and K.A. Müller received a Nobel prize for making a ceramic material that became a superconductor at 30 K. This opened a new field of research to find "high-temperature" superconductors that become superconducting above the temperature of liquid nitrogen (77 K), which is much cheaper than liquid helium. New materials have been developed that are superconducting at more than 130 K.

Increasing Efficiency

Because superconductors have no electrical resistance, a current can flow indefinitely through them without losing energy. In one experiment, a current traveled through a superconducting loop for more than two years without losing energy. How can we make use of such efficient conductors?

The use of superconductors could eliminate much of the electrical energy waste we experience today. Ten percent of the energy transmitted through electrical power lines is lost as heat. Superconductors could also make electric motors, generators, computer

Program Resources

 Reproducible Masters
Study Guide, p. 95 L1
Reinforcement, p. 95 L1
Enrichment, p. 95 L3
Cross Curricular Integration, p. 28 L1

Transparencies
Section Focus Transparency 91 L1

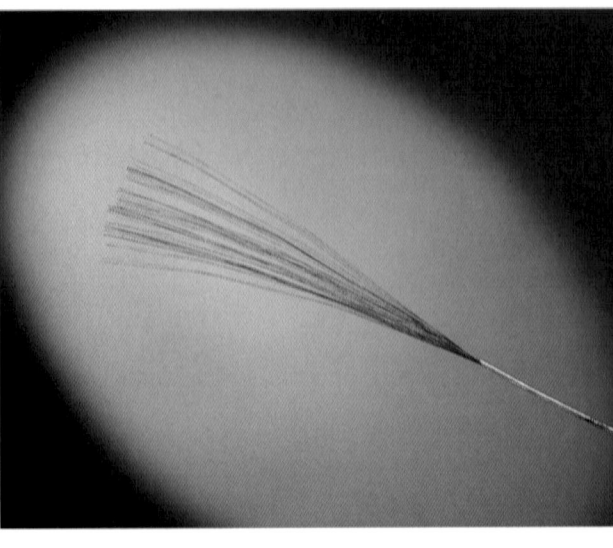

Figure 22-15
One of the most studied high-temperature superconductors is a compound of yttrium, barium, copper, and oxygen.

parts, and magnetic resonance imaging instruments more efficient, as well. One problem to be resolved is that superconducting materials are often brittle, and therefore hard to shape into wires, although researchers have had some success in developing thin films of superconducting materials.

Magnetic Levitation

Figure 22-14 shows the spectacular levitation effects of magnets over superconductors. Recall that a magnet moving toward a conductor induces a current in the conductor. The current, in turn, produces a magnetic field. If the conductor is a superconductor cooled to its critical temperature, the current will move continuously through it. The magnetic forces induced by these currents repel the magnet and cause the magnet to float, or levitate, above the superconductor. If a train had a powerful magnet beneath it and the rails were made of a superconducting material, the train would move above the rails. The only friction would be with the air. The train would lose little energy to the environment and would not give off pollutants.

Section Wrap-up

Review

1. How do superconductors differ from ordinary conductors?

2. What physical property of superconducting materials currently prevents them from being used in devices such as computers or motors?

Explore the Technology

What would be the advantages and disadvantages of developing superconducting trains? Use information in this section to determine what the difficulties of perfecting this system might be. In what ways might superconducting trains be a part of the solution to energy shortages?

CONNECT TO

CHEMISTRY
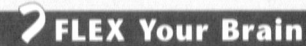

In 1987, a new compound of yttrium, barium, copper, and oxygen was discovered to be superconducting at 90 K. Find the symbols and atomic numbers of these four elements and *identify* each as a metal, a nonmetal, or a metalloid.

SCIENCE & SOCIETY

22-4 Superconductivity   **641**

? FLEX Your Brain

Use the Flex Your Brain activity to have students explore SUPERCONDUCTIVITY.

📁 **Activity Worksheets,** p. 5

Reteach

LS **Logical-Mathematical** Have the students calculate how cold the "high-temperature" superconductors actually are. Convert 120 K to °C. *120 – 273 = –153°C*. This should help students realize how inconvenient it would be to keep superconductors cool.

Extension

📁 For students who have mastered this section, use the **Reinforcement** and **Enrichment** masters.

4 Close

Try to borrow a kit that shows superconductivity and perform the demonstration in **Figure 22-14.**

Section Wrap-up

Review

1. Superconductors have no resistance to electrical currents.

2. Superconductors are quite brittle and break easily. Superconductors need to be very cold to work.

Explore the Technology

Answers will vary. Make sure students can support their answers.

Science & ART

Sources

- Chapman, Richard. *The Complete Guitarist.* Dorling Kindersley, Inc., New York, 1995.
- Denyer, Ralph. *The Guitar Handbook.* Alfred A. Knopf, New York, 1995.
- Trynka, Paul (ed.). *The Electric Guitar—an Illustrated History.* Chronicle Books, San Francisco, 1995.

Background

Most electric guitars are made of hardwoods including maple, alder, ash, walnut, and mahogany. The denser the wood, the greater the natural sustain the instrument will have. The tone of the guitar can be changed by using a combination of woods in the body and the neck of the instrument.

Teaching Strategies

Auditory-Musical Invite some local guitar players to perform for the class. Perhaps some of your students are such musicians. If possible, have students compare and contrast the sounds made by the acoustic guitar with those of the electric guitar.

Auditory-Musical If an instrument is available, allow students to take turns strumming and/or picking the strings of any guitar, having them note differences in sounds and the reasons for the differences.

- Make sure students understand how an electric guitar works by posing the following questions. **How are sounds produced with an acoustic guitar?** *Sounds are produced when the strings vibrate and resonate within the soundbox or body of the guitar.* **How are sounds made with an electric guitar?** *The pick-up converts the vibrations of the strings into alternating current, which travels*

Guitars—Plugged or Unplugged

Strumming Along

Historians have found evidence of a guitar-like instrument dating back to the Renaissance period of history. Since then, the guitar has become the world's most popular music maker. How does an acoustic guitar differ from its electrical counterpart?

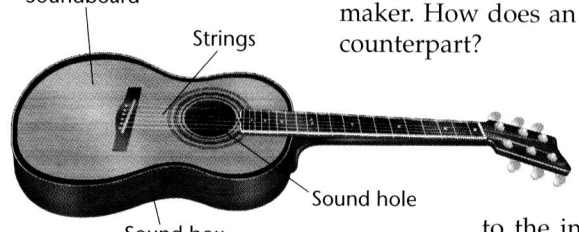

Soundboard
Strings
Sound hole
Sound box

The Acoustic Guitar

Acoustic guitars produce sound when the strings are plucked. This transfer of energy from the musician to the instrument causes the strings to vibrate. The energy is then transferred from the strings to the hollow sound box, or body, of the guitar. The sound box then amplifies the sound waves so that they can be heard. The sounds made by this type of guitar are bright and clear, but early guitarists strained to be heard above other instruments. This desire to be heard led to the development of the electric guitar.

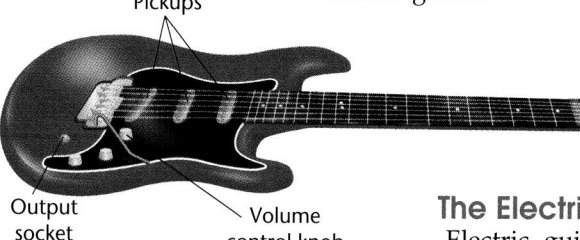

Pickups
Output socket
Volume control knob

The Electric Guitar

Electric guitars vary widely. But the one thing these generally solid-bodied guitars have in common is one or more magnetic pickups. A pickup is a device that converts the vibrations of the strings into alternating current. These electrical signals travel to an amplifier that magnifies them. The signals are then converted back into sound waves by the loudspeakers.

The single-coil pickup consists of a small, permanent magnet wrapped with copper wiring. The magnet generates a magnetic field that allows the vibrating steel strings of an electric guitar to interact with the field to generate small pulses of electrical energy in the wire. The twin-coil pickup consists of two coils wired in series that allow electrical currents to flow out of phase. Sound distortion, often referred to as humming, is eliminated in twin-coil pickups.

Science Journal

In your Science Journal, explain why you like or don't like the sounds made with electric guitars. How do they compare to sounds made by acoustic guitars?

642   Chapter 22 Magnetism and Its Uses

to an amplifier that magnifies it. The signals are then converted back into sound waves by loudspeakers.

Classics

Among the well-known electric guitar players are Chuck Berry, Stanley Clarke, Eric Clapton, Bo Diddley, Jimi Hendrix, Chrissie Hynde, Jaco Pastorius, Les Paul, Bonnie Raitt, Patti Smith, Stevie Ray Vaughn, Eddie van Halen, and Steve Vai.

Summary

22-1: Characteristics of Magnets
1. Opposite poles of magnets attract; like poles repel.
2. The magnetic field is the region around the magnet where magnetic forces act.
3. Groups of atoms with aligned magnetic poles are called magnetic domains.

22-2: Uses of Magnetic Fields
1. An electric current passing through a coil of wire can produce a magnetic field around the wire. The coil becomes an electromagnet; one end of the coil forms a north pole, and the other end forms a south pole.
2. Ammeters measure electrical current in amperes and should be connected in series. Voltmeters measure the potential difference in volts and should be connected in parallel.
3. An electric motor contains a rotating electromagnet that converts electrical energy to mechanical energy.

22-3: Producing Electric Current
1. A generator produces electric current by rotating a loop of wire in a magnetic field.
2. Direct current flows in one direction through a wire; alternating current reverses its direction in a regular pattern.
3. The number of turns of wire in the primary and secondary coils of a transformer determines whether it increases or decreases voltage.

22-4: Science and Society: Superconductivity
1. Superconductors are materials that have no electrical resistance.
2. The use of superconductors can eliminate electrical energy waste in the form of heat.

Key Science Words

a. alternating current (AC)
b. ammeter
c. direct current (DC)
d. electric motor
e. electromagnet
f. electromagnetic induction
g. generator
h. magnetic domain
i. magnetic field
j. magnetic pole
k. magnetism
l. superconductor
m. transformer
n. voltmeter

Reviewing Vocabulary

Match each phrase with the correct term from the list of Key Science Words.

1. region around a magnet in which a magnetic force acts
2. a property of some materials in which there is a force of repulsion or attraction between like or unlike poles
3. a temporary magnet made of a wire coil through which an electric current passes
4. measures electric current
5. a device that converts electrical energy into mechanical energy
6. the production of an electric current by moving a wire through a magnetic field
7. current that flows in only one direction
8. a device that changes the voltage of an alternating current
9. material with no electrical resistance
10. region of a magnet where magnetic lines of force are most dense

Chapter 22 Review 643

Review

Summary

Have students read the summary statements to review the major concepts of the chapter.

Reviewing Vocabulary

1. i	**6.** f
2. k	**7.** c
3. e	**8.** m
4. b	**9.** l
5. d	**10.** j

✔ Assessment

Portfolio Encourage students to place in their portfolios one or two items of what they consider to be their best work. Examples include:
- Using Technology answers, p. 629
- Connect to Earth Science answers, p. 634
- Activity 22-2 results and answers, pp. 638-639 [P]

Performance Additional performance assessments may be found in **Performance Assessment** and **Science Integration Activities.** Performance Task Assessment Lists and rubrics for evaluating these activities can be found in Glencoe's **Performance Assessment in the Science Classroom.**

GLENCOE TECHNOLOGY

Videodisc

Glencoe Physical Science

Interactive Videodisc

Use the videodisc segment *Comparing Motors and Generators* from Lesson 8 to review the principles of motors and generators.

MindJogger Videoquiz

Chapter 22 Have students work in groups as they play the Videoquiz game to review key chapter concepts.

Checking Concepts

1. a		6. d	
2. c		7. c	
3. d		8. b	
4. b		9. b	
5. d		10. a	

Understanding Concepts

11. The magnetic domain model assumes that metals such as iron contain groups of atoms that behave as miniature magnets. The domains are randomly oriented in nonmagnetic materials, but will rearrange themselves in the presence of a strong magnetic field.

12. Galvanometers and DC motors involve coils of wire rotating between the poles of permanent magnets. A motor also has a commutator that changes the direction of electric flow in the coil every half turn.

13. When a coil is moved so that it cuts across magnetic fields, an electric flow is produced. This process is called electromagnetic induction.

14. It contains two coils of wire wrapped around an iron core. An alternating current passes through the first coil and produces a changing magnetic field, which induces current in the second coil. If the second coil has more turns than the primary coil, the voltage in the second coil is greater. This type of transformer is a step-up transformer. In a step-down transformer, the secondary coil has fewer turns than the primary coil and the voltage of the current is reduced.

15. Superconducting power lines would conduct electricity with no energy loss due to electrical resistance.

Checking Concepts

Choose the word or phrase that completes the sentence or answers the question.

1. A magnet's force is strongest at its _____.
 a. north and south poles
 b. south pole
 c. north pole
 d. center

2. As the distance between two magnetic poles decreases, the magnetic force _____.
 a. remains constant
 b. changes unpredictably
 c. increases
 d. decreases

3. Domains at the north pole of a bar magnet have _____.
 a. north magnetic poles only
 b. south magnetic poles only
 c. no magnetic poles
 d. both north and south magnetic poles

4. Which of the following would not change the strength of an electromagnet?
 a. increasing the amount of current
 b. changing the current's direction
 c. inserting an iron core inside the loop
 d. increasing the number of loops

5. Ammeters should be _____.
 a. designed to have high resistance
 b. designed without magnets
 c. connected in parallel
 d. calibrated in amperes

6. A device containing a wire coil suspended in a magnetic field that measures potential difference is called a(n) _____.
 a. transformer
 b. electromagnet
 c. ammeter
 d. voltmeter

7. The direction of the electric current in an AC circuit _____.
 a. remains constant
 b. is direct
 c. changes regularly
 d. changes irregularly

8. Before current in power lines can enter your home, it must pass through a _____.
 a. step-up transformer
 b. step-down transformer
 c. commutator
 d. voltmeter

9. Some materials lose all electrical resistance when they are _____.
 a. at room temperature
 b. at their critical temperature
 c. below absolute zero
 d. heated

10. Superconductors can levitate magnets because the continuous electric current through the superconductor produces _____.
 a. a repulsive magnetic force
 b. an attractive magnetic force
 c. an antigravity force
 d. a mechanical force

Understanding Concepts

Answer the following questions in your Science Journal using complete sentences.

11. Describe the magnetic domain model.

12. Describe the basic device used in both galvanometers and DC electric motors. What additional device does an electric motor have? Why is this device necessary?

13. Explain how a generator produces an electric current.

14. Explain how a transformer works. How do step-up and step-down transformers differ?

15. How might superconductors help conserve energy resources?

Thinking Critically

16. If a magnetic compass needle points north, what is the actual polarity of Earth's northern magnetic pole? Explain.

17. Explain why dropping or heating a permanent magnet causes its magnetic domains to shift out of alignment.

18. Why must an ammeter be connected in series with a circuit? Why must a voltmeter be connected in parallel across a circuit?

Thinking Critically

16. The magnetic pole at Earth's north pole must actually be a south magnetic pole, because it attracts the north magnetic pole of a compass needle.

17. Magnetic domains are formed when the magnetic fields of atoms having unpaired electrons are lined up. When a magnet is dropped or heated, the alignment of magnetic domains is destroyed.

18. Connecting the ammeter in parallel will change the flow in the rest of the circuit. A voltmeter is connected in parallel because the voltage is the same across all branches.

19. 1200 V is ten times 120 V, so the secondary coil has 1/10 the turns, or ten turns.

20. You must find a substance that is not too brittle and exhibits superconductivity at normal temperatures.

19. A step-down transformer reduces a 1200-V current to 120 V. If the primary coil has 100 turns, how many must its secondary coil have?

20. You are developing a superconducting train. On what two problems will you concentrate your research? Why?

Developing Skills

If you need help, refer to the **Skill Handbook**.

21. Comparing and Contrasting: Compare and contrast electric and magnetic forces.

22. Comparing and Contrasting: Compare and contrast AC generators and DC motors.

23. Recognizing Cause and Effect: Earth's magnetic field has reversed itself many times in the past. What would be the likely effect of this switching of Earth's magnetic poles on the alignment of magnetic minerals deposited in Earth?

24. Hypothesizing: A compass needle will point north due to the magnetic field of Earth. When a bar magnet is brought near the compass, the needle is attracted or repelled by the bar magnet. Propose a hypothesis about the relative strengths of a bar magnet and Earth's magnetic field.

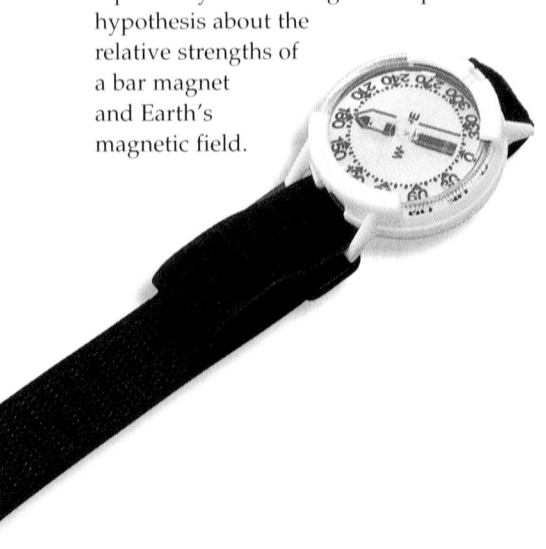

25. Concept Mapping: Complete the following events chain map by supplying the name and function of devices used to convert the mechanical energy of a turbine at an electrical power plant into the mechanical energy of an electric fan in your home.

Initiating Step

Generator: Converts mechanical energy of turbine into electrical current

↓

Step-up transformer: Raises voltage of current to minimize loss in power lines

↓

Step-down transformer: Lowers voltage to standard 120V

↓

Final Outcome

Electric motor: Turns fan blades by converting electrical energy back into mechanical energy

Performance Assessment

1. Writing a Newspaper Article: Research the most recent developments in superconductors. Write a newspaper editorial stating whether or not the government should spend more money on superconductor research.

2. Poster: In Activity 22-2 on pages 638-639, you built and experimented with a simple transformer. Make a poster showing what transformers have in common with AC generators.

3. Invention: Invent a new device that uses an electric motor. Demonstrate your device for the class.

Developing Skills

21. Comparing and Contrasting Both electric and magnetic forces are produced by electrons. Charged objects produce electric fields. Magnetic fields are produced by the motion of electrons. Both types of forces become weaker as the distance from the source of the force is increased. Both forces are attractive when unlike charges or poles are brought together and repulsive when like charges or poles are brought together.

22. Comparing and Contrasting Both have wire coils suspended between two fixed magnets. In a generator, the coil mechanically rotates in the magnetic field. This induces an electric current in the coil. In an electric motor, a current passes through the coil and makes it an electromagnet that rotates in the field of the fixed magnets.

23. Recognizing Cause and Effect Some layers in Earth have magnetic minerals aligned to the north, whereas other layers have minerals aligned to the south. Such alternating layers are strong evidence that Earth's magnetic field has often reversed its polarity.

24. Hypothesizing Near the compass, the magnetic field of the bar magnet is stronger than Earth's magnetic field.

25. Concept Mapping See student page.

Performance Assessment

1. Accept either position. Use the Performance Task Assessment List for Newspaper Article in **PASC,** p. 69. **P**

2. Both transformers and AC generators have coils and magnets. Use the Performance Task Assessment List for Poster in **PASC,** p. 73. **P**

3. The only requirement should be that the invention works. Use the Performance Task Assessment List for Invention in **PASC,** p. 45. **P**

Chapter Organizer

Section	Objectives/Standards	Activities/Features
Chapter Opener		**Explore Activity:** Observe the electronics inside a musical greeting card. p. 647
23-1 Semiconductor Devices (1 session, 1 block)*	1. **Describe** how the two types of doped semiconductors conduct a current. 2. **Explain** the device that changes AC into DC. 3. **Describe** the practical benefits of an integrated circuit. **National Content Standards:** (5-8) UCP1-UCP3, UCP5, E2, G1; (9-12) UCP1-UCP3, UCP5, E2, G1	**Connect to Chemistry,** p. 649 **Activity 23-1:** Semiconductors, p. 650 **Science Journal,** p. 653 **Skill Builder:** Interpreting Scientific Illustrations, p. 653
23-2 Radio and Television (1 session, ½ block)*	1. **Describe** how radio and television programs are transmitted. 2. **Explain** the operation of a cathode-ray tube. **National Content Standards:** (5-8) UCP1-UCP3, UCP5, A2, E2, G1; (9-12) UCP1-UCP3, UCP5, A2, E2, G1	**MiniLAB:** What influences the reception of radio signals? p. 655 **Problem Solving:** Sending Signals with Remote Control, p. 656 **Using Math,** p. 657 **Skill Builder:** Concept Mapping, p. 657
23-3 Microcomputers (1 session, ½ block)*	1. **Identify** the basic parts of a microcomputer. 2. **Describe** the role of the microprocessor. 3. **Distinguish** between RAM and ROM. **National Content Standards:** (5-8) UCP1-UCP3, UCP5, A1, A2, E1, E2, F5, G1, G3; (9-12) UCP1-UCP3, UCP5, A1, A2, E1, E2, F6, G1, G3	**MiniLAB:** What makes a calculator user friendly? p. 661 **Using Technology:** Navigating the Net, p. 662 **Using Computers,** p. 663 **Skill Builder:** Comparing and Contrasting, p. 663 **Activity 23-2:** Magnetic Messages, pp. 664-665
23-4 Science and Society (1 session, ½ block)*	1. **Discuss** the types of crimes that can be committed by computer misuse. 2. **Predict** the possible consequences of computer crimes. **National Content Standards:** (5-8) E2, F5, G1; (9-12) E2, F1, F6, G1	**Science Journal,** p. 667 **Explore the Technology,** p. 667 **People and Science:** Dr. Tridib Banerjee, Urban Planner, p. 668

* A complete Planning Guide that includes block scheduling is provided on pages 32T-35T.

Activity Materials

Explore	Activities	MiniLABs
page 647 musical greeting card, powerful hand lens	page 650 diode, 2 different-colored LEDs, hook-up wire, 1000-ohm resistor, 6-V battery, 6-V AC power supply pages 664-665 16 disk magnets, 2 sheets and a thin strip of cardstock paper, binder-hole reinforcing stickers or other symbols, transparent tape, metric ruler	page 655 2 AM/FM radios—one with a tall antenna and one without a visible antenna page 661 calculator

Need Materials? Call Science Kit (1-800-828-7777).

Teacher Classroom Resources

Reproducible Masters	Transparencies	Teaching Resources
Study Guide, p. 96 **Reinforcement**, p. 96 **Enrichment**, p. 96 **Activity Worksheets**, pp. 136-137	**Section Focus Transparency 92,** Computers: Key to the Future **Teaching Transparency 45,** Semiconductors	**Glencoe Physical Science Interactive Videodisc** **Physical Science CD-ROM** **Spanish Resources** **English/Spanish Audiocassettes** **Cooperative Learning Resource Guide** **Lab Partner** **Lab and Safety Skills** **Lesson Plans**
Study Guide, p. 97 **Reinforcement**, p. 97 **Enrichment**, p. 97 **Activity Worksheets**, p. 140	**Section Focus Transparency 93,** Then and Now **Teaching Transparency 46,** Public Address System	
Study Guide, p. 98 **Reinforcement**, p. 98 **Enrichment**, p. 98 **Activity Worksheets**, pp. 138-139, 141 **Multicultural Connections**, p. 49 **Science and Society Integration**, p. 27 **Concept Mapping**, pp. 51-52 **Cross-Curricular Integration**, p. 29 **Critical Thinking/Problem Solving**, p. 29	**Section Focus Transparency 94,** Virtual World **Science Integration Transparency 23,** The Semiconductor Computer Chip	**Assessment Resources** **Chapter Review**, pp. 49-50 **Assessment**, pp. 151-154 **Performance Assessment in the Science Classroom (PASC)** **MindJogger Videoquiz** **Alternate Assessment in the Science Classroom** **Performance Assessment** **Chapter Review Software** **Computer Test Bank**
Study Guide, p. 99 **Reinforcement**, p. 99 **Enrichment**, p. 99 **Science Integration Activity 23,** My Computer Has the Flu!	**Section Focus Transparency 95,** Mousetrap!	

GLENCOE TECHNOLOGY

The following multimedia resources are available from Glencoe.

Science and Technology Videodisc Series (STVS)
Physics
  Making Integrated Circuits
  Computer Graphics
  Computerized Star Imaging
  Factory of the Future
  Computerized Apple
Chemistry
  High-Tech Ceramics
Earth & Space
  Modeling Black Holes
  Computerized Weather Forecasting

The Infinite Voyage Series
Miracles by Design

Glencoe Physical Science Interactive Videodisc
Electricity and Magnetism

Physical Science CD-RO

Key to Teaching Strategies

The following designations will help you decide which activities are appropriate for your students.

L1 Level 1 activities should be within the ability range of all students, including those with learning difficulties.

L2 Level 2 activities should be within the ability range of the average to above-average student.

L3 Level 3 activities are designed for the ability range of above-average students.

LEP LEP activities should be within the ability range of Limited English Proficiency students.

LS These activities are designed to address different learning styles.

COOP LEARN Cooperative Learning activities are designed for small group work.

P These strategies represent student products that can be placed into a best-work portfolio.

Teacher Classroom Resources

This is a representation of key blackline masters available in the Teacher Classroom Resources.

Teaching Aids

Section Focus Transparencies

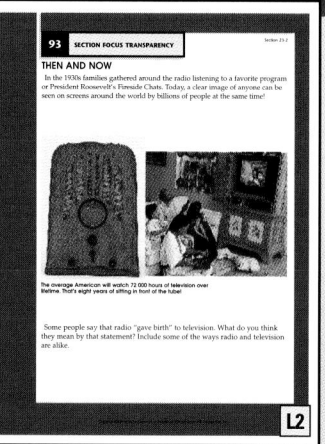

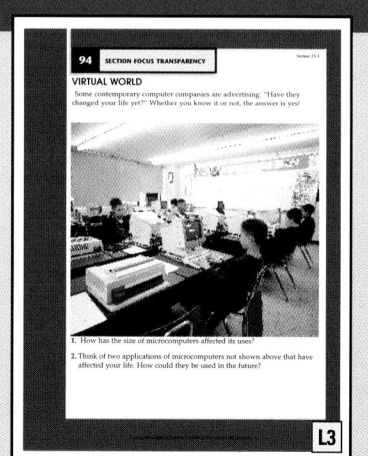

Science Integration Transparencies

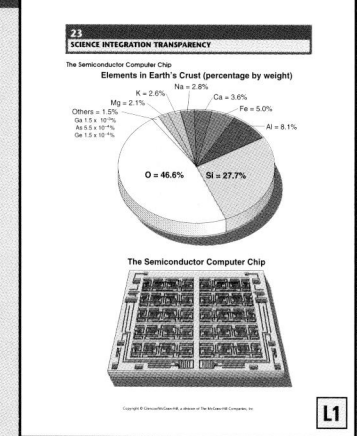

Teaching Transparencies

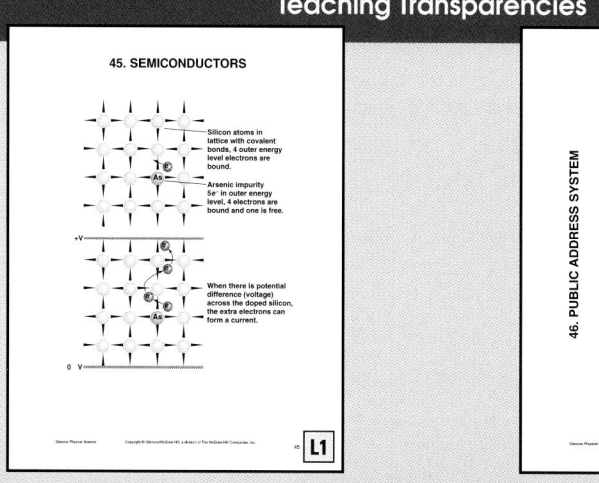

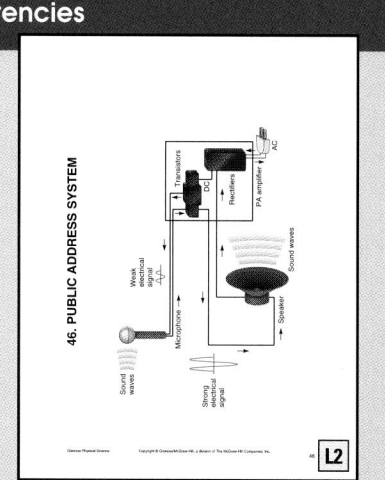

Meeting Different Ability Levels

Study Guide

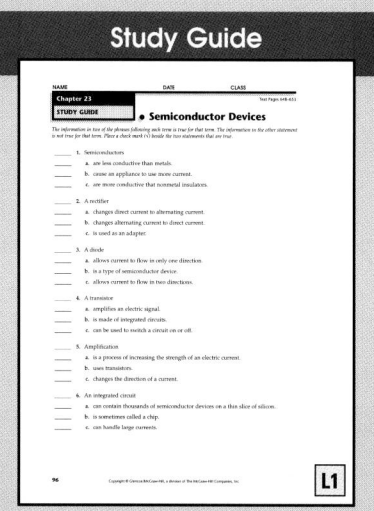

Reinforcement

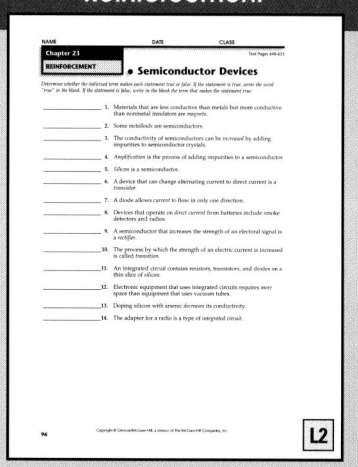

Enrichment Worksheets

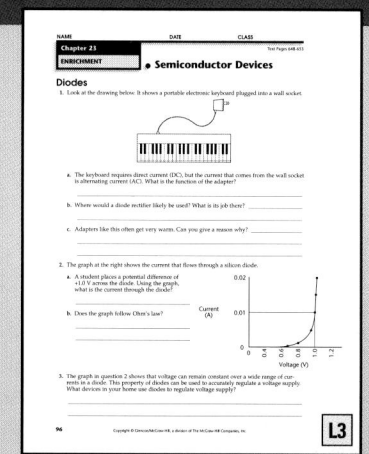

Hands-On Activities

Science Integration Activity

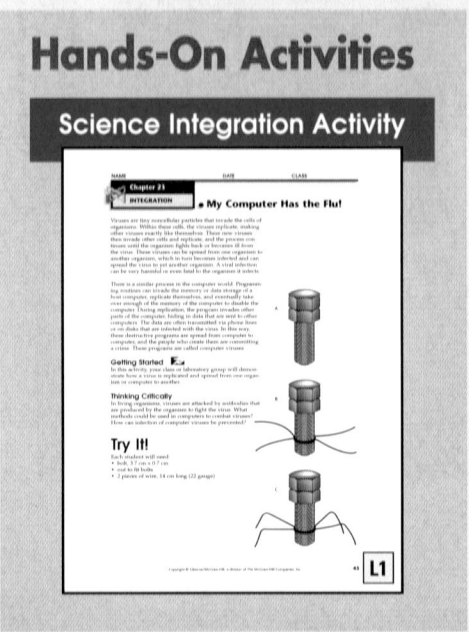

Assessment

Performance Assessment

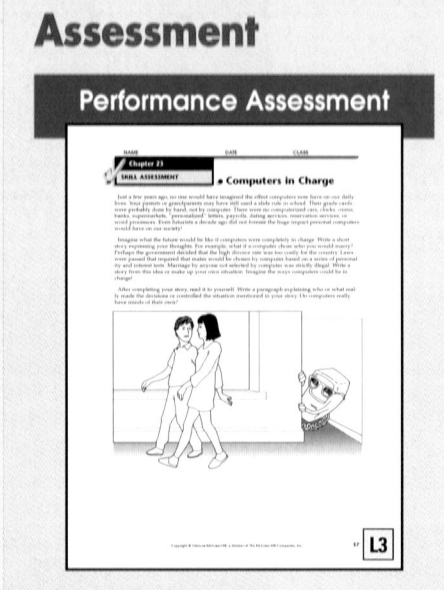

Enrichment and Application

Critical Thinking/Problem Solving

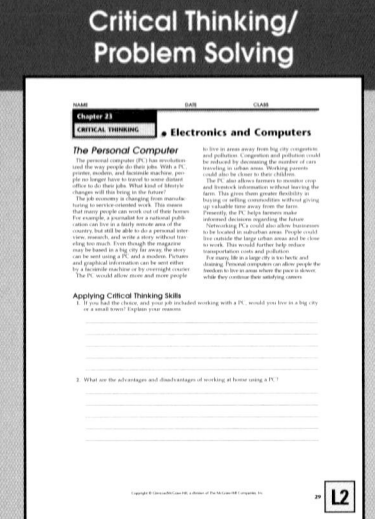

Cross-Curricular Integration

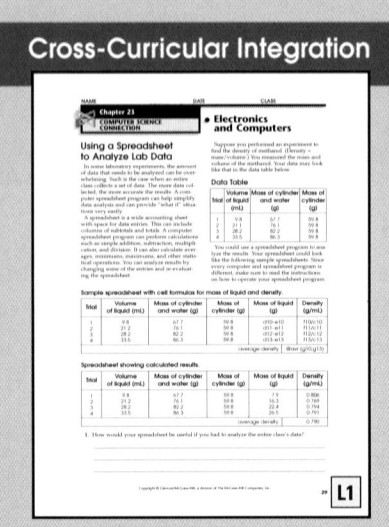

Science and Society Integration

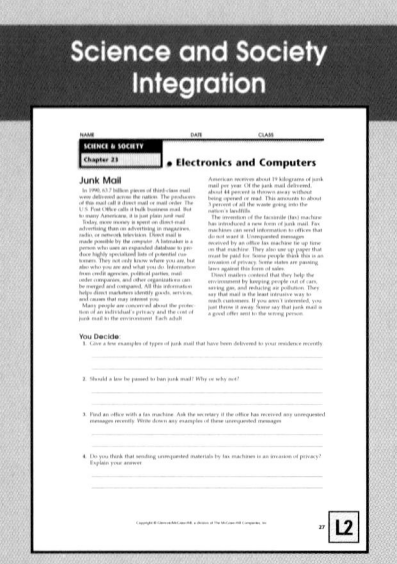

Multicultural Connections

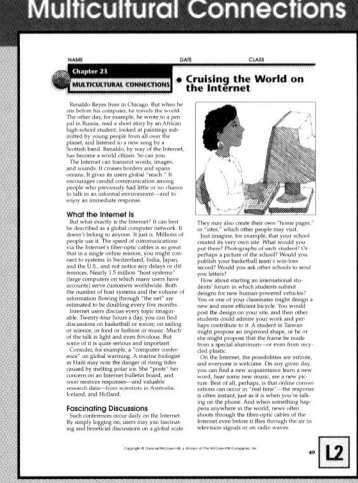

Concept Mapping

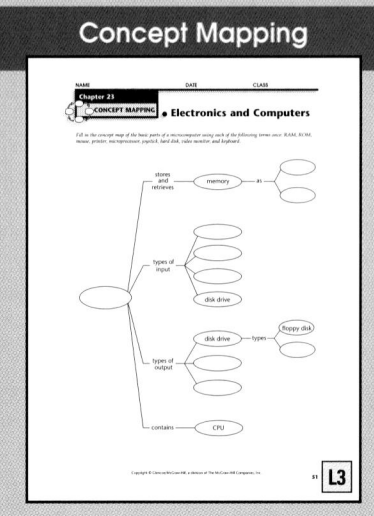

Electronics and Computers

CHAPTER OVERVIEW

Section 23-1 This section introduces the theory of semiconductors. Semiconductor applications including diodes, transistors, amplification, and integrated circuits are illustrated.

Section 23-2 This section discusses radio and television transmission and shows how a cathode-ray tube produces an image on the screen.

Section 23-3 This section identifies the components of a microcomputer, discusses the microprocessor, and presents the function of computer memory and methods of data storage.

Section 23-4 Science and Society This section features a discussion of computer crimes, including viruses and hacking. The Explore the Issue question asks students to evaluate how computer crimes should be treated by our legal system.

Chapter Vocabulary

rectifier	micro-
diode	processor
transistor	RAM
amplification	ROM
integrated	computer
circuit	virus
cathode-ray tube (CRT)	

Theme Connection

Systems and Interaction In this chapter emphasize how electronic and magnetic systems interact to help us create useful devices. The previous chapter covered the connection between electricity and magnetism. Numerous applications will illustrate it here.

Previewing the Chapter

646

Learning Styles   Look for the following logo for strategies that emphasize different learning modalities. **LS**

Kinesthetic	Reteach, p. 656; MiniLAB, p. 661
Visual-Spatial	Explore, p. 647; Demonstration, p. 649; Activity, p. 652; Reteach, p. 652; Visual Learning, pp. 649, 651, 655, 659, 662
Interpersonal	Activity 23-1, p. 650; Activity 23-2, pp. 664-665; Close, p. 667
Intrapersonal	MiniLAB, p. 655
Logical-Mathematical	Science Journal, p. 667
Linguistic	Science Journal, p. 659; Reteach, p. 667; Science Journal, p. 667

Chapter 23

Electronics and Computers

What ways do people have today of communicating with each other over great distances? The telephone has been around for many years, but advances in electronics and computers have made it possible for people to communicate through methods such as portable cellular phones, and computer networks run games and educational programs on CD-ROM. Have you ever examined the parts inside an electronic device like a calculator or radio?

EXPLORE ACTIVITY

Observe the electronics inside a musical greeting card.

1. Obtain a musical greeting card and listen to the greeting.
2. Remove the electronic device from the card. Examine it with a powerful hand lens, and in your Science Journal, sketch and label the parts you see.
3. Observe the card again. Can you determine what turns the musical card on and off?

Observe: Along with the diagram in your Science Journal, predict the function of any of the parts you see. Which part is the speaker?

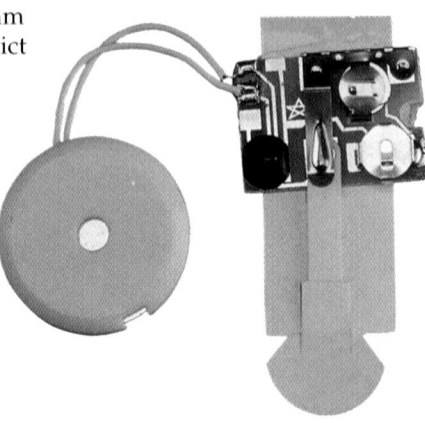

Previewing Science Skills

▶ In the Skill Builders, you will **interpret scientific illustrations, map concepts,** and **compare and contrast.**

▶ In the Activities, you will **observe, infer, classify,** and **communicate.**

▶ In the MiniLABs, you will **observe, infer,** and **interpret.**

647

EXPLORE ACTIVITY

Purpose

IS **Visual-Spatial** Use the Explore Activity to help students notice the parts and details of electronic devices. Explain that they will be learning the functions of some of the parts they see in the device. L1

Preparation

Obtain several musical greeting cards from card shops.

Materials

musical greeting cards, hand lenses

Teaching Strategies

Have students try to draw the parts of the electronic device and label them descriptively. Have them speculate where the music originates and where the sound comes out. Tell them they will be learning about some of the parts used in electronic devices.

Observe

Students will probably be able to identify the flat speaker membrane and the connecting wires.

✓ Assessment

Oral Have students describe what type of switch triggers the sound. Use the Performance Task Assessment List for Making Observations and Inferences in **PASC,** p. 17. P

Assessment Planner

Portfolio
Refer to page 669 for suggested items that students might select for their portfolios.

Performance Assessment
See page 669 for additional Performance Assessment options.
Skill Builders, pp. 653, 657, 663
MiniLABS, pp. 655, 661
Activities 23-1, p. 650; 23-2, pp. 664-665

Content Assessment
Section Wrap-ups, pp. 653, 657, 663, 667
Chapter Review, pp. 669-671
Mini Quizzes, pp. 653, 657, 663

Group Assessment
Opportunities for group assessment occur with Cooperative Learning Strategies and Flex Your Brain Activities.

Prepare

Section Background

- The science behind semi-conductors is included in the vast field of solid-state physics, which is the study of the structure of solids. An explanation of the properties and behavior of semiconductors at the atomic level can be found in electron band theory, based on quantum mechanics.
- Pure, undoped semiconductors are said to be intrinsic.

Preplanning

- Obtain examples of transistors, diodes, resistors, and microchips from a local electronics store. They may even have some spare parts they would be willing to donate to your school.
- Gather materials for Activity 23-1.

1 Motivate

Bellringer

 Before presenting the lesson, display **Section Focus Transparency 92** on the overhead projector. Assign the accompanying **Focus Activity** worksheet. L1 LEP

INTEGRATION
Chemistry

Have students research how pure semiconductors, such as silicon or germanium, are obtained. Pure silicon crystals often have less than 1 part in 1 billion as impurities.

23•1 Semiconductor Devices

Science Words
- rectifier
- diode
- transistor
- amplification
- integrated circuit

Objectives
- Describe how the two types of doped semiconductors conduct a current.
- Explain the device that changes AC into DC.
- Describe the practical benefits of an integrated circuit.

INTEGRATION
Chemistry

Semiconductors

Do you use a calculator when you do your homework, average your grades, or add up the amount of money you've saved? Your calculator is only one of thousands of complex electronic devices made possible by advances in the applications of electricity and magnetism. If you carefully removed the front of your calculator, as shown in **Figure 23-1,** you would see tiny circuits with all sorts of unusual parts inside. You might see several kinds of semiconductor devices: a diode, a rectifier, and a transistor.

Chemistry in Your Calculator

What elements do you think might act as semiconductors? To answer this question, you must first think about the periodic table you studied in Chapter 10. Recall that the elements on the left side and center of the table are metals and conduct electricity. Nonmetals—poor conductors of electricity—are found on the right side of the table. They are electrical insulators. How would you classify the elements found along the staircase-shaped border between the metals and nonmetals? Would you call these conductors or insulators?

The elements located between the metals and nonmetals on the periodic table are metalloids. Some metalloids, such as silicon and germanium, are semiconductors. Semiconductors are less conductive than metals but more conductive than nonmetal insulators. Your calculator and other electronic devices in

Figure 23-1

A calculator contains several different types of semiconductor devices. It contains tiny diodes, rectifiers, and transistors.

648

 Reproducible Masters
Study Guide, p. 96 L1
Reinforcement, p. 96 L1
Enrichment, p. 96 L3
Activity Worksheets, pp. 136-137 L1

Transparencies
Section Focus Transparency 92
Teaching Transparency 45

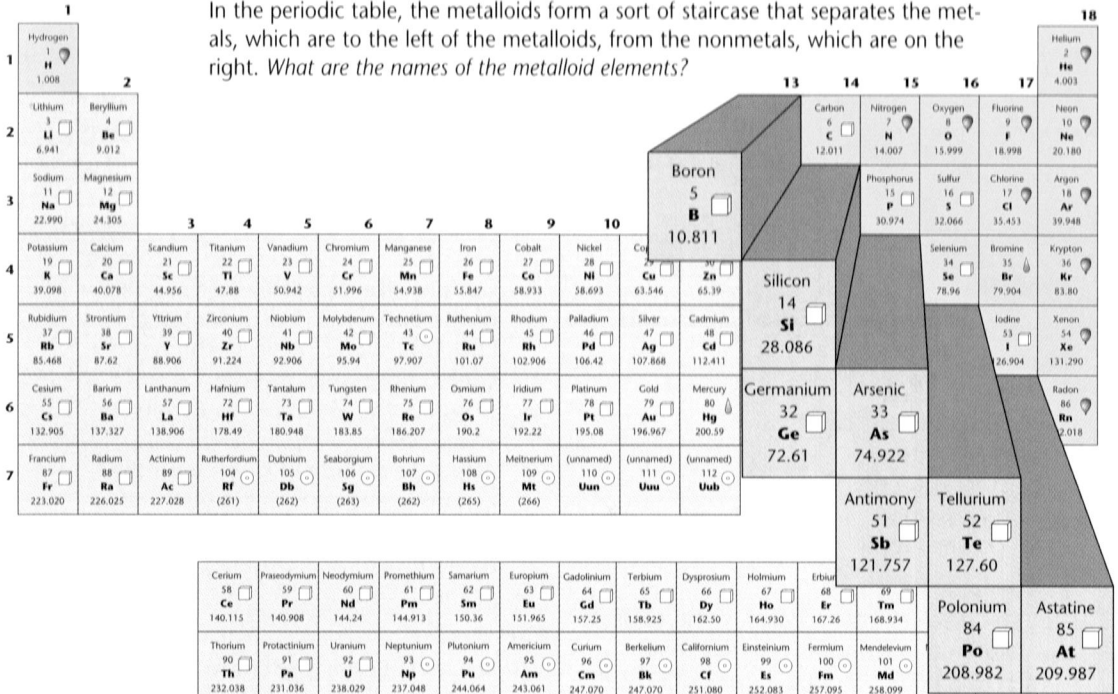

Figure 23-2

In the periodic table, the metalloids form a sort of staircase that separates the metals, which are to the left of the metalloids, from the nonmetals, which are on the right. *What are the names of the metalloid elements?*

your home that use semiconductors use less current to operate than similar devices that don't use semiconductors. **Figure 23-2** shows the location of the materials on the periodic table.

Adding Impurities

The conductivity of semiconductor crystals can be increased by adding impurities. This process is called doping. Doped silicon is a commonly used semiconductor. Silicon atoms have four electrons in their outer energy level. This electron arrangement stabilizes the crystal lattice structure. If small amounts of another element with either more or fewer than four outer energy level electrons are added, a few high-energy electrons will be able to move more easily through the material. So, the conductivity of the compound is higher than that of pure silicon.

Silicon is often doped with arsenic or gallium. Locate these elements on the periodic table. How many electrons does each of these elements have in its outer energy level? Even one other type of atom among a million silicon atoms will significantly change the conducting properties of the material. By controlling the type and amount of doping, semiconductors with a variety of conducting properties can be created.

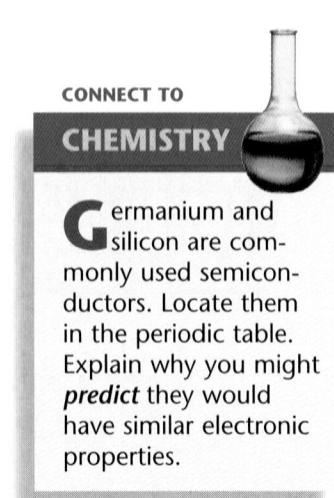

CONNECT TO

CHEMISTRY

Germanium and silicon are commonly used semiconductors. Locate them in the periodic table. Explain why you might *predict* they would have similar electronic properties.

23-1 Semiconductor Devices **649**

2 Teach

Demonstration

Visual-Spatial If possible, bring a transparent phone to class. The wiring in a phone appears to be complex, but it is actually made of just a few kinds of electrical devices. Ask students whether they can identify any of the parts.

CONNECT TO

CHEMISTRY

Answer Ge and Si are in the same group on the periodic table, so they would likely have the same number and arrangement of outer level electrons. **P**

Student Text Questions

• **How would you classify the elements found along the staircase-shaped border between the metals and nonmetals?** *The elements are classified as metalloids.*

• **Would you call these conductors or insulators?** *Their ability to conduct is between conductors and insulators.*

Visual Learning

Figure 23-2 **What are the names of the metalloid elements?** *The elements are boron, silicon, germanium, arsenic, antimony, tellurium, astatine, and polonium.* **LEP** **IS**

Activity 23-1

Semiconductors

Have you ever passed through a turnstile as you entered a store? You and other people could go through in only one direction. Electrical devices called diodes change AC to DC current by allowing current in a circuit to flow in only one direction.

Problem

How do diodes affect AC and DC circuits?

Materials

• diode
• 2 LEDs, different colors
• hook-up wire
• resistor, 1000 ohms
• battery, 6-volt
• AC power supply, 6 volts

Procedure

1. Attach the resistor to one of the LEDs before connecting it to the battery. Reverse battery connections until the LED lights.
2. Mark the positive side of the LED.
3. Repeat steps 1 and 2 with the second LED.
4. Twist the positive wire of one LED with the negative wire of the other. Attach the resistor to the other two wires.

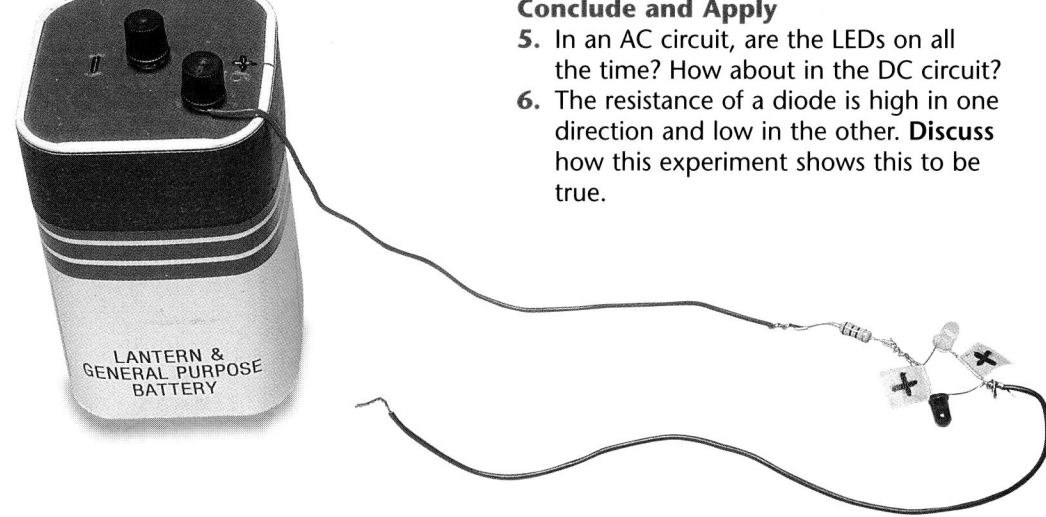

5. Connect the assembly to the poles of the battery and observe which LED light works. Reverse the battery connections and observe.
6. Connect the assembly to 6 volts AC and observe.
7. Attach the diode in series with the assembly and connect it to 6 volts AC.
8. Reverse the direction of the diode and reconnect to AC. Observe the response of the LEDs.

Analyze

1. A battery sends current in one direction. How does the LED assembly show this?
2. How does the LED assembly detect an alternating current?
3. **Infer** what the addition of a diode does to an alternating current.
4. **Infer** what reversing the direction of the diode accomplishes.

Conclude and Apply

5. In an AC circuit, are the LEDs on all the time? How about in the DC circuit?
6. The resistance of a diode is high in one direction and low in the other. **Discuss** how this experiment shows this to be true.

Electronic Gates— Rectifiers and Diodes

Have you noticed that some portable radios can operate on batteries, but they may also have adapters that plug into wall sockets? Why do they need an adapter? Radios operate on low-voltage direct current. For the radio to operate with household alternating current, the voltage of the current must be lowered and the current must be changed to flow in one direction. A radio adapter contains a transformer, which reduces the voltage, and a rectifier. A **rectifier** is a device that changes alternating current into direct current.

Many household devices are built to operate on direct current. Some, such as smoke detectors and clocks, can easily be supplied with direct current from batteries. Other appliances that use large amounts of electricity, such as your television or computer, would not be practical to operate with direct current from batteries. Therefore, transformers and rectifiers are wired into the circuits inside of radios, televisions, computers, and other similar appliances so that they can be supplied with low-voltage direct current from a source of alternating current.

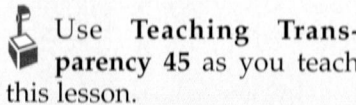

Diodes

A diode is one type of rectifier. A **diode** can be thought of as a type of valve or gate that allows current to flow only in one direction. A diode can be made by doping the ends of a crystal with different elements. Electrons can easily flow through the diode in one direction only. This device, the portable radio's adapter plug, shown in **Figure 23-3,** probably contains a diode rectifier.

Transistors

Have you wondered why radios are sometimes called transistor radios? They contain electrical devices called transistors. A **transistor** is a semiconductor that amplifies or strengthens an electric signal or acts as a tiny electric on-off switch. The signal in an electronic device is the varying electric current that represents a sound, picture, or some other piece of information. Transistors like those shown in **Figure 23-4C** on page 652 amplify a weak electric signal until it is strong enough to be useful. **Amplification** is the process of increasing the strength of an electric signal.

Figure 23-3

Diodes and rectifiers, above, change alternating current into direct current. *Why are these useful in a radio?*

Visual Learning

Figure 23-3 Why are these useful in a radio? *A radio needs DC to operate.* **LEP** **LS**

Teacher F.Y.I.

A zener [za-ner] diode is a diode used to create a steady voltage in a circuit over a range of currents.

Use **Teaching Transparency 45** as you teach this lesson.

Content Background

A graph of current versus voltage for resistors in a circuit produces a straight line. However, a graph of current versus voltage for diodes and transistors in a circuit does not produce a straight line; they are nonlinear devices.

Enrichment

Have students investigate how integrated circuits are constructed in such miniature size. Note that radios will now even fit in a wristwatch. They used to be the size of modern televisions. This decrease in size is the result of integrated circuits. **L3**

GLENCOE TECHNOLOGY

Videodisc
STVS: Physics
Disc 1, Side 2
Making Integrated Circuits (Ch. 15)

CD-ROM
Physical Science CD-ROM
Have students perform the interactive exploration for Chapter 23 to reinforce important chapter concepts and thinking processes.

Cultural Diversity

A Well-Trained Scientist Shirley Jackson's interest in science began with stories her mother read her about Benjamin Banneker. When she entered MIT in 1964, she was one of two African American women in a group of no more than 30 to 40 women in the school—a pioneer for both women and African Americans. During her college years, she also worked to help others by becoming a co-founder of the Black Student Union. This group was instrumental in initiating MIT's active recruitment of minority students. In 1973, she became the first African American to receive a doctoral degree in physics from MIT. Her current area, theoretical explanations for the behavior of semiconductors, makes her a member of a select group in physics. Only 20% of the people with doctorates in physics are working as theorists.

Explaining how a transistor can amplify an electrical signal can be confusing. Emphasize that the small current received as an input signal controls the large current from the power source.

Activity

IS **Visual-Spatial** Have some microchips with integrated circuits on them available for students to observe. Examine one through a microscope. They should realize how tiny the components can be and still be effective. You might want to bring in an old calculator to show them the difference in size and price between early and modern calculators.

3 Assess

Check for Understanding

? FLEX Your Brain

Use the Flex Your Brain activity to have students explore RECTIFIERS.

🗁 **Activity Worksheets,** p. 5

Reteach

IS **Visual-Spatial** Have students interpret and explain the diagram in Figure 23-2 on page 649. Have them explain why some semiconductors are doped.

Extension

🗁 For students who have mastered this section, use the **Reinforcement** and **Enrichment** masters.

Figure 23-4

Portable stereos (A) contain circuit boards (B), which contain miniaturized transistors (C).

Amplifying Signals

The radio wave that travels many kilometers from a broadcast station to your stereo and television receivers is converted into an electric signal that is too weak to reproduce the information it carries into picture or sound. In other words, it is not strong enough to cause vibration of the loudspeaker and create sound that can be heard. As seen in **Figure 23-4,** a transistor in the electric circuit of the stereo uses a small current from a weak incoming signal to control a large current provided by a power source. Thus, the small input signal supplied to the transistor results in a large, varying output current that can vibrate the speaker.

Your school public address (PA) system contains several transistors. When someone speaks over the system, the sound wave is converted to an electrical signal by a microphone and amplified by transistors within the circuit of the system. The amplified signal is then changed back into sound waves by a speaker. Transistors and their ability to amplify signals make it possible for you to use tape players, bullhorns, hearing aids, and televisions. None of these electronic devices or their applications would be possible without the transistor. It is considered to be the most fundamental part of an electronic circuit.

Integrated Circuits

Look at **Figure 23-5.** Before transistors were invented, devices called vacuum tubes were used to amplify electric signals. These low-pressure glass or metal tubes regulated the electron flow of a circuit's current as it went through each tube. Your grandparents probably owned a television or radio that used these tubes to amplify signals. Televisions and radios today have semiconductor components instead of vacuum tubes. After the development of the transistor, integrated circuits became a reality. As a result of this electronics break-through, today's televisions, radios, calculators, and other electronic devices can be much smaller than the older ones.

An **integrated circuit** can contain thousands of resistors, diodes, and transistors on a thin slice of silicon. These thin silicon slices, called chips, can be smaller than 1 cm on a side. The miniature circuit components in an integrated circuit are made by doping the silicon chip with small amounts of impurities. The conductors between the components are made by using photosensitive chemicals to etch circuits on the silicon chip. Having circuit components so close together reduces the time required for a current to travel through a circuit. As a result, the integrated circuit is an effective design for rapid information processing, which is essential in devices such as microcomputers.

Figure 23-5

A Before the invention of transistors, vacuum tubes were used to amplify signals in early electronic devices.

B Thousands of resistors, diodes, transistors, and other devices exist in an integrated circuit on a silicon chip a little larger than your thumbnail.

Section Wrap-up

Review

1. Describe how the conductivity of silicon as a semiconductor can be improved.

2. What device can you use to change AC to DC? How does it work?

3. How are integrated circuits made?

4. **Think Critically:** Would carbon atoms be useful in doping a silicon semiconductor? Explain.

Skill Builder
Interpreting Scientific Illustrations
Look at the periodic table of the elements on page 649. List at least four examples each of metals, nonmetals, and metalloids. Classify each of these as a conductor, an insulator, or a semiconductor. If you need help, refer to Interpreting Scientific Illustrations in the **Skill Handbook.**

Science Journal
Radio and television are relatively recent inventions that make use of such devices as diodes and transistors. In your Science Journal, write a paragraph describing how your life would be different without these technologies.

•MINI•QUIZ•

Use the Mini Quiz to check students' recall of chapter content.

1. _____ are less conductive than conductors but more conductive than insulators. *Semiconductors*

2. A common method of increasing the conductivity of a semiconductor is the process of _____ . *doping*

3. An adapter in a radio that changes AC into DC probably contains a(n) _____ . *rectifier*

4. What is a major use for transistors? *to amplify an electrical signal*

Section Wrap-up

Review

1. by doping it with gallium or arsenic

2. a rectifier or diode—it allows current to flow in only one direction

3. Integrated circuits are made by doping silicon chips with impurities and photo-etching circuits on the chips.

4. **Think Critically** No, carbon atoms have the same number of electrons in their outer energy level as does silicon.

Skill Builder
The elements listed for metals should come from the left side and center of the periodic table; nonmetals are on the right side; and metalloids border the staircase between the metals and nonmetals. The metals should be listed as conductors, the nonmetals as insulators, and the metalloids as semiconductors.

Assessment

Content Locate silicon, germanium, arsenic, and gallium in the periodic table. List their symbols and atomic numbers. Use the Performance Task Assessment List for Science Journal in **PASC,** p. 103. **P**

Science Journal
Accept all logical answers. Students might include the effect on social life, how they get news, and how they are notified of emergencies. **P**

Prepare

Section Background

- Recall that electromagnetic waves were first used to transmit messages over long distances after the invention of the wireless telegraph in the late 1800s.

- The electrical signal carrying sound information is called an AF, or audio frequency, signal. The frequencies correspond to the range of hearing (20 to 20 000 Hz). The carrier waves are called RF, or radio frequency, waves. They have much higher frequencies.

Preplanning

Gather radios for the Mini-LAB, or have students bring them to class.

1 Motivate

Bellringer

Before presenting the lesson, display **Section Focus Transparency 93** on the overhead projector. Assign the accompanying **Focus Activity** worksheet. **L1** **LEP**

Tying to Previous Knowledge

Ask students what radio waves are. Remind them that radio waves make up the longest wavelengths in the electromagnetic spectrum. They are hitting each student right now. Students will find out what it takes to change these silent, invisible waves to sounds they can hear and images they can see.

Student Text Question

Do AM or FM stations operate at a higher frequency? *FM stations*

23•2 Radio and Television

Science Words

cathode-ray tube (CRT)

Objectives

- Describe how radio and television programs are transmitted.
- Explain the operation of a cathode-ray tube.

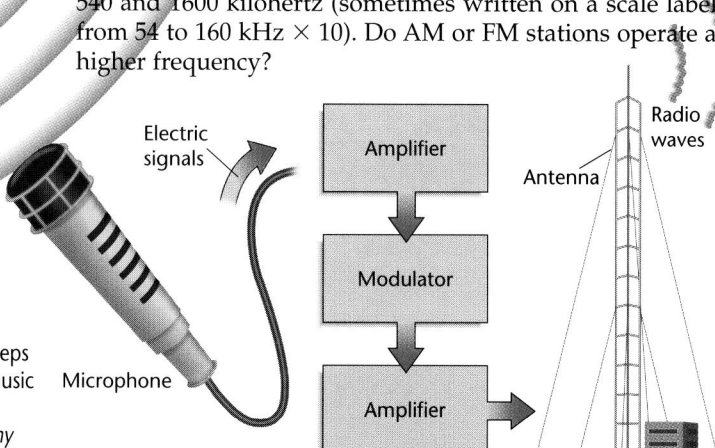

Figure 23-6

This diagram illustrates the steps that must occur to change music at a radio station into radio waves and back to music. *Why are radio waves used to transmit the signal over long distances rather than sound waves?*

Radio Transmission

Now that you've studied sound, electromagnetic waves, and electricity, you should be comfortable with most of the concepts needed to explain how a radio works. Radios operate by changing electromagnetic radio waves into vibrations that produce compressional sound waves. Recall that radio waves are the kind of electromagnetic radiation with the longest wavelength and shortest frequency. Like all electromagnetic radiation, radio waves travel at the speed of light. However, the conversion to sound waves is not possible without first converting the radio waves to electrical signals.

Carrier Waves

On a radio dial, a set of numbers on the top of the dial corresponds to the FM frequency range, and a separate set of numbers on the bottom corresponds to the AM frequency range. Every radio station, AM or FM, operates at a specific frequency and wavelength called a carrier wave. FM stations broadcast at carrier frequencies ranging from 88 to 108 megahertz. AM stations broadcast at carrier frequencies between 540 and 1600 kilohertz (sometimes written on a scale labeled from 54 to 160 kHz × 10). Do AM or FM stations operate at a higher frequency?

Sending Signals—From Sound Waves to Radio Waves

At the radio station, a microphone collects the compressional waves created by sounds, as shown in **Figure 23-6,** and the microphone changes the sound waves into electrical signals. The electric current vibrations vary according to the sound

654 Chapter 23 Electronics and Computers

---

vibrations. These signals are then amplified and passed through the modulator, as discussed in Chapter 19. The type of modulation varies in AM and FM radio stations. When the electrical signals are used to produce variations in the amplitude of the carrier waves, they are called *amplitude-modulated,* or AM, waves. When the electrical signals are used to produce variations in the frequency of the carrier waves, they are called *frequency-modulated,* or FM, waves. The amplified, modulated electric currents are sent to an antenna, where they are transformed into radio waves.

Receiving Signals—From Radio Waves to Sound Waves

Figure 23-6 shows that your receiving radio has an antenna that collects these radio waves and transforms them into electric currents. If the radio is tuned to the same frequency as the carrier waves, the signal will be amplified again. The carrier wave is then removed, leaving only the electric signal that corresponds to the original sound waves. The radio's loudspeaker vibrates to cause sound waves for you to hear.

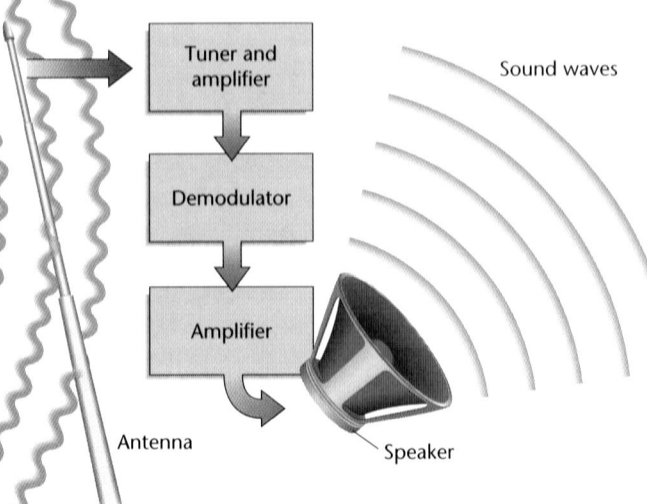

Antenna   Speaker

Television Transmission

What do radios have in common with television sets? Both have tuners that tune the circuits to the frequency of the carrier waves. Both also have loudspeakers to vibrate and produce sound and an antenna to collect radio signals. The antennas on modern radios and television sets are often hidden in the circuits of these devices. Televisions not only turn the radio waves into sound, but they also use these signals to create a visual image. Television is like a radio with pictures.

23-2 Radio and Television   **655**

What influences the reception of radio signals?

Procedure
1. Obtain an AM/FM radio with a tall antenna and one without a visible antenna, such as on a clock radio.
2. Using the FM broadcast band of one radio, count the stations you can clearly receive as you slowly scan from the highest to the lowest frequency. Record this information in your Science Journal.
3. Repeat step 2 with the other radio and record your data.
4. Using the radio with the best reception, listen for the locations of the broadcasting stations as you scan both AM and FM bands. Record the locations of stations from each band.

Analysis
1. Which radio clearly received signals from the most radio stations? How does this relate to the antenna system in each radio?
2. Do you have reception from greater distances on AM or FM bands?

Inclusion Strategies

Gifted Students can test the effectiveness of various radio antennas that they design and construct from different materials. Vary the size and shape of the antenna, or connect it to available items such as pipes, phone lines, and so on. They will test the antenna's effectiveness by recording how many radio stations can be received. Predictions and results should be charted and shared. **L3**

MiniLAB

Purpose
IS **Intrapersonal** Students will compare and contrast the AM and FM reception from two different radios. **L1** **LEP**

Materials
two radios, one with a tall antenna and one without a visible antenna

Teaching Strategies
Troubleshooting You may wish to conduct this as a demonstration.

Analysis
1. Answers will vary. Students will realize that invisible antennae in power cords also do a reasonable job.
2. They will likely observe that AM bands usually provide reception from greater distance.

✓ Assessment

Oral Have students compare and contrast AM and FM transmission based on their observations in this MiniLAB and content in this section. Use the Performance Task Assessment List for Making Observations and Inferences in **PASC**, p. 17. **P**

📁 **Activity Worksheets,** pp. 5, 140

Visual Learning

Figure 23-6 Why are radio waves used to transmit the signal over long distance rather than sound waves? *Radio waves travel faster, are more penetrating and more easily controlled, and lose less energy during transfer.* **LEP** **IS**

655

Problem Solving

Solve the Problem

1. Pulses of infrared radiation. Yes, you could block it with your hand.
2. The signal selects the electronic device that should respond and tells it what to do, such as changing a channel, raising the volume, or rewinding a tape.

Think Critically

1. The remote did not produce a signal that commanded your VCR or stereo to respond.
2. Manufacturers agree to a common signal sensor that is used in different brands.

3 Assess

Check for Understanding

? FLEX Your Brain

Use the Flex Your Brain activity to have students explore RADIO TRANSMISSION.

 Activity Worksheets, p. 5

Reteach

LS **Kinesthetic** Bring a radio or television into your classroom. Have students demonstrate the use of each device with proper physics explanations. They should include explanations such as how the radio waves are picked up and transformed to sound and what is happening when you change the channel. ⌐L1⌐

Extension

For students who have mastered this section, use the **Reinforcement** and **Enrichment** masters.

Problem Solving

Sending Signals with Remote Control

Do you use a remote control to change a television channel, control a VCR, or change the volume on a stereo? Remote controls are now available that can communicate with up to 16 different electronic devices in a home entertainment system. How do remote controls work?

Most remote controls emit pulses of infrared radiation at a frequency that can be detected by the electronic products designed for remote operation. Pressing different keys causes unique groups of pulses to be produced. Part of the infrared code is used to select which electronic product should respond to the remote command. The other part of each code tells the responding device what to do, such as change a channel, raise the volume, or rewind a tape.

Sometimes, people become frustrated when a remote included with a new television set will not control their VCR or other remote-operated devices. What do you think might cause this problem?

Solve the Problem:

1. **What carries the signal from a remote control to a television set? Do you think it would be possible to block this signal with your hand?**
2. **What kinds of information are included in the signal emitted from a remote control?**

Think Critically:

Suppose you purchased a new television that came with a remote control, but the remote control would not operate your VCR or stereo.
1. What possible explanations for this problem can you suggest?
2. How is it possible for remote controls made by different manufacturers to control other products?

The audio and video signals for television programs are sent and received like an FM radio signal. Just as a microphone is used to change sound waves into varying electric currents, a television camera is used to change light into electric currents that represent the images. This electric video signal then modulates a carrier wave. A television station simultaneously transmits the audio and video signals from its antenna.

Tuning into a television station is much like tuning a radio. When you select a channel, only a certain carrier frequency is picked up and amplified within the set. The audio portion of the television program is changed to sound by a loudspeaker. The picture you see on the screen is created by a more complex process.

Making Images with Cathode-Ray Tubes

You may have heard that if the picture tube inside your television goes out, you are faced with a major repair bill or should probably buy a new television. That's because this picture tube is the most expensive component in your TV set. The picture tube is a type of cathode-ray tube. A **cathode-ray tube (CRT)** uses electrons and fluorescent materials to produce images on a screen.

Look at the CRT in the television set in **Figure 23-7.** A CRT is a sealed glass vacuum tube. When power is applied to the components in the tube, electrons come off a negative cathode and are focused into a beam. They move toward the screen, which is coated on the inside with materials that glow when struck by the electrons. The direction of the beam is changed by electromagnets outside of the CRT. This allows the beam of electrons to sweep the surface of the screen many times each second.

The cathode-ray tube of a color TV contains a screen lined with different materials that give off light in one of the three primary colors: red, blue, or green. The radio waves that carry the picture signal are changed into electrical signals in the television. These electrical signals control the intensity and position of the electron beams, which determine the colors and patterns of the image that forms on the screen. How is this process similar to the way your eye forms color images?

Figure 23-7

The screen of a television set is really the end of a cathode-ray tube that is coated with fluorescent materials.

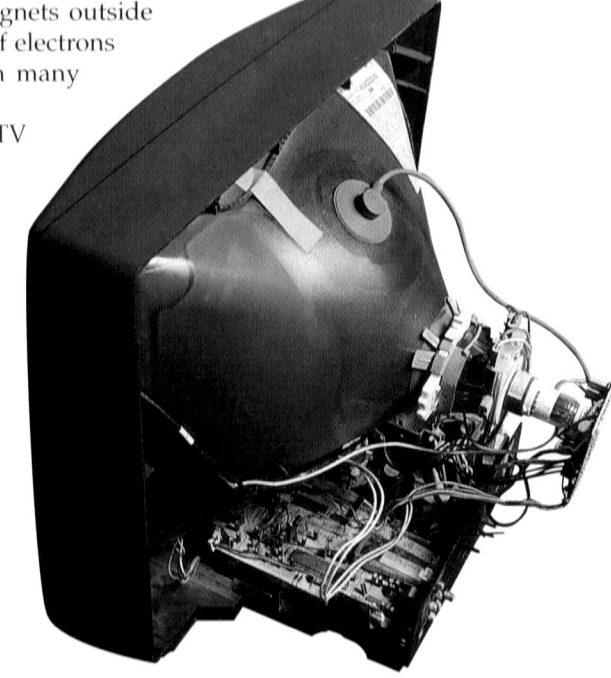

Section Wrap-up

Review

1. Explain what happens inside your radio when you tune in to a particular radio station.

2. How does a cathode-ray tube produce an image?

3. **Think Critically:** Explain the similarities in how eyes and CRTs in televisions produce color images.

Skill Builder
Concept Mapping
Make a concept map that shows the steps involved in radio broadcasts. Begin with the voice at the microphone and end with the radio waves leaving the transmitter. If you need help, refer to Concept Mapping in the **Skill Handbook.**

USING MATH

Radio waves travel at the speed of light, about 3.0×10^8 m/s. Calculate the wavelength of the radio waves if you are listening to an AM station with a frequency of 1000 kHz. (Hint: Recall that the velocity of a wave = wavelength × frequency.)

Skill Builder

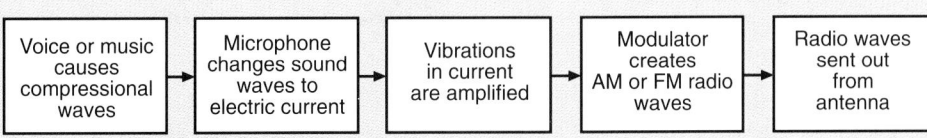

| Voice or music causes compressional waves | → | Microphone changes sound waves to electric current | → | Vibrations in current are amplified | → | Modulator creates AM or FM radio waves | → | Radio waves sent out from antenna |

---

4 Close

•MINI•QUIZ•

Use the Mini Quiz to check students' recall of chapter content.

1. **What is the role of the loudspeaker in a radio?** *to vibrate in response to an electrical signal, producing sound waves*

2. **What kind of signal transmits TV video images?** *amplitude-modulated*

Section Wrap-up

Review

1. The radio picks up and amplifies radio carrier waves of the frequency to which the receiver is tuned.

2. Electron beams cause specific materials on the screen of the tube to fluoresce. The intensity and pattern of the beams determine the image.

3. **Think Critically** Both the picture tube and the retina are sensitive to red, green, and blue. With the eye, there are red-, blue-, and green-sensitive cones. With the TV there are red-, blue-, and green-fluorescing materials.

USING MATH

$$\lambda = \frac{v}{f}$$

$1000 \text{ kHz} = 10^6 \text{ Hz}$

$$\lambda = \frac{3.0 \times 10^8 \text{ m/s}}{1 \times 10^6 \text{ Hz}} =$$

3.0×10^2 m

✓ Assessment

Performance Have students add the steps showing how radio waves become audible sounds. Use the Performance Task Assessment List for Scientific Drawing in **PASC,** p. 55. [P]

Prepare

Section Background

Magnetic disks are made by coating the disk with a thin liquid suspension of tiny magnetized iron oxide particles. These permanent magnets are aligned by a strong magnetic field as the liquid is dried, making them resistant to reasonably rough treatment.

Preplanning

- Make sure you have at least one computer available for demonstration purposes to use with instruction in this section.
- Locate calculators to use in the MiniLAB. Collect disk magnets for Activity 23-2.

1 Motivate

Bellringer

Before presenting the lesson, display **Section Focus Transparency 94** on the overhead projector. Assign the accompanying **Focus Activity** worksheet. `L1` `LEP`

Tying to Previous Knowledge

Take a poll: **How many students used a computer of some kind in the last week?** *Answers will vary. Most new cars, some calculators, grocery store cash registers, and even some appliances use microcomputers.* In this section students will learn about the structure and function of microcomputers.

23•3 Microcomputers

Science Words

microprocessor
RAM
ROM

Objectives

- Identify the basic parts of a microcomputer.
- Describe the role of the microprocessor.
- Distinguish between RAM and ROM.

Figure 23-8

When you think of a computer, you probably picture the input and output devices.

Microcomputer Components

How often do you use microcomputers? You may use them to play computer games, practice math problems, or locate a book in the library. A computer is a device you can program to carry out calculations and make logical decisions. How many simple addition problems can you do in 10 seconds? Computers can perform up to several billion calculations each second. This ability makes them efficient at processing and storing large amounts of information.

Microcomputers must have three main capabilities. First, they must be capable of receiving and storing the data or information needed to solve a problem. Next, microcomputers must be able to follow instructions to perform tasks in a logical way. Finally, microcomputers must communicate their information to the outside world. These requirements can be fulfilled with a combination of hardware and software components. Computer hardware refers to the major permanent components of the microcomputer, as shown in **Figure 23-8.**

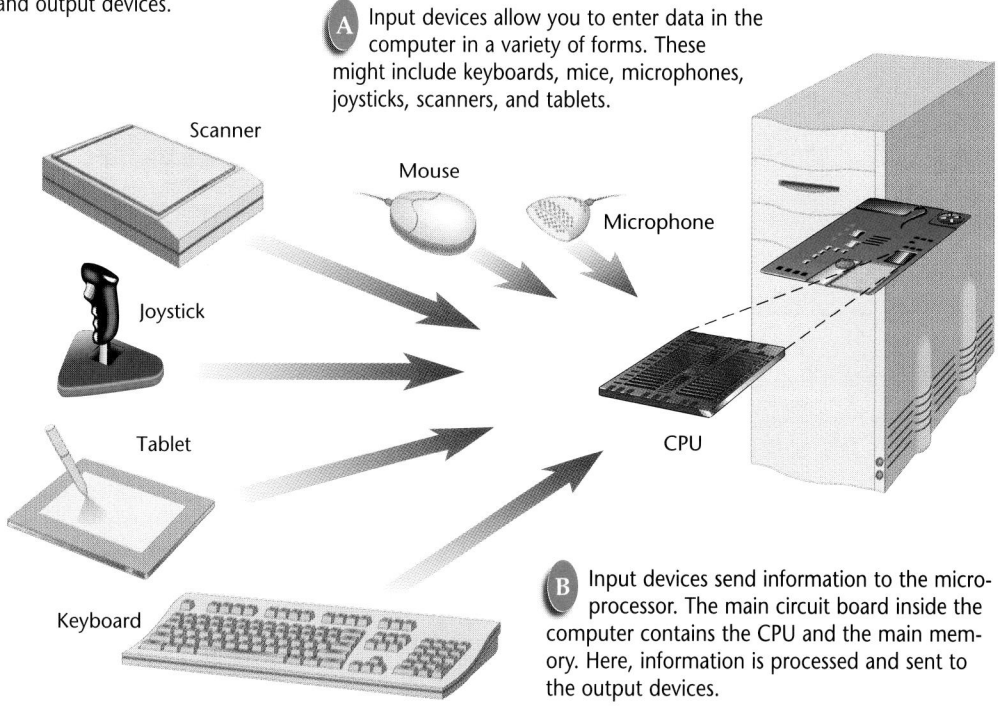

A Input devices allow you to enter data in the computer in a variety of forms. These might include keyboards, mice, microphones, joysticks, scanners, and tablets.

Scanner
Mouse
Microphone
Joystick
Tablet
CPU
Keyboard

B Input devices send information to the microprocessor. The main circuit board inside the computer contains the CPU and the main memory. Here, information is processed and sent to the output devices.

658 Chapter 23 Electronics and Computers

Program Resources

📁 **Reproducible Masters**
Study Guide, p. 98 `L1`
Reinforcement, p. 98 `L1`
Enrichment, p. 98 `L3`
Cross-Curricular Integration, p. 29
Science and Society Integration, p. 27
Activity Worksheets, pp. 138-139, 141 `L1`
Concept Mapping, p. 51

Multicultural Connections, pp. 49-50
Critical Thinking/Problem Solving, p. 29

📦 **Transparencies**
Section Focus Transparency 94 `L1`
Science Integration Transparency 23

Microprocessors

The first computers were very large and slow compared to the computers you are familiar with. The United States built several of the first computers shortly after World War II, between 1946 and 1951. These early computers operated on complex circuits composed of thousands of vacuum tubes. They used a lot of energy and were large enough to fill entire rooms. The invention of the integrated circuit rapidly improved the efficiency of computers. **Figure 23-9** illustrates a computer using vacuum tubes.

A **microprocessor** serves as the brain of the computer. A microprocessor is an integrated circuit on the main circuit board. It receives electrical input from the user and tells other parts of the computer how to respond, just as your brain tells your hand to move when you touch a hot pan. It might store information using a disk drive, change a video display, or make a sound. The microprocessor contains the central processing unit, or CPU, which is the circuitry that actually carries out the arithmetic and logical operations. CPUs are different among various kinds of microcomputers.

Figure 23-9

A modern home video game has more memory and performs more operations than this early computer that used vacuum tubes. *What inventions made this decrease in the size of computers possible?*

 Output devices, such as monitors, speakers, and printers, communicate information to the user.

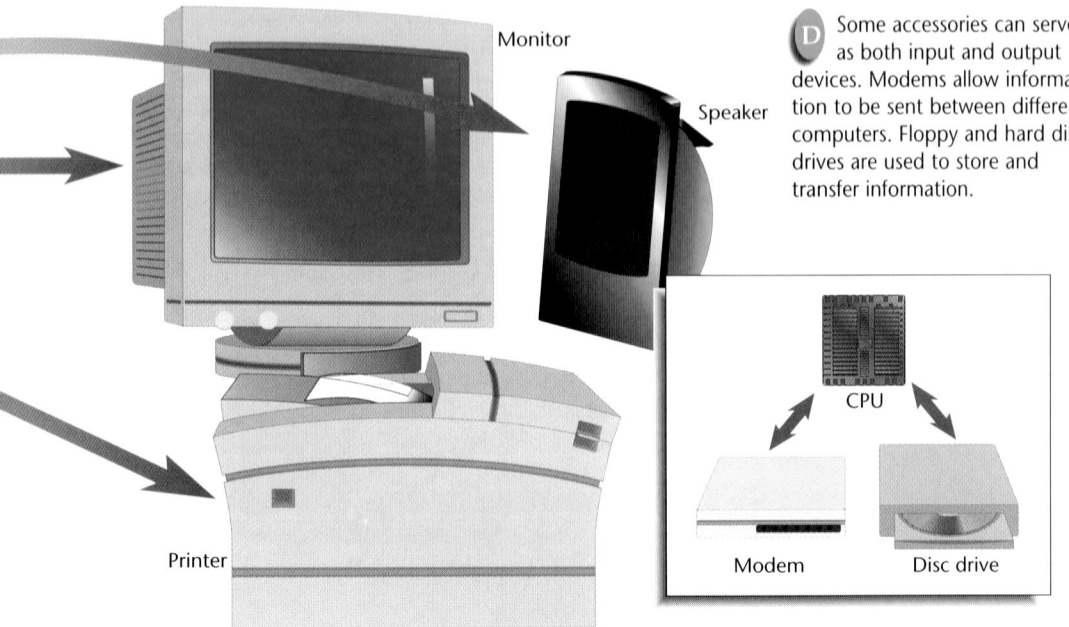

Monitor

Speaker

 Some accessories can serve as both input and output devices. Modems allow information to be sent between different computers. Floppy and hard disk drives are used to store and transfer information.

CPU

Printer

Modem

Disc drive

23-3 Microcomputers 659

Visual Learning

Figure 23-8 Have a computer set up in the room so students can see the various parts. Ask them what kinds of things the computer might be useful for. They may mention games, quiz questions, typing papers, or programming.

Figure 23-9 What inventions made this decrease in the size of computers possible? *The invention of transistors and then integrated circuits made miniaturization possible.*

LEP **LS**

GLENCOE TECHNOLOGY

Videodisc

The Infinite Voyage Series: Miracles by Design
Chapter 8
Ceramic Superconductors: Effects on the Computer

STVS: Chemistry
Disc 2, Side 1
High-Tech Ceramics
(Ch. 13)

STVS: Physics
Disc 1, Side 2
Computer Graphics (Ch. 16)

Computerized Star Imaging (Ch. 17)

Factory of the Future (Ch. 18)

Computerized Apple (Ch. 20)

Theme Connection

Systems and Interactions Use **Figure 23-8** to emphasize the theme of systems and interactions. Ask students how the capabilities of a computer system would be changed if any one element (input, output) were not present.

Science Journal

What has it done for me lately? Before beginning this section, have students write a paragraph describing all of the tasks or activities they have used a computer to do. Have them discuss other uses for computers they may have seen or heard about. *Students might mention video games, word processing, running programs, creating diagrams, or monitoring heart signals.* **L1** **LS**

Computer Memory

Information is collected and stored in the memory of the computer. The memory contains thousands of tiny circuits that have switches with two positions: open (off) or closed (on). Recall that a switch must be closed for current to flow. All computer information is processed with combinations of just two numbers, zero and one, to represent these situations. This is called a binary number system. Each 0 (off) or 1 (on) represents one binary digit and is called a bit. Numbers, letters, and symbols are represented in your computer by arrangements of eight bits called bytes. **Table 23-1** shows a commonly used binary code for numerical data.

Table 23-1

Binary Code					
Number	Code	Switch	Number	Code	Switch
0	0011 0000	○○●● ○○○○	5	0011 0101	○○●● ○●○●
1	0011 0001	○○●● ○○○●	6	0011 0110	○○●● ○●●○
2	0011 0010	○○●● ○○●○	7	0011 0111	○○●● ○●●●
3	0011 0011	○○●● ○○●●	8	0011 1000	○○●● ●○○○
4	0011 0100	○○●● ○●○○	9	0011 1001	○○●● ●○○●

Random Access Memory

Microcomputers have several kinds of memory. Temporary memory stores documents, programs, and data while they are being used. This temporary memory is called random access memory, or **RAM,** because any bit can be used in storing information. Because information is electronically stored in RAM, the information is lost when the computer is turned off.

Read Only Memory

When you purchase a computer, it comes with some information already stored in permanent memory. This information contains instructions required by the microprocessor to operate the computer. The computer can read this memory, but information can't be added to it. For this reason, it is called read only memory, or **ROM.** Information in ROM is permanently stored inside the computer and, therefore, isn't lost when the computer is turned off. Neither RAM nor ROM is useful in helping you store your work between uses.

Floppy Disks and Hard Drives

One of the main advantages of using computers is that you can save information and then come back later to make changes to it. Most computers have at least one disk drive for storing information on a floppy disk, **Figure 23-10B.** A floppy disk is a thin, round, plastic disk coated with a magnetic material, such as iron oxide. The disk is encased in a thicker plastic case for protection. Information can be saved on the disk, and the disk can be removed from the disk drive for storage. Floppy disks can be used to transfer data or programs from one computer to another computer.

Hard disk drives, **Figure 23-10A,** are found inside the main part of most computers. Hard disks are rigid metal disks that stay inside the computer and spin continuously when the computer is on. Hard drives are useful because they hold many times more information than floppy disks can hold. They also retrieve information much faster.

CD-ROM

Have you listened to music recorded on a compact disk, or CD? If so, you have used an optical disk, on which information is written and read by lasers. You probably know that although you can listen to it, you cannot record music yourself onto a CD. Computers can also read text, sound, and graphics recorded on a similar product called a CD-ROM, **Figure 23-10C.** What do you think *CD-ROM* stands for? This abbreviation represents the term *Compact Disk—Read Only Memory.* Remember that ROM means you cannot save information to the CD. CD-ROMs can hold much greater amounts

Figure 23-10

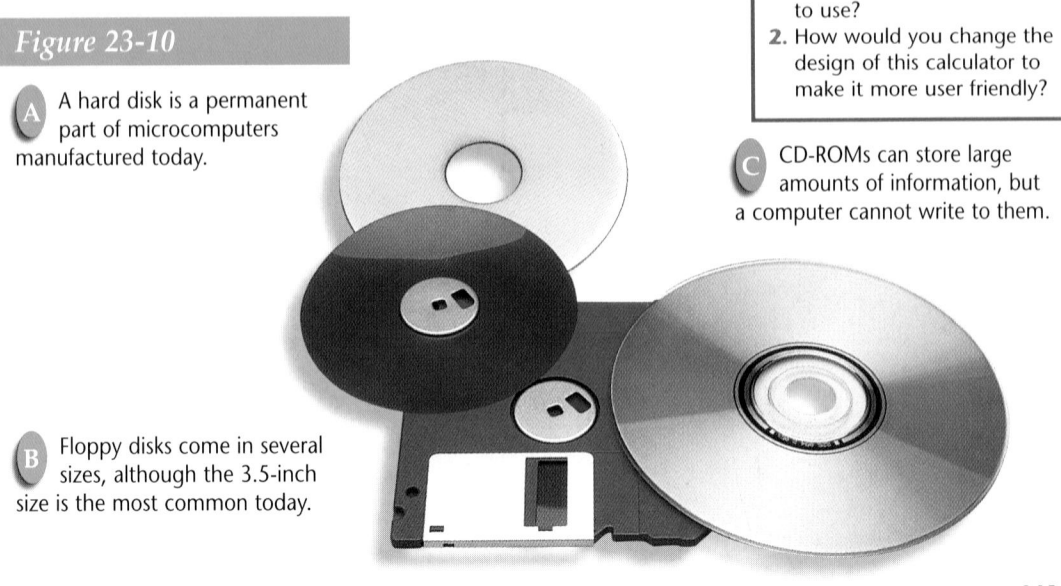

A A hard disk is a permanent part of microcomputers manufactured today.

B Floppy disks come in several sizes, although the 3.5-inch size is the most common today.

C CD-ROMs can store large amounts of information, but a computer cannot write to them.

Integrating the Sciences

Chemistry Floppy disks are coated with materials having strong magnetic properties. Elements with these properties, such as iron, cobalt, and nickel, are called ferromagnetic. **Find out what the prefix ferro- means.** Ferro- *comes from the Latin word* ferrum, *meaning iron.*

Try to initiate communication with students in a different part of the country or world.

Think Critically

You could use the Internet to contact people and library references in Kenya. By interacting with people in Kenya, you could ask specific questions and obtain examples not found in your local library.

Visual Learning

Figure 23-11 What advantages do computers provide for desktop publishing? *Editors can arrange and rearrange copy and illustrations many times before the page is actually produced.* **LEP** **LS**

3 Assess

Check for Understanding

? FLEX Your Brain

Use the Flex Your Brain activity to have students explore MEMORY STORAGE.

 **Activity Worksheets,** p. 5

Reteach

Draw an analogy between taking an open-note exam and the RAM/ROM difference. The material you do not know is written in front of you in the notes (RAM). If you take the notes away, the information is gone. Some answers are stored permanently in your brain (ROM). **L1**

Extension

 For students who have mastered this section, use the **Reinforcement** and **Enrichment** masters.

Navigating the Net ▼ ▲

Imagine sitting down at a computer in your home or school and being able to access information from computers around the world within seconds! This is possible with use of the Internet, an international system of thousands of smaller networks linked together by wired or wireless connections. Many companies and schools provide Internet access free of charge to employees or students, but you can also pay a monthly fee for access through an Internet service. All you need is a computer, a modem with communications software, and a phone line.

The most popular service on the Internet is electronic mail, or E-mail, which enables users to send and receive messages between computer terminals around the world in just seconds. People can even "talk" over Internet by typing conversation back and forth. Bulletin board discussion groups in a system called Usenet allow people to post messages and read comments from around the world about several thousand different topics of interest.

Because the information available on the Internet changes daily, a directory is not available.

Think Critically: How might the Internet be helpful if you were assigned to write a paper about the government of Kenya in your social studies class?

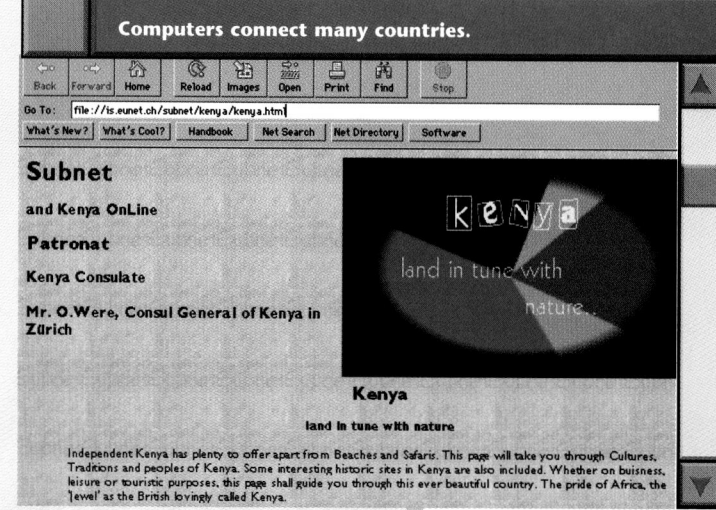

Computers connect many countries.

Across the Curriculum

Computer Science Write a BASIC program to carry out a simple mathematical function, such as adding or multiplying two numbers. Some students will be able to do this with little difficulty, so you can make the program more challenging.

---

of information than floppy disks and even more than some hard disks. They are commonly used to hold reference materials such as encyclopedias, magazines, and digitized video and sound.

Uses for Computers

Computers can't complete a task without instructions for carrying out a series of operations. A program is a group of instructions that tell a computer what to do. Programs are sometimes referred to as software. Computer programmers use special languages that convert your language into instructions the CPU can understand. BASIC, Pascal, Fortran, and COBOL are examples of computer languages. Whenever you play a computer game, use a word processor, or solve mathematical problems, a computer program is instructing the computer to perform in a certain way.

Hidden Computers

As shown in **Figure 23-11,** microcomputers are becoming useful in more and more applications, such as desktop publishing. You often use them in situations where a computer screen is not visible. For example, computers are regularly used to regulate mechanical processes in cars, to monitor heating and cooling systems in buildings, and to enter inventory bar codes in grocery stores. Many calculators can now be programmed, as well. Even some kitchen appliances contain simple, small computers to enhance their operations. Microcomputers would not exist without advances in electronics, such as the transistor and the integrated circuit. How do you think we'll use microcomputers of the future?

Figure 23-11

Look closely at the computer screen to the right and you will see a two-page spread from this book. *What advantages do computers provide for desktop publishing?*

Section Wrap-up

Review

1. What are the functions of the three main parts of a computer?

2. What is the difference between RAM and ROM?

3. **Think Critically:** How would you transfer a computer program from your computer to a friend's computer?

Skill Builder
Comparing and Contrasting
Compare and contrast the advantages and disadvantages of floppy disk drives and hard disk drives. If you need help, refer to Comparing and Contrasting in the **Skill Handbook.**

Using Computers

Word Processing
Locate a microcomputer in your school. Sketch the system and label any parts you can identify as input, output, or processing devices. Next, examine the software on the machine. Using a word processor, write a report about the hardware and software telling what programs you find and what the purpose of each one is.

23-3 Microcomputers **663**

Skill Builder
Floppy disk drives are useful because floppy disks can be used to transport information from one computer to another and are relatively inexpensive. However, hard disk drives can store much more information and can rapidly access information. The main disadvantage of hard disk drives is that they are expensive.

 **Assessment**

Process Have students use their comparisons to explain why both hard and floppy disk drives are commonly used in a computer. Use the Performance Task Assessment List for Making Observations and Inferences in **PASC,** p. 17. **P**

4 Close

•MINI•QUIZ•

Use the Mini Quiz to check students' recall of chapter content.

1. **What does CPU stand for? What does the CPU do?** *Central processing unit; it is the circuitry that performs arithmetic and logical operations.*

2. **Which kind of memory can be readily accessed and used to store temporary information?** *RAM*

3. **Which holds more information, hard disks or floppy disks?** *hard disks*

4. **A(n) _____ is a set of instructions that tells a computer what to do.** *program*

Section Wrap-up

Review

1. They must receive and store data, decide what operations to perform next, and communicate with the outside world.

2. Information can be added to RAM by the user and read by the computer. Information is temporarily stored in RAM and is lost when the computer is turned off. ROM is permanent memory that the computer can only read, giving it instructions for operating the microprocessor.

3. **Think Critically** Transfer the program to a floppy disk, or use modems or a computer network.

Using Computers

Accept all reasonable, supported answers.

Design an Experiment
Magnetic Messages

Computers represent, process, and store information using a binary number system. In this system, coded combinations of the numbers zero and one represent all characters and symbols entered into a computer. Can you design a binary code and store a message magnetically?

PREPARATION

Purpose
LS Interpersonal Students will invent a system of communication that stores information in magnetic fields. **L2**

COOP LEARN

Process Skills
classifying, communicating, making models, making and using tables, observing and inferring, forming hypotheses, using numbers

Time
one 45-minute class period

Materials
Provide each team with a thin strip and two sheets of card stock and at least 16 ceramic magnets. Small stars or stickers may be used to indicate the poles of magnets.

Alternate Materials Any shape of magnets with poles on the faces will work, but the magnets should be of equal strength.

Possible Hypotheses
Students may hypothesize that each letter of the alphabet can be coded by a sequence of magnets with specific poles up or down. By knowing the polarity of a reading magnet and the code, the message could be "read."

📁 **Activity Worksheets,** pp. 5, 138-139

PLAN THE EXPERIMENT

Possible Procedures
Students should devise a binary code for the alphabet and label their magnets as instructed. They should recall that opposite poles attract. By knowing the polarity (as a 1 or 0) of the down side of the test magnet, they should be able to determine the hidden message.

Problem
How can you invent a binary code for the letters of the alphabet and use magnets to write a word in this code?

Form a Hypothesis
Develop a hypothesis about how a binary code could be used to represent all letters of the alphabet. In addition, hypothesize how the attraction and repulsion between magnets could be used to encode a message with your binary code.

Objectives
• Invent a code for each letter of the alphabet using a sequence of five binary digits.

• Use magnets to write and read a message using the binary code.

Possible Materials
• disk magnets, 16
• card stock paper, 2 sheets and a thin strip
• binder-hole reinforcing stickers or other symbols
• transparent tape
• metric ruler

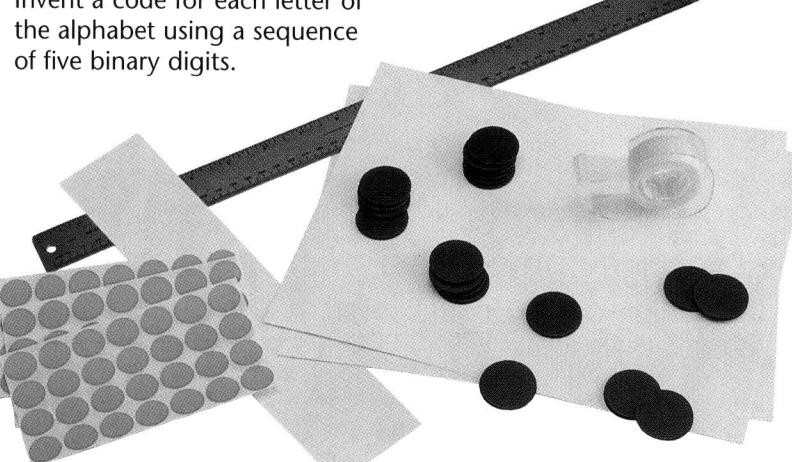

PLAN THE EXPERIMENT

1. As a group, agree upon and write out the hypothesis statement.
2. Invent and record your binary code for each letter of the alphabet. Each letter should be represented by five binary digits, with each digit having the value 0 or 1. For example, the code for *A* might be 00001, *B* might be 00010, and so on.

Check the Plan
1. Put all of the magnets in a single stack. Use small stickers or tape to label the top side of each magnet with the number 1 and the bottom with the number 0.
2. Use a ruler to divide one sheet of the card stock into three rows of five squares as shown. Mark the second sheet in exactly the same way.

3. Choose a three-letter word to write in your magnetic code. Place the code for one letter on each line, and tape the magnets to the cardboard with the appropriate code number right-side up.
4. Cover the magnets with the second sheet marked with the locations of the squares.
5. How will you test and decode the hidden message? Devise a way to use the remaining magnet to test the digit under each square.
6. *Make sure your teacher approves your plan and that you have included any changes suggested in the plan.*

DO THE EXPERIMENT

1. Carry out the magnetic testing of the message as planned, recording in your Science Journal which side of each hidden magnet you believe is facing up.
2. Use the results of your magnetic testing to decode the message.

Analyze and Apply
1. **Explain** how you know which side of the magnet is facing up underneath your testing magnet.

2. Which side, 0 or 1, of your testing magnet was facing the code? If you were to repeat the experiment while holding your test magnet the opposite way, **hypothesize** how your interpretations would differ.
3. **Infer** why there are only two numbers in this magnetic code. Could you think of each digit as being like an off/on switch? Explain.

Go Further

List your operating rules for reading your word. Communicate your rules, letter code, and magnetic message with another team, and see whether they can decode your message.

23-3 Microcomputers   **665**

✓ Assessment

Oral Ask students if it would be possible to invent a letter code based on three digits (0, 1, 2) using the same method of magnetic decoding. Discuss the fact that there would be no way to have three possible answers if only two options occur: attraction and repulsion. Use the Performance Task Assessment List for Making Observations and Inferences in **PASC**, p. 17. **P**

Teaching Strategies
- This is a good activity with which to use cooperative learning strategies. If students are working in groups, you may wish to divide the tasks of labeling the magnets, designing the code, and testing the code.
- Follow-up discussion can further develop the lab analogy to the writing and reading of information on the magnetic material of a computer disk.

DO THE EXPERIMENT

Expected Outcome
Students will be able to decode the messages by observing attraction and repulsion with the reading magnet.

Analyze and Apply
1. The reading magnet will be either attracted or repelled depending on which pole of the message magnet is facing up.
2. Students will answer 0 or 1. The observations of attraction and repulsion would be reversed if the reading magnet were reversed.
3. The magnet has only two poles so the only possible responses are attraction and repulsion. In this sense, you could say attraction was "on" and repulsion was "off."

Go Further
Team operating rules will vary. The reader must know which way to scan, right to left, left to right, top to bottom, and so on. The reader must also know if attraction to the reader magnet means the message magnet is 1 or 0.

Prepare

Section Background

A distinction is sometimes made between a computer virus and a worm program. A virus replicates itself and attaches itself to other programs to produce sometimes bizarre effects. A worm loads a system with information that replicates itself and fills the memory and files in the computer. This slows or stops the computer from functioning.

1 Motivate

Bellringer

Before presenting the lesson, display **Section Focus Transparency 95** on the overhead projector. Assign the accompanying **Focus Activity** worksheet. **L1** **LEP**

Tying to Previous Knowledge

In this chapter, students have studied microcomputers. With modems it is possible for computer systems in different places to communicate with each other through phone lines. Explain how this is a possible problem, as it provides a door that can be broken into.

2 Teach

Discussion

Photocopy or make an overhead transparency of a software copyright agreement. Discuss how easy it is to copy software, even with the copyright.

TECHNOLOGY:
23•4 Computer Crimes

Science Words

computer virus

Objectives

- Discuss the types of crimes that can be committed by computer misuse.
- Predict the possible consequences of computer crimes.

How serious are computer crimes?

What do you think of when you hear the word *crime?* You probably think of stealing, vandalism, or violence. An increasingly common type of crime is being committed by the misuse of computers.

You may have copied a friend's cassette tape onto a blank tape for your own personal use. Computer programs can also be copied from one floppy disk to another. When you purchase a program, the software company usually intends for you to make one backup copy in case the original one becomes damaged. The companies don't intend for you to copy the program to give to your friends. Cassette tapes and many computer programs are protected by copyright laws, as shown in **Figure 23-12.** However, some programs are free and can be legally shared. Sharing of illegal copies of software is probably the most common of computer crimes, as well as the most difficult computer crime to safeguard against.

Figure 23-12

Most purchased computer programs are copyrighted, making it illegal to share copies with other computer owners. You can protect a disk from contracting a virus by keeping the window open when loading files from the disk. This prevents any new material from being written to the disk.

Computer Viruses—A Contagious Problem

In recent years, computer viruses have become a problem. A **computer virus** is a type of program that can multiply inside a computer and use so much memory that it harms the system. Like a virus in your body, it can remain inactive in the computer until something causes it to spread. As a result of the virus attack, data in memory might be lost or a strange message might appear on the screen. The infected computer might even just stop functioning. Viruses can be sent through phone lines linking networked computers or can spread from infected

---

Program Resources

 Reproducible Masters
Study Guide, p. 99 **L1**
Reinforcement, p. 99 **L1**
Enrichment, p. 99 **L3**
Science Integration Activities, pp. 45-46

Transparencies
Section Focus Transparency 95 **L1**

software shared between computers. Sometimes, antivirus programs can find and destroy viruses before they spread or even fix the damage already done. Why do you think it is a crime to purposefully spread a computer virus?

Hacking—Finding Security Flaws

Perhaps the most controversial computer crime is hacking, or entering closed computer or telecommunications systems without permission. Hackers may have several motives. Some hackers claim they try to gain unauthorized access to a system for the intellectual challenge. They argue that breaking into computer systems helps improve security measures by pointing out their flaws.

Hacking is an invasion of privacy. Those who break into systems for malicious purposes are sometimes called crackers. Crackers may be seeking financial information, stealing software, trying to destroy data or computer hardware, or even altering records. This type of crime is so new that until recently law enforcement personnel have been hampered by a lack of laws defining computer crime. Now that the extent and seriousness of the activities of hackers and crackers have been recognized, steps are being taken to pass specific legislation dealing with these crimes. Many believe hacking and cracking are like physically breaking into a home or business and stealing goods or information. They argue that those who enter systems illegally should be prosecuted as criminals. What do you think?

Section Wrap-up

Review

1. How does a computer virus interfere with a computer?

2. What is computer cracking?

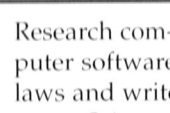
interNET CONNECTION How can a company or organization protect their computer systems from hacking or computer viruses? Are there requirements or standards that help define how safe a system is? Visit the Chapter 23 Internet Connection at Glencoe Online Science, **www.glencoe.com/sec/science/physical,** for a link to more information about computer crimes.

SCIENCE & SOCIETY

23-4 Computer Crimes   **667**

Science Journal

Research computer software laws and write a report in your Science Journal. Distinguish between a felony and a misdemeanor.

Science Journal

Felonies and Misdemeanors Felonies and misdemeanors are distinguished by the severity of the punishment, misdemeanors being the less severe. Federally, a crime is a felony if the punishment is death or imprisonment for more than one year. **LS**

3 Assess

Check for Understanding

? FLEX Your Brain

Use the Flex Your Brain activity to have students explore COMPUTER CRIMES.

📁 **Activity Worksheets,** p. 5

Reteach

LS Linguistic For students who have had life science, you could draw a useful analogy between viruses that cause human diseases and computer viruses. For example, the virus that causes AIDS can be inactive for years before it spreads and attacks the body. **L1**

Extension

📁 For students who have mastered this section, use the **Reinforcement** and **Enrichment** masters.

4 Close

LS Interpersonal Have students work in groups to make a list of different types of computer crimes. **L1** **COOP LEARN**

Section Wrap-up

Review

1. A virus multiplies and expands to fill the memory of a computer. It may prevent the computer from functioning.

2. Cracking is breaking into a computer system to gain information without permission.

Background

- Urban planners like Dr. Banerjee foresee that the widespread use of computers may have an enormous impact on the traditional structure of cities. Dr. Banerjee poses this question: What will be the future of the human-made environments, especially huge office buildings, when fewer people have to report physically at their place of employment? He also wonders how the growth of shopping by computers will affect shopping centers. Since he feels that people go to malls to see and be seen as well as to shop, shopping centers fulfill a basic human need to come together. He continues by saying that conventional public spaces, such as parks and plazas, are declining in popularity, partly because of crime, and malls are taking over their function.

- GIS has long been used by energy companies, utilities, railroads, and governments. With appropriate computer programs, these agencies can perform such tasks as plotting the route of pipelines, tracking the movement of populations, and mapping minerals in geologic formations. Here are some other applications of GIS: mapping the location of public-assistance services that deal with disasters like fires, floods, and earthquakes; establishing locations and patterns of crime; monitoring pest infestation of crops; planning tree-cutting in a forest.

- GIS is particularly useful in planning wildfire-protection programs. The maps involved could pinpoint the location of fire hydrants, escape trails, and fire-engine turnarounds.

People and Science

DR. TRIDIB BANERJEE, *Urban Planner*

On the Job

Q In what specialized ways are computers used in the Urban Planning Department at your university?

A Geographic Information Systems— GIS for short—are made up of digitalized maps and computer software that allow manipulation of demographic data on the computer screen. This complicated-sounding equipment lets planners play the old game of "What If?" For instance, they might ask themselves, "What if there were a new freeway in this part of the city?" With GIS, planners can look ahead at what could happen and make decisions based on a simulation of reality.

Q Can GIS be helpful to the average citizen?

A Our department is placing computers with GIS capability in community centers and libraries. The aim is to assist in job creation and economic development. Say, for instance, that a person wants to open a laundromat. He or she needs to learn what the potential market is for that business. GIS can show the demographics of the

Personal Insights

proposed laundromat's location—who lives there and what the income level is. It could also show what similar, competing businesses are already in the area. A person starting a business needs this kind of information to qualify for a bank loan.

Q Why did you choose your career?

A Cities—like Calcutta, where I grew up—are fascinating phenomena. They are becoming the natural habitat of humanity, since 80 percent of Americans now live in cities. Making cities good places to live, work, and play requires careful planning.

Career Connection

Gather some data about your city or town that could be used by urban planners. On a map, mark the areas where young people gather, such as schools, playgrounds, and entertainment centers. Use your data to show where it might be most logical to construct community centers with programs for young people.

Investigate the following careers:

- **Data Analyst**
- **Data Entry Operator**

Teaching Strategies

If possible, make available to students computer simulation games that involve planning for the growth and problems of imaginary cities.

Career Connection

Urban planners deal with the problems and potentials of large cities. A few of the areas they study include urban design, urban history, social policies, transportation, and demographic and population trends. Careers for urban planners exist in the public, private, and volunteer sectors. College undergraduate and graduate degrees are usually required in this field.

Summary

23-1: Semiconductor Devices

1. The conductivity of semiconductors can be increased by adding impurities to them.
2. A rectifier changes alternating current into direct current.
3. Electronic devices containing integrated circuits are much smaller than those that don't contain these circuits.

23-2: Radio and Television

1. Radio and television signals are transmitted by changing electrical signals into electromagnetic radiation.
2. A cathode-ray tube uses electrons and fluorescent materials to produce pictures on a television screen.

23-3: Microcomputers

1. The basic parts of a microcomputer include the input devices, the output devices, disk drives, and the main circuit board.
2. The microprocessor receives input and tells the computer how to respond.
3. RAM is the temporary memory in a computer. ROM is memory that is permanently stored inside the computer.

23-4: Science and Society: Computer Crimes

1. Computer crimes include illegally copying software, planting computer viruses, and hacking.
2. Some computer crimes can involve stealing restricted information or altering records.

Key Science Words

a. amplification
b. cathode-ray tube (CRT)
c. computer virus
d. diode
e. integrated circuit
f. microprocessor
g. RAM
h. rectifier
i. ROM
j. transistor

Reviewing Vocabulary

Match each phrase with the correct term from the list of Key Science Words.

1. miniature components on a slice of silicon
2. any device that converts AC to DC
3. amplifies an electric signal
4. the process of making an electric current stronger
5. a TV picture tube
6. receives input and tells the computer how to respond
7. a computer's temporary memory
8. a computer program that multiplies and affects a computer's memory
9. contains stored information needed by a microprocessor to operate a computer
10. a type of rectifier

Summary

Have students read the summary statements to review the major concepts of the chapter.

Reviewing Vocabulary

1. e	6. f
2. h	7. g
3. j	8. c
4. a	9. i
5. b	10. d

✔ Assessment

Portfolio Encourage students to place in their portfolios one or two items of what they consider to be their best work. Examples include:

- Connect to Chemistry explanation, p. 649
- Science Journal paragraph, p. 653
- Explanation paragraph for Activity 23-2 code and message, pp. 664-665 **P**

Performance Additional performance assessments may be found in **Performance Assessment** and **Science Integration Activities**. Performance Task Assessment Lists and rubrics for evaluating these activities can be found in Glencoe's **Performance Assessment in the Science Classroom**.

GLENCOE TECHNOLOGY

▣ MindJogger Videoquiz

Chapter 23 Have students work in groups as they play the Videoquiz game to review key chapter concepts.

Checking Concepts

1. a   **6.** a
2. a   **7.** a
3. d   **8.** b
4. a   **9.** a
5. b   **10.** c

Understanding Concepts

11. In diodes, semiconductors are used to convert an alternating current into a direct current. In transistors, they amplify the electric signal or act as on-off switches.

12. The advantages of integrated circuits are that they have little electrical resistance and minimize energy loss, and a current can travel faster through the circuit. Integrated circuits cannot handle large currents, and they are too small to be easily manipulated.

13. Computers must be able to store data, perform tasks, and have some way to communicate with the outside world.

14. Computer crime includes copying and distributing copyrighted computer software, sending a computer virus to grow and wipe out a computer's memory, and breaking into closed computer systems. Answers will vary, but students should support their answers.

15. Answers may include such activities as playing computer games, writing papers on word processors, buying groceries, and driving a car.

Thinking Critically

16. CD-ROMs store huge amounts of information, but you cannot save information to the CD yourself.

17. It is not necessary to take the CD player back be-

Checking Concepts

Choose the word or phrase that completes the sentence.

1. Elements that are semiconductors are located on the periodic table _____.
   a. between metals and nonmetals
   b. on the right side
   c. at the bottom
   d. on the left side

2. Solid-state electronic devices use _____.
   a. low current   c. no current
   b. high current   d. vacuum tubes

3. Rectifiers are used to _____.
   a. change DC to AC
   b. make sure something is right side up
   c. amplify radio signals
   d. change AC to DC

4. The signal in an electronic device that represents sound and images is a _____.
   a. varying current   c. radio wave
   b. constant current   d. sound wave

5. Transistors are used in all of the following except _____.
   a. compact discs   c. hearing aids
   b. TV vacuum tubes   d. tape players

6. Integrated circuits consist of _____.
   a. doped silicon chips with photographed wires
   b. generators and appliances
   c. televisions and radios
   d. vacuum tubes and transistors

7. The signals from an FM radio station are transmitted to your home as _____.
   a. frequency-modulated radio waves
   b. sound waves
   c. cathode rays
   d. electric current

8. Computer hardware consists of all of the following except _____.
   a. video display terminal
   b. programs
   c. disk drives
   d. keyboards

9. A computer represents information using _____.
   a. a binary number system
   b. sequences of numbers
   c. arrangements of four bits
   d. the numbers 1 through 10

10. Computer data and programs both can be stored permanently and transferred to other computers using _____.
   a. RAM   c. floppy disks
   b. ROM   d. hard disks

Understanding Concepts

Answer the following questions in your Science Journal using complete sentences.

11. How are semiconductors used in diodes and transistors?

12. Describe the advantages and disadvantages of integrated circuits.

13. Identify three functions of a microcomputer.

14. Describe three types of computer crimes. Which do you think is the most serious? Why?

15. Describe how computers affect you in everyday life.

Thinking Critically

16. What are the advantages and disadvantages of using a CD-ROM as a data-storage method?

17. You have connected your new compact disc player directly to your stereo speakers, but when you turn on the CD, you hear nothing. Should you return the CD player to the store? Why or why not?

18. Which would do more harm to the operation of a computer, the loss of its random access memory or the loss of its read only memory? Explain your answer.

cause the only problem is that an amplifier was never hooked up to the player. The CD player's signal is too weak to vibrate the speakers and requires a larger outside current provided by the amplifier current.

18. The loss of read only memory would be much more serious because the information stored here is absolutely necessary for the operation of the computer by the microprocessor.

19. The best way to store such information

would be to store it on the computer's hard disk because this disk can hold much more information, and it can be retrieved much more rapidly than from a floppy disk.

20. Because of problems such as user error and computer viruses, computer records could be irretrievably lost. Written records are necessary for these reasons, as well as in case a hacker invades the school's computer system and changes grades.

19. You need to store information on your computer that you can retrieve rapidly whenever you use your computer. What is the best way to store this information?

20. Why might your school keep written records even when it has computers that can store these records much more efficiently?

Developing Skills

If you need help, refer to the **Skill Handbook.**

21. **Recognizing Cause and Effect:** Complete the table below by identifying either the cause or effect of problems associated with cathode-ray tubes.

Cathode-Ray Tubes

Cause	Effect
1. Inside of tube coated with a nonfluorescent material.	1. Fluorescent light is not produced—no picture.
2. Electromagnets outside CRT fail to function properly.	2. Electron beam doesn't sweep the entire screen.
3. Incoming electrical signals are garbled.	3. Patterns and colors of images are garbled.

22. **Comparing and Contrasting:** Compare and contrast the arrangement of electrons in conductors and insulators. Discuss why the arrangement in conductors makes them suitable to conduct electric currents.

23. **Making and Using Tables:** Construct a table to organize what you know about the parts of the microcomputer and the function of each.

24. **Concept Mapping:** Make a concept map showing the transfer of radio waves from the receiving antenna to your ear as sound waves.

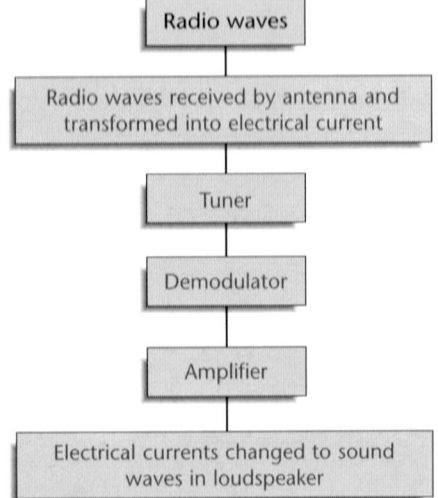

25. **Formulating a Hypothesis:** While using a word processor to write a paper, suppose you notice smiley faces appearing throughout your text. Suggest a possible explanation for this problem.

Performance Assessment

1. **Booklet:** If you are computer-wise, help novices by developing a dictionary of computer terms. Start with the terms in computer ads.

2. **Formulating a Hypothesis:** In Activity 23-2 on pages 664-665, you communicated information magnetically. Hypothesize why your "reading" magnet might no longer interact with your code magnets after being dropped.

3. **Poster:** Make a poster showing the passage of information through a computer from an input device to an output device.

Developing Skills

21. **Recognizing Cause and Effect**
   Effect: 1. Fluorescent light is not produced. No picture.
   Cause: 2. Electromagnets outside the CRT fail to function properly.
   Effect: 3. Patterns and colors of image are garbled.

22. **Comparing and Contrasting** In conductors, the outer electrons are loosely bound; in insulators they are tightly bound. Conductors are suitable for antennas because when radio waves reach an antenna, they exert a force on the loosely bound electrons that can generate an electric current.

23. **Making and Using Tables** See table below.

24. **Concept Mapping** See student page.

25. **Formulating a Hypothesis** Your computer may have picked up a computer virus.

Performance Assessment

1. Answers will vary. Be sure all terms are correctly defined. Use the Performance Task Assessment List for Booklet or Pamphlet in **PASC,** p. 57. **P**

2. Dropping a magnet can cause the magnetic domains to change alignment and reduce the magnetic effect. Use the Performance Task Assessment List for Formulating a Hypothesis in **PASC,** p. 21. **P**

3. Posters should be similar to the illustration in Figure 23-8 on pp. 658-659. Use the Performance Task Assessment List for Poster in **PASC,** p. 73. **P**

Component	Function
Input devices (keyboards, etc.)	Allow the user to communicate with the computer
Output devices (screen, etc.)	Allow computer to communicate with the user
Microprocessor	Carries out the computer's operations; receives and translates the user's commands to the computer
RAM	Temporarily stores information while the computer is being used
ROM	Stores the information needed by the microprocessor
Floppy disk	Contains information and programs to be stored and transferred to other computers
Hard disk	Contains information and programs to be stored permanently in the computer

Section	Objectives/Standards	Activities/Features
Chapter Opener		Explore Activity: Model a nuclear reaction. p. 673
24-1 Radioactivity (1 session, ½ block)*	**1. Discuss** the discovery of radioactivity. **2. Contrast** properties of radioactive versus stable nuclides. National Content Standards: (5-8) UCP1-UCP3, A2, E2, F1, F5, G1, G3; (9-12) UCP1-UCP3, A2, B2, E2, F1, F5, F6, G1, G3	MiniLAB: How are neutrons and protons related to atomic stability? p. 675 Using Technology: Irradiated Food, p. 676 Science Journal, p. 677 Skill Builder: Comparing and Contrasting, p. 677 Science and Art: Is it real or is it fake? p. 678 Science Journal, p. 678
24-2 Nuclear Decay (1 session, ½ block)*	**1. Distinguish** among alpha, beta, and gamma radiation. **2. Calculate** the amount of a radioactive substance remaining after a time based on its half-life. **3. Relate** half-life to the process of radioactive dating. National Content Standards: (5-8) UCP2, UCP3, UCP5, A1, A2, B1, D2, E1; (9-12) UCP2, UCP3, UCP5, A1, A2, E1	Using Math, p. 680 Problem Solving: Invisible Invader, p. 681 Using Math: Calculating the Quantity of a Radioactive Element, p. 682 Using Math, p. 683 Skill Builders: Interpreting Scientific Illustrations, p. 683 Activity 24-1: Investigating Half-Life, pp. 684-685
24-3 Detecting Radioactivity (1 session)*	**1. Describe** how radioactivity can be detected. **2. Explain** how a Geiger counter can determine the quantity of nuclear radiation present. National Content Standards: (5-8) UCP2, UCP3, UCP5; (9-12) UCP2, UCP3, UCP5	Connect to Earth Science, p. 687 Using Computers, p. 688 Skill Builder: Observing and Inferring, p. 688
24-4 Nuclear Reactions (2 sessions, 1 block)*	**1. Distinguish** between nuclear fission and fusion. **2. Explain** how nuclear fission can begin a chain reaction. **3. Discuss** how nuclear fusion occurs in the sun. National Content Standards: (5-8) UCP2, UCP3, UCP5, D3, G1, G3; (9-12) UCP2, UCP3, UCP5, B2, D3, D4, G1, G3	MiniLAB: How does a chain reaction start? p. 690 Connect to Chemistry, p. 691 Science Journal, p. 691 Skill Builder: Concept Mapping, p. 691
24-5 Science and Society (1 session, ½ block)*	**1. Describe** how radioactive tracers and PET can be used to diagnose medical problems. **2. Discuss** how radioactive isotopes can aid in the treatment of cancers. National Content Standards: (5-8) UCP2, UCP3, UCP5, C1, C2, E2, F1, F5; (9-12) UCP2, UCP3, UCP5, C1-C3, E2, F1, F6	Explore the Technology, p. 693 Activity 24-2: Figuring Out Fusion, p. 694

Activity Materials

Explore	Activities	MiniLABs
page 673 32 marbles, clay, large beaker, watch or clock with second hand	pages 684-685 several pennies, graph paper page 694 6 balls of green clay, 6 smaller balls of white clay	page 675 pencil, paper page 690 dominoes

Need Materials? Call Science Kit (1-800-828-7777).   * A complete Planning Guide that includes block scheduling is provided on pages 32T-35T.

Teacher Classroom Resources

Reproducible Masters	Transparencies	Teaching Resources
Study Guide, p. 100 **Reinforcement**, p. 100 **Enrichment**, p. 100 **Activity Worksheets**, p. 146 **Lab Manual 47**, The Effect of Radiation on Seeds **Critical Thinking/Problem Solving**, p. 30	**Section Focus Transparency 96,** Changes of a Third Kind	**Glencoe Physical Science Interactive Videodisc** **Physical Science CD-ROM** **Spanish Resources** **English/Spanish Audiocassettes** **Cooperative Learning Resource Guide** **Lab Partner** **Lab and Safety Skills** **Lesson Plans**
Study Guide, p. 101 **Reinforcement**, p. 101 **Enrichment**, p. 101 **Activity Worksheets**, pp. 142-143 **Lab Manual 48**, Radioactive Decay— A Simulation **Multicultural Connections**, pp. 51-52 **Concept Mapping**, pp. 53-54	**Section Focus Transparency 97,** The Shroud of Turin—How Old? **Science Integration Transparency 24,** Isotopes and Dating Techniques	
		Assessment Resources
Study Guide, p. 102 **Reinforcement**, p. 102 **Enrichment**, p. 102 **Science Integration Activity 24,** Absorption of Beta Radiation **Science and Society Integration**, p. 28	**Section Focus Transparency 98,** Detecting Overexposure to Radiation	**Chapter Review**, pp. 51-52 **Assessment**, pp. 155-158 **Performance Assessment in the Science Classroom (PASC)** **MindJogger Videoquiz** **Alternate Assessment in the Science Classroom** **Performance Assessment** **Chapter Review Software** **Computer Test Bank**
Study Guide, p. 103 **Reinforcement**, p. 103 **Enrichment**, p. 103 **Activity Worksheets**, p. 147	**Section Focus Transparency 99,** Nuclear Reactors in the United States **Teaching Transparency 47,** Types of Nuclear Radiation **Teaching Transparency 48,** Nuclear Chain Reaction	
		Key to Teaching Strategies
Study Guide, p. 104 **Reinforcement**, p. 104 **Enrichment**, p. 104 **Activity Worksheets**, pp. 144-145 **Cross-Curricular Integration**, p. 30	**Section Focus Transparency 100,** Using Radioisotopes	The following designations will help you decide which activities are appropriate for your students.

GLENCOE TECHNOLOGY

The following multimedia resources are available from Glencoe.

Science and Technology Videodisc Series (STVS)
Chemistry
  Carbon-14 Dating
Neutron Activation
  Analysis of Paintings
Dating by Thermoluminescence

Earth & Space
  Fossil CAT Scan
Human Biology
  PET Scanner

The Infinite Voyage Series
Unseen Worlds
Fires of the Mind
The Dawn of Humankind

Glencoe Physical Science Interactive Videodisc
Chemical Detectives

Physical Science CD-ROM

Key to Teaching Strategies

The following designations will help you decide which activities are appropriate for your students.

L1 Level 1 activities should be within the ability range of all students, including those with learning difficulties.

L2 Level 2 activities should be within the ability range of the average to above-average student.

L3 Level 3 activities are designed for the ability range of above-average students.

LEP LEP activities should be within the ability range of Limited English Proficiency students.

LS These activities are designed to address different learning styles.

COOP LEARN Cooperative Learning activities are designed for small group work.

P These strategies represent student products that can be placed into a best-work portfolio.

Teacher Classroom Resources

This is a representation of key blackline masters available in the Teacher Classroom Resources.

Teaching Aids

Section Focus Transparencies

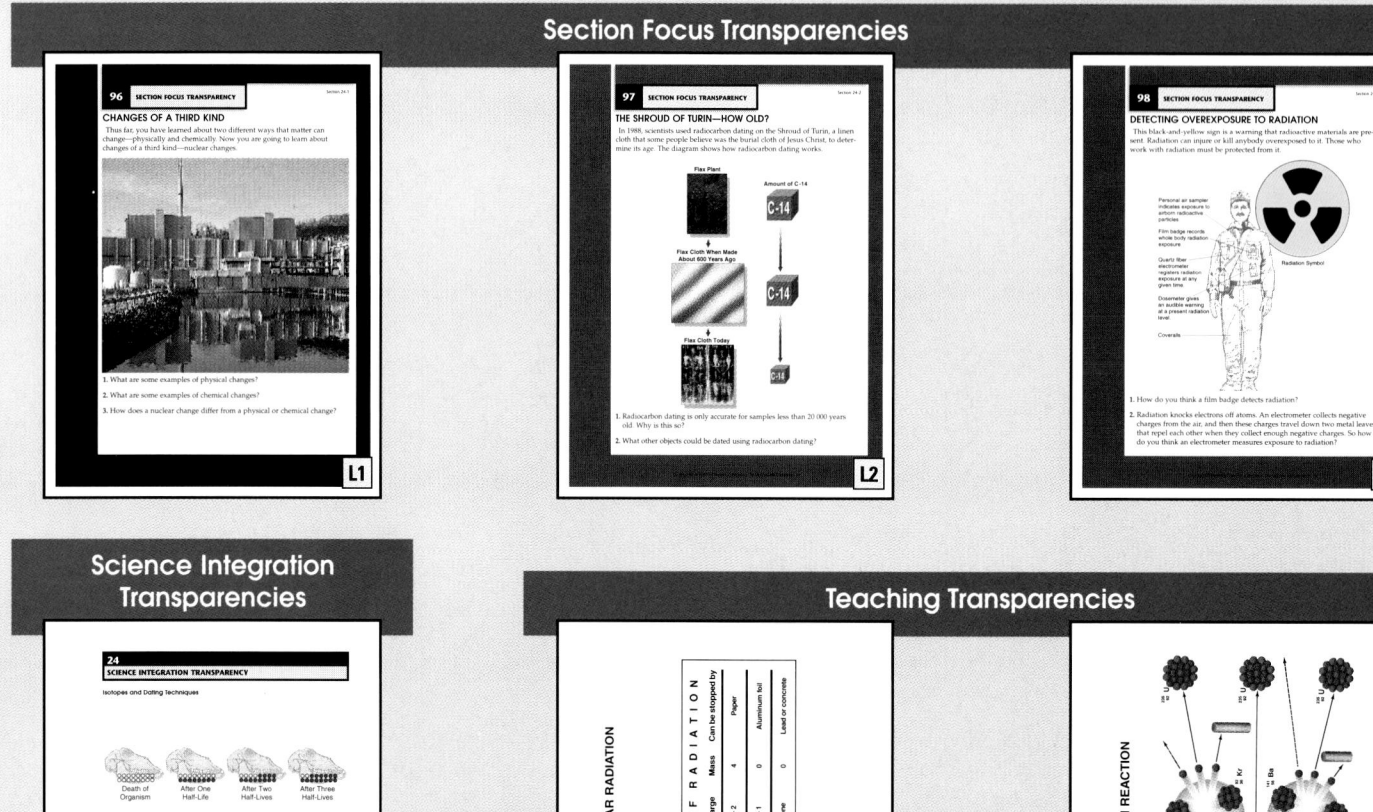

96 SECTION FOCUS TRANSPARENCY

CHANGES OF A THIRD KIND

Thus far, you have learned about two different ways that matter can change—physically and chemically. Now you are going to learn about changes of a third kind—nuclear changes.

1. What are some examples of physical changes?
2. What are some examples of chemical changes?
3. How does a nuclear change differ from a physical or chemical change?

L1

97 SECTION FOCUS TRANSPARENCY

THE SHROUD OF TURIN—HOW OLD?

In 1988, scientists used radiocarbon dating on the Shroud of Turin, a linen cloth that some people believe was the burial cloth of Jesus Christ, to determine its age. The diagram shows how radiocarbon dating works.

Flax Plant

Amount of C-14

C-14

Flax Cloth When Made About 600 Years Ago

C-14

Flax Cloth Today

C-14

1. Radiocarbon dating is only accurate for samples less than 20 000 years old. Why is this so?
2. What other objects could be dated using radiocarbon dating?

L2

98 SECTION FOCUS TRANSPARENCY

DETECTING OVEREXPOSURE TO RADIATION

This black-and-yellow sign is a warning that radioactive materials are present. Radiation can injure or kill anybody overexposed to it. Those who work with radiation must be protected from it.

Radiation Symbol

1. How do you think a film badge detects radiation?
2. Radiation knocks electrons off atoms. An electrometer collects negative charges from the air, and then these charges travel down two metal leaves that repel each other when they collect enough negative charges. So how do you think an electrometer measures exposure to radiation?

L3

Science Integration Transparencies

24 SCIENCE INTEGRATION TRANSPARENCY

Isotopes and Dating Techniques

Death of Organism | After One Half-Life | After Two Half-Lives | After Three Half-Lives

Isotope	Half-Life	Use
Thorium-232	14 Billion Years	Old rocks
Potassium-40	1300 Million Years	Fossils older than 50,000 years
Carbon-14	5730 Million Years	Recent fossils and remains 50 000 years or less

L1

Teaching Transparencies

47. TYPES OF NUCLEAR RADIATION

THREE TYPES OF RADIATION

Type	Symbol	Charge	Mass	Can be stopped by
alpha	$^4_2He, \alpha$	+2	4	Paper
beta	e, β	−1	0	Aluminum foil
gamma	γ	none	0	Lead or concrete

L1

48. NUCLEAR CHAIN REACTION

L2

Meeting Different Ability Levels

Study Guide

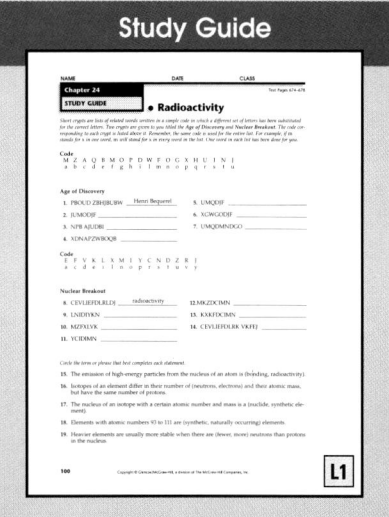

NAME   DATE   CLASS

Chapter 24

STUDY GUIDE   Text Pages 674-678

• Radioactivity

Secret crypts are lists of related words written in a simple code to solve a different set of letters has been substituted for the correct letters. True crypts are given to you titled the Age of Discovery and Nuclear Breakout. The code corresponding to each crypt is listed above it. Remember, the same code is used for the entire list. For example, if in stands for a in one word, m will stand for a in every word in the list. One word in each list has been done for you.

Code
M Z A Q B M O F D W E T O G X H U I N J
a b c d e f g h i l m n o p q r s t u

Age of Discovery
1. PROUD ZBHJBCBW   Henri Becquerel
2. JUMOOJF
3. NFB AJUDBI
4. XONAFZWBQJB

Code
E F V K L X M I Y C N D Z R J
a d e i l n o p q r s t u v y

Nuclear Breakout
8. CEVLEEPDJLDI   radioactivity
9. LNJDEYKN
10. MZFVLVK
11. VCJDMIV

12. MKJZDCIMN
13. KAKFDCIMN
14. CEVLEEPDJK VKJFEJ

Circle the term or phrase that best completes each statement.

15. The emission of high-energy particles from the nucleus of an atom is (bonding, radioactivity).
16. Isotopes of an element differ in their number of (neutrons, electrons) and their atomic mass, but have the same number of protons.
17. The nucleus of an isotope with a certain atomic number and mass is a (nuclide, synthetic element).
18. Elements with atomic numbers 93 to 111 are (synthetic, naturally occurring) elements.
19. Heavier elements are usually more stable when there are (fewer, more) neutrons than protons in the nucleus.

L1

Reinforcement

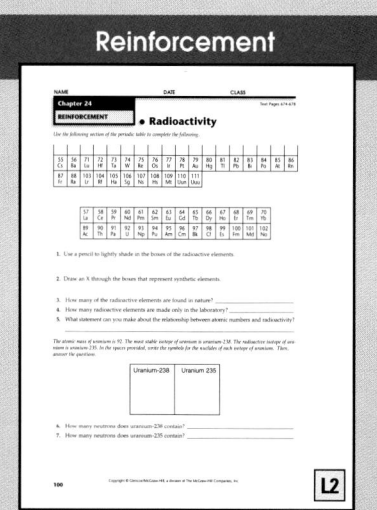

NAME   DATE   CLASS

Chapter 24

REINFORCEMENT   Text Pages 674-678

• Radioactivity

Use the following section of the periodic table to complete the following.

1. Use a pencil to lightly shade in the boxes of the radioactive elements.
2. Draw an X through the boxes that represent synthetic elements.
3. How many of the radioactive elements are found in nature?
4. How many radioactive elements are made only in the laboratory?
5. What statement can you make about the relationship between atomic numbers and radioactivity?

The atomic mass of uranium is 92. The most stable isotope of uranium is uranium 238. The radioactive isotope of uranium is uranium 235. In the spaces provided, write the symbols for the nuclides of each isotope of uranium. Then, answer the questions.

Uranium-238 | Uranium 235

6. How many neutrons does uranium-238 contain?
7. How many neutrons does uranium-235 contain?

L2

Enrichment Worksheets

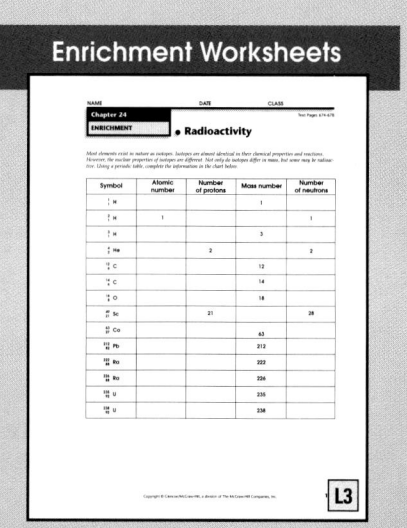

NAME   DATE   CLASS

Chapter 24

ENRICHMENT   Text Pages 674-678

• Radioactivity

Most elements exist in nature as isotopes. Isotopes are almost identical in their chemical properties and reactions. However, the nuclear properties of isotopes are different. Not only do isotopes differ in mass, but some may be radioactive. Using a periodic table, complete the information in the chart below.

Symbol	Atomic number	Number of protons	Mass number	Number of neutrons
^1_1H		1		1
^2_1H			3	
^3_1H			3	
^4_2He		2		2
$^{12}_6C$			12	
$^{14}_6C$			14	
$^{18}_8O$			18	
$^{45}_{21}Sc$		21		24
$^{59}_{27}Co$			63	
$^{82}_{82}Pb$			212	
$^{88}_{88}Ra$			223	
$^{88}_{88}Ra$			226	
$^{92}_{92}U$			235	
$^{92}_{92}U$			238	

L3

Chapter 24 Radioactivity and Nuclear Reactions

Hands-On Activities

Science Integration Activity

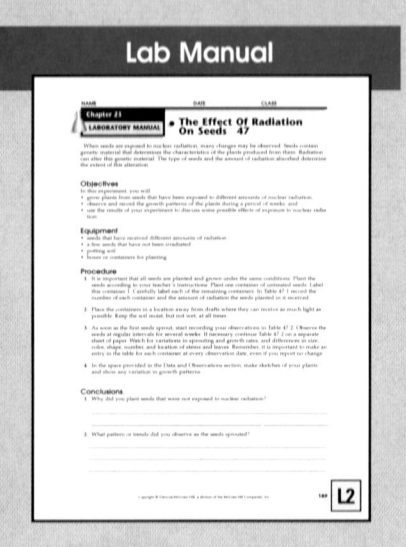

Absorption of Beta Radiation

L1

Lab Manual

The Effect Of Radiation On Seeds 47

L2

Assessment

Performance Assessment

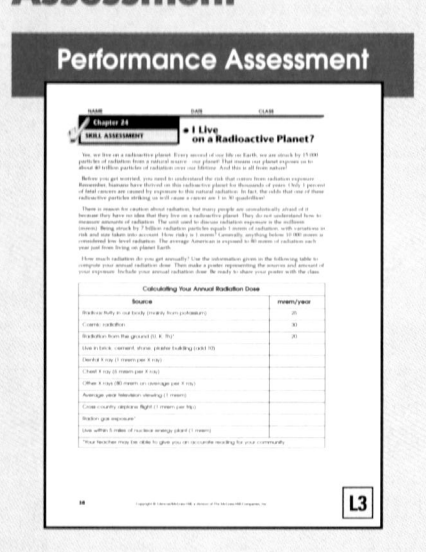

I Live on a Radioactive Planet?

L3

Enrichment and Application

Critical Thinking/Problem Solving

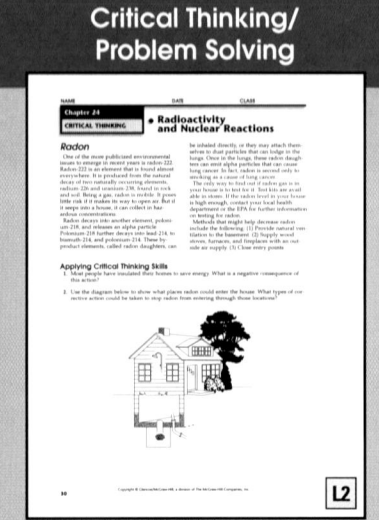

Radioactivity and Nuclear Reactions

Radon

L2

Cross-Curricular Integration

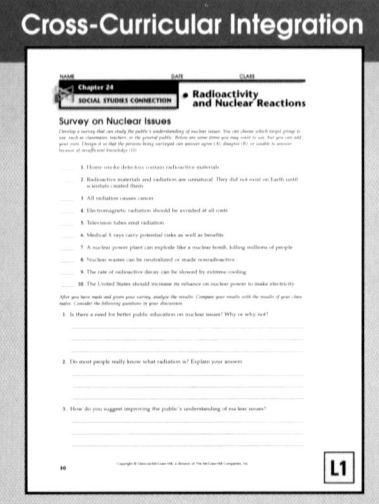

Radioactivity and Nuclear Reactions

Survey on Nuclear Issues

L1

Science and Society Integration

Radioactivity and Nuclear Reactions

Low-Level Radiation

L2

Multicultural Connections

Dr. Wu and Particles that Defied the Law

L2

Concept Mapping

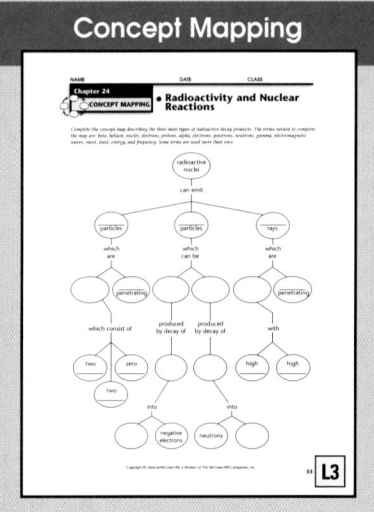

Radioactivity and Nuclear Reactions

L3

Radioactivity and Nuclear Reactions

CHAPTER OVERVIEW

Section 24-1 The discovery and current applications of radioactivity illustrate the nature of radioactive elements.

Section 24-2 Three types of radioactive decay and half-life are introduced.

Section 24-3 Methods of detecting and counting radiation are illustrated.

Section 24-4 Processes and applications of nuclear fission are presented and compared with those of nuclear fusion.

Section 24-5 Science and Society This section features a discussion of the use of radioisotopes in medicine.

Chapter Vocabulary

radioactivity	half-life
cloud chamber	tracer
nuclide	PET
bubble	gamma ray
chamber	beta particle
alpha particle	nuclear
chain reaction	fission
transmutation	
nuclear fusion	

Theme Connection

Stability and Change Throughout this book, the nucleus of the atom has been modeled as essentially stable. In this chapter, students learn that the nucleus can undergo patterns of change and transformation.

Previewing the Chapter

672

Learning Styles

Look for the following logo for strategies that emphasize different learning modalities. **LS**

Kinesthetic	Explore, p. 673; Reteach, p. 676; MiniLAB, p. 690; Activity 24-2, p. 694
Visual-Spatial	Science and Art, p. 678; Demonstration, p. 686; Reteach, p. 687; Inclusion Strategies, p. 687
Interpersonal	MiniLAB, p. 675; Reteach, p. 691
Logical-Mathematical	Activity 24-1, p. 684
Linguistic	Science and Art, p. 678; Science Journal, pp. 681, 691

LS

Chapter 24

Radioactivity and Nuclear Reactions

Have you ever seen a nuclear power station such as the one in the photo? Inside a reactor in this power station, a change in the nucleus of unstable atoms releases energy that is converted to electrical power. Waste is also produced during the nuclear reaction. How is nuclear waste different from other wastes?

EXPLORE ACTIVITY

Model a nuclear reaction.

1. Put 32 marbles, each with an attached lump of clay, into a large beaker. These marbles with clay represent unstable atoms.
2. During a 45-second period, remove 16 marbles and pull off the clay. Place the clay into a pile and the marbles into another beaker. Marbles without clay represent stable atoms. The clay represents by-products or wastes of the reaction—smaller atoms that may still be capable of giving off additional energy.
3. Repeat this procedure four more times, each time removing half the marbles remaining in the original beaker.

Observe: In your Science Journal, explain how this model points out one of the main problems associated with using nuclear power to make electricity.

Previewing Science Skills

► In the Skill Builders, you will **compare and contrast, interpret scientific illustrations, observe and infer,** and **map concepts.**

► In the Activities, you will **predict, interpret,** and **formulate models.**

► In the MiniLABs, you will **observe and infer.**

673

Assessment Planner

Portfolio
Refer to page 695 for suggested items that students might select for their portfolios.

Performance Assessment
See page 695 for additional Performance Assessment options.
Skill Builders, pp. 677, 683, 688, 691
MiniLABS, pp. 675, 690
Activities 24-1, pp. 684, 685; 24-2, p. 694

Content Assessment
Section Wrap-ups, pp. 677, 683, 688, 691, 693
Chapter Review, pp. 695-697
Mini Quizzes, pp. 677, 683, 688

Group Assessment
Opportunities for group assessment occur with Cooperative Learning Strategies and Flex Your Brain Activities.

Prepare

Section Background

Elements 93 and beyond can only be produced synthetically.

Preplanning

For the MiniLAB on page 675, divide students into groups.

1 Motivate

Bellringer

 Before presenting the lesson, display **Section Focus Transparency 96** on the overhead projector. Assign the accompanying **Focus Activity** worksheet. L1 LEP

Activity

Tell students they are being exposed to radiation. See how many possible sources they can list. Sources may include cosmic radiation, smoke detectors, radon gas in their basements, and medical/dental X rays. L1

2 Teach

NATIONAL GEOGRAPHIC SOCIETY

 **Videodisc**

GTV: Planetary Manager
Energy

47474

47945

Science Words

radioactivity
nuclide

Objectives

- Discuss the discovery of radioactivity.
- Contrast properties of radioactive versus stable nuclides.

Figure 24-1

Many smoke detectors, such as this one, contain a small amount of a radioactive element.

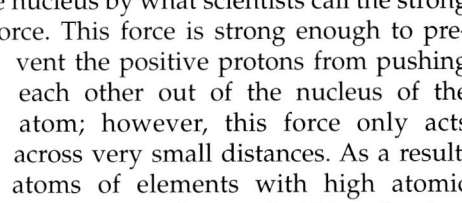

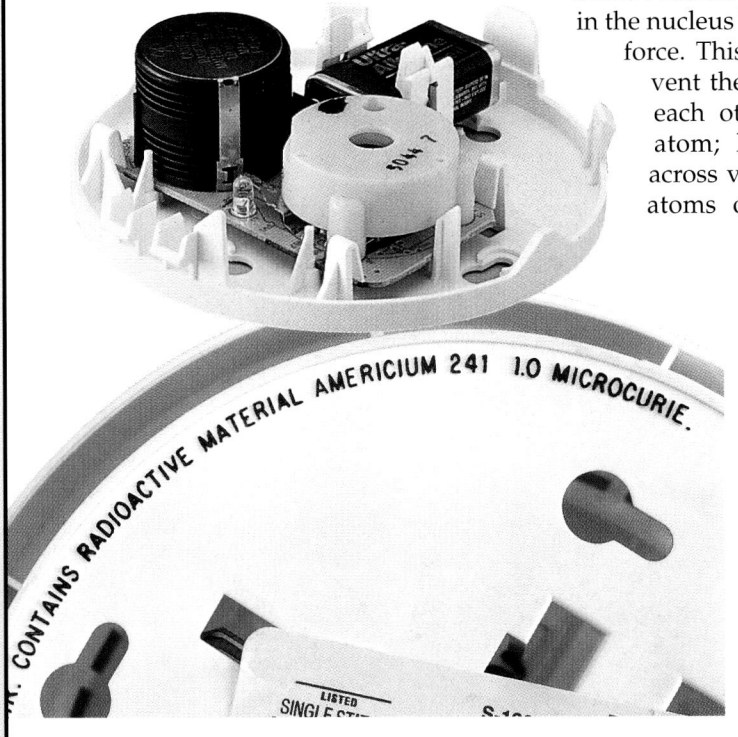

Radioactive Elements

Do you have a smoke detector, such as the one shown in **Figure 24-1**, in your home to alert you if there's a fire? If it is an ionizing smoke detector, it probably contains a small amount of a radioactive element called americium-241.

Look around the room. Can you detect any evidence of radioactivity? You can't see, hear, taste, touch, or smell radioactivity. Do you realize that there are small amounts of radioactivity all around you? You even have small amounts of radioactive material inside your body.

The discovery of radioactivity about 100 years ago has led to major advances in medical diagnoses and treatments, the use of nuclear energy, and the designing of nuclear weapons. It's certainly easy to see why some uses of radiation and nuclear energy are so controversial.

What is radioactivity?

You know that all elements are made of atoms, and that atoms are made up of protons, neutrons, and electrons. Protons and neutrons are held together in the nucleus by what scientists call the strong force. This force is strong enough to prevent the positive protons from pushing each other out of the nucleus of the atom; however, this force only acts across very small distances. As a result, atoms of elements with high atomic numbers are held together less securely than an oxygen atom, for example, which has relatively few protons and neutrons in its nucleus. Particles or energy can escape from all nuclei with atomic numbers of 84 or higher. This process is called radioactive decay. These nuclei are unstable. **Radioactivity,** therefore, is the emission of high-energy radiation or particles from the nucleus of a radioactive atom.

674 Chapter 24 Radioactivity and Nuclear Reactions

Program Resources

 Reproducible Masters
Study Guide, p. 100 L1
Reinforcement, p. 100 L1
Enrichment, p. 100 L3
Activity Worksheets, p. 146 L1
Lab Manual 47
Critical Thinking/Problem Solving, p. 30

Transparencies
Section Focus Transparency 96 L1

Discovery

How was radiation first discovered if it can't be detected by your senses? Henri Becquerel accidentally discovered radioactivity in 1896 when he left uranium salt in a desk drawer with a photographic plate. Later, when he removed the plate and developed it, he found an outline of the uranium salt. He hypothesized that the uranium had given off some invisible energy and exposed the film. This process is called radiation. Two years after Becquerel's discovery, Marie and Pierre Curie discovered the elements polonium and radium. These elements are even more radioactive than uranium.

Notice where these three elements are found on the periodic table. They all have high atomic numbers and, therefore, are located near the bottom of the chart. Each element that has an atomic number greater than 83 is radioactive. Some other elements also have common radioactive isotopes.

Find the elements with atomic numbers 93 to 111 on the periodic table. Elements with these atomic numbers don't exist naturally on Earth. They have been produced only in laboratories and are called synthetic elements. These synthetic elements are unstable. Why might these elements be difficult to study?

Nuclides

Recall from Chapter 10 that most elements have isotopes. Isotopes of an element have the same number of protons but differ in their number of neutrons and in their atomic mass. Some isotopes are radioactive and others are not. Why is this so? The nucleus of an isotope with a certain atomic number and mass is called a **nuclide.** In many nuclides, the strong force is enough to keep the nucleus permanently together, creating a stable nuclide. When the strong force is not sufficient to hold unstable nuclides together permanently, they decay, giving off matter and energy.

Isotopes of elements differ in the ratio of neutrons to protons, shown in **Figure 24-2.** This ratio is related to the stability of the nucleus. An isotope of a less massive element is stable if the ratio is about 1 to 1. Isotopes of the heavier elements are

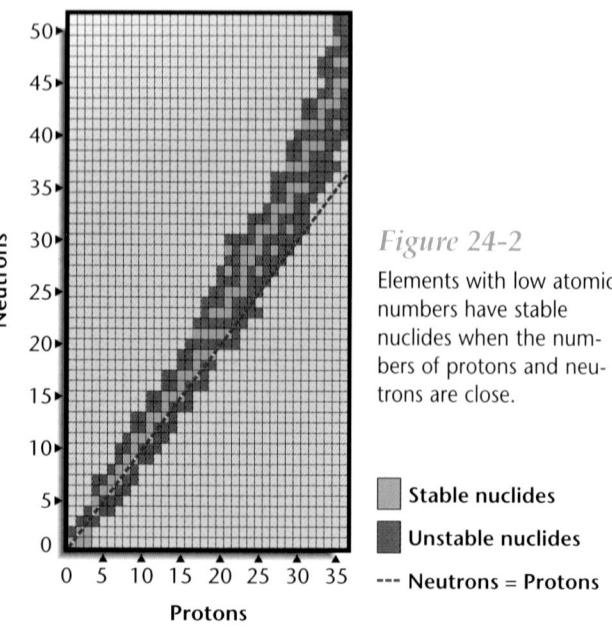

Figure 24-2

Elements with low atomic numbers have stable nuclides when the numbers of protons and neutrons are close.

- ■ Stable nuclides
- ■ Unstable nuclides
- --- Neutrons = Protons

24-1 Radioactivity   **675**

To illustrate how ionizing radiation can disrupt molecular functions, have two pieces of paper with paper clips on each. Join the papers at the paper clips by sticking the clips into each side of a thin piece of clay. This setup represents a particular molecular structure—paper, clip, clay, clip, paper. While holding the pieces of paper, snap the clay. The papers are no longer attached. In the same manner, the energy of gamma rays may break a protein structure at a key point.

Think Critically

Accept answers supported by evidence.

Student Text Question

See p. 677

How many neutrons does potassium-40 have? *21*

3 Assess

Check for Understanding

? FLEX Your Brain

Use the Flex Your Brain activity to have students explore NUCLIDES.

📁 **Activity Worksheets,** p. 5

Reteach

Kinesthetic Write each of the following nuclides on a piece of paper and have a student hold each one. Instruct them to find other nuclides that have the same atomic number. Some may be left alone. Emphasize that the groups represent different isotopes of the same element.

$^{234}_{91}Pa$ $^{214}_{82}Pb$ $^{214}_{83}Bi$ $^{214}_{84}Po$ $^{210}_{82}Pb$

$^{210}_{83}Bi$ $^{206}_{82}Pb$ $^{210}_{84}Po$ $^{218}_{85}At$

USING TECHNOLOGY

Irradiated Food ▼ ▲

Even well-planned food storage and preparation do not prevent thousands of food-related deaths and millions of food poisoning cases each year. The battle to control organisms that cause these problems has a weapon in nuclear technology.

Controlled doses of gamma rays can be used to ionize atoms in important molecules in bacteria. When the atoms lose electrons, the way they bond to other atoms changes. These changes in bonding can also change the function of key molecules, which could kill the bacteria. The United Nations has approved irradiation of food. Cobalt-60 and cesium-137 are gamma-producing isotopes used to irradiate potatoes, fruits, pork, poultry, and spices.

Possible Problems

Exposure to radiation does not make the food radioactive, and it reduces the amount of chemicals needed to keep food safe. However, critics are concerned about loss of nutritional value, possible flavor changes, and possible unknown reactions involving newly unbonded atoms.

Think Critically:

Would you support the use of radiation to preserve foods? Justify your answer.

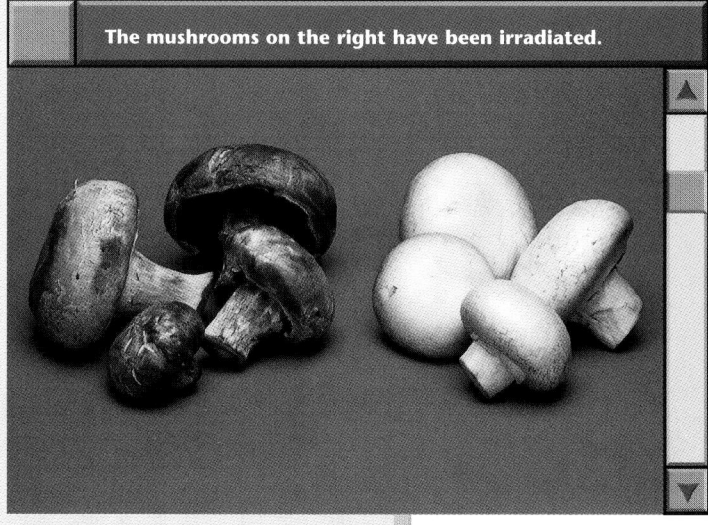

The mushrooms on the right have been irradiated.

Theme Connection

Systems and Interactions Systems and interactions are emphasized in this section. As the system of bonded atoms with chemical bonds interacts with ionizing radiation they break apart, often changing the system to operate with a different function.

stable when the ratio of neutrons to protons is about 3 to 2. However, the nuclei of isotopes of both lighter and heavier elements that differ much from these ratios are unstable; that is, nuclei with too many or too few neutrons compared to the number of protons are radioactive.

Nuclide Numbers

How can you distinguish one isotope from another of the same element? A nuclide can be represented by a symbol that gives the atomic number, mass number, and element symbol, as shown in **Table 24-1.** You know the atomic number is the same as the number of protons in the element, and the mass number is the total number of protons and neutrons in the nucleus of that atom. The symbol for the nucleus of the stable isotope of potassium is shown as an example. Examine what each number and symbol represents.

Table 24-1

Radioactive Nuclides of Some Elements

Element	Nuclide	Atomic mass number	Protons (atomic number)	Neutrons
Hydrogen	$^{3}_{1}H$	3	1	2
Helium	$^{5}_{2}He$	5	2	3
Lithium	$^{8}_{3}Li$	8	3	5
Carbon	$^{14}_{6}C$	14	6	8
Nitrogen	$^{16}_{7}N$	16	7	9
Potassium	$^{40}_{19}K$	40	19	21

mass number → $^{39}_{19}K$ ← element symbol
atomic number →

Now, compare the stable isotope of potassium to the radioactive isotope below.

mass number → $^{40}_{19}K$ ← element symbol
atomic number →

The stable isotope is called potassium-39. This isotope has 19 protons and 20 neutrons. The radioactive isotope is potassium-40. How many neutrons does potassium-40 have?

Section Wrap-up

Review

1. Identify the contributions of the three scientists who discovered the first radioactive elements.

2. What is the range of atomic numbers in which all isotopes are radioactive? Which of these are synthetic?

3. **Think Critically:** What is the ratio of protons to neutrons in lead-214? Explain whether you would expect this isotope to be radioactive or stable.

Skill Builder
Comparing and Contrasting
Compare and contrast stable and unstable isotopes of potassium. What do they have in common? How do they differ? If you need help, refer to Comparing and Contrasting in the **Skill Handbook.**

Science Journal

In your Science Journal, make a list of the first things you think of when you hear the word *radiation.* Write one paragraph describing your positive thoughts about radiation and another describing your negative thoughts.

24-1 Radioactivity   **677**

Skill Builder
The stable and unstable isotopes of potassium have the same number of protons and the same number of electrons. However, the unstable isotope of potassium has a higher mass number because it has one more neutron than does the stable nuclide.

✓ Assessment

Performance Have students organize their information about potassium isotopes into a chart that shows comparisons and contrasts. Use the Performance Task Assessment List for Venn Diagram and Pro/Con Issue in **PASC,** p. 95. **P**

Extension

For students who have mastered this section, use the **Reinforcement** and **Enrichment** masters.

4 Close

•MINI•QUIZ•

Use the Mini Quiz to check students' recall of chapter content.

1. All elements that have atomic numbers greater than _____ are radioactive. *83*

2. What is radioactivity? *the emission of high-energy radiation or particles from the nucleus of a radioactive atom*

3. When are isotopes of heavier elements most stable? *when they have a neutron-to-proton ratio of about 3 to 2*

Section Wrap-up

Review

1. Becquerel found uranium salt gave off radioactivity. Marie and Pierre Curie isolated polonium and radium from pitchblende.

2. Elements 84-111 are radioactive. Elements 93-111 are synthetic.

3. **Think Critically** 82/132 = 0.62. It is likely to be radioactive because the ratio of protons to neutrons differs from the 2-to-3 stable ratio for heavy elements.

Science Journal

Accept all reasonable answers. Positive responses may include medical examples and energy resources. Negative responses may include nuclear waste. **L1**

Science & ART

Source

Moché, Dinah. *Radiation: Benefits/Dangers.* Franklin Watts, 1979.

Background

Physicist Maurice Cotter and chemist Kathleen Taylor used neutron activation analysis on oil paintings by U.S. landscape artist Ralph Albert Blakelock (1847-1919) to determine the chemical composition of the paint. The paintings became radioactive, but only for a short time. They were able to be rehung in a museum about two months later.

Teaching Strategies

 Linguistic Have students work in small groups to think of other objects that would be useful to analyze and possibly date using neutron activation analysis. **COOP LEARN**

GLENCOE TECHNOLOGY

 **Videodisc**

STVS: Chemistry
Disc 2, Side 2
Neutron Activation Analysis of Paintings (Ch. 6)

Dating by Thermoluminescence (Ch. 7)

CD-ROM

Physical Science CD-ROM
Have students perform the interactive exploration for Chapter 24 to reinforce important chapter concepts and thinking processes.

Portrait of a Boy, by Sonfonisba Anguissolla

*inter*NET
CONNECTION

Visit the Chapter 24 Internet Connection at Glencoe Online Science, **www.glencoe.com/sec/ science/physical**, for a link to more information about art forgeries.

Is it real or is it fake?

Paintings by great artists are sold for thousands and even millions of dollars. That is why some talented (but dishonest) people paint copies of great works of art and try to pass them off as the real thing. That is also why methods to detect such forgeries have been developed. One method of detecting art forgeries relies on a type of radioactive dating. This method helps date the paint used on the painting.

The chemical makeup of paints has changed over the years. Lead pigments used in oil paints today are not the same as those used hundreds of years ago. Lead that was refined before the year 1650 contains traces of chromium. Lead that was refined after the year 1950 contains traces of zinc and antimony. So the presence or absence of these trace elements in oil paint can show the general time period in which a painting was done. If the paint contains zinc and antimony, it must have been applied within the past few decades and not, for example, by a master artist of the 1700s or 1800s.

Analysis by Neutron

How can traces of an element in a paint be detected? One way is to bombard the painting with neutrons. Stable atoms within the paint then start to decay. Gamma rays are emitted and are observed with radiation detectors. Because each element produces its own pattern of gamma radiation, studying the patterns can reveal which elements are present in the paint. The age of the paint can then be generally determined. This method of discovering the chemical makeup of an object is called neutron activation analysis.

During the bombardment of the paint with neutrons, beta particles are also released. The trails of these particles can be captured on film, producing a picture called an autoradiograph. An autoradiograph reveals what layers of materials lie under the surface oil paint. Together with X rays, autoradiographs can be used to see the stages of a master artist's work, paintings that were painted over, and parts of a painting that may have been done by students of the artist rather than the artist. Autoradiographs are, therefore, extremely useful to art historians.

 Kinesthetic Invite interested students to paint in oils. Then ask them to compare the way they use and view the paint with the way a scientist would view it.

 Visual-Spatial Display an oil painting to the class. Ask students to compare the way they view the painting with the way a scientist would view it.

Nuclear Decay 24•2

Nuclear Radiation

You've heard the terms *radioactivity* and *radioactive* ever since you were a young child. Maybe you didn't know what these words meant then, but now you know that they refer to unstable nuclei that emit radiation composed of small particles or energy called nuclear radiation. There are three common types of nuclear radiation—alpha, beta, and gamma radiation. **Figure 24-3** shows the path that each type of nuclear radiation follows as it moves through a magnetic field. As you can see, each type of radiation is affected differently. Which type of radiation is most affected by a magnetic field?

Radiation in the form of **alpha particles** is given off when a nucleus releases two protons and two neutrons. Notice that the alpha particle is the same as a helium nucleus, as shown in **Table 24-2**. An alpha particle has a charge of 2+ and an atomic mass of 4. It is the largest and slowest form of radiation, so it is also the least penetrating. Alpha particles can be

Science Words

alpha particle
beta particle
transmutation
gamma ray
half-life

Objectives

- Distinguish among alpha, beta, and gamma radiation.
- Calculate the amount of a radioactive substance remaining after a time based on its half-life.
- Relate half-life to the process of radioactive dating.

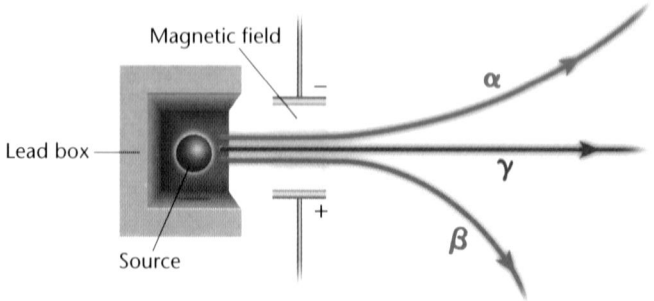

Figure 24-3

Notice how positive alpha and negative beta particles are deflected by a magnetic field. *Why aren't gamma rays deflected?*

Table 24-2

Three Types of Nuclear Radiation

	Type	Symbol	Charge	Mass
	alpha	$^4_2 He$, a	2+	4
	beta	$^{\ 0}_{-1} e$, b	1–	0
	gamma	g	none	0

24-2 Nuclear Decay   679

Program Resources

Reproducible Masters
Study Guide, p. 101 [L1]
Reinforcement, p. 101 [L1]
Enrichment, p. 101 [L3]
Activity Worksheets, pp. 142-143 [L1]
Concept Mapping, p. 53
Lab Manual 48
Multicultural Connections, pp. 51-52

Transparencies
Section Focus Transparency 97
Science Integration Transparency 24

Section 24•2

Prepare

Section Background

Both alpha and beta particles cause transmutation because the nucleus loses or gains protons, causing a change in atomic number. Emitting gamma radiation does not change the identity of a nuclide.

Preplanning

Obtain pennies and graph paper for Activity 24-1.

1 Motivate

Bellringer

Before presenting the lesson, display **Section Focus Transparency 97** on the overhead projector. Assign the accompanying **Focus Activity** worksheet. [L1] [LEP]

Tying to Previous Knowledge

Ask the students to recall the most energetic kinds of electromagnetic radiation. Gamma radiation has the highest frequency and shortest wavelength. Explain that this is one kind of radiation emitted during radioactive decay.

2 Teach

Visual Learning

Figure 24-3 Why aren't gamma rays deflected? *They have no charge.* [LEP] [IS]

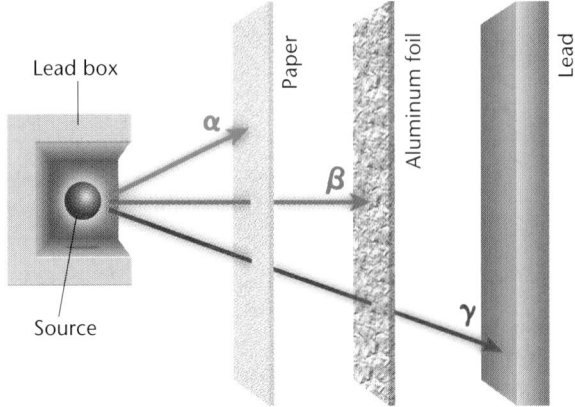

Figure 24-4

Alpha particles can be stopped by paper, beta particles by aluminum foil, and gamma rays by lead.

USING MATH

The most common isotope of uranium, $^{238}_{92}$U, transmutes to thorium by emitting an alpha particle with a mass of 4 and a charge of 2+. What is the symbol for the nucleus of this isotope of thorium?

Figure 24-5

In a transmutation, as in any nuclear reaction, the atomic mass number of the decayed nuclide equals the sum of the mass numbers of the newly formed nuclide and the emitted particles.

stopped by a sheet of paper, **Figure 24-4,** but they can be dangerous if they are released by atoms inside your body.

The smoke detector mentioned earlier in this chapter gives off alpha particles that ionize the surrounding air. An electric current flows through the ionized air, but the circuit is broken when smoke particles enter the ionized air. When the circuit is broken, the alarm is set off.

Decaying Neutrons and Protons

Sometimes, a neutron inside a nucleus decays spontaneously into a proton and an electron. The electron is emitted from the nucleus at high speed and is called a **beta particle.** The proton formed by the decaying neutron stays in the nucleus, forming an element with an atomic number one greater than that of the original element. Because the total number of protons and neutrons has not changed, the atomic mass number of the new element is the same as that of the original element. A proton can also decay into a neutron and a positron. A positron is similar to an electron, but has a positive charge. Positrons are also considered to be beta particles. Beta particles are much faster and more penetrating than alpha particles. However, they can be stopped by a sheet of aluminum foil.

Isotopes that give off alpha or beta particles undergo transmutation. **Transmutation** is the process of changing one element to another through nuclear decay. The nuclear equations in **Figure 24-5** show a nuclear transmutation. Because two protons are lost from the nucleus in alpha decay, the new element formed has an atomic number two less than that of the original element. The atomic mass number of the new element is four less than the original element. These two examples of

$$^{218}_{84}\text{Po} \longrightarrow ^{214}_{82}\text{Pb} + ^{4}_{2}\text{He}$$

$$^{214}_{82}\text{Pb} \longrightarrow ^{214}_{83}\text{Bi} + ^{0}_{-1}e$$

Cultural Diversity

Hideki Yukawa Hideki Yukawa was the first Japanese person to receive the Nobel Prize. He accomplished this by developing a theory about what holds the nucleus of the atom together. He reasoned that there must be a nuclear force that involves the transfer of some entity other than a proton or neutron. His theory also stated that this force must be stronger over short distances than electromagnetic forces in order to overcome the repulsion between protons. When he published his theory in 1935, no particle had been discovered that matched the exchange particle his ideas required. It was not until 1947 that the particle Yukawa predicted was actually found. As a result of the verification of his theory, Yukawa was awarded the Nobel Prize in 1949.

transmutation confirm that the charge of the original nuclide equals the sum of the charges of the nuclide and particle formed.

The most penetrating form of radiation is not made of particles at all. **Gamma rays** are electromagnetic waves with very high frequency and energy. They have no mass, no charge, and they travel at the speed of light. They are usually released along with alpha or beta particles. Gamma rays are the most penetrating form of radiation, and thick blocks of materials with dense nuclei are required to stop them. Lead and concrete are commonly used barriers for gamma rays. The three kinds of radiation are summarized in **Table 24-2** on page 679.

Half-Life

Safe disposal of some radioisotopes is not a problem because their nuclides decay to stable nuclides in less than a second. However, nuclides of certain radioactive isotopes may require millions of years to decay. A measure of the time required by the nuclides of an isotope to decay is called half-life. The **half-life** of an isotope is the amount of time it takes for half the nuclides in a sample of a given radioactive isotope to decay. Half-lives vary widely among the radioactive isotopes. For example, polonium-214 has a half-life of less than a thousandth of a second, but uranium-238 has a half-life of 4.5 billion years.

Problem Solving

Invisible Invader

In a typical day, everyone comes into contact with many forms of nuclear radiation. Some of the sources are from commercial products or the result of human activity. However, in recent years, there has been increasing awareness of another source of exposure.

Depending on the location, type of soil, and types of building materials, residents may be exposed to unusual amounts of alpha radiation from radon gas. Radon can seep in around building foundations that are not properly sealed. It can also decay into other radioactive elements that accumulate in poorly ventilated areas. Homeowners can purchase kits that can be used to monitor amounts of radon in their homes. Of the total average amount of radiation exposure in the United States, more than 50 percent can be attributed to radon. How can this potentially harmful exposure be reduced?

Think Critically:

1. If you purchased a radon testing kit, where would you most likely place it?
2. If a home tested high in radon levels, why would sealing help solve the problem in some cases and increased ventilation help in other situations?
3. Radon and its by-products emit alpha particles. Because alpha particles are relatively easy to stop, why is radon exposure considered such a risk for humans?

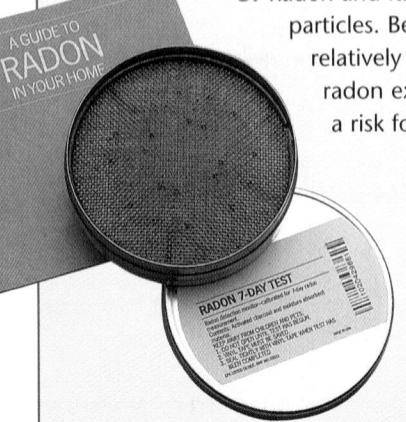

Think Critically

1. The manufacturer will supply specific instructions. The test kit should be placed in areas where there is dead air space and in foundation areas such as basements.
2. Sealing may be needed if the ground around the house contains higher levels of radioactivity. Ventilation may be needed if some building materials have high levels of radioactivity.
3. If the alpha source is inhaled, the high-ionizing alpha particles will be inside the lung tissue, where extensive damage could result.

USING MATH

$^{234}_{90}\text{Th}$

Teacher F.Y.I.

Note that the rate of radioactive decay is nearly impossible to change. Attempts to change this rate by altering temperature and pressure conditions, other chemical reactions, or exposure to electric and magnetic fields have been unsuccessful.

Content Background

Radiocarbon dating was developed in the 1940s by Willard Libby. Cosmic rays from space interact with atoms in Earth's atmosphere and liberate neutrons. The presence of carbon-14 in our atmosphere results from neutron bombardment of nitrogen, which causes a proton to be emitted.

$$^1_0n + {}^{14}_7\text{N} \rightarrow {}^{14}_6\text{C} + {}^1_1\text{H}$$

 Use **Science Integration Transparency 24** as you teach this lesson.

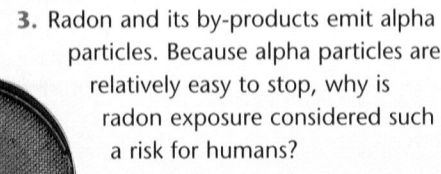 **Science Journal**

LS **Linguistic** Have students compare and contrast the amount of radiation they could detect in a set time span from several varying situations. For example, compare a large sample with a short half-life with a large sample with a long half-life. Also, compare a small sample with a long half-life with a large sample with a short half-life.

Table 24-3

Sample Half-Lives			
Isotope	**Half-Life**	**Isotope**	**Half-Life**
$^{3}_{1}\text{H}$	12.3 years	$^{212}_{82}\text{Pb}$	10.6 hours
$^{14}_{6}\text{C}$	5730 years	$^{194}_{84}\text{Po}$	0.7 second
$^{131}_{53}\text{I}$	8.04 days	$^{235}_{92}\text{U}$	7.04×10^{8} years

1. Amount of original
   sample = 20 g
   Time Elapsed = 45 h
   Half-life of Na-24 = 15 h
   n = number of half-lives
   elapsed = 3
   Amount Remaining =

   $$\frac{\text{Amount of Original}}{2^{n}}$$

   $$= \frac{20\text{ g}}{2^{3}} = 2.5\text{ g}$$

Visual Learning

Figure 24-6 Do you think carbon-14 dating could be used to determine the age of a dinosaur skeleton? *no, because they are too old* **LEP**
LS

3 Assess

Check for Understanding

Use the Flex Your Brain activity to have students explore RADIOACTIVE DATING.

 **Activity Worksheets,** p. 5

Reteach

Emphasize that the relatively large size and positive charge of alpha particles makes them easily captured by other atoms or molecules. Beta particles are many times smaller, but their charge causes attractions and repulsions with other particles. Gamma radiation has no mass or charge, so it is difficult to interact with.

Extension

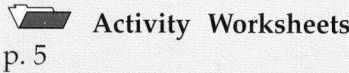 For students who have mastered this section, use the **Reinforcement** and **Enrichment** masters.

You can determine the amount of a radioactive sample that will remain after a given amount of time if you know the half-life of the sample. The half-lives of some radioactive elements are listed in **Table 24-3.** Look at the problem below to see how to calculate the amount of remaining sodium-24, which has a half-life of 15 hours.

USING MATH

Calculating the Quantity of a Radioactive Element

Example Problem:
How much of a 20-g sample of sodium-24 would remain after decaying for 30 hours?

Problem-Solving Steps:
1. Identify the known information:
   Amount of original sample of sodium-24 = 20 g
   Time elapsed = 30 hours
   Half-life of sodium-24 = 15 hours
2. Identify the unknown information.
   amount of remaining sodium-24
3. Solve for the unknown information.

   $$\text{amount remaining} = \frac{\text{amount of original sample}}{2^{n}}$$

   n = number of half-lives elapsed = 30 hours/15 hours = 2
4. **Solution:**

   $$\frac{20\text{ g}}{2^{2}} = \frac{20\text{ g}}{4} = 5\text{ g}$$

   The sample will contain 5 g of sodium-24 and 15 g of the decay product.

Practice Problem
How much of a 20-g sample of sodium-24 would remain after 45 hours?
Strategy Hint: Determine how many half-lives of sodium-24 elapse in 45 hours.

Across the Curriculum

Archaeology Have students do research to learn how radioactive dating has been used to determine the ages of artifacts found in archaeological digs.

Radioactive Dating

It was mentioned earlier that you have some radioactive materials inside your body. One of these is a radioactive isotope of carbon, carbon-14. This isotope is in some of the carbon dioxide molecules plants take in as they respire. As a result, you have a small amount of carbon-14 inside of you from the plants and animals you eat. Carbon-14 emits a beta particle and decays into nitrogen-14.

All living things contain a somewhat constant amount of carbon-14. Decaying carbon-14 is constantly replaced by new carbon-14 in a living organism, but when the organism dies, its carbon-14 decays without replacement. The half-life of carbon-14 is 5730 years. It will take 5730 years for half of the carbon-14 atoms in a dead organism to decay to nitrogen-14. In another 5730 years, half of the remaining carbon will decay. By measuring the amount of carbon-14 remaining in a fossil or skeleton, such as those shown in **Figure 24-6,** scientists can determine the approximate age of the material. This process is called carbon-14 dating. Only remains of plants and animals that lived within the last 50 000 years contain enough carbon-14 to measure.

Figure 24-6

Some scientists consider carbon-14 dating to be accurate to only 20 000 years because of the difficulty of detecting lesser amounts of carbon-14. *Do you think carbon-14 dating could be used to determine the age of a dinosaur skeleton? Explain.*

Section Wrap-up

Review

1. Describe each of the three types of radiation, and compare their penetrating power.

2. Write a nuclear equation to show how radon-222 decays to give off an alpha particle and another element. What is the other element?

3. Think Critically: Is it possible for an isotope to decay to an element with a higher atomic number? Explain.

Skill Builder

Interpreting Scientific Illustrations
Use **Figure 24-3** on page 679 to explain why the alpha particle curves up, the beta particle curves down, and the gamma particle travels straight through the magnetic field. If you need help, refer to Interpreting Scientific Illustrations in the **Skill Handbook.**

USING MATH

The half-life of iodine-131 is about eight days. Calculate how much of a 40-g sample will be left after eight days. After 16 days? After 32 days?

24-2 Nuclear Decay   **683**

Skill Builder

The gamma ray has no charge and will not be affected by the magnetic field. The oppositely charged alpha and beta particles will be deflected in opposite directions.

Performance   Ask students to sketch what would happen if the magnetic field in the diagram were reversed. Use the Performance Task Assessment List for Scientific Drawing in **PASC,** p. 55. P

4 Close

•MINI•QUIZ•

Use the Mini Quiz to check students' recall of chapter content.

1. **What kind of radiation is identical to a helium nucleus?** *alpha particles*

2. **Arrange the three kinds of nuclear radiation in order of increasing penetrating power.** *alpha particles, beta particles, gamma rays*

3. **What is transmutation?** *the process by which one element changes to another by nuclear decay involving alpha or beta particles*

4. **What isotope is used in radioactive dating some artifacts found in archaeological digs?** *carbon-14*

Section Wrap-up

Review

1. Alpha particles, helium nuclei, are the least penetrating and can be stopped by paper. Beta particles are electrons and can be stopped by aluminum foil. Gamma rays are the most penetrating kind of nuclear radiation and are stopped by lead or concrete.

2. $^{222}_{86}\text{Rn} \rightarrow ^{218}_{84}\text{Po} + ^{4}_{2}\text{He}$; polonium

3. **Think Critically** Yes. In beta decay, a neutron decays into an electron and a proton, thereby increasing the atomic number of the element by one.

USING MATH

20 g after 8 days, 10 g after 16 days, 2.5 g after 32 days

Activity 24-1

PREPARATION

Purpose

IS **Logical-Mathematical** Use a model of the half-life of an isotope to predict nuclear decay and construct a graph to show the decay pattern. L1

COOP LEARN

Process Skills

predicting, interpreting data, classifying, formulating operational definitions, formulating models

Time

40 minutes

Materials

Gather enough pennies for each student to have one.

Possible Hypotheses

If a radioactive sample has a known half-life, then the amount of material remaining after that half-life can be predicted.

Activity Worksheets, pp. 5, 142-143

PLAN THE EXPERIMENT

Possible Procedures

One possible procedure is to gather all the pennies and record the number. Declare one half-life, and give the coins an identical toss. Record the number that land heads as "not yet decayed." Repeat this procedure (using only the heads, each time) to see how many remained undecayed after each half-life.

Teaching Strategies

- Establish some rules for the tossing of the coins to avoid hazards of flying objects. Stress that the method of data collection should be consistent for all persons.

Design Your Own Experiment

Investigating Half-Life

The decay rate of radioactive isotopes can vary from fractions of a second to several billions of years. If the half-life of an isotope is known, can you predict the amount of a radioactive sample that will remain after a known amount of time? Can you predict when a specific atom will decay?

PREPARATION

Problem

How can a model be used to demonstrate the usefulness of half-life in predicting the remaining amounts of a radioactive isotope?

Form a Hypothesis

Based on what you know about half-life, state a hypothesis about how the half-life of a radioactive isotope can be used to calculate the amounts of a radioactive isotope remaining.

Objectives

- Make a model to represent atoms in a radioactive sample.

- Calculate the amount of change in the objects that represent isotopes for each half-life.
- Design an experiment that tests how useful half-life is in predicting the amount of radioactive material remaining after a known amount of time.

Possible Materials

- several pennies
- graph paper

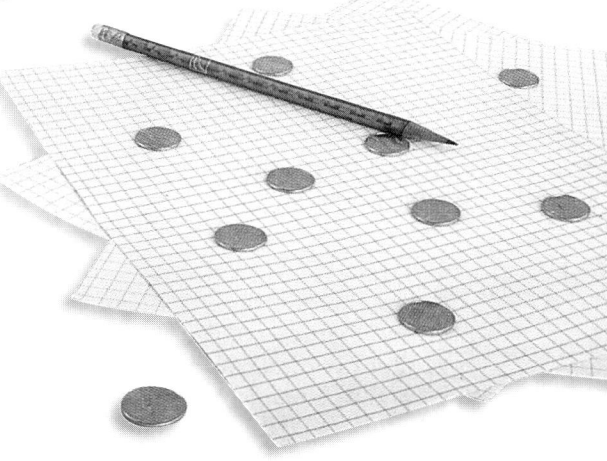

Across the Curriculum

Math The half-life of radium-226 is 1600 years. **Predict how much of a 10-g sample will remain after 1600 years. After 3200 years.** *5 g; 2.5 g* **How can we know the half-life of uranium-238 if it has a half-life of 4.5 billion years?** *By looking at decay rates during known amounts of time, mathematical relationships allow us to calculate the actual half-life.* L2 IS

PLAN THE EXPERIMENT

1. As a group, agree upon and write out the hypothesis statement.
2. As a group, list the steps that you need to perform to test your hypothesis. Assume that each penny represents an atom in a radioactive sample and that each coin that lands heads up after flipping has not decayed.
3. Make a list of the materials that you will need to complete your experiment.
4. In your Science Journal, design a data table containing two columns—*Half-Life* and *Atoms Remaining*.

Check the Plan

1. In your model of radioactive decay, what can you do with pennies to represent the radioactive decay of an isotope?

2. What will you consider to be one half-life for your model?
3. How many half-lives do you want to investigate?
4. What are the variables and controls in your model?
5. What will be on each axis of your graph?
6. *Make sure your teacher approves your plan and that you have included any changes suggested in the plan.*

DO THE EXPERIMENT

1. Carry out the approved experiment as planned.
2. Record your results for several half-lives in your data table.

Analyze and Apply

1. **Compare** your results with those of other groups in the class.
2.  The relationship among the starting number of pennies, the number of pennies remaining (*Y*), and the number

of half-lives (*X*) is given by $Y = $ starting number of pennies$/2^X$. Use your graphing calculator to graph this equation. Use your graph to find the number of pennies remaining after 2.5 half-lives.
3. Can you use your model to **predict** which individual atoms will decay during one half-life? Can you **predict** the total number that will decay in one half-life? Explain.

24-2 Nuclear Decay   685

Go Further

Use the original number of pennies and the method described on page 673 to calculate the expected number of atoms left after three half-lives. Compare this number to your experimental results. Explain any differences.

- This activity should statistically provide an excellent model for nuclear decay.
- Students may need assistance in setting up their graphs, plotting the points, and drawing the curve.

DO THE EXPERIMENT

Expected Outcome
When tossed, approximately half of the coins should land heads up. If the tails-up coins are removed, about half the remaining coins should land heads up. At the end of each toss, approximately half the number of coins remain.

Analyze and Apply

1. Each group should find that approximately half the number of pennies would land heads up at the end of each toss.
2. Student answers will vary depending on the original number of pennies used. If 50 pennies were originally used, the trace function can be used to determine that 9 pennies remain after 2.5 half-lives.
3. No; yes, because the sample size is larger.

Go Further
The greatest number of atoms probably decayed in the first half-life. The calculation will vary depending upon the original sample size, but the following equation should be used.
amount (number) remaining = amount (number) of original sample$/2^n$
The result should be close to the actual value.

Prepare

Section Background

There are several units for measuring radiation. The curie and the becquerel (the SI unit) are based on the number of disintegrations per second. The roentgen, the rad, and the gray (SI unit) measure the effect the radiation has on the absorbing material. The rem measures biological damage most effectively.

1 Motivate

Bellringer

Before presenting the lesson, display **Section Focus Transparency 98** on the overhead projector. Assign the accompanying **Focus Activity** worksheet. L1 LEP

Demonstration

LS **Visual-Spatial** Bring in a smoke detector that uses americium-241. Light some flash paper or a candle to produce a controlled amount of smoke and use it to make the smoke detector go off.

2 Teach

Visual Learning

Figure 24-7 How do you think the trail of a beta particle might differ from the trail of an alpha particle? *Because they have different masses and charges, alpha trails are wider and less curved.* LEP LS

Science Words

cloud chamber
bubble chamber

Objectives

- Describe how radioactivity can be detected.
- Explain how a Geiger counter can determine the quantity of nuclear radiation present.

Radiation Detectors

Because you can't feel a single particle or gamma ray, you must use instruments to detect their presence. Some methods of detecting radioactivity use the fact that radiation forms ions by removing electrons from matter it passes through. The newly formed ions can be detected in several ways.

A **cloud chamber,** shown in **Figure 24-7A,** can be used to detect charged nuclear particles as they leave cloud tracks. A cloud chamber contains supersaturated water or ethanol vapor. When a charged particle from a radioactive sample moves through the chamber, it leaves a path of ions behind as it knocks electrons off the atoms in the air. The vapor condenses around these ions to provide a visible path of droplets along the track of the particle. Beta particles leave long, thin trails, and alpha particles leave shorter and thicker trails.

Another way to detect and monitor the paths of nuclear particles is by using a bubble chamber. A **bubble chamber** holds a superheated liquid, which doesn't boil because of increased pressure in the system. When a moving particle

Figure 24-7

Several different instruments can be used to detect radiation. *How do you think the trail of a beta particle might differ from the trail of an alpha particle?*

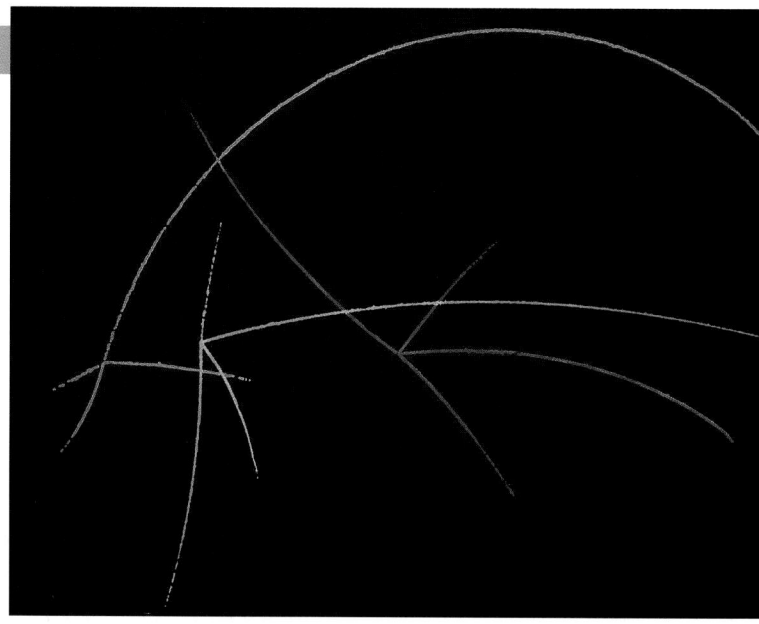

A A cloud chamber is a device used to detect particles of nuclear radiation.

Program Resources

📁 **Reproducible Masters**
Study Guide, p. 102 L1
Reinforcement, p. 102 L1
Enrichment, p. 102 L3
Science and Society Integration, p. 28
Science Integration Activities, pp. 47-48

Transparencies
Section Focus Transparency 98

leaves ions behind, the liquid boils along the trail. The path shows up as tracks of bubbles, as you can see in **Figure 24-7B.**

Do you remember how an electroscope was used to detect electric charges in Chapter 21? When an electroscope is given a negative charge, the leaves repel each other and spread apart. They will remain apart until the extra electrons have somewhere to go and discharge the electroscope. A nuclear particle moving through the air will remove electrons from molecules in the air, leaving them positively charged. When this occurs near an electroscope, some positively charged air molecules will come in contact with the electroscope and attract the electrons from the leaves, as shown in **Figure 24-7C.** As these negatively charged leaves lose their charges, they move together.

Measuring Radiation

You constantly receive small doses of radiation from your environment. It is not known whether this radiation is harmful to your body tissues. However, larger doses of radiation can be harmful to living tissue. If you worked or lived in an environment that had potential for exposure to high levels of radiation—a nuclear testing facility, for example—you might want to know exactly how much radiation you were being exposed to. A simple method of counting radioactivity is to use a Geiger counter. A Geiger counter is a device that produces an electric current when radiation is present.

B Particles of nuclear radiation can be detected as they leave trails of bubbles in a bubble chamber.

CONNECT TO

EARTH SCIENCE

Compare the process of forming a visible path of droplets in a cloud chamber with the formation of a cloud in the sky.

C As negatively charged leaves lose their charges to positively charged molecules in the air, the leaves no longer repel.

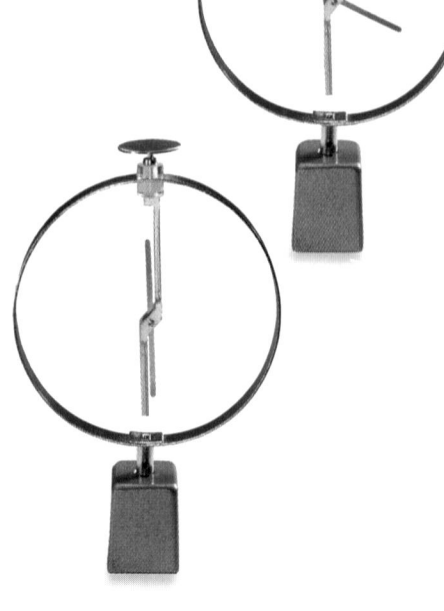

24-3 Detecting Radioactivity   **687**

CONNECT TO

EARTH SCIENCE

Answer In a cloud chamber, water or ethanol vapor condenses around ions to form visible droplets. Clouds form around dust particles in the atmosphere.

GLENCOE **TECHNOLOGY**

⊙ **Videodisc**
STVS: Chemistry
Disc 2, Side 2
Radon Danger (Ch. 3)

3 Assess

Check for Understanding

? FLEX Your Brain

Use the Flex Your Brain activity to have students explore DETECTING RADIATION.

🗁   **Activity Worksheets,** p. 5

Reteach

LS **Visual-Spatial** Have students explain the operation of a Geiger counter by examining the one pictured in **Figure 24-8.**

Extension

🗁   For students who have mastered this section, use the **Reinforcement** and **Enrichment** masters.

Across the Curriculum

Geology Geiger counters are used at job sites to determine radiation levels. Find out how much additional radiation workers in a hospital nuclear medicine lab or nuclear power plant would be exposed to.

Inclusion Strategies

Learning Disabled Students can make a time line of the important discoveries and events in the area of nuclear power beginning with Henri Becquerel's discovery. **LS**

4 Close

•MINI•QUIZ•

Use the Mini Quiz to check students' recall of chapter content.

1. **Which device for detecting radioactivity uses condensation trails?** *cloud chamber*

2. **A(n) _____ is a device commonly used to measure amounts of radiation.** *Geiger counter*

3. **Which device for detecting radioactivity produces a path of boiling liquid?** *bubble chamber*

Section Wrap-up

Review

1. Geiger counter, cloud chamber, bubble chamber, electroscope, and photographic film

2. Radiation enters the tube, causing a current that is amplified with a speaker to click.

3. **Think Critically** A Geiger counter, if it is small and portable; it can precisely measure the intensity of the radiation.

Using Computers

Sample Program:
10 REM HALF-LIFE
20 PRINT "ENTER THE MASS OF THE RADIONU-CLIDE IN GRAMS"
30 INPUT M
40 PRINT
50 PRINT "ENTER THE NUMBER OF HALF-LIVES TO ELAPSE"
60 INPUT H
70 PRINT
80 FOR C=0 TO H
90 PRINT "HALF-LIVES", H–C; "MASS REMAINING"; M; "g"
100 PRINT
110 LET M=M/2
120 NEXT C
130 END

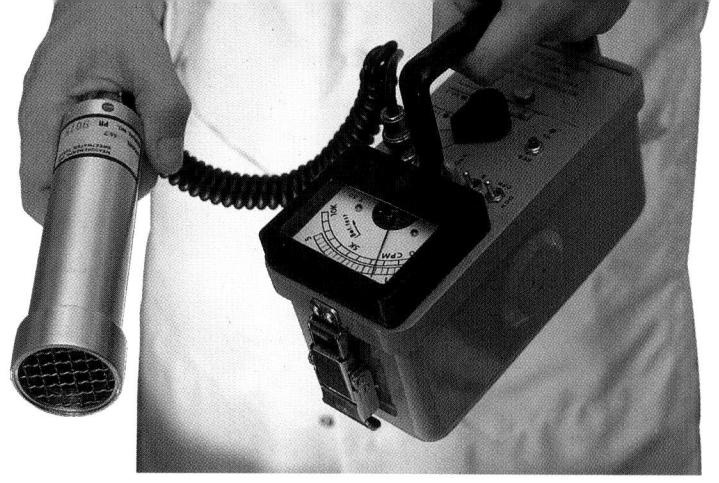

Figure 24-8

A Geiger counter can be used to measure the intensity of nuclear radiation.

Figure 24-8 shows a Geiger counter. The tube is filled with gas at a low pressure. A positively charged wire runs through the center of a negatively charged copper cylinder. This tube is connected to a voltage source. Radiation enters the tube at one end, stripping electrons from the gaseous atoms. The electrons are attracted to the positive wire. They knock more electrons off the atoms in the gas, and an "electron avalanche" is produced. A large number of electrons reaches the wire, producing a short, intense current in the wire. This current is amplified to produce a clicking sound or flashing light. The intensity of radiation present is determined by the number of clicks in each second.

Geiger counters can be made small and portable. They are often used to test the radioactivity at job sites where workers can be exposed to radioactive materials.

Section Wrap-up

Review

1. What are four ways radioactivity can be detected?

2. Briefly explain how a Geiger counter operates.

3. **Think Critically:** Suppose you needed to check the level of radioactivity in the laboratory of a hospital. Which method would you use? Explain your choice.

Skill Builder

Observing and Inferring

You are observing the presence of nuclear radiation with a bubble chamber and see two kinds of trails. Some trails are short and thick, and others are long and thin. What type of nuclear radiation might have caused each trail? If you need help, refer to Observing and Inferring in the **Skill Handbook**.

Using Computers

Word Processing
Write a short program that allows you to input the mass and the half-life of a radioactive sample and calculate what mass remains after a certain number of half-lives.

Skill Builder

Alpha particles have a higher charge and lower penetrating power and would produce the shorter trails. Beta particles are much smaller, and have higher penetrating ability, and would produce the long, thin trails.

✔ Assessment

Oral Have students explain why beta particles would likely leave longer trails. Use the Performance Task Assessment List for Making Observations and Inferences in **PASC**, p. 17. **P**

Nuclear Reactions 24 • 4

Nuclear Fission

Have you ever played a game of pool? What happens when you shoot the cue ball into an area of densely packed balls? They spread apart, or "break" the pack. In 1938, physicists Otto Hahn and Fritz Strassmann found that a similar result occurs when a neutron is shot into the large nucleus of a uranium-235 atom. Earlier, physicist Enrico Fermi had tried to bombard large nuclei with neutrons in an effort to make nuclei that would be larger than uranium. Splitting an atomic nucleus wasn't the expected outcome of this process.

Explain the Process

Lise Meitner was the first to offer a theory to explain the splitting of a nucleus. She concluded that the neutron fired into the nucleus disturbs the already unstable nucleus and causes it to split into two nuclei of nearly equal mass. The process of splitting a nucleus into two nuclei with smaller masses is called **nuclear fission**. The term *fission* means "to divide."

Only large nuclei with atomic numbers above 90 can undergo nuclear fission, an example of which is shown in **Figure 24-9**. The products of a fission reaction usually include two or three individual neutrons. The total mass of the products is somewhat less than the mass of the uranium-235 nucleus and the neutron. Some of the mass has been converted to a tremendous amount of energy.

$$\,_0^1 n + \,_{92}^{235}U \longrightarrow \,_{56}^{141}Ba + \,_{36}^{92}Kr + 3\,_0^1 n$$

Figure 24-9

The uranium-235 nucleus is split by bombarding it with a neutron, and there can be many possible fission products. Sometimes, it forms a barium-141 nucleus and a krypton-92 nucleus and releases three neutrons.

Do you think the energy released in the fission of a single nucleus would be significant? Probably not by itself. However, the three neutrons produced in the fission reaction can bombard other nuclei in the sample to split more nuclei.

Science Words

nuclear fission
chain reaction
nuclear fusion

Objectives

- Distinguish between nuclear fission and fusion.
- Explain how nuclear fission can begin a chain reaction.
- Discuss how nuclear fusion occurs in the sun.

24-4 Nuclear Reactions   689

Section 24•4

Prepare

Section Background

When a neutron is absorbed into a nucleus, the added internal energy causes the nucleus to take on an elongated form. The strong nuclear force is weakened over this increased distance, and the repulsive electric forces dominate. The nucleus splits into two fission fragments and several neutrons.

Preplanning

Obtain several sets of dominoes for the MiniLAB.

1 Motivate

Bellringer

 Before presenting the lesson, display **Section Focus Transparency 99** on the overhead projector. Assign the accompanying **Focus Activity** worksheet. **L1** **LEP**

Tying to Previous Knowledge

Tie nuclear science to students' previous knowledge of the use of nuclear energy, especially in generating electricity and in nuclear bombs, and the tremendous amounts of energy involved.

Program Resources

Reproducible Masters
Study Guide, p. 103 **L1**
Reinforcement, p. 103 **L1**
Enrichment, p. 103 **L3**
Activity Worksheets, p. 147 **L1**

Transparencies
Section Focus Transparency 99 **L1**
Teaching Transparencies 47, 48

689

2 Teach

MiniLAB

Purpose
 Kinesthetic To illustrate the concept of a chain reaction. L1

Materials
packages of dominoes

Teaching Strategies
Set up one set of dominoes to fall from the tumbling of one. Set up another set to purposely fail. Compare the results.

Analysis
1. Many reactions beginning from only a few neutron-emitting fission reactions can be seen.
2. The tumbling sequence stops. This can happen in a chain reaction if released neutrons do not find other fissionable target nuclei.

✓ Assessment

Oral Ask students to explain why it is difficult to control nuclear fission reactions. Use the Performance Task Assessment List for Making Observations and Inferences in **PASC**, p. 17. P

INTEGRATION
Earth Science

Stars synthesize lighter-weight atoms up to iron. After that, huge explosions assist the production of other, heavier elements.

CONNECT TO
CHEMISTRY

Answer The periodic table shows an atomic mass of 238.029, representing the average of all isotopes. Uranium-235 is relatively rare.

690

Figure 24-10

A chain reaction occurs when neutrons produced by nuclear fission bombard other nuclei, releasing more neutrons. Some of these neutrons are absorbed by control rods in the reactor.

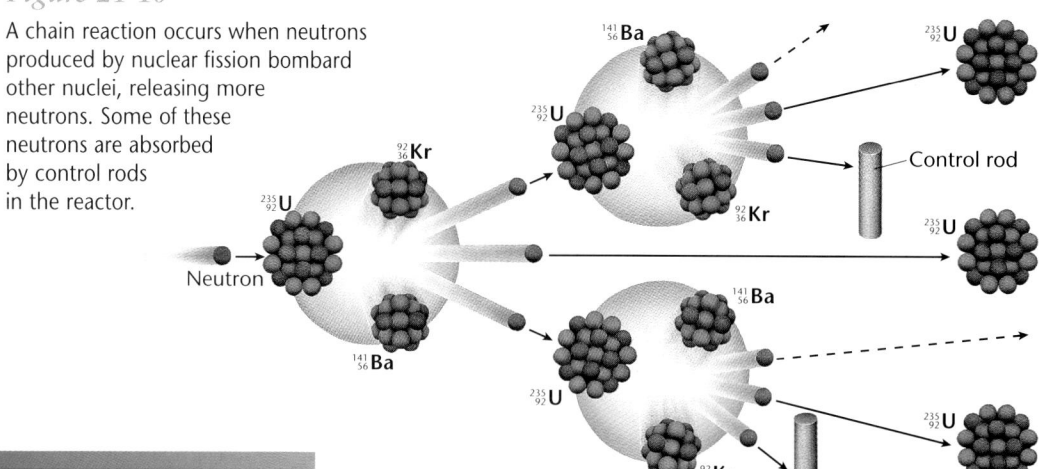

MiniLAB

How does a chain reaction start?

Find out how a chain reaction starts and then continues.

Procedure
1. Place dominoes close together on their edges in a straight line.
2. When all the dominoes are set up, gently knock over the first domino.

Analysis
1. In what ways is your model similar to the chain reaction that can take place in a sample of uranium-235?
2. What would happen if one domino were not positioned so that it struck the next domino when it fell? How is this situation similar to what could happen in a chain reaction involving uranium?

 INTEGRATION
Earth Science

These reactions will each release more neutrons, and so on. If there is not some other material present to absorb these neutrons, an uncontrolled chain reaction can result. A **chain reaction,** shown in **Figure 24-10,** is an ongoing series of fission reactions. This can cause billions of reactions to occur each second, resulting in the release of tremendous amounts of energy. In the next chapter, you will read about ways scientists control nuclear fission and use the energy to produce electricity.

Nuclear Fusion

You read in the last section how tremendous amounts of energy can be released in nuclear fission. In fact, splitting one uranium-235 nucleus produces several million times more energy than chemically reacting one molecule of dynamite. Even more energy can be released in another type of nuclear reaction. This process is caused by the fusing together of nuclei, so it is the opposite of fission. **Nuclear fusion** is the combining of two nuclei with low masses to form one nucleus of larger mass.

Natural Fusion

Controlled nuclear fusion is difficult to induce because extremely high temperatures are required for this process to occur. For this reason, fusion reactions are also known as thermonuclear reactions. However, these extremely high temperatures do exist in the stars, including the sun, shown in **Figure 24-11.** At these high temperatures, atoms have so much kinetic energy that positive nuclei and electrons exist separately. Matter in this state is called plasma.

690 Chapter 24 Radioactivity and Nuclear Reactions

 Use these program resources as you teach this lesson.
Study Guide, p. 103

Use **Teaching Transparencies 47** and **48** as you teach this lesson.

Scientists hypothesize that when the sun first formed, most of its nuclei were hydrogen nuclei. As thermonuclear fusion occurred, four hydrogen nuclei fused to become a helium nucleus. This process is still occurring, and about one percent of the mass of the sun's original nuclei is changed to energy and released in the form of electromagnetic radiation. As the star ages, the percentage of helium nuclei gradually increases and the number of hydrogen nuclei decreases until they are used up, and the reaction can no longer occur. Scientists estimate that the sun has enough hydrogen to keep this reaction going for another 5 billion years.

Scientists are searching for a way to induce and control nuclear fusion. If they achieve this goal, fusion could provide an almost unlimited supply of energy to produce electricity.

CONNECT TO

CHEMISTRY

Use the periodic table to find the atomic mass of uranium. *Infer* why this value differs from the mass number 235 given for fissionable uranium.

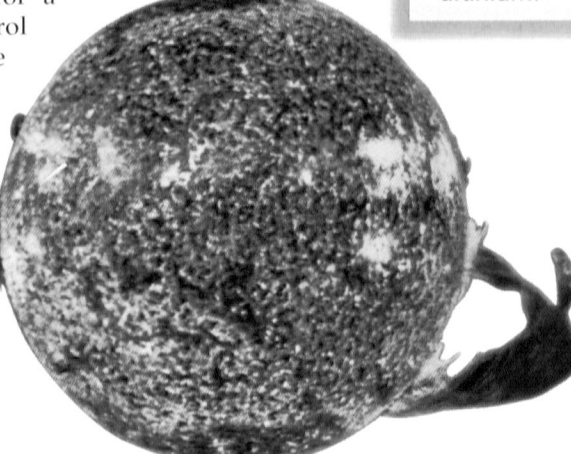

Figure 24-11

The sun is a star of medium internal temperature, averaging about 20 million degrees Celsius.

Section Wrap-up

Review

1. Compare and contrast nuclear fission and fusion.

2. Why does a chain reaction often occur when a uranium-235 nucleus is split?

3. **Think Critically:** Explain why the hydrogen fusion reaction that takes place in the sun can't currently produce energy for electricity on Earth.

Skill Builder
Concept Mapping
Make a concept map to show how a chain reaction occurs when U-235 is bombarded with a neutron. Show how each of the three neutrons given as products begins another fission reaction. If you need help, refer to Concept Mapping in the **Skill Handbook.**

Science Journal

In your Science Journal, explain why finding a practical way to use fusion as an energy source would be an improvement over using fission.

24-4 Nuclear Reactions   691

Skill Builder

```
        Neutron
           │
          hits
           │
      U-235 nucleus
       ╱           ╲
  and forms      and releases
    ╱    ╲        ╱    │    ╲
Ba-141  Kr-92  Neutron Neutron Neutron
                  │       │       │
              which hits which hits which hits
                  │       │       │
               U-235    U-235    U-235
               nucleus  nucleus  nucleus
```

✓ Assessment

Performance Ask students to redraw their Skill Builder concept maps to show how a chain reaction may be prevented. Use the Performance Task Assessment List for Scientific Drawing in **PASC**, p. 55. **P**

---

3 Assess

Check for Understanding

? FLEX Your Brain

Use the Flex Your Brain activity to have students explore FUSION.

📁 **Activity Worksheets,** p. 5

Reteach

LS **Interpersonal** Have one student tell another how to model fission with clay and have him or her do it.

Extension

📁 For students who have mastered this section, use the **Reinforcement** and **Enrichment** masters.

4 Close

Activity

For fusion, show how two small balls of clay can fuse to become a larger nucleus. **LEP**

COOP LEARN

Section Wrap-up

Review

1. Fission is splitting large nuclei, and fusion is joining of small nuclei.

2. The reaction produces 3 neutrons that can bombard and cause fission in other uranium-235 nuclei.

3. **Think Critically** Fusion requires extremely high temperatures that cannot be maintained on Earth.

Science Journal

LS **Linguistic** Nuclear waste would be nearly eliminated; reactants are easier to obtain.

691

Prepare

Section Background

Immature or rapidly dividing cells are most susceptible to damage by radiation. For this reason, cancer cells can often be destroyed with radiotherapy.

1 Motivate

Bellringer

 Before presenting the lesson, display **Section Focus Transparency 100** on the overhead projector. Assign the accompanying **Focus Activity** worksheet. L1 LEP

2 Teach

GLENCOE TECHNOLOGY

 **Videodisc**

The Infinite Voyage Series: Unseen Worlds

Chapter 8

Digital X Rays, 3-D X Rays: Detection Made Easy

Fires of the Mind

Chapter 4

Positron Emission Tomography (PET)

STVS: Human Biology

Disc 7, Side 1

PET Scanner (Ch. 19)

Science Words

tracer
PET

Objectives

* Describe how radioactive tracers and PET can be used to diagnose medical problems.
* Discuss how radioactive isotopes can aid in the treatment of cancers.

Figure 24-12

Iodine-131 accumulates in the thyroid gland. The amount of radiation detected is related to how well the thyroid gland functions.

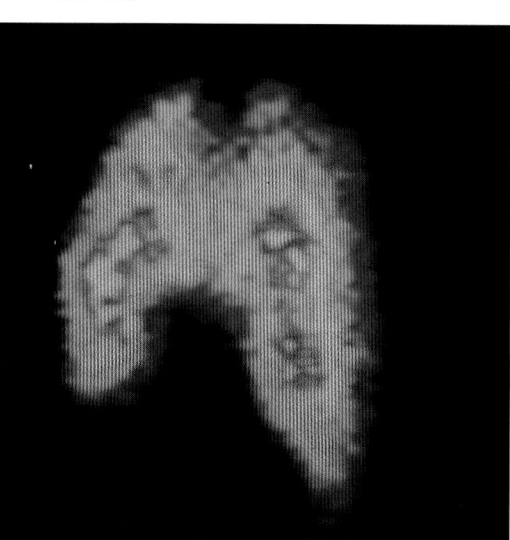

TECHNOLOGY:

24•5 Using Nuclear Reactions in Medicine

Tracers—Identified Isotopes

If you were going to meet a friend in a crowded area, it would be easier to find you if you told your friend what you would be wearing. If scientists want to follow where a particular molecule goes in your body or how a particular organ functions, they do some identifying of their own. Just as a bright red hat might set you apart from hundreds of other people, a radioactive atom differs from other surrounding atoms. These radioisotopes that are used inside of a body to monitor body processes are called **tracers.** Examples of tracers include carbon-11, iodine-131, and sodium-24, all of which are used as tracers because of their relatively short half-lives.

Thyroid—The Iodine Connection

The thyroid gland helps regulate several body processes, including growth. Iodine from food is transported to the thyroid, where it is incorporated into hormones that the thyroid produces. If a patient is suspected of having a thyroid problem, he or she may be given a mixture that contains iodine-131, which emits beta particles. The thyroid gland absorbs the iodine-131. The energy from emitted beta particles can be detected and used to evaluate the patient's thyroid, as shown in **Figure 24-12.** If little radiation shows up, the doctor may suspect a tumor.

Cancer Treatment

The radiation from nuclear decay is strong enough to ionize nearby atoms. This ionization process can be a way to destroy the advance of some cancerous tumors. Cancerous cells grow quickly, making them more susceptible to absorbing radiation than healthy cells. If a critical molecule, such as the DNA or RNA of a cancerous cell, had an atom that was ionized, then

692 Chapter 24 Radioactivity and Nuclear Reactions

Program Resources

 Reproducible Masters
Study Guide, p. 104 L1
Reinforcement, p. 104 L1
Enrichment, p. 104 L3
Cross-Curricular Integration, p. 30
Activity Worksheets, pp. 144-145 L1

 **Transparencies**
Section Focus Transparency 100

the molecule might no longer function in its intended way. This could cause death or mutation in the cancerous cell. When possible, a radioactive isotope such as gold-198 or iridium-192 is implanted within or near the tumor. If a surgical implant is impractical, a tumor is often treated from outside the body. Typically, an intense gamma ray from cobalt-60 is focused on the tumor for a specific time period. The gamma rays pass through the body and into the tumor.

A Practical Use for Positrons

To study brain functions and disorders, scientists attach radioactive fluorine-18 to molecules that are circulated to the brain. A patient is then placed inside a large tube with an array of detectors. When positrons are emitted from decaying fluorine that is in the blood supply of the brain, they hit electrons from other molecules and form two gamma rays. Analyzing the emitter gamma rays can reveal tumors and various activity levels in the brain. This technological advancement in medicine, shown in **Figure 24-13,** is called Positron Emission Tomography, or **PET.**

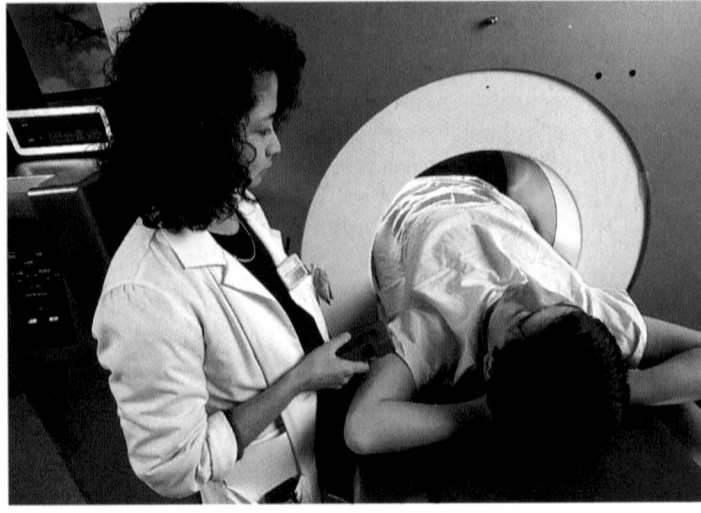

Figure 24-13

When detectors are placed around a person's head, PET scans reveal brain function and can reveal brain disorders, such as tumors.

Section Wrap-up

Review

1. What is a radioactive tracer? Give an example of a tracer.

2. Why is it important that tracers have relatively short half-lives?

Explore the Technology

Although many benefits have resulted from nuclear applications in medicine, there are some disadvantages. During cancer treatment, for example, healthy cells are also damaged. In fact, high levels of radiation have been linked with causing cancer. Explain how radiation can both cause and treat cancer.

SCIENCE & SOCIETY

24-5 Using Nuclear Reactions in Medicine   **693**

Community Connection

Nuclear Medicine If possible, have a medical doctor or radiation technician visit the classroom to discuss nuclear medicine examples. As an alternative, write to or visit the nearest facility to obtain up-to-date brochures that deal with nuclear medicine.

Figuring Out Fusion

The energy that warms Earth is produced in thermonuclear reactions in the sun, which is about 1.5×10^8 km away. We can model this process of nuclear fusion.

Purpose

LS **Kinesthetic** Illustrate the hydrogen fusion process through the use of physical models. **L1** **LEP** **COOP LEARN**

Process Skills

formulating models, interpreting data, inferring, sequencing, recognizing cause and effect, interpreting scientific illustrations

Time

30 minutes

📁 **Activity Worksheets,** pp. 5, 144-145

Teaching Strategies

Alternate Materials Any combination of large and small objects can be used to represent neutrons and positrons. Clay requires no pins to hold parts together.

Troubleshooting Be sure students construct the correct nuclei at each step. If they do not, they will not be able to draw a correct conclusion.

- The proton model shown here is useful for this activity but otherwise not representative. The released positron is a short-lived particle that reacts with an electron to become gamma radiation.

Answers to Questions

1. Two protons fuse as a positron is released, and a proton changes to a neutron. Then a hydrogen-1 nucleus fuses with hydrogen-2 to make helium-3. Finally, two helium-3 nuclei fuse to make one helium-4 and two hydrogen-1 nuclei.

2. Two positrons are lost. Two protons are left over.

3. It takes six hydrogen-1 nuclei to make one helium-4 nucleus. However, two hydrogens are regained in the process. Therefore, the net usage is four hydrogen-1 atoms.

Problem

How can you model nuclear fusion?

Materials

- 6 balls of green clay
- 6 smaller balls of white clay

Procedure

1. Make six hydrogen nuclei (protons) by sticking one small white positron to each large green neutron.
2. Refer to step 1 of the table and make two particles of $^2_1$H.
3. Refer to step 2 of the table and make two particles of $^3_2$He.
4. Refer to step 3 of the table and make one particle of $^4_2$He.

Analyze

1. Sequence the steps necessary to change $^1_1$H into $^4_2$He.
2. In this process, what is lost and what is left over?
3. How many atoms of hydrogen does it take to make one atom of helium?

Conclude and Apply

4. What can the leftover hydrogen be used for?
5. The mass of a hydrogen-1 nucleus is 1.0078 u, while the mass of a helium-4 nucleus is 4.0026 u. **Compare** the mass of one helium nucleus to the mass of four hydrogen nuclei. If mass changes to energy, where does the energy of the sun come from?

Fusion Reaction

Step 1	$^1_1$H + $^1_1$H → $^2_1$H + positron
Step 2	$^1_1$H + $^2_1$H → $^3_2$He
Step 3	$^3_2$He + $^3_2$He → $^4_2$He + 2$^1_1$H

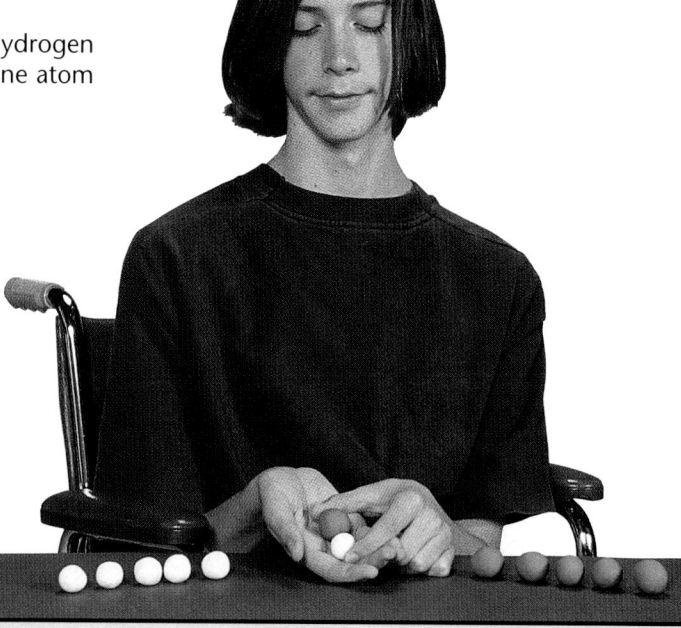

4. Hydrogen that is produced in the fusion reaction can be used for the next cycle of fusion.

5. He = 4.0026 u; 4H = (4)(1.0078 u)
   = 4.0312 u

   mass loss = 4.0312 u – 4.0026 u
   = 0.0286 u

This mass is converted to energy in the fusion process. The energy of the sun comes from the conversion of mass during fusion according to the relationship $E = mc^2$.

✔ Assessment

Oral Ask students to explain how many other hydrogen nuclei would be needed to join the two remaining nuclei to form the next helium nucleus. Use the Performance Task Assessment List for Making Observations and Inferences in **PASC,** p. 17. **P**

Summary

24-1: Radioactivity

1. Radioactivity was accidentally discovered by Henri Becquerel about 100 years ago.
2. Radioactive nuclides are unstable and therefore decay. Stable nuclides don't decay.

24-2: Nuclear Decay

1. The three common types of radiation emitted from a decaying nucleus are alpha particles, beta particles, and gamma rays.
2. The amount of a remaining radioactive substance can be determined if you know the amount of the original sample and the number of half-lives that have elapsed.
3. The age of some previously living materials can be determined by a process called carbon-14 dating.

24-3: Detecting Radioactivity

1. Radioactivity can be detected with a cloud chamber, a bubble chamber, an electroscope, or a Geiger counter.
2. A Geiger counter indicates the intensity of radiation present by producing a clicking sound or a flashing light that increases in frequency as more radiation is present.

24-4: Nuclear Reactions

1. Atomic nuclei are split during fission and combined during fusion.
2. Neutrons released from a nucleus during fission can split other nuclei.

24-5: Science and Society: Using Nuclear Reactions in Medicine

1. Radioactive tracers can go to certain areas of the body to indicate abnormalities.

2. Radiation from some radioactive isotopes can kill cancer cells.

Key Science Words

a. alpha particle
b. beta particle
c. bubble chamber
d. chain reaction
e. cloud chamber
f. gamma ray
g. half-life
h. nuclear fission
i. nuclear fusion
j. nuclide
k. PET
l. radioactivity
m. tracer
n. transmutation

Reviewing Vocabulary

Match each phrase with the correct term from the list of Key Science Words.

1. emission of high-energy particles or radiation from an unstable nucleus
2. a particle of nuclear radiation with a charge of 2+ and an atomic mass of 4
3. nuclear decay that results in the formation of a new element
4. the time for one-half of a sample of a radioactive isotope to decay
5. detects nuclear particles as they leave a trail of condensed water or ethanol vapor
6. splitting a nucleus into two smaller masses
7. radioactive isotopes that are used to monitor human body functions
8. the nucleus of a specific isotope
9. radiation in the form of an electromagnetic wave
10. two nuclei combining into a larger nucleus

Summary

Have students read the summary statements to review the major concepts of the chapter.

Reviewing Vocabulary

1. l	**6.** h
2. a	**7.** m
3. n	**8.** j
4. g	**9.** f
5. e	**10.** i

✔ Assessment

Portfolio Encourage students to place in their portfolios one or two items of what they consider to be their best work. Examples include:

- Explore Assessment, p. 673
- Skill Builder Assessment, p. 683
- Activity 24-1 Assessment, p. 685 P

Performance Additional performance assessments may be found in **Performance Assessment** and **Science Integration Activities.** Performance Task Assessment Lists and rubrics for evaluating these activities can be found in Glencoe's **Performance Assessment in the Science Classroom.**

GLENCOE TECHNOLOGY

Videodisc

Glencoe Physical Science

Interactive Videodisc

Use the videodisc lesson 6, *Chemical Detectives,* to review the principles of half-life and radioactive dating methods .

MindJogger Videoquiz

Chapter 24 Have students work in groups as they play the Videoquiz game to review key chapter concepts.

Checking Concepts

1. a   **6.** b

2. a   **7.** a

3. b   **8.** c

4. a   **9.** a

5. a   **10.** d

Understanding Concepts

11. Radioactivity was discovered by Henri Becquerel in 1896. Marie and Pierre Curie discovered polonium and radium. Nuclear fission was produced in 1938 by Otto Hahn and Fritz Strassman and explained by Lise Meitner.

12. The strong nuclear force is less effective when the ratio of protons to neutrons in the nucleus is not a stable ratio.

13. There is a fixed percentage of carbon-14 while an organism is alive. By measuring the percentage of carbon-14 in a sample of once-living material and comparing this with the percentage of carbon-14 in a living sample, scientists can approximate the age of the sample.

14. In a cloud chamber, ions cause supersaturated vapor to condense in a cloud trail. In a bubble chamber, ionized atoms of a superheated liquid boil, leaving a trail of bubbles. In an electroscope, radiation strips electrons from air molecules, giving the molecules a positive charge. These ions cause the electroscope to discharge.

15. Cancer cells have accelerated physiological processes. Thus, most of the tracer will accumulate at the tumor and can be detected.

Thinking Critically

16. Yes. Other than being radioactive, the salt would

Checking Concepts

Choose the word or phrase that completes the sentence.

1. Radioactivity can be _____.
  a. found in your classroom   c. seen
  b. prevented by a tracer   d. smelled

2. Elements that have been produced artificially through nuclear reactions have atomic numbers _____.
  a. greater than 92   c. between 83 and 92
  b. greater than 83   d. between 90 and 111

3. When a neutron decays, the electron produced is called _____.
  a. an alpha particle   c. gamma radiation
  b. a beta particle   d. both a and c

4. The time for half the nuclides in a sample of a radioactive isotope to decay _____.
  a. is constant   c. increases with time
  b. varies   d. decreases with time

5. Carbon-14 dating could be used to date _____.
  a. an ancient Roman scroll
  b. an ancient marble column
  c. dinosaur fossils
  d. Earth's oldest rocks

6. Radiation in a nuclear laboratory could best be measured with _____.
  a. a cloud chamber   c. an electroscope
  b. a Geiger counter   d. a bubble chamber

7. A _____ is an ongoing series of fission reactions.
  a. chain reaction
  b. decay reaction
  c. positron emission
  d. fusion reaction

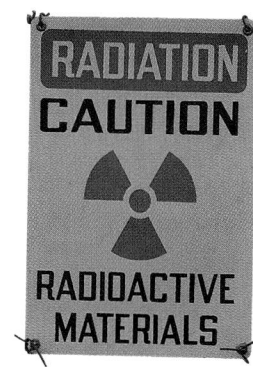

8. The sun is powered by _____.
  a. nuclear decay   c. thermonuclear fusion
  b. nuclear fission   d. combustion

9. A radioisotope that acts as an external source of ionizing radiation in the treatment of cancer is _____.
  a. cobalt-60   c. gold-198
  b. carbon-14   d. technetium-99

10. A major medical use of radiation is to _____.
  a. assist breathing   c. heal broken bones
  b. ease pain   d. treat cancers

Understanding Concepts

Answer the following questions in your Science Journal using complete sentences.

11. Briefly discuss the history of nuclear science.

12. Explain why certain nuclides decay.

13. Describe the basic principles of carbon-14 dating.

14. Discuss the similarities and differences in the devices that detect radioactivity.

15. Discuss the use of radioactive tracers. How do they detect cancer cells?

Thinking Critically

16. Can a radioactive isotope of sodium combine with chlorine to form table salt? If so, would its chemical and physical properties be the same as those of ordinary table salt? Explain your answer.

17. Copper-66 releases a beta particle as it decays, and radium-226 releases an alpha particle. Write a nuclear equation for each of these transmutations.

18. Explain the presence of relatively heavy elements such as carbon, oxygen, magnesium, and iron in stars.

19. Explain how radioisotopes are used to preserve foods and to study how plants take up nutrients from the soil.

696   Chapter 24 Radioactivity and Nuclear Reactions

be chemically the same as normal salt.

17. $^{66}_{29}\text{Cu} \rightarrow {}^{66}_{30}\text{Zn} + {}^{0}_{-1}e$

$^{226}_{88}\text{Ra} \rightarrow {}^{222}_{86}\text{Rn} + {}^{4}_{2}\text{He}$

18. The temperatures and pressures in the interior of a star are sufficiently high for atoms in the plasma state to undergo fusion and form successively heavier atoms as lighter nuclei collide.

19. Radiation can be used to kill bacteria in food and give it longer shelf-life. Radioisotopes can be placed in the soil of

lab plants. Using radiation detectors, scientists can monitor how they are taken up by the plant.

20. Four half-lives elapse in 40 minutes. remaining mass = $320 \text{ g}/2^4 = 20$ grams

Developing Skills

21. **Concept Mapping** See student page.

22. **Making and Using Tables** See sample table at the bottom of page 697.

20. Nitrogen-13 has a half-life of about 10 minutes. How much of a 320-g sample of nitrogen-13 would remain after decaying for 40 minutes?

Developing Skills

If you need help, refer to the **Skill Handbook**.

21. **Concept Mapping:** Complete the following concept map summarizing how a Geiger counter works.

Initiating event

| Radiation enters negatively charged copper tube. |

↓

| Radiation strips electrons from gaseous atoms. |

↓

| Free electrons move to positively charged wire in tube. |

↓

Final outcome

| Resulting current is amplified to produce a click. |

22. **Making and Using Tables:** Construct a table summarizing the characteristics of each of the three common types of radiation.

23. **Observing and Inferring:** Another type of particle of nuclear radiation is a positron, which has a charge of 1+ and the same mass as an electron. This particle is given off when a proton spontaneously changes into a neutron. Infer what type of radiation will be emitted from each of the following radioisotopes:
   a. Boron-8, which is unstable due to an extra proton.
   b. Thorium-232, which is unstable due to its large nucleus containing too many protons and neutrons.
   c. Potassium-40, which is unstable due to an extra neutron.

24. **Making and Using Graphs:** Using the data below, construct a graph plotting the mass numbers vs. the half-lives of radioisotopes. Plot mass numbers to the nearest ten. Is it possible to use your graph to predict the half-life of a radioisotope given its mass number?

Isotope Half-Lives

Radio-isotope	Mass Number	Half-Life
Radon	222	4 days
Thorium	234	24 days
Iodine	131	8 days
Bismuth	210	5 days
Polonium	210	138 days

25. **Interpreting Data:** Review your graph of decay from Activity 24-1 on pages 684-685. What would the half-life curve look like if your sample contained twice as many coins? Predict the new data and sketch the corresponding graph. Then try it and see.

Performance Assessment

1. **Booklet:** Write a biography of a person who made an important contribution to nuclear science. Include sketches or photos of the person and his or her work.
2. **Oral Presentation:** Research the causes and effects of radon pollution in the home. Report your findings to the class.
3. **Model:** Use the clay models from Activity 24-2 on page 694 to determine the product of a simple fusion reaction between a proton ($_1^1 H$) and a neutron ($_0^1 n$). Complete the following nuclear equation.
$$_1^1 H + _0^1 n \rightarrow \underline{\hspace{1cm}} + \text{gamma radiation}$$

Chapter 24 Review

23. **Observing and Inferring**
   a) emission of positron
   b) emission of alpha particle
   c) emission of beta particle
24. **Making and Using Graphs** The graph shows no correlation between mass number and greatly differing half-lives.

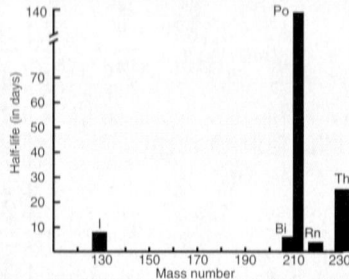

25. **Interpreting Data** The curve would have the same shape, but the data points would have higher values and one more half-life would be required for the entire sample to decay.

Performance Assessment

1. Have students use reference materials. Use the Performance Task Assessment List for Writer's Guide to Nonfiction in **PASC,** p. 85. **P**
2. Student reports should include sources of radon, detection, and solutions to the problem. Use the Performance Task Assessment List for Oral Presentation in **PASC,** p. 71. **P**
3. $_1^2 H$, deuterium; Use the Performance Task Assessment List for Model in **PASC,** p. 51. **P**

22.

Type of Radiation	Charge	Mass	Rel. Speed	Penetrating Power	Symbol
Alpha	2+	4	Slow	Low; stopped by paper	$_2^4 He$; α
Beta	1–	0	Fast	Moderate; stopped by aluminum foil	$_{-1}^{0} e \, _{+1}^{0} e$; β
Gamma	0	0	Speed of light	Great; stopped by thick lead or concrete	γ

Chapter Organizer

Section	Objectives/Standards	Activities/Features
Chapter Opener		**Explore Activity:** Use the energy from the wind to provide turning power. p. 699
25-1 Fossil Fuels (1 session)*	1. **Discuss** the origin and characteristics of the three main types of fossil fuels. 2. **Describe** the need for and methods of energy conservation. National Content Standards: (5-8) UCP2, UCP3, UCP5, A2, B3, E2, F4, F5; (9-12) UCP2, UCP3, UCP5, A2, B5, E2, F6	**MiniLAB:** How can an updated design improve efficiency? p. 701 **Using Technology:** No More Gas Tanks? p. 702 **Science Journal,** p. 703 **Skill Builder:** Comparing and Contrasting, p. 703
25-2 Nuclear Energy (1 session)*	1. **Outline** the operation of a nuclear reactor. 2. **Describe** the problems and methods associated with nuclear waste disposal. 3. **Discuss** nuclear fusion as a possible energy source. National Content Standards: (5-8) UCP1-UCP3, UCP5, A2, B3, D1; (9-12) UCP1-UCP3, UCP5, A2	**Using Math,** p. 706 **Connect to Earth Science,** p. 708 **Using Computers,** p. 708 **Skill Builder:** Concept Mapping, p. 708 **Activity 25-1:** Nuclear Waste Disposal, p. 709
25-3 Science and Society (1 session)*	1. **Discuss** problems associated with storing nuclear waste. 2. **Compare** long-term nuclear waste-storage options. National Content Standards: (5-8) UCP2, UCP5, F1-F5, G1; (9-12) UCP2, UCP5, F1, F3-F6, G1	**Explore the Issue,** p. 711
25-4 Alternative Energy Sources (2 sessions)*	1. **Analyze** the need for alternative energy sources. 2. **Discuss** the methods of generating electricity with several energy sources. 3. **Describe** the advantages and disadvantages of several alternative energy sources. National Content Standards: (5-8) UCP1-UCP3, UCP5, A1, A2, B3, D3, E1, G1; (9-12) UCP1-UCP3, UCP5, A1, A2, D1, E1, G1	**Connect to Chemistry,** p. 713 **MiniLAB:** How do hydroelectric power plants use water to drive turbines? p. 715 **Problem Solving:** Energy Exercises, p. 716 **Using Math,** p. 717 **Skill Builder:** Classifying, p. 717 **Activity 25-2:** Energy Alternatives—A Local Plan, pp. 718-719 **People and Science:** Yvonne Ho Cardinale, Hydroelectric Engineer, p. 720

Activity Materials

Explore	Activities	MiniLABs
page 699 several sheets of paper, scissors, metric ruler, straight pins, window fan	page 709 4 sodium hydroxide pellets, four 100-mL beakers, phenolphthalein solution, plastic food wrap, aluminum foil, 3 rubber bands, 2 twist ties, modeling clay, forceps, safety goggles, apron pages 718-719 large poster board, colored markers, miscellaneous construction objects such as empty paper towel or toilet paper rolls, glue, cellophane tape, aluminum foil, empty cardboard milk cartons	page 701 balance, candle, water, graduated cylinder, beaker, thermometer, aluminum foil page 715 clay, unsharpened pencil, flat wooden sticks, large plastic container, water

Need Materials? Call Science Kit (1-800-828-7777).   * A complete Planning Guide that includes block scheduling is provided on pages 32T-35T.

Teacher Classroom Resources

Reproducible Masters	Transparencies	Teaching Resources
Study Guide, p. 105 Reinforcement, p. 105 Enrichment, p. 105 Activity Worksheets, p. 152 Critical Thinking/Problem Solving, p. 31	Section Focus Transparency 101, Fossil Fuels Teaching Transparency 50, Petroleum Distillation	**Glencoe Physical Science Interactive Videodisc** **Physical Science CD-ROM** **Spanish Resources** **English/Spanish Audiocassettes** **Cooperative Learning Resource Guide** **Lab Partner** **Lab and Safety Skills** **Lesson Plans**
Study Guide, p. 106 Reinforcement, p. 106 Enrichment, p. 106 Activity Worksheets, pp. 148-149	Section Focus Transparency 102, Nuclear Power Plants Worldwide Teaching Transparency 49, Nuclear Power Plant	**Assessment Resources** Chapter Review, pp. 53-54 Assessment, pp. 159-162 Performance Assessment in the Science Classroom (PASC) MindJogger Videoquiz Alternate Assessment in the Science Classroom Performance Assessment Chapter Review Software Computer Test Bank
Study Guide, p. 107 Reinforcement, p. 107 Enrichment, p. 107 Cross-Curricular Integration, p. 31	Section Focus Transparency 103, What Do You Do with an Old Nuclear Power Plant?	
Study Guide, p. 108 Reinforcement, p. 108 Enrichment, p. 108 Activity Worksheets, pp. 150-151, 153 Lab Manual 49, Solar Cells Lab Manual 50, Using the Sun's Energy Multicultural Connections, p. 53 Science Integration Activity 25, "Free" Hot Water Science and Society Integration, p. 29 Concept Mapping, pp. 55-56	Section Focus Transparency 104, Energy Use in the Future Science Integration Transparency 25, Wind Power Works!	**Key to Teaching Strategies** The following designations will help you decide which activities are appropriate for your students. L1 Level 1 activities should be within the ability range of all students, including those with learning difficulties. L2 Level 2 activities should be within the ability range of the average to above-average student. L3 Level 3 activities are designed for the ability range of above-average students. LEP LEP activities should be within the ability range of Limited English Proficiency students. LS These activities are designed to address different learning styles. COOP LEARN Cooperative Learning activities are designed for small group work. P These strategies represent student products that can be placed into a best-work portfolio.

GLENCOE TECHNOLOGY

The following multimedia resources are available from Glencoe.

Science and Technology Videodisc Series (STVS)
Chemistry
 Hydrogen Sponge
 Dealing with Hazardous
 Materials
 Solar House
 Spiral Solar Concentrator
 Solar Tower
 Hydroelectric Power
 Mini-Hydroelectric Power Plants
 Geothermal Wells

National Geographic Society Series
GTV: Planetary Manager

Glencoe Physical Science Interactive Videodisc
Carbohydrates and Hydrocarbons

Physical Science CD-ROM

Teacher Classroom Resources

This is a representation of key blackline masters available in the Teacher Classroom Resources.

Teaching Aids

Section Focus Transparencies

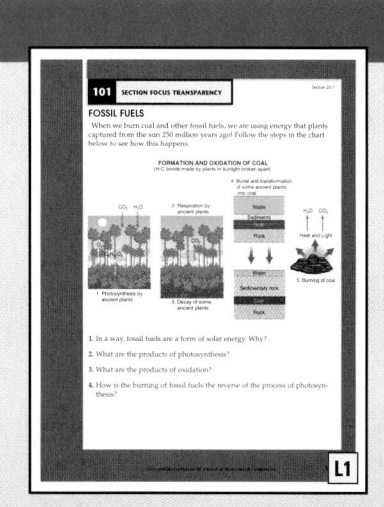

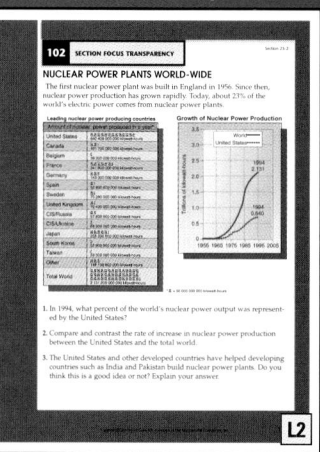

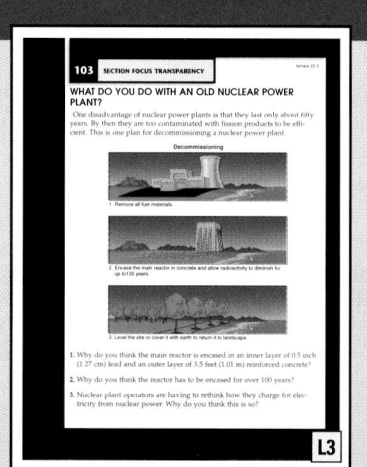

Science Integration Transparencies

Teaching Transparencies

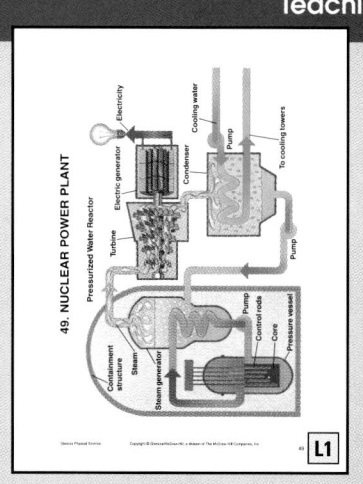

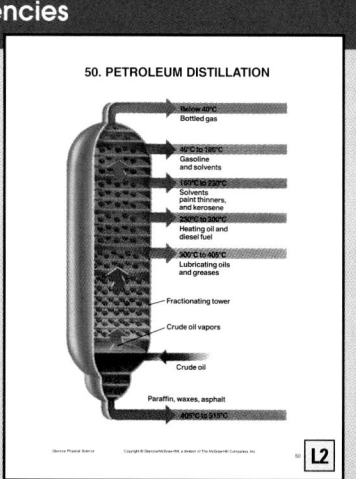

Meeting Different Ability Levels

Study Guide

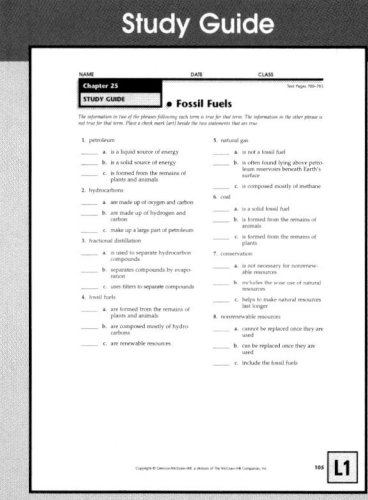

Reinforcement

Enrichment Worksheets

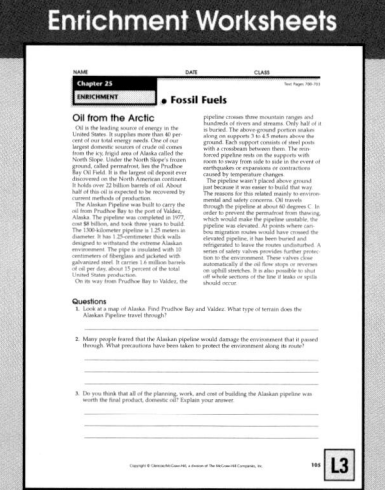

Chapter 25 Energy Sources

Hands-On Activities

Science Integration Activity

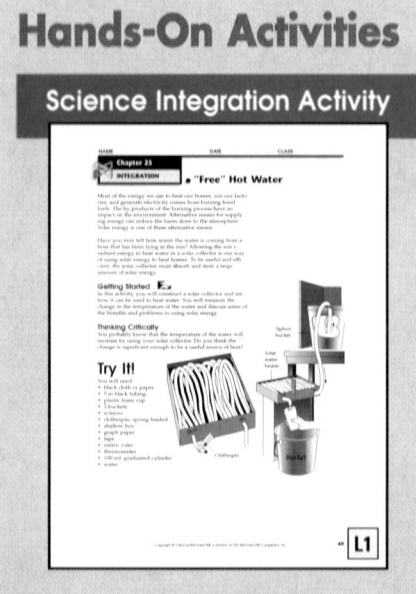

"Free" Hot Water

L1

Lab Manual

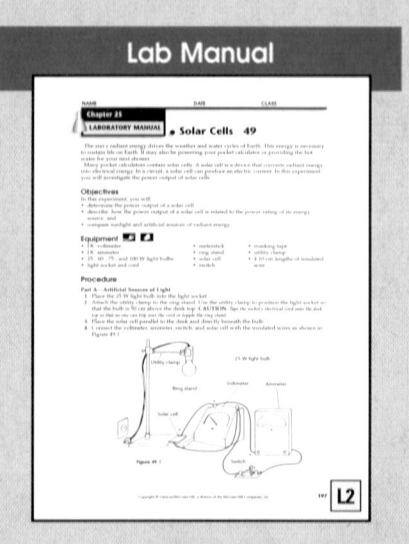

Solar Cells 49

L2

Assessment

Performance Assessment

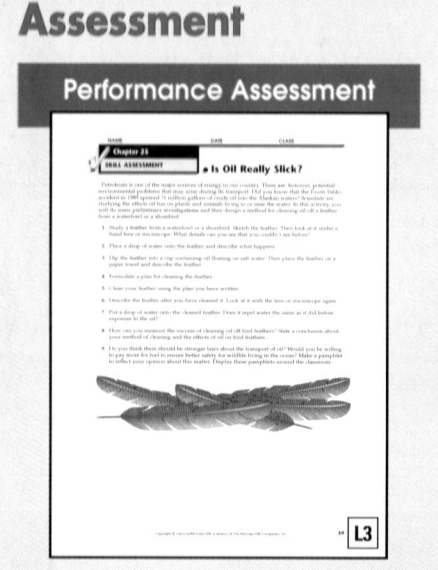

Is Oil Really Slick?

L3

Enrichment and Application

Critical Thinking/Problem Solving

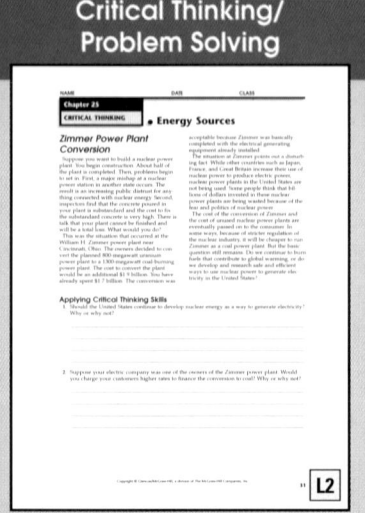

Energy Sources

L2

Cross-Curricular Integration

Energy Sources

L1

Science and Society Integration

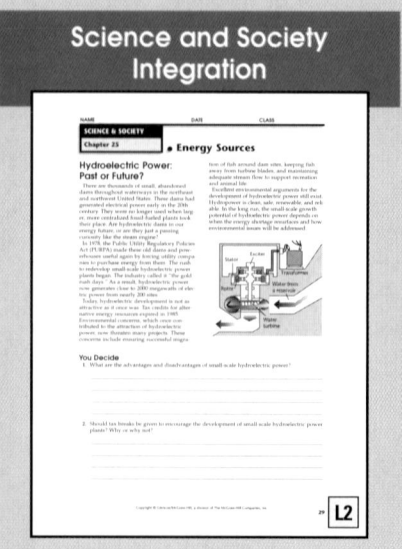

Energy Sources

L2

Multicultural Connections

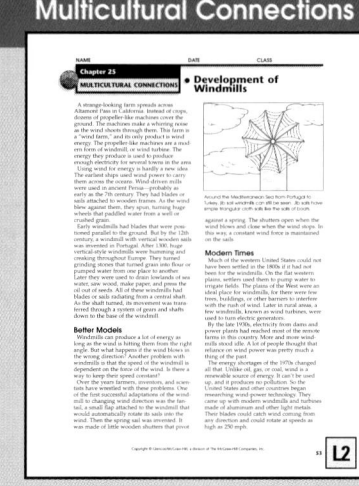

Development of Windmills

L2

Concept Mapping

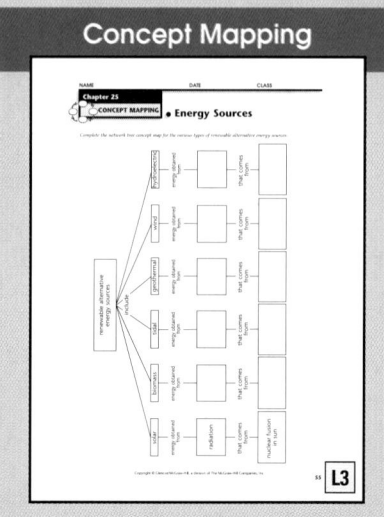

Energy Sources

L3

698D

Energy Sources

CHAPTER OVERVIEW

Section 25-1 This section focuses on methods of using fossil fuels as our most common energy resource. Fossil fuels are discussed as a nonrenewable resource.

Section 25-2 The process of using nuclear reactors to generate electricity is illustrated. Nuclear waste disposal is discussed. The challenges of developing fusion power are also presented.

Section 25-3 Science and Society This section provides a description of nuclear waste disposal and storage options.

Section 25-4 Alternate energy sources are compared. These include biomass, solar energy, hydroelectricity, tidal energy, wind energy, and geothermal energy.

Chapter Vocabulary

petroleum
fractional
 distillation
nonrenewable
 resource
nuclear reactor
nuclear waste
NIMBY

photovoltaic
 cell
hydroelec-
 tricity
tidal energy
geothermal
 energy

Theme Connection

Energy Energy is an obvious theme throughout this chapter. The sections compare the use of fossil fuels, nuclear energy, solar energy, and other alternative energy sources as methods of supplying useful energy to large communities. The environmental effects of generating usable energy should also be discussed.

Previewing the Chapter

698

Learning Styles

Look for the following logo for strategies that emphasize different learning modalities.

Kinesthetic	MiniLAB, pp. 701, 715; People and Science, p. 720
Visual-Spatial	Cultural Diversity, p. 702; Visual Learning, pp. 700, 703, 705, 707, 713, 714
Interpersonal	Reteach, p. 702; Discussion, p. 706; Activity 25-1, p. 709
Intrapersonal	Explore, p. 699
Linguistic	Activity, pp. 706, 707; Activity 25-2, pp. 718-719

Chapter

25

Energy Sources

People need energy daily for almost everything we do. When we need it, where do we get energy? What energy sources and uses do you notice in the photo on the facing page?

In the activity that follows, you will see how straight-line movement of air can be converted into circular motion. Other forces can be used the same way.

Consider some of the alternative ways other forces can be used to generate electricity through moving turbines as you read this chapter.

EXPLORE ACTIVITY

Use the energy from the wind to provide turning power.

1. Cut several different sizes of squares from sheets of paper. Make a cut from each corner of the square toward the center, stopping when you are within 1-2 cm from the center.
2. Fold each point toward the center, running a straight pin through all four tips and the center of the square.
3. Hold the wind catcher by the pointed end of the pin in front of a window fan, and note which size spins fastest.

Observe: In areas of consistent winds, wind power has been the source of energy to generate electricity. In your Science Journal, describe an advantage wind power has over fossil fuels.

Previewing Science Skills

▶ In the Skill Builders, you will compare and contrast, map concepts, and outline.

▶ In the Activities, you will diagram, research, and make observations.

▶ In the MiniLABs, you will measure, observe, and draw conclusions.

699

EXPLORE ACTIVITY

Purpose

LS Intrapersonal Use the Explore Activity to help students think about the nature of our energy resources. Point out that alternatives to fossil fuels are necessary to fulfill the energy demands of society. L1 LEP

Preparation

You may want to prepare one wind catcher before class to use as an example.

Materials

paper samples, scissors, straight pins, window fan

Teaching Strategies

Try the wind catchers with a variable speed fan to determine if one size wind catcher works equally well at slow and fast speeds.

Observe

Wind power does not have the air pollution by-products of fossil fuels. In addition, obtaining fossil fuels often involves environmentally sensitive situations.

✔ Assessment

Oral Have students explain how the direction of force from the wind is turned into a circular motion. Use the Performance Task Assessment List for Making Observations and Inferences in **PASC,** p. 17. P

Assessment Planner

Portfolio
Refer to page 721 for suggested items that students might select for their portfolios.

Performance Assessment
See page 721 for additional Performance Assessment options.
Skill Builders, pp. 703, 708, 717
MiniLABS, pp. 701, 715
Activities 25-1, p. 709; 25-2, pp. 718-719

Content Assessment
Section Wrap-ups, pp. 703, 708, 711, 717
Chapter Review, pp. 721-723
Mini Quizzes, pp. 703, 717

Group Assessment
Opportunities for group assessment occur with Cooperative Learning Strategies and Flex Your Brain Activities.

Prepare

Section Background

More than 50% of the world's oil reserves are located in the Persian Gulf region.

Preplanning

The MiniLAB requires votive candles and aluminum foil.

1 Motivate

Bellringer

Before presenting the lesson, display **Section Focus Transparency 101** on the overhead projector. Assign the accompanying **Focus Activity** worksheet. L1 LEP

Tying to Previous Knowledge

Students have studied hydrocarbons in Chapter 13. Petroleum and natural gas, both fossil fuels, are made of hydrocarbon molecules.

Brainstorming

Have students list the ways in which they have encountered products made from petroleum. They may think of gasoline, petroleum jelly, baby oil, kerosene, or motor oil. Point out that many plastics, cosmetics, and detergents are also derived from oil.

Visual Learning

Figure 25-1 **How might the fossils have ended up under these layers?** *Answers will vary. layers of mud and dirt collected over many years, or a cataclysmic event like a flood or volcanic eruption* LEP

LS

Science Words

petroleum
fractional distillation
nonrenewable resource

Objectives

* Discuss the origin and characteristics of the three main types of fossil fuels.
* Describe the need for and methods of energy conservation.

Petroleum

Do you own a sweater made of synthetic fibers such as polyester or nylon? If so, are you aware that your sweater has something in common with the gasoline used for fuel in cars and buses? Synthetic fibers are made with chemicals that come from crude oil, and gasoline is refined from crude oil.

Crude oil, or **petroleum**, is a liquid source of energy made from the decayed remains of plants and animals. Petroleum, along with natural gas and coal, is called a fossil fuel. When plants and animals died, they were covered with layers of sand, mud, volcanic ash, and other matter that collected on the surface of Earth over many years. Great pressure, heat, and bacterial action acted on these buried organisms, forming fossil fuels.

Petroleum is pumped to the surface from wells drilled deep into the ground, as shown in **Figure 25-1.** The crude oil that comes from these wells contains a variety of chemicals. Because plants and animals contain large amounts of hydrogen and carbon, petroleum is composed mostly of compounds containing these elements, called hydrocarbons.

Figure 25-1

Oil wells are used to find fossil fuels beneath layers of earth. *How might the fossils have ended up under these layers?*

Sorting out Hydrocarbons

These various hydrocarbon compounds must be separated to make use of the chemical energy stored in petroleum. If you have ever seen an oil-refining plant, you may have noticed several tall towers. These towers are called fractionating towers. Fractionating towers use a process called **fractional distillation** to separate hydrocarbon compounds. First, crude oil is pumped into the bottom of the tower and heated. The chemical compounds in the crude oil boil and evaporate according to their individual boiling points. Those materials with the lowest boiling points rise to the top of the column as vapor and are separated and collected. When each vapor cools below its boiling point, it again turns to a liquid. Hydrocarbons with high boiling

700 Chapter 25 Energy Sources

Program Resources

Reproducible Masters
Study Guide, p. 105 L1
Reinforcement, p. 105 L1
Enrichment, p. 105 L3
Activity Worksheets, p. 152 L1
Critical Thinking/Problem Solving, p. 31

Transparencies
Section Focus Transparency 101
Teaching Transparency 50

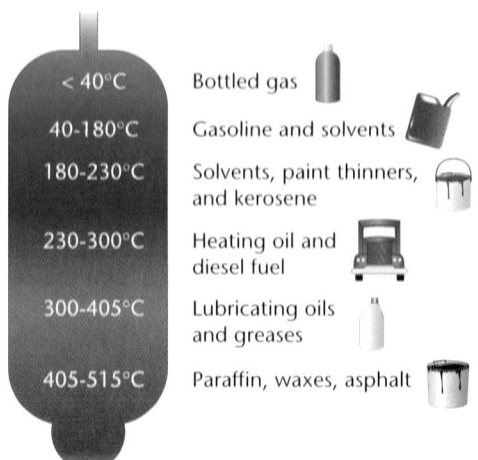

< 40°C	Bottled gas
40-180°C	Gasoline and solvents
180-230°C	Solvents, paint thinners, and kerosene
230-300°C	Heating oil and diesel fuel
300-405°C	Lubricating oils and greases
405-515°C	Paraffin, waxes, asphalt

Figure 25-2

Fractional distillation is used to separate the hydrocarbon components in petroleum.

points may remain as liquids and be drained off through the bottom of the tower. Notice the different products made from the petroleum fractions that are shown in **Figure 25-2.**

Oil must travel from the oil fields to the refineries. Oil can be carried through pipelines, by ship, or in tanker trucks. You use petroleum every day for electricity and transportation, but not all the effects of petroleum usage are helpful to you or the environment. When petroleum is burned in cars and at electric power plants, it gives off smoke as well as carbon monoxide and other chemical compounds. These particles and compounds affect the quality of the air you breathe.

Other Fossil Fuels

Do you cook on a gas stove at home? If so, you burn natural gas to get energy to cook your food. Natural gas, like petroleum, is a fossil fuel. It is often found trapped above liquid petroleum reservoirs below Earth's surface; it is extracted with the petroleum.

Natural gas is composed mostly of methane, CH_4, but it also contains smaller amounts of hydrocarbon gases such as propane, C_3H_8, and butane, C_4H_{10}. When natural gas or petroleum is burned, it combines with oxygen, and heat is given off. During this process, carbon dioxide and water are formed as chemical by-products. This is a combustion reaction. Natural gas is burned to provide energy for cooking, heating, and manufacturing. When natural gas is used properly, it doesn't pollute the environment with the compounds that result from burning petroleum. However, the carbon dioxide that it produces can cause Earth's atmospheric temperature to rise.

MiniLAB

How can an updated design improve efficiency?

Procedure

1. Measure and record the mass of a candle.
2. Measure 50 mL of water into a container. Record the temperature of the water.
3. Use the lighted candle to heat the water 10°C. Put out the candle and measure its mass again.
4. Repeat steps 1 to 3 with an aluminum chimney surrounding the candle to help direct the heat upward.

Analysis

1. Compare the mass change in the two trials. Does a smaller or larger mass change in the candle show greater efficiency?
2. Gas burners are used to heat hot-water tanks. What must be considered in the design of these heaters?

Across the Curriculum

History Investigate the political situation that led to the oil embargo in 1973. Another oil shortage also occurred in the late 1970s. How did these shortages affect consumers and our approach to energy resources?

Science Journal

Temperature's Rising Have students write out an explanation for the following question. **How does carbon dioxide given off in the combustion of fossil fuels cause the temperature of Earth to rise?** *It causes the atmosphere to reflect radiation trying to escape Earth. This locks more heat within our atmosphere.* **L1**

GLENCOE TECHNOLOGY

 Videodisc

STVS: Chemistry
Disc 2, Side 1
Hydrogen Sponge (Ch. 10)

3 Assess

Check for Understanding

Reteach

IS Interpersonal Divide students into teams of three to consider the advantages and disadvantages of using petroleum, natural gas, and coal. Have them combine their lists and report their conclusions. **COOP LEARN**

Extension

For students who have mastered this section, use the **Reinforcement** and **Enrichment** masters.

USING TECHNOLOGY

No More Gas Tanks?

Someday, the familiar gasoline pump may be found only in museums. Instead of gasoline, we may be using natural gas, electricity, or hydrogen for our cars. Earth has a limited supply of petroleum to make gasoline. To make existing petroleum supplies last longer, automobile designers are using different technologies to power automobiles.

Natural gas vehicles (NGVs) are designed to run on methane. They produce less air pollution than cars that burn gasoline. Methane can be obtained from naturally decaying plant material. Electrically powered cars produce less exhaust pollution, but the energy used to generate the electricity may offset any savings in petroleum. Solar power could be used to make electricity. This would improve the conservation and pollution advantages of an electrically powered car. Hydrogen is another possible automobile fuel. However, carrying hydrogen in a car would be dangerous. To solve this problem, metal sponges are being made that absorb hydrogen, releasing it only when needed.

Think Critically: These technologies are not being implemented quickly. How would tripling the current price of gasoline affect the rate of implementation?

Car of the future?

Burning Coal for Energy

Coal is a solid fossil fuel made from the remains of plants. It is sometimes mined near the surface but is more commonly mined about one hundred meters underground. The quality of a coal sample depends on its age and on the type of plant life from which it formed.

When used as a fuel, coal is cleaned to remove impurities that would be released into the air when the coal is burned. Sulfur is removed to prevent the formation of compounds that cause acid rain. Harmful nitrogen oxides form when coal is burned at high temperatures. Therefore, it is desirable to find ways to burn coal at lower temperatures.

Because coal is mined, digging coal mines can disturb the environment. Today, laws require that the environment be returned to its original state when a coal mine is closed.

Fuel Conservation

Think of how many times you ride in a car, turn on a light, or use an electric appliance. Each time you do one of these things, you probably use some type of fossil fuel as a source of energy. Fossil fuel reserves are decreasing as our population and industrial demands are increasing. At our current rate of consumption, the United States may be out of oil in less than one century. Coal is more plentiful, but like petroleum, it is a nonrenewable resource. All fossil fuels are **nonrenewable resources**—they cannot be replaced after they are used.

We must conserve the nonrenewable energy resources we do have. Conservation can be as simple as turning off a light when you leave a room, avoiding excessive speeds when driving a car, or riding a bike instead of driving a car at all. Using energy-efficient appliances can reduce energy use in your home. Trees around your home can keep it cooler in summer and reduce the use of air conditioning. Replacing old windows with new, energy-efficient windows, as shown in **Figure 25-3**, means buildings will use less energy for heating and cooling. We need to develop some alternative energy sources to help us conserve now and to prepare for the time when fossil fuels may not be readily available.

Figure 25-3

Does your school have energy-efficient windows? How could replacing windows save energy?

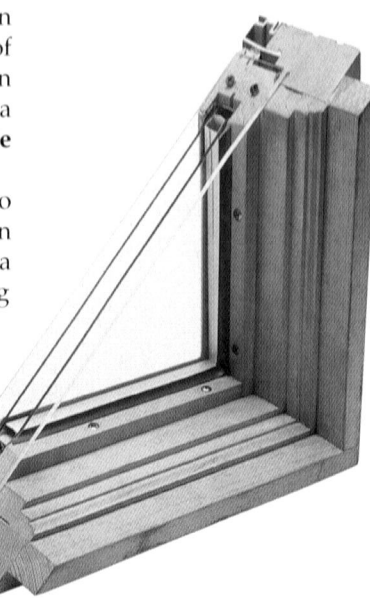

Section Wrap-up

Review

1. Describe the three main types of fossil fuels.

2. Which type of fossil fuel is the most abundant in the United States? What are the advantages and disadvantages of using this fuel?

3. **Think Critically:** As the plants and animals living on Earth today die, they will decay and be buried to form the fossils of tomorrow. If the formation of fossil fuels is a continuous cycle, why are they considered to be *nonrenewable* resources?

Skill Builder
Comparing and Contrasting
Compare and contrast the different fossil fuels. Include the advantages and disadvantages of using each as a source of energy. If you need help, refer to Comparing and Contrasting in the **Skill Handbook**.

Science Journal

In your Science Journal, make a list of areas in your school where existing energy could be conserved.

Skill Builder
Students should compare the fossil fuels based on the points discussed in this section. **P**

✓ Assessment

Performance Have students organize the advantages and disadvantages into a table. Use the Performance Task Assessment List for Data Table in **PASC**, p. 37. **P**

Visual Learning

Figure 25-3 How could replacing windows save energy? *You use less energy for heating and cooling because of less leakage.* **LEP** **IS**

4 Close

•MINI•QUIZ•

Use the Mini Quiz to check students' recall of chapter content.

1. **Petroleum, natural gas, and coal are all _____ .** *fossil fuels*
2. **What is fractional distillation?** *a process used to separate hydrocarbons into usable forms*
3. **All fossil fuels are _____ resources.** *nonrenewable*

Section Wrap-up

Review

1. Petroleum is a liquid composed mostly of hydrocarbons. Natural gas is composed mostly of methane. Coal is a solid fossil fuel.
2. Coal is the most abundant fossil fuel, and it gives off large amounts of energy. Burning impure coal pollutes the air, mining deforms the environment, and it is a nonrenewable resource.
3. **Think Critically** We use up fossil fuels faster than they are replaced.

Science Journal

Encourage students to support their opinions. They may suggest newer insulated windows, insulation around doorways, etc.

25•2 Nuclear Energy

Prepare

Section Background

• Compared to fossil fuels, nuclear energy has only recently been developed as an energy resource. The atom was first split in 1939.

• Countries differ widely in their opinion on the use of nuclear energy. France is aiming to eventually produce up to 90 percent of its electricity from nuclear energy, while Sweden and some other nations have decided to decrease or omit reliance on nuclear energy.

Preplanning

A number of items are needed for Activity 25-1 on page 709. Begin assembling these materials.

1 Motivate

Bellringer

Before presenting the lesson, display **Section Focus Transparency 102** on the overhead projector. Assign the accompanying **Focus Activity** worksheet. [L1] **LEP**

Tying to Previous Knowledge

In the last chapter students learned that nuclear fission and fusion release huge amounts of energy. In this section, they will study how this energy can be used to generate electricity.

Use **Teaching Transparency 49** as you teach this lesson.

Science Words

nuclear reactor
nuclear waste

Objectives

• Outline the operation of a nuclear reactor.
• Describe the problems and methods associated with nuclear waste disposal.
• Discuss nuclear fusion as a possible energy source.

Nuclear Reactors

In the last chapter, you learned that nuclear fission chain reactions give off a great deal of energy. A **nuclear reactor** uses the energy from a controlled nuclear fission chain reaction to generate electricity. Nuclear reactors can vary in design. Most fission reactors have several parts in common, including fuel, control rods, and cooling systems. The actual fission of the radioactive fuel occurs in a relatively small part of the reactor—the core.

In the reactor core, where the uranium oxide fuel is found, neutrons are being absorbed by U-235. When this happens, U-235 nuclei undergo fission and release more neutrons. These neutrons can strike other U-235 nuclei, causing them to also release two or three neutrons as they undergo fission. This process continues quickly, in the frame of milliseconds, releasing tremendous amounts of energy from the tiny fuel pellets shown in **Figure 25-5.** What keeps the process from continuing out of control, with each millisecond producing more energy than the one before?

To control the reaction, rods containing boron or cadmium are used to absorb some of the neutrons. Moving these control rods deeper into the reactor allows them to capture more neutrons and slow down the chain reaction. Eventually, only one neutron per fission is able to react with a U-235 atom to produce another fission, and energy is released at a constant rate. **Figure 25-4** explains how nuclear fission is used to produce electricity.

Figure 25-4

Nuclear reactors actually use water to turn turbines and generate electricity. The nuclear reaction simply gives the water the energy to move the turbines.

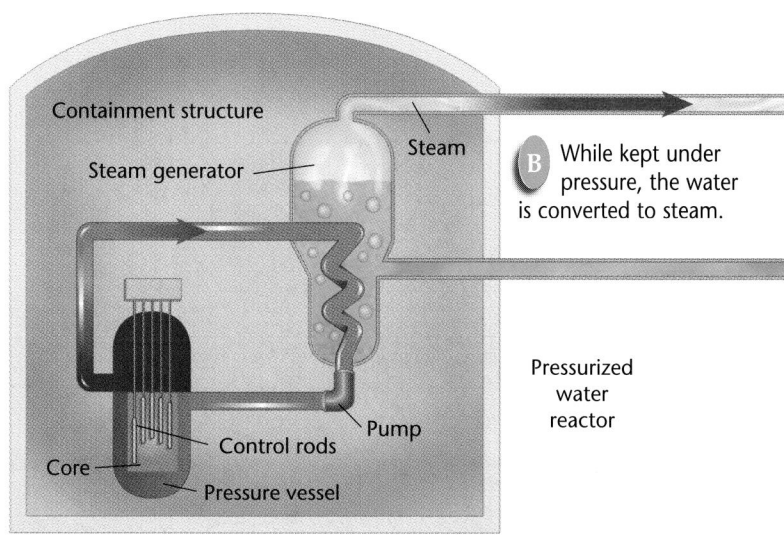

Containment structure
Steam generator
Steam
B While kept under pressure, the water is converted to steam.
Pressurized water reactor
Pump
Control rods
Core
Pressure vessel

A The core of the reactor uses energy from fission to heat contained water. This water is not released into the main water supply.

704

Program Resources

Reproducible Masters
Study Guide, p. 106 [L1]
Reinforcement, p. 106 [L1]
Enrichment, p. 106 [L3]
Activity Worksheets, pp. 148-149 [L1]

Transparencies
Section Focus Transparency 102
Teaching Transparency 49

Nuclear Generation of Electricity

Nuclear fission reactors currently supply some of our nation's electricity. One advantage of using nuclear energy is that it's less harmful to the environment than the use of fossil fuels. The fission process produces no air pollution, whereas the burning of coal and petroleum creates nearly 20 000 metric tons of pollutants each day. Nuclear fission doesn't produce carbon dioxide that escapes into the atmosphere and contributes to the problem of global warming. On the other hand, the mining of uranium and extraction of U-235 does cause environmental damage.

Using nuclear fission to generate electricity also has other disadvantages. The water that circulates around the core of the reactor must cool before it goes back into streams and rivers. If it is released into those waterways while still warm, the excess heat could harm fish and other animals and plants in the water. **Figure 25-6** on page 706 shows the large, hollow towers used to cool the warm water. The most serious risk of nuclear fission is the escape of harmful radiation from the power plants. Nuclear reactors have elaborate systems of safeguards, strict safety precautions, and highly trained workers who can prevent most accidents.

However, after the chain reaction has occurred, the fuel rods contain fission products that are highly radioactive. These products must be contained in heavily shielded surroundings while they decay so that no radiation escapes

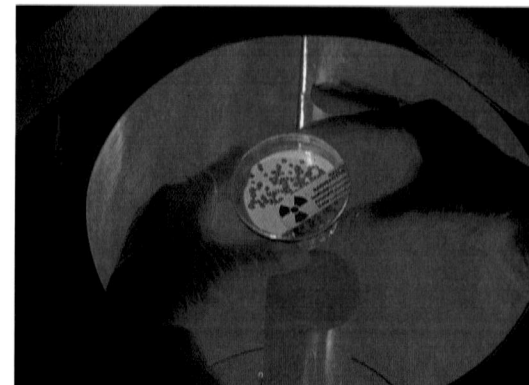

Figure 25-5
Nuclear fuel pellets are assembled into fuel rods. The rods may be up to about 4 m in length.

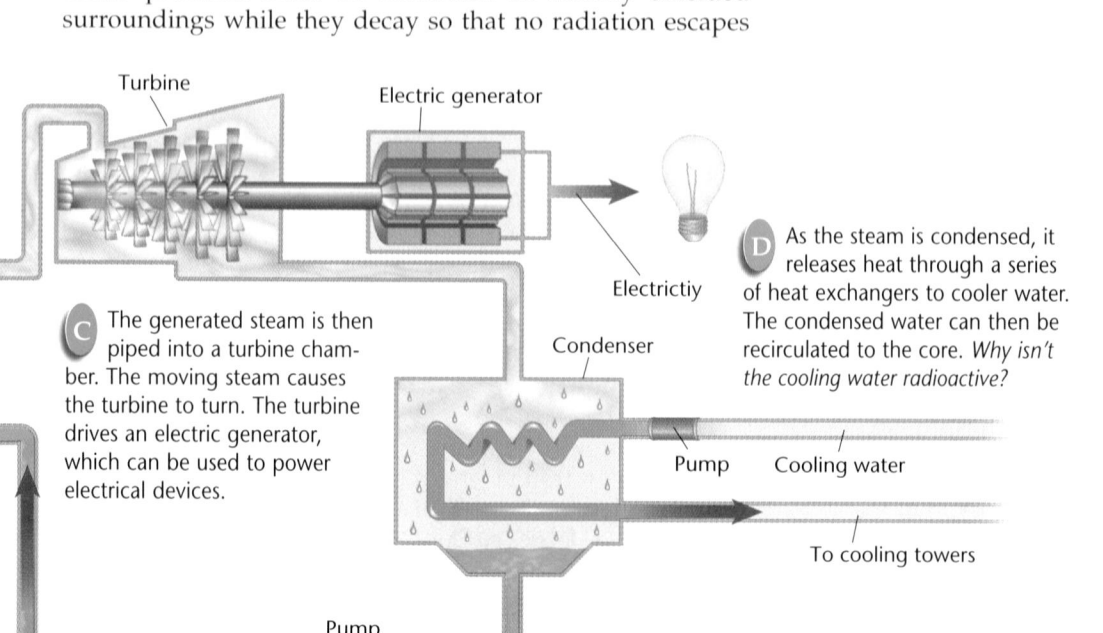

Turbine

Electric generator

Electrictiy

C The generated steam is then piped into a turbine chamber. The moving steam causes the turbine to turn. The turbine drives an electric generator, which can be used to power electrical devices.

Condenser

D As the steam is condensed, it releases heat through a series of heat exchangers to cooler water. The condensed water can then be recirculated to the core. *Why isn't the cooling water radioactive?*

Pump   Cooling water

To cooling towers

Pump

25-2 Nuclear Energy   **705**

Figure 25-6
The cooling towers, often the most visible part of a nuclear reactor, help to get rid of extra heat from water that has been heated in the power plant.

USING MATH

In 1945, the sources of energy in the United States were: hydroelectric and other —5%, natural gas—12%, petroleum—32%, and coal—51%. In 1995, the sources were: hydroelectric and other—4% nuclear—7%, coal—23%, natural gas—24%, and petroleum—42%. Use graphs to show how these energy sources have changed.

into the environment. Because these fission products have long half-lives, they must be stored in containers that will last the duration of the period of radioactive decay, which can be as long as tens of thousands of years.

Another problem is the disposal of the reactor itself when it no longer runs as it should. Every nuclear reactor has a limited useful life span of a few decades. After that time, the reactor's efficiency is severely reduced by the presence of large amounts of fission products that can't be cleaned up. The reactor is then shut down or decommissioned.

Nuclear Waste Disposal

Has your watch ever run down because the battery was too old? The battery wasn't able to produce enough energy to allow your watch to keep correct time. Once you replaced the battery, your watch worked. A similar thing happens to the fuel source in a nuclear reactor. After about three years, there is not enough fissionable U-235 left in the fuel pellets to sustain the chain reaction. These used fuel pellets are called *spent fuel.* The spent fuel contains the radioactive fission products in addition to the remaining uranium. This is an example of nuclear waste. **Nuclear wastes** are radioactive by-products that result when radioactive materials are used. They are usually classified as high-level or low-level wastes for disposal.

Radiation is around you all the time. Low-level nuclear waste usually contains a small amount of radioactive material diluted by a large amount of nonradioactive material. Products of some medical and industrial processes are low-level wastes. They may include items of clothing that are used in handling radioactive materials, as shown in **Figure 25-7.**

Use these program resources as you teach this lesson.
Critical Thinking/Problem Solving, p. 31
Study Guide, p. 106

Low-level wastes are usually buried in sealed containers in locations licensed by the federal government. When dilute enough, they are sometimes released into the air or water.

High-level nuclear waste is generated in nuclear power plants and by defense research. After spent fuel is removed from a reactor, it is stored in a deep, heavily insulated pool of water. Many of the radioactive materials in high-level nuclear waste have short half-lives. But the spent fuel also contains materials that will continue to decay for tens of thousands of years. For this reason, the waste must be disposed of in extremely durable and stable containers.

Figure 25-7

Which type of radiation is most easily shielded against by the clothing and gloves of these workers?

Fusion Power

Imagine the amount of energy the sun must give off to heat Earth from 93 million miles away. In Chapter 24, thermonuclear fusion was explained as the process that releases this energy. Recall that thermonuclear fusion is the joining together of small nuclei at high temperatures. Fusing the nuclei in 1 g of heavy hydrogen gives off about the same amount of energy as burning more than 8 million grams of coal. If we could control thermonuclear fusion in a laboratory, we would likely have the answer to Earth's energy problems.

Let's examine the benefits of nuclear fusion as a source of energy. Hydrogen nuclei are the source of energy for nuclear fusion. Hydrogen is the most abundant element in the universe. Unlike those in nuclear fission, the products of nuclear fusion are not radioactive. Helium is the main product of hydrogen fusion. Someday, nuclear fusion may provide a permanent and economical way to generate electricity.

25-2 Nuclear Energy   **707**

707

Answer The location has a stable rock structure that would withstand the forces of erosion and earth movement.

4 Close

Show the students the calculations of how much radioactive material would remain of a 200-g sample with a half-life of 6 years, after 24 years. Use the blackboard or an overhead projector.

Section Wrap-up

Review

1. Control rods and circulating water absorb excess neutrons.

2. It is difficult to maintain the high temperatures required and contain the fusion.

3. **Think Critically** It is low-level waste and should be sealed in an insulated container and buried in a repository.

Using Computers

Advantages include less pollution and greater efficiency. Disadvantages include waste disposal and the risk of thermal pollution.

In 1987, Congress voted to build the first national permanent nuclear repository site at Yucca Mountain, Nevada. Burial of waste is scheduled to start in 2003, although the plan has changed several times. *Classify* the geological features this location has that make it suitable for a nuclear repository.

The challenge lies in creating and containing nuclear fusion. The temperature needed to carry out a nuclear fusion reaction is more than 1 million degrees Celsius. The plasma containing the hydrogen nuclei can't be contained by any material at this temperature. However, it can be contained for a short time in a magnetic bottle, which uses a magnetic field to keep the particles in a small volume. An experimental fusion reactor is shown in **Figure 25-8**. At the present time, the energy required to maintain the high temperatures needed for the fusion reaction is greater than the energy output from the fusion of the nuclei.

Figure 25-8

This experimental fusion reactor is used to house a nuclear fusion reaction.

Section Wrap-up

Review

1. How is the rate of fission in a nuclear reactor controlled?

2. Explain the major obstacles in controlling nuclear fusion.

3. **Think Critically:** Suppose that in a research project, you have generated a 10-g sample of nuclear waste. Some of the materials have a short half-life and some will decay for thousands of years. How would you classify this waste, and how will it likely be disposed of?

Skill Builder
Concept Mapping

Design an events chain concept map for the generation of electricity in a nuclear fission reactor. Begin with the bombarding neutron and end with electricity in overhead lines. If you need help, refer to Concept Mapping in the **Skill Handbook.**

Using Computers

Word Processing
Create a table with two divisions for the advantages and disadvantages of nuclear power. Type in as many entries under each as you can list.

Skill Builder

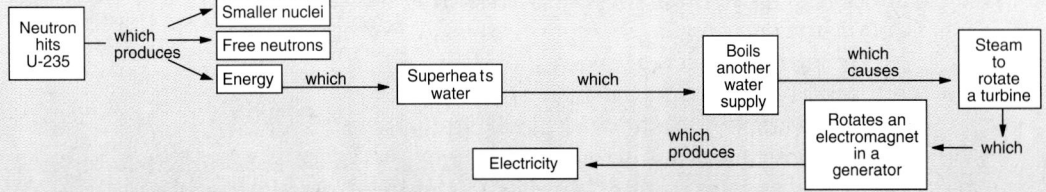

✓ Assessment

Performance Have students complete the concept map beyond where smaller nuclei and free neutrons are shown. Use the Performance Task Assessment List for Concept Map in **PASC,** p. 89.

P

Activity 25-1

Nuclear Waste Disposal

Because many radioactive wastes will continue to decay for hundreds of years, it is important to test containment procedures that will prevent leakage. What kinds of materials do you think could effectively hold radioactive wastes while they decay?

Problem

How can you build a model to represent the concerns about containment of radioactive waste?

Materials

- sodium hydroxide pellets (4)
- 100-mL beakers (4)
- phenolphthalein solution
- modeling clay
- plastic food wrap
- aluminum foil
- rubber bands (3)
- twist ties (2)
- forceps
- goggles

Procedure

1. Let the sodium hydroxide pellets represent pellets of nuclear waste. Your job is to test materials that may be able to keep unwanted chemicals from leaking into the environment. **CAUTION:** *Do not touch the sodium hydroxide pellets or solution with your hands. Always wear goggles when handling sodium hydroxide. Wipe up all spills.*
2. Fill all four beakers with water and add 3 or 4 drops of phenolphthalein to each.
3. Using the forceps, place one pellet in one of the beakers. Record your observations.
4. Wrap each of the remaining pellets: one in aluminum foil, one in plastic wrap, and one in clay. Secure the plastic with twist ties.

5. Drop the wrapped pellets into the remaining three beakers, and cover them with food wrap.
6. Make observations each day for three days to determine whether there is leakage from the containers. Record your observations in the data table.
7. After the last observations, pour the solutions into a container provided by your teacher.

Data and Observations   Sample Data

Beaker Containing Pellets	Observations		
	Day 1	Day 2	Day 3
Unwrapped	red color	indicates	leakage
Wrapped in aluminum foil	leakage	leakage	leakage
Wrapped in plastic wrap	no leakage	leakage	leakage
Embedded in clay	no leakage	no leakage	no leakage

Analyze

1. Was red color seen in any of the beakers after the first day?
2. After three days, which, if any, of the beakers show no red color?

Conclude and Apply

3. **Infer** which kind of wrapping is the most leakproof.
4. **Hypothesize** why the storage and disposal of nuclear waste is a major concern of society.

25-2 Nuclear Energy   **709**

plastic containers may show leakage and turn red.

2. Clay, if properly applied, will continue to contain the chemical and the solution will remain colorless.
3. the clay
4. Nuclear generation of electricity is a reliable energy source if plant safety can be maintained and a secure method of disposal can be found. Some nuclear waste will be radioactive for thousands of years.

✓ Assessment

Oral Have students discuss how this model fails to reflect some of the problems associated with nuclear waste disposal. Use the Performance Task Assessment List for Analyzing the Data in **PASC**, p. 27. **P**

Activity 25-1

Purpose

IS Interpersonal Students will simulate the problems of nuclear waste disposal.

Process Skills

observing, interpreting, recording data, communicating, recognizing cause and effect, making models, predicting

Time

3 days, 20 minutes/day

Safety Precautions

- This activity would work well as a demonstration. Sodium hydroxide is caustic. It must not come in contact with skin or other tissues. Eye safety is essential and goggles must be worn. Make eyewash equipment available. Keep track of every pellet. Wipe up all spills. Have students wash their hands if they come in contact with the pellets and before they leave the lab.
- Before disposing of solutions, neutralize them by adding just enough dilute hydrochloric acid to make the red color disappear.

📁 **Activity Worksheets,** pp. 5, 148, 149

Teaching Strategies

Troubleshooting When sealing the pellet in clay, start with a patty of clay, pull the clay up around the pellet, and roll the material gently into a smooth ball.

- The need for following strict handling precautions helps make the point of the activity.
- Provide small squares (5 cm × 5 cm) of aluminum foil and plastic. Provide enough modeling clay to enclose a pellet.

Answers to Questions

1. All of the beakers containing aluminum samples will turn red. Some of the

Prepare

Section Background

Radioactive waste storage is a problem. Two points of view about how to solve this problem are presented.

1 Motivate

Bellringer

Before presenting the lesson, display **Section Focus Transparency 103** on the overhead projector. Assign the accompanying **Focus Activity** worksheet. [L1] [LEP]

Discussion

Lead students in a risk-benefit discussion about a current choice in their lives; i.e., riding in a car, food choices, playing sports. Are the benefits of nuclear power worth the risks?

2 Teach

Inquiry Question

What problems must be addressed with a national storage site? *transportation safety from other sites, possible terrorist plans for theft, political changes on policies and funding*

Activity

Have students contact a nuclear medicine facility to find out what is done with their nuclear waste. [L1]

Science Words
NIMBY

Objectives

- Discuss problems associated with storing nuclear waste.
- Compare long-term nuclear waste-storage options.

Figure 25-9

This map shows where the different kinds of waste are now being stored.

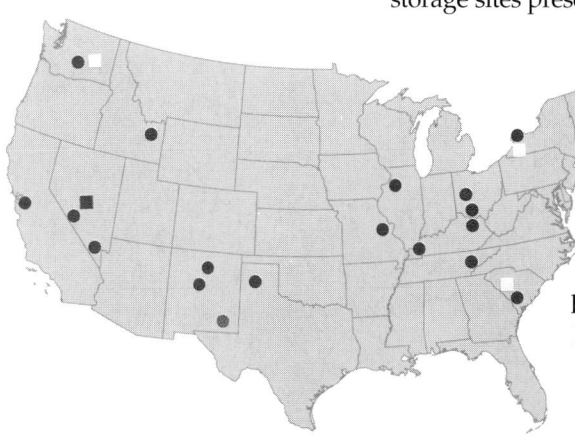

- Low-level military and commercial waste site
- Temporary high-level military waste site
- Permanent high-level military waste site
- ■ Proposed high-level commercial waste site

710   Chapter 25 Energy Sources

NIMBY

Using nuclear fission to generate electrical power offers several advantages. In the United States, a portion of our electrical output can be attributed to this technology. However, you have also read about the problem of nuclear waste disposal. When it comes to trying to find a suitable location for storage of waste or transport of nuclear products, many people support the **NIMBY,** Not in My Backyard, point of view. In other words, nuclear power may be agreeable as long as it is not too close to home.

Classifying Nuclear Waste

In the previous section, you learned that nuclear waste is radioactive and is classified as low level or high level. As you read, low-level waste is disposed of in special landfills licensed by the federal government. However, high-level wastes are not disposed of nearly as easily. **Figure 25-9** shows where nuclear storage sites presently are located in the United States.

The radioactive products of uranium fission pose major problems with storage. What should be done with the radioactive fuel rods? The rods contain high-level waste. High-level waste is made up of such fission products as Sr-90 and Cs-137, with half-lives in the 30-year range. Some I-129 and Tc-99, with half-lives on the order of hundreds of thousands of years, are also in high-level waste. As you have read, a national storage site for this type of waste has been proposed near Yucca Mountain, Nevada. Sealing the waste in ceramic glass, placing the glass globules in protective containers, then storing the containers hundreds of meters below ground in stable rock or salt deposits may keep the material safely away from human activity for hundreds of years.

Exposure to excessive nuclear radiation can cause extreme health problems, and even death. Nuclear waste has already been generated. With approximately

---

Program Resources

📁 **Reproducible Masters**
Study Guide, p. 107 [L1]
Reinforcement, p. 107 [L1]
Enrichment, p. 107 [L3]
Cross-Curricular Integration, p. 31

🔦 **Transparencies**
Section Focus Transparency 103

100 active nuclear power plants and many medical applications, we will continue to generate radioactive waste. How can we minimize our exposure to radiation from nuclear waste?

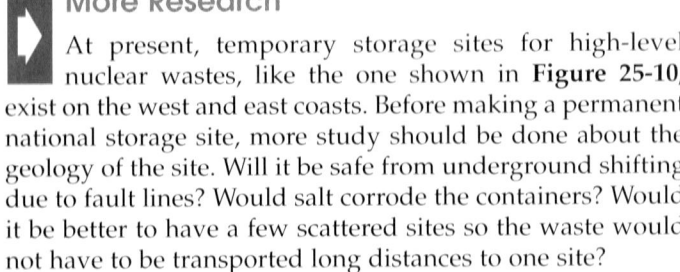

2 Points of View

More Research

At present, temporary storage sites for high-level nuclear wastes, like the one shown in **Figure 25-10,** exist on the west and east coasts. Before making a permanent national storage site, more study should be done about the geology of the site. Will it be safe from underground shifting due to fault lines? Would salt corrode the containers? Would it be better to have a few scattered sites so the waste would not have to be transported long distances to one site?

Safely Below Ground

Research has already shown that deep, stable rock deposits offer the best hope for safely containing radioactive waste. By having one national storage site, federal monitoring of strict design rules and safety standards could be maintained. This would be easier than monitoring several sites. Some of the temporary sites have already shown some leaking. The Nuclear Waste Policy Act of 1982 charges the federal government with establishing a national permanent site by 2010.

Figure 25-10
These temporary sites will be empty when a national site is chosen.

Section Wrap-up

Review

1. What problems are associated with storing nuclear waste?

2. Cite one advantage and one disadvantage to the proposed Yucca Mountain site for permanent nuclear storage.

 Visit the Chapter 25 Internet Connection at Glencoe Online Science, **www.glencoe.com/sec/science/physical,** for a link to more information about storing nuclear wastes.

SCIENCE & SOCIETY

3 Assess

Check for Understanding

? FLEX Your Brain

Use the Flex Your Brain activity to have students explore NUCLEAR WASTE.

📁 **Activity Worksheets,** p. 5

Reteach

Ask students to consider what happens to their trash from school. Explain why nuclear "trash" cannot be taken care of in the same way.

Extension

📁 For students who have mastered this section, use the **Reinforcement** and **Enrichment** masters.

4 Close

Remind students that each form of power generation has unique problems. We must consider other options. This will help bridge to the next section where other alternatives are discussed.

Section Wrap-up

Review

1. Transportation to site, leakage, and future safety of area.

2. Advantage—deep stable rock deposits Disadvantage—the high levels of salt

Prepare

Section Background

- During the Arab Oil Embargo of 1973, President Nixon declared that the U.S. would meet its own energy needs by the end of the decade. Years later, we still rely heavily on imported energy.
- In Hawaii, the sugar industry produces bagasse, dry fiber similar to wood. More than one-third of the energy on two Hawaiian islands is supplied by this biomass.

Preplanning

- Gather materials needed for Activity 25-2. The earlier students begin to prepare for this activity, the better.
- The MiniLAB requires plastic gallon jugs and modeling clay.

1 Motivate

Bellringer

Before presenting the lesson, display **Section Focus Transparency 104** on the overhead projector. Assign the accompanying **Focus Activity** worksheet. L1 LEP

Tying to Previous Knowledge

Mention that students have learned about fossil fuels and nuclear fission, which supply the majority of our energy requirements. But those are nonrenewable resources and they will run out.

Use these program resources as you teach this lesson.

Lab Manual 49, 50

Science Integration Activity, pp. 49, 50

25•4 Alternative Energy Sources

Science Words

photovoltaic cell
hydroelectricity
tidal energy
geothermal energy

Objectives

- Analyze the need for alternative energy sources.
- Discuss the methods of generating electricity with several energy sources.
- Describe the advantages and disadvantages of several alternative energy sources.

The Need for Alternatives

Can you name any sources of energy other than fossil fuels and nuclear energy? Although we have enough of these energy sources to fill our energy demands today, there is a great need to develop alternative sources of energy for the future. As you have already discovered, depending on fossil fuels and nuclear fission for our energy needs has many disadvantages.

Nuclear fission and the burning of fossil fuels are both processes used to boil water to produce steam. However, other materials can be burned to give off energy, as well. Biomass is renewable organic matter, such as wood, sugarcane fibers, rice hulls, and animal manure. It can be burned in the presence of oxygen to convert the stored chemical energy to thermal energy. Burning biomass is probably the oldest use of natural resources for human needs.

You may have seen gasohol advertised at a gas station. How does this fuel differ from normal gasoline? Corn and other plant fibers can be fermented to convert the sugar and starch in the grain to ethanol. The ethanol is combined with gasoline to produce gasohol for use in your car engine.

Biomass and gasohol are just two examples of many energy alternatives that reduce our consumption of nonrenewable fuels. The energy alternatives discussed in this section make use of processes that occur naturally on Earth. What natural processes do you see around you that could be used to generate electricity or provide heat?

Solar Energy

The sun is Earth's only source of new energy. There are now automobiles powered by sunlight. Solar panels on the car collect and use solar energy for power. Methods of collecting and using solar energy are categorized as passive or active.

Passive solar heating is the direct use of the sun's energy in maintaining comfortable indoor temperatures. Passive solar heating was used centuries ago by the Romans to heat their bathhouses. Efforts in energy conservation have renewed interest in this method of heating. Buildings constructed with strategically placed windows can be heated by the sun. On warm days, these windows can be covered with blinds to prevent excessive heating.

Program Resources

 Reproducible Masters
Study Guide, p. 108 L1
Reinforcement, p. 108 L1
Enrichment, p. 108 L3
Concept Mapping, p. 55
Lab Manual 49, 50
Multicultural Connections, pp. 53, 54
Activity Worksheets, pp. 150, 151, 153

Science Integration Activity, pp. 49, 50
Science and Society Integration, p. 29

Transparencies
Section Focus Transparency 104
Science Integration Transparency 25

In active solar heating, solar panels collect and store solar energy. Solar panels are made of large, dark-colored trays covered with transparent glass or plastic. Large mirrors are sometimes used to focus the sun's radiation into these solar collectors. The panels absorb the sun's energy and use it to heat water. The heated water can be used directly, or it can be stored to give off thermal energy. Solar energy can even be used to drive electric power generators in solar thermal power plants.

Electricity Directly from Light

A device used to convert solar energy into electricity is the **photovoltaic cell,** also called the solar cell. Do you own a solar-powered calculator like the one in **Figure 25-11?** It contains a solar cell. Photovoltaic cells are made of a semiconductor lined on both surfaces with a conducting metal. As light strikes the surface of the cell, electrons flow between the two metal layers, creating a current. Many cells connected in a circuit can provide significant amounts of electricity. In order to be a useful technology, photovoltaic cells must be capable of producing in their lifetime more energy than is used in producing the cells. The cells must also be economical and practical to manufacture.

This method of producing electricity is more expensive on a large scale than the use of non-renewable fuels, but it can be less expensive in isolated areas when the cost of building transmission lines to those areas is considered. Solar energy is a pollution-free resource that is becoming more economical as our solar technology develops.

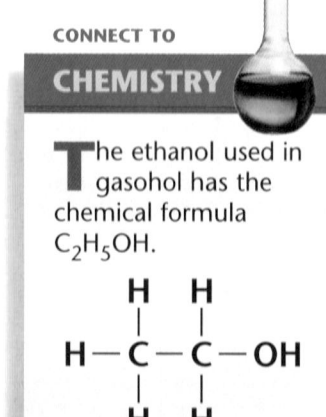

Figure 25-11

In what way does this solar-powered calculator allow our existing energy resources to last longer?

2 Teach

Visual Learning

Figure 25-11   In what way does this solar-powered calculator allow our existing energy resources to last longer? *It does not use electricity generated by fossil fuels.*
LEP LS

GLENCOE TECHNOLOGY

 Videodisc
STVS: Chemistry
Disc 2, Side 2
Solar House (Ch. 10)

Spiral Solar Concentrator (Ch. 12)

Solar Tower (Ch. 13)

 **CD-ROM**

Physical Science CD-ROM
Have students perform the interactive exploration for Chapter 25 to reinforce important chapter concepts and thinking processes.

Across the Curriculum

Engineering   Have students build small solar water heaters. Students should analyze and compare various designs and build several types of models to test them for effectiveness. Have them time how long it takes each model to heat a constant amount of water on the same day, so weather conditions are equal. A great many books on homemade solar heaters were published in the 1970s and 1980s.

Integrating the Sciences

Life Science Have the students research the possible ecological hazards of building tidal power facilities. Some shallow-water organisms depend on unrestricted cycles of tides to bring food and remove waste. L2

Hydroelectricity

One way to produce electricity is with water. Water flowing in rivers carries tremendous amounts of kinetic and potential energy. Dams are built to store vast amounts of water. Behind a dam, the water is deep. Near the base of a hydroelectric dam, water is allowed to rush out through tunnels. The rushing water spins a turbine, rotating the shaft of an electric generator to produce electricity.

Hydroelectricity is electricity produced by the energy of moving water. Hydroelectric power plants are an efficient way to produce electricity. They produce almost no pollution. The bodies of water held back by dams can provide lakes for recreational uses and irrigation. The ongoing natural water cycle makes hydroelectric power a permanent resource. After the initial cost of building a dam and power plant, the electricity is relatively cheap. However, artificial dams can disturb the balance of natural ecosystems.

Tidal Energy

The gravitational forces of the moon and the sun cause bulges in Earth's oceans. As Earth rotates, the two bulges of ocean water move westward. Each day, the level of the ocean on a coast rises and falls continually. A kind of hydroelectric power can be generated by these ocean tides. The moving water can be trapped at high tide by building a dam at the opening of a river or bay. **Figure 25-12** shows the forces that can be harnessed in tides and the water level changes associated with them. The flowing water at low tide spins a turbine, which operates an electric generator. Energy generated by tidal motion is called **tidal energy.**

Figure 25-12

Waves caused by the gravitational pull of the sun and moon cause these churning tides in the photograph. Spring tides occur when the sun, moon, and Earth are aligned (at new and full moon). Neap tides occur when the sun, Earth, and moon form a right angle (at the first and third quarters of the moon). *Do you think tidal energy would be a reliable form of energy? Why or why not?*

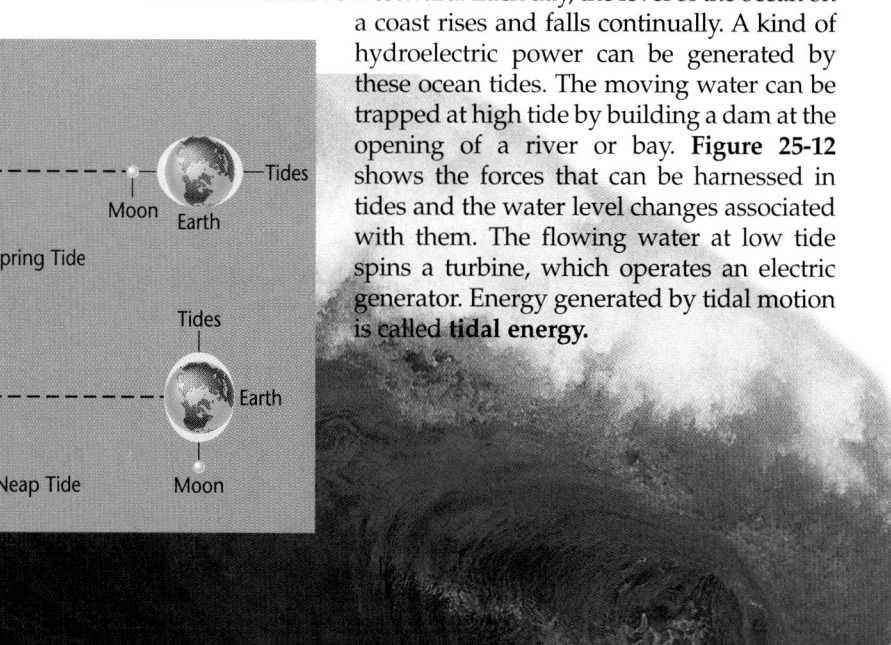

Figure 25-13

This is not a propeller, but a windmill. You don't see many windmills, but in the right place they are powerful producers of energy.

Like producing hydroelectric power, generating electricity from tidal motion is nearly pollution-free. But, there are only a few places on Earth where the difference between high and low tide is large enough to be an efficient source of energy. In the United States, the use of tidal energy is being explored in the Cook Inlet in Alaska and Passamaquoddy Bay in Maine. The ocean environment will possibly make construction and maintenance of these plants difficult. The only functioning tidal power station in North America is at Annapolis Royal, Nova Scotia. Tidal energy probably will be a limited, but useful, source of energy in the future.

Wind Energy

You may have seen a windmill on a farm. Or, in a windy region, you may have seen several hundred windmills like the ones in **Figure 25-13.** A windmill is a turbine that is turned by the wind instead of steam or water. The windmill spins and rotates an electric generator to produce electricity.

Only a few places on Earth consistently have enough wind to rely on wind power to meet energy needs. Improved design of wind generators can increase their efficiency, so new methods of using the wind's energy are being researched. Wind generators do not actually use up any resources. They do not pollute the atmosphere or water. However, they do change the appearance of a landscape.

25-4 Alternative Energy Sources   715

Mini LAB

How do hydroelectric power plants use water to drive turbines?

Procedure
1. Mold a clay collar around the middle of a pencil.
2. Insert flat wooden sticks into the clay to resemble the spokes of a wheel.
3. Punch a small hole near the bottom of a large plastic container.
4. Fill the container with water, holding your finger over the hole.
5. Release your finger, holding the pencil so that a stream of water strikes the wooden sticks.

Analysis
1. How do you know that the kinetic energy of the water has been transferred to your model turbine?

Cultural Diversity

The Cost of Electricity What would life be like without electric lights, television, radios, or video games? Would you choose to live without them? There is a group of people in several states in the United States who have chosen to do so—the Amish. They approach what we consider modern necessities based on strongly held beliefs about the value of family, church, and community. To their way of thinking, the chance to work together is more important than having technological conveniences. Machines that make it possible for only one person to do a job change the ways families and communities relate to each other. The same is true of entertainment technologies. The Amish are very careful about what new technology they allow into their lives. These positions are a reflection of their strong stance to protect and preserve their families.

Problem Solving

Energy Exercises

After reading about several alternative energy sources, you may be thinking that some of them would not easily apply to your community. Certainly, ocean tides will not be much help in states without a coastline. Solar-powered generators may not work in northern areas. However, increased efficiency and better conservation methods can help make our existing energy supplies last longer. *Efficiency* refers to the amount of benefit obtained compared to the amount of energy used. For example, automobiles now obtain more miles per gallon of gasoline used than they did ten years ago. Conservation methods include practices such as recycling. By recycling aluminum cans, coal deposits can be saved. It takes less coal to make a can from recycled aluminum than from mined aluminum.

Solve the Problem:

1. **Compare and contrast the terms** *conservation* **and** *efficiency.*
2. **Is riding a bus to school, rather than riding in a passenger car, an example of a conservation method or an increase in the efficiency of technology?**

Think Critically:

Fluorescent lightbulbs last longer than typical incandescent bulbs. A lower-wattage fluorescent bulb can provide the same light as a higher-wattage incandescent bulb. Discuss the use of each bulb as it relates to efficiency and conservation.

 **INTEGRATION**
Earth Science

Geothermal Energy

Look down at the ground. What do you see and feel underneath your feet? Although you may think of Earth as a solid sphere, hot gases and molten rock lie far beneath the surface. The inner parts of Earth contain a great deal of thermal energy, called **geothermal energy.** You do not usually notice this energy because most of it is far below Earth's crust.

In some places, Earth's crust has cracks or thin spots in it. These areas allow some of the geothermal energy to rise up near the surface of Earth. Active volcanoes permit hot gases and molten lava from deep within Earth to escape. Perhaps you have seen a geyser shoot steam and hot water from Earth. Have you ever visited or seen pictures of the famous geyser, Old Faithful, in Yellowstone National Park? This water is heated by geothermal energy. Wells can be drilled deep within Earth to pump out this hot water, which ranges in temperature from 150°C to 350°C. The steam can be used to rotate turbines and turn electric generators. Use of geothermal energy can release some sulfur compounds from gases within Earth. This pollution can be controlled by pumping the water and steam back into Earth.

Where on Earth?

Only certain places on Earth have the proper geological characteristics to use geothermal energy. Iceland makes use of geothermal plants, as shown in **Figure 25-14**, because it is located near the boundary of two continental plates. Volcanic activity and hot springs abound in the area. Lardello, Italy, began using geothermal energy in the early 1900s. Geothermal applications are limited because of the geological characteristics needed. However, the United States, particularly in the western areas, has some potential for the use of heat from Earth to make electricity.

Figure 25-14

Geothermal energy may be used in areas like Iceland. *Why isn't this form of energy the answer for other areas on Earth?*

Section Wrap-up

Review

1. Why is there a need for developing and using alternative energy sources?

2. Describe three main ways to use direct solar energy.

3. **Think Critically:** What single resource do most of the energy alternatives discussed in this section depend on, either directly or indirectly?

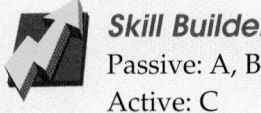

Skill Builder

Classifying

Classify the following examples of solar heating as passive or active. (A) The interior of an enclosed automobile on a sunny day. (B) Air in a room of a house near a window warms. (C) Hot water in a pipe is pumped from a rooftop to a radiator device in a bedroom.

USING MATH

The sources of energy in the United States are: 42 percent petroleum, 24 percent natural gas, 23 percent coal, 7 percent nuclear, and 4 percent other. If the percent of nuclear energy were represented with a 1-m strip of paper, how long would the other strips be?

25-4 Alternative Energy Sources   **717**

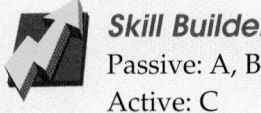

Skill Builder
Passive: A, B
Active: C

Assessment

Oral Have students explain the role of a water pump in active solar heating. Use the Performance Task Assessment List for Making Observations and Inferences in **PASC**, p. 17. **P**

INTEGRATION
Earth Science

Based on current calculations, geothermal energy is a nonrenewable resource. From the places where it is used, it cannot be replaced as fast as it is being withdrawn.

4 Close

•MINI•QUIZ•

Use the Mini Quiz to check students' recall of chapter content.

1. **A device that is used to convert solar energy into electricity is called a(n)** _____ . *photovoltaic cell*

2. **How do windmills produce electricity?** *They spin and rotate an electric generator.*

3. **Where does geothermal energy come from?** *the hot gases and molten rock beneath the surface of Earth*

Section Wrap-up

Review

1. Many present fuels are nonrenewable resources. Fossil fuels are in limited supply and cause pollution. Nuclear waste disposal is another problem.

2. Passive solar heating: Let the sun warm and light rooms. Active solar heating: Use solar panels and mirrors for heating water or air. Photovoltaic cells: Produce an electric current from the sun's energy.

3. **Think Critically** the sun

USING MATH

Petroleum	$= 42/7 \times 1m = 6$ m
Coal	$= 23/7 \times 1m = 3.3$ m
Natural Gas	$= 24/7 \times 1m = 3.4$ m
Others	$= 4/7 \times 1m = 0.6$ m

717

Activity 25-2

PREPARATION

Purpose
IS Linguistic Research and present arguments for decision making regarding the use of various sources of energy.
L1 COOP LEARN

Process Skills
communicating, researching, diagramming, comparing and contrasting, recognizing cause and effect, predicting

Time
several class periods

Materials
butcher paper, markers, cardboard, tape, scissors, written resources on alternative energy sources

Possible Hypothesis
Accept reasonable and supported answers. An example, for an ocean based community, could be as follows: If tidal energy could be harnessed, then it could provide significant amounts of electrical power.

📁 **Activity Worksheets,** pp. 5, 150, 151

PLAN THE EXPERIMENT

Possible Procedures
Divide the class into energy problem solving teams. Assign each team an energy source such as coal, geothermal, hydro, nuclear, petroleum, solar, tidal, or wind. Allow one period for research, planning, and poster making; one day for model making and report preparation; and one day for reporting and voting.

Teaching Strategies
Troubleshooting If a team wants to take the position that no plant is necessary, they must provide ways of reducing energy needs and defend the possibility of a reduced standard of living.

718

Activity 25-2

Design Your Own Experiment
Energy Alternatives— A Local Plan

Electrical power is a basic need for a community. What power source should be considered when generating electricity? In this activity, you will investigate the main problems associated with building a power plant.

PREPARATION

Problem
Your class has been selected to prepare a recommendation for the construction of a new power plant in your community. What variables should your team consider for your community?

Form a Hypothesis
Based on what you have just read and a knowledge of your community's resources, form a hypothesis about the best type of power plant for your community.

Objectives
• Consider available resources when planning a power plant.
• Consider environmental advantages and disadvantages of your suggested power plant.
• Prepare an oral report with a poster or model of your design for a power plant.

Possible Materials
• large poster board
• colored markers
• miscellaneous construction objects

Inclusion Strategies

Learning Disabled Students can create mobiles illustrating different energy sources, conservation models, and examples of waste. They can make drawings or find them in other media. Illustrations should be labeled and grouped according to category.

PLAN THE EXPERIMENT

1. As a group, agree upon and write out your hypothesis statement about your planned power plant.
2. Decide together on the components in your design, such as resources available, possible environmental aspects, and economic factors.
3. Make a list of materials you will need to complete your oral report when you present your findings to the class.

Check the Plan

1. Read over your design to make sure you have considered all variables.
2. Will you have to visit a resource center to obtain more information about the type of power plant you are suggesting?

3. Will your presentation include a scale model, a collage, or a poster?
4. *Make sure your teacher approves your plan.*

POWER STATION

DO THE EXPERIMENT

1. Decide how your team will make its presentation.
2. Obtain the materials for your presentation props.

Analyze and Apply

1. Listen to the reports of other groups. **Compare and contrast** the advantages and disadvantages each power plant design offers.

2. Will the plant designs presented be used primarily as supplements for the existing population, or will they provide for anticipated expanding populations?
3. **Analyze** how economic factors compare to environmental factors when deciding which plant to use.

Go Further

Contact your local power plant authority. Ask what alternatives the community might consider if the present plant were going to be replaced.

Inclusion Strategies

Gifted Students can write a story about how they survived an imaginary week-long loss of energy sources (electric and fossil fuel), describing their daily routine. They should try to come up with ideas for an alternate energy source. **L3**

Gifted Have students brainstorm ways energy is wasted and then ways to turn the waste to conservation. **L3**

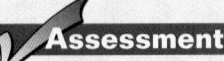 Assessment

Performance Have each student summarize the team's report in paragraph or outline form, and sketch a diagram that represents some aspect of the construction or operation of their power plant. Use the Performance Task Assessment List for Investigating an Issue Controversy in **PASC**, p. 65. **P**

DO THE EXPERIMENT

Expected Outcome
Student teams should have reports that include the resources, environmental pros and cons, and economic aspects of the type of power plant they design.

Analyze and Apply

1. Accept all reasonable answers. Coal offers a plentiful supply, but can increase air pollution. Geothermal causes little pollution, but can only be used in limited areas. Hydro offers low pollution levels, but often has environmental problems due to damming rivers. Nuclear power offers a long supply of fuel, but causes problems with nuclear waste storage. Petroleum is relatively cheap, but has a limited supply of raw material. Solar power is clean, but the technology for use is still not well developed. Tidal power is a low-pollution source of power, but is limited to coastal areas. Wind power has the potential to supply clean power, but is limited to areas of lower populations and high, consistent winds.
2. Accept well-defended answers.
3. Answers should include considerations of both sides. As environmental concerns are minimized, inexpensive electricity may be the result.

Go Further

Answers will vary. The contact could be done as a class letter or class video that summarizes the class reports, followed by the Go Further question.

Background

- Several dams are sometimes constructed on a single river, making it possible to use the same water to generate electricity more than once. Some dams are constructed for storage purposes only.

- Most hydroelectric projects in the United States were built early in the 20th century. As a result many of these facilities are more than 50 years old and most construction activity today involves repairing and maintaining existing structures.

- Water resources used to produce electricity must be shared among many parties. Local fish and game and wildlife agencies require that a constant flow is maintained. A consistent flow protects the habitat of wildlife that depend on the stream for food or shelter. Other water users include cities and irrigation districts. In some cases different utility companies have power plants on the same river, making it necessary to communicate and make cooperative decisions regarding water storage and flow rates.

Teaching Strategies

LS Kinesthetic Have students demonstrate how the force of water flowing downhill can be used to do work by holding a plastic pinwheel under a stream of water flowing from the tap or from a garden hose. Have students discuss ways to experiment with the shapes of pinwheel blades to get the most work out of a given water flow. Ask students to discuss experimenting with the rate of flow to determine what is most efficient for a given blade design.
L1

720

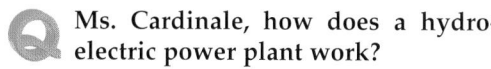

People and Science

YVONNE HO CARDINALE, *Hydroelectric Engineer*

On the Job

Q Ms. Cardinale, how does a hydroelectric power plant work?

A The power plant is located inside a dam built to hold back the water in a river, creating a reservoir. A large tunnel, called the intake structure, opens into the lowest part of the reservoir. Water flows through the tunnel and across the blades of a turbine. The turbine generates electricity.

Q How is a hydro plant built?

A For a large project here in California several years ago, I was responsible for quality control in the construction of the main tunnel, penstock, and surge chamber. We used a tunnel-boring machine to carve out the main tunnel, and drilling and blasting for smaller access tunnels.

Q What is a surge chamber?

A It's a vertical shaft that serves as a pressure release if a turbine has to be shut down. A huge amount of water—representing a lot of kinetic energy—rushes through the tunnel and penstock, so you have to have somewhere for it to go if it can't move on through the turbine.

Personal Insights

Q What is a typical workday like ?

A I help predict how much electricity some of our hydro plants can generate, based on weather factors. I go out to monitor weather stations and measure snow depth. I use computer models to help decide how much water to save behind the dam to produce power when it's needed.

Career Connection

Civil engineers are involved in the design, planning, and construction of power plants, buildings, roads, and underground cable and pipe systems. Interview a civil engineer who works in your community. Ask about the training needed to become an engineer, as well as for descriptions of a typical workday.

Career Connection

- Civil engineers earn at least a bachelor's degree at a four-year college. Some engineers earn a bachelor's degree in physics or another scientific discipline, then go on to earn a graduate degree in applied engineering.

- Power plant operators are usually union workers who may or may not have a college degree. Many learn their skills through on-the-job training.

For More Information

- The Media Relations or Education Department of your local electric utility will be able to supply information about career opportunities. They can also supply information about how different types of power plants operate.

- American Society of Civil Engineers (ASCE)
  ATTN: Student Services
  1015 15th Street, N.W.
  Washington, DC 20005

Summary

25-1: Fossil Fuels

1. The three types of fossil fuels are petroleum, natural gas, and coal. They formed from the buried remains of plants and animals.
2. All fossil fuels are nonrenewable energy resources. Supplies will run out soon if they aren't used wisely.

25-2: Nuclear Energy

1. A nuclear reactor uses the energy from a controlled nuclear chain fission reaction to generate electricity.
2. Nuclear wastes must be carefully contained and disposed of so radiation from nuclear decay will not leak into the environment.
3. Nuclear fusion releases greater amounts of energy than does nuclear fission, but fusion must occur at temperatures that are too high to be contained in a laboratory.

25-3: Science and Society: Nuclear Waste and NIMBY

1. Due to long half-lives and radioactivity levels that can be harmful, storing nuclear waste is an environmental and political problem.
2. A national permanent storage area has been proposed in Nevada.

25-4: Alternative Energy Sources

1. Alternate energy resources are needed to supplement or replace nonrenewable energy resources.
2. Other sources of energy for generating electricity include hydroelectricity and solar, wind, tidal, and geothermal energy.

3. Although some alternative energy sources don't pollute the environment and are renewable, their use is often limited to the regions where the energy source is available.

Key Science Words

a. fractional distillation
b. geothermal energy
c. hydroelectricity
d. NIMBY
e. nonrenewable resource
f. nuclear reactor
g. nuclear waste
h. petroleum
i. photovoltaic cell
j. tidal energy

Reviewing Vocabulary

Match each phrase with the correct term from the list of Key Science Words.

1. separates the hydrocarbons in crude oil
2. liquid remains of dead organisms
3. generates electricity from a controlled fission reaction
4. thermal energy inside Earth
5. source of energy that can't be replaced
6. converts solar energy directly into electricity
7. electricity produced by the energy of moving water
8. energy produced by the rise and fall of ocean levels
9. an attitude that suggests that easy answers to problems can be found as long as the solution is not too close to home
10. by-product of fission reactions

Summary

Have students read the summary statements to review the major concepts of the chapter.

Reviewing Vocabulary

1. a	**6.** i
2. h	**7.** c
3. f	**8.** j
4. b	**9.** d
5. e	**10.** g

Assessment

Portfolio Encourage students to place in their portfolios one or two items of what they consider to be their best work. Examples include:

• Skill Builder answers, p. 703
• Activity 25-1 results and answers, p. 709
• Solve the Problem answers, p. 716 **P**

Performance Additional performance assessments may be found in **Performance Assessment** and **Science Integration Activities.** Performance Task Assessment Lists and rubrics for evaluating these activities can be found in Glencoe's **Performance Assessment in the Science Classroom.**

GLENCOE TECHNOLOGY

 Videodisc

Glencoe Physical Science

Interactive Videodisc

Use the videodisc lesson *Carbohydrates and Hydrocarbons* to review the principles of this chapter.

MindJogger Videoquiz

Chapter 25 Have students work in groups as they play the Videoquiz game to review key chapter concepts.

Checking Concepts

1. d   6. d
2. c   7. b
3. a   8. b
4. a   9. c
5. c   10. d

Understanding Concepts

11. Fossil fuels and biomass are burned to boil water and produce steam. Steam is produced naturally by geothermal energy. Nuclear reactors produce steam with heat from fission reactions. The steam produced by these sources turns a turbine. Hydroelectric and tidal energy use moving water to turn a turbine. Wind turns a turbine-like device. Turbines turn generators, which produce electricity.

12. Our current major energy resource—fossil fuels—is nonrenewable and will soon be depleted. Energy is conserved by turning off lights, adjusting the thermostat, and so on.

13. It is not possible because the concentration of fissionable U-235 in the reactor fuel rods is far below that needed for a nuclear explosion.

14. The reaction cannot be contained with present technology.

15. Fossil fuels are nonrenewable and produce pollutants. Nuclear reactors produce radioactive wastes that aren't easily disposed of. Alternative energy resources are not used because the necessary technology does not yet exist, is too expensive, or has limited applicability.

Thinking Critically

16. Answers will vary; be sure support is given.

Chapter 25 Review

Checking Concepts

Choose the word or phrase that completes the sentence.

1. Plant and animal remains that are buried under sediments are not acted upon by _____ to form fossil fuels.
   a. bacteria   c. heat
   b. pressure   d. radiation

2. Hydrocarbons react with _____ during the combustion of fossil fuels.
   a. carbon dioxide   c. oxygen
   b. carbon monoxide   d. water

3. Fossil fuels are becoming more scarce as industrial demands increase and _____.
   a. the population increases
   b. the population decreases
   c. the number of nuclear reactors increases
   d. fewer plants and animals die

4. Both burning fossil fuels and nuclear fission must first be used to produce _____ in order to produce electricity.
   a. steam   c. plutonium
   b. carbon dioxide   d. water

5. A major disadvantage of using nuclear fusion reactors is that they _____.
   a. use hydrogen from water as fuel
   b. produce less radioactivity
   c. require extremely high temperatures
   d. use only small nuclei

6. Nuclear wastes include _____.
   a. products of fission reactors
   b. materials with short half-lives
   c. products from medical and industrial processes
   d. all of these

7. High-level nuclear wastes are currently disposed of by _____.
   a. releasing them into water
   b. storing them in a deep, insulated pool of water
   c. burying them in unstable areas
   d. releasing them into the air

---

8. All of Earth's energy resources can ultimately be traced back to _____.
   a. plants   c. geothermal resources
   b. the sun   d. fossil fuels

9. Photovoltaic cells must be made _____ before they can be more widely used to produce electricity.
   a. pollution-free   c. less expensive
   b. nonrenewable   d. larger

10. _____ is an alternate source of energy that uses water heated naturally by Earth's internal heat.
   a. Hydroelectricity   c. Tidal energy
   b. Nuclear fission   d. Geothermal energy

Understanding Concepts

Answer the following questions in your Science Journal using complete sentences.

11. Most of the energy resources discussed in this chapter produce electricity by means of an electric generator. Briefly discuss how each resource is used to do this.

12. Why is energy conservation important? What are some ways to conserve energy?

13. How great is the possibility of a nuclear explosion in a nuclear reactor?

14. Why isn't fusion being used today as a source of energy?

15. What specific problems have created the need for alternative energy resources? Why aren't these resources more widely used today?

---

17. (a) wind energy, hydroelectricity, tidal energy; (b) geothermal energy; (c) nuclear fission and fusion; (d) fossil fuels, biomass; (e) photovoltaic cells (solar energy)

18. (a) Not a good way because containment canisters might rupture in geologically unstable areas and release radiation. (b) This removes wastes from human populations; but corrosion by ocean water, pressure, and geological activity could result in leakage. (c) The possibility of an accidental rocket explosion should be considered.

19. renewable: solar energy, hydroelectricity, tidal energy, geothermal energy, wind energy; nonrenewable: fossil fuels, nuclear fusion, fission reactors

20. The burning of oil as a source of energy should be curtailed so that more of it could be used to produce important chemicals, plastics, and medicines.

Thinking Critically

16. Which fossil fuel do you think we should use to generate electricity? What are the pros and cons of using this fossil fuel?

17. Match each of the energy resources described in the chapter with the proper type of energy conversion listed below.
    a. kinetic energy to electricity
    b. thermal energy to electricity
    c. nuclear energy to electricity
    d. chemical energy to electricity
    e. light energy to electricity

18. Evaluate the following disposal methods of high-level nuclear wastes.
    a. Bury the wastes in an area of high earthquake and volcanic activity.
    b. Place the wastes on the ocean floor.
    c. Rocket the wastes into space.

19. Classify the energy resources discussed in this chapter as renewable or nonrenewable.

20. Suppose that new reserves of petroleum were discovered and that a nonpolluting way to burn them for energy were found. Why would it still be a good idea to decrease our use of petroleum as a source of energy? (HINT: Consider fractional distillation.)

Reactor Problems

Cause	Effect
1. The cooling H_2O is released hot	1.
2. Control rods are removed	2.
3.	3. Reactor core overheats; meltdown

23. **Making and Using Tables:** Construct a table to summarize the advantages and disadvantages of each of the energy resources discussed in this chapter.

24. **Classifying:** The following concepts relate to nuclear power as an energy source. Classify each as connected to fission or fusion. (a) Hydrogen nuclei are the main fuel source. (b) Products are not radioactive. (c) Boron and cadmium can be used to help control the process. (d) Magnetic forces can be used to help control the process.

25. **Using Numbers:** Two radioactive samples have the following half-lives. X has a half-life of 30 years, and Y has a half-life of 10 years. If you have 10.0 g of X and 100 g of Y initially, how much will remain after 90 years?

Developing Skills

If you need help, refer to the **Skill Handbook.**

21. **Sequencing:** The ultimate source of energy for cooking food on an electric stove is the sun. List in order the steps that must occur before you can use the sun's energy in this way.

22. **Recognizing Cause and Effect:** Complete the following table that describes changes in the normal operation of a nuclear reactor and the possible effects of these changes.

Performance Assessment

1. **Oral Presentation:** Research the ways in which scientists are currently trying to contain and control fusion. Write a report and present it to your classmates.

2. **Newspaper Article:** Write a newspaper article to raise public awareness of current energy problems and solutions.

3. **Asking Questions:** Review your power plant design from Activity 25-2 as if you were a state representative. What questions would you want answered?

Developing Skills

21. **Sequencing** Solar energy is: (1) stored as chemical energy in plants and animals; (2) released as thermal energy when fossil fuels are combusted; (3) converted to the kinetic energy of steam in a power plant; (4) converted to the mechanical energy of a turbine; (5) converted to electricity by a generator; (6) converted to thermal energy for cooking by a stove.

22. **Recognizing Cause and Effect** (1) Thermal pollution. (2) The reaction will continue on uncontrolled and will overheat the reactor core. (3) The reactor's cooling system fails or the control rods fail.

23. **Making and Using Tables** Tables will vary; look for proper placement of advantages and disadvantages.

24. **Classifying** Fission: c; Fusion: a, b, d

25. **Using Numbers** X: 1.25 g, Y: 0.19 g

Performance Assessment

1. Look for current references and mention of the magnetic bottle. Use the Performance Task Assessment List for Oral Presentation in **PASC,** p. 71. **P**

2. Problems may include pollution and the exhaustion of nonrenewable resources. Use the Performance Task Assessment List for Newspaper Article in **PASC,** p. 69. **P**

3. Questions may include: What pollution will be created? How will this be funded? Will this solve our energy problem? Use the Performance Task Assessment List for Asking Questions in **PASC,** p. 19. **P**

Assessment Resources

📁 Reproducible Masters

Chapter Review, pp. 53, 54
Assessment, pp. 159-162
Performance Assessment, p. 39

Glencoe Technology

⊙ **Chapter Review Software**
⊙ **Computer Test Bank**
📼 **MindJogger Videoquiz**

Objectives

LS Intrapersonal Students will search for, gather, and analyze information from the Internet about problems of their choosing. They will make presentations to the class using the information they obtained in their searches. **L1**

Summary

The Internet is a collection of data bases stored in hundreds of thousands of computers that are loosely linked together. Accessing the Internet requires a computer, modem, and a subscription with one of the commercial servers. Using the Internet, students can access scientists and universities, as well as private corporations throughout the world. After deciding on a topic about which they would like to learn more, students will collect data and prepare a report that contains text and graphics, and perhaps sound and photography.

Time Required

One-half class period will be needed to discuss the requirements of the project and to decide if students will work independently or with a partner. Each student will need approximately 3 hours on-line at the computer and an additional 3 to 5 hours to prepare a report. Allow an additional 2 to 3 days for student presentations.

Preparation

- If students will be using a computer lab or computers in the library, arrange access with the person in charge.

- Prepare a letter to parents telling them that students will be using Internet resources.

724

UNIT PROJECT 7

Researching Via the Internet

How do you use resources to find out more about a topic in which you are interested? In the past, you might have gone to the library and used a card catalog or periodical listing to find books, magazine and newspaper articles that were related to your topic. Now, at most libraries, you can find these same book and article listings on a computer. Many libraries have put their holdings in one large database to make research less time consuming and more successful.

Within the last few years a new, and more powerful source for information retrieval has become widely available. This source of information is called the Internet. If you have a computer or have access to a school computer, you can do research on the Internet without ever leaving your chair.

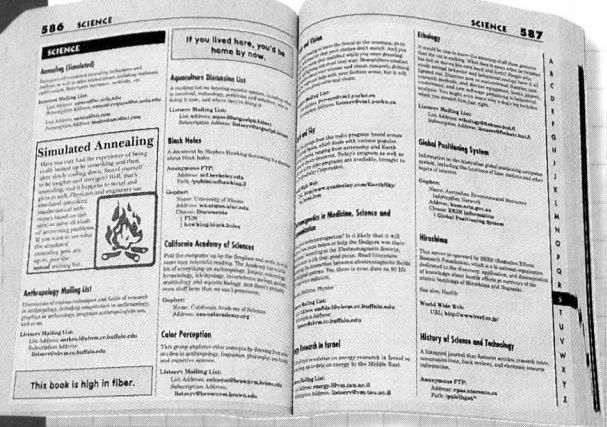

What is the Internet?

The Internet is made up of millions of computers all around the world that are linked to each other. The information available in these computers is varied and vast. You can find information on almost any imaginable topic. Using electronic mail, or e-mail, you can speak directly to an expert in a particular field or to a person in another country. Some computers store articles or entire books about particular subjects. Others contain thousands of photographs and illustrations.

Getting Started

To use the Internet you must have a way to access it. Many schools have direct access to the Internet or have a connection through a local college or university. A phone line can also give you access by way of a commercial server. Each server differs slightly in the way it works. Time, practice, and perhaps some cooperative work with a few friends will help you locate the information you need. Reference books on using the Internet will be helpful as will consulting with persons already proficient at research using the Internet.

Beginning Your Search

For this project you will choose a new technology that you wish to research and report on to your class. Look at the Using Technology features in this book for an idea. If you have an idea for a topic from another source, check it out with your teacher before beginning. Pick a topic that isn't too broad or narrow.

Internet Yellow Pages

Internet yellow pages contain the Internet addresses for locating information on any imaginable topic. However, because no one is in charge of the Internet, an Internet address may be available for a time, then the people managing it may decide to remove it.

Internet Search Tips

One important method for finding information is a search tool. A search tool allows you to type in a key word or words. The tool then searches thousands of sites on the Internet for those that contain your key words. After a few seconds, the search tool returns many addresses for other links related to your topic. The *Gopher* is one type

of search tool. There are thousands of Gopher servers available on the Internet.

Making Your Presentation

After you've completed your search for information and gathered all of your data, you should be ready to prepare a presentation for your class. Your presentation should be put together using a multimedia computer program. Your presentation should consist of at least five different screens with text, graphics, and perhaps sound and photos. Design each screen so that it is appealing to the audience. Your report should include information from at least five sites on the Internet. Identify your sites and explain to the class how you got to them.

A word of caution. Not everything on the Internet is valid. Much of the information has not been edited, reviewed, or even verified as true. If you have questions about anything you find, ask a parent, teacher or other adult to help you determine if it is useful or not.

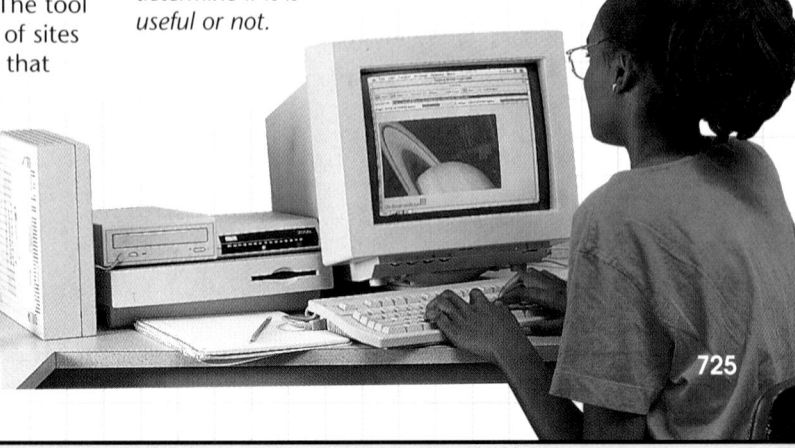

- Make a list of URLs (Internet addresses) that students can use as a starting point. Since URLs change often, it is especially important that you check the addresses before you provide them to students.
- If you are not familiar with the Internet, ask another faculty member for background help.

Teaching Strategies

- Remind students to take notes while they are online, just as they would for any other research project. URLs are important for navigating back to an area of interest.
- If students work in pairs, it is best to have a student who does not have Internet access at home work with one who does.
- Remind students that community libraries and university libraries provide Internet access. There may be a fee for logging onto the Internet.
- Explain that search engines may provide useful starting points. The Web Crawler is one example.
- Provide students with a rubric that explains what portion of their grade will be determined by each part of their Internet project.

Go Further

Once students have begun to use the Internet, they can use it as a vehicle to answer many different questions and complete other classroom exercises. You may wish to assign another problem that challenges them even further.

References

The following books are published by Ventana Press, 4911 S. Alston Avenue, Durham, NC 27713.
- *HTML Publishing on the Internet for Mac*
- *Internet Roadside Attractions*
- *Walking the World Wide Web*

Appendices

Appendix A

Appendix B

Appendix C

Appendix A

SI Units of Measurement

Table A-1

SI Base Units

Measurement	Unit	Symbol	Measurement	Unit	Symbol
length	meter	m	temperature	kelvin	K
mass	kilogram	kg	amount of substance	mole	mol
time	second	s	intensity of light	candela	cd
electric current	ampere	A			

Table A-2

Units Derived from SI Base Units

Measurement	Unit	Symbol	Expressed in Base Units
energy	joule	J	$kg \cdot m^2/s^2$ or $N \cdot m$
force	newton	N	$kg \cdot m/s^2$
frequency	hertz	Hz	$1/s$
potential difference	volt	V	$kg \cdot m^2/(A \cdot s^3)$ or W/A
power	watt	W	$kg \cdot m^2/s^3$ or J/s
pressure	pascal	Pa	$kg/(m \cdot s^2)$ or N/m^2
quality of electric charge	coulomb	C	$A \cdot s$

Table A-3

Common SI Prefixes

Prefix	Symbol	Multiplier	Prefix	Symbol	Multiplier
Greater than 1			*Less than 1*		
mega-	M	1 000 000	*deci-*	d	0.1
kilo-	k	1 000	*centi-*	c	0.01
hecto-	h	100	*milli-*	m	0.001
deka-	d	10	*micro-*	µ	0.000 001

Table A-4

SI/Metric to English Conversions

	When you want to convert:	Multiply by:	To find:
Length	inches	2.54	centimeters
	centimeters	0.39	inches
	feet	0.30	meters
	meters	3.28	feet
	yards	0.91	meters
	meters	1.09	yards
	miles	1.61	kilometers
	kilometers	0.62	miles
*Mass and Weight	ounces	28.35	grams
	grams	0.04	ounces
	pounds	0.45	kilograms
	kilograms	2.20	pounds
	tons	0.91	metric tons
	metric tons	1.10	tons
	pounds	4.45	newtons
	newtons	0.23	pounds
Volume	cubic inches	16.39	cubic centimeters
	cubic centimeters	0.06	cubic inches
	cubic feet	0.03	cubic meters
	cubic meters	35.31	cubic feet
	liters	1.06	quarts
	liters	0.26	gallons
	gallons	3.78	liters
Area	square inches	6.45	square centimeters
	square centimeters	0.16	square inches
	square feet	0.09	square meters
	square meters	10.76	square feet
	square miles	2.59	square kilometers
	square kilometers	0.39	square miles

* Weight as measured in standard Earth gravity

SI/Temperature Scale Conversions

Table A-5

SI/Temperature Scale Conversions

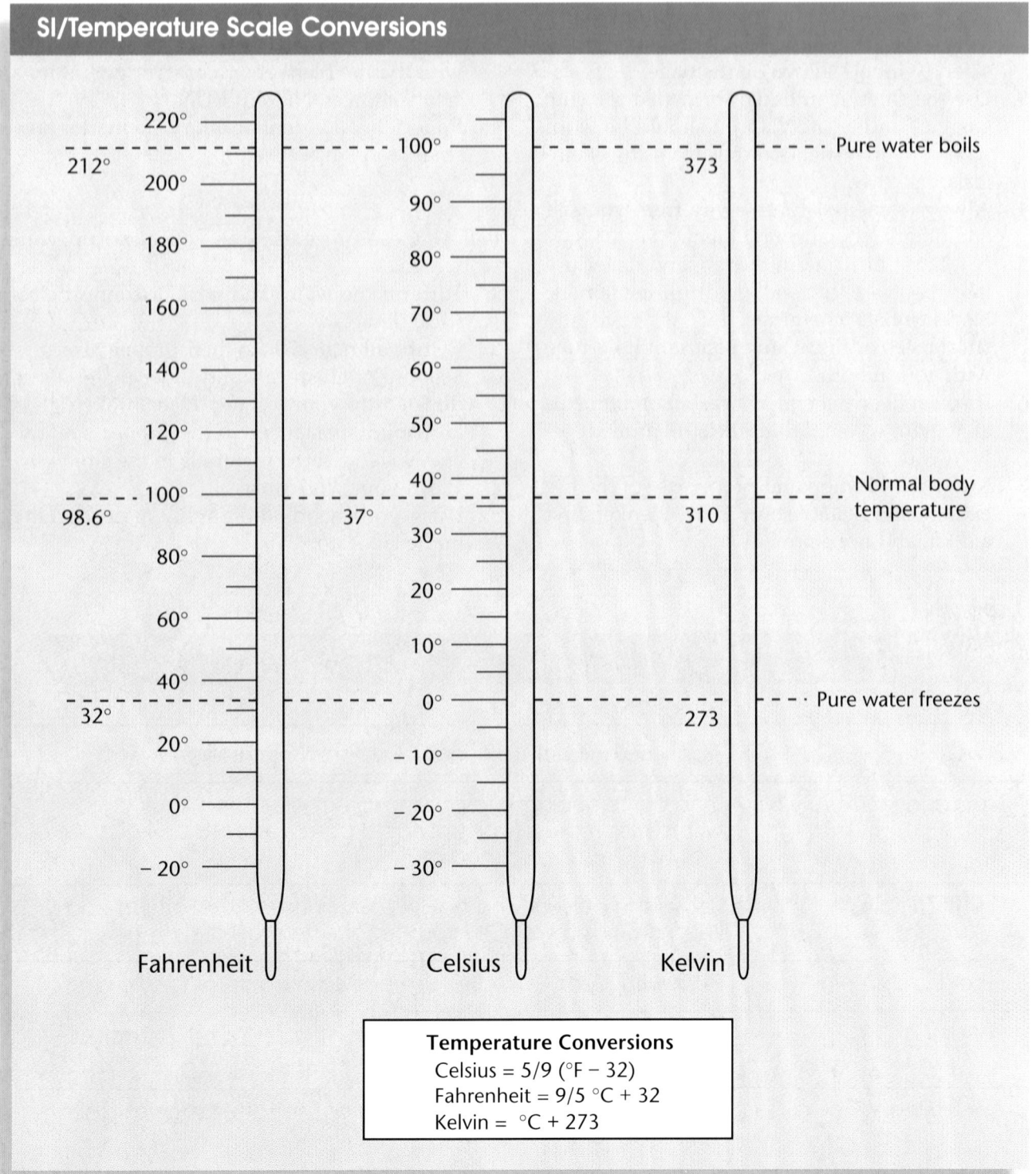

Temperature Conversions
Celsius = 5/9 (°F − 32)
Fahrenheit = 9/5 °C + 32
Kelvin = °C + 273

Safety in the Classroom

1. Always obtain your teacher's permission to begin an investigation.
2. Study the procedure. If you have questions, ask your teacher. Be sure you understand any safety symbols shown on the page.
3. Use the safety equipment provided for you. Goggles and a safety apron should be worn when any investigation calls for using chemicals.
4. Always slant test tubes away from yourself and others when heating them.
5. Never eat or drink in the lab, and never use lab glassware as food or drink containers. Never inhale chemicals. Do not taste any substances or draw any material into a tube with your mouth.
6. If you spill any chemical, wash it off immediately with water. Report the spill immediately to your teacher.
7. Know the location and proper use of the fire extinguisher, safety shower, fire blanket, first aid kit, and fire alarm.
8. Keep all materials away from open flames. Tie back long hair and loose clothing.
9. If a fire should break out in the classroom, or if your clothing should catch fire, smother it with the fire blanket or a coat, or get under a safety shower. NEVER RUN.
10. Report any accident or injury, no matter how small, to your teacher.

Follow these procedures as you clean up your work area.

1. Turn off the water and gas. Disconnect electrical devices.
2. Return all materials to their proper places.
3. Dispose of chemicals and other materials as directed by your teacher. Place broken glass and solid substances in the proper containers. Never discard materials in the sink.
4. Clean your work area.
5. Wash your hands thoroughly after working in the laboratory.

Table B-1

First Aid	
Injury	**Safe Response**
Burns	Apply cold water. Call your teacher immediately.
Cuts and bruises	Stop any bleeding by applying direct pressure. Cover cuts with a clean dressing. Apply cold compresses to bruises. Call your teacher immediately.
Fainting	Leave the person lying down. Loosen any tight clothing and keep crowds away. Call your teacher immediately.
Foreign matter in eye	Flush with plenty of water. Use eyewash bottle or fountain.
Poisoning	Note the suspected poisoning agent and call your teacher immediately.
Any spills on skin	Flush with large amounts of water or use safety shower. Call your teacher immediately.

Safety Symbols

This textbook uses the safety symbols in **Table B-2** below to alert you to possible laboratory dangers.

Table B-2

Safety Symbols

 **DISPOSAL ALERT**
This symbol appears when care must be taken to dispose of materials properly.

 **ANIMAL SAFETY**
This symbol appears whenever live animals are studied and the safety of the animals and the students must be ensured.

 **BIOLOGICAL HAZARD**
This symbol appears when there is danger involving bacteria, fungi, or protists.

 **RADIOACTIVE SAFETY**
This symbol appears when radioactive materials are used.

 **OPEN FLAME ALERT**
This symbol appears when use of an open flame could cause a fire or an explosion.

 **CLOTHING PROTECTION SAFETY**
This symbol appears when substances used could stain or burn clothing.

 **THERMAL SAFETY**
This symbol appears as a reminder to use caution when handling hot objects.

 **FIRE SAFETY**
This symbol appears when care should be taken around open flames.

 **SHARP OBJECT SAFETY**
This symbol appears when a danger of cuts or punctures caused by the use of sharp objects exists.

 **EXPLOSION SAFETY**
This symbol appears when the misuse of chemicals could cause an explosion.

 **FUME SAFETY**
This symbol appears when chemicals or chemical reactions could cause dangerous fumes.

 **EYE SAFETY**
This symbol appears when a danger to the eyes exists. Safety goggles should be worn when this symbol appears.

 **ELECTRICAL SAFETY**
This symbol appears when care should be taken when using electrical equipment.

 **POISON SAFETY**
This symbol appears when poisonous substances are used.

 **SKIN PROTECTION SAFETY**
This symbol appears when use of caustic chemicals might irritate the skin or when contact with microorganisms might transmit infection.

 **CHEMICAL SAFETY**
This symbol appears when chemicals used can cause burns or are poisonous if absorbed through the skin.

Appendix C

PERIODIC TABLE OF THE ELEMENTS

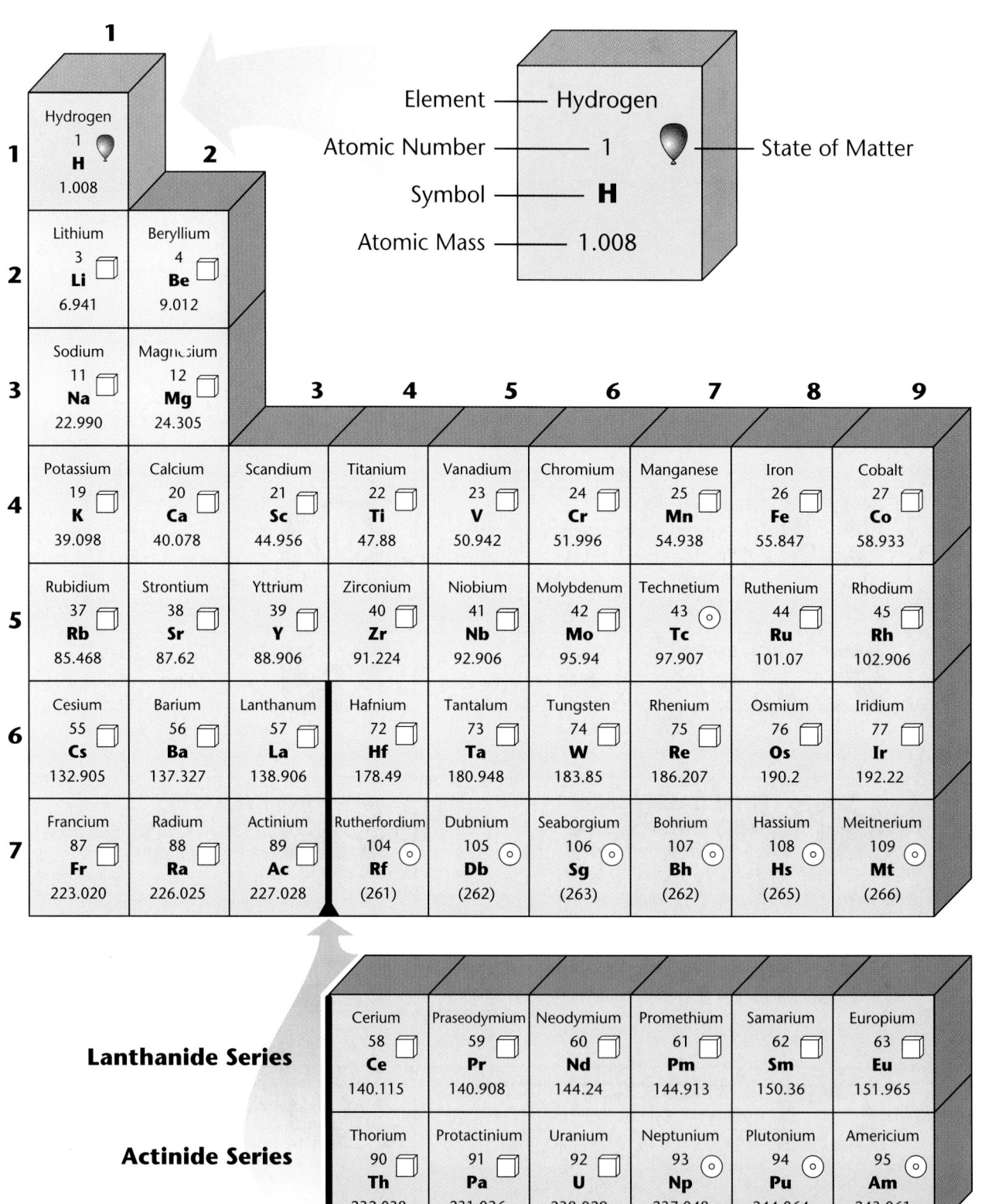

	Element	Hydrogen	State of Matter
	Atomic Number	1	
	Symbol	**H**	
	Atomic Mass	1.008	

1

1 Hydrogen 1 **H** 1.008

2

2 Lithium 3 **Li** 6.941 — Beryllium 4 **Be** 9.012

3 Sodium 11 **Na** 22.990 — Magnesium 12 **Mg** 24.305

	3	**4**	**5**	**6**	**7**	**8**	**9**

4 Potassium 19 **K** 39.098 — Calcium 20 **Ca** 40.078 — Scandium 21 **Sc** 44.956 — Titanium 22 **Ti** 47.88 — Vanadium 23 **V** 50.942 — Chromium 24 **Cr** 51.996 — Manganese 25 **Mn** 54.938 — Iron 26 **Fe** 55.847 — Cobalt 27 **Co** 58.933

5 Rubidium 37 **Rb** 85.468 — Strontium 38 **Sr** 87.62 — Yttrium 39 **Y** 88.906 — Zirconium 40 **Zr** 91.224 — Niobium 41 **Nb** 92.906 — Molybdenum 42 **Mo** 95.94 — Technetium 43 **Tc** 97.907 — Ruthenium 44 **Ru** 101.07 — Rhodium 45 **Rh** 102.906

6 Cesium 55 **Cs** 132.905 — Barium 56 **Ba** 137.327 — Lanthanum 57 **La** 138.906 — Hafnium 72 **Hf** 178.49 — Tantalum 73 **Ta** 180.948 — Tungsten 74 **W** 183.85 — Rhenium 75 **Re** 186.207 — Osmium 76 **Os** 190.2 — Iridium 77 **Ir** 192.22

7 Francium 87 **Fr** 223.020 — Radium 88 **Ra** 226.025 — Actinium 89 **Ac** 227.028 — Rutherfordium 104 **Rf** (261) — Dubnium 105 **Db** (262) — Seaborgium 106 **Sg** (263) — Bohrium 107 **Bh** (262) — Hassium 108 **Hs** (265) — Meitnerium 109 **Mt** (266)

Lanthanide Series — Cerium 58 **Ce** 140.115 — Praseodymium 59 **Pr** 140.908 — Neodymium 60 **Nd** 144.24 — Promethium 61 **Pm** 144.913 — Samarium 62 **Sm** 150.36 — Europium 63 **Eu** 151.965

Actinide Series — Thorium 90 **Th** 232.038 — Protactinium 91 **Pa** 231.036 — Uranium 92 **U** 238.029 — Neptunium 93 **Np** 237.048 — Plutonium 94 **Pu** 244.064 — Americium 95 **Am** 243.061

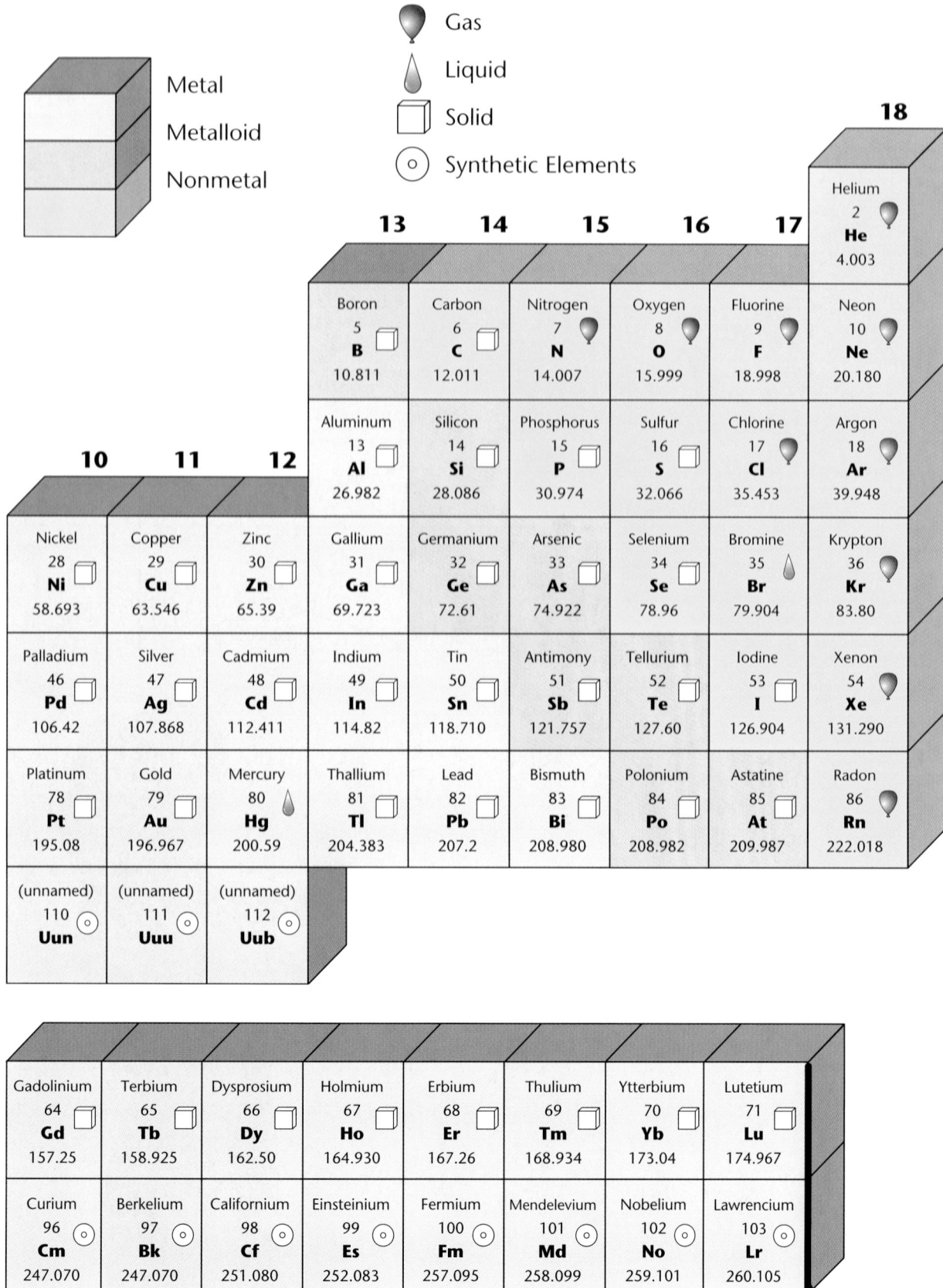

Metal

Metalloid

Nonmetal

Gas

Liquid

Solid

Synthetic Elements

13	14	15	16	17	18
					Helium 2 He 4.003
Boron 5 B 10.811	Carbon 6 C 12.011	Nitrogen 7 N 14.007	Oxygen 8 O 15.999	Fluorine 9 F 18.998	Neon 10 Ne 20.180
Aluminum 13 Al 26.982	Silicon 14 Si 28.086	Phosphorus 15 P 30.974	Sulfur 16 S 32.066	Chlorine 17 Cl 35.453	Argon 18 Ar 39.948

10	11	12						
Nickel 28 Ni 58.693	Copper 29 Cu 63.546	Zinc 30 Zn 65.39	Gallium 31 Ga 69.723	Germanium 32 Ge 72.61	Arsenic 33 As 74.922	Selenium 34 Se 78.96	Bromine 35 Br 79.904	Krypton 36 Kr 83.80
Palladium 46 Pd 106.42	Silver 47 Ag 107.868	Cadmium 48 Cd 112.411	Indium 49 In 114.82	Tin 50 Sn 118.710	Antimony 51 Sb 121.757	Tellurium 52 Te 127.60	Iodine 53 I 126.904	Xenon 54 Xe 131.290
Platinum 78 Pt 195.08	Gold 79 Au 196.967	Mercury 80 Hg 200.59	Thallium 81 Tl 204.383	Lead 82 Pb 207.2	Bismuth 83 Bi 208.980	Polonium 84 Po 208.982	Astatine 85 At 209.987	Radon 86 Rn 222.018
(unnamed) 110 Uun	(unnamed) 111 Uuu	(unnamed) 112 Uub						

Gadolinium 64 Gd 157.25	Terbium 65 Tb 158.925	Dysprosium 66 Dy 162.50	Holmium 67 Ho 164.930	Erbium 68 Er 167.26	Thulium 69 Tm 168.934	Ytterbium 70 Yb 173.04	Lutetium 71 Lu 174.967
Curium 96 Cm 247.070	Berkelium 97 Bk 247.070	Californium 98 Cf 251.080	Einsteinium 99 Es 252.083	Fermium 100 Fm 257.095	Mendelevium 101 Md 258.099	Nobelium 102 No 259.101	Lawrencium 103 Lr 260.105

Table of Contents

Organizing Information

Communicating

The communication of ideas is an important part of our everyday lives. Whether reading a book, writing a letter, or watching a television program, people everywhere are expressing opinions and sharing information with one another. Writing in your Science Journal allows you to express your opinions and demonstrate your knowledge of the information presented on a subject. When writing, keep in mind the purpose of the assignment and the audience with which you are communicating.

Examples Science Journal assignments vary greatly. They may ask you to take a viewpoint other than your own; perhaps you will be a scientist, a TV reporter, or a committee member of a local environmental group. Maybe you will be expressing your opinions to a member of Congress, a doctor, or to the editor of your local newspaper, as shown in **Figure 1.** Sometimes, Science Journal writing may allow you to summarize information in the form of an outline, a letter, or in a paragraph.

Figure 1

A Science Journal entry.

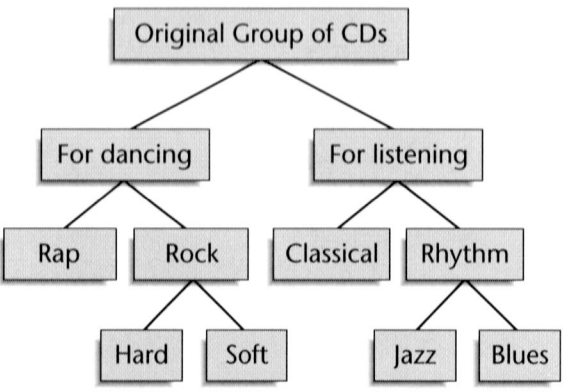

Figure 2

Classifying CDs.

Classifying

You may not realize it, but you make things orderly in the world around you. If you hang your shirts together in the closet or if your favorite CDs are stacked together, you have used the skill of classifying.

Classifying is the process of sorting objects or events into groups based on common features. When classifying, first observe the objects or events to be classified. Then, select one feature that is shared by some members in the group but not by all. Place those members that share that feature into a subgroup. You can classify members into smaller and smaller subgroups based on characteristics.

Remember, when you classify, you are grouping objects or events for a purpose. Keep your purpose in mind as you select the features to form groups and subgroups.

Example How would you classify a collection of CDs? As shown in **Figure 2,** you might classify those you like to dance to in one subgroup and CDs you like to listen to in the next column. The CDs you like to dance to could be subdivided into a rap subgroup and a rock subgroup. Note that for each feature selected, each CD fits into only one subgroup. You would keep selecting features until all the CDs are classified. **Figure 2** shows one possible classification.

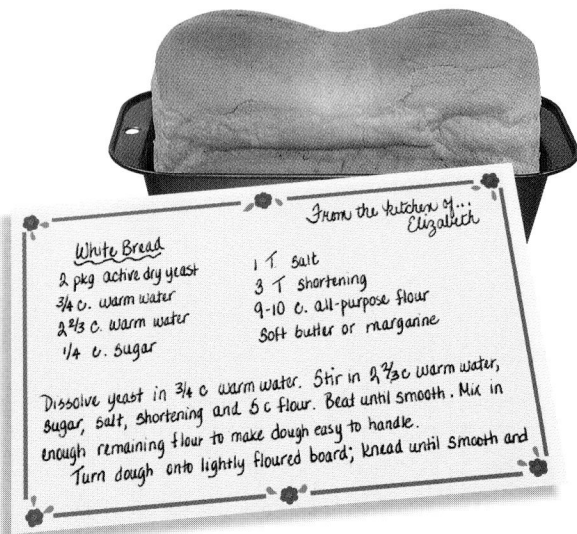

Figure 3

A recipe for bread contains sequenced instructions.

Sequencing

A sequence is an arrangement of things or events in a particular order. When you are asked to sequence objects or events within a group, figure out what comes first, then think about what should come second. Continue to choose objects or events until all of the objects you started out with are in order. Then, go back over the sequence to make sure each thing or event in your sequence logically leads to the next.

Example   A sequence with which you are most familiar is the use of alphabetical order. Another example of sequence would be the steps in a recipe, as shown in **Figure 3.** Think about baking bread. Steps in the recipe have to be followed in order for the bread to turn out right.

Concept Mapping

If you were taking an automobile trip, you would probably take along a road map. The road map shows your location, your destination, and other places along the way. By looking at the map and finding where you are, you can begin to understand where you are in relation to other locations on the map.

A concept map is similar to a road map. But, a concept map shows relationships among ideas (or concepts) rather than places. A concept map is a diagram that visually shows how concepts are related. Because the concept map shows relationships among ideas, it can make the meanings of ideas and terms clear, and help you understand better what you are studying.

There is usually not one correct way to create a concept map. As you construct one type of map, you may discover other

Figure 4

Network tree describing U.S. currency.

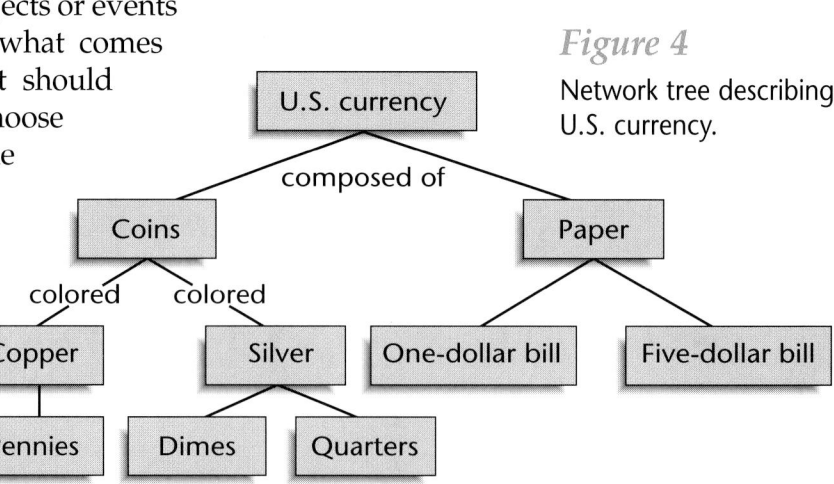

ways to construct the map that show the relationships between concepts in a better way. If you do discover what you think is a better way to create a concept map, go ahead and use the new one. Overall, concept maps are useful for breaking a big concept down into smaller parts, making learning easier.

Examples

Network Tree Look at the concept map about U.S. currency in **Figure 4.** This is called a network tree. Notice how some words are in rectangles while others are written across connecting lines. The words inside the rectangles are science concepts. The lines in the map show related concepts. The words written on the lines describe the relationships between concepts.

When you are asked to construct a network tree, write down the topic and list the major concepts related to that topic on a piece of paper. Then look at your list and begin to put them in order from general to specific. Branch the related concepts from the major concept and describe the relationships on the lines. Continue to write the more specific concepts. Write the relationships between the concepts on the lines until all concepts are mapped. Examine the concept map for relationships that cross branches, and add them to the concept map.

Events Chain An events chain is another type of concept map. An events chain map, such as the one describing a typical morning routine in **Figure 5,** is used to describe ideas in order. In science, an events chain can be used to describe a sequence of events, the steps in a procedure, or the stages of a process.

When making an events chain, first find the one event that starts the chain. This event is called the initiating event. Then,

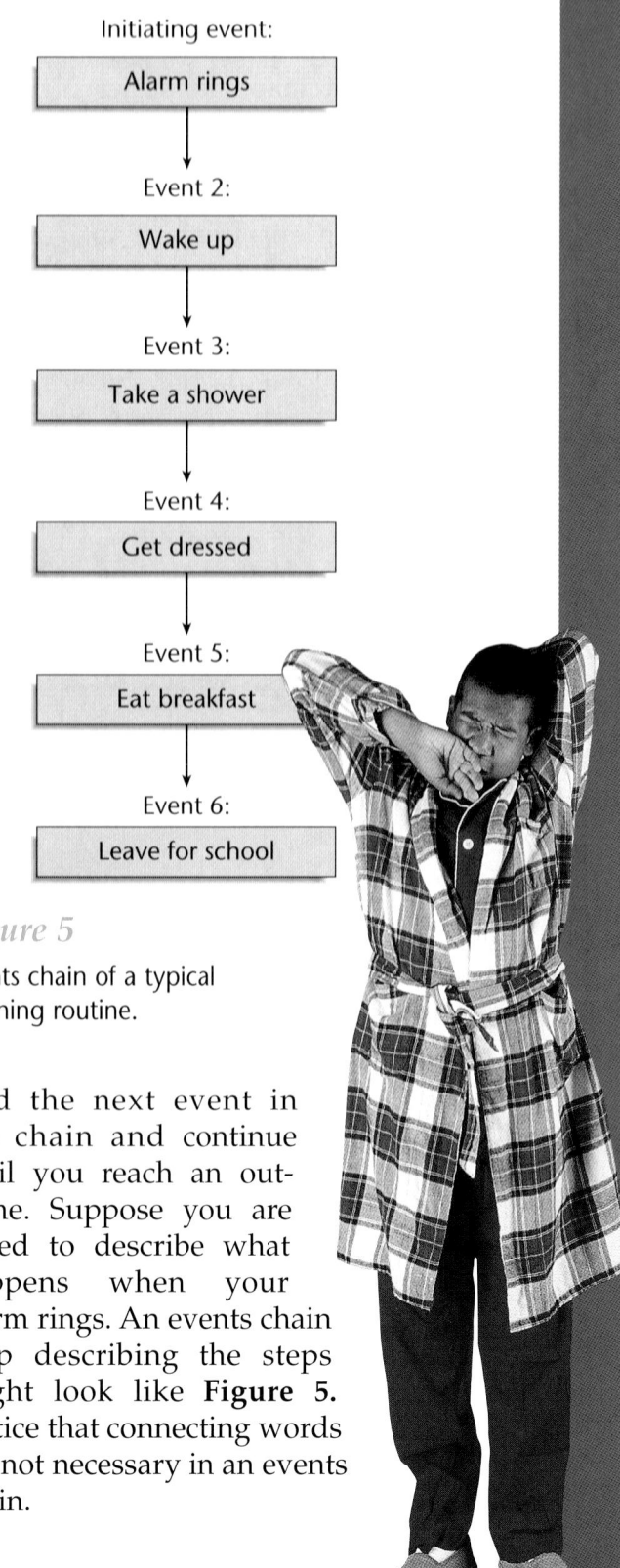

Initiating event:

Alarm rings

Event 2:

Wake up

Event 3:

Take a shower

Event 4:

Get dressed

Event 5:

Eat breakfast

Event 6:

Leave for school

Figure 5

Events chain of a typical morning routine.

find the next event in the chain and continue until you reach an outcome. Suppose you are asked to describe what happens when your alarm rings. An events chain map describing the steps might look like **Figure 5.** Notice that connecting words are not necessary in an events chain.

Cycle Map A cycle concept map is a special type of events chain map. In a cycle concept map, the series of events does not produce a final outcome. Instead, the last event in the chain relates back to the initiating event.

As in the events chain map, you first decide on an initiating event and then list each event in order. Because there is no outcome and the last event relates back to the initiating event, the cycle repeats itself. Look at the cycle map describing the relationship between day and night in **Figure 6.**

Figure 6
Cycle map of day and night.

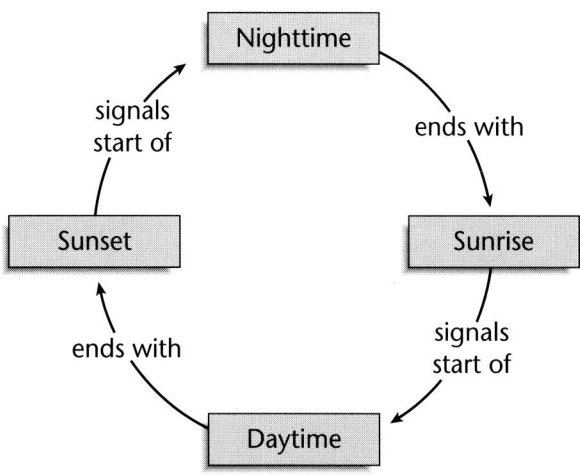

Spider Map A fourth type of concept map is the spider map. This is a map that you can use for brainstorming. Once you have a central idea, you may find you have a jumble of ideas that relate to it, but are not necessarily clearly related to each other. As illustrated by the homework spider map in **Figure 7,** by writing these ideas outside the main concept, you may begin to separate and group unrelated terms so that they become more useful.

Figure 7
Spider map about homework.

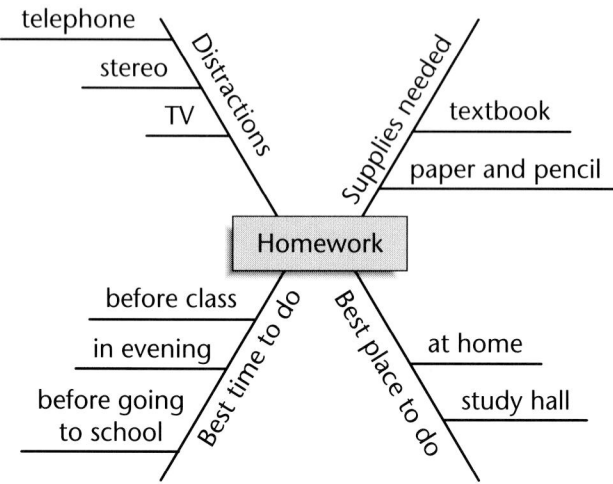

Making and Using Tables

Browse through your textbook and you will notice tables in the text and in the activities. In a table, data or information is arranged in a way that makes it easier for you to understand. Activity tables help organize the data you collect during an activity so that results can be interpreted more easily.

Examples Most tables have a title. At a glance, the title tells you what the table is about. A table is divided into columns and rows. The first column lists items to be compared. In **Figure 8,** the collection of recyclable materials is being compared in a table. The row across the top lists the specific characteristics being compared. Within the grid of the table, the collected data are recorded.

What is the title of the table in **Figure 8?** The title is "Recycled Materials." What is being compared? The different materials being recycled and on which days they are recycled.

Making Tables To make a table, list the items to be compared down in columns

Figure 8

Table of recycled materials.

Recycled Materials			
Day of Week	Paper (kg)	Aluminum (kg)	Plastic (kg)
Mon.	4.0	2.0	0.5
Wed.	3.5	1.5	0.5
Fri.	3.0	1.0	1.5

and the characteristics to be compared across in rows. The table in **Figure 8** compares the mass of recycled materials collected by a class. On Monday, students turned in 4.0 kg of paper, 2.0 kg of aluminum, and 0.5 kg of plastic. On Wednesday, they turned in 3.5 kg of paper, 1.5 kg of aluminum, and 0.5 kg of plastic. On Friday, the totals were 3.0 kg of paper, 1.0 kg of aluminum, and 1.5 kg of plastic.

Using Tables How much plastic, in kilograms, is being recycled on Wednesday? Locate the column labeled "Plastic (kg)" and the row "Wed." The data in the box where the column and row intersect are the answer. Did you answer "0.5"? How much aluminum, in kilograms, is being recycled on Friday? If you answered "1.0," you understand how to use the parts of the table.

Making and Using Graphs

After scientists organize data in tables, they may display the data in a graph. A graph is a diagram that shows the relationship of one variable to another. A graph makes interpretation and analysis of data easier. There are three basic types of graphs used in science—the line graph, the bar graph, and the circle graph.

Examples

Line Graphs A line graph is used to show the relationship between two variables. The variables being compared go on two axes of the graph. The independent variable always goes on the horizontal axis, called the x-axis. The dependent variable always goes on the vertical axis, called the y-axis.

Suppose your class started to record the amount of materials they collected in one week for their school to recycle. The collected information is shown in **Figure 9.**

You could make a graph of the materials collected over the three days of the school week. The three week days are the independent variables and are placed on the x-axis of your graph. The amount of materials collected is the dependent variable and would go on the y-axis.

After drawing your axes, label each with a scale. The x-axis lists the three weekdays. To make a scale of the amount of materials collected on the y-axis, look at the data values. Because the lowest amount collected was 1.0 and the highest

Figure 9

Amount of recyclable materials collected during one week.

Materials Collected During Week		
Day of Week	Paper (kg)	Aluminum (kg)
Mon.	5.0	4.0
Wed.	4.0	1.0
Fri.	2.5	2.0

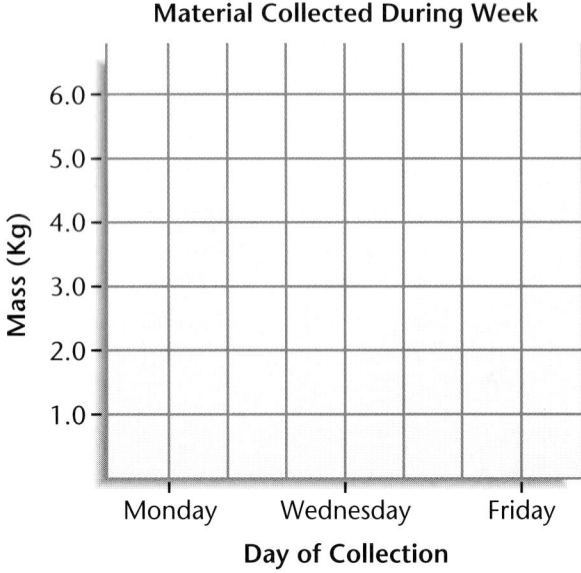

Material Collected During Week

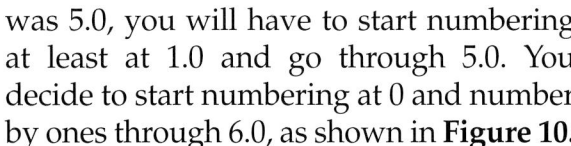

Figure 10

Graph outline for material collected during week.

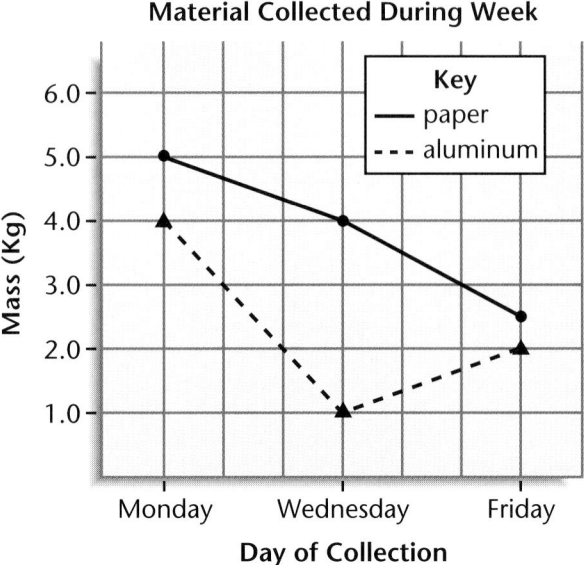

Material Collected During Week

Figure 11

Line graph of materials collected during week.

was 5.0, you will have to start numbering at least at 1.0 and go through 5.0. You decide to start numbering at 0 and number by ones through 6.0, as shown in **Figure 10.**

Next, plot the data points for collected paper. The first pair of data you want to plot is Monday and 5.0 kg paper. Locate "Monday" on the x-axis and locate "5.0" on the y-axis. Where an imaginary vertical line from the x-axis and an imaginary horizontal line from the y-axis would meet, place the first data point. Place the other data points the same way. After all the points are plotted, connect them with the best smooth curve. Repeat this procedure for the data points for aluminum. Use continuous and dashed lines to distinguish the two line graphs. The resulting graph should look like **Figure 11.**

Bar Graphs Bar graphs are similar to line graphs. They compare data that do not continuously change. In a bar graph, vertical bars show the relationships among data.

To make a bar graph, set up the x-axis and y-axis as you did for the line graph.

The data are plotted by drawing vertical bars from the x-axis up to a point where the y-axis would meet the bar if it were extended.

Look at the bar graph in **Figure 12** comparing the mass of aluminum collected over three weekdays. The x-axis is the days on which the aluminum was collected. The y-axis is the mass of aluminum collected, in kilograms.

Circle Graphs A circle graph uses a circle divided into sections to display data. Each section represents part of the whole. All the sections together equal 100 percent.

Suppose you wanted to make a circle graph to show the number of seeds that germinated in a package. You would count the total number of seeds. You find that there are 143 seeds in the package. This represents 100 percent, the whole circle.

You plant the seeds, and 129 seeds germinate. The seeds that germinated will make up one section of the circle graph, and the seeds that did not germinate will make up the remaining section.

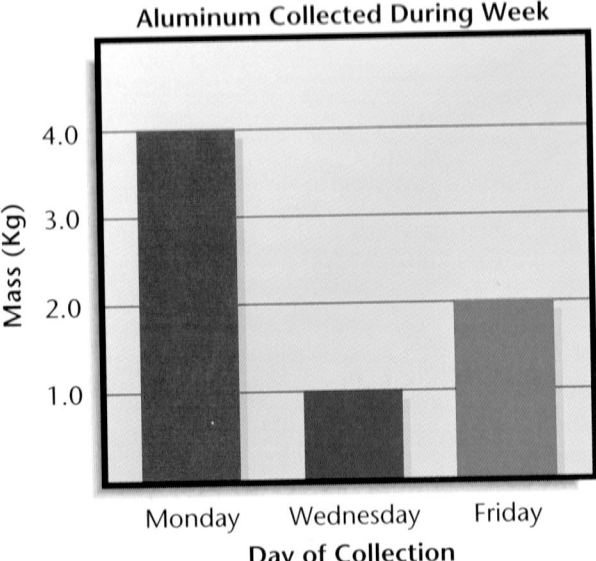

Aluminum Collected During Week

Figure 12

Bar graph of aluminum collected during week.

To find out how much of the circle each section should take, divide the number of seeds in each section by the total number of seeds. Then multiply your answer by 360, the number of degrees in a circle, and round to the nearest whole number. The section of the circle graph in degrees that represents the seeds germinated is figured below.

$$\frac{129}{143} \times 360 = 324.75 \text{ or } 325 \text{ degrees (or } 325°)$$

Plot this group on the circle graph using a compass and a protractor. Use the compass to draw a circle. It will be easier to measure the part of the circle representing the nongerminating seeds, so subtract 325° from 360° to get 35°. Draw a straight line from the center to the edge of the circle. Place your protractor on this line and use it to mark a point at 35°. Use this point to draw a straight line from the center of the circle to the edge. This is the section for the group of seeds that did not germinate. The other section represents the group of 129 seeds that did germinate. Label the sections of your graph and title the graph as shown in **Figure 13.**

Figure 13

Circle graph of germinated seeds.

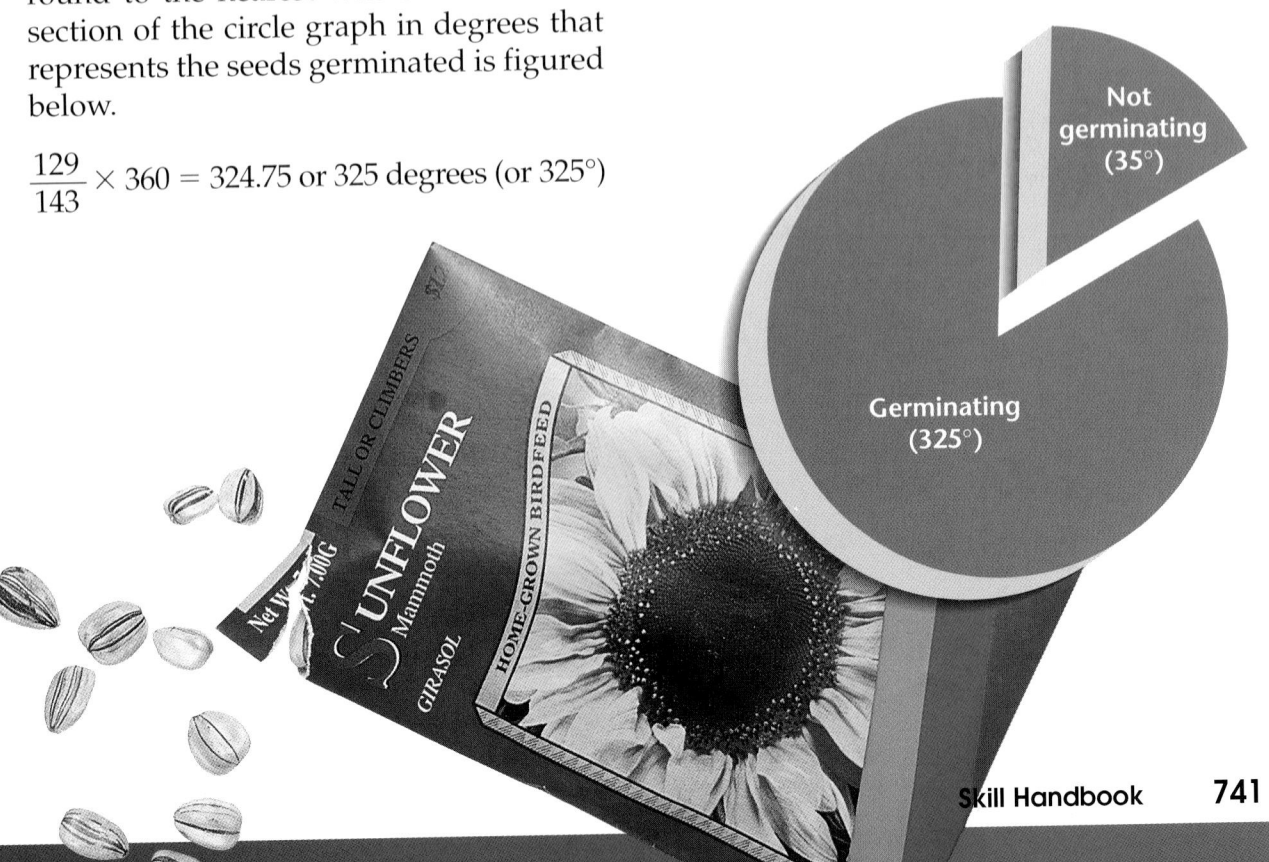

Seeds Germinated

Thinking Critically

Observing and Inferring

Observing Scientists try to make careful and accurate observations. When possible, they use instruments such as microscopes, thermometers, and balances to make observations. Measurements with a balance or thermometer provide numerical data that can be checked and repeated.

When you make observations in science, you'll find it helpful to examine the entire object or situation first. Then, look carefully for details. Write down everything you observe.

Example Imagine that you have just finished a volleyball game. At home, you open the refrigerator and see a jug of orange juice on the back of the top shelf. The jug, shown in **Figure 14,** feels cold as you grasp it. Then you drink the juice, smell the oranges, and enjoy the tart taste in your mouth.

Figure 14

Why is this jug of orange juice cold?

As you imagined yourself in the story, you used your senses to make observations. You used your sense of sight to find the jug in the refrigerator, your sense of touch when you felt the coldness of the jug, your sense of hearing to listen as the liquid filled the glass, and your senses of smell and taste to enjoy the odor and tartness of the juice. The basis of all scientific investigation is observation.

Inferring Scientists often make inferences based on their observations. An inference is an attempt to explain or interpret observations or to say what caused what you observed.

When making an inference, be certain to use accurate data and observations. Analyze all of the data that you've collected. Then, based on everything you know, explain or interpret what you've observed.

Example When you drank a glass of orange juice after the volleyball game, you observed that the orange juice was cold as well as refreshing. You might infer that the juice was cold because it had been made much earlier in the day and had been kept in the refrigerator or you might infer that it had just been made, using both cold water and ice. The only way to be sure which inference is correct is to investigate further.

Comparing and Contrasting

Observations can be analyzed by noting the similarities and differences between two or more objects or events that you observe. When you look at objects or events to see how they are similar, you are comparing them. Contrasting is looking for differences in similar objects or events.

Figure 15

Table comparing the nutritional value of *Candy A* and *Candy B*.

Nutritional Value		
	Candy A	**Candy B**
Serving size	103 g	105 g
Calories	220	160
Total Fat	10 g	10 g
Protein	2.5 g	2.6 g
Total Carbohydrate	30 g	15 g

Example Suppose you were asked to compare and contrast the nutritional value of two candy bars, *Candy A* and *Candy B.* You would start by looking at what is known about these candy bars. Arrange this information in a table, like the one in **Figure 15.**

Similarities you might point out are that both candy bars have similar serving sizes, amounts of total fat, and protein. Differences include *Candy A* having a higher calorie value and containing more total carbohydrates than *Candy B.*

Recognizing Cause and Effect

Have you ever watched something happen and then made suggestions about why it happened? If so, you have observed an effect and inferred a cause. The event is an effect, and the reason for the event is the cause.

Example Suppose that every time your teacher fed the fish in a classroom aquarium, she or he tapped the food container on the edge of the aquarium. Then, one day your teacher just happened to tap the edge of the aquarium with a pencil while making a point. You observed the fish swim to the surface of the aquarium to feed, as shown in **Figure 16.** What is the effect, and what would you infer to be the cause? The effect is the fish swimming to the surface of the aquarium. You might infer the cause to be the teacher tapping on the edge of the aquarium. In determining cause and effect, you have made a logical inference based on your observations.

Perhaps the fish swam to the surface because they reacted to the teacher's waving hand or for some other reason. When scientists are unsure of the cause of a certain event, they design controlled experiments to determine what causes the event. Although you have made a logical conclusion about the behavior of the fish, you would have to perform an experiment to be certain that it was the tapping that caused the effect you observed.

Figure 16

What cause and effect situations are occurring in this aquarium?

Practicing Scientific Processes

You might say that the work of a scientist is to solve problems. But when you decide how to dress on a particular day, you are doing problem solving, too. You may observe what the weather looks like through a window. You may go outside and see whether what you are wearing is warm or cool enough.

Scientists use an orderly approach to learn new information and to solve problems. The methods scientists may use include observing to form a hypothesis, designing an experiment to test a hypothesis, separating and controlling variables, and interpreting data.

Forming Operational Definitions

Operational definitions define an object by showing how it functions, works, or behaves. Such definitions are written in terms of how an object works or how it can be used; that is, what is its job or purpose?

Figure 17

What observations can be made about this dog?

Example Some operational definitions explain how an object can be used.
- A ruler is a tool that measures the size of an object.
- An automobile can move things from one place to another.

Or such a definition may explain how an object works.
- A ruler contains a series of marks that can be used as a standard when measuring.
- An automobile is a vehicle that can move from place to place.

Forming a Hypothesis

Observations You observe all the time. Scientists try to observe as much as possible about the things and events they study so they know that what they say about their observations is reliable.

Some observations describe something using only words. These observations are called qualitative observations. Other observations describe how much of something there is. These are quantitative observations and use numbers, as well as words, in the description. Tools or equipment are used to measure the characteristic being described.

Example If you were making qualitative observations of the dog in **Figure 17,** you might use words such as *furry, yellow,* and *short-haired.* Quantitative observations of this dog might include a mass of 14 kg, a height of 46 cm, ear length of 10 cm, and an age of 150 days.

Hypotheses Hypotheses are tested to help explain observations that have been made. They are often stated as *if* and *then* statements.

Examples Suppose you want to make a perfect score on a spelling test. Begin by thinking of several ways to accomplish this. Base these possibilities on past observations. If you put each of these possibilities into sentence form, using the words *if* and *then,* you can form a hypothesis. All of the following are hypotheses you might consider to explain how you could score 100 percent on your test:

If the test is easy, then I will get a perfect score.

If I am intelligent, then I will get a perfect score.

If I study hard, then I will get a perfect score.

Perhaps a scientist has observed that plants that receive fertilizer grow taller than plants that do not. A scientist may form a hypothesis that says: If plants are fertilized, then their growth will increase.

Designing an Experiment to Test a Hypothesis

In order to test a hypothesis, it's best to write out a procedure. A procedure is the plan that you follow in your experiment. A procedure tells you what materials to use and how to use them. After following the procedure, data are generated. From this generated data, you can then draw a conclusion and make a statement about your results.

If the conclusion you draw from the data supports your hypothesis, then you can say that your hypothesis is reliable. *Reliable* means that you can trust your conclusion. If it did not support your hypothesis, then you would have to make new observations and state a new hypothesis—just make sure that it is one that you can test.

Example Super premium gasoline costs more than regular gasoline. Does super premium gasoline increase the efficiency or fuel mileage of your family car? Let's figure out how to conduct an experiment to test the hypothesis, "*if* premium gas is more efficient, *then* it should increase the fuel mileage of our family car." Then a procedure similar to **Figure 18** must be written to generate data presented in **Figure 19** on page 746.

These data show that premium gasoline is less efficient than regular gasoline. It took more gasoline to travel one mile (0.064) using premium gasoline than it does to travel one mile using regular gasoline (0.059). This conclusion does not support the original hypothesis made.

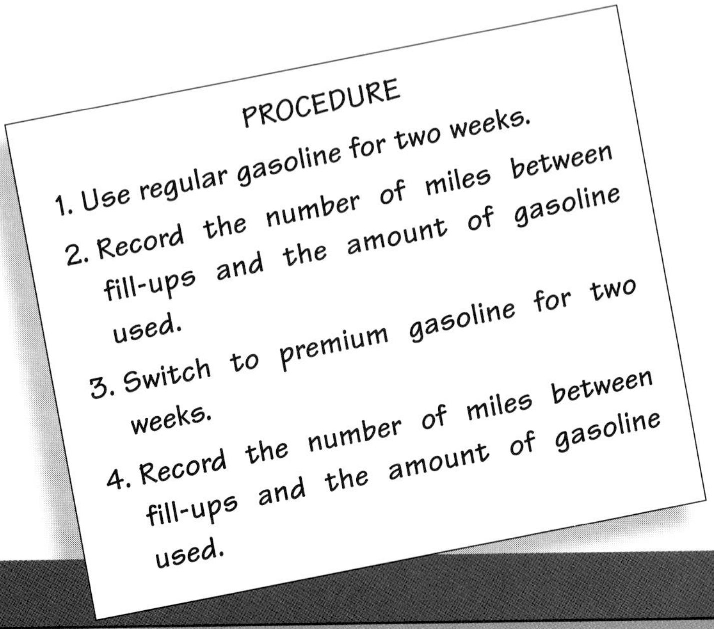

PROCEDURE

1. Use regular gasoline for two weeks.
2. Record the number of miles between fill-ups and the amount of gasoline used.
3. Switch to premium gasoline for two weeks.
4. Record the number of miles between fill-ups and the amount of gasoline used.

Figure 18
Possible procedural steps.

Figure 19

Data generated from procedure steps.

	Miles traveled	Gallons used	Gallons per mile
Regular gasoline	762	45.34	0.059
Premium gasoline	661	42.30	0.064

Separating and Controlling Variables

In any experiment, it is important to keep everything the same except for the item you are testing. The one factor that you change is called the *independent variable.* The factor that changes as a result of the independent variable is called the *dependent variable.* Always make sure that there is only one independent variable. If you allow more than one, you will not know what causes the changes you observe in the independent variable. Many experiments have *controls*—a treatment or an experiment that you can compare with the results of your test groups.

Example In the experiment with the gasoline, you made everything the same except the type of gasoline being used. The driver, the type of automobile, and the weather conditions should remain the same throughout. The gasoline should also be purchased from the same service station. By doing so, you made sure that at the end of the experiment, any differences were the result of the type of fuel being used—regular or premium. The type of gasoline was the *independent factor* and the gas mileage achieved was the *dependent factor.* The use of regular gasoline was the *control.*

Interpreting Data

The word *interpret* means "to explain the meaning of something." Look at the problem originally being explored in the gasoline experiment and find out what the data show. Identify the control group and the test group so you can see whether or not the variable has had an effect. Then you need to check differences between the control and test groups.

Figure 20

Which gasoline type is most efficient?

These differences may be qualitative or quantitative. A qualitative difference would be a difference that you could observe and describe, while a quantitative difference would be a difference you can measure using numbers. If there are differences, the variable being tested may have had an effect. If there is no difference between the control and the test groups, the variable being tested apparently has had no effect.

Example Perhaps you are looking at a table from an experiment designed to test the hypothesis: If premium gas is more efficient, then it should increase the fuel mileage of our family car. Look back at **Figure 19** showing the results of this experiment. In this example, the use of regular gasoline in the family car was the control, while the car being fueled by premium gasoline was the test group.

Data showed a quantitative difference in efficiency for gasoline consumption. It took 0.059 gallons of regular gasoline to travel one mile, while it took 0.064 gallons of the premium gasoline to travel the same distance. The regular gasoline was more efficient; it increased the fuel mileage of the family car.

What are data? In the experiment described on these pages, measurements were taken so that at the end of the experiment, you had something concrete to interpret. You had numbers to work with. Not every experiment that you do will give you data in the form of numbers. Sometimes, data will be in the form of a description. At the end of a chemistry experiment, you might have noted that one solution turned yellow when treated with a particular chemical, and another remained clear, like

Figure 21

Data.

water, when treated with the same chemical. Data, therefore, are stated in different forms for different types of scientific experiments.

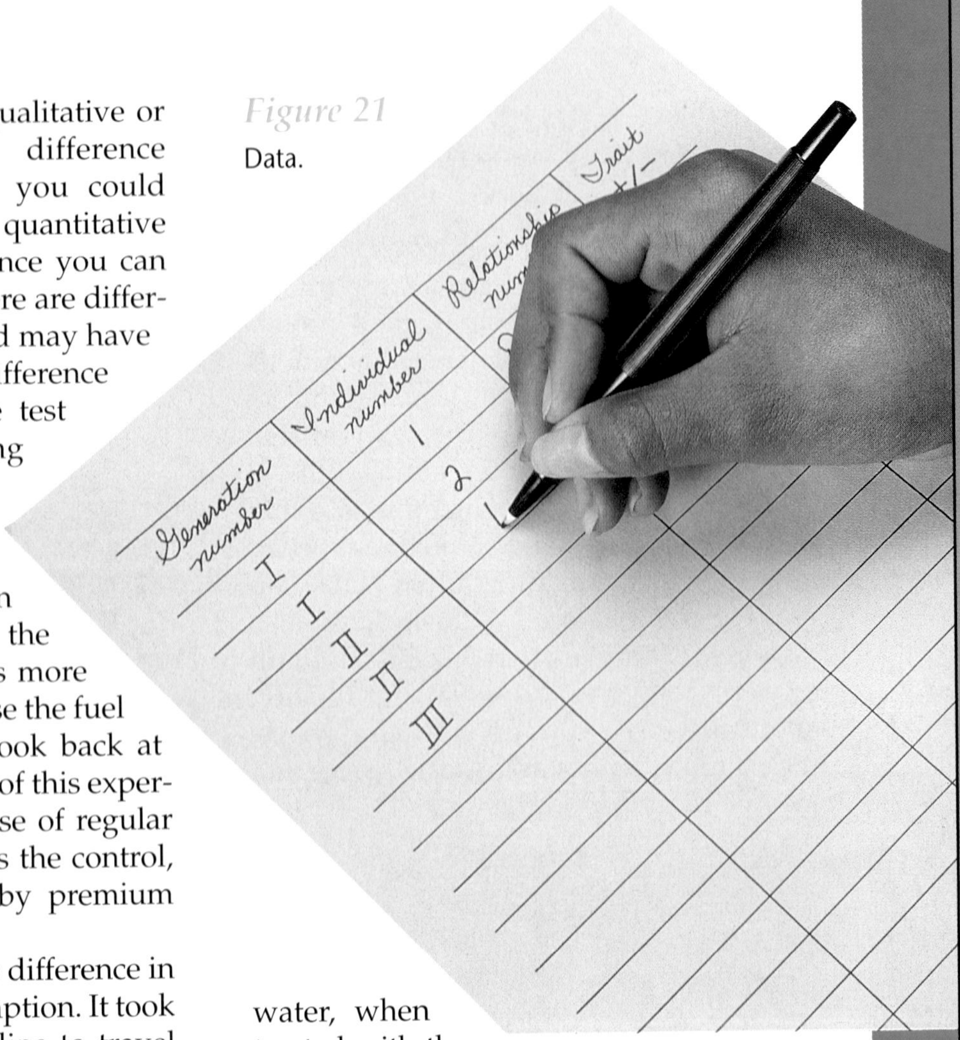

Are all experiments alike? Keep in mind as you perform experiments in science that not every experiment makes use of all of the parts that have been described on these pages. For some, it may be difficult to design an experiment that will always have a control. Other experiments are complex enough that it may be hard to have only one dependent variable. Real scientists encounter many variations in the methods that they use when they perform experiments. The skills in this handbook are here for you to use and practice. In real situations, their uses will vary.

Representing and Applying Data

Interpreting Scientific Illustrations

As you read a science textbook, you will see many drawings, diagrams, and photographs. Illustrations help you to understand what you read. Some illustrations are included to help you understand an idea that you can't see easily by yourself. For instance, we can't see atoms, but we can look at a diagram of an atom and that helps us to understand some things about atoms. Seeing something often helps you remember more easily. Illustrations also provide examples that clarify difficult concepts or give additional information about the topic you are studying. Maps, for example, help you to locate places that may be described in the text.

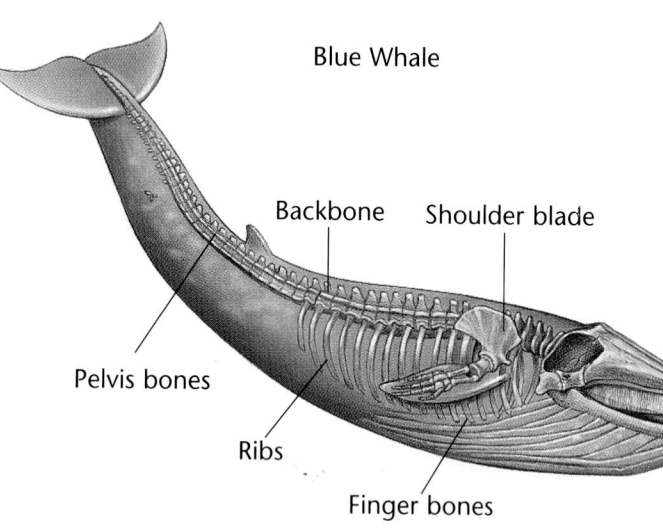

Blue Whale

Figure 22

A labeled diagram of a blue whale.

Figure 23

If an inanimate object had dorsal and ventral surfaces, its various sides would be marked for orientation like the one in this drawing.

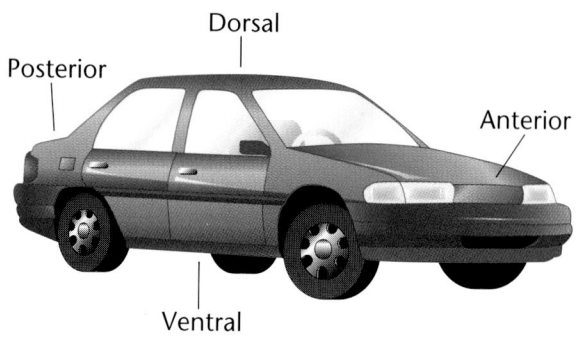

Examples

Captions and Labels Most illustrations have captions. A caption is a comment that identifies or explains the illustration. Diagrams, such as **Figure 22,** often have labels that identify parts of the organism or the order of steps in a process.

Learning with Illustrations An illustration of an organism shows that organism from a particular view or orientation. In order to understand the illustration, you may need to identify the front (anterior) end, tail (posterior) end, the underside (ventral), and the back (dorsal) side of the object shown in **Figure 23.**

You might also check for symmetry. A shark in **Figure 24** has bilateral symmetry. This means that drawing an imaginary line through the center of the animal from the anterior to posterior end forms two mirror images.

Radial symmetry is the arrangement of similar parts around a central point. An

Figure 24

A shark (A) illustrating bilateral symmetry and a pear (B) illustrating a longitudinal section and a cross section.

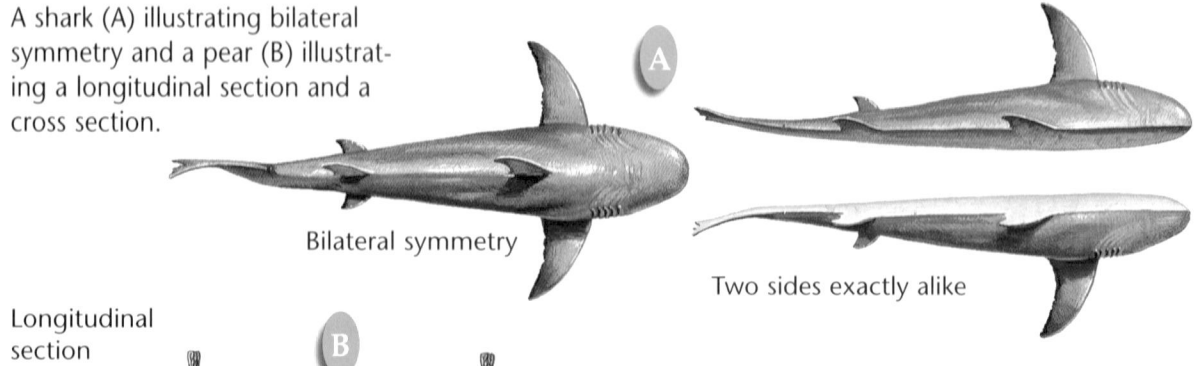

Bilateral symmetry

Two sides exactly alike

Longitudinal section

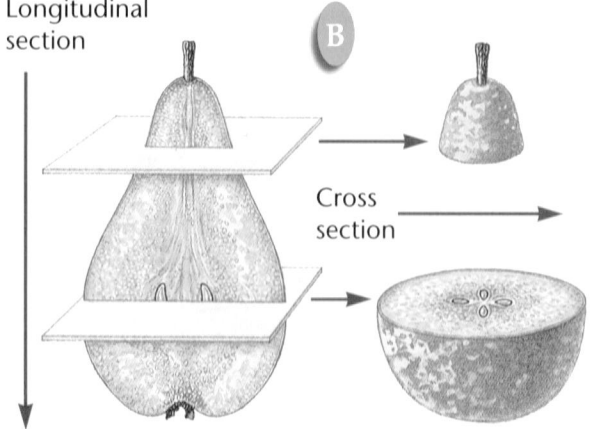

Cross section

object or organism such as a hydra can be divided anywhere through the center into similar parts.

Some organisms and objects cannot be divided into two similar parts. If an organism or object cannot be divided, it is asymmetrical. Regardless of how you try to divide a natural sponge, you cannot divide it into two parts that look alike.

Some illustrations enable you to see the inside of an organism or object. These illustrations are called sections. **Figure 24** also illustrates some common sections.

Look at all illustrations carefully. Read captions and labels so that you understand exactly what the illustration is showing you.

Making Models

Have you ever worked on a model car or plane or rocket? These models look, and sometimes work, much like the real thing, but they are often on a different scale than the real thing. In science, models are used to help simplify large or small processes or structures that otherwise would be difficult to see and understand. Your understanding of a structure or process is enhanced when you work with materials to make a model that shows the basic features of the structure or process.

Example In order to make a model, you first have to get a basic idea about the structure or process involved. You decide to make a model to show the differences in size of arteries, veins, and capillaries. First, read about these structures. All three are hollow tubes. Arteries are round and thick. Veins are flat and have thinner walls than arteries. Capillaries are small.

Now, decide what you can use for your model. Common materials are often best and cheapest to work with when making models. As illustrated in **Figure 25** on page 750, different kinds and sizes of pasta might work for these models. Different sizes of rubber tubing might do just as well. Cut and glue the different noodles or tubing onto thick paper so the openings can be seen. Then label each. Now you have a simple, easy-to-understand model showing the differences in size of arteries, veins, and capillaries.

What other scientific ideas might a model help you to understand? A model

Figure 25

Different types of pasta may be used to model blood vessels.

of a molecule can be made from gumdrops (using different colors for the different elements present) and toothpicks (to show different chemical bonds). A working model of a volcano can be made from clay, a small amount of baking soda, vinegar, and a bottle cap. Other models can be devised on a computer. Some models are mathematical and are represented by equations.

Measuring in SI

The metric system is a system of measurement developed by a group of scientists in 1795. It helps scientists avoid problems by providing standard measurements that all scientists around the world can understand. A modern form of the metric system, called the International System, or SI, was adopted for worldwide use in 1960.

The metric system is convenient because unit sizes vary by multiples of 10. When changing from smaller units to larger units, divide by 10. When changing from larger units to smaller, multiply by 10. For example, to convert millimeters to centimeters, divide the millimeters by 10. To convert 30 millimeters to centimeters, divide 30 by 10 (30 millimeters equal 3 centimeters).

Prefixes are used to name units. Look at **Figure 26** for some common metric prefixes and their meanings. Do you see how the prefix *kilo-* attached to the unit *gram* is *kilogram,* or 1000 grams? The prefix *deci-* attached to the unit *meter* is *decimeter,* or one-tenth (0.1) of a meter.

Examples

Length   You have probably measured lengths or distances many times. The meter is the SI unit used to measure length. A baseball bat is about one meter long. When measuring smaller lengths, the meter is divided into smaller units called centimeters and millimeters. A centimeter is one-hundredth (0.01) of a meter, which is about the size of the width

Figure 26

Common metric prefixes.

Metric Prefixes			
Prefix	**Symbol**	**Meaning**	
kilo-	k	1000	thousand
hecto-	h	100	hundred
deka-	da	10	ten
deci-	d	0.1	tenth
centi-	c	0.01	hundredth
milli-	m	0.001	thousandth

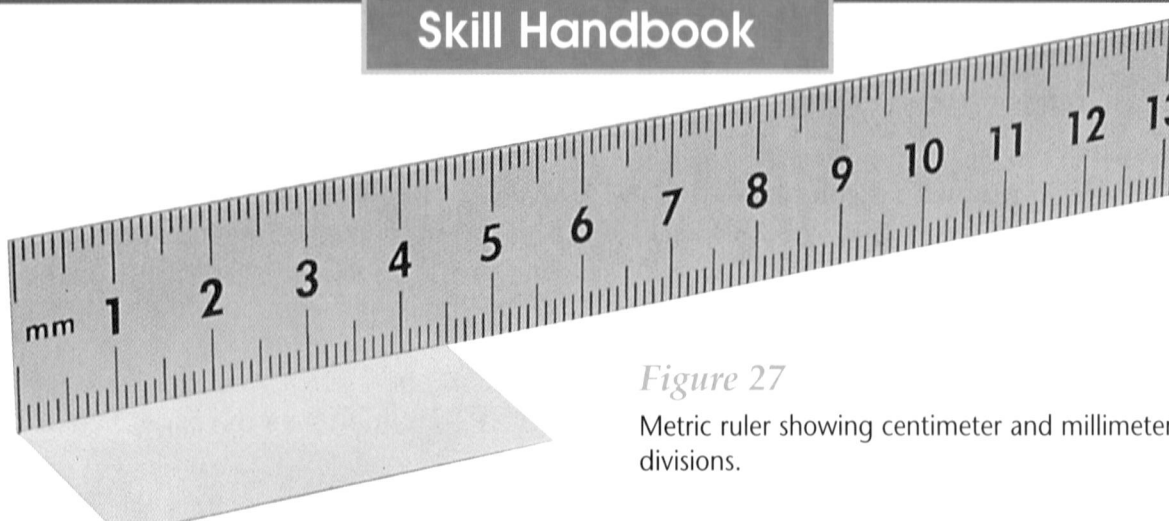

Figure 27

Metric ruler showing centimeter and millimeter divisions.

of the fingernail on your ring finger. A millimeter is one-thousandth of a meter (0.001), about the thickness of a dime.

Most metric rulers have lines indicating centimeters and millimeters, as shown in **Figure 27.** The centimeter lines are the longer, numbered lines; the shorter lines are millimeter lines. When using a metric ruler, line up the 0-centimeter mark with the end of the object being measured, and read the number of the unit where the object ends, in this instance 4.5 cm.

Surface Area Units of length are also used to measure surface area. The standard unit of area is the square meter (m^2). A square that's one meter long on each side has a surface area of one square meter. Similarly, a square centimeter, (cm^2), shown in **Figure 28,** is one centimeter long on each side. The surface area of an object is determined by multiplying the length times the width.

Volume The volume of a rectangular solid is also calculated using units of length. The cubic meter (m^3) is the standard SI unit of volume. A cubic meter is a cube one meter on each side. You can determine the volume of rectangular solids by multiplying length times width times height.

Liquid Volume During science activities, you will measure liquids using beakers and graduated cylinders marked in milliliters, as illustrated in **Figure 29.** A graduated cylinder is a cylindrical container marked with lines from bottom to top.

Liquid volume is measured using a unit called a liter. A liter has the volume of 1000 cubic centimeters. Because the prefix *milli-* means thousandth (0.001), a milliliter equals one cubic centimeter. One

Figure 28

A square centimeter.

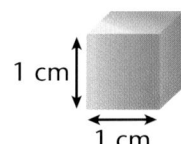

1 cm

1 cm

Figure 29

A volume of 79 mL is measured by reading at the lowest point of the curve.

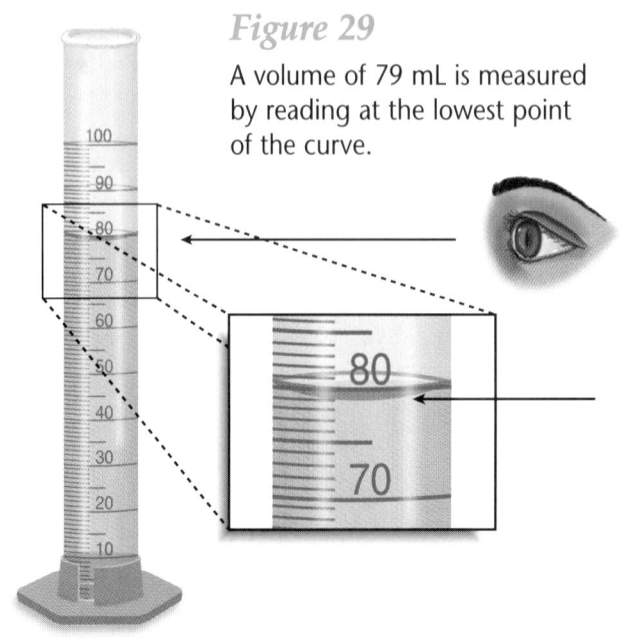

milliliter of liquid would completely fill a cube measuring one centimeter on each side.

Mass Scientists use balances to find the mass of objects in grams. You will use a beam balance similar to **Figure 30.** Notice that on one side of the balance is a pan and on the other side is a set of beams. Each beam has an object of a known mass called a *rider* that slides on the beam.

Before you find the mass of an object, set the balance to zero by sliding all the riders back to the zero point. Check the pointer on the right to make sure it swings an equal distance above and below the zero point on the scale. If the swing is unequal, find and turn the adjusting screw until you have an equal swing.

Place an object on the pan. Slide the rider with the largest mass along its beam until the pointer drops below zero. Then move it back one notch. Repeat the process on each beam until the pointer swings an equal distance above and below the zero point. Add the masses on each beam to find the mass of the object.

You should never place a hot object or pour chemicals directly onto the pan.

Instead, find the mass of a clean beaker or a glass jar. Place the dry or liquid chemicals in the container. Then find the combined mass of the container and the chemicals. Calculate the mass of the chemicals by subtracting the mass of the empty container from the combined mass.

Predicting

When you apply a hypothesis, or general explanation, to a specific situation, you predict something about that situation. First, you must identify which hypothesis fits the situation you are considering.

Examples People use prediction to make everyday decisions. Based on previous observations and experiences, you may form a hypothesis that if it is wintertime, then temperatures will be lower. From past experience in your area, temperatures are lowest in February. You may then use this hypothesis to predict specific temperatures and weather for the month of February in advance. Someone could use these predictions to plan to set aside more money for heating bills during that month.

Figure 30

A beam balance is used to measure mass.

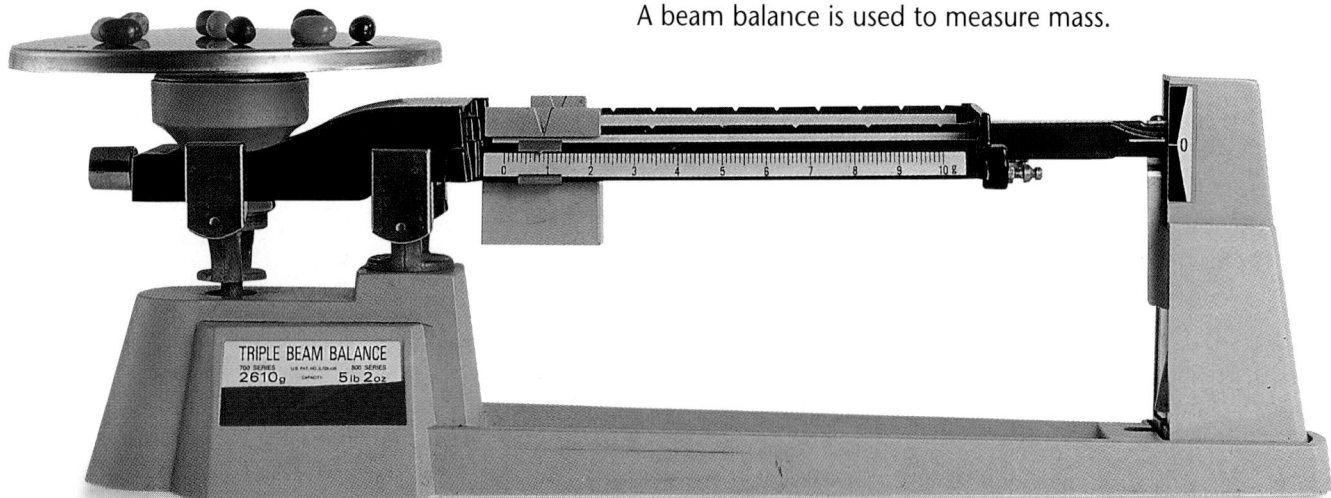

Using Numbers

When working with large populations of organisms, scientists usually cannot observe or study every organism in the population. Instead, they use a sample or a portion of the population. To sample is to take a small representative portion of organisms of a population for research. By making careful observations or manipulating variables within a portion of a group, information is discovered and conclusions are drawn that might then be applied to the whole population.

Scientific work also involves estimating. To estimate is to make a judgment about the size of something or the number of something without actually measuring or counting every member of a population.

Examples Suppose you are trying to determine the effect of a specific nutrient on the growth of black-eyed Susans. It would be impossible to test the entire population of black-eyed Susans, so you would select part of the population for your experiment. Through careful experimentation and observation on a sample of the population, you could generalize the effect of the chemical on the entire population.

Here is a more familiar example. Have you ever tried to guess how many beans

were in a sealed estimating. What i beans held one liter knew that 30 beans w 100-milliliter jar, how many be you estimate to be in the one-liter you said about 300 beans, your estim would be close to the actual number of beans. Can you estimate how many jelly beans are on the cookie sheet in **Figure 31?**

Scientists use a similar process to estimate populations of organisms from bacteria to buffalo. Scientists count the actual number of organisms in a small sample and then estimate the number of organisms in a larger area. For example, if a scientist wanted to count the number of bacterial colonies in a petri dish, a microscope could be used to count the number of organisms in a one-square-centimeter sample. To determine the total population of the culture, the number of organisms in the square-centimeter sample is multiplied by the total number of square centimeters in the culture.

Figure 31

Sampling a group of jelly beans allows for an estimation of the total number of jelly beans in the group.

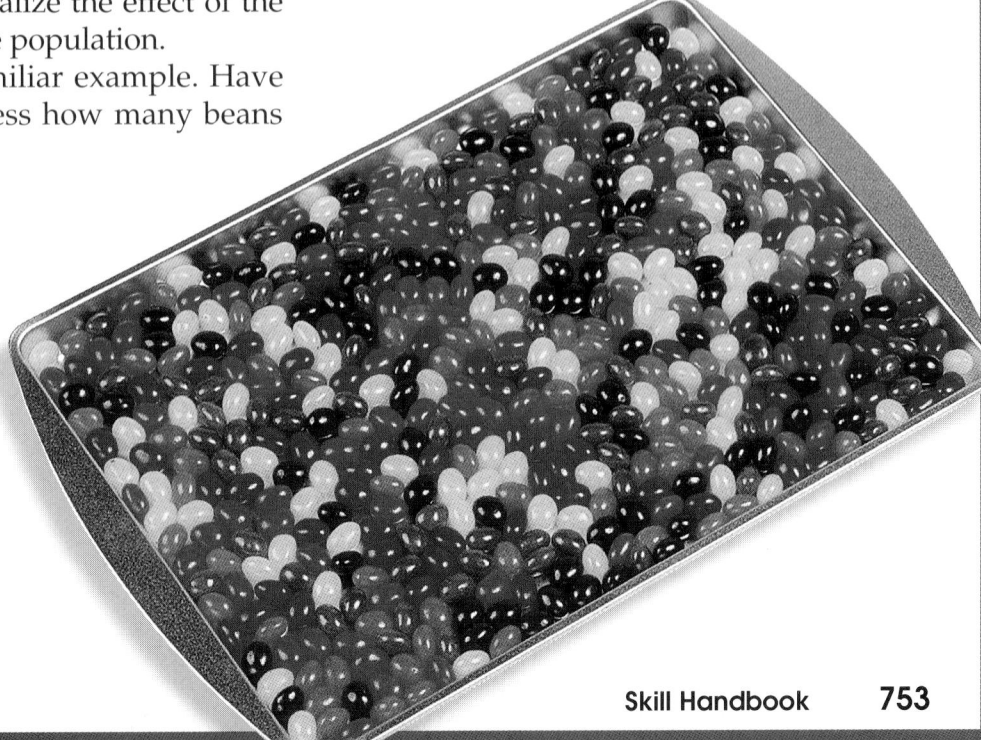

...change
...ount of
...nd time

acid: ...ions (H^+)
in soluti... ...less than 7.
(Chap. 17, p. 4...)

acid rain: Rain with a p... ...produced by
substances in the air reac... ...rainwater to
make it acidic. Acid rain is harmf... to animals and
plants and increases the rate of weathering. (Chap.
17, p. 478)

acoustics: The study of sound. (Chap. 18, p. 520)

actinide: Any of the 14 radioactive elements having
atomic numbers 90-103; used in nuclear power generation and nuclear weapons.

active solar heating: Collecting the sun's energy with
solar panels, heating water with that energy, and
storing the heated water to use the energy later.

aerosol: A liquid sprayed from a pressurized container; for example, a can of insect spray.

air resistance: Frictional force air exerts on a moving
object; acts opposite in direction to the object's
motion. The amount of air resistance depends on
an object's shape, density, speed, and size. (Chap.
4, p. 97)

alchemist: A medieval version of the modern chemist;
a practitioner who blended primitive chemistry
with magic, seeking to turn ordinary metals into
gold.

alcohol: Type of compound formed when $-OH$
groups replace one or more hydrogen atoms in a
hydrocarbon. (Chap. 13, p. 368)

allotropes: Different structural forms of the same element; for example, some carbon molecules form
soft graphite, whereas others form hard diamonds.
(Chap. 12, p. 347)

alloy: A mixture consisting of a metal and one or more
elements. (Chap. 14, p. 388)

...pha particle: A particle of nuclear radiation emitted
from a decaying atomic nucleus; has a charge of 2+,
an atomic mass of 4, and is the largest, slowest, and
least penetrating form of radiation. (Chap. 24,
p. 679)

alternating current (AC): Electric current that reverses its direction in a regular pattern; the 60-Hz AC
in our homes changes direction 120 times each
second. (Chap. 22, p. 635)

amalgam: An alloy containing the element mercury;
for example, dental fillings are amalgams. (Chap.
14, p. 391)

ammeter: A galvanometer that measures electrical
current passing through in amperes; connected in
series with the circuit. (Chap. 22, p. 630)

amorphous: Something that has no specific shape; for
example, a liquid or gas.

ampere: The unit for measuring current, the rate of
flow of electrons in a circuit; abbreviated A;
1 ampere = 1 coulomb of charge flowing past a
point in 1 second.

amplification: The process of increasing the strength
of an electric signal. (Chap. 23, p. 651)

amplitude: In a wave, the distance from the rest position of the medium to either the crest or trough.
(Chap. 18, p. 499)

amplitude-modulated (AM) waves: Radio waves
whose amplitude is varied with voice, music,
video, or data for transmission over long distances.

angle of incidence: In waves, the angle formed by the
incident wave and the normal (perpendicular);
labeled i.

angle of reflection: In waves, the angle formed by the
reflected wave and the normal (perpendicular);
labeled r.

anhydrous: A chemical compound that normally has
water molecules attached to its ions but from which
the water has been removed.

antacid: An "anti-acid," or a chemical that changes an
acid substance to a neutral substance.

antifreeze: A solute added to a solvent to lower the
temperature at which the solvent will freeze; for
example, the antifreeze added to water in a car's
cooling system.

aqueous: Describes a solution made with water; for
example, ammonia water is an aqueous solution of
ammonia gas.

Archimedes' principle: The Greek mathematician
Archimedes stated that the buoyant force on an
object in a fluid is equal to the weight of the fluid
displaced by the object. (Chap. 8, p. 235)

aromatic compounds: Chemical compounds that contain the benzene ring structure; most have distinctive odors. (Chap. 13, p. 366)

artificial satellite: Human-made device that orbits Earth; used for communication, weather-monitoring, military, and scientific purposes. (Chap. 4, p. 108)

atomic number: The number of protons in an atom's nucleus. (Chap. 10, p. 271)

average atomic mass: The average mass of an element is the average mass of the mixture of its isotopes. (Chap. 10, p. 280)

average speed: A rate of motion determined by dividing the total distance traveled by the total travel time. (Chap. 3, p. 65)

balance: A device used in laboratories to measure mass; it works by balancing a mass to be determined with a standard mass that is known.

balanced chemical equation: A chemical equation that has the same number of atoms of each element on both sides of the equation. (Chap. 16, p. 448)

balanced forces: Forces that are equal in size and opposite in direction; balanced forces acting on a body do not change the motion of the body. (Chap. 3, p. 78)

bar graph: A type of graph used to show information collected by counting; uses vertical or horizontal bars of different lengths to help people compare quantities.

base: A substance that produces hydroxide ions (OH^-) in solution; basic solutions have a pH over 7. (Chap. 17, p. 470)

Bernoulli's principle: The Swiss scientist Daniel Bernoulli stated that as the velocity of a fluid increases, the pressure exerted by the fluid decreases. (Chap. 8, p. 237)

beta particle: A negatively charged electron or positively charged positron emitted from a decaying atomic nucleus; beta particles are faster moving and more penetrating than alpha particles. (Chap. 24, p. 680)

BHA: A food preservative; it is a chemical inhibitor that slows the chemical reactions that spoil food (*Butylated Hydroxy Anisole*).

BHT: A food preservative; it is a chemical inhibitor that slows the chemical reactions that spoil food (*Butylated Hydroxy Toluene*).

binary compound: A chemical compound composed of two elements; for example, sodium chloride. (Chap. 11, p. 315)

biodegradable: Describes a material that can be decomposed into a harmless form by the biological action of organisms such as bacteria.

biogas: Mixture of gases, mostly methane, produced when biomass is allowed to rot in the absence of air. (Chap. 13, p. 370)

biomass: Organic material from such sources as wood, corn, and wastes from animals and crops; biomass is the source of biogas and the ethanol used in gasohol. (Chap. 13, p. 370)

bionics: The science of designing artificial replacements (prostheses) for parts of the human body that are not working properly. (Chap. 7, p. 196)

boiling point: The temperature at which vapor bubbles form in a liquid and rise to the surface, increasing evaporation.

Boyle's law: British scientist Robert Boyle stated that the volume of a gas decreases when the pressure increases, provided the temperature stays the same. (Chap. 8, p. 229)

brittleness: A characteristic of certain materials that makes them break easily when stressed.

bubble chamber: Device filled with superheated liquid; used to detect and monitor the path of charged nuclear particles, which leave a trail of bubbles as they pass through the chamber. (Chap. 24, p. 686)

buoyant force: Ability of a fluid to exert an upward force on an object immersed in the fluid. (Chap. 8, p. 234)

butane: A flammable gas (C_4H_{10}); a part of natural gas.

byte: A basic unit of computer memory that represents a character (number, symbol, or alphabet letter); consists of 8 bits.

calorimeter: An instrument used to measure changes in thermal energy.

carbohydrate: An organic compound having twice as many hydrogen atoms as oxygen atoms. (Chap. 13, p. 376)

carbon-14 dating: Age-determining method for carbon-containing objects up to 50 000 years old; works by calculating the amount of carbon-14 remaining in the object.

catalyst: A substance that speeds a chemical reaction without itself being permanently changed. (Chap. 16, p. 456)

cathode-ray tube (CRT): Sealed glass vacuum tube that uses electrons and fluorescent material to produce images on a screen. (Chap. 23, p. 656)

central processing unit (CPU): The main circuit board inside a computer that performs the calculating and holds the main memory.

centripetal acceleration: Acceleration toward the center of a circle by an object moving along a circular path. (Chap. 4, p. 102)

centripetal force: The force that causes an object moving along a circular path to move toward the center of the path.(Chap. 4, p. 102)

ceramic: A material made from dried clay or claylike mixtures. (Chap. 14, p. 392)

cermet: A tough, heat-resistant material that has the properties of both a ceramic and an alloy; ceramic-metal (*cer*amic + *met*al). (Chap. 14, p. 394)

chain reaction: A continuing series of fission reactions in which neutrons from fissioning nuclei cause other nuclei to split, releasing more neutrons, which split more nuclei, and so on. (Chap. 24, p. 690)

Charles's law: French scientist Jacques Charles stated that the volume of a gas increases when temperature increases, provided the pressure stays the same. (Chap. 8, p. 230)

chemical bond: The force that holds together the atoms in a compound; chemical bonding occurs because atoms of most elements become more stable by losing, gaining, and sharing electrons. (Chap. 11, p. 301)

chemical change: The change of substances to different substances. (Chap. 9, p. 257)

chemical formula: A precise statement that tells which elements are in a compound and their ratios. (Chap. 11, p. 300)

chemical property: A characteristic of a substance that indicates whether it can undergo a specific chemical change. (Chap. 9, p. 259)

chemical reaction: A change in which one or more substances are converted to different substances (the reactants are changed into the products). (Chap. 16, p. 442)

chemical symbol: A shorthand way to write the name of an element; for example: C for carbon, Ag for silver. (Chap. 10, p. 270)

chemically stable: Describes an atom whose outer energy level is completely filled with electrons. (Chap. 11, p. 300)

chemically unstable: Describes an atom whose outermost energy level is not filled with electrons so it seeks electrons from other atoms and thus forms compounds; for example: sodium, chlorine.

chloro-: Prefix that indicates presence of chlorine, as in tetrachloroethylene or chlorofluorocarbon.

chlorofluorocarbon (CFC): A group of compounds whose decomposition releases chlorine atoms that destroy ozone molecules in the upper atmosphere; used, for example, in air-conditioning systems. (Chap. 16, p. 447)

circuit: A closed path through which electrons (electricity) flow. (Chap. 21, p. 601)

circuit breaker: A device that protects an electrical circuit; if too much current flows, the circuit breaker opens the circuit, stopping the current.

cloud chamber: Device filled with water- or ethanol-saturated air; used to detect charged nuclear particles, which leave a trail as they pass through. (Chap. 24, p. 686)

coagulation: Process that destroys colloid structure; can be used to reduce a colloidal form of air pollution. (Chap. 9, p. 253)

coal: A rock formed of ancient decayed plants; burned as a fossil fuel.

coefficient: In a chemical equation, the number that represents the number of units of each substance taking part in a chemical reaction. (Chap. 16, p. 445)

coherent light: A beam of light in which all the electromagnetic waves travel with the crests and troughs aligned; thus, the beam does not spread out. (Chap. 20, p. 575)

colloid: A heterogeneous mixture containing tiny particles that never settle out; for example, milk and gelatin. (Chap. 9, p. 249)

combustion: Rapid burning. (Chap. 6, p. 166)

composite: A mixture of two materials, one of which is embedded in the other. (Chap. 14, p. 401)

compound: Substance made of the combined atoms of two or more elements. (Chap. 9, p. 246)

compound machine: A combination of two or more simple machines. (Chap. 7, p. 198)

compression: In compressional waves, the dense area of the wave.

compressional wave: A type of wave where matter vibrates in the same direction as the wave travels. (Chap. 18, p. 505)

computer: A device you can program to do calculations, make logical decisions, and manipulate data.

computer virus: Type of program designed to infect a computer, erase data, scramble other programs, or fill up so much memory that the system is harmed. (Chap. 23, p. 666)

concave lenses: Lenses that are thinner in the middle and thicker at the edges and thus curve inward; form virtual, upright, smaller images of an object. (Chap. 20, p. 565)

concave mirror: A mirror whose surface curves inward; produces real images. (Chap. 20, p. 558)

concentrated solution: A solution in which the amount of solute is near the maximum the solvent can hold at that temperature.

concentration: Generally, the proportion of a solute dissolved in a solvent.

condensation: The change of a substance from a gas to a liquid, which usually takes place when a gas is cooled to or below its boiling point. (Chap. 8, p. 225)

condense: To go from the gas state to the liquid state, due to a loss of heat.

conduction: The transfer of energy through matter in which energy moves from particle to particle; conduction takes place more easily in solids than in liquids and gases. (Chap. 6, p. 152)

conductor: (electrical) A material that allows electrons to move easily through it. (Chap. 21, p. 594)

constant: In an experiment, a factor that doesn't change. (Chap. 1, p. 21)

constant speed: Speed that does not change. (Chap. 3, p. 65)

contraction: Movement of molecules toward one another, so that they occupy a smaller space.

control: In an experiment, a standard for comparison that is often needed to draw a meaningful conclusion. (Chap. 1, p. 21)

convection: The transfer of energy by the bulk movement of matter in which particles move from place to place in a fluid, carrying the energy with them. (Chap. 6, p. 153)

convex lenses: Lenses that are thicker in the middle than at the edges; can produce both real and virtual images. (Chap. 20, p. 564)

convex mirror: A mirror with a surface that curves outward; produces upright, smaller, virtual images of an object. (Chap. 20, p. 561)

corrosive: Hazardous compound that attacks and alters metals, human tissue, or other materials; for example, oven cleaners and battery acid. (Chap. 11, p. 312)

covalent bond: A type of chemical bond formed by atoms when they share electrons; covalent bonding produces molecules. (Chap. 11, p. 306)

coulomb: The charge carried by 6.24 billion billion electrons.

crest: The highest point of a wave. (Chap. 18, p. 499)

critical temperature: In superconductors, the very low temperature at which a material ceases to have any electrical resistance.

crystals: In most solids, the arrangements of particles in repeating geometric patterns. (Chap. 8, p. 215)

current: The flow of electrons through a wire or any conductor; measured in amperes (A) with an ammeter. (Chap. 21, p. 601)

deceleration: The rate of change in velocity (speed and/or direction) when velocity is decreasing; also called negative acceleration.

decibel: The unit of measure for sound intensity, abbreviated dB. The faintest sound most people can hear = 0 dB.

decomposition reaction: A chemical reaction in which a substance breaks down (decomposes) into two or more simpler substances. (Chap. 16, p. 452)

dehydrating agent: A substance that can remove water from materials. (Chap. 17, p. 468)

density: The mass per unit volume of a material (g/cm^3); describes how tightly packed a sub-

stance's molecules are. (Chap. 2, p. 43)

dependent variable: In an experiment, the factor whose value changes because of a change in the independent variable. (Chap. 1, p. 22)

derived unit: Unit of measurement obtained by combining SI units; for example, density is a derived unit formed by dividing an object's mass by its volume. (Chap. 2, p. 42).

detergent: An organic salt similar to soap, except that detergents do not form soap scum in hard water. (Chap. 17, p. 483)

diatomic molecule: A molecule composed of two atoms of the same element; for example, hydrogen (H_2) or oxygen (O_2). (Chap. 12, p. 341)

diesel engine: An internal combustion engine that compresses a fuel-air mixture so much that it ignites from the heat of compression without a spark.

diffraction: The bending of waves around a barrier. (Chap. 19, p. 548)

diffraction grating: A piece of glass or plastic with many parallel slits that acts like a prism, causing white light that passes through it to separate into its component colors (the visible spectrum). (Chap. 19, p. 550)

dilute solution: A solution in which the amount of solute is much less than the maximum the solvent can hold at that temperature.

diode: A type of rectifier that allows electric current to flow in only one direction. (Chap. 23, p. 651)

direct current (DC): Electrical current that flows in only one direction through a wire. (Chap. 22, p. 635)

disinfectant: A chemical that kills bacteria, such as alcohol.

dissociation: The breaking apart of an ionic compound (such as salt) into positive and negative ions when dissolved in water. (Chap. 15, p. 433)

DNA: *DeoxyriboNucleic Acid*; an acid in the nuclei of cells that codes and stores genetic information.

doping: Adding an impurity to a semiconductor to increase its electrical conductivity.

Doppler effect: An increase or decrease in wave frequency, caused by motion of the source and/or motion of the observer; applies to all waves (including light, heat, sound, and so on).

dot diagram: A diagram to represent electrons in the outer energy level of an atom; uses the element symbol and dots. (Chap. 10, p. 288)

double-displacement reaction: A chemical reaction in which two ionic compounds in solution react, forming a precipitate, gas, or water. (Chap. 16, p. 453)

dry cell: A power source that acts as an electron pump and generates electric current by a chemical reaction; "dry" because it uses a thick, pasty electrolyte. (Chap. 21, p. 602)

ductile: Ability of metals to be pulled into wires. (Chap. 12, p. 330)

efficiency: The ratio of the useful work put out by a machine to the work put into that machine; efficiency is always less than 100 percent. (Chap. 7, p. 200)

effort arm: The part of a lever on which an effort force (F_e) is applied. (Chap. 7, p. 186)

effort force: The force applied to a machine when a machine is used to do work. (Chap. 7, p. 181)

electric field: An area surrounding an electron that exerts a force on anything nearby with an electric charge; is strongest nearest the electron and weakens with distance. (Chap. 21, p. 594)

electric motor: A device that contains a rotating electromagnet that changes electrical energy to mechanical energy. (Chap. 22, p. 630)

electrical power: The rate at which electrical energy is converted to another form of energy; electrical power is expressed in watts (W) or kilowatts (kW). (Chap. 21, p. 614)

electrolyte: A substance that separates or forms ions in a water solution, making the solution an electrical conductor. (Chap. 15, p. 433)

electromagnet: Strong temporary magnet made by inserting an iron core into a wire coil and passing an electric current through the coil. (Chap. 22, p. 628)

electromagnetic induction: Process by which electrical current is induced in a wire when it is moved through a magnetic field. (Chap. 22, p. 633)

electromagnetic radiation: Transverse waves that transfer energy by radiation; vary in length from very long radio waves to extremely short gamma waves. (Chap. 19, p. 528)

electron arrangement: In an atom, how the electrons are distributed in the atom's various energy levels.

electron cloud: Region where electrons most probably are found surrounding the nucleus of an atom. (Chap. 10, p. 272)

electrons: Negatively charged particles that move around the nucleus of an atom. (Chap. 10, p. 270)

electroscope: A device containing two suspended metal leaves in a jar that move apart when charged; used to detect the presence of electric charges. (Chap. 21, p. 596)

element: Substance in which all the atoms in a sample are alike. (Chap. 9, p. 246)

endothermic reaction: A chemical reaction in which energy is absorbed. (Chap. 16, p. 458)

energy: The ability to cause change. (Chap. 5, p. 124)

energy farming: The growing of plants for use as fuel. (Chap. 13, p. 370)

energy transfer: The movement of energy from one object to another; for example, thermal energy flowing as heat from a heated stove to a skillet.

ester: An organic compound formed by reacting an organic acid with an alcohol; some esters provide scents to flowers and fruits and flavors to gelatin and candy. (Chap. 17, p. 484)

evaporation: The gradual change of a substance from a liquid to a gas at temperatures below the boiling point. (Chap. 8, p. 224)

exothermic reaction: A chemical reaction in which energy is released. (Chap. 16, p. 456)

expansion: Outward movement of molecules away from one another so that they occupy a larger space.

experiment: An organized procedure for testing a hypothesis. An experiment typically has a control and dependent and independent variables. (Chap. 1, p. 20)

external combustion engine: An engine in which the fuel is burned outside the engine; for example, an old-fashioned locomotive steam engine. (Chap. 6, p. 167)

factor: A condition that affects the outcome of an event.

farsighted: Describes a person who sees faraway things clearly, but has trouble focusing on nearby objects.

fiberglass: Hairlike strands of glass that make a good insulator when arranged in puffy layers.

filter: In working with light, a device that allows one or more colors to be transmitted while others are absorbed or blocked.

flame test: A laboratory test to identify elements by heating a substance in a flame and observing the flame color.

flammable: A chemical characteristic of a substance that allows it to oxidize rapidly; also called burnable.

flash distillation: A desalination process where seawater is pumped into a vacuum chamber so it boils quickly and at a lower temperature (because the normal air pressure is absent).

fluid: Any material that flows; liquids and gases are fluids. (Chap. 6, p. 153)

fluorescence: Occurs when a material absorbs ultraviolet radiation that stimulates it to radiate visible light.

fluorescent light: Light produced when ultraviolet radiation inside a fluorescent light bulb causes its fluorescent coating to glow; uses less energy than incandescent lights. (Chap. 19, p. 544)

focal length: The distance from the center of a lens or mirror to its focal point. (Chap. 20, p. 559)

focal point: A point on the optical axis of a concave mirror or convex lens where the light rays come together. (Chap. 20, p. 559)

force: A push or a pull one body exerts on another. Unbalanced forces always change the motion of the body. (Chap. 3, p. 78)

fractional distillation: A process based on boiling points used in oil refineries to separate the chemical compounds in crude oil into gasoline, kerosene, and other products. (Chap. 25, p. 700)

fractionating towers: Towers at oil refineries used for fractional distillation of petroleum.

free-fall: How an object moves in space when it is influenced only by gravity.

Freon: A refrigerant gas used in refrigerators and air conditioners.

frequency: The number of wave crests that pass a point during one second; expressed in hertz (Hz). (Chap. 18, p. 499)

frequency-modulated (FM) waves: Radio waves whose frequency is varied with voice, music, video, or data for transmission over long distances.

friction: The force that opposes motion between two surfaces that are touching each other. The amount of friction depends on the types of surfaces and the kinds of force pressing the surfaces together. (Chap. 3, p. 82)

fuel rod: A metal rod filled with uranium pellets, used as the fuel in a nuclear reactor.

fulcrum: The fixed point around which a lever pivots. (Chap. 7, p. 186)

fuse: A device that protects an electrical circuit. If too much current flows, the fuse melts, breaking the circuit to stop current flow.

galvanometer: An instrument (meter) used to detect electric currents; most electrical meters are galvanometers.

gamma rays: High-frequency electromagnetic waves that travel at the speed of light, have no mass or charge, and are the most penetrating form of radiation. (Chap. 19, p. 534; Chap. 24, p. 681)

gaseous solution: A homogeneous gas that is composed of two or more gases.

gasohol: A mixture of ethanol and gasoline that is a useful substitute for gasoline, but whose production may be damaging to the environment; a biomass fuel. (Chap. 13, p. 371)

gear: A wheel with teeth around its edge designed to mesh with teeth on another gear so as to transfer force and motion.

gelatin: A substance obtained by boiling animal bones; used in glues and goods.

generator: A device that uses electromagnetic induction to induce electrical current by rotating loops of wire through a magnetic field. (Chap. 22, p. 633)

geothermal energy: Thermal energy source located far below Earth's crust. (Chap. 25, p. 716)

glass: A ceramic mixture with no regular crystal structure. (Chap. 14, p. 394)

graduated cylinder: A cylinder marked with a volume scale, used in laboratories for measuring liquid volumes.

granite: An igneous rock that is a mixture of crystals of several different compounds.

graph: A visual display of information or data—such as line graphs, bar graphs, and circle graphs—organized to help people interpret, understand, or quickly find information. (Chap. 2, p. 48)

graphite: A mineral made of carbon atoms arranged in layers that easily slide past one another, forming a dry lubricant.

gravity: Force exerted by every object in the universe on every other object. The amount of gravitational force depends on the masses of the objects and the distance between them. (Chap. 3, p. 83)

grounded: Electrically connected to Earth (the ground), either directly or through a wire or other metal object.

groundwater: Water in the ground that comes from rain and melting snow. Although often pure enough to drink, it is easily polluted by dumps, sewers, and chemical spills.

group: In the periodic table, each of the 18 vertical columns of elements; each group (family) is made up of elements with similar properties. (Chap. 10, p. 288)

hacker: A person who uses a computer to break into other computer systems without permission.

half-life: The amount of time required for one-half of the nuclides in a sample of radioactive isotope to decay. (Chap. 24, p. 681)

halogens: Highly active elements in periodic table Group 17 (fluorine, chlorine, bromine, iodine, astatine); halogens have seven electrons in their outer shells and readily combine with Group 1 elements such as sodium.

heat: Thermal energy that flows from a warmer material to a cooler material. (Chap. 5, p. 136)

heat engine: A device that converts thermal energy that is produced by burning fuel into mechanical energy; for example, an automobile engine. (Chap. 6, p. 166)

heat mover: A device that moves thermal energy from one location and releases it in another location having a different temperature; for example, a refrigerator, air conditioner, or heat pump. (Chap. 6, p. 168)

heat of fusion: The amount of energy needed to change a material from the solid state to the liquid state. (Chap. 8, p. 226)

heat of vaporization: The amount of energy needed to change a material from a liquid to a gas. (Chap. 8, p. 226)

herbicide: A chemical poison that kills undesirable plants.

hertz: The unit of measure for frequency, abbreviated Hz; for example, 440 waves per second = 440 Hz.

heterogeneous mixture: A mixture in which different parts can be easily distinguished. (Chap. 9, p. 248)

high-speed photography: A photographic method that takes clear (not blurred) pictures of fast-moving objects.

homogeneous mixture: A mixture in which different materials are blended evenly so that the mixture is the same throughout; also called a solution. (Chap. 9, p. 248)

horizontal: A direction parallel to Earth's surface.

hydrate: A compound that has water molecules chemically attached to its ions and written into its formula. (Chap. 11, p. 319)

hydraulic: Describes a system operated by applying pressure to a liquid.

hydrocarbon: A compound containing only carbon and hydrogen atoms. (Chap. 13, p. 361)

hydroelectricity: Electricity produced by the energy of moving water. (Chap. 25, p. 714)

hydronium ion: The ion (H_3O^+) that makes a solution acidic; formed when a polar molecule such as HCl is dissolved in water and the H^+ is removed from the polar molecule. (Chap. 17, p. 472)

hypothesis: A testable prediction used to see how something works or to solve a problem. (Chap. 1, p. 16)

ideal machine: A machine in which work input equals work output; such a "perfect" machine would be frictionless and 100 percent efficient. (Chap. 7, p. 181)

incandescent light: Light produced by a thin tungsten wire, or filament, that is heated in an incandescent bulb until it glows. (Chap. 19, p. 544)

inclined plane: A simple machine consisting of a sloping surface (ramp) used to raise objects. (Chap. 7, p. 193)

incoherent light: Light rays that are nearly parallel, but spread out because their electromagnetic waves do not travel in the same direction; crests and troughs do not coincide. (Chap. 20, p. 576)

independent variable: In an experiment, the factor adjusted to a different value by the experimenter to see what effect it will have on the dependent variable. (Chap. 1, p. 22)

indicator: An organic compound that changes color in an acidic solution or a basic solution. (Chap. 17, p. 466)

induction: Electrically charging an object or creating an electrical current in it, without physically touching it.

inertia: The tendency of an object to resist any change in its motion. If motionless, it tends to remain at rest; if moving, it tends to keep moving at the same speed and in the same direction. (Chap. 3, p. 79)

infrared radiation: Electromagnetic waves that have a wavelength slightly longer than visible light; indicates the presence of heat. (Chap. 19, p. 531)

infrasonic waves: Waves at frequencies below the limit of human hearing (below 20 Hz).

inhibitor: A substance that slows or prohibits a chemical reaction. (Chap. 16, p. 457)

instantaneous speed: The rate of motion at a given instant in time, such as that given on a car's speedometer. (Chap. 3, p. 64)

insulator: A material that does not allow heat or electrons to move through it easily. (Chap. 6, p. 155; Chap. 21, p. 595)

integrated circuit: A thin slice of silicon, often less than 1 cm on a side, which can contain thousands of resistors, diodes, and transistors; used in computers and electronic equipment. (Chap. 23, p. 653)

intensity: In sound waves, the amount of energy in each wave; the intensity level of a sound is measured in decibels (dB). (Chap. 18, p. 510)

interference: The ability of two or more waves to combine and form a new wave. (Chap. 18, p. 519)

internal combustion engine: An engine in which fuel is burned inside the engine in chambers (cylinders); for example, an automobile engine. (Chap. 6, p. 166)

ion: A positively or negatively charged atom. (Chap. 11, p. 304)

ionic bond: A type of chemical bond formed by the attraction between opposite charges of the ions in an ionic compound. (Chap. 11, p. 305)

ionization: The breaking apart of certain polar substances to form ions when dissolved in water. (Chap. 15, p. 433)

isomers: Compounds that have identical chemical formulas but different molecular structures and shapes. (Chap. 13, p. 364)

isotopes: Atoms of the same element that have different numbers of neutrons; for example, boron-10 and boron-11. (Chap. 10, p. 279)

J

joule (JEWL): The basic unit of energy and work, named for English scientist James Prescott Joule; 1 joule (J) = a force of 1 newton moving through 1 meter.

K

kelvin: The SI unit of temperature; zero K = absolute zero (the coldest possible temperature, $-273°C$). (Chap. 2, p. 45)

kilogram: The SI unit of mass; 1 kilogram (kg) = 1000 grams (g). (Chap 2, p. 43)

kilowatt-hour: The unit of electrical energy; 1 kilowatt-hour (kWh) = 1000 watts (W) of power used for 1 hour. (Chap. 21, p. 616)

kinetic energy: Energy in the form of motion, as in a moving car or truck; the greater the mass and velocity of a moving object, the more kinetic energy it has. (Chap. 5, p. 125)

kinetic theory of matter: The idea that all matter is made up of constantly moving, tiny particles. (Chap. 8, p. 215)

L

lanthanide: Any of the 14 metallic elements having atomic numbers 58-71; used in magnets, ceramics, and television picture tubes.

laser: A device that emits a beam of photons that travel in the same direction and phase, producing a beam of coherent light. (Chap. 20, p. 575)

law of conservation of energy: A law stating that energy can change form but cannot be created or destroyed under ordinary conditions. This law applies to systems in which energy cannot enter or exit the system. (Chap. 5, p. 129)

law of conservation of mass: A law stating that matter is neither created nor destroyed during a chemical change. (Chap. 9, p. 260)

law of conservation of momentum: A law stating that the total momentum of a group of objects is conserved unless a net force acts on the objects. (Chap. 4, p. 115)

lever: A simple machine consisting of a bar that is free to pivot (rotate) around a fixed point (fulcrum). (Chap. 7, p. 186)

line graph: A type of graph used to show trends or continuous change by drawing a line that connects data points.

lipids: Fats, oils, and related organic compounds. (Chap. 13, p. 377)

liquid solution: A liquid solvent that has dissolved in it a gas, liquid, or solid.

liter: The unit of liquid volume (L) that occupies the same volume as a cubic decimeter (dm^3) and is slightly larger than a quart. (Chap. 2, p. 42)

loudness: The human perception of sound intensity. (Chap. 18, p. 510)

lubricant: A substance used to reduce the friction between two surfaces that move together; for example, oil, grease, or graphite.

M

machine: A device that makes work easier by changing the size of the force applied to it and/or the direction of that force. (Chap. 7, p. 180)

magnetic bottle: A powerful magnetic field that creates a container ("bottle") to hold the hydrogen plasma needed for a nuclear fusion reaction.

magnetic domains: Groups of atoms with aligned magnetic poles. (Chap. 22, p. 626)

magnetic field: The region around a magnet where magnetic forces act. (Chap. 22, p. 624)

magnetic poles: The two ends of a piece of magnetic material where the magnetic forces are strongest, labeled north pole (N) and south pole (S). (Chap. 22, p. 624)

magnetism: A property of some materials in which there is a force of repulsion or attraction between certain like and unlike poles. (Chap. 22, p. 624)

magnifier: A device that makes things appear larger so that more detail can be seen; for example, a magnifying glass or microscope.

malleable: Ability of metals to be hammered or rolled into thin sheets. (Chap. 12, p. 330)

mass: A measurement of the amount of matter in a object; its SI unit is the kilogram (kg). (Chap. 2, p. 43)

mass number: Sum of the number of protons and neutrons in an atom's nucleus. (Chap. 10, p. 279)

mechanical advantage: The number of times a machine multiplies the effort force applied to it. (Chap. 7, p. 182)

mechanical energy: The total amount of kinetic energy and potential energy in a system. (Chap. 5, p. 129)

medium: A material (solid, liquid, or gas, or a combination of these) through which a wave travels. (Chap. 18, p. 498)

melt: The changing of a substance from a solid state to a liquid state when heated above the substance's freezing/melting point.

melting point: The temperature at which a solid changes to a liquid.

metallic bonding: The type of chemical bond in which positively charged ions are surrounded by freely moving electrons; metallic bonding holds metals together. (Chap. 12, p. 331)

metalloids: Elements having properties of both metals and nonmetals. In the periodic table, metalloids are found along the stair-step line that separates metals from nonmetals. (Chap. 10, p. 291)

metals: Elements usually having these common properties: shiny, good conductors of heat and electricity, are solids at room temperature. In the periodic table, metals are found at the left of the stair-step line. (Chap. 10, p. 290)

meter: The SI base unit of length (m); 100 cm = 1 m. (Chap. 2, p. 38)

microprocessor: The computer's "brain." It receives input and tells the computer how to respond. (Chap. 23, p. 659)

microscope: An optical instrument that uses two convex lenses with relatively short focal lengths to magnify small, close-up objects. (Chap. 20, p. 570)

microwaves: Radio waves with the highest frequency and energy; used in communications and microwave ovens. (Chap. 19, p. 530)

mixture: A material made of elements or compounds stirred together but not combined chemically.

model: A symbolic representation of an idea, system, or structure to make something understandable. Models help us solve problems and deal with things difficult to see because they are too large or too small. (Chap. 1, p. 12)

modulation: Process of adding voice, music, video, computer information, or other data to radio waves by using electrical currents to vary either amplitude or frequency. (Chap. 19, p. 530)

momentum: A property of any moving object; the momentum (p) of an object is the product of its mass and velocity ($p = mv$). (Chap. 4, p. 112)

monomers: Organic molecules that are strung together to form polymers.

music: Sound created using specific pitches, sound quality, and regular patterns. (Chap. 18, p. 516)

N

nearsighted: Describes a person who sees nearby things clearly, but has trouble focusing on distant objects.

net force: The sum of the forces on an object when unbalanced forces are applied to it. A net force changes the object's speed or direction or both. (Chap. 3, p. 78)

neutralization: A chemical reaction between an acid and a base; the acid's hydronium ions (H_3O^+) combine with the base's hydroxide ions (OH^-) to produce water (H_2O), and the acid's negative ions combine with the base's positive ions to produce a salt. (Chap. 17, p. 480)

neutralize: To change an acidic solution or a basic solution so that it is neutral.

neutron: Atomic particle with no charge (neutral) that is part of an atom's nucleus. (Chap. 10, p. 271)

Newton's first law of motion: Describes the relationship between velocity and forces. An object moving at a constant velocity keeps moving at that velocity unless a net force acts on it. An object at rest stays at rest unless a net force acts on it. (Chap. 3, p. 81)

Newton's second law of motion: Describes the acceleration of an object in the direction of the net force applied to it. Expressed in equation form as *force = mass × acceleration,* or *F = ma.* (Chap. 4, p. 94)

Newton's third law of motion: Describes action-reaction pairs; for every action force, there is an equal and opposite reaction force. (Chap. 4, p. 110)

NIMBY: *Not in My Backyard;* a point of view that supports an issue such as nuclear generation of electricity as long as it does not take place close to home. (Chap. 25, p. 710)

noise: Sound that has no regular pattern or definite pitch. (Chap. 18, p. 516)

nonelectrolyte: A substance, such as pure water, that does not conduct electricity. (Chap. 15, p. 433)

nonmetals: Elements that usually are gases or brittle solids at room temperature; most do not conduct heat or electricity well. In the periodic table, nonmetals are found at the right of the stair-step line. (Chap. 10, p. 291)

nonpolar molecule: A molecule that does not have oppositely charged ends. (Chap. 11, p. 307)

nonrenewable resources: Resources such as coal, oil, and natural gas, which cannot be replaced after they are used up. (Chap. 25, p. 703)

normal: In the study of light, an imaginary line drawn perpendicular to a reflecting surface or perpendicular to a medium that light is entering.

nuclear fission: Process in which an atom's nucleus is split into two nuclei with smaller masses. (Chap. 24, p. 689)

nuclear fusion: Process in which two atomic nuclei with low masses are fused into a single nucleus of larger mass; also known as a thermonuclear reaction. (Chap. 24, p. 690)

nuclear reactor: A device that generates electricity from a controlled nuclear fission chain reaction. (Chap. 25, p. 704)

nuclear waste: Radioactive by-products from nuclear power generation, nuclear medicine, and so on; storing nuclear waste is a political and environmental problem. (Chap. 25, p. 706)

nucleic acid: An organic polymer that controls the activities and reproduction of cells. (Chap. 13, p. 375)

nucleus: The positively charged center of an atom that contains protons and neutrons. (Chap. 10, p. 270)

nuclide: The nucleus of a specific isotope having a certain specific atomic number and atomic mass. (Chap. 24, p. 675)

observation: Using your senses to gather information; in science we use instruments such as microscopes and balances to carefully gather accurate and detailed observations. (Chap. 1, p. 16)

ocean thermal energy conversion: Process that uses heat engines to convert differences in ocean temperatures into mechanical energy to drive turbines. (Chap. 6, p. 173)

ohm: The unit for measuring resistance.

Ohm's law: States that potential difference = current × resistance, or V (volts) = I (amperes) × R (ohms). (Chap. 21, p. 606)

opaque materials: Materials you can't see through because they absorb or reflect all light. (Chap. 19, p. 536)

optical axis: A line perpendicular to the center of a mirror or lens.

optical fiber: Transparent glass fiber that can transmit light from one place to another. (Chap. 20, p. 578)

organic compounds: Most chemical compounds that contain the element carbon are organic compounds; more than 90 percent of all compounds are organic. (Chap. 13, p. 360)

organic solvent: A type of liquid often found in building materials; contains potentially harmful chemicals. (Chap. 15, p. 422)

oxidation number: A positive or negative number that indicates how many electrons an atom has lost, gained, or shared when bonding with other atoms. (Chap. 11, p. 314)

parallel circuit: An electrical circuit where the current flows through more than one path. If one path is interrupted, current will still flow through the other paths. (Chap. 21, p. 609)

pascal: The SI unit of pressure; 1 pascal (Pa) of pressure = 1 newton/square meter (N/m^2). (Chap. 8, p. 228)

Pascal's principle: French scientist Blaise Pascal stated that pressure applied to the fluid is transmitted unchanged through the fluid. (Chap. 8, p. 236)

passive solar heating: Direct use of the sun's energy to heat something, without using fans or mechanical devices to transfer heat from one area to another.

period: A horizontal row in the periodic table. (Chap. 10, p. 290)

periodic table: A table of the elements arranged according to repeated changes in properties. (Chap. 10, p. 284)

PET: *Positron-Emission Tomography*; PET scans are used in medicine to reveal brain function and certain brain disorders such as tumors. (Chap. 24, p. 693)

petroleum: Crude oil, formed by decayed remains of plants and animals; a fossil fuel that is burned and used to make lubricants and plastics. (Chap. 25, p. 700)

pH: A measure of hydronium ion (H$_3$O$^+$) concentration in solution; expressed on the pH scale from 0 to 14. From 7 down to 0, a solution is increasingly acidic; at 7 it is neutral (pure water); from 7 to 14, a solution is increasingly basic (pH = *p*otential of *H*ydrogen). (Chap. 17, p. 475)

phenolphthalein (feen ul THAYL een): A chemical used as a color indicator in titration; colorless in an acidic solution, but turns pink in a basic solution.

photon: A tiny particlelike bundle of radiation. (Chap. 19, p. 529)

photovoltaic cell: A device used to convert solar energy directly into electrical energy; also called a solar cell. (Chap. 25, p. 713)

physical change: A change in the size, shape, color, or state of matter. (Chap. 9, p. 256)

physical property: Any characteristic of a material that can be observed without changing the identity of the material itself. (Chap. 9, p. 254)

physical science: The study of matter and energy; topics for study include chemistry and physics. (Chap. 1, p. 8)

pickling: A process that removes impurities from the surfaces of steel and other metals by dipping them in hydrochloric acid. (Chap. 17, p. 469)

pigment: Colored material that absorbs some colors and reflects others; the colors of pigments are determined by the colors they reflect. (Chap. 19, p. 540)

pitch: The highness or lowness of a sound, which is determined by the frequency of the sound waves. (Chap. 18, p. 508)

plane mirror: A mirror with a flat surface that produces a virtual image. (Chap. 20, p. 558)

plankton: Tiny plants and animals that live in water and are food for small fish; they are easily killed by acid rain. (Chap. 17, p. 478)

plasma: A gaslike mixture of charged particles that

exists at extremely high temperatures. (Chap. 8, p. 218)

plastic: Polymer-based material than can be easily molded. (Chap. 14, p. 396)

polar molecule: A molecule with opposite charges on each end. (Chap. 11, p. 307)

polarized light: Light in which the transverse waves vibrate only along one plane. (Chap. 20, p. 574)

polarizing filter: A filter made of chains of molecules in parallel rows that will transmit only light waves vibrating in the same direction as the molecular chains.

polluted water: Water that contains high levels of unwanted substances that may be harmful to living things. (Chap. 8, p. 222)

polyatomic ion: A group of covalently bonded atoms in which the whole group is positively or negatively charged. (Chap. 11, p. 318)

polymer: A huge molecule made up of many smaller organic molecules, called monomers, that have formed new bonds and are linked together. (Chap. 13, p. 373)

positron: A positively charged particle similar to an electron.

potential difference: The difference in electric potential energy between two different points; measured in volts (V). (Chap. 21, p. 600)

potential energy: Stored energy. The amount of potential energy of an object depends on its position or condition. (Chap. 5, p. 125)

power: The measure of the amount of work done in a specific amount of time; power is measured in watts and equals 1 joule/second. (Chap. 7, p. 202)

precipitate: An insoluble compound formed during a double-displacement reaction. (Chap. 16, p. 453)

pressure: The amount of force exerted per unit of area. (Chap. 8, p. 228)

principle: A basic rule or law describing how something always works in the natural world; for example, "gravity always pulls objects toward each other."

products: In a chemical reaction, the substances produced by the reaction. (Chap. 16, p. 442)

projectile: Any object shot or thrown through the air; all projectiles have both horizontal and vertical velocities. (Chap. 4, p. 100)

propane: A flammable gas (C_3H_8); a part of natural gas.

protein: Organic polymer formed from amino acids. (Chap. 13, p. 373)

proton: Atomic particle with a positive charge that is part of an atom's nucleus. (Chap. 10, p. 271)

pulley: A simple machine consisting of a grooved wheel with a rope or a chain running along the groove. (Chap. 7, p. 190)

quality: In sound, the differences among sounds of the same pitch and loudness. (Chap. 18, p. 517)

quarks: Very small particles of matter that make up protons and neutrons; presently, six different types of quarks are known. (Chap. 10, p. 276)

radiation: The transfer of energy in the form of waves. Radiation is a type of energy transfer that does not require matter. (Chap. 6, p. 154)

radiator: A device with a large surface area that transfers heat to surrounding air by conduction. (Chap. 6, p. 162)

radio waves: Electromagnetic waves that have long wavelengths (low frequencies); radio waves are used in communications. (Chap. 19, p. 529)

radioactive element: An unstable element whose nucleus breaks down and gives off particles, radiation, and energy. (Chap. 12, p. 332)

radioactivity: The emission of high-energy radiation or particles from the nucleus of a radioactive atom. (Chap. 24, p. 674)

RAM: *Random-Access Memory* in a computer; temporary memory that is lost when the computer is turned off. (Chap. 23, p. 660)

rarefaction: In compressional waves, the less dense area of the wave.

reactants: The starting substances in a chemical reaction. (Chap. 16, p. 442)

real image: An image produced where light rays converge, as with a concave mirror or convex lens; a real image is real, enlarged, upside-down, and can be projected onto a screen. (Chap. 20, p. 559)

rectifier: Any device that converts alternating current (AC) into direct current (DC). (Chap. 23, p. 651)

reflecting telescope: An optical instrument that uses a concave mirror, a plane mirror, and a convex lens to magnify distant objects. (Chap. 20, p. 569)

reflection: Occurs when a wave strikes an object and bounces off. (Chap. 19, p. 546)

refracting telescope: An optical instrument that uses two convex lenses to magnify distant objects. (Chap. 20, p. 568)

refraction: The bending of waves, caused by changing their speed. (Chap. 19, p. 547)

resistance: The opposition to the flow of electrons through a conductor; measured in ohms. (Chap. 21, p. 603)

resistance arm: The part of a lever that exerts the resistance force (F_r). (Chap. 7, p. 186)

resistance force: The force exerted by a machine to overcome resistance to gravity or friction. (Chap. 7, p. 181)

resonance: The tendency of an object to vibrate at the same frequency as another vibrating source. (Chap. 18, p. 517)

reverberation: The echoing effect produced by multiple reflections of sound. (Chap. 18, p. 520)

RNA: *RiboNucleic Acid*; a nucleic acid that controls production of proteins that make new cells.

ROM: *Read-Only Memory* in a computer; it is permanent memory stored inside the computer, even when the power is turned off. (Chap. 23, p. 660)

salt: A compound formed during a neutralization reaction when negative ions from an acid combine with positive ions from a base. (Chap. 17, p. 480)

saponification: The process of making soap. (Chap. 17, p. 483)

saturated hydrocarbon: A hydrocarbon that contains only single-bonded carbon atoms; an example is propane. (Chap. 13, p. 362)

saturated solution: A solution that has dissolved all the solute it can normally hold at a specific temperature. (Chap. 15, p. 426)

scientific law: A rule that describes, but doesn't explain, a pattern in nature and predicts what will happen under specific conditions. (Chap. 1, p. 17)

screw: A simple machine consisting of a special type of inclined plane wrapped in a spiral around a cylindrical post. (Chap. 7, p. 194)

second: The SI unit for time. (Chap. 2, p. 44)

semiconductor: An element that conducts electricity under certain conditions. (Chap. 12, p. 346)

series circuit: An electrical circuit where the current has only one path. If the path is interrupted in any point, it stops current flow in the entire circuit. (Chap. 21, p. 608)

SI: International System of Units; standard, easy to use, worldwide system of measurement based on powers of 10. The SI standards are used by all scientists, and are a modern version of the metric system. (Chap. 2, p. 35)

simple machine: A device that accomplishes work with only one movement. The lever, pulley, wheel and axle, inclined plane, screw, and wedge are all simple machines. (Chap. 7, p. 180)

single-displacement reaction: A chemical reaction in which one element replaces another element in a compound. (Chap. 16, p. 452)

soap: An organic salt made by reacting fats or oils with a strong base such as sodium hydroxide. (Chap. 17, p. 483)

solar collector: A device that absorbs radiant energy from the sun that can be converted to thermal energy and used to heat buildings. (Chap. 6, p. 164)

solar energy: Energy from the sun. Solar energy is free and can be converted into thermal energy to heat homes and other buildings. (Chap. 6, p. 163)

solubility: The amount of a substance (solute) that will dissolve in a solvent; generally expressed as the maximum number of grams of solute that will dissolve in 100 g of a solvent at a specific temperature. (Chap. 15, p. 424)

solute: The substance being dissolved in a solvent. (Chap. 15, p. 417)

solution: A homogeneous mixture containing particles so tiny that they cannot be seen even with a microscope; particles in a solution don't settle and don't scatter light. (Chap. 9, p. 248)

solvent: The substance that dissolves a solute. (Chap. 15, p. 417)

specific heat: The amount of energy needed to raise the temperature of 1 kg of a material 1 K; it is measured in joules per kilogram per kelvin. (Chap. 5, p. 141)

speed: The rate of motion, or the rate at which a body changes position. (Chap. 3, p. 64)

standard: In measurement, an exact quantity that everyone agrees to use as a basis of comparison; for example, a centimeter, meter, kilogram, cubic decimeter, milliliter, liter, kelvin, joule, and so on. (Chap. 2, p. 34)

state of matter: Any of the four conditions in which matter can exist: solid, liquid, gas, or plasma; the state of a sample of matter depends on its temperature. (Chap. 8, p. 214)

static electricity: The net buildup of electric charges on an object. (Chap. 21, p. 592)

step-down transformer: An electrical transformer that decreases (steps down) the voltage of a power line.

step-up transformer: An electrical transformer that increases (steps up) the voltage of a power line.

strong acid: An acid that ionizes almost completely in solution; for example, hydrochloric acid and sulfuric acid. (Chap. 17, p. 474)

strong base: A base that dissociates completely in solution; for example, sodium hydroxide. (Chap. 17, p. 474)

sublimation: The process in which a solid changes directly to a vapor without forming a liquid. (Chap. 12, p. 343)

submerge: to fully immerse (completely cover) something in a fluid.

substance: Matter that is an element or a compound. (Chap. 9, p. 247)

substituted hydrocarbon: A hydrocarbon in which one or more hydrogen atoms have been replaced by atoms of other elements. (Chap. 13, p. 367)

supercollider: A device to make protons collide at high speed so they break apart into quarks.

superconductor: A supercooled material that has no electrical resistance; a current can flow indefinitely through a superconductor without losing energy. (Chap. 22, p. 640)

supersaturated solution: An unstable solution that contains more solute than a saturated solution can at that same specific temperature. (Chap. 15, p. 428)

suspension: A heterogeneous mixture containing a liquid in which larger particles eventually settle out. (Chap. 9, p. 250)

synthesis reaction: A chemical reaction in which two or more substances combine to form a different substance. (Chap. 16, p. 451)

synthetic fiber: A thin strand of synthetic polymer that can be woven into fabrics; examples include nylon and Kevlar fibers. (Chap. 14, p. 399)

technology: The practical use of scientific information to improve the quality of human life. (Chap. 1, p. 8)

telephoto lens: A lens having a long focal length and producing an enlarged, close-up image of an object. (Chap. 20, p. 571)

temperature: A measure of the average kinetic energy of the particles that make up a sample of matter. As an object's particles move faster, its temperature rises, and as an object's particles move slower, its temperature falls. (Chap. 5, p. 134)

terminal velocity: The greatest velocity reached by a falling object. Terminal velocity is achieved when the force of gravity is balanced by air resistance. (Chap. 4, p. 99)

theory: The most logical explanation of why things work the way they do. A theory is a former hypothesis that has been tested with repeated experiments and observations and found always to work. (Chap. 1, p. 17)

thermal energy: Total energy of a material's particles, including both kinetic energy (vibrations and movement within and between particles) and potential energy (resulting from forces that act within or between particles). (Chap. 5, p. 135)

thermal expansion: A characteristic of almost all matter that causes it to expand when heated and contract when cooled. (Chap. 8, p. 219)

thermal pollution: Pollution caused when waste heat raises the temperature of the environment. Thermal pollution in bodies of water can damage or destroy plants and animals that live there. (Chap. 5, p. 138; Chap. 8, p. 223)

thermonuclear fusion: Nuclear fusion that occurs under conditions of enormous heat (millions of degrees), as in a star or our sun.

tidal energy: Energy produced by the rise and fall of ocean levels; used to generate electricity. (Chap. 25, p. 714)

time: The interval between two events; the SI unit for time is the second. (Chap. 2, p. 44)

titration: Process in which a solution of known concentration is used to determine the concentration of an acidic or basic solution. (Chap. 17, p. 481)

total internal reflection: Occurs when all the light striking a surface between two materials is reflected totally back into the first material. (Chap. 20, p. 577)

toxic: Hazardous substance than can injure living tissue; toxic compounds are poisonous. (Chap. 11, p. 312)

tracer: A radioisotope used in medical diagnosis to allow doctors to monitor human body functions, locate tumors, detect fluid movement, and so on. (Chap. 24, p. 692)

transformer: A device that can increase or decrease the voltage of an alternating current. (Chap. 22, p. 635)

transistor: A semiconductor that amplifies or strengthens an electrical signal or acts as a tiny on-off switch. (Chap. 23, p. 651)

transition element: An element in Groups 3-12 of the periodic table; typically, these are metals with one or two electrons in their outer energy level. (Chap. 12, p. 334)

translucent materials: Materials that can be hazily seen through because they allow some light to pass through them, but not enough for a clear image. (Chap. 19, p. 536)

transmutation: Changing one element to another through radioactive decay. (Chap. 24, p. 680)

transparent materials: Materials that can be clearly seen through because they allow light to pass through them. (Chap. 19, p. 536)

transuranium element: Any element having more than 92 protons, the atomic number of uranium. (Chap. 12, p. 338)

transverse wave: A type of wave where the medium moves at right angles to the direction the wave is traveling. (Chap. 18, p. 499)

trough: The lowest point of a wave. (Chap. 18, p. 499)

Tyndall effect: The scattering of light by particles in a mixture; this effect can be seen in all colloids. (Chap. 9, p. 249)

ultrasonic technology: Technology using high-frequency sound waves for many different purposes such as medical diagnoses, sonar, and jewelry cleaning. (Chap. 18, p. 514)

ultraviolet radiation: Electromagnetic waves that have a higher frequency than visible light; abbreviated UV. (Chap. 19, p. 532)

unsaturated hydrocarbons: Hydrocarbons that contain at least one double or triple bond between carbon atoms; an example is ethyne. (Chap. 13, p. 364)

unsaturated solution: A solution that is capable of dissolving more solute at a specific temperature. (Chap. 15, p. 427)

velocity: The speed and direction of a moving body, such as a storm or a basketball thrown across a court. (Chap. 3, p. 72)

Venturi effect: Reduction in pressure of a fluid resulting from the speed increase as fluids are forced to flow faster through narrow spaces. (Chap. 8, p. 238)

virtual image: An image formed of diverging light rays, as in a plane or convex mirror, or seen through a concave lens. A virtual image can't be projected onto a screen. (Chap. 20, p. 558)

visible radiation: Electromagnetic waves in the only part of the electromagnetic spectrum we can see—light. (Chap. 19, p. 532)

volt: The unit for measuring electrical potential energy; abbreviated V.

voltage: A difference in electrical potential, measured in volts with a voltmeter.

voltmeter: A galvanometer that measures potential differences in volts and is placed in parallel across a part of the circuit. (Chap. 22, p. 630)

volume: The amount of space occupied by an object; volume units are found by combining other SI units of length. (Chap. 2, p. 41)

wave: A rhythmic disturbance that carries energy through matter or space. (Chap. 18, p. 498)

wavelength: The distance between identical points on two adjacent waves; for example, the distance between two crests or two troughs. (Chap. 18, p. 499)

weak acid: An acid that partially ionizes in solution; for example, carbonic acid. (Chap. 17, p. 474)

weak base: A base that partially dissociates in solution; for example, magnesium hydroxide. (Chap. 17, p. 474)

wedge: A simple machine consisting of a moving inclined plane with one or two sloping sides; examples are knives and chisels. (Chap. 7, p. 194)

weight: The measure of the force of gravity on an object. Weight changes with changes in gravity and is measured in newtons (N). (Chap. 3, p. 84)

wet cell: A power source that generates electric current by a chemical reaction; "wet" because it uses a liquid electrolyte. (Chap. 21, p. 602)

wheel and axle: A simple machine consisting of two different-sized wheels that rotate together, such as a doorknob or wheel-handled faucet. (Chap. 7, p. 191)

wide-angle lens: A lens with a short focal length that produces a relatively small image of an object, but includes much of the object's surroundings. (Chap. 20, p. 571)

work: The transfer of energy through motion; *work = force × distance*. Work is measured in joules and is accomplished only when force produces motion in the direction of the force. (Chap. 5, p. 126)

X rays: Electromagnetic waves having a wavelength shorter (higher frequency) than ultraviolet radiation; often used in medical diagnosis and photography because they can penetrate human tissue. (Chap. 19, p. 534)

This glossary defines each key term that appears in **bold type** in the text. It also shows the page number where you can find the word used. Some additional helpful terms are also provided.

acceleration/aceleración: índice del cambio en la velocidad (un cambio en la dirección o en la rapidez). La cantidad de aceleración depende del cambio en la velocidad y del intervalo de tiempo. (Cap. 3, pág. 73)

acid/ácido: sustancia que produce iones de hidrógeno H^+, en solución. La presencia de los iones de H^+ es lo que les da a los ácidos sus propiedades características. (Cap. 17, pág. 466)

acid rain/lluvia ácida: cualquier lluvia cuyo pH es menor de 5.6. Se produce cuando las sustancias en el aire reaccionan con el agua de lluvia y la vuelven ácida. La lluvia ácida es dañina para los animales y las plantas, además de aumentar el ritmo de la meteorización. (Cap. 17, pág. 478)

acoustics/acústica: estudio del sonido. (Cap. 18, pág. 520)

actinide/actínido: cualquiera de los 14 elementos radiactivos cuyos números atómicos van desde 90 a 103 y los cuales se usan en la producción de potencia nuclear y de armas nucleares.

active solar heating/calefacción solar activa: la recolección de la energía solar con paneles nucleares para calentar agua con esa energía. Esta agua calentada se almacena para ser usada más tarde como fuente energética.

aerosol/aerosol: líquido que se rocía de un recipiente a presión. Por ejemplo, una lata de insecticida.

air resistance/resistencia del aire: fuerza que el aire ejerce sobre un objeto en movimiento. (Cap. 4, pág. 97)

alchemist/alquimista: una versión medieval del químico moderno; un practicante que combinaba la química primitiva con magia, en busca de convertir metales común y corrientes en oro.

alcohol/alcohol: tipo de compuesto que se forma cuando los grupos –OH reemplazan uno o más átomos de hidrógeno en un hidrocarburo. (Cap. 13, pág. 368)

allotrope/alotropo: elemento que exhibe diferentes formas teniendo además estructuras moleculares diferentes. El carbono exhibe alotropía en sus distintas formas cristalinas, como por ejemplo, el grafito y el diamante. (Cap. 12, pág. 347)

alloy/aleación: mezcla de un metal y uno o más elementos. Una aleación conserva las propiedades comunes de los metales, pero no es un metal puro. (Cap. 14, pág. 388)

alpha particle/partícula alfa: tipo de radiación despedido cuando un núcleo libera dos protones y dos neutrones. Es la forma más extensa de radiación y la más lenta, de modo que es la menos penetrante. (Cap. 24, pág. 679)

alternating current (AC)/corriente alterna: corriente que cambia su dirección en un patrón regular. La corriente alterna de 60 Hz de nuestros hogares cambia dirección 120 veces cada segundo. (Cap. 22, pág. 635)

amalgam/amalgama: aleación que contiene mercurio. Las amalgamas dentales consisten de mercurio, plata y cinc. Los dentistas utilizan las amalgamas para rellenar los huecos de las caries dentales. (Cap. 14, pág. 391)

ammeter/amperímetro: instrumento que mide, en amperios, la corriente eléctrica que pasa por un circuito, el cual debe estar conectado en serie. (Cap. 22, pág. 630)

amorphous/amorfo: algo que no tiene forma específica; por ejemplo, un líquido o un gas.

ampere/amperio: la unidad para medir la corriente, la razón del flujo de electrones en un circuito; abreviado A; 1 amperio = 1 culombio de carga que pasa por un punto dado en un segundo.

amplification/amplificación: proceso de incrementar la fortaleza de una señal eléctrica. (Cap. 23, pág. 651)

amplitude/amplitud: distancia desde la cresta (o valle) de una onda hasta la posición de descanso del medio. Corresponde a la cantidad de energía transportada por una onda. Las ondas que transportan grandes cantidades de energía tienen alturas o amplitudes altas y las ondas que transportan menos energía tienen amplitudes más bajas. (Cap. 18, pág. 499)

amplitude-modulated (AM) waves/ondas de amplitud modulada: ondas radiales cuya amplitud varía con la voz, la música, el video o los datos al ser transmitidas a través de largas distancias.

angle of incidence/ángulo de incidencia: en ondas, el ángulo formado por la onda de incidencia y la normal (perpendicular); se representa como *i*.

angle of reflection/ángulo de reflexión: en ondas, el ángulo formado por la onda reflejada y la normal (perpendicular); se representa como *r*.

anhydrous/anhidro: un compuesto químico que normalmente contiene moléculas de agua adheridas a sus iones, pero al cual se le ha secado el agua.

antiacid/antiácido: un "anti-ácido" o una sustancia química que cambia una sustancia ácida en una sustancia neutra.

antifreeze/anticongelante: un soluto que se le agrega a un disolvente para reducir la temperatura a la cual se congela el disolvente; por ejemplo, el anticongelante que se le añade al agua del sistema de enfriamiento de un carro.

aqueous/acuoso: describe una solución hecha con agua; por ejemplo, el agua de amoníaco es una solución acuosa de gas amoniacal.

Archimedes' principle/principio de Arquímedes: dice que la fuerza boyante de un objeto en un fluido es igual al peso del fluido desplazado por el objeto. (Cap. 8, pág. 235)

aromatic compound/compuesto aromático: tipo de compuesto que contiene la estructura de anillo bencénico. Un ejemplo de un compuesto aromático es la naftalina. (Cap. 13, pág. 366)

artificial satellites/satélites artificiales: dispositivos hechos por los seres humanos que giran en órbitas alrededor de la Tierra, por lo general, con un propósito específico. (Cap. 4, pág. 108)

atomic number/número atómico: número de protones en el núcleo de un átomo. (Cap. 10, pág. 271)

average atomic mass/masa atómica promedio: masa promedio de la mezcla de isótopos de un elemento. (Cap. 10, pág. 280)

average speed/rapidez promedio: se halla dividiendo el total de la distancia viajada entre el total del tiempo viajado. (Cap. 3, pág. 65)

balance/balanza: un dispositivo que se usa en laboratorios para medir la masa; funciona equilibrando la masa que se está determinando con una masa estándar conocida.

balanced chemical equation/ecuación química equilibrada: ecuación que tiene el mismo número de átomos de cada elemento en ambos lados de la ecuación. (Cap. 16, pág. 448)

balanced forces/fuerzas equilibradas: fuerzas del mismo tamaño pero opuestas en dirección. (Cap. 3, pág. 78)

bar graph/gráfica de barras: un tipo de gráfica que se usa para mostrar información recogida por conteo; utiliza barras horizontales o verticales de diferentes longitudes para ayudarnos en la comparación de cantidades.

base/base: sustancia que produce iones de hidróxido, OH^- en solución. En estado puro y sin disolver, las bases son sólidos cristalinos, pero en solución se sienten resbalosas y tienen sabor amargo. Sin embargo, las bases no se deben nunca tocar ni probar porque pueden causar quemaduras graves. (Cap. 17, pág. 470)

Bernoulli's principle/principio de Bernoulli: dice que a medida que aumenta la velocidad de un fluido, disminuye la presión ejercida por el fluido. (Cap. 8, pág. 237)

beta particle/partícula beta: electrón emitido del núcleo, a alta velocidad. (Cap. 24, pág. 680)

BHA/BHA: una sustancia para preservar alimentos; es un inhibidor químico que demora las reacciones químicas que descomponen los alimentos (*B*util-*h*idroxi*a*nisol).

BHT/BHT: una sustancia para preservar alimentos; es un inhibidor químico que demora las reacciones químicas que descomponen los alimentos (*B*util-*h*idroxi*t*olueno).

binary compound/compuesto binario: compuesto químico formado de dos elementos. Por ejemplo, el cloruro de sodio es un compuesto binario hecho de cloro y sodio. (Cap. 11, pág. 315)

biodegradable/biodegradable: describe un material que se puede descomponer en sustancias no tóxicas mediante la acción biológica de organismos como las bacterias.

biogas/biogas: mezcla de gases, principalmente, metano que se produce de la biomasa. (Cap. 13, pág. 370)

biomass/biomasa: material orgánico proveniente de la madera, la caña de azúcar, el maíz, las cosechas y los desperdicios de los animales. (Cap. 13, pág. 370)

bionics/biónica: ciencia que diseña repuestos artificiales para las partes del cuerpo humano. Los repuestos artificiales a menudo se llaman prótesis. (Cap. 7, pág. 196)

boiling point/punto de ebullición: la temperatura a la cual se forman burbujas de vapor en un líquido y se elevan hasta la superficie, aumentando la evaporación.

Boyle's law/ley de Boyle: dice que si se disminuye el volumen de un recipiente de gas, la presión del gas aumenta, siempre y cuando no se altere la temperatura. Al aumentar el volumen, la presión disminuye. (Cap. 8, pág. 229)

brittleness/quebradizo: una característica de ciertos materiales que hace que se rompan fácilmente bajo estrés.

bubble chamber/cámara de burbuja: cámara llena de un líquido supercalentado, el cual no hierve debido al aumento de presión en el sistema. (Cap. 24, pág. 686)

buoyant force/fuerza boyante: capacidad de un fluido—líquido o gas—de ejercer una fuerza ascendente sobre un objeto sumergido en él. La cantidad de fuerza boyante determina si un objeto se hunde o flota en un fluido. (Cap. 8, pág. 234)

butane/butano: un gas inflamable (C_4H_{10}); una parte del gas natural.

byte/byte: una unidad básica de memoria de computadora que representa un carácter (número, símbolo o letra del alfabeto); consiste de 8 bitios.

calorimeter/calorímetro: un instrumento que se usa para medir cambios en la energía térmica.

carbohydrate/carbohidrato: compuesto orgánico que posee el doble de átomos de hidrógeno que de oxígeno. Un grupo de carbohidratos son los azúcares. (Cap. 13, pág. 376)

carbon-14 dating/datación con carbono 14: método para determinar la edad de objetos que contienen carbono 14 y que datan de hasta 50 000 años atrás; funciona calculando la cantidad de carbono 14 que aún contiene el objeto.

catalyst/catalizador: sustancia que acelera una reacción química, sin cambiar permanentemente la sustancia misma. Cuando se le añade un catalizador a una reacción, se termina con la misma cantidad de catalizador con que se comenzó. (Cap. 16, pág. 456)

cathode–ray tube (CRT)/tubo de rayos catódicos: dispositivo que usa electrones y materiales fluorescentes para producir imágenes sobre una pantalla. (Cap. 23, pág. 656)

central processing unit (CPU)/unidad de procesamiento central: el tablero principal de circuitos dentro de una computadora que ejecuta los cálculos y almacena la memoria principal.

centripetal acceleration/aceleración centrípeta: aceleración que se efectúa hacia el centro de una trayectoria circular o curva. (Cap. 4, pág. 102)

centripetal force/fuerza centrípeta: fuerza que actúa hacia el centro de una trayectoria circular o curva. (Cap. 4, pág. 102)

ceramic/cerámica: material hecho de arcilla seca y de mezclas de arcilla seca. Hace miles de años que los seres humanos comenzaron a fabricar recipientes de cerámica para almacenar alimentos y transportar agua. (Cap. 14, pág. 392)

cermet/cermet: o *cer*ámica–*met*al, es un material que posee propiedades tanto de las cerámicas como de las aleaciones. (Cap. 14, pág. 394)

chain reaction/reacción en cadena: serie de reacciones de fisión en curso. Puede causar billones de reacciones cada segundo, dando como resultado la liberación de tremendas cantidades de energía. (Cap. 24, pág. 690)

Charles's law/ley de Charles: dice que el volumen de un gas aumenta cuando se aumenta la temperatura, siempre y cuando no se altere la presión. Lo contrario también es cierto. (Cap. 8, pág. 230)

chemical bond/enlace químico: fuerza que mantiene unidos los átomos en una sustancia. (Cap. 11, pág. 301)

chemical change/cambio químico: cambio de una sustancia de un material en una sustancia diferente. (Cap. 9, pág. 257)

chemical formula/fórmula química: presentación precisa de los elementos que contiene un compuesto y las proporciones de los átomos de esos elementos. (Cap. 11, pág. 300)

chemical property/propiedad química: característica de una sustancia que indica si puede sufrir cierto cambio químico. (Cap. 9, pág. 259)

chemical reaction/reacción química: un ejemplo de un cambio químico. Cuando ocurre una reacción química, una o más sustancias se convierten en nuevas sustancias. (Cap. 16, pág. 442)

chemical symbol/símbolo químico: forma abreviada de escribir el nombre de un elemento. Por ejemplo,:

C para el carbono, Ag para la plata. (Cap. 10, pág. 270)

chemically stable/estable químicamente: describe un átomo cuyo nivel de energía externo está completamente lleno de electrones. El nivel de energía externo, de la mayoría de los átomos de los elementos, está lleno cuando contiene ocho electrones. (Cap. 11, pág. 300)

chemically unstable/químicamente inestable: describe un átomo cuyo nivel de energía externo no está lleno de electrones de manera que busca electrones de otros átomos y así forma compuestos; por ejemplo, el cloruro sódico.

choloro-/cloro: prefijo que indica la presencia de cloro, como en tetracloroetileno o clorofluorocarbono.

chlorofluorocarbon (CFC)/clorofluorocarbono: grupo de compuestos cuya descomposición libera átomos de cloro que destruyen las moléculas de ozono en la atmósfera superior. Se usan en los sistemas de aire acondicionado y en la fabricación de algunos tipos de espumas de polímeros. (Cap. 16, pág. 447)

circuit/circuito: trayectoria cerrada a través de la cual pueden fluir los electrones. (Cap. 21, pág. 601)

circuit breaker/cortacircuitos: un dispositivo que protege un circuito eléctrico; si fluye demasiada corriente, el cortacircuitos abre el circuito, deteniendo el flujo de corriente.

cloud chamber/cámara de niebla: cámara que contiene agua supersaturada o vapor de etanol y la cual se usa para detectar partículas nucleares cargadas, las cuales dejan rastros de vapor al pasar. (Cap. 24, pág. 686)

coagulation/coagulación: destrucción de la estructura de coloides, debido a la atracción mutua de pequeñas partículas suspendidas. (Cap. 9, pág. 253)

coal/carbón: una roca formada de plantas antiguas descompuestas; se quema como un combustible fósil.

coefficient/coeficiente: representa el número de unidades de cada sustancia que forman parte de una reacción química. (Cap. 16, pág. 445)

coherent light/luz coherente: energía electromagnética de una sola longitud de onda que viaja con sus crestas y valles alineados. El haz no se dispersa porque todas las ondas viajan en la misma dirección. (Cap. 20, pág. 575)

colloid/coloide: mezcla heterogénea que, así como una solución, nunca se asienta. (Cap. 9, pág. 249)

combustion/combustión: quiere decir combustión rápida y es el proceso mediante el cual las máquinas de calor queman combustible. (Cap. 6, pág. 166)

composite/compuesto: mezcla de dos o más materiales, uno de ellos incrustado en el otro. (Cap. 14, pág. 401)

compound/compuesto: material formado de la combinación de átomos de dos o más elementos. La proporción de los diferentes átomos en un compuesto es siempre la misma. Por ejemplo, los elementos hidrógeno y oxígeno se pueden combinar para formar el compuesto agua. Los átomos de los elementos en el agua siempre se encuentran en la proporción de dos átomos de hidrógeno por cada átomo de oxígeno. (Cap. 9, pág. 246)

compound machine/máquina compuesta: combinación de dos o más máquinas simples. (Cap. 7, pág. 198)

compression/compresión: el área densa de la onda, en ondas de compresión.

compressional wave/onda de compresión: onda en la cual la materia vibra en la misma dirección en que viaja la onda. En las ondas de compresión, el medio se desplaza en la misma dirección en que se mueve la energía. (Cap. 18, pág. 505)

computer/computadora: un dispositivo que se programa para ejecutar cálculos, tomar decisiones lógicas y manipular datos.

computer virus/virus de computadora: tipo de programa que se puede multiplicar dentro de una computadora y llegar a usar tanta memoria que perjudica el sistema. (Cap. 23, pág. 666)

concave lenses/lente cóncava: lente más delgada en el centro y más gruesa en las orillas y por consiguiente se curvan hacia adentro, forman una imagen virtual, al derecho y más pequeña que el objeto en sí. (Cap. 20, pág. 565)

concave mirror/espejo cóncavo: espejo cuya superficie se curva hacia adentro, como la parte interior de una cuchara. (Cap. 20, pág. 558)

concentrated solution/solución concentrada: una solución en que la cantidad de soluto está casi al máximo de lo que el disolvente puede retener a una temperatura dada.

concentration/concentración: Por lo general, la proporción de soluto disuelto en un disolvente.

condensation/condensación: sucede cuando un gas se convierte en líquido. Por lo general, un gas se condensa cuando se enfría a su punto de ebullición o por debajo de este. (Cap. 8, pág. 225)

conduction/conducción: transferencia de energía a través de la materia por contacto directo de las partículas. (Cap. 6, pág. 152)

conductor/conductor: material que permite que los electrones se muevan fácilmente a través de él. Los metales como el cobre y la plata están hechos de átomos que no sostienen sus electrones muy apretados, de modo que los electrones pueden moverse fácilmente a través de materiales hechos de estas clases de átomos. (Cap. 21, pág. 594)

constant/constante: factor que no varía en un experimento. (Cap. 1, pág. 21)

constant speed/rapidez constante: rapidez que no varía. (Cap. 3, pág. 65)

contraction/contracción: movimiento en que las moléculas se condensan para ocupar un área más pequeña.

control/control: un estándar para la comparación. (Cap. 1, pág. 21)

convection/convección: transferencia de energía debido al movimiento de materia. (Cap. 6, pág. 153)

convex lenses/lente convexa: lente más gruesa en el centro que en las orillas. Los rayos luminosos que se acercan a la lente paralelos al eje óptico son refractados hacia el centro de la lente. Estos rayos convergen en el centro focal, de modo que pueden formar imágenes reales, las cuales pueden proyectarse sobre una pantalla. (Cap. 20, pág. 564)

convex mirror/espejo convexo: espejo en que los rayos nunca se juntan, de modo que la imagen es siempre virtual, al derecho y más pequeña que el objeto en sí. (Cap. 20, pág. 561)

corrosive/corrosivo: término proveniente de la raíz en latín que quiere decir "desprender, separar". Este tipo de compuesto es útil para destapar cañerías. (Cap. 11, pág. 312)

covalent bond/enlace covalente: atracción que se forma entre los átomos cuando comparten electrones. (Cap. 11, pág. 306)

coulomb/culombio: la carga que transportan 6.24 billones de billones de electrones.

crest/cresta: punto más alto de una onda. (Cap. 18, pág. 499)

crystal/cristal: patrón en que se encuentran arregladas las partículas de la mayoría de los sólidos. (Cap. 8, pág. 215)

current/corriente: flujo de electrones a través de un alambre u otro conductor. La cantidad de corriente eléctrica depende del número de electrones que pasan por un punto en un momento dado. La corriente en un circuito se mide en amperios, (A). (Cap. 21, pág. 601)

deceleration/deceleración: La razón del cambio en velocidad (rapidez o dirección) cuando la velocidad disminuye; también llamada aceleración negativa.

decibel/decibel: la unidad de medida de la intensidad sonora, abreviada dB. El sonido más débil que la mayoría de la gente puede oír es = 0 dB.

decomposition reaction/reacción de descomposición: reacción en la cual una sustancia se descompone en dos o más sustancias más simples. (Cap. 16, pág. 452)

dehydrating agent/agente deshidratador: sustancia que puede extraer agua de los materiales. (Cap. 17, pág. 468)

density/densidad: masa por unidad de volumen de un material. (Cap. 2, pág. 43)

dependent variable/variable dependiente: factor cuyo valor depende del valor de la variable independiente. (Cap. 1, pág. 22)

derived unit/unidad derivada: unidad que se obtiene de la combinación de unidades del SI. (Cap. 2, pág. 42)

detergent/detergente: sal orgánica cuya estructura se parece a la del jabón. (Cap. 17, pág. 483)

diatomic molecule/molécula diatómica: molécula que consiste en dos átomos del mismo elemento. (Cap. 12, pág. 341)

diesel engine/motor diesel: un motor de combustión interna que comprime una mezcla de aire y combustible que se enciende debido al calor de compresión sin necesidad de una chispa.

diffraction/difracción: desviación de las ondas alrededor de una barrera, en la cual las ondas rodean el obstáculo. Todas las ondas electromagnéticas, de sonido y de agua pueden ser difractadas. Por ejemplo, las ondas se desvían alrededor de una persona que nada en el agua. (Cap. 19, pág. 548)

diffraction grating/retícula de difracción: pedazo de vidrio o de plástico hecho de muchas hendiduras paralelas. Cuando se hace brillar una luz blanca a través de una retícula de difracción, los colores de la luz se separan. (Cap. 19, pág. 550)

dilute solution/solución diluida: una solución en que la cantidad de soluto es mucho menor que el máximo que el disolvente puede retener a una temperatura dada.

diode/diodo: parece como un tipo de válvula o de puerta que permite que una corriente fluya en una sola dirección. (Cap. 23, pág. 651)

direct current (DC)/corriente directa: corriente que fluye en una sola dirección a través de un alambre. (Cap. 22, pág. 635)

disinfectant/desinfectante: una sustancia química que mata bacterias. Por ejemplo, el alcohol es un desinfectante.

dissociation/disociación: proceso en que un sólido iónico se disuelve en agua y sus iones positivos y negativos se separan unos de otros. Por ejemplo, el cloruro de sodio se disocia cuando se disuelve en agua. (Cap. 15, pág. 433)

DNA:DNA: *Ácido Desoxirribonucleico*; un ácido en el núcleo de las células que codifica y almacena información genética.

doping/adulteración: la adición de un impureza a un semiconductor con el propósito de aumentar su conductividad eléctrica.

Doppler effect/efecto Dopler: un aumento o disminución en la frecuencia de ondas, causado por movimiento de la fuente o movimiento del observador. Se aplica a todas las ondas (inclusive a las de la luz, de calor, del sonido, etc.).

dot diagram/diagrama de puntos: diagrama que utiliza el símbolo del elemento junto con unos puntos para representar los electrones en el nivel de energía externo. (Cap. 10, pág. 288)

double–displacement reaction/reacción de desplazamiento doble: reacción que ocurre cuando un precipitado, agua o gas se forma de la combinación de dos compuestos iónicos en solución. (Cap. 16, pág. 453)

dry cell/pila seca: es la que puede actuar como una bomba de electrones porque posee una diferencia de potencial entre los terminales positivo y negativo. (Cap. 21, pág. 602)

ductile/ductil: propiedad de los metales que permite que puedan ser enrollados en alambres. (Cap. 12, pág. 330)

efficiency/eficiencia: medida que se usa para calcular la cantidad de trabajo útil que se ha producido, del trabajo inicial aplicado a una máquina. (Cap. 7, pág. 200)

effort arm/brazo de esfuerzo: parte de la palanca donde se aplica la fuerza de esfuerzo. (Cap. 7, pág. 186)

effort force (F_e)/fuerza de esfuerzo: la fuerza que se le aplica a la máquina. (Cap. 7, pág. 181)

electric field/campo eléctrico: espacio que ejerce una fuerza sobre cualquier cosa que tiene una carga eléctrica. El campo eléctrico es más fuerte cerca de un electrón y se debilita al aumentar su distancia del mismo. (Cap. 21, pág. 594)

electric motor/motor eléctrico: dispositivo que cambia la energía eléctrica en energía mecánica, para hacer funcionar un artefacto como por ejemplo, un ventilador. (Cap. 22, pág. 630)

electrical power/potencia eléctrica: proporción a la cual la energía eléctrica se convierte en otra forma de energía. (Cap. 21, pág. 614)

electrolyte/electrolítica: sustancia que se separa en iones o que forma iones en una solución acuosa. (Cap. 15, pág. 433)

electromagnet/electroimán: imán temporal que se forma pasando una corriente por un eje de hierro que ha sido introducido en un espiral. (Cap. 22, pág. 628)

electromagnetic induction/inducción electromagnética: proceso en que al mover un alambre por un campo magnético se produce una corriente. También se produce una corriente moviendo un imán por dentro y por fuera de un espiral de alambre. (Cap. 22, pág. 633)

electromagnetic radiation/radiación electromagnética: ondas electromagnéticas transversales producidas por el movimiento de partículas eléctricamente cargadas. Se llama radiación electromagnética porque las partículas irradian las ondas. (Cap. 19, pág. 528)

electron/electrón: partícula con carga negativa que rodea el núcleo. (Cap. 10, pág. 270)

electron arrangement/disposición de electrones: la manera en que están distribuidos los electrones en los varios niveles de energía de un átomo.

electron cloud/nube de electrones: modelo del átomo que describe la región en donde un electrón se puede encontrar en un momento dado. (Cap. 10, pág. 272)

electroscope/electroscopio: instrumento hecho de dos láminas metálicas delgadas adheridas a una barra metálica que tiene una perilla en la cúspide. Se usa para detectar cargas de electricidad estática. (Cap. 21, pág. 596)

element/elemento: tipo de materia cuyos átomos son todos iguales. (Cap. 9, pág. 246)

endothermic reaction/reacción endotérmica: reacción que requiere energía para llevarse a cabo. Frecuentemente se usa para obtener un metal de su mena. (Cap. 16, pág. 458)

energy/energía: capacidad de causar cambio. Los científicos han tenido problemas definiendo lo que es la energía porque existe en tantas formas diferentes. Algunas de estas formas incluyen la energía radiante, la energía eléctrica, la energía química, la energía térmica y la energía nuclear. (Cap. 5, pág. 124)

energy farming/cultivo de energía: cultivo de plantas con el propósito de usarlas para proporcionar energía. El cultivo y el uso controlados de plantas puede convertir esta fuente energética en una fuente renovable. (Cap. 13, pág. 370)

energy transfer/transferencia de energía: el movimiento de energía de un objeto a otro; por ejemplo, la energía térmica que fluye, en forma de calor, desde una estufa caliente a un sartén.

ester/éster: compuesto orgánico que se forma de la reacción de un ácido orgánico con un alcohol. Los ésteres son los responsables de los maravillosos olores y sabores de flores, frutas y otros alimentos. (Cap. 17, pág. 484)

evaporation/evaporación: cambio gradual de un líquido a un gas, el cual ocurre a temperaturas inferiores al punto de ebullición. (Cap. 8, pág. 224)

exothermic reaction/reacción exotérmica: tipo de reacción en la cual se libera alguna forma de energía. Por ejemplo, la quema de leña o la explosión de la dinamita. (Cap. 16, pág. 456)

expansion/expansión: movimiento hacia afuera de las moléculas alejándose unas de otras, para ocupar un espacio mayor.

experiment/experimento: procedimiento organizado que se hace con el propósito de probar una predicción. (Cap. 1, pág. 20)

external combustion engine/motor de combustión externa: máquina en que el combustible se quema fuera del motor. (Cap. 6, pág. 167)

factor/factor: una condición que afecta el resultado de un evento.

farsighted/présbita: describe a una persona que ve bien los objetos lejanos, pero que tiene problemas para enfocar los objetos cercanos.

fiberglass/vitrofibra: hilos de vidrio que parecen cabellos. Son buenos aisladores cuando se arreglan en capas espumosas.

filter/filtro: un dispositivo que permite que uno o más colores se transmitan mientras que otros son absorbidos o bloqueados, en el trabajo con la luz.

flame test/prueba de la llama: una prueba de laboratorio para identificar elementos mediante el calentamiento de la sustancia en una llama y luego observando el color de la llama.

flammable/inflamable: una característica química de una sustancia que le permite ser oxidada rápidamente; también denominada combustible.

flash distillation/evaporación relámpago: un proceso de desalinización en que el agua de mar se bombea hacia una cámara al vacío, para hierva rápidamente y a una temperatura más baja (debido a la ausencia de presión atmosférica).

fluid/fluido: cualquier material que fluye. El aire y el agua son dos ejemplos de fluidos. (Cap. 6, pág. 153)

fluorescence/fluorescente: cuando un material absorbe radiación ultravioleta, y es estimulado a irradiar luz visible.

fluorescent light/luz fluorescente: tipo de iluminación que produce luz sin pérdida excesiva de energía calórica. Para producir la misma cantidad de luz, este tipo de bombillas usa solo una quinta

parte de la electricidad que la que usa una bombilla común y corriente. (Cap. 19, pág. 544)

focal length/longitud focal: distancia desde el centro de un espejo hasta el punto focal. (Cap. 20, pág. 559)

focal point/punto focal: punto en el eje óptico a través del cual todos los rayos de luz paralelos a dicho eje son reflejados. (Cap. 20, pág. 559)

force/fuerza: empuje o fuerza de atracción que un cuerpo ejerce sobre otro. (Cap. 3, pág. 78)

fractional distillation/destilación fraccional: proceso que se usa en la refinación del petróleo para separar los compuestos hidrocarburos. (Cap. 25, pág. 700)

fractionating towers/torres de fraccionamiento: torres en las refinerías petroleras que se usan para destilar el petróleo por fracciones.

free-fall/caída libre: la manera en que un objeto se mueve por el espacio cuando está bajo la influencia de la gravedad solamente.

Freon/freón: un gas refrigerante que se utiliza en refrigeradores y acondicionadores de aire.

frequency/frecuencia: número de crestas que pasan por un punto fijo cada segundo. Se expresa en hertz (Hz). Un hertz equivale a una onda por segundo. (Cap. 18, pág. 499)

frequency-modulated (FM) waves/ondas de frecuencia modulada: ondas radiales cuya frecuencia varía con la voz, la música, el video o los datos al ser transmitidas a través de largas distancias.

friction/fricción: fuerza que se opone al movimiento entre dos superficies que están en contacto una con otra. (Cap. 3, pág. 82)

fuel rod/barra combustible: una barra metálica llena de perdigones de uranio, la cual se usa como combustible en los reactores nucleares.

fulcrum/fulcro: punto fijo de una palanca. (Cap. 7, pág. 186)

fuse/fusible: un dispositivo que protege un circuito eléctrico. Si fluye mucha corriente, el fusible se derrite interrumpiendo el circuito y detiene así el flujo de corriente.

galvanometer/galvanómetro: un instrumento (medidor) que se usa para detectar corrientes eléctricas; la mayoría de los medidores eléctricos son galvanómetros.

gamma ray/rayo gamma: la más penetrante forma de radiación, la cual no consiste en partículas, sino en ondas electromagnéticas con frecuencia y energía muy altas. Los rayos gamma no tienen ni masa, ni carga y viajan a la velocidad de la luz. (Cap. 24, pág. 681)

gamma ray/rayo gamma: tipo de radiación con la más alta frecuencia y las más penetrantes de todas las ondas electromagnéticas, ubicado en el extremo opuesto al las ondas radiales en el espectro electromagnético. Los núcleos de los átomos radiactivos emiten rayos gamma. (Cap. 19, pág. 534)

gaseous solution/solución gaseosa: un gas homogéneo compuesto de dos o más gases.

gasohol/gasohol: fermentación del maíz para producir alcohol y mezclarlo con gasolina. Este tipo de mezcla permitirá que los abastecimientos actuales de gasolina duren más tiempo. (Cap. 13, pág. 371)

gear/engranaje: una rueda dentada en el borde diseñada para encajar en los dientes de otro engranaje y así transferir la fuerza y el movimiento.

gelatin/gelatina: una sustancia que se obtiene de hervir los huesos de animales. Se usa en las pegas y otros artículos.

generator/generador: dispositivo que produce corriente eléctrica mediante la rotación de una bobina de alambre por un campo magnético. La bobina de alambre se conecta a una fuente de energía mecánica y se coloca en medio de los polos de un imán. (Cap. 22, pág. 633)

geothermal energy/energía geotérmica: tipo de energía que contiene el núcleo terrestre en grandes cantidades. Esta energía se puede utilizar en lugares en donde la corteza terrestre presenta grietas que permiten que parte de la energía geotérmica se escape por ellas. (Cap. 25, pág. 716)

glass/vidrio: cerámica sin la estructura cristalina regular. Existen miles de variedades de vidrios, el más común es el de las ventanas, que está hecho principalmente de oxígeno y silicio, con pequeñas cantidades de sodio y calcio y una pizca de aluminio. (Cap. 14, pág. 394)

graduated cylinder/cilindro graduado: un cilindro marcado con una escala de volumen que se usa en laboratorios para medir volúmenes líquidos.

granite/granito: una roca ígnea que es una mezcla de cristales de diferentes compuestos.

graph/gráfica: despliegue visual de información o datos. (Cap. 2, pág. 48)

graphite/grafito: un mineral hecho de átomos de carbono arreglados en capas que se deslizan fácilmente una sobre otra, formando un lubricante seco.

gravity/gravedad: fuerza que cada objeto en el universo ejerce sobre todos los otros objetos. (Cap. 3, pág. 83)

grounded/soterrado: conectado eléctricamente a tierra, ya sea de forma directa o a través de alambres u otros objetos.

groundwater/agua subterránea: agua que proviene de lluvia y nieve derretida depositada en reservorios bajo tierra. Aunque a menudo es lo suficientemente pura para beber, puede ser contaminada fácilmente por depósitos de basura, cloacas y derrames químicos.

group/grupo: columna vertical en la tabla periódica. (Cap. 10, pág. 288)

hacker/hacker: una persona que usa una computadora para tener acceso ilegalmente a otros sistemas de computadoras.

half–life/período de media vida: cantidad de tiempo que tarda en desintegrarse la mitad de los nucleidos en una muestra de un isótopo radiactivo dado. Los períodos de medias vidas de los isótopos radiactivos varían de un elemento a otro. Por ejemplo, el plutonio–214 tiene un período de media vida de una milésima de segundo, mientras que el período de media vida del uranio–238 es de 4.5 billones de años. (Cap. 24, pág. 681)

halogens/halógenos: elementos altamente activos en el Grupo 17 de la tabla periódica (flúor, cloro, bromo, yodo, astato); los halógenos poseen siete electrones en su exterior y se combinan fácilmente con los elementos del Grupo 1, como el sodio.

heat/calor: energía térmica o interna que fluye desde algo que tiene una temperatura más alta hasta algo con temperatura más baja. El calor siempre fluye de materiales más calientes a materiales más fríos. (Cap. 5, pág. 136)

heat engine/máquina de calor: dispositivo que convierte la energía térmica en energía mecánica. (Cap. 6, pág. 166)

heat mover/movedor de calor: dispositivo que traslada energía térmica de un lugar a otro lugar con una temperatura diferente. (Cap. 6, pág. 168)

heat of fusion/calor de fusión: cantidad de energía necesaria para cambiar un material del estado sólido al líquido. (Cap. 8, pág. 226)

heat of vaporization/calor de evaporación: cantidad de energía necesaria para cambiar un material del estado líquido al gaseoso. (Cap. 8, pág. 226)

herbicide/herbicida: un veneno químico que mata plantas no deseadas.

hertz/hertz: la unidad de medida de la frecuencia, se abrevia Hz. Por ejemplo, 440 ondas por segundo = 440 Hz.

heterogeneous mixture/mezcla heterogénea: mezcla en la que los diferentes materiales que la forman pueden ser distinguidos fácilmente. (Cap. 9, pág. 248)

high-speed photography/fotografía ultrarrápida: un método fotográfico que toma fotografías claras (no borrosas) de objetos que se mueven rápidamente.

homogeneous mixture/mezcla homogénea: mezcla en que dos o más sustancias están esparcidas uniformemente por toda su extensión, como por ejemplo, el agua salada. (Cap. 9, pág. 248)

horizontal/horizontal: una dirección paralela a la superficie terrestre.

hydrate/hidrato: compuesto que tiene agua adherida químicamente a sus iones. Hidrato proviene de la palabra que significa "agua". (Cap. 11, pág. 319)

hydraulic/hidráulico: describe un sistema operado mediante la aplicación de presión a un líquido.

hydrocarbon/hidrocarburo: compuesto formado exclusivamente de átomos de carbono y de hidrógeno. (Cap. 13, pág. 361)

hydroelectricity/hidroelectricidad: electricidad que se produce mediante la utilización del agua en movimiento. Las plantas de energía hidroeléctrica son una manera eficiente de producir electricidad. (Cap. 25, pág. 714)

hydronium ion/ion de hidronio: ion que se forma de la disolución de una molécula polar, como el HCl, en agua. La carga negativa de las moléculas de agua atraen la carga positiva de la molécula polar, extrayendo el H^+ de la molécula polar. (Cap. 17, pág. 472)

hypothesis/hipótesis: predicción científica que se puede medir. (Cap. 1, pág. 16)

ideal machine/máquina ideal: es aquella en que el trabajo de entrada es igual al trabajo de salida. (Cap. 7, pág. 181)

incandescent light/luz incandescente: luz producida por un alambre delgado llamado filamento, el cual por lo general está hecho de tungsteno. (Cap. 19, pág. 544)

inclined plane/plano inclinado: superficie inclinada que se usa para levantar objetos. (Cap. 7, pág. 193)

incoherent light/luz incoherente: luz que puede contener más de una longitud de onda y cuyas ondas electromagnéticas no viajan en la misma dirección, haciendo que el haz se esparza. La luz de una bombilla común y corriente es luz incoherente. (Cap. 20, pág. 576)

independent variable/variable independiente: es el factor que el experimentador modifica. (Cap. 1, pág. 22)

indicator/indicador: compuesto orgánico que cambia de color en un ácido o en una base. (Cap. 17, pág. 466)

induction/inducción: cargar un objeto eléctricamente o crear una corriente eléctrica en el objeto, sin tocarlo físicamente.

inertia/inercia: tendencia de un objeto a resistir cualquier cambio en su movimiento. (Cap. 3, pág. 79)

infrared radiation/radiación infrarroja: radiación con una longitud de onda un poco más larga que la luz visible. La razón por cual la piel se siente más caliente al estar expuesta al sol es porque está absorbiendo parte de la radiación infrarroja proveniente del sol. (Cap. 19, pág. 531)

infrasonic waves/ondas infrasónicas: ondas con frecuencias inferiores a las perceptibles por el oído humano (menos de 20 Hz).

inhibitor/inhibidor: sustancia que impide que ocurra cierta reacción química, como por ejemplo las sustancias preservativas de alimentos BHT y BHA que previenen el deterioro de ciertos alimentos. (Cap. 16, pág. 457)

instantaneous speed/rapidez instantánea: razón de movimiento en cualquier instante dado. (Cap. 3, pág. 64)

insulator/aislador: material que no permite que el calor circule fácilmente a través de él. (Cap. 6, pág. 155)

insulator/aislador: material que no permite que los electrones se muevan fácilmente a través de él. Además del plástico, la madera, el caucho y el vidrio son buenos aisladores. (Cap. 21, pág. 595)

integrated circuit/circuito integrado: circuito que puede contener miles de resistores, diodos y transistores en una lámina delgada de silicio. (Cap. 23, pág. 653)

intensity/intensidad: depende de la cantidad de energía de cada onda sonora. (Cap. 18, pág. 510)

interference/interferencia: capacidad de dos o más ondas de combinarse y formar una nueva onda. (Cap. 18, pág. 519)

internal combustion engine/motor de combustión interna: máquina en que el combustible se quema dentro de unas cámaras llamadas cilindros. (Cap. 6, pág. 166)

ion/ion: átomo con carga negativa o positiva. (Cap. 11, pág. 304)

ionic bond/enlace iónico: fuerza de atracción entre los iones de carga opuesta en un compuesto iónico. (Cap. 11, pág. 305)

ionization/ionización: proceso en que el agua separa las moléculas de ciertas sustancias polares al ser éstas disueltas en agua. (Cap. 15, pág. 433)

isomers/isómeros: compuestos que tienen fórmulas químicas idénticas, pero diferentes estructuras moleculares y formas. (Cap. 13, pág. 364)

isotope/isótopo: átomo del mismo elemento que posee diferentes números de neutrones. (Cap. 10, pág. 279)

joule/julio: la unidad básica de energía y trabajo, nombrada por el científico inglés James Prescott Joule; 1 julio (J) = una fuerza de 1 newton que se mueve a través de una distancia de 1 metro.

kelvin (K)/kelvin (K): es la unidad de temperatura del SI. (Cap. 2, pág. 45)

kilogram/kilogramo: es la unidad de masa del SI. (Cap. 2, pág. 43)

kilowatt-hour/kilovatio-hora: unidad de energía eléctrica. Un kilovatio–hora equivale a 1000 vatios de potencia usada en una hora. (Cap. 21, pág. 616)

kinetic energy/energía cinética: energía en forma de movimiento. (Cap. 5, pág. 125)

kinetic theory of matter/teoría cinética de la materia: ley que dice que toda la materia está compuesta por unas partículas pequeñísimas en constante movimiento. (Cap. 8, pág. 215)

laser/láser: dispositivo que produce rayos de luz coherente. El haz del láser es angosto y enfocado y no se esparce a medida que viaja a través de distancias largas. (Cap. 20, pág. 575)

lanthanide/lantánido: cualquiera de los 14 elementos metálicos con números atómicos del 58 al 71; se usan en imanes, cerámicas e ionoscopios de televisión.

law of conservation of energy/ley de conservación de la energía: enuncia que la energía puede cambiar de forma, pero no puede ser creada ni destrui-

da bajo condiciones común y corrientes. Esta ley se aplica a sistemas cerrados en donde la energía no puede ni entrar ni salir del sistema. (Cap. 5, pág. 129)

law of conservation of mass/ley de conservación de la masa: ley que dice que la materia no puede ser creada ni destruida durante un cambio químico. De acuerdo a esta ley, la masa de todas las sustancias antes de un cambio químico es igual a la masa de todas las sustancias después de dicho cambio químico. (Cap. 9, pág. 260)

law of conservation of momentum/ley de conservación del momento: ley que dice que la cantidad total de momento de un grupo de objetos no cambia a menos que fuerzas exteriores actúen sobre los objetos. (Cap. 4, pág. 115)

lever/palanca: barra que está en libertad de girar, o moverse alrededor de un punto fijo. (Cap. 7, pág. 186)

line graph/gráfica lineal: un tipo de gráfica que se usa para mostrar tendencias haciendo uso de una recta que conecta puntos en la gráfica.

lipid/lípido: compuesto orgánico que contiene los mismos elementos que los carbohidratos, pero en diferentes proporciones y combinaciones, por ejemplo, las grasas y los aceites. (Cap. 13, pág. 377)

liquid solution/solución líquida: un disolvente líquido que ha disuelto un gas, un líquido o un sólido.

liter/litro: ocupa el mismo espacio que un decímetro cúbico. (Cap. 2, pág. 42)

loudness/volumen: percepción humana de la intensidad del sonido. Entre más altas son la intensidad y la amplitud, más alto es el sonido. (Cap. 18, pág. 510)

lubricant/lubricante: sustancia que se usa para disminuir la fricción entre dos superficies que se mueven juntas; por ejemplo, el aceite, la grasa o el grafito.

machine/máquina: dispositivo que facilita el trabajo. (Cap. 7, pág. 180)

magnetic bottle/botella magnética: un poderoso campo magnético que crea un recipiente (botella) para retener el plasma de hidrógeno necesario para causar una reacción de fusión nuclear.

magnetic domain/dominio magnético: grupos de átomos con polos magnéticos alineados en elementos como el cobalto, el níquel y otros materiales magnéticos, en los que el campo magnético creado por cada átomo ejerce una fuerza sobre los otros átomos. (Cap. 22, pág. 626)

magnetic field/campo magnético: región alrededor de un imán en donde actúan las fuerzas magnéticas. (Cap. 22, pág. 624)

magnetic pole/polo magnético: extremo de un imán en donde las fuerzas magnéticas son más fuertes. Todos los imanes tienen dos polos magnéticos opuestos—el norte (N) y el sur (S). Si un imán de barra se suspende de manera que pueda moverse libremente, el extremo norte señalará hacia el norte. (Cap. 22, pág. 624)

magnetism/magnetismo: propiedad de la materia de ejercer una fuerza de atracción entre polos distintos o de repulsión entre polos iguales. Las fuerzas magnéticas son más fuertes cerca de los extremos, o polos magnéticos de un imán. (Cap. 22, pág. 624)

magnifier/amplificador: un dispositivo que hace que las cosas se vean más grandes y con más detalles; por ejemplo, una lupa o un microscopio.

malleable/maleable: propiedad de los metales que hace posible que puedan ser martillados o enrollados en láminas. (Cap. 12, pág. 330)

mass/masa: es una medida de la materia de un objeto. (Cap. 2, pág. 43)

mass number/número de masa: suma del número de protones y de neutrones en el núcleo de un átomo. (Cap. 10, pág. 279)

mechanical advantage (MA)/ventaja mecánica: número de veces en que una máquina multiplica la fuerza de esfuerzo. (Cap. 7, pág. 182)

mechanical energy/energía mecánica: cantidad total de energía cinética y energía potencial en un sistema. (Cap. 5, pág. 129)

medium/medio: material a través del cual una onda transfiere energía. Puede ser sólido, líquido, gas o una combinación de estos. Las ondas radiales y las ondas luminosas son dos tipos de ondas que pueden viajar sin necesidad de un medio. (Cap. 18, pág. 498)

melt/derretir: cambio de una sustancia del estado sólido al líquido cuando se le calienta a más del punto de fusión o de congelación.

melting point/punto de fusión: la temperatura a la cual un sólido se convierte en líquido.

metal/metal: elemento cuyas propiedades comunes son que existe en forma sólida a temperatura ambiental, es brillante y buen conductor de calor y electricidad. (Cap. 10, pág. 290)

metallic bonding/enlace metálico: tipo de enlace en que los iones metálicos positivos están rodeados por un "mar de electrones". (Cap. 12, pág. 331)

metalloid/metaloide: elemento ubicado a los lados de la línea divisoria escalonada de la tabla periódica. (Cap. 10, pág. 291)

meter/metro: es la unidad básica de longitud del SI. La longitud de un bate de béisbol es de más o menos un metro. Las reglas métricas y las varas se usan para medir la longitud. (Cap. 2, pág. 38)

microprocessor/microprocesador: dispositivo que funciona como el cerebro de la computadora. Consiste de un circuito integrado en el tablero principal de circuitos. Recibe datos del usuario y ordena a otras partes de la computadora cómo responder, así como el cerebro de una persona ordena a una mano que se retire de algo caliente. (Cap. 23, pág. 659)

microscope/microscopio: instrumento que usa dos lentes convexas con longitudes focales relativamente cortas para ampliar objetos pequeños y cercanos. (Cap. 20, pág. 570)

microwave/microonda: onda radial con la más altas frecuencia y energía. Algunos usos de las microondas son en la comunicación y en los hornos de microondas. (Cap. 19, pág. 530)

mixture/mezcla: material hecho de elementos o compuestos revueltos, pero que no están combinados químicamente.

model/modelo: idea, sistema o estructura que representa cualquier cosa que uno trate de explicar. Un modelo nunca es exactamente igual a lo que uno trata de explicar, pero se parece lo suficiente para permitir comparaciones. (Cap. 1, pág. 12)

modulation/modulación: proceso de variar las ondas radiales. Además de la voz, otras cosas como las imágenes luminosas, la información de computadoras y la música también se pueden usar para modular las ondas radiales. (Cap. 19, pág. 530)

momentum/momento: una propiedad que posee un objeto en movimiento debido a su masa y a su velocidad. (Cap. 4, pág. 112)

monomers/monómeros: moléculas orgánicas enhebradas para formar polímeros.

music/música: sonido que se crea usando tonos específicos y calidad sonora, siguiendo un patrón regular. (Cap. 18, pág. 516)

nearsighted/miope: describe a una persona que ve claramente los objetos cercanos, pero que tiene dificultad enfocando los objetos lejanos.

net force/fuerza neta: fuerza impuesta sobre un objeto, la cual siempre hace que cambie su velocidad. (Cap. 3, pág. 78)

neutralization/neutralización: reacción química entre un ácido y una base. Durante una reacción de neutralización, los iones de hidronio del ácido se combinan con los iones de hidronio de la base para producir agua. A medida que los iones de hidróxido y de hidronio reactivos son extraídos de la solución, las propiedades ácidas y básicas de los reactantes se cancelan o neutralizan. Las reacciones de neutralización son iónicas. (Cap. 17, pág. 480)

neutralize/neutralizar: proceso de cambiar una solución ácida o una básica, de manera que sea neutra.

neutron/neutrón: uno de dos tipos principales de partículas que componen el núcleo de un átomo. Tiene casi la misma masa que un protón. (Cap. 10, pág. 271)

Newton's first law of motion/primera ley de movimiento de Newton: esta ley enuncia que un objeto que se mueve a una rapidez constante continúa moviéndose a esa rapidez a menos que una fuerza neta actúe sobre él. (Cap. 3, pág. 81)

Newton's second law of motion/segunda ley de movimiento de Newton: ley que dice que una fuerza neta que actúa sobre un objeto hace que el objeto acelere en la dirección de la fuerza. (Cap. 4, pág. 94)

Newton's third law of motion/tercera ley de movimiento de Newton: ley que describe pares de fuerzas de acción–reacción de la siguiente manera: cuando un objeto ejerce una fuerza sobre un segundo objeto, el segundo, ejerce sobre el primero una fuerza igual en tamaño y opuesta en dirección. En otras palabras, por cada fuerza de acción existe una fuerza de reacción igual y opuesta. (Cap. 4, pág. 110)

NIMBY (Not in My Backyard)/NIMBY (No en mi vecindario): punto de vista apoyado por mucha gente que está de acuerdo con el uso de las plantas de energía nuclear, siempre y cuando no se construyan muy cerca a sus vecindarios. (Cap. 25, pág. 710)

noise/ruido: sonido que carece de patrón regular y de tono. (Cap. 18, pág. 516)

nonelectrolyte/no electrolítica: sustancia que no conduce electricidad, como el agua pura. (Cap. 15, pág. 433)

nonmetal/no metal: elemento ubicado a la derecha de la línea divisoria en forma de escalera de la tabla periódica. (Cap. 10, pág. 291)

nonpolar molecule/molécula no polar: molécula que carece de extremos con carga opuesta. (Cap. 11, pág. 307)

nonrenewable resource/recurso no renovable: fuente energética, como los combustibles fósiles, que no se puede reemplazar después de ser usada. (Cap. 25, pág. 703)

normal/normal: una línea imaginaria perpendicular a una superficie reflectora o perpendicular a la superficie de un medio al cual entra la luz.

nuclear fission/fisión nuclear: proceso de romper un núcleo en dos núcleos más pequeños. La palabra fisión quiere decir "dividir". Los núcleos grandes, con números atómicos mayores de 90 pueden ser sometidos a la fisión nuclear. (Cap. 24, pág. 689)

nuclear fusion/fusión nuclear: combinación (o unión) de dos núcleos de masas bajas para formar un núcleo de mayor masa. (Cap. 24, pág. 690)

nuclear reactor/reactor nuclear: dispositivo que usa la energía de una reacción de fisión nuclear en cadena controlada para generar electricidad. (Cap. 25, pág. 704)

nuclear waste/desperdicio nuclear: subproducto radiactivo que resulta del uso de materiales radiactivos. (Cap. 25, pág. 706)

nucleic acid/ácido nucleico: polímero que controla las actividades y la reproducción de las células. Un tipo de ácido nucleico es el ácido desoxirribonucleico, DNA. (Cap. 13, pág. 375)

nucleus/núcleo: centro con carga positiva del átomo. Contiene la mayor parte de la carga de un átomo. (Cap. 10, pág. 270)

nuclide/nucleido: núcleo de un isótopo con cierto número y masa atómicos. En muchos nucleidos, la fuerza nuclear fuerte es suficiente para mantener el núcleo unido permanentemente, creando así un nucleido estable. (Cap. 24, pág. 675)

observation/observación: primer paso en un método científico, en el cual se usan los sentidos para reunir información. En la ciencia, usamos instrumentos como microscopios y balanzas para reunir cuidadosamente observaciones precisas y detalladas. (Cap. 1, pág. 16)

ocean thermal energy conversion (OTEC)/conversión de energía térmica del océano: proceso que utiliza máquinas de calor para convertir las diferencias en temperatura del agua del océano en energía mecánica para accionar turbinas y producir electricidad. (Cap. 6, pág. 173)

ohm/ohmio: la unidad para medir la resistencia.

Ohm's law/ley de Ohm: esta ley enuncia que la diferencia de potencial = corriente × resistencia, o $V = I$ (amperios) $× R$ (ohmios). (Cap. 21, pág. 606)

opaque material/material opaco: material que absorbe o refleja toda la luz y no permite ver ningún objeto a través de él. (Cap. 18, pág. 536)

optical axis/eje óptico: una línea perpendicular al centro de un espejo o una lente.

optical fiber/fibra óptica: fibra de vidrio transparente que puede transmitir la luz de un lugar a otro. (Cap. 20, pág. 578)

organic compound/compuesto orgánico: nombre con que se denominan la mayoría de los compuestos que contienen carbono. Existen algunas excepciones como el monóxido de carbono y los carbonatos. (Cap. 13, pág. 360)

organic solvent/disolvente orgánico: tipo de disolvente que se encuentra en el pegamento, las pinturas, el quitapinturas, el diluyente de pintura o barniz y algunos compuestos de calafateo y alfombrado. Se sabe que muchos disolventes orgánicos afectan el sistema nervioso y ocasionan problemas de crecimiento y desarrollo. En algunos casos, se ha sabido que varios de ellos causan cáncer. (Cap. 15, pág. 422)

oxidation number/número de oxidación: número positivo o negativo asignado a cada elemento para mostrar su capacidad de enlace en un compuesto. (Cap. 11, pág. 314)

parallel circuit/circuito en paralelo: circuito que tiene bifurcaciones a lo largo de las cuales se mueve la corriente. (Cap. 21, pág. 609)

pascal (Pa)/pascal: es la unidad de presión del SI. (Cap. 8, pág. 228)

Pascal's principle/principio de Pascal: el científico francés Blaise Pascal enunció que la presión que se le aplica a un fluido es transmitida sin cambio a través del fluido. (Cap. 8, pág. 236)

passive solar heating/calefacción solar pasiva: uso directo de la energía solar para calentar algo, sin usar ventiladores ni dispositivos mecánicos para transferir el calor de un área a otra.

period/período: hilera horizontal de elementos en la tabla periódica. (Cap. 10, pág. 290)

periodic table/tabla periódica: tabla en que los elementos se clasifican en orden de masa atómica ascendente. Está organizada en hileras y columnas para mostrar las propiedades repetitivas (periódicas) de los elementos. (Cap. 10, pág. 284)

PET/PET: (Positron Emission Tomography o tomografía de emisión de positrones) avance tecnológico en la medicina, en que mediante el análisis de rayos gama emisores se pueden revelar tumores en el encéfalo. (Cap. 24, pág. 693)

petroleum/petróleo: fuente líquida de energía que se formó de la descomposición de plantas y animales hace millones de años. El petróleo, junto con el gas natural y el carbón, se llaman combustibles fósiles. (Cap. 25, pág. 700)

pH/pH: medida de la concentración de iones de hidronio en una solución. (Cap. 17, pág. 475)

phenolphthalein/fenolftaleína: Una sustancia química que se usa como un indicador del color en valoración. Es incolora en una solución ácida, pero se vuelve rosada en una solución básica.

photon/fotón: cuanto diminuto de radiación que se parece a una partícula. Los fotones con la más alta energía corresponden a la luz con la frecuencia más alta. Los fotones con muy alta energía pueden en realidad causar daños a la materia, como por ejemplo, las células del cuerpo humano. (Cap. 19, pág. 529)

photovoltaic cell/pila fotovoltaica: también llamada pila solar, este es un dispositivo que se usa para convertir energía solar en electricidad. (Cap. 25, pág. 713)

physical change/cambio físico: cambio en tamaño, forma o estado de la materia. (Cap. 9, pág. 256)

physical property/propiedad física: cualquier característica que se puede observar en un material sin cambiar las sustancias que lo componen. Por ejemplo: color, forma, tamaño, densidad, punto de ebullición y punto de fusión. (Cap. 9, pág. 254)

physical science/ciencia física: el estudio de la materia y la energía. En el universo, todas las cosas que se pueden medir son materia o energía. Las plantas y los animales, las rocas y las nubes, los huevos y los elefantes son ejemplos de materia. Los relámpagos, los rayos, el movimiento y la luz solar son ejemplos de energía. (Cap. 1, pág. 8)

pickling/decapado: proceso en que se eliminan los óxidos y otras impurezas de las superficies de los metales, mediante la inmersión en ácido clorhídrico. (Cap. 17, pág. 469)

pigment/pigmento: material de color que absorbe algunos colores y refleja otros. Los artistas utilizan los pigmentos para hacer varios colores de pinturas. (Cap. 19, pág. 540)

pitch/tono: la agudeza o gravedad de un sonido, dependiendo de la frecuencia de las ondas sonoras. (Cap. 18, pág. 508)

plane mirror/espejo plano: espejo con una superficie plana. (Cap. 20, pág. 558)

plankton/plancton: plantas y animales acuáticos que forman la base de la cadena alimenticia para los peces pequeños. (Cap. 17, pág. 478)

plasma/plasma: mezcla parecida a un gas de partículas con cargas positivas y negativas. (Cap. 8, pág. 218)

plastic/plástico: material con base de polímeros que puede ser moldeado en diferentes formas. (Cap. 14, pág. 396)

polar molecule/molécula polar: molécula con un extremo positivo y otro negativo. (Cap. 11, pág. 307)

polarized light/luz polarizada: luz en la cual las ondas transversales vibran en un solo plano. (Cap. 20, pág. 574)

polarizing filter/filtro polarizador: un filtro formado de cadenas de moléculas, ubicadas en hileras paralelas que transmiten solo las ondas luminosas que vibran en la misma dirección de las cadenas moleculares.

polluted water/agua contaminada: se refiere al agua que contiene niveles tan altos de materiales residuales que esto la hace inaceptable para beber y otros propósitos específicos. (Cap. 8, pág. 222)

polyatomic ion/ion poliatómico: grupo de átomos unidos covalentemente y cargados positiva o negativamente, de manera que el compuesto en su totalidad contiene tres o más elementos. Muchos compuestos, incluyendo la soda de hornear, están compuestos de dos o más elementos porque contienen iones poliatómicos. El prefijo *poli* significa "-muchos", por consiguiente la palabra poliatómico quiere decir "muchos átomos". (Cap. 11, pág. 318)

polymer/polímero: molécula enorme compuesta de muchas moléculas orgánicas más pequeñas que han formado nuevos enlaces y se han unido unas a otras. (Cap. 13, pág. 373)

positron/positrón: una partícula con carga positiva parecida a un electrón.

potential difference/diferencia de potencial: diferencia en potencial entre dos lugares diferentes. Se mide en voltios, (V). (Cap. 21, pág. 600)

potential energy/energía potencial: es la energía almacenada. (Cap. 5, pág. 125)

power/potencia: proporción a la cual se realiza trabajo. (Cap. 7, pág. 202)

precipitate/precipitado: compuesto insoluble que se forma durante una reacción de desplazamiento doble. (Cap. 16, pág. 453)

pressure/presión: cantidad de fuerza ejercida por unidad de área. (Cap. 8, pág. 228)

principle/principio: una regla o ley básica que describe cómo funciona siempre algo en la naturaleza; por ejemplo, "la gravedad siempre atrae los objetos unos hacia otros".

product/producto: sustancia producida en una reacción química. (Cap. 16, pág. 442)

projectile/proyectil: cualquier cosa que se dispara o se lanza al aire. (Cap. 4, pág. 100)

propane/propano: un gas inflamable (C_3H_8); una parte del gas natural.

protein/proteína: polímero formado de compuestos orgánicos llamados aminoácidos. Existen millones de proteínas, pero solo 20 aminoácidos comunes.

(Cap. 13, pág. 373)

proton/protón: uno de dos tipos principales de partículas que componen el núcleo de un átomo. Tiene casi la misma masa que un neutrón. (Cap. 10, pág. 271)

pulley/polea: rueda que contiene una ranura por donde se puede pasar una cuerda o cadena. (Cap. 7, pág. 190)

quality/calidad: describe las diferencias entre los sonidos del mismo tono y volumen. (Cap. 18, pág. 517)

quark/quark: partícula pequeñísima que compone el protón y el neutrón. (Cap. 10, pág. 276)

radiation/radiación: transferencia de energía en forma de ondas. (Cap. 6, pág. 154)

radiator/radiador: dispositivo que tiene un área de superficie grande, el cual está diseñado para calentar el aire a su alrededor, por medio de la conducción. (Cap. 6, pág. 162)

radio wave/onda radial: onda con longitud de onda larga y frecuencia baja. Las ondas radiales poseen la energía más baja de los fotones. (Cap. 19, pág. 529)

radioactive element/elemento radiactivo: elemento cuyo núcleo se descompone e irradia partículas de energía. (Cap. 12, pág. 332)

radioactivity/radiactividad: emisión de radiación de alta energía o de partículas del núcleo de un átomo radiactivo. (Cap. 24, pág. 674)

RAM/RAM: (*Random Access Memory* o memoria de acceso aleatorio) tipo de memoria temporal de microcomputadora que almacena datos, documentos y programas mientras están en uso. La información almacenada en RAM se pierde cuando se apaga la computadora. (Cap. 23, pág. 660)

rarefaction/rarefacción: el área menos densa de la onda, en ondas de compresión.

reactant/reactante: cualquier sustancia que está a punto de reaccionar en una reacción química. (Cap. 16, pág. 442)

real image/imagen real: imagen verdadera ampliada

y patas arriba en la cual los rayos luminosos se encuentran en la imagen, de modo que uno puede ver una pantalla sostenida allí. (Cap. 20, pág. 559)

rectifier/rectificador: dispositivo que cambia la corriente alterna en corriente directa. (Cap. 23, pág. 651)

reflecting telescope/telescopio reflector: telescopio que usa un espejo cóncavo, un espejo plano y una lente convexa para ampliar objetos distantes. (Cap. 20, pág. 569)

reflection/reflexión: ocurre cuando una onda choca contra un objeto y rebota del mismo. La reflexión ocurre con todo tipo de ondas—electromagnéticas, de sonido y de agua. (Cap. 19, pág. 546)

refracting telescope/telescopio refractor: telescopio que usa dos lentes convexas para recoger y enfocar la luz de objetos distantes. (Cap. 20, pág. 568)

refraction/refracción: cambio en la dirección de propagación de las ondas causado por un cambio en velocidad al moverse de un medio a otro. La cantidad de refracción depende de la velocidad de la luz en ambos materiales. Entre mayor sea la diferencia entre las velocidades en los dos medios, más se dobla o refracta la luz al pasar, haciendo un ángulo de un medio al otro. (Cap. 19, pág. 547)

resistance/resistencia: tendencia de los materiales a oponerse al flujo de electrones, cambiando la energía eléctrica en energía térmica y luz. La resistencia se mide en ohms. (Cap. 21, pág. 603)

resistance arm/brazo de resistencia: parte de la palanca que ejerce la fuerza de resistencia. (Cap. 7, pág. 186)

resistance force (F_r)/fuerza de resistencia: la fuerza aplicada por la máquina para superar la resistencia. (Cap. 7, pág. 181)

resonance/resonancia: tipo de vibración que sucede cuando el sonido que llega hasta un objeto tiene la misma frecuencia natural del objeto, haciendo que el objeto comience a vibrar a esa frecuencia. (Cap. 18, pág. 517)

reverberation/reverberación: efecto producido por múltiples reflexiones del sonido. (Cap. 18, pág. 520)

RNA/RNA: Ácido RiboNucleico; un ácido nucleico que controla la producción de las proteínas que forman nuevas células.

ROM/ROM: (*Read Only Memory* o memoria de solo lectura) tipo de información almacenada permanentemente en una computadora y la cual contiene las instrucciones necesarias para que el microprocesador pueda hacer funcionar la computadora. La computadora puede leer este tipo de información, pero el usuario no puede añadirle más ni cambiar este tipo de información. La información en ROM queda almacenada permanentemente, es decir, no se pierde cuando se apaga la computadora. (Cap. 23, pág. 660)

salt/sal: compuesto que se forma cuando los iones negativos de un ácido se combinan con los iones positivos de una base. (Cap. 17, pág. 480)

saponification/saponificación: proceso de fabricación del jabón. (Cap. 17, pág. 483)

saturated hydrocarbon/hidrocarburo saturado: hidrocarburo que solo contiene átomos de carbono de enlace sencillo. (Cap. 13, pág. 362)

saturated solution/solución saturada: solución que ha disuelto todo el disolvente que era capaz de disolver normalmente, a una temperatura dada. (Cap. 15, pág. 426)

scientific law/ley científica: regla de la naturaleza que resume observaciones relacionadas y resultados experimentales con el propósito de describir patrones en la naturaleza. (Cap. 1, pág. 17)

screw/tornillo: plano inclinado enrollado en forma de espiral alrededor de un poste cilíndrico. (Cap. 7, pág. 194)

second/segundo: es la unidad de tiempo del SI. (Cap. 2, pág. 44)

semiconductor/semiconductor: material que conduce una corriente eléctrica bajo ciertas circunstancias. (Cap. 12, pág. 346)

series circuit/circuito en serie: circuito en el que la corriente solo tiene una ruta a través de la cual puede fluir. (Cap. 21, pág. 608)

SI/SI: o sistema internacional de unidades, es el sistema estándar de medidas que se usa a través del mundo. En el SI, cada tipo de medida tiene una unidad básica, como por ejemplo, el metro, que es la unidad básica de longitud. (Cap. 2, pág. 35)

simple machine/máquina simple: dispositivo que realiza trabajo con un solo movimiento. (Cap. 7, pág. 180)

single–displacement reaction/reacción de desplazamiento simple: ocurre cuando un elemento reemplaza a otro en un compuesto. (Cap. 16, pág. 452)

soap/jabón: sal orgánica. Se forma de la reacción de grasas o aceites con hidróxido de sodio o de potasio. (Cap. 17, pág. 483)

solar collector/colector solar: dispositivo que absorbe la energía radiante del Sol. (Cap. 6, pág. 164)

solar energy/energía solar: energía proveniente del sol. (Cap. 6, pág. 163)

solubility/solubilidad: máximo número de gramos de una sustancia que se pueden disolver en 100 gramos de disolvente a cierta temperatura. (Cap. 15, pág. 424)

solute/soluto: sustancia que se disuelve en el disolvente. (Cap. 15, pág. 417)

solution/solución: es otro nombre para una mezcla homogénea. (Cap. 9, pág. 248)

solvent/disolvente: sustancia que disuelve el soluto. (Cap. 15, pág. 417)

specific heat/calor específico: es la cantidad de energía que se necesita para subir un kelvin, en la temperatura de un kilogramo de material. (Cap. 5, pág. 141)

speed/rapidez: razón de cambio en posición. (Cap. 3, pág. 64)

standard/estándar: una cantidad exacta que la gente acuerda usar para hacer comparaciones. (Cap. 2, pág. 34)

state of matter/estado de la materia: cada una de las cuatro formas en que puede existir la materia: sólido, líquido, gas y plasma. (Cap. 8, pág. 214)

static electricity/electricidad estática: acumulación de cargas eléctricas en un objeto. (Cap. 21, pág. 592)

step-down transformer/transformador rebajador: un transformador eléctrico que disminuye (rebaja) el voltaje de una línea de fuerza eléctrica.

step-up transformer/transformador elevador: un transformador eléctrico que aumenta (eleva) el voltaje de una línea de fuerza eléctrica.

strong acid/ácido fuerte: ácido que se ioniza casi completamente en solución. (Cap. 17, pág. 474)

strong base/base fuerte: es la que se disocia completamente en solución. (Cap. 17, pág. 474)

sublimation/sublimación: proceso en que un sólido cambia directamente a vapor sin pasar por el estado líquido. (Cap. 12, pág. 343)

submerge/sumergir: introducir y cubrir completamente algo en un líquido.

substance/sustancia: puede ser un elemento o un compuesto. (Cap. 9, pág. 247)

substituted hydrocarbon/hidrocarburo de sustitución: hidrocarburo en que uno más de sus átomos de hidrógeno han sido reemplazados por átomos de otros elementos. (Cap. 13, pág. 367)

supercollider/superchocador: un dispositivo que hace que los protones choquen a altas velocidades y se conviertan en quarks.

superconductors/superconductores: materiales que no presentan resistencia eléctrica. Cuando estos materiales se sobreenfrían a una temperatura de casi cero absoluto (0K), –273°C, pierden toda resistencia a la electricidad. (Cap. 22, pág. 640)

supersaturated solution/solución supersaturada: solución que contiene más soluto que una solución saturada a una temperatura dada. (Cap. 15, pág. 428)

suspension/suspensión: mezcla heterogénea que contiene un líquido en el cual se han asentado las partículas visibles. (Cap. 9, pág. 250)

synthesis reaction/reacción de síntesis: reacción en la cual dos o más sustancias se combinan para formar otra sustancia. (Cap. 16, pág. 451)

synthetic fiber/fibra sintética: filamento de un polímero sintético. Un filamento de una fibra sintética llamada Kevlar es cinco veces más fuerte que un filamento similar de acero. (Cap. 14, pág. 399)

technology/tecnología: uso práctico de la información tecnológica. Los descubrimientos científicos pueden conducir a innovaciones tecnológicas, las cuales a su vez, pueden conducir a más descubrimientos. Tanto la ciencia pura, para el avance de los conocimientos, como la ciencia aplicada, o tecnología, son partes importantes de la ciencia. (Cap. 1, pág. 8)

telephoto lens/teleobjetivo: lente con distancia focal más larga y ubicado más lejos de la película que un objetivo granangular. Son fáciles de reconocer porque sobresalen de la cámara para aumentar la distancia entre la lente y la película. La imagen se ve ampliada y el objeto parece estar más cerca de lo que en realidad está. (Cap. 20, pág. 571)

temperature/temperatura: es una medida del promedio de energía cinética de las partículas en una muestra de materia. (Cap. 5, pág. 134)

terminal velocity/velocidad terminal: es la máxima velocidad alcanzada por un objeto que cae. (Cap. 4, pág. 99)

theory/teoría: explicación basada en muchas observaciones, las cuales se basan en resultados experimentales. Es la manera más lógica de explicar el funcionamiento de las cosas. (Cap. 1, pág. 17)

thermal energy/energía térmica: es la energía total de las partículas en un material. Este total incluye tanto energía cinética como potencial. (Cap. 5, pág. 135)

thermal expansion/expansión térmica: cualidad que tiene la materia de expandirse cuando se calienta y de contraerse cuando se enfría. (Cap. 8, pág. 219)

thermal pollution/contaminación térmica: agua calentada excesivamente por las plantas de energía eléctrica o las industrias y que luego se vierte en los ríos o fuentes de agua. (Cap. 8, pág. 223)

thermal pollution/contaminación térmica: ocurre cuando el calor residual cambia la temperatura del ambiente de modo significativo. (Cap. 5, pág. 138) thermonuclear fusion/fusión termonuclear: fusión nuclear que ocurre bajo condiciones de intenso calor (millones de grados), como en una estrella o en nuestro sol.

tidal energy/energía de las mareas: energía generada por el movimiento de las mareas del océano. (Cap. 25, pág. 714)

time/tiempo: intervalo entre dos eventos. (Cap. 2, pág. 44)

titration/valoración: proceso en que una solución de concentración conocida se usa para determinar la concentración de otra solución. (Cap. 17, pág. 481)

total internal reflection/reflexión interna total: ocurre cuando la luz que choca contra una superficie entre dos materiales se refleja completamente en el primer material. (Cap. 20, pág. 577)

toxic/tóxico: compuestos venenosos. (Cap. 11, pág. 312)

tracer/indicador radiactivo: uso de radioisótopos dentro del cuerpo humano para inspeccionar los procesos corporales. El carbono–11, el yodo–131 y el sodio–123 son indicadores radiactivos porque poseen medias vidas relativamente cortas. (Cap. 24, pág. 692)

transformer/transformador: dispositivo que puede aumentar o disminuir el voltaje de una corriente alterna. Su funcionamiento involucra principios tanto de electromagnetismo como de inducción electromagnética. (Cap. 22, pág. 635)

transistor/transistor: semiconductor que amplifica o fortalece una señal eléctrica o que actúa como un diminuto interruptor eléctrico. (Cap. 23, pág. 651)

transition element/elemento de transición: elementos pertenecientes a los grupos del 3 al 12 de la tabla periódica. (Cap. 12, pág. 334)

translucent material/material translúcido: material que permite que cierta luz lo atraviese, pero no deja ver claramente los objetos a través de él. (Cap. 19, pág. 536)

transmutation/transmutación: proceso de cambiar un elemento en otro mediante la desintegración nuclear. (Cap. 24, pág. 680)

transparent material/material transparente: material que permite que la luz los atraviese y por consiguiente deja ver los objetos a través de él. (Cap. 19, pág. 536)

transuranium elements/elementos transuránicos: elementos que tienen más de 92 protones, que es el número atómico del uranio. Todos estos elementos sintéticos son inestables y se descomponen rápidamente. (Cap. 12, pág. 338)

transverse wave/onda transversal: tipo de onda en el cual el medio se mueve formando ángulos rectos con la dirección en que viaja la onda. (Cap. 18, pág. 499)

trough/valle: punto más bajo de una onda. (Cap. 18, pág. 499)

Tyndall effect/efecto de Tyndall: el esparcimiento de la luz por las partículas de una mezcla. (Cap. 9, pág. 249)

ultrasonic technology/tecnología ultrasónica: tecnología que hace uso del ultrasonido para una variedad de propósitos, como en la joyería, en la medicina y en la industria química. (Cap. 18, pág. 514)

ultraviolet radiation/radiación ultravioleta: radiación con frecuencia más alta que la luz visible, de modo que sus fotones son más energéticos y tienen más potencia de penetración que los fotones de la luz visible. (Cap. 18, pág. 532)

unsaturated hydrocarbon/hidrocarburo no saturado: hidrocarburo, como el etino o el eteno, que contiene por lo menos un enlace doble o triple entre los átomos de carbono. (Cap. 13, pág. 364)

unsaturated solution/solución no saturada: cualquier solución que puede disolver más soluto, a cierta temperatura dada. (Cap. 15, pág. 427)

velocity/velocidad: describe tanto la rapidez como la dirección de un objeto. (Cap. 3, pág. 72)

Venturi effect/efecto de Venturi: aumento en la velocidad de un fluido, que al ser forzado a fluir a través de espacios estrechos, disminuye su presión. Una demostración dramática del efecto de venturi ha ocurrido en ciudades donde el viento es forzado a fluir entre hileras de rascacielos—la reducción en presión fuera del edificio durante vientos fuertes ha hecho que las ventanas se desprendan debido a que la presión dentro del edificio es mayor. (Cap. 8, pág. 238)

virtual image/imagen virtual: imagen en la cual los rayos luminosos no pasan a través de ella. La imagen virtual que forma un espejo plano está erguida y aparece tan atrás del espejo como el objeto esté en frente del mismo. (Cap. 20, pág. 558)

visible radiation/radiación visible: o luz, es la única parte del espectro electromagnético que podemos ver. Cubre solo un pequeño alcance del espectro, comparada con otros tipos de energía. (Cap. 19, pág. 532)

volt/voltio: la unidad para medir el potencial de energía eléctrica. Se abrevia V.

voltage/voltaje: una diferencia en potencial eléctrico, medida en voltios con un voltímetro.

voltmeter/voltímetro: instrumento que mide, en voltios, la diferencia de potencial en un circuito. (Cap. 22, pág. 630)

volume/volumen: cantidad de espacio que ocupa un objeto. (Cap. 2, pág. 41)

wave/onda: perturbación rítmica que transporta energía a través de la materia o del espacio. (Cap. 18, pág. 498)

wavelength/longitud de onda: distancia entre un punto de una onda y otro punto idéntico en la siguiente onda, como por ejemplo, de una cresta a la siguiente o de un valle al siguiente. (Cap. 18, pág. 499)

weak acid/ácido débil: ácido que solo se ioniza parcialmente en solución. (Cap. 17, pág. 474)

weak base/base débil: es la que no se disocia completamente en solución. (Cap. 17, pág. 474)

wedge/cuña: plano inclinado con uno o dos lados que se inclinan. (Cap. 7, pág. 194)

weight/peso: medida de la fuerza de gravedad sobre un objeto. (Cap. 3, pág. 84)

wet cell/pila húmeda: es la que tiene dos placas conectadas, las cuales están hechas de diferentes metales o compuestos metálicos, en una solución electrolítica. La mayoría de las baterías de los carros, llamadas baterías de almacenaje de plomo, contienen una serie de seis pilas húmedas hechas de placas de plomo y de dióxido de plomo en una solución de ácido sulfúrico. (Cap. 21, pág. 602)

wheel and axle/rueda y eje: máquina simple que consiste en dos ruedas de diferentes tamaños las cuales rotan juntas. (Cap. 7, pág. 191)

wide-angle lens/objetivo granangular: lente con distancia focal corta que produce una imagen relativamente pequeña del objeto pero que incluye la mayor parte de sus alrededores. Este tipo de lente debe colocarse cerca de la película para enfocar la imagen con su corta longitud focal. (Cap. 20, pág. 571)

work/trabajo: transferencia de energía a través del movimiento. (Cap. 5, pág. 126)

X ray/rayo X: tipo de radiación que posee una longitud de onda más corta y una frecuencia más alta que la radiación ultravioleta. Los fotones de los rayos X transportan energía más alta y poseen una mayor potencia de penetración que cualquier otro tipo de radiación electromagnética, a excepción de los rayos gamma. (Cap. 19, pág. 534)

Index

The Index for *Glencoe Physical Science* will help you locate major topics in the book quickly and easily. Each entry in the Index is followed by the numbers of the pages on which the entry is discussed. A page number given in **boldface type** indicates the page on which that entry is defined. A page number given in *italic type* indicates a page on which the entry is used in an illustration or photograph. The abbreviation *act.* indicates a page on which the entry is used in an activity.

15, 31, 42, 45, 48, Morgan-Cain and Associates; 48, David Ashby; 49, 50, 51, 57, 65, 67, Morgan-Cain and Associates; 68, Bill Singleton/John Edwards; 69, Precision Graphics; 70, 73, 74, Morgan-Cain and Associates; 86, Thomas Gagliano; 91, Morgan-Cain and Associates; 98, Antonio Castro; 101, 108, 109, Morgan-Cain and Associates; 110, Antonio Castro; 112, Jim Shough; 121, Morgan-Cain and Associates; 125, George Bucktell; 128, Dick Smith; 132, 133, 134, 139, Morgan-Cain and Associates; 143, Henry Hill; 149, Morgan-Cain and Associates; 152, Sarah Woodward; 154, 155, Morgan-Cain and Associates; 157, George Bucktell; 158, Jacque Auger; 162, 163, 164, 165, Morgan-Cain and Associates; 166, Thomas Gagliano; 167, Jacque Auger; 168, Morgan-Cain and Associates; 173, Thomas Gagliano; 177, 181, 183, 186, 188, Morgan-Cain and Associates; 189, Thomas Gagliano; 190, 191, Morgan-Cain and Associates; 192, Henry Hill/John Edwards; 197, 199, Thomas Gagliano; 207, (top right) Morgan-Cain and Associates; 207, (bottom left) Dick Smith; 208, Tom Kennedy/Romark Illustration; 215, 216, Morgan-Cain and Associates; 217, Jim Shough; 226, Henry Hill; 226, 228, 231, 235, Morgan-Cain and Associates; 236, Henry Hill/John Edwards; 243, 248, 249, 252 Morgan-Cain and Associates; 272, Henry Hill; 273, 276, 277, 279, Morgan-Cain and Associates; 281, Henry Hill; 285, (t) & (b) Morgan-Cain and Associates; 286-287, Dave Reed; 289 (t) & (b), 295, 299 (t) & (b), 301 (t) & (b), 302, 304, 305 (t) & (b), 306 (t) & (b), 308, 314, 315, 323 (t) & (b), 325, 327, 331-334, 336 (t) & (b), 338-340, 341 (t) & (b), 342, 343, 346, Morgan-Cain and Associates; 347 (t), Tom Kennedy/Romark Illustrations; 347 (b), 348 (t), Morgan-Cain and Associates; 348 (b), George Bucktell; 349, 350, 357, 361 (t) & (b), 362, 363, (t) & (b), 366 (t) & (b), 367, 373, 375, 390, Morgan-Cain and Associates; 398, 400, Jacque Auger; 409, 411, 418, 419, 427, 432, 433 (t) & (b), 445, 446, 463, 469, 472, 473, 475, Morgan-Cain and Associates; 478, 479, Chris Forsey; 481, 483, Morgan-Cain and Associates; 493, Tom Kennedy/Romark Illustration; 499, 500, 505, 506, Morgan-Cain and Associates; 508, Jim Jobst; 510 (t), Morgan-Cain and Associates; 510 (b), Chris Forsey; 517, 518, Morgan-Cain and Associates; 519, Glen Wasserman; 529, 530, 533, 538, Morgan-Cain and Associates; 540, Bill Watterson; 541, 542, 544, 546, 547, 550, 555, Morgan-Cain and Associates; 558, 559, 562, 564, 565, 566, 567, Thomas Gagliano; 569, Jim Shough; 570, David Fischer; 573, Jacque Auger; 574, 576, Thomas Gagliano; 579, Jacque Auger; 556, 592, 593, Morgan-Cain and Associates; 597, Henry Hill/John Edwards; 600, Sarah Woodward; 601, Morgan-Cain and Associates; 602, Jacque Auger; 604, 610, 621, Morgan-Cain and Associates; 634, Jacque Auger; 646, 649, 658-659, 671, 654-655, 675, 679, 680, 689, Morgan-Cain and Associates; 690, Henry Hill; 697, 701, 704-705, 710, Morgan-Cain and Associates; 714, Tom Kennedy/Romark Illustration; 735, 736 (t) & (b), 737, 738 (t) & (b), 740 (t) & (b), 741(t) & (b), 748 (t), Rolin Graphics, 748 (b), Morgan-Cain and Associates; 749 (t), Barbara Hoopes-Ambler, 749 (b), David Ashby; 751, (t) & (b), Morgan-Cain and Associates

Photo Credits

Cover Ken Ross/FPG International; **Connect to Chemistry** (flask)K&C Fischer; **vi,** Jeff Smith/Fotosmith; **vii,** (t)Rick Doyle, (b)David W. Hamilton/The Image Bank; **viii,** (t)Timothy Fuller, (b)Peter Skinner/Photo Researchers; **viii-ix,** Jeff Smith/Fotosmith; **ix,** Richard Megna/Fundamental Photographs, (b)Jeff Smith/Fotosmith; **x,** Wayne Sproul/International Stock Photo; **xi,** (t)Jeff Smith/Fotosmith, (b)Dominic Oldershaw; **xii,** (t) Timothy Fuller, (b)Alfred Pasieka/Peter Arnold, Inc.; **xiii,** (t)Timothy Fuller, (c)Yoav Levy/PhotoTake NYC, (b)Richard Megna/Fundamental Photographs; **xiv,** (l)Sean Justice, (r)Doug Martin; **xv,** (t)Michael Tamborrino/FPG International, (b)Gordon Gahar/Lawrence Livermore Laboratory; **xvi,** Russell D. Curtis/Photo Researchers; **xvii, xviii,** Timothy Fuller; **xix,** Dominic Oldershaw; **xx,** Jeff Smith/The Image Bank; **xxi,** Timothy Fuller; **xxii,** (t)GCA/CNRI/PhotoTake NYC, (b)courtesy Insurance Institute for Highway Safety; **xxiii,** Jeff Smith/Fotosmith; **xxiv** David M. Allen; **xxv,** (t)SuperStock, used by permission of the artist's estate, (b)David Sutherland/Tony Stone Images; **2-3,** Jeff Smith/Fotosmith; **3,** Richard T. Nowitz/Photo Researchers; **4-5,** Sean Justice; **6,** Don Smetzer/Tony Stone Images; **7,** Jeff Smith/Fotosmith; **8,** T. Wilfit/The Image Bank; **9,** Timothy Fuller; **10,** Jeff Smith/Fotosmith; **11,** Franklin Over; **12,** Telegraph Colour Library/FPG International; **13,** Dominic Oldershaw; **14,** (tl tr)Franklin Over, (b)Visuals Unlimited/Charlie Heidecker; **16,** Sean Justice; **17,** Telegraph Colour Library/FPG International; **18, 19,** The Kobal Collection; **20,** KS Studios; **21,** Sean Justice; **22,** Thomas Veneklasen; **23,** Sean Justice; **24,** (bc)Thomas Veneklasen, (others)Timothy Fuller; **25,** Timothy Fuller; **26, 27,** Dominic Oldershaw; **28,** Patricia Clancey; **30,** Dominic Oldershaw; **32, 33,** Timothy Fuller; **34,** Jeff Smith/Fotosmith; **35,** National Bureau of Weights and Measures; **37,** Jeff Smith/Fotosmith; **39,** Timothy Fuller; **40,** Jeff Smith/Fotosmith; **41,** NASA; **43,** Jeff Smith/Fotosmith; **44,** (l)courtesy The Children's Museum, (tc)V.A.P Neal/Photo Researchers, (tr bc)Timothy Fuller, (br)Visuals Unlimited/National Bureau of Standards; **46,** Jeff Smith/Fotosmith; **52,** Timothy Fuller; **53,** (t)Dr. E.R. Degginger/Color-Pic, (b)Eunice Harris/Photo Researchers; **54,** SuperStock, used by permission of the artist's estate; **58,** Dr. E.R. Degginger/Color-Pic; **59,** KS Studios; **60-61,** Harold Lee Miller; **61,** Russell D. Curtis/Photo Researchers; **62-63,** Liaison International; **63,** Jeff Smith/Fotosmith; **64,** Yann Guichaoua/Vandystadt/Photo Researchers; **65,** Timothy Fuller; **66,** Jeff Smith/Fotosmith; **70,** StudiOhio; **71,** Timothy Fuller; **72,** Warren Faidley/International Stock Photo; **73,** NHRA Photo; **74,** Bettmann Newsphotos; **76,** courtesy Insurance Institute for Highway Safety; **77,** Bud Fowle; **78,** Tim DeFrisco/Allsport USA; **79,** (t)Timothy Fuller, (b)Harry Skull/Allsport USA; **80,** (t)Ben Rose/The Image Bank, (b)Romilly Lockyer/The Image Bank; **81, 82,** Jeff Smith/Fotosmith; **83,** Jonathan Daniel/Allsport USA; **84, 85,** NASA; **86,** Timothy Fuller; **87,** Doug Martin; **88,** David M. Allen; **92-93,** NASA/Peter Arnold, Inc.; **93,** Jeff Smith/Fotosmith; **94, 95,** Timothy Fuller; **96,** Peticolas-Megna/Fundamental Photographs; **97,** Timothy Fuller; **98,** (l)Bob Firth/International Stock Photo, (r)Warren Faidley/International Stock

Fuller; **249**, (t c)Jeff Smith/Fotosmith, (b)Timothy Fuller; **251**, Doug Martin; **253**, James Stevenson/Science Photo Library/Photo Researchers; **254**, Jeff Smith/Fotosmith; **255**, (t)Jeff Smith/Fotosmith, (bl)Visuals Unlimited/Arthur R. Hill, (br)Doug Martin; **256**, P&G Bowater/The Image Bank; **257**, John Brennels ©1990, DISCOVER Publications; **258**, (t)Visuals Unlimited/Richard Thorn, (b)David W. Hamilton/The Image Bank; **259**, **260**, **261**, Timothy Fuller; **262**, **263**, Jeff Smith/Fotosmith; **264**, David Noble/FPG International; **266**, Jeff Smith/Fotosmith; **267**, Tom McHugh/Photo Researchers; **268**, Mark Marten/NASA/Photo Reserachers; **268-269**, AP/Wide World Photos; **269**, Timothy Fuller; **270**, **272**, Jeff Smith/Fotosmith; **273**, Timothy Fuller; **275**, Doug Martin; **277**, Hank Morgan/Science Source/Photo Researchers; **278**, Timothy Fuller; **279**, **280**, Doug Martin; **283**, Jeff Smith/Fotosmith; **284**, Stamp from the Collection of Prof. C.M. Lang, Photography by Gary Shulfer, University of WI-Steven's Point: Russia 3608; **285**, Dr. E.R. Degginger/Color-Pic; **288**, (t)Jeff Smith/Fotosmith, (b)Timothy Fuller; **290**, Visuals Unlimited/IBMRL; **291**, (t)Manfred Kage/Peter Arnold, Inc., (b)Richard Megna/Fundamental Photographs; **292**, Fermilab Visual Media Services; **295**, Jeff Smith/Fotosmith; **296-297**, Tony Wiles/Tony Stone Images; **297**, **298**, Timothy Fuller; **299**, (l)Charles E. Zirkle, (c)Richard Megna/Fundamental Photographs, (r)Jeff Smith/Fotosmith; **300**, Timothy Fuller; **302**, First Image; **303**, StudiOhio; **304**, Richard Megna/Fundamental Photographs; **307**, Manfred Kage/Peter Arnold, Inc.; **309**, Stuart Brinin; **310**, **311**, **313**, Timothy Fuller; **314**, (l)Mercury Archives/The Image Bank, (r)H.R. Bramaz/Peter Arnold, Inc.; **319**, Jeff Smith/Fotosmith; **320**, Timothy Fuller; **322**, **323**, Jeff Smith/Fotosmith; **324**, (l)FPG International, (tr)Bettmann Archive, (br)UPI/Bettmann Newsphotos; **326-327**, Tim Courlas; **328-329**, SuperStock; **329**, Richard Megna/Fundamental Photographs; **330**, (l)Diane Schiumo/Fundamental Photographs, (r)Visuals Unlimited/Charles Sykes; **331**, Timothy Fuller; **332**, Richard Megna/Fundamental Photographs; **333**, (l)Doug Martin, (c)ZEFA/H. Armstrong Roberts, (r)StudiOhio; **334**, Andy Caulfield/The Image Bank; **335**, Mugshots/The Stock Market; **336**, (t)Jeff Smith/Fotosmith, (b)Timothy Fuller; **338**, EPRI/Science Source/Photo Researchers; **339**, Timothy Fuller; **340**, Mark Burnett/Photo Researchers; **341**, Annie Griffiths/Westlight; **342**, (t)Timothy Fuller, (b)Ben Simmons/The Stock Market; **343**, (t)Jeff Smith/Fotosmith, (b)Jerry Schad/Photo Researchers; **344**, Charles Steiner/International Stock Photo; **345**, Matt Meadows; **346**, Timothy Fuller; **347**, (t)Timothy Fuller, (b)Visuals Unlimited/Smithsonian Institution, A.J. Copley; **349**, (l)Timothy Fuller, (r)Doug Martin; **350**, (l)E.R. Degginger/Photo Researchers, (r)Gregory G. Dimijian/Photo Researchers; **351**, Ken Frick; **352**, **through 360**, Timothy Fuller; **362**, Ken Frick; **364**, Timothy Fuller; **365**, (l)Doug Martin, (r)Russell D. Curtis/Photo Researchers; **366**, Timothy Fuller; **367**, Ken Frick; **368**, **369**, Timothy Fuller; **370**, Jean Higgins/Unicorn Stock Photo; **371**, Visuals Unlimited/Link; **372**, Doug Martin; **373**, (l)Visuals Unlimited/David M. Phillips, (r)Ken Frick; **374**, Aaron Haupt; **376**, Timothy Fuller; **377**, (t)Ken Frick, (b)Bill Denison/Uniphoto; **378**, Wayne Sproul/International Stock Photo; **379**, (l)John M. Brunley/Photo Researchers, (r)Visuals Unlimited/William J. Weber; **380**, **381**, Timothy Fuller; **382**, Stuart Brinin; **384**, Aaron Haupt; **386-387**, courtesy Raytheon Aircraft; **387**, Jeff Smith/Fotosmith; **388**, Timothy

Fuller; **389,** (l r)Jeff Smith/Fotosmith, (c)Timothy Fuller; **390,** (tl)Dr. E.R. Degginger/Color-Pic, (others)Doug Martin; **391,** Visuals Unlimited/SIU; **392,** Smithsonian Institution; **393,** (t)Ken Frick, (bl)Brent Turner/BLT Productions, (br)Jeff Smith/Fotosmith; **394,** Timothy Fuller; **395,** Doug Martin; **396, 399, 400,** Timothy Fuller; **401,** Jeff Smith/Fotosmith; **402,** Timothy Fuller; **403,** Doug Martin; **404,** (t)David Madison/DUOMO, (b)Marvin E. Newman/The Image Bank; **405,** (t)Mark D. Phillips/Photo Researchers, (b)Timothy Fuller; **406,** Cory Warner; **409,** Dominic Oldershaw; **410,** Timothy Fuller; **412-413,** Laurence Parent; **413,** Simon Fraser/Science Photo Library/Photo Researchers; **414-415,** Wiley & Wales/Adventure Photo; **415,** Thomas Veneklasen; **416,** (l)Tate Gallery, London/Art Resource NY, (r)Norbert Wu; **417,** Mark Burnett; **418,** Sean Justice; **420,** Mark Burnett; **421, 422,** Matt Meadows; **423,** Ed Taylor/FPG International; **424,** Sean Justice; **425,** Mark Burnett; **428,** Custom Medical Stock Photo; **429,** Richard Megna/Fundamental Photographs; **430,** Thomas Veneklasen; **431,** Matt Meadows; **432,** Jeff Smith/Fotosmith; **433,** Norbert Wu; **435,** Pictor/Uniphoto; **436,** Thomas Veneklasen; **438,** Norbert Wu; **439,** Leonard Lessin/Peter Arnold, Inc.; **440-441,** Visuals Unlimited/Jeff Greenberg; **441,** Timothy Fuller; **442,** Bettmann Archive; **443, 444,** Timothy Fuller; **447,** NASA; **448,** Timothy Fuller; **451,** TSADO/NASA/Tom Stack & Associates; **452,** Peticolas-Megna/Fundamental Photographs; **454-455,** Timothy Fuller; **456,** Kent & Donna Dennen/Photo Researchers; **457,** AC Delco, General Motors/Peter Arnold, Inc.; **458,** Visuals Unlimited/Bernd Wittich; **459,** Timothy Fuller; **460,** SuperStock; **462,** Jeff Smith/Fotosmith; **463,** Timothy Fuller; **464-465,** Uniphoto; **465, 466, 467,** Dominic Oldershaw; **468,** Doug Martin; **470,** Dominic Oldershaw; **471,** Grant Heilman/Grant Heilman Photography; **474,** Dr. E.R. Degginger/Color-Pic; **476,** Dominic Oldershaw; **477,** Doug Martin; **482, 486, 487,** Timothy Fuller; **488,** Cory Warner; **492, 493,** Thomas Veneklasen; **494-495,** Steve Jennings/LGI; **495,** Will & Deni McIntyre/Photo Researchers; **496-497,** Susan McCartney/Photo Researchers; **497,** Timothy Fuller; **498,** Guy Marche/FPG International; **502, 503,** Timothy Fuller; **504, 507,** Francois Gohier/Photo Researchers; **509,** NASA/Science Source/Photo Researchers; **511,** Dr. E.R. Degginger/Color-Pic; **512,** Timothy Fuller; **513,** StudiOhio; **514,** Robert Goldstein/Photo Researchers; **515,** PhotoTake NYC; **516, 517,** Timothy Fuller; **518,** (l)Wide World Photos, (r)Timothy Fuller; **520,** courtesy CAPA, photo by D.R. Goff; **521,** SuperStock; **522,** Runk/Schoenberger from Grant Heilman; **526,** Pete Saloutos/The Stock Market; **527,** Dominic Oldershaw; **528,** Telegraph Colour Library/FPG International; **529, 530,** Dominic Oldershaw; **531,** (t)Howard Sochurek/The Stock Market, (b)Alfred Pasieka/Peter Arnold, Inc.; **532,** Dominic Oldersahw; **533,** Richard Hutchings/Photo Researchers; **534,** (t)Timothy Fuller, (b)Scott Camazine/Photo Researchers; **535,** Alvis Upitis/The Image Bank; **536,** Thomas Veneklasen; **537,** Gabriel Covian/The Image Bank; **538,** Jeff Smith/Fotosmith; **539,** Timothy Fuller; **540,** CALVIN AND HOBBES ©1993, Bill Watterson./Dist. by UNIVERSAL PRESS SYNDICATE. Reprinted with permission. All rights reserved; **543-547,** Dominic Oldershaw; **548,** David Parker/Science Photo Library/Photo Researchers; **549,** Hank Morgan from Rainbow; **550,** Richard Laird/FPG International; **551,** Doug Martin; **552,** Dominic Oldershaw; **556, 557,** Timothy Fuller; **559,**

PERIODIC TABLE OF THE ELEMENTS

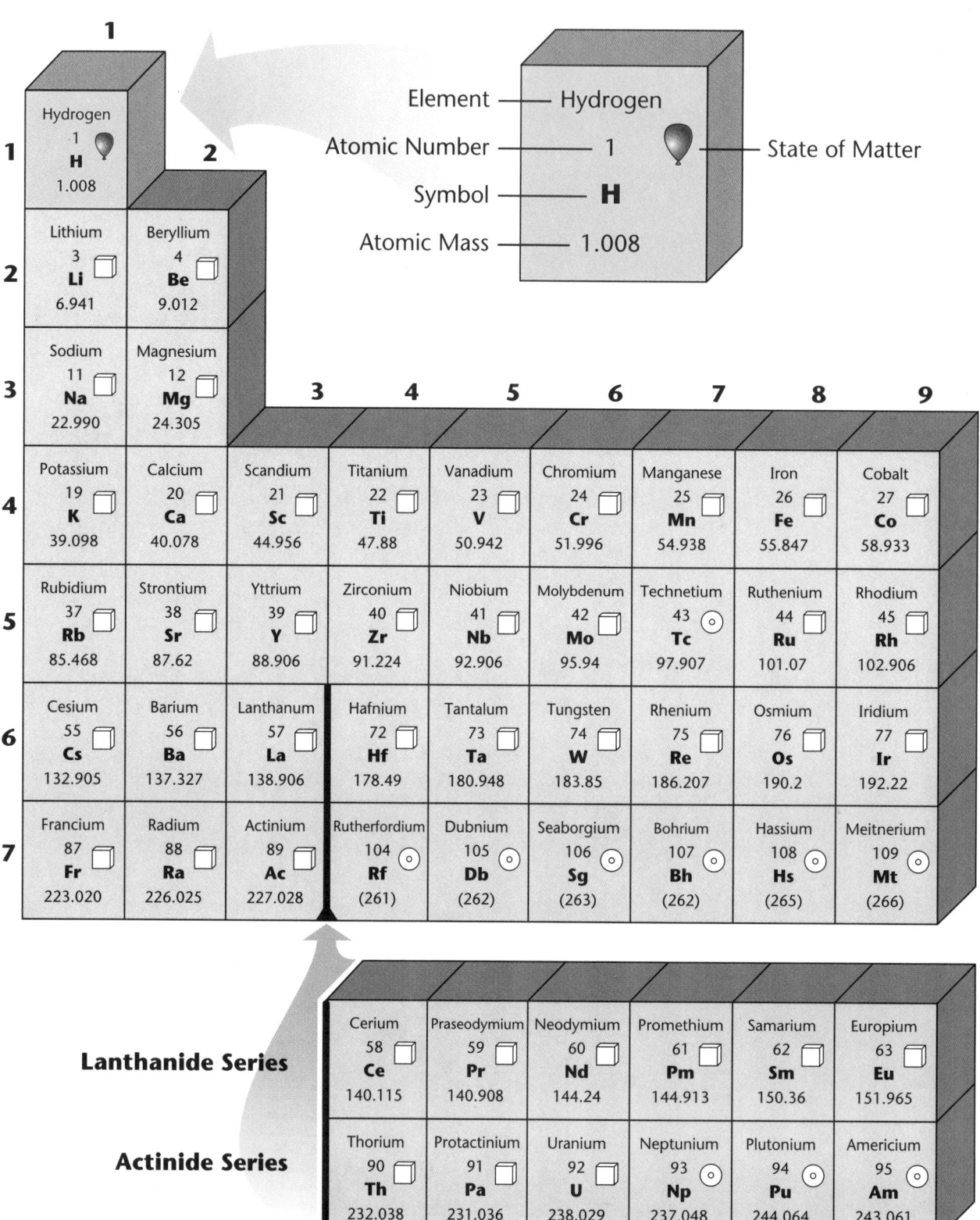

Element — Hydrogen
Atomic Number — 1 — State of Matter
Symbol — H
Atomic Mass — 1.008

1	2	3	4	5	6	7	8	9
Hydrogen 1 **H** 1.008								
Lithium 3 **Li** 6.941	Beryllium 4 **Be** 9.012							
Sodium 11 **Na** 22.990	Magnesium 12 **Mg** 24.305							
Potassium 19 **K** 39.098	Calcium 20 **Ca** 40.078	Scandium 21 **Sc** 44.956	Titanium 22 **Ti** 47.88	Vanadium 23 **V** 50.942	Chromium 24 **Cr** 51.996	Manganese 25 **Mn** 54.938	Iron 26 **Fe** 55.847	Cobalt 27 **Co** 58.933
Rubidium 37 **Rb** 85.468	Strontium 38 **Sr** 87.62	Yttrium 39 **Y** 88.906	Zirconium 40 **Zr** 91.224	Niobium 41 **Nb** 92.906	Molybdenum 42 **Mo** 95.94	Technetium 43 **Tc** 97.907	Ruthenium 44 **Ru** 101.07	Rhodium 45 **Rh** 102.906
Cesium 55 **Cs** 132.905	Barium 56 **Ba** 137.327	Lanthanum 57 **La** 138.906	Hafnium 72 **Hf** 178.49	Tantalum 73 **Ta** 180.948	Tungsten 74 **W** 183.85	Rhenium 75 **Re** 186.207	Osmium 76 **Os** 190.2	Iridium 77 **Ir** 192.22
Francium 87 **Fr** 223.020	Radium 88 **Ra** 226.025	Actinium 89 **Ac** 227.028	Rutherfordium 104 **Rf** (261)	Dubnium 105 **Db** (262)	Seaborgium 106 **Sg** (263)	Bohrium 107 **Bh** (262)	Hassium 108 **Hs** (265)	Meitnerium 109 **Mt** (266)

Lanthanide Series

Cerium 58 **Ce** 140.115	Praseodymium 59 **Pr** 140.908	Neodymium 60 **Nd** 144.24	Promethium 61 **Pm** 144.913	Samarium 62 **Sm** 150.36	Europium 63 **Eu** 151.965

Actinide Series

Thorium 90 **Th** 232.038	Protactinium 91 **Pa** 231.036	Uranium 92 **U** 238.029	Neptunium 93 **Np** 237.048	Plutonium 94 **Pu** 244.064	Americium 95 **Am** 243.061